T0133402

TURFGRASS HISTORY AND LITERATURE

TURFGRASS HISTORY AND LITERATURE

Lawns, Sports, and Golf

by

James B Beard

Harriet J. Beard

James C Beard

Michigan State University Press

East Lansing, Michigan, USA

∞ The paper used in this publication meets the minimum requirements of ANSI/NISO Z39.48-1992 (R 1997) (Permanence of Paper).

Michigan State University Press
East Lansing, Michigan 48823-5245

Printed and bound in the United States of America.

20 19 18 17 16 15 14   1 2 3 4 5 6 7 8 9 10

LIBRARY OF CONGRESS CATALOGING-IN-PUBLICATION DATA

Beard, James B, 1935–
    Turfgrass history and literature : lawns, sports, and golf / James B, Harriet J., and James C Beard.
      pages cm
    Includes bibliographical references.
    ISBN 978-1-61186-103-7 (cloth : alk. paper) 1. Turfgrasses—History. 2. Turfgrasses—Bibliography. I. Beard, Harriet J. II. Beard, James C. III. Title.

    SB433.B269 2013
    635.9'642—dc23   2012049435

Book and cover design by Scribe Inc.

green press INITIATIVE Michigan State University Press is a member of the Green Press Initiative and is committed to developing and encouraging ecologically responsible publishing practices. For more information about the Green Press Initiative and the use of recycled paper in book publishing, please visit www.greenpressinitiative.org.

In appreciation to the United States Golf Association Green Section, whose generous financial contribution made the publication of this book possible

Contents

Preface

ONE CAN LEARN MUCH FROM PAST HISTORY AND EXPERIENCES. THE TURFGRASS BOOKS, CONFERence proceedings, periodicals, and reports are our most valuable resource in documenting the evolutionary development and heritage of turfgrass science and culture. The objective of this book is to chronicle the evolution and history of turfgrass use on a worldwide basis as documented by early publications and photographs. This book should serve as an important reference and background resource for scholars, collectors, and those interested in the evolution and history of turfgrass art and science.

The first chapter focuses on documentation of the historical evolution of turfgrasses through the available, worldwide turfgrass literature. Subsequent chapters encompass a bibliography of (a) turfgrass book authors and publishers, (b) turfgrass research development, and (c) turfgrass educational programs via reviews of the range in books, scientific journals, research reports, and trade publications related to turfgrasses. Of special interest is the inclusion of numerous selected quotes and original photographs from the older publications that depict the early activities, equipment, and conditions of the time.

The stimulus for this book evolved naturally as a result of extensive literature searches while we were authoring seven earlier books, including the **Turfgrass: Science and Culture** and **Turfgrass Bibliography— From 1672 to 1972.** The compilation of titles, citations, and descriptions for this book was an arduous, time-consuming task spanning more than 40 years. Many of the publications cited herein were not available in public libraries at the time. This book, titled **Turfgrass History and Literature: Lawns, Sports, and Golf** documents the evolution of books, proceedings, periodicals, scientific journals, and research publications concerning the science and culture of turfgrasses as well as the originating authors. The erratic initiation and termination of certain turfgrass publications, especially conference proceedings, trade periodicals, research summaries, and field day reports, make the information in this book of special value.

Perspective. The true heritage of turfgrass science and culture is best represented by the principles, cultural practices, grasses, materials, and equipment that have been developed by turfgrass scientists, private companies, professional turfgrass managers, and amateur practitioners. There have been many improved turfgrass cultivars, fertilizers, pest controls, cultural approaches, and maintenance equipment, which in their time were significant improvements that served a valuable role. However, most of this equipment no longer exists, except for a few pieces of equipment preserved by collectors. Thus the historical record of evolutionary advances in the art and science of turfgrass culture is best preserved through writings and pictures in publications.

Books provide a limited, but the only, historical record for the 1700s and 1800s. In these early years, advances were based on trial-and-error approaches under field conditions. The turfgrass conference proceedings and trade periodicals became a significant source of historical records during the early 1900s. These publications provide early documentation of turfgrass advances and techniques developed primarily by turfgrass practitioners and gardeners.

The 1960s, 1970s, and 1980s were when the major advances in the science of turfgrass culture and quality occurred. These turfgrass advances are recorded in publications such as scientific periodicals, research summaries, and field day reports. The science of turfgrass culture became more important than the art during this period. Still, the art aspects learned only through actual field experience remain important in the development of turfgrass practitioners, especially for greens and sports fields.

Research Sources. The authors have made an effort to personally review the publications cited in the book. In accomplishing this, visitations were required to the Boston Public Library, Boston, Massachusetts; British Library, London, England; British Museum Library, London, England; Brooklyn Botanical Garden, Brooklyn, New York; Denver Public Library, Denver, Colorado; Golf House Library, US Golf Association, Far Hills, New Jersey; Kew Botanical Gardens Herbarium/Library, Richmond, England; Institute of Groundsmanship Library, Milton Keynes, England; Lindley Library, Royal Horticultural Society, London, England; Michael J. Hurdzan Collection, Columbus, Ohio; Museum of English Rural Life, University of Reading, Reading, England; National Archives, Richmond, England; National Playing Fields Association Library, London, England; New York Botanical Garden Library, Bronx, New York; Melvin B. Lucas Collection, South Dartmouth, Massachusetts; Michigan State University Library, East Lansing, Michigan; New York City Public Library, New York; Purdue University Library, West Lafayette, Indiana; Rijks Museum, Amsterdam, The Netherlands; Sports Turf Research Institute Library, Bingley, England; St Andrews University Library, St Andrews, Scotland; Football Association Library, London, England; US Department of Agriculture National Agricultural Library, Beltsville, Maryland; US Library of Congress, Washington, DC; and Victoria and Albert Museum Library, London, England. A grant from Exeter Bentgrass Association partially funded the travel expenses.

The publications cited herein are based primarily on the very comprehensive James B Beard Turfgrass Collection that was donated to the Michigan State University Turfgrass Collection in 2003. The holdings at this library represent the most complete collection of turfgrass related publications available today. Nevertheless, there are still missing titles that need to be secured in order to obtain a complete historical record. Peter Cookingham at Michigan State University, cooking1@mail.lib.msu.edu, should be contacted in this regard. It is hoped that this book will stimulate others to become involved in the collection and preservation of the remaining turfgrass books, conference proceedings, periodicals, research reports, and field day reports before they are discarded by individuals without realizing their true significance.

A great deal of time was required to obtain the author resumes presented in part 2. The resume of authors not deceased were obtained via a one page resume form that each of them completed. In some cases, the resumes are minimal or lacking in spite of three to eight written requests to potential sources. A few individuals chose not to respond to these multiple requests, and consequently their resumes are absent or brief. Deceased authors were especially challenging; even some employers (organizations, etc.) no longer retained records.

Alas, privacy laws in certain countries, such as the United Kingdom, presented an unyielding barrier. Thus the information presented in this book represents the best available known documentation.

Acknowledgments. This book would not have been as comprehensive without the cooperation of turfgrass scientists, scholars, and collectors from throughout the United States and the world during the past 40 years. Special acknowledgments for their inputs in searching citations and descriptions for numerous titles are given to David E. Aldous, Gary W. Beehag, Bill G. Casimaty, Donald S. Lock, Peter E. McMaugh, and Brett Robinson for Australia; Pam Charbonneau for Canada; David M. Gilstrap for China; Kate Daniels, Robert Daniels, Rémy Dorbeau, Francis Lemaire, and Paul Mansat for France; Klaus Muller-Beck for Germany; Aidan O'Hara for Ireland; Hisakazu Kihara, Yoshisuke Maki, Hideaki Tonogi, and Susumu Yoshida for Japan; Megan Cushnahan, D. Keith McAuliffe, and G.S. Robinson for New Zealand; Massimo Mocioni, Francesco Modestini, and Paolo Croce for Italy; Douglas Lee for Malaysia; C.A.R. Schulenburg for South Africa; Ki Sun Kim for South Korea; Javier Garcia Ircio for Spain; Sven-Ove Dahlsson and Peder Weibull for Sweden; Stig Persson for Sweden and Denmark; William A. Adams, John C. Campbell, John R. Escritt, Roger D.C. Evans, Graeme Forbes, Gordon McKillop, and Jeffrey Perris for the United Kingdom; and Lee C. Dieter, D. Thomas Duff, Harry C. Eckhoff, Michael J. Hurdzan, James M. Latham, John A. Long, Melvin B. Lucas Jr., Thomas C. Mascaro, Morris Brown, James G. Prusa, Alexander M. Radko, James B. Ricci, William Rose, Janet M. Seagle, James T. Snow, Charles Tadge, Joseph Troll, and Alfred J. Turgeon for the United States. There are hundreds of other individuals who helped with one to a few author and book entries.

The encouragement and cooperation of the O.J. Noer Research Foundation and Charles G. Wilson while the authors assembled the original Turfgrass Collection at Michigan State University deserves a very special acknowledgment, along with the assistance of Richard E. Chapin, who was then Director of Libraries. The worldwide book search and loan acquisition assistance of Peter O. Cookingham, Sue Depoorter, and Mike Schury during our extensive research work at the Michigan State University Library and the cooperation of Director Clifford H. Haka were invaluable and greatly appreciated. Finally, as parents, James B and Harriet acknowledge the digitization and photo editing of the many old photographs presented in the book accomplished via the expertise of James C Beard.

Summation. The authors hope readers of this text will be lenient concerning errors of omission or commission. It was not for lack of effort over the past 40 years of collecting and documenting turfgrass history. Some early historical turfgrass activities may never have been recorded or, if recorded, have since been lost or have not been found. Certainly this historical documentary is more advanced than anything that has preceded it in terms of an organized presentation in a single edition.

James B, Harriet J., and James C Beard

A PERSONAL PHILOSOPHY OF TURFGRASS
ART, CULTURE, AND SCIENCE

Professional turfgrass culture is one of the few areas of applied plant
    science where:
The ultimate goal is perfection,
Which seldom is achieved,
When reached, it is fleeting,
Therein lies the ultimate challenge.

Further

Upon approaching perfection in turfgrass,
The more evident the imperfections become,
The more difficult and costly they are to correct.
Therein lies the intrigue.

JAMES B BEARD

Common and Scientific Names of Turfgrasses

bentgrass, colonial	*Agrostis capillaris*
bentgrass, creeping	*Agrostis stolonifera*
bentgrass, velvet	*Agrostis canina*
bermudagrass, African	*Cynodon transvaalensis*
bermudagrass, dactylon	*Cynodon dactylon* ssp. *dactylon*
bluegrass, annual	*Poa annua* var. *annua*
bluegrass, Canada	*Poa compressa*
bluegrass, creeping	*Poa annua* var. *reptans*
bluegrass, Kentucky	*Poa pratensis*
bluegrass, rough	*Poa trivialis*
bluegrass, Texas	*Poa arachnifera*
buffalograss, American	*Buchloe dactyloides*
carpetgrass, common	*Axonopus fissifolius*
carpetgrass, broadleaf	*Axonopus compressus*
centipedegrass	*Eremochloa ophiuroides*
fescue, Chewings	*Festuca rubra* ssp. *commutata*
fescue, hard	*Festuca trachyphylla*
fescue, sheep	*Festuca ovina*
fescue, slender-creeping red	*Festuca rubra* ssp. *littoralis*
fescue, strong-creeping red	*Festuca rubra* ssp. *rubra*
fescue, meadow	*Schedonorus pratensis*
fescue, tall	*Schedonorus arundinaceus*
gramagrass, blue	*Bouteloua gracilis*
gramagrass, sideoats	*Bouteloua curtipendula*
paspalum, seashore	*Paspalum vaginatum*
ryegrass, annual	*Lolium multiflorum*

ryegrass, perennial	*Lolium perenne*
St. Augustinegrass	*Stenotaphrum secundatum*
wheatgrass, crested	*Agropyron cristatum*
wheatgrass, western	*Pascopyrum smithii*
zoysiagrass, Japanese	*Zoysia japonica*
zoysiagrass, manila	*Zoysia matrella* ssp. *matrella*
zoysiagrass, mascarene	*Zoysia pacifica*

PART I

History of Turfgrass Evolution and Development

CHAPTER 1 IS BASED PRIMARILY ON RELATIVELY RECENT technology advances and allied research. The authors fully recognize that the grass evolutionary events and dating may be adjusted even more in the future if the advances of the past decade are any indicator. The turfgrass history addressed in chapters 2 through 10 covers primarily the period up to 1950 and is focused on the United States.

Turfgrass Origins, Migrations, and Diversifications

AN UNDERSTANDING OF THE RELATIVELY RECENT ADVANCES IN GRASS SPECIES EVOLUTION AND development has dictated revision of how native and naturalized turfgrass species are viewed, especially in North America. The supporting research for these updates is reviewed herein.

Native Turfgrass Species. *Webster's Third New International Dictionary* defines "native" as something indigenous to a particular locality. "Indigenous" is defined as being native to and as originating or developing or being produced naturally in a particular land, region, or environment. The associated emotive terminologies relative to the term "native" that have been used include naturalized, alien, exotic, and nonnative. Criteria for classifying native species that have been utilized or proposed include historical, geographical, economical, behavioral, sociological, and/or political considerations. The latter four involve value judgments based on an arbitrary time frame that results in controversy due to the diversity of human preferences and needs (53).* Thus the native aspects of turfgrasses addressed herein will rely on temporal and geographic dimensions documented by sound hard science.

There are proponents who consider the plant species present in America when Europeans arrived as the native species that have persisted from time immemorial, free from human disturbances. However, recent research indicates there are woodland areas that have developed subsequent to anthropological disturbances during the late Holocene, at least 4,000 years ago (80). The Mound Builders and early American populations were active in domesticated crop propagation involving land use by fire to remove forested areas. Such anthropological disturbance of plant species and reinvasion by intermediate succession species have been recently documented for regions in southeastern and in northeast central North America (33, 80), and others probably will be found.

One can only speculate as to the grass species that would have been naturally introduced and would have persisted in recent times if human disturbances had not occurred. Disturbed environments can have long-term or short-term effects on plant community composition depending on the degree and duration of the disturbance event. Disturbance of plant ecosystems can be a natural aspect caused by drought, fire, disease, insect, glaciation, heat stress, freeze stress, volcanic smoke, soil deposition, soil salt level, animal pressures, and so on. Defining a native plant based on a fixed point in time—such as pre-Holocene, pre-Neolithic, or pre–European colonial—is unrealistic, as it assumes no nonhuman disturbances would have occurred subsequently. It also fails to acknowledge humans as a component of the ecosystem. Historically, the species composition of most plant communities has been a continuous, dynamic process of transitional

---

* The numbers in parentheses correspond to the reference list found at the end of the chapter.

diversification and recovery. It is likely the natural distribution of plant communities would still be different even if human intervention had not occurred. Similarly, species move geographically in response to environmental changes caused both naturally and by human disturbances. Thus native sites should not be viewed as fixed geographical areas that existed prior to human activity. Also, the following premise stated by some individuals is not valid: native turfgrasses inherently possess lower cultural requirements, resource inputs, and/or water use rates, plus better human benefits, than nonnative species (53).

Application of the native terminology for grasses has varied over the past century and previously emphasized the effects of tectonic processes that resulted in assumed geographic barriers such as oceanic separation of continents. Some grass taxonomy and flora books include a geographic description that proposes a continent(s) or region(s) where each species is native. In the early 1900s, there were North American texts on grass taxonomy that referred to grass species such as the *Agrostis*, *Festuca*, and *Poa* as natives. Then by the mid-1900s, most grass taxonomy, flora, and other texts listed such species as native to Eurasia. This change in the center-of-origination concept was based on plant exploration that indicated the geographic region thought to possess the greatest morphological diversity. Unfortunately, the true center-of-origination for a species may no longer exist botanically due to severe, extended drought (as had occurred in Africa) or due to ice age glacial presence (e.g., the Northern Hemisphere). Also, the oceanic separation of continents is not proving to be the migration barrier it was once assumed to be (48, 85).

The development of molecular phylogenetic techniques during the past two decades, involving DNA sequencing instrumentation and unique phenetic and cladistic analyses via computer programs for large data sets, has drastically changed our understanding of when and where diversifications of grass genera, subgenera, species, and subspecies occurred. Thus an update of the native or naturalized status of the major turfgrass species is addressed herein.

Naturalized Turfgrass Species. The term "naturalized turfgrass species" is defined herein as a plant that is introduced, becomes established, and persists without direct human inputs. The introduction may be unintended or intended. An introduced, heterozygous species may diversify and naturalize as a local ecotype that is well adapted, functionally valuable, and noninvasive. Also, multiple independent origins of similar genotypes have occurred in widely separated but similar environments, as has been documented for *Agrostis capillaris*, *Agrostis stolonifera*, and *Poa annua* (57). The plant community composition is inherently dynamic, with preservation of the status quo on a long-term basis unlikely. It is natural to have both migration and introduction and naturalization of species. Anthropologically related naturalization contributes to the species diversity of plant communities.

Taxonomy. The grass taxonomy employed herein is that of M.E. Barkworth and others (6, 96, 97). The classification system for grasses is kingdom Plantae, phylum Magnoliophyta (syn. division), class Liliopsida (syn. Monocotyledonae), subclass Commelinidae, superorder Poanae, order Poales, and family Poaceae (syn. Gramineae). Subfamily names end in *oidea*, tribes in *eae*, and subtribes in *inae*.

The traditional grass taxonomic system is based on morphological similarities, especially the inflorescence. A more recent classification system is cladistics or phylogenetic systematics, in which groups or clades are organized based on relative dating of common ancestors with shared diversity of anatomical, physiological, biochemical, and morphological features. Basically, a clade may encompass all descendant species of an ancestor. Two major grass clades are BEP (encompassing the Bambusoideae, Ehrhartoideae, and Pooideae subfamilies) and PACCMAD (encompassing the Panicoideae, Arundinoideae, Chloridoideae, Centothecoideae, Micrairoideae, Aristidoideae, and Danthonioideae subfamilies) (15, 32, 42, 43, 77).

The BEP clade contains exclusively C_3 grass species, while the PACCMAD clade includes C_3 species, as well as all known C_4 grasses. The latter are mainly distributed within the Chloridoideae, Aristidoideae, and Panicoideae subfamilies. The crown node for the BEP clade is estimated to be ~10mya older than that of the

PACCMAD clade (17). Numerous subclades become evident from the cladistic applications (2, 15, 25, 26, 39, 48, 51, 58, 62, 67, 85, 86), resulting in taxonomic classifications that more clearly depict evolution of the grass family. Cladistic classification of the monocotyledons indicates the Joinvilleaceae are a sister group to the Poaceae within the Commelinidae subclass (35, 51, 86, 89).

Plant Evolution Phases. The geophysical aspects of grass species evolution involve (a) a primitive ancestral center-of-origination, (b) early diversification, (c) migration or dispersal, and (d) secondary-centers-of-origin or diversification in order to adapt and survive changes. These include environmental (temperature, water, irradiation, and carbon dioxide), edaphic (soil texture, pH, nutrients, and salts), and biotic (animal grazing, human activities, and pests) changes. There also are continuing intraspecies diversification regions in response to current natural and human environmental disruptions. Accordingly, microcenters-of-diversification occur in small areas where mutations and hybridization are significant, as represented by the presence of diploid, triploid, tetraploid, hexaploid, and even octoploid plant species (101).

Previously, a sometimes ill-defined center-of-origination had been considered phenotypically to be a region where the greatest diversity of a grass species occurs. More recently, confirmation of or major adjustments in dating and location have been made possible by more sophisticated paleobotany using ultra-structural electron microscopic techniques plus stable carbon isotope dating instrumentation and research procedures. Actually, the species' secondary-centers-of-origin and/or diversification may be more important than the center-of-origination (48), especially if those ancestors in the center-of-origination became extinct as a result of adverse climatic changes, such as the severe drought that occurred in Africa. Also, key secondary-centers-of-diversification for temperate C_3 grasses were formed after subsidence of the last glacial ice age. Current data indicate that certain morphological diversifications of grasses have occurred at least 23 mya after the origin of a particular character (51). Diversification is a dynamic, ongoing process with multiple microsecondary-centers-of-origin. A grass population has the potential for further differentiation that leads to better genetic adaptation to a specific environment by natural selection.

Early Plate Tectonics. Grass-dominated ecosystems now cover more than one-third of the earth's land surface area. This was not always the case. The early evolutionary history of primitive ancestral grasses and their broad migration patterns are thought to have been influenced by shifting positions of land masses (27) and seas that created and even destroyed environments where the ancestral grasses originated or migrated. Plate tectonic theory has been advanced by new geological evidence and contributes to concepts of angiosperm-Poaceae distributions. The supercontinent Pangaea formed ~500 million years ago (mya). Its configuration as it existed in the medial Cretaceous prior to continental breakup is shown in the following figure 1.1. Precontinental separation of Pangaea and possible ancestral grass migrations are also shown.

Pangaea split into Laurasia (now North America and Eurasia) and Gondwana (now South America, Africa, Antarctica, and Australia) during the late Mesozoic era, ~160 mya. A land connection between Gondwanan Africa and Laurasian Europe was present in the western Mediterranean region. The early Laurasian land connection between Europe and North America is termed the "North Atlantic Land Bridge." Gondwana was characterized by mild temperatures and a huge range of flora and fauna for millions of years. The early West Gondwana formation may have allowed South America to be more directly land accessible to primitive ancestral plant migration from Africa than from Laurasian North America.

Intercontinental Plant Migration. The movement of plant species between continents and directionally across large continental areas is referred to as migration. Transcontinental dispersals have been found to occur earlier and more frequently than was once assumed (48, 85). Wind, birds, animals, and/or floating debris may have contributed to such dispersals. Species distributions and multiple centers-of-diversification tend to be influenced by conditions existing since the Pleistocene. Direct intercontinental plant migration

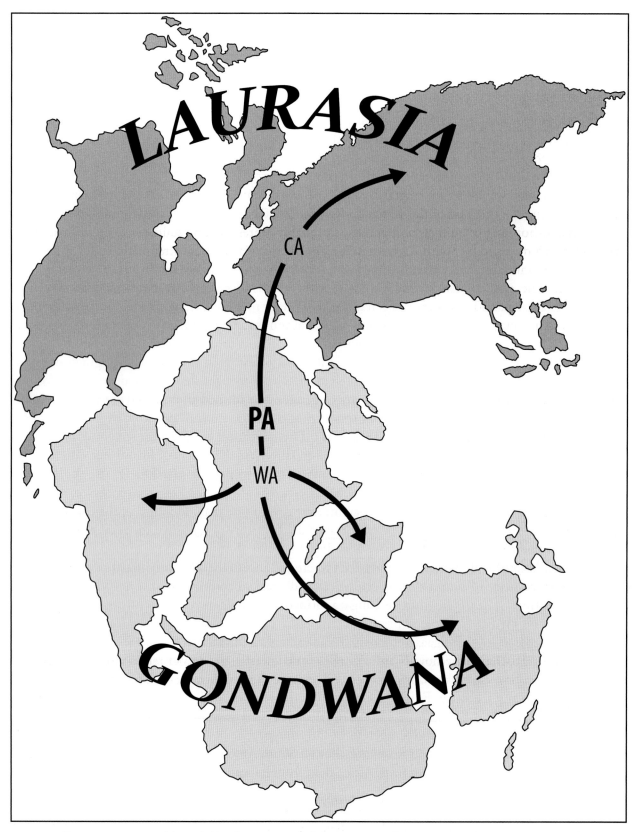

Figure 1.1. Depiction of the probable relative positions of the Laurasian and Gondwanan land masses in the medial Cretaceous ~100 mya (adapted from Smith et al., 1973) and possible broad migration patterns of primitive ancestral grasses (PA), plus later migration of cool-season ancestral grasses (CA) and warm-season ancestral grasses (WA). [26]*

---

* The numbers in brackets correspond to the figure source acknowledgments listed in the appendix.

of primitive ancestral grasses was possible up to or following specified dating, as extensively reviewed by P.H. Raven and D.I. Axelrod (72; table 1.1). Patterns of potential intermittent ancestral plant migration to North America (72) may have involved a Laurasian direct land route from Europe, an indirect land route from Asia via the Bering Land Bridge, and a more recently formed land route from South America via the Panamanian land bridge (table 1.1).

Note that moderate biogeographic plant migration by oceanic island-hopping probably occurred via birds, wind, and/or flotation for a significant time following geologically dated continent separation. A narrow gap interspersed with volcanic islands between Africa and South America presumably could have allowed primitive flora migrations up into the Paleocene (58, 72). It has been proposed that tropical grass subfamilies may have continued to migrate until continental separation reached a water gap of 745 miles (1,200 km) during the first half of the Tertiary, ~50 mya (28, 72).

Major island-hopping of animals and grasses is probably represented by southeast Asia to Australia via New Guinea ~15 mya and by South America to North America via an intermittent land connection ~10 to 12 mya (72). Migration from Asia-Europe to North America via the Bering Land Bridge has occurred since ~55 mya (72). The ancestral Bovidae migrated to North America from Asia-Europe via the Bering Land Bridge (73). There also was a land uprising to form the Gibraltar Strait Bridge between Europe and Africa ~5 mya, by which Bovidae migrated to Africa. Oceanic island-hopping between South America and North America may have occurred from ~45 mya. Mountain-hopping from Europe to Asia to North America to South America occurred starting at least 10 mya for *Agrostis*, *Festuca*, and *Poa* (39, 48, 72).

Primitive Ancestral Grasses. Definitive evidence is limited concerning the continental center-of-origination for primitive ancestral grasses. Fossil records for the grass family Poaceae within the angiosperms and their ancestors are very incomplete. Based on paleobotanical evidence, including silicified plant tissues termed *phytoliths* and stable carbon isotope studies, primitive ancestral grasses appeared rather late in the earth's history—somewhere between 55 and 70 mya (43, 48, 51). It has been proposed, and supported by phylogenetic analyses, that the Arundinoideae was the ancestor from which the cool-season monophyletic Pooideae arose (28, 52).

Available evidence suggests the primitive ancestors of Poaceae occurred in the tropical forests and fringes of Gondwanan Africa (17, 28, 58, 86). An extensive review of pollen fossil records by Muller (64) indicates basal Poaceae were present in the Paleocene epoch ~55 to 60 mya in central to eastern Africa and east central South America. Macrofossil plant remains from a formation in western Tennessee suggest the presence of a common ancestor for the Poaceae under tropical, seasonally dry conditions ~55 mya (30). This is a significant finding relative to native grass species evolution in North America.

TABLE 1.1 SUMMARY OF WHEN POSSIBLE EARLY INTERCONTINENTAL PLANT MIGRATION
PATTERNS FOR PRIMITIVE ANCESTRAL GRASSES COULD HAVE OCCURRED (72).

Event occurrence	*Specific continental contact event*	*Event dating (millions of years ago)*
Last	Africa and India	~100 late Lower Cretaceous
	Africa and South America	~90 early Upper Cretaceous
	Africa and Eurasia, prior to Miocene event	~63 early Paleocene
	Europe and North America	~49 mid-Eocene
	South America to Australia via Antarctica	~38 mid-Eocene
First	India contacts Asia	~45 early Eocene
	Reestablished contact between Africa and Eurasia	~17 mid-Miocene
	South America and North America	~5.7 late Miocene

The crown node of the BEP and PACCMAD clades has been estimated phylogenetically at ~57 mya (17). Phytoliths preserved in coprolites from central India revealed five extant Poaceae subclades present during the Upper Cretaceous ~65 to 70 mya (70). Molecular phylogenetic and cladistic biogeographic analyses propose a dating range of Poaceae at the Late Cretaceous ~65 to 96 mya (16, 17, 26, 35, 58).

Both diversifications and migrations of grasses have been influenced by changing regional environmental, edaphic, and biotic influences, including changes in tectonic plate processes, ice age glaciation, volcanic activity, tropical dominance, sea level, fire occurrence, water availability, favorable temperatures, seasonality, and competitive flora and fauna. Note that ice age climatic change during the Pleistocene was particularly disruptive in terms of grass species diversification and migration in the Northern Hemisphere.

Diversification. Phytogeographical analyses suggest that early differentiation of the lineages for tribes of the Poaceae may have started by the end of the Upper Cretaceous (26, 58). The Pooideae probably diversified initially in temperate environments at high latitudes and altitudes in the tropics (58). From ancestral origination in fringe-forested regions, there appears to have been further evolution to savannas and then to temperate steppes in the case of the Pooideae. The C_3 cool-season ancestral grasses are estimated to have arisen during the Eocene ~44 to 55 mya (30, 95). The Pooideae are estimated to have migrated to the steppes of Eurasia ~38 to 47 mya (17). Grass-pollen fossils are more abundant during this epoch (64).

The C_4 physiological characteristics of warm-season ancestral grasses evidently appeared as at least 12 distinct events and locations starting in the Oligocene ~30 mya and continuing through 15 mya (16, 17, 43, 51, 54, 86). The C_4 warm-season ancestral grasses of the Panicoideae and the Chloridoideae are estimated to have originated ~26 mya and 29 mya, respectively (17). The emergence of C_4 grasses was associated with a drop in atmospheric carbon dioxide (CO_2) levels (16, 37) and/or seasonally drier climates (28, 56). Macrofossils of Chloridoideae grass leaves from Kansas indicate that C_4 Kranz leaf anatomy existed in the Miocene ~five to seven mya (97). Also, evidence indicates that C_4 grasses expanded in the Americas during the late Miocene (56).

Early grasses eventually became significant components of plant communities in the dryer savannas, steppes, and prairies. The increasing dominance of grasses was probably aided by periodic lightning-induced fires that destroyed most tree and shrub competitions, while the grasses survived via basal tillering. Actually, the dry grass biomass provided fuel for the fires.

It is postulated that since Gondwana phytogeographically separated into the present major continents, numerous distinct grass diversifications have occurred on four separate continents. The original ancestral grasses have now evolved into a large family of angiosperms with ~40 tribes, ~800 genera, and ~11,000 species (26).

Migration. Grasses had eventually dispersed to all continents ~60 mya after their ancestral Gondwanan Africa origin (17). Phylogenetic studies suggest that the first migration within the Pooideae, from Eurasia to North America, occurred ~38 mya via the North Atlantic Land Bridge when there was a global warming trend (17). Subsequently, there is evidenced that other BEP taxa migrated from Europe via Asia to North America by the Bering Land Bridge (48).

Modes of seed dispersal away from the parent plant can occur by animals, gravity, wind, and water. Certain grass species possess characteristics that allow either internal or external transport by animals. Internal mechanisms can occur via the digestive tract, primarily of large grazing animals (66, 71). The ingestion and internal passage of viable seeds through the digestive systems of animals have been documented for C_3 grasses *Agrostis stolonifera, Festuca rubra, Poa annua, Poa compressa, Poa pratensis, and Schedonorus arundinaceus* (34) and for C_4 grasses *Axonopus fissifolius* (19, 65), *Buchloe dactyloides, Cynodon dactylon,* and *Paspalum notatum* (19).

Certain chaff-like grass seeds facilitate external transport on the fur coats and hairy hides of mobile animals. Seeds with distinct awns, hairy lemmas, and/or sharp-pointed florets allow penetration in animal hair and resist easy withdrawal (24).

Many grasses adapted for turfgrass purposes have very small, lightweight seeds that can migrate via wind. For example the approximate number of seeds per pound (0.45 kg) for *Agrostis capillaris* is greater than 5 million, for *Cynodon dactylon* is greater than 1.5 million, for *Poa pratensis* is greater than 1 million, and for fine-leaf *Festuca* is greater than 400,000 (10). Large dust storms from Africa to the Americas may have furthered intercontinental dispersal of certain grass seeds.

The same chaff-like, lightweight seed characteristics also aid migration in flowing water down streams and rivers, especially in association with eroded soil. *Stenotaphrum secundatum* seed has an inflorescence structure that enhances flotation on ocean currents between islands and along coastal shorelines (79).

The presence of a chemical germination inhibitor in the seed is another mechanism by which some species of grass seeds extend the duration for potential migration. *Poa pratensis* is an example with modern cultivars having seed germination delays of three to four months (10) following maturation. Certain earlier ancestral grasses probably had longer germination inhibition periods due to a higher inhibitor content and/or greater resistance to inhibitor degradation.

Some grass species have what is termed "hard seed," in which the seed coat is impermeable to water needed to initiate germination. Typical turfgrass species with hard seeds include *Cynodon dactylon*, *Eremochloa ophiuroides*, and *Zoysia japonica* (10). Germination can be improved by mechanical or acid scarification to open the seed coat barrier. Either the presence of a hard coat or an inhibitor in the seed could facilitate distance dispersal of such grass species in the digestive systems of birds and animals, especially herbaceous grass–feeding animals (24).

Grazing Mammals and Turfgrasses. The ancestral species for some turfgrass species evolved in association with herbivorous grazing mammals. Early, now extinct browsing ungulates known as Condylarths evolved while feeding on ancestral C_3 grass terrestrial ecosystems during the early Paleocene ~65 mya (8, 42). Subsequently, during the early Eocene, ~55 mya, early herbivorous ruminant Artiodactyls arose. The Bovid subfamilies exhibited a rapid diversification ~23 mya (63). The Bovidae family, with a distinct tooth structure, arose in the late Miocene ~five to nine mya. Evolving in Eurasia were cattle, antelope, and bison of the subfamily Bovinae, plus goats, sheep, water buffalo, and musk-ox of the subfamily Caprinae (63, 95).

Early grasses persisted as a minor component of forested plant communities for a long time. Eventually, ancient grass–dominated ecosystems developed in certain open regions ~five to six mya as a result of significant changes to a drier environment that caused a decline in tree populations (91). During this time, certain herbivorous mammals also were evolving with mouth parts adapted to grazing more closely and also with high-crowned teeth and/or constantly growing teeth needed for grazing on grasses with high phytoliths or silicified fibers (61). Consequently, there was natural selection toward grass species in certain regions that were structurally adapted to survive severe defoliation (88).

While open-habitat grasses exhibited taxonomic diversification in North America at least 34 mya, phylogenic studies suggest they did not become ecologically important until formation of the savanna-like Great Plains around 23 to 27 mya (90). A major expansion of C_4 grasses occurred during the late Miocene ~six to eight mya during a major climate change to a dryer environment and lower atmospheric CO_2 levels (23, 37, 56). Intercontinental distribution of both C_3 and C_4 grasses stabilized during the late Miocene ~six mya (61).

Carbon dating of herbivorous mammalian skeletons and attached grass remnants by paleobotanical studies and DNA sequencing confirm a comigration among a number of grazing mammals and certain perennial grasses (16, 24). The interrelationship of certain low-growing grasses and herbivorous mammals continued over an estimated period of ~10 to 20 mya (8). The result was a number of grass species having (a) leaves with basal meristems, (b) shoots with short basal internodes, and (c) prostrate, creeping growth habits via lateral stems termed *stolons* and *rhizomes*. Evolution of these morphological adaptations in some grasses allowed subsequent use for turfgrass purposes involving frequent, low mowing.

Some centers-of-diversification for temperate grasses are of more recent post-Pleistocene development ~two to one mya due to glacial effects from the last ice age. Examples are the North American Northeast and Midwest regions and the higher altitude meadowlands of central Europe.

Human Dispersal. Intercontinental human grass dispersals may have occurred even earlier than New World colonial times. More than 600 years ago, world voyagers and global explorers may have dumped from their ships at various worldwide locations sweepings containing straw and hay bedding composed primarily of grasses. Later, colonists traveling via boat to the New World, primarily from Europe, brought grasses for agrarian forage and pasture purposes. These introduced grasses had been selected by Europeans for more than 400 years based on a higher yield or biomass production needed for feeding herbivorous domesticated animals (81). The early colonists preferred the taller-growing, high-yielding introduced grasses for agrarian purposes. European forage and pasture grasses were less desirable as turfgrasses, especially in early times of laborious manual mowing. Many European-introduced grass genotypes naturalized in North America through intraspecies diversification and natural selection over a period of 300 to 400 years. These grasses became interspersed with precolonial native grass communities, especially in eastern North America. For example, a good germplasm source for turfgrass selection is in old cemeteries that have been mowed regularly but typically not irrigated or fertilized. Clarification as to the specific genetic sources from which many contemporary turfgrass cultivars were derived will require molecular phylogenetic research.

Modern Diversification. Globally, there are ~40 grass species utilized as turfgrasses to varying degrees. Most are heterozygous with inflorescence fertilization by wind-aided cross-pollination. The result is a wide range in diversity with many grass ecotypes that have specific adaptation potentials to the numerous global variations in the environmental, edaphic, and biotic components contributing to the broad array of ecosystems. Furthermore, the diversification process for individual grass species is ongoing with new genotypes emerging and some genotypes possibly becoming extinct if poorly adapted to the specific environment in certain ecosystems (81).

To the untrained observer, a turfgrass community, such as a lawn, may appear to lack diversity. This is because the basic leaf characteristics are more similar in shape, orientation, and color compared to the broad-leaf dicotyledonous species such as trees, shrubs, and flowers. Actually the genetic diversity is substantial in most mature turfgrass lawn communities. It is a key to the capability of grasses to recover and survive as perennials under wide seasonal variations in temperature, water, shade, traffic, and pest stresses.

COOL-SEASON TURFGRASSES

The primitive ancestral center-of-origination for cool-season grasses (Pooideae) is thought to have been from the forest margins of Africa at higher altitudes (28). The unique evolutionary feature possessed by Pooideae was adaptation to cold climates that allowed migration to the temperate steppes of Eurasia (28). Subsequently, the Pooideae may have emerged under a tree-shrub dominant, temperate ecosystem with sufficient precipitation for sustained growth and possibly survival during occasional short periods of droughty conditions. Taxonomic divergence of the base Pooideae group appears to have been initiated after a global supercooling period ~26 to 33.5 mya (78). Diversification also was probably characterized by a cool temperate climate plus reasonably good precipitation and soil fertility. Cool-season grass diversification in Eurasia may have been greater in the mountainous areas where shade from trees was reduced. Plant explorers

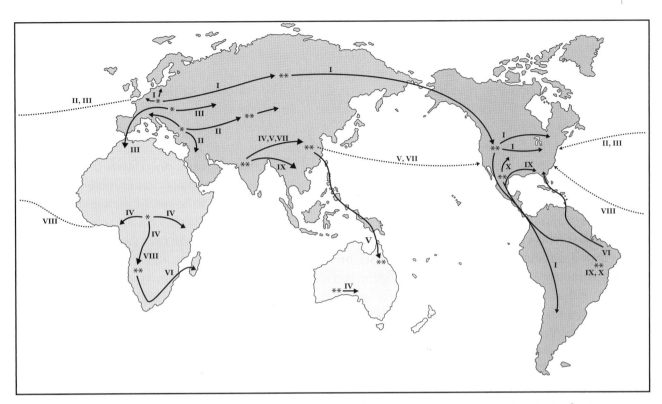

Figure 1.2. Broad global depiction of the current understanding and probable post-Pangaea centers-of-origination,* major migration routes, and key secondary-centers-of-origin** or centers-of-diversification of early grasses from which extant turfgrasses evolved. Designations are as follows: I. *Festuca* (fine-leaf), *Poa*, and *Agrostis*; II. *Lolium*; III. *Schedonorus arundinaceus*; IV. *Cynodon*; V. *Zoysia*; VI. *Stenotaphrum*; VII. *Eremochloa*; VIII. *Paspalum* (fine-leaf); IX. *Axonopus*; X. *Buchloe* and *Bouteloua*. Solid lines depict preanthropological migration and dash lines represent human introduction. [26]

(W.A. Meyers, Rutgers University, personal communication, 2011) report greater diversity within the *Festuca*, *Poa*, and *Agrostis* species in Europe and Asia at altitudes above 3,000 feet (915 m).

The cool-season turfgrasses discussed herein are taxonomically classified in the Pooideae, with C_3 physiology (96). Within the Pooideae there are two supertribes, including Pooideae, and seven tribes, including Poeae. The Pooideae includes ~3,300 species. The Poeae encompass ~115 genera and ~2,500 species (6) and occur with a high species frequency percentage in the higher latitudes and higher altitudes where summer and winter temperatures are colder (92). The cool-season turfgrasses occur in the temperate climates of both the northern and southern hemisphere and in mountains of the tropics, with an optimum temperature range of 60 to 75 °F (16 to 24 °C; 10, 11). Migration of certain C_3 cool-season perennial grasses evidently occurred by mountain-hopping. The likely migration route for the freeze tolerant *Agrostis*, *Festuca*, and *Poa* species into the South American Andes was via the North American, Asian, and European mountains (18, 48). In North America, these three genera typically have their largest range of species diversity in the western mountainous regions. Geophysical aspects of extant cool-season turfgrass will be addressed for the major species utilized in the United States. Included are the fine-leaf fescues, bluegrasses, bentgrasses, ryegrasses, and broad-leaf tall fescues.

Fine-Leaf Fescue. The early species center-of-origin for fine-leaf fescues (*Festuca* species) has been reported to be in central Europe during the mid-Miocene ~13 mya, based on phenotypic and biogeographical analyses (48). *Festuca* is reported to be an ancient group and a main evolutionary line in diversification of the Poeae tribe (19). Two main *Festuca* clades are evident from phenetic and phylogenetic analyses, being fine-leaf and broad-leaf, which had split into diverging lineages (29, 31, 93, 94). Within the fine-leaf *Festuca* clade, the *F. ovina* and *F. rubra* groups arose (94). The latter is smaller in number of subgroups (69) and is yet to be fully defined due to the paraphyletic nature (2, 6, 10). The *F. ovina* complex is the oldest of the turfgrass types (31), followed in lineage by *F. trachyphylla* and *F. rubra* ssp. *rubra*. Based on phenetic distance analysis, *F. ovina* and *F. trachyphylla* are distinct but closely related (2).

Fine-leaf fescues probably diversified in cool, continental, open forest regions, and possibly in mountainous areas and are included in the *Festuca* subgenus (1, 2, 6, 7). Diversification of the *Festuca* genus resulted in a major expansion, with ~400 known species (6). Their extant occurrence ranges from temperate to alpine to polar regions of both hemispheres (2). Chewing's fescue (*F. rubra* ssp. *commutata*), hard fescue (*F. trachyphylla*), sheep fescue (*F. ovina*), slender-creeping red fescue (*F. rubra* ssp. *littoralis*), and strong-creeping red fescue (*F. rubra* ssp. *rubra*) are the fine-leaf *Festuca* species now used as turfgrasses in the United States (10, 11). These five fine-leaf *Festuca* are seed propagated and largely cross-pollinated. The taxonomic classification of Chewings fescue as traditionally used is open to question (2). The fine-leaf fescues migrated outward throughout much of Europe, into central and eastern temperate Asia (25), and into Africa. Eventually they moved across the Bering Land Bridge into North America and migrated southward into Central America and over the Panamanian Land Bridge into South America (the previous figure). The basic route in the Americas was north to south via the western mountains of North America to the Andes to Patagonia over a period ~3.8 to 10 mya. This migration route has been documented among the *Festuca* species by biogeographical analyses (48). There are at least 37 *Festuca* species that are native to North America (6, 7), including the red, hard, and sheep fescues now used as turfgrasses. There also were outward migrations into the temperate climatic regions of the continent.

Most turfgrass-type fine-leaf fescue species (*Festuca*) are native to North America, having persisted for more than three centuries. They are used as a turfgrass in the cooler portions of the cool climatic regions of the United States, being especially adapted to shaded and low-nitrogen fertility conditions (10, 11).

Bluegrass. The early species center-of-origination for the bluegrasses (*Poa*) based on phenotypic and cladistic analyses (85) is temperate, open Eurasia. The *Poa* genus is classified in the *Poeae* tribe (96), with diversification being expansive to ~500 known species. Global distribution studies reveal the *Poa* species have the highest relative specific differentiation in regions of high latitude and high altitude (47). *Poa pratensis* evolved and diversified in geographical regions characterized by a cool continental climate and probably in fringe open meadow areas interspersed in forests on relatively fertile soils. *Poa trivialis* is thought to have evolved in the cooler, northerly, forest regions of Europe under relatively wet climatic and soil conditions on clayey soils. The diploid *P. trivialis* ssp. *trivialis* has naturalized in North America (6). The bluegrass species are propagated by seed and many vegetatively by lateral stems such as rhizomes or stolons.

Poa annua is a ubiquitous, cosmopolitan grass that is very well adapted to anthropological habitats. It is considered an annual weed in some turfgrass ecosystems, especially the bunch-type annual bluegrass (*P. annua* var. *annua*). The stoloniferous perennial creeping bluegrass (*P. annua* var. *reptans*) tends to dominate when under a long-term managed component in certain turfgrass situations (11). *P. annua* is thought to have arisen from hybridization between *P. infirma* and *P. supina* (6).

A biogeographic migration pattern similar to the fine-leaf fescues through Asia via the Bering Land Bridge to North America (72) is probable for the bluegrasses (the previous figure), especially as many *Poa* species have similar temperature responses including good to excellent freeze stress tolerance (2, 9). *P. pratensis* has a circumpolar distribution (85), with a capability to develop numerous ecotypes to thrive in distinctly different habitats (93). More than 60 species of *Poa* are now considered native to North America (6). *Poa* has a high species frequency percentage in regions with a July midsummer isotherm below 75 °F (24 °C; 47). Chloroplast-DNA phylogenetics and cladistic analyses for biogeographical events revealed at least six groups of *Poa* species had independently colonized secondary-centers-of-diversification in North America, including Kentucky bluegrass and Texas bluegrass. The *Poa* are found extensively in the western mountain ranges of North America (41), and the *P. pratensis* complex has been termed native to the Rocky Mountains (50). The extant Kentucky bluegrasses of the eastern United States may be of an integrated ancestry with postglacial reinvasion from the west or south and with European colonist introductions (22).

The *Poa* species utilized in the United States as turfgrasses resulting from diversification or migration include Kentucky bluegrass (*P. pratensis*), rough bluegrass (*P. trivialis*), and Texas bluegrass (*P. arachnifera*) (10, 11). *P. pratensis* is a facultative apomictic and highly polyploid. Kentucky bluegrass is widely used as a turfgrass in the cool climatic regions of the United States under minimum shade, while rough bluegrass is best adapted in moist-to-wet, shaded conditions in the cooler portions of the cool-humid climatic region (10, 11).

Texas bluegrass is unique among the turfgrass-type *Poa* species in adaptation to warm climates and is native to the southern Great Plains (85). It is best adapted to moist, fertile soils and bottom lands. Available evidence suggests *P. arachnifera* or its ancestor migrated north from South America, as two very close sister species occur there (85). Recently, hybrid cultivars of *P. pratensis* × *P. arachnifera* have been developed for turfgrass use.

To conclude, a number of *Poa* species are North American natives, including *P. arachnifera* and possibly *P. pratensis*. Biogeographical analyses and molecular phylogenetic research with a range of *Poa* species similar to that for the fine-leaf fescues is needed for proper documentation, which should include *P. arachnifera*, *P. annua*, *P. pratensis*, and *P. trivialis*.

Bentgrass. Based on phenotypic evidence, the early diverged species center-of-origination for the bentgrasses (*Agrostis* species) was in south central Mediterranean Eurasia (84). The *Agrostis* genus is of ancient origin and has been morphologically classified in the Poeae, with ~200 known species (6, 96, 97), including at least 21 native North American species. The *Agrostis* probably diversified in an environment characterized by fringe forest zones of partial shade and minimal moisture stress on poorly drained clayey soils. Mountainous regions favored the origination of the polyploid forms of *Agrostis* (84), with most adapted to narrow habitats (7). Diversification resulted in three bentgrass turfgrasses, which are now utilized in the United States, including colonial bentgrass (*A. capillaris*, syn. *A. tenuis*), creeping bentgrass (*A. stolonifera*, syn. *A. palustris*), and velvet bentgrass (*A. canina*) (10, 11).

Low levels of interspecific hybridization may occur between *A. stolonifera* and both *A. capillaris* and *A. castellana* but are not likely for *A. canina* (13). Allotetraploids *A. capillaris* and *A. stolonifera* have the A_2 subgenome in common (49, 85). Phylogenetic analyses and divergence time estimates indicate these two species diverged from a common ancestor ~2.2 mya (76). Cluster analysis shows the primary cultivated species *A. capillaris* and *A. stolonifera* are more similar to each other than the diploid *A. canina* is to either (4). *Agrostis clavata* has been suggested as a progenitor of *A. capillaris* and *A. stolonifera* based on MITE-display marker clustering and ploidy prediction (3). Also genetic marker studies suggest *A. canina* may have contributed to the evolution of *A. stolonifera* (74). Turfgrass-type *Agrostis* species are propagated primarily by seed, being largely cross-pollinated. *A. stolonifera* also can be propagated vegetatively by stolons.

A mountainous migration pattern of *Agrostis* species from Europe to North America (72) probably occurred similar to that documented for the fine-leaf fescues (the previous figure). The turfgrass-type *Agrostis* have good-to-excellent freeze stress tolerance (9, 11). Northern salt marsh and lakeside populations of *A. stolonifera* are considered native to North America (6). The species is well adapted to temporary flooding and wet soil sites, probably reflecting the conditions under which creeping bentgrass originated in Eurasia. The extant bentgrass cultivars tolerate close, frequent mowing while sustaining good shoot density (10, 11).

The turfgrass-type *Agrostis* species have been viewed by some as being introduced to the United States due to oceanic separation from the early center-of-origination. By inference from recently available evidence, specifically from the fine-leaf fescue research, the turfgrass-type bentgrasses would be considered possible natives to the United States. It is hoped that needed, clarifying biogeographical analyses and molecular phylogenetic investigations will be accomplished.

Ryegrass. The early species center-of-origination for the ryegrasses (*Lolium* species) based on genetic diversity was the open, temperate, southern Eurasian region along the Mediterranean Sea. The *Lolium* genus is classified in the Poeae (96), with diversification limited to ~five known species that occurred starting ~two mya (6, 48). It has been proposed that the *Lolium* genus and *Schedonorus pratensis* evolved from a common ancestor (74, 79) and split ~2.8 mya (48). Extant *Lolium* turfgrass cultivars tend to have limited shade adaptation, and their center-of-origin and diversification probably were characterized by open areas.

Propagation is by seed being cross-pollinated, with most mowed turfgrass cultivars being bunch-types. The turfgrass-type species used in the United States include perennial ryegrass (*L. perenne*) and annual ryegrass (*L. multiflorum*) (10, 11). They are interfertile and naturally intergrade (6). The turfgrass-type perennial ryegrass is utilized in the warmer portion of the cool climatic region of the United States and is utilized for winter overseeding of warm-season turfgrasses. They are best adapted to mild winters and cool-moist summers and to fertile soils (10, 11).

Migration of the *Lolium* was south into northern Africa, especially in mountainous regions; into northern Europe; and also eastward into Asia and India (the previous figure). Transcontinental pre-Holocene migration research is lacking. Lack of freeze stress tolerance (9, 10, 11) may have prevented migration to North America via the Bering Land Bridge.

Thus based on the available phenotypic evidence the ryegrasses utilized for turfgrass purposes in the United States are probably naturalized grasses first introduced by human transatlantic movement in the mid-1600s for forage and pasture utilization in domestic animal agriculture. Confirmation is needed via biogeographical analyses and molecular phylogenetic research.

Tall Fescue. The early species center-of-origination for broad-leaf *Schedonorus* was in mild-temperate Eurasia, separating from fine-leaf *Festuca* during the Miocene ~15 mya (48, 93). The *Schedonorus* genus is classified in the *Schedonorus* subgenus (96, 97), with the turfgrass-type being allopolyploid tall fescue (*Schedonorus arundinaceus*, syn. *Festuca arundinacea*; 6). Tall fescue probably diversified under summers characterized by higher temperatures and less rainfall. *S. arundinaceus* is younger in lineage than the turfgrass-type fine-leaf *Festuca* and based on cluster analysis is closely related to *Schedonorus pratensis* (6, 99). Biological clock estimates indicate the Asian American clade of broad-leaf *Schedonorus* apparently diverged from the fine-leaf *Festuca* clade ~12 mya (48). Certain hexaploid turfgrass-type tall fescues, such as Rebel II and Bonanza, are clustered in a subgroup according to cluster analysis (99). Propagation is by seed being primarily cross-pollinated, with tall fescue usually behaving as a bunch-type when regularly mowed as a turfgrass.

Migration of *S. arundinaceus* from Europe extended eastward into Asia and south into Africa, especially via the mountainous regions (the previous figure). Potential transcontinental migration to the United States prior to Holocene human activities has not been properly investigated. Northward dispersal in Europe may have been limited due to a lack of freeze stress tolerance (9, 10, 11). Tall fescue is best adapted for turfgrass use in the transitional and warmer-cool climatic regions of the United States under regular mowing at moderately high heights (10, 11).

Based on the limited available information, tall fescue is considered a naturalized species thought to have been intentionally introduced into the United States by transcontinental human activities in the mid-1800s for forage and pasture use in domestic animal production. Biogeographical analyses and molecular phylogenetic research are needed for confirmation.

WARM-SEASON TURFGRASSES

The primitive ancestral center-of-origination for the warm-season grasses is thought to have been Gondwanan Africa (28, 72), with West Gondwanan migration to South America and East Gondwanan migration to India and to Australia via Antarctica. Key synapomorphies include Kranz leaf anatomy and chloridoid bicellular microhairs on the leaf epidermis (55). The C_4 photosynthetic pathway characteristic of warm-season grasses evolved in Africa from C_3 ancestral grasses at a later time in the earth's history and involved multiple

TABLE 1.2. A REPRESENTATION OF KEY EVENTS IN GRASS ORIGIN AND EVOLUTION DURING THE LATTER PORTION OF EARTH GEOLOGICAL HISTORY, BASED ON CURRENT KNOWLEDGE.

Epoch	*Cumulative time* (millions of years)	*Approximate duration* (millions of years)	*Major plant/grass events*	*Related events*
Lower/Early Cretaceous	99.6–145.5	45.9	• Extinction of many species of African flora	• South America isolated from Africa • India, Antarctica, and Australia separate from Africa
Upper/Late Cretaceous	65.5–99.6	34.1	• Evolution of nongrass flowering plants • Reports of grass pollen fossils	• Climate becomes mild to tropical
Paleocene	55.8–65.5	9.7	• Definitive reports of grass pollen fossils • Flowering plants become more successful • Extant plants appear	• North America–Greenland isolated from Europe • Seas retreat and climate tropical • Rapid spread of mammals • Early herbivorous mammals • India contacts Asia
Eocene	33.9–55.8	21.9	• Primitive C_3 cool-season grasses arise (*Poaceae*) • Increasing evidence of grass pollen fossils and macrofossils	• Climate warms and seas flood the land • Australia separates from Antarctica • Alps and Rocky mountains form • Modern mammals emerge
Oligocene	23.0–33.9	10.9	• Abundant grass pollen fossils • Diversification of C_3 grass species • Trees and grasses cover much of land • Migration of C_3 grasses to Africa • Primitive C_4 warm-season grasses evolve	• Climate cools and becomes dryer • Widespread forests • Atmospheric CO_2 levels lowered • Increased fire frequency
Miocene	5.3–23.0	17.7	• Fine-leaf fescues emerge in Europe (*Festuca*) • Grasses become ubiquitous globally • Large temperate grasslands form globally	• Sea level falls, Himalayas and Andes rise • *Bovidae* evolve in Europe and Asia • Golden age of mammals, with most diversity

TABLE 1.2. *(continued)*

Epoch	Cumulative time* (millions of years)	Approximate duration (millions of years)	Major plant/grass events	Related events
Pliocene	2.6–5.3	2.7	• Expansion of grasslands and grasses emerge on US Great Plains • Savannas replace many forests in Africa	• Climate cools and becomes dryer • Continents in present position • *Bovidae* invade Africa
Pleistocene	0.01–2.6	2.59	• Shift in grass species of US Great Plains	• Panamanian land bridge forms • Last ice age covers northern lands and sea level falls • Humans evolve & develop hunting skills • Withdrawal of ice sheets/glaciers of ice age
Holocene (ongoing)	0–0.01	0.01	• Present flora and fauna • Turfgrass use developed in last 800 years • Natural diversification/selection of cultivars in mowed turfgrass areas	• Climate warms and sea level rises • Development of agriculture by humans • Rise of human civilization

*The geologic time scale is from the Geological Society of America (2009).

biochemical and histological C_4 events (16, 17, 32, 51, 82). The first evidence of C_4 photosynthesis in the Chloridoideae occurred ~33 to 30 mya (15, 17).

Fossil records indicate the ancestral species of warm-season Poaceae originated in West Gondwana, now Africa, mainly in savanna ecosystems (72). Phylogenetic analyses suggest the monophyletic Panicoideae and Chloridoideae are derived from the ancestral polyphyletic Arundinoideae (32, 52). The PACCMAD clade exhibited phylogenetic diversifications during the mid-Miocene through late Quaternary (16). Molecular phylogenetic studies reveal the Chloridoideae may date as early as the late Oligocene ~31 mya (16, 17), but major global migration may not have occurred until the Pliocene (67). Species of C_4 turfgrass types tend to have multiple secondary-centers-of-origination in geographical regions closer to the equator under warm climatic conditions. Twenty-four shifts in phylogenetic diversification rates have been identified among the C_4 grass lineages, with 19 during the Miocene and 5 during the Pliocene (16).

Most warm-season turfgrasses are taxonomically classified within either the Chloridoideae or Panicoideae (96). The Chloridoideae has eight tribes, including Cynodonteae and Zoysieae, while the Panicoideae encompasses two tribes, including the Paniceae. Extant warm-season turfgrasses occur mostly in the warm, tropical, and subtropical climates, with an optimum temperature range of 80 to 95 °F (27 to 35 °C). Most are prone to chill stress in the 52 to 58 °F (11 to 15 °C) range, with resultant low temperature discoloration and winter dormancy (10, 11, 12). Extant Chloridoideae typically are adapted to survive at mean temperatures above 50 °F (10 °C; 64) for the coldest months. Locations and dating for species centers-of-origin and geophysical migrations are increasingly based on phenotypic, molecular phylogenetic, and allied cladistic assessments. Fossil types of evidence are limited as fossilization was restricted in dry savanna climates where warm-season grasses diversified. Geophysical characteristics of extant warm-season turfgrasses are addressed for the major species utilized in the United States, including bermudagrass, zoysiagrass, St Augustinegrass, centipedegrass, seashore paspalum (fine-leaf), carpetgrass, American buffalograss, and gramagrass.

Bermudagrass. The primary center-of-origination for the bermudagrasses (*Cynodon* species) is in open, southeastern Africa (38). Subsequent early diversification was associated with large herds of herbivorous, African, hoofed mammals. Based on biosystematic and molecular phylogenetic research, the *Cynodon* secondary-centers-of-origin or centers-of-diversification include South Africa, India, Afghanistan, and Australia (45).

The ubiquitous, cosmopolitan *C. dactylon* is now found throughout the warm-temperate, subtropical, and tropical climatic regions of the world (45). The *Cynodon* genus is classified in the Cynodonteae and Chloridinae, with limited diversification of ~9 known species and 10 varieties (6, 46, 67, 96). Cluster analysis of DNA amplification products has confirmed this earlier taxonomic classification. Evidently the genetic diversity of *C. dactylon* ssp. *dactylon* was generated internally (45). It thrives under soil disturbance on sunny sites. Low-growing *Cynodon* species used as turfgrasses in the United States include dactylon bermudagrass (*C. dactylon* ssp. *dactylon*), a sterile hybrid bermudagrass (*C. dactylon* ×*C. transvaalensis*), and a natural sterile triploid hybrid (*Cynodon* ×*magennisii*) (10, 11). Molecular phylogenetic analysis has identified cluster groups broadly based on relative leaf width (20).

The presumed geophysical migration of the ancestral grass was via the East Gondwanan land connection. Subsequent pre-Holocene migration was outward in Africa and into southern Europe, Asia, and Australia (the previous figure). Propagation is vegetatively by stolons and rhizomes and by seed for *C. dactylon* being cross-pollinated. The climate under which bermudagrass species evolved was subtropical with typically dry summers that resulted in a deep extensive root system, low evapotranspiration rate, extensive lateral stem development, and superior drought resistance. The modern *Cynodon* turfgrass cultivars have retained these water-conserving characteristics, but most lack shade adaptation (10, 11).

Unintended transatlantic human movement of *C. dactylon* via ships of the Spanish conquistadors probably occurred in the 1600s to both North and South America (100), followed by further dispersal and naturalization and intraspecies diversification in the warm climatic regions. The dactylon bermudagrass (*C. dactylon* ssp. *dactylon*) introduced into the United States was readily dispersed and became naturalized. The questions are when was it introduced, by what means, and were extant native *C. dactylon* already present in the United States as a result of earlier natural migration from Africa, possibly via South America. In contrast, African bermudagrass (*C. transvaalensis*) was probably a human-introduced species from South Africa to the United States, with minimal dispersal and naturalization. Biogeographical analyses and molecular phylogenetic research are needed to clarify these dispersal aspects of both species.

Zoysiagrass. Phenotypically, the zoysiagrasses (*Zoysia* species) are recognized to have a postancestral center-of-origin under warm, tropical, open conditions in southeast China. Their primitive ancestral grass center-of-origination probably occurred in Africa, where it is assumed they subsequently became extinct during a severe arid period (72). The *Zoysia* genus is classified in the Cynodonteae and Zoysiinae (68), with diversification being relatively small consisting of ~11 known species (7, 67, 96). Shoots of the turfgrass-type *Zoysia* species possess a high silica content, which suggests that herbivorous grazing mammals may not have been as significant in their diversification and migration. Based on molecular phylogenetic analyses, the *Zoysia* genus arose later than the *Cynodon* (15). *Zoysia* species now utilized for turfgrass purposes in the United States include Japanese zoysiagrass (*Z. japonica*), manila zoysiagrass (*Z. matrella* var. *matrella*), and, to a lesser extent, mascarene zoysiagrass (*Z. pacifica*, syn. *Z. matrella* var. *tenuifolia*) (7, 10, 11). These three readily intercross via pollination suggesting a close genetic relationship. Also used as a turfgrass is hybrid zoysiagrass (*Z. japonica* × *Z. pacifica*), cultivar Emerald. Cluster analyses of RAPD revealed groupings that differ from the classical taxonomic species classification (98).

East Gondwanan migration of *Zoysia* primitive ancestors probably occurred from Africa to India, although a West Gondwanan migration of the Zoysiinae to South America cannot be ruled out entirely without additional investigations (67). Following the separation of India from Africa and the eventual contact of India with Asia, the *Zoysia* ancestors migrated to Southeast Asia (72). Pre-Holocene migration and secondary-centers-of-origin or centers-of-diversification for *Zoysia* species appear phenotypically to have been limited to the Australasian region, extending north up into Japan and Korea and south into northeast Australia and to the south Pacific Islands, extending to Polynesia (the previous figure). The sandy coastal regions and salt marshes of the southwest Pacific have concentrations of *Z. matrella*, including the Malay Peninsula, New Guinea, the Philippines, and southeast China (39). Propagation has been primarily vegetative by rhizomes and stolons, with limited seed availability for *Z. japonica*. Seed germination among *Zoysia* species ranges from poor to nonexistent, which may have impaired their migration and diversification.

Zoysiagrasses are used in the United States primarily for turfgrass purposes in the humid-subhumid transitional zone between the cool and warm climatic regions. The *Zoysia* species diversified in open habitats under a tropical climate characterized by the hot, humid summers of southeast China with a large amount of precipitation distributed throughout the year. As a consequence, the zoysiagrasses evolved with a short root system that continues to be expressed in modern turfgrass-type cultivars and contributes to a significant reduction in dehydration avoidance and results in less drought resistance than bermudagrass (10, 11). Zoysiagrass cultivars used as turfgrasses are characterized by a very slow establishment rate.

Japanese zoysiagrass was present in the United States by the late 1800s. It probably was intentionally introduced by people from southeast China and, subsequently, has naturalized. Biogeographical analyses and molecular phylogenetic research are needed to clarify the migratory or introduced status.

St Augustinegrass. Evidently, the *Stenotaphrum* genus is a paleotropic offshoot of *Paspalidium* (79). The postancestral center-of-origination for the St Augustinegrasses (*Stenotaphrum* species), based on phenotypic diversity, suggests coastal southeastern Africa, since six of the seven known species occur in that region (79). The extensive occurrence of *S. secundatum* in the tropical and subtropical Gulf of Mexico-Caribbean regions of the Americas suggests a secondary-center-of-origin or secondary-center-of-diversification, possibly via the West Gondwanan bridge from Africa and via South America. The *Stenotaphrum* genus is classified in the Paniceae and Setariinae, with diversification limited to ~seven known species (6, 96).

The *Stenotaphrum* species most probably migrated outward into tropical regions around the primary and secondary species centers-of-origin or centers-of-diversification, especially in sandy coastal and stream swale regions (79). Most propagation is vegetative by stolons, although seed of some genotypes are viable being cross-pollinated, while others are self-sterile (79). The inflorescence is adapted for subregional island dispersal of seed by ocean currents (79). Transcontinental migration to Southeast Asia and northeast Australia probably occurred prior to human movements via the coastal areas along the Indian Ocean and then southwest via tropical oceanic island-hopping (79; the previous figure).

One species dominates turfgrass usage in the United States, *S. secundatum* (10, 11). Floral stigmas and stolon internodes of the Florida-centered, coarse-textured *S. secundatum* group are purple, while the narrower-leafed, Texas-centered Gulf Coast group have yellow stigmas and green stolon internodes (60). The purple stigma types are triploids and tetraploids both forming seeds that are highly sterile (60). This would have limited their migration primarily to Florida. The yellow stigma types are diploids that form seeds with a reasonable germination capability and considerable heterozygosity. This allowed selection for improved freeze stress tolerance that resulted in wider adaptation and migration along the Gulf Coast and northward.

The *S. secundatum* species of St Augustinegrass is used for turfgrass purposes in the warm, tropical, and subtropical southeastern and south central United States, extending from Florida to the southern coastal region of Texas (40). It is best adapted to sandy coastal sites and to moist, well-drained, fertile soils plus a relatively high mowing height (10, 11). The yellow-stigma Gulf Coast group has good shade adaptation, whereas the purple-stigma group typically has both poor shade adaptation and poor freeze stress tolerance.

Phenotypic evidence indicates St. Augustinegrass is a naturally diversified species that has become native to the United States with two secondary-centers-of-origin or secondary-centers-of-diversification. Biogeographical analyses and molecular phylogenetic research are needed to determine the early intercontinental migration routing and timing.

Centipedegrass. An East Gondwanan migration of the primitive ancestors to the *Eremochloa* species may have occurred from Africa to India then from India to Asia, following the separation of India from Africa and its eventual contact with Asia (72). The postancestral center-of-origination for centipedegrass (*Eremochloa* species), based on phenotypic genetic diversity, is open, temperate, and subtropical Southeast Asia (40). The *Eremochloa* genus is classified in the Paniceae and Setariinae, with diversification limited to ~11 known species (6, 96). *E. ophiuroides* is the only species of centipedegrass now utilized as a low-growing turfgrass species in the United States (10, 11).

Migration of *Eremochloa* species appears to have been outward into tropical southeast and central China (44) and to Southeast Asia and to Australia (the previous figure). Propagation is by seed, being asexually reproducing diploids with some self-incompatibility and vegetatively by stolons (44).

The *Eremochloa* species evolved in an open tropical climate with hot, humid summers and a large amount of precipitation distributed throughout much of the year. As a consequence, centipedegrasses evolved with a short root system that continues to be expressed in modern turfgrass-type cultivars (11). Centipedegrass as utilized for turfgrass purposes has a very low nitrogen requirement (10, 11).

Government records indicate one species of centipedegrass was intentionally introduced into the United States in 1916. Whether there were natural intercontinental migrations to the Americas before Holocene human activities need investigation via biogeographical analyses and molecular phylogenetic research. Until needed research is accomplished, *E. ophiuroides* should be considered as introduced into the United States. It probably became naturalized to the humid and subhumid southeastern and south central regions of the United States (40).

Seashore Paspalum. The primitive ancestral center-of-origination for the *Paspalums* is thought to be West Gondwanan African, with migration to South America prior to the continental plate separation of Pangaea. Thus South America was a key center-of-origin, typically in brackish areas and salt marshes. However, the species center-of-origin for certain fine-leaf, turfgrass-type *Paspalum* species developed in America is tropical and subtropical southern Africa, based on phenotypic evidence. Subsequent secondary-diversification-centers occured in coastal South America and southeastern North America, based on molecular phylogenetic research and cluster analysis (59). The *Paspalum* genus is classified in the Paniceae and Setariinae, with ~320 known species (6, 96) and 7 cluster groups identified (59). There are very coarse, intermediate, and fine-leaf species and ecotypes. The fine-leaf *P. vaginatum* is the only species group utilized as a turfgrass in the United States (36).

Pre-Holocene migration of the fine-leaf *P. vaginatum* expanded primarily around the tropical and subtropical coastal and stream swales of southern Africa, which were characterized by brackish marsh ecosystems that included tidal flooding (the previous figure). Propagation is vegetative via rhizomes and stolons and also can be by seed being diploids with cross-pollination and considerable self-sterility (21). It is most valuable as a turfgrass for use on soils with high salt levels.

Early unintended introduction of fine-leaf *P. vaginatum* into the United States probably was by human activity via ships to southeastern Atlantic coastal ports during the 1700s (36), as well as into the coastal West Indies. This ecotype has become a naturalized species, persisting primarily in the brackish, sandy coastal regions of southeastern and south central United States.

Carpetgrass. A postancestral, distinct secondary-center-of-origin for the carpetgrasses (*Axonopus* species) is phenotypically considered to have been tropical Brazil and to have extended into the Caribbean region of the Americas (14, 40, 67). Their primitive ancestor grass and related geophysical aspects may have involved origination in Africa and migration to east and central Brazil in South America (14). The *Axonopus* genus is classified in the Paniceae and Setariinae, with significant diversification indicated by ~100 species (6). The two low-growing, turfgrass-type species in use are common carpetgrass (*A. fissifolius*, syn. *A. affinis*) and tropical or broadleaf carpetgrass (*A. compressus*; 10, 11). Both are adapted to open sites and wet soils that are periodically flooded.

Pre-Holocene migration of the *Axonopus* species presumably was outward regionally from the African ancestral center-of-origination and then from secondary-centers-of-origin in South America, eventually reaching Southeast Asia and the southeastern United States (67; the previous figure). Propagation is vegetatively by stolons and by seed being diploids with cross-pollination and some self-sterility (21). Phenotypic observations reveal a widespread, extant global distribution in the tropical-subtropical climatic regions of the Americas, Southeast Asia and the Pacific Islands, and Africa. In terms of turfgrass use, common carpetgrass is found in the southeastern United States with limited use, while tropical carpetgrass is widely used in Southeast Asia. Common carpetgrass is best adapted to acidic, moist-to-wet soils (10, 11).

Carpetgrass is native to the southeastern United States and the American tropics and subtropics (6, 41). Biogeographical analyses and molecular phylogenetic research would elucidate the early diversification and migration of carpetgrass.

SEMIARID WARM- AND COOL-SEASON TURFGRASSES

The US Great Plains extends from the Rocky Mountains eastward to the forested region and from the Gulf of Mexico northward to the coniferous forests of central Montana. This extensive, temperate grassland has exhibited radical, multiple changes in flora. Fossil evidence indicates species in the Stipeae, such as the now infrequently found *Piptochaetium*, occurred in the Pleiocene ~23 to 27 mya (89), followed by species of *Hordeum* and *Bromus* (87).

The present grasslands of the North American Great Plains are relatively recent ecosystems that emerged like a savanna during the Miocene-Pliocene transition ~5 to 7 mya (5). Aridity increased during the Pliocene, including a distinct drought season that resulted in a dry, flammable biomass of grass. This natural fuel, combined with the smooth to rolling topography of the plains and active winds, favored fires. Lightning ignited the fires in earlier times, and later, humans employed fire to harvest food, drive animals, reduce pests, clear land, provide openings for travel, and employ both defensive and offensive strategies. An additional biotic factor was the emergence of numerous browsing and grazing mammals on the Great Plains during the late Pleistocene.

Wheatgrass. Subsequently, the tall-growing wheatgrasses (*Agropyron* and *Pascopyrum* species) migrated from the western mountain areas into the temperate, dry northern Great Plains (87). Previously, they had migrated via the Bering Land Bridge from the open Euro-Mediterranean region to Asiatic China and eastward. They were key components of what has been called the tallgrass prairie (5, 40). Both genera are C_3 cool-season grasses classified in the Triticeae (96). There are ~15 *Agropyron* species and 1 *Pascopyrum* species (6, 96, 97). Included were crested wheatgrass (*A. cristatum*) and western wheatgrass (*P. smithii*), which have been used as minimal maintenance, unirrigated turfgrasses in the northern portion of the semiarid region (10, 11). Western wheatgrass is best adapted to alkaline, mesic, meadow-like conditions (7).

American Buffalograss and Gramagrass. Another major shift in grass species dominating the Great Plains occurred during the Pleistocene (87). The environmental causes of the shift to *Buchloe* and *Bouteloua* species probably were related to the northern Great Plains being covered with glacial ice plus soil changes in the southern portions related to moraine sands and silts deposited by flooding from large rivers flowing from the melting glaciers. Also there was a bison invasion ~0.5 to 1.2 mya (61), an expansion that formed large dominating grazing herds where horses had formerly been prominent (24).

Several key warm-season, low-growing grasses of *Buchloe* and *Bouteloua* species emerged during the Pleistocene. Their ancestral grass origin may have been West Gondwanan Africa, with geophysical aspects involving continental migration to South America prior to the significant separation of Pangaea (72). Subsequent migration to Central America, where the two low-growing semiarid genera evolved within a relatively comparable time frame (15), presumably occurred. The two genera contributed to what has been termed the shortgrass prairie of the Great Plains (5, 40).

American buffalograss (*Buchloe dactyloides*) evolved in a secondary-diversification-center in Mexico relatively recently (87) and then migrated northward into the US Great Plains (the previous figure). *Buchloe* has only one known species and is classified in the Chlorideae (6, 96). It has C_4 physiology, is usually dioecious, and is best adapted to the open, upland, semiarid western third of the Great Plains. Propagation is by seed and vegetatively by stolons.

The gramagrasses (*Bouteloua* species) evolved at a secondary-diversification-center in open northern Mexico. Included were blue gramagrass (*B. gracilis*) and the apomictic side-oats gramagrass (*B. curtipendula*), which became major components of the shortgrass prairie. The *Bouteloua* genus is classified in the Cynodonteae and Boutelouinae, with ~40 known species and a C_4 physiology (6, 67, 68, 96). They also exhibited a northward migration into the United States (the previous figure). Propagation is by seed and vegetatively by rhizomes.

The three grass species have been utilized as turfgrasses primarily under unirrigated conditions in the Great Plains region of the United States (10, 11). They are best adapted to semiarid climates where the precipitation is not sufficient to support most tree and shrub species (10, 11). They tend to become dormant during extensive summer droughts. Their drought survival is achieved primarily by a dormancy escape mechanism rather than an inherent dehydration tolerance component of drought resistance. These semiarid, warm-season grasses have a relatively moderate to low shoot density.

Perennial turfgrasses with a higher water requirement are less favored in the unirrigated, semiarid environments of the Great Plains. However, the high density, more aggressive turfgrass species, such as Kentucky bluegrass and bermudagrass, tend to dominate over these three species in higher precipitation regions (10, 11). Also, the semiarid environments in which these genera evolved resulted in genotypes that were not subjected to pathogen pressures common to humid regions and consequently did not result in strong selection for resistance to such diseases.

American buffalograss and blue and side-oat gramagrasses were originally naturalized species to the United States that may be termed native species.

SUMMARY

The characterization of native or naturalized status for certain turfgrasses in North America is now possibly based on paleobotanical and stable carbon isotopes studies plus molecular phylogenetic and biographical analyses. The current state of knowledge is summarized in table 1.3. Where confirming modern research is lacking for some turfgrass species, the species dating placement have been based on the more thoroughly researched species possessing similar centers-of-origin or centers-of-diversification, geophysical-environmental adaptation, and taxonomy. It is hoped that the needed molecular phylogenetic research for a number of turfgrass species will be completed in the near future.

Certain groups advocate that only plants and grasses native to a particular region should be grown in that landscape. A premise for this concept is that native grasses have superior adaptation compared to any introduced, naturalized grasses. However, there are naturalized grasses as well adapted or even better adapted to certain climatic-soil regions of the world than native grasses. And, actually, neither category is genetically stable—rather, they continue to diversify.

The term native has been used as an arbitrary concept to access whether a species belongs naturally at a specific site. However, it is a questionable approach to separate a natural process from the presence and influence of human activity. While it is appropriate and fortunate that some individuals are interested in preserving and propagating native grass stands in various climatic regions, it is unwise and even detrimental to promote laws or other means that force all individuals to use only native grasses in landscape plantings throughout a city or region. Advocating only the use of native plants in what is essentially an artificial nonnative environment of urban structures, concrete, asphalt, and disturbed lands is not logical. Many plants arbitrarily identified as native to a local natural environment may not be adapted to a nearby urban ecosystem, while certain naturalized plant species may be better adapted and more capable of enhancing the functional human quality of life in urban areas.

TABLE 1.3. SUMMARY OF NATIVE, MOSTLY SECONDARY CENTERS, AND NATURALIZED CATEGORIES OF 25 TURFGRASSES FOUND IN NORTH AMERICA, BASED ON CURRENT KNOWLEDGE.

Native/naturalized categories	Grass genus	Turfgrasses common name
Native, based on biogeographical, phenotypic and phylogenetic studies	*Festuca rubra*	Red fescues
	" *trachyphylla*	Hard fescue
	" *ovina*	Sheep fescue
Native, based on phenotypic and genetic assessments*	*Axonopus compressus*	Broadleaf carpetgrass
	" *fissifolius*	Common carpetgrass
	Bouteloua curtipendula	Side-oats gramagrass
	" *gracilis*	Blue gramagrass
	Buchloe dactyloides	American buffalograss
	Poa arachnifera	Texas bluegrass
	Stenotaphrum secundatum	St Augustinegrass
Possible native, based on phenotypic assessments and allied-species biogeographical, phylogenetic studies*	*Agrostis capillaris*	Colonial bentgrass
	" *stolonifera*	Creeping bentgrass
	" *canina*	Velvet bentgrass
	Poa pratensis	Kentucky bluegrass
Possible naturalized, following anthropological introduction*	*Eremochloa ophiuroides*	Centipedegrass
	Lolium multiflorum	Annual ryegrass
	" *perenne*	Perennial ryegrass
	Paspalum vaginatum	Seashore paspalum
	Poa trivialis	Rough bluegrass
	Schedonorus arundinaceus	Tall fescue
	Zoysia japonica	Japanese zoysiagrass
	" *matrella* ssp. *matrella*	Manila zoysiagrass
	" *pacifica*	Mascarenegrass
Undocumented, needs research	*Cynodon dactylon* ssp. *dactylon*	Dactylon bermudagrass
	Poa annua	Annual bluegrass

*Needs further biogeographical and molecular phylogenetic research to confirm.

REFERENCES

1. Aiken, S.G., M.J. Dallwitz, C.L. McJannet, and L.L. Consaul. 1996 (onward). *Festuca* of North America: Descriptions, illustrations, identification and information retrieval. Inst. of Bot., Chinese Acad. of Sci. http://delta-intkey.com/festuca.

2. Aiken, S.G., M.J. Dallwitz, C.L. McJannet, and L.L. Consaul. 1997. Biodiversity among *Festuca* (Poaceae) in North America: Diagnostic evidence from DELTA and clustering programs, and an INTKEY package for interactive, illustrated identification and information retrieval. Can. J. of Bot. 75:1527–1555.

3. Amundsen, K.L. 2009. Origin and evolution of cultivated *Agrostis* spp. Ph.D. diss., George Mason Univ., Fairfax, Virginia, USA.

4. Amundsen, K., and S. Warnke. 2011. Species relationships in the genus *Agrostis* based on flow cytometry and MITE-display molecular markers. Crop Sci. 51:1224–1231.

5. Axelrod, D.I. 1985. Rise of the grassland biome, central North America. Bot. Rev. 5(2):163–201.

6. Barkworth, M.E., K.M. Capels, S. Long, and M.B. Piep, editors. 2003. Flora of North America, Magnoliophyta: Commelinidae (in part): Poaceae, Part 2, vol. 25. Oxford Univ. Press, New York, USA.

7. Barkworth, M.E., L.K. Anderton, K.M. Capels, S. Long, and M.B. Piep, editors. 2007. Manual of Grasses for North America. Intermountain Herbarium and Utah State Univ. Press, Logan, Utah, USA.

8. Barnard, C., and O.H. Frankel. 1964. Grass, grazing animals, and man in historic perspective. In: C. Barnard, editor, Grasses & Grasslands. Macmillan and Co. Ltd., London, UK. p. 1–12.

9. Beard, J.B. 1966. Direct low temperature injury of nineteen turfgrasses. Quarterly Bull. of the Michigan Agric. Experiment Stn. 48(3):377–383.

10. Beard, J.B. 1973. Turfgrass: Science and Culture. Prentice-Hall Inc., Englewood Cliffs, New Jersey, USA.

11. Beard, J.B, and H.J. Beard. 2005. Beard's Turfgrass Encyclopedia. Michigan State Univ. Press, East Lansing, Michigan, USA.

12. Beard, J.B, and P.O. Cookingham. 2007. William J. Beal—Pioneer applied botanical scientist and research society builder. Agron. J. 99:1180–1187.

13. Belanger, F.C., T.R. Meagher, P.R. Day, K. Plumley, and W.A. Meyer. 2003. Interspecific hybridization between *Agrostis stolonifera* and related *Agrostis* under field conditions. Crop Sci. 43:240–246.

14. Black, G.A. 1963. Grasses of the genus *Axonopus*, a taxonomic treatment. Advancing Frontiers of Plant Sci. 5:1–186.

15. Bouchenak-Khelladi, Y., N. Salamin, V. Savolainen, F. Forest, M. van der Bank, M.W. Chase, and T.R. Hodkinson. 2008. Large multi-gene phylogenetic trees of the grasses (Poaceae): Progress toward complete tribal and generic level sampling. Mol. Phylogenetics and Evolution 47:488–505.

16. Bouchenak-Khelladi, Y., G.A. Verboom, T.R. Hodkinson, N. Salamin, O. Francois, G.N. Chonghaile, and V. Savolainen. 2009. The origins and diversification of C_4 grasses and savanna-adapted ungulates. Global Change Biology 15:2397–2417.

17. Bouchenak-Khelladi, Y., G.A. Verboom, V. Savalainen, and T.R. Hodkinson. 2010. Biogeography of the grasses (Poaceae): A phylogenetic approach to reveal evolutionary history in geographical space and geological time. Bot. J. of the Linnean Soc. 162:543–557.

18. Burkart, A. 1975. Evolution of grasses and grasslands in South America. Taxon 24(1):53–66.

19. Burton, G.W., and J.S. Andrews. 1948. Recovery and viability of seed of certain southern grasses and lespedezas passed through the Bovine digestive tract. J. of Agric. Res. 76:95–103.

20. Caetano-Anollés, G., L.M. Callahan, P.E. Williams, K.R. Weaver, and P.M. Gresshoff. 1995. DNA amplification fingerprinting analysis of bermudagrass (*Cynodon*): Genetic relationships between species and interspecific crosses. Theoretical Applied Genet. 91:228–235.

21. Carpenter, J.A. 1958. Production and use of seed in seashore paspalum. J. of the Australian Inst. of Agric. Sci. 24:252–256.

22. Carrier, L., and K.S. Bort. 1916. The history of Kentucky bluegrass and white clover in the United States. J. of the Am. Soc. of Agron. 8(4):256–266.

23. Cerling, T.E., Y. Wang, and J. Quade. 1993. Expansion of C_4 ecosystems as an indicator of global ecological change in the late Miocene. Nat. 361:344–345.

24. Chapman, G.P. 1996. The Biology of Grasses. CAB International, Wallingford, Oxon, UK.

25. Chen, X., S.G. Aiken, M.J. Dallwitz, and P. Bouchard. 2003. Systematic studies of *Festuca* (Poaceae) occurring in China compared with taxa in North America. Can. J. of Bot. 81:1008–1028.

26. Clark, L.G., W. Zhang, and J.F. Wendel. 1995. A phylogeny of the grass family (Poaceae) based on *ndhF* sequence data. Systematic Bot. 20(4):436–460.

27. Clayton, W.D. 1975. Chorology of the genera of *Gramineae*. Kew Bull. 30:111–132.

28. Clayton, W.D. 1981. Evolution and distribution of grasses. Ann. of the Missouri Bot. Garden 68:5–14.

29. Clayton, W.D., R. Cerros-Tlatilpa, M.S. Kinney, M.E. Siqueiros-Delgado, H.L. Bell, M.P. Griffith, and N.F. Refulio-Rodiguez. 2007. Phylogenetics of Chloridoideae (Gramineae): A preliminary study based on nuclear ribosomal internal transcribed spacer and chloroplast *trn*L-F sequences. In: J.T. Columbus, E.A. Fariar, J.M. Porter, L.M. Prince, and M.G. Simpson, editors, Monocots: Comparative Biology and Evolution—Poales. Rancho Santa Ana Bot. Garden, Claremont, California, USA. p. 565–579.

30. Crepet, W.L., and G.D. Feldman. 1991. The earliest remains of grasses in the fossil record. Am. J. of Bot. 78:1010–1014.

31. Darbyshire, S.J., and S.I. Warwick. 1992. Phylogeny of North American *Festuca* (Poaceae) and related genera using chloroplast DNA restriction site variation. Can. J. of Bot. 70:2415–2429.

32. Davis, J.I., and R.J. Soreng. 1993. Phylogenetic structure in the grass family (Poaceae), as determined from chloroplast DNA restriction site variation. Am. J. of Bot. 80:1444–1454.

33. Delcourt, P.A., H.R. Delcourt, C.R. Ison, W.E. Sharp, and K.J. Gremillion. 1998. Prehistoric human use of fire, the eastern agricultural complex, and Appalachian oak-chestnut forests: Paleoecology of Cliff Palace Pond, Kentucky. Am. Antiquity 63(2):263–278.

34. Dore, W.G., and L.C. Raymond. 1942. Pasture studies XXIV: Viable seeds in pasture soil and manure. Sci. Agric. 23(2):69–79.

35. Doyle, J.J., J.I. Davis, R.J. Soreng, D. Garvin, and M.J. Anderson. 1992. Chloroplast DNA inversions and the origin of the grass family (Poaceae). Proc. of the Natl. Acad. of Sci. 89:7722–7726.

36. Duncan, R.R., and R.N. Carrow. 2000. Seashore Paspalum: The Environmental Turfgrass. Ann Arbor Press, Chelsea, Michigan, USA.

37. Ehleringer, J.R., R.F. Sage, L.B. Flanagan, and R.W. Pearcy. 1991. Climate change and the evolution of C_4 photosynthesis. Trends Ecological Evolution 6(3):95–99.

38. Forbes, I., Jr., and G.W. Burton. 1963. Chromosome numbers and meiosis in some *Cynodon* species and hybrids. Crop Sci. 3:75–79.

39. Goudswaard, P.C. 1980. The genus Zoysia (Gramineae) in Malesia. Blemea 26:169–175.

40. Gould, F.W. 1975. The Grasses of Texas. Texas A&M Univ. Press, College Station, Texas, USA.

41. Gould, F.W., and R.B. Shaw. 1983. Grass Systematics. Texas A&M Univ. Press, College Station, Texas, USA.

42. Grass Phylogeny Working Group. 2000. A phylogeny of the grass family (Poaceae) as inferred from eight character sets. In: S.W.L. Jacobs and J.E. Everett, editors, Grasses: Systematics and Evolution. CSIRO, Melbourne, Australia. p. 3–7.

43. Grass Phylogeny Working Group. 2001. Phylogeny and subfamilial classification of the grasses (Poaceae). Ann. of the Missouri Bot. Garden 88(3):373–457.

44. Hanna, W.W., and G.W. Burton. 1978. Cytology, reproductive behavior, and fertility characteristics of centipedegrass. Crop Sci. 18:835–837.

45. Harlan, J.R., and J.M.J. de Wet. 1969. Sources of variation in *Cynodon dactylon* (L.) Pers. Crop Sci. 9:774–778.

46. Harlan, J.R., J.M.J. de Wet, K.M. Rawal, M.R. Felder, and W.L. Richardson. 1970. Cytogenetic studies in *Cynodon* L.C. Rich. (Gramineae). Crop Sci. 10:288–291.

47. Hartley, W. 1961. Studies on the origin, evolution, and distribution of the Gramineae: The genus *Poa* L, vol. 4. Australian J. of Bot. 9:152–161.

48. Inda, L.A., J.G. Segarra-Moragues, J. Müller, P.M. Peterson, and P. Catalán. 2008. Dated historical biogeography of the temperate Loliinae (Poaceae, Pooideae) grasses in the northern and southern hemispheres. Mol. Phylogenetics and Evolution 46:932–957.

49. Jones, K. 1956. Species differentiation in *Agrostis*: The significance of chromosome pairing in the tetraploid hybrids of *Agrostis canina* subsp. *Montana* Hartmn., *A. tenuis* Sibth. and *A. stolonifera* L, part 2. J. of Genet. 54:377–393.

50. Keck, D.D. 1965. Poa. In: A. Munz, editor, California Flora. Univ. of California Press, Berkeley, California, USA. p. 1482–1490.

51. Kellogg, E.A. 2000. The grasses: A case study in macroevolution. Annu. Rev. of Ecology and Systematics 31:217–238.

52. Kellogg, E.A., and C.S. Campbell. 1987. Phylogenetic analyses of the Gramineae. In: T.R. Soderstron, K.W. Hilu, C.S. Campbell, and M.E. Barkworth, editors, Grass Systematics and Evolution. Smithsonian Inst. Press, Washington, DC, USA. p. 310–322.

53. Kendle, A.D., and J.E. Rose. 2000. The aliens have landed! What are the justifications for "native only" policies in landscape plantings? Landscape and Urban Planning 47:19–31.

54. Kingston, J.D., B.D. Marino, and A. Hill. 1994. Isotopic evidence for Neogene hominid paleoenvironments in the Kenya Rift Valley. Sci. 264:955–959.

55. Krans, J.V., J B Beard, and J.F. Wilkinson. 1979. Classification of C_3 and C_4 turfgrass species based on CO_2 compensation concentration and leaf anatomy. HortScience 14:183–185.

56. Latorre, C., J. Quade, and W.C. McIntosh. 1997. The expansion of C_4 grasses and global change in the late Miocene: Stable isotope evidence from the Americas. Earth and Planetary Sci. Letters 146:83–96.

57. Levin, D.A. 2001. The recurrent origin of plant races and species. Systemic Bot. 26(2):197–204.

58. Linder, H.P. 1987. The evolutionary history of the Poales/Restionales—a hypothesis. Kew Bull. 42(2):297–318.

59. Liu, Z.W., R.L. Jarret, R.R. Duncan, and S. Kresovich. 1994. Genetic relationships and variation among ecotypes of seashore paspalum (*Paspalum vaginatum* Swartz) determined by random amplified polymorphic DNA (RAPD) markers. Genome 37:1011–1017

60. Long, J.A., and E.C. Bashaw. 1961. Microsporogenesis and chromosome numbers in St. Augustinegrass. Crop Sci. 1:41–43.

61. MacFadden, B.J. 2000. Origin and evolution of the grazing guild in Cenozoic New World terrestrial mammals. In: H. Sues, editor, Evolution of Herbivory in Terrestrial Vertebrates: Perspectives from the Fossil Record. Cambridge Univ. Press, Cambridge, England, UK. p. 223–244.

62. Mathews, S., R.C. Tsai, and E. Kellogg. Phylogenetic structure in the grass family (Poaceae): Evidence from the nuclear gene phytochrome B. Am. J. of Bot. 87:96–107.

63. Matthes, C.A., and S.K. Davis. 2001. Molecular insights into the evolution of the Family Bovidae: A nuclear DNA perspective. Mol. Biology and Evolution 18(7):1220–1230.

64. Muller, J. 1981. Fossil pollen records of extant angiosperms. Bot. Rev. 47:1–140.

65. Neto, M.S., R.M. Jones, and D. Ratcliff. 1987. Recover of pasture seed ingested by ruminants: Seed of six tropical pasture species fed to cattle, sheep and goats, vol. 1. Australian J. of Experimental Agric. 27:237–246.

66. Ortmann, J., W.H. Schacht, and J. Stubberndieck. 1998. The "foliage is the fruit" hypothesis: Complex adaptations in buffalograss (*Buchloe dactyloides*). Am. Midland Naturalist 140(2):252–263.

67. Peterson, P.M., J.T. Columbus, and S.J. Pennington. 2007. Classification and biogeography of new world grasses: Chloridoideae. Aliso 23:580–594.

68. Peterson, P.M., K. Romaschenko, and G. Johnson. 2010. A classification of the Chloridoideae (Poaceae) based on multigene phylogenetic trees. Mol. Phylogenetics and Evolution 55:580–598.

69. Piper, C.V. 1906. North American species of *Festuca*. Contrib. from the US Natl. Herbarium 10:1–48.

70. Prasad, V., C.A.E. Strömberg, H. Alimohammadian, and A. Sahni. 2005. Dinosaur coprolites and the early evolution of grasses and grazers. Sci. 310:1177–1180.

71. Quinn, J.A., D.P. Mowrey, S.M. Emanuele, and R.D.B. Whalley. 1994. The "foliage is the fruit" hypothesis: *Buchloe dactyloides* (Poaceae) and the shortgrass prairie of North America. Am. J. of Bot. 81(12):1545–1554.

72. Raven, P.H., and D.I. Axelrod. 1974. Angiosperm biogeography and past continental movements. Ann. of the Missouri Bot. Garden 61:539–673.

73. Rezaei, H.R., S. Naderi, I.C. Chintauan-Marquier, P. Taberlet, A.T. Virk, H.R. Naghash, D. Rioux, M. Koboli, and F. Pompanon. 2010. Evolution and taxonomy of the wild species of the genus *Ovis* (Mammalia, Artiodactyla, Bovidae). Mol. Phylogenetics and Evolution 54:315–326.

74. Rotter, D., K.V. Ambrose, and F.C. Belanger. 2010. Velvet bentgrass (*Agrostis canina* L.) is the likely ancestral diploid maternal parent of allotetraploid creeping bentgrass (*Agrostis stolonifera* L.). Genet. Resource Crop Evolution 57:1065–1077.

75. Rotter, D., K. Amundsen, S.A. Bonos, W.A. Meyer, S.E. Warnke, and F.C. Belanger. 2009. Molecular genetic linkage map for allotetraploid colonial bentgrass. Crop Sci. 49:1609–1619.

76. Rotter, D., A.K. Bharti, H.M. Li, C. Luo, S.A. Bonos, S. Bughrara, G. Jung, J. Messing, W.A. Meyer, S. Rudd, S.E. Warnke, and F.C. Belanger. 2007. Analysis of EST sequences suggests recent origin of allotetraploid colonial and creeping bentgrass. Mol. Genet. and Genomics 278:197–209.

77. Sanchez-Ken, J.G., L.G. Clark, E.A. Kellogg, and E.E. Kay. 2007. Reinstatement and emendation of subfamily Micrairoideae (Poaceae). Systematic Bot. 32:71–80.

78. Sandve, S.R., and S. Fiellheim. 2010. Did gene family expansions during the Eocene-Oligocene boundary climate cooling play a role in Pooideae adaptation to cool climates? Mol. Ecology 19:2075–2088.

79. Sauer, J.D. 1972. Revision of *Stenotaphrum* (Gramineae: Paniceae) with attention to its historical geography. Brittonia 24:202–222.

80. Scharf, E. 2010. Archaeology, land use, pollen and restoration in the Yazoo Basin. Vegetation History and Archaeobotany 19:159–175.

81. Scholz, H. 1975. Grassland evolution in Europe. Taxon 24(1):81–90.

82. Sinha, N.R., and E.A. Kellogg. 1996. Parallelism and diversity in multiple origins of C_4 photosynthesis in grasses. Am. J. of Bot. 83:1458–1470.

83. Smith, A.G., J.C. Briden, and G.E. Drewry. 1973. Phanerozoic world maps. In: N.F. Hughes, editor, Organisms and Continents through Time. Paleontological Association, London. p. 1–43.

84. Sokolovskaya, A.P. 1938. A caryo-geographical study of the genus *Agrostis*. Cytologia (Tokyo) 8:452–467.

85. Soreng, R.J. 1990. Chloroplast-DNA phylogenetics and biogeography in a reticulating group: Study in Poa (Poaceae). Am. J. Bot. 77(11):1383–1400.

86. Soreng, R.J., and J.I. Davis. 1998. Phylogenetics and character evolution in the grass family (Poaceae): Simultaneous analysis of morphological and chloroplast DNA restriction site character sets. Bot. Rev. 64:1–85.

87. Stebbins, G.L. 1975. The role of polyploid complexes in the evolution of North American grasslands. Taxon 24(1):91–106.

88. Stebbins, G.L. 1981. Coevolution of grasses and herbivores. Ann. of the Missouri Bot. Garden 68:75–86.

89. Stevenson, D.W., and H. Loconte. 1995. Cladistic analysis of monocot families. In: P.J. Rudall, P.J. Cribb, D.F. Cutler, and C.J. Humphries, editors, Monocotyledons: Systemics and Evolution. Royal Bot. Gardens, Kew, England, UK. p. 543–578.

90. Strömberg, C.A.E. 2005. Decoupled taxonomic radiation and ecological expansion of open-habitat grasses in the Cenozoic of North America. Proc. of the Natl. Acad. of Sci. 102:11980–11984.

91. Strömberg, C.A.E. 2011. Evolution of grasses and grassland ecosystems. Annu. Rev. of Earth and Planetary Sci. 39:517–544.

92. Thomasson, J.R., M.E. Nelson, and R.J. Zakrzewski. 1986. A fossil grass (Gramineae: Chloridoideae) from the Miocene with Kranz anatomy. Sci. 233:876–878.

93. Torrecilla, P., and P. Catalán. 2002. Phylogeny of broad-leaved and fine-leaved *Festuca* lineages (Poaceae) based on nuclear ITS sequences. Systematic Bot. 27:241–251.

94. Torrecilla, P., J.A. Rodriguez, D. Stanick, and P. Catalán. 2003. Systematics of *Festuca* L. Sects. *Eskia* Willk., *Pseudatropis* Kriv., *Amphigenes* (Janka) Tzvelev., *Pseudoscariosa* Kriv., and Scariosae Hack. based on analyses of morphological characters and DNA sequences. Plant Systematics and Evolution 239:113–139.

95. Vislobokova, I.A. 2008. The major stages in the evolution of Artiodactyl communities from the Pliocene-early middle Pleistocene of northern Eurasia, part 1. Paleontological J. 42(3):297–312.

96. Watson, L., and M.J. Dallwitz. 1992. The Grass Genera of the World. CAB International, Wallingford, Oxon, UK.

97. Watson, L., and M.J. Dallwitz. 1992 (onward). The grass genera of the world: Descriptions, illustrations, identification, and information retrieval; including synonyms, morphology, anatomy, physiology, phytochemistry, cytology, classification, pathogens, world and local distribution, and references. Inst. of Bot., Chinese Acad. of Sci. http://delta-intkey.com/grass (23 April 2010).

98. Weng, J.H., M.J. Fan, C.Y. Lin, Y.H. Liu, and S.Y. Huang. 2007. Genetic variation of *Zoysia* as revealed by random amplified polymorphic DNA (RAPD) and isozyme pattern. Plant Production Sci. 10(1):80–85.

99. Xu, W.W., and D.A. Sleper. 1994. Phylogeny of tall fescue and related species using RFLPs. Theoretical and Applied Genet. 88:685–690.

100. Wu, Y. 2011. Cynodon. In: C. Kobe, editor, Wild Crop Relatives: Genomic and Breeding Resources, Millets and Grasses. Springer-Verlag, Berlin, Germany. p. 53–71.

101. Zhukovsky, P.M. 1968. New centres of the origin and new gene centres of cultivated plants including specifically endemic micro-centres of species closely allied to cultivated species. Botanicheskii Zhurnal 53:430–460.

CHAPTER 2

Early European and American Lawns

Ancient gardens are represented by the Persian pleasure gardens and later by the Arabian gardens (23).* The basic plant and hard-surface landscape constituents of these gardens remain unknown but are thought to have involved low-growing flowering plants of dicotyledonous species. Subsequently, the Greeks and then the Romans adapted the pleasure gardens of Persia and Arabia to their civilizations. These garden landscapes may have been quite different from the mowed turfgrass lawns of the twentieth century. The specific species present in these garden landscapes are not known. Also, there probably were ancient grazed park-like grasslands or wildlife-hunting areas on the grounds of emperors and kings and their aristocratic associates. These were not the managed grass areas focused on functional, recreational, and quality-of-life activities for the general populous, as known during the modern times of the past three centuries.

ORIGINS OF EUROPEAN LAWNS

Twelfth Century. The concept of a landscape dominated by low-growing, green vegetation may have emerged in monastic gardens in feudal Europe during the twelfth century. Translators of non-English writings of this century have chosen to use the word turf or lawn without documentation as to whether the twelfth century usage relates to our modern lexicon for these words (10). The poems and writings of the period emphasized the fragrance of vegetative surfaces, suggesting the presence of herbs. Perhaps these small garden areas within cloisters consisted of low-growing species with numerous aromatic, flowering plants. Common dwarf herbs were thought to include chamomile, wild thyme, pennyroyal, and yarrow. Some grasses may have been present as what are now termed weeds.

Thirteenth Century. There are indications that "medieval gardens" existed in the courtyards of Crown palaces in the thirteenth century (1). Indications are that a dominant component of the low-growing garden areas was chamomile (23). Outdoor seats with a green flowering surface were a unique feature (12). Whether these low-growing flowery areas contained grasses is speculative (10).

Fourteenth Century. There is a void in historical documentation concerning lawns or turfs. This was a period of hard times in Europe that included the Black Death and the Hundred Years' War. Any gardens

---

* The numbers in parentheses correspond to the reference list found at the end of the chapter.

were probably limited to the walled holdings of a few Crown rulers and to church cloisters. Records exist that classical imperial gardens were being manually cut by sickle in Japan in 1320 (18). These probably were three-dimensional surfaces of zoysiagrass over a base of works.

Fifteenth Century. Emergence of the Renaissance period brought development of commerce, travel, cities, and industry. However, emphasis was still placed on fortifying castles for war. Landscape gardening during this century remained unchanged and only for the wealthy few and the monks. Written or artistic representations of lawns or turfs were seriously deficient.

Sixteenth Century. Formal gardens were being constructed during the height of the Renaissance (16) but were primarily found in ecclesiastical centers and royal pleasure grounds. Villages established central parks, termed a "village green or common," that functioned as a recreation and meeting area for townspeople. The term "lawn" first appeared in the mid-1500s lexicon as an open space among trees, with no reference to grasses (10). Landscapes including expansive pastoral areas came into use on the residential estates of very large landowners in Great Britain, Austria, Germany, France, Italy, and the Netherlands and in Scandinavian countries during the late 1500s.

Seventeenth Century. Country houses flourished in the 1600s. They included relatively large open areas with walks of green vegetation among the extensive, formally patterned flower beds. The term "lawn" was used in a general sense to involve a large expanse of green vegetation that may have been grazed by animals so that it would appear low growing (10). The concept of a lawn composed primarily of grass was starting to emerge. In a 1676 book by Moses Cook, the design of landscapes surrounding houses was specified to include a spacious lawn of one acre (0.4 ha) or more (4). The first significant gardening books published appeared during the latter half of the seventeenth century in England (7, 15). Preplant weed control was by treatment with boiling water. In 1665, John Rea published suggestions on preparing a planting bed and on the selecting, harvesting, and transplanting of "turfs" (22).

Eighteenth Century. In England, grasses became a focus for use in lawn gardens, flower gardens, pleasure gardens, and greens during the eighteenth century. The great houses constructed during this period were planned with large grassy vistas promoted by Lancelot "Capability" Brown. His naturalistic designs featured large, gently undulating grass areas in the foreground of the house with scattered clumps of trees and distant serpentine lakes. A low-growing appearance was sustained by grazing sheep. Grass vegetative covers came to the fore in parks and church cemeteries of Great Britain during the eighteenth century (12). The first Japanese gardening book that addressed the planting and care of grasses also was published in the early 1700s.

Miller's **The Gardeners Kalendar** of 1745 included specific instructions on lawn care (21). This is one of the earliest books to use the word "lawn" to mean an area of grass laid down by design to enhance the visual appearance around a house. Book authors noted the importance of mowing, rolling, edging, weeding, and obtaining good seed in producing quality lawns. John Evelyn's writings instructed gardeners that the turfgrass walks and bowling greens should be mowed and rolled at 15-day intervals, with a turf beater being used during wet seasons to smooth uneven areas (7, 8).

Nineteenth Century. Gardening books focused more attention on the culture of lawns during the nineteenth century. Guidelines from 1805 included mowing "grass-plats" whenever there was the least hold for the scythe, removing leaf clippings, and edging grass walks periodically (19). Writings reveal that lawns of grass were being cut in China during this time.

An extensive 1807 presentation of turfgrass establishment and culture is found in McDonald's **A Complete Dictionary of Practical Gardening** published in England (20). His observations concerning lawn

Figure 2.1. Drawing of the Weston Burt estate and grounds in Gloucestershire, England, UK, in the late 1700s. [26]*

establishment were that sodding was preferred to seeding of lawns; that September through April was the best time to sod or seed in England; that the grass seed must be free of weed seeds; that the grass selected should be a deep-rooted, prostrate growing, permanent, heat-resistant species; and that poling of lawns be accomplished regularly to break up worm casts.

Prior to 1892, most published information on turfgrass culture was included in gardening books. The earliest known nonscientific publication of over 20 pages devoted strictly to turfgrasses was titled **The Formation of Lawns, Tennis and Cricket Grounds from Seed** published by Sutton & Sons of Reading, England, UK, in 1892 (26). This was the first of seventeen revised editions published from 1892 to 1962, an extraordinary span of 70 years (see chapter 12).

Twentieth Century. In the European countries, land was traditionally owned by a few wealthy individuals and was won by warfare and/or through royal allies and through major service to the king and his lords. Opportunities were rare for lesser-connected individuals to own a residence with space for landscape gardening. This long history in Europe influenced the degree to which an emerging middle class could secure landownership. Also, the middle class evolved primarily in old, densely populated urban areas where space for private gardening was limited.

In contrast, there were large areas in the United States that were settled by pioneering, agrarian families. They acquired or homesteaded small land areas by clearing openings, plowing among stumps and stone

---

* The numbers in brackets correspond to the figure source acknowledgments listed in the appendix.

piles, and planting a diversity of crops and vegetables needed to sustain the family and their domesticated animals. They usually were the only workers on the land they owned or would own. From these hardworking farmers emerged a significant middle class of private landowners. On this private space around the house, these individuals chose to develop lawns that improved the environment in which they lived. Within several generations, many sons and daughters of these rural homesteaders moved to urban centers to pursue employment opportunities. These individuals retained an appreciation for the rural life that they had known and expressed it via urban lawns and gardens.

The use of perennial grasses for lawns began to increase in the early 1900s. This growth was aided by the availability of mechanical mowers at a more reasonable cost. Emergence of a large middle class in North America enabled ownership of separate houses on a small but private piece of land that accommodated development of a lawn and garden, which offered an improved personal quality of life. Accordingly, the following sections will focus on the United States. Grass seed production from fields grown specifically for turfgrass use started to develop in the 1920s, with major expansion after WWII (see chapter 8). Allied with a major expansion of suburban residential areas and associated lawns was an explosion in commercial sod production (see chapter 9). In support of the substantial growth of lawn care, an expansion in turfgrass research (see chapter 6) and turfgrass education (see chapter 7) grew.

THE AMERICAN LAWN

Village Green. Multiple, usually attached family houses evolved in Europe centuries earlier than in the United States. Small groupings of houses plus barns in certain countries, with the farmers traveling outward onto nearby lands to conduct their cropping and pasturing activities, were typical of rural areas. This was based on a long history during which human security was a priority. Also, it was the model used by large landowners for housing the families of their farm workers. The land available around the houses in these small communities typically was very limited, especially in terms of lawn installations. The alternative to a turfgrass lawn for these small communities evolved during the 1500s in the form of a "village green or common." This European model was employed for some early settlements and colonies along what became the Atlantic coastal states of the United States.

Garden Lawns. Large aesthetic-based garden lawns began to be designed for the great houses of large landowners in Britain during the 1700s. Many of the plantations of the southeastern and mid-Atlantic colonies or states of the United States were characterized by more formal gardens. These were modeled similar to the estate gardens of the United Kingdom and Europe discussed earlier. Early US presidents George Washington and Thomas Jefferson emulated these large garden lawn designs on their own estates.

Evolution of the American Lawn

An evolutionary model for the American lawn in the United States is presented herein. Land was available to pioneering rural settlers, especially in the central and western areas of the United States. Government agencies developed low-cost land sales and various types of land grant and homestead programs whereby a family was awarded up to a section (640 ac or 260 ha) of land if they developed and lived on this land for a specified number of years. The American lawn evolved on these modest rural homesteads to accommodate their functional needs.

Many early rural settlers to the United States, especially in the Midwestern region, focused their initial efforts in securing a reliable food source and shelter. They were not from wealthy European families with large land holdings. Being oriented to survival via agrarian approaches, they sought soils suited for sustainable food crop and animal production. Construction of simple, single-room homesteads adapted to the local climate followed. A horse or ox was desired for power in agrarian activities and for transportation but was costly to acquire for many.

The first plant species propagated were vegetables, fruits, and grains, the latter for both human and domestic animal consumption. Storable foods were vital and included potatoes, pumpkins, onions, squashes, and sweet potatoes, plus grains such as corn, wheat, barley, and cereal rye. Gardens to grow vegetables and fruits typically were sited near the homestead. Grain crops were planted in nearby clearings among the tree stumps. Horses and cattle stomping through the garden and pigs, sheep, and goats feeding on the vegetables eventually necessitated the construction of a protective barrier.

The immediate area surrounding the homestead was gradually subjected to increased treading by a growing number of family members and an enlarged population of domesticated animals, such as chickens, ducks, geese, cows, sheep, hogs, goats, and/or horses. The consequences were major problems with mud and dust intermixed with animal excrement and urine. When the sun dried trafficked ground areas, the resultant dust would be inches deep, and when it rained, one could hardly walk due to suction of the mud. Thus a garden fence was eventually extended around the homestead to exclude the domesticated animals.

The protected ground inside this newly installed fence was released to support growth for a diversity of desirable and undesirable naturalized plant species. The undesirable plants included pollen-producing dicotyledonous species, such as ragweed, which causes allergic reactions in humans. Poison ivy and poison oak also may have invaded. The tall vegetative growth provided a favorable habitat for biting insects such as mosquitos, chiggers, and ticks, which are vectors for certain serious human diseases. Unwanted poisonous snakes and rodents, such as rats and mice that also are carriers of human diseases, were harbored within the tall vegetation. Due to these negative factors affecting human health and quality of life, it was logical that the area inside the homestead fence would be cut back periodically by manual scythe.

One or two cuttings annually would have involved vegetation of one to six feet (0.3–1.8 m) in height, with the taller plants usually being broadleaf species. Under two or more scythe-cuttings per year, the populations of larger dicotyledonous species would progressively decline. As grasses became more dominant, the cut vegetation was probably cured, raked, and transported to a storage barn to use as winter feed for cattle and horses. What remained would have been clumps of brown stems with a few green leaves and numerous scattered areas of bare soil.

In most ecosystems that were periodically defoliated by grazing or mowing, the monocotyledons or grasses would have concomitantly increased to eventually become dominant. An experienced scytheman could cut grasses to a relatively smooth level. Maintaining a sharp cutting blade was key. Grasses were most effectively cut when wet, such as following a rain or intense morning dew. More intense defoliation in terms of frequent, closer cutting would result in a population shift from tall, scattered, bunch-type grass species to lower-growing, dense, creeping-type, perennial grass species. This more dense cover resulted in a further reduction in mud and dust problems and increased competition against many weedy, unreliable dicotyledonous species. Also, as rooms and lean-tos were added to the homestead, the fence was probably moved outward, and the emerging turfgrass lawn was enlarged.

Thus the American home lawn evolved for the benefit of families of very modest and even poor means.* This also was the model that was applied as rural settlers moved to early small towns because of the key functional benefits such as mud, dust, pest, and human disease control. Similarly, the reason Dr. William J.

---

* One of the authors, James B Beard, was raised on a farm in western Ohio that was homesteaded in 1832 by his great, great grandfather, William Beard (1789–1840), who traveled from Adams County, Pennsylvania, via oxen and covered wagon and settled near a flowing spring on the south bank of Greenville Creek with his wife and four children. This author experienced some of the latter stages in evolution of the American lawn described herein.

Beal of the State Agricultural College in Michigan initiated the first published turfgrass research in the United States in 1880 was to solve the severe mud problem on the college campus and resultant tracking of mud into campus buildings (2). Most rural areas had dirt roads and walkways, as concrete sidewalks and stone or brick roads were only found in wealthy larger cities, if at all.

Early Turfgrasses. The sole cultural practice commonly utilized during this early evolution was defoliation by scythe-cutting. Compared to lawns in the post-1950 period, these early mowed grasses tended to be more thin and prone to periodic loss due to diseases and insect pests plus environmental stresses such as drought, heat, and winterkill. Recovery and survival was most likely to occur from perennial grass species possessing lateral stems—that is, rhizomes and stolons—in contrast to annual, bunch-type grass species. Many of these perennial species existed in North America before Europeans became interested in the Americas as a place for permanent habitation. Included would have been the fine-leaf fescues (*Festuca*) in the cool-humid woodlands, bluegrasses (*Poa*) in the cool-humid meadows interspersed within wooded areas, and bentgrasses (*Agrostis*) in the cool-humid, open, poorly drained wetlands along seas and rivers and along fringe forest areas. American buffalograss (*Buchloe*) was common in the open, semiarid plains of both the warm and cool climatic regions. St Augustinegrass (*Stenotaphrum*) occurred in the warm-humid climate on soils along rivers and coastal lowlands. Carpetgrass (*Axonopus*) also was found in the warm-humid climates of sandy coastal areas. These grass species adapted to associations with wild grazing mammals and, subsequently, to human agrarian activities such as grazing of domesticated animals and to periodic cutting by the scythe.

Landscaping. The ladies of the house probably were a key driving force in the evolution of the American lawn, since they were the primary individuals faced with cleaning the mud and dust problems within the homestead. Also, it was probably the lady of the house who led the propagation of flowers and ornamental plants within the fenced yard area. Placement of larger woody shrubs and trees tended to be around the perimeter of the yard rather than immediately adjacent to the house. There was a key functional reason for this landscape strategy. Specifically, it was fire prevention for the wooden house structures, especially during dry summer periods. The mowed lawn around the house functioned as a barrier that slowed the rate at which fires could spread laterally, as there were no water mains, pump stations, fire hydrants, or lawn irrigation systems.

During the evolution of the American lawn, there were three key developments: the push reel mower, manuring, and clean grass seed. Each will now be addressed.

Lawn Mower. For centuries turfgrass had been cut by grazing animals, primarily sheep and rabbits, and by human manual efforts with the hand sickle and scythe on formally groomed lawns and greens. The relatively smooth cutting of lawns with a hand scythe required experience and attention to detail plus a properly sharpened blade. Otherwise, circular inequities in turfgrass height, scars, and even bare areas resulted. Cutting turfgrass with a sickle or scythe was best done in the early morning when the turfgrass leaves were still wet with dew.

The reel lawn mower, also known as a cylinder mower in England, was invented and patented in October 1830 by Edwin Beard Budding (1796–1846) of Stroud, Gloucestershire, England, UK (10, 13, 14). It was a hand lawn mower—that is, it was manually pushed, with a 19-inch (0.48 m) width of cut but lacked a mechanical height control roller to prevent scalping. The concept was derived from a mechanical shearing machine designed and patented by John Lewis of nearby Brimscombe in 1815 to remove nap from woolen cloth as used in the numerous textile mills in the area (10). Between Stroud and Brimscombe were the village of Thrupp and another wool mill. It eventually was leased to John Ferrabee who reorganized the facility to manufacture machines under the name Phoenix Ironworks (10). Edwin Budding became associated with the Thrupp firm circa the 1820s. Actually, Edwin Budding and John Ferrabee lived in adjacent residences.

Figure 2.2. Budding manual-push, roller-drive, reel mower as patented in 1830. [26]

In May of 1830, they signed an agreement acknowledging that Budding had invented the mower and that Ferrabee would invest the funds needed to manufacture and market this new, innovative machine. The original Budding mower was tested outdoors at night to ensure secrecy. In 1832, J.R. and A. Ransomes of Ipswich, England, UK, began manufacturing the original Budding design under a license agreement with John Ferrabee (10, 11, 14, 27).

The basic design style is known as a "rear roller mower," and also as a "drum mower." The rear roller causes the cutting cylinder or reel to turn by means of a series of gears. It is interesting to note that the original design of Budding's 1830 lawn mower is still recognizable in a modern greensmower—that is, drive drum, cutting cylinder, and bedknife. The initial concept for a reel mower was a back-roller-drive type that was followed by the side-wheel-drive type. In 1869, the British firm of Follows & Bates in Gorton, Manchester, England, introduced the Climax side-wheel-drive hand lawn mower with cutting widths of 6, 7, 8, and 10 inches (15.2, 17.8, 20.3, and 25.4 cm).

In 1842, the first horse-drawn mower was registered in Scotland by Alexander Shanks of Arbroath, Scotland, UK (13). The reel unit had a 42-inch (106.7 cm) cutting width. Leather horse boots then became available to minimize turfgrass damage. The next major concept change was the horse-drawn, large-wheel, riding, reel mower with chain side-wheel drive of the early 1900s.

The first reel lawn mowers to be manufactured in the United States were made by H.N. Swift of Beacon, New York, beginning in 1855.* His "roller" model lawn mower was based on an imported English mower. It wasn't until after the Civil War that other companies entered this newly emerging equipment market. The Hills Archimedean Lawn Mower Company of Hartford, Connecticut, began production in 1867, with 12- and 14-inch (30.5 & 35.6 cm) cutting width hand-push models available by 1872. The following year saw the establishment of Chadborn & Coldwell Manufacturing Company of Newburgh, New York. Interestingly the towns of Beacon and Newburgh are across the Hudson River from each other. In addition, Thomas Coldwell had previously worked for Swift. In 1871, they were marketing 14- and 18-inch (35.6

---

* Information sourced from: James B. Ricci, The Reel Lawn Mower History and Preservation Project, Haydenville, Massachusetts, US.

Figure 2.3. Drawing of an 1881 Philadelphia reel lawn mower with side-wheel-drive design. [36]

& 45.7 cm) cutting width hand-push mowers and a 30-inch (76.2 cm) cutting width horse-drawn mower. The Chadborn & Coldwell Mfg. Co. grew rapidly and, by 1872, bought out the lawn mower operations of H.N. Swift.

The second style of reel lawn mower architecture is the "side wheel." With this type of lawn mower the rear roller is split and moved to the sides. It takes the form of two narrow wheels that seem to be attached to the ends of the cutting cylinder. Graham, Emlin, and Passmore of Philadelphia, Pennsylvania, introduced their Philadelphia side-wheel lawn mower in 1869 with cutting widths of 15 and 20 inches (38.1 & 50.8 cm). This type of mower eventually dominated the American market, while the roller mower was preferred in England.

Typically, instructions from these early companies would correlate the size of the mower with the motive power. Lawn mowers up to about 16-inches (0.41 m) wide would be for one person. A 20-inch (0.5 m) mower would be for two people, one pushing and one pulling. The 24-inch (0.6 m) mower would be pulled by a pony, with 30-inches (0.76 m) and wider models being horse powered. The usage of hand lawn mowers in the United States corresponds with the spread of urbanization. The reel lawn mower dominated the market until the mid-1950s when sales of the rotary lawn mower surpassed the reel lawn mower.

To sustain healthy turfgrass, no more than one-third of the leaf area should be removed at any one cutting. Regular use dictated by the reel mower greatly improved the quality of turfgrass lawns compared to the scythe. This was because the reel type of cut necessitated a regular mowing frequency. Failure to do so resulted in an excessive shoot biomass that made it physically difficult to manually push a functioning reel mower through the grass, with the only option being two to four passes of the mower to reestablish the normal cutting height.

In 1893, the first mechanically powered mower was patented by James Sumer of Leyland, Lancashire, England, UK. The steam-driven, reel mower weighed approximately 1.5 tons (1,360 kg). Both 25-inch (63.5 cm) and 30-inch (76.2 cm) cutting widths were marketed in 1897. Then a petrol motor–powered mower with a 40-inch (1 m) cutting width was sold by Coldwell Mfg. Co. in Newberry, New York, USA, in 1899 (13). Shortly thereafter, the first production-line, internal combustion engine–powered, reel mower was introduced to Great Britain in 1902 by Ransomes, Sims, and Jefferies of Ipswich, England,

Figure 2.4. First horse-drawn, reel mower with roller-drive, invented in 1842 by Alexander Shanks. [38]

Figure 2.5. Coldwell improved horse lawn mower, mid-1890s. [36]

UK (10, 12, 27). The cutting width was 42 inches (106.7 cm), and it had a front grass box with a chain-crank emptying device.

Manuring. Animal agriculture had a long history of applying manure to croplands for improved productivity. Thus a second cultural practice to be used in early American lawn culture was nutrient supplementation, then known as "manuring." It probably became a notable practice in the 1800s. In urban areas, the horse was the main power source for transportation. Thus horse manure was plentiful, with great mounds accumulating within cities. Manure disposal was a major problem. High-volume transport devices had not been developed, and long-haul roadway systems did not yet exist. Some manure was used by local farmers involved in fresh produce production near cities. Thus a practical solution for use of this byproduct was spreading the composted horse manure on nearby gardens and lawns. Manure street sweepings were particularly desirable for lawns as this fractionated material was more easily spread.

Figure 2.6. Coldwell motor lawn mower, 1907. [27]

By 1920, the automobile powered by an internal combustion engine was in ascendency, and horse-drawn vehicles were becoming obsolete, especially in urban areas. Consequently supplies of manure became inadequate for lawns and gardens. In addition, as street paving with petroleum-based materials increased, it resulted in the manure street sweepings becoming contaminated with organic compounds that were toxic to the turfgrass.

What were then termed "artificial manures" came to the fore. These included animal and plant byproducts and mined or synthesized minerals. Key among the former were bone meal, dried blood, guano, fish meal, and soybean meal. The latter included ammonium sulfate, calcium nitrate, superphosphate, and potash, which were promoted as odorless lawn fertilizers. Nationally marketed products included Vigaro, introduced in 1922 by the Swift Company of Chicago, which contained their agricultural-based fertilizer with certain micronutrients added. Then in 1928, the O.M. Scott and Son Company of Marysville, Ohio, introduced Turf Builder nationally, which was a 10-6-4 controlled-release fertilizer consisting of soybean and cotton seed meals, other organic nitrogen sources, and essential phosphorus and potassium.

Weed-Free Grass Seed. The third cultural improvement for turfed lawns in the United States prior to 1950 was the introduction of weed-free grass seed in the late 1890s. Typically grass seed was harvested from pastures contaminated with numerous weed species. Thus the only option in establishing a quality lawn was by sodding sourced from a weed-free stand. The higher cost limited use of this option.

A pioneer in mechanical seed cleaning and subsequent national marketing and promoting of weed-free grass seed was Orlando M. Scott (1837–1923) of Marysville, Ohio. This was a major advance that may be difficult to appreciate in a modern context. No significant advances were accomplished up to 1950 in terms of breeding turfgrass cultivars for lawns that were improved over the forage and pasture grass cultivars traditionally used.

Historical Documentation. In 1906, the first major book devoted to lawn establishment and culture titled **Lawns and How to Make Them** was authored by Leonard Barron and published in New York, NY, USA (3). Also published in 1906 was **The Lawn** by L.C. Corbett (5). Shortly thereafter **Les Gazons** by J.C.N. Forestier of France was published in 1908 (9). Subsequently, somewhat smaller lawn books were published including **Lawn Soils** by Oswald Schreiner and J.J. Skinner in 1911 (24), **Making A Lawn** by Luke J. Doogue in 1912 (6), and **Lawns and Soils** by Oswald Schreiner and others in 1912 (25). Turfgrass books oriented primarily to US

Figure 2.7. Cloth bag and label for Turf Builder® as first marketed in 1928 by O.M. Scott & Sons Co. [32]

home owners did not appear in significant numbers until the 1930s. These lawn books and many others are described in chapter 12. A major increase in books devoted to turfgrasses and their culture occurred after 1950. Post-1950 lawn books practically oriented to lawn care and planting are described in chapter 19, section A, and books covering both lawns and ornamental landscapes are described in chapter 19, section B.

REFERENCES

1. Albertus, M. 1867. De Vegetabilibus Lilori VII. Critical edition began by Ernest Meyer and completed by Karl Jessen. G. Reimer, Berlin, Germany.
2. Baker, R.S., and J.B. Baker. 1925. An American Pioneer in Science: The Life and Service of William James Beal. Privately printed, Amherst, Massachusetts, USA.
3. Barron, L. 1906. Lawns and How To Make Them. Doubleday, Page & Co., New York, New York, USA.
4. Cook, M. 1676. The Manner of Raising, Ordering and Improving Forest Trees: With Directions How to Plant, Make and Keep Woods, Walks, Avenues, Lawns, Hedges, etc. Peter Parker, Leg and Star in Cornhil, London, England, UK.
5. Corbett, L.C. 1906. The Lawn. Farmers' Bull. 248. United States Dep. of Agric., Washington, DC, USA.
6. Doogue, L.J. 1912. Making A Lawn. McBride, Nast and Co., New York, New York, USA.
7. Evelyn, J. 1669. Kalendarium Hortense. John Martin & James Allestry, London, England, UK.
8. Eveyln, J. 1932. Directions for the Gardiner at Says-Court. Geoffrey Veynes, editor, Nonesuch Press, London, England, UK.
9. Forestier, J.C.N. 1908. Les Gazons. Lucien Laveur, Paris, France.
10. Fort, T. 2001. The Grass Is Greener. HarperCollins Publ., London, England, UK.

11. Grace, D.P., and D.C. Phillips. 1975. Ransomes of Ipswich: A History of the Firm and Guide to Records. Inst. of Agric. History, Univ. of Reading, Reading, England, UK.

12. Haldane, E.S. 1934. Scots Gardens in Old Times (1200–1800). Alexander Maclehose & Co., London, England, UK.

13. Halford, D.G. 1982. Shire Album No. 91: Old Lawn Mowers. Shire Publ. Ltd., Aylesbury, Buckinghamshire/Botley, Oxford, England, UK.

14. Hall, A. 1984. The Hall and Duck Collection: A Concise History of Lawn Mower Development. Andrew Hall, Sheffield, England, UK.

15. Henrey, B. 1975. British Botanical and Horticultural Literature, vols. 1, 2, and 3. Oxford Univ. Press, London, England, UK.

16. Hill, T. 1568. The Profitable Art of Gardening. Thomas Marshe, London, England, UK.

17. Langley, B. 1728. New Principles of Gardening. A. Bettesworth et al., London, England, UK.

18. Maki, Y. 1976. Utilization, research activities, and problems of turfgrass in Japan. Rasen Grüflächen Begrünungen 2:24–28.

19. Marshall, C. 1805. Introduction to the Knowledge and Practice of Gardening. Bye & Law, Clerkenwell, England, UK.

20. McDonald, A. 1807. A Complete Dictionary of Practical Gardening, vol. 1 and 2. R. Taylor & Co., London, England, UK.

21. Miller, A. 1745. The Gardeners Kalendar. J. Rivington, London, England, UK.

22. Rea, J. 1665. Flora, Ceres and Pomona. Richard Marriot, London, England, UK.

23. Rohde, E.S. 1928. The Garden: Lawns: Nineteenth Century and After, Part 2. 104:200–209.

24. Schreiner, O., and J.J. Skinner. 1911. Bull. 75. Bureau of Soils, United States Dep. of Agric., Washington, DC, USA.

25. Schreiner, O., J.J. Skinner, L.C. Corbett, and M.L. Mulford. 1912. Farmers Bull. 494, United States Dep. of Agric., Washington, DC, USA.

26. Sutton & Sons, 1892. The Formation of Lawns, Tennis and Cricket Grounds from Seed. Simpkin, Marshall, Hamilton, Kent & Co. Ltd., London, England, UK.

27. Weaver, C., and M. Weaver. 1989. Ransomes 1789–1989: 200 Years of Excellence—A Bicentennial Celebration. Ransomes Sims & Jefferies PLC, Ipswich, England, UK.

CHAPTER 3

History of Sports Field Turfgrass Development

OUTDOOR GAMES IN WHICH TEAMS ATTEMPT TO HIT, CARRY, KICK, OR THROW SOME FORM OF A variably shaped ball into, over, or across the opponents designated goal have a long history in most parts of the world. Early "team" games frequently involved no rules; an unlimited number of "players," usually male; and a playing field that extended from one village to the next. This combination resulted in dangerous body contact of all sorts and serious injuries. Eventually a few competition rules of these ancient games may have been agreed to by the competing teams on game day. The playing surface could be composed of varying types of vegetation and landscapes, ranging from trees to rivers, with the location decided by the local host team. The first truly nationwide rules for many sports were developed during the 1850 to 1900 period in the United Kingdom. This resulted in sports within other countries responding similarly, including the United States.

Proper documentation as to the true origins of most outdoor sports played on grass has been lost in the mists of antiquity. While recognizing this limitation, attempts are made herein to suggest a possible time frame based on scattered bits of historical records. Many of the team sports surfaces eventually evolved to a grassy surface kept short by grazing animals, especially sheep. The evenly distributed rainfall and intensive sheep husbandry in Britain contributed to this country being a pioneer in turfgrass sports surfaces. These play areas were being "marked off" by the late 1440s. During the 1500s, areas of land were being designated and reserved for sports activities.

Rolling became especially important for bowls and cricket as clubs were formed and a permanent ground acquired. Eventually, patches of excessive grass growth not removed by grazing sheep would have been cut by manual scything. Manuring was avoided or infrequent, especially on larger sports fields so that laborious manual scything would be minimized. This was a reasonable approach in the early days of relatively light usage and low turfgrass quality expectations of those who recalled playing on pasture areas. Seasonal shoot growth could vary significantly depending on the rainfall pattern, as supplemental irrigation was not an option. Team sports ground construction and cultural practices utilizing practitioner groundsmen probably did not come into common usage until the 1850 to 1900 period, which coincides with the nationalization of rules and extensive organization of sports clubs (table 3.1). Just how many of the groundsmen were employed part time versus full time is unknown. Dual use of sports fields was practiced in the formative years from 1850 to 1900, with cricket outfields being used in the winter for soccer, field hockey, and/or even rugby.

Early Turfgrass Cultural Innovations. One may be surprised that a number of pioneering basic cultural practices originated for use on sports surfaces other than golf courses. Sodding was practiced on bowling greens by at least 1663. This is documented by a bill for construction of a bowling green at Windsor Castle in England stating "for cutting Turfe for ye green" (4). However, it was in Scotland that sodding came to the fore in the early 1700s via the use of natural "salt faill" (sea-marsh) turfgrasses. The sod was cut at a thickness of 1.3 to 1.5 inches (3.3–3.8 cm).

TABLE 3.1. CHRONOLOGICAL STAGES IN THE INITIATION OF TURFGRASS CULTURAL PRACTICES ON VARIOUS SPORTS FIELD SURFACES. YEAR LISTED IS WHEN WRITTEN DOCUMENTATION IS AVAILABLE, BUT IT MAY HAVE BEEN UTILIZED EARLIER.

Known year	*Cultural practice*
1660s	Sodding of bowling greens
1700s	Rolling of bowling greens
1700s	Sheep grazing of team sports fields in summer
Mid-1700s	Rolling of cricket wickets
Late 1700s	Sand patching of thinned and damaged surfaces
1800s	Systematic method of root-zone construction for bowling greens
1870s	Construction of surface contoured sports fields for drainage
1880s	Field surface markings for team sports fields
Late 1800s	Manual forking
Early 1900s	Horse-drawn rolling of sports fields
1920s	Lightweight sports field motorized power units
1930s	Horse/tractor-drawn spiking and slicing units for sports fields

Figure 3.1. Manual-push, 26-inch-wide, cast iron Shanks roller. [38]**

* The numbers in parentheses correspond to the reference list found at the end of the chapter.

** The numbers in brackets correspond to the figure source acknowledgments listed in the appendix.

Figure 3.2. Boxing turfgrass cut from a sea-washed marsh on the UK coast. [30]

A second pioneering, early turfgrass cultural practice on bowling greens in the United Kingdom was rolling. It was being practiced by at least the year 1700 using heavy, carved stones. Rolling of cricket wickets followed by at least the mid-1700s and was widely practiced in the early 1800s.

What evolved to the practice of topdressing an entire green or cricket wicket probably originated as what was termed "patching." When this practice was introduced lacks documentation. However, the need for smooth bowling greens probably led to patching by at least the late 1700s, when rolling use expanded. Organic composting for use in patching bare areas was beneficial in stimulating turfgrass recovery, as it was a primary source of key nutrients.

During the 1800s, the first known widely adopted systematic method for construction of root zones evolved in Scotland for bowling greens (4). In 1662, it is documented that authorization was given in rainy Glasgow, Scotland, for "the laying out of one bowling green on the sand." The "Scotch Greens" were flat in contrast to the crown greens of northern England, and thus internal drainage of excess water was desired to minimize cancellation of play days. This indicated a need for construction on sand sites. However, many bowling greens sited in villages and estates around Scotland were on poorly drained native soils. The solution developed in Scotland was construction of a root zone, thought to have improved internal drainage of water. This method consisted in variations of the following "soilless" profile: digging a flat subbase, placing clay pipe drains at a 9- to 21-foot (2.7–6.4 m) spacing; spreading a 6- to 12-inch-deep (15–30 cm) layer of coarse clinker, broken stone, or brick-bats; adding a 1- to 3-inch-thick (2.5–7.6 cm) layer of fine ash; and topping it with 1 to 3 inches (2.5–7.6 cm) of sand upon which the sea-marsh turf is laid. This method eventually was widely used in bowling green construction throughout the United Kingdom. One further note is that the early crown bowling greens may have been constructed due to the problems with flat greens, such as lack of water drainage causing loss of turfgrass and objectionable soil settling.

Historical Documentation. The early developments of turfgrass culture for sports greens and fields in England are documented in two unique series of booklets of 20 or more pages. The earliest was published by Sutton & Sons of Reading, England, starting in 1892 and continuing via 15 revisions through 1937. Only tennis and cricket were addressed in the 1892 and 1896 issues. Putting greens were added in 1899 and bowling greens and croquet in 1902. Perhaps this reflects the order in which sports surfaces adapted improved cultural practices and/or budgets, as the company probably would emphasize those sport surfaces that were purchasing seed and related turfgrass maintenance products (see Sutton & Sons in chapter 12).

The other early series of booklets was published by James Carter & Co. as authored by Reginald Beale, starting in 1906 and continuing through 1939 for 25 revised editions. Golf putting greens were their focus in 1906, and then croquet, tennis, and bowls were added in 1908. Team field sports of soccer and field hockey were mentioned in 1910, and rugby was mentioned in 1931 (see Beale, Reginald, in chapter 12).

Many of the books described in chapter 12 address the principles and practices of turfgrass culture that are basic to maintenance of sports fields and greens. A number of books that discuss the overall aspects of sports field construction and/or maintenance are described in chapter 12 (see the summary that follows).

The general construction aspects of sports fields are addressed specifically via books in chapter 12 that are listed by authors and organizations such as Baker in 1990 & 2006 (UK), Brown in 2005 (UK), Matthias & Bartz in 2002 & 2008 (Germany), Smith in 1950 (UK), and Stewart in 1994 (UK).

In addition, books on general turfgrass sports field construction and/or maintenance are found in chapter 13 under authors and organizations such as Bandeville in 1942 & 1952 (France); Critchley in 1987 (Australia); Dury in 1998 (UK); Fried in 2005 (USA); National Playing Fields Association in 196x, 1970, & 1976 (UK); Phillips in 1996 (USA); Sportsmark Group in 1980 and 1993 (UK); and Webster in 1940 (UK).

Books focused primarily on sports field contracts are described in chapter 13 under authors and organizations such as Chadwick in 1990 (UK); Dury in 2004 (UK); Glazner in 2001 (USA); Gooch in 1959, 1963, 1965, & 1975 (UK); Sayers in 1991 (UK); The Scottish Sport Council in 1992 (UK); and Sports Council in 198x (UK).

Construction and maintenance books for turfgrass surfaces are referenced within the following sections for 14 sports. These individual sport discussions are presented in approximately the order they became managed turfgrass surfaces. The outdoor sports discussed herein are mainly those most commonly played on turfgrass surfaces in the United States. Space limitations do not permit a comprehensive, worldwide discussion of sports played on grassy surfaces either historically or in modern times.

Summary of Sports Surface Construction and Culture Books Described in Chapter 12, as Listed by Author or Organization

Abramashvili in 1970, 1979, & 1988 (USSR)	Daniels in 1990 (Canada)
Adams & Gibbs in 1994 (UK & New Zealand)	Department of Education and Science in 1966 (UK)
American Sports Builders Association in 2010 (USA)	Dury in 1998 (UK)
Army Sport Control Board in 1940 (UK)	English Sports Council in 2000 (UK)
Association Française pour le Développement des Equipements de Sportifs et de Loisira in 1980 (France)	Evans in 1994 (UK)
	Goatley et al. in 2008 (USA)
Beale in 1924 (UK)	Goss & Cook in 1983 (USA)
Bošković in 1975 (Yugoslavia)	Gruttadaurio in 2008 (USA)
Bouhana in 1946 (France)	Harper in 1967 & 1986 (USA)
Bowles in 1967 & 1978 (UK)	Harradine in 1955 & 1961 (Germany)
Bureš & Blažek in 1975 (Czechoslovakia)	Hooper circa 1967 (UK)
Cettour in 1996 (France)	Hope in 1978 (UK)
Cockerham, Gibeault, & Silva in 2004 (USA)	Liffman in 1984 (Australia)
	Macself in 1924 (UK)

Mascaro in 1955 (USA)
McCarty et al. in 2005 (USA)
McNaughton circa 1934 (Malaysia)
Merino & Miner in 1998 (Spain)
Ministère de L'Education Nationale in 1960 (France)
Ministère de la Jeunesse et des Sports in 1992 (France)
Ministry of Education in 1955 (France)
Nakahara in 1994 (Japan)
New Zealand Institute for Turf Culture in 1961 & 1995 (New Zealand)
New Zealand Sports Turf Institute in 2000 (NZ)
O.M. Scott & Sons Company in 1932 (USA)
Perry in 1995 (USA)
Peterson in 1975 & 1978 (Denmark)
Pettigrew in 1937 (UK)

Phillips & Hardiman circa 195x (UK)
Puhalla, Krans, & Goatley in 1999 (USA)
Sanders in 1910 & 1920 (UK)
Sheard in 2000, 2005, & 2008 (Canada)
Singapore Parks & Recreation Department in 1977 (Singapore)
Skirde et al. in 1980 & 1983 (Germany)
Snook in 1978 (UK)
Spencer in 2008 (Australia)
Sport England in 2000 (UK)
Sudell in 1957 (UK)
Sutton & Sons of Reading in 1948 (UK)
Sutton in 1962 (UK)
Turfgrass Society of Korea in 2002 (South Korea)
Turner in 1962 (UK)
White & Bowles in 1952 (UK)

LAWN BOWLS

The rolling or tossing of a round object for a short distance toward a marker or a jack is an ancient game from Egyptian, Roman, and Polynesian times. The objects varied from stones to steel balls to wood to sown leather–covered balls filled with grain, chaff, etc. The marker may have been a fixed stone or a thrown jack. These early games were probably played on a diversity of mainly grassless grounds of varied slopes or levelness. They have been documented as being played in European countries from 1,000 to 2,000 years ago, including the British Isles (bowls), France (*boule*), Germany (*kegel*), and Italy (*boccie*).

Figure 3.3. View of turfgrass on a lawn bowling green in Canada, circa 1930s. [41]

In France and Scotland, "bowls" became a preferred recreational activity during the 1200 to 1500 period (4). During the 1400s through the early 1600s, edicts prohibited the lower classes from playing bowls. Originally, kings issued the bans to protect and promote archery that was a needed expertise for military battles. Later, local governors issued similar repressive bans to control drunkenness and gambling around taverns that typically had a bowling green nearby.

By the 1800s, lawn bowls had spread to Ireland, Australia, Canada, India, Malaysia, New Zealand, South Africa, and parts of the United States. Lawn bowls rules emerged during the 1860s. The first national flat-rink governing associations were formed in the Australian states of New South Wales and Victoria circa 1890, followed by the Scottish Bowling Association (SBA) in 1892 and the English Bowling Association (EBA) in 1903, which is now known as Bowls England. The American Lawn Bowls Association (ALBA) was formed in 1915 and the American Women Lawn Bowls Association (AWLBA) was organized in 1970. In 2000, these two merged into the United States Lawn Bowls Association. Modern bowls is played by both men and women of many age groups. Bowls competitions are primarily organized within facilities at private and public clubs.

Green Layout. There were English garden designs of the 1500s that included what were termed "bowling alleys." They functioned as garden walks within the floral-shrub plantings. Most probably these alleys were composed of grass-herb mixtures, with chamomile being a primary or even dominant herb component (4). Then in the 1600s, bowling greens of turfgrass were being constructed as part of the garden design on the private grounds of the wealthy and aristocrats.

Organized construction of grassed bowling greens finally came to the fore during the 1700s and 1800s in Scotland. These early turfgrass bowling greens probably were the forerunners of modern fine turfgrass. Three developments were keys to this advance in grassed, level bowls surfaces (4). Included were the use of sea-marsh sod, a silt-based root zone with underdrainage, and the invention of the mechanical reel mower. The late 1800s also were characterized by the construction of public bowling greens in village parks throughout Britain. Bowling greens are primarily constructed as flat, square turfgrass areas with 126-foot (38.4 m) side dimensions, although crown bowling greens are found in northern England. Intensively used bowling greens should be constructed with a well-drained root-zone profile such as a perched-hydration, high-sand system with a closely spaced under-drain-line network similar to that used on golf putting greens.

Turfgrass Surface. Early bowls may have been played on selected turfgrass areas where rabbits grazed on fine-leaf fescue. Modern turfgrass cultural and construction practices of the twentieth century provided quality bowls playing surfaces. New Zealand is unique in that lawn bowls is played on cotula (*Leptinella* spp.) rather than the commonly used turfgrasses such as bentgrass or hybrid bermudagrass, depending on the climate. Typically, mowing happens daily at a very close height of 0.1 to 0.16 inch (2.5–4.1 mm). Rolling also is routinely practiced prior to each playing day. Specific books on turfgrass culture for bowling greens are described in chapter 12 by authors Evans in 1988 and 1992 (UK); Haley in 196x, 1972, and 1979 (USA); Howard in 1993 (New Zealand); Howell in 1968 (UK); Levy circa 1949 (New Zealand); Liffman in 1984 (Australia); New Zealand Sports Turf Institute in 2008 (New Zealand); Perris in 2008 (UK); Pierce & Rigney in 1951 (Australia); Powell in 1947 (South Africa); and Sutton in 1962 (UK). Also see chapter 13 under author Connell in 1962 (USA). Other books with small sections concerning bowling green turfgrass culture and construction are described in chapter 12 under authors and organizations such as Beale in 1924, 1931, & 1952 (UK); Boyce in 194x (Canada); Dawson in 1939 & 1968 (UK); DeThabrew in 1973 (UK); Dury in 198x (UK); Hawthorn in 1977 (UK); Holborn in 1989 (UK); Lewis in 1948 (UK); MacDonald in 1923 (UK); Reed in 1950 (UK); Rees in 1962 (Australia); Sanders in 1910 & 1920 (UK); Sutton in 1962 (UK); Victoria Department of Agriculture in 19xx & 1971 (Australia); and Walker in 1971 (New Zealand).

Figure 3.4. View of turfgrass on cricket ground in Rye, New York, USA, circa 1928. [34]

CRICKET

This outdoor sport contributed a key evolutionary role in the early development of turfgrass for sport surfaces. Various club-and-ball games were being played during the thirteenth and fourteenth centuries in western Europe and the United Kingdom. Names for these ancient games included club-ball, stool-ball, and trap-ball. An early version of cricket was played in England in the 1500s called "creckett." The game of cricket grew substantially in southeast England during the 1600s and expanded nationwide in the 1700s. The Marylebone Cricket Club (MCC) at Lord's Old Ground was formed in 1787 and became the focal point for the game's countrywide rules. County level clubs were organized in England in the early 1800s. Cricket also expanded in popularity across the British Commonwealth countries during the 1800s, with international matches being played by the mid-1800s. The Imperial Cricket Conference was formed in 1909, renamed the International Cricket Conference in 1956, and renamed again as the International Cricket Council (ICC) in 1989. Teams are composed of 11 active players per side. Competitions are primarily among men at facilities maintained by colleges, schools, clubs, and parks, and at the professional level.

Ground Layout. The cricket ground may be a variable shape ranging from oval to somewhat circular. The turfgrass outfield typically extends 120 to 160 feet (37–49 m) around the central cricket table. The rectangular-shaped turfgrass cricket table for major competitions has dimensions of 100 by 66 feet (30 × 20 m), with 10 wicket positions, each being 10 feet (3 m) wide on the table. The wicket table is the turfgrass surface onto which the ball is bowled and where the batsman attempts to strike the ball into the 360 surrounding outfield. Selection of the proper clayey soil profile for the cricket table is a critical aspect that affects the action of a bowled cricket ball. Placed at each end of the wicket are bowling creases 8 feet and 8 inches (2.64 m) wide, and in the center of each are placed three vertical, pointed rods that are stuck in the turfgrass soil and topped with two bails or crosspieces placed end-to-end.

Turfgrass Surface. Cricket is the oldest team sport with documentation of grasses being intentionally used as the playing surface since the 1700s (3). Churchyards were a popular site, although discouraged by the clergy. When the location of Lord's cricket ground was moved in 1811, the turfgrass cricket table also was moved. The turfgrass at Lord's was maintained by a flock of sheep that were penned along the edge of the outfield when not required for maintenance of the grass. In addition, rolling was practiced by at least the late 1700s.

The turfgrass species used on the cricket outfield are bentgrass, hybrid-bermudagrass, fine-leaf fescue, perennial ryegrass, or polystand composites, depending on the climate. A typical mowing height is in the range of 0.25 to 0.75 inch (6.4–19 mm). Rolling is practiced regularly. Turfgrass species used on the cricket table are bentgrasses or hybrid bermudagrass, depending on the climate. Intense, multiple rolling passes and appropriate watering is an art required in the preparation of an individual wicket for an upcoming event.

Books focused on turfgrass culture and construction of cricket grounds are described in chapter 12 under authors and organizations such as Dury in 198x (UK); Evans in 1991 & 1996 (UK); Gibbs in 1895 (UK); Liffman in 1984 (Australia); Holborn in 1989 (UK); Lock in 1957, 1972, & 1987 (UK); Mansfield circa 1980 (UK); McIntyre & McIntyre in 2001 (Australia); Middlesex County Cricket Club circa 1956 (UK); New South Wales Cricket Association in 1981 (Australia); and Tainton & Klug in 2002 (South Africa).

CROQUET

Ancestral games of croquet date back to at least the 1300s. They involved hitting a ball with a wooden mallet. However, these games did not utilize the modern croquet stroke. A game similar to croquet was played in Ireland by at least the 1830s and was introduced to England ~1851 (9). The croquet game as known today emerged in England during the mid-1800s and became popular during the 1860s. A set of rules was published in 1864 and was most probably played on turfgrass by that time. The game spread through the British Commonwealth countries and to the United States. It has remained a minor sport but not without enthusiasm for the men and women who participate. The Croquet Association was formed as the United All England Croquet Association in 1897 in England and governs association croquet. American Six-Wicket Croquet is also popular and is governed by the United States Croquet Association (USCA), formed in 1977. A variation termed "backyard croquet" is played with nine wickets on higher cut residential lawns. The World Croquet Federation (WCF) was organized in 1986 as an umbrella governing body for international competitions. Croquet is played by men and women across a wide range of ages at facilities maintained by colleges, clubs, and parks.

Court Layout. The modern playing surface is a rectangle with dimensions of 84 feet by 105 feet (25.6 × 32 m) plus a turfgrass surround that is 5 feet (1.5 m) wide. The flat court surface used for high-level competitions necessitates a well-drained soil profile over closely spaced drain lines similar to the perched-hydration, high-sand constructions used for golf putting greens.

Figure 3.5. View of turfgrass on Union Croquet Club, New York, USA, circa 1924. [34]

Playing Surface. Lawn croquet has probably been played on turfgrass since the mid-1800s. A uniform, flat, smooth playing surface is desired. The main turfgrasses used depending on the climatic conditions are bentgrasses, hybrid bermudagrass, or perennial ryegrass plus some fine-leaf fescue. Occasional residential lawn play also involves higher-cut Kentucky bluegrass in North America. For competitive croquet, the turfgrass cultural practices are similar to those for golf putting greens. The typical mowing height for championship play is in the 0.25 inch (6.4 cm) range. Books devoted to croquet turfgrass culture and construction are described in chapter 12 under authors and organizations such as Mabee in 1991 (USA) and the New Zealand Sports Turf Institute in 2009 (New Zealand). Turfgrass culture and construction of croquet courts are addressed in sections of certain books reviewed in chapter 12 that are listed under authors such as Beale in 1924, 1931, & 1952 (UK); Dawson in 1939 & 1968 (UK); Hawthorn in 1977 (UK); Puhalla, Krans, and Goatley in 1999 (USA); and Sanders in 1910 & 1920 (UK).

FIELD HOCKEY

Ancient stick-and-ball games similar to field hockey were played throughout Europe in the Middle Ages. Included were *beikou* in Inner Mongolia, *cambuca* in England, *het kolven* in the Netherlands, *hurling* in Gaelic Ireland, *jeu de mail* in France, and *shinty* in Scotland. Modern field hockey emerged in England in the mid-1800s. At first its popularity evolved in public schools, with organized clubs forming in the late 1800s for both men and women. The game also spread abroad to the British Empire, Europe, and the United States during this same period. It is now the second largest team sport in the world. Field hockey is played by 11 active players per team. Both men and women participate at facilities provided by colleges, schools, clubs, and parks. In England, the Hockey Association was formed in 1886, and the Fédération Internationale de Hockey (FIH) was founded in 1924 in France, with both functioning as governing bodies. The United States Field Hockey Association (USFHA) was organized in 1922, and the Field Hockey Association of America (FHAA) was formed in 1928 for men. Then in 1993, the two governing bodies were merged to form US Field Hockey (USFH).

Field Layout. Field hockey is played on a rectangular area of 100 by 60 yards (91.4 × 54.8 m), with a goal opening of 12 feet wide by 7 feet high (3.7 × 2.1 m) on each goal line. There also is a turfgrass surround of varying width.

Playing Surface. The early British field hockey playing surfaces were grassy, with cricket outfields offering a dual-use facility. Managed turfgrass surfaces became more common by the late 1800s. Turfgrasses commonly used, depending on the climate, are Kentucky bluegrass, perennial ryegrass, or bermudagrass. As field hockey involves rolling a ball on the field surface, turfgrass maintenance is similar to that for soccer. The typical mowing height is in the 0.5 to 1.0 inch (1.3–2.5 cm) range during the playing season. Maintenance of turfgrass for field hockey surfaces is included in sections of certain books described in chapter 12 under authors such as Beale circa 1914, in 1924, & in 1931 (UK); Dawson in 1939 & 1968 (UK); Evans in 1994 (UK); Hawthorn in 1977 (UK); Puhalla, Krans, and Goatley in 1999 (USA); and Reed in 1950 (UK).

Synthetic surfaces emerged in the 1980s and have become mandatory for international and most national field hockey competitions. In England, field hockey typically was played during the winter on cricket fields. This situation was a factor in the change to a synthetic surface. Turfgrass surfaces are still in use for some local hockey fields.

Figure 3.6. View of turfgrass on soccer field in England, UK, circa 1923. [15]

SOCCER

Running and kicking an object or ball has been a popular recreation in most parts of the world since at least the third century BC. Ancient versions of what became football evolved in Europe and England starting in medieval times. There were few and varying localized rules, with opposing teams being unlimited and quite large in number of participants. Due to its violent nature, these activities have been referred to as "medieval or mob football."

Early development of modern soccer is thought to have occurred in the United Kingdom. Edicts banning "fute-ball" appeared in England in the mid-1300s and increased in the 1400s. This game had numerous versions of running, body contact, and kicking a ball. The rules for play varied among regions and even villages. Then in 1815, rules for English football separated games based on carrying the ball, or rugby football, and no-hands contact with the ball, or association football, into two groups. Subsequently, Association Football was officially organized in England in 1863, with handling of the ball strictly banned by specific rules. In 1904, the Fédération Internationale de Football Association (FIFA) was formed to govern the sport. Soccer continued to grow into the most popular team sport worldwide. It is now commonly known as football in most parts of the world, association football in England, and soccer in the United States. Teams are composed of 11 active players per side. Competition involves both men and women at facilities maintained by colleges, schools, youth leagues, clubs, and parks and at the professional level.

Field Layout. Soccer is played on a rectangular field with dimensions varying in the range of 110 to 120 yards (100.6–109.7 m) long by 70 to 80 yards (64–73.2 m) wide for international competitions. For noninternational events, the dimensions vary in the range of 100 to 130 yards (91.4–118.9 m) long by 50 to 100 yards (45.7–91.4 m) wide. The longer lines are termed touch lines. A rectangular goal centered on each goal line has posts 8 yards (7.3 m) apart, with an upper crossbar positioned 8 feet (2.44 m) above the ground. A turfgrass surround is maintained for a varying width outside the field of play.

Playing Surface. Since the early village host team selected the site on which football (soccer) was to be played, it was likely that a nearby grassy pasture grazed by sheep would have been selected. Football has been played in England on grassy surfaces since at least the 1600s and possibly during the sixteenth century.

Apparently some forms of turfgrass maintenance evolved on these areas in the 1700s. The commonly used turfgrass, depending on the climate, are bermudagrass, Kentucky bluegrass, or perennial ryegrass. A level, firm, fast surface with ball bounce between the knee and shoulder are desired. The preferred mowing height is in the range of 0.5 to 1.5 inch (13–38 mm), depending on the turfgrass species and cultivar plus the caliber of competitors. Books devoted to culture and construction of turfgrass soccer fields are described in chapter 12 under authors and organizations such as Bošković in 1971, 1975, 1984, 2002, & 2004 (Yugoslavia); Bures in 1985 (Czechoslovakia); the Football Association in 1951 (UK); Gionet in 2005 (Canada); and Hawthorn in 1972 (UK). There also are sections in certain books in chapter 12 that address turfgrass soccer field culture and construction under authors and organizations such as Beale in 1924, 1931, & 1952 (UK); Bouhana in 1946 (France); Brown in 1995 (Canada); Dawson in 1939 & 1968 (UK); DeThabrew in 1973 (UK); Evans in 1994 (UK); Hawthorn in 1977 (UK); Lewis in 1948 (UK); Liffman in 1984 (Australia); MacDonald in 1923 (UK); Puhalla, Krans, & Goatley in 1999 (USA); Reed in 1950 (UK); and Walker in 1971 (New Zealand).

RUGBY

The ancient origins of rugby are probably from medieval football described in the first paragraph under soccer. Rugby is a form of football developed from an earlier version thought to have been played at the Rugby School in England between 1750 and 1859. This was a rough game with limited rules that involved a large, variable number of players per team. Then by 1865, advancement of the ball via running and carrying became accepted. Two main forms of what is now modern rugby evolved in the late 1800s in the United Kingdom—rugby union and rugby league. Union rugby football is governed in England by the Rugby Football Union formed in 1871 and is administered globally by the International Rugby Board (IRB), renamed in 1998 from the International Rugby Football Board (IRFB) and originally formed in 1887. League rugby football is governed by the Rugby League formed in 1922 from the former Northern Rugby Football Union organized in 1895. Global administration is by the Rugby League International Federation (RLIF) organized in 1948. Rugby popularity spread mainly into western Europe, Australia, New Zealand, Polynesia, the United States, South Africa, and Southeast Asia. Rugby teams involve 13 active players for league and 15 for union. The game is played primarily by men in facilities provided at colleges, schools, clubs, and parks and at the professional level.

Field Layout. Rugby league is played on a rectangular field with dimensions not exceeding 74 by 109 yards (68 × 100 m), plus an in-goal turfgrass area extending 6.6 to 12 yards (6–11 m) beyond each try (goal) line. At each goal line, a set of two goal posts spaced 6 yards (5.5 m) apart is placed, with a crossbar 9.84 feet (3 m) high, forming an H-shape. Rugby union also has a rectangular field but of 77 by 109 yards (70 × 100 m), plus an in-goal turfgrass area extending 24 yards (22 m) outward to the dead ball line from each try (goal) line. The sides are known as touch lines. There is a turfgrass surround for a varying width outside the formal field of play.

Playing Surface. Both rugby union and league have probably been played on turfgrass since their modern origins in the 1860s. The main turfgrass species utilized, depending on the climate, are bermudagrass, Kentucky bluegrass, or perennial ryegrass. When played as a winter sport in wet climates, rugby play can be very destructive to turfgrass fields. Typical mowing height is 2 to 3 inches (5.1–7.6 cm), depending on the turfgrass species. Small sections on turfgrass cultural practices utilized on rugby fields are addressed in chapter 12 by authors such as Beale in 1924, 1931, & 1952 (UK); Dawson in 1939 & 1968 (UK); DeThabrew in 1973 (UK); Evans in 1994 (UK); Hawthorn in 1977 (UK); Liffman in 1984 (Australia); Puhalla, Krans, & Goatley in 1999 (USA); Reed in 1950 (UK); Stewart in 1994 (UK); and Sutton in 1962 (UK).

LACROSSE

As one of the oldest team games in North America, possibly dating back to the fifth century, lacrosse had its origins among the eastern woodland and plains Indian tribes (7). Included were the Choctaw, Eastern Cherokee, Iroquois, Mohawk, Ojibwe, Onondaga, Ottawa, Sioux, and Winnebago. Many different forms of stick-and-ball games were played by these tribes with few rules. Involved was very rough play by large teams over an extensive area of miles and of irregular topography and low-growing vegetation. The equipment consisted of a ball composed of wood or deerskin stuffed with hair, plus a stick originally carved like a giant wood spoon, which evolved to a wood stick with a circle on one end filled with netting. The lack of rules led to violent clashes at times, as the running and contact prepared participants for actual tribal wars.

European settlers in eastern Canada from Quebec to Toronto began to adapt a more organized version of field lacrosse in the 1800s. Lacrosse clubs were formed in Canada during the mid-1800s, with the game eventually spreading to Australia and England. Formal rules for lacrosse were published in the 1860s. The National Lacrosse Association (NLA) was formed in 1867 as the national governing body for Canada. The game had grown substantially in the eastern United States by the early 1900s, especially at colleges and amateur sports clubs. Early governing bodies for field lacrosse were the men's Federation of International Lacrosse (FIL) in 1974 and the International Federation of Women's Lacrosse Association (IFWLA) in 1972. These two were merged to form the Federation of International Lacrosse (FIL) in 2008. Modern lacrosse involves 10 active players competing for each team. Competition is primarily among men at facilities provided by colleges, schools, youth leagues, clubs, and parks.

Field Layout. Lacrosse field dimension involve a rectangular shape of 110 by 60 yards (101 × 55 m) with a goal 15 yards (13.7 m) in from each end line that has a 6 by 6 foot (1.8 × 1.8 m) opening. A turfgrass surround of varying widths is maintained outside the formal playing field. Actually, many men's high school and college lacrosse competitions are played on dual-use sports fields with American football, while most little-league lacrosse is played on municipal or county dual-use fields with soccer.

Playing Surface. Turfgrass was the playing surface of choice for lacrosse by the late 1800s. Maintenance practices of turfgrasses on lacrosse fields would have been similar to that of soccer. Thus the turfgrass cultural practices are basically similar to what are used for those sports. The turfgrass species commonly used are hybrid bermudagrass, Kentucky bluegrass, or perennial ryegrass, depending on the climate. Typical mowing heights are in the range of 0.75 to 1.5 inch (1.9–3.8 cm). A section on the culture and construction of turfgrass lacrosse fields is found in books described in chapter 12 under Liffman in 1984 (Australia) and Puhalla, Krans, and Goatley in 1999 (USA).

LAWN TENNIS

In the thirteenth century, rudimentary versions of tennis involved striking a ball with the hand, followed by use of a glove, then a paddle, and, eventually, a racquet. An indoor game involving a racket and ball called court tennis emerged during the sixteenth century in France. Subsequently, outdoor lawn tennis was played in England in 1870. Thereafter, lawn tennis spread to the United States in 1874 and grew rapidly in popularity in many parts of the world. Many of the basic modern rules were established at the All England Club in 1877. The United States National Lawn Tennis Association (USNLTA) was formed in 1881 and is now known as the United States Tennis Association (USTA). The British Lawn Tennis Association was organized in 1888. The International Lawn Tennis Federation (ILTF) was founded in 1913 and the name changed to International Tennis Federation (ITF) in 1977. Competitions involve both men and women plus mixed doubles and are available at facilities maintained at colleges, schools, clubs, and public parks.

Court Layout. Lawn tennis is played on a rectangular-shaped, flat surface with court dimensions of 78 feet (23.8 m) in length by 27 feet (8.2 m) in width for singles matches and 36 feet (11 m) in width for doubles matches. A turfgrass surround extends outward from both base lines ~21 feet (6.4 m) and ~12 feet (3.7 m) from each sideline. Placed halfway between the base lines is a net that is 42 inches (106.7 cm) high at the posts and 36 inches (91.4 cm) high in the center. Tennis court surfaces consisted of either turfgrass or clay up until the mid-1900s, when hard courts such as concrete or asphalt options came into use. The United States championships were played on a turfgrass surface until 1975. The 1970s led to the introduction of synthetic tennis surfaces. Clay courts remain popular in certain countries such as Spain and Italy, while synthetic courts are widely used in North America. Turfgrass tennis courts continue to be used in some countries such as the United Kingdom and New Zealand.

Turfgrass Surface. Lawn tennis has been played on turfgrass from its origin in the early 1870s, with the earliest supposedly being on a croquet court. The surface characteristics desired for turfgrass tennis courts are uniformity in the level, pace, and ball bounce. The main turfgrass utilized, depending on the climate, are bentgrass, perennial ryegrass, or hybrid bermudagrass. In the early 1900s, both mowing and rolling of quality turfgrass tennis courts were accomplished with horse-drawn units. By the 1950s, quality lawn tennis courts were mowed two to three times per week at between 0.13 and 0.5 inch (3.3–12.7 mm) during the growing season. Rolling and irrigation were routine practices, with coring or spiking or piercing being practiced in the spring and autumn. A close mowing height in the range of 0.13 to 0.25 inch (3.3–6.4 mm) is desired for modern lawn tennis. Rolling also is practiced regularly during periods of play. Use of a pony roller dates back to at least 1877.

Sections on maintenance and construction of turfgrass lawn tennis courts are discussed in chapter 12 under authors such as Beale in 1924, 1931, & 1952 (UK); Bošković in 1975 (Yugoslavia); Cunningham in 1914 (UK); Dawson in 1939 & 1968 (UK); DeThabrew in 1973 (UK); Dury in 198x (UK); Hawthorn in 1977 (UK); Holborn in 1989 (UK); Lewis in 1948 (UK); MacDonald in 1923 (UK); Perris in 2000 (UK); Puhalla, Krans, & Goatley in 1999 (USA); Reed in 1950 (UK); Sanders in 1910 & 1920 (UK); Stevens in 1909 (UK); Stewart in 1994 (UK); Sutton in 1962 (UK); Thorn in 1991 (UK); and Walker in 1971 (New Zealand). Also see a book described in chapter 13 under Haake in 2000 (UK).

BASEBALL

A baseball-like game called "rounders" and also baseball was played by youngsters in England and Ireland in the early 1700s. It was and is played in innings with a batter and base runner opposing nine defensive fielders on a diamond with four posts or bases. The first known use of the term baseball in the United States was in 1791. Various ball-and-bat baseball-like games were being played with varying local rules in the eastern region of the United States for the next five decades. Then in 1845, a New York club played baseball utilizing what are considered modern rules. In 1857, the National Association of Baseball Players (NABBP) was formed to govern baseball. Baseball expanded throughout the United States during the 1860s. Eventually the National Association of Professional Baseball Players (NAPBP) split off from the NABBP in 1870 but was dissolved in 1875 and the National League was formed. These tumultuous times for baseball organizations steadied when the National Association of Professional Baseball Leagues (NAPBL) was formed in 1901. The game of baseball spread across America. There are nine active players per team. Competitions are played primarily by men at facilities provided by colleges, schools, youth leagues, clubs, and parks and at the professional level.

Figure 3.7. View of turfgrass on baseball ground at Ebbits Field, Brooklyn, New York, USA, 1920s. [34]

Field Layout. Baseball is played on grounds of variable outer dimensions. The infield or diamond consists of a clayey, bare perimeter area with bases on each corner designated as home, first, second, and third that form a square of 90 feet (27.4 m) per side. Within the "skinned" infield is a turfgrass area. There also is a bare, clay, raised pitcher's mound of 18 feet (5.5 m) in diameter within the turfgrass infield. The outfield surface is turfgrass and extends in a 90 degree arc 95 feet (29 m) from the front edge of the pitcher's mound. The turfgrass outfield extends from home plate for a varying distance in the order of 300 to 325 feet (91.4–99 m) down each foul line and 350 to 400 feet (106.7–121.9 m) in center field. A nonturfed warning zone is included at the outer perimeter of many baseball outfields. There also is a turfgrass surround of varying width outside both foul lines.

Playing Surface. Baseball play on at least the outfield has probably involved a grassy surface since at least the 1870s. The turfgrass species used on baseball fields, depending on the climate, are bermudagrass, Kentucky bluegrass, or perennial ryegrass. The mowing height during the playing season is in the range of 0.5 to 1.7 inches (1.3–4.3 cm), depending on the turfgrass species and cultivar. Books concerning the culture and construction of turfgrass baseball fields are described in chapter 12 under authors and organizations that include Brown in 1995 (Canada); Perry in 1997 (USA); Puhalla, Krans, & Goatley in 1999 & 2003 (USA); and United States Baseball Federation in 1980 & 1992 (USA).

SOFTBALL

The game of softball descended from baseball in the late 1800s, with varying versions played until the development of standard countrywide rules for the United States in 1934. Both slow pitch and fast pitch versions continue to be played, with the latter becoming dominant in the 1940s. Softball has a number of administrative groups within the United States and worldwide. It is the most popular participant sport in the United States and has spread to a number of other countries, including Australia, China, Japan, and New Zealand. Softball is played among nine players from each team. Both men and women across a wide range of ages participate at facilities provided by colleges, schools, clubs, and parks. Women are particularly active in playing softball.

Field Layout. Only the softball outfield has a turfgrass surface that ranges from 150 to 275 feet (46–84 m) down both foul lines and 175 to 300 feet (53–91 m) into center field from home plate, depending on the competitive level. The entire softball infield is a bare, clayey soil. This skinned area includes a square area of bases ranging from 55 to 65 feet (17–20 m) apart and extends outward in an arc of 55 to 65 feet (17–20 m) from the front center of the pitcher's plate. A turfgrass surround of varying width usually is utilized.

Playing Surface. Modern softball outfield surfaces probably have been composed of turfgrass since the 1930s. The commonly used turfgrass species are bermudagrass, Kentucky bluegrass, or perennial ryegrass, depending on the climate. The mowing height during the playing season ranges from 0.5 to 1.7 inches (1.3–4.3 cm), depending on the species. Books addressing the culture and construction of softball fields are described in chapter 12 under authors Brown in 1995 (Canada); Perry in 1994 & 1997 (USA); and Puhalla, Krans, & Goatley in 1999 & 2003 (USA).

AMERICAN FOOTBALL

Early versions of rugby and soccer served as the basis for the evolution of American football. Uniform rule changes introduced in the 1880s formed the basis for the team sport, which was distinctly different from its ancestral sports previously mentioned. The sport emerged primarily in eastern United States colleges. Significant expansion had occurred by 1900, especially among colleges in the industrial Midwest region. The popularity of American football continued to grow into a national sport in the early 1930s. The Intercollegiate Athletic Association of the United States was formed in 1905 to address football injury problems. It changed its name to the National Collegiate Athletic Association (NCAA) in 1910. The American Professional Football Association was formed in 1920, with the name being changed to the National Football League (NFL) in 1922. There are 11 active players for each team. The game is played primarily by men at facilities maintained by colleges, schools, youth leagues, and parks, and at the professional level.

Field Layout. This game is played on a rectangular shaped field with dimensions of 120 by 53.3 yards (109.7 × 48.7 m), not including the outer sideline area. Goal posts are placed on each back line of the end zones at a spacing of 70.8 feet (21.6 m) in from the sideline. There is also a turfgrass surround of varying width. The increasing intensity of football field use has resulted in soil profile construction involving a perched-hydration, high-sand root zone above a closely spaced drain-line network.

Princeton-Yale Football Game, showing the perfect turf at the Palmer Stadium, Princeton, N. J.

Figure 3.8. View of turfgrass on an American football field at Princeton, New Jersey, USA, circa 1924. [34]

Playing Surface. American football has probably been played on a grassy surface since its modern origin in the late 1800s. Commonly used turfgrasses are bermudagrass, Kentucky bluegrass, perennial ryegrass, or turfgrass-type tall fescue, depending on the climate. The mowing height ranges from one to two inches (25–51 mm). Books addressing American football field turfgrass culture and construction are described in chapter 12 under authors such as Puhalla, Krans, & Goatley in 1999 and 2003 (USA).

AUSTRALIAN RULES FOOTBALL

The game of Australian rules football originated in Melbourne, Victoria, Australia, in 1858. It was developed by cricket players as an exercise during the winter off-season. There are 18 active players per team, wearing studded shoes. The players may carry, kick, or "hand pass" the ball in which the ball is held in one hand and punched with the other. Forward passing is illegal. The game has grown to become a major competitive sport that has expanded outward from the Victoria region nationally. The Australian Football League (AFL) was formed in 1990.

Field Layout. Australian rules football has been played on turfgrass since it originated and typically is sited on dual-use fields with cricket. Thus the turfgrass playing field is oval-shaped with dimensions being a maximum of 202 yards long by 170 yards wide (185 × 155 m) and a minimum of 148 yards long by 120 yards wide (135 × 110 m). A turfgrass surround outside the playing field is minimally 4.4 yards (4 m) wide. There are four goal posts on each end longwise and spaced 7 yards (6.4 m) apart, with the goal posts 6.6 yards (6 m) high and the behind posts 3.3 yards (3 m) high.

Playing Surface. The commonly used turfgrass species are bermudagrass, Kentucky bluegrass, or perennial ryegrass, depending on the climate. The mowing height range is 0.8 to 1.6 inches (2–4 cm). Turfgrass culture is discussed in a section of a book described in chapter 12 under Liffman in 1984 (Australia).

TURFED HORSE RACECOURSE TRACKS

Humans domesticated horses starting about 4,000 BC. Some forms of local racing probably followed by 3,000 BC. Just when permanent horse racecourses on formally organized grassy routes occurred is unclear. The British Isles is a likely pioneering location. Formal turfgrass racecourse routes at permanent sites were in use by 1700, with running distances in the order of 4 miles (6.4 km). Thoroughbred horse racing on grass emerged in the United States during the mid-1800s. However, nongrass racetracks have retained their dominance in the United States over the years. In contrast, most countries around the world conduct horse racing on turfgrass surfaces. Also, a variation in horse racing on turfgrass is the steeplechase racecourse that includes artificial obstacles such as hedges. There also are specialty horse jumping courses that involve competition on a turfgrass surface.

Racecourse Layout. Turfgrass racecourse shapes vary widely around the world. The interlocking mesh element system has proven very effective for stabilizing root zones plus enhancing water and air movement on racecourses. Also, the system reduces injuries to horses and improves race times.

Racecourse Surface. Turfgrass is a safer surface for running horses in terms of the possibility of injuries. Quality turfgrass racecourses will have a well-rooted, uniform, dense turfgrass biomass for cushioning the impact of horses' hooves plus a soil profile characterized by rapid drainage of excess water to minimize

Figure 3.9. Turfgrass on horse racecourse in England, UK, circa 1923. [34]

cancellation of race meets. For thoroughbred horse racetracks, perennial ryegrass and/or Kentucky bluegrass are used in cool-climatic regions and bermudagrass in warm-climates. For transitional climates, zoysiagrass is used in Japan, kikuyugrass in South Africa, and tall fescue-Kentucky bluegrass in the United States. The mowing is higher than for most sports field turfgrasses, with a height in the range of 4 to 6 inches (10–15 cm), except for bermudagrass and zoysiagrass that are mowed in the range of 2 to 3 inches (5.1–7.6 cm). Hooves of running horses are very destructive. Thus appropriate fertilization is a key to turfgrass recovery, with irrigation also valuable under drier conditions during the racing season. A brief section concerning maintenance of turfgrass racecourses is found in chapter 12, listed under authors such as Beale in 1924 (UK), Dawson in 1939 & 1968 (UK), Hawthorn in 1977 (UK), and Walker in 1971 (New Zealand).

POLO

A medieval game *chaughan* was played in Persia (Iran) and central Asia by the first century AD and earlier (1). There were at least six early versions of what became polo. By the tenth century, polo was a relatively organized team sport with basic playing concepts similar to those now practiced. It eventually spread to central India, especially the Manipur region, and to China during the fifteenth and sixteenth centuries. Modern polo is thought to have evolved in Tibet where it was known as *pulu*. Polo became a popular equestrian team sport in India during the 1600s. The grounds' perimeters were marked at about 300 by 170 yards (274 × 155 m), leveled and dusty due to a lack of grass.

Polo was introduced to England in the 1860s by army cavalry units and from there to other countries in western Europe, Argentina, and the United States during the 1870 to 1890 period. Also, formal national rules for polo competitions originated during this time frame. Four team members compete, with each riding on a string of polo ponies. Long-handled mallets ranging from 50 to 53 inches (1.27–1.35 m) are used to strike a wooden or plastic ball, with the ultimate objective to hit the ball between two vertical goal posts of the opposing team. The ball is struck with the longer sides of the mallet head.

The Federation of International Polo (FIP) was formed in 1982 to govern world competitions, while the United States Polo Association (USPA), founded in 1890 as the Polo Association, oversees US activities.

Field Layout. The polo playing field is rectangular with dimensions of 300 by 200 or 160 yards (274 × 183 or 146 m). The shorter 160 yard width may be used if there are side boards that generally are 6 inches (15.2 cm) high. A set of vertical, collapsible goal posts are placed 8 yards (7.3 m) apart on each end line. There also is a 30-yard-wide (27.4-meter-wide) turfgrass safety area behind each goal line, plus 10-yard (9.1 m) turfgrass surrounds outside each sideline.

Playing Surface. Just when grass became the preferred surface for field polo is unclear. Selection of a grass species and cultivar with less slippery leaves is especially important for polo competitions plus a dense shoot biomass for cushioning horses' hooves and a deep, well-rooted subbase for turfgrass stability. Bermudagrass is preferred but limited to warm climates. Kentucky bluegrass is another rhizomatous species used in cool regions. The rhizomes provide an underground matrix and survival structure that is important under the destructive force of horse hooves turning, stopping, and starting. Uniformity in ball roll and bounce is desired, which dictates the cultural practices. A mowing height of 2 to 3 inches (5.1–7.6 cm) is common, with the lower height employed on bermudagrass. Manual replacement of divots during and after each competition is commonly practiced. Sections concerning polo field turfgrass culture are found in a few books described in chapter 12 under authors such as Beale in 1924 (UK), Cunningham in 1914 (UK), Dawson in 1939 & 1968 (UK), and Hawthorn in 1977 (UK).

GROUNDSMAN/SPORTS TURF MANAGER

The first rudimentary groundsman probably emerged in the United Kingdom in the late 1600s. During the early years in maintenance of sports grounds, part-time workers were employed to conduct the minimal maintenance of grassy sports fields and greens. These workers usually had experience in agricultural plant and soil husbandry. The main activities probably would have been herding sheep over the play area, sweeping or brushing off debris and excrement, and replacing divots for certain sports.

Manual harvesting and transplant sodding were practiced on bowling greens as early as the 1660s in Scotland. Records from Sterling, Scotland, in 1738 concerning sodding of a bowling green refer "to William Dawson, gardener, and keeper of the said Green" (4). It is unclear whether the worker was employed full time or part time.

As expectations of better quality playing surfaces emerged among the sports participants, the need for a full-time groundskeeper became apparent, and appropriate budgets were allocated. Just when this actually occurred is unclear. The Marylebone Cricket Club hired a full-time groundsman in 1864 (3). As larger, heavier rollers were introduced in the 1880s for cricket outfields and similar closely cut turfgrass sports surfaces, a groundsman experienced in handling horses was an added requirement. Introduction of the mechanical reel mower added to the need for a knowledgeable groundsman, especially for closely mowed bowling greens and cricket tables. However, widespread adoption of reel mowers did not occur until the late 1800s. Some traditionalist in England clung to sheep grazing of sports fields into the early 1900s. The wider, horse-drawn, reel mower was introduced in 1842 in Scotland, which greatly reduced labor costs for mowing larger sports fields. Large horse-drawn mowers increased in acceptance and continued to be used into the early 1900s on many sports fields. The groundsman/sports turfgrass manager was developing into a specialist profession as a result of these innovations. After WWII, full-time sports field groundsman positions become more common.

Professional Organization

One of the earliest national groups formed was originally known in England as the National Association of Groundsman in 1934, with the name changed to the Institute of Groundsmanship (IOG) in 1969. The organization published a periodical known as the **Journal of the National Association of Groundsman** from 1948 to 1952 and the **Groundsman** since 1952. Each summer since 1938, they also have sponsored a major equipment and product exhibition (SALTEX).

It was not until 1981 that the Sports Turf Managers Association (STMA) was formed in the United States. The organization published a newsletter called **Sports Turf Newsletter** from 1985 to 1986, **Sports Turf Manager** from 1987 to 2003, and, since 2004, **STMA News Online**. STMA also has cooperatively published the periodicals **SportsTURF** from 1988 to 1989, **Golf & sportsTURF** from 1990 to 1992, and **SportsTURF** since 1992. An annual winter educational conference has been held by the STMA since 1986.

REFERENCES

1. Dale, T.F. 1905. Polo: Past and Present. Country Life Library of Sport, George Newnes, London, England, UK.
2. Dingley, H.J. 1893. Touchers and Rubs on Ye Anciente Royal Game of Bowles. Simpkin, Marshall, Hamilton, Kent & Co. Ltd., London, England, UK.
3. Evans, R.D.C. 1991. Cricket Grounds: The Evolution, Maintenance and Construction of Natural Turf Cricket Tables and Outfields. Sports Turf Research Institute, Bingley, West Yorkshire, England, UK.
4. Evans, R.D.C. 1992. Bowling Greens: Their History Construction and Maintenance. 2nd rev. ed. Sports Turf Research Institute, Bingley, West Yorkshire, England, UK.
5. Evans, R.D.C. 1994. Winter Games Pitches. Sports Turf Research Institute, Bingley, West Yorkshire, England, UK.
6. Godfree, L.A., and H.B.T. Wakelam. 1937. Lawn Tennis. David McKay Co., Philadelphia, Pennsylvania, USA.
7. Fisher, D.M. 2002. Lacrosse: A History of the Game. Johns Hopkins Univ. Press, Baltimore, Maryland, USA.
8. Perris, J., editor. 2000. Grass Tennis Courts: How to Construct and Maintain Them. Sports Turf Research Institute, Bingley, West Yorkshire, England, UK.
9. Prichard, D.M.C. 1981. The History of Croquet. Cassell Ltd., London, England, UK.

Figure 3.10. View of turfgrass on polo field in Westbury, New York, USA, circa 1930. [41]

Figure 4.1. View of turfgrass on putting green at Arcola Country Club, New Jersey, USA, 1915. [41][*]

* The numbers in brackets correspond to the figure source acknowledgments listed in the appendix.

CHAPTER 4

History of Golf Course Turfgrass Development

NUMEROUS ANCIENT GAMES WERE PLAYED WITH A STICK USED TO HIT AN OBJECT FROM A HARD surface such as bare ground, stone streets, and ice. Examples are *paganica* of Roman times, *jeu de mail* or *á la chicane* in France during late-Roman times, *chole* and *croisse* in Flanders during the 1100s, *cambuca* in England from the 1200s, and *colf* in the Dutch Netherlands in the 1300s. Apparently, golf evolved from one or more of the games played in the lowland countries of western Europe, possibly from chole. Most of these games used a carved, wooden ball. The basic concepts of modern golf emerged in eastern Scotland in the sandy coastal ecosystem. These concepts included the following:

- Hitting a ball from a stationary position
- Preset target holes
- Location on an extensive playing area of grass
- Long playing distance
- Multiple types of clubs used

Records for the term "golf" first appeared in the Acts of the Scottish Parliament of 1457, 1471, and 1491, but this game may have been a transitional version in terms of the criteria just listed. These early references involved efforts to discourage golf in favor of archery, which was needed as a military expertise for war. Documentation of golf play during the 1500s and 1600s are lacking but probably occurred in eastern Scotland in some advanced transitional version. Records of more modern-style golf play emerged in the mid-1700s, with rudimentary clubs for golfers being formed. A golf competition played on the St Andrews public linksland in 1754 involved the Noblemen and Gentlemen of Fife and the Gentlemen Golfers of Edinburgh. Now known as the St Andrews Old Course, this site was one of the earliest golf courses and still is in play. The narrow strip of low-growing turfgrass bordered by masses of dense vegetation leading to the harbor became known as the Green. A legal title was acquired in 1552 when the town charter reserved for the citizens of St Andrews the right of using the links for "golfe, fute-ball, shuting and all games, as well as casting divots, gathering turfs, and for pasturing of their livestock." This charter legally confirmed the citizens' right to play golf over the linksland. Note that the infertile, fine-sand soil was not desired for agricultural operations, being poorly supportive of even sheep grazing thus the granting of recreational rights to the townspeople. Early evolution of golf courses may be interpreted in relation to actual detailed records from the Royal and Ancient Golf Club of St Andrews as chronologically presented in table 4.1.

TABLE 4.1. CHRONOLOGY OF KEY TURFGRASS FEATURE DEVELOPMENTS ON THE ST ANDREWS OLD COURSE DURING THE EIGHTEENTH AND NINETEENTH CENTURIES.

1754	Competitive round of golf set at 22 holes.
1754	Teeing of a ball must be within 1 club length of putting hole.
1764	Standard round of golf stabilized at 18 putting holes, from 22.
1777	Teeing area extended to between 1 and 4 club lengths from putting hole.
1789	Term hole-green first mentioned.
1812	Teeing area no less than 2 club lengths nor more than 4 from putting hole.
1830s	Fair greens widened to between 30 and 40 yards (27–37 m).
1850s	Putting greens enlarged.
1851	Teeing area extended to between 4 and 6 club lengths from putting hole.
1858	Teeing area extended to between 6 and 8 club lengths from putting hole.
1860s	Fair greens widened to between 50 and 60 yards (46–55 m).
1875	Teeing area extended to between 8 and 12 club lengths from putting hole.
1877	Option to have teeing ground site separate from putting green.
1880s	Fair greens widened to between 80 and 100 yards (73–91 m).
1882	Turfgrass teeing areas fully separated from putting greens.
1891	Putting hole diameter standardized at 4 ¼ inches (108 mm).

Golf progressively expanded in the United Kingdom and eventually spread worldwide. In North America, records of golf play appeared in the 1780s in South Carolina but had disappeared by the War of 1812. Then, five golf organizations were formed in Quebec and Ontario, Canada, during the 1870s, followed by additional clubs for golfers in four states of the United States during the 1880s as well as in other English speaking countries. The game grew more rapidly, especially in North America. Golf enjoyed an unprecedented growth in player numbers and golf course facilities following WWII. Communities were developed with a golf course as the central attraction. Resort hotels around the world found that a golf course added to their popularity.

Golf Course Layout. The standard golf course layout consists of 18 golf holes, each with a teeing ground(s), usually a fairway, and putting green. The total area for a golf course typically is in the range of 130 to 225 acres (53–91 ha). The playing length of an 18-hole golf course is variable but normally ranges from 6,200 to 6,600 yards (5,669–6,035 m), and championships range from 6,600 to 7,200 yards (6,035–6,584 m). Designs that blend with the existing topography are preferred.

Modern putting greens under intense play are best constructed of a high-sand root zone with a perched-hydration system over a closely spaced drain-line network (1, 2).* Putting greens normally range in size from 5,000 to 7,500 square feet (465–697 m²), with a resultant total area of 2 to 4 acres (0.8–1.6 ha). Golf tees typically have the same total area. The fairway width normally ranges from 25 to 60 yards (23–55 m), with a total area of 25 to 50 acres (10–20 ha). Surrounding each golf hole are two to three differentiated turfgrass areas of progressively higher mowing heights referred to as intermediate, primary, and secondary roughs. Other design components that may be included and vary in both number and placement include sand bunkers, water features, trees, shrubs, practice putting greens, a practice range, a turfgrass nursery, cart paths, and an operations center.

Books focused on golf course architecture and planning are described in chapter 13. Also, specialty books on irrigation, mowing, and other equipment are discussed in chapter 13.

---

* The numbers in parentheses correspond to the reference list found at the end of the chapter.

EARLY HISTORY OF TURFGRASS SURFACES

Little is known about early turfgrass culture on golf courses. The linksland along the Scottish coast where a number of golf courses evolved was formed at the close of the Ice Age as the sandy floor rose above the sea. The resulting sand dunes and marshlands formed an undulating terrain. Regular rains favored grasses, and close-grazing rabbits resulted in patches of low-growing grass species. The linksland also had natural hazards such as heathers, gorses, reeds, pools, and pits of sand where sheep huddled for shelter from the wind.

On Scottish linksland, the closely grazed grasses were composed of fine-leaf fescue plus some colonial bentgrass. A rabbit grazing population of 60 to 80 per acre (150–200 per ha) can maintain a height of grass ~0.4 inch (10 mm), typically in irregular patches near their warrens. Also a mature sheep grazing population of ~5 per acre (12 per ha) could produce a grass height of 1.2 to 2.0 inches (30–51 mm). Sweeping was practiced prior to play days to remove the rabbit and sheep "purls" (excrement). The biological cutting of grass on hole greens with rabbits was at no direct cost, but there was the labor expense to remove the rabbit byproduct that interfered with ball movement.

By 1820, rudimentary maintenance on the Old Course at St Andrews included the following:

- Cunnigers (rabbits) grazing the hole greens (putting green sites)
- Sheep grazing the fair greens (fairways)
- Putting holes made or repaired and switching at 3 to 4 week intervals
- No mechanical mowing
- Cunniger burrows and scrapes filling in play areas periodically

By 1885, the Old Course golf culture involved the following:

- No irrigation
- Putting holes changed and swept (switching) weekly as of 1851
- Putting greens mowed occasionally with a push-reel mower
- Sheep grazing the fair greens
- Cunniger burrows and scraps filled in play areas
- Composting for sand patching
- Sand patching practiced mainly on putting green cuts and selected fair green areas
- Rolling practiced periodically

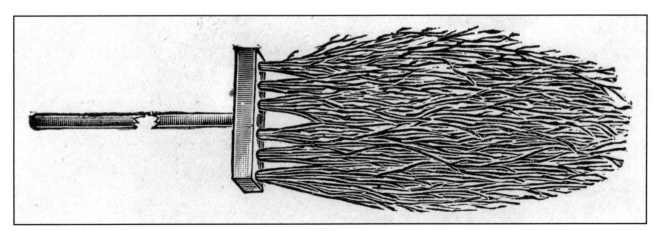

Figure 4.2. Broom used in sweeping greens to remove rabbit droppings, earthworm casts, and debris. The broom tips were made of natural birch or bamboo, circa 1920s. [14]

- Manual graiping (forking) practiced
- "Returfing" (sodding) practiced, including stabilizing bunker perimeters

The "feathery" golf ball consisted of a sown leather cover tightly filled with feathers. It was commonly used in Scotland for 200 years up until the 1850s, as the "gutty" (*gutta percha*) ball was introduced in the late 1840s. The gutty ball tended to roll over the turfgrass surface, especially when struck with a metal putter. In contrast, "scloffing" (a stroke) from a wooden club on a feathery ball lifted the ball toward the hole. This ball roll technique change eventually led to the implementation of cultural practices needed to produce smoother putting greens including rolling, sand patching, and occasional mowing.

Longnose wooden golf clubs were used from the fifteenth into the nineteenth centuries. Use of this club usually involved a stroke with less divoting of the turfgrass. Their use declined in the 1850s after the gutta percha ball was developed. During the 1880s, cleek clubs with iron heads became popular, causing greater turfgrass "cuts" (divoting) damage. This change in golf clubs probably resulted in separation of the hole green (putting green) and teeing ground in the 1880s in order to protect the turfgrass putting surface.

MIDTERM HISTORY OF TURFGRASS SURFACES

Up to the 1920s, much of the golf course work involved in turfgrass culture was accomplished manually, with some assistance from horses. For example, a row of workers on their hands and knees aligned across a fairway was used to remove weeds. During the 1920s, some courses could afford the new tractors with gasoline-powered engines and wide steel wheels.

The traditional approach to putting green construction during the early 1900s involved elevating a small plateau using the adjacent soil; they were termed "pushup" greens. Then some constructions utilized a thick base of large clinker or cinders, followed by a shallow layer of pea gravel, and topped by a 1.5 to 2 foot depth of fine soil. Note the very general characterizations. There was a lack of quantitative specifications in terms of particle size, range, and depth. It was not until the 1960s that detailed science-based construction specifications came into use.

To illustrate the relative timing in probable adoption of various turfgrass cultural practices on golf courses, a summary of practices implemented at the St Andrews Old Course during the 1890 to 1940 period is presented chronologically in table 4.2.

Figure 4.3. A July 1858 calotype of turfgrass on the eighteenth putting
green at the St Andrews Old Course, Scotland, UK. [47]

TABLE 4.2. CHRONOLOGY OF KEY TURFGRASS CULTURAL DEVELOPMENTS ON THE ST ANDREWS OLD COURSE FROM 1890 TO 1940.

1887	Shallow wells dug near putting greens for watering
1891	Mowing fairway with horse-drawn, roller-drive, reel mower
c. 1899	Manual "graiping" (forking)
Late 1800s	Manure composting/application
Early 1900s	Broad-scale topdressing
1911	Motor-powered, reel mower utilized
1912	Nine cisterns were dug deeper for putting green watering
1912	Rabbit control authorized
1919	Earthworm control tests initiated on putting greens
1920s	Manual weed removal on putting greens initiated
1923	Horse-drawn, three-gang, reel mower utilized on fairways
1927	Tractor-drawn, gang, reel mowers
Late 1920	Sweeping twice weekly
1928	Earthworm irritant mowrah meal applied
c. 1930	Mechanical spiking
1933	Manual core cultivation of greens and fairways
1930s	Artificial manuring (mineral fertilizers) used
1934	Light, frequent topdressing of putting greens
1936	Sheep removed from links except for WWII

Smoothing and Rolling. Mechanical smoothing by rolling is one of the earliest turfgrass cultural practices, dating back to the 1700s. It predates even mowing. These heavy manually pushed rollers were hand carved from stone. Just when rolling was first practiced on putting greens is unclear. Rolling with manually pushed, lightweight, wooden rollers was being used on putting greens by the mid-1800s, and metal, water-ballast rollers were in use by the 1890s. Larger horse-drawn rollers also were used during the late 1800s on fairways. The main function was to smooth out surfaces after sand patching or filling of rabbit burrows and scrapes as well as to smooth out hoof marks from animals. Rolling as a regular practice declined in the early 1900s, as potential problems with soil compaction were recognized.

Grazing and Mowing. Biological cutting by grazing rabbits on hole greens and by sheep on fair greens were the original standard. However, large populations of rabbits were not available in North America and many other parts of the world. In these countries, the approach on putting greens was to keep the soil nutrient levels impoverished and dry. When rains occurred that stimulated too much leaf growth, these areas were cut manually by scything.

Mechanical mowing of putting greens became common in the 1870s, using manual-push, side-wheel-drive, reel mowers. It was not until the early 1900s that motor-powered, reel mowers for putting greens became available. The three-gang motor-powered "Overgreen" was introduced in the 1930s. Developmental stages in mower use advances were similar but delayed somewhat.

Fairway mowing did not become feasible until the development of the horse-drawn, 42-inch-wide (4.3-meter-wide), roller-drive, reel mower by Alexander Shanks of Scotland in 1842. Then in the late 1800s, the horse-drawn, side-wheel-drive, reel mower was introduced. A breakthrough innovation in 1912 for

Figure 4.4. Motor-powered Worthington Overgreen with three, attached, 36-inch-wide, reel mowers, Worthington Mower Co., Pennsylvania, USA, circa 1931. [2]

mowing extensive fairway areas was the three-gang, reel mower developed by Charles Worthington in the United States. Multigangs used horse-drawn hitches initially, but most had changed to tractor-drawn power units by the late 1920s.

Historically the mowing heights of putting greens, tees, and fairways have been gradually reduced as represented in table 4.3. The degree of closeness has accelerated since 1980. A key factor facilitating close mowing was the development of cultivars with the capability to sustain the desired leaf density under the severe defoliation. As dictated by the lowered cutting height, the mowing frequency was increased. For putting greens, the frequency has progressively changed from weekly in the 1920s to three times per week to four to five times per week to daily.

Sand Patching and Topdressing. Early sand patching was used on the linksland in Scotland by the 1840s as sand was readily available nearby. It was used to fill cunniger scrapes and burrows. Eventually, sand was composted with animal manures, butcher scraps, and/or seaweed. The compost contained nutrients that

TABLE 4.3. TYPICAL TRENDS IN MOWING HEIGHTS (INCH[ES]) OF TURFGRASS
FEATURES ON US GOLF COURSES FROM 1920 TO 2010.

Year	Putting greens	Tees	Fairway	Primary rough
1920	0.3–0.5	0.5–1.0	1.5–2.5	—
1950	0.2–0.35	0.4–0.75	0.6–1.2	2.0–4.0
1980	0.2–0.3	0.3–0.6	0.5–1.0	1.5–4.0
2010	0.1–0.2	0.25–0.5	0.4–1.0	1.5–4.0

Figure 4.5. Manually transporting via wheelbarrow, dumping in piles, and scattering shredded manure over a turfgrass, circa 1905. [10]

stimulated grass growth and recovery of weak, thinned turfgrass patches on greens and fairways. Composted topdressing was used over putting greens by the late 1800s to smooth over cuts and ball marks, with several annual topdressings used by the 1920s. The application method involved hauling to the site by horse-drawn wagon; filling of wheelbarrows; manual pushing/tipping in piles around the area to be treated; manual spreading by slinging in an arc via pitch-fork or shovel, depending on the degree of decomposition; and dragging or brushing to further aid uniform spreading.

Manuring and Fertilization. The use of animal excrements as a plant nutrient source was referred to as manuring. It was not used in earlier times over broad areas of fairways, as it created excessive leaf growth that necessitated laborious manual scything. However, by the 1880s, raw manure was being applied to thinned or bare areas from wear, sheep scalds, and other stresses in the turfgrass to stimulate revegetation. Shortly thereafter, composting of manure for turfgrass use became commonplace, especially as many sources were intermixed with straw, stover bedding, and road sweepings.

Organic materials encompassing manure, slaughterhouse waste, seaweed-kelp, peat moss, and leaf mold were being added to the compost piles by 1900. Also there were composting operations at golf courses in which sand was being added in layers and intermixed. As this practice increased, the term "composting" was used to represent both the process and also the application. Eventually the term top-dressing was used for the latter.

Some processed commercial manures, mostly produced from rural agricultural operations, were being marketed into urban areas in the early 1900s. Included were bone meal, dried blood, hoof-and-horn meal, guano, and seed meals. A new term "artificial manure" came into use in the 1920s. During this time, the age of the automobile had arrived and horse-drawn transportation began a long decline, as did the supply of manure. The artificial manures were actually mineral-based fertilizers. Included were ammonium nitrate, ferrous sulfate, potassium chloride, potassium nitrate, potassium sulfate, rock phosphate, and sodium nitrate.

After WWII, the organic, petroleum-based nitrogen carriers came into use. One of the earliest was water-soluble, quick-release urea. Controlled-release types also became available such as the ureaformaldehyde (UF), methylene ureas (MU), and isobutylidene diurea (IBDU). Another form of controlled-release carrier that came into common use on turfgrasses is the coated type such as sulfur coated and plastic resin coated.

Early application of mineral-based fertilizers was by hand slinging onto patches of thinned turf. Also, wide-mouthed shovels were used with a throwing motion in an arc. More uniform particle distribution was achieved when gravity-drop, manually pushed spreaders became available in the 1910s. Also, a manually pushed centrifugal type spreader was available.

Watering and Irrigation. The removal of sheep in favor of mechanical mowing occurred at the St Andrews Old Course decades later than most golf courses in North America. Perhaps the last several decades were symbolic, as tradition is a vital heritage at this historic site. Also, irrigation was minimized and long delayed in comparison to other parts of the world where summer droughts are more commonly experienced.

Early attempts at watering putting greens involved manual pumping from cisterns (shallow dug wells) into barrels hauled by a horse-drawn wagon or by a large wooden tank mounted on wheels with horse-drawn hitch. Water delivery onto the putting greens was by manual pumping with a "water bowser" combined with gravity flow. Eventually in the early 1900s, underground piping systems were being installed on golf courses, extending from a water source to each putting green and terminating in an aboveground water hydrant. A moveable hose would be connected to the hydrant and manually extended onto various positions on the putting green with an attached hose-end sprinkler. Similar systems were installed along some fairways in the United States in the 1920s with hydraulic-powered traveling sprinklers pulled over the fairways. The next improvement was a network of fixed underground piping with regular spaced risers connected to coupler valves positioned at the soil surface for manual attachment of quick-coupler impact sprinkler heads. It came into widespread use around putting greens and tees after WWII and shortly thereafter on fairways.

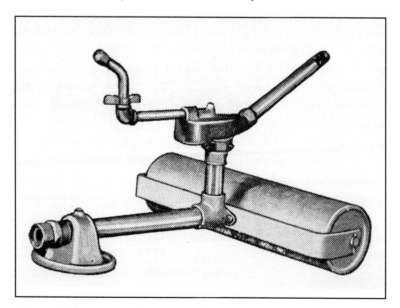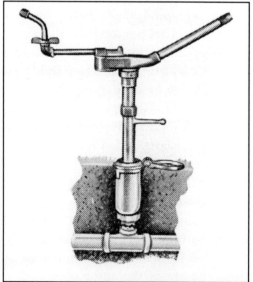

Figure 4.6. Rotary spray above-ground sprinkler on roller base for hose-end attachment (left), and "hoseless" underground sprinkler system consisting of self-closing snap valve and rotary spray sprinkler head with manual coupler (right), both by Skinner Irrigation, 1930s. [41]

Graiping and Turfgrass Cultivation. Problems with soil compaction eventually occurred as the intensity of golf play increased. Early correction efforts involved "graiping," which was a downward action that forced holes into soil by manual piercing as by a solid, four-pronged fork. It became a common practice in Scotland by the 1880s. A significant advance in turfgrass cultivation was the development of Paul's Process Patent Turf Fork in 1925 in Scotland. It was a manual, foot-applied, three-tined coring device with core ejection that continued to be used into the 1950s.

Various manual-push, shallow spike units mounted on rollers or rotating reels became available during the 1920s. Wider spiking and slicing units mounted on rotating reels became available in the 1930s for use on fairways.

Figure 4.7. "Paul's Process" for manual turfgrass core cultivation, involving a hollow three-tined fork as used in 1919 in Scotland, UK. [33]

Figure 4.8. Motor-powered core cultivation machine invented by Thomas Mascaro of Pennsylvania, USA. [4]

The first truly effective motor-powered, mechanical coring machine was invented by Thomas Mascaro of Pennsylvania, USA, in 1946. This was another breakthrough innovation that came into widespread use, including wider tractor-pulled units for fairways. Then in 1980, a tractor-mounted, solid-tine, punch-type cultivation device was developed in the Netherlands. It had the key capability to penetrate soils two to three times deeper than earlier turfgrass cultivation devices.

Drainage. Drainage of excess surface water has always been a concern, especially since most early golf courses were positioned on the natural contours of a site. One of the earliest corrective approaches was to fill in small depressions where standing water accumulated. Another approach was to dig shallow drainage ways or even deep ditches where appropriate and where a lower outlet was available. Eventually all these reasonable options were accomplished. There remained level, fine-textured soil sites that could only be drained of excess standing water by internal drainage arrangements. "Pipes" were being installed in the 1880s as well as "tarred shoots." Whether these were comparable is unclear. Presumably tarred shoots were elongated, wooden, covered troughs. Clay drain tiles were in common usage by the late 1800s. They were the standard used until the late 1960s, when perforated polyethylene drainage pipe became available.

MODERN TURFGRASS SURFACES

Modern golf courses are covered with turfgrass of specific species and mowing heights, except for sand bunkers, water features, cart paths, and woodlands. Putting green surfaces are composed of creeping bentgrass in cool climates or low-growing hybrid bermudagrass, such as Champion, in warm climates. Some colonial bentgrass plus fine-leaf fescue is used in moist-cool climates such as the United Kingdom. Vegetatively propagated creeping bentgrass cultivars dominated on putting greens in North America from 1920 to 1960, when the seeded cultivar Penncross was successfully introduced. Actually, stemmy dactylon bermudagrass was used on putting greens up to 1956 when the first hybrid-bermudagrass cultivar Tifgreen was developed.

Fairways and tees consist of creeping bentgrass, colonial bentgrass plus fine-leaf fescue, hybrid-bermudagrass, Kentucky bluegrass, perennial ryegrass, seashore paspalum, or zoysiagrass, depending on the climate and soil. Roughs are basically seeded dactylon bermudagrass, fine-leaf fescue, Kentucky bluegrass, tall fescue, tropical carpetgrass, or Japanese zoysiagrass, depending on the climate.

The modern mowing height for putting greens is in the range of 0.1 to 0.2 inch (2.5–5.1 mm), and it usually occurs daily during the playing season. Tees are mowed at 0.25 to 0.5 inch (6.4–12.7 mm), depending on the turfgrass species. The fairways may be mowed in the range of 0.4 to 1.0 inch (10–25 mm), depending on the species. Mowing of roughs varies in width, height, and combinations, with the intermediate rough at 0.75 to 2 inches (1.9–5.1 cm) and the primary rough at 1.5 to 4 inches (3.8–10.2 cm), depending on the turfgrass species. The secondary rough is commonly unmowed.

Historical Documentation. The first major book on golf course turfgrass culture, **Golf Greens and Green-Keeping**, was published in 1906 in London, England, UK, under the editorship of Horace G. Hutchinson (5). Subsequently, **Turf for Golf Courses**, a technically oriented book on the maintenance of turfgrasses for golf courses, was published in 1917 under the authorship of Drs. Charles V. Piper and Russell A. Oakley (8); this was followed by **Turf Management** by H.B. Musser in 1950 and 1962 (6, 7) and by **Turf Management for Golf Courses** by Dr. James B Beard in 1982 and 2001 (1, 2). These latter three books were sponsored by the United States Golf Association. Books devoted fully to golf course culture and construction described in chapter 12 are listed in the summary that follows.

*Summary of Golf Course Turfgrass Culture and Construction Books
Described in Chapter 12 as Listed by Author or Organization*

Arthur in 1997 & 2003 (UK)
Bandeville in 1928 (France)
Beale in 1905 & 1908 (UK)
Beard in 1982, 1988, 1990, 1991, 1992, 1993, & 2001 (USA)
Bengeyfield in 1989 (USA)
Campbell in 1982 (UK)
Casati in 2004 (Italy)
Colt in 1900 (UK)
Crockford in 1993 (Australia)
Cubbon & Markuson in 1933 (USA)
Davis in 1990 (USA)
Dawson in 1932 (UK)
Ehara in 1982 (Japan)
Farley in 1931 (USA)
Federazione Italiana Golf in 1990 & 1994 (Italy)
Ferguson in 1968 (USA)
Gault circa 1912 (UK)
Graves in 1998 (USA)
Hacker in 1992 (UK)
Hatsukade in 1995 (Japan)
Hawtree in 1983 (UK)
Hayes in 1992 (UK)
Hotchkin circa 1938 (UK)
Hurdzan in 1985, 1996, 2004, & 2006 (USA)
Hutchinson in 1906 (UK)
Joo in 1992 & 2002 (South Korea)
Kim in 1992 & 1999 (South Korea)
Kim in 1993 (South Korea)
Kim in 2005 (South Korea)
Knoop in 1977 & 1980 (USA)
Lee in 2005 & 2007 (South Korea)
Lees in 1918 (USA)
Levy et al. circa 1950 (New Zealand)
Luff in 2002 (USA)
Mackenzie in 1920 & 1995 (UK)
Maki in 1991 & 1992 (Japan)
Malaysian Golf Association in 1984
Mascaro in 1958, 1960, 1964, 1974, & 1992 (USA)

McCarty & Elliot in 1993 (USA)
McCarty et al. in 2001 & 2005 (USA)
McIntyre et al. in 2007 (Australia)
McMaugh in 1970 (Australia)
Metsker in 1996 (USA)
Monje-Jiménez in 1997 & 2002 (Spain)
Murdock in 1968, 1978, & 1991 (USA)
Murray in 1932 (South Africa)
Musser in 1950, 1962, & 1964 (USA)
Nakamura in 1993 (Japan)
National Golf Foundation in 1947 & 1995 (USA)
Nikolai in 2005 (USA)
Noer in 1928, 1937, 1946, 1947, 1950, & 1959 (USA)
O.M. Scott & Sons in 1921, 1922, 1923, 1928, & 1931 (USA)
Park in 1990 (UK)
Perris in 1996 (UK)
Piper & Oakley in 1917 (USA)
Price in 1989 & 2002 (UK)
Quast in 2004 (USA)
Sachs in 2002 (USA)
Schumann et al. in 1998 (USA)
Shim in 1992 (South Korea)
Snow in 1993 (USA)
Souma in 1937 (Japan)
Stanley Thompson circa 193x (Canada)
Stinson in 1981 (USA)
Sutton & Sons in 1896 & 1948 (UK)
Sutton in 1933 & 1950 (UK)
Sutton circa 1912 (UK)
Thorburn and Co. in 1899 & 1908 (USA)
United States Golf Association in 1968, 1994, & 2002 (USA)
Viergever in 1970 (USA)
White in 2000 (USA)
Wilson in 1966, 1970, & 1971 (USA)
Witteveen & Bavier in 1998 & 2005 (USA)
Yanagi in 2002 (Japan)

There are other books that have small sections on golf putting greens and fairways described in chapter 12 under authors and organizations such as Barron in 1906 & 1923 (USA); Beale in 1924, 1931, & 1952 (UK); Boyce in 194x (Canada); Collins in 192x (USA); Cunningham in 1914 (USA); Dawson in 1939 &

1968 (UK); Greenfield in 1962 (UK); Hawthorn in 1977 (UK); Hope in 1983 & 1990 (UK); Kansai Golf Union in 1983 & 1991 (Japan); Lewis in 1948 (UK); Macself in 1924 & circa 1930 (UK); McIntyre in 2004 (Australia); New Zealand Institute of Turf Culture in 1961; Palmer circa 193x (UK); Panella in 1972, 1981, & 2000 (Italy); Pettigrew in 1973 (UK); Reed in 1950 (UK); Rees in 1962 (Australia); Sanders in 1910 & 1920 (UK); Smith in 1948 & 1950 (UK); Tucker in 1915 (USA); and Walker in 1971 (New Zealand).

GREENKEEPERS AND GOLF COURSE SUPERINTENDENTS

Just when early greenkeepers were employed for maintenance of golf courses is unclear. The earliest greenkeeper positions were part time and were established in the late 1700s. Note that the correct term is "greenkeeper," not greenskeeper. Historically, the term "green" referred to the whole golf course. The Society of St Andrews Golfers (later to become the Royal and Ancient Golf Club of St Andrews) requested special playing privileges on the Old Course in 1754. In return, the society agreed to pay for maintenance of the Old Course. This is a very significant historical event because it indicates the society was concerned with "golf course maintenance" in 1754. The question is what did maintenance mean at that time? Many early greenkeeper positions involved much time devoted to what are now golf club manager and golf professional responsibilities. It was not until after 1900 that true full-time golf greenkeeper or superintendent positions became common at golf courses.

ORGANIZATIONS

The United States Golf Association (USGA) was formed in 1894 to administer the game and rules nationally. Then in 1920, the USGA became active in turfgrass research and education by organizing the USGA Green Section. Key early educational periodicals were **The Bulletin** published from 1921 to 1933, **Turf Culture** from 1939 to 1942, **USGA Journal and Turf Management** from 1948 to 1963, and **USGA Green Section Record** from 1963 to 2010.

The Regional Turfgrass Advisory Service was formed by the USGA Green Section in 1953. Annual consultation visits to individual golf courses were made available. In addition, the USGA Green Section has provided substantial financial support of turfgrass research at many state universities throughout the United States. Since inception of the Green Section in 1920 and through 2010, the USGA has invested more than $36 million in turfgrass research activities. This funding and encouragement have been significant forces behind the advances in turfgrass research and education programs at many state universities, with the resultant technology benefitting all segments of turfgrass use, including home lawns. A major increase in USGA funding occurred in 1983, through the efforts of J. B Beard, with a focus on water conservation (3).

The National Association of Greenkeepers of America (NAGA) was formed in 1926 to further the professional recognition and competence of greenkeepers through a continuing education program. The organization continued to grow and is today known as the Golf Course Superintendents Association of America (GCSAA). The association sponsored a formal golf course educational conference starting in 1928, and it has continued this annual event. A golf course equipment and maintenance products show was subsequently added. In 1971, a certification program (the Certified Golf Course Superintendent program; CGCS) was initiated. The association began publishing the **National Greenkeeper** in 1927 through 1933, followed by the **National Greenkeeper and Turf Culture** in 1933, **The Greenskeepers' Bulletin** in 1933, **The Greenkeepers' Reporter** from 1933 to 1951, **The Golf Course Reporter** from 1951 to 1965, **Golf Superintendent** from 1966 to 1978, **Golf Course Management** from 1976 to 2006, and **GCM** since 2007.

REFERENCES

1. Beard, J.B. 1982. Turf Management for Golf Courses. 1st ed. Burgess Publ. Co., Minneapolis, Minnesota, USA.
2. Beard, J.B. 2001. Turf Management for Golf Courses. 2nd rev. ed. Ann Arbor Press, Chelsea, Michigan, USA.
3. Beard, J.B. 2011. A brief history of the science of turfgrass water use and conservation. In: S.T. Cockerham and B. Leinauer, editors, Turfgrass Water Conservation, Publication 3523, Univ. of California Agric. and Nat. Resources. p. 5–14.
4. Hardt, F.M. 1939. Introducing "turf culture." Turf Culture 1(1):1–5.
5. Hutchinson, H.G. 1904. Golf Greens and Green-Keeping. Charles Schribners' Sons, London, England, UK.
6. Musser, H.B. 1950. Turf Management. 1st ed. McGraw-Hill Book Co., New York, New York, USA.
7. Musser, H.B. 1962. Turf Management. 2nd rev. ed. McGraw-Hill Book Co., New York, New York, USA.
8. Piper, C.V., and R.A. Oakley. 1917. Turf for Golf Courses. Macmillan Co., New York, NY, USA.

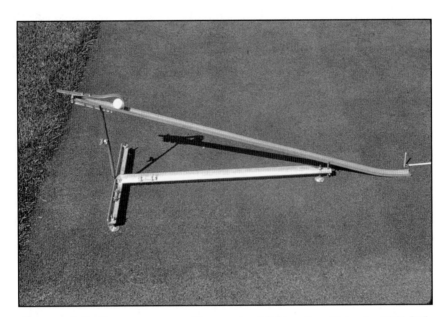

Figure 4.9. Ball roll test apparatus used in research at Michigan State University, 1962. [26]

Figure 4.10. View of turfgrass in front of the Saint Andrews Golf Clubhouse, the first permanent golf club organized in the United States, in Mt. Hope, New York, USA, in the early 1920s. [34]

CHAPTER 5

Early Turfgrass Innovations

PRIOR TO THE TWENTIETH CENTURY, THE "ART" OF TURFGRASS CULTURE WAS DEVELOPED THROUGH trial-and-error experiences. Typically, unprocessed manures were used for fertilizer, Bordeaux mixture for disease control, nicotine solution for insect control, and sulfuric acid or knives for weed removal. The height of the grass was originally controlled by the intense grazing of sheep and rabbits. Hand scything was employed on some turfgrass areas.

The true heritage of turfgrass science and culture is best represented by the principles, cultural practices, grasses, materials, and advances in equipment that have been developed by turfgrass scientists, private companies, professional turfgrass managers, and amateur practitioners. There have been many improved turfgrass cultivars, fertilizers, pest controls, cultural approaches, and advances in maintenance equipment, which, in their time, were significant improvements that served a valuable role. However, most of these no longer exist, except for a few pieces of equipment preserved by collectors. Thus the historical record of evolutionary advances in the art and science of turfgrass culture is best preserved through writings and pictures in books (see chapter 12).

Books provide the best historical record, especially in the 1700s and 1800s. In these earlier years, advances were based on trial-and-error approaches under field conditions, when the art of turfgrass culture was far more advanced than the science. Other significant publications in terms of a historical record include the turfgrass conference proceedings (see chapter 16) and trade periodicals (see chapter 17) of the 1900s. These publications provide early documentation of turfgrass advances and techniques developed primarily by turfgrass practitioners and gardeners.

The 1960s, 70s, and 80s were the periods when the major advances in the science of turfgrass culture and quality occurred. These turfgrass advances are recorded in publications such as scientific periodicals, research summaries, and field day reports (see chapter 15). The science of turfgrass culture became more important than the art during this period. Still, the art aspects, learned only through actual field experience, remained important in the development of turfgrass practitioners, especially for greens and sports fields.

The evolution of turfgrass use as practiced today occurred in association with animal agriculture in climates favorable for grass growth, especially rainfall and temperature. The earliest significant uses of managed turfgrass for lawns, parks, and sports fields were in the United Kingdom, where the rainfall distribution throughout the year was reasonably good and where the moderate temperatures favored the growth of cool-season turfgrasses, such as *Agrostis, Festuca, Lolium,* and *Poa.* In addition, the grazing of sheep was a significant agricultural activity throughout the countryside from 1800 to 1950.

EARLY PIONEERING INNOVATIONS

Key advances that furthered the use of turfgrasses involved inventions and developments achieved through trial-and-error activities that are termed "the art of turfgrass culture" (1).* There were many significant contributions made over the past two centuries, with a few key events presented herein. Ten developments that highlight the turfgrass discovery and invention era are summarized in table 5.1 and are discussed in the following sections.

Reel Mower. For years turfgrass areas were cut to a relatively uniform height, either by a hand scythe or by a hand sickle in the case of closely maintained turfgrasses that were cut more frequently. The leaves of grasses were best cut by this method when the grass was wet, such as during early morning dews or after rains. This

TABLE 5.1. KEY EVENTS IN THE TURFGRASS DISCOVERY AND INVENTION ERA FROM 1830 TO 1950.

Year (circa)	Contribution/invention	Inventor/contributor
1830	Reel mower, mechanical and hand pushed	E.B. Budding, England, UK
c. 1840s	Cylindrical clay tile drains	England, UK
c. 1880s	Weed-free grass seed processing and testing	O.M. Scott, Ohio, USA
c. 1890s	Irritant for earthworm management	P.W. Lees, England, UK
1912	Side-wheel-drive mowers on multigang frame	C.C. Worthington, Pennsylvania, USA
1928	Slow-release (organic) turfgrass fertilizer	O.M. Scott & Sons Co., Ohio, USA
1928	Powered rotary mower	J.M. Miller, Kentucky, USA
1930–32	Turfgrass fungicide development	J.L. Monteith & A.S. Dahl, Washington, DC, USA
c. 1930–35	Pop-up sprinkler head	W.V.E. Thompson, California, USA
1945	Selective broadleaf weed control—2,4-D	G.F.F. Davis, Washington, DC, USA

Figure 5.1. Trenching and installing clay-tile subsurface drainage lines prior to planting. [44]**

---

* The numbers in parentheses correspond to the reference list found at the end of the chapter.

** The numbers in brackets correspond to the figure source acknowledgments listed in the appendix.

Figure 5.2. Wood-frame, manual-crank grass seed cleaning apparatus as used by O.M. Scott in the 1880s. [32]

was a very laborious, time-consuming activity. This started to change in 1830 with the invention of the reel mower by Edward Beard Budding (2, 4, 6). The first manual-push, reel mower was more cost effective, which allowed the opportunity for middle-class residents to maintain residential and village green turfgrass areas that enhanced their functional quality of life.

Clay Drain Tile. Fired clay tiles for subsurface drains were developed in the United Kingdom circa the 1840s. This was the standard worldwide technique for subsurface drainage of soils for over 100 years. During most of that period, the clay tiles were installed by manual digging of the trenches. Thus subsurface drains did not come into widespread use until the development of petrol-powered, mechanical, wheel-trenching machines in the early 1900s.

Weed-Free Grass Seed. The next major advance occurred in the 1880s when the marketing of weed-free grass seed based on proper seed cleaning, processing, and testing techniques was pioneered by Orlando M. Scott (see chapter 12) of Ohio, USA. Initially, he utilized a manually cranked, wooden seed cleaning machine that he had modified. Prior to that time, grass seed was harvested from pastures that were typically contaminated with weeds, and the resultant seed was sold directly to turfgrass users. There were no effective, selective controls for the weeds in harvest fields or in the home lawns and turfgrass areas planted with the weed-contaminated grass seed. Thus the solution was the development of procedures to clean weed seeds out of the grass seed. In addition, O.M. Scott pioneered seed testing procedures long before state governmental agencies enacted laws requiring seed testing and labeling.

Earthworm Management. A major advance in the 1890s was the development of an earthworm management control by Peter W. Lees of England, UK (see chapter 12). Prior to this event, the two main practices discussed in gardening books were rolling and mowing of the turfgrass, with rolling listed first. This can be attributed to the disruption of the surface by extensive earthworm populations, particularly

Figure 5.3. First gang mower, circa 1913, by Charles C. Worthington, with three, side-wheel-drive, reel units. [2]

in England, where early turfgrass culture evolved. Thus development of an effective earthworm irritant resulted in rolling being substantially less important as a turfgrass cultural practice. In fact, rolling was eventually recognized in the 1920s as having negative effects in terms of soil compaction, especially on clayey soils.

The earthworm management procedure involved applying an irritant, mowrah meal, to the soil surface and drenching it in with excess quantities of water (see Beale, Reginald, in chapter 12). As a result, the earthworms came to the surface, were raked into piles, shoveled onto wheelbarrows, and physically hauled off the turfgrass area (7).

It also should be noted that prior to this innovation the game of golf and golf courses had been limited principally to the coastal areas of Scotland and northern England, called linksland or seaside courses. Earthworms are not well adapted to sandy soils. Attempts to develop upland golf courses were relatively unsuccessful, principally because of the unplayable, clayey putting green surfaces caused by earthworm castings. The emergence and major expansion of golf courses on upland soils occurred coincidently and could be attributed to the earthworm management procedure developed by Peter Lees.

Multiple-Gang Mower. The first side-wheel-drive mower on a wooden multiple-gang frame with an 84-inch (2.13 m) mowing width was developed in circa 1912 by Charles Campbell Worthington.* A patent was filed in 1913 by C.C. Worthington and was granted in January 1917. First production by the Shawnee Mower Company was probably in Hoboken, New Jersey and in 1915 moved to Stroudsburg, Pennsylvania, USA. The name was changed to the Worthington Mower Company in 1920. The gang mower was a major advancement and opened the way for even wider multigang mowers to economically mow extensive turfgrass areas of parks, golf course fairways, sports fields, recreational areas, and other large open areas (see chapter 13).

Slow-Release Turfgrass Fertilizer. The first commercially produced slow-release turfgrass fertilizer was marketed in 1928 by the O.M. Scott and Sons Company in Ohio, USA. It was a natural organic product developed at Ohio State University through research funded by the O.M. Scott and Son Company of

---

* Information sourced from James B. Ricci, The Reel Lawn Mower History and Preservation Project, Haydenville, Massachusetts, US.

nearby Marysville, Ohio. It was marketed in a large cloth bag under the name of Scotts Turf Builder. This branded slow-release fertilizer continues to be sold today—some 85-plus years later.

Turfgrass Disease Control. During the late 1920s and early 1930s, two fungicides for controlling of a number of major turfgrass diseases were developed by Drs. John L. Monteith and Arnold S. Dahl of the USDA-USGA Arlington Turf Research Center in Washington, DC, USA (see Monteith, John, in chapter 12). These were the first effective fungicides for the control of *Microdochium* patch, *Rhizoctonia* brown patch, *Sclerotinia* dollar spot, and *Typhula* blight.

Rotary Mower. A number of patents were filed for various types of rotary mowers dating back to circa 1900. However, current records indicate* the earliest successful patent in terms of marketing and distribution of a motor-powered rotary mower was a 1928 US patent filing granted in 1931 to Josephus Miller of Louisville, Kentucky, and manufactured by the Louisville Electric Company. Then William Beazley of St. Petersburg, Florida, filed a US patent in 1929; it was granted in 1931 and the mower was manufactured by the Beazley Power Mower Company. David Cockburn of Iver, England, filed a Great Britain patent, which was granted in 1932 for the Rotoscythe that was manufactured by Power Specialities Limited, Maidenhead, England (3). Development of the powered rotary mower resulted in the capability to mow minimal maintenance turfgrasses growing at a higher height and at less frequent intervals. These are conditions in which reel-type mowers are less effective.

Popup Sprinkler Head. In the early 1930s, the first underground pop-up sprinkler head for irrigation was developed by Walter Van E. Thompson of the Thompson Manufacturing Company in California, USA. This was a major advance compared to the numerous types of individual, fixed, hose-end sprinklers of the oscillating or rotating type previously available, as they had to be manually moved frequently for effective irrigation.

Selective Broad-Leaved Weed Control. In the mid-1940s, the first truly effective herbicide for the selective removal of broad-leaved weeds from perennial grasses was developed by Dr. Gretchen Fannie-Fern

Figure 5.4. Large-scale herbicide sprayer with 400-gallon tank and 23-foot-wide spray boom, used for 2,4-D experiments, as operated on the National Capital Parks, Washington, DC, USA. [19]

---

* Information sourced from James B. Ricci, The Reel Lawn Mower History and Preservation Project, Haydenville, Massachusetts, US.

Davis in Washington, DC, USA. Some of the earliest turfgrass tests were conducted on the turfgrass mall area between the US Capitol and the Washington Monument (see Davis, Gretchen, in chapter 12). The development-use strategy for 2,4-D on turfgrasses was a major event. It remains a key herbicide in the management of turfgrass areas more than 50 years later.

Summary. In the modern times of the twenty-first century, some of these developments seem of minimal significance. However, at the time they were developed or invented, these contributions were very major advances in improving the quality and functional benefits of turfgrass. Modern turfgrass science evolved gradually based on these early inventions and art-dominated, trial-and-error developments between 1800 and 1950. These pioneering individuals and companies need our utmost respect for their very important contributions.

REFERENCES

1. Beard, J.B. 2001. The Art and Invention Era in the Evolution of Turfs: 1830 to 1952. Intl. Turfgrass Res. Conf., Toronto, Ontario, Canada.
2. Grace, D.P., and D.C. Phillips. 1975. Ransomes of Ipswich: A History of the Firm and Guide to Records. Inst. of Agric. History, Univ. of Reading, Reading, England, UK.
3. Halford, D.G. 1982. Shire Album No. 91: Old Lawn Mowers. Shire Publ. Ltd., Aylesbury, Buckinghamshire/Botley, Oxford, England, UK.
4. Hall, A. 1984. The Hall and Duck Collection: A Concise History of Lawn Mower Development. Andrew Hall, Sheffield, England, UK.
5. Lees, P.W. 1918. Care of the Greens. C.W. Wilcox, New York, New York, USA.
6. Weaver, C., and M. Weaver. 1989. Ransomes 1789–1989: 200 Years of Excellence—A Bicentennial Celebration. Ransomes Sims & Jefferies PLC, Ipswich, England, UK.

Figure 5.5. A portion of the O.M. Scott and Sons turfgrass research plots at Marysville, Ohio, USA, 1947. [32]

CHAPTER 6

History of Turfgrass Research

FROM THE TWELFTH THROUGH EIGHTEENTH CENTURIES, MUCH OF THE EARLY KNOWLEDGE OF turfgrass culture slowly evolved from practices originating in the United Kingdom and Europe. Turfgrass culture was an art attained through apprenticeship experience, keen observation, and trial-and-error methods. In 1805, William Curtis described his observations of the following five perennial grasses in terms of adaptation for use on lawns and bowling greens in England (10).[*]

Colonial bentgrass (*Agrostis capillaris* L.)—fine bent-grass.
For grass-plats and lawns, it seems likely to be the best of all our English species, being of ready growth, bearing the scythe well, producing fine foliage, and resisting drought better than most.

Kentucky bluegrass (*Poa pratensis* L.)—smooth-stalked meadow grass.
We shall just mention one striking character of this grass; it never throws up any flowering stems or bents, but once in a season (May), while many other grasses, especially the Ray-Grass and Dwarf-Meadow, are putting them forth perpetually; from this peculiarity, joined to its hardiness and verdure, it would appear to be a good grass for lawns or grass plats.

Hard fescue (*Festuca longifolia* Thuill.)—hard fescue grass.
It is early and productive, its foliage is fine, and of a beautiful green; hence we have thought it was of all grasses the fittest for a grass-plat, or bowling-green: but we have found, that though it thrives very much, when first sown or planted, it is apt to become thin, and almost disappear, after a while; from its natural place of growth.

Sheep's fescue (*Festuca ovina* L. var. *ovina*)—sheep's fescue-grass.
It appears to be applicable only to the purpose of making a fine-leaved grass-plat, that shall require little or no mowing. For this purpose it must be sown about the middle of August, in an open, not too dry, situation, broadcast, and that thickly, on ground very nicely prepared and levelled; when it has once got possession of the soil, it will form so thick a turf, as to suffer few intruding weeds, and may be kept in order with little trouble.

---

[*] The numbers in parentheses correspond to the reference list found at the end of the chapter.

Perennial ryegrass (*Lolium perenne* L.)—ray-grass or rye-grass.
As its foliage is of rapid growth, and its flowering stems are continually shooting forth, it should never be sown to form a lawn, grass-plat, or bowling-green.

The United States pioneered turfgrass research initially and has continued to provide leadership in the development of knowledge concerning the science of turfgrass culture. Most of the early turfgrass research was initiated through the efforts and monetary support of the golf interests in English speaking countries and in Japan. Manufacturers developed new, efficient, greatly improved turfgrass maintenance equipment by the early 1950s. Chemical companies developed improved, turfgrass-specific pesticide formulations for disease, insect, and weed control. New turfgrass culture findings became available from the state agricultural experiment stations of the major land-grant universities.

United States. Turfgrass species and polystand studies for lawns were initiated in 1880 at the Michigan Agricultural Experiment Station of Michigan A&M College, East Lansing, Michigan, USA, by the noted botanist Dr. William J. Beal. Three scientific papers were published from these studies (1, 2, 3). Thus 1892 marked the formal introduction of documented turfgrass science (table 6.1). It was not until 25 years later in 1917 that the next scientific paper on turfgrass research was published by the Rhode Island Agricultural Experiment Station in Rhode Island, USA.

TABLE 6.1. CHRONOLOGICAL LIST OF NINE EARLY SCIENTIFIC RESEARCH PAPERS DEVOTED SPECIFICALLY TO TURFGRASS CULTURE THAT WERE PUBLISHED UP TO 1930. NOTE THAT INVESTIGATIONS CONCERNING THE CONTROL OF WEEDS , DISEASES, AND INSECTS ARE NOT INCLUDED.

Year	Author	Title & citation	Location
1892	W.J. Beal	Some selections of grasses promising for field and lawn (1).	Michigan
1893	W.J. Beal	Mixtures of grasses for lawns (2).	Michigan
1898	W.J. Beal	Lawn-grass mixtures as purchased in markets compared with a few of the best (3).	Michigan
1917	B.L. Hartwell & S.C. Damon	The persistence of lawn and other grasses as influenced by the effect of manures on the degree of soil acidity (27).	Rhode Island
1927	J.D. Wilson	The measurement and interpretation of the water-supplying power of the soil with special reference to lawn grasses and some other plants (45).	Ohio
1928	B.E. Gilbert	An analytical study of the putting greens of Rhode Island golf courses (18).	Rhode Island
1929	E.S. Garner & S.C. Damon	The persistence of certain lawn grasses as affected by fertilization and competition (17).	Rhode Island
1930	H.B. Sprague & E.E. Evaul	Experiments with turf grasses in New Jersey (39).	New Jersey
1930	L.S. Dickinson	The effect of air temperature on the pathogenicity of *Rhizoctoniasolani* parasitizing grasses on putting-green turf (14).	Massachusetts

Subsequently, evaluations of grasses for lawn use were started in New Haven, Connecticut, USA, in 1886 (26). Then a formal turfgrass experimental garden was established by J.B. Olcott in association with the Connecticut Agricultural Experiment Station at South Manchester, Connecticut, in 1890. The objectives were to describe, evaluate, and select improved turfgrass species for lawns and seed production (36). No data or scientific research papers were published concerning the results from these efforts in Connecticut.

Turfgrass experiments were initiated in 1890 by L.F. Kinny at Kingston, Rhode Island, by the Rhode Island Agricultural Experiment Station (8, 28). The first scientific paper published from this research appeared in 1917 under the authorship of Burt L. Hartwell and S.C. Damon (26). In 1910, Lyman Carrier initiated turfgrass research at Virginia Polytechnic College, Blacksburg, Virginia.

Under the leadership of the United States Department of Agriculture (USDA) researchers Drs. Charles V. Piper and Russell A. Oakley, the Arlington Turf Garden was established at Arlington, Virginia, in 1916. A cooperative agreement was drawn up between the USGA Green Section and the USDA in 1921 to jointly finance turfgrass research at Arlington (24). The pioneering investigations at Arlington emphasized turfgrass cultivar

Figure 6.1. Turfgrass cultivar plots for greens at Rhode Island State College, USA, established in 1892. [8]*

Figure 6.2. Creeping bentgrass planting made in 1917 at the USDA Arlington Farm, Virginia, USA. [7]

* The numbers in brackets correspond to the figure source acknowledgments listed in the appendix.

evaluation, fertilization, and disease control. Most of these research findings were published in summary form in **The Bulletin** of the United States Golf Association Green Section starting in 1926. The turfgrass experimental plots were moved from Arlington to the USDA Plant Industry Station at Beltsville, Maryland, in 1942.

The year when turfgrass research (including unreplicated demonstration test plots) was initiated in each state in the United States is shown in table 6.2. Based on the pioneering turfgrass research conducted during the late 1800s at the Agricultural Experiment Station in Michigan, followed by the stations in Connecticut and Rhode Island, plus the encouragement of Drs. Piper and Oakley, turfgrass field plots were initiated at the agricultural experiment stations (AES) in the following states: California by C.M. Haring in 1921 (25), Florida by L.E. Stokes and C.R. Enlow in 1922 (16, 40), Massachusetts by L.S. Dickinson in 1923 (13), Kansas by J.W. Zahnley and L. Quinlan in 1924 (7, 46), Nebraska by F.D. Keim in 1925 (18), New Jersey by G.W. Musgrave and H.B. Sprague in 1925 (33, 39), Ohio by C.G. Williams and J.D. Wilson in 1926 (44), and Pennsylvania by H.B. Musser and A.D. Lasker in 1929 (34). Many of these field tests initiated in the various states during the 1920s were not replicated and thus were more in the mode of demonstration

TABLE 6.2. KNOWN STATE HISTORICAL RECORDS FOR YEAR TURFGRASS RESEARCH OR TESTING WAS INITIATED.

State	*Turfgrass research initiated*	*State*	*Turfgrass research initiated*
Michigan	1880	Colorado	1948
Connecticut	1885	Kentucky	1948
Rhode Island	1890	Oklahoma	1948
Virginia	1910	Arizona	1949
Wisconsin	1920	Alaska	1950
California	1921	New Mexico	1954
		South Dakota	1955
Florida	1922		
Massachusetts	1923	Mississippi	1956
Kansas	1924	Maine	1958
Nebraska	1925	Missouri	1958
New Jersey	1925	Utah	1958
		Arkansas	1959
Ohio	1926	South Carolina	1959
Alabama	1927	Vermont	1959
Pennsylvania	1929		
		Louisiana	1960
Iowa	1931	Oregon	1960
Maryland	1931	New Hampshire	1961
Illinois	1934	North Carolina	1961
Minnesota	1936	Wyoming	1962
Tennessee	1938	Hawaii	1963
Texas	1940	Nevada	1965
Indiana	1942	Delaware	1965
Washington	1942	West Virginia	1965
Georgia	1946	Idaho	1970
New York	1947	Montana	1978
		North Dakota	1986

test plots. Actually, some were unmowed rows of various grass species for initial evaluation as to adaptation. This approach allowed the harvest of seed for future plantings. There was a substantial delay between installation of the first turfgrass test plots in many states and when the first peer-reviewed publication of replicated research findings was available for distribution to other turfgrass researchers and to turfgrass practitioners. The early official university faculty appointments in turfgrass research are presented in table 6.3.

United Kingdom. Early turfgrass research plots were established by Martin H. Sutton of Sutton & Sons, Reading, England, circa the 1880s (15). Two other United Kingdom companies active in early turfgrass research around the 1890s were James Carter & Company of Raynes Park, England, led by Reginald Beale

TABLE 6.3. KNOWN EARLY STATE TURFGRASS RESEARCH HISTORY.

	Official turfgrass research appointment		*Forerunner who unofficially pursued turfgrass research*	*Year*
Year	Institution	Name		
1934	Penn State	H.B. Musser	H.B. Musser + A.D. Lasker	1929
1936	Iowa State	S.W. Edgecombe	V. Stoutemeyer	1932
1946	Purdue	G.O. Mott	G.O. Mott	1942
1946	Tifton (Georgia) USDA	G.W. Burton	None	—
1947	Cornell (New York)	J.E. Cornman	None	—
1950	Texas A&M	J.R. Watson	None	—
1950	Rutgers (New Jersey)	R.E. Engle	G.W. Musgrave	1923
1951	Florida	G.C. Nutter	L.E. Stokes + C.R. Enlow	1922
1953	Univ. of California, Davis	J.H. Madison	C.M. Haring	1921
1953	Oklahoma State	W.W. Huffine	W.C. Elder	1948
1953	Virginia PI	R.E. Blaser	L. Carrier	1910

Figure 6.3. Turfgrass experimental plots of Sutton & Sons, Reading, England, UK, in 1900. [44]

(see chapter 12) and shortly thereafter by James C. MacDonald of Harpenden, England (see chapter 12). Then in 1929, a Joint Advisory Council of the four national golf unions formed the Board of Greenkeeping Research and established a turfgrass research station on the St Ives Estate near Bingley, Yorkshire, England, UK (11, 12). Robert B. Dawson served as the first Director and continued in that role until his retirement in 1963. The original board was reorganized in 1951 and since has been known as the Sports Turf Research Institute, with offices and research plots continuing at Bingley.

Japan. The first investigations of Japanese turfgrasses and their culture were initiated around 1910 by H. Hora at the Faculty of Agriculture, University of Tokyo. Subsequently, the Kansai Green Section Institute was established in 1961 and the Nishi-Nippon Green Research Institute in 1962. The turfgrass research effort, as evidenced by the scientific publications generated, has ranked as one of the strongest in the Pacific Rim.

South Africa. Observational assessments of turfgrasses were pioneered privately by Dr. Charles M. Murray starting in 1914. Then research concerned with the fertility requirements of bermudagrass were initiated in 1929 under the direction of T.D. Hall and D. Meredith (23). More detailed turfgrass investigations were started in 1933 at the Frankenwald Botanical Research Station of the University of Witwatersrand located near Johannesburg.

Canada. Successful efforts of the United States in initiating turfgrass research stimulated the Royal Canadian Golf Association to request federal funds for turfgrass research (30). As a result, turfgrass experimental plots were established by G.P. McRostie at the Central Experimental Farm near Ottawa, Ontario, Canada, in 1924. Subsequently, turfgrass research was initiated at Ontario Agricultural College by R. Goodwin-Wilson in the late 1940s.

New Zealand. Fertilizer experiments on turfgrasses were initiated in 1931 by George Holford and Ernest A. Madden at the Grassland Division facilities of the Department of Agriculture, Palmerston North.

Figure 6.4. Turfgrass research plots at the Frankenwald Botanical Research Station, University of Witwatersrand, South Africa, in 1938. [1]

Figure 6.5. Experimental plots for a worm control study in New Zealand, circa 1933. [23]

Subsequently, the New Zealand Golf Association established a Greenkeeping Research Committee in 1932 that conducted research at Hokowhitu on the Manawatu Golf Club property (9). The research work was transferred to the Department of Scientific and Industrial Research in 1936. This committee was reorganized in 1949 as the New Zealand Institute for Turf Culture (42).

Sweden. Turfgrass experiments were initiated in Sweden in 1931 by the Weibullsholm Plant Breeding Institute of Landskrona (35). These studies of seed mixtures were conducted under the guidance of Frederick Nielsson. Subsequently, an enlarged set of research plots were established at the Weibullsholm Research Station by G. Weibull at Landskrona during the 1950s (43) and were expanded by Bjarne Langvad in the 1960s.

Australia. In 1931, a study of soil suitability for turfgrass cricket wickets was conducted by C.S. Piper of the Waite Agricultural Research Institute, University of Adelaide, South Australia (37). Research was initiated at golf courses by the New South Wales Golf Council Greens Research Committee in 1932 as described in their bulletin. The Victorian Inland Golf Club Research Committee conducted bentgrass cultivar evaluations for putting greens in 1934. Then the Queensland Board of Greenkeeping Research was active in turfgrass research in 1935 and 1936 as described in their bulletin. The Australian Board of Greenkeeping Research was set up just before WWII but ceased activity due to the war. The Grass Research Bureau (NSW) Ltd. (GRB) was established in Sydney, New South Wales, in 1954 under joint sponsorship by the New South Wales Golf Association and the Department of Technical and Further Education at the Ryde School of Horticulture. The first full-time staff member was established in 1964, titled Research Officer-in-Charge with Peter McMaugh selected to the position. It was renamed the Australian Turf Grass Research Institute Ltd. (ATRI) in 1970 and continued turfgrass research and educational activities until closure in 1998. The Turf Research and Advisory Institute located in Frankston, Victoria, was organized by the Victoria Department of Agriculture in 1973 and operated through 1991. Then in 2000, a turfgrass research program was initiated by Dr. Donald Loch as Redlands Turf Research and sponsored by the Queensland Department of Primary Industries.

Summary. The 1950s was the decade when the demand and use of turfgrasses expanded far beyond the expectations of most individuals associated with the turfgrass industry. Modern turfgrass culture continues to become more technically oriented. However, a certain degree of art is still retained and will remain, especially in the maintenance of greens. The increasing demand for professional turfgrass managers who have college training indicates the important role of science in modern turfgrass culture. Further evidence is represented by growth of turfgrass research and educational programs. The United States now has turfgrass research programs in most states. Varying intensities of turfgrass research have been developed in other countries during the 1950s, including France, Germany, Italy, the Netherlands, and South Korea in addition to the previously mentioned United States, Japan, Canada, Great Britain, South Africa, New Zealand, Sweden, and Australia.

TURFGRASS RESEARCH CONTRIBUTIONS

There have been many problems solved by research that contributed significantly to our understanding of turfgrass culture and to advances in cost-efficient, effective, science-based turfgrass practices. Most of the major research-based advances have occurred since 1945 (6). The demand for improved turfgrass quality and functionality emerged as the intensity of use rocketed higher. There was an important transition in the 1970s from a trial-and-error art era in turfgrass culture to a dominance of science-based technology.

While a relatively young science, the turfgrass research findings published in the past 50 years are extensive. Accordingly, a historical overview of 25 key scientific research contributions that furthered practical turfgrass culture is summarized in table 6.4. Extensive related research has complimented these key advances. These advances were made possible through the combined efforts of turfgrass researchers at the state agricultural experiment stations (AES), research conducted by private firms, and individual professional turfgrass managers.

TABLE 6.4. TURFGRASS PROGRESS THROUGH RESEARCH FROM 1945 TO 2000.

Time period	Key areas of turfgrass research emphases and achievements
1945–55	• Selective broadleaf weed control: phenoxy-herbicides • Insecticide development for efficacy & persistence
1950–60	• Equipment: powered coring, slicing, spiking, & vertical cutting machines • Mowing practices: height & frequency physiological/species aspects • Culture of turfgrass communities: mixtures & blends/polystands
1955–65	• Post- and preemergence weed control through selective herbicides • Turf-type fertilizer ratios, analyses, & formulations, including slow-release types • Warm-season turfgrass cultivar development: bermudagrasses & zoysiagrasses
1960–70	• High-sand root zone modification: USGA-Texas high-sand method • Roadside establishment techniques: improved seeding & mulching methods
1965–75	• Cool-season turfgrass cultivar development through breeding: Kentucky bluegrasses, perennial ryegrasses, & tall fescues • Sod production expansion: cultural practices, harvesting equipment, & shipping • Transition from an art-dominated to science-dominated turfgrass culture
1970–80	• Expansion of turfgrass science and technology around the world • Turfgrass disease characterizations plus systemic fungicide programs • New turfgrass nutritional practices, emphasizing potassium (K) & iron (Fe) • Lawn care industry technology emergence
1975–85	• Growth and physiology of turfgrasses in relation to environmental stress tolerances: cold, heat, drought, wear, & shade • Plant growth regulator advances across species • Advances in fully automated irrigation systems

1980–90	• Development of turfgrass cultural practices & cultivars that conserve water, energy, & nutrient resources
	• Environmentally based prediction models for pests & evapotranspiration
	• Development of root-zone stabilization & enhancement techniques
1985–95	• Deep turfgrass cultivation & broad-area topdressing equipment
	• Environmental protection & quality-of-life enhancement through turfgrasses, especially in urban areas
	• Development of more effective growth regulators
1990–2000	• Pest-specific growth regulators for insect control
	• Cultural enhancement of turfgrass sports field playing quality & safety
	• Documentation of turfgrass functional benefits to human quality of life
	• Emergence of metabolic enhancers

REFERENCES

1. Beal, W.J. 1892. Some selections of grasses promising for field and lawn. Proc. Soc. Prom. Agric. Sci. 13:135–137.

2. Beal, W.J. 1893. Mixtures of grasses for lawns. Proc. Soc. Prom. Agric. Sci. 14:28–33.

3. Beal, W.J. 1898. Lawn-grass mixtures as purchased in the markets compared with a few of the best. Proc. Soc. Prom. Agric. Sci. 19:59–63.

4. Beard, J.B. 1973. Turfgrass: Science and Culture. Prentice-Hall Inc., Englewood Cliffs, New Jersey, USA.

5. Beard, J.B. 1989. The role of Gramineae in enhancing men's quality of life. Japanese Sci. Counc., Natl. Com. Agric. Sci., Tokyo, Japan.

6. Beard, J.B. 2004. Evolution of turfgrass research contributions. Turfgrass Bull. 226:51–56.

7. Call, L.E. 1926. Turf and lawn grass experiments. Director's Rep. Kansas Agric. Exp. Stn., 1924–1926. p. 47–48.

8. Card, F.W., M.A. Blake, and H.L. Barnes. 1907. Lawn experiment. Rep. Rhode Island Agric. Exp. Stn., 1905–1906, part 2, 19. p. 162–166.

9. Corkill, L. 1967. Turf improvement: The future. New Zealand Inst. Turf Culture Newsletter 52:3–4.

10. Curtis, W. 1805. Practical Observation on the British Grasses. 3rd ed. H.D. Symonds, London, England, UK.

11. Dawson, R.B. 1929. St Ives Research Station, its surroundings and historical associations. J. Board Greenkeeping Res 1(1):9–11.

12. Dawson, R.B. 1929. Some greenkeeping problems and practices, the programme of the research station and the factors influencing growth of turf. J. Board Greenkeeping Res. 1(1):12–23.

13. Dickinson, L.S. 1928. Turf experiments at the Massachusetts Agricultural College. Bull. USGA Green Section 8(12):248–249.

14. Dickinson, L.S. 1930. The effect of air temperature on the pathogenicity of Rhizoctonia solani parasitizing grasses on putting-green turf. Phytopath. 20:597–608.

15. Early Local History Group. 2006. Sutton Seeds: A History 1806–2006. Univ. of Reading, Reading, England, UK.

16. Enlow, C.R. 1928. Turf studies at the Florida Experiment Station. Bull. USGA Green Section 8(12):246–247.

17. Garner, E.S., and S.C. Damon. 1929. The persistence of certain lawn grasses as affected by fertilization and competition. Agric. Exp. Stn. Rhode Island State College Bull. 217.

18. Gilbert, B.E. 1928. An analytical study of the putting greens of Rhode Island golf courses. Rhode Island Agric. Exp. Stn. Bull. 212.

19. Grace, D.P., and D.C. Phillips. 1975. Ransomes of Ipswich: A History of the Firm and Guide to Records. Inst. of Agric. History, Univ. of Reading, Reading, England, UK.

20. Grau, F.V. 1928. Turf experiments at Nebraska College of Agriculture. Bull. USGA Green Section 8(12):253–254.

21. Halford, D.G. 2008. Old Lawnmowers. Shire Library Classics, Shire Publ., Botley, Oxford, England, UK.

22. Hall, A. 1984. The Hall and Duck Collection: A Concise History of Lawn Mower Development. Andrew Hall, Sheffield, England, UK.

23. Hall, T.D. 1948. Introduction. Experiments with *Cynodon dactylon* and other species at the South African Turf Research Station, African Explosives and Chemical Industries Limited, Johannesburg, South Africa. p. 5–7.

24. Hardt, F.M. 1939. Introducing "Turf Culture." Turf Culture 1(1):1–5.

25. Haring, C.M. 1922. Grass and forage plant investigations. Rep. College Agric. & Agric. Exp. Stn., Univ. of California, 1921.

26. Hartwell, B.L., and S.C. Damon. 1917. The persistence of lawn and other grasses as influenced especially by the effect of manures on the degree of soil acidity. Agric. Exp. Stn. Rhode Island State College Bull. 170.

27. Johnson, S.W. 1888. The station forage garden. Rep. Connecticut Agric. Exp. Stn., 1887. 11: 163–176.

28. Kinney, L.F. 1891. Trial of lawn grasses. Rep. Rhode Island State Agric. Exp. Stn., part 2, 1890. 3:156.

29. MacDonald, J.C. 1923. Lawns, Links, & Sportsfields. Country Life Ltd. & Georges Newnes Ltd., London, England, UK.

30. McRostie, G.P. 1928. Turf studies at the Central Experimental Farm, Ottawa. Bull. USGA Green Section 8(12):250–251.

31. Monteith, J. 1928. Demonstration turf gardens on golf courses. Bull. USGA Green Section 8(12):250–251.

32. Monteith, J. 1928. The Arlington turf garden. Bull. USGA Green Section 8(12):244–245.

33. Musgrave, G.W. 1926. Rep. Dep. of Agron. Miscell. Rep. New Jersey State Agric. Exp. Stn., 1925. 46:275.

34. Musser, H.B. 1929. Unprojected research. Rep. Pennsylvania Agric. Exp. Stn. Bull. 243, 42: 33–34.

35. Nilsson, F. 1931. Om gräsmattor. Weibull's Yearbook 26:26–29.

36. Olcott, J.B. 1891. Grass-gardening. Rep. Connecticut Agric. Exp. Stn., 1890. 14:162–174.

37. Piper, C.S. 1932. Some characteristics of soils used for turf wickets in Australia. Trans. Proc. Royal Soc. South Australia 56:15–18.

38. Sprague, H.B. 1928. Turf experiments at the New Jersey state station. Bull. USGA Green Section 8(12):251–252.

39. Sprague, H.B., and E.E. Evaul. 1930. Experiments with turf grasses in New Jersey. New Jersey Agric. Exp. Stn. Bull. 497.

40. Stokes, W.E. 1922. Lawn grass studies. Rep. Univ. Florida Agric. Exp. Stn.

41. Sutton & Sons. 1906. Garden Lawns, Tennis Lawns, Bowling Greens, Croquet Grounds, Putting Greens, Cricket Grounds. Simpkin, Marshall, Hamilton, Kent & Co. Ltd., London, England, UK.

42. Walker, C. 1957. Report on greenkeeping research. New Zealand Inst. Turf Culture 7:1.

43. Weibull, G. 1958. Experiments with turf-grasses 1948–1955. Agric. Hortique Genetica 16:209–220.

44. Williams, C.G. 1928. Lawn and golf course problems being studied. Rep. Ohio Agric. Exp. Stn., 1926–1927, Bull. 417. 46:18–20.

45. Wilson, J.D. 1927. The measurement and interpretation of the water-supplying power of the soil with special reference to lawn grasses and some other plants. Plant Physiol. 2:385–440.

46. Zahnley, J.W. 1928. Experiments with turfgrasses in Kansas. Bull. USGA Green Section 8(12):249–250.

CHAPTER 7

History of Turfgrass Science Education

THE EARLY TURFGRASS PRACTITIONERS LEARNED THE ART OF TURFGRASS MAINTENANCE VIA TRIAL-and-error, on-the-job experiences. This accumulated art was then passed on to the younger generation via an apprenticeship arrangement. As the science of turfgrass culture evolved through research, the need for associated formal turfgrass courses became apparent.

UNDERGRADUATE TURFGRASS EDUCATION

The earliest of the undergraduate turfgrass course offerings were organized in 1927 at Massachusetts State College, taught by Lawrence Dickinson; in 1932 at Rhode Island State College, taught by Leslie Keegan; and in 1935 at Oregon State Agricultural College, taught by Dr. H.A. Schoth.

Figure 7.1. Greenkeeping class at Massachusetts Agricultural College with Professor Lawrence Dickinson, third from the right, circa 1928. [2]*

---

* The numbers in brackets correspond to the figure source acknowledgments listed in the appendix.

Further expansion followed WWII, led by Purdue University in 1950 and followed by Penn State University and Washington State University, both in 1955 (table 7.1). Many other state universities followed in the next two decades. The science of turfgrass culture became the dominant dimension for modern turfgrass professionals, while the art aspects gained through experience remained a valued asset.

TWO-YEAR ASSOCIATE PROGRAMS AND WINTER SHORT COURSE

The first 2-Year Turfgrass Course was developed in the Stockbridge School at the Massachusetts State Agriculture School in 1927, also organized by Lawrence Dickinson. This program continues to this day. After WWII, a few more two-year turfgrass associate programs evolved at other universities. Two-year educational programs include Golf Turf Management at Penn State University in 1958, initiated by H. Burton Musser; Turfgrass and Golf Course Management at the University of Maryland in 1965, initiated by Drs. Charles W. Laughlin and Douglas T. Hawes; Golf Course Management at Michigan State University in 1966, initiated by Dr. John W. King; Golf Course Management at North Carolina State University in 196x, initiated by Dr. William B. Gilbert; Lawn Care Technology at Michigan State University from 1989 to 1991, first coordinated by Dr. Eric Miltner; Lawn Care & Athletic Fields Technology from 1991 to 1992, first coordinated by Dr. Paul E. Rieke; Sports and Commercial Turf Management at Michigan State University in 1993, first coordinated by Dr. David M. Gilstrap; and Turfgrass Management at the University of Guelph in 2003, first led by Rob Witherspoon.

The associate programs require a summer internship period at an operational turfgrass facility. Most of these major university turfgrass associate degree programs allow internal transfer of individual courses to an undergraduate BSc degree if a grade above a specified level is achieved. For those unable to pursue an undergraduate education for various reasons, the two-year associate program has been an important alternative. Actually, many modern two-year turfgrass programs have a significant number of students enrolled who previously had completed two to four years of undergraduate courses in another subject area. There also are independent two-year schools and community colleges that offer one or more turfgrass courses. Some offer a legitimate curriculum in turfgrass science, while others have a broader course of study across landscaping.

TABLE 7.1. KNOWN EARLY FORMAL UNDERGRADUATE TURFGRASS EDUCATION IN THE UNITED STATES FROM 1927 THROUGH 1955. BASED ON OFFICIAL PRINTED UNIVERSITY/COLLEGE COURSE CATALOGUES.

Year of first official turfgrass course(s)	Institution	Title of course(s)	Instructor	Teaching full-time equivalent	Year of official turfgrass course of study
1927	Massachusetts State College	Greenkeeping	L.S. Dickinson	1.0	—
1932	Rhode Island State College	(a) Agrostology (b) Golf courses	L.A. Keegan	1.0	1932
1935	Oregon State Agricultural College	Lawns & Turf	Dr. H.A. Schoth	0.1	1977
1950	Purdue	Turfgrass	Dr. W.H. Daniel	0.25	1960
1955	Penn State	Turf Management	H.B. Musser	—	1992
1955	Washington State	Turf Management	Dr. A.G. Law	0.15	1955

Winter Short Course. A pioneering turfgrass educational effort was initiated in 1925 by Lawrence Dickinson at the Massachusetts State Agriculture School when a single 8-Week Winter School for Turfgrass Managers was initiated. This evolved to a 12-Week Winter Turfgrass Course in 1927 and is now a 7-Week Turfgrass School. Penn State College offered a 4-Week Winter Course titled Greenkeepers Short Course from 1931 to 1940. A 10-Week 2-Year Professional Golf Turf Management Winter Short Course was available at Rutgers University in 1962, first led by Dr. Ralph Engel. Also offered is a 3-Week Golf Course Management Preparatory Short Course, initiated in 1992 and coordinated by James Morris. Since 1970, the University of Guelph in Ontario, Canada, has offered a 4-Week Turf Managers' Short Course, first organized by Dr. Jack Eggens.

ADVANCED GRADUATE EDUCATION

As there were no turfgrass scientists in the early years, technical questions regarding turfgrass problems usually were routed to those individuals with the best knowledge concerning grasses. These were the early educators and researchers formally assigned to the forage, pasture, and/or grassland subject areas at agricultural colleges. After WWII, the ever increasing volume of questions about turfgrass culture resulted in a number of the forage scientists being assigned part time to the turfgrass subject area. From this formative phase, full-time positions addressing the turfgrass subject needs evolved. Subsequently, academic faculty positions increased significantly during the 1955 to 1965 period, with most being joint appointments involving teaching, research, and/or extension.

The opportunity for academic positions in the emerging turfgrass science field led a significant number of graduate students to pursue PhD degrees in the turfgrass subject area (table 7.2). Many had served in

Figure 7.2. Teaching diagnostic techniques to a group of graduate students with Dr. Jim Beard, fifth from the right, 1960s. Can you recognize future Michigan State University turfgrass science PhD recipients Bob Carrow, Keith Karnok, John Kaufmann, John King, David Martin, Bob Shearman, and Al Turgeon? [26]

TABLE 7.2. DOCTORATE (PhD) DEGREES EARNED IN THE TURFGRASS AREA
FROM 1931 THROUGH 1966 IN THE UNITED STATES.

Year	Graduate institution	PhD recipient	Advisor/major professor	Turfgrass thesis subject
1931	Wisconsin	K. Arnold S. Dahl	L.R. Jones	Disease/plant pathology
1942	Pennsylvania	Walter S. Lapp	J.H. Schramm	Cultural practices
1949	Purdue	Willis Skrda	G.O. Mott	Airfield root zones
1950	Michigan State	William H. Daniel	J. Tyson	Cultural practices
1950	Maryland	Marvin H. Ferguson	H.G. Gauch	Soil fertility
1950	Purdue	Richard R. Davis	G.O. Mott	Root zones
1950	Penn State	James R. Watson	H.B. Musser	Soil cultural practices
1951	Rutgers	Ralph E. Engel	H.B. Sprague	Cultural practices
1951	Penn State	Neal Wright Linlie	H.B. Musser	Breeding
1952	Cornell	Gene C. Nutter	J.E. Cornman	Weed control
1952	Penn State	John C. Harper	H.B. Musser	Cultural practices
1955	Michigan State	Felix V. Juska	J. Tyson	Cultural practices
1957	Penn State	Joseph M. Duich	H.B. Musser	Breeding
1958	Purdue	Michael P. Britton	H. Cumings	Disease/plant pathology
1960	Purdue	Norman R. Goetze	W.H. Daniel	Soil fertility
1960	Washington State	Roy L. Goss	A.G. Law	Cultural practices
1961	Purdue	James B Beard	W.H. Daniel	Physiology/biochemistry
1965	Texas A&M	John A. Long	E.C. Bashaw	Genetics/cytology
1965	Massachusetts	Joseph Troll	R. Rhoade	Nematology
1965	Virginia Polytech	Richard E. Schmidt	R.E. Blaser	Physiology
1966	Virginia Polytech	Andrew J. Powell	R.E. Blaser	Physiology

WWII, which created a major increase in turfgrass PhD recipients during the early 1950s. Ten universities dominated turfgrass graduate education, with Purdue University (G.O. Mott and W.H. Daniel), Penn State University (H.B. Musser), and Michigan State University (J. Tyson) being especially active pre-1960.

The same graduate college institutions that were active in PhD education during the pioneering 1950s also were most active in master of science (MSc) degree training, with Rhode Island University also being active. By the 1980s, an MSc turfgrass degree program was offered within most state agricultural colleges in the United States.

DISTANCE EDUCATION

A more recent innovation in formal turfgrass education programs is via electronic online means, termed "distance education." A course of study leading to an undergraduate degree is now possible while continuing to live in one's normal place of residence and associated employment. A pioneer in turfgrass distance education has been Penn State University under the leadership of Dr. Alfred J. Turgeon. The Online Basic Certificate in Turfgrass Management through Penn State's World Campus became available in 1998 and the Online Advanced Certificate in Turfgrass Management began in 2000. The Online Turfgrass Science BSc degree program began in 2006, the Online Turfgrass Science Associate degree program began in 2009, and the Online Master of Professional Studies in Turfgrass Management (MPS-TM) degree program started in 2010.

TURFGRASS EXTENSION EDUCATION

The education of turfgrass professionals is an ongoing lifelong need, especially as the field is driven by technical advances and science-based research. However, in terms of requests, the largest audience by far is the homeowner. The state-based Extension Service fills this turfgrass information need for individuals after completing their formal education. The personnel involve both county-based agents and statewide turfgrass specialists. The early state turfgrass extension specialists are summarized in table 7.3. Information dissemination involves lectures, bulletins and circulars, newsletters, telephone calls, radio and television broadcasts, electronic transmissions, field days, and conferences.

State Extension Lawn Publications. For the homeowner, a popular source of lawn care and planting information has been the state extension bulletin or circular. A historical summary of these early state publications from 1906 through 1935 is presented in table 7.4. The earliest of the publications were produced by Purdue University in 1913, Michigan State College in 1913, and Cornell University in 1916. The United States Department of Agriculture, in Washington, DC, published lawn bulletins in 1906 and 1912.

State Turfgrasses Conferences. State turfgrass conferences were a key early means for education of professionals and practitioners. Six were organized primarily at state agricultural colleges during the 1928 to 1938 period. Turfgrass conferences evolved as major educational events after WWII as represented in table 7.5. More detailed information concerning published proceedings of the conferences is provided in chapter 16. The published conference proceedings were a key source of up-to-date technical information concerning turfgrasses and their culture. Eight states started publishing proceedings of their turfgrass conferences in the late 1940s, twelve in the 1950s, and thirteen in the 1960s. Proceeding publications waned in the 1980s and most were terminated by the 1990s. Technical, association, and trade periodicals became more timely and economical information sources. These are described in chapter 17.

TABLE 7.3. KNOWN EARLY STATE TURFGRASS EXTENSION SPECIALIST ACTIVE FROM 1935 THROUGH 1950.[*]

		First Official State Turfgrass Extension Specialist		Earlier Active Unofficial Turfgrass Extension Person	
Year	State	Name	Full-Time Equivalent (FTE)	Year	Name
1935	Pennsylvania (Penn State)	Dr. Fred V. Grau	0.5	c. 1930	Nicholar Schmitz
1947	New Jersey (Rutgers)	Dr. Ralph E. Engel	0.5	1936	Dr. T.C. Longnecker
1947	New York (Cornell)	Dr. John F. Cornman	~0.2	c. 1915	M.E. Montgomery
1948	Iowa (Iowa State)	Arthur Edward Cott	0.1	1939	Dr. Sam W. Edgecomb
1950	Indiana (Purdue)	Dr. William H. Daniel	0.25	c. 1937	Dr. Gerald O. Mott

[*]See Table 7.4 for other individuals involved in early state turfgrass extension publications.

TABLE 7.4. KNOWN STATE AND USDA TURFGRASS LAWN PUBLICATIONS FROM 1906 THROUGH 1935.

Date	State	Publication no.	Title	Author(s)	Page
1906	Bureau of Plant Industry	Farmers' Bulletin No. 248	The Lawn	L.C. Corbett	20
1912	USDA	Farmers' Bulletin 494	Lawn Soils and Lawns	O. Schreiner, J.J. Skinner, L.C. Corbett & F.L. Mulford	48
Apr. 1913	Indiana (Purdue)	Leaflet No. 41	How to Make and Maintain a Lawn	S.D. Conner & M.L. Fisher**	6
Apr. 1913	Michigan	Circular No. 20	Starting a Lawn*	C.P. Halligan**	4
Jan. 1916	New York (Cornell)	Bulletin No. 7	Lawns and Lawn Making	M.E. Montgomery	4
June 1926	New Jersey (Rutgers)	Extension Bulletin 54	Better Lawns	H.R. Cox	8
May 1927	Rhode Island	Bulletin No. 48	The Making and Care of Lawns	S.C. Damon**	13
June 1929	Ohio	Special Circular 18	Better Lawns	F.A. Welton & R.M. Salter*	16
Sept. 1929	Colorado	Bulletin 308A	Lawns	G. Beach	8
Dec. 1929	Florida	Bulletin 209	Lawns In Florida	C.R. Enlow & W.E. Stokes**	20
Jan. 1930	New Jersey (Rutgers)	Bulletin No. 76	Better Lawns	H.R. Cox	8
July 1930	Minnesota	Special Bulletin 130	Making the Home Lawn	L.E. Longley**	11
July 1931	Missouri	Circular No. 274	Development and Care of Lawns	H.F. Major	12
Oct. 1931	North Carolina	Circular No. 1889	Lawn Seeding and Care	C.B. Williams**	8
Oct. 1931	Pennsylvania	Circular 143	Better Lawns In North Carolina	N. Schmitz	10
Oct. 1931	USDA	Farmers' Bulletin No. 1677	Lawns	H.L. Westover & C.R. Enlow**	18
Mar. 1932	Connecticut	Circular 83	Planting and Care of Lawns	Anonymous	4
Mar. 1932	Kentucky	Circular No. 256	The Lawn	N.R. Elliott	8
Apr. 1932	Indiana (Purdue)	Leaflet No. 41 (rev.)	How To Make and Maintain a Lawn	S.D. Conner & M.L. Fisher**	6
Apr. 1932	Minnesota	Special Bulletin 130 (rev.)	Making the Home Lawn	L.E. Longley**	12
June 1932	Michigan	Extension Bulletin No. 125	Insects Infesting Golf Courses and Lawns	R.H. Pettit	11
Sept. 1933	Massachusetts	Leaflet No. 85	Facts on Lawn Management	L.S. Dickinson	12
May 1934	Kansas	Bulletin 267	Lawns In Kansas**	J.W. Zahnley & L.R. Quinlan**	32
June 1934	New York (Cornell)	Bulletin 296	Lawns: Construction & Maintenance	R.W. Curtis & J.A. DeFrance	52
Apr. 1935	Massachusetts	Leaflet No. 95 (rev.)	Lawn Management	L.S. Dickinson	12
May 1935	Illinois	No. 1 (rev.)	Lawns and Their Care	F.F. Weinard	7
Oct. 1935	USDA	Farmers' Bulletin No. 1677 (rev.)	Planting and Care of Lawns	H.L. Westover & C.R. Enlow**	20

* Revised four times through 1935.
** Published and/or authored with Agricultural Experiment Station.

State Turfgrass Field Days. The state turfgrass field day provided practitioners the opportunity to actually observe the ongoing investigations as presented by researchers involved. Typically, there is the opportunity to ask questions during or immediately after the formal tour. State turfgrass field day events are held annually or biannually. Some also include equipment exhibits. Three state agricultural colleges held turfgrass field days prior to 1940, including Rutgers of New Jersey in 1928, Michigan State in 1932, and Pennsylvania State in 1937. Most state turfgrass field days were initiated in the 1950s or later (table 7.5). Some states also publish turfgrass field day reports that are described in chapter 15.

TABLE 7.5. KNOWN EARLY STATE TURFGRASS CONFERENCES AND FIELD DAYS INITIATED IN THE UNITED STATES.

State	Turfgrass conference		Year of first turfgrass field day
	First held	*Proceedings first published*	
New Jersey (Rutgers)	1928	1949	1928
Pennsylvania	1929	1949	1937
Michigan	1931	1972	1932
Iowa	1932	None	1957
Indiana (Purdue; Midwest)	1937	1948	1951
Ohio	1938	1967	1968
Washington (Northwest)	1941	1946	1956
New Mexico (Southwest)	1947	1947	2004
New York (Cornell)	1948	1977	1980
California	1949	1950	1951
Florida	1949	1946	1952
Maryland (Mid-Atlantic)	1949	1949	197x
Texas	1949	1949	1958
Kansas (Central Plains)	1950	1950	1950
Georgia (Southeast)	1951	1951	1946
Oklahoma	1952	1953	195x
Arizona	1953	1953	1954
Colorado (Rocky Mountain)	1954	1954	195x
Utah (Intermountain)	1954	1954	1999
Massachusetts	1956	1956	195x

Figure 7.3. Student laboratory exercise concerning turfgrass morphology studies. [26]

Figure 7.4. Student field trip to study turfgrass *in-vivo* diagnostic techniques. [26]

CHAPTER 8

History of Turfgrass Seed Production

UP TO THE TWENTIETH CENTURY, THE PLANTING OF TURFGRASSES BY HAY SEEDS WAS TYPICALLY unacceptable because of severe contamination of tall, off-type grasses and broadleaf weeds. Consequently, sodding was the preferred establishment approach for quality turfgrass uses.

Seed production of turfgrass species evolved in the United States, principally in Rhode Island, on native bentgrass stands first described in the mid-1700s (2).* Bentgrasses composed the principal turfgrass seed produced in southern New England, with production increasing in the early 1880s. A key aspect involved intensive grazing by sheep that reduced the off-type, coarser grass species population. It was described in 1887:

> There are thousands of acres of rough pasture sward in every county of the State, largely composed of this old English "Fine Bent." Sheep, if well managed, improve a dwarf *Agrostis* sod, while eating most other kinds of grass out. The connection between fine mutton and Fine Bent seed-growing is natural and practically absolute. Since flocks have disappeared, the fine grass has been overgrown by coarser ones, or with weeds and bushes. Formerly hundreds of bushels of fine *Agrostis* seed were saved from sheep pastures in this State. Now we learn of none being grown for market. (6)

Bentgrass. Importing of bentgrass seed into the United States from Germany was significant up to WWII. It typically was a mixture of 75 to 85% colonial, 10 to 20% velvet, and 1 to 5% creeping bentgrass. This mixture, referred to as "South German Mixed Bentgrass," was used extensively on putting greens. Subsequent segregation into distinct patches served as the genetic source for bentgrass cultivars developed in the United States in the latter half of the twentieth century. Some early bentgrass seed production also occurred in the Netherlands, Belgium, and England. Seed production advanced to a specialty crop with the "Rhode Island" colonial bentgrass seeded into a prepared soil. By 1930, production of velvet and creeping bentgrass seed was on the rise and colonial bentgrass was declining in Rhode Island (5). At this time, substantial quantities of colonial bentgrass seed were being imported from Germany and New Zealand. Eventually, seed production shifted westward from Rhode Island to Michigan to Oregon, where bentgrasses could be produced with much higher yields and less harvesting problems due to dry summer conditions.

A native monostand of what was locally called "bermuda grass" around Coos Bay, Oregon, was identified as creeping bentgrass in 1923 (7). By 1926, seed was being harvested from this area, which became known as Seaside creeping bentgrass. Harvesting of native bentgrass expanded on marshland along the Pacific

---

* The numbers in parentheses correspond to the reference list found at the end of the chapter.

coast up to Canada. Native monostands in the lower Columbia River became known as Astoria colonial bentgrass. Inland farmers also had seeded colonial bentgrass that was first harvested in 1930. Eventually, a major bentgrass seed production industry emerged in Oregon during the 1950s, especially after the release of improved cultivars, such as Penncross in 1954. Seed production of colonial and creeping bentgrasses has continued primarily in northwest Oregon.

Fine-Leaf Fescue. Early seed production of fine-leaf fescue seed occurred from native stands in northern Europe, such as The Netherlands. Significant quantities were imported to the United States, mostly dominated by creeping red fescue. Another significant early imported source to North America was "browntop" or Chewings fescue from New Zealand. The latter was associated with a breakthrough in seed germination research that revealed the importance of sealed containers during shipping to maintain a low atmospheric humidity (4). Fine-leaf fescue seed production in Oregon was initiated in the early 1930s. Seed production increased dramatically in western North America during the 1950s. Principal production regions now include western and eastern Oregon, eastern Washington, and western Canada.

Redtop. Early North American redtop grass seed production was centered in southern Illinois and nearby Missouri. By the 1950s, redtop seed production had declined substantially due to its inferior turfgrass characteristics.

Perennial Ryegrass. Successful seed production of perennial ryegrass was developed in western Oregon in the early 1930s. Within the decade, imports of perennial ryegrass from southern Europe, Britain, New Zealand, and western Victoria, Australia, began an extended decline. Some of the earliest cultivars planted for seed production were Norlea, Linn, and NK-100. A major stimulant in quality seed production of perennial ryegrass was marked by the release of a true turf-type cultivar, Manhattan, in 1967, as developed by Dr. C. Reed Funk at the New Jersey Agricultural Experiment Station, Rutgers.

Kentucky Bluegrass. Early production of "common" Kentucky bluegrass seed came from naturalized pasture stands in Kentucky up to 1900 and then from Missouri. Seed production expanded rapidly up to WWII, extending to North Dakota, Minnesota, and Ohio (9). Minnesota has continued to produce

Figure 8.1. A gang of three bluegrass seed strippers in operation near Lexington, Kentucky, USA, in the 1940s. There was a seed box in front of each machine attendant; when these were full, the rough seed heads were manually transferred to bags for transport to the curing yard. [25]*

---

* The numbers in brackets correspond to the figure source acknowledgments listed in the appendix.

Kentucky bluegrass seed. Seed production of improved Kentucky bluegrass cultivars developed in the Pacific Northwest states of Oregon, eastern Washington, and northern Idaho during the 1950s. The first major breakthrough was the cultivar Merion, released in 1947. Some Kentucky bluegrass seed has been produced in Canada, the Netherlands, Denmark, Poland, and central Europe.

Rough Bluegrass. The source of rough bluegrass seed up until 1980 was primarily from naturalized stands in Denmark. Then development of proprietary cultivars served as the focus for rough bluegrass seed production in Oregon. The first cultivar breakthrough was Sabre, developed in 1977 by Dr. C. Reed Funk of the New Jersey Agricultural Experiment Station, Rutgers.

Tall Fescue. Seed production of Alta tall fescue was initiated in Oregon in 1935. Forage-type tall fescue was planted extensively in the western and southeastern United States in the late 1940s for animal production. Forage-type cultivars, such as Alta and Kentucky 31, were also being used on roadsides and minimally used, low-maintenance areas. It was not until true turf-type cultivars of tall fescue were released in the 1980s that proprietary seed production was stimulated.

Bermudagrass. Early seed of "common" dactylon bermudagrass was derived from the cleaning of alfalfa seed produced in western Arizona and adjacent southeast California, USA, from 1906 to the late 1930s. During the 1940s and 1950s, bermudagrass seed was produced in the Gila and Wellton-Mohawk Valleys in southwest Arizona (1). The Wellton-Mohawk irrigation project, with its high salinity, resulted in bermudagrass seed production dominating cropping starting in 1952. Certified "Arizona common" dactylon bermudagrass was not available until 1963. By 1980, significant bermudagrass seed production had developed in the Imperial Valley of California. The main seed harvest is in the summer with a fall harvest possible but with a smaller seed yield. Dehulling is an important phase in post-harvest processing. With development of improved seeded dactylon bermudagrass cultivars starting in the late 1980s, production fields are now being seeded, maintained, and harvested. Key early turfgrass-type cultivars of dactylon bermudagrass were NuMex SAHARA in 1987 and Sonesta in 1991, both developed by Dr. Arden Baltensperger of New Mexico State University. Subsequently, seed of certain cultivars, such as Riviera, released by Oklahoma State University, is being produced in Oklahoma.

Figure 8.2. Bermudagrass seed threshing operation in Yuma County, Arizona, USA, circa 1960. [26]

Zoysiagrass. Limited seed quantities have been manually harvested from native Japanese zoysiagrass stands in South Korea, Japan, and China since the mid-1960s. Cultivars of improved seeded Japanese zoysiagrass were developed starting in the 1990s, with seed production fields located in the Lakeland area in south-central Georgia, USA.

Centipedegrass. The short-stalked seed heads of centipedegrass impaired traditional mechanical seed harvesting methods for a long time. An innovative suction head technique made mechanical seed harvesting of centipedegrass from old naturalized stands feasible in the 1950s. This led to continuing seed production in south Georgia, USA.

Seashore Paspalum. Seed production of seashore paspalum has evolved only recently in the Willamette Valley of northwestern Oregon.

Carpetgrass. Limited quantities of common carpetgrass seed were harvested from old, naturalized stands in the southern areas of Louisiana and Mississippi, USA, up until the 1960s. Harvesting consisted of cutting, raking into windrows, and stationary threshing or threshing by direct combining. Seed is now produced along the north coast area of New South Wales, Australia.

Kikuyugrass. Seed production of kikuyugrass is highly specialized and located near Quirindi in northeast New South Wales, Australia. A suction head pickup from windrows for combines is utilized along with multiple passes.

American Buffalograss. Since the late 1930s, seed of American buffalograss has been harvested from old, native stands in the central Great Plains states. However, seed yields ranged from erratic to total failures. The release of improved turfgrass-type cultivars of American buffalograss in the 1990s resulted in the specialty planting, culture, and harvesting of seed. Specialized harvesting techniques involve sickle-bar cutting and collection of the burs by suction, brush, or beater-type equipment. Seed germination is enhanced significantly by a post-harvest dehulling process.

TURFGRASS SEED PRODUCTION METHODS

The origins of turfgrass seed harvesting involved naturalized stands of old, grazed perennial grasses. Early seed harvesting was by hand stripping. Subsequently, species with long-stemmed grass seed heads were manually cut by scythe or sickle, tied in sheaves, and stacked to dry, and consequently, the seeds were manually thrashed usually by flail methods. The seed heads were cut when semi-mature, as many perennial grass species are prone to seed shattering and loss if thrashed when fully mature.

Early rudiments of commercial turfgrass seed production were present in central Germany in the late 1800s. The development of specialty turfgrass seed production in Germany is described in 1909 as follows:

> During the past quarter of a century the growing of fine grasses for seed has been developed in Germany until it has become, like the bulb-growing in Holland, a business upon which whole districts of people are supported. The grasses are grown by peasant cultivators, who own their small areas of land and religiously preserve each field for one kind of grass from year to year, and generation to generation. Sometimes whole villages will unite to ensure that only one kind of fine grass is grown in the district and no admixture is admitted. By this means, by systematic cleaning of the land from weeds,

and by the admirable system of constant testing made possible by Government establishing local stations for the purpose, remarkable degree of purity in grass seeds has been arrived at. (8)

Manual cutting was necessary where the mature plants were scattered about, as was the situation for bentgrass in Germany during the 1880s (2). This early processed seed had a purity of 70 to 80 percent.

Early commercial turfgrass seed production fields were planted by broadcasting that produced solid stands. By the late 1950s, planting in rows was adopted as weed control was facilitated and seed yields increased for most turfgrass species.

Eventually mechanical harvesting—developed involving horse-drawn binders—was utilized to cut and tie the stemmed seed heads of taller species into sheaves; the sheaves were then manually placed in shocks for drying, followed by thrashing with a stationary, belt-powered unit. Then tractor-drawn binders came into use during the 1920s. This technique was replaced by a comb-type seed head stripper that was used on cool-season turfgrasses grown in the Midwest, especially Kentucky bluegrass. Subsequently, two types of a more efficient stripper came into widespread use in the Midwest (9). One consisted of three sickle bars mounted on a revolving frame, and the second was a revolving cylinder with protruding spikes. Both threw the seed heads with upper stems into boxes that were manually emptied into large bags.

The "green" seed with chaff was immediately hauled to a curing yard or shed and spread out in narrow windrows of 15 to 18 inches (38–46 cm) high for drying. These windrows were turned 3 to 4 times daily to prevent seed death by heating. When dry, the seed was bagged and processed through a cleaner or multiple types of cleaners with appropriately sized screens relative to the seed size of the specific turfgrass species.

Modern seed harvesting for many turfgrasses consists of a self-propelled sickle bar cutter-swather, air drying, and self-propelled combining via a pickup device. Direct combining has been practiced, but the problem with seed shattering for many turfgrass species resulted in greater seed loss than when a swathing step is practiced. The earlier grain-type concaves in combines were converted to more effective rubber concaves and bars in the 1960s. Also, early combines had a platform where the grass seed was manually bagged, and the filled bags were tied and allowed to slide down a shoot onto the ground to be picked up. More recently, the seed handling has changed from a large bag system to a system in which trucks are filled

Figure 8.3. Manual harvesting of bentgrass seed in Germany, mid-1920s. [31]

and seed transported to the drying and cleaning facility. Also, mechanical dryers came into use to cure the "green" seed before seed cleaning.

Grass Seed Processing. One of the early pioneers in the marketing of quality grass seed in the United States was Orlando M. Scott of Marysville, Ohio, founder of the Scott Company (see O.M. Scott in Chapter 12). He approached the problem by utilizing specially designed, manually powered, mechanical seed cleaning devices for removing unwanted contaminants from the grass seed, with special emphasis on cool-season turfgrasses. The seed cleaning machine was built of wood and contained a series of interchangeable screens with different mesh sizes that allowed the removal of unwanted crop and weed seeds based on their differential sizes. O.M. Scott also pioneered practical methods for testing seed in terms of viability that predated government regulations requiring this procedure. These seed improvement practices by mechanical cleaning were initiated in the 1880s by O.M. Scott. This quality grass seed was initially marketed principally to farms for use in pastures in the Midwest. One marketed species was Kentucky bluegrass that also was being purchased for turfgrass areas. In 1906, the O.M. Scott Company began to market quality seed by mail. In 1916, an order was received for 5,000 pounds (2,268 kg) of Kentucky bluegrass seed for Brentwood in the Pines Golf Club on Long Island, New York. This order marked a turning point for the O.M. Scott Company, resulting in major emphasis placed on marketing quality grass seeds for the turfgrass industry in North America.

Quality commercial turfgrass seed availability started to evolve in England around the 1900s. The pioneer leaders were Martin F.H. Sutton (1875–1930) of Sutton & Sons, Reading, England; Reginald E.C. Beale (1877–1952) of James Carter & Company, Raynes Park, England; and John L. Forbes (1883–1967) of Stewart and Company, Edinburgh, Scotland (see Chapter 12). Each was actively involved in improving the weed-free aspects of turfgrass seeds, and each was the author-publisher of an early series of booklets concerning turfgrass establishment from seed and the subsequent culture.

Plant Variety Protection Act. The adoption of the United States Plant Variety Protection Act of 1970 provided a stimulus for proprietary plant breeding activities by private firms. Prior to the act, only limited

Figure 8.4. Group of grass seed cleaning machines at Sutton & Sons, late 1950s. [20]

cultivar breeding work was being done by university grass breeders. These public cultivars were released to any grower requesting the seed. New cultivars released by these public tax-supported entities periodically resulted in surplus production and an associated reduction in grower prices. In contrast, seed grown on a proprietary contract allowed for a controlled supply of seed more coordinated with the demand.

Cultivar Seed Associations. Grass seed associations were readily formed as new turfgrass cultivars became available. The concept was inaugurated by formation of the Merion Bluegrass Association in 1953, which was composed of seed growers for that cultivar. Quality growing was the major purpose of the association that worked closely with Pennsylvania State University to maintain high-quality certified seed. As new cultivars and species of grass came on the market, more associations were formed under the leadership of Bill L. Rose to grow seed, including the Exeter Bentgrass Association in 1966, the Manhattan Ryegrass Growers Association in 1970, and the Penncross Bentgrass Association in 1971. Cultivar seed associations soon recognized the possibilities of managing overproduction and maintaining seed quality through such grower groups. The associations recruited top seed growers to participate in the production and set standards of quality that were not easy to maintain. Also, marketers usually were selected for the grass seed through a contract arrangement.

Cultivar Royalties. As the cultivar seed associations proliferated and growers realized the protection that associations offered, they continued to offer new provisions. In the 1960s, Bill Rose and Dick Bailey began a program to offer royalties to university researchers who developed new, improved cultivars of turfgrass. The association assessment on sales of a new cultivar was made to the association, which in turn was paid to the university for development of the cultivar.

The consortium has worked well, and the spectacular improvement in turfgrasses has been accepted by the seed merchants and especially by the customers of the grass seed industry. Grass seed growers have found that growing new cultivars to specified high standards has caused prices to stabilize. Growers are less prone to huge seed surpluses that were once prevalent in the turfgrass seed industry.

OREGON TURFGRASS SEED PRODUCTION

This section is adapted from a document written by W. Scott Lamb titled "A History of the Oregon Turfgrass Seed Industry." It provides an overview concerning development of turfgrass seed production in the Willamette Valley. The Oregon Agricultural Experiment Station Bulletin No. 4 of 1890 contained a research report by E. Grimm (3) that predicted, "The indications are that this is a most wonderful grass country, and when farmers once become fully awakened to the value of their land for grasses and clovers, wheat growing will take a secondary position in the Willamette Valley, which in time will be the home of the grasses and clovers and that in their highest perfection." Bentgrass seed harvesting from native stands began along the Pacific coast in Oregon during the 1920s. Then formal turfgrass production evolved and became centered in the Willamette Valley of northwest Oregon. This region produced seed of turfgrass species such as turf-type bentgrass, perennial ryegrass, fine-leaf fescue, Kentucky bluegrass, tall fescue, and rough bluegrass.

Grass seed is now produced to varying extent throughout the state. Northeast Oregon is a sizeable growing area for Kentucky bluegrass and fine-leaf fescue. Seed growers in the area were perhaps the first to plant grass seed in rows, which facilitated cultivation, enabling unwanted species to be rogued out. Central Oregon also has produced quality Kentucky bluegrass and fine-leaf fescue seed. This is high-mountain country, where growers had to learn seed production practices that were different from those used by the valley farmers.

Grass Seed Certification. The objective of seed certification by a state is to provide assurance that a species or cultivar is true to the specified type. Grass seed certification in Oregon began in 1926 with Astoria colonial bentgrass in Clatsop County in northwest Oregon (7), followed in 1927 by Seaside creeping bentgrass in Coos County in southwest Oregon. Perennial ryegrass was included in 1932, with Highland colonial bentgrass added in 1934. Eventually, a system evolved in which breeder seed from the eastern United States was sent to Oregon, where certified seed of genetic integrity was grown and returned to the eastern areas for turfgrass use. A similar program resulted in cultivars of European grass seed also being sent to Oregon for production.

Seed Grower Promotion Organized. In the 1950s, growers began to consider the promotion of Oregon grass seed. Grass seed dealers sold seed to companies across the nation, where the seeds were mixed and sold under a specific company brand name. Typically, no mention was made of the quality produced by Oregon grass seed growers. Some growers of fine-leaf fescue seed met to investigate the possibility of forming a promotional organization. The enabling act to authorize formation of a commission was passed by the Oregon legislature, and in 1955 the Chewings and Creeping Red Fescue Commission was formed. An assessment on the first purchaser of fine-leaf fescue was taken out of the seed payment to each grower and turned over to the commission for promotional work. Subsequently, the Highland Bentgrass Commission was formed in 1959, and the Ryegrass Commission came into being in 1965.

Field Burning Issues. As the ryegrass production expanded in the 1940s, so did the diseases that hampered production. Blind seed disease caused by *Gloeotinia granigena* became particularly devastating, reducing crop yields. During the late 1940s, researchers had found that field burning was effective in controlling the disease, and also growers noticed that burning stimulated plants into greater seed production the following year. Field burning became a common practice, and some 100,000 acres (40,500 ha) were burned in 1950. As more people moved into the Willamette Valley, complaints of smoke from the field burning escalated in the 1960s. The new Oregon Department of Environmental Quality began to question the practice and legislators from affected districts began introducing bills to curtail open burning of fields. The Oregon Seed Council began experimenting with straw densification programs and mechanical field burners to eliminate the smoke. Millions of dollars were invested in research programs focused on both approaches. Metal field burners tended to collapse under the intense heat of burning straw. There has been success in densifying straw into large bales for animal feed, with a key market developed in Asia. The Oregon legislature eventually passed a bill aimed at phasing out field burning. Consequently, grass residue management has been a key to future turfgrass seed production in western Oregon. Also, cool-season turfgrass seed production has become more dispersed into Canada, Idaho, Oregon, and Washington.

WASHINGTON AND IDAHO TURFGRASS SEED PRODUCTION

Major turfgrass seed production areas developed in eastern Washington and in northern Idaho. Seed production locations were mainly Spokane County in Washington and the Rothdrum Prairie area near Post Falls, Idaho. These seed production areas were specifically planted for this purpose. There were both unirrigated and irrigated production fields. Early emphasis was placed on Kentucky bluegrass. Jacklin Seed Company was an early leader in production and marketing of turfgrass seed from this region, especially Merion Kentucky bluegrass. Other key turfgrass processors were Dye Seed and Seeds Inc. Post-harvest burning of grass seed fields in Washington was phased out during the 1996 through 1998 period. It is still being practiced in Idaho.

The Merion Bluegrass Association formed in 1953 and was followed by the Intermountain Grass Grower's Association (IGGA) in the 1970s. Subsequently, in the late 1990s, the latter organization split into the IGGA and the North Idaho Farmer's Association. The IGGA is now known as the Washington Turfgrass Seed Commission.

The following description is adapted from a document written by Arden W. Jacklin titled "Evolution of Jacklin Seed Company." It provides a personal perspective of turfgrass seed production history in Washington and Idaho, represented by a key company directly involved.

The Jacklin organization initiated turfgrass seed production by contracting for a planting of an Olds creeping red fescue seed production field in 1944. Subsequently the company began purchasing land for seed production to compliment the contract growing, seed processing, and marketing operations. In 1945, seed production of two Kentucky bluegrass cultivars was initiated, specifically Delta and Geary. The planting of a Merion Kentucky bluegrass seed field followed in 1947. Demand for Merion grew rapidly, with the Jacklin Seed Company becoming the largest grower-processor in the United States. Other early Kentucky bluegrass cultivars added to Jacklin Seed Company field production were Windsor, 0217, and S-21. Then, in 1956, the first seed production field of Penncross creeping bentgrass was planted by the Jacklin Seed Company with production continuing until 1987. Perennial ryegrass seed production was eventually added with cultivars Pelo and NK100 around 1963.

REFERENCES

1. Baltensperger, A., B. Dossey, L. Taylor, and J. Klingenberg. 1993. Bermudagrass, *Cynodon dactylon* (L.) Pers., seed production and variety development. International Turfgrass Society Res. J. 7:829–838.

2. Edler, G.C. 1930. Bent seed production in Germany. Bull. USGA Green Section 10(11):205.

3. Grimm, E. 1890. Notes on Farm Crops. Oregon Agric. Exp. Sta. Bull. 4, p. 3–10.

4. Kearns, V., and E.H. Toole. 1939. Relation of temperature and moisture content to longevity of Chewings fescue seed. Technical Bull. No. 670, US Dep. of Agric.

5. Odland, T.E. 1930. Bent grass seed production in Rhode Island. Bull. USGA Green Section 19(11):201–204.

6. Olcott, J.B. 1888. Fine versus coarse *Agrostis*. Connecticut Agric. Exp. Sta. Annual Rep, 1887. New Haven, Connecticut, USA, p. 177–181.

7 Schoth, H.A. 1930. Bent grass seed production in the Pacific Northwest. Bull. US Golf Assoc. Green Section 10(11):206–211.

8. Stevens, W.J. 1909. Lawn. Agric. & Horticultural Assoc. Ltd., London, England, UK.

9. Wheeler, W.A. 1950. Forage and Pasture Crops. D. Van Nostrand Co. Inc., Princeton, New Jersey, USA.

Figure 8.5. A fine-leaf fescue clonal grass breeding nursery at Michigan State University in the early 1960s. [26]

Figure 9.1. Turfgrass sod production planting experiments at the Michigan State University Muck Experimental Farm, Rose Lake, Michigan, USA, 1965. [26]*

---

* The numbers in brackets correspond to the figure source acknowledgments listed in the appendix.

CHAPTER 9

History of Sod Production

SODDING WAS THE PRINCIPLE METHOD FOR ESTABLISHING QUALITY TURFGRASS FOR MOST OF THE evolutionary period in turfgrass development from the twelfth through nineteenth centuries. This was because most of the perennial grass seeds available were of a very erect, open, course-textured forage type and the seed was heavily contaminated with seeds of unwanted broadleaf weeds and weedy grasses. Thus sodding was the principle means of achieving a quality turfgrass.

EARLY SODDING PRACTICES

The first written literature describing sodding appeared in 1159 in the first Japanese book on gardening titled **Satu-tei-kai** (20).[*] This sodding probably involved unmowed, low-growing ornamental zoysiagrass plantings. In 1665, the Englishman J. Rae authored a book wherein sod harvesting and transplanting was discussed (19). The preferred turfgrass was colonial bentgrass harvested from sea marshes. An interesting excerpt from his book follows:

> The best turfs for this purpose are had in the most hungry Common, and where the grass is thick and short, prick down a line eight to ten feet long, and with a spade cut the turfs thereby, then shift the line a foot or fifteen inches further, and so proceed until you have cut so far as you desire, then cross the line to the same breadth, that the turfs may be square, and cut them thereby; then with a straight bitted spade, or turving-iron (which many for that purpose provide) and a short cord tied to it near the bit, and the other end to the middle of a strong staff, whereby on thrusting the spade forward under the turfs, and another by the staff pulling backward, they will easily be flaid and taken up, but not too many at a time for drying, but as they are layed; which must be done by a line, and a long level, placing them close together, and beating them down with a mallet; having covered the quarter, or place intended, let it be well watered, and beaten all over with a heavy broad beater.

For the most part, sods were not harvested from commercial production fields as is commonly practiced today. Rather, the sod was harvested from old stands of turfgrass that had been grazed for an extended period of time under specific soil and climatic conditions. Sodding was referred to as "returfing" in the

---

[*] The numbers in parentheses correspond to the reference list found at the end of the chapter.

United Kingdom. Salt fail or sea-marsh turf was a key source for sodding bowling and putting greens in Scotland during the early 1700s and continued in the United Kingdom into the 1920s. The commercial sod harvest sites were on natural seaside environments, specifically marshes, with a minimal weed content and fine-textured silts that allowed more easy turfgrass lifting and trimming by boxing. An entire double green of 20,776 square feet (1,930 m²) was returfed at the St Andrews Old Course in the late 1800s.

Specialty sod production of turfgrass was recorded in Japan in 1701 (16). Another Englishman, A. McDonald, published a book titled **The Complete Dictionary of Practical Gardening** in 1807 (18). In it he states that "sodding is preferred over seeding." In 1906, the first book on lawns was published in the United States (1). The author Leonard Barron made the following statement on sodding:

> Sods are generally cut for convenience's sake three feet long and one foot wide, and in quantity can usually be bought at twenty cents a turf. I know of one man who has developed the regular business of growing sods for sale. His trade is in a city that is famous for its well kept gardens. He uses only the highest grade lawn mixture for raising of his crop which is given careful attention from first to last and he gets ten cents per square foot, thus realizing a profit of between five and six hundred dollars to the acre every three years. No turfs are cut and sold under that age. This is necessary, in fact, because the Kentucky blue grass will not have made a proper growth before this time.
>
> The average workman can lay in a day and do it perfectly about five or six hundred square feet of sod, giving thorough attention to levelling and making complete union. An expert can cover as much as eight hundred square feet or more. This is not work that can ordinarily be done by a common day labourer, and will generally cost two dollars a day. The cost for laying an acre at this rate would therefore be about one hundred and sixty dollars.
>
> After the turf is laid as evenly as possible, and the unions filled with fresh soil, there comes the very essential work of beating. This is really hard work. The turves must be beaten and pounded down to ensure intimate contact with the soil below. If this is not done the roots fail to take hold and the grasses die after a few days of dry, hot weather. Watering will help a great deal, and should be done all summer on a newly laid turf lawn.

Many of the early sod operations were simply pasture stripping activities cut from short-term rented sites. It consisted of harvesting sod from a grassy pasture area after one to several mowings, removing the cut vegetation, and cutting out the sod. It was rarely weed-free.

Commercial sod production is an agriculturally oriented operation in which turfgrasses are specifically planted and maintained for the purpose of harvesting for shipment to consumer sites to be transplanted onto lawns, sports fields, and commercial sites. This was a drastic change from the previous pasture stripping operations as it became the main source of sod due to a major improvement in quality. Several commercial sod production operations were initiated during the 1920s in Michigan, New England, and Ohio. However, it was not until the 1960s that the commercial sod production industry expanded rapidly in acreage and economic value, especially in North America (3, 4, 8).

STOLON AND SPRIG PRODUCTION

A number of commercial sod operations originally produced vegetatively propagated stolons of creeping bentgrass in cool climates or sprigs of bermudagrass or St Augustinegrass in warm climates.

Stolonizing. The 1920s and 1930s was a time when vegetative stolon production of creeping bentgrass for lawns and golf courses peaked in North America. The main species utilized was creeping bentgrass. During that time, the better quality turfgrasses were produced from genetically uniform, weed-free, vegetatively

Figure 9.2. Creeping bentgrass vegetative stolon production at O.M. Scott & Sons, Marysville, Ohio, USA, circa 1928. [31]

propagated creeping bentgrass. Planting of stolons in rows 3 to 5 feet (0.9–1.5 m) apart was a typical method. Soil fumigation prior to planting was practiced to eliminate off-types and weedy species. Early harvesting consisted of manually cutting off the stolons and chopping into ~6 inch (15 cm) lengths by use of large knives and a wooden cutting board. After WWII, powered mechanical sod cutters were used to harvest the vegetative rows, followed by chopping in a modified soil shredder or silage chopper developed by growers and then bagging for shipment. Stolons also have been harvested from full turfgrass sods by cutting the sod just at the soil surface and passing it through a shredder.

Early bentgrass stolon planting on user sites involved manually spreading over a prepared soil, manually covering with a screened topdressing or soil material by sling shoveling, rolling, and irrigation where needed. Stolonizing of bentgrass essentially ceased in the 1960s as seeding became dominate.

Sprigging. Sprigging is practiced for the warm-season grasses such as bermudagrass, zoysiagrass, centipede-grass, St Augustinegrass, and carpetgrass. Early techniques involved manual cutting from a sod field at the soil surface, manual chopping via large knives on a wooden chopping board, and bagging. Actually, vegetative propagation of bermudagrass did not become widely practiced until the 1960s when turfgrass-type cultivars were developed. Key early hybrid bermudagrass cultivars produced for sprigging were Tifgreen, released in 1956, and Tifdwarf, released in 1965, both by the Georgia Agricultural Experiment Station (AES) and Crops Research Division—Agricultural Research Service (CRD-ARS). These sprigs were cut from mature sod with a powered mechanical sod cutter and cut up into short sprigs via modified mechanical soil shredders or silage choppers. Eventually, mechanical harvesters were developed by growers that cut and shredded sod, screened out soil, and conveyed the sprigs into holding bins or bags.

Early planting of sprigs utilized a manual hoe to cut shallow trenches into which sprigs were placed; the soil was then pulled back over the trench and manually tamped. Mechanical planters were developed by growers in the 1960s with rolling cutters that cut a shallow furrow into which sprigs were dropped; the furrow was pushed back and press rolled to firm soil around the sprigs. The sprig rows are spaced 10 to 18 inches (25–46 cm) apart. A version of stolonizing is used for planting bermudagrass on putting greens involving mechanical devices for broadcast spreading, press rolling, topdressing, and rolling.

Plugging. A method of vegetative planting 2 to 4 inch (5–10 cm) full-turfgrass pieces as circular or square shaped plugs is used for planting primarily zoysiagrasses, especially in residential lawns. The 4-inch size is preferred. Historically, plugging was a manual process with cutting of both the plug and hole opening by means of an appropriately shaped metal cutter positioned on the end of a handle. Specialized mechanical plug harvest and plug planting machines were developed in the 1970s by growers. Plug planting typically is on 12 to 16 inch (30–41 cm) centers. The key zoysiagrass cultivars produced for plugging include Meyer Japanese zoysiagrass, released in 1951 by the CRD-ARS and USGA Green Section, and, to a lesser extent, Emerald manila zoysiagrass, released in 1955 by the Georgia AES, CRD-ARS, and USGA Green Section.

SOD PRODUCTION

The time required to form a sod of proper density and strength during the early years was three or more years. This caused sod to be increasingly expensive for use on residential lawns. Obviously, a cultural system to meet the unique requirements of commercial sod production was needed. Certain technologies from intensive golf course culture were adaptable. Other aspects were learned by observant sod growers via trial-and-error experiences. Still, some innovative technologies only emerged from detailed research, as outlined in a later section. Eventually, knowledgeable sod producers could produce on-average three crops in two years with the improved cultivars available.

The desired production site consists of a relative level landform with a stone-free soil root zone. Typical early sod operations failed to provide any site preparation activities prior to seeding, which resulted in the inability to harvest numerous void areas scattered throughout a field, weedy grass stands requiring costly control measures, poor sod strength that delayed harvesting, and low-quality sod, resulting in a reduced market price. Drainage of excess water was critical and has been provided via open ditches. Initial development of sod production land now involves two or more years. This entails clearing brush, roots, and debris; land leveling; installing surface and subsurface drainage; and fallowing to control weedy species.

The principle species utilized in commercial sod production have been and are the cool-season Kentucky bluegrass and polystands with perennial ryegrass or tall fescue and warm-season St Augustinegrass, bermudagrass, and zoysiagrass. Early plantings for cool-season turfgrasses frequently were attempted by broadcasting seed, followed by mechanical mixing into the soil, with rolling sometimes done. A resultant problem was erratic seedling stand densities that extended the time to form harvestable sod strength. The causes were seed placed too deeply and poor seed-soil contact needed for moisture transfer. The solution was planting with a cultipacker seeder with a gravity seed box placed between the two off-set rollers. It also became apparent that uniform, timely irrigation was key to achieving the desired stand density.

Following seedling establishment, the priority in early sod growing practices after WWII was solely to produce a dense, dark-green shoot biomass. This was accomplished by high rates and frequent applications of nitrogen. However, the lack of adequate sod strength needed for harvesting and handling was a common problem that extended the production time. Subsequent research revealed a lack of rhizomes caused by the high nitrogen levels forcing aboveground shoot growth that robbed the carbohydrates needed for rhizome growth. The excessively high nitrogen fertilization also reduced the rooting rate of transplanted sod.

Another early improper culture practice was mowing at very infrequent intervals, which also extended the sod production time. The resultant excessive leaf blade removal at any one mowing tended to restrict the carbohydrate supply for rhizome growth needed to achieve proper sod strength.

Agricultural tractor tires were commonly employed in early sod production operations. However, a negative consequence was depressed tracks across the field. The depressions impaired maximum yield in sod harvesting, as surface soil smoothness was vital. Thus rolling of production fields became an important culture practice. Also, sod growers developed flat metal wheels of an extended width that minimized disruption of the smooth soil surface.

Much of the 1960s sod production expansion was located on muck soils (21). These organic soils facilitated shorter production times and a lighter mass that aided ease in harvest and handling plus shipping larger loads to distant markets. In the 1960s, Michigan sod growers were shipping sod for distances up to 400 miles (644 km).

Net Sod. Sod production with net reinforcement was targeted to increase sod strength for earlier harvesting and for turfgrass species with a bunch-type or weak lateral stem growth, such as perennial ryegrass and tall fescue. The concept was originated in 1970 by F. Brian Mercer of Blackburn, England, researched positively at Michigan State University (8), and successfully field tested in California (11) and Michigan. Net sod techniques can shorten the production by up to 75%. It continues to be used primarily in specialty situations and under intensive production.

Clippings Pelletizing and Utilization. Techniques for pelletizing leaf clippings of Kentucky bluegrass were initiated in 1967 by Drs. James B Beard and Michael B. Tessar at Michigan State University and were eventually successful after attempting various procedures (4). Two tons of pelletized clippings were processed. Analyses conducted were element content, protein content, total digestible nutrients, and rate of digestibility. Then utilization studies were successfully conducted by comparative feeding experiments with horses and chickens. The high xanthophyll was of particular benefit for chickens. However, the most promising use tested was as laboratory animal litter for rats and rabbits. Superiority was noted in odor suppression while causing no skin rashes. Subsequently, clippings utilization facilities for chicken feed were constructed and successfully operated by the Warren Turf (22) and Calturf sod production operations.

SOD HARVESTING

During the twelfth through nineteenth centuries, sods were harvested manually by individuals using a "turving iron," which was a rudimentary form of what is now known as a sod knife. In order to lay level sods on formal turfgrass areas like greens, the sod was manually harvested by a turving iron and then placed upside down in a "gauge box" where a soil "sheering knife" was drawn across the underlying soil at a uniform depth to produce a level underside for the sod. Advanced versions of the latter were called "trimming boxes." Eventually, a sled-like device was designed and successfully used for harvesting sod to a somewhat controlled width and depth. It was laboriously pushed or kicked along by an individual and eventually was operated by the pulling force of a horse.

The 1960s was a period of rapid expansion for the commercial sod production industry (9). As part of this expansion, various mechanical cutting, rolling or folding, conveyer loading, and palletized loading machines were developed by growers. The basic reason behind these mechanical harvesting advances was to achieve more competitive costs that reduced what had been a very costly, labor intensive industry.

A powered mechanical sod cutter was invented by Frank Phillips in 1944 and marketed in the late 1940s (10). This marked a major advancement in cutting quality and labor efficiency. Eventually, sod cutters were designed that could cut widths of 12, 16, 18, and 24 inches (30, 41, 46, and 61 cm) plus a mechanical blade attachment in the back cut the sod in controlled lengths. Also, a number of different types of sod rolling devices were designed, constructed, and tested by inventors such as Daymon and Ryan in the late 1960s. Originally, these rolling devices operated as separate machines but, subsequently, were combined in single sod cutting and rolling units. A popular, simple device was the Sulky-Ryder Roller, originating in Michigan, which could be hitched directly to the back of a sod cutter.

Initially, sod was manually loaded onto trucks for delivery to the user site. Then conveyor sod loading onto semitrailer trucks was developed in the late 1960s, followed in the early 1970s by loading onto pallets and use of forklifts to place the loaded sod pallets onto trucks. At least six designs of an integrated sod cutting, rolling or folding, and palletized loading machine mounted on a tractor or a self-propelled platform were constructed and successfully tested by growers. Sod harvesters were developed and marketed by Big J, Brouwer, Nunes, Princeton, Ryan and Wessel, all in North America. Several of the integrated mechanical sod harvesters came into common use by the late 1970s.

Figure 9.3. A horse-drawn "Champion" sled sod cutter with 12-inch (30.5-cm) cutting width, 1920s. [41]

Big Roll. Sod transplanting has long been and continues to be a labor intensive practice on small residential properties. A key innovation during the 1970s was what is called the "big roll," developed for extensive turfgrass areas. It consists of a mechanical cutter and rolling apparatus that cuts the sod into 48 inch (1.2 m) widths and rolls the sod in up to 60 foot (18 m) lengths. Then a transplant rolling device is used to position and lay the sod onto the installation site, followed by rolling.

Soilless Sod. Another innovation is "washed sod," developed in 1974 by Ben O. Warren of Illinois, USA. A basic principle in sod transplanting of putting greens, teeing grounds, and sports fields is that the soil attached to the sod must be of the same texture as the underlying soil. Otherwise, internal water drainage is seriously impaired and divot damage greatly increased. Thus an apparatus that employs a flexible conveyor mat with perpendicular directed, pressurized water sprays to wash soil from the cut sod was developed, successfully tested, and used commercially. The soilless sod roots faster than unwashed sod in addition to minimizing the problems indicated earlier.

SOD PRODUCTION RESEARCH

Much early sod production methodology was developed by trial-and-error experiences of individual growers (13). During initial development of the pioneering sod production research program at Michigan State University, innovative devices were designed, tested, and successfully developed for the assessment of sod strength (7, 20, 24) and transplant sod rooting (12, 13). They were essential in order to study the relative sod performance characteristics of turfgrass species and cultivars and for mixtures and blends as well as for

Figure 9.4. Motor-powered, reciprocating-knife sod cutter with cut-off knife and sulky sod roller attachments. [26]

testing the effects of mowing height and frequency, nitrogen fertility level, seeding rate, and pesticide effects. Major innovations arising out of this unique sod production research program included the following:

- Substantially reduced time to grow a commercial sod crop (4, 5)
- Enhanced rate of transplant sod rooting on customer sites (13)
- Development of net sod production techniques (8)
- Development of the first pelletized clippings and utilization assessments (4, 5, 22)
- Cultural techniques for the prevention of sod heating during transport (4, 14, 15)
- Shallow root-zone production of turfgrasses over plastic within two weeks (8)
- Comparative soil subsidence (5, 21)

Paralleling this sod production research was the development of a major turfgrass breeding program by Dr. Reed Funk at Rutgers University. The main cool-season species grown on commercial sod farms was rhizomatous Kentucky bluegrass, which is apomictic. Dr. Funk developed an interspecific hybridization procedure that was a major breakthrough. His outstanding research led to a number of improved turfgrass cultivars released in the 1970s, such as Adelphi, Bonnieblue, and Majestic Kentucky bluegrass, that were used in sod production. Key early warm-season turfgrass cultivars aiding sod production in the southern United States include Tifway hybrid bermudagrass, released in 1960 from the Georgia AES and CRD-ARS, and Floratam St Augustinegrass, released in 1973 from the Florida AES and Texas AES.

The Michigan State University and Rutgers University research efforts plus mechanical innovations by sod growers culminated in (a) major advances in sod production technology, (b) development of numerous improved turfgrass cultivars used in sod production, and (c) mechanization of commercial sod production.

This eventually led to the development of a mature sod production industry that offered quality, affordable sod to a broad market of consumers in North America and around the world.

SOD PRODUCERS ORGANIZATION

In response to this rapid sod industry growth, a unique research and educational program for commercial sod production was developed at Michigan State University under the leadership of Drs. James B Beard, Paul E. Rieke, and Joseph M. Vargas. From this major sod production research effort evolved the summer Michigan State University Sod Producers Field Day and a Sod Production Educational Section held during the annual winter Michigan Turfgrass Conference in East Lansing (2, 3, 5). These field days and winter educational sessions drew 200 to 300 attendees representing sod production operations from throughout the United States and Canada as well as a few growers from other parts of the world. As a result of these activities, there was a growing cohesiveness among sod producers that also served as a focal point in the final formation and start-up of the American Sod Producers Association.

After a number of failed efforts, progress was made in forming a national sod organization. On July 11, 1967, in conjunction with the Michigan State University (MSU) Turfgrass Field Day (2), the American Sod Producers Association (ASPA) was officially organized (17). It was at an evening meeting in Anthony Hall on the MSU campus that growers from across the country began to sign up as charter members. The ASPA officers elected at this first ASPA meeting were President Ben Warren, Illinois; Vice President Robert Daymon, Michigan; Treasurer Louis DeLea, New York; Secretary Richard Horner, Wisconsin; and board members Tobias Grether, California; J.E. Ousley, Florida; and Wiley Miner, New Jersey.

On the second day, the first ASPA field day was held. Through the efforts of Dr. James B Beard, Dr. Paul E. Rieke, and Don Juchartz of Michigan State University, an all-day field trip via buses was arranged. The morning tour was to Emerald Valley Turf, where 700 acres (283 ha) of Merion Kentucky bluegrass was in production. Those in attendance viewed the on-site sod production operation, ranging from specialized harvesting equipment to modern cultural practices. The afternoon was spent at Halmick Sod Nursery near East Lansing, Michigan, where attendees watched demonstrations of all the latest sod harvesting equipment. This innovative equipment was shown by grower-innovators and manufacturers from throughout North America. There were 124 commercial sod growers from the United States, Canada, and Europe on this tour. Both days were a success.

The ASPA was born and started to grow from the first day. The name was changed to Turfgrass Producers International (TPI) in 1994. An annual winter sod producers conference has been held by TPI since 1967 as well as an annual summer sod field day and equipment show. TPI also has published the bimonthly periodical **Turf News** since 1977. Eventually, sod grower organizations were formed in other countries such as the United Kingdom and Australia.

REFERENCES

1. Barron, L. 1906. Lawns and How to Make Them. Doubleday, Page & Co., New York, USA.
2. Beard, J.B, editor. 1967. Michigan State University Turfgrass Field Day Report and Sod Producers Field Day. Michigan Agric. Exp. Sta. Publ., East Lansing, Michigan, USA.
3. Beard, J.B, editor. 1969. Michigan State University Sod Producers Field Day Report. Michigan Agric. Exp. Sta. Publ., East Lansing, Michigan, USA.
4. Beard, J.B, editor. 1969. Sod and Turf—Michigan's $350 Million Carpet. Michigan Sci. in Action, East Lansing, Michigan, USA.
5. Beard, J.B, editor. 1971. Michigan State University Sod Producers Field Day Report. Michigan Agric. Exp. Sta. Publ., East Lansing, Michigan, USA.

6. Beard, J.B. 1972. Comparative sod strengths and transplant sod rooting of Kentucky bluegrass cultivars and blends. 42nd Michigan Turfgrass Conf. Proc. Michigan State University, East Lansing, Michigan, USA. 1:123–127.

7. Beard, J.B. 1973. Turfgrass: Science and Culture. Prentice-Hall Inc., Englewood Cliffs, New Jersey, USA.

8. Beard, J. B, D.P. Martin, and F.B. Mercer. 1980. Investigation of net-sod production as a new technique. Proc. Intl. Turfgrass Res. Conf. 3:353–360.

9. Beard, J.B, and P.E. Rieke. 1969. Producing quality sod. In: A.A. Hansen and F.V. Juska, editors, Turfgrass Science. Am. Soc. Agron. Ser. 14. Am. Soc. Agron., Madison, Wisconsin, USA.

10. Betts, J.L., and W.G. Mathews, editors. 1992. Turfgrass: Nature's Constant Benediction—The History of the American Sod Producer's Association 1967–1992. Am. Sod Producers Assoc., Rolling Hills, Illinois, USA.

11. Cockerham, S.T. 1988. Turfgrass Sod Production. Publication 21451, Division of Agric. and Nat. Resources, Oakland, California, USA.

12. King, J.W., and J. B Beard. 1969. Measuring rooting of sodded turfs. Agron. J. 61:497–498.

13. King, J.W., and J. B Beard. 1972. Post harvest cultural practices affecting the rooting of Kentucky bluegrass sods grown on organic and mineral soils. Agron. J. 64:259–262.

14. King, J.W., J. B Beard, and P.E. Rieke. 1982. Factors affecting survival of Kentucky bluegrass sod under simulated shipping conditions. J. Am. Soc. Hort. Sci. 107(4):634–637.

15. King, J.W., J. B Beard, and P.E. Rieke. 1982. Effects of carbon dioxide, oxygen, and ethylene levels on Kentucky bluegrass sod. J. Am. Soc. Hort. Sci. 107(4):638–640.

16. Maki, J. 1976. Utilization, research activities, and problems of turfgrass in Japan. Rasen Gruflachen Bergrunungen 2:24–28.

17. Mathews, W., and W. Pemrick, editors. 2007. History of Turfgrass Producers International: 40th Anniversary 1967–2007. Turfgrass Producers Int., East Dundee, Illinois, USA.

18. McDonald, A. 1807. A Complete Dictionary of Practical Gardening, vols. 1 and 2. R. Taylor and Co., London, England, UK.

19. Rae, J. 1665. Flora, Ceres and Pomona. Richard Marriot, London, England, UK.

20. Rieke, P.E., J. B Beard, and C.M. Hansen. 1968. A technique to measure sod strength for use in sod production studies. In: Agronomy Abstracts. American Society of Agronomy, Madison, Wisconsin, USA. p. 60.

21. Rieke, P.E., J. B Beard, and R.E. Lucas. 1968. Grass sod production on organic soils in Michigan. Proc. Third Intl. Peat Congr. 3:350–354.

22. Warren, B. 1992. Grass clipping utilization. Proc. Illinois Turfgrass Conf. Univ. of Illinois, Urbana, Illinois, USA. 13:57.

Figure 9.5. The original sod strength test apparatus first developed at Michigan State University in 1966. [26]

Figure 10.1. A group of Midwest university turfgrass researchers during a meeting at Michigan State University in 1967. [26]*

* The numbers in brackets correspond to the figure source acknowledgments listed in the appendix.

CHAPTER 10

Organization of United States Turfgrass Scientists

TURFGRASS SCIENCE DEVELOPED VIA RESEARCH WAS VERY SLOW TO EVOLVE COMPARED TO MOST field and horticulture crops. Prior to 1946, there had been very few peer-reviewed research papers published on the subject of turfgrass. This began to change after WWII. Actually, a number of forage scientists with some turfgrass activities served as members of the United States Army for the duration of WWII with responsibility for grass vegetation maintenance on airplane runways and camp installations. In this capacity, they were interacting regularly. This has contributed to the subsequent formation of turfgrass scientists discussed herein. Certainly a national organization was needed in the United States where emerging turfgrass researchers and educators could meet to exchange information and discuss approaches to solving problems facing turfgrass culture.

At a leadership meeting of the American Society of Agronomy (ASA) in Cincinnati, Ohio, in March 1946, a proposal to hold an oral paper presentation session on "turf" was approved. It was assigned to the "Special Crops" Section along with weed control. Finally, the stage was set for turfgrass scholars to participate in a face-to-face forum where their individual research and education activities could be formally presented and discussed. The early paper presentation sessions and appointed session chairs for the 1946 through 1952 formative period are summarized in table 10.1. The situation at the first paper presentation session in 1947 was tenuous at best, according to the following description:

> Leaders of the many turfgrass groups throughout the country are invited and urged to take membership in the American Society of Agronomy through the secretary-treasurer, Dr. G.G. Pohlman, Morgantown, West Virginia, and to attend the meetings of the Society which are announced in the official journal. In September, 1946, titles and abstracts of papers for the first Turf Management Program were received and approved.
>
> The first Turf meeting held within the Society at Omaha, Nebraska, in November, 1946, very nearly proved to be the last. The space allotted to the Turf Group was on the Mezzanine, just inside the railing above the hotel lobby. The noise and confusion of the lobby were magnified and intensified to the point where it seemed, at times, foolish to continue the meeting. The fact that those present persisted and did present their papers while those trying to listen kept on trying was something of a miracle. That ordeal having been met and the test passed, plans were made for the second meeting to be held in November, 1947, in Cincinnati.

Basically, the newly renamed Crop Science Division of the American Society of Agronomy (ASA) was expanded to five subsections in 1946. One subsection was designated V, Special Topics. Turfgrass presentations were approved by the ASA for inclusion as a session within the Subsection V, Special Topics. By the next year, the Special Topics Section 5 was divided into Section 5 "Turf" and Section 6 "Weed Control" within the Crop Science Division of the ASA. In 1953, a name change was made to Division XI, Turf Management, within the Crop Science Division of the ASA.

A Turfgrass Committee, with activities distinctly separate from the preceding formal ASA paper presentations also was formed. It functioned from 1946 through 1955, with F.V. Grau as chair. An annual committee report was presented and published. This led to a "turf" session held within the Agronomic Application Division of the ASA, wherein F.V. Grau became active. A joint session with the Turf Section 5 was held in 1949 with five papers presented. Then a session separate from the Turf Section 5 was organized, with four turfgrass papers in 1950 and four turfgrass papers in 1951. The session was chaired by H.B. Sprague in 1949, G.W. Burton in 1950, and C.K. Hallowell in 1951. The consequence of this separation was a reduction of paper presentations in Crop Section 5, Turf, of the Crop Science Division during this period. The Turfgrass Committee and separate sessions were dissolved after 1951.

Official CSSA Designation. The Crop Science Society of America (CSSA) was formed in 1955 with G.O. Mott elected as the first president for 1956. The question of including a turfgrass division in the new CSSA was argued extensively by the organizing committee, with Dr. Gerald O. Mott successfully defending inclusion. Thus the designation "Division XI, Turf Management," was retained, with a Division XI member elected to serve on the CSSA Board of Directors by the CSSA membership. Also, the first listing of an elected vice chair appeared in 1955. This individual was the chair-elect for the upcoming year and was assigned the upcoming program planning. The first election of a CSSA Division XI member as a Representative to the ASA Board of Directors was implemented in 1961. The division name was changed to Turfgrass Management Division C-5 in 1962, to Turfgrass Division C-5 in 1973, and to Turfgrass Science Division C-5 in 1987.

The Turfgrass Science Division C-5 meets annually at locations around the United States, during which turfgrass research, education, extension, and symposium papers are presented. In addition, turfgrass research of members can be peer-reviewed and published in the technical journals of the CSSA and

TABLE 10.1. SUMMARY BY YEAR OF THE SECTION 5 CHAIR AND NUMBER
OF PAPER PRESENTATIONS, 1946 THROUGH 1952.

Year	Appointed section chair	State	Number of papers presented[*]		
			T	R	D
1946	F.V. Grau[**]	Maryland	6	0	6
1947	H.R. Albrecht[**]	Indiana	16	3	13
1948	H.B. Sprague[**]	Texas	17	1	16
1949	H.B. Musser	Pennsylvania	11	8	3
1950	G.O. Mott[**]	Indiana	10	5	5
1951	R.E. Engel	New Jersey	6	3	3
1952	R.R. Davis[**]	Ohio	6	3	3

[*] T: total papers; R: research papers; D: descriptive papers.
[**] Part-time turfgrass responsibilities, usually with forages.

ASA, known as **Crop Science** and **Agronomy Journal**, respectively. The turfgrass scientists who have been elected Division Chair by their CSSA peers in education, extension, and research are summarized in table 10.2a and table 10.2b.

TABLE 10.2A. SUMMARY BY YEAR OF THE XI/C-5 DIVISION CHAIR, ASA BOARD REPRESENTATIVE, AND NUMBER OF PAPERS PRESENTED BY CATEGORY FROM 1953 THROUGH 1984.

Year	Elected chair	State	Elected ASA Board Rep.	Number of papers presented[*]			
				T	R	D	S
1953	M.H. Ferguson	Texas		6	4	2	
1954	W.H. Daniel	Indiana		6	6		
1955	J.R. Watson	Minnesota		15	10	5	
1956	C.G. Wilson	California		16	13	3	
1957	G.C. Nutter	Florida		16	15	1	
1958	W.W. Huffine	Oklahoma		17	17		
1959	H.B. Musser	Pennsylvania		12	12		
1960	R.E. Engel	New Jersey		17	17		
1961	V.B. Youngner	California	E.C. Roberts	11	11		
1962	R.R. Davis[**]	Ohio	E.C. Roberts	21	21		
1963	F.V. Juska	Maryland	E.C. Roberts	14	14		
1964	N.R. Goetze[**]	Oregon	W.H. Daniel	36	21		15
1965	E.C. Roberts	Iowa	W.H. Daniel	23	18		5
1966	J. B Beard	Michigan	W.H. Daniel	25	19	2	4
1967	C.Y. Ward[**]	Mississippi	R.R. Davis	32	27		5
1968	J.A. Simmons	Ohio	R.R. Davis	40	29		11
1969	R.E. Schmidt	Virginia	R.R. Davis	25	20		5
1970	W.B. Gilbert	North Carolina	J. B Beard	25	25		
1971	G.C. Horn	Florida	J. B Beard	33	33		
1972	R.W. Miller	Ohio	J. B Beard	29	25		4
1973	P.E. Rieke	Michigan	W.W. Huffine	33	29		4
1974	L.M. Callahan	Tennessee	W.W. Huffine	42	37		5
1975	J.F. Shoulders	Virginia	W.W. Huffine	38	34		4
1976	A.E. Dudeck	Florida	K.T. Payne	40	31		9
1977	J.H. Dunn	Missouri	K.T. Payne	30	30		
1978	W.R. Kneebone	Arizona	K.T. Payne	45	32	2	11
1979	A.J. Powell Jr.	Kentucky	P.E. Rieke	45	44	1	
1980	D.B. White	Minnesota	P.E. Rieke	47	38	3	6
1981	A.J. Turgeon	Illinois	P.E. Rieke	41	39	2	
1982	R.W. Duell	New Jersey	J.H. Dunn	44	43	1	
1983	R.C. Shearman	Nebraska	J.H. Dunn	59	56	3	
1984	D.P. Martin	Ohio	J.H. Dunn	48	42		6

[*] T: total papers; R: research papers; D: descriptive papers; S: invited symposium papers.
[**] Part-time turfgrass responsibilities, usually with forages.

TABLE 10.2B. SUMMARY BY YEAR OF THE C-5 DIVISION CHAIR, ASA BOARD REPRESENTATIVE, AND NUMBER OF PAPERS PRESENTED BY CATEGORY FROM 1985 THROUGH 2010.

Year	Elected chair	State	Elected ASA Board Rep. CSSA* Board Rep.	Number of papers presented**					
				T	O	P	S	Ex	Ed
1985	R.N. Carrow	Georgia	A.E. Dudeck	72	65		7		
1986	D.V. Waddington	Pennsylvania	A.E. Dudeck	75	58	5	7	5	
1987	B.J. Johnson	Georgia	A.E. Dudeck	66	30	32		4	
1988	T.P. Riordan	Nebraska	R.C. Shearman	77	31	30	10		6
1989	N.E. Christians	Iowa	R.C. Shearman	97	25	52	8	8	4
1990	T.L. Watschke	Pennsylvania	R.C. Shearman	109	54	45	10		
1991	W.A. Meyer	Oregon	R.N. Carrow	94	28	53	13		
1992	J.M. DiPaola	North Carolina	R.N. Carrow	94	43	45	6		
1993	A.M. Petrovic	New York	R.N. Carrow	95	41	34	20		
1994	K.J. Karnok	Georgia	N.E. Christians	101	41	44	16		
1995	J.V. Krans	Mississippi	N.E. Christians	112	52	52	6	1	1
1996	C.S. Throssell	Indiana	N.E. Christians	98	41	52	3	1	1
1997	R.J. Hull	Rhode Island	T.P. Riordan	119	60	44	11	1	3
1998	G.H. Snyder	Florida	T.P. Riordan	168	86	58	14	4	6
1999	L.A. Brilman	Georgia	T.P. Riordan	157	72	76	5	4	
2000	J.L. Nus	Kansas	J.L. Cisar	143	48	79	11	4	1
2001	M.C. Engelke	Texas	J.L. Cisar	198	63	109	22	4	
2002	T.K. Danneberger	Ohio	J.L. Cisar	185	99	75	4	4	3
2003	G.K. Stahnke	Washington	R.E. Gaussoin	205	83	105	12	4	1
2004	J.D. Fry	Kansas	R.E. Gaussoin	215	90	103	12	4	6
2005	T.W. Fermanian	Illinois	R.E. Gaussoin	179	67	91	7	6	8
2006	E.A. Guertal	Alabama	T.J. Koski	188	61	102	19	6	
2007	R.E. Gaussoin	Nebraska	T.J. Koski	234	89	117	28		
2008	J.C. Stier	Wisconsin	T.J. Koski	163	67	85	11		
2009	D.W. Williams	Kentucky	M. Richardson	186	80	97	9		
2010	T.J. Koski	Colorado	M. Richardson	174	76	93	3	1	1

* ASA Board Rep position terminated in 2005, and CSSA Board Rep continued.

** T: total; O: oral research; P: poster research; S: invited symposia; Ex: extension; Ed: education.

The summary paper presentation numbers may vary from earlier annual C-5 Historian Reports. Those numbers were based on phone conversations with each chair made months before each scheduled Annual Meeting, while the paper number listings herein are based on actual counts from the printed programs.

A very high recognition by one's peers is to be elected President of the CSSA or ASA. Scientists with at least some turfgrass activities who have been elected President of the CSSA total five and President of ASA total six (table 10.3). Only James B Beard and Robert C. Shearman for CSSA and Arden A. Baltensperger for ASA were full-time turfgrass scientists and educators when elected to office.

TABLE 10.3. ELECTED SOCIETY PRESIDENTS INVOLVED IN TURFGRASS SCIENCE.

CSSA President		ASA President	
1956	Gerald O. Mott (first President)	1914	Charles V. Piper
1960	Howard B. Sprague	1943	Frank D. Keim
1974	Richard R. Davis	1962	Glen W. Burton
1986	James B Beard	1964	Howard B. Sprague
1995	Robert C. Shearman	1970	Roy E. Blaser
		1990	Arden A. Baltensperger

Chronological Summary of Turfgrass Science Division C-5 Activities

1960	First time the **Division XI Turfgrass Field Tour** was listed in the ASA Annual Meeting Program, as organized by V.B. Youngner and R.E. Engel.
1961	Division XI published **Nomenclature of Some Plants Associated with Turfgrass Management** of 21 pages, as written by the XI Turfgrass Nomenclature Committee composed of F.V. Juska, W.H. Daniel, E.C. Holt, and V.B. Youngner.
1964	First **Division C-5 Turfgrass Symposium** listed in the ASA Annual Meetings Program, as organized by N.R. Goetze.
1966	A Division C-5 **Operations Manual** was written by J. B Beard.
1966	The organization of an **International Turfgrass Research Conference** was proposed by J. B Beard and was formally encouraged by the Division C-5 members.
1969	The ASA Agronomy Monograph No. 14 titled **Turfgrass Science** of 715 pages was published, as written by 33 Division C-5 members with A.A. Hanson and F.V. Juska serving as editors.
1969	The Division **C-5 Historian** position was established, with C.G. Wilson appointed.
1973	First Division **C-5 By-Laws** were approved by the CSSA, with J. B Beard chair of the By-Laws Committee.
1975	Guidelines for **Turfgrass Metric Equivalents** were adopted by Division C-5, based on a draft proposal by the Division C-5 Terminology Committee chaired by J. B Beard.
1976	Division C-5 published **Post 1976 Turfgrass Industry Challenges In Research, Teaching And Continuing Education** of 33 pages, as organized by the C-5 Bicentennial Committee chaired with J.F. Shoulders as chair.
1979	A comprehensive **Glossary of Turfgrass Terms** was adopted by Division C-5, as drafted by a committee with J. B Beard as chair.
1984	First **Division C-5 Graduate Student Outstanding Paper Award** competition was implemented, based on the concept promoted by J.F. Shoulders, with the award criteria and procedures developed by a committee chaired by D.P. Martin.
1986	First Division C-5 paper session on **Turfgrass Extension** listed in the ASA Annual Meetings Program, as organized by P.E. Rieke.
1986	A major revision of the **Editorial Process for Crop Science** was instituted under the leadership of CSSA President J. B Beard, in which turfgrass scientists became part of the formally appointed editorial decision-making process for publication of turfgrass related papers.
1987	The **Fred V. Grau Turfgrass Science Award** was implemented by the CSSA, with the selection criteria and procedures developed by a committee chaired by D.V. Waddington.

1988	First Division C-5 paper session for **Turfgrass Education** listed in the ASA annual meeting program, as organized by J.M. DiPaola.
1989	A revision of the Division **C-5 By-Laws** was approved, as developed by a committee chaired by J. B Beard.
1992	The ASA, CSSA, and Soil Science Society of America (SSSA) Agronomy Monograph No. 32 titled **Turfgrass** of 805 pages was published, as written by 40 Division C-5 members with D.V. Waddington, R.N. Carrow, and R.C. Shearman serving as editors.
1994	A revision of the Division **C-5 By-Laws** was approved, with K.J. Karnok as chair.
1994	The ASA and CSSA published a book titled **Turf Weeds and Their Control** of 259 pages, as written by 14 Division C-5 members with A.J. Turgeon serving as editor.
1995	A **Division C-5 Newsletter** was initiated by J.V. Krans.
1996	The **Division C-5 Web Site** was implemented by a committee with T.W. Fermanian as chair.
2003	The Division **C- 5 Operations Manual** was updated by a committee chaired by J. B Beard.
2008	Division C-5 had the most paper presentations of any CSSA division during the 2008 meetings in Houston.
2008	Division C-5 was the first tri-society division to initiate a review process for annual meeting abstracts.

National Society Honors. The achievement of Fellow of the CSSA, ASA, or SSSA is the highest professional honor a member can receive. The Fellow Awards are presented to a maximum of 0.3% of the active membership annually. Through 2010, the Turfgrass Science Division C-5 members honored as Fellows total 34 for the CSSA, 58 for the ASA and 5 for the SSSA, with 20, 33, and 1, respectively, that have devoted all of their time to turfgrass science (table 10.4a and table 10.4b).

TABLE 10.4A. SUMMARY OF C-5 TURFGRASS SCIENCE MEMBERS WHO ARE RECIPIENTS OF THE FELLOW AWARD FROM ONE OF THE TRI-SOCIETIES FROM 1907 THROUGH 1979.

Year	Recipient	State	CSSA	ASA	SSSA
1925	C.V. Piper*	Maryland		X	
1927	R.A. Oakley*	Maryland		X	
1937	F.D. Keim*	Nebraska		X	
1941	H.B. Sprague*	New Jersey		X	
1947	T.E. Odland*	Rhode Island		X	
1949	G.W. Burton*	Georgia	X	X	
1951	H.R. Albrecht*	Pennsylvania	X	X	
1953	R.E. Blaser*	Virginia	X	X	X
1955	G.O. Mott*	Indiana		X	
1955	H.B. Musser	Pennsylvania		X	
1958	D.G. Sturkie*	Alabama	X	X	
1962	A.G. Law*	Washington		X	
1963	W.H. Daniel	Indiana	X	X	
1963	A.A. Hanson*	Maryland	X	X	
1964	R.M. Hagan*	California		X	X

TABLE 10.4A. (*continued*).

1966	E.C. Holt*	Texas	X	X	
1966	F.V. Juska	Maryland		X	
1967	M.E. Bloodworth*	Texas		X	X
1967	R.R. Davis*	Ohio	X	X	
1967	K.T. Payne*	Michigan	X	X	
1968	V.B. Youngner	California		X	
1969	R.E. Engel	New Jersey		X	
1969	F.V. Grau	Maryland	X	X	
1969	H.G. Hodgson*	Alaska		X	X
1969	W.W. Huffine	Oklahoma	X	X	
1969	C.Y. Ward*	Mississippi	X	X	
1971	J. B Beard	Michigan	X	X	
1971	E.C. Roberts	Iowa	X	X	
1977	J.F. Shoulders	Virginia	X	X	
1979	J.R. Watson	Minnesota	X	X	
1979	L.N. Wise*	Mississippi		X	

* Part-time turfgrass responsibilities, usually with forages or administration.

TABLE 10.4B. SUMMARY OF C-5 TURFGRASS SCIENCE MEMBERS WHO ARE RECIPIENTS OF THE FELLOW AWARD FROM ONE OF THE TRI-SOCIETIES FROM 1980 THROUGH 2010.

Year	Recipient	State	CSSA	ASA	SSSA
1980	A.A. Baltensperger*	New Mexico	X	X	
1984	C.R. Funk	New Jersey	X	X	
1985	W.W. Hanna*	Georgia	X	X	
1986	M.H. Niehaus*	Colorado	X	X(89)**	
1986	P.E. Rieke	Michigan	X		
1986	R.E. Schmidt	Virginia	X	X(97)	
1986	D.V. Waddington	Pennsylvania		X	
1988	B.J. Johnson	Georgia	X(89)	X	
1988	R.C. Shearman	Nebraska	X(89)	X	
1988	A.J. Turgeon	Pennsylvania	X	X(89)	
1991	T.L. Watschke	Pennsylvania	X(92)	X	
1992	R.R. Duncan*	Georgia	X	X	
1993	R.N. Carrow	Georgia		X	
1993	W.A. Meyer	Oregon	X	X(94)	
1994	T.P. Riordan	Nebraska	X(95)	X	
1994	D.B. White	Minnesota		X	
1995	N.E. Christians	Iowa	X(97)	X	
1995	K.J. Karnok	Georgia	X	X	
1995	A.J. Powell Jr.	Kentucky		X	
1998	G.H. Snyder	Florida		X	X(99)
1999	C.M. Taliaferro*	Oklahoma	X(01)	X	

TABLE 10.4B. (*continued*).

Year	Recipient	State	CSSA	ASA	SSSA
2001	J.V. Krans	Mississippi		X	
2003	J.L. Cisar	Florida		X	
2003	P.H. Dernoeden	Maryland	X(07)	X	
2003	B. Huang	New Jersey	X(04)	X	
2004	A.D. Brede	Idaho	X(05)	X	
2004	B.B. Clarke	New Jersey	X(06)	X	
2005	R.E. Gaussoin	Nebraska	X(06)	X	
2006	M.C. Engelke	Texas		X	
2007	A.L. Brilman	Oregon	X		
2007	P.H. Dernoeden	Maryland	X		
2009	E.A. Guertal	Alabama		X	
2009	J.L. Cisar	Florida	X		

* Part-time turfgrass responsibilities, usually with forages.
** Numbers in parentheses indicate the year that the award was received.

There also are other CSSA and ASA awards that involve a single recipient each year for which Turfgrass Science Division C-5 members have been recognized. Included are society-wide competitions in research, teaching, and breeding plus the Turfgrass Science Award as summarized in table 10.5.

TABLE 10.5. INDIVIDUAL CSSA AND ASA AWARDS GIVEN TO TURFGRASS SCIENCE DIVISION C-5 MEMBERS.

CSSA Crop Science Research Award	1991	James B Beard
CSSA Crop Science Teaching Award	1997	Keith J. Karnok
CSSA NCCPA Genetics and Plant Breeding Award for Industry	1991	William A. Meyer
	2005	A. Douglas Brede
	2007	A. Leah Brilman
	2008	C. Rose-Fricker
ASA Agronomic Resident Education Award	1994	Keith J. Karnok
ASA Agronomic Service Award	1973	William H. Daniel
	1977	James R. Watson
CSSA Fred V. Grau Turfgrass Science Award	1987	James R. Watson
	1988	James B Beard
	1989	Jack J. Murray
	1990	C. Reed Funk

TABLE 10.5. (*continued*).

1991	Glen W. Burton
1992	Robert C. Shearman
1993	Donald V. Waddington
1994	William A. Meyer
1995	B.J. Johnson
1996	Terrance P. Riordan
1997	Keith J. Karnok
1998	A.J. Powell Jr.
1999	Nick E. Christians
2000	Richard E. Schmidt
2001	Wayne W. Hanna
2002	Alfred J. Turgeon
2003	Paul E. Rieke
2004	T. Karl Danneberger
2005	A. Douglas Brede
2006	Arden A. Baltensperger
2007	Peter H. Dernoeden
2008	Roch E. Gaussoin
2009	Thomas L. Watschke

REFERENCES

Beard, J.B. 2005. History Records of Division C-5, Turfgrass Science—1946 thru 2004. Consists of a four-volume bound set held in the Turfgrass Collection at the Michigan State University Library:
  Volume I. Annual Business Meeting Minutes.
  Volume II. Annual Scientific Program Turfgrass Paper Presentations.
  Volume III. By-Laws and Publications.
  Volume IV. Annual Historian Reports.

Figure 11.1. Group photo of attendees at the first International Turfgrass Research Conference in 1969 at Harrogate, England, UK. [26]*

Front row (sitting) (left to right)	Second row (left to right)	Third row (left to right)	Back row (left to right)
Dr. H.W. Daniel	Mr. J.P. Shildrick	Dr. R.W. Schery	Dr. J.H. Madison
Dr. S.W. Bingham	Dr. P.E. Rieke	Mr. E. Helmbring	Dr. L.E. Janson
Dr. G.M. Wood	Ir. H. Vos	Dr. J. Troll	Dr. R.W. Miller
Dr. E. Ebert-Jehle	Dr. V.I. Stewart	Mr. K. Potter	Dr. R.A. Keen
Mr. R.L. Morris	Mr. R.W. Palin	Mr. J.C. Knolle	Mr. P. Bowen
Prof. P. Boeker	Mr. D. Soper	Mr. G. Akesson	Dipl. Ing. E.W. Schweizer
Mr. B. Langvad	Dr. C.E. Wright	Dr. L.H.J. Korsten	Dr. C.M. Switzer
Dr. J.R. Watson	Mr. D.T.A. Aldrich	Dr. B. Werminghausen	Dr. C.W.H.M. Schaepman
Dr. J.B Beard	Ir. J.P. van der Horst	Dr. E.L. Entrup	Mr. J. Andringa
Mr. J.R. Escritt	Dr. H.H. Williams	Dr. P.R. Henderlong	Mr. C. Eisele
Ir. G.J. Ruychaver	Dr. R.R. Davis	Dr. W.A. Adams	Mr. R.C. O'Knefski
Dr. Z.B. - Kuninska	Mr. A.V. Bogdan	Dr. A. Pap	Dr. I. Yoshikawa
Miss M.-L. Denecke	Mr. J.F. Shoulders	Dr. L.E. Moser	Mr. W.H. Bengeyfield
Mr. W.C. Morgan	Mr. J.L. Kidwell	Mr. M.A. Wood	Mr. D.J. Glas
Mr. G.S. Robinson	Dr. R.L. Goss	Dr. J. Stubbs	Dr. R.E. Schmidt
Ir. W.A. Eschauzier	Dr. W.B. Gilbert	Mr. M. Peterson	Mr. J.A. Simmons
	Dr. D.B. White	Mr. G.G. Fisher	Dr. T. Eggers
	Dr. T.E. Freeman	Ir. M. Kamps	Mr. R. Vijn
	Dr. G.C. Horn	Mr. J.L. Dawson	Dr. W.W. Huffine
	Dr. K. Ehara	Dr. R.E. Engel	Dr. J.E. Howland

* The numbers in brackets correspond to the figure source acknowledgments listed in the appendix.

CHAPTER 11

Organization of International Turfgrass Scientists

International communication among most turfgrass educators and researchers was almost nonexistent prior to 1969, even within Europe. In 1965, Dr. James B Beard initiated inquires to several international scientific societies regarding the addition of a turfgrass science section. The International Society for Horticultural Science was clearly not interested. The United States representatives to the International Grassland Society were very supportive but the European and Australian contingents rejected the proposal. Thus a decision was made to organize an independent group.

In 1966, Dr. J. B Beard hosted Bjarne Langvad of Sweden at his home in Okemos, Michigan, USA. The concept of an international turfgrass research organization and conference was proposed by J. Beard. B. Langvad was very encouraging as to the need and potential for such an organization in Europe. It was indicated that the conference should be held in Europe in order to attract the most participants.

At the August 1966 annual meeting of the Turfgrass Management Division C-5 of the Crop Science Society of America in Stillwater, Oklahoma, USA, Dr. J. B Beard presented a proposal regarding the need for enhanced communications among international turfgrass scientists via an international turfgrass research conference. Considerable discussion followed, with the C-5 members deciding to pursue organization of a nonaffiliated international conference. A motion was made and passed charging Dr. Beard to lead a committee in "forming an international turfgrass organization." This was a major challenge as there were no start-up funds available.

Subsequently, Dr. Beard contacted John Escritt of England and Dr. James Watson of the United States to serve with Bjarne Langvad on an Organizing Committee for the First International Turfgrass Research Conference. Dr. Beard expended over $700 of personal funds in the many activities needed to organize the first conference.

The first ITRC was held in 1969 in Harrogate, England, UK, with the Sports Turf Research Institute as host. The initial organization officers were Dr. James B Beard of the United States, serving as President; Bjarne Langvad of Sweden, serving as Vice President; Dr. James R. Watson of the United States, serving as Secretary; and John R. Escritt of England, serving as Treasurer. The International Turfgrass Society (ITS) was organized at this conference. Its main activity is sponsoring the International Turfgrass Research Conference every four years plus publishing the **ITRC Proceedings/Journal**.

The initial projection was if 20 turfgrass scientists participated in the First International Turfgrass Research Conference, it would be deemed a successful start. The response was impressive with 77 participants from 12 countries. These original ITRC participants are listed as follows. Certificates of Appreciation were given to four members who had participated in all ten International Turfgrass

Research Conferences from 1969 through 2005. They were Dr. William A. Adams of the United Kingdom, Dr. James B Beard of the United States, Dr. Richard E. Schmidt of the United States, and Martin Peterson of Denmark.

Charter Members of International Turfgrass Society (1969)

Dr. William A. Adams	Aberystwyth, Wales, UK	Mr. D.J. Glas	Rilland-Bath, the Netherlands
Mr. G. Akesson	Hammenhog, Sweden	Dr. Roy L. Goss	Washington, USA
Mr. D.T.A. Aldrich	Cambridge, England, UK	Mr. E. Helmbring	Malmoe, Sweden
Mr. J. Andringa	Scheemda, the Netherlands	Dr. Paul R. Henderlong	Ohio, USA
		Dr. Granville C. Horn	Florida, USA
Dr. James B Beard	Michigan, USA	Ir. J.P. van der Horst	Gravenhage, the Netherlands
Mr. William H. Bengeyfield	California, USA	Dr. Joseph E. Howland	Nevada, USA
Dr. S. Wayne Bingham	Virginia, USA	Dr. Wayne W. Huffine	Oklahoma, USA
Prof. Peter Boeker	Bonn, West Germany	Dr. Lars-Eric Janson	Stockholm, Sweden
Mr. A.V. Bogdan	Maidenhead, England, UK	Ir. M. Kamps	Vlijmen, the Netherlands
Mr. P. Bowen	Ipswich, Suffolk, UK	Dr. Ray A. Keen	Kansas, USA
		Mr. Jack L. Kidwell	Virginia, USA
Dr. John F. Cornman	New York, USA	Mr. Jurgen C. Knolle	Postfach, West Germany
Dr. William H. Daniel	Indiana, USA	Dipl. Ing. Bretislav Komicek	Kormornicka, Czechoslovakia
Dr. Richard R. Davis	Ohio, USA		
Mr. J.L. Dawson	Newbridge, Scotland, UK	Dr. L.H.J. Korsten	Rotterdam, the Netherlands
Miss M.L. Denecke	Munchen, West Germany	Mr. Bjarne Langvad	Landskrona, Sweden
Dr. Edith Ebert-Jehle	Basle, Switzerland	Dr. John H. Madison	California, USA
Dr. T. Eggers	Hamburg, West Germany	Dr. Robert W. Miller	Ohio, USA
		Mr. Wayne C. Morgan	California, USA
Dr. Kaoru Ehara	Fukuoka, Japan	Mr. Robert L. Morris	Ipswich, Suffolk, UK
Mr. C. Eisele	Darmstadt, West Germany	Dr. Lowell E. Moser	Ohio, USA
Dr. Ralph E. Engel	New Jersey, USA	Mr. Robert C. O'Knefski	New York, USA
Dr. Ernst L. Entrup	Lippstadt, West Germany	Mr. R.W. Palin	Reading, UK
		Dr. A. Pap	Witham, England, UK
Ir. W.A. Eschauzier	Vlijmen, the Netherlands	Mr. Martin Petersen	Odense, Denmark
Mr. John R. Escritt	Bingley, England, UK	Mr. K. Potter	Witham, England, UK
Mr. G.G. Fisher	Cambridge, England, UK	Dr. Paul E. Rieke	Michigan, USA
		Mr. G.S. Robinson	Palmerston North, New Zealand
Dr. T. Edward Freeman	Florida, USA		
Dr. William B. Gilbert	North Carolina, USA	Ir. G.J. Ruychaver	Gravenhage, the Netherlands

Dr. C.W.H.M. Schaepman	Arnhem, the Netherlands	Dr. Joseph Troll	Massachusetts, USA
Dr. Robert W. Schery	Ohio, USA	Mr. R. Vijn	Vlijmen, the Netherlands
Dr. Richard E. Schmidt	Virginia, USA		
Dipl. Ing. Edgar W. Schweizer	Thun, Switzerland	Ir. H. Vos	Wageningen, the Netherlands
Mr. John P. Shildrick	Bingley, England, UK	Dr. James R. Watson	Minnesota, USA
		Dr. Berhard Werminghausen	Limburgerhof, West Germany
Mr. John F. Shoulders	Virginia, USA		
Mr. James A. Simmons	Ohio, USA	Dr. Donald B. White	Minnesota, USA
Mr. Derek Soper	Ongar, England, UK	Dr. H. Hamilton Williams	California, USA
		Dr. Glen M. Wood	Vermont, USA
Dr. V.I. Stewart	Aberystwyth, Wales, UK	Mr. M.A. Wood	Belfast, Northern Ireland, UK
Dr. J. Stubbs	Nr. Haslemere, England, UK	Dr. Charles E. Wright	Armagh, Northern Ireland, UK
Dr. Clayton M. Switzer	Guelph, Ontario, Canada	Dr. Isao Yoshikawa	Takarazuka, Japan

At a business meeting of the first ITRC in Harrogate, England, UK, the participants discussed and voted to form the International Turfgrass Society and to hold an International Turfgrass Research Conference at four-year intervals. A chronological summary of ITRC and ITS history is presented as follows. The progress, participation, and improvements in the ITRC have been significant. All members have gained knowledge through this shared learning experience.

There were 99 papers presented at the First International Turfgrass Research Conference. Many of the presentations at this first conference were information oriented regarding turfgrass research and education activities in the represented countries. This approach had the objective of furthering an understanding of the turfgrass industry, problems, research, and education activities in the various countries. The first ITRC resulted in the European representatives organizing a regional research conference. Subsequent ITRC events have focused on an exchange of research findings plus a set of invited keynote presentations on current topics. The length of the **ITRC Proceedings/Journals** has ranged from 530 to 1,432 pages. A total of 175 peer reviewed papers were presented and published through 2009 (table 11.1).

Chronology of ITRC and ITS History

July 1969

- First International Turfgrass Research Conference held at Old Swan Hotel in Harrogate, England, UK. The host organization was the Sports Turf Research Institute. The postconference turfgrass tour was to Scotland, England, Sweden, West Germany, and the Netherlands.
- Attendees voted to form an International Turfgrass Society and to hold a quadrennial conference.

July 1973

- Second International Turfgrass Research Conference held on the campus of Virginia Polytechnic Institute and State University in Blacksburg, Virginia, USA. The host organization was the College of Agriculture and Life Sciences. Both preconference (east) and postconference (Midwest) turfgrass tours were conducted.

TABLE 11.1. SUMMARY OF PARTICIPATION IN THE INTERNATIONAL TURFGRASS RESEARCH CONFERENCES FROM 1969 THROUGH 2009.

Year and place of conference	Number of registrants	Number of countries represented	Number of papers presented	Number of peer-reviewed papers published in Journal	Number of pages in Journal	ITRC Supplement pages[*]
1969, Harrogate, England, UK	78	12	99	99	610	—
1973, Blacksburg, Virginia, USA	248	15	80	71	602	—
1977, Munich, West Germany	265	19	91	59	530	—
1981, Guelph, Ontario, Canada	240	18	81	62	564	61
1985, Avignon, France	266	21	86 (76 oral; 10 poster)	84	870	110
1989, Tokyo, Japan	814	18	98	98	458	163
1993, Palm Beach, Florida, USA	368	23	149 (109 oral; 40 poster)	149	1,035	83
1997, Sydney, New South Wales, Australia	242	22	168 (76 oral; 92 poster)	144[**]	1,432	—
2001, Toronto, Ontario, Canada	340	26	224 (118 oral; 105 poster)	164[**]	1,056	—
2005, Llandudno, Wales, UK	262	25	174 (88 oral; 86 poster)	174[**]	1,261	109
2009, Santiago, Chile	155[***]	19	112 (53 oral; 59 poster)	111[**]	1,252	70

[*] Nonreviewed papers.
[**] Published in two parts.
[***] Includes accompanying persons.

- A Constitution and Bylaws for the International Turfgrass Society were presented, amended, and adopted by vote of the participants in attendance.
- ITS Distinguished Service Awards were presented to Dr. James B Beard and Mr. John R. Escritt.
- As funds were lacking, ITS sought a sponsor to publish the papers presented at ITRC. The American Society of Agronomy, the Crop Science Society of America, and the Soil Science Society of

America agreed to jointly publish the Proceedings of the Second and Third ITRC, with the understanding that the publication schedule would depend on when their staff had free time.

July 1977

- Third International Turfgrass Research Conference held at the Penta Hotel in Munich, West Germany. A postconference turfgrass tour was organized through southern Europe.
- This was the first ITRC at which translation between two languages was organized.
- ITS membership dues were initiated.

July 1981

- Fourth International Turfgrass Research Conference held on the campus of the University of Guelph in Guelph, Ontario, Canada. The host organization was the Ontario Agricultural College.
- A new feature of the ITRC was a series of keynote papers that headline each subject area. The invited presentations were by Drs. W.A. Adams, J. B Beard, C.R. Funk, R.L. Goss, and A.J. Turgeon.
- This was the first ITRC at which poster presentations were offered.
- Revisions to the ITS Constitution and Bylaws were presented, amended, and adopted by vote of the members in attendance.

July 1984

- Fifth International Turfgrass Research Conference held in the Palais de Papes inside the walled city of Avignon, France. The host organization was the Institut National de la Recherche Agronomique (INRA).
- A new approach at this ITRC was a series of six special topic workshops organized as a discussion group interaction.
- Invited keynote papers were presented by J. B Beard, B. Bourgoin, P.M. Canaway, P. Daget, M. Masson, N. O'Neill, and T.L. Watschke.

August 1989

- Sixth International Turfgrass Research Conference held in the Takanawa Prince Hotel in Tokyo, Japan. The host organization was the Japanese Society of Turfgrass Science.
- A format for selecting host sites was established involving an eight-year advance plan.
- Initiated a procedure for the ITS Executive Committee to meet midway or two years before the upcoming ITRC at the host site.

July 1993

- Seventh International Turfgrass Research Conference held at The Breakers in Palm Beach, Florida, USA. The host organization was the Turfgrass Science Division C-5 of the Crop Science Society of America.
- New ITS Articles of Incorporation and Bylaws were proposed and adopted by vote of the members present.
- The International Turfgrass Society was incorporated in Minnesota, USA, and received a 501(c)(3) tax exemption.

- An ITS Newsletter titled **International Turfgrass** was initiated with eight issues published in the past four years. Dr. J.R. Hall III served as editor.
- A computerized data base of members was established.
- The Proceedings name was changed to the **International Turfgrass Society Research Journal**.
- A format and procedure document for the peer-review of scientific papers to be published in the **ITS Research Journal** was prepared.

July 1997

- The Eighth International Turfgrass Research Conference was held at the University of Sydney in Sydney, New South Wales, Australia.
- The **ITS Research Journal** was published in two separate parts.
- The preconference tour was focused on New Zealand.
- Dr. John Cisar assumed editorship of **International Turfgrass**, the Newsletter of ITS.

July 2001

- The Ninth International Turfgrass Research Conference was held at the Westin Harbour Castle Hotel in Toronto, Ontario, Canada. The host organization was the Guelph Turfgrass Institute.
- An ITRC website was established by the Institute.
- **ITS NETNOTES** was established by Dr. Joseph DiPaola.

July 2005

- The Tenth International Turfgrass Research Conference was held at the North Wales Conference Center in Llandudno, Wales, United Kingdom. The host organization was the Institute of Biological Sciences, University College of Wales.

July 2009

- The Eleventh International Turfgrass Research Conference was held in Santiago, Chile. The host organization was Prados-Floricultura, Universidad de Chile.

ITS Officers and Directors. A summary of the individual members who have served as officers and/or directors of the International Turfgrass Society for the first 11 four-year terms from 1969 through 2009 are summarized in table 11.2a and table 11.2b. They represent 16 different countries from throughout the world. In most cases, the President of the ITS has been from the country scheduled to host the International Turfgrass Research Conference (ITRC), although that is not a specific requirement in the ITS Bylaws.

ITRC Participation. Sponsorship of the International Turfgrass Research Conference at four-year intervals is the primary function and activity of the International Turfgrass Society. A total of 77 participants from 12 countries were represented at the first conference in 1969. During the second through eleventh conferences, from 1973 to 2009, the attendance has been in the range of 155 to 814 participants (table 11.3). A maximum of 26 countries has been represented at any one conference, with a total of 44 countries represented over the period of the 11 International Turfgrass Research Conferences. Seven countries have had representatives at all 11 International Turfgrass Research Conferences, including Canada, Germany, Japan, Sweden, Switzerland, the United Kingdom, and the United States. A total of 3,277 participants have attended the first 11 International Turfgrass Research Conferences.

TABLE 11.2A. SUMMARY OF EXECUTIVE COMMITTEE AND OFFICERS SERVING THE INTERNATIONAL TURFGRASS SOCIETY FROM 1969 THROUGH 1989.

Executive committee position	Conference, year, city, and country					
	First, 1969, Harrogate, UK	Second, 1973, Blacksburg, USA	Third, 1977, Munich, West Germany	Fourth, 1981, Guelph, Canada	Fifth, 1985, Avignon, France	Sixth, 1989, Tokyo, Japan
President	J. B Beard	R.R. Davis	P. Boeker	C.M. Switzer	P. Mansat	Y. Maki
Vice president	B. Langvad	B. Langvad	J.P. van der Horst	H. Vos	H. Vos	W.A. Adams
Secretary	J.R. Watson	J.F. Shoulders	F.B. Ledeboer	F.B. Ledeboer	J.F. Shoulders	J.F. Shoulders
Treasurer	J.R. Escritt	R.E. Schmidt	R.E. Schmidt	R.E. Schmidt	R.E. Schmidt	R.E. Schmidt
Past president	—	J. B Beard	R.R. Davis	P. Boeker	C.M. Switzer	P. Monsat
Directors	—	W.H. Daniel C.M. Switzer J.P. van der Horst	K. Ehara R.E. Engel A.C. Ferguson R.L. Morris W. Skirde	W.A. Adams K. Ehara W.W. Huffine P. Mansat D.K. Taylor	W.A. Adams Y. Maki R.W. Sheard T.R. Siviour J.R. Watson	F. Lemaire P. McMaugh R.W. Sheard A.J.P. van Wijk J.R. Watson
Historian	—	—	J. B Beard	J. B Beard	J. B Beard	J. B Beard

TABLE 11.2B. SUMMARY OF EXECUTIVE COMMITTEE MEMBERS AND OFFICERS SERVING THE INTERNATIONAL TURFGRASS SOCIETY FROM 1993 TO 2009.

Executive committee position	Conference, year, and country				
	Seventh, 1993, USA	Eighth, 1997, Australia	Ninth, 2001, Canada	Tenth, 2005, Wales, UK	Eleventh, 2009, Chile
President	J.R. Watson	P. McMaugh	P. Charbonneau	W.A. Adams	C.P. Muller
Vice president	W.A. Adams	R.N. Carrow	R.N. Carrow	R.N. Carrow	R.J. Gibbs
Secretary	J.R. Hall III	J.R. Hall III	T. Yamada	K. Carey	K. Carey
Treasurer	R.E. Schmidt	G.H. Snyder	G.H. Snyder	J.L. Cisar	J.L. Cisar
Past president	Y. Maki	J.R. Watson	P. McMaugh	P. Carbonneau	W.A. Adams

TABLE 11.2B. (*continued*).

Executive committee position	Conference, year, and country				
	Seventh, 1993, USA	Eighth, 1997, Australia	Ninth, 2001, Canada	Tenth, 2005, Wales, UK	Eleventh, 2009, Chile
Elected directors	F. Lemaire	B. Bourgoin	W.A. Adams	W.A. Adams	I. Chivers
	P. McMaugh	P.M. Canaway	B. Bourgoin	S.W. Baker	B.B. Clarke
	W.A. Meyer	P. Charbonneau	M.P. Canaway	K. Carey	T.K. Danneberger
	A.J.P. van Wijk	J.L. Cisar	J. Choi	R.N. Carrow	R. Dorberau
	H. Yanagi	K.W. McAuliffe	J.L. Cisar	I. Chivers	M. Engelsjord
		W.A. Meyer	S. Dahlsson	J.S. Choi	S. Kimura
		M. Peterson	J.M. DiPaola	J.L. Cisar	L. Han
		H. Yanagi	B. Holl	B.B. Clarke	H.G. Mueller-Beck
			K. McAuliffe	S.O. Dahlsson	P. Nektarios
			I.C. McIver	J.M. DiPaola	A. Richter
			Y. Noma	M. Engelsjord	D. Thorogood
			M. Peterson	R.J. Gibbs	M. Volterrani
			H. Richter	J.P. Lebourcher	
				H. Liebao	
				K.G. Mueller-Beck	
				C.P. Muller	
				J. Neylan	
				Y. Noma	
				H. Richter	
				D. Thorogood	
Historian	J. B Beard	J. B Beard	J. B Beard	J. B Beard	S.W. Baker

TABLE 11.3. SUMMARY OF PARTICIPATION AT THE INTERNATIONAL TURFGRASS RESEARCH CONFERENCE FROM 1969 THROUGH 2009.

Country	Conference and year											
	First, 1969	Second, 1973	Third, 1977	Fourth, 1981	Fifth, 1985	Sixth, 1989	Seventh, 1993	Eighth, 1997	Ninth, 2001*	Tenth, 2005*	Eleventh, 2009**	Total
Argentina	—	—	—	—	—	—	1	—	—	—	2	3
Australia	—	1	2	3	5	6	15	67	14	15	12	140
Austria	—	1	3	3	2	2	—	1	2	1	1	16
Barbados	—	—	—	—	—	—	—	—	—	—	1	1
Belgium	—	2	1	—	3	—	3	—	2	1	1	13
Canada	1	12	5	35	4	1	5	7	34	13	7	124
Chile	—	—	—	—	—	—	—	1	2	1	6	10

TABLE 11.3. (*continued*).

Country	First, 1969	Second, 1973	Third, 1977	Fourth, 1981	Fifth, 1985	Sixth, 1989	Seventh, 1993	Eighth, 1997	Ninth, 2001*	Tenth, 2005*	Eleventh, 2009**	Total
China	—	—	—	—	—	3	2	3	5	6	5	19
Czecho-slovakia	1	—	1	—	—	1	—	—	—	2	—	5
Denmark	1	1	2	1	2	1	2	1	3	6	—	20
Ecuador	—	—	—	—	—	—	—	—	1	—	1	2
Egypt	—	—	—	—	—	—	—	—	—	—	2	1
Finland	—	1	—	1	—	—	—	—	1	—	—	3
France	—	12	29	17	95	8	20	4	7	5	—	197
Germany	7	11	75	9	8	2	9	4	6	2	2	135
Greece	—	—	1	—	1	—	—	—	—	2	—	4
Hong Kong	—	—	—	—	—	—	—	3	—	—	—	3
India	—	—	—	—	—	—	—	—	—	1	—	1
Ireland	—	—	—	—	—	—	—	—	—	3	—	3
Israel	—	—	—	1	—	—	1	—	—	—	—	2
Italy	—	—	2	2	5	—	2	3	7	11	2	34
Japan	2	8	20	16	21	711	23	11	7	9	1	829
Nether-lands	10	12	21	11	16	2	4	—	—	5	—	81
New Zealand	1	1	1	—	—	4	3	11	4	5	1	30
Norway	—	—	1	1	—	—	1	—	2	4	2	11
Philip-pines	—	—	—	—	—	—	1	—	—	—	—	1
Poland	—	—	—	—	1	—	—	—	—	—	—	1
Portugal	—	—	—	—	—	—	—	—	—	1	—	1
Russia	—	—	—	2	—	2	1	—	1	—	—	6

TABLE 11.3. (*continued*).

Country	First, 1969	Second, 1973	Third, 1977	Fourth, 1981	Fifth, 1985	Sixth, 1989	Seventh, 1993	Eighth, 1997	Ninth, 2001*	Tenth, 2005*	Eleventh, 2009**	Total
Saudi Arabia	—	—	—	—	—	—	—	—	1	—	—	1
South Africa	—	—	—	—	1	—	—	4	1	1	2	9
South Korea	—	—	—	1	—	13	6	3	6	2	—	31
Spain	—	—	—	—	2	—	—	—	—	—	—	2
Sweden	4	2	4	4	3	2	2	2	4	6	2	35
Switzer-land	2	1	7	1	5	1	1	1	2	4	3	28
Taiwan	—	—	—	—	—	1	—	1	1	—	—	3
Thailand	—	—	—	—	—	—	1	—	—	—	—	1
Tunisia	—	—	—	—	1	—	—	—	—	—	—	1
Turkey	—	—	—	—	—	—	—	—	1	—	—	1
Turks and Caicos	—	—	—	—	—	—	2	—	—	—	—	2
United Kingdom	17	8	13	6	13	4	14	4	11	30	1	120
United States	31	175	76	125	77	30	249	111	215	126	103	1,438
Venezuela	—	—	—	1	—	—	—	—	—	—	—	1
Yugoslavia	—	—	1	—	1	—	—	—	—	—	—	2
Grand Total	77	248	265	240	266	814	368	242	339	262	156	3,277

* Preregistered.
** Includes accompanying persons.

REFERENCE

Beard, J.B. 2005. Tenth Anniversary Historian Report. Int. Turfgrass Soc.

PART II

Literature of Turfgrass

Figure 12.1. The James B Beard Turfgrass Library Collection at the Michigan State University Library, Overview (above) and Periodical Section (below). [6]*

---

* The numbers in brackets correspond to the figure source acknowledgments listed in the appendix.

CHAPTER 12

Bibliography of Books/Monographs on Turfgrass Culture

A. ALPHABETICAL PRESENTATION BY AUTHOR WITH DETAILED PHYSICAL AND SUBJECT DESCRIPTIONS OF THE BOOKS PLUS AUTHOR RESUMES (1892 TO 2010)

THESE BOOKS REPRESENT THE STATUS OF TURFGRASS SCIENCE AND CULTURE AT A GIVEN TIME DURing the evolution of turfgrass maintenance from an art-dominated approach to a science-based approach. This documentation of progress in turfgrass art and science through time via books is primarily a phenomenon of the twentieth century. For most of the 1800s, the limited written information available on turfgrass maintenance appeared as short sections in gardening books and focused on ornamental lawns.

Pioneering Series Publications by Private Companies

The first publication of 20 or more pages devoted strictly to turfgrass culture appeared in 1892 in England under the title **Formation of Lawns, Tennis & Cricket Grounds from Seed**. Note that the first turfgrass research paper, also published in 1892, was titled **Some Selections of Grasses for Field and Lawn,** authored by Dr. William James Beal of Michigan, USA. This was a time when the first significant books about the games of field sports and golf also were being published.

The original 1892 publication was followed through 1912 by eight revisions of this semisoft cover series sponsored by Sutton & Sons of Reading, England. It is of interest that from 1892 to 1899 the subject emphasis was on lawns, tennis courts, and cricket grounds. In 1899, the subject of putting greens was added, with bowling greens subsequently added in 1904. Authorship of these pioneering educational booklets, such as **Garden Lawns, Tennis Lawns, Bowling Greens, Croquet Courts, Putting Greens, Cricket Grounds** in 1904, is attributed to Martin H.F. Sutton. Under his early leadership and the sponsorship of Sutton & Sons, a series of 17 revised editions was published concerning turfgrasses, cultural practices, pest controls, and equipment used in England during the first half of the twentieth century. This series from 1892 to 1962 offers a unique historical perspective because of the continuity of revisions over a 70-year period.

A second soft cover series of interesting turfgrass publications titled **The Practical Greenkeeper** was initiated by Reginald Beale in 1907 under the sponsorship of James Carter & Co. of London, England, UK. This series of 31 editions through 1939 also provides an interesting early written and visual documentation of turfgrass cultural practices in England.

In the United States, the O.M. Scott & Sons Company initiated a series of pioneering turfgrass books starting in 1921. Their books represent the oldest and largest private educational effort in the United States.

Pioneering Authors of Turfgrass Books

Pioneering individuals in articulating the development of turfgrass cultural practices are summarized in table 12.1. They were the first to devote the time and effort to develop concepts and philosophies of turfgrass culture and, subsequently, to present this information in an organized, written form from which others could learn. Please note that the author of an original text offers a permanent written record for others to criticize. All too often the critics of such venturesome authors only criticize but seldom present their own written philosophies for critical review.

In 1906, a major book devoted strictly to turfgrass establishment and culture for golf courses was published in England titled **Golf Greens and Green-Keeping**. This 219-page book was an assemblage of 14 papers by 12 authors under the editorship of Horace G. Hutchinson. Each author discussed the establishment and culture of golf course turfgrasses on the different types of soil-climate-climax vegetation regions in the United Kingdom such as (a) seaside links, (b) light inland soil, (c) heath land, (d) pine forests, (e) medium soil, (f) chalk down, (g) heavy soil, and (h) park areas. The need for adjustments in turfgrass culture as affected by soil type and climate was being recognized.

Reginald Beale's 1924 book titled **Lawns for Sports—Their Construction and Upkeep,** plus a **Supplement Issued With Lawns for Sports Materials, Tools And Fittings,** is particularly interesting and very well illustrated. This book is the most extensive and comprehensive written to that date on sports fields. The supplement contains by far the largest number of photos and drawings of turfgrass equipment that was published in that era.

It is interesting to note that the early turfgrass books published in the United Kingdom focused on sport turfgrasses, while most early turfgrass books published in the United States addressed primarily lawn establishment and culture. Also, all the turfgrass books published prior to 1925 were from the United Kingdom and the United States, except for one 1908 book from France.

Of the books published or revised through 1930, the only turfgrass book written by scholars-researchers holding PhDs was **Turf for Golf Courses** by Drs. Charles V. Piper and Russell A. Oakley of the United States, published in 1917. This book represented a major advance in that the science-based technology of the time was brought more to the fore.

Interspersed in this chapter are excerpts from many of the early books. They represent the state of knowledge at that point in time and illustrate to modern turfgrass scientists and practitioners just how far the science and technology of turfgrass culture have evolved in 100-plus years. Even now, the basic pool of knowledge concerning the science of turfgrass culture is rather modest compared to many areas of applied plant science devoted to field and horticultural crops.

Also included are a number of photographs and drawings that depict the turfgrass conditions, tools, equipment, and techniques used in turfgrass establishment and culture during the early part of the twentieth century. Again, the advances made in the latter half of the twentieth century have been impressive.

The turfgrass books by senior authors listed in this chapter are presented alphabetically by author, editor, or sponsor organization. The criteria for books included in this chapter are (a) those devoted for the most part to the subjects of turfgrass science, establishment, and culture, and (b) those of 20 pages or more in length.

TABLE 12.1. YEAR, AUTHOR, AND TITLE WHEN THE EARLY TURFGRASS BOOKS* WERE PUBLISHED, INCLUDES FIRST EDITION PUBLICATIONS OF 20 PAGES OR MORE IN LENGTH FROM 1892 THROUGH 1924.

Country	Year	Author (editor)	Pages	Title
United Kingdom	1892*	Sutton & Sons	21	Formation of Lawns, Tennis & Cricket Grounds from Seed
	1895	J.A. Gibbs	42	The Improvement of Cricket Grounds on Economical Principles
	1900	E.K. Toogood	20	Toogood's Treatise on Lawns, Cricket Grounds, &c.
	1906	H.G. Hutchison (ed)	219	Golf Greens and Green-Keeping
	1906	R. Beale	32	Golf
	1908*	R.E.C. Beale	69	The Practical Greenkeeper
	1908	Stewart and Co.	26	Links and Lawn Grasses
	1908	R. Beale	20	Manures and Composts for Golf Greens, &c.
	1909	W.J. Stevens	20	Lawns
	1909	T.W. Sanders	148	Lawns and Greens: Their Formation and Management
	191x	M.H.F. Sutton	47	The Laying Out and Upkeep of Golf Courses and Putting Greens
	1910	T.W. Sanders	137	Lawns and Greens: Their Formation and Management
	1912	M.H.F. Sutton (ed)	212	The Book Of The Links
	1913	W.K. Gault	101	Practical Golf Greenkeeping
	1923	J. MacDonald	78	Lawns, Links, & Sportsfields
	1924	R.E.C. Beale	276	Lawns for Sports—Their Contraction and Upkeep
	1924	R. E. C. Beale	67	Supplement Issued With Lawns for Sports—Materials, Tools, And Fittings
	1924	A.J. Macself	204	Grass
	1924	A. Key	21	Hints on Lawns
United States	1892	J.B. Olcott	26	Talk About A Grass Garden
	1899	J.M. Thornburn & Co.	24	The Seeding and Preservation of Golf Links
	1906	L.W. Barron	174	Lawns and How to Make Them
	1906	L.C. Corbett	20	The Lawn
	1911	O. Schreiner and J. Skinner	55	Lawn Soils
	1912	L.J. Doogue	51	Making A Lawn
	1912	O. Schreiner et al.	48	Lawn Soils and Lawns
	1914	S.A. Cunningham et al.	40	Lawns, Golf Courses, Polo Fields, Turf Courts, and How to Treat Them
	1914	L.J. Doogue	22	The Proper Care of Lawns
	1915	W. Tucker	36	Practical Illustrations of Turf Production
	1916	Odorless Plant Food Co.	28	The Maintenance of Lawns: Applicable Also to Golf Courses
	1917	C.V. Piper and R.A. Oakley	267	Turf for Golf Courses
	1918	P.W. Lees	93	Care of the Green
	1918	Commission for Beautifying the City	30	Norfolk Lawns and How to Grow Them
	1920	O.M. Scott & Sons	31	Weedless Lawns and Golf Courses
	1920	R. E. C. Beale	40	Lawns and Sports Grounds
	1922	O.M. Scott & Sons	54	The Seeding and Care of Golf Courses
France	1908	J.C.N. Forestier	128	Les Gazons (Turf)

* The first booklet in a series.

RESOURCE LIST FORMAT

Citation Style Used:
- Name of senior author or sponsoring organization
- Year of publication
- Edition, if applicable
- Title of book
- Name of author(s) or editor(s) as it appears in book, with contributing authors listed as appropriate
- Series title and number, if applicable
- Name and location of publisher—city, state, and country
- Approximate numbers of pages, illustrations, and tables
- Whether indexed and extent of the index
- Type of binding, color, and color of lettering
- Size of pages in inches (mm) by width and height
- Language in which the book is printed, if not in English

Abbreviations Used:
illus. Number of illustrations in book
n.d. No publication date shown in book
circa An approximation
pp. Number of pages in book
† Books held in the Beard Turfgrass Collection at Michigan State University Library
* Book contains chapter(s) or section(s) on turfgrass but addresses a broader array of subjects than just turfgrass
✳ Books that are old (most published before 1955) and rare (based on over 30 years of experience by the senior author examining book stores, dealer catalogues, libraries, etc.)
Dr. Doctor of philosophy (PhD) degree holder. In the case of dual titles, the title doctor will be used rather than professor, as the former is a higher ranking in the United States
MSc Master of science
BSc Bachelor of science

[1] Author formally asked that resume not be included.
[2] Author did not respond to multiple written mailings.
[3] Efforts to obtain a proper resume were unsuccessful even though multiple communications were sent to former employers and organizations where active. Additionally, research at national archives was done. Also, laws in certain countries are very restrictive (e.g., United Kingdom).

APS	American Phytopathological Society		NGF	National Golf Foundation
ASA	American Society of Agronomy		NPFA	National Playing Fields Association
ASGCA	American Society of Golf Course Architects		OBE	Officer of the Order of the British Empire
ASHA	American Society of Horticultural Science		RHS	Royal Horticultural Society
ASPA	American Sod Producers Association		SSSA	Soil Science Society of America
ATRI	Australian Turfgrass Research Institute		STMA	Sports Turf Managers Association
BIGGA	British and International Golf Greenkeepers Association		STRI	Sports Turf Research Institute
CSSA	Crop Science Society of America		TPI	Turfgrass Producers International
EPA	Environmental Protection Agency		UK	United Kingdom
GCSAA	Golf Course Superintendents Association of America		US	United States
IOG	Institute of Groundsmanship		USDA	United States Department of Agriculture
ITS	International Turfgrass Society		USGA	United States Golf Association

A

Abramashvili, Guivi Georgierich

Guivi Abramashvili was born in Tbilissi, Georgia, USSR, and earned a PhD in Agricultural Sciences from the Sport-Project Institute in 1957. Starting in 1962, he served as Chief of the Sector for Sports Turfgrasses at the Sport-Project Institute. He provided leadership and research in sports field construction and culture in Russia, including the Moscow Olympics, and retired in 1989. Abramashvili was awarded *Honoured Fellow* for Sports Constructions and Agronomy of Soccer Pitches by the USSR in 1997.

1970† **Long-Lasting Turfed Areas for Sport and Leisure.** G.G. Abramashvili. Publication House of Building, Moscow, Russia, USSR. 104 pp., 46 illus., 29 tables, no index, glossy semisoft bound in green and white with black lettering, 6.6 by 8.3 inches (167 × 212 mm). Published entirely in the Russian language.

This small book addresses the safety and effectiveness of turfgrasses as used on sport fields. The existing research is reviewed, along with the author's experience in the art and science of turfgrass culture. The impacts of soils, nutrient regimes, and other cultural practices on the development of turfgrasses are discussed. It is oriented to professionals responsible for the culture of turfgrasses for sport and recreation in Russia.

1979† **Town and Sporting Grass Fields.** G.G. Abramashvili. Moscow Worker, Moscow, Russia, USSR. 104 pp., 48 illus., 17 tables, no index, soft bound in green and white with black lettering, 4.8 by 7.9 inches (121 × 200 mm). Published entirely in the Russian language.

A practical book on site preparation, establishment, and culture of turfgrasses for sports fields under the conditions in the USSR. Emphasis is placed on root-zone construction. Also, methods for electrical soil warming are addressed, with an example being the Zapolyar soccer field in the Russian territory north of the Arctic Circle.

1980† **Turfgrass Maintenance, Foreign Experience.** G.G. Abramashvili. Soviet Sport, Moscow, Russia, USSR. 120 pp., 50 illus., 4 tables, soft bound in green with black lettering, 4.8 by 5.9 inches (140 × 150 mm). Published entirely in the Russian language.

A book of observations by the author concerning turfgrass cultural practices utilized in countries outside of the USSR.

1988† **Turfgrass for Sport Fields.** G.G. Abramashvili. Soviet Sport, Moscow, Russia, USSR. 160 pp., 18 illus., 29 tables, no index, soft bound in green and white with black lettering, 5.5 by 8.3 inches (140 × 212 mm). Published entirely in the Russian language.

A book on the physical-water properties of soils and modified root zones required for sport fields in Russia. A new system is proposed involving water-air tubes that enhance air ventilation and drainage of excess water.

Adams, William Arthur (Dr. and Professor)

Bill Adams was born in Ashby-de-la-Zouch, Leicestershire, England, UK, and earned a BSc degree in Agricultural Chemistry with a specialization in Soil Science and Plant Nutrition from Nottingham University in 1962 with *first class honours* and a PhD from Nottingham University in 1965. He has served on the faculty at the Institute of Biological Sciences at the University of Wales, Aberystwyth, Wales, since 1966, progressing from Lecturer to Senior Lecturer in 1983, to Reader in 1994, and to Professor in 1999. He retired in 2006. He served as a teacher and researcher in grassland soil chemistry and microbiology and in turfgrass-soil science. Drs. Adams and Vic Stewart conducted the original research on a slit drainage system involving vertical minislits at right angles and connected with main slits with permeable fill that are oriented over drain lines. He was elected International Turfgrass Society (ITS) Vice President from 1981 to 1985 and from 1989 to 1993 and ITS President from 2001 to 2005 while serving on the ITS Board of Directors from 1977 to 1993 and from 1997 to 2009. Dr. Adams has been honored with the Award for Research of the National Turfgrass Council in 1990, Friendship Award from the People's Republic of China in 1994, Honorary Advisor of the National Playing Fields Association in 1996, and Lifetime Achievement Award of the Institute of Groundsmanship in 1999.

1994† **Natural Turf for Sport and Amenity: Science and Practice.** W.A. Adams and R.F. Gibbs. CAB International, Oxon, UK. 404 pp., 177 illus. with 9 in color, 75 tables, small index, glossy semisoft bound in greens and brown with dark brown and green lettering, 6.1 by 9.2 inches (155 × 233 mm).

A technical book on the construction and turfgrass establishment of greens, tennis courts, soccer and rugby fields, cricket grounds, and horse racetracks. Emphasis is on the construction approaches. It is oriented to the United Kingdom and New Zealand conditions. Three chapters were written by contributing authors: "Horse Racing Tracks" by T.R.O. Field, "Amenity Grass for Non-sport Use" by E.J.P. Marshall, and "Warm-Season Turfgrasses" by Dr. J.R. Watson. Each chapter is referenced with a selected list of citations.

Agnew, Michael Lewis (Dr.)

Mike Agnew was born in Atchison, Kansas, USA, and earned BSc, MSc, and PhD degrees majoring in Horticulture from Kansas State University in 1979, 1981, and 1984, respectively. He served as Extension Turfgrass Specialist and Assistant and Associate Professor at Iowa State University from 1984 to 1994. Since then he has been Field Technical Manager for Syngenta Professional Products and has been involved in turfgrass fungicide, herbicide, and insecticide development. In 2004, he became Senior Field Technical Manager. Dr. Agnew has been honored with the President's Citation in 1989 and Certificate of Merit in 1991 from the Iowa Horticultural Society, and the Meritorious Service Award in 1992 and the Educational Service Award in 1994 from the Iowa Turfgrass Institute.

1992†   **The Mathematics of Turfgrass Maintenance.** First revised edition. Michael L. Agnew and Nick E. Christians. Golf Course Superintendents Association of America, Lawrence, Kansas, USA. 77 pp., 18 illus., 3 tables, no index, textured semisoft bound in cream with black lettering, 8.4 by 11.0 inches (214 × 279 mm).

A revision of the instructional booklet on mathematical calculations typically encountered in golf course operations. Sections on topdressing and irrigation have been added. It was prepared in coordination with a one-day seminar presentation.

Originally published circa 1977. See Beard, James B—1977.
Revised in 1997.

1997†   **The Mathematics of Turfgrass Maintenance.** Second revised edition. Coauthor.
See Christians, Nick Edward—1997.

Ahn, Yong Tae (Dr.)

Yong Tae Ahn was born in Gyeongsang-Nom-Do, South Korea. He earned a BA degree in Commercial Science from Kookim University in 1971 and an MSc degree in Financial Affairs from the Korea Graduate School of Business in 1974, completed the Chief Executive Offices Course at Seoul Graduate School of Business in 1996, and earned a PhD degree in Physical Education from Kyunghee University, South Korea, in 2006. He formed GMI Inc. in 1995 and serves as President. Dr. Ahn is active in golf course design and management consulting.

1999   **System and Leadership for Golf Courses.** Coauthor.
See Kim, In Seob—1999.

Air Force, United States

1949†   **Grounds Construction and Maintenance.** Department of the Air Force, Washington, D.C., USA. 77 pp., 50 illus. plus 3 world climatic maps in color, 19 tables, no index, glossy hard board covers in tan with blue lettering in a 3-hole binder, 7.7 by 10.2 inches (196 × 260 mm).

Figure 12.2. Budding manual-push, roller-drive, reel mower as manufactured by Ransomes in the 1830s. [26]

∗   A small, quality manual on the construction and establishment of turfgrasses on military air field installations under a wide diversity of climatic and soil conditions. The focus is on the use of turfgrasses for dust control, soil erosion control, safety zones, and traffic areas, with an emphasis on turfgrass airfield landing strips. The need for turfgrass areas on military installations was summarized as follows:

> The establishment of adequate grounds treatments is an integral part of the construction of the facilities at an air installation. Construction is not complete until all grounds involved in the project or group of projects have been treated for necessary dust and erosion control, any affected safety zones or traffic areas in the flying field have been given adequate treatment, authorized athletic and recreation grounds have been effectively treated to make them serviceable, and until hospital grounds, installation cemeteries, headquarters areas, and other prominent features of developed areas have had grounds treated to create a pleasing, orderly appearance, compatible with the proper functioning of the facility, and with economical maintenance.

Alderson, James Stillman

James Alderson was born in Lampasas, Texas, USA, and earned a BSc degree in Range Science from Texas A&M University in 1971. He joined the United States Department of Agriculture (USDA) Soil Conservation Service as a Range Conservationist and progressed to District Conservationist in 1981 to Area Plant Scientist in 1985, to Plant Materials Center Manager in 1991, and, finally, to Plant Materials Specialist of the USDA National Resources Conservation Service (NRCS) in Temple, Texas, in 1993; he retired in 2005. Alderson received the Superior Service Award from the USDA in 1993.

1994†   **Grass Varieties in the United States.**∗ Fourth revised edition. James Alderson and W. Curtis Sharp. Agricultural Handbook No. 170, Soil Conservation Service, United States Department of Agriculture, Washington, D.C., USA. 296 pp., 2 illus., 2 tables, indexed by cultivar name or number, semisoft bound in greens with black lettering, 8.5 by 11.0 inches (216 × 279 mm).

The revision consists of a major update on the status of named cultivars, newly-released cultivars, and experimental lines of grasses available in the United States and Canada.

†   *Reprinted in 1995 by CRC Press Inc., Boca Raton, Florida, USA.*

Aldous, David Ernest (Dr. and Honorary Principal Fellow and Associate Professor)

David Aldous was born in Sydney, New South Wales, Australia, and earned a BSc degree with *honours* in Agriculture from the University of Sydney in 1971, an MSc degree from Cornell University in 1976, and a PhD degree in Crop Science and Turfgrass from Michigan State University in 1978, studying with Dr. John Kaufmann. He served as Lecturer at the University of Queensland from 1972 to 1974, as Senior Lecturer in amenity horticulture at Massey University in New Zealand from 1978 to 1981, and as Principal Lecturer in the Department of Resource Management, Forestry and Amenity Horticulture at the Institute of Land and Food Resources, University of Melbourne, Burnley Campus, since 1981. He was involved in teaching and research of turfgrasses and amenity grasslands, retiring in 2008. He was elected President of the International Federation of Park and Recreation Administration in 1998. Dr. Aldous received Fellow from Parks and Leisure Australia in 1986, the National Award in Park Management from the Royal Australia Institute of Parks and Recreation in 1995, the Innovation and Best Practice Award from Parks and Leisure Australia in 2002, and the Frank Stewart Award from Parks and Leisure Australia in 2005.

1999†   **International Turf Management Handbook.** David Aldous (editor) with 24 contributing authors. Butterworth-Heinemann, Port Melbourne, Victoria, Australia. 430 pp., 72 illus., 66 tables, indexed, glossy hardbound in multicolor with white lettering, 7.4 by 9.6 inches (188 × 245 mm).

A book of 20 chapters that addresses four subject areas: turfgrass culture, facility management, performance assessment, and environmental issues. There are selected references at the end of each chapter.

2002†   **Sports turf & amenity grasses: a manual for use and identification.** D.E. Aldous and I.H. Chivers. Landlinks Press, Collingwood, Victoria, Australia. 120 pp.; 167 illus. with 147 in color; 30 tables; small index; glossy hardbound in multicolor with white, black, and yellow lettering; 7.9 by 10.2 inches (200 × 260 mm).

A book on the vegetative identification, description, adaptation, and use of 57 turfgrasses that are presented in 26 major and 31 minor use groups. A visual vegetative key is presented in the front.

Alison, Charles Hugh (OBE)

Charles Alison (1882–1952) was born in Preston, Lancashire, England, UK, and studied at New College, Oxford University from 1900 to 1902. He was a quality golfer and cricketer. After working as a journalist, he then became the Club Secretary at the new Stoke Poges Golf Club in London. It was there that he met and became a protégé of golf course architect

Harry S. Colt. He first worked as a construction superintendent and then as an associate. They eventually formed a partnership that continued for 20 years. He was active worldwide in golf course architecture, with a concentration of efforts in North America and Far East.

1920† **Some Essays on Golf-Course Architecture.** Coauthor.
>> See *Colt, Harry Shapland—1920.*

   * Alison commented on inland golf course playing conditions prior to 1900 in the United Kingdom:

> Through the winter months the ball frequently stuck where it pitched, and was almost invariably covered with mud. As regards the putting greens, the greenkeeper relied mainly on the heavy roller for obtaining an even surface. He had not yet discovered the art of creating a porous soil in which the fine grasses could flourish, and which would remain firm in wet weather. The use of sand, except on the teeing grounds, was practically unknown, and no worm-killer had been invented.

Alma, Alberto (Dr. and Professor)

Alberto Alma was born in Strambino, Italy, and earned BSc degree in Agricultural Science in 1982 and a PhD degree in Entomology in 1990, both from the University of Turino. He joined the faculty at the University of Turino in 1995 and rose to Full Professor in 2001. He has been involved in teaching and research concerning agricultural entomology, systemic entomology, integrated pest management (IPM), and biological control of pest insects.

2000† **La difesa dei tappeti erbosi: Malattie fungine, nemici animali e infestanti (Pest control on turfgrasses: Diseases, animal enemies and weeds).** Coauthor.
>> See *Gullino, Maria Lodovica—2000.*

American Sod Producers Association (ASPA)

An organization composed primarily of members who are growers and marketers of vegetative turfgrasses in the form of sod, sprigs, and/or plugs. It was formed in 1967, with the first president being Ben O. Warren. The name was changed to Turfgrass Producers International (TPI) in 1994.

1992† **Turfgrass: Nature's Constant Benediction: The History of the American Sod Producers Association.**
>> See *Betts, Janice Lake—1992.*
>> Also see *Mathews, Wendell Glen—2007.*

American Sports Builders Association (ASBA)

The American Sports Builders Association (ASBA) was founded by a group of contractors in 1965 as the US Tennis Court and Track Builders Association (USTC and TBA). The name was changed in 2004 to ASBA in order to represent the broader range of work contracted by its members. The ASBA holds a winter meeting of its members and a technical meeting with trade show.

2010† **Sports Fields—A Construction and Maintenance Manual.** American Sports Builders Association, Ellicott City, Maryland, USA. 286 pp., 109 illus. with 65 in color, 2 tables, indexed, glossy semisoft covers in multicolor with black and blue lettering in a spiral ring binder, 8.5 by 11.0 inches (215 × 279 mm).

   * A book concerning the planning, design, specifications, construction, and maintenance of sports fields in the United States. Both natural turfgrass and artificial surfaces are addressed. ASTMA International Standards and a glossary of terms are included in the back.

Ansorena Miner, Javier (Dr.)

Javier Ansorena was born in San Sebastian, Pais Vasco, Spain, and earned a Chemical Technician degree from Valladolid University in 1970; he also earned a BSc in Chemical Science in 1975 and a PhD in 1980, both from País Vasco University. He worked as Associate Professor for País Vasco University from 1976 to 1982 and as Laboratory Director of Agriculture and Environment for the Diputación Foral de Guipúzcoa from 1982 to 1997. Since 1997 he has served as Chief of the Environmental Department for the Diputación Foral de Guipuzcoa. Dr. Ansorena is President of ANORCADE (Association for Normalization and Quality Standards for Turfgrass Sportfields Construction in Spain).

1998† **Césped Deportivo: Construcción y Mantenimiento (Sport Turfgrass: Construction and Maintenance).** Coauthor.
>> See *Merino Merino, Domingo—1998.*

Aoki, Kohichi

Kohichi Aoki was born in Nishinomiya City, Hyogo Prefecture, Japan, and earned a BSc in Agriculture from Kyoto University in 1974. He served as Researcher in Ornamental Plants at the Landscaping Lab from 1980 to 1998, as General Manager of

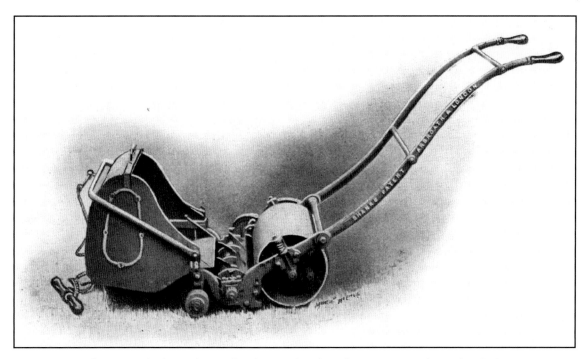

Figure 12.3. Shanks 1904 manual-push, 12-inch-wide, reel green mower with grass box. [38]

the Sandy Soil Vegetable Crops Lab from 1998 to 2002, as General Manager of the Ornamental Plants Lab from 2002 to 2004, and as General Manager of the Genetic Engineering Laboratory from 2004 to 2008, all at Chiba Prefectural Agriculture Research Center. Aoki received the Award of Technology for Group from the Japanese Society of Turfgrass Science in 2000.

1998   **Turfgrasses and Their Cultivars.** Coauthor.
         See *Asano, Yoshito—1998.*

Army Sport Control Board, Great Britain

1940   **Recreation Grounds.** Army Sport Control Board, London, England, UK. 44 pp., no illus., 2 tables, no index, soft bound in red with black lettering, 4.8 by 7.1 inches (122 × 181 mm).

  * A pioneering booklet on the establishment and culture of turfgrasses for sport and recreational uses for the military in the United Kingdom. It encompasses cricket pitches and grounds, tennis courts, bowling greens, golf courses, and track & field facilities. The approach then used on soccer fields was as follows:

> Football pitches are a problem at this time, especially those situated on heavy ground, and badly drained. Every attempt should be made to avoid overplay under wet conditions; if it is necessary to roll for further play after the ground has been badly cut up, and before it has had time to dry out thoroughly, then it is advisable to spike-roll and lightly harrow, before rolling. If available, a dressing of sand should be put on the goal areas, or wet patches, before rolling.

Army, United States Corps of Engineers

1947†   **Planting Turf.** E.M. Part XIV, Engineering Manual for War Department Construction, Corps of Engineers, Washington, D.C., USA. 25 pp., 12 illus., 1 table, no index, stapled semihard covers in tan with brown lettering, 7.7 by 10.2 inches (195 × 260 mm).

         A small manual concerning the soils, site preparation, turfgrass adaptation, planting, and establishment of turfgrasses on military installations in the United States.

1961†   **Planting Turf—Engineering and Design Manual.** EM 1110-1-321, Corps of Engineers, United States Army, Washington, D.C., USA. 45 pp., 20 illus., 8 tables, no index, semisoft bound in black and white with black lettering, 7.3 by 9.9 inches (184 × 251 mm).

         A small booklet on site preparation, planting, and establishment of turfgrasses designed for use by the Corps of Engineers. Guidelines are given for turfgrass species selection based on specific soil and climatic conditions in the United States. There is a short bibliography in the back.

Army, United States

1958† **Repairs and Utilities—Grounds Maintenance and Land Management.** * Technical Manual No. 5-630, Department of the Army, Washington, D.C., USA. 126 pp., 75 illus., 28 tables, indexed, semisoft bound in tan with brown lettering, 7.8 by 10.1 inches (198 × 256 mm).

A manual that addresses the management of military lands, with emphasis on turfgrass and landscape plant utilization. Soil and climatic influences within the United States are discussed. The outline for a land management plan and a selected bibliography are presented in the back.

Arthur, James Herbert

Jim Arthur (1920–2005) was born in London, England, UK, and earned a BSc degree in Agriculture from Reading University in 1946. During WWII, he was in military service from 1939 to 1945. He worked as an Advisory Officer at the Sports Turf Research Institute from 1946 to 1952 then as a Technical Advisor and eventually Managing Director of Twyford Pedigree Seed from 1952 to 1970. Subsequently, Arthur worked as a private consultant in the United Kingdom, with an emphasis on golf courses.

1997† **Practical Greenkeeping.** First edition. Jim Arthur. The Royal and Ancient Golf Club of St Andrews, Fife, St Andrews, Scotland, UK. 271 pp., 192 illus. with 136 in color, 4 tables, small index, cloth hardbound in green with gold lettering and a glossy jacket in multicolor with white and black lettering, 6.8 by 9.3 inches (173 × 235 mm).

A book on golf course maintenance under the environmental conditions in the United Kingdom. As stated by the author, most of the material presented is based on his 50-plus years of field experience. Emphasis is on the assumed art of turfgrass culture, with an apparent disdain for the scientific aspects based on sound research. The approach is very provincial, with potential success based on the absence of significant environmental stresses, including traffic. Numerous claims of successes or failures on golf courses from a particular cultural practices are discussed, but there is no actual documentation. The approaches espoused would fail in many regions of the world.

2003† **Practical Greenkeeping.** Second revised edition. Jim Arthur. The Royal and Ancient Golf Club of St Andrews, Fife, Scotland, UK. 312 pp., 192 illus. with 16 in color, 4 tables, indexed, cloth hardbound in green with gold lettering, and a jacket in multicolor with white and black lettering, 6.8 by 9.3 inches (173 × 235 mm).

A reprinting of the 1997 first edition, with additions organized as appendices for each chapter placed in the back of the book.

Asano, Yoshito (Dr. and *Professor*)

Yoshito Asano (1949–2007) was born in Ichikawa City, Chiba Prefecture, Japan, and earned a BSc degree in Horticulture from Chiba University in 1973 and a DrAgri in Horticulture from Hokkaido University in 1979. He was Chief Researcher at the National Institute of Agrobiological Sciences from 1982 to 1989. He then served on the Faculty of Horticulture at Chiba University, starting as an Assistant Professor in 1989 and becoming a Full Professor in 2002. His research focused on gene recombination and transfer in turfgrasses. Dr. Asano received the Encouragement Award of the Japanese Society for Horticultural Science in 1981 and the Scientific Award of the Japanese Society of Turfgrass Science in 1998.

1998 **Turfgrasses and Their Cultivars.** Yoshito Asano and Kohichi Aoki (editors). Soft Science Inc., Tokyo, Japan. 405 pp.; 386 illus. with 151 in color; 90 tables, indexed, cloth hardbound in green with gold lettering and a jacket in multicolor with green, white, and black lettering; 7.1 by 10.0 inches (180 × 256 mm). Printed in the Japanese language.

A basic book on turfgrass science, with emphasis on breeding improved cultivars through biotechnology. Lists of selected references are included at the ends of certain chapters.

Asay, Kay Harris (Dr.)

Kay Asay was born in Lovell, Wyoming, USA, and earned a BSc degree in Agronomy from the University of Wyoming in 1957, an MSc degree in Agronomy and Crop Breeding from the University of Wyoming in 1959, and a PhD degree in Crop Breeding from Iowa State University in 1965. He has led an active research program in the breeding and genetics of grasses, including low-maintenance turfgrasses, at the University of Missouri from 1965 to 1974 and at the United States Department of Agriculture-Agriculture Research Service (USDA-ARS) laboratory at Logan, Utah, from 1974 to 2002, when he retired. Dr. Asay was awarded Fellow of the Crop Science Society of America in 1982, Fellow of the American Society of Agronomy in 1985, USDA-ARS Outstanding Scientist of the Year in 1988, and USDA-ARS Certificate of Merit in 1995.

1989† **Contributions from Breeding Forage and Turf Grasses.** * Coeditor.
See *Sleper, David Allen—1989.*

Ashman & Associates

1999† **Sports Turf Management Program.** Ashman & Associates, Arlington Heights, Illinois, USA. 130 pp., 27 illus., 12 tables, no index, semisoft covers in multicolor with black lettering and a clear overlay front in a comb binder, 8.5 by 11.0 inches (216 × 279 mm).

   A guide book on the culture and renovation of baseball fields and associated equipment plus maintenance of the dirt infield areas.

Association Francaise pour le Développement des Équipements de Sportifs et de Loisirs (AFDESL; French Association for the Development of Facilities for Sports and Recreation)

1980 **Les Terrains de Sports: Concevoir, Réaliser, Exploiter (Sports Grounds: Planning, Construction, Maintenance).** Association Francaise pour le Développement des Équipements de Sportifs et de Loisirs. Éditions du Moniteur, Paris, France. 244 pp. plus 40 pp. covering producers and distributors, 464 illus. with 70 in color, 42 tables, small index, glossy semisoft bound in multicolor with white lettering, 8.3 by 11.6 inches (210 × 295 mm). Printed entirely in the French language.

   A book on the design, construction, and maintenance of sports fields for France, including field layouts. There is a bibliography in the back.

Australian Turfgrass Research Institute Limited (ATRI)

ATRI was first organized in 1954 as the Grass Research Bureau (NSW) Ltd. funded by the New South Wales Golf Association (NSWGA), located at the Ryde School of Horticulture, and staffed from 1960 to 1961 by Douglas Corbett. In 1956, the Royal New South Wales Bowling Association (RNSWBA) became a cocontributor to the Grass Research Bureau (GRB). Then after a period of inactivity, Peter McMaugh was employed in 1964 to lead the fledgling Grass Research Bureau with the title of Research Officer-in-Charge. The original objectives were to provide a scientific-based turfgrass service to bowling and golf clubs through advisory activities and research so as to solve problems in greens and to provide improved playing surfaces. The name was changed to the Australian Turfgrass Research Institute Limited (ATRI) in 1970. In 1971, the Bureau was moved to purpose-built laboratories in Concord West, New South Wales, with Sam Sutton appointed Director on a part-time basis. Peter McMaugh was named full-time Director in 1976 and served till 1978. Successor Directors were Trevior Siviour, from 1978 to 1982; Gerald Schultz, from 1982 to 1987; and Ian McIver, from 1987 to 1998. The GRB first published their official journal called **Grass Research** in 1955. It was replaced in 1958 by **The Bulletin** and subsequently a new format was published in 1965 as **Grass Research**. ATRI organized a series of 24 Topic Seminars of 1 to 2 days duration held from 1982 to 1995. From 1900 to 1997, the ATRI sponsored 14 conferences and symposia of 2 days in duration with published proceedings. The **ATRI Turf Notes** was an information newsletter published periodically. The ATRI was terminated in 1998 when the golf organizations withdrew financial support.

1987† **Disease, Insect & Weed Control in Turf.** First edition. David Worrad. Australian Turfgrass Research Institute, Concord West, New South Wales, Australia. 34 pp., no illus., 26 tables, no index, fold-stapled semisoft covers in yellow with black and green lettering, 8.1 by 11.6 inches (207 × 294 mm).

   A guide to injury symptoms on turfgrasses and the controls for the major disease, insect, and weed pests that occur in Australia.

   Revised in 1989.

1988† **Turf Market.** The Australian Turfgrass Research Institute Limited, Concord West, New South Wales, Australia. 46 pp., no illus., 29 tables, no index, semisoft covers in yellow and green with yellow and green lettering in a spiral binder, 8.5 by 11.5 inches (215 × 292 mm).

   A comprehensive review of statistics concerning the Australian turfgrass market to identify and quantify the major turfgrass diseases, insect pests, and weeds by the type of grassed playing surface and by state. The objective was to provide the value effect of each pest.

1989† **Disease, Insect & Weed Control in Turf.** Second revised edition. Australian Turfgrass Research Institute, Concord West, New South Wales, Australia. 54 pp., no illus., 34 tables, no index, fold-stapled glossy semisoft covers in white and green with black and green lettering, 8.1 by 11.7 inches (205 × 297 mm).

   An updated, expanded guide to turfgrass pest diagnosis and control for recreational and sports fields in Australia.

   Reprinted in 1989 with minor changes.

   Revised in 1993.

1991† **Sportsturf Water Management Manual.** First edition. E.W. Barlow, Peter McMaugh, Neill Rathbone, Don Ensign, Phillip Kesterton, Maria Zannetides, Tony Fogarty, and Mark Parker. The Australian Turfgrass Research Institute, Concord West, New South Wales, Australia. 53 pp., 17 illus., 8 tables, no index, fold-stapled textured semisoft covers in green with black lettering, 8.5 by 11.8 inches (215 × 300 mm).

A manual concerning turfgrass water-use conservation under the conditions in Australia, including plant-water relationships, turfgrass water-use efficiency, and irrigation strategies. It was designed for use with ATRI training courses.

Revised in 1992.

1992† **Soil Manual I.** Murray G. Wallis. The Australian Turfgrass Research Institute, Concord West, New South Wales, Australia. 22 pp., 11 illus., 4 tables, no index, fold-stapled textured semisoft covers in maroon with black lettering, 8.2 by 11.75 inches (208 × 298 mm).

A basic manual concerning the physical characteristics of soils plus water movement and retention. There are three short sections on root zone modification. It was designed for use with ATRI training courses.

Reprinted in 1993 with revisions.

1992† **Sportsturf Water Management Manual.** Second revised edition. David Yates, Peter McMaugh, David G. Bedingfield, John Chambers, Phillip L. Kesterton, Murray G. Wallis, Ian C. McIver, Phillip Allan, and Greg Mathews. The Australian Turfgrass Research Institute, Concord West, New South Wales, Australia. 66 pp., 22 illus., 17 tables, no index, fold-stapled textured semisoft covers in blue with black lettering, 8.5 by 11.7 inches (215 × 297 mm).

An expanded revision that includes two case studies concerning water management planning. Selected references are included at the back of each section. It was designed for use with ATRI training courses.

1992† **Turfgrass Variety Manual.** Martin Eade, Trevor Siviour, Peter McMaugh, Michael Robinson, John Neylan, John D. Clark, D.K. McIntyre, Rod Riley, John Odell, and Peter Hickman. The Australian Turfgrass Research Institute, Concord West, New South Wales, Australia. 50 pp., no illus., 12 tables, no index, fold-stapled textured semisoft covers in mustard with black lettering, 8.2 by 11.7 inches (208 × 297 mm).

A manual concerning cultivars of both cool-season and warm-season turfgrass for use in Australia. Included are cultivar assessments conducted in Australia plus sections on requirements for turfgrass performance on bowling greens, golf courses, and sports fields. It was designed for use with ATRI training courses.

1993† **Disease, Insect & Weed Control in Turf.** Third revised edition. David S. Drane (editor) with contributing authors Gary W. Beehag, Ian C. McIver, Murray G. Wallis, and David E. Westall. Australian Turfgrass Research Institute, Concord West, New South Wales, Australia. 85 pp., 1 illus., 45 tables, no index, fold-stapled glossy semisoft covers in white, green, and black with black lettering, 8.1 by 11.7 inches (206 × 296 mm).

A revised and updated guide to turfgrass pest and disease management and control for sports and recreational turfgrasses in Australia. A section on environmental quality aspects has been added.

Reprinted in 1993 with minor corrections.
Revised in 1996.

1993† **Turf Disease Manual.** Ian C. McIver, Melissa Haslam, and Frank Dempsey. The Australia Turfgrass Research Institute, Concord West, New South Wales, Australia. 45 pp., 25 illus., 2 tables, no index, fold-stapled semisoft covers in cream with black lettering, 8.2 by 11.75 inches (208 × 298 mm).

A basic manual concerning the diseases of turfgrasses encompassing pathology, identification, host infection, and control. There are specific sections on fairy rings, Helminthosporium diseases, nematodes, and spring dead spot. It was designed to use with the ATRI turfgrass workshops.

1993† **Turfgrass Weed Manual.** Gary W. Beehag and Dr. Peter Michael (editors). The Australian Turfgrass Research Institute Limited, Concord West, New South Wales, Australia. 23 pp., 49 illus., no tables, no index, semisoft bound in green with black lettering, 8.5 by 11.5 inches (215 × 292 mm).

A basic manual concerning identification of the common species of broadleaf weeds and grassy weeds and the significance of plant form on herbicide control. It was designed for use with ATRI training workshops.

1994† **Turf Tips Compendium—Volume I.** First edition. Angelene Smith (editor). The Australian Turfgrass Research Institute Limited, Concord West, New South Wales, Australia. 88 pp., 21 illus., 24 tables, no index, semisoft covers in blue with black lettering in a spiral binder, 8.5 by 11.5 inches (215 × 292 mm).

An assemblage of 42 short articles on turfgrass culture with emphasis on diseases, insects, weeds, grasses, and fertilizers.

Reprinted as a second edition in 1997.

1994† **Turf Nutrition and Fertilisers.** Jyri Kaapro (editor). The Australian Turfgrass Research Institute Limited, Concord West, New South Wales, Australia. 48 pp., 1 illus., 34 tables, no index, semisoft bound in fawn with black and white lettering, 8.5 by 11.5 inches (215 × 292 mm).

A basic manual concerning the measurement and significance of soil nutrition and of fertilizer, lime, and gypsum selection. It was designed for use with ATRI training workshops.

1994† **Soil Chemistry Manual.** Jyri Kaapro (editor). The Australian Turfgrass Research Institute Limited, Concord West, New South Wales, Australia. 38 pp., no illus., 34 tables, no index, semisoft bound in blue with black lettering, 8.5 by 11.5 inches (215 × 292 mm).

Figure 12.4. Shanks 1914 Caledonia manual-push, 14-inch-wide, reel green mower. [38]

A basic manual concerning the measurement and significance of soil nutrition and of fertilizer, lime, and gypsum selection. It was designed for use with ATRI training workshops.

1995† **Turfgrass Identification Manual.** Gary W. Beehag (editor). The Australian Turfgrass Research Institute Limited, Concord West, New South Wales, Australia. 32 pp., 41 illus., 1 table, no index, semisoft bound in fawn with black lettering, 8.5 by 11.5 inches (215 × 292 mm).

A basic manual concerning the vegetative identification of the common species of cool-season and warm-season turfgrasses using a plant key. It includes a plant key and a glossary. It was designed for use with ATRI training workshops.

1996† **Disease, Insect & Weed Control in Turf.** Fourth revised edition. Ric Wozniak (editor) with five contributing authors: Gary W. Beehag, Maria F. Guerrera, Jyri Kaapro, Ian C. McIver, and David E. Westall. Australian Turfgrass Research Institute Limited, Concord West, New South Wales, Australia. 108 pp. plus 8 pp. of advertising, 40 illus. with 36 in color, 37 tables, no index, glossy semisoft bound in multicolor with white and pink lettering, 8.4 by 11.6 inches (213 × 295 mm).

A major revision of the 1993 guide. New chapters have been added concerning parasitic nematodes, mites, and growth regulators.

1997† **Turf Tips Compendium—Volume I.** Second revised edition. Angelina Smith (editor) with eight contributing authors. The Australian Turfgrass Research Institute Limited, Concord West, New South Wales, Australia. 71 pp., 28 illus., 9 tables, no index, semisoft covers in purple with black lettering in a comb binder, 8.25 by 11.6 inches (209 × 295 mm).

A reprint of the 1994 edition with minor updates.

1997† **Turf Tips Compendium—Volume II.** Gary W. Beehag and Angelene Smith (editors). The Australian Turfgrass Research Institute Limited, Concord West, New South Wales, Australia. 70 pp., 29 illus., 27 tables, no index, semisoft covers in orange with black lettering in a spiral binder, 8.5 by 11.5 inches (215 × 292 mm).

An assemblage of 40 short articles on turfgrass culture covering turfgrass construction, environmental issues, grasses, insects, nematodes, pesticides, soils, and water.

B

Bahir, Zvi

Zvi Bahir (1910–2002) was born in Pinczowska County, Poland, and immigrated to Israel in 1934. He was Chief Gardener at Kibbutz Nir-David from 1940 to 1957 and the Director of National Park Gan-Hoshloska from 1957 to 1987. He pioneered turfgrass development in Israel and the development of the Kibbutz Public Gardens concept in the new state of Israel. He served as Director of the Department of Irrigation Efficiency at the Israel Ministry of Agriculture from 1966 to 1976. From 1987 to 1995, he was Chair of the Committee for Grassed Sporting Fields, Physical Education & Sport Authority for the Israel Ministry of Education & Culture, and he was on the Editorial Board of **Garden & Landscape**. He was elected to the Board of Directors of the Israel Gardeners Association. Bahir received the Person of the Year Award from the Israel Minister of Agriculture in 2000.

1963†   **The Lawn and Its Care.** Zvi Bahir and Jakob Shoor. Agricultural Extension Service No. 54, Jewish Agency Settlement Department, Ministry of Agriculture, Tel Aviv, Israel. 64 pp.; 43 illus.; 9 tables; no index; fold-stapled semisoft covers in green, white, and black with black lettering; 6.2 by 9.1 inches (158 × 231 mm). Printed entirely in the Hebrew language.

     A **rare**, small handbook on the establishment and culture of lawns under the arid conditions in Israel. Major emphasis is placed on irrigation. It is the first turfgrass book published in the Middle East.

Baker, Stephen William (Dr.)

Steve Baker was born in Bushey, Hertfordshire, England, UK, and earned a BSc degree with *first class honours* in Geography from the University College of London in 1977 and a PhD degree in Geography from the University of Bristol in 1982. He served as Soil Physicist from 1980 to 1988, then as Senior Research Officer from 1988 to 1996, and since as Head of Soils and Sports Surface Science, all with the Sports Turf Research Institute. He was elected to the International Turfgrass Society Board of Directors from 2001 to 2009.

1990†   **Sands for Sports Turf Construction and Maintenance.** S.W. Baker. The Sports Turf Research Institute, Bingley, West Yorkshire, England, UK. 67 pp., 36 illus., 15 tables, no index, glossy semisoft bound in yellow with brown lettering, 5.7 by 8.2 inches (145 × 210 mm).

     A small manual summarizing research on sand materials for turfgrass root zones used for sports field, with recommendations made for winter sports surfaces in the United Kingdom. It contains a selected list of references in the back.

2005†   **STRI Guidelines for Golf Green Construction in the United Kingdom.** Steven Baker. The Sports Turf Research Institute, Bingley, West Yorkshire, England, UK. 20 pp., 14 illus. with 9 in color, 2 tables, no index, fold-stapled glossy semisoft covers in greens and white with white and black lettering, 8.2 by 11.7 inches (209 × 297 mm).

     A small guide to materials selection and construction methods for building putting greens in the United Kingdom. It is based on the USGA Method, with some modifications.

2006†   **Handboek Greenonderhoud (Golf Green Maintenance Handbook).** Stephen Baker, as translated by the Nederlandse Golf Federatie. The Nederlandse Golf Federatie, De Meern, the Netherlands. 71 pp.; 51 illus. in color; 1 table; no index; glossy semisoft bound in multicolor with green, white, and black lettering; 5.8 by 8.2 inches (147 × 209 mm). Printed in the Dutch language.

     A handbook concerning the root zones, construction, turfgrass selection, culture, pest management, and surface-quality monitoring of putting greens under conditions in the Netherlands. There is a list of selected references in the back.

2006†   **Rootzones, Sands and Top Dressing Materials for Sports Turf.** Dr. Stephen Baker. The Sports Turf Research Institute, Bingley, West Yorkshire, England, UK. 104 pp. with 8 pp. of advertising, 56 illus. with 19 in color, 21 tables, no index, glossy semisoft bound in multicolor with black lettering, 5.8 by 8.2 inches (148 × 209 mm).

     A small book concerning construction guidelines for the root-zone profile and the specifications to use on sports fields under conditions in the United Kingdom. There is a list of selected references in the back.

Bandeville, Maurice Louis

Maurice Bandeville (1877–1953) was born in Arras, France. Bandeville was President of the following: Société des Sports du Touquet, from 1905 to 1914; Cannes CLTC; Golf Club de Fourqueux; and Conseil du Parc des Princes. He was also Fondateur du Golf Club de Lys-Chantilly and Ancien Président de l'Athletisme à l'USFSA.

1928   **Le Parcours de Golf: Sa Construction-son Entretien-les Gazons en Général (The Golf Course—Construction, Maintenance, Turfs in General).** Maurice L. Bandeville. Guides Plumon, Paris, France. 147 pp., 32 illus., no tables, no index, hardbound in beige with red and black lettering, 4.3 by 6.7 inches (110 × 171 mm). Printed entirely in the French language.

  *   The second turfgrass book published in France and the first to address the construction, establishment, and culture of golf course turfgrasses under conditions in France. This is a **very rare** book that is a must for collectors of historical turfgrass books.

1942   **Les Pelouses et Terrains de Sport, Construction et Entretien: Les Gazons en Général (Lawns and Sports Grounds, Construction and Maintenance: Turfs in General).** First edition. Maurice L. Bandeville. Editions Berger-Levrault, Paris, France. 159 pp., 29 illus., 2 tables, no index, semisoft bound in brown with black lettering and a front photograph insert in greens, 4.3 by 6.7 inches (110 × 171 mm). Printed entirely in the French language.

* A book on the construction, establishment, and culture of turfgrasses on lawns and sports fields in France. There are sections on turfgrasses for rugby, field hockey, soccer, track & field, and lawn tennis, including field layouts. A **very rare** book. It is oriented to conditions in France.

    Revised in 1952.

1952   **Pelouses et Terrains de Sport (Lawns and Sport Grounds).** Second revised edition. Maurice L. Bandeville. Editions Berger-Levrault, Paris, France. 158 pp.; 21 illus.; 2 tables; no index; semisoft bound in green, black, and white with black lettering; 4.5 by 6.9 inches (115 × 175 mm). Printed entirely in the French language.

* An update of a book on the construction, establishment, and culture of turfgrasses on lawns and sports fields. Included are sections on turfgrasses for rugby, field hockey, soccer, track & field facilities, and lawn tennis courts. It is oriented to conditions in France. Preface by Henri Delauncy.

Barkworth, Mary Elizabeth (Dr. and *Professor)*

Mary Barkworth was born in Marlborough, Wiltshire, England, UK, and earned a Bachelor of Letters degree in Mathematics and Physics from the University of British Columbia in 1960, a Master of Education degree in General Science from Western Washington University in 1968, and a PhD degree in Botany from Washington State University in 1975. She served as a Research Scientist for the Biosystematics Research Unit at Agriculture Canada from 1975 to 1978; she joined the faculty at Utah State University in 1979 as Assistant Professor of Biology and served as Director of the Intermountain Herbarium from 1980 to 1983 and from 1989 to the present. Dr. Barkworth has specialized in the systematics of North American grasses, especially the *Stipeae* and *Triticeae*.

1986†   **Grass Systematics and Evolution.*** Coeditor.

    See *Soderstrom, Thomas Robert—1986.*

2003   **Flora of North America, Volume 25, Magnoliophyta: Commelinidae (in part): Poaceae, Part 2.** Mary E. Barkworth, Kathleen M. Capels, Sandy Long, and Michael B. Piep (editors). Oxford University Press, New York City, New York, USA. 809 pp., 1,567 illus., 2 tables, indexed, cloth hardbound in green and blue with gold lettering, 8.4 by 10.9 inches (213 × 276 mm).

    A major scholarly compilation concerning the taxonomy of the grass family (Poaceae) of North America, north of Mexico. The order of presentation is grass tribes, genera, and species, with the latter totaling 733. It is extensively illustrated as to anatomical characters and distribution maps. There are 54 contributing authors.

2007   **Flora of North America, Volume 24, Magnoliophyta: Commelinidae (in part): Poaceae, Part 1.** Mary E. Barkworth, Kathleen M. Capels, Sandy Long, Laurel K. Anderton, and Michael B. Piep (editors). Oxford University Press, New York City, New York, USA. 939 pp., 1,451 illus., 2 tables, indexed, cloth hardbound in green and blue with gold lettering, 8.5 by 11.0 inches (216 × 279 mm).

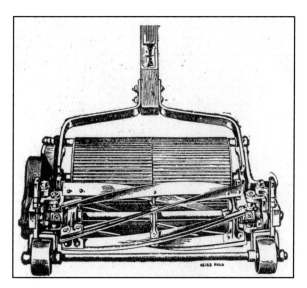

Figure 12.5. Manual-push, 18-inch-wide, Pennsylvania putting green mower with 6 steel blades, adjustable cutting height of ³⁄₁₆ to 1¼ inch and 7-inch halved roller drive, circa 1924. [34]

A major scholarly compilation concerning the taxonomy of the grass family (*Poaceae*) of North America, north of Mexico. The order of presentation is 25 grass tribes, 23 genera, and 1,373 species. It is extensively illustrated as to anatomical characters and distribution maps. There are 63 contributing authors.

2007† **Manual of Grasses for North America.** Mary E. Barkworth, Laurel K. Anderton, Kathleen M. Capels, Sandy Long, and Michael B. Piep (editors). Utah State University Press, Logan, Utah, USA. 639 pp., 1,983 illus., no tables, large index, glossy semisoft covers in multicolor with white and gray lettering, 8.3 by 10.9 inches (212 × 278 mm).

A condensed version of **Flora of North America,** volumes 24 and 25, with a few updates.

Barnard, Colin (Dr.; OBE)

"Bernie" Barnard (1904–1974) was born in Kalgoorlie, Western Australia, Australia, and earned a BSc degree with *honours* in 1925, an MSc degree in 1928, and a DSc in 1936, all in Botany from the University of Sydney. He joined the Council for Scientific and Industrial Research as a Botanical Assistant at Mildura, Australia, in 1928 and move to Canberra, Australia, in 1932 where he rose to Senior Principal Research Scientist in 1962, retiring in 1969. During WWII he initiated a phytochemical survey of native flora in Australia. He served as Head of the Plant Introduction Section of the Commonwealth Scientific and Industrial Research Organization (CSIRO) from 1962 to 1969. Dr. Barnard conducted research on the anatomy of inflorescence development in grasses. He was elected President of the Royal Society of Australia for 1946 through 1949 and Secretary of the Australian & New Zealand Association for the Advancement of Science from 1939 through 1958. Dr. Barnard was awarded Member of the Order of the British Empire (OBE) in 1970.

1964† **Grasses & Grasslands.** C. Barnard (editor) with 15 contributing Australian authors who were active in grass research. Macmillan & Company Limited, London, England, UK. 281 pp., 63 illus., 13 tables, indexed, textured cloth hardbound in green with gold lettering and a jacket in dark blue and green with green and black lettering, 6.2 by 9.7 inches (157 × 247 mm).

An important, scholarly book devoted to the basic biology of grasses and grasslands. It is extensively referenced. The chapters on grass systematics, distribution, form and structure, germination, growth, and environmental control relate to turfgrasses.

Barron, Leonard W.

Leonard Barron (1868–1938) was born in Chiswick, Middlesex, England, UK. He was a horticultural writer and pioneering author of the first turfgrass book published in the United States. His father, Archibald F. Barron (1835–1903), was a Royal Horticultural Society Veitch Memorial Medal recipient in 1889 and Victoria Medal of Honour recipient in 1897 and was Superintendent of the Royal Horticultural Society Gardens at Chiswick. This was where Leonard Barron acquired his early experience in gardening. He studied Botany at Gunnersbury College from 1880 to 1884 and took special courses at Birkbeck Institute in London from 1884 to 1892. He was engaged as Assistant Editor of **The English Gardeners' Chronicle** from 1885 to 1893. In 1894, he immigrated to the United States and eventually to Rockville Center, New York, USA. During this period he devoted his efforts to horticultural journalism. His first US position was Managing Editor of **American Gardening** from 1894 to 1904, succeeding Dr. L.H. Baily. He subsequently initiated **American Gardeners Chronicle.** Barron then joined the firm of Doubleday, Page and Company, where he became Horticultural Editor of a magazine from 1928 to 1936 that eventually became the **American Home and Country Life.** He was Managing Editor of the **Garden Magazine** from 1905 to 1911 and Editor from 1911 to 1928 and also served as Editor of **Garden and Home Builder** from 1924 to 1928 and Horticulture Editor of **American Home and Country Life** from 1928 to 1936. He was a charter member of the New York Horticultural Society, served as Secretary from 1900 to 1909, and was elected to the Board of Directors from 1928 to 1938. Barron was awarded the Gold Achievement Medal of the Horticultural Society of New York in 1938 and Fellow of the American Association for the Advancement of Science in 1931.

1906† **Lawns and How to Make Them—Together with the Proper Keeping of Putting Greens.** First edition. Leonard Barron. The Garden Library, Volume III, Doubleday, Page and Company, New York City, New York, USA. 174 pp., 31 illus., 11 tables, indexed, cloth hardbound in green with gold lettering, 5.0 by 7.5 inches (127 × 190 mm).

This is the first major book written specifically about the establishment and culture of lawns in the United States. It is a must for all collectors of historical turfgrass books. Surprisingly, this earliest of lawn books addresses both cool-season and warm-season turfgrasses of the United States. It includes a short chapter on putting greens written by golfer Walter J. Travis. The chapter subject titles include the following:

I. Renovating the Old Lawn	IV. Which is Better: Turf or Seed?
II. How to Make a Lawn Once for All	V. The Fine Art of Mowing, Rolling, and Watering
III. Economical Grading	VI. How to Feed a Lawn

VII.	Solving the Weed Problem: Insects, etc.	XI.	The Best Lawn Tools and Their Use
VIII.	The Truth about "Lawn Mixtures"	XII.	How to Make Lawn Pictures
IX.	Seed Mixtures for Special Purposes	XIII.	The Peculiar Requirements of Putting Greens
X.	Lawns for Subtropical Regions	XIV.	Guide to the Best Lawn Grasses

In the Chapter IV title, the term "turf" refers to sodding as compared to seeding. This pioneering, practically oriented book must have made an important educational contribution in its time, as evidenced by the number of reprintings. It contains many interesting photographs that represent some of the earliest pictorials published, including a unique photograph of a steam-driven, riding lawn mower weighing more than 1,000 pounds (454 kg).

The philosophies of turfgrass lawn culture, especially the use of rolling and hand scythes, in the United States during the early 1900s were represented by the following:

> Rolling, mowing, and watering are the three essential details which require attention every year, and they must be carried out thoroughly year after year, without any sort of lessened vigore because the lawn is established. The whole object and aim of the after attention is to secure a uniform sod and even texture over the entire surface of the lawn.
>
> Nothing conduces more to the maintenance of perfect condition in the lawn than persistent and early rolling each year; not that rolling should be omitted any time during the season, but it is especially necessary in the early spring. Just as soon as the ground becomes workable and the grass starts into growth the whole surface should be thoroughly rolled again and again to effectively overcome the loosening effects of the freezing and thawing of winter.
>
> Top dressings of good garden loam—rich fertiliser is not indicated—and rolling again and again will accomplish wonders. It is hardly possible to make the surface of the lawn too compact by this process. A roller which will exert a pressure of a thousand or even fifteen hundred pounds will not be too heavy.
>
> Just because so few people nowadays understand its manipulation, the scythe—the ideal cutting instrument—has fallen into disuse. For newly made lawns it is infinitely superior to the lawn mower as it cuts without tearing, without pulling. And a lawn mowed by an expert shows no signs whatever of having been cut. The inevitable streakiness which follows the use of the lawn mower operated in different directions is not seen after the scythe is used. Another advantage of this instrument is that the depth at which it will cut can be graded to a nicety, and there is no necessity to follow it up with trimming shears because it can be used in the sharpest corners.

† *Reprinted in the Garden Library series in 1909, in 1910 in light green with dark green lettering, and in 1914 in gray with black lettering with different bindings.*
  Revised in 1923.

> Sixteen years ago the little book on Lawns was submitted to the public, and it has seemingly filled a niche in our gardening handbooks that justifies the present revision. Not much that is startling has occurred in that elapsed period, but we have better knowledge of some troubles and their control, and such new facts are now included. This little treatise still remains the only one of its kind. Its purpose is to enable any one to establish a respectable and adequate greensward in any sort of soil where grass can be made to grow.

1923† **Lawn Making—Together with the Proper Keeping of Putting Greens.** Second revised edition. Leonard Barron. The Amateur's Book of the Garden Series, Doubleday, Page and Company, New York City, New York, USA. 176 pp., 31 illus., 11 tables, indexed, cloth hardbound in black with gold lettering and a jacket in multicolor with cream lettering, 5.4 by 8.0 inches (137 × 203 mm).

There are only minor revisions of this pioneering book. In the introduction Barron stated:

> THE PERFECT LAWN IN AN IDEAL SETTING
> Forming the foreground of some nature picture, the lawn fulfils its greatest mission, and is the greatest achievement of the landscape gardener's art. The smooth velvety greensward is a harmonious accent in the general colour scheme, unifying the nearby trees and shrubbery with the distant view. Should we not, therefore, make it with solicitous care?

† *Reprinted in the Garden Library series for the National Garden Association in 1928 and 1929 by Doubleday, Doran, and Company Inc.*

Batten, Steven Michael

Steve Batten was born in Oklahoma City, Oklahoma, USA. He served the US Navy in Arctic Submarine Service from 1966 to 1970. Batten earned both BSc and MSc degrees in Agronomy and Turfgrass from Oklahoma State University in 1975 and 1977, respectively, studying with Dr. Wayne Huffine. He was Turfgrass Research Associate at Texas A&M University from 1977 to 1981, working with Dr. Jim Beard; Southeastern Regional Agronomist for the USGA Green Section from 1981 to 1984; Turfgrass Agronomist for Golden Bear International from 1984 to 1989; and, since then, President of Steve Batten and Associates Inc. of Lake Worth, Florida. Batten has been active in golf course turfgrass consulting and as an illustrator, including drawings for **Turf Management for Golf Courses.**

1979† **Introduction To Turfgrass Science And Culture Laboratory Exercises.** Coauthor.
　　　See *Beard, James B—1979.*

Bavier, Michael Robert

Mike Bavier was born in Wilmar, Minnesota, USA, and earned an Associate degree from Pennsylvania State College in 1963. He served in the US Marine Corps from 1964 to 1970. Bavier then became Assistant Superintendent at Oak Ridge Country Club in Hopkins, Minnesota. In 1965, he moved to Calumet Country Club in Homewood, Illinois, as Superintendent and in 1969, became Superintendent at Inverness Golf Club near Chicago where he continues to serve. Bavier was elected President of the Midwest Golf Course Superintendents Association (GCSA) in 1978 and President of the Golf Course Superintendents Association of America in 1981. He was honored twice in 1990 and 2001 with the Charles Bartlett Award from the Midwest GCSA. Bavier was honored with the GCSAA Distinguished Service Award in 2000 and, in 2002, was recognized by the Village of Inverness as Arbor Day Citizen of the Year for his 20-plus years as a volunteer arborist for the Village.

1998† **Practical Golf Course Maintenance—The Magic of Greenkeeping.** First edition. Coauthor.
　　　See *Witteveen, Gordon Cornelis—1998.*

2003 **Guía Pràctica para Manejo de Pastos en Campos de Golf (Practical Golf Course Maintenance).** Coauthor.
　　　See *Witteveen, Gordon Cornelis—2003.*

2005† **Practical Golf Course Maintenance—The Magic of Greenkeeping.** Revised second edition. Coauthor.
　　　See *Witteveen, Gordon Cornelis—2005.*

Beal, William James (*Dr.* and *Professor* Emeritus)

William Beal (1833–1924) was born near Adrian in the Michigan Territory. Young Beal experienced living in a log cabin with an open stone fireplace, crops planted among tree stumps on land tilled by wooden plows, and Native Americans living nearby. He earned AB and AM degrees in Classical Studies from the University of Michigan in 1859 and 1861, respectively. Then an SB degree in Natural History from Harvard University in 1965 and an MSc degree from Chicago University in 1875. William Beal was one of the visionary educators and researchers of his time while serving as a Professor on the faculty of the new State Agricultural College in Michigan from 1871 to 1910. He was a key leader in the experimental movement of agricultural botany. He conducted (a) the first demonstration of hybrid vigor by controlled crossing of corn lines in 1879, (b) the initiation of the oldest ongoing US botanical experiment involving the vitality of buried seeds in 1879, and (c) the first turfgrass experiments, including polystand compatibility, in 1880. He initiated early, extensive purity and viability testing of seeds in 1877 and also organized the oldest, continuously operating US botanical garden in 1877. These works resulted in more than 1,200 papers plus seven extensive texts. Professor Beal was a proponent of scholarly communications among the few, isolated scientists active in applied botany and in agricultural research. His visionary efforts contributed to the formation of several important national agricultural science organization. Dr. Beal was a key founder and the first President of (a) the Society for the Promotion of Agricultural Science (SPAS) in 1880, (b) the Association of Botanists in the United States Agricultural Experiment Stations in 1888, (c) the Botanical Club of the American Association for the Advancement of Science in 1888, and (d) the Michigan Academy of Sciences in 1884. Dr. Beal received an Honorary PhD from the University of Michigan in 1880, an Honorary ScD from the Michigan Agricultural College in 1905, and an Honorary DAgr from Syracuse University in 1916, as well as the Wilder Silver Medal from the American Pomology Society in 1885 and Fellow of the American Association for the Advancement of Science.

1887† **Grasses of North America for Farmers and Students.** First edition. W.J. Beal. Published by the author, Agricultural College, Michigan, USA. 473 pp., 175 illus., 19 tables, large index, textured cloth bound in brown with imprinted gold lettering, 5.8 by 8.7 inches (147 × 221 mm).

Figure 12.6. Dr. William Beal, circa 1900. [6]   **Figure 12.7.** Reginald Beale, circa 1923. [15]

An early, widely used, practical book concerning grasses and their use, primarily for agriculture purposes. It includes the anatomy, morphology, physiology, growth, taxonomy, description, culture, and pests. There is a chapter on "Grasses for the lawn, the garden, and for decoration," plus a bibliography in the back.

Reprinted in 1896 by Henry Holt and Company as volume I.

1896† **Grasses of North America.** Volume II. William Beal. Henry Holt and Company, New York City, New York, USA. 715 pp., 274 illus., 5 tables, large index, textured cloth hardbound in burgundy with imprinted gold lettering, 5.8 by 8.7 inches (147 × 221 mm).

＊ A major text of that time concerning the taxonomy of grasses in North America. There are 12 tribes and 146 genera described and illustrated. Their geographical distribution and a bibliography are included in the back.

Beale, Reginald Evelyn Child

Reginald Beale (1877–1952) was born in Kingston, Surrey, England, UK, and joined James Carter & Company in 1897 as a Horticulturist, serving for 55 years. He became the Turfgrass Specialist and eventually Manager of the Sports and Grass Department at James Carter & Company. He provided leadership for some of the early turfgrass research in England and was a very active writer on the subject of turfgrass establishment and culture. Reginald Beale emphasized turfgrasses for golf courses and sports fields. He became a Director of Carters Tested Seeds Limited. Reginald Beale was selected as a Vice President of the Golf Greenkeepers Association at the time it was formed in 1912.

Edward John Beale (1835–1902), the father of Reginald Beale, was a turfgrass specialist and the Senior Partner of James Carter & Company, a seed firm in the United Kingdom. He received the Victoria Medal of Honour in Horticulture from the Royal Horticultural Society in 1897.

~

James Carter & Company was founded circa 1836 by seedsman James Carter (1797–1855) at 238 High Holborn in London, England, UK. Later the company acquired land and developed a nursery at Raynes Park, England, UK. The company name was changed to Carters Tested Seeds in 1927. In 1985, Carters Tested Seeds, then owned by the Horticultural Botanical Association, was sold to Sutton & Sons.

From 1837, the company regularly published an annual list of seeds offered for sale. Mention of turfgrass seed mixtures for lawns appeared in the 1860s, with a full page or more of turfgrass seed listings appearing from 1869 onward. Then small 4- to 6-page sections on turfgrass establishment and culture appeared annually in the seed booklets published by James Carter & Company after 1884. In 1908, James Carter & Co. initiated a series of 31 turfgrass publications titled **The Practical Greenkeeper** by Reginald Beale. Twenty-five are described herein, including the first, third, fourth, fifth, sixth, seventh, eighth, tenth (autumn), fourteenth (autumn), sixteenth, seventeenth, and eighteenth editions from circa 1908 through 1927 published under the name James Carter & Company plus 13 editions for 1927 through 1939 published under the name Carters Tested Seeds. Also the company published four American editions titled **Carters Practical Greenkeeper**, with the second, third, and fourth editions included herein.

~

1906   **Golf.** Reginald Beale. James Carter & Co., High Holborn, London, England, UK. 32 pp., 27 illus., no tables, no index, fold-stapled semisoft covers in green with embossed red lettering, 9.6 by 7.2 inches (245 × 182 mm).

*   A **very rare**, small book concerning methods for site preparation and seeding a golf course under conditions in England. There also is a section by greenkeeper Peter W. Lees titled "Worms in putting and other greens" that includes the series of eight photographs illustrating the original procedure for the use of mowrah meal as an irritant in earthworm control.

1908   **Manures and Composts for Golf Greens, & c.** Reginald Beale. James Carter & Co., High Holborn, London, England, UK. 20 pp., no illus., 3 tables, no index, fold-stapled semisoft covers in green with black lettering, 7.3 by 9.5 inches (185 × 242 mm).

*   A **very rare**, small early guide booklet of manure sources for fertilization of greens, manuring practices, and composting procedures. There is a preface by Horace G. Hutchinson.

n.d.   **The Practical Greenkeeper.** First edition. Reginald Beale. James Carter & Co., London, England, UK. 69 pp. includ-
circa   ing 2 pp. of advertising, 34 illus., 3 tables, no index, fold-stitched semisoft covers in green with embossed orange
1908   lettering, 7.0 by 9.5 inches (178 × 241 mm).

*   A **truly rare** booklet that is a must for collectors of historical turfgrass books, as is the entire series of revised editions. It is an early booklet on the construction, establishment, and culture of turfgrasses for lawns, putting greens, croquet courts, tennis courts, bowling greens, and floral hazards. It is oriented to English conditions. The booklet contains sections on the pioneering worm control work for greens by greenkeeper Peter W. Lees and on manures and composts. The preface is by Horace G. Hutchinson. There also is a section on weed control. The book contains a listing of costs for various turfgrass establishment and cultural practices as they existed in the first decade of the twentieth century in England. There is a pictorial description of the original earthworm irritant, mowrah meal, used on greens as a control method, which was developed by greenkeeper Peter Lees.

In 1908 Reginald Beale took a clear position concerning the use of sheep on golf courses in comparison to mechanical reel mowers:

> Sheep, under certain conditions, will manure the grass, keep it short, and, by constantly moving about, help to give the turf a firm surface; this assists in saving expense of cutting, rolling and manuring.
>
> Sheep may be grazed in the spring and early summer, providing they are cake or artificially fed; breeding ewes are seldom given much cake or roots till they have lambed. If not cake or artificially fed, sheep do little good other than keeping the grass short.
>
> Sheep, even on old pastures, are apt to pull out the small grass rootlets. They should not be allowed on new grass under any circumstances until it is ascertained that the roots are strong enough to resist being lifted.
>
> The droppings from the sheep are always more or less an inconvenience to those playing on a golf green, and in the dry weather sheep scalds are very frequent.
>
> To sum the matter up briefly, if proprietors do not mind the expense of cutting, rolling and manuring, we see no reason why sheep should be introduced. We have explained the only good they can possibly do under the best circumstances; while, if not cake fed, they may impoverish the land, and in any case are of more or less a nuisance on grass that is being played upon, particularly on a putting green. On a garden lawn or similarly confined space their presence would be impossible.

n.d.†   **The Practical Greenkeeper.** Third revised edition. Reginald Beale. James Carter & Company, London, England, UK.
circa   103 pp., 59 illus., 8 tables, small index, fold-stitched textured semisoft covers in tan with embossed gold lettering, 7.2
1910   × 9.5 inches (184 × 243 mm).

*   A **very rare**, revised, expanded manual on the establishment and culture of lawns, greens, cricket fields, croquet lawns, soccer fields, hockey fields, and tennis courts in England. It includes a preface by Horace G. Hutchinson and a section on manures and composts. It also includes a section by greenkeeper Peter W. Lees concerning worm control on greens. It contains listings of costs for products. The manual is well illustrated with representative photographs of the time.

Figure 12.8. (a) Manually spreading mowrah meal. (b) Drenching mowrah meal irritant into soil. (c) Earthworms moving to turfgrass surface. (d) Earthworms raked into pile. (e) Earthworms shoveled into wheelbarrow for removal. [11]

n.d.   **The Practical Greenkeeper.** Fourth revised edition. Reginald Beale. James Carter & Co., London, England, UK. 104
circa   pp., 116 illus., 13 tables, small index, fold-stitched textured semisoft covers in red with embossed gold lettering, 7.2 by
1911   9.5 inches (184 × 243 mm).
  *   A **very rare**, small book that has some revisions, especially the illustrations and tables. In the early 1900s, the merits of
       a newly developed gravity drop spreader were compared to the traditional manual method of planting grass seeds:

> Seed Sowing Machines.
>   These machines, which were invented by our firm, are specially constructed to sow grass seeds over large areas at rates varying from 6 to 12 bushels per acre. An experienced man can, with the help of a machine, sow from 3 to 4 acres a day, sowing the seed evenly and at the desired rate per acre without waste.
>   The advantage of using a machine over hand sowing is this: the machine saves time, sows the seed evenly and well, does not waste seed, and can be used with success on days when the wind would make it impossible to sow by hand. We are always pleased to rent a machine when large areas are to be sown, charging carriage on the machine only, unless an expert is sent with the machine to supervise the sowing, when his time and expenses will be charged also.

Revised circa 1913.

n.d.† **The Practical Greenkeeper.** Fifth revised edition. Reginald Beale. James Carter & Co., London, England, UK. 103
circa pp., 66 illus., 8 tables, indexed, fold-stitched semisoft covers in blue with embossed gold lettering, 7.2 by 9.5 inches
1913 (184 × 243 mm).

＊ A **very rare**, small book. It contains some revisions of the earlier edition, primarily the photographs. Reginald Beale
described a 1913 method for keeping cattle, horses, and sheep off putting greens in England:

> Horses and cattle can be kept off putting greens if a few shovelfuls of finely sifted coal-fire cinders are scat-
> tered over the greens every now and again after mowing.
>   The cinders will also keep off sheep to a very large extent, but a few will stray on now and again in
> spite of them.
>   The cinders do not hurt the live-stock in any way, they simply make it uncomfortable for them to feed
> where they have been thrown, with the result that they move on to clean herbage and leave the gritting
> greens alone.

n.d. **The Practical Greenkeeper.** Sixth revised edition. Reginald Beale. James Carter & Co., London, England, UK. 103
circa pp. including 1 p. of advertising, 105 illus., 14 tables, small index, stitched semisoft covers in blue with embossed gold
1914 lettering, 7.2 by 9.5 inches (184 × 243 mm).

＊ A **very rare**, small book that includes a section by Horace C. Hutchinson. Addressed are turfgrass culture for lawn bowling greens, golf courses, cricket fields and wickets, croquet lawns, soccer fields, hockey fields, and tennis lawns. There is
a unique section on floral hazards for golf courses.

n.d. **Carters Practical Greenkeeper.** Second revised American edition. Reginald Beale. Carters Tested Seeds Inc., Boston,
circa Massachusetts, USA. 100 pp., 73 illus., 5 tables, no index, semisoft bound in purple with gold lettering, 7.2 by 9.6
1914 inches (183 × 243 mm).

＊ A revision of an interesting early manual on the Carter system of construction, establishment, and culture of turfgrasses
for golf courses and bowling greens. Also addressed was the ongoing debate at that time concerned with design of the
bunker face. It is a **very rare** publication.
Revised in 1916 and 1921.

1916† **Carters Practical Greenkeeper.** Third revised American edition. Reginald Beale. Carters Tested Seeds Inc., Boston,
Massachusetts, USA. 64 pp., 61 illus., 9 tables, small index, fold-stapled semisoft covers in light brown with an embossed
gold crest and black lettering, 7.2 by 9.5 inches (184 × 242 mm).

＊ A **very rare,** revised booklet on site construction, planting, and culture of turfgrasses for golf courses, bowling greens,
tennis courts, and lawns. There are sections on "Modern golf architecture" and "Lilliput Links" by A.W. Tillinghast and
"Worms in putting and other greens" by greenkeeper Peter W. Lees. There is a "Preface on manures and composts" by
Horace G. Hutchinson.
Revised in 1921.

1920 **The Practical Greenkeeper.** Seventh revised edition. Reginald Beale. James Carter & Co., Raynes Park, London, England, UK. 48 pp., 35 illus., 5 tables, no index, exterior string stitching with semisoft covers in dark blue with embossed
gold lettering, 7.3 by 9.6 inches (185 × 244 mm).

＊ A **very rare**, small booklet that includes a section on golf course architecture and the construction of putting greens by
T. Simpson. Also addressed are manures and composts, pest control, and floral hazards.
Revised in 1921.

n.d. **Lawns and Sports Grounds.** Reginald Beale. James Carter & Co., Raynes Park, London, England, UK. 40 pp. includ-
circa ing 4 pp. of advertising, 19 illus., 3 tables, no index, fold-stapled semisoft covers in browns and cream with brown
1920 lettering, 7.2 by 4.8 inches (184 × 121 mm).

＊ A **very rare**, small book concerning the planting and culture of lawns and sports fields, including bowling greens, croquet lawns, cricket fields and wickets, and tennis lawns.

1921 **Carters Practical Greenkeeper.** Fourth revised American edition. Reginald Beale. Carters Tested Seeds Inc., New York
City, New York, USA. 48 pp., 27 illus., 1 table, indexed, semisoft bound in green with black and silver lettering, 7.5 by
10.2 inches (191 × 257 mm).

＊ A **very rare**, interesting, early manual on the Carter system of construction, establishment, and culture of turfgrasses for
golf courses. Emphasis is placed on pest control, fertilization, and composting.

1921 **The Practical Greenkeeper.** Eighth revised edition. Reginald Beale. James Carter & Co., London, England, UK. 48 pp.
including 3 pp. of advertising, 76 illus., 8 tables, small index, fold-stapled semisoft covers in blue with embossed gold
lettering, 7.3 by 9.6 inches (185 × 244 mm).

Figure 12.9. Clock arrangement for experimental testing of turfgrasses at James Carter & Co., circa 1913. [13]

* A **very rare**, small book that includes a section on golf course construction by T. Simpson. It contains minor revisions, primarily illustrations.

1922 **The Practical Greenkeeper.** Tenth revised edition, autumn. Reginald Beale. James Carter & Co., Raynes Park, London, England, UK. 50 pp. including 5 pp. of advertising, 77 illus., 11 tables, small index, fold-stapled semisoft covers in blue with embossed gold lettering, 7.3 by 9.6 inches (185 × 244 mm).

* A **very rare**, small book with minor revisions.

1923 **Lawns and Sports Grounds.** Autumn edition. Reginald Beale. James Carter & Co., Raynes Park, London, England, UK. 40 pp. including 4 pp. of advertising, 20 illus., 3 tables, no index, fold-stapled semisoft covers in multicolor with brown lettering, 7.1 by 4.6 inches (181 × 118 mm).

* A **very rare**, small book with minor revisions, primarily in the price lists.

1924 **The Practical Greenkeeper.** Fourteenth revised edition, autumn. Reginald Beale. James Carter & Co., Raynes Park, London, England, UK. 52 pp. including 7 pp. of advertising, 90 illus., 16 tables, small index, fold-stapled semisoft covers in blue with embossed gold lettering, 7.2 by 9.6 inches (184 × 245 mm).

* A **very rare**, small book with some revisions, primarily in the illustrations and tables.
  Revised in 1925.

1924† **Lawns for Sports: Their Construction and Upkeep.** Reginald Beale. Simpkin, Marshall, Hamilton, Kent, and Company Limited, London, England, U.K. 276 pp., 116 illus. with 5 color plates, no tables, indexed, cloth hardbound in cream with gold lettering and a jacket in multicolor with white and brown lettering, 6.5 by 9.7 inches (165 × 246 mm).

* This is the most comprehensive book on the establishment and culture of sports turfgrasses that had been written up to 1924. It is a **relatively rare** book that is a must for collectors of historical turfgrass books. It addresses turfgrass culture for golf fairways, putting greens, turfed tees, bunkers, tennis courts, croquet courts, bowling greens, cricket, football and hockey fields, polo grounds, and racecourses under conditions in the United Kingdom. The chapter subject titles include the following:

I.	Golf	XI.	Football and Hockey
II.	The Course through the Green	XII.	Polo Fields and Race Courses
III.	Putting Greens	XIII.	Sowing
IV.	Tees	XIV.	Turfing
V.	Bunkers	XV.	Weeds
VI.	The Upkeep of Putting Greens	XVI.	Lime and the Soil
VII.	Upkeep of the Course through the Green	XVII.	Fertilisers, Top-Dressings and Composts
VIII.	Lawn Tennis and Croquet Courts	XVIII.	Sand-Charcoal-Breeze
IX.	Bowling Greens	XIX.	Quantities of Fertiliser and Other Materials for Proper Upkeep of Sports Grounds
X.	Cricket Fields		

XX.	Worms		XXVI.	Rolling
XXI.	Pests		XXVII.	Rollers
XXII.	Drainage		XXVIII.	Hard Courts
XXIII.	Water		XXIX.	Cost of Construction of Sports Grounds
XXIV.	Mowing		XXX.	A Few Facts in Brief
XXV.	Mowing Machines, Description and Prices		XXXI.	Miscellaneous

The book is well illustrated, including 22 plates of weeds and extensive illustrations of turfgrass equipment and hand implements available in 1924. Specifications, photographs, and prices for the equipment also are presented. It is a pioneering book that includes a photograph of Reginald Beale. Regarding the major problem of earthworms in England, Reginald Beale stated the following:

The Action of Worms

Worms are the worst pest that attacks turf; they riddle the soil and turf to such an extent that it becomes unnaturally muddy and soft, which cannot be corrected by rolling, and otherwise strong turf is made so tender that it soon wears out.

They cover the surface with slimy casts, which not only foul the turf and balls, but when trodden or rolled down actually smother out the fine grasses, and the damage they do to mowing machines is beyond belief.

They work hardest in low, damp areas, which sink into little pot-holes and so destroy the accuracy of the surface. Although there is no apparent connection with worms and weeds, they are closely allied inasmuch as their slimy casts make wonderful seed-beds for weeds, and in consequence they are always found together, and more or less in ratio.

How to Destroy the Worms

Leave the turf unrolled for several days, and so allow the worms to open up their runs.

Select a dull, misty, muggy, warm day with the wind in the south or west when the ground is moist, and the worms are working actively. Broadcast Carters Powdered Wormkiller over the surface at the rate of half a pound per square yard.

Water the Wormkiller in immediately with a hose, water cart or can, and use as much water as possible. The effect is instantaneous, and the worms, large and small, struggle to the surface in thousands to die.

The Wormkiller is absolutely infallible, provided that the worms are working close to the surface, and plenty of water is used.

The best time of the year to kill them is during the breeding seasons, when they work quite close to the surface, which are roughly from the end of August to the beginning of December, and from the end of March to the end of May.

It should not be forgotten that the invention and introduction of Carters Wormkiller in 1902 made Winter Golf possible on inland courses, particularly on those standing on heavy soil; and that its use has increased the standard of play of other games played on turf.

See also *Supplement Issued With Lawns for Sports—Materials, Tools, and Fittings. Beale, Reginald Evelyn Child—1924.*

1924† **Supplement Issued With Lawns for Sports—Materials, Tools and Fittings.** Reginald Beale. James Carter & Co., Raynes Park, England, UK. 67 pp., extensively illustrated, no tables, no index, fold-stapled semisoft covers in light brown with brown lettering, 6.5 by 9.7 inches (165 × 246 mm).

* A **very rare**, interesting catalogue of descriptions, illustrations, and prices for the wide range of seeds, materials, tools, and equipment used in sports turfgrass maintenance operations in England in the 1920s. The categories presented include the following:

· Grass Seed for Lawns	· Croquet Fittings
· Fertilizers	· Cricket Fittings
· Wormkillers	· Association Football Fittings
· Top Dressings	· Hockey Field Fittings
· Weeding Tools and Weed Killers	· Hand Tools
· Golf Course Requisites	· Turfing Tools
· Tennis Lawn and Hard Court Fittings	· Drainage Tools

• Watering Appliances	• Horse Boots
• Sieves and Screens	• Harrows
• Brooms	• Plants
• Earth Augers	

The supplement contains instructions for ordering horse boots:

> Rubber Soles. Fig. 1 or Fig. 2. A quality Boot made with Solid Rubber Outer Soles instead of Leather, 5s. per set extra, for sizes up to No. 1. Sizes: Nos. 2, 2a, 3, 3a, 7s 6d. per set; Nos. 4 and 5, 10s extra.
> *Important.*—In ordering Boots, pencil outlines on strong paper, on one side only, showing shape and exact length of a fore and hind foot, should be sent. These patterns should be taken by placing a piece of cardboard or stiff paper under foot and drawing a pencil line close round outside the hoof or iron shoe, carefully marking each where the heels end, as per woodcut; also particulars of any peculiarity about the feet (such as "Drop Sole," etc.), so that this may be arranged for. If the shoes have calks, their depth over all, including thickness of shoe, should be given (this is very important). Special Boots are supplied for very hilly ground. The pattern required should be mentioned.

1924   **Lawns and Sports Grounds.** Spring edition. Reginald Beale, James Carter & Co., Raynes Park, London, UK. 40 pp., 19 illus., 2 tables, no index, fold-stapled semisoft covers in brown plus a multicolor plate on the front with dark brown lettering, 4.9 by 7.2 inches (124 × 183 mm).

*   A **very rare**, small book with some revisions.

1925   **The Practical Greenkeeper.** Sixteenth revised edition. Reginald Beale. James Carter & Co, Raynes Park, London, England, UK. 52 pp. including 7 pp. of advertising, 69 illus., 5 tables, no index, fold-stitched semisoft covers in blue with embossed gold lettering, 7.2 by 9.6 inches (183 × 244 mm).

*   Contains minor revisions of this **very rare** booklet.
     Revised in 1926.

1926   **The Practical Greenkeeper.** Seventeenth revised edition. Reginald Beale. James Carter & Co, Raynes Park, London, England, UK. 52 pp. including 7 pp. of advertising, 96 illus., 16 tables, small index, fold-stapled semisoft covers in blue with embossed gold lettering, 7.3 by 9.6 inches (185 × 244 mm).

*   A **very rare**, small book with minor revisions, primarily the illustrations.
     Revised in 1927.

1927†   **The Practical Greenkeeper.** Eighteenth revised edition. Reginald Beale. James Carter & Co, Raynes Park, London, England, UK. 52 pp. including 7 pp. of advertising, 64 illus., 17 tables, small index, fold-stapled semisoft covers in blue with embossed gold lettering, 7.2 by 9.6 inches (183 × 244 mm).

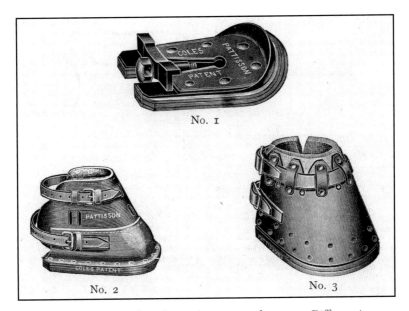

Figure 12.10. Horse boots for reducing damage to turfgrass areas. Different sizes are needed for each horse, as well as for the fore and hind feet. [16]

Figure 12.11. Turfgrass experimental plots at James Carter & Co., 1913. [13]

* Contains minor revisions of a **very rare**, small book. The following procedure was recommended for mole control in 1927:

> Take a long darning needle, nip off half the eye and fix the point into a cork or other suitable handle. Take some wool (Scotch fingering) and cut it into 2-in. lengths. Soak these pieces in a strong solution of strychnine, about double the strength of the British Pharmacopoeia Liquid Strychnine; the solution will then be as strong as it can be made. Allow the pieces of wool to dry.
> Procure a quantity of good-sized worms. Catch the middle of a length of wool in the truncated eye of the needle and thrust it into the worm near the head; insert all the wool and then withdraw the needle.
> Take a pointed stick, make one or two holes in each mole run and drop a treated worm in each. Level the mole-hills so that you can see at a glance if any moles have been missed.

See also *The Practical Greenkeeper, New Edition—1927N to 1939N.*

1927N **The Practical Greenkeeper.** Revised new edition. Reginald Beale. Carters Tested Seeds, Raynes Park, London, England, UK. 56 pp. including 8 pp. of advertising, 73 illus., 12 tables, no index, fold-stapled textured semisoft covers in tans with embossed gold and black lettering, 7.3 by 9.6 inches (185 × 243 mm).

* A **very rare**, small manual on the construction, establishment, and culture of turfgrasses on putting greens and fairways in the United Kingdom. It is organized in six parts with the last two addressing turfgrass pest controls. There are five pages of turfgrass equipment photographs and prices in the back. Reginald Beale had the following comments on bentgrass (*Agrostis* spp.) greens:

> The diseases which attack the bents are known as large and small brown patch, they are caused by moulds or fungi for which no remedy is known.
> Whether the Acidity Theory will prove to be a success or not in this Country time alone can tell but it is no use adopting it unless one has command of an adequate supply of water, compost and labour, because without these, satisfactory results cannot be hoped for.

1927 **Lawns & Weeds.** Reginald Beale. Carters Tested Seeds. Raynes Park, London, England, UK. 32 pp. including 9 pp. of advertising, 17 illus., 21 tables, no index, fold-stapled semisoft covers in cream with a circular multicolor insert and brown lettering, 6.5 by 8,.0 inches (165 × 202 mm).

* A **very rare** booklet concerning the planting, fertilization, weed control, and worming of lawns and sports fields in the United Kingdom.

1928N **The Practical Greenkeeper.** Revised new edition. Reginald Beale. Carters Tested Seeds, Raynes Park, London, England, UK. 56 pp. including 8 pp. of advertising, 74 illus., 13 tables, no index, fold-stapled textured semisoft covers in tans with embossed gold and black lettering, 7.2 by 9.4 inches (183 × 239 mm).

    ✳   A **very rare**, small book with minor revisions.

1928   **Lawns & Weeds.** Reginald Beale. Carters Tested Seeds, Raynes Park, London, England, UK. 32 pp., 17 illus., 21 tables, no index, fold-stapled textured semisoft covers in brown with a circular multicolor insert and dark brown lettering, 6.5 by 8.0 inches (165 × 202 mm).

    ✳   A **very rare** booklet with minor revisions.

1929N   **The Practical Greenkeeper.** Reginald Beale. Carters Tested Seeds, Raynes Park, London, England, UK. 56 pp. including 9 pp. of advertising; 74 illus.; 15 tables; no index; fold-stapled textured semisoft covers in greens, blues, and black with green and black lettering; 7.3 by 9.6 inches (185 × 244 mm).

    ✳   A **very rare**, small book with minor revisions.

1930N   **The Practical Greenkeeper.** Reginald Beale. Carters Tested Seeds Inc., Raynes Park, London, England, UK. 56 pp. including 8 pp. of advertising, 74 illus., 15 tables, no index, fold-stapled textured semisoft covers in greens and black with green and black lettering, 7.3 by 9.6 inches (185 × 244 mm).

    ✳   A **very rare**, small book with minor revisions.

1931   **The Book Of The Lawn—A Complete Guide to the Making and Maintenance of Lawns and Greens for all purposes.** First edition. Reginald Beale. Cassell and Company Limited, London, England, UK. 151 pp. plus 7 pp. of advertising, 27 illus., no tables, indexed, cloth hardbound in green with black lettering and a jacket in multicolor with magenta lettering, 4.8 by 7.2 inches (122 × 184 mm).

    ✳   The basics of turfgrass establishment and culture are covered in this book. It is practically oriented to the amateur home lawn enthusiasts of England but also contains sections on tennis courts, croquet courts, bowling greens, cricket fields, soccer fields, rugby fields, hockey fields, and clock golf. In 1931, Reginald Beale observed the effectiveness in mowing seedling turfgrasses:

> The modern mowing machine is now so perfect that it can be used on young grass with absolute safety as soon as it grows about an inch high, having previously rolled it.

    Revised in 1952.

1931   **Lawns & Weeds.** Reginald Beale. Carters Tested Seeds Limited, Raynes Park, London, England, UK. 32 pp., 17 illus., 3 tables, no index, fold-stapled glossy semisoft covers in multicolor, 6.4 by 8.0 inches (163 × 202 mm).

    ✳   A **very rare**, small booklet on the construction and establishment of turfgrasses for lawns and greens, with emphasis on weed control. It is practically oriented to lawn enthusiasts and cool-season turfgrass conditions in the United Kingdom.

1931N†   **The Practical Greenkeeper.** Revised new edition. Reginald Beale. Carters Tested Seeds Limited, Raynes Park, London, England, UK. 56 pp. including 8 pp. of advertising, 75 illus., 7 tables, no index, fold-stapled textured semisoft covers in greens with embossed dark green lettering, 7.2 by 9.5 inches (184 × 242 mm).

    ✳   A **very rare**, small book with minor revisions.

1932N   **The Practical Greenkeeper.** Revised new edition. Reginald Beale. Carters Tested Seeds Limited, Raynes Park, London, UK. 56 pp. including 8 pp. of advertising, 45 illus., 4 tables, no index, fold-stapled textured semisoft covers in greens and black with embossed green lettering on a black background, 7.2 by 9.6 inches (183 × 244 mm).

    ✳   A minor revision of the 1931 new edition. It is a **very rare**, small book.

1933N   **The Practical Greenkeeper.** Reginald Beale. Carters Tested Seeds Limited, Raynes Park, London, England, UK. 56 pp.; 78 illus.; 19 tables; no index; fold-stapled textured semisoft covers in cream, green, and blue with embossed blue and light green lettering; 7.3 by 9.5 inches (185 × 242 mm).

    ✳   A **very rare**, small book that contains minor revisions.

1933   **Lawns & Weeds.** Reginald Beale. Carters Tested Seeds Ltd., Raynes Park, London, England, UK. 32 pp. including 12 pp. of advertising, 24 illus., 13 tables, no index, fold-stapled glossy semisoft covers in multicolor and white with black lettering, 6.5 by 8.0 inches (165 × 203 mm).

    ✳   A **very rare** booklet with minor revisions in the advertising and pricing.

1934N   **The Practical Greenkeeper.** Reginald Beale. Carters Tested Seeds Limited, Raynes Park, London, England, UK. 56 pp. including 6 pp. of advertising, 83 illus., 21 tables, no index, fold-stapled textured semisoft covers in greens with embossed dark green lettering, 7.2 by 9.6 inches (183 × 243 mm).

    ✳   A **very rare**, small book containing significant revisions. There are sections on golf course maintenance by Sir Gay Campbell and T. Simpson.

1935N   **The Practical Greenkeeper.** Revised new edition. Reginald Beale. Carters Tested Seeds Limited, Raynes Park, London, England, UK. 56 pp. including 6 pp. of advertising, 80 illus., 10 tables, no index, fold-stapled textured semisoft covers in browns and gold with embossed dark green lettering, 7.3 by 9.6 inches (185 × 244 mm).

* A **very rare**, small book containing minor revisions. There are five pages of turfgrass equipment photographs with prices in the back.

1935 **Lawns & Weeds.** Reginald Beale. Carters Tested Seeds Ltd., Raynes Park, London, England, UK. 32 pp. including 12 pp. of advertising, 61 illus., 13 tables, no index, fold-stapled glossy semisoft covers in multicolor and white with black lettering, 6.5 by 8.0 inches (165 × 202 mm).

* A **very rare** booklet with minor changes in advertising and pricing.

1936N† **The Practical Greenkeeper.** Revised new edition. Reginald Beale. Carters Tested Seeds Limited, Raynes Park, London, England, UK. 56 pp. including 2 pp. of advertising, 80 illus., 1 table, no index, fold-stapled textured semisoft covers in greens with embossed light green lettering, 7.4 by 9.6 inches (188 × 244 mm).

* A **very rare** book with major revisions. It addresses the planning, construction, establishment, and culture of golf courses in the United Kingdom. There is a major section on weeds with 22 photographs and a major section on pests, including worms, moles, horses, cattle, rabbits, and sheep. There are two pages of turfgrass equipment photographs with prices in the back.

1937N **The Practical Greenkeeper.** Revised new edition. Reginald Beale. Carters Tested Seeds Limited, Raynes Park, London, England, UK. 56 pp. including 2 pp. of advertising, 78 illus., 1 table, no index, fold-stapled semisoft covers in dark green with embossed gold trim and gold lettering, 7.6 by 9.7 inches (193 × 246 mm).

* A minor revision of the 1936 new edition. It is a **very rare**, small book.

1938N **The Practical Greenkeeper.** Revised new edition. Reginald Beale. Carters Tested Seeds Limited, Raynes Park, London, England, UK. 56 pp. including 2 pp. of advertising, 88 illus., 22 tables, no index, fold-stapled semisoft covers in green with impressed gold lettering, 7.2 by 9.6 inches (183 × 243 mm).

* A **very rare**, small book with minor revisions.

1939N† **The Practical Greenkeeper.** Revised new edition. Reginald Beale. Carters Tested Seeds Limited, Raynes Park, London, England, UK. 56 pp. including 2 pp. of advertising, 75 illus., 3 tables, no index, fold-stapled semisoft covers in green with impressed gold lettering, 7.2 by 9.5 inches (184 × 242 mm).

* A **very rare**, small book that contains major revisions, especially in the planning and construction of golf courses in the United Kingdom. Major sections are devoted to pest control in turfgrasses. Reginald Beale had the following observations on materials used for topdressing turfgrasses in England in the 1930s:

> Pure sand is composed of minute fragments of stone; charcoal is calcined wood or bone; and breeze, cinder, clinker and coke dust partly burnt coal. None of these materials is of any real manurial value and they are primarily used to open up stiff soils, improve the surface drainage and make the putting surface firm and true.

> *Revised circa 1949.*
> See also *Carters Tested Seeds Limited—1949 to 1969.*

1952† **The Book Of The Lawn—A Complete Guide to the Making and Maintenance of Lawns and Greens for all purposes.** Second revised edition. Reginald Beale. Cassell & Company Limited, London, England, UK. 151 pp., 33 illus., 1 table, indexed, cloth hardbound in green with gold lettering and a jacket in greens and white with red and black lettering, 4.8 by 7.2 inches (123 × 184 mm).

There are only small revisions of this second edition.

Beard, Harriet Jean

Harriet Beard was born on a farm near Pleasant Hill, Ohio, USA, and married Jim Beard in 1955. She worked side-by-side with her husband in the development of his publications, as well as in raising two sons. She received the Oberly Award of the American Library Association in 1979.

1977† **Turfgrass Bibliography—From 1672 to 1972.** Coauthor.
See *Beard, James B—1977.*

2005† **Beard's Turfgrass Encyclopedia—for Golf Courses, Grounds, Lawns, Sports Fields.** Coauthor.
See *Beard, James B—2003.*

2005† **History Records of the Division C-5—Turfgrass Science 1946–2004. Volumes I through IV.** Coeditor.
See *Beard, James B—2005.*

Beard, James B (Dr., DA, VMM, and *Professor* Emeritus *of Turfgrass Science*)

Jim Beard was born on a farm near Bradford, Ohio, USA, and earned a BSc degree, graduating *summa cum laude* and majoring in Agronomy and Crops from Ohio State University in 1957; he also earned an MSc degree in Crop Ecology and Statistics in 1959 and a PhD degree in Plant Physiology and Biochemistry in 1961 as a National Science Foundation

Figure 12.12. Shanks 1914 Triumph horse-drawn, riding, side-wheel-drive, 36-inch-wide, reel mower. [38]

Graduate Fellow, both from Purdue University, studying with Dr. Bill Daniel. He was awarded a National Science Foundation Postdoctoral Fellowship to study plant biochemistry at the Life Sciences Department, University of California, Riverside, from 1969 to 1970. Dr. Beard was an honored recipient of a Doctorate of Agriculture *honoris causa* from Purdue University in 2004. He was active in turfgrass research and education while on the faculty at Michigan State University from 1961 to 1975 and at Texas A&M University from 1975 to 1992. He served as Interim Head of the Soil and Crop Science Department in 1980. Dr. Beard focused on environmental stress physiology and prevention, root ecology, plant water relations, and sod production. He initiated the 2-Year Turfgrass Technical Education Program at Michigan State University, with Dr. John W. King as the Instructor. Dr. Beard also developed major undergraduate and graduate turfgrass education programs at both Michigan State University and Texas A&M University. He founded the International Sports Turf Institute Inc. in 1992 through which he pioneered turfgrass education worldwide, lecturing in 30-plus countries, with most on two to four occasions. Dr. Beard was elected to the Crop Science Society of America (CSSA) Board of Directors from 1964 to 1966 and from 1985 to 1987, to the Board of Directors of the American Society of Agronomy from 1969 to 1971 and from 1984 to 1987, and to the Council for Agricultural Science and Technology from 1989 to 1991; he was also elected as Turfgrass Science Division C-5 Chair in 1965 and CSSA President in 1986. He was elected and served on the Board of Directors of the Scotts Company from 1987 to 1999. Dr. Beard was founder and first President of the International Turfgrass Society (ITS) in 1969 and served on the ITS Board of Directors from 1969 to 2007. He received the Veitch Memorial Medal of the Royal Horticultural Society, London, in 2008; the Donald Rossi Award of the Golf Course Builders Association in 1999; the Founders Award of the Sports Turf Managers Association in 1998; the Distinguished Alumni Award of Ohio State University in 1995; the Presidents Outstanding Service Award of the Nebraska Turfgrass Association in 1991; an Honorary Membership to the OJ Noer Research Foundation in 1995; the first Wilkie Lectureship at Michigan State University in 1994; the Presidential Award of the Turfgrass Council of North Carolina in 1994; the Crop Science Research Award of the CSSA in 1991; a Fellow of the American Association for the Advancement of Science in 1990; the Green Section Award of the United States Golf Association in 1989; the Distinguished Service Award of the Massachusetts Turfgrass Association in 1989; the CSSA Turfgrass Science Award in 1988; the Faculty Distinguished Achievement Award in Research from Texas A&M University in 1987; the Oberly Award of the American Library Association in 1979; the Meritorious Service Award of the Michigan Turfgrass Foundation in 1978; the Service Recognition and Honorary Member Award of the American Sod Producers Association in 1975; the Distinguished Service Award of the International Turfgrass Society in 1973; a Fellow of the American Society of Agronomy in 1971; and a Fellow of the Crop Science Society of America (CSSA) in 1971. He has been presented various honorary memberships from numerous states and countries. The James and Harriet Beard Endowed Turfgrass Stress Physiology Fellowship was established in 2002 at Ohio State University. Then the James B Beard Turfgrass Collection and Room were dedicated at the Michigan State University Library in 2003.

1973† **Turfgrass: Science and Culture.** James B Beard. Prentice-Hall Inc., Englewood Cliffs, New Jersey, USA. 658 pp., 206 illus., 75 tables, detailed index, textured cloth hardbound in dark green with gold lettering and a glossy semisoft jacket in greens with white and black lettering plus a clear overjacket, 7.0 by 9.2 inches (178 × 233 mm).

A comprehensive, authoritative basic treatise on turfgrass science. This scholarly text led the transition of turfgrass culture from an art-oriented craft to a science-based profession in applied plant science. It is widely used throughout the world, including universities. The construction, establishment, and culture of turfgrass areas are addressed. A unique feature of the book involves four chapters on turfgrass microclimate, environmental stress, and cultural modifications of the turfgrass microenvironment. It is the first text fully referenced and based on a comprehensive review of the scientific literature. This text remains the original, scholarly source of information on turfgrass science worldwide. The book is a must for collectors of historical turfgrass books.

† *Over 25 reprintings.*

1975† **How to have a Beautiful Lawn.** First edition. James B Beard. Intertec Publishing Corporation, Kansas City, Missouri, USA. 113 pp., 157 illus. with 68 in color, 32 tables, indexed, glossy semisoft bound in multicolor with black and white lettering, 8.1 by 11.1 inches (205 × 282 mm).

A small book on the basics of lawn establishment and culture presented in a how-to manner. It is well illustrated and contains the most extensive characterizations of turfgrass cultivars and pesticides registered for use in the United States. This practical book has been used in statewide County Extension Offices and in Master Gardener education programs.

† *Revised in 1979.*

n.d.† **The Mathematics of Turfgrass Maintenance.** James B Beard, Paul E. Rieke, W.E. Knoop, and Tom J. Rogers. Golf circa Course Superintendents Association of America, Lawrence, Kansas, USA. 58 pp., 8 illus., 2 tables, no index, fold-1977 stapled textured semisoft covers in yellow with black lettering, 8.4 by 11.0 inches (214 × 279 mm).

The first instructional booklet on mathematical calculations typically encountered in golf course operations. Included are areas; rates of pesticide, fertilizer, and seeding; volumes; and metric conversions. Sets of problem exercises are included. It was prepared in coordination with a one-day seminar presentation.

Revised in 1992.

See *Agnew, Michael Lewis—1992.*

1977† **Turfgrass Bibliography—From 1672 to 1972.** James B Beard, Harriet J. Beard, and David P. Martin. Michigan State University Press, East Lansing, Michigan, USA. 730 pp., no illus., no tables, contains both author and subject indexes, cloth hardbound in charcoal gray with gold lettering, 6.2 by 9.2 inches (157 × 235 mm).

This extensive book is a must reference source for collectors of turfgrass books. It represents a compilation of turf-grass scientific, technical, and popular literature published over the previous 300 years. The Bibliography contains over 16,000 references listed alphabetically by author and is cross indexed on a subject basis containing more than 40,000 entries. It is a unique, one-of-kind book. In the process of assembling the publications required for references in the Bibliography, the Noer Memorial Turfgrass Library was initiated by Dr. Beard at Michigan State University, with the support of Director of Libraries Dr. Richard Chapin. This Bibliography won the Oberly Award of the American Library Association in 1979, a seldom awarded honor.

1979† **How to have a Beautiful Lawn.** Second revised edition. James B Beard. Beard Books, College Station, Texas, USA. 113 pp., 169 illus. with 71 in color, 40 tables, indexed, glossy semisoft bound in multicolor with white and black lettering, 8.1 by 11.1 inches (205 × 282 mm).

This edition contains substantial revisions, including an update of the turfgrass cultivar, pesticide, and pest control descriptions registered for use in the United States.

† *Revised in 1983.*

1979† **Introduction To Turfgrass Science And Culture Laboratory Exercises.** James B Beard, Joseph M. DiPaola, Don Johns Jr., Keith J. Karnok, and Steven M. Batten. Burgess Publishing Company, Minneapolis, Minnesota, USA. 217 pp., 112 illus., 50 tables, no index, glossy semisoft bound in multicolor with white lettering, 8.3 by 11.0 inches (210 × 279 mm).

The first laboratory instructional manual published for turfgrass students. It consists of 14 exercises. At the end of each exercise, there are a set of activities for the student to complete, a set of references, and guidelines for the instructor.

1982† **Turf Management for Golf Courses.** First edition. James B Beard. Burgess Publishing Company, Minneapolis, Minnesota, USA. 641 pp., 513 illus. with 29 in color, 55 tables, detailed index, glossy hardbound in multicolor with black lettering, 7.3 by 9.2 inches (185 × 233 mm).

The major, comprehensive text on both the basic concepts and how-to dimensions of turfgrass culture and management for golf courses. There are specific chapters addressing the construction, establishment, and culture of turfgrasses on greens, turfed tees, fairways, roughs, and bunkers, including summary cultural systems by grass species. Also, it contains chapters on architecture, construction, equipment, irrigation systems, and personnel management, records, and budgets. A chapter on pest management contains excellent drawings of weeds and color photographs of diseases and insects. An extensive glossary and appendix are in the back. The book was sponsored by the United States Golf

Figure 12.13. Observing root growth on an *in-vivo* glass-faced rhizotron at Purdue University, 1957. [26]

Association, as were the earlier books by C.V. Piper and R.A. Oakley (1917) and by H.B. Musser (1950). Members of the Editorial Board were A.M. Radko, W.H. Bengeyfield, W.G. Buchanan, J.B. Moncrief, C.H. Schwartzkopf, and S.J. Zontek, while the Review Board members were W.J. Carson, L.A. Eggleston, P. Maples Jr., S.A. Sherwood, A.A. Snyder, R.M. Williams, and J.A. Zoller. The text is illustrated by Steven M. Batten. This book is a must for collectors of historical turfgrass and golf books.

† *Numerous reprintings.*
  Revised in 2001.

† *A Chinese translation also was published in 1999* consisting of 444 pages, glossy semisoft bound with multicolor covers.

1983† **Better Turfgrass Nutrition.** James B Beard. Beard Books and Estech Inc., Winter Haven, Florida, USA. 22 pp., 23 illus. in color, 17 tables, no index, glossy semisoft bound in multicolor with white lettering, 9.0 by 12.0 inches (228 × 305 mm).

A concise booklet that profiles the major turfgrass species in North America, with emphasis on their nutritional requirements. It contains nine unique, large color illustrations of these turfgrasses by Steven M. Batten.

1983† **How to have a Beautiful Lawn.** Third revised edition. James B Beard. Beard Books, College Station, Texas, USA. 113 pp., 169 illus. with 71 in color, 40 tables, index, glossy semisoft bound in multicolor with white and black lettering, 8.1 by 11.1 inches (205 × 282 mm).

This edition contains a revised update of turfgrass cultivar, pesticide, and pest control descriptions of materials registered for use in the United States.
  Revised in 1988 and 1993.

1984† **Turfgrass Plant Physiology and Botany.** James B Beard. International Turfgrass Manual Number 1, Golf Course Superintendents Association of America, Lawrence, Kansas, USA. 20 pp., 2 illus., 3 tables, no index, soft covers in white with black lettering in a 3-hole notebook, 8.5 by 10.9 inches (216 × 303 mm).

A seminar booklet in outline format. The fundamentals of turfgrass growth, development, and physiology are addressed. There is a list of selected references in the back. It was developed in coordination with a two-day seminar presentation.

† *Reprinted in 1984 and 1985.*
† *Revised in 1986.*

1984† **Golf Course Management.** James B Beard. International Turfgrass Manual Number 2, Beard Books and Club Management Institute Workshop, Club Managers Association of America, Alexandria, Virginia, USA. 71 pp., 22 illus., 10 tables, no index, semisoft bound in white with black lettering in a 3-hole notebook, 8.5 by 11.0 inches (216 × 279 mm).

An introductory manual on turfgrass growth, soil root zones, feature construction, turfgrass establishment, and cultural systems for golf course putting greens, turfed tees, fairways, and roughs, plus bunkers. There is a list of selected references in the back. It was prepared to coordinate with a one-day seminar presentation at various locations in the United States.

1986† **Basic Turfgrass Botany and Physiology.** First edition. James B Beard and Jeffrey V. Krans. Golf Course Superintendents Association of America, Lawrence, Kansas, USA. 44 pp., 67 illus., 3 tables, no index, soft bound in white with black lettering, 8.9 by 11.0 inches (226 × 279 mm).

A lecture notebook on the fundamentals of cell biology, physiology, development, and growth of perennial turfgrasses. It encompasses the shoots, roots, seeds, and inflorescences. The book codes are referenced to **Turfgrass: Science and Culture**. It was prepared to coordinate with a two-day seminar presentation at various locations in the United States.

† *Annual reprintings through 1992.*
Revised in 1993.

1988† **Manuale per la Conduzione dei Tappeti Erbosi dei Campi di Golf Italiani (Manual on Italian Golf Course Turf Management)** James B Beard. Beard Books and Federazione Italiana Golf, Roma, Italy. 241 pp., 19 illus., 21 tables, no index, semisoft bound in light gray with black lettering, 6.5 by 9.5 inches (165 × 240 mm). Printed entirely in the Italian language.

A manual concerning the construction, turfgrass establishment, and culture of golf course greens, turfed tees, fairways, and roughs. It is oriented to Italian conditions. The manual was originally developed to coordinate with a three-day seminar presentation in Milan, Italy. It was the first organized turfgrass educational effort in Italy and part of a five-year series.

1988† **How to have a Beautiful Lawn.** Fourth revised edition. James B Beard. Beard Books, College Station, Texas, USA. 117 pp., 169 illus. with 71 in color, 40 tables, indexed, glossy semisoft bound in multicolor with white and black lettering, 8.1 by 11.1 inches (205 × 282 mm).

This edition contains a major revised update of the turfgrass cultivar descriptions plus updates of pest descriptions and pesticide materials registered for use in the United States.
Revised in 1993.

1988† **Turfgrass: Identification, Description, Culture.** James B Beard. International Turfgrass Manual Number 8, Beard Books, College Station, Texas, USA. 49 pp., 22 illus., 14 tables, no index, semisoft bound in white with black lettering, 8.5 by 11.0 inches (216 × 279 mm).

A manual on the identification of perennial turfgrasses by vegetative characteristics and by seeds. Also outlined are the cultural strategies for six cool-season turfgrasses and three warm-season turfgrasses. It was prepared to coordinate with a one-day seminar presentation at various locations in the United States.

1989† **Malattie Parassitarie, Attacchi degli Insetti e le Infestanti dei Tappeti Erbosi (Diseases, Insects and Weeds of Turfgrasses).** James B Beard, Beard Books and Federazione Italiana Golf, Roma, Italy. 20 pp., no illus., 6 tables, no index, soft covers in white with black letters in a comb binder, 8.3 by 11.7 inches (210 × 297 mm). Printed entirely in the Italian language.

A guide to the identification and control of turfgrass pests common to golf courses in Italy. This manual was originally developed to coordinate with a two-day seminar presentation in Rome, Italy.

1990† **Guida per la Costruzione e la Manutenzione dei Greens nei Campi di Golf Italiani (Italian Guide to Putting Green Construction and Maintenance).** James B Beard with translation to Italian by Paolo Croce and Francesco Saverio Modestini. Beard Books and Federazione Italiana Golf, Roma, Italy. 90 pp., 5 illus., 10 tables, no index, clear semisoft covers over white pages with black lettering in a comb binder, 8.3 by 11.7 inches (210 × 297 mm). Printed entirely in the Italian language.

A small practical guide to the construction and culture of creeping bentgrass (*Agrostis stolonifera*) putting greens in Italy. The appendix is devoted to the major pests and their control. The guide was originally developed to coordinate with a three-day seminar presentation in Milan, Italy.

1990† **A Manual For Construction and Maintenance of Bentgrass Putting Greens.** James B Beard. Beard Books and Nichino Ryokka Company Limited, Tokyo, Japan. 44 pp., 3 illus., 10 tables, no index, textured semisoft bound in cream with black lettering, 8.2 by 11.7 inches (208 × 298 mm). Printed entirely in the Japanese language.

A small manual on the root-zone construction, establishment, and culture of creeping bentgrass (*Agrostis stolonifera*) putting greens under Japanese climatic conditions. The manual was originally developed to coordinate with a two-day seminar presentation in Tokyo, Japan.

1990† **Svenska Golfförbundets Skötselmanual för Golfgräs (A Manual on Swedish Golf Course Turf).** James B Beard. Beard Books, Svenska Golfförbundets, and Swedish Greenkeepers Association, Farsta, Sweden. 116 pp., 18 illus., 22 tables, no index, soft covers in light blue with dark blue lettering in a 2-hole binder, 8.4 by 11.7 inches (213 × 297 mm). Printed entirely in the Swedish language.

A manual on grass growth structures, cool-season turfgrasses, soil modification, site development, turfgrass establishment, and cultural practices for putting greens, turfed tees, fairways, roughs, and bunkers of golf courses in Sweden. The manual was originally developed to coordinate with a two-day seminar presentation in Norrköping, Sweden.

1991† **A Manual on Malaysian Golf Course Turf Construction, Establishment and Maintenance.** James B Beard. Beard Books and Malaysian Golf Association, Kuala Lumpur, Malaysia. 125 pp., 17 illus., 25 tables, no index, semisoft covers in white with green lettering in a comb binder, 8.2 by 11.7 inches (208 × 298 mm).

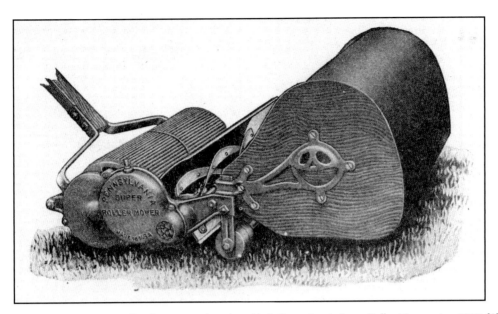

Figure 12.14. Manual-push, roller-drive, 18-inch-wide, 7-blade Pennsylvania Super Roller Mower, circa 1927. [2]

A manual on grass botany, warm-season turfgrass species, soils, root-zone modification, basic cultural practices, and cultural systems for turfgrasses on putting greens, turfed tees, fairways, roughs, and bunkers of golf courses in Malaysia. The manual was originally developed to coordinate with a two-day seminar presentation in Kuala Lumpur, Malaysia.

1991†　**A Manual on Golf Course Construction, Turf Establishment and Cultural Practices.** James B Beard. Beard Books and British & International Golf Greenkeepers Association, Aldwark, York, UK. 83 pp., 19 illus., 22 tables, no index, semisoft covers in white with black lettering in a comb binder, 8.2 by 11.7 inches (208 × 296 mm).

A small manual on grass botany, cool-season turfgrass species, soils, root-zone modification, establishment, and basic cultural practices for golf course turfgrasses in the United Kingdom. The manual was originally developed to coordinate with a two-day seminar presentation in Harrogate, England, UK.

1991†　**Guida per la Costruzione e la Manutenzione di Fairway e Tee nei Campi di Golf Italiani (Italian Guide to Fairway and Tee Construction and Maintenance).** James B Beard. Beard Books and Federazione Italiana Golf, Roma, Italy. 107 pp., 14 illus., 20 tables, no index, soft bound in white with black lettering, 8.5 by 11.7 inches (216 × 297 mm). Printed entirely in the Italian language.

A guide to the turfgrass species and cultivars, and their establishment and culture for fairways and turfed tees in Italy. The guide was originally developed to coordinate with a three-day seminar presentation in Milan, Italy.

1991†　**Turfgrasses and Soils for the Southeastern United States.** James B Beard. International Turfgrass Manual Number 14, Beard Books, College Station, Texas, USA. 71 pp., 30 illus., 19 tables, no index, semisoft bound in white with black lettering, 8.3 by 10.8 inches (211 × 274 mm).

A basic manual on the morphology, taxonomy, adaptation, and identification of perennial turfgrasses plus soil characteristics, root-zone modification, and fertilizers, as applied to the southeastern United States. It was originally prepared to coordinate with a two-day seminar presentation at various locations in the southeastern and south central United States.

1991†　**Turfgrass Culture and Pest Management for the Southeastern United States.** James B Beard. International Turfgrass Manual Number 15, Beard Books, College Station, Texas, USA. 96 pp., 1 illus., 22 tables, no index, semisoft bound in white with black lettering, 8.5 by 11.0 inches (216 × 279 mm).

A basic manual on the establishment procedures, cultural practices, and cultural systems of perennial turfgrasses plus pest diagnosis and management, as applied to the southeastern United States. It was originally prepared to coordinate with a two-day seminar presentation at various locations in the southeastern and south central United States.

1992†　**Criteri Generali per la Pianificazione della Costruzione di un Percorso di Golf (Guidelines for Planning Construction of a Golf Course).** James B Beard with translation to Italian by Paolo Croce, Francesco S. Modestini, and Alessandro De Luca. Beard Books and Federazione Italiana Golf, Roma, Italy. 48 pp., 3 illus., no tables, no index, clear semisoft covers over white pages with black lettering in a comb binder, 8.2 by 11.7 inches (209 × 297 mm). Printed entirely in the Italian language.

A manual on golf course site selection, architect selection, permitting, architecture, plans, contracts, specifications, bidding, and contractor selection for Italian conditions. The manual was originally developed to coordinate with a three-day seminar presentation in Milan, Italy.

1992† **Golfbanesseminarium—Handbook för Anläggning och skötsel an Golfbanor i Sverige (A Manual on Swedish Golf Course Turf Construction, Establishment and Maintenance).** James B Beard. Beard Books, Svenska Golf Fobundet, and Swedish Greenkeepers Association, Farsta, Sweden. 118 pp., 19 illus., 24 tables, no index, semisoft covers in blue and yellow with black lettering in a 4-hole binder, 8.3 by 11.7 inches (211 × 297 mm). Printed entirely in the Swedish language.

A manual on grass botany, cool-season turfgrass species, soils, root-zone modification, basic cultural practices, and cultural systems for putting greens, turfed tees, fairways, roughs, and bunkers on golf courses in Sweden. This manual was originally developed to coordinate with a two-day seminar presentation in Stockholm, Sweden.

1992† **Turfgrasses and Soils for the Midwestern United States.** James B Beard. International Turfgrass Manual Number 20, Beard Books, College Station, Texas, USA. 72 pp., 31 illus., 17 tables, no index, soft covers in white with black lettering in a 3-hole notebook, 8.5 by 11.0 inches (216 × 279).

A basic manual on the morphology taxonomy, adaptation, and identification of perennial turfgrasses plus soil characteristics, root-zone modification, and fertilizers, as applied to the Midwestern United States. It was originally prepared to coordinate with a two-day seminar presentation at various locations in the Midwestern and central United States.

1992† **Turfgrass Culture and Pest Management for the Midwestern United States.** James B Beard. International Turfgrass Manual Number 21, Beard Books, College Station, Texas, USA. 95 pp., 1 illus., 21 tables, no index, semisoft covers in white with black lettering in a 3-hole notebook, 8.5 by 11.0 inches (216 × 279 mm).

A basic manual on the establishment procedures, cultural practices, and cultural systems of perennial turfgrasses plus pest diagnosis and management, as applied to the Midwestern United States. It was originally prepared to coordinate with a two-day seminar presentation at various locations in the Midwestern and central United States.

1992† **A Manual on Turfgrasses, Root Zone Construction, Turf Establishment and Cultural Practices for Japan.** James B Beard. International Turfgrass Manual Number 22, Beard Books and Nichino Ryokka Company Limited, Tokyo, Japan. 85 pp., 29 illus., 19 tables, no index, semisoft bound in white with black lettering, 8.5 by 11.0 inches (216 × 279 mm). Printed entirely in the Japanese language.

A basic manual on the morphology, taxonomy, establishment, and culture of perennial turfgrasses plus soil characteristics and root-zone modification. There is a short list of selected references in the back. The manual was originally prepared to coordinate with a two-day seminar presentation in Tokyo, Japan.

1993† **How to have a Beautiful Lawn.** Fifth revised edition. James B Beard. Beard Books, College Station, Texas, USA. 117 pp., 169 illus. with 71 in color, 39 tables, indexed, glossy semisoft bound in multicolor with white and black lettering, 8.1 by 11.1 inches (205 × 282 mm).

This edition contains a major revised update for the pest descriptions and pesticides registered for use in the United States.

1993† **Turfgrass Handbook for Southeast Asian Golf Course and Sport Facilities.** James B Beard. Beard Books, College Station, Texas, USA. 137 pp., 9 illus., 34 tables, no index, soft covers in white with black lettering in a comb binder, 8.5 by 10.9 inches (215 × 277 mm).

A handbook on turfgrass botany, root-zone modification, cultural systems for golf courses and sports fields, pest management, fertilization, irrigation, and turfgrass benefits. The handbook was originally developed to coordinate with a three-day seminar presentation in Kuala Lumpur, Malaysia.

1993† **Basic Turfgrass Botany and Physiology.** Second revised edition. James B Beard and Jeffrey V. Krans. Golf Course Superintendents Association of America, Lawrence, Kansas, USA. 77 pp., 100-plus illus., 7 tables, no index, semisoft bound in grays with black lettering, 8.5 by 11.0 inches (216 × 279 mm).

A lecture notebook on the fundamentals of cell biology, physiology, development, and growth of perennial turfgrasses. It encompasses the shoots, roots, seeds, and inflorescences with numerous illustrations added. There is a list of selected references in the back. The book codes are referenced to **Turfgrass: Science and Culture.** It was prepared to coordinate with a two-day seminar presentation at various locations in the United States.

† *Annual reprintings through 1998.*

1993† **A Manual for Construction and Maintenance of Creeping Bentgrass Putting Greens.** James B Beard. International Turfgrass Manual Number 23, Beard Books and PGA European Tour Greenkeepers Conference, College Station, Texas, USA. 56 pp., 4 illus., 11 tables, no index, semisoft bound in white with black lettering, 8.3 by 10.8 inches (211 × 274 mm).

A manual concerning the construction, establishment, culture, and hole changing of creeping bentgrass (*Agrostis stolonifera*) putting greens under the conditions in Europe. There is a list of selected references in the back. The manual was originally developed to coordinate with a three-day seminar presentation in Malága, Spain.

1995† **Manual on Diagnosing Turfgrass Problems and Corrective Actions.** James B Beard. International Turfgrass Manual Number 26. Beard Books, College Station, Texas, USA. 78 pp., 6 illus., 21 tables, no index, soft covers in white with black lettering in a comb binder, 8.5 by 10.9 inches (215 × 278 mm).

Figure 12.15. Shanks 1904 horse-drawn, 24-inch-wide, reel mower with grass box and mechanical discharge. [38]

A manual on the turfgrass-soil problems of southeast Asia, their diagnoses, and appropriate corrective actions. The manual was originally developed to coordinate with a three-day seminar presentation in Kuala Lumpur, Malaysia.

1997† **Color Atlas of Turfgrass Diseases.** First English edition. Coauthor.
See *Tani, Toshikazu—1997.*

1998† **Turfgrass Trends.** James B Beard. International Turfgrass Manual Number 28, Beard Books and Workshop Book, Club Managers Association of America, Alexandria, Virginia, USA. 55 pp., no illus., 12 tables, no index, semisoft covers in white with black lettering in a 3-hole notebook, 9.3 by 10.6 inches (286 × 269 mm).

A booklet on trends and current research in turfgrass culture for golf courses. Included are new low-growing cultivars for putting greens, rolling putting greens, winter overseeding transition, heat stress, low-temperature freezing, and environmental issues. The booklet was originally prepared to coordinate with a one-day seminar presentation at various locations in the United States.

2002† **Turf Management for Golf Courses.** Second revised edition. James B Beard. Ann Arbor Press, Chelsea, Michigan, USA, and John Wiley & Sons, Hoboken, New Jersey, USA, since 2003. 793 pp., 617 illus. with 522 in color, 82 tables, detailed index, glossy hardbound in multicolor with white lettering, 8.0 by 10.0 inches (203 × 253 mm).

An extensive revision and update of the major book on both the basic concepts and how-to dimensions of turfgrass culture and management for golf courses. It is an authoritative reference that is widely used around the world. Most photographs are in color. It includes selected references. There is a glossary of terms in the back.

2003† **La Pianificazione della Costruzione di un Percorso di Golf (Planning the Construction of a Golf Course).** James B Beard, as revised and translated by Paolo Croce, Alessandro De Luca, and Massimo Mocioni. Federazione Italiana Golf, Roma, Italy. 52 pp., 27 illus. in color, no tables, no index, glossy semisoft bound in multicolor with black lettering, 8.3 by 11.7 inches (210 × 297 mm). Printed entirely in the Italian language.

A guide book on the planning, design, and construction of golf courses in Italy. There is a bibliography in the back.

2005† **Beard's Turfgrass Encyclopedia—for Golf Courses, Grounds, Lawns, Sports Fields.** James B Beard and Harriet J. Beard. Michigan State University Press, East Lansing, Michigan, USA. 511 pp., 327 illus., 102 tables, textured hardbound in dark green with light green lettering, 8.4 by 10.9 inches (214 × 276 mm).

This major work encompasses both the practical and scientific turfgrass terminology in use. Each term is defined, followed by an expanded discussion as appropriate. The text is illustrated by James Converse, with emphasis on the grasses, weeds, and insects. It is the first turfgrass encyclopedia and a must for the reference library of both turfgrass practitioners and scholars.

2005† **History Records of the Division C-5—Turfgrass Science 1946–2004. Volumes I through IV.** James B Beard and Harriet J. Beard (editors). Turfgrass Science Division C-5, Crop Science Society of America, Madison, Wisconsin, USA.

1,157 pp., 2 illus., 88 tables, no index, textured cloth hardbound (each of four volumes) in light green with gold lettering, 8.5 by 11.0 inches (216 × 279 mm).

A compilation of the history records of what is now known as the Turfgrass Science Division C-5 of CSSA, since it was formed in 1946. It is assembled in four separate bound volumes as follows: Volume I, Annual Business Meeting Minutes, 283 pages; Volume II, Annual Scientific Program, 368 pages; Volume III, By-Laws and Publications, 193 pages; and Volume IV, Annual Historian Reports, 313 pages.

2008† **Water Quality and Quantity Issues for Turfgrasses in Urban Landscapes.** James B Beard and Michael P. Kenna (editors) with 25 contributing authors. Council of Agricultural Science and Technology, Ames, Iowa, USA. 318 pp., 46 illus. with 25 in color, 29 tables, no index, glossy hardbound in multicolor with white and green lettering, 5.7 by 8.6 inches (146 × 218 mm).

A science-based presentation that represents the culmination of presentations and in-depth discussions among 48 researchers, environmental specialists, and water industry managers during a three-day workshop. It is a major resource for government officials and policymakers.

Bechelet, Henry Charles

Henry Bechelet was born in Gosport, Hampshire, England, UK, and earned a BSc with *honours* in Geography and Geology from Keele University in 1989 and a Postgraduate Diploma in Crop Protection from Harper Adams College in 1990. He was a Turfgrass Agronomist for the Sports Turf Research Institute (STRI) from 1990 to 1992, a Technical Sales Representative for Vitax Limited from 1992 to 2000, and, since 2000, a Turfgrass Agronomist for STRI.

2007† **STRI Disturbance Theory.** Henry Bechelet and Richard Windows. The Sports Turf Research Institute, Bingley, West Yorkshire, England, UK. 80 pp., 11 illus., 2 tables, no index, semisoft covers in dark green and gold with gold lettering in a ring binder, 5.8 by 8.3 inches (147 × 210 mm).

This booklet presents a theory concerning the use of certain cultural practices to select for two turfgrass species-colonial bentgrass (*Agrostis capillaris*) and creeping red fescue (*Festuca rubra* ssp. *rubra*), especially under putting green conditions. This theory has specific application to certain cool-humid climatic conditions that occur in the United Kingdom. The authors point out that this proposed approach needs documentation via research.

Bengeyfield, William Henry

Bill Bengeyfield was born in East Williston, New York, USA, and earned a BSc degree in Floriculture and Ornamental Horticulture from Cornell University in 1948. From 1942 to 1945 and from 1951 to 1953, he served five years in the US Army and the US Air Force as a navigator on B-25 and B-29 bombers. He served as New York County Agricultural Agent for the USDA Extension Service from 1948 to 1951, as Western Director for the Green Section of the United States Golf Association (USGA) from 1953 to 1978, as Director of Grounds Maintenance for Industry Hills Golf Course from 1978 to 1981, and as National Director of the USGA Green Section from 1981 to 1993. He became owner-operator of the Frankfort Golf Club near Frankfort, Michigan, USA, until 2006. Bengeyfield received the Southern California Golf Association Service Award in 1989, the GCSAA Distinguished Service Award in 1990, and the Service Recognition Award of the European PGA Tour in 2000.

1989† **Specifications for a Method of Putting Green Construction.** William H. Bengeyfield (editor) with the contributing authors being the United States Golf Association (USGA) Green Section Staff Agronomists. United States Golf Association, Far Hills, New Jersey, USA. 25 pp., 12 illus. in color, 1 table, no index, fold-stapled glossy semisoft covers in multicolor with blue lettering, 6.0 by 9.0 inches (152 × 228 mm).

A small booklet of revised USGA specifications for putting green construction in the United States. The changes were based primarily on field observations of the USGA Green Section Agronomists. The original science-based USGA specifications were published in 1968 under the editorship of Dr. Marvin H Ferguson.
Revised in 1993.
See *Snow, James Taft—1993.*

Bennett, Jesse Merle

Jesse Bennett (1886–1943) was born in Camden, Michigan, USA, and earned a BSc degree in Forestry from Michigan State College in 1919. In 1922, he was first employed and, in 1925, became Superintendent of Parks and Forestry for the Board of County Road Commissioners of Wayne County (Detroit), Michigan. He was a national pioneer in the forestry, nursery, and landscape aspects of roadside construction. He was responsible for establishment of the first public arboretum in Michigan located in Hines Park, and subsequently, it was named Bennett Arboretum.

1929† **Roadside Development.**˙ J.M. Bennett. Land Economics Series, The Macmillan Company, New York City, New York, USA. 265 pp., 101 illus., 5 tables, extensive index, textured cloth hardbound in blue with gold lettering and a jacket in black and white with black lettering, 5.2 by 7.7 inches (133 × 196 mm).

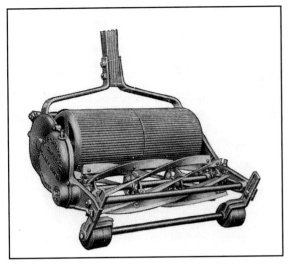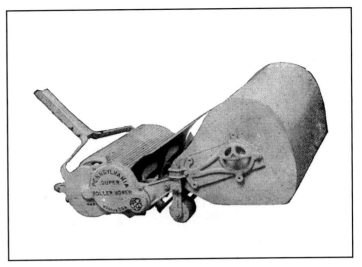

Figure 12.16. New Pennsylvania Super manual-push, 7-bladed, roller-drive, reel mower, without grass box (left) and with grass box (right), 1920s. [41]

* This **very rare**, pioneering, comprehensive book on roadside development discusses design, construction, and maintenance of roadsides. It encompasses both turfgrasses and landscape plantings plus chapters on public utilities, comfort stations, bridges, and lighting. There is an appendix on state laws relating to roadsides. It is a unique book and the first one devoted to roadsides. It contains a detailed bibliography. Bennett's philosophy concerning roadside development was as follows:

> The work as described in this book is not a matter of beautifying the highways as it is so often referred to, but is the business of developing the highways for the purpose of safely increasing their practical use and improving the roadsides to a degree where Nature can step in and make them beautiful.

1936† **Roadsides—The Front Yard of the Nation.*** J.M. Bennett. The Stratford Company Publishers, Boston, Massachusetts, USA. 250 pp., 17 illus., no tables, no index, cloth hardbound in dark green with gold lettering, 5.3 by 7.7 inches (134 × 196 mm).

* A **very rare**, pioneering book on roadside development organized in a different manner than the author's 1929 book titled **Roadside Development**. There is more emphasis placed on the planting materials. It contains a bibliography in the back.

Bidwell, Warren Alvis

Warren Bidwell (1913–1999) was born in Owensboro, Kentucky, USA. He began his apprenticeship in golf course maintenance at Homestead Golf Club in Cincinnati, Ohio, at the age of 14. He held golf course superintendent positions at Seaview Golf Club in Absecon, New Jersey, in 1948; Olympia Fields Country Club in Chicago in 1959; Philadelphia Country Club in Gladwyne, Pennsylvania, in 1965; Congressional Country Club in Maryland in 1975; and again Olympia Fields Country Club from 1978 to 1983. He subsequently worked as a consultant and speaker for the Tee-2-Green Corporation. He was elected President of the Philadelphia Association of Golf Course Superintendents, President of the Midwest Golf Course Superintendents Association in 1963, and to the Board of Directors of the Golf Course Superintendents Association of America for 1969 through 1971. Bidwell received the GCSAA Distinguished Service Award in 1984.

1974† **Turfgrass Course.** Coauthor.
　　See *Mascaro, Thomas Charles—1974.*

Billing, Bernhard

Bernhard Billing was born in Wintefhur, Switzerland, and earned a *Diploma Landwirt* from Farmers College of Zürich in 1992, a Dipl.-Ing. Agr. FH from Swiss College of Agriculture SHL, and a BSc BUAS in Agriculture in Zollikofen, Bern, Switzerland, in 1997. After two years in marketing farm products, he became Assistant Turf Extension Officer in the Turf Department at Otto Hauenstein Seeds AG, serving from 1999 to 2008. Since 2009, he has been owner-operator of a seed production business in wild flowers.

2004† **Die Rasenfibel (The Turf Primer).** Bernhard Billing. Otto Hauenstein Seeds, Rafz, Switzerland. 190 pp. plus 1 p. of advertising, 213 illus. with 200 in color, 34 tables, small index, glossy semisoft bound in multicolor with white and black lettering, 5.8 by 8.2 inches (147 × 209 mm). Printed entirely in the German language.

A book on the basics of turfgrass planting and culture organized as to distinct uses, such as residential lawns, sports turfgrasses, golf turfgrasses, and green roofs. It is oriented to cool-season turfgrasses under the conditions in Switzerland. The foreword is by Otto Weilenmann and Antoine Berger.

Black, Neil Duncan

Neil Black was born in Griffith, New South Wales, Australia, and earned a BScAgr degree in Agricultural Science from the University of Sydney in 1971, a Horticultural Certificate with *honours* from Sydney Technical College in 1974, and a Diploma in Education (Technology) from the University of Technology-Sydney in 1986. He has served as Head of the Division of Horticulture for the New South Wales (NSW) Technical & Further Education (TAFE) Commission from 1979 to 1989, as State Manager for the Rural and Mining Training Division of NSW-TAFE Commission from 1989 to 1994, and as Institute Director for both the Western Institute of TAFE, from 1991 to 2000, and the North Coast Institute of TAFE, since then. He was elected President of the NSW Branch of the Australian Institute of Horticulture in 1985. Black received the Award of Excellence for Outstanding Contributions to Horticulture from the Australian Institute of Horticulture in 1988.

1984† **Growing Media for Ornamental Plants and Turf.** Coauthor.
      See *Handreck, Kevin Arthur—1984.*

1994† **Growing Media for Ornamental Plants and Turf.** Coauthor.
      See *Handreck, Kevin Arthur—1994.*

2002† **Growing Media for Ornamental Plants and Turf.** Coauthor.
      See *Handreck, Kevin Arthur—2002.*

Black, Robert John (Dr.)

Bob Black was born in Houma, Louisiana, USA, and earned BSc and MSc degrees in Horticulture from Louisiana State University in 1966 and 1968 respectively, and a PhD degree in Plant and Soil Sciences from the University of Tennessee in 1972. He served as a Louisiana Extension Horticulture Agent in New Orleans from 1973 to 1975 and then joined the University of Florida in the Department of Environmental Horticulture as a Consumer Horticulture Specialist serving from 1975 to 2002.

1990† **Florida Lawn Handbook.** First edition. Coeditor.
      See *McCarty, Lambert Blanchard—1990.*

1997† **Florida Lawn Handbook.** Second revised edition. Coeditor.
      See *Ruppert, Kathleen Carlton—1997.*

Blažek, Otakar

Otakar Blažek (1929–2006) was born in Dobronin, Jihlava District, Czechoslovakia. He completed studies at the Business Academy in 1949 and at Tyrš Institute of Physical Education in Nymburk in 1953 and then graduated from the Faculty of Physical Education within Charles University in Prague in 1958. He became Secretary in 1953 and Director in 1962 of the Sport Center in Nymburk (a town near Prague) and was involved in the construction and turfgrass establishment of experimental soccer fields and tennis courts at the Center. Blažek initiated a formal study program in turfgrass management in 1975 and developed an expanded turfgrass experimental area in 1992. He retired in 1995.

1975† **Travnatá Hřiště (Turf Playing Fields).** Coauthor.
      See *Bureš, František—1975.*

Board of Greenkeeping Research (BGR)

The Board of Greenkeeping Research (BGR) was formed in 1929 at the St Ives Estate just west of Bingley, West Yorkshire, England, UK, with R.B. Dawson employed as Director from 1929 to 1963. It was originally an organization of UK golf clubs formed to support turfgrass research and provide advisory services for golf courses. The name was changed to the Sports Turf Research Institute (STRI) in 1951, and the role was expanded to address a broad range of turfgrass facilities. The STRI is now organized as two separate entities: research and advisory.

1931† **Guide To The Experiments of the St Ives Research Station.** Board of Greenkeeping Research, Bingley, Yorkshire, England, UK. 46 pp., 1 illus., 15 tables, no index, fold-stapled semisoft covers in cream and green with green lettering, 5.5 by 8.4 inches (140 × 213 mm).

  * A **rare** booklet describing the newly established field plot turfgrass research experiments of the Board of Greenkeeping Research, with lists of specific treatments. The St Ives Research Station was officially founded in May 1929.

Bošković, Boris Petar̄

Boris Bošković was born in Tuzla, Bosnia and Herzegovina, the son of Petar̄ Bošković. He earned a Civil Engineering degree from the Faculty of Technical Sciences in 1991. He is licensed by the state as a supervisor of construction technical

Figure 12.17. First experimental plots being seeded in 1929 at the St Ives research site of the Board of Greenkeeping Research. [17]

standards and regulations as well as works design and performance. Bošković wrote the sections on standards for building a stadium.

2004† **Sve o Savremenim Fudbalskim i Golf Terenima (All about Football and Golf Fields).** Coauthor.
  See *Bošković, Petař Milóvan—2004.*

Bošković, Petař Milóvan (Dr.)

Petař Bošković was born in Derventa, Bosnia and Herzegovina, Yugoslavia, and earned a Candidate of Science (CSc) degree from the University of Brno, Czechoslovakia, in 1975 and a PhD from the University of Zagreb, Croatia, Yugoslavia, in 1979. He served as a turfgrass specialist for sports and soccer (football) organizations, including Manager of Stadia in Novi Sad and then advisor of field facilities for the Yugoslav Football Association from 1970 to 1993 and SIGIS Tunis from 1993 to 1998. He also has consulted on soccer (football) fields in Greece, Cypress, and North Africa. Dr. Bošković retired in 1993.

1971† **Podizanje i Održavanje Travnatih Terena (Construction and Culture of Turfgrass Fields).** Petař Bošković and František Bureš. Zavod za Unapredenje Fizicke Kulture "Vojvodine," Novi Sad, Yugoslavia. 84 pp., 34 illus., 20 tables, no index, semisoft bound in white and black with green and black lettering, 6.5 by 9.5 inches (165 × 241 mm). Printed entirely in the Yugoslav language.
  A small book on the construction, establishment, and culture of turfgrasses for sports fields, with emphasis on procedures for soccer (football) fields under the conditions in Yugoslavia. A total of 1,000 copies were printed. It was the first turfgrass book published in Yugoslavia.

1975† **Travnate Površine—Izgradnja i Održavanje (Turfgrass Areas—Construction and Culture).** Petař Bošković. Sportska Tribina, Zagreb, Yugoslavia. 91 pp. with 2 pp. of advertising; 50 illus.; 11 tables; no index; glossy semisoft bound in green, white, and black with white and black lettering; 5.6 by 7.8 inches (141 × 199 mm). Printed entirely in the Yugoslav language.
  A small book on the construction, establishment, and culture of turfgrasses for sports fields under the conditions in Yugoslavia. It is a practically oriented book that encompasses turfgrass areas for soccer, tennis, and track & field. It contains a short list of selected references in the back. A total of 5,000 copies were printed.

1975† **Biološki Problemi Travnjaka na Terenima Jugoslavije (Turf Biological Problems on Yugoslavian Playing Fields).** Petař Bošković. Pokrajinska Zajednica za Naučni Rad u Novom Sadu, Novi Sad, Yugoslavia. 83 pp., 11 illus., 26 tables, no index, semisoft bound in white with brown lettering, 6.6 by 9.4 inches (167 × 238 mm). Printed entirely in the Yugoslav language.

A small publication on the research findings from the author's dissertation for the CSc degree from the University of Brno. A total of 100 copies were printed.

1984† **Football Grounds—Their Construction and Standards of Maintenance.** Petař Bošković. Yugoslav Federation of Physical Culture, Belgrade, Yugoslavia. 80 pp., 21 illus., 18 tables, no index, glossy semisoft bound in multicolor with black lettering, 5.5 by 8.1 inches (139 × 205 mm).

A small book on the construction, establishment, and culture of turfgrasses for soccer fields in Yugoslavia. The main focus is on site selection, design, drainage, and construction. The printing totaled 500 copies.

Originally published under the title **Fudbalski Tereni** in the Yugoslav language.

2001 **Fudbalski Stadioni i Igrališta—Tehnički Kriteriji FSJ, UEFE i FIFE—Projektovanje, Izgradnja i Bezbednost-Evropska Igrališta i Održavanje (Football Stadium and Playground: Technical Requirement, Regulation, Safety on the FSJ, UEFA, FIFA—Design, Construction and Maintenance).** Petař Bošković. FSJ, Belgrade, Yugoslavia. 135 pp., 16 illus., 19 tables, no index, glossy semisoft bound in green with black lettering, 6.5 by 9.8 inches (165 × 250 mm). Printed in the Yugoslav language.

A book on the development and maintenance of soccer facilities under Yugoslavian conditions. There are chapters on the construction of football stadia, FIFA requirements for stadia, construction of soccer pitches, grass varieties for football fields, and maintenance of turfgrass football fields. A total of 1,000 copies were printed.

2002† **Football Stadium and Playground For Hot Arid, Semiarid and Wet Regions.** Petař Bošković. Published by the author, Novi Sad, Yugoslavia. 117 pp., 37 illus. plus 13 in color, 17 tables, no index, glossy semisoft bound in yellow with green lettering, 5.9 by 8.2 inches (150 × 208 mm).

A book on the development and maintenance of soccer facilities under African conditions. There are chapters on construction of football stadia, FIFA requirements for stadia, construction of soccer pitches, grass varieties for football fields, and maintenance of turfed football fields. A small list of selected references is presented in the back. Total of 500 copies printed for members of FIFA.

2003† **Moj Travnjak (My Lawn).** Dr. Petař Bošković. "Dobro Jutro" Dnevnik, Novi Sad, Serbia. 72 pp. plus 10 pp. of advertising, 63 illus. in color, 4 tables, no index, glossy semisoft bound in multicolor with green and white lettering, 5.5 by 8.2 inches (140 × 207 mm). Printed entirely in the Yugoslav language.

A small book on the construction and culture of lawns. It is practically oriented to lawn enthusiasts under the conditions in Serbia. There is a list of selected references in the back. A total of 1,500 copies were printed.

2004† **Sve o Savremenim Fudbalskim i Golf Terenima (All about Football and Golf Fields).** Dr. Petař Bošković and Boris Bošković. FSS GC, Novi Sad, Serbia. 134 pp., 52 illus. in color, 16 tables, no index, glossy semisoft bound in multicolor with white lettering, 5.4 by 8.2 inches (138 × 209 mm). Printed entirely in the Yugoslav language.

A book on turfgrass culture of soccer (football) fields under conditions in Serbia. Soil and water analyses and interpretation are emphasized. There is a list of selected references in the back. A total of 1,000 copies were printed. Also has been published in the Hungarian language.

2007 **Ukrasni Travnjak (Decorative Turfgrass).**
See *Čižek, Jan Josip—2007.*

Bossard, Bruce J.[2]

1980† **The Bossard Sportsfields Landscape Groundskeeping Pedology Handbook.** B.J. Bossard. Bossard Sportsfields, Euclid, Ohio, USA. 63 pp., 3 illus., 2 tables, no index, fold-stapled semisoft covers in green with black lettering, 5.3 by 8.3 inches (135 × 211 mm).

A small book in which the author relates his personal philosophies of turfgrass establishment and culture in the northern United States, with an intermix of comments on lawns and sports fields.

Boufford, Robert William

Bob Boufford was born in Detroit, Michigan, USA, and earned an Associate of Applied Science from Oakland Community College, Michigan, in 1973, a BSc degree from Michigan State University in 1975, and an MSc degree from Kansas State University in 1979. He served as Instructor of Agriculture and Turf Management at Dodge City Community College, Kansas, from 1975 to 1977, taught turfgrass subjects in the Professional Golf Management Program at Ferris State University in Big Rapids, Michigan, from 1979 to 1988, was Instructor in Landscape and Grounds Management at the Agricultural and Technical Institute, Ohio State University, in Wooster, Ohio, from 1989 to 1994, was Assistant and Associate Professor of Horticulture at Clark State Community College in Springfield, Ohio, from 1994 to 1999, was Senior Instructor at WebCT Inc. in Boston, Massachusetts, from 2000 to 2002, and has been Team Lead of E-Learning Support and Technical Development at the Academic Information and Communications Technologies (AICT) Department at the University of Alberta, Canada, since 2002. He has specialized in computer applications for turfgrass and landscape operations and, more recently, in E-Learning services. He was founding publisher of the **Ohio Lawn Care**

Association Newsletter in 1991 and coeditor through 1999. Boufford received the Faculty Award for Professional Excellence from Clark State Community College in 1999.

1995† **Turfgrass Science and Management Lab Manual.** Coauthor.

See *Emmons, Robert David—1995.*

Bouhana, Charles[3]

1946 **Terrains de sports—Stades, gazons, drainage, irrigation (Sports grounds—Stadiums, turfs, drainage, irrigation).** Charles Bouhana. E. Baudelot & Cie., Paris, France. 517 pp., 131 illus., 6 tables, no index, soft bound in red and dark green with gilded lettering, 6.3 by 8.7 inches (160 × 220 mm). Printed entirely in the French language.

* A **relatively rare** book on the construction of sports fields and stadiums and the subsequent establishment and culture of turfgrasses. It is practically oriented to the conditions in France. Included are turfgrasses for soccer, rugby, field hockey, cricket, baseball, golf, polo, and airfields.

Bowles, William Henry George (OBE)

Bill Bowles (1900–1989) was born in Bromley, Kent, England, UK. He was the son of a groundsman and, starting at 12 years of age, apprenticed under his father at Forest Hill Cricket Club in London, England, UK. At age 18, he began service in the Army Kings Royal Rifle Corps during WWI. After being demobilized, he returned to a groundsman career, initially at the Commercial Union Assurance Company Athletics Club in London from 1919 to 1921, Holdron's Department Store Ground from 1922 to 1934, and Westminster Bank Sports Club Ground from 1934 to 1936. He then became Head Groundsman at Eton College from 1936 to 1984, with responsibility for 20 cricket wickets, 50-plus soccer pitches, 36 rugby fields, lawn tennis courts, athletic tracks, and a 9-hole golf course. He was a key pioneer of modern sports field culture in England. In 1934, Bowles founded the Institute of Groundsmanship, originally named the National Association of Groundsmen until 1955. He served as Honorary Secretary from 1934 to 1936, Chairman from 1936 to 1953, Vice President from 1954 to 1958, and President from 1958 to 1989. Bowles was initiated as an OBE in 1973.

1952† **Practical Groundsmanship.** Coauthor.

See *White, Lawrence Walter—1952.*

* Coauthors White and Bowles commented on groundsmanship in 1952:

> As Great Britain possess the greatest area per square mile devoted to outdoor sport in the world, and at the same time has the most unpredictable climate known, a fixed working programme for the groundsman is obviously impossible. Yet throughout the centuries during which the cultivation of sports turf has been so important a part of the recreational life of the country, there have always been men who were devoted to this work.
>
> Inevitably, they have all acquired, over a period of time, a knowledge born of their experience in combating the problems that confronted them. Whatever they have learned, most are at least aware that a subject so complex as groundsmanship can never be completely learned.

1967† **Sports Ground Maintenance—an elementary guide to club committees.** First edition. W.H. Bowles. National Playing Fields Association, London, England, UK. 40 pp.; 29 illus.; 2 tables; no index; fold-stapled semisoft covers in green, black, and white with black lettering; 5.4 by 8.5 inches (138 × 215 mm).

* An abbreviated handbook on the maintenance of turfgrasses for sports fields, cricket wickets, bowling greens, and tennis courts. It is practically oriented to cool-season turfgrass conditions in England.

Revised in 1978 and 1989.

1978† **Sports Ground Maintenance—an elementary guide to club committees.** Second revised edition. W.H. Bowles. National Playing Fields Association, London, England, UK. 31 pp.; 21 illus.; no tables; no index; glossy semisoft bound in green, black, and white with black lettering; 5.9 by 8.2 inches (149 × 209 mm).

Minor revisions, with the sections on turfgrass species and seed mixtures deleted. There is a significant updating of the photographs.

Revised in 1989.

See *Holborn, John Charles—1989.*

Bowman, Daniel Clark (Dr. and Associate Professor)

Dan Bowman was born in Milwaukee, Wisconsin, USA, and earned a BSc degree in Horticulture from the University of Wisconsin in 1976 and a MSc degree in Environmental Horticulture and Turfgrass and a PhD degree in Plant Physiology and Turfgrass both from the University of California, Davis, in 1985 and 1987, respectively. He served as an Assistant and Associate Professor in the Plant Science Department at the University of Nevada, Reno, from 1988 to 1994 and as Associate Professor at North Carolina State University since 1994. Dr. Bowman is involved in research and teaching of turfgrass physiology and soils related aspects.

1990†   **The Sand Putting Green—Construction and Management.** Coauthor.

See *Davis, William Boyce—1990.*

Boyce, James Henderson

Jim Boyce (1910–2000) was born in Stamford Centre, Ontario, Canada, and earned a Bachelor of Science in Agriculture (BScAgr) degree in Field Husbandry from the Ontario Agricultural College, Guelph, in 1932. He then earned an MSc degree from Rutgers University in 1939. He accepted a position as Agricultural Scientist in the Division of Forage Crops with the Dominion Experimental Farms in Ottawa in 1932, where he managed the turfgrass research plots under the direction of R.I. Hamilton. Upon retirement circa 1962 he became involved in soil testing in liaison with Brookside Laboratories of Ohio. Then in 1970, he was appointed Executive Director of the Canadian Golf Superintendents Association officed in Ottawa. His main focus was monthly production of **The Green Master** for three years until late 1972. Boyce was awarded Honorary Life Membership in the Sports Turf Association in 1989.

n.d.   **The Establishment and Maintenance of Turf on Putting and Bowling Greens.** J.H. Boyce. Division of Forage Crops,
circa   Dominion Experimental Farms Service, Central Experimental Farms Service, Central Experimental Farm, Ottawa, Can-
194x   ada. 21 pp., no illus., no tables, no index, stapled soft covers in black with dark blue lettering, 8.0 by 9.0 inches (217 × 250 mm).

*   A small mimeograph concerning the cool-season turfgrass for greens, along with their establishment and culture.

1950   **The Construction and Care of Lawns.** James H. Boyce. Canada Department of Agriculture, Ottawa, Canada. 24 pp., no illus., 1 table, no index, fold-stapled soft covers in white and green with black lettering, 6.0 by 9.0 inches (157 × 241 mm).

A booklet concerning the characteristics of cool-season turfgrasses for lawns as grown in Canada, plus their planting and care. This is a revision of the earlier 1939 and 1948 editions.

Brandenburg, Rick Lynn (Dr. and Professor)

Rick Brandenburg was born in Wabash, Indiana, USA, and earned a BSc degree in Entomology from Purdue University in 1977 and a PhD degree in Entomology from North Carolina State University in 1981. He served as Assistant Professor at the University of Missouri from 1981 to 1985 and then joined the Entomology faculty at North Carolina State University (NCSU) where he progressed to Full Professor in 1993. He conducts research and adult extension in turfgrass and peanut entomology. In 2001, he became Co-Director of the Center for Turfgrass Environmental Research and Education at NCSU. He was named a William Neal Reynolds Professor in 2007. Dr. Brandenburg received the State Early Career Award from Epsilon Sigma Phi in 1990; the Outstanding Service Award from the Turfgrass Council of North Carolina in 1994; the Extension Education Award from the North Carolina Association of Cooperative Extension Specialists in 1995; the Distinguished Achievement Award in Extension Entomology from the Entomology Society of America Southeastern Branch in 1995; the Certificate of Excellence in Educational Materials from the American Society of Agronomy in 1998; the State Midcareer Award from Epsilon Sigma Phi in 2001; the Recognition Award in Urban Entomology from the Southeastern Branch of the Entomological Society of America in 2002; the John V. Osmum Alumni Professional Achievement Award from Purdue University in 2002; the Outstanding Extension Service Award from the College of Agricultural and Life Sciences, NCSU, in 2002; the Alumni Outstanding Extension and Outreach Award from NCSU in 2003; the Alumni Distinguished Professor of Extension and Outreach from NCSU in 2003; the Certificate for Outstanding Faculty Scholarship from NCSU in 2003; a Fellow from the American Peanut Research and Education Society in 2003; a Fellow from the Entomological Society of America in 2004; the Recognition Award in Urban Entomology from the Entomological Society of America in 2004; the Extension Communication Award from American Society for Horticultural Science in 2004; the Outstanding Book Award from the American Society for Horticultural Science in 2004; and the Outstanding Contributions to Entomology Award from the North Carolina Entomological Society in 2005.

1995†   **Handbook of Turfgrass Insect Pests.** Rick L. Brandenburg and Michael G. Villani (editors). Entomological Society of America, Lanham, Maryland, USA. 141 pp., 117 illus. with 55 in color, 4 tables, indexed, glossy semisoft bound in multicolor with black lettering, 8.4 by 10.8 inches (213 × 274 mm).

A handbook on the insect and mite pests of turfgrasses, including identification, life history, and pest management practices. There are 31 insect and mite groupings described. A list of selected references is presented at the back of most sections. A glossary of selected terms is included in the back.

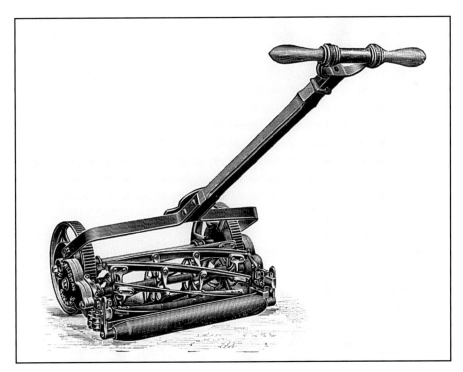

Figure 12.18. Shanks 1904 Talisman manual-push, high-wheel-drive, 16-inch-wide, reel mower. [38]

Brede, Andrew Douglas (Dr.)

Doug Brede was born in Pittsburg, Pennsylvania, USA, and earned a BSc in Agronomy and Turfgrass from Pennsylvania State University in 1975. He worked two years as an Assistant Golf Course Superintendent. Subsequently, he earned MSc and PhD degrees both in Agronomy and Turfgrass from Pennsylvania State University in 1978 and 1982, respectively, studying with Dr. Joe Duich. He then joined the faculty at Oklahoma State University working in turfgrass management research and teaching. In 1986, Dr. Brede was employed as Research Director for Jacklin Seed Company with major responsibilities in turfgrass breeding. Since 1997, he has been Director of Research and Executive Vice President of the Jacklin Seed Division, J.R. Simplot Company. Dr. Brede was awarded Fellow of the American Society of Agronomy in 2004 and of the Crop Science Society of America (CSSA) in 2005. Also in 2005, he received the CSSA Turfgrass Science Award and the CSSA and NCCPA Genetics and Plant Breeding Award for Industry.

2000† **Turfgrass Maintenance Reduction Handbook—Sports, Lawns, and Golf.** Doug Brede. Ann Arbor Press, Chelsea, Michigan, USA. 388 pp., 93 illus., 43 tables, indexed, glossy hardbound in greens with white and light green lettering, 8.5 by 10.9 inches (216 × 277 mm).

Approaches for lowering cultural inputs in turfgrass culture are addressed. One-half of the book is devoted to turfgrass and cultivar selection, with nontraditional species receiving much attention. A catalogue of unconventional grasses and a list of selected references are included in the back.

Brett, Walter Percivall

Walter Brett was born in England, UK. He became a gardening journalist and served as editor of **Home Gardening**, **The Home Gardening Encyclopaedia**, **The New Book of Gardening**, and **The Rose Encyclopedia**, plus numerous gardening books.

n.d.† **The Lawn—How to Make and Maintain It.** Walter Brett. The Pearson Practical Handbook Series, C. Arthur Pearson circa Limited, London, England, UK. 144 pp., 22 illus., 4 tables, small index, both cloth hardbound in light green with dark 1937 green lettering and also semisoft bound in white and black with green and black lettering, 4.8 by 7.1 inches (122 × 181 mm).

* A **rare** book on the basics of turfgrass establishment and culture, with emphasis on lawns. It is oriented to amateur home lawn enthusiasts and cool-season turfgrass conditions in England. There also is a chapter on grass banks, verges, and paths. The last chapter briefly addresses tennis courts, badminton courts, croquet courts, bowling greens, and garden golf greens. In 1930, Walter Brett had the following comment on power mowers:

> Power mowers are surely a real luxury. You can run over a "vast" stretch of lawn at an amazing speed and finish up as fresh as though you'd gone for a gentle amble in the park.

British Seed Houses Limited (BSH)

Samuel McCausland founded a seed business in 1825 in Belfast, Northern Ireland. He specialized in the sale of Irish ryegrass seed. In 1963, Germinal Holdings Ltd. was formed through the merger of Samuel McCausland Ltd. and Joseph Morton Ltd., Banbridge, Northern Ireland. Then in 1969, British Seed Houses Ltd. was formed following the merger of McCausland and Morton's interests in England and the acquisition of David Bell Ltd. in Scotland. Germinal Holdings Ltd., who own British Seed House of Lincoln, England, are now involved in the agricultural, amenity grass, and clover seed sectors. Their activities include breeding, assessment, production, and marketing.

n.d.† circa 1978 **Turfgrass Manual.*** First edition. R. Jeremy C. Hawarth. British Seed Houses Limited, Warrington, Cheshire, England, UK. 47 pp., 14 illus., 27 tables, no index, fold-stapled glossy semisoft bound in white with blue lettering, 6.7 by 9.5 inches (171 × 241 mm).

A small manual for the turfgrass professionals involved in the design, specifications, establishment, and maintenance of sports grounds and amenity grassland areas in the United Kingdom. It encompasses lawns, bowling greens, cricket grounds, hockey fields, soccer fields, rugby fields, horse racecourses, roadsides, and sand dunes.
Revised in 1982, 1988, and 1994.

n.d.† circa 1982 **Turfgrass Manual.*** Second revised edition. R. Jeremy C. Hawarth. British Seed Houses Limited, Warrington, Cheshire, England, UK. 47 pp., 14 illus., 27 tables, no index, fold-stapled glossy semisoft bound in white with blue lettering, 6.7 by 9.5 inches (171 × 241 mm).

This second edition has only minor revisions.
Revised in 1988 and 1994.

n.d.† circa 1988 **Turfgrass Manual.*** Third revised edition. British Seed Houses Limited, Warrington, Cheshire, England, UK. 60 pp., 14 illus., 32 tables, no index, fold-stapled glossy semisoft covers in greens and white with blue lettering, 6.7 by 9.5 inches (170 × 242 mm).

This third edition has significant revisions. The foreword is by John Shildrick.
Revised in 1993.

n.d.† circa 1993 **Turfgrass Manual.*** Fourth revised edition. British Seed Houses Limited, Warrington, Cheshire, England, UK. 67 pp.; 14 illus.; 28 tables; no index; fold-stapled glossy semisoft covers in green, yellow, and blue with white lettering; 6.6 by 9.4 inches (168 × 241 mm).

This fourth edition has been revised and reorganized. There is a foreword by Dr. P. Michael Canaway.

Brown, Denis[2]

1995 **Guide d'Aménagement des Terrains Extérieurs: Baseball, Soccer, Softball (Guide for Installation of Outdoor Fields: Baseball, Soccer, Softball).** Denis Brown. Régie de la Sécurité dans les Sports du Québec, Trois-Rivéres, Québec, Canada. 40 pp., 48 illus., 7 tables, no index, glossy semisoft bound in multicolor with black lettering, 8.3 by 9.8 inches (212 × 250 mm). Printed in the French language.

A booklet on the design and construction of turfgrass fields for baseball, soccer, and softball under conditions in Quebec Provence, Canada.

2005 **Guide d'Aménagement et d'Entretien desTerrains de Soccer Extérieurs (Guide for Installation and Maintenance of Outdoor Soccer Fields).** Coauthor.
See *Gionet, Luc—2005.*

Brown, Stewart[2]

2005† **Sports Turf & Amenity Grassland Management.** Stewart Brown. The Crowood Press Limited, Marlborough, Wiltshire, England, UK. 192 pp., 143 illus. with 24 in color, 30 tables, small index, glossy semisoft bound in multicolor with white and green lettering, 6.4 by 9.2 inches (163 × 234 mm).

A book on turfgrasses and their culture, establishment, and renovation for sports, lawns, and functional uses in the United Kingdom. It is practically oriented to both public and private turfgrass areas. There is a small glossary in the back.

Bunnell, Brian Todd

Todd Bunnell was born in Lexington, Kentucky, USA, and earned a BSc degree in Plant and Soil Science from the University of Kentucky in 1997; he also earned an MSc degree in Horticulture in 2000 and a PhD degree in Plant Physiology and Turfgrass in 2003 both from Clemson University. In 2003, Dr. Bunnell became Manager of Turfgrass and Ornamental Research at SePRO Corporation in Carmel, Indiana.

2003† **Fundamentals of Turfgrass and Agricultural Chemis**try. Coauthor.
See *McCarty, Lambert Blanchard—2003.*

Bureš, František *(Dr.* and *Associate Professor)*

František Bureš (1921–1999) was born in Lhotice na Moravě, Czechoslovakia, and earned a Dipl. Ing. degree with a major in Agronomy in 1949, a DrTech degree in 1950, and a CSc (now PhD) degree in 1962, all from Mendel University of Agriculture and Forestry in Brno, Czechoslovakia. He served in the Department of Grassland Management & Forage Production at Mendel University in Brno, Czechoslovakia, from 1950 to 1970, rising from Lecturer in 1963 to Assistant Professor in 1964 and Associate Professor in 1968. He then focused on the development of a turfgrass research facility at the Research Institute of Pastures in Banská Bystrica and at its branch Research Institute of Pastures and Turf in Liberec. He was editor of the periodicals **Trávniky, Sport, Zeleň, Ekologie (Turfgrass, Sports, Verdure, Ecology)** from 1991 through 1992 and **Zahrada, Park, Krajina, Trávniky (Garden, Park, Landscape, Turfgrass)** from 1993 through 1994. Dr. Bureš was active in turfgrass research and in education for sports field groundsman and golf course greenkeepers, and he also expanded the turfgrass management course of study at Mendel University in Brno. He was the first Czechoslovakian scholar to devote his full efforts to turfgrass science.

1971† **Podizanje i Održavanje Travnatih Terena (Construction and Culture of Turfgrass Fields).** Coauthor.
    See Bošković, Petař—1971.

1975 **Travnatá Hřiště (Turf Playing Fields).** František Bureš and Otakar Blažek. Tělovýchovná Škola ČÚV, Praha, Czechoslovakia. 224 pp., no illus., no tables, no index, soft bound in red, white and black with black lettering, 8.1 by 11.2 inches (207 × 295 mm). Printed in the Czechoslovak, Russian, German, and English languages.

    This publication is an annotated bibliography that reviews the turfgrass problems encountered on sports fields and recreation areas. Interpretations are applied to Czechoslovakian conditions.

1985 **Fotbalová Travnatá Hřiště (Grassy Football Playgrounds).** František Bureš. ÚV ČSTV Vedeckometodické Oddělení, Prague, Czechoslovakia. 104 pp., 41 illus., 10 tables, no index, semisoft bound in blue and white with black lettering, 8.2 by 11.5 inches (208 × 292 mm). Printed in the Czechoslovak language.

    A manual concerning the construction, establishment, and turfgrass culture of soccer fields under the conditions in Czechoslovakia.

1991 **Handbuch Rasen: Grundlagen—Anlage—Pflege (Handbook for Turfgrass: Basics, Installation, Maintenance).** Coauthor.
    See Gandert, Klaus-Dietrich—1991.

Burpee, Lee Loun *(Dr.* and *Professor)*

Lee Burpee was born in Brooklyn, New York, USA, and earned a BA degree in Biology from Gettysburg College in 1971, and both a MSc and a PhD degrees in Plant Pathology from Pennsylvania State University in 1974 and 1979, respectively, studying with Dr. Herb Cole. He has served as Plant Pathologist for the Bermuda Department of Agriculture from 1979 to 1982, as Assistant and Associate Professor at the University of Guelph, Canada, from 1982 to 1989, and as Full Professor at the University of Georgia since 1989. He conducts research on the etiology and management of turfgrass diseases in the Department of Plant Pathology at the University of Georgia, Griffin Research Center.

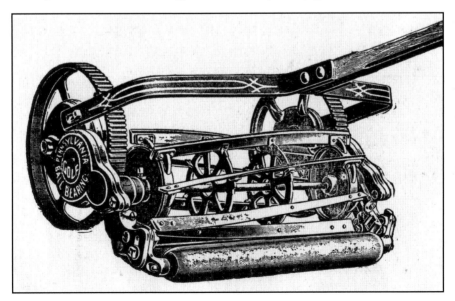

Figure 12.19. Manual-push, 19-inch-wide, side-wheel-drive, Pennsylvania Aristocrat reel mower with seven blades and grass catcher, circa 1924. [34]

1993† **A Guide to Integrated Control of Turfgrass Diseases. Volume I. Cool Season Turfgrasses.** L.L. Burpee. GCSAA Press, Golf Course Superintendents Association of America, Lawrence, Kansas, USA. 242 pp., no illus., no tables, no index, textured glossy semisoft covers in burgundy with gold lettering in a ring binder, 6.0 by 8.9 inches (152 × 227 mm).

A guide to the diseases of C_3, perennial, cool-season turfgrasses, with emphasis on control rather than diagnosis. It is organized by turfgrass species with two-page subheadings for each disease on each host.

1995† **A Guide to Integrated Control of Turfgrass Diseases. Volume II. Warm Season Turfgrasses.** L.L. Burpee. GCSAA Press, Golf Course Superintendents Association of America, Lawrence, Kansas, USA. 175 pp., no illus., no tables, no index, textured glossy semisoft covers in burgundy with gold lettering in a ring binder, 6.0 by 9.0 inches (152 × 228 mm).

A guide to the diseases of C_4, perennial, warm-season turfgrasses, with emphasis on control rather than diagnosis. It is organized by turfgrass species with two-page subheadings for each disease on each host.

C

Campbell, Christopher Stuart (Dr. and Professor)

Christopher Campbell was born in Boston, Massachusetts, USA, and earned a BA degree from Harvard University in 1968, an MSc degree from the University of Maine in 1975, and a PhD from Harvard University in 1980. He served as Assistant Professor at Rutgers University from 1980 to 1983 and, since, as Assistant Professor through Full Professor in Plant Systemics in the School of Biology and Ecology at the University of Maine.

1986† **Grass Systematics and Evolution.**[*] Coauthor.
See *Soderstrom, Thomas Robert—1986.*

Campbell, John Kirkwood

John Campbell (1919–2003) was born in Milngavie, Scotland, UK, and started apprenticeship with his father in golf course turfgrass maintenance at Hilton Park Golf Club in the west of Scotland at the age of 15 while also studying commercial art via night school. John had joined the Gordon Highlanders prior to the outbreak of WWII and was captured along with most of the 51st Highland Division at St Valery France in 1940. He was held as a prisoner of war in a German prison camp in Poland for five years. With the Russians advancing, his camp was evacuated and he was marched 1,000 miles (more than 1600 km) with surviving fellow prisoners until finally liberated by American military troops near Berlin. John Campbell served as Head Greenkeeper at Royal North Devon Golf Club from 1947 to 1953, at East Herts Golf Club till 1958, and then at Longniddry Golf Club. He became the first Links Supervisor at St Andrews in Scotland in 1961, where he was responsible for four golf courses, Old, New, Jubilee, and Eden, and also laid out and constructed the Balgove course. He served in that capacity until 1974 and, subsequently, was involved in construction and maintenance of a 36-hole golf course at Foxhills, Surrey, England, and also did some golf course design work in the United Kingdom and abroad. John Campbell was elected Chairman of the Scottish

Figure 12.20. Mowing a lawn with a manual-push, side-wheel-drive, reel mower, circa 1905. [10]

Greenkeepers Association from 1965 to 1968. He developed a periodical called **Turf Craft** that led a renaissance in the Scottish Greenkeepers Association. John Campbell was an enthusiastic person who elevated the professional image of greenkeepers in the United Kingdom.

1982† **Greenkeeping.** John Campbell. A. Quick & Company Limited, Harwich, Essex, UK. 88 pp., 9 illus., no tables, no index, glossy semisoft bound in black and white with blue lettering, 5.9 by 8.5 inches (150 × 216 mm).

A small book on the culture and management of golf courses under the conditions found in the United Kingdom. It also contains chapters on architecture, history of bunkers, and management of the greenkeeping staff and budgets. This is the first book written by a greenkeeper of the Old Course at St Andrews, Scotland. John Campbell's biography in the front of the book states the following:

> After demob from the army, he continued to work as a member of his father's staff. Then, in 1947, he was appointed head greenkeeper to the Royal North Devon Golf Club, whose links are the famous Westward Ho! He was 27.
> Looking after Westward Ho! which is laid out on common land, where sheep, cattle and ponies are allowed to graze without restriction, was a tough assignment and an unforgettable experience just after the war. The damage inflicted on the course by these roving animals would drive the staff to the depths of despair at times, and they had to work hard against all odds to keep the course trim.

Carrow, Robert Norris (*Dr.* and *Professor*)

Bob Carrow was born in Mt. Pleasant, Michigan, USA, and earned a BSc degree in 1968 and a PhD in Soil Fertility and Turfgrass in 1972, both from Michigan State University, studying with Dr. Paul Rieke. He served as a faculty member at the University of Massachusetts from 1972 to 1976 and at Kansas State University from 1976 to 1984. He then accepted a research position as Professor of Turfgrass Science in the Department of Crop and Soil Sciences, University of Georgia, at the Georgia Experiment Station in Griffin, Georgia. Dr. Carrow's research focus has been in the areas of (a) climatic and soil stresses, (b) water conservation and irrigation water quality issues, (c) management of salt-affected sites, (d) soil fertility and plant nutrition, and (e) traffic stresses on recreational sites. He was elected to the Crop Science Society of America (CSSA) Board of Directors from 1984 to 1986 and to the Turfgrass Science Division C-5 Chair in 1985, to the Board of Directors of the American Society of Agronomy from 1991 to 1993, and to Vice President of the International Turfgrass Society (ITS) from 1993 to 2005 while on the ITS Board of Directors from 1993 to 2005. Dr. Carrow was honored with the Outstanding Professor Award from the Stockbridge School, University of Massachusetts, in 1975 and as Fellow of the American Society of Agronomy in 1993.

1992† **Turfgrass.** Coeditor.
See *Waddington, Donald VanPelt—1992.*

1998† **Salt-Affected Turfgrass Sites: Assessment and Management.** R.N. Carrow and R.R. Duncan. Ann Arbor Press, Chelsea, Michigan, USA. 185 pp., 36 illus. with 26 in color, 31 tables, small index, glossy hardbound in multicolor with gold lettering, 6.2 by 9.3 inches (157 × 236 mm).

A specialty book on the assessment, principles, and treatment of high-salt conditions in terms of the soil, turfgrass, and cultural aspects. A set of case studies is included. Selected references are listed in the back.

2001† **Turfgrass Soil Fertility and Chemical Problems—Assessment and Management.** R.N. Carrow, D.V. Waddington, and P.E. Rieke. Ann Arbor Press, Chelsea, Michigan, USA. 412 pp., 99 illus. with 55 in color, 113 tables, indexed, glossy hardbound in multicolor with white lettering, 7.0 by 10.0 inches (177 × 254 mm).

A text on soil chemistry and plant nutrition as related to turfgrass culture. It includes soil chemical properties, essential plant nutrients, soil chemical problems, fertilizers, and fertilization programs. There is a list of selected references at the end of each chapter.

2009† **Turfgrass and Landscape Irrigation Water Quality: Assessment and Management.** Coauthor.
See *Duncan, Ronny Rush—2009.*

Carters Tested Seeds Limited

A series of booklets of more than 20 pages and up to 50 pages were published by the company after WWII. Specific years identified included 1949, 1952, 1956, 1957, 1958, 1961, 1966, and 1969. An example is shown of the 1966 edition.

1966† **Treatment of Golf Courses and Sports Grounds.** Carters Tested Seeds Limited, Raynes Park, London, England, UK. 31 pp.; 50 illus.; 3 tables; no index; fold-stapled glossy semisoft covers in black, white, and light green with black and white lettering; 5.8 by 9.3 inches (147 × 236 mm).

A small booklet concerning the planning, construction, establishment, and culture of turfgrasses for golf courses. It also contains sections on sports fields. It is oriented to cool-season turfgrasses and conditions in the United Kingdom.
See also *Beale, Reginald Evelyn Child.*

Casati, Maria Paolo

Maria Casati was born in Padova, Italy, and earned a Doctor in Law degree from the University of Ferrara in 1998. She became a Golf Professional in 2000 serving as Golf Teacher at Golf Club della Montecchia in Padova. Casati has won numerous Italian and European golf championships.

2004† **Il campo pratica per il gioco del golf—Costruzione, manutenzione e gestione (The Practice golf range—Construction, maintenance and management).** Mario Paolo Casati, Paolo Croce, Alessandro De Luca, and Massimo Mocioni. Associazione Italiana Tecnici Di Golf, Castelnuovo Rangone, Italy. 82 pp. plus 4 pp. of advertising, 40 illus. in color, 3 tables, no index, glossy semisoft bound in multicolor with black and white lettering, 6.5 by 9.3 inches (165 × 235 mm). Printed entirely in the Italian language.

A unique, small book concerning the layout, construction, turfgrass establishment, and culture of a practice golf range, including pitching area, bunkers, and putting green. It is oriented to conditions in Italy. There is a list of selected references in the back.

Casler, Michael Darwin (Dr.)

Mike Casler was born in Green Bay, Wisconsin, USA, and earned a BSc degree in Agronomy from the University of Illinois in 1976; he also earned an MSc in 1979 and PhD in 1980, both in Plant Breeding from the University of Minnesota. He joined the faculty at the University of Wisconsin in 1980, conducting research in the breeding and genetics of perennial cool-season grasses. He rose to Associate Professor in 1985 and Full Professor in 1986. Then in 2002, he joined the USDA-ARS US Dairy Forage Research Center in Madison, Wisconsin. Dr. Casler was elected to the Crop Science Society of America (CSSA) and the American Society of Agronomy (ASA) Boards of Directors for 2000 to 2003. Honors he has received include the CSSA Young Crop Scientist Award in 1991, the American Forage and Grassland Council Merit Certificate in 1992, CSSA Fellow in 1993, ASA Fellow in 1997, and Wisconsin Agricultural Research Stations' Researcher of the Year Award in 1998.

2003† **Turfgrass Biology, Genetics, and Breeding.** Michael D. Casler and Ronny R. Duncan (editors) with 24 contributing authors. John Wiley & Sons, Hoboken, New Jersey, USA. 377 pp.; 61 illus.; 19 tables; large index; glossy hardbound in green and black with yellow, green, and white lettering; 7.0 by 9.9 inches (177 × 251 mm).

A text on the basics of turfgrass improvement. There are 14 chapters on cool-season turfgrasses and 7 chapters on warm-season turfgrasses. Each is addressed in terms of distribution and adaptation, morphology, cytotaxonomy, and breeding and genetic improvement.

Cave, Lewis Walter

Lewis Cave (1901–1996) was born in Bedford, Bedfordshire, England, UK. He pursued apprentice horticulture training at several private estates and then completed a two-year course of study at the Royal Horticultural Gardens at Wisley. From 1927 to 1930, he was in charge of the Experimental Gardens of Joseph Fison & Company and then was appointed their Horticultural Consultant. He was Turf Advisor to the Advisory Department of Fisons Limited from 1945 to 1955 and, subsequently, was Turf Consultant to Fison Horticulture Limited. Cave was especially active on the European continent as a specialist in turfgrass culture for lawn bowling greens.

1967† **Cave's Guide to Turf Culture.** Lewis W. Cave. Pelham Books Limited, London, England, UK. 188 pp., 96 illus. with 16 in color, 49 tables, no index, cloth hardbound in brown with silver lettering and a jacket in green and white with white and black lettering, 5.2 by 8.3 inches (131 × 212 mm).

A guide book in which the technical terms used in turfgrass culture in the United Kingdom are arranged in alphabetical order, followed by a basic definition and/or practical description.

† *Reprinted in 1972 for Readers Union Limited, Newton Abbot, England, UK, in cloth hardbound in light green with gold lettering and a jacket in green and red with white, red, and black lettering.*

Cettour, Henri[2]

1996 **Stades et Terrains de Sports: Guide Technique et Réglementaire (Stadiums and Sports Fields: Technical and Regulatory Guide).** Second revised edition. Henri Cettour. Éditions Le Moniteur, Paris, France. 367 pp.; 337 illus.; 21 tables; indexed; glossy semisoft bound in multicolor with red, black, white, and blue lettering; 11.0 by 11.6 inches (230 × 295 mm). Printed in the French language.

A book emphasizing the design, planning, construction, and maintenance of sports facilities. There is a preface by Edward Landrain. A list of selected references is in the back plus a small glossary.

Chapman, Geoffrey Peter (Dr.)

Geoff Chapman was born in England, UK, and earned a BSc degree with *first class honors* in Botany in 1955 and a PhD degree in Cytogenetics in 1958, both from the University of Birmingham. He served as Senior Research Fellow in the Botany

Department at the University of West Indies, Jamaica, from 1961 to 1969, as Lecturer at St Johns College, York, UK, from 1970 to 1971, and as Senior Lecturer in the Biological Science Department at Wye College, University of London, UK, from 1971 to 1997, when he retired. Dr. Chapman worked in the area of reproductive biology of grasses.

1996† **The Biology of Grasses.** G.P. Chapman. CAB International, Oxon, England, UK. 273 pp.; 63 illus.; 9 tables; indexed by author, by taxonomy, and by subject; glossy hardbound in multicolor with white lettering; 6.1 by 9.2 inches (156 × 234 mm).

A scholarly text on the evolution, domestication, and taxonomy of grasses worldwide. There is an extensive list of references in the back and also a detailed glossary of terms related to grasses.

Chase, Mary Agnes (Dr.)

Agnes (Meara) Chase (1869–1963) was born in rural Iroquois County, Illinois, USA, with her early years spent in Chicago. She obtained a grammar school education before employment in clerical positions. During the 1890s she worked nights as a proofreader for the **Inter Ocean** newspaper, which allowed her to collect and illustrate plants during the day and take courses at the Lewis Institute and at the University of Chicago. From 1901 to 1903, she was an Assistant in Botany at the Field Museum of Natural History in Chicago where she continued to perfect techniques for artistic line drawings of plants. Chase was appointed a Botanical Illustrator with the United States Department of Agriculture in Washington, DC, in 1903. She subsequently served as Scientific Assistant in Agrostology from 1907 to 1923, working with Dr. Albert Hitchcock; as Assistant Botanist from 1923 to 1925; as Associate Botanist from 1925 to 1936; and then as Senior Botanist in charge of all systemic agrostology. When officially retired from the USDA in 1939, Chase was appointed Research Associate for the Smithsonian Institution where she worked full time in the US National Herbarium. Following Dr. Hitchcock's death in 1935, she continued his research activities at the Washington Grass Herbarium and wrote and illustrated the revision of Dr. Hitchcock's 1950 **Manual of Grasses of the United States**. Dr. Chase was awarded Honorary Fellow of the Smithsonian Institute in 1958 and an Honorary Doctor of Science degree from the University of Illinois in 1958. She continued her studies at the Herbarium until her death at age 94.

1922† **First Book of Grasses: The Structure of Grasses Explained for Beginners.** First edition. Agnes Chase. The Macmillan Company, New York City, New York, USA. 129 pp., 215 illus., no tables, small index, cloth hardbound in green with gold and black lettering, 4.8 by 7.2 inches (123 × 184 mm).

A book delineating a basic method for teaching the identification of grasses via the structures of the inflorescences. It is organized with twelve lessons formatted by genera from the simplest to most complex seed head.

Revised in 1937 and 1977.

1937† **First Book of Grasses: The Structure of Grasses Explained for Beginners.** Second revised edition. Agnes Chase. W.A. Silveus, San Antonio, Texas, USA. 140 pp., 225 illus., no tables, small index, cloth hardbound in green with black lettering, 4.7 by 7.2 inches (120 × 184 mm).

An update of a book delineating a basic method for teaching the identification of grasses via the structures of the inflorescence.

Revised in 1959.

Figure 12.21. Horse-drawn, riding, reciprocating sickle bar cutter. It was best used on tall coarse-leaved grasses, circa 1905. [10]

1950† **Manual of the Grasses of the United States.**˙ Second revised edition. Coauthor.
　　　See *Hitchcock, Albert Spear—1950.*

1959　**First Book of Grasses: The Structure of Grasses Explained for Beginners.** Third revised edition. Agnes Chase. Smithsonian Institution Press, Washington, D.C., USA. 149 pp., 229 illus., no tables, small index, cloth hardbound in green with black lettering plus a jacket in greens with black lettering, 5.2 by 8.0 inches (131 × 202 mm).

　　　A minor revision of a book delineating a basic method for teaching the identification of grasses via structures of the inflorescence. There is an interesting foreword about Dr. Chase by Leonard Carmichael.
　　　Reprinted in 1964, 1968, and 1977.
　　　Revised in 1996.

1996　**Agnes Chase's First Book of Grasses: The Structure of Grasses Explained for Beginners.** Coauthor.
　　　See *Clark, Lynn Gail—1996.*

Chivers, Ian Harold

Ian Chivers was born in Yallourn, Victoria, Australia, and earned a BSc degree in Plant Science in 1978 and a Master of Applied Science degree in Agronomy in 1988, both from the University of Melbourne. Chivers has been involved as a private turfgrass consultant and a grass breeder and seed producer serving as CEO of Racing Solutions since 1982 and as CEO of Native Seeds since 1988. He served on the Board of Directors of the International Turfgrass Society from 2001 to 2009.

2002†　**Sports turf & amenity grasses: a manual for use and identification.** Coauthor.
　　　See *Aldous, David Ernest—2002.*

Christians, Nick Edward (Dr. and *Professor*)

Nick Christians was born in Mason City, Iowa, USA, and earned a BSc degree in Outdoor Recreation and Forestry from Colorado State University in 1972, and both an MSc degree in 1977 and a PhD degree in 1979 in Agronomy and Turfgrass from Ohio State University, studying with Dr. Dave Martin. He joined the faculty as an Assistant Professor at Iowa State University (ISU) in 1979 and rose to Full Professor in 2000. His responsibilities have been in turfgrass teaching and research. He was elected to the Crop Science Society of America (CSSA) Board of Directors and Turfgrass Science Division C-5 Chair in 1989; he was also elected to the Board of Directors of the American Society of Agronomy (ASA) from 1994 to 1996. Dr. Christians received the Iowa Legislative Teaching Excellence Award in 1986, the ISU College of Agriculture Louis Thompson Advisor Award in 1889, the American Society of Horticulture Science (ASHS) Outstanding Undergraduate Educator Award in 1991, the ASHS Extension Education Aids Award in 1995, Fellow of ASA in 1995 and of CSSA in 1996, the Nebraska Turfgrass Foundation President's Award in 1995, the Iowa Inventor of the Year Award in 1998, the CSSA Turfgrass Science Award in 1999, the ISU Agriculture College Faculty Member of the Year in 2002, and the ISU Foundation Award for Excellence in Academic Advising in 2004.

1992†　**The Mathematics of Turfgrass Maintenance.** First edition. Coauthor.
　　　See *Agnew, Michael Lewis—1992.*

1997†　**The Mathematics of Turfgrass Maintenance.** Second revised edition. Nick E. Christians and Michael L. Agnew. Ann Arbor Press, Chelsea, Michigan, USA. 149 pp., 43 illus., no tables, very small index, glossy hardbound in light lavender with black lettering, 6.1 by 9.3 inches (155 × 236 mm).

　　　A revision of a small book on mathematical procedures involved to implement cultural practices on extensive turfgrass areas. Included are calculations for area, volume, fertilization, pesticide applications, spreader and sprayer calibrations, irrigation, and seeding rates. Practice problems are included at the back of each chapter.
　　　For the first edition, see *Agnew, Michael Lewis—1992.*
　　　Revised in 2000 and 2008.

1998†　**Fundamentals of Turfgrass Management.** First edition. Nick Christians. Ann Arbor Press, Chelsea, Michigan, USA. 301 pp., 135 illus., 36 tables, small index, glossy hardbound in multicolor with white lettering, 7.0 by 10.0 inches (177 × 254 mm).

　　　An introductory book on the principles of turfgrass culture. There are chapters on athletic field management, sod production, professional lawn care, and golf course maintenance. Selected references are included for each chapter.
　　　Revised in 2004 and 2007.

2000†　**The Mathematics of Turfgrass Maintenance.** Third revised edition. Nick E. Christians and Michael L. Agnew. Ann Arbor Press Inc., Chelsea, Michigan, USA. 176 pp., 51 illus., 6 tables, very small index, glossy hardbound in yellow with black lettering, 6.0 by 9.3 inches (253 × 235 mm).

　　　Contains minor revisions to the 1997 edition plus the addition of chapters on the metric system and the application of small amounts of materials to glasshouse pots.
　　　Revised in 2008.

2002† **Scotts Lawns—Your Guide to a Beautiful Lawn.** Nick Christians and Ashton Ritchie. Meredith Books, Des Moines, Iowa, USA. 192 pp.; 426 illus. in color; 7 tables; large index; glossy semisoft bound in multicolor with white, yellow, and black lettering; 8.1 by 10.9 inches (206 × 277 mm).

An extensively illustrated book on the culture and pest management of lawns plus site preparation and establishment of turfgrasses.

2004† **Fundamentals of Turfgrass Management.** Second revised edition. Nick Christians. John Wiley & Sons Inc., Hoboken, New Jersey, USA. 365 pp.; 169 illus. with 52 in color; 41 tables; small index; glossy hardbound in multicolor with white, black, and yellow lettering; 6.9 by 9.9 inches (175 × 251 mm).

A revised update with color photographs added.

Revised in 2007.

2007† **Fundamentals of Turfgrass Management.** Third revised edition. Nick Christians. John Wiley & Sons Inc., Hoboken, New Jersey, USA. 380 pp., 237 illus. with 91 in color, 47 tables, indexed, glossy hardbound in multicolor with green and black lettering, 7.5 by 9.2 inches (191 × 233 mm).

This is an updated version of the second edition.

2008† **The Mathematics of Turfgrass Maintenance.** Fourth revised edition. Nick E. Christians and Michael L. Agnew. John Wiley & Sons Inc., Hoboken, New Jersey, USA. 168 pp., 59 illus., 2 tables, very small index, glossy hardbound in blues and white with white and dark blue lettering, 6.0 by 9.0 inches (152 × 228 mm).

A revised edition of the 2000 version.

Civil Aeronautics Administration, United States

1949† **Airport Turfing.** Civil Aeronautics Administration, United States Department of Commerce, Washington, D.C., USA. 40 pp., 9 illus., 4 tables, no index, fold-stapled semisoft covers in dark green with white lettering, 7.7 by 9.9 inches (195 × 251 mm).

∗ A **rare**, small, pioneering guide on the development, establishment, and maintenance of turfgrasses for airports, with emphasis on runways. It is oriented to owners, operators, and builders of airports in the United States. The evolution of airfields was summarized as follows:

One of the first problems encountered in making the aircraft a practical vehicle was that of taking off and landing. The Wrights, from 1901 until 1910, used a monorail car for launching their machine—both the gliders and the powered versions. Landings were made on skids on the sands of Kitty Hawk and the grassy meadows of the Miami Valley near Dayton, Ohio. In 1910 the Wrights abandoned the monorail car and attached wheels directly to the skids of their aircraft.

Experimenters in Europe in the early 1900's attempted to launch their would-be flying machines from floats, but were unsuccessful. In 1914, the French Army offered prizes and held a competition for aircraft equipped with landing gears that would take off and land on a plowed field. Results were not too satisfactory.

During World War I, the so-called military airfields were simply tracts of fairly level and smooth ground, which had been used for grazing or farming. About the only work done on them was some ditching for drainage purposes. These fields were generally satisfactory for the type of airship used at that time.

The municipal and private airports for many years following the First World War were turf fields, without hard-surfaced runways. The type of turf used was given little attention, except in isolated cases. Since the planes were light in weight and few in number, a fairly smooth meadow proved satisfactory as an airport.

As aircraft increased in weight and speed, it became necessary on major airports to provide paved runways capable of supporting heavier planes in all kinds of weather and of withstanding wear from increased frequency of traffic. A good turf landing strip, however, is still favored by the majority of operators of the lighter weight planes. It is cheaper to construct and maintain, and its use results in more economical aircraft operation through reduced tire wear and damage to the landing gear.

With the increased knowledge and experience in turf culture, soil characteristics, grading, compaction, and drainage, gained during and since World War II, more satisfactory and durable turfed landing areas are now being developed. Properly established and maintained turf strips have much greater resistance to wear and recover more rapidly from minor damage than the old-type "pasture" field. Such strips are now being used by relatively heavier planes through greater extremes of weather. Turf also performs several other important functions on all types of airports.

Figure 12.22. Tractor-drawn, three-gang, side-wheel-drive, reel Worthington Triple Mower, USA, circa 1926. [2]

Čižek, Jan Josip (Dr. and Professor)

Jan Čižek (1925–2006) was born in Valpovo, Croatia, Yugoslavia, and earned a PhD from the University of Zagreb, Croatia, in 1956. He became a Professor "Provizvodnja Krmnog Bilja" at the University of Zagreb, specializing in forage crops.

2007† **Ukrasni Travnjak (Decorative Turfgrass).** Jan Čižek, Petař Bošković, and Nikola Samardžija. Školska Knjiga, Zagreb, Croatia. 88 pp., 93 illus. with 76 in color, 10 tables, no index, glossy hardbound in greens with white lettering, 6.6 by 9.5 inches (167 × 240 mm). Printed entirely in the Yugoslav language.

A small book on the establishment and culture of decorative lawns. It is practically oriented to lawn enthusiasts under conditions in Croatia. A total of 1,000 copies were printed.

Clare, Kenneth Edward[3]

1961† **The Use and Control of Vegetation on Roads and Airfields Overseas.** K.E. Clare. Road Research Technical Paper No. 52, Department of Scientific and Industrial Research, London, England, UK. 48 pp., 29 illus., 9 tables, no index, semisoft bound in tan with black lettering, 5.9 by 9.6 inches (150 × 245 mm).

A bulletin concerning early 1960s practices in the construction and culture of vegetation on airfields outside England, including Africa, Cyprus, Fiji, Hong Kong, Malta, Singapore, and the West Indies. There is emphasis on airfield runways.

1963† **The Management of Vegetation on Airfields Overseas.** K.E. Clare. Road Research Technical Paper No. 65, Road Research Laboratory, Department of Scientific and Industrial Research, London, England, UK. 35 pp., 12 illus., 1 table, no index, semisoft bound in tan with black lettering, 5.9 by 9.5 inches (150 × 241 mm).

There are some revisions to the 1961 version.

Clark, Lynn Gail (Dr. and Professor)

Lynn Clark was born in West Orange, New Jersey, USA, and earned BSc degrees in Botany and in Horticulture from Michigan State University in 1979 and a PhD degree in Systemic Botany from Iowa State University (ISU) in 1986. She then joined the faculty in the Department of Botany at ISU and, subsequently, became a member of the Department of Ecology, Evolution and Organismal Biology at ISU in 2003. Her research focus is on the systemics, morphology, and evolution of grasses, especially woody bamboos. She is now Director of the Ada Hayden Herbarium and the Biological and Premedical Illustration Program.

1996 **Agnes Chase's First Book of Grasses: The Structure of Grasses Explained for Beginners.** Fourth revised edition. Lynn G. Clark, Richard Walter Pohl, and Agnes Chase. Smithsonian Institution Press, Washington, D.C., USA. 127 pp.; 124 illus.; no tables; small index; glossary semisoft bound in taupe, teal, and white with brown lettering; 5.5 by 9.0 inches (140 × 230 mm).

An update of the classic book delineating a basic method for teaching the identification of grasses via the structures of the inflorescence. Lessons 1 and 2 have been combined, and a lesson on bamboos has been added. Some illustrations have been added.

Clarke, Bruce Bennett (Dr. and *Professor*)

Bruce Clarke was born in Englewood, New Jersey, USA, and earned both BSc and PhD degrees from Rutgers University in 1977 and 1982, respectively. He joined the Plant Pathology faculty at Rutgers University in 1982 and became an Associate Professor in 1988 and Full Professor in 1997. Dr. Clark has served as Director of the Center for Turfgrass Science since 1993 and Chair of the Plant Pathology Department from 1999 to 2001. He has been active in teaching, extension, and research in turfgrass pathology, emphasizing ectotrophic root and crown infecting fungi. In 2005, Dr. Clarke was awarded the Ralph Geiger Endowed Chair in Turfgrass Science. He was elected President of the North East Division of the American Phytopathological Society for 2001 and appointed to the International Turfgrass Society Board of Directors from 2001 to 2009. Dr. Clarke has received both the Team Research Award in 1965 and the Weissblat Distinguished Service Award in 2001 from Rutgers University, the Distinguished Service Award of the North East Division of the American Phytopathological Society in 2001, and Fellow of the American Society of Agronomy in 2004.

1992† **Compendium of Turfgrass Diseases.** Coauthor.
> See *Smiley, Richard Wayne—1992.*

1993† **Turfgrass Patch Diseases Caused by Ectotrophic Root-Infecting Fungi.** Bruce B. Clarke and Ann B. Gould (editors) with nine contributing authors: B.B. Clarke, A.B. Gould, R.W. Smiley, N. Jackson, P.J. Landschoot, W.W. Shane, J.C. Stier, S.T. Nameth, and P.H. Dernoeden. APS Press, The American Phytopathological Society, St. Paul, Minnesota, USA. 161 pp., 77 illus. with 70 in color, 10 tables, no index, glossy semisoft bound in multicolor with green and white lettering, 6.0 by 9.0 inches (152 × 228 mm).
> A scholarly book on the current status of research on patch diseases of turfgrasses, including the biology, taxonomy, epidemiology, and detection. It contains a list of selected references at the end of each chapter.

1996 **Plagas y Enfermedades de los Céspedes (Compendium of Turfgrass Diseases).** Coauthor.
> See *Smiley, Richard Wayne—1996.*

Clifford, Harold Trevor (Dr. and *Professor* Emeritus)

Trevor Clifford was born in Melbourne, Victoria, Australia, and earned a BSc degree from the University of Melbourne in 1948, an MSc degree with *first class honours* from the University of Melbourne in 1951, a PhD degree from Durham University in England in 1955, and a Doctor of Science (DSc) degree from the University of Melbourne in 1974. He has served as a Lecturer in Agricultural Botany at the University College, Ibadan, Nigeria, from 1955 to 1958, became a Lecturer in Botany at the University of Queensland in 1958, and, subsequently, became a Senior Lecturer, Reader, and then Professor. Dr. Clifford retired in 1992. He was awarded Fellow of the Australian Institute of Biology. Dr. Clifford has been Honorary Research Associate of the Queensland Herbarium and Honorary Fellow (Geoscience) of the Queensland Museum, working with fossils on botanical taxonomy and geological problems related to evolution.

1977† **Identifying Grasses: data, methods, and illustrations.** H.T. Clifford and L. Watson. University of Queensland Press, St. Lucia, Queensland, Australia. 146 pp., 357 illus., 8 tables, indexed, cloth hardbound in green with white and black lettering and a glossy jacket in multicolor with white lettering, 9.9 by 10.9 inches (252 × 278 mm).
> A book that employs a computer approach to the scientific identification of grasses, including vegetative characteristics. It contains excellent close-up photographs of the leaf blade surface and leaf cross sections for a wide array of grass species.

Clouston, David W.A. (Dr.)

David Clouston (1901–1965) was born in Orkney, Scotland, UK, and earned a BSc degree in 1923, an MA degree in 1924, and a DSc degree in 1935, all in Agriculture from Edinburgh University. He was employed by the Ministry of Agriculture at Belfast University, Northern Ireland, until 1929, then was Lecturer of Agricultural Botany, and, subsequently, became Deputy Principal and Head of the Departments of Agricultural Botany, Mycology, and Grassland Husbandry at the North of Scotland College of Agriculture at Aberdeen until he retired in 1962.

1937† **The Establishment and Care of Fine Turf for Lawns and Sports Grounds.** First edition. David Clouston. D. Wyllie & Son, Aberdeen, Scotland, UK. 121 pp., 50 illus., 1 table, no index, cloth hardbound in aqua with black lettering, 5.5 by 8.6 inches (139 × 218 mm).

∗ A **very rare**, small book on the establishment and care of turfgrasses for greens, lawns, and sport fields. It is oriented to the cool-season turfgrass conditions in the United Kingdom. Major emphasis is placed on descriptions of species of turfgrasses and weeds plus some insects and diseases. The publisher D. Wyllie & Son was a seedsmen and nurserymen company in Aberdeen. All drawings were prepared by John G. Taylor. Dr. Clouston discussed the use of ammonium sulfate (NH_4SO_4) in weed control as follows:

> When the acidifying effects of sulphate of ammonia on the soil were first established, the tendency was to attribute the improvement to acid intoxication. An acid soil is undoubtedly advantageous in the production

of a turf, since grasses such as *Agrostis, Festuca,* and others grow well naturally under acid conditions and can withstand a degree of acidity in the soil which is harmful to many of the broad-leaved weeds such as daisies, plantains and clover. Sulphate of ammonia has also a manurial effect which stimulates the growth of grasses and enables them to compete more successfully with weeds. There is also the further point that many of the common weeds in lawns are very low in carbohydrates and appear to be more susceptible than grasses to injury from ammonia compounds in the soil. These factors doubtless all participate to some extent in making sulphate of ammonia an efficient lawn or sports ground improver. The selective corrosive action of sulphate of ammonia on broad-leaved weeds appear, however, to be one of the most important assets of sulphate of ammonia.

Sulphate of ammonia treatment by itself may not be sufficient to eradicate all the weeds in a lawn; daisies, plantains, clover, selfheal and mouse-ear chickweed generally yield to sulphate of ammonia treatment. Others such as pearlwort and moss may require treatment with sulphate of iron, which has also a corrosive action, and generally it may be said that weed eradication is hastened when a small proportion of sulphate of iron is added along with the sulphate of ammonia.

In the treatment of lawns and sports grounds with sulphate of ammonia particular attention should be paid to points such as the following if good results are to be obtained:

(1) Sulphate of ammonia should not be applied immediately before or after mowing; if possible, it should not be applied during the period one to two days before and after mowing. If sulphate of ammonia is applied to freshly cut grass it may have a harmful scorching effect on the freshly cut surface and thus injure the grass as well as the weeds.

Revised in 1939.

1939† **The Establishment and Care of Fine Turf for Lawns and Sports Grounds.** Second revised edition. David Clouston. D. Wyllie & Son, Aberdeen, Scotland, UK. 126 pp. plus 2 pp. of advertising; 50 illus., mostly photographs; 2 tables; no index; cloth hardbound in aqua with black lettering; 5.5 by 8.5 inches (139 × 216 mm).

* This second edition contains some updated changes. It is a **very rare** book. Lawn sand has been widely used for weed control in turfgrasses in the United Kingdom, as described by Dr. Clouston in the following:

LAWN SAND.—The use of lawn sand has done much to maintain sports turves in relatively weed-free condition. The important ingredients are sulphate of ammonia and calcined iron sulphate made up as follows:—

Sulphate of ammonia   3 parts;
Calcined iron sulphate, 1 part;
Sand or soil,   20 parts.

The mixture is usually applied at the rate of 3–4 oz. per sq. yd. every 10–14 days until the weeds are eradicated. Several dressings may be necessary. It is essential that the lawn sand should be applied during a dry spell, for, if rain ensues, the value of the chemical as a weed killer is largely lost. On the other hand do not apply during a period of drought or severe scorching will occur. Lawn sand will cause damage to the grass if applied immediately after cutting. It is desirable to apply it two to three days before mowing or two to three days after mowing, where possible.

Once weed eradication has been successfully accomplished applications should cease until again required.

The deep rooted weeds such as dandelion, cat's ear, etc., which are resistant to lawn sand, are best treated with 10% arsenic acid or 10% sodium chlorate, by means of a weed injector. The latter is inserted near the crown and a small quantity of the weed killer solution injected. Arsenic is poisonous, and this must be borne in mind if sheep are grazing, as difficulties might ensue. Care should be exercised also if drinking water is involved in any way. The chlorate would be preferable under these circumstances. There is a certain amount of danger of fire with chlorate. Waterproof clothing should be worn when this chemical is being applied, and it should be washed from the clothes afterwards.

Cobb, Patricia Ann Powell (Dr. and *Professor* Emeritus)*

Pat Cobb was born in Montgomery, Alabama, USA, and earned a BSc degree in Biology and Chemistry from Huntingdon College in Montgomery in 1961. She taught high school chemistry and physics for five years. Subsequently, she completed both MSc and PhD degrees in Entomology from Auburn University in 1967 and 1973, respectively. She joined the faculty at Auburn University in 1978, retiring as a Full Professor in 1999. She was active in adult extension programs involving insect problems of turfgrasses and ornamentals, with an emphasis on mole crickets and fire ants. Dr. Cobb

Figure 12.23. Worthington Quintuplex side-wheel-drive, reel gang mower. Frame designed to be pulled as a single 30-inch-wide mower, a 3-gang triple 84-inch-wide mower, or a 5-gang 132-inch-wide mower, circa 1924. [34]

received the Distinguished Achievement for Extension Excellence from the Southeastern Branch of the Entomological Society of America in 1989, the Award of Merit from the National Association of County Agricultural Agents in 1998, and the USGA Green Section Award in 2001.

1998† **IPM Handbook for Golf Courses.** Coauthor.
See *Schumann, Gail Lynn—1998.*

Cockerham, Stephen T

Steve Cockerham was born in Elwood, Indiana, USA, and earned a BSc degree in Agronomy and Turfgrass from Purdue University in 1961, studying with Dr. Bill Daniel; an MSc degree in Agronomy and Turfgrass from New Mexico State University in 1966; and an MBA degree from Southern Illinois University, Edwardsville, in 1985. He served as Superintendent at the Country Club of Decatur, Illinois, from 1961 to 1962. From 1962 to 1964, he was in the fourth US Peace Corps group formed and served in El Salvador as a surveyor for the Salvadoran Soil Conservation Service. Then, he became Assistant Lackawanna County Agent in Scranton, Pennsylvania, from 1964 to 1968, Research Specialist for the Chevron Chemical Company from 1968 to 1970, Research Director at Cal-Turf Inc. from 1970 to 1975, and Managing General Partner of Rancho Verde Turf Farms from 1977 to 1983. Steve Cockerham then became Superintendent of Agricultural Operations at the University of California, Riverside, in 1983. He was elected President of the Southern California Turfgrass Council for 1979 to 1980, President of the American Sod Producers Association for 1981 to 1982, President of the Sports Turf Managers Association in 1989, and Chair of the A-7 Experiment Station Managers Division of the American Society of Agronomy in 1998.

1985† **Turfgrass Water Conservation.** Coeditor.
See *Gibeault, Victor Andrew—1985.*

1988† **Turfgrass Sod Production.** Stephen T Cockerham. Publication 21451, Division of Agricultural and Natural Resources, University of California, Oakland, California, USA. 84 pp.; 15 illus.; 10 tables; no index; glossy semisoft bound in green, brown, and white with white lettering; 6.0 by 9.0 inches (152 × 228 mm).

A small book that addresses the business of commercial sod production in the United States, using a hypothetical sod company. It is a one-of-a-kind book that is based on the author's experiences in this industry.

2004† **Establishing and Maintaining the Natural Turf Athletic Field.** Stephen T Cockerham, Victor A. Gibeault, and Deborah B. Silva. Publication 21617, University of California Agriculture and Natural Resources, Oakland, California, USA. 60 pp., 20 illus. plus 8 in color, 8 tables, no index, fold-stapled glossy semisoft covers in multicolor with white lettering, 5.9 by 9.1 inches (150 × 231 mm).

A manual on the design, construction, turfgrass planting, and cultural practices for sports fields. Small specific sections on community fields, major league quality, and problem diagnosis are included in the back.

2011† **Turfgrass Water Conservation.** Second revised edition. Stephen T Cockerham and Bernd Leinauer (editors) with 18 contributing authors. Publication 3523, University of California, Agriculture and Natural Resources, Richmond, California, USA. 168 pp., 26 illus. with 19 in color, 6 tables, indexed, glossy semisoft bound in multicolor with white and blue lettering, 8.5 by 11.0 inches (217 × 279 mm).

A complete update of the 1985 first edition, based on the Victor A. Gibeault Symposium held at the University of California, Riverside. There are twelve chapters with supporting research cited at the end of each chapter. There also is a glossary of selected terms in the back.

Collen, Paul[2]

1997 **Référence gazon: Avec applications sur tableur (Lawn References: With Application on Spreadsheets).** Paul Collen. Editions Synthése Agricole, Bordeauz, France. 198 pp., numerous illus., 5 tables, no index, glossy semisoft bound in light green with purple and white lettering, 6.0 by 9.0 inches (163 × 244 mm). Printed in the French language.

A book concerning the establishment and culture of turfgrasses for ornamental, amenity, and sports field areas. It is practically oriented to turfgrass managers under the conditions in France.

Collins, William J.[3]

n.d. **Lawns, Golf and Sport Turf.** William J. Collins. Collin's Seed Service Co., Boston, Massachusetts, USA. 29 pp., no
circa illus., no tables, no index, fold-stapled textured semisoft bound in light green and orange with orange lettering, 5.4 by
192x 7.8 inches (133 × 191 mm).

* A **rare**, small booklet concerning turfgrass species descriptions, planting procedures, renovations with seeds, brief sections on the culture of putting greens and fairways, and earthworm problems.

Colt, Harry Shapland

Harry Colt (1869–1951) was born in St Amanda, England, UK, and studied law at Clare College, Cambridge University, earning a BA degree in 1890. He was a practicing lawyer and fine amateur golfer. While serving as secretary of the new Sunningdale Golf Club from 1901 to 1913, he evolved an interest in the design of golf courses. Subsequently, he essentially led development of the full-time profession of golf course architecture during the first half of the twentieth century. He pioneered the construction of golf courses in forested areas of England and Europe and utilized attractive, formal renderings. His protégés and eventual partners included such notable golf course architects as Charles H. Alison, John F.S. Morrison, and Dr. Alister Mackenzie.

1920† **Some Essays on Golf-Course Architecture.** H.S. Colt and C.H. Alison. "Country Life" and George Newnes Limited, London, England, UK and Charles Scribner's Sons, New York City, New York, USA. 69 pp., 18 illus., no tables, no index, quarter cloth hardbound in blues with black lettering with a jacket in light green with black lettering, 4.8 by 7.2 inches (123 × 183 mm).

* A **rare**, small, early book on golf course architecture and construction written by two widely recognized, pioneering golf architects, with four chapters by Harry S. Colt. It is a classic of its time. The chapter subject titles and listed contributing authors are as follows:

I.	Golf in the Nineties—C.H. Alison	VI.	Labour-Saving Machinery and the Cost of Construction
II.	The Modern Course–Framework		
III.	The Placing of Bunkers	VII.	Golf in Belgium
IV.	Construction	VIII.	Other Opinions
V.	Financial Considerations		

English golf courses and cool-season turfgrass conditions are emphasized. The book contains numerous interesting early photographs. Included is a photograph of Harry S. Colt. It is a must for collectors of historical golf and/or turfgrass books. Coauthors Harry Colt and Charles Alison commented on a 1920s trend to construct golf courses near urban centers:

> When the game of Golf began to achieve its widespread popularity in England, a demand naturally arose for golf-courses in the immediate neighbourhood of the large towns. Prior to 1890 the game had been played, for the most part, either on seaside links or on common land. But the new generation of golfers, many of whom were busy men, were unwilling or unable to waste precious time in travelling, and therefore tackled the problem of creating golf-courses on any land which was available in the suburbs.

1990† **Some Essays on Golf-Course Architecture.** First reprint edition. H.S. Colt and C.H. Alison. Grant Books Limited, Droitwich, Worcestershire, England, UK. 78 pp., 35 illus., no tables, cloth hardbound in green with gold lettering, 6.5 by 9.3 inches (166 × 237 mm).

A reprinting of a classic 1920 book by golf course architects H.S. Colt and C.H. Alison. Included are an introduction by Fred Hawtree and a commentary by Geoffrey Cornish titled "The Pervasive Influence of H.S. Colt in the History of Golf Architecture." Sixteen additional illustrations have been added. It was a limited first reprinting of 700 copies.

See original printing *Colt, Harry Shapland—1920.*

† *Reprinted a second time in 1993 by Grant Books* totaling 1,000 copies.

Conover, Herbert Spencer, Jr.

Herbert Conover (1908–1982) was born in Northampton, Pennsylvania, USA, and earned a BSc degree in Landscape Architecture from Pennsylvania State University in 1930. He was a Landscape Architect with the Division of Reservoir Properties of the Tennessee Valley Authority for 18 years. Subsequently, in 1956, he joined Chas. T. Main Inc. as Resident Landscape Architect for park and recreation areas at the St. Lawrence Power Project, Massena, New York, and then at the Niagara Power Project, Niagara Falls, New York. In 1963, Herbert Conover became an Associate Member of Chas. T. Main Inc. in Boston where he supervised the design and construction of over 50 state and municipal park and recreation areas. He retired in 1974.

1953   **Public Grounds Maintenance Handbook.** First edition. Herbert S. Conover. Tennessee Valley Authority, Knoxville, Tennessee, USA. 495 pp.; 273 illus.; 60 tables; detailed index; textured semisoft bound in salmon, grays, and black with black lettering; 7.9 by 10.4 inches (200 × 264 mm).

＊   This **very rare** handbook was originally published by the Tennessee Valley Authority (TVA) as an operating manual that focused on the specific grounds maintenance problems of the TVA reservations. It was developed originally for internal use. As a result of numerous requests from other agencies, it was then published on a limited distribution basis. A list of references is included in the back.

Revised in 1958 and 1977.

1958†   **Grounds Maintenance Handbook.** Second revised edition. Herbert S. Conover. F.W. Dodge Corporation, New York City, New York, USA. 501 pp., 195 illus., 102 tables, indexed, cloth hardbound in green with gold lettering and a jacket in light green with black lettering, 5.9 by 8.9 inches (150 × 226 mm).

A book on the establishment and maintenance of public, industrial, and institutional grounds. This revision was broadened in applicability beyond the TVA problems and also updated. There are chapters on turfgrasses, trees, and landscape plants. It encompasses the array of facilities found in parks and public grounds including roadways, parking areas, picnic areas, and sanitary facilities. There is a glossary of selected terms in the back.

Revised in 1977.

1977†   **Grounds Maintenance Handbook.** Third revised edition. Herbert S. Conover. McGraw-Hill Book Company Inc., New York City, New York, USA. 631 pp., 247 illus., 92 tables, indexed, textured cloth hardbound in black with gold lettering and a jacket in green and black with white lettering, 5.9 by 8.9 inches (151 × 226 mm).

This edition represents a major update and revision of design methods and equipment advances. Additions include maintenance of golf courses, tennis courts, and artificial turfs. An extensive appendix has been added.

Figure 12.24. Tractor-drawn, 12-foot-wide, 5-gang, side-wheel-drive, 7-blade, reel Worthington mower, USA, circa 1926. [2]

Converse, James Thayer

Jim Converse was born in Columbus, Ohio, USA. He was a golf course superintendent at Marysville Country Club from 1955 to 1956 and then joined the O.M. Scott and Sons Company, working in the Turf Research Group from 1957 to 1965, then as Manager of the Scotts Technical Institute from 1965 to 1970, and then as Director of Technical Services of the Pro Turf Division from 1970 to 1980. He is a noted illustrator and photographer of turfgrass plants and ecosystems. Jim Converse was responsible for the drawings in **Beard's Turfgrass Encyclopedia.**

1979† **Scotts Guide to the Identification of Grasses.** Jim Converse. The O.M. Scott and Sons Company, Marysville, Ohio, USA. 82 pp., 73 illus., no tables, index in front, glossy semisoft covers in green with gold lettering in a ring binder, 6.3 by 3.3 inches (159 × 83 mm).

A field guide-book for the identification of 60 cool- and warm-season turfgrasses of North America. It includes both written descriptions and drawings of the vegetative plant, flower, and seed. Included is a key to vegetative identification. It is a quality reference guide.

1979† **Scotts Guide to the Identification of Dicot Turf Weeds.** Jim Converse. The O.M. Scott & Sons Company, Marysville, Ohio, USA. 119 pp., 100 illus., no tables, index in front, textured semisoft covers in black with gold lettering in a ring binder, 6.0 by 3.2 inches (152 × 82 mm).

A field guidebook for the identification of 100 broadleaf-dicotyledonous weeds commonly found in turfgrasses in North America. It includes both written descriptions and drawings of the vegetative plant, the flower, and the seed. It is a quality reference guide.

n.d.† **Proturf Guide to the Identification of Grasses.** Jim Converse. The O.M. Scott and Sons Company, Marysville, Ohio, circa USA. 82 pp., 73 illus., no tables, index in front, textured semisoft covers in black with orange and green lettering in a early ring binder, 6.3 by 3.3 inches (159 × 83 mm).
198x This field guide book is essentially a reprint of the 1979 guide.
  † *Reprinted in 1985.*
    See *O.M. Scott and Sons Company.*

n.d.† **Proturf Guide to the Identification of Dicot Turf Weeds.** Jim Converse. The O.M. Scott & Sons Company, Marysville, Ohio, USA. 119 pp., 100 illus., no tables, index in front, textured semisoft covers in black with orange and green 198x lettering in a ring binder, 6.0 by 3.2 inches (152 × 82 mm).
This field guide book is essentially a reprint of the 1979 guide.
  † *Reprinted in 1985.*
    See *O.M. Scott and Sons Company.*

1982† **Scotts Guide To The Identification of Turfgrass Diseases and Insects.** Jim Converse. O.M. Scott & Sons Company, Marysville, Ohio, USA. 105 pp., 160 illus. in color, no tables, no index, glossy semisoft bound in multicolor with green and white lettering, 5.5 by 8.2 inches (140 × 209 mm).

A practical reference book on the identification and symptom diagnosis of turfgrass diseases, viruses, nematodes, insects, and mites. It includes a pest description, turfgrass injury symptoms, and life cycle. This comprehensive book is well illustrated in color.
  † *Reprinted in 1987.*

Cook, Thomas William (Associate Professor)

Tom Cook was born in Omak, Washington, USA, and earned a BSc in Agronomy and Soils from Washington State University in 1972 and an MSc in Crop and Soil Science from the University of Rhode Island in 1975. He served as a Turfgrass Research Associate at Washington State University at Puyallup from 1975 to 1977, then as an Assistant Professor from 1977 to 1983, and as an Associate Professor from 1983 to 2008 at Oregon State University (OSU). His key focus was teaching. Cook received the Teaching Merit Award from the North American College Teachers of Agriculture in 1985, the R.M. Wade Award for Excellence in Teaching from the OSU College of Agricultural Sciences in 1985, the Distinguished Professor Award of the OSU Agriculture Executive Council in 1988, the Distinguished Service Award of the Oregon Golf Course Superintendents Association (OGCSA) in 2006, and the Distinguished Service Award of the Golf Course Superintendents Association of America in 2006; he was inducted to the OGCSA Hall of Fame in 2010.

1983† **Construction and Maintenance of Natural Grass Athletic Fields.** Coauthor.
    See Goss, Roy Leon—1983.

Corbett, Lee Cleveland (Dr.)

Lee Corbett (1867–1940) was born in Watkins, New York, USA, and earned both BSc and MSc degrees from Cornell University in 1890 and 1896, respectively. He served as Assistant Horticulturalist at Cornell University from 1891 to

1893, as Professor of Horticulture and Forestry at South Dakota State College from 1893 to 1895, and as Professor of Horticulture at West Virginia University from 1895 to 1901. Then in 1901, he joined the United States Department of Agriculture (USDA) Bureau of Plant Industry in Washington, DC. He was a USDA Horticulturalist from 1901 to 1913, an Assistant Chief of the Bureau of Plant Industry from 1913 to 1915, a Senior Horticulturist in Charge of Horticultural and Pomological Investigations from 1915 to 1929, and as Principal Horticulturalist from 1930 to 1938, when he retired. He was elected President of the American Society of Horticultural Science for 1914. Dr. Corbett received the Silver Medal at the Paris Exposition in 1901, the Bronze Medal of the Louisiana Purchase Expedition in 1904, and an Honorary Doctor of Agriculture (DA) from the University of Maryland in 1921.

1906† **The Lawn.** L.C. Corbett. Farmers' Bulletin 248, United States Department of Agriculture, Washington, D.C., USA. 20 pp., 5 illus., no tables, no index, fold-stapled soft covers in white with black lettering, 5.5 by 8.5 inches (139 × 217 mm).

   * A **very rare** booklet on the site preparation and establishment of lawns in the United States, with a short section on lawn care in the back. It is oriented to the lawn enthusiasts. This bulletin is a must for collectors of historical turfgrass books. This booklet is one of the earliest in the United States.

1912† **Lawn Soils and Lawns.** Coauthor.
    See *Schreiner, Oswald—1912.*

Couch, Houston Brown (Dr. and Professor Emeritus)

Houston Couch (1924–2004) was born in Estill Springs, Tennessee, USA. After high school in June 1943, he became part of the first military draft of 18-year-olds in the United States. He was placed in a unit that was the youngest US Infantry Unit in the army—the 517th Parachute Infantry Combat Team, one of the army's first elite combat units. During WWII, the combat team endured some of the heaviest fighting of the European campaigns. Houston Couch was with the 517th as it began its campaign in Italy in June 1944. He was part of an early morning parachute jump into southern France and saw action at the Battle of the Bulge. He was wounded in action in Belgium in January 1945, for which he was awarded the Purple Heart. Subsequently, he earned a BSc degree in Agronomy from Tennessee Technological University in Cookeville in 1950 and a PhD degree in Vegetable Plant Pathology from the University of California, Davis, in 1954. He accepted an Assistant Professor position as a Forage Pathologist at Pennsylvania State University in 1954 and began a conversion to Turfgrass Pathology one and a half years later. Then in 1965, he became Head of the Department of Plant Pathology and Physiology at Virginia Polytechnic University, serving till 1974. He then refocused on research and teaching related to environmental causes of turfgrass diseases and their control, with emphasis on root disease ecology. Dr. Couch retired in 2003. He was elected President of the Northeastern Division of the American Phytopathological Society (APS) in 1964 and President of the Potomac Division of APS in 1972; he was also elected to the Governing Council of APS from 1974 to 1976. Dr. Couch received the Meritorious Achievement Award in 1976 and the R.D. Cake Award in 1990, both from the Virginia Turfgrass Council, and the USGA Green Section Award in 2003.

1962† **Diseases of Turfgrasses.** First edition. Houston B. Couch. Reinhold Publishing Corporation, New York City, New York, USA. 289 pp., 45 illus. with 13 in color, 5 tables, good index, cloth hardbound in yellow and green with yellow and green lettering and a jacket in green and white with black lettering, 5.9 by 8.9 inches (150 × 226 mm).

    A treatise on the causal pathogens of turfgrass diseases. Each disease is discussed in terms of symptoms, causal fungus, turfgrass hosts, disease cycle, and control. The text is documented by extensive references. A chapter on nematodes is included.
    Revised in 1973 and 1995.

1973† **Diseases of Turfgrasses.** Second revised edition. Houston B. Couch. Robert E. Krieger Publishing Company Inc., Huntington, New York, USA. 348 pp., 43 illus. with 13 in color, 7 tables, indexed, cloth hardbound in light greens with black lettering, 6.0 by 8.9 inches (152 × 226 mm).

    This book is a revised, expanded update. Documentation of the text has been expanded, several newly identified diseases of turfgrasses have been added, and the disease control chapter has been updated. There is a new chapter on viruses and mycoplasma-like organisms.
   † *Reprinted in 1976,* as a cloth hardbound in green with silver lettering version.
    Revised in 1995.

1995† **Diseases of Turfgrasses.** Third revised edition. Houston B. Couch. Robert E. Krieger Publishing Company, Malabar, Florida, USA. 421 pp., 325 illus. with 180 in color, 52 tables, detailed index, glossy hardbound in multicolor with gold lettering and a jacket in multicolor with white and black lettering, 8.9 by 11.9 inches (226 × 302 mm).

    This large, expanded book encompasses significant revisions that update information on newly recognized diseases of turfgrasses, improved diagnostic procedures, and recently developed cultural practices, resistant cultivars, and fungicides. A glossary and extensive list of references are included in the back.

Figure 12.25. Tractor-drawn, five-gang, side-wheel-drive, five-blade, reel F. & N. Quintet Fairway mower, Indiana, USA, circa 1928. [2]

2000† **The Turfgrass Disease Handbook.** Houston B. Couch. Krieger Publishing Company, Malabar, Florida, USA. 209 pp., 173 illus. with 126 in color, 4 tables, no index, glossy hard covers in multicolor with black lettering in a ring binder with backing, 5.5 by 8.5 inches (140 × 216 mm).

    A handbook on the description, diagnosis, and control of 44 diseases of cool- and warm-season turfgrasses. Also included are nematodes, algae, and black-layer problems. Integrated disease management is discussed.

Croce, Paolo

Paolo Croce was born in Rome, Italy, and earned a BSc degree in Crop Production from the University of Torino in 1981. He then studied Turfgrass Science at Texas A&M University in 1988 with Dr. Jim Beard. He was Coordinator of the Green Section of the Italian Golf Federation from 1989 to 2004; he was involved in teaching, research, and advising on golf course turfgrasses. Paolo Croce then formed a private turfgrass consulting firm, Verde Golf, advising throughout Italy via an office in Turin. He also has been teaching a short course in Golf Course Management at the University of Pisa since 2002 and serves on the Golf Environment Europe Steering Committee.

2000† **Tappeti Erbosi (Turfgrasses).** Coauthor.
    See *Panella, Adelmo—2000.*

2003† **La Pianificazine della Costruzione di un Percorso di Golf (Planning the Construction of a Golf Course).** Coauthor.
    See *Beard, James B—2003.*

2004† **Il campo pratica per il gioco del golf—Costruzione, manutenzione e gestione (The practice golf range—Construction, maintenance and management).** Coauthor.
    See *Casati, Maria Paolo—2004.*

2006† **Tappeti Erbosi—Cura, gestione e manutenzione, delle aree verdi pubbliche e private (Turfgrasses—Care, management and maintenance of public and private green areas).** Second revised edition. P. Croce, A. DeLuca, M. Falcinelli, F.S. Modestini, and F. Veronesi. L'Edagricole, Bologna, Italy. 340 pp., 231 illus. with 192 in color, 54 tables, indexed, semisoft bound in multicolor with green lettering, 7.7 by 10.2 inches (195 × 260 mm). Printed in the Italian language.

    A revision and update of the 2000 first edition. A key addition addresses vegetative propagation and planting techniques.

Crockford, Claude

Claude Crockford (1904–1995) was born in the rural Surrey Hills area of Victoria, Australia. He aspired to become an architect and studied several years at Swinburne College until the Depression. Starting in 1927, he apprenticed on the green staff at Yarra Bend Public Golf Course and then, in 1935, accepted the Assistant Greenkeeper position at the renowned Royal Melbourne Golf Club in Australia. Crockford was elevated shortly thereafter to Course Manager, serving from 1935 to 1975, when he retired.

1993† **The Complete Golf Course: Turf and Design.** Claude Crockford. Thomson Wolveridge & Associates Pty. Limited, Mount Eliza, Victoria, Australia. 150 pp., 58 illus. with 40 in color, 8 tables, no index, textured cloth hardbound in multicolor with gold lettering, 6.7 by 9.6 inches (169 × 244 mm).

A small book organized in three parts: Part I, Design and construction of golf course features; Part II, Turf production and maintenance; and Part III, Special problems in turfgrass maintenance. It represents the author's experiences gained over 40 years as golf course superintendent at the Royal Melbourne Golf Club in Melbourne, Australia. It was a limited edition printing of 1,000 copies.

Cubbon, Miles Hugo (Dr.)

Miles Cubbon (1896–1958) was born in Grand Valley, Pennsylvania, USA, and earned a BSc degree in 1921 and a PhD degree in 1925, both from Cornell University. He served on the faculty as Assistant Professor of Agronomy at Massachusetts State Agricultural College from 1926 to 1933, specializing in turfgrass-soil science, and then was associated with the United States Soil Conservation Service in Upper Darby, Pennsylvania.

1933† **Soil Management For Greenkeepers.** M.H. Cubbon and M.J. Markuson. W.F. Humphrey Press, Geneva, New York, USA. 152 pp., 10 illus., 2 tables, no index, cloth hardbound in greens with black lettering, 4.9 by 7.7 inches (125 × 197 mm).

✳ This is a **truly rare** book that is a must for collectors of historical turfgrass books. It is one of the earliest books devoted solely to soils, soil chemistry, drainage, fertilization, irrigation, and pesticides as related to turfgrasses for golf course greens and fairways. It was a unique book in its time and a first of its kind in terms of being science based. It was written in two parts, with the back one-third of the book devoted to golf course drainage, as authored by Miner J. Markuson. The chapter subject titles include the following:

Part I	
1. General Make–Up of Soils	8. Fertilization of Fairways
2. Fundamentals of Chemistry	9. Watering Greens
3. Plant Nutrients and Soil Acidity	10. Use of Weed Killers and Other Poisons
4. Effects of Organic Matter on Soils	Part II
5. Nitrogen Changes in Soils	1. Golf Course Drainage
6. General Considerations in Fertilizers	2. Engineering Methods
7. Fertilization of Golf Greens	3. Profile Leveling

Dr. Cubbon's observations on two ideas being promoted for use on golf courses in the United States in the 1930s were as follows:

Heating loam to kill weed seeds apparently gives satisfactory results in more cases than it fails. Whether the practice is economical can best be answered by the individual greenkeeper. I can see no real good reason for heating loam as a general plan. Weed seeds may be killed by the process, and some probably are. Heating soil generally increases the amount of soluble plant food in that soil. So when heated loam is used with good results, the question at once comes up, is the better growth of grass due to fewer weeds or to more soluble nitrogen and other plant food. If the good effects from heated loam are due to plant food, then why not add plant food directly? In most cases plant food in the form of fertilizers is much cheaper than in heated loam. Usually if grass grows thickly enough no weeds can get in. Better concentrate on the grass and let the weeds die a natural death.

Beware of all culture and preparations put on the market for the purpose of inoculating soils. First of all, the numbers of active organisms in such cultures are often too low to be worth while. Secondly, why try to grow legumes, or to induce them to get started, when nitrogen in ordinary fertilizers costs less? And thirdly, soils receive all the inoculation they need from dust blowing, from players tramping the course, and from additions of compost or topdressing. Soils are usually well supplied with organisms, but have too little energy material for the organisms to work on. Instead of buying inoculation, better buy food for the bacteria that are already there. Such food consists of organic matter and mineral elements, the same as plants need.

Cunningham, Samuel Alfred

Samuel Cunningham (1860–c. 1917) was born in Maryland, USA. His professional involvement was as a banker in New York City and as a worker in a safe deposit company in 1910. He served as the Chairman of the Green Committee of the Englewood Golf Club in Englewood, New Jersey, which hosted the USGA Open Championship in 1904. This may have stimulated his interest in turfgrass culture. A 1914 advertisement by The Coe-Mortimer Company in the **American Golfer** stated "Mr. S.A. Cunningham perfected Country Club Brands after many years of experimental work."

~

The **E. Frank Coe Company** was established in 1857. Evidently, E. Frank Coe started to make superphosphate of lime in 1860, which was primarily dicalcium phosphate. This precipitated bone fertilizer was then a byproduct of glue-stock manufacture from the skeletons of vertebrates. The $CaHPO_4$ was readily available to plants via the soil solution. The E. Frank Coe Company of New York was marketing "High Grade Animal Bone Fertilizers" by 1905, including American Lawn Fertilizer. Then in 1908, The Coe-Mortimer Company started to manufacture and market "Country Club Golf and Lawn Fertilizers" for a select private trade. This is one of the earliest records in the United States of branded fertilizer being marketed for a specific turfgrass use.

Elmore Frank Coe was born circa 1807 in Connecticut, USA. At age 30, in 1837, he purchased a company in New York City that manufactured glue and precipitated bone meal from animal bones. By 1870, he was living in Brooklyn, New York, with his occupation being listed in the census as a guano manufacturer. At that time, guano not only referred to bird excrement but also referred to fertilizer derived from other animal products such as bone and meat and fish scraps. By 1880, Frank Coe was listed in the census as a fertilizer manufacturer.

~

1914† **Lawns, Golf Courses, Polo Fields, Turf Courts, and How to Treat Them.** S.A. Cunningham and George D. Leavens. Coe-Mortimer Company, New York City, New York, USA. 40 pp.; 23 illus.; 11 tables, no index, fold-stapled semisoft covers in gray, green, orange and tan with green and orange lettering; 6.5 by 8.6 inches (165 × 218 mm).

* A **very rare**, early booklet on turfgrass selection, seed establishment, renovation, and culture of turfgrass for lawns, golf greens and fairways, polo fields, and tennis courts. It is oriented to conditions in the northeastern United States. There are scattered pages in the text concerning the fertilizers and worm controls available from The Coe-Mortimer Company. Soil preparation for putting greens using dynamite was described in 1914 as follows:

> If, however, the sub-soil is compact it is best to remove the entire top-soil and work first with the sub-soil itself. Sub-surface drains if needed should be put in place while the top-soil is removed. If the hard-pan or compact sub-surface is not too deep it may be broken up by exploding dynamite in holes bored from two to two and one-half feet deep and from ten to twenty feet apart. This shattering of the sub-soil often makes sub-surface drains unnecessary, but in this matter be guided by the advice of your engineer or turf expert.

D

Dahl, Karl Arnold Sixten (Dr.)

Arnold Dahl was born in Superior, Wisconsin, USA. He served in the US military in 1916 and 1917 and then earned a BSc degree, studying diseases of lawns and golf greens in 1924; an MSc degree, studying *Rhizoctonia* spp. of turfgrasses in 1925; and a PhD degree, investigating snow molds of turfgrasses in 1931, all in Plant Pathology from the University of Wisconsin, studying with Professor L.R. Jones. He was one of the earliest to receive both formal MSc and PhD education in the turfgrass pathology area. During this time, he also worked as a Disease Specialist for the USGA Green Section and the United States Department of Agriculture (USDA) under the leadership of Dr. John Monteith from 1927 to 1934. He also taught in the Biological Sciences Department at Des Moines University and at Kansas City University. He was subsequently an Agronomist for the US Soil Conservation Service in Washington, DC, from 1934 to 1942, the War Food Administration from 1943 to 1944, the Commodity Credit Administration from 1945 to 1947, the Production and Marketing Administration from 1947 to 1953, the Commodity Stabilization Service from 1954 to 1961, and, subsequently, the Agricultural Stabilization and Conservation Service.

1932† **Turf Diseases and Their Control.** Coauthor.
    See *Monteith, John L., Jr.—1932.*

* In 1932, coauthors Monteith and Dahl commented on dollar spot disease of turfgrasses, which was thought to be a species of *Rhizoctonia* prior to that time:

> *Cause.*—Dollarspot is a fungus disease caused by a species of Rhizoctonia, which in some respects is similar to the fungus causing brownpatch. The fungus lives in the soil and attacks the grass in a manner

similar to brownpatch (page 91 to 97). This fungus is generally more destructive at a somewhat lower temperature than is favorable for brownpatch. It, too, requires abundant moisture. An adequate supply of readily-available nitrogen, when other conditions are favorable, enables grass to quickly recover from an attack of the disease. Abundant sulphur and a decided deficiency of lime tend to encourage dollarspot on turf.

Dahlsson, Sven-Ove (Dr.)

Sven Dahlsson was born in Flen, Sweden, and earned a BSc degree in Plant Husbandry and a PhD degree in Agronomy, both from the Agricultural University of Uppsla, Sweden in 1967 and 1973, respectively. He served in turfgrass research and extension at W. Weibull AB from 1970 to 1979; and since 1980, he has been active as a consultant in turfgrass research and education, with emphasis on sports fields, for S-O Dahlsson Gräskonult HB. Dr. Dahlsson is the first PhD-educated scientist in Sweden and other Scandinavian countries to work full-time in the turfgrass area. He served on the Board of Directors of the International Turfgrass Society from 1997 to 2005.

1987† **Klippta Gräsytor (Mown Grass Areas).** Sven-Ove Dahlsson. Stad & Land, Nr 61, Alnarp, Techomatorp, Sweden. 165 pp., 92 illus., 42 tables, no index, semisoft bound in white and orange with black lettering, 6.8 by 9.6 inches (174 × 244 mm). Printed entirely in the Swedish language.

A book on the construction, establishment, and culture of turfgrasses under the cool-season turfgrass conditions in Sweden. Specific sections address diseases and weeds plus sod production. It contains an extensive list of references in the back. The book was written on behalf of the Swedish Agricultural University.

† *Reprinted in 1987* as a soft bound version in green and white.

Dallwitz, Michael John (Dr.)

Mike Dallwitz was born in Adelaide, South Australia, Australia, and earned a BA degree with *honors*, majoring in Pure Mathematics in 1964; a BSc degree with *honors*, majoring in Physics in 1965; and a PhD degree, majoring in Physics in 1969, all from the Australian National University. He served in the Division of Entomology of the Commonwealth Scientific and Industrial Research Organization (CSIRO) in Canberra, Australian Capital Territory, Australia, from 1970 to 2000, rising to Senior Principal Research Scientist. He was investigating ecological modeling and then computer-software work in biology. He won the University Prize in Pure Mathematics in 1965. Dr. Dallwitz was awarded the CSIRO Medal in 1996 for development of the DELTA taxonomic computer programs, which involve a format for coding taxonomic descriptions.

1992 **The Grass Genera of the World.** First edition. Coauthor.
See *Watson, Leslie—1992.*

1994 **The Grass Genera of the World.** Revised second edition. Coauthor.
See *Watson, Leslie—1994.*

Daniel, William Hugh (Dr. and *Professor* Emeritus)

Bill Daniel (1919–1995) was born in Sparkman, Arkansas, USA, and earned a BA degree in Social Science from Ouchita College in 1941. After serving as a bomber pilot in the US Air Force during WWII, he received a BSc degree in Agriculture from the University of Arkansas in 1947 and both MSc and PhD degrees in 1948 and 1950, respectively, from Michigan State University, specializing in Soil Science while studying with Dr. Jim Tyson. He served on the faculty in the Agronomy Department at Purdue University from 1950 to 1985, where he was active in turfgrass extension, research, and teaching, with emphasis on weed control and root-zone management. His development of these activities through the Midwest Regional Turf Foundation and Midwest Regional Turf Conference was the model emulated by the surrounding eight states in developing their own university turfgrass science programs. Under his guidance, Purdue University developed one of the earliest undergraduate majors in Turfgrass Management. He was elected American Society of Agronomy (ASA) Turf Management Division XI Chair in 1954, to the ASA Board of Directors for 1964 through 1967, and to the International Turfgrass Society Board of Directors from 1969 to 1973. Dr. Daniel was awarded Fellow of the ASA in 1963, Fellow of the Crop Science Society of America in 1963, the ASA Agronomic Service Award in 1973, the GCSAA Distinguished Service Award in 1975, an Honorary Member of the American Sod Producers Association in 1981, the United States Golf Association Green Section Award in 1984, the Council of the Sagamores of the Wabash Award from the Governor of Indiana in 1985, and the Distinguished Service Award of the Midwest Turf Foundation in 1994.

1979† **Turf Managers' Handbook.** First edition. W.H. Daniel and R.P. Freeborg. Business Publications Division, Harvest Publishing Company, Cleveland, Ohio, USA. 424 pp., 246 illus. with 96 in color, 194 tables, no index, cloth hardbound in salmon with gold lettering, 5.8 by 8.9 inches (148 × 226 mm).

A handbook on the construction, establishment, and culture of turfgrasses, as well as the management of turfgrass facilities. In the back, there are chapters covering the diverse types of turfgrass use. It also contains a glossary in the back.

Revised in 1987.

1980† **Turf Managers' Handbook.** Same 1979 first edition in a semisoft bound version with a light green cover and black and white lettering, 5.8 by 8.9 inches (148 × 226 mm).

1987† **Turf Managers' Handbook.** Second revised edition. W.H. Daniel and R.P. Freeborg. Business Publications Division, Harvest Publishing Company, Cleveland, Ohio, USA. 437 pp., 219 illus. with 32 in color, 211 tables, indexed, cloth hardbound in white with green lettering, 5.9 by 8.9 inches (151 × 227 mm).

A slight revision of the 1979 edition, primarily in the back sections. An index has been added.

Daniels, Robert Wesley (Dr. and Professor)

Bob Daniels was born in Windsor, Nova Scotia, Canada, and earned a BSc degree in Horticulture from McGill University in 1966, an MSc degree in Horticulture from Michigan State University in 1970, and a PhD degree in Turfgrass Science from Pennsylvania State University in 1989. Since 1975, he has been active in turfgrass teaching and research in the Department of Environmental Sciences at Nova Scotia Agricultural College, becoming an Associate Professor in 1981 and a Full Professor in 1989. He also is coordinator of the NSAC Turfgrass Research Center.

1989 **Golf Courses: A Guide to the Construction and Maintenance of Low to Medium Budget Facilities.** Robert W. Daniels. Nova Scotia Agricultural College, Truro, Nova Scotia, Canada. 66 pp., 16 illus., 6 tables, no index, glossy semisoft bound in brown with black lettering, 8.2 by 10.7 inches (208 × 272 mm).

A guide to construction, turfgrass establishment, and culture of modestly maintained turfgrasses for golf courses under conditions in Canada.

1990 **Recreational Fields—A Guide to Site Selection, Construction and Maintenance.** Robert W. Daniels. Nova Scotia Agricultural College, Truro, Nova Scotia, Canada. 72 pp., 37 illus., 8 tables, no index, glossy semisoft bound in cream with black and green lettering, 8.2 by 10.7 inches (208 × 272 mm).

A guide to the site selection, construction, turfgrass establishment, and culture of sports fields under conditions in Canada.

Danneberger, Tom Karl (Dr. and Professor)

Karl Danneberger was born in Champaign, Illinois, USA, and earned a BSc degree in Agronomy and Turfgrass from Purdue University in 1977, studying with Dr. Bill Daniel; an MSc degree in Horticulture and Turfgrass from the University of Illinois in 1979, studying with Dr. Al Turgeon; and a PhD degree in Plant Pathology and Turfgrass from Michigan State University in 1983, studying with Dr. Joe Vargas. He then joined the faculty at Ohio State University where he has been active in turfgrass teaching and research, with emphasis on physiological and ecological aspects. Dr. Danneberger was elected to the Crop Science Society of America Board of Directors for 2001 through 2003, the Turfgrass Science Division C-5 Chair in 2002, and the International Turfgrass Society Board of Directors for 2005 through 2009.

1993† **Turfgrass Ecology & Management.** T. Karl Danneberger. Franzak and Foster, G.I.E. Publishers, Cleveland, Ohio, USA. 201 pp.; 103 illus. in color; 39 tables; indexed; glossy hardbound in multicolor with white, green, and brown lettering; 7.0 by 9.2 inches (178 × 234 mm).

Figure 12.26. Tractor-drawn, 12-foot-cutting-width (3.7 m), 5-gang, 5-blade, side-wheel-drive, reel New Pennsylvania Super Mower, 1920s. [41]

A small book written for the advanced turfgrass undergraduate student. Topics addressed are climate, turfgrass environment, soil ecology, and population dynamics of turfgrass communities. It contains a glossary and a selected set of references in the back.

Davis, Fanny-Fern Gretchen (Dr. and Chair Emeritus of Biology)

Fanny-Fern Smith (1902–1989) was born in Red Bud, Illinois, USA. She received an AB degree in Botany and Zoology in 1923, an MSc degree in Plant Genetics in 1924, and a PhD degree in Plant Physiology in 1926, all from Washington University of St. Louis. She first taught botany at Lindenwood College for Women in St. Charles, and then, with new husband Dr. Everett Davis, moved to Blacksburg, Virginia. She was the first woman faculty member at Virginia Polytechnic Institute and State University (VPI&SU), teaching for one year before starting a family. She then joined the United States Golf Association Green Section at Arlington in 1938, where she served as a research botanist, advisory specialist, and ghost writer and editor of **Timely Turf Topics,** signing as "FFD." In 1943, she became Acting Director of the USGA Green Section, where she led the first developmental research program that resulted in 2,4-D becoming a key selective herbicide for broadleaf weed control in turfgrasses. In 1945, she resigned to devote more time to her family of three children. She accepted a part-time position as Turf Consultant for the National Capital Parks, which allowed her to continue the 2,4-D research on turfgrasses. Subsequently, in 1957, she became Associate Professor and Head of the Biology Department at George Mason College. Then she established the Biology Department at Okaloosa-Walton Community College in Niceville, Florida, in 1965, where she retired in 1968 as Chair *Emeritus* of Biology. Dr. Davis was honored with the USGA Green Section Award in 1975.

1947†   **Turf Weed Control With 2,4-D.** Fanny-Fern Davis. National Park Service, United States Department of the Interior, Washington, D.C., USA. 59 pp., 32 illus., 5 tables, no index, fold-stapled soft covers in black and white with black lettering, 10.5 by 8.0 inches (267 × 203 mm).

  ✻   The **very rare**, pioneering publication on broadleaf weed control in turfgrasses. It is a publication that summarizes the initial research on 2,4-D as a herbicide for the control of weeds in turfgrasses. The status of broadleaf weed control options prior to 1947 were summarized by Dr. Davis as follows:

> It has long been felt that the balance could be shifted in favor of the grass at the expense of the weeds much more rapidly, however, if the weeds could be killed directly, while simultaneously giving encouragement to the grass, making it possible for the grass to fill in promptly the areas originally occupied by the weedy species. Many different chemicals, notably the arsenicals and sodium chlorate, have been used for this purpose. However, the selectivity of these chemicals has not been entirely satisfactory. Even when relatively small amounts have been used, the turf grasses have suffered some temporary injury due to the burning action of the herbicides on the foliage. Although the recovery of the grass usually has been more or less complete, the initial setback resulting from the temporary burn has prevented the grass from filling in the areas occupied by the weeds promptly as the latter die. Moreover, in many instances it has been necessary to reduce the amount of herbicide used so much that the weeds as well as the grass have recovered, and a second or even a third application—each with its temporary burn to the grass—has been necessary to effect satisfactory control.

Davis, William Boyce (Extension Environmental Horticulturalist Emeritus)

Bill Davis was born in Clearfield, Pennsylvania, USA, and earned a BSc degree in Horticulture in 1950, a Master in Education degree in Agricultural Education in 1951, and an MSc degree in Landscape Horticulture in 1960, all from the University of California, Davis. From 1960 to 1987, he served as the Turfgrass and Landscape Specialist for the University of California, Davis. He was active in turfgrass adult extension and research, specializing in high-use turfgrass areas. Bill Davis received the Golf Course Superintendents Association of America (GCSAA) Distinguished Service Award in 1988.

1990†   **The Sand Putting Green—Construction and Management.** William B. Davis, Jack L. Paul, and Daniel Bowman. Publication 21448, Cooperative Extension, Division of Agriculture and Natural Resources, University of California, Oakland, California, USA. 22 pp., 19 illus. with 6 in color, 6 tables, no index, fold-stapled glossy semisoft covers in multicolor with black and white lettering, 8.4 by 10.9 inches (214 × 278 mm).

     A booklet in which the specifications and procedures for the construction of 100-percent sand putting greens under California conditions are discussed. The interpretations of the authors are presented. Also addressed are guidelines for the establishment and culture of turfgrasses on 100-percent sand greens.

Dawson, Robert Brian (OBE)

"RB" Dawson (1903–1969) was born in Belfast, Northern Ireland, UK, and earned a BSc degree with *honours* from the University of Manchester in 1925 specializing in General Science. He served on the staff of the Field and Chemical Department

Figure 12.27. Robert Dawson, circa 1960. [3]

of Rothamsted Experimental Station in Hertfordshire, England, from 1926 to 1929. During this time he was awarded an MSc degree in 1928 from the University of Manchester. He then became the first Director of the Board of Green-keeping Research (later named the Sports Turf Research Institute in 1951) at the time of its formation in 1929 and continued in that capacity for 34 years until his retirement in 1963. It was during this period that he was actively involved in turfgrass research and advisory activities. R.B. Dawson was elected President of the Horticultural Education Association in 1949. He was initiated as an OBE in 1956.

1932†   **Greenkeeping for Golf Courses.** R.B. Dawson and T.W. Evans. Scottish Golf Union (Western District), Glasgow, Scotland, UK. 60 pp., 3 illus., 6 tables, no index, fold-stapled textured semisoft covers in green with black lettering, 5.6 by 8.7 inches (141 × 222 mm).

    A compilation of four lectures on greenkeeping presented in 1931 and 1932 by two STRI turfgrass specialists. R.B. Dawson addressed sodding and seeding procedures, while Dr. T.W. Evans discussed composting, fertilizers, and fertilization that was then termed manuring.

1939†   **Practical Lawn Craft.** First edition. R.B. Dawson. Agricultural and Horticultural Handbooks, Crosby Lockwood & Son Limited, London, England, UK. 300 pp. plus 16 pp. of advertising, 46 illus., 14 tables, indexed, cloth hardbound in green with white lettering and a jacket in greens and white with green and white lettering, 5.4 by 8.5 inches (138 × 217 mm).

∗   A **relatively rare** book on the basics of turfgrass establishment and culture as they evolved in England in the 1930s. A set of chapters in the back discusses golf courses, bowling greens, lawn tennis courts, croquet courts, soccer fields, rugby fields, hockey fields, cricket grounds, polo fields, horse racecourses, cemeteries, and airfields. Some references are cited. Also discussed is the development and maintenance of turfgrasses in other portions of the world. This book is a must for collectors of historical turfgrass books. There is a foreword by Professor R. George Stapledon. R.B. Dawson proposed the following compost preparation procedure for material to be used in topdressing greens:

> In practice a good compost heap is best built up with alternate layers of soil (9 in. deep) and organic mat-ter (3 to 4 in. deep) the heap being allowed to stand for a period of about 12 to 18 months to ensure good decomposition. The material is then cut down and screened. Sharp sand may then be added. For most purposes a 3/16-in. mesh is adequate, but for finer turf a 1/8-in. mesh should be used, and when large quantities have to be handled a rotary screen is essential. It is advisable to test the material for weed seeds by placing some in a box and keeping it moist and warm.
>
> If dung is used as the source of organic matter it is important to realize its variability, due to the type of food fed to the animals, the kind of animal, the age of the dung, the storage, and the nature of the

> litter. Horse manure on peat moss-bedding is generally accepted as the best for this purpose, since more of the liquid excrement is retained in the peat and there is less likelihood of weed impurities than with straw litter.
>
> Compost heaps and prepared material should always be protected from the elements. A heat exposed to rain loses nitrogen, while screening is made difficult if the material is saturated with moisture. It is a good plan for a new heap to be in course of construction at the time the old one is being brought into use so that a continuous supply is available.
>
> All the dangers and anxieties of using soil and manure containing weed seeds in compost piles can readily be overcome by the process of sterilization by heat. There are three main methods by which this may be carried out, namely by baking, by steaming, and by electricity.

Revised in 1945, 1947, 1949, 1954, 1959, and 1968.

1945† **Practical Lawn Craft.** Second revised edition. R.B. Dawson. Crosby Lockwood & Son Limited, London, England, UK. 306 pp. plus 15 pp. of advertising, 51 illus., 17 tables, indexed, cloth hardbound in green with gold lettering and a jacket in greens and white with green and white lettering, 5.4 by 8.4 inches (138 × 214 mm).

✻ This **rare** second revision contains small changes, including additional information on mowing practices and the adjustment and maintenance of mowers. Note that pitch is an alternate word used in the United Kingdom for sport field. Foreword by Sir R. George Stapledon. R.B. Dawson had the following suggestions regarding pitch maintenance:

> During the winter months occasional rolling becomes necessary and the surface often becomes wet and greasy through the puddling action of the feet. Under such conditions the moisture is retained on the surface even though the ground may be well drained. Regular routine spiking and sanding will do much to prolong the life of the pitch and give better winter conditions. Rolling in winter often follows immediately play finishes for the day. This must be done to guard against a sudden fall in temperature causing freezing in the uneven state, thus possibly jeopardizing the next game. After sanding the surface, important association matches are played even if the turf be frozen.

Revised in 1947, 1949, 1954, 1959, and 1968.

1947† **Practical Lawn Craft.** Revised and reprinted unnumbered edition. R.B. Dawson. Agricultural and Horticultural Handbooks, Crosby Lockwood & Son Limited, London, England, UK. 306 pp., 52 illus., 22 tables, indexed, cloth hardbound in green with gold lettering and a jacket in greens and white with green and white lettering, 5.4 by 8.4 inches (138 × 214 mm).

✻ The book contains minor revisions. The foreword is by Sir R. George Stapledon.

Revised in 1949, 1954, 1959, and 1968.

1949† **Practical Lawn Craft.** Third revised edition. R.B. Dawson. Crosby Lockwood & Son Limited, London, England, UK. 315 pp. plus 43 pp. of advertising, 53 illus., 16 tables, indexed, cloth hardbound in green with gold lettering and a jacket in green and white with green and white lettering, 5.4 by 8.4 inches (138 × 214 mm).

✻ The revisions involved additions to the appendix, primarily weed and insect controls. The foreword is by Sir R. George Stapledon. In 1949, Dawson offered the following rationale for use of the acid theory in turfgrass culture:

> Much of the turf found on old-established sports grounds, golf courses, park lands and garden lawns is of this acid matted type. Such acid turf contains the desirable grasses (bent and fescue), some Yorkshire fog, scarcely any crested dog's-tail or perennial rye-grass, and very few weeds. It is characterized by low fertility and low pH. The miscellaneous plants which exist, for example heath bedstraw, wood-rush, cat's-ear, and creeping hawkweed, do not appreciably detract from the turf unless their proportions become very great. Under acid matted conditions worm activity in turf is either eliminated or very much reduced. Further, such turf provides a dry resilient sward during the winter months, though it has the disadvantage of being susceptible to drought and when once dried out re-wetting by rain is often very slow.

Revised in 1954, 1959, and 1968.

1954† **Practical Lawn Craft and Management of Sports Turf.** Fourth revised edition. R.B. Dawson. Agricultural and Horticultural Series, Crosby Lockwood & Son Limited, London, England, UK. 320 pp. plus 31 pp. of advertising; 73 illus.; 16 tables; indexed; cloth hardbound in green with gold lettering and a jacket in green, black, and white with white and black lettering; 5.4 by 8.3 inches (135 × 212 mm).

This edition contains significant revisions within the same chapter format. The photographs have been updated substantially, especially those depicting equipment. The foreword is by Sir R. George Stapledon.
Reprinted in 1955.
Revised in 1959 and 1968.

1957   **Practical Lawn Craft and Management of Sports Turf.** Fourth revised edition. R.B. Dawson, as translated by Boris Sigalov. Main Botanical Garden, USSR Academy of Sciences, Moscow, Russia, USSR. 221 pp.; 37 illus.; 9 tables; indexed; cloth hardbound in brown, tan, and white with brown, green, and white lettering; 5.7 by 8.7 inches (145 × 220 mm). Printed entirely in the Russian language.
    A Russian translation of the 1954 fourth revised edition.

1959†   **Practical Lawn Craft and Management of Sports Turf.** Fifth revised edition. R.B. Dawson. Agricultural and Horticultural Series, Crosby Lockwood & Son Limited, London, England, UK. 320 pp. plus 27 pp. of advertising; 73 illus.; 16 tables; indexed; cloth hardbound in green with gold lettering and a glossy jacket in green, white, and black with white and black lettering; 5.4 by 8.4 inches (138 × 215 mm).
    There are only minor revisions.
    Revised in 1968.

1960†   **Lawns—For garden and playing field.** First edition. R.B. Dawson. Penguin Handbook PH56, Penguin Books Limited, Harmondsworth, England, UK. 176 pp., 129 illus., 10 tables, indexed, glossy semisoft bound in multicolor with black lettering, 4.3 by 7.1 inches (110 × 180 mm).
    A book on the establishment and culture of lawns in the United Kingdom. It is practically oriented to the home lawn enthusiasts. There is a short chapter on tennis courts, croquet lawns, putting lawns, park lawns, and playing fields plus a small glossary in the back. This book was prepared in collaboration with the Royal Horticultural Society.
    Revised in 1963.

1963†   **Lawns—For garden and playing field.** Second revised edition. R.B. Dawson. Penguin Books Limited, Harmondsworth, England, UK. 176 pp., 126 illus., 10 tables, indexed, glossy semisoft bound in multicolor with black lettering, 4.4 by 7.1 inches (111 × 180 mm).
    There are only minor revisions.

1965   **Gazons en Sportvelden (Lawns and Sportsfields).** R.B. Dawson, with translation by J.F. Kliphuis. Van der Have, Kapelle-Biezelinge, the Netherlands. 168 pp., 128 illus., no tables, no index, semisoft bound in multicolor with green and black lettering, 4.5 by 7.0 inches (113 × 179 mm). Printed in the Dutch language.
    A Dutch translation with minor revisions of the 1963 English edition of **Lawns—For gardens and playing fields.**

1968†   **Practical Lawn Craft and Management of Sports Turf.** Sixth revised edition. R.B. Dawson. Crosby Lockwood & Son Limited, London, England, UK. 320 pp. plus 7 pp. of advertising; 73 illus.; 17 tables; indexed; cloth hardbound in dark green with gold lettering and a jacket in green, black, and white with white and black lettering; 5.4 by 8.4 inches (137 × 214 mm).
    There are only minor revisions.
    The book was revised under the title **Dawson's Practical Lawncraft.**
    See *Hawthorn, Robert—1977.*

Decker, Henry Fleming (Dr.)

Henry Decker was born in Camden, New Jersey, USA, and earned a BA degree in English in 1953 and an MSc degree in Agronomy in 1958, both from Rutgers University; he earned a PhD degree in Botany and Grass Biosystematics from Yale University in 1962 and then pursued Postdoctoral study in grass genetics at the University of Florida from 1964 to 1965. He served as Assistant Professor through Full Professor of Botany at Ohio Wesleyan University from 1962 to 1972 and, since, as Adjunct Professor. He has studied the grasses of Africa, the Amazon, the Arctic, and Central America under grants from the Byrd Institute, the United States Department of Agriculture, and the Smithsonian Institute. He formed Buckeye Bluegrass Farms, A Research Group of Ostrander, Ohio, and served as the President from 1972 to 2005.

1988†   **Lawn Care—A Handbook for Professionals.** Henry F. Decker and Jane M. Decker. Prentice-Hall Inc., Englewood Cliffs, New Jersey, USA. 270 pp., 88 illus., 26 tables, indexed, glossy hardbound in greens with green and white lettering, 6.9 by 9.2 inches (176 × 233 mm).
    An introductory book on the culture and establishment of turfgrasses in the United States. Major emphasis is placed on weed, disease, and insect problems of turfgrasses. There is a small glossary in the back.

Decker, Jane Cynthia (Dr. and Professor)

Jane (McLaughlin) Decker (1935–1988) was born in Cleveland, Ohio, USA, and earned an AB degree from Mt. Holyoke College in 1957 and both an MSc degree in 1958 and a PhD degree in 1961 in Plant Anatomy from Yale University.

She served as a Botany Instructor at Ohio State University from 1962 to 1964 and then joined the faculty at Ohio Wesleyan University, rising to Assistant Professor in 1973, Associate Professor in 1978, and to Full Professor in 1982, specializing in wood anatomy. She was awarded the Allen Trimble Chair in Botany in 1983. She was a respected teacher of botany and also conducted research concerning the biochemical classification of plants. She was elected President of the Ohio Academy of Science from 1983 to 1984. Dr. Decker was honored with the Bishop Herbert Welsh Meritorious Teaching Award from Ohio Wesleyan University in 1988 and the university also established the Jane M. Decker Arboretum.

1988†   **Lawn Care—A Handbook for Professionals.** Coauthor.
      See *Decker, Henry Fleming—1988.*

Dehaye, Alain

Alain Dehaye was born in Vouziers, France, and earned a Diplôme D'études Supérieur Appliqué (DESA) degree in Horticulture from the University of Reims in 1980. He served as Enseignant PCETA-Neuvic de (Professor Certified for Agronomy and Horticulture) from 1980 to 1989. From 1989 to 1992, he was responsible for construction of the Chateau de Taulane golf course. He has been a consultant involved in golf course construction, culture, and management since 1993. Alain Dehaye also has served as a consultant for AGREF (Golf Course Intendant Association of France) since 1986.

1992†   **Les Maladies Cryptogamiques des Gazons: Biologie et Contrôle (Fungal Diseases of Turfgrasses: Biology and Control).** First edition. Alain Dehaye. A.G.R.E.F., Egerie-Golf, Mareil sur Mauldre, France. 128 pp., 158 illus. with 48 in color, 14 tables, no index, cloth hardbound in multicolor with white lettering, 5.6 by 8.2 inches (150 × 216 mm). Printed entirely in the French language.
      A book concerning causal pathogens, hosts, and control of diseases on turfgrasses that occur under the conditions in France.

2007†   **Les Maladies Cryptogamiques des Gazons: Biologie et Contrôl (Fungal Diseases of Turfgrasses: Biology and Control).** Second revised edition. Alain Dehaye. A.G.R.E.F. Biarrize, France. 161 pp., 141 illus. with 96 in color, 51 tables, no index, glossy hardbound in lime green with white lettering, 6.1 by 8.9 inches (155 × 226 mm). Printed in the French language.
      A revision of the 1992 first edition.

DeLuca, Alessandro

Sandro DeLuca was born in Osimo, Italy, and earned a LAUREA degree from the University of Bologna in 1987; he studied Turfgrass Science at Texas A&M University with Dr. Jim Beard. He has since served as an Instructor at the Federazione Italiana Golf (FIG) Italian Turfgrass Technical Training School and as an Advisory Turfgrass Agronomist with the Green Section of the FIG.

2000†   **Tappeti Erbosi (Turfgrasses).** Coauthor.
      See *Panella, Adelmo—2000.*

2003†   **La Pianificazione della Costruzione di un Percorso di Golf (Planning the Construction of a Golf Course).** Coauthor.
      See *Beard, James B—2003.*

Figure 12.28. Powered tractor and front-mounted, side-wheel-drive, reel gang mower, late 1920s. [6]

2004† **Il campo pratica per il gioco del golf—Costruzione, manutenzione e gestione (The Practice golf range—Construction, maintenance and management).** Coauthor.
    See *Casati, Maria Paolo—2004.*

2006† **Tappeti Erbosi—Cura, gestione e manutenzione, delle aree verdi pubbliche e private (Turfgrasses—Care, management and maintenance of public and private green areas).** Coauthor.
    See *Croce, Paolo—2006.*

Department of Education and *Science, England*

1966 **Playing Fields and Hard Surface Areas.** Second revised edition. Building Bulletin 28, Department of Education and Science, Her Majesty's Stationery Office, London, England, UK. 92 pp., 29 illus. with 10 in color, 16 tables, no index, glossy semisoft bound in multicolor with black lettering, 8.3 by 11.9 inches (211 × 301 mm).
    A manual concerning the building and care of playing fields, primarily for schools. Included are site selection, planning, layout, construction, planting, turfgrass establishment, and maintenance.

Dernoeden, Peter Hamilton (Dr. and *Professor)*

Peter Dernoeden was born in Philadelphia, Pennsylvania, USA, and earned a BSc degree in 1970 from Colorado State University, specializing in Landscape Horticulture. He served in the US Army as a field artillery surveyor from 1970 to 1973. Then he received an MSc degree in Turfgrass and Horticulture, studying with Dr. Jack Butler and a PhD degree in Plant Pathology and Turfgrass from the University of Rhode Island in 1980, studying with Dr. Noel Jackson. Dr. Dernoeden joined the Agronomy faculty at the University of Maryland in 1980 and became involved in turfgrass research, adult extension, and teaching. He rose to Full Professor in the Department of Plant Sciences and Landscape Architecture in 1994. His professional expertise emphasizes the interactions among turfgrass pathology, weed science, and culture of low-maintenance turfgrasses. Dr. Dernoeden received the Extension and Industry Award from the Northeast Branch of the American Society of Agronomy in 1990, Fellow of the American Society of Agronomy in 2003, the Outstanding Researcher Award of the Northeastern Weed Science Society in 2007, and Fellow of the Crop Science Society of America in 2007.

1992† **Compendium of Turfgrass Diseases.** Coauthor.
    See *Smiley, Richard Wayne—1992.*

1995† **Managing Turfgrass Pests.** Coauthor.
    See *Watschke, Thomas Lee—1995.*

1996† **Plagas y Enfermedades de los Céspedes (Compendium of Turfgrass Diseases).** Coauthor.
    See *Smiley, Richard Wayne—1996.*

2000† **Creeping Bentgrass Management: Summer Stresses, Weeds, and Selected Maladies.** Peter H. Dernoeden. Ann Arbor Press, Chelsea, Michigan, USA. 144 pp., 67 illus. with 47 in color, 24 tables, indexed, glossy hardbound in multicolor with black and white lettering, 6.9 by 9.9 inches (175 × 251 mm).
    A specialty book on the environmental and biological stresses that may occur in the summer on bentgrass (*Agrostis* spp.) as grown on golf courses. Emphasis is on diseases and their management, weed control, and plant growth regulators. There is a bibliography in the back.

De Thabrew, Wijewarnakula Vivian

Vivian De Thabrew was born in Balapitiya, Sri Lanka, and earned a BA degree in 1962 from London University. His apprentice turfgrass training was with the London County Council and then with the Swanley Council where he played cricket for a county club. He subsequently became Head Groundsman for the Swanley Council, Kent, serving from 1962 to 1968 and earned a National Diploma in Turfculture from the Institute of Groundsmanship in 1966. Then in 1969, Vivian De Thabrew changed professions after earning a Diploma in Library Science from the University of North London.

n.d.† **Lawns—A manual of Lawn care.** W.V. De Thabrew. Thornhill Press, Gloucester, England, UK. 184 pp., 43 illus.,
circa 5 tables, indexed, both cloth hardbound in light green with black lettering, 5.6 by 7.9 inches (141 × 200 mm) and also
1972 glossy semisoft bound in multicolor with white and black lettering, 5.7 by 8.1 inches (146 × 206 mm).
    This manual addresses the establishment and culture of lawns. It is practically oriented to the home lawn enthusiasts and cool-season turfgrass conditions in England. There is a chapter on nongrass lawns plus a glossary and a list of selected references in the back.

1973† **Lawns, Sportsgrounds and Playing Areas—Including Non-Grass Surfaces.** W.V. De Thabrew. Thornhill Press, Gloucester, England, UK. 279 pp., 71 illus., 16 tables, indexed, cloth hardbound in blue with black lettering and a jacket in greens with black lettering, 5.6 by 8.0 inches (141 × 204 mm).

A book concerning construction, establishment, and culture of turfgrasses on lawns and sports fields in England. The first part on lawns is a reprinting of Vivian De Thabrew's 1972 book titled **Lawns**. The second part encompasses turfgrasses for cricket, tennis, bowling, soccer, rugby, hockey, golf courses, and playing areas. Chapters also are included on cinder running tracks, hard-surface tennis courts, and equipment. There is a list of selected references in the back.

Dickinson, Lawrence Sumner (*Professor* Emeritus *of Agrostology*)

Professor Lawrence Dickinson (1888–1965) was born in Amherst, Massachusetts, USA. He earned a BSc degree from Massachusetts Agricultural College and Boston University in 1910 and an MSc degree from Massachusetts Agricultural College in 1936. His first employment at Massachusetts Agricultural College in 1913 was as Foreman of Grounds and Coach of Track (and unofficial security officer). He was a pioneering instructor of turfgrass culture at what is now known as the University of Massachusetts. He initiated the 10-week Stockbridge Winter School in golf course turfgrass culture at the Massachusetts Agricultural College in 1927. A two-year course specializing in turfgrass studies was initiated in 1946. He became an Assistant Professor of Horticulture at the Massachusetts Agricultural College in 1928, an Associate Professor in 1946, and a Full Professor in 1954. He retired in 1958. Professor Dickinson received the GCSAA Distinguished Service Award in 1958 and the second USGA Green Section Award in 1962.

1930† **The Lawn—The Culture of Turf in Park, Golfing and Home Areas.** First edition. Lawrence S. Dickinson. Farm and Garden Library Series, Orange Judd Publishing Company Inc., New York City, New York, USA. 128 pp., 13 illus., 14 tables, indexed, cloth hardbound in blue with orange lettering and a jacket in greens and white with black lettering, 5.0 by 7.2 inches (127 × 184 mm).

* This book is a must for collectors of historical turfgrass books as the author was one of the earliest college teachers on the subject of turfgrasses and their culture. It is a small book on the basics of turfgrass establishment and culture that was widely used in its time. It is a **relatively rare** book. This is one of the first turfgrass books to include a chapter on maintenance of cemeteries. It contains 10 pages of tables in the back. Lawrence Dickinson noted that 23 different grass species and one legume were being sold in lawn seed mixtures in 1930. The chapter subject titles include the following:

Figure 12.29. Lawrence Dickinson, circa 1932. [2]

I.	The General View	VIII.	Controlling Pests
II.	Molding the Lawn	IX.	General Maintenance
III.	Preparation of the Seed Bed	X.	Lawn Mowers
IV.	Seed Selection and Planting	XI.	Park Turf
V.	Important Turf Plants	XII.	Cemetery Turf
VI.	Planting a Lawn with Stolons	XIII.	Useful Tables
VII.	Fertilizing		

Dickinson described the status of cemetery turfgrasses in 1930 as follows:

A cemetery turf is the least worked of any public or semi-public turf. With the exception of Memorial Day, traffic is very light, and it is never violent. Few divots are taken and it is seldom disturbed except for a new interment. Yet with so little use cemetery turf is not as a rule good, and like the park turf the main reason for failure is too close clipping and too little fertilizer or wrong fertilization.

The old time cemeteries and those of the 90's present a real problem because they are usually situated on thin worn out sandy soil, overlying a gravel subsoil, and exposed to wind and the sun rays. In addition many have the lots raised from eight inches to a foot above the drives and paths, and are maintained with small endowments or smaller municipal appropriations.

The sites for the more modern cemeteries were chosen because of their landscape beauty and appropriateness for the dignified and religious atmosphere. These cemeteries have reasonably good soil, trees to break the drying winds and are void of raised lots or graves. The endowments are sufficient for a good standard of maintenance.

Revised in 1931.

1931† **The Lawn—The Culture of Turf in Park, Golfing, and Home Areas.** Second revised edition. Lawrence S. Dickinson. Orange Judd Publishing Company Inc., New York City, New York, USA. 128 pp., 13 illus., 14 tables, indexed, cloth hardbound in blue with orange lettering and a jacket in green and white with black lettering, 5.0 by 7.2 inches (127 × 184 mm).

  * A slight revision of a **relatively rare** book. Lawrence Dickinson suggested the following procedure for mole control:

An inexpensive and effective method of eliminating the mole is to cut an old inner tube, fasten one end to the exhaust pipe of an automobile (any make) and the other end to a hose. Carefully insert the hose in the central part of runway. Be sure the run is not blocked by falling earth so as to prevent a free flow of the exhaust gas (carbon monoxide). Let the automobile engine run slowly for twenty minutes and Mr. Mole is painlessly killed. Mole runs should be tamped (after the mole has been killed) and thoroughly watered so as to hasten the rooting of the lifted grass plants.

  † *Reprinted in 1936.*

Diesburg, Kenneth Lynn (Dr. and Assistant Professor)

Ken Diesburg was born in New Hampton, Iowa, USA, and earned a BSc degree in Botany in 1974, an MSc degree in Plant Breeding in 1980, and a PhD in Horticulture and Turfgrass in 1987, studying with Dr. Nick Christians, all from Iowa State University. He worked as Assistant Breeder for Northrup-King Seed Company from 1978 to 1982 and Breeder for International Seeds in 1988. Dr. Diesburg joined the Faculty at Southern Illinois University in 1989. His responsibilities include research, teaching, and adult extension working with turfgrasses, with a focus on zoysiagrass (*Zoysia* spp.) and growth regulators.

2004† **Turf Management In The Transition Zone.** Coauthor.
    See *Dunn, John Henry—2004.*

DiPaolo, Joseph Michael (Dr.)

Joe DiPaolo was born in in New Brunswick, New Jersey, USA, and earned a BSc degree in Plant Science from Rutgers University in 1975, and both MSc and PhD degrees in Turfgrass Physiology from Texas A&M University in 1977 and 1979, respectively, studying with Dr. Jim Beard. He served on the faculty at North Carolina State University from 1979 to 1994, rising to Full Professor involved in turfgrass research and teaching. He then joined Ciba Turf &

Ornamentals in 1994 as Regional R&D Manager; he later became Lawns Market Manager of Novartis Crop Protection in 1997, and Golf Market Manager of Syngenta Crop Protection in 2000, and Senior Marketing Manager of Professional Products at Syngenta Lawn and Garden Group in 2006. He has served as Adjunct Professor at Ohio State University. He was elected President of Gamma Sigma Delta NCSU Chapter from 1992 to 1993; he was also elected to the Crop Science Society of America Board of Directors from 1991 to 1993, to the Turfgrass Science Division C-5 Chair in 1992, and to the Board of Directors of the International Turfgrass Society from 1997 to 2005.

1979† **Introduction To Turfgrass Science And Culture Laboratory Exercises.** Coauthor.
  See *Beard, James B—1979.*

Dittmer, Howard James (Dr. and *Professor* Emeritus *of Biology*)

Howard Dittmer (1910–2001) was born in Pekin, Illinois, USA, and earned both BA and EMA degrees from the University of New Mexico in 1933 and 1934, respectively, and a PhD degree in Plant Morphology from the University of Iowa in 1938. He served on the faculty at Dubuque University and Chicago Teachers College from 1938 to 1943 and then joined the faculty at the University of New Mexico in 1943, rising from Assistant Professor to Full Professor of Biology, plus being Assistant Dean and Associate Dean of the College of Arts and Sciences from 1956 to 1970 and from 1970 to 1975, respectively. He conducted research on root systems as related to soil physics, including studies on turfgrasses. Dr. Dittmer was elected President of the Southwest and Rocky Mountain Division of the American Association for the Advancement of Science in 1968 and President of the New Mexico Academy of Science in 1972. He was awarded Fellow of the American Association for the Advancement of Science in 1949.

1950† **Lawn Problems of the Southwest.**˙ Howard J. Dittmer. University of New Mexico Publications in Biology Number Four, The University of New Mexico Press, Albuquerque, New Mexico, USA. 76 pp., no illus., 12 tables, no index, semisoft bound in green with black lettering, 5.9 by 8.9 inches (151 × 226 mm).

 * A **very rare**, small book on the establishment and culture of lawns in New Mexico. It is practically oriented to the home lawn enthusiasts. It includes a brief discussion of trees and shrubs. It is a pioneering, early book devoted specifically to lawn turfgrasses grown on alkaline soils in hot, arid climates.

Doogue, Luke Joseph

Luke Doogue (1866–1925) was born in Boston, Massachusetts, USA, to parents Elisabeth and William Doogue who had emigrated from Ireland. His father, William Doogue (1828–1906), was a pioneering Superintendent of the Boston Public Grounds Department from 1878 to 1906. After schooling, Luke Doogue entered the florist business in Boston. He then joined the Boston Public Grounds Department in 1886 as an Assistant Forester and a Landscape Gardener serving through 1906. He became a well-known landscape gardener, specializing in artistic floral designs for public areas. Luke Doogue became a Clerk and Droughtsman for the Boston Park and Recreation Department in 1913, serving until his retirement in 1924.

1912† **Making A Lawn.** First edition. Luke J. Doogue. House and Garden Making Books, McBride, Nast and Company, New York City, New York, USA. 51 pp., 7 illus., no tables, no index, cloth hardbound in green with white lettering and a jacket in white with green lettering, 4.2 by 6.2 inches (108 × 157 mm).

 * One of the oldest books published on lawns in the United States. Its popularity at that time is evidenced by the numerous reprintings. It is a **rare** book that is a must for collectors of historical turfgrass books. This small book is devoted primarily to site preparation and establishment of cool-season turfgrass lawns in the northeastern regions of the United States, with short chapters on the culture of lawns. The chapter subject titles include the following:

• The Small Lawn, Old and New	• Sodding
• The Treatment of Large Areas	• Good Loam and Fertilizers
• Grass Seed	• Lawn-Mower, Roller, and Hose
• Sowing the Seed	• Weeds and Other Pests

Concerning the problem of earthworms in lawns and suggested irrigation techniques, the author indicated the following:

> . . . an application of lime-water will drive many (earthworms) to the surface, where they can be swept up; or a heavy rolling with a 1,500-lb roller will do much to discourage them.

† *Reprinted as the second edition in April 1913; the third edition in May 1913; the fourth edition in January 1914, in light green; and the fifth edition in July 1917, in green cloth and hardbound.*

1914† **The Proper Care of Lawns.** L.J. Doogue. Booklet No. 38, The Dunham Company, Berea, Ohio, USA. 22 pp. plus 9 pp. of advertising, 23 illus., 6 tables, no index, fold-stapled textured semisoft covers in multicolor with black lettering, 5.4 by 6.8 inches (137 × 173 mm).

✳ A **rare**, small booklet concerning the establishment and care of lawns. It is practically oriented to lawn enthusiasts and cool-season turfgrass conditions in the cool northern United States. There is a small section on lawn tennis courts. The Dunham Company manufactured and marketed various turfgrass rollers for use on turfgrasses.

Duble, Richard Lee (Dr.)

Richard Duble was born in Galveston, Texas, USA, and earned BSc, MSc, and PhD degrees in Agronomy, all from Texas A&M University (TAMU) in 1962, 1965, and 1967, respectively. He worked in forages at the Texas Agricultural Experiment Station (TAES) Overton Center from 1967 to 1970 and in turfgrass from 1970 to 1973. Since 1974, he has been an Extension Turfgrass Specialist at TAMU, retiring in 1996.

1989† **Southern Turfgrasses: Their Management and Use.** First edition. Richard L. Duble. TexScape, College Station, Texas, USA. 351 pp., 149 illus. with 113 in color, 23 tables, small index, cloth hardbound in light green with green lettering and a glossy jacket in greens and white with white and black lettering, 7.5 by 9.5 inches (190 × 241 mm).

A book on turfgrasses of the southern United States and their establishment and culture plus the associated pest problems and management.

1996† **Turfgrasses: Their Management and Use in the Southern Zone.** Second revised edition. Richard L. Duble. Texas A&M University Press, College Station, Texas, USA. 331 pp., 135 illus. with 51 in color, 52 tables, small index, cloth hardbound in green with black lettering and a glossy jacket in greens with white and black lettering, 7.0 by 9.0 inches (177 × 230 mm).

An updated version of the 1989 edition. A list of selected references has been added in the back.

† Also printed as a semisoft bound version in multicolor covers with white and black lettering.

Duncan, Ronny Rush (Dr.)

Ronny Duncan was born in Hereford, Texas, USA, and earned a BSc degree in agronomy from Texas Tech University in 1969, and both an MSc in Crop Science and a PhD in Plant Breeding from Texas A&M University in 1974 and 1977, respectively. He joined the University of Georgia conducting research in breeding of grain sorghum from 1977 to 1992 and of turfgrasses from 1992 to 2003. He rose to Associate Professor in 1982 and Full Professor in 1988. Since 2004, he has been Vice President of Turf Ecosystems. Dr. Duncan received Fellow in both the Crop Science Society of America and American Society of Agronomy in 1992.

1998† **Salt-Affected Turfgrass Sites: Assessment and Management.** Coauthor.
See *Carrow, Robert Norris—1998.*

2003† **Turfgrass Biology, Genetics, and Breeding.** Coeditor.
See *Casler, Michael Darwin—2003.*

2009† **Turfgrass and Landscape Irrigation Water Quality: Assessment and Management.** Ronny R. Duncan, Robert N. Carrow, and Michael T. Huck. CRC Press, Boca Raton, Florida, USA. 485 pp., 37 illus. with 20 in color, 59 tables, large index, glossy hardbound in multicolor with white lettering, 6.1 by 9.3 inches (155 × 235 mm).

A specialty book concerning water quality use in turfgrass and ornamental plant irrigation. Included are constituents of concern in water quality, sources of poor quality water, turfgrass and soil management for sites irrigated with poor quality water, and environmental issues associated with irrigation from poor quality water sources. There is a selected list of references in the book.

Dunn, John Harvey McHenry (Dr. and Professor Emeritus)

John Dunn (1937–2012) was born in Upper Darby, Pennsylvania, USA, and earned a BSc degree in Agronomy from Pennsylvania State University in 1960 and both MSc and PhD degrees from Rutgers University in 1966 and 1968, respectively, studying with Dr. Ralph Engel. Dr. Dunn joined the faculty in the Department of Horticulture at the University of Missouri in 1968 and rose to Full Professor in 1978. He initiated the turfgrass research program at the University of Missouri, with emphasis on the adaptation of warm-season turfgrasses in the transition climatic zone. Dr. Dunn also taught turfgrass and plant nutrition courses, as well as serving as Turfgrass Extension Specialist from 1977 to 1984 and from 1997 to 2003, when he retired. He was elected to the Crop Science Society of America Board of Directors from 1976 to 1978, and Turfgrass Division C-5 Chair in 1977, and to the American Society of Agronomy Board of Directors from 1982 through 1984.

2004† **Turf Management In The Transition Zone.** John H. Dunn and Kenneth Diesburg. John Wiley & Sons Inc., Hoboken, New Jersey, USA. 288 pp.; 51 illus.; 11 tables; small index; glossy hardbound in greens and white with black, white, green, and yellow lettering; 6.1 by 9.2 inches (155 × 233 mm).

A speciality book focused on the turfgrasses and cultural practices specific to the transitional climatic zone, primarily in the United States. There is a list of selected references at the back of each chapter.

Dury, Peter Leslie Kennan (Dr.)

Peter Dury was born in Nottingham, Nottinghamshire, England, UK. He was in national military service from 1954 to 1956 and, subsequently, worked as Groundsman and then Foreman for the Derby Parks Department from 1956 to 1961 and Head Groundsman for Southport and Birkdale Cricket Club from 1961 to 1965. He earned a National Diploma in Turfculture and Sports Ground Management from the Institute of Groundsmanship in 1964. Then he served as Playing Fields Supervisor for Nuneaton Borough Council from 1965 to 1967, as Playing Fields Superintendent for Huntingdon and Peterborough County Council from 1967 to 1968, and as both Deputy and Head County Playing Fields Officer for the Nottinghamshire County Council Education Department from 1968 to 1988. Following retirement he has been active as a private consultant on sports facilities. His experiences have encompassed sport and landscape provisions from the initial site survey through construction to the actual management of the facilities. He pioneered performance quality standards for sports facilities, such as for the British Standards Institute. Dury was Pitches Consultant to the Irish Cricket Union from 1978 to 1988, Consultant to the Nottinghamshire Playing Fields Association from 1968 to 1988, Consultant to the National Cricket Association from 1976 to 1996, and Consultant to the England and Wales Cricket Board from 1996 to 2001. He was elected as National Executive of the Association of Landscape Management from 1974 to 1994 and of the Institute of Groundsmanship from 1968 to 1982, as well as President of the Association of Landscape Management (formerly the Association of Playing Fields Officers) in 1980 and Chairman of the Institute of Groundsmanship Education Committee in 1982. Peter Dury received the National Turf Grass Council Award in 1991, an Honorary Doctorate degree from the University of Essex (Writtle College) in 2002, the Life Time Achievement Award from the Institute of Groundsmanship in 2003, and the Presidents Certificate for Services to the National Playing Fields Association in 2003.

198x **Sportsground Construction & Management.** Peter L.K. Dury. Playing Fields Service. Nottinghamshire County Council, Nottingham, England, UK. 84 pp., 152 illus., 4 tables, no index, soft bound in greens and white with black lettering, 8.3 by 11.6 inches (211 × 295 mm).

A manual on the design, planning, construction, and culture of turfgrasses for sports fields, bowling greens, cricket pitches, tennis courts, and golf courses. It is oriented to the cool-season turfgrass conditions in England.

n.d. **Cricket Pitch Management.** Peter L.K. Dury. Part I of In the World of Cricket Pitches, Nottinghamshire Council, Nottingham, England, UK. 57 pp., no index, soft bound in light green and black with black lettering, 8.3 by 11.6 inches circa (210 × 295 mm).
mid-
198x A booklet on the construction and maintenance of a cricket pitch or square under cool-season conditions in England. The desired pitch characteristics, soil, and subsequent preparation are discussed.

1987† **Cricket—Take Care of Your Square.** Coauthor.
See *Lock, Herbert Christmas—1987.*

1997 **Grounds Maintenance: Managing Outdoor Sport and Landscape Facilities.** Peter L.K. Dury. Thorogood Limited, London, England, UK. 150 pp., 26 illus., 40 tables, no index, glossy semisoft bound in dark teal with white lettering, 8.1 by 11.5 inches (205 × 292 mm).

A book concerning the planning, construction management, and maintenance of recreational facilities and landscapes, with emphasis on sport and physical recreation. Included are the assessment of facility and surface quality, cost effectiveness, safety, legislative effects, and future changes.

2004 **Design and Maintenance of Outdoor Sports Facilities.** Peter L.K. Dury. National Playing Fields Association, London, England, UK. 104 pp., 80 illus., 80 tables, no index, glossy semisoft bound in multicolor with white lettering, 8.2 by 11.7 inches (209 × 296 mm).

A book concerning the principles, performance criteria tests, selection of a contractor for construction, and monitoring the construction of outdoor sports facilities. The orientation is to cool-season conditions in the UK. There is a glossary of selected terms in the back.

E

Eggens, Jack Lambert (Dr. and Professor Emeritus)

Jack Eggens was born in Manotick Station, Ontario, Canada, and earned a BSc degree from the Royal Military College in 1960, a Bachelor of Science in Agriculture degree (BScAgr) in Horticultural Science from the University of Toronto in 1965, and both MSc and PhD degrees in Horticultural Science from the University of Guelph in 1966 and 1970, respectively. From 1966 to 1970, Jack Eggens served as Lecturer at the University of Guelph and rose to Assistant Professor from 1970 to 1973, Associate Professor from 1975 to 1985, and Full Professor from 1985 to 1995, when he retired. He was very active in teaching turfgrass science and also conducted research concerning cultivar characterization, thatch management, nitrogen fertilization, and annual bluegrass (*Poa annua*) culture and control. He was Deputy Chair of Horticulture from 1989 to 1992 and Director of the Guelph Turfgrass Institute from 1993 to 1994. Honors received by Dr. Eggens include Honorary Class President of the Ontario Agricultural College Class of 1988 and 1992, Ontario Agricultural College Waghorne Teaching Fellowship in 1994, Canadian Golf Superintendents Association John B. Steel Distinguished Service Award in 1995, Honorary Life Member of the Nursery Sod Growers Association in 1996, and establishment of the "J.L. Eggens Scholarship" for the Turf Manager's Short Course in 1996.

1998   **Turf Management: Principles and Practices—Study Guide.** J.L. Eggens. Department of Horticulture, University of Guelph, Guelph, Ontario, Canada. 244 pp., 11 illus., 129 tables, indexed, semisoft covers in green with black lettering in a comb-binder with a clear cover overlay, 8.5 by 11.0 inches (217 × 279 mm).

An instructional manual on turfgrass including their characteristics, root-zone construction, turfgrass establishment, cultural practices, environmental stresses, and pest management. At the end of each of the eight chapters there are a set of selected references and multiple choice questions. There is a glossary in the back.

There are ten earlier editions dated 1985, 1986, 1987W, 1987F, 1988, 1989, 1991, 1992, 1993, and 1995.

Ehara, Kaoru (Dr. and Professor Emeritus)

Kaoru Ehara (1907–1998) was born in Hokkaido, Japan, and earned a Bachelor of Agriculture (BAgr) degree from Tokyo University in 1934, and a Doctorate of Agriculture (DrAgr) degree in Crop Science from Kyusyu University in 1948. Dr. Ehara served as Associate Professor from 1944 to 1959 and Professor from 1959 to 1970 at Kyusyu University. He then was Vice President of the Japanese Society of Grassland Science from 1971 to 1975. He was elected Director of the Nishinihon Green Research Institute from 1970 to 1972, elected to the International Turfgrass Society Board of Directors from 1973 to 1981, and elected Director of the Higasinihon Green Research Institute from 1985 to 1990. Dr. Ehara has been honored with the Award of Grassland Science from the Japanese Society of Grassland Science in 1962 and the Award of Agronomy from the Japanese Society of Agronomy in 1966.

Figure 12.30. Front-mounted, five-gang, side-wheel-drive Toro mower with hitch travelline and Toro tractor, USA, 1926. [2]

1967† **Turfgrass Establishment and Care of the Home Lawn—Including Home Garden Design.**ˣ Kaoru Ehara, Yoshisuke Maki, and Hidesato Chono (editors). Seibundo Shink-osha Inc., Tokyo, Japan. 237 pp., 349 illus. with 55 designs of lawn gardens, 3 tables, no index, cloth hardbound in scarlet and black with scarlet lettering and a glossy jacket in multicolor with black lettering, 5.8 by 7.9 inches (147 × 201 mm). Printed entirely in the Japanese language.

    A manual on lawn establishment and lawn garden development oriented to home owners in Japan. Subjects include lawn culture, lawn landscape gardening, twenty-eight models of lawn garden layouts, arrangement of flowers and trees in the lawn garden, flower garden development, and ground cover plants.

1968† **Turfgrass and Turf—Establishment and Maintenance.** First edition. Kaoru Ehara. Yokendo Limited, Tokyo, Japan. 500 pp., 215 illus. with 8 in color, 89 tables, indexed in Japanese and English, cloth hardbound in green with gold lettering and a slip case in white and green with dark blue and white lettering, 5.8 by 8.2 inches (147 × 208 mm). Printed in the Japanese language.

    An extensive book on the basics of turfgrasses and their culture. It is a key early Japanese book on turfgrasses. Emphasis is placed on golf course maintenance. Included are chapters on the history of turfgrass utilization and research, climate, soils, water, fertilizers, turfgrass characteristics, breeding, propagation, culture, pest control, and weeds and their control. There is a list of selected references in the back.

    Revised in 1971, 1976, 1984, and 1993.

1971† **Turf and Turfgrasses—Establishment and Maintenance.** Second revised edition. Kaoru Ehara. Yokendo Limited, Tokyo, Japan. 550 pp., 215 illus. with 8 in color, 89 tables, indexed in English and Japanese, cloth hardbound in green with gold lettering, 5.8 by 8.2 inches (147 × 208 mm). Printed in the Japanese language.

    There are small revisions.

    Revised in 1976, 1984, and 1993.

1976† **Turf and Turfgrasses—Establishment and Maintenance.** Third revised edition. Kaoru Ehara. Yokendo Limited, Tokyo, Japan. 550 pp., 278 illus. with 24 in color, 112 tables, indexed in English and Japanese, cloth hardbound in green with gold lettering, 5.8 by 8.2 inches (147 × 208 mm). Printed in the Japanese language.

    There are only minor revisions.

    Revised in 1984 and 1993.

1977† **Monograph of Turf Culture and Turfgrasses.** Kaoru Ehara and Fumio Kitamura (editors) with 45 contributing authors. Soft Science Company Limited, Tokyo, Japan. 480 pp., 155 illus. with 61 in color, 147 tables, indexed, cloth hardbound in green with gold lettering and a slip case in light green with dark green lettering, 7.5 by 10.4 inches (190 × 265 mm). Printed entirely in the Japanese language.

    The first integrated book written specifically about turfgrass culture in Japan. Contains chapters on the history of turfgrass utilization and research, environment, sod industry, turfgrass taxonomy, establishment, culture, pest control, and turfgrass use and culture on gardens, parks, golf courses, athletic fields, horse racetracks, and playgrounds, plus equipment, chemicals, and soil amendments.

    Reprinted in 1978 and 1980.

1980† **Development of Beautiful Lawn Garden—Understandable at a Glance.** Kaoru Ehara. Chikuma Shuhansha Inc., Tokyo, Japan. 158 pp.; 136 illus. with 68 in color; 21 tables; no index; glossy semisoft bound in multicolor with yellow, white, and black lettering; 5.0 by 7.0 inches (127 × 179 mm). Printed entirely in the Japanese language.

    A guide book on lawn garden development, including design. It is practically oriented to the lawn enthusiasts and turfgrass conditions in Japan.

1982† **Knowledge of Course Management at Golf Club.** Kaoru Ehara. Golf Nippon, Tokyo, Japan. 561 pp., 70 illus., 76 tables, no index, textured cloth hardbound in dark blue with gold lettering and a slip case in multicolor with black lettering, 5.7 by 8.1 inches (146 × 207 mm). Printed entirely in the Japanese language.

    An extensive manual on course management oriented to golf course owners of Japan. Topics addressed are golf course construction, turfgrass culture, diseases, insects, weeds, and equipment.

1984† **Turf and Turfgrasses—Establishment and Maintenance.** Fourth revised edition. Kaoru Ehara. Yokendo Limited, Tokyo, Japan. 572 pp., 219 illus. with 8 in color, 111 tables, indexed in Japanese and English, cloth hardbound in green with gold lettering plus a slip case in white and green with dark blue and white lettering, 5.8 by 8.2 inches (147 × 208 mm). Printed entirely in the Japanese language.

    There are only minor revisions.

    Revised in 1993.

1993† **Turf and Turfgrass—Establishment and Maintenance.** Fifth revised edition. Karoru Ehara. Yokendo Limited, Tokyo, Japan. 572 pp., 283 illus. with 28 in color, 114 tables, extensive index in Japanese and English, cloth hardbound in green with gold lettering and a jacket in white and green with dark blue and white lettering plus a slip case in white and green with dark blue and white lettering, 5.8 × 8.2 inches (147 × 209 mm). Printed in the Japanese language.

    There are only minor revisions.

Eisele, Christoph[3]

1962† **Rasen, Gras und Grünflächen (Turf, Grasses and Green Spaces).** First edition. Christoph Eisele. Verlag Paul Parey, Berlin and Hamburg, West Germany. 135 pp., 28 illus., 6 tables, indexed, glossy hardbound in multicolor with white and orange lettering, 4.8 by 7.6 inches (122 × 193 mm). Printed entirely in the German language.

   A book on the construction, establishment, and culture of turfgrass areas, especially in relation to conditions in Germany. It includes sections on turfgrasses for soccer, lawn tennis, golf, and lawns.
   Revised in 1973.

1973† **Rasen, Gras und Grünflächen (Turf, Grasses and Green Spaces).** Second revised edition. Christoph Eisele. Verlag Paul Parey, Berlin and Hamburg, West Germany. 144 pp., 20 illus., 7 tables, indexed, glossy hardbound in green with white and green lettering, 4.8 by 7.6 inches (122 × 192 mm). Printed entirely in the German language.

   A revised, updated edition.

Elliott, Monica Lynn (*Dr.* and *Professor*)

Monica Elliott was born in Mattoon, Illinois, USA, and earned both BSc degrees in Environmental Biology and Botany and a BA degree in History from Eastern Illinois University in 1978; she also earned MSc and PhD degrees in Plant Pathology from Montana State University in 1983 and 1987, respectively. She became Assistant Professor at the Fort Lauderdale Research and Education Center, University of Florida, in 1987; Associate Professor in 1993; and Full Professor in 2002. Dr. Elliott conducts research on the diseases of warm-season turfgrasses and their control.

1993† **Best Management Practices for Florida Golf Courses.** Coauthor.
   See *McCarty, Lambert Blanchard—1993.*

1998† **IPM Handbook for Golf Courses.** Coauthor.
   See *Schumann, Gail Lynn—1998.*

1999† **Best Management Practices for Florida Golf Courses.** Coauthor.
   See *Unruh, Joseph Bryan—1999.*

Emmons, Robert David (*Professor* Emeritus)

Robert Emmons was born in Camden, New Jersey, USA. He received the Bronze Star and the Purple Heart while serving in the US Army from 1968 to 1970. He earned a BA degree in Prelaw from Boston University in 1968 and an MSc degree in Plant Science from the University of New Hampshire in 1975. Emmons was an Assistant through Full Professor in the Plant Science Department at the State University of New York (SUNY), Cobleskill, from 1975 to 2006 and was involved in the turfgrass teaching program. Professor Emmons received the Certificate of Merit from the United States Department of Agriculture in 1981, the Citation of Merit from the New York State Turfgrass Association in 1991, the Excellence in Teaching Award from SUNY in 1997, and the GCSAA Distinguished Service Award in 2003.

Figure 12.31. Motor-powered, reel mower with grass box and clippings side delivery, as first manufactured by Ransomes in 1902. [26]

1984† **Turfgrass Science and Management.** First edition. Robert D. Emmons. Delmar Publishers Inc., Albany, New York, USA. 451 pp., 373 illus., 56 tables, small index, glossy hardbound in multicolor with white lettering, 7.9 by 9.1 inches (200 × 231 mm).

    An introductory book concerning construction, establishment, and culture of turfgrasses. Emphasis is placed on pest problems. There are 31 problem sets of turfgrass calculations in the back, along with a glossary and selected bibliography of general books.

    Translated and published in 1992 with the Chinese language title Cao Ping Ke Xu Geian Li.

    Revised in 1995, 2000, and 2008.

1995† **Turfgrass Science and Management.** Second revised edition. Robert D. Emmons. Delmar Publishers, Albany, New York, USA. 512 pp., 406 illus. with 48 in color, 43 tables, small index, glossy hardbound in multicolor with yellow and blue lettering, 8.0 by 9.1 inches (203 × 232 mm).

    A revision of an introductory book on turfgrass culture and management. A short glossary is found in the back plus a selected bibliography of general books.

    Revised in 2000 and 2008.

1995† **Turfgrass Science and Management Lab Manual.** Second revised edition. Robert Emmons and Robert Boufford. Delmar Publishers, Albany, New York, USA. 163 pp., 48 illus., 30 tables, no index, glossy semisoft bound in multicolor with yellow and light blue lettering, 8.5 by 10.9 inches (216 × 278 mm).

    A laboratory manual of 15 turfgrass-related exercises. The exercises can be accomplished via coordinated software; a supplemental files disk is attached in the back. A short glossary of terms and a list of selected references are included in the back.

    Revised in 2008.

2000† **Turfgrass Science and Management.** Third revised edition. Robert D. Emmons. Delmar Thomson Learning, Albany, New York, USA. 528 pp.; 337 illus. with 48 in color; 59 tables; small index; glossy hardbound in multicolor with white, yellow, and green lettering; 8.0 by 9.2 inches (203 × 234 mm).

    A slight revision of an introductory book on turfgrass culture and management. A short glossary and a list of selected general books are included in the back.

2008† **Turfgrass Science and Management.** Fourth revised edition. Robert D. Emmons. Thomson Delmar Learning, Clifton Park, New York, USA. 584 pp., 468 illus. with 370 in color, 68 tables, indexed, glossy hardbound in multicolor with yellow and white lettering, 8.0 by 10.0 inches (204 × 253 mm).

    An update of the third edition with major additions to the illustrations.

English Sports Council

Established in 1965, the English Sports Council functioned as an advisory body to the government. Then in 1972, it was granted independent status by the Royal Charter and a permanent staff was funded. The staff implements policies and decisions made by the Sports Council and provides technical and advisory services to local authorities and other sports organizations.

2000 **Natural Turf for Sport.** Sport England Publications, London, England, UK. 35 pp.; 45 illus. with 23 in color; 1 table; no index; fold-stapled semisoft covers in blue and white with blue, red, and black lettering; 8.25 by 11.7 inches (209 × 297 mm).

    A small guide providing an overview of the design, construction, planting, and culture of sports turfgrass surfaces.

Escritt, John Robert

John Escritt (1916–2003) was born at Stockton-on-Tees, England, UK, and earned a BSc degree with *honours* in Chemistry from the University College of North Wales-Bangor in 1937 and an MSc degree in Chemistry and Weed Control from the University of Wales in 1948. John Escritt served as Chemist for the Board of Greenkeeping Research at St Ives in 1938 and 1939. Following service during WWII, he returned in 1945 as Staff Chemist involved in turfgrass research such as weed and insect controls, growth regulators, and fertilizer assessments. He became Assistant Director in 1955 and then served as the second Director of the Sports Turf Research Institute from 1964 to 1981, when he retired.

1965† **Sports Ground Construction—Specifications For Playing Facilities.** First edition. Coauthor.

    See *Gooch, Robert Butler—1965.*

1975† **Sports Ground Construction Specifications.** Second revised edition. Coauthor.

    See *Gooch, Robert Butler—1975.*

1978† **ABC of Turf Culture.** J.R. Escritt. Kaye & Ward Limited, London, England, UK. 248 pp., 47 illus., 12 tables, indexed, cloth hardbound in green with silver lettering and a jacket in multicolor with green and black lettering, 5.3 by 8.5 inches (134 × 215 mm).

    A reference book of the technical terms used in turfgrass culture in the UK. The terms are arranged in alphabetical order, followed by a basic definition and/or practical description.

Evans, Roger David Calder

Roger Evans was born in Carmarthen, South Wales, UK, and earned a BSc degree in Agricultural Botany with *honours* from the University College of Wales, Aberystwyth, in 1964. He served as a Turfgrass Advisory Agronomist and Librarian with the Sports Turf Research Institute at Bingley from 1964 to 1998. Roger Evans had a strong interest in and has studied the history of turfgrass development in the United Kingdom.

1988† **Bowling Greens—Their History, Construction and Maintenance.** First edition. R.D.C. Evans. The Sports Turf Research Institute, Bingley, West Yorkshire, England, UK. 196 pp.; 142 illus.; 13 tables; no index; glossy hardbound in green, white, and black with white, yellow, and black lettering; 8.3 by 11.6 inches (210 × 295 mm).

A book on the construction and culture of lawn bowling greens, with emphasis on cool-season turfgrass conditions in the United Kingdom. There is a major section on the history of lawn bowling. It includes reprints of two research papers in the appendix plus an extensive bibliography in the back. The most comprehensive book on the turfgrass cultural dimensions of lawn bowling greens at that time.

† *Revised in 1992 and 2008.*

1991† **Cricket Grounds—The Evolution, Construction and Maintenance of Natural Turf Cricket Tables and Outfields.** R.D.C. Evans. The Sports Turf Research Institute, Bingley, West Yorkshire, England, UK. 280 pp.; 133 illus.; 35 tables; no index; glossy hardbound in red, white, and black with white and yellow lettering; 8.2 by 11.6 inches (208 × 295 mm).

A book on the maintenance and construction of cricket grounds and cricket tables. It is the first major book published on this subject, with emphasis on cool-season turfgrass conditions in the United Kingdom. There are eight articles on special turfgrass-soil topics related to cricket grounds included in the back.

1992† **Bowling Greens—Their History, Construction and Maintenance.** Second revised edition. R.D.C. Evans. The Sports Turf Research Institute, Bingley, West Yorkshire, England, UK. 211 pp.; 205 illus. with 16 in color; 21 tables; no index; glossy hardbound in green, white, and black with white, yellow, green, and black lettering; 8.3 by 11.6 inches (210 × 295 mm).

A significant revision, including the addition of a third research paper in the appendix. The number of illustrations has been increased significantly.

Revised in 2008.

See *Perris, Jeffrey—2008.*

1992† **The Care of the Golf Course.** First edition. Coeditor.

See *Hayes, Peter—1992.*

1994† **Winter Games Pitches—The Construction and Maintenance of Natural Turf Pitches for Team Games.** R.D.C. Evans (editor) with eight contributing authors. The Sports Turf Research Institute, Bingley, West Yorkshire, England, UK. 209 pp., 146 illus., 17 tables, no index, glossy hardbound in multicolor with white lettering, 6.7 by 9.1 inches (170 × 230 mm).

A book on the design, construction, turfgrass selection, culture, and renovation of sport fields, with emphasis on cool-season turfgrasses. There also is a section on soil warming plus an extensive bibliography in the back. It is oriented to the cool-season climatic and soil conditions in the United Kingdom.

1996† **The Care of the Golf Course.** Second revised edition. Coeditor.

See *Perris, Jeffrey—1996.*

1996† **The Cricket Groundsman's Companion—A Basic Guide To The Maintenance of Cricket Tables & Outfields.** R.D.C. Evans. The Sports Turf Research Institute, Bingley, West Yorkshire, England, UK. 21 pp. plus 1 p. advertising; 11 illus. with 5 in color; no tables; no index; fold-stapled glossy semisoft covers in dark green with yellow, white, and light green lettering; 4.5 by 9.4 inches (114 × 238 mm).

A booklet outlining the maintenance and marking practices for cricket grounds under the conditions in England.

1996† **The STRI Golf Greenkeepers Training Course Craft Level Theory.** R.D.C. Evans (editor) including thirteen contributing STRI authors. The Sports Turf Research Institute, Bingley, West Yorkshire, England, UK. 198 pp., numerous illus. plus 20 color photographs, no tables, no index, semisoft covers in black and green with white lettering in a ring binder, 10.6 by 12.4 inches (270 × 315 mm).

A training manual for students in the Greenkeepers Training Course, National Vocational Qualification (NVQ) or Scottish Vocational Qualification (SVQ) at craft level 2 of Amenity Horticulture (Greenkeeping).

1997† **On Course Field Guide to the Identification of Golf Course Grasses.** R.D.C. Evans. Greenkeepers's Training Committee, British and International Golf Greenkeepers Association, Aldwark Manor, Alne, York, UK. 24 pp., 1 illus. plus 10 color photographs, no tables, no index, waterproof semisoft covers in multicolor with black lettering in a ring binder, 4.3 by 7.3 inches (110 × 185 mm).

A small guide book to the identification of the major turfgrasses found in the United Kingdom.

Evans, Thomas Watcyn (Dr.)

Thomas Evans was born in Mountain Ash, Glamorgan, Wales, UK, and earned a BSc degree in Science in 1927 and a PhD in Agricultural Biochemistry in 1930, both from the University College of Wales, Aberystwyth. He then served as Scientist and Chief Assistant to the Director of Research at the St Ives Research Station, Board of Greenkeeping Research, at Bingley from 1930 to 1936, specializing in soil and fertilizer aspects. Dr. Evans then joined with I.G. Lewis, also formerly with BGR, to become the Editors and Publishers of **The Agrostologist,** first issued in the autumn of 1936 from their offices in London. They also were active as Consulting Agrostologists to turfgrass facilities throughout the United Kingdom. **The Agrostologist** was a quality bulletin on turfgrass management that was of a high standard for those times in the United Kingdom and Europe. WWII brought these operations to a halt in the summer of 1940. Dr. Evans was an associate of the Institute of Chemistry, which became the Royal Society of Chemistry.

1932† **Greenkeeping for Golf Courses.** Coauthor.
      See *Dawson, Robert Brian—1932.*

Everest, John W.[2] (Dr.)

2001† **Color Atlas of Turfgrass Weeds.** Coauthor.
      See *McCarty, Lambert Blanchard—2001.*
2008† **Color Atlas of Turfgrass Weeds.** Second revised edition. Coauthor.
      See *McCarty, Lambert Blanchard—2008.*

F

Fagerness, Matthew James (Dr.)

Matt Fagerness was born in Chehalis, Washington, USA, and earned a BSc in Crop and Soil Sciences from Washington State University in 1995, an MSc degree in Weed Science from Michigan State University in 1997, and a PhD degree from North Carolina State University in 2000, studying weed science and plant growth regulators with Dr. Fred Yelverton. He served as an Assistant Professor and Extension Turfgrass Specialist at Kansas State University from 2000 to 2004.

Figure 12.32. Shanks 1910 motor-powered, riding, 42-inch-wide, reel mower with grass box and mechanical discharge. [38]

2004† **Turfgrass Chemicals and Pesticides—A Practitioner's Guide.** Matt Fagerness and Rodney Johns. The McGraw-Hill Companies Inc., New York City, New York, USA. 530 pp.; 111 illus.; 32 tables; small index; glossy hardbound in blue, green, and red with white and green lettering; 5.9 by 8.9 inches (150 × 226 mm).

    A book concerning pest problems and pesticides for their control on turfgrasses. The chapter on turfgrass diseases is authored by Ross Brown.

Falcinelli, Mario (Professor)

Mario Falcinelli was born in Torgiano, Italy, and earned a BSc degree in Agricultural Science from the University of Perugia in 1967. He has served on the faculty at the University of Perugia since 1973 and became a Full Professor in 1990. He has been Director of the Department of Plant Biology and Biotechnology Agro-Environmental since 2000 and has been involved in teaching and research related to the genetics and breeding of grasses. Professor Falcinelli was elected President of the International Seed Production Group for 1997.

2000† **Tappeti Erbosi (Turfgrasses).** Coauthor.
    See *Panella, Adelmo—2000.*
2006† **Tappeti Erbosi—Cura, gestione e manutenzione, delle aree verdi pubbliche e private (Turfgrasses—Care, management and maintenance of public and private green areas).** Coauthor.
    See *Croce, Paolo—2006.*

Farley, Gertrude Adelia

Gertrude MacGill (1885–1964) was born in Bartlett, New Hampshire, USA. She was raised on a farm in east central New Hampshire and, subsequently, earned a two-year degree from Boston's Burdette College. Three years after the death of her husband, Otis Farley, who died due to gas wounds during WWI, she moved to Cleveland, Ohio. Farley initially worked for the Cleveland District Golf Association and was appointed the first of the local Green Section secretaries in 1921, charged to conduct a service bureau under the tutelage and supervision of Drs. C.V. Piper and R.A. Oakley. This was a little more than a year after the Green Section of the United States Golf Association became operational in Washington, DC. For six years, she was their Secretary, handling the buying for about 40 golf clubs in the Cleveland District Association of Greenkeepers and also managing and directing a clearinghouse of information. She was instrumental in forming of the National Association of Greenkeepers of America (NAGA) and contributed to the early writing and production of **The National Greenkeeper**. Her writings and unique poems were attributed to GAF so that she would not be identified as a lady in a male-dominant readership of greenkeepers. A 1926 poem,

Figure 12.33. Gertrude Farley, 1927. [2]

"Reflections of a Green Section Secretary," by GAF is presented herein. She was the first NAGA salaried officer and clerk as Assistant Secretary and Treasurer, responsible for the clerical work, accounting records, bookkeeping, dues collection, and payment of expenses. She subsequently resigned for unstated reasons in February 1928. In 1929, Farley joined the Davey Tree Expert Company in Kent, Ohio, specializing in tree surgery for golf courses. Later, Farley published **Golf Course Commonsense**, which was the first turfgrass book authored by a woman. In March 1936, Gertrude Farley initiated publication of **The Turf Survey: A Topical Textbook of Turf**, a monthly periodical that was short lived. Then in the mid-1930s, she was employed by the Addressograph-Multigraph Company in Cleveland until her retirement in 1954.

Reflections of a Green Section Secretary

I don't know how to keep a green,
Nor how to run a tractor;
My knowledge isn't very keen
On any golfing matter.
All day they ask me why are moles
And cuppy lies and ants.
And why the golfers dig such holes,
And wear such funny pants.
I tell 'em how to mow a tee,
And what to buy for seed;
The information's full and free,
It's all in what I read.
And when an engine stalls and quits,
They ask me what's the matter,

I sympathize and feed 'em bits
Of borrowed motor patter.
But when I step down off the deck,
Give up my frail command,
I'd not advise a nervous wreck
To take the work in hand.
Let him be strong and versed in low
Of greens and other things;
Let him be two of me and more,
Equipped with spreading wings.
Let him have courage, might to see
The thing to do—and do it,
No matter what the odds may be
The job! There's nothing to it!

—G.A.F.—

1931†   **Golf Course Commonsense.** G.A. Farley. Farley Libraries, Cleveland Heights, Ohio, USA. 256 pp., 59 illus., 7 tables, no index, cloth hardbound in dark green with gold lettering, 6.0 by 8.9 inches (152 × 226 mm).

   *   This is a **very rare** book that is a must for collectors of historical turfgrass books. It is the first turfgrass book written by a woman. It is a key book on both agronomic turfgrass culture and management of golf courses. It was the most advanced work of its time, with both the fundamentals and how-to aspects presented. Note the inclusion of chapters on Growing Choice Flowers and Golf in Community Welfare. These were truly modern topics for golf course operations that were seldom addressed in the 1930s. The chapter subject titles include the following:

I.	A Short History of Golf	XI.	Top-Dressing and Turf Repair
II.	Description Standard Eighteen Holes	XII.	Weeds and Diseases
III.	Greenkeeper in the Making	XIII.	Birds, Animals and Insects
IV.	Soils, Fertilization and Growth	XIV.	Equipment and Supplies
V.	Drainage and Water Systems	XV.	Keeping Course Records
VI.	Grasses	XVI.	Greenkeeper in the South
VII.	Teeing Grounds	XVII.	Concrete Construction.
VIII.	Fairways	XVIII.	Golf Course Tees
IX.	Putting Greens	XIX.	Growing Choice Flowers
X.	Hazards	XX.	Golf in Community Welfare

Faulkner, Reginald Percy [3]

1950†   **The Science of Turf Cultivation.** R.P. Faulkner. The Technical Press Limited, Kingston Hill, Surrey, England, UK. 64 pp.; 31 illus.; 1 table; no index; textured cloth hardbound in dark green and yellow with silver lettering and a jacket in cream, black, and white with brown lettering; 5.3 × 8.5 inches (135 × 216 mm).

   *   A small, practically oriented book on the construction, establishment, and culture of cool-season turfgrasses under the conditions in the United Kingdom. It is a **very rare** book. It represents a series of articles originally published in a monthly magazine. The foreword is by Martin A.F. Sutton. Faulkner's views on potassium requirements in 1950 were as follows:

Turning next to the element potassium, there is little evidence that this is required in any quantity by turf. Indeed much excellent turf exists that has never had a dressing of a potassium-containing fertilizer. Yet we know full well that potassium is needed by grass as it is needed by all plants, and grass mowings when analyzed are found to contain potassium, albeit only in very small quantity.

Federazione Italiana Golf (FIG)

FIG is an organization of golf clubs located in Italy; it is headquartered in Rome. It was formed in 1928 with the objective to further the development of golf. Activities include sponsorship of the Italian Open and other amateur championship competitions. FIG owns and operates a golf course, Le Querce, and also supports the FIG Green Section and Turfgrass Technical School. The Green Section provides turfgrass technical assistance for golf courses, conducts turfgrass research, and presents educational seminars and lectures. The Turfgrass Technical School was formed in 1989 under the leadership of Board Member Roberto Rivetti, with Dr. Jim Beard serving as Faculty Advisor. It is headquartered at the Federal Technical Center near Sutri, which is 28 mi (45 km) northeast of Rome. The Center has 3 instructional rooms, a library, 3 offices, and an 18-hole golf course—Le Querce.

n.d.†
circa
1990

Scuola Nazionale di Golf (National School of Golf). Federazione Italiana Golf, Roma, Italy. 32 pp., no illus., no tables, no index, semisoft bound in multicolor with black lettering, 6.7 by 9.2 inches (169 × 233 mm). Printed entirely in the Italian language.

An information booklet concerning the educational opportunities offered by the National School of Golf under sponsorship by the Federazione Italiana Golf. Included are specific course offerings, instructional facilities, teaching faculty, and summer work-study program.

1994†

Campi di Golf e Ambiente: Localizzazione, Progettazione e Gestione (Golf Courses and Environment—Site, Design and Management). Federazione Italiana Golf, Roma, Italy. 83 pp., 7 illus., 3 tables, no index, glossy semisoft bound in dark green with white lettering, 6.7 × 9.4 inches (169 × 238 mm). Printed entirely in the Italian language.

A small book on the relationships between golf and the environment. The design, construction, and turfgrass cultural practices that can be used to ensure a quality environment under Italian conditions are discussed.

1999†

Linee guida generali per una manutenzione ecocompatibile dei percorsi di golf italiani (Guidelines for the environment-friendly maintenance of Italian golf courses). National School of Golf, Green Section, Federazione Italiana Golf, Sutri, Viterbo, Italy. 31 pp., no illus., 10 tables, no index, fold-stapled semisoft covers in multicolor with black lettering, 8.3 by 11.6 inches (210 × 296 mm). Printed in the Italian and English languages.

A booklet on the turfgrass establishment and cultural practices contributing to protection of the environment on golf courses in Italy. There is a list of selected references in the back.

Ferguson, Marvin H (Dr.)

Marvin Ferguson (1918–1985) was born in Buda, Texas, USA, and earned a BSc degree in Agronomy from Texas A&M College in 1940 and a PhD degree from the University of Maryland in 1950. Dr. Ferguson joined the United States Golf Association (USGA) Green Section working at the Turf Gardens of the Arlington Research Station from 1940 to 1942 during the move of key plant materials to the new United States Department of Agriculture (USDA) Plant Industry Station in Beltsville, Maryland. He then served in the Navy as a Medical Corpsman during WWII. He rejoined the USGA in Beltsville as a Research Agronomist from 1947 to 1951, when he took a one-year leave to serve as Chief of the Grounds Maintenance and Erosion Control Section of the Military Air Transport Service. From 1952 through 1968, he was USGA Green Section National Research Coordinator and Director of the Midcontinent Region, officed at Texas A&M University, College Station, Texas. Dr. Ferguson played a key leadership role in educational efforts concerning the newly-developed USGA Root-Zone Modification System. It was a major challenge at that time, which he pursued in a respectful, scholarly manner to those questioning the new technique. He also served as Editor of the **Green Section Record**. He left the Green Section in 1958 to form AgriSystems of Texas, officed in Bryan, Texas. He was elected to the American Society of Agronomy Board of Directors and Turfgrass Management Division XI Chair in 1953. Dr. Ferguson received Fellow of the American Association for the Advancement of Science in 1964 and the USGA Green Section Award in 1973.

1968†

Building Golf Holes for Good Turf Management.˙ Marvin H Ferguson (editor) with six contributing authors: William H. Bengeyfield, James L. Holmes, Holman Griffin, Alexander M. Radko, James B. Moncrief, and Lee Record. United States Golf Association, New York City, New York, USA. 55 pp., 3 illus., 11 tables, no index, fold-stapled semisoft covers in light blue with black lettering, 6.0 by 9.0 inches (152 × 228 mm).

A handbook on the design, specifications, construction, and planting of golf courses, with specific chapters on turfgrasses for putting greens, turfed tees, fairways, roughs, and bunkers. There are small sections on trees and on equipment for golf courses. It is a unique how-to manual of its time.

Fermanian, Thomas Walter (Dr. and *Associate Professor* Emeritus)

Tom Fermanian was born in Milwaukee, Wisconsin, USA, and earned a BSc degree in Biology from the University of Wisconsin-Whitewater in 1972. He was Assistant Director of Quality Assurance for the Beatrice Foods Company from 1972 to 1974. He then received an MSc degree in Agronomy and Turfgrass in 1978 and a PhD degree in Crop Science and Turfgrass in 1980, both from Oklahoma State University, studying with Dr. Wayne Huffine. Since 1980, he has served in turfgrass research, teaching, and adult extension in the Department of Natural Resources and Environmental Sciences at the University of Illinois at Urbana-Champaign, specializing in weed management, plant growth regulation, and computer-based turfgrass management modeling. He retired from the University of Illinois in 2007 and became active in private consulting. He was elected to the Crop Science Society of America Board of Directors from 2002 through 2004 and Turfgrass Science Division C-5 Chair in 2003. Dr. Fermanian has received the Karl E. Gardner Outstanding Undergraduate Advisor Award in 1999, the Distinguished Service Award of the Illinois Turfgrasses Foundation in 1999, and an induction into the Academy of Teaching Excellence of the ACES College, University of Illinois, for 1999 to 2002.

1987† **Controlling Turfgrass Pests.** First edition. Coauthor.
    See *Shurtleff, Malcolm C.—1987.*

1997† **Controlling Turfgrass Pests.** Second revised edition. Thomas W. Fermanian, Malcolm C. Shurtleff, Roscoe Randall, Henry T. Wilkinson, and Philip L. Nixon. Prentice-Hall Inc., Upper Saddle River, New Jersey, USA. 655 pp., 939 illus. with 60 in color, 71 tables, large index, glossy hardbound in multicolor with dark green lettering, 7.0 by 9.2 inches (178 × 234 mm).

    A major revision and update that addresses the biology, management and control of weed, disease, and insect problems of both cool- and warm-season turfgrasses. Emphasis is placed on integrated pest management. There is an extensive glossary in the back.

2003† **Controlling Turfgrass Pests.** Third revised edition. Thomas W. Fermanian, Malcolm C. Shurtleff, Roscoe Randall, Henry T. Wilkinson, and Philip L. Nixon. Prentice-Hall Inc., Upper Saddle River, New Jersey, USA. 668 pp.; 763 illus. with 60 in color; 63 tables; extensive index; glossy hardbound in multicolor with green, blue, white, and yellow lettering; 7.0 by 0.2 inches (178 × 233 mm).

    An update of the 1997 second revised edition.

Fertiliser Journal

1954† **The Handbook of Organic Fertilisers—Their Properties and Uses.*** Fertiliser Journal Limited, London, England, UK. 84 pp., no illus., 23 tables, no index, cloth hardbound in green with black lettering, 5.4 by 8.5 inches (138 × 215 mm).

    A small book that describes the organic carrier sources for nitrogen, phosphorus, and potassium fertilizers available in the United Kingdom.

The Football Association (FA)

The FA was formed in 1863 by representatives of twelve soccer clubs and schools to frame a set of rules to facilitate uniformity in competitions. Currently the members are from affiliated clubs, leagues, and associations in England. The FA organizes eleven different competitions including the FA Cup, FA Youth Cup, and FA Women's Cup. Their publications includes two periodicals—**Insight** (Coaching) and **Refereeing**—plus annuals such as **The FA Handbook**, **The Official FA and England Yearbook**, and **Laws of Association Football**.

1951† **The Soccer Club Groundsman.** First edition. The Football Association. Naldrett Press, London, England, UK. 108 pp., 20 illus., 3 tables, no index, cloth hardbound in light green with dark green lettering and a jacket in greens and white with white and green lettering, 4.7 by 7.2 inches (119 × 182 mm).

   ＊   A key, small book for collectors of historical turfgrass books. An early, quality book devoted solely to the construction, establishment, and culture of turfgrasses on soccer fields. It is oriented to cool-season turfgrasses and conditions in England. Also included is a chapter on stands and spectators. The book was written in collaboration with the Sports Turf Research Institute, with a foreword by Sir Stanley Rous. The suggested seed mixture for English soccer fields in 1951 was as follows:

> In practice it is best to arrange a blend of several kinds which will intermingle and support one another in a suitable mixture. The mixture should contain Rye Grass and Crested Dog's Tail as the background together with a proportion of Timothy, Fescue, Bent and possibly Rough Stalked Meadow Grass. It should always be borne in mind, however, that good preparation of the land to be sown is just as important as the choice of the right seed. If the ground is very wet and therefore more at the mercy of the players' boots, it may be necessary to increase the proportion of the coarser grasses. In hard-pressed parts of the playing-area, such as the goal-mouth, it may be advisable to use only a mixture of Perennial Rye Grass and Crested Dog's-tail.

    Revised in 1972.
    See *Hawthorn, Robert—1972.*

Forestier, Jean Claude Nicholas

Jean Forestier (1861–1930) was born in Aix-les-Bains, France, and was originally trained for the engineering profession having graduated from the Ecole Polytechnique but, subsequently, attended the Ecole Forestiére at Nancy, France. He then worked on reforestation projects in the Pyrenees and the Alps. He became a landscape architect and urban planner of gardens and parks. Jean Forestier worked as an assistant to R. Alphand who was Director-General of Public Works in Paris. Initially Jean Forestier worked in the Bois de Vincennes and then in eastern Paris. He also was Director of the Municipal Horticulture School. In 1897, he became Director of the Bois de Boulogue in the western section of Paris where he served for thirty years as forester and garden architect. He was responsible for the urban transformations of Paris during the late nineteenth and early twentieth centuries, including the role of free spaces within an urban setting. Jean Forestier was one of the most prominent French landscape architects of the early twentieth century. This was accomplished via his actual designs and constructions of parks in diverse locations worldwide plus through his writings and lectures. Jean Forestier's honors included Officer de la Légion d'Honneur, Commandeur du Mérite Civil d'Espagne, and Officier d'Académie and du Mérite Agricole.

1908† **Les Gazons (Turfs).*** J.C.N. Forestier. Lucien Laveur, Paris, France. 136 pp., 29 illus. with 12 plates, 19 tables, no index, cloth bound in red, 5.4 by 7.2 inches (137 × 184 mm). Printed entirely in the French language.

* A **truly rare** book that is a must for collectors of historical turfgrass books. It is a pioneering book on the construction, establishment, and culture of lawns in France. The book is oriented to the field practitioner and contains sections on tennis and croquet courts. Much of his advice was based on English lawn culture. It is the first turfgrass book published in France and one of the earliest in the world. There is one chapter on flowers.

Franklin, Stuart Bruce

Stuart Franklin was born in New York City, New York, USA, and earned a BA degree from Syracuse University in 1973. He has been the owner-operator of a lawn fertilizing company, Nature's Lawn and Garden Inc., in western New York. Stuart Franklin was elected Treasurer of the Western New York Society for Organic Horticulture for 2002 through 2006.

1988† **Building a Healthy Lawn: A Safe and Natural Approach.** Stuart Franklin. A Garden Way Publishing Book, Storey Communications Inc., Pownal, Vermont, USA. 168 pp., 83 illus. plus 16 in color, 5 tables, small index, glossy semisoft bound in multicolor with white and black lettering, 6.0 by 9.0 inches (152 × 228 mm).

A book on lawn care, with a small section on turfgrass establishment. The philosophy focuses on organic approaches. It is oriented to conditions in the United States. There is a small glossary of selected terms in the back.

There are multiple reprintings.

Freeborg, Raymond Paul (Dr.)

Ray Freeborg (1924–1989) was born in St. Louis, Missouri, USA, and earned a BSc degree in Agronomy and Turfgrass from Purdue University in 1959; an MA degree in Botany from Washington University, St. Louis, in 1966; and a PhD degree in Agronomy and Turfgrass from Purdue University in 1971, studying with Dr. Bill Daniel. Dr. Freeborg served as Executive Secretary of the St. Louis Turf Research Association from 1959 to 1963 and as Research Agronomist at Purdue University from 1973 to 1985. He was involved in turfgrass teaching, adult extension, and research with emphasis on herbicides and plant growth regulators.

1979† **Turf Managers' Handbook.** First edition. Coauthor.
&
1980†

    See *Daniel, William Hugh—1979 and 1980.*

1987† **Turf Managers' Handbook.** Second revised edition. Coauthor.
    See *Daniel, William Hugh—1987.*

Fry, Jack Douglas (Dr. and *Professor*)

Jack Fry was born in Kansas City, Kansas, USA, and earned a BSc in Horticulture from Kansas State University in 1982; an MSc in Agronomy from the University of Maryland in 1984, studying with Dr. Peter Dernoeden; and a PhD in Horticulture and Turfgrass from Colorado State University in 1987, studying with Dr. Jack Butler. From 1987 to 1991, he served as an Assistant Professor of Turfgrass Science at Louisiana State University and, since, has been on the faculty at the Department of Horticulture, Forestry and Recreation Resources at Kansas State University (KSU), becoming an Associate Professor in 1994 and a Full Professor in 2000. He has been active in both teaching and research. He was elected to the Crop Science Society of America Board of Directors from 2003 through 2005 and the Turfgrass Science Division C-5 Chair in 2004. Dr. Fry was awarded Outstanding Faculty Mentor from the KSU College of Agriculture in 2001 and the Outstanding Research Award from KSU Gamma Sigma Delta in 2003.

Figure 12.34. Shanks 1914 motor-powered, walk-behind, 30-inch-wide, reel mower with grass box. [38]

2004† **Applied Turfgrass Science and Physiology.** Jack Fry and Bingru Huang. John Wiley & Sons Inc., Hoboken, New Jersey, USA. 320 pp., 97 illus., 103 tables, indexed, glossy hardbound in multicolor with white and green lettering, 8.5 by 9.2 inches (190 × 233 mm).

A speciality text on the physiology and culture of turfgrasses. The presentation is in three parts: carbohydrate metabolism and turfgrasses, environmental stress and pests, and cultural practices. There is a list of selected references at the end of each chapter.

G

Gandert, Klaus-Dietrich *(Dr.* and *Professor)*

Dietrich Gandert was born in Berlin, Germany, and earned a Diplom-Gärtner in 1953, a Doctorate of Agronomy (DrAgro) in 1957, and a Doctorate of Agricultural Science (DrScAgr) in 1980, all from Humboldt University. Dr. Gandert served in the Sektion Pflanzenproduktion at Humboldt University of Berlin where his main focus was teaching. He was named Associate Professor of Village and Landscape Design in 1980 and Professor for Landschafts- und Grünanlagengestaltung (Landscape and Green Space Design) in 1990. Dr. Gandert initiated the first turfgrass research in Germany in 1958.

1960 **Rasen: Bedeutung, Anlage, Pflege (Turf: Importance, Installation, Care).** Klaus-Dietrich Gandert. VEB Deutscher Landwirtschaftsverlag, Berlin, East Germany. 156 pp., 86 illus., 16 tables, no index, glossy semisoft bound in greens with red and black lettering, 5.9 by 8.1 inches (149 × 207 mm). Printed in the German language.

An early German book concerning the characteristics of cool-season turfgrasses, site and soil preparation, turfgrass establishment, and culture under cool-season conditions in Germany. There is a list of selected references in the back.

1976† **Rasen für Sport und Spiel (Turf for Sports and Games).** Klaus-Dietrich Gandert and Astulf Schnabel. VEB Deutscher Landwirtschaftsverlag, Berlin, East Germany. 171 pp.; 63 illus.; 18 tables; small index; glossy hardbound in multicolor with white, black, and red lettering; 5.7 by 8.3 inches (145 × 212 mm). Printed entirely in the German language.

A book on soil problems, construction, turfgrass species, culture, and renovation of sports fields under the conditions in East Germany. There is a short list of selected European references in the back.

Reprinted in 1977.

1991 **Handbuch Rasen: Grundlagen-Anlege-Pflege (Handbook for Turfgrass: Basics, Installation, Maintenance).** Klaus-Dietrich Gandert and František Bureš. Deutscher Landwirtschaftsverlag GmbH, Berlin, Germany. 364 pp.; 167 illus.

with 31 in color; 60 tables; small index; glossy hardbound in multicolor with yellow, white, and black lettering, 5.7 by 8.4 inches (145 × 214 mm). Printed in the German language.

A basic text concerning turfgrass science as applied to turfgrass culture in Germany. There are sections on construction and culture for a diverse range of turfgrass uses, including sports fields, gymnastic turfgrasses, horse sports grounds, lanes & parking areas, airfields, roofs, roadsides, high altitudes, and shaded areas. There is an extensive list of references in the back.

Garcia-Ircio, Francisco Javier

Javier Ircio was born in Zaragoza, Spain, and earned a BSc (Agronomist Engineer) degree in Pytotechnics from Madrid Polytechnic University in 2003, an MSc degree in Technology and Quality Control of Foods from Centro de Estudios Superiores de la Industria Farmaceutica, Madrid, in 2005, and an Associate Certificate in Turfgrass Management from Michigan State University (MSU) in 2009. He worked as Laboratory Director of Food Hygiene and Safety for the Agriculture Department of Angola from 2005 to 2006. Javier Garcia-Ircio received a scholarship from the Royal Spanish Golf Association-GCSAA to study at MSU in 2007 and 2008.

2008† **El Swing Del Agua (Swing of the Water).** Francisco Javier Garcia-Ircio. Castilla La Mancha Golf Federation, Guadalajara, Spain. 92 pp., 35 illus. in color, 5 tables, no index, glossy semisoft bound in blues and white with white and black lettering, 6.3 by 9.0 inches (160 × 229 mm). Printed in the Spanish language.

A small book on water use and conservation that is focused primarily on the sociological and political aspects in Spain. Turfgrasses as employed on golf courses and agricultural crops are the primary examples represented.

Gault, William Kerr

William Gault (1876–1916) was born in Scotland, UK. Following apprenticeship as a greenkeeper, he became Head Greenkeeper at Glasgow Golf Club at Blackhill, Scotland, and, from 1903 to 1905, at Glasgow Golf Club at Killermont, which he constructed in six months from 1903 to 1904 based on a centerline fixed by Tom Morris. From 1905 to 1910, he was Head Greenkeeper at Portmarnock Golf Club in County Dublin, Ireland, and then at the Bruntsfield Links Golfing Society, Cramond, in Edinburgh, Scotland, from 1910 to 1916. William Gault was a contributing writer on greenkeeping for **The Irish Golfer**, was active as a greenkeeping consultant, and produced and sold his own fertilizer under the name "Gault's Patent Manure."

n.d.   **Practical Golf Greenkeeping.** W.K. Gault. The Golf Printing and Publishing Company, London, England, UK. 101
circa   pp. with 9 pp. of advertising, 3 illus., no tables, no index, cloth hard-bound in dark green with gold reverse embossed
1912   lettering, 5.5 by 8.5 inches (140 × 216 mm).

*   A **truly rare** book that is a must for collectors of historical turfgrass books. It is the first book written by a working greenkeeper. The book is organized in a calendar format of cultural practices for golf courses, with emphasis on adverse climatic conditions. Each chapter represents a month of the year. This book was written after an adverse, dry year in 1911. Also included is a photograph of William Gault. The lines of play, sheep for roughs, and motor mowers plus a description of a 1912 golf exhibition that included turfgrass equipment are discussed as follows:

Lines of Play
The closer we cut the lines of play the better for the turf, and the good golfer will appreciate the close-lying ball in a position similar to that which he finds at the seaside; and where sheep are kept, the best way to remedy the evil of their constant presence on the fairway is to have it close cut. If sheep cannot find enough to eat on the proper course, then, and only then, will they thin out the rough, and keep the bottom so that a ball may be found. When the latter is accomplished, it is better than cutting the rough with a reaper, which plan is so commonly practised. The cutting away of the long grass means the removal of a useful hazard for a pulled or sliced ball, and the obliteration of the true line of play. I think the traditions of the Royal and Ancient game should be maintained, regardless of petty annoyances or losses, at least so far as allowing those natural hazards to remain. And long grass, of whatever kind, is a natural hazard, which calls for more science in playing from than do most of our present-day artificial sand bunkers. We seldom see the rough trimmed on seaside links. Bents, bracken, whins, and other rank growing plants are allowed to remain as hazards and ornaments. Why not leave the rough on inland courses?

Top Dressing
The sand is of no value as a manure, but when judiciously applied, it counteracts and reduces the organic matter which rich soil contains in excess of the amount required for growing fine grasses. Sand makes a firm, porous surface on a green, and encourages the finer grasses to sprout up, and it has also a refining effect on the coarser growing kinds, as well as making the plants of a hardier substance; yet, after all these attainments, it is necessary to use sand with discretion, because, as in other things, one

Figure 12.35. William Gault, circa 1910. [22]

must be careful not to overdo the right amount. Far better to use too little and be on the safe side, than to put on too much. Just apply it in small quantities; allow the turf time to absorb one dressing before putting on another.

Motors
With the increased popularity of golf, and the improvements achieved in all other departments of the game, the motor mower is the only additional implement which has been added to the greenkeeper's equipment that is worth speaking about. The present advanced state in which the motor for golf links is presented, shows a decided improvement, both in construction and in simplicity of working, as compared with the same class of machine only a few years ago. In 1912, the Golf Exhibition was held at Murifield during the week of the Open Championship, and the Scottish greenkeepers had a good opportunity of studying the latest in machinery as applicable to golf links, a great number of greenkeepers being present. The various stalls presented an animated appearance. It is needless to mention the number of exhibitors, suffice to say they were all there.

Gibbs, Joseph Arthur[3]

1895   **The Improvement of Cricket Grounds on Economical Principles.** J.A. Gibbs. Horace Cox, London, England, UK. 42 pp. plus 35 pp. of advertising, no illus., 1 table, small index, cloth hardbound in dark green with gold impressed lettering, 4.8 by 7.1 inches (123 × 181 mm).

✳   A **truly rare**, pioneering booklet on soil selection, construction, culture, and costs of turfed cricket wickets and grounds in the late 1800s in England. It is the second oldest turfgrass book published in the United Kingdom.

Gibbs, Richard John (Dr.)

Richard Gibbs was born in Bristol, England, UK, and earned a BSc degree with *first class honours* in Agricultural Chemistry from the University of Leeds in 1981 and a PhD degree in Soil Science from Lincoln College, New Zealand, in 1986. He worked as a Postdoctoral Research Assistant with the soils-turfgrass program at the University College of Wales, Aberystwyth, from 1986 to 1990 and then as Senior Lecturer in Turfgrass Science and Culture at Myerscough College in Lancashire, England. Dr. Gibbs immigrated to New Zealand in 1991 and became a Scientific Officer at the New Zealand Sports Turf Institute at the Fitzherbert Science Centre, Palmerston North, New Zealand, where he administered the construction and research activities. Then in 2003, Dr. Gibbs joined Recreational Services Limited as Manager of the Sports Surface Design and Management Division. Dr. Gibbs was elected to the Board of Directors of the International Turfgrass Society (ITS) for 2001 to 2009 and as Vice President of ITS for 2005 to 2009.

1994† **Natural Turf for Sport and Amenity: Science and Practice.** Coauthor.
See *Adams, William Arthur—1994.*

Gibeault, Victor Andrew (Dr. and Professor Emeritus in Botany and Plant Sciences)

Vic Gibeault was born in Pawtucket, Rhode Island, USA, and earned both BSc and MSc degrees in Agronomy from the University of Rhode Island in 1963 and 1965, respectively, studying with Dr. Richard Skogley, and a PhD degree in Crop Physiology and Taxonomy from Oregon State University in 1971, studying with Dr. Norm Goetze. From 1969 to 2004, he served as Extension Environmental Horticulturalist at the University of California, Riverside. His emphasis was on turfgrass species and cultivar performance and cultural practices. Dr. Gibeault received the GCSSA Distinguished Service Award in 1993.

1985† **Turfgrass Water Conservation.** First edition. Victor A. Gibeault and Stephen T Cockerham (editors) with fourteen contributing authors: Joseph P. Rossillon, James R. Watson, Victor B. Youngner, James B Beard, Robert C. Shearman, Jack D. Butler, Paul E. Rieke, David D. Minner, Robert N. Carrow, Jewell L. Meyer, Bruce C. Camenga, Phillip F. Colbaugh, Clyde L. Edmore, and Cal O. Olson. Publication 21405, Cooperative Extension, Division of Agriculture and Natural Resources, University of California, Oakland, California, USA. 155 pp., 77 illus., 43 tables, no index, glossy semisoft bound in multicolor with green, blue, and black lettering, 8.4 by 10.9 inches (213 × 276 mm).

The book consists of 11 technical review papers based on a turfgrass water conservation symposium sponsored by the American Sod Producers Association. Each section is supported by a set of scientific literature citations and a practicum. A small glossary of selected terms is in the back. This was an important early assemblage of information concerning turfgrass water conservation.

2004† **Establishing and Maintaining the Natural Turf Athletic Field.** Coauthor.
See *Cockerham, Stephen T—2004.*

2011† **Turfgrass Water Conservation.** Second revised edition. Coeditor.
See *Cockerham, Stephen T—2011.*

Gionet, Luc

Luc Gionet was born in Montreal, Quebec, Canada, and earned a Diplôme D'études Collégiales (DEC) degree in Science from Bois-de-Boulogne College in 1984 and a BAU degree in Landscape Architecture from Montreal University in 1988. He served as a Landscape Architect for Pluram Inc. from 1987 to 1990 and as a Landscape Architect for the City of Montreal since 1990 plus as Teaching Assistant at Quebec University of Trois-Rivieres from 1991 to 1994. He is responsible for technical assistance and research in sports fields and equipment for public park and playgrounds. Luc Gionet has been involved in facilities for a number of national and international competitions such as soccer (FIFA), tennis (Canada Cup), and track & field (Canada Championships).

2005 **Guide d'Aménagement et d'Entretien des Terrains de Soccer Extérieurs (Guide for Installation and Maintenance of Outdoor Soccer Fields).** Luc Gionet and Denis Brown. Ministére de l'Éducation, du Loisir et du Sport, Québec, Canada. 77 pp., 50 illus., 15 tables, no index, glossy hardbound in greens and blue with white and green lettering, 8.2 by 10.9 inches (208 × 277 mm). Printed in the French language.

A booklet on surfaces for soccer fields, both turfgrass and artificial. Specifics of root-zone profile construction and both subsurface and surface drainage techniques are addressed for conditions in the Province of Quebec. Other aspects included field layout, irrigation facilities, and safety relative to fixed structures.

Goatley, James Michael, Jr. (Dr. and Professor)

Mike Goatley was born in Springfield, Kentucky, USA, and earned both a BSc degree in Agronomy and an MSc degree in Crop Science from the University of Kentucky in 1983 and 1986, respectively. He then earned a PhD degree majoring in Agronomy and Turfgrass from Virginia Tech University in 1988, studying with Dr. Richard Schmidt. He joined the Agronomy Faculty at Mississippi State University (MSU) in 1988 conducting turfgrass teaching and advising and research in the Golf and Sports Turf Management Program. He rose to Full Professor in 1998 and, subsequently, accepted a Turfgrass Extension Specialist position at Virginia Tech University in 2004. Dr. Goatley has received the Outstanding Turfgrass Professional Award from the Mississippi Turfgrass Association in 2001, the Outstanding Undergraduate Advisor from MSU in 2001, the Excellence in Teaching Award from MSU in 2002, and the Outstanding Faculty Advisor Award from the Academic Advising Association in 2002.

1999† **Sports Fields—A Manual for Design, Construction, and Maintenance.** Coauthor.
See *Puhalla, James—1999.*

2003† **Baseball and Softball Fields: Design, Construction, Renovation, and Maintenance.** Coauthor.
See *Puhalla, James—2003.*

2008† **Sports Turf Management in the Transition Zone.** Michael Goatley, Shawn Askew, Erik Ervin, David McCall, Bob Studholme, Peter Schultz, and Brandon Horvath. Pocahontas Press Inc., Blacksburg, Virginia, USA. 203 pp., 100 illus.

with 96 in color, 4 tables, indexed, glossy hardbound in multicolor with white and black lettering, 6.0 by 9.0 inches (152 × 228 mm).

A speciality book concerning turfgrasses and soils of the transition climatic zone as utilized in the establishment and culture of sports fields. It also includes maintenance of skinned infields and field marking.

Goit, Whitney[3]

1932   **Grass–Golf Clubs Can Do It, So Can You.** Whitney Goit and Randolph Shaffer. Golf Course Equipment Company, Chicago, Illinois, USA. 30 pp., no illus., no tables, no index, fold-stapled textured semisoft covers in green and black with black lettering, 5.7 by 8.8 inches (146 × 223 mm).

*   A **very rare**, small book on the establishment and culture of lawns. It is oriented to lawn enthusiasts and cool-season turfgrass conditions in the northern United States.

Gooch, Robert Butler

Bob Gooch (1913–1987) was born in Ramsgate, Kent, England, UK. His early apprenticeship was for the Kent County Council starting in 1929. He then joined the National Playing Fields Association (NPFA) in 1937 and, except for military service in the Territorial Army during WWII, he worked for the Association until his retirement in 1978. After WWII, he served the NPFA as Deputy to P.W. Smith, and then as Chief Technical Officer and as Technical Director. Bob Gooch received the NPFA President's Certificate from Prince Philip in 1979.

1956†   **Selection and Layout of Land for Playing Fields and Playgrounds.** First edition. R.B. Gooch. The National Playing Fields Association, London, England, UK. 95 pp., 42 illus., 4 tables, no index, semisoft bound in yellow with green lettering, 5.4 by 8.3 inches (138 × 212 mm).

A manual on site selection, design, and layout of sports and recreation facilities under the conditions in the United Kingdom. Included are the dimensional requirements for most sports, the layout of playing fields on from 3 to 32 acres, and children's playground layouts.

Revised in 1959 and 1963.

1959†   **Selection and Layout of Land for Playing Fields and Playgrounds.** Second revised edition. R.B. Gooch. The National Playing Fields Association, London, England, UK. 92 pp., 44 illus., 4 tables, no index, semisoft bound in light blue with dark blue lettering, 5.4 by 8.3 inches (138 × 210 mm).

A minor revision of the 1956 edition.

Revised in 1963.

1963†   **Selection and Layout of Land for Playing Fields and Playgrounds.** Third revised edition. R.B. Gooch. The National Playing Fields Association, London, England, UK. 107 pp., 43 illus., 4 tables, no index, fold-stapled semisoft covers in yellow with black lettering, 5.4 by 8.2 inches (138 × 208 mm).

An update and revision of the 1959 edition.

1965†   **Sports Ground Construction—Specifications For Playing Facilities.** First edition. R.B. Gooch and J.R. Escritt. The National Playing Fields Association, London, England, UK. 104 pp., 16 illus., 1 table, no index, cloth hardbound in green with black lettering, 5.4 by 8.5 inches (138 × 215 mm).

A small handbook of procedures, contracts, and specifications for outdoor sports field construction and establishment in England. Included are such sports as lawn bowling, croquet, golf greens, lawn tennis, and cricket. Hard, nonturf surfaces also are included, such as tennis courts and running tracks. An appendix in the back includes a glossary of terms and explanations plus a list of British standards.

There was an earlier version published by the National Playing Fields Association titled **Specifications of Playing Facilities**.

See *Smith, Percy White—1948.*

Revised in 1975.

1975†   **Sports Ground Construction—Specifications.** Second revised edition. R.B. Gooch and J.R. Escritt. National Playing Fields Association, London, England, UK. 126 pp., 8 illus., 7 tables, indexed, cloth hardbound in light green with black lettering, 5.4 by 8.5 inches (137 × 216 mm).

An updated and slightly expanded third revision, presented in metric units.

Goss, Roy Leon (Dr.)

Roy Goss was born in Weslaco, Texas, USA. He earned a BSc degree in Agriculture in 1950, a Bachelor of Education (BEd) degree in 1951, and a PhD degree in Agronomy in 1960, all from Washington State University. From 1958 to 1988, he served as the Turfgrass Research and Extension Agronomy Specialist for Washington State University, officed in Puyallup, Washington. He also was Executive Secretary of the Northwest Turfgrass Association from 1960 to 1988 and was elected to the Washington State University Faculty Senate for 1970 to 1972 and to the position of

Parks Board Commissioner for Anderson Island Washington Park for 1968 to 1972. Dr. Goss received the GCSAA Distinguished Service Award in 1978, USGA Green Section Award in 1988, an induction to the Oregon Golf Superintendents Hall of Fame in 2003, and a Lifetime Member of both the Northwest Turfgrass Association in 1987 and the Western Canada Turfgrass Association in 1988.

1983†   **Construction and Maintenance of Natural Grass Athletic Fields.** Roy L. Goss and Thomas Cook. Pacific Northwest Cooperative Extension Publication PNW 0240, Puyallup, Washington, USA. 27 pp., 24 illus. with 13 in color, 4 tables, no index, fold-stapled glossy covers in multicolor with white and black lettering, 8.4 by 10.9 inches (213 × 277 mm).

A small guide to the construction and culture of sport fields under the cool conditions of the Pacific Northwest. Drainage system designs for various types of sports fields are diagramed, and suggested seasonal maintenance calendars are presented.

Gould, Ann Brooks (Dr.)

Ann Gould was born in Chicago, Illinois, USA, and earned a BSc degree in Chemistry from Illinois State University in 1978, an MSc degree in Biology from Illinois State University in 1980, and a PhD in Plant Pathology from the University of Kentucky in 1988. She was a Postdoctoral Associate at the University of Florida from 1996 to 1997 and 1998 to 1999. She has served as an Extension Specialist in the Diseases of Ornamentals at Rutgers University since 1990. Dr. Gould received the Merle V. Adams Award for Academic Excellence in Extension in 1995 and the Rutgers Cooperative Extension Award for Diversity in 1995.

1993†   **Turfgrass Patch Diseases Caused by Ectotrophic Root-Infecting Fungi.** Coeditor.
      See *Clarke, Bruce Bennett—1993.*

Gould, Frank Walton (Dr. and Distinguished Professor Emeritus)

Frank Gould (1913–1981) was born in Mayville, North Dakota, USA, and earned a Bachelor of Engineering (BE) degree from Northern Illinois State Teachers College in 1935, an MA degree in Botany from the University of Wisconsin in 1937, and a PhD degree in Botany from the University of California, Berkeley, in 1941. He served as Instructor at Compton Junior College from 1942 to 1944, as Assistant Professor at the University of Arizona from 1944 to 1949, and as Associate Professor through Distinguished Professor at Texas A&M University in the Range Science Department from 1949 to 1979. He was a noted grass taxonomist and Curator of the S.M. Tracy Herbarium grass collection. Dr. Gould received the Faculty Distinguished Achievement Award in Research from Texas A&M University in 1969.

1968†   **Grass Systematics.*** First edition. Frank W. Gould. McGraw-Hill Book Company, New York City, New York, USA. 382 pp., 188 illus., 7 tables, indexed, cloth hardbound in light blue and green with green lettering, 5.9 by 8.9 inches (150 × 227 mm).

An important scholarly book on grass classification and systematics, including the structure and growth of the grass plant. It is oriented to the grass species of North America. A key to the grass genera of the United States is presented. Certain subject chapters include a list of selected references. There is a glossary of terms in the back.
      Revised in 1983.

Figure 12.36. Steam-powered, roller-drive, reel mower, weighing 1,000 or more pounds, circa 1905. [10]

1983† **Grass Systematics.** Second revised edition. Frank W. Gould and Robert B. Shaw. Texas A&M University Press, College Station, Texas, USA. 397 pp., 205 illus., 9 tables, index, both cloth hardbound in dark green with gold lettering and also glossy semisoft bound in white, green, and black with green and black lettering, 6.0 by 9.0 inches (152 × 228 mm).

A minor revision completed after Dr. Gould's death and based on his work. Included are updated subfamily names, incorporating his most recent concept of the Paniceae, and new information on the importance of the photosynthetic pathway and C_4 subtype in grass systematics.

Graves, Robert Muir

Bob Graves (1930–2003) was born in Trenton, Michigan, USA, and earned a BSc degree in Landscape Architecture from the University of California, Berkeley, in 1953. He served in the US Navy during the Korean conflict and served 22 years in the Naval Reserves achieving the rank of Commander. He entered private practice as a licensed landscape architect in 1955, officed in Walnut Creek, California. Within five years, he became focused on golf course architecture. He also was active as a lecturer on golf course design at a number of universities, including the Harvard Graduate School of Design, starting in 1989 with colleague Geoff Cornish. Bob Graves was elected President of the American Society of Golf Course Architects in 1974 and was named Fellow of the Society in 2003.

1998† **Golf Course Design.** Robert Muir Graves and Geoffrey S. Cornish (editors), with 21 contributing authors: Thomas A. Marzolf, Ronald G. Dodson, Desmond Muirhead, Kenneth De May, William W. Amick, James McC. Barrett, John H. Foy, James Fransis Moore, Norman Hummel Jr., James T. Snow, W. Gary Paumen, Richard H. Elyea, Virgil Meier, Dean Mosdell, Christine Faulks, Philip Arnold, Richard L. Norton, S. Jeffrey Anthony, Barbara B. Beal, Keven J. Franke, and Mark A. Mungeam. John Wiley & Sons Inc., New York City, New York, USA. 446 pp., 329 illus., 31 tables, small index, cloth hardbound in dark green with silver lettering, 7.5 by 9.8 inches (191 × 248 mm).

A book based on a compilation of seminars presented at Harvard University as part of the Professional Development Summer Continuing Education Program. The subjects addressed include the history of golf and golf course design, site selection, course routing, golf hole design, construction, drainage, irrigation system, turfgrass selection, planting, financing, securing permits, and training of golf course architects. The appendix includes design exercises and symbols in golf course design. Included in the back is a selected bibliography and a glossary of terms.

Greenfield, Ian Ronald

Ian Greenfield (1930–2004) was born in Godstone, Kent, England, UK, and earned a BSc degree in Agriculture from Wye Agricultural College, and an MI in Biology. He served as Chief Agricultural Advisor to F.W. Berk & Company Limited and was involved in research and development of pest control chemicals during his early career. In 1963, he then became Managing Director of Cayford Technical Service and Cayford Chemicals Limited of Bishops Stortford, Hertfordshire, England (specialists in golf course and sports ground development and construction), serving into the early 1990s.

1962† **Turf Culture.** Ian Greenfield. Leonard Hill (Books) Limited, London, England, UK. 364 pp. plus 5 pp. of advertising, 135 illus. with 1 color plate, 59 tables, detailed index, textured cloth hardbound in green with gold lettering and a jacket in multicolor with white and black lettering, 5.9 by 9.5 inches (150 × 243 mm).

A basic book on soils, grasses, construction, establishment, and culture of turfgrasses oriented primarily for the United Kingdom. It contains sections on equipment and contracts. Types of turfgrass maintenance discussed were soccer, rugby, field hockey, cricket, tennis courts, croquet, lawn bowling, polo, horse racecourses, parks, golf courses, lawns, and roadsides. There is one chapter summarizing recent turfgrass developments in other parts of the world, especially pest control. The foreword is by Martin Sutton.

Gruttadaurio, Joann[2]

2008† **Sports Field Management.** Joann Gruttadaurio. New York State Turfgrass Association, Latham, New York, USA. 42 pp., 15 illus., 19 tables, no index, glossy semisoft covers in multicolor with white lettering in a spiral ring binder, 11.0 by 8.5 inches (279 × 215 mm).

A manual concerning the culture and management of cool-season turfgrasses on sport fields under the conditions in New York state. A seasonal maintenance calendar is presented. There is a sport field assessment form in the back.

Guérin, Jean-Paul (Dr.)

Jean-Paul Guérin was born in Cholet, France, and earned an Agricultural Engineering degree in 1968 and a PhD degree in Agronomy in 1970, both from Rennes University. He served as Ingénier for the Ministere of Sports from 1974 to 1986, as Ingénier for Groupe Jeanjean from 1986 to 1989, and Ingénier of Research and Development for BASF Compo from 1989 to 2007, when he retired. He was elected Secrétaire for the Société Francaise des Gazons for 1989 to 2007. Dr. Guérin was awarded the Chevalier Mérite Agricole by the Ministere de l'Agriculture in 1996.

1985† **Los Cespedes (Turfs).** Coauthor.
See *Thomas, Robert—1985.*

Gullino, Maria Lodovica (Dr. and Professor)

Maria Gullino was born in Saluzzo, Italy, and earned a BSc in Biology in 1979 and a PhD degree in Biological Sciences in 1975, both from the Botany Institute, University di Turino. She became a National Research Council Fellow in Plant Pathology in 1978, a Researcher in 1980, an Associate Professor in 1986, and a Full Professor in Biological and Integrated Plant Disease Management (Phytopathological Biotechnologies) in 2000, all from the University di Turino. The author of five books, she has been active in research concerning plant disease management, biological control of diseases, and sustainable plant protection. She was elected Chair of the Italian Society for Crop Protection from 1999 to 2007, Chair of the Italian Association of Agricultural Scientific Societies from 2003 to 2004, and Vice President of the International Society for Plant Pathology in 2003. Dr. Gillino received the International Novamont Award in 2002 and the Van Den Brande Award from the University of Gent, Belgium, in 2005.

2000† **La difesa dei tappeti erbosi: Malattie fungine, nemici animali e infestanti (Pest control on turfgrasses: Diseases, animal enemies and weeds).** M. Lodovica Gullino, Massimo Mocioni, Guiseppe Zanin, and Alberto Alma. Edizioni L'Informatore Agrario, Verona, Italy. 196 pp., 176 illus. with 158 in color, 28 tables, indexed, cloth hardbound in gray with gold lettering and a glossy jacket in multicolor with black lettering, 6.6 by 9.4 inches (167 × 239 mm). Printed entirely in the Italian language.

A book concerning the pests of turfgrasses common to Italy, including diseases, insects, and weeds. Individual pests are described, along with the turfgrass injury symptoms and the appropriate management practices. There is list of selected references and a glossary of terms in the back.

H

Hacker, John William

John Hacker was born in Devizes, Wiltshire, UK, and earned a Diploma in Horticulture from the Royal Botanic Gardens, Kew, in 1975 and an MSc degree from Rutgers University in 1977, studying rose culture. He has served as Demonstrator in Horticulture at the University of Bath from 1977 to 1982, as Senior Lecturer in Turfgrass Technology at Lancashire College of Agriculture and Horticulture from 1982 to 1989, and, then, as Director of Professional Sportsturf Design (NW) Limited.

1992† **Golf Course Presentation.** John Hacker and George Shiels. Professional Sportsturf Design (N.W.) Limited, Preston, England, UK. 45 pp., 16 illus. in color, 3 tables, no index, fold-stapled glossy semisoft covers in multicolor with white and yellow lettering, 5.8 by 8.2 inches (147 × 209 mm).

The surface turfgrass appearance or presentation of a golf course is addressed in this small booklet. Emphasis is placed on highlighting the individual or unique aspects of each golf course. It is oriented to cool-season turfgrass conditions in the United Kingdom.

Hackett, Norman

Norman Hackett (1882–1957) was born in Accrington, Lancashire, England, UK. He apprenticed in the leather tanning and currying business of his family and emerged as a West Riding industrialist. He also was an amateur golfer and member at Keighley Golf Club in Yorkshire, England, UK. He was one of the pioneering founders of the Board of Greenkeeping Research (BGR) at Bingley in 1929, later renamed the Sports Turf Research Institute (STRI). He served as the Board's Honorary Secretary from 1929 to 1951. He conducted turfgrass experiments at Keighley Golf Club well before the STRI was founded and presented the findings in publications such as **Golfing** in the early 1920s. Based on these findings, Norman Hackett also acted as a turfgrass advisor at golf courses in the United Kingdom.

1928† **Soil Acidity: The Vital Importance of Top-Dressing And Other Notes.** Norman Hackett. British Golf Union Joint Advisory Committee, Glasgow, Scotland, UK. 37 pp., 4 illus., 1 table, no index, fold-stapled semisoft covers in green with black lettering, 5.5 by 8.5 inches (140 × 218 mm).

 * A **truly rare** booklet that is a must for collectors of historical turfgrass books. This small booklet focuses on the philosophy of acidic soil management and topdressing for putting greens. The author cites supporting turfgrass research from the United States and other locations. It contains an introduction by Sir Robert Greig. This booklet represents the first written introduction of the acid theory of turfgrass culture into the United Kingdom from the United States. In the front, the author emphasized the need for a turfgrass research and educational organization in England. Norman Hackett summarized his philosophies on turfgrass fertilization and topdressing as follows:

Figure 12.37. Oxen teams pulling 36-inch-wide, reel mowers on the fairways at Durban Golf Club, South Africa, 1928. [38]

Out of all the field of investigations and experience, including those at Arlington and elsewhere, has emerged what appears to be an extremely simple course of procedure for the fertilising of Greens. As the case stands today, it seems sure that the most satisfactory fertiliser treatment from the standpoint of turf and economy involves the combined use of *ammonium sulphate and suitable compost.*

Just a few reasons by it is thought that this combination is what should be used.

(1) Nitrogen is the outstanding fertiliser element for turf grasses. To be a good grass fertiliser a substance must be rich in nitrogen in a form available to the plant. Ammonium sulphate and ammonium phosphate are high in available nitrogen.

(2) It seems to be amply proved that the use of acid-reacting fertilisers goes far towards reducing the expense of weeding Greens. Ammonium sulphate and ammonium phosphate appear to be the best acid-reacting nitrogenous fertilisers now available, price, and other factors considered. Not only are they capable of acidifying soils, but they promote the growth of grass which is satisfactory alike in vigour and texture.

(3) *The addition of phosphorus and potassium does not appear to be necessary at all other than when added by the application of compost.*

If nitrogen is the important fertiliser element to be supplied and the creation and maintainance of a rather highly acid condition in the soil is desirable in solving the weed problem, then it would appear that ammonium sulphate is the most effective and economical fertiliser to use.

As for the use of compost, in addition to its many other functions, some of which are well-known, it supplies suitable organic matter, which is probably much needed and tends to counteract any evil effect the long-continued use of acid-reacting inorganic fertiliser may have on the soil.

The function of top-dressing may be stated briefly as follows:—(1) It fills small depressions in the Green, thus making the surface smooth and true to putt on. With bent Greens especially, it keeps the creeping roots covered and prevents the development of the loose and fluffy growth above the soil surface which is always undesirable. (2) It provides and maintains a satisfactory mechanical condition of the soil which is essential to the continuous health and vigour of the turf, and to the maintenance of sufficient resiliency in the Green to hold a properly pitched shot, and yet retain that moderate degree of firmness necessary to permit the putted ball to roll true. (3) It supplies plant food, other than that in the fertiliser, which is conducive to the growth of the best turf. (4) It helps in a way as yet unexplained. When such inert substances as ground cork, charcoal, or brush are scattered over the turf, the growth of grass is stimulated.

Haley, Edgar Raymond (MD)

Edgar Haley (1910–1998) was born in St John, New Brunswick, Canada, and earned a BSc degree in engineering from UCLA in 1932 and received his Doctor of Medicine and Master of Surgery (MDCM) degree from McGill University, Montreal,

Canada, in 1936. His residency was at Johns Hopkins as a medical intern from 1936 to 1941. During WWII, he served as a Captain in the US Army. Dr. Haley became a Diplomat of the American Board of Surgery in 1950 and designed the first incubator for premature babies while at Johns Hopkins in Baltimore, Maryland. Dr. Haley then pursued an active medical practice in California as a General Surgeon for ~26 years, retiring in 1978. A lifelong lawn bowling enthusiast, he was a member of the Escondido, Santa Monica, Beverly Hills, and Brentwood Lawn Bowling Clubs. For three decades, he conducted tests at his own expense on various root-zone constructions, turfgrasses, and cultural practices on four bowling greens built at his Escondido, California, property. These studies were pursued after consultations with Drs. Bill Davis, Jim Beard, and Vic Gibeault. Based on these activities, he pioneered sand-based lawn bowling green construction. He also invented a greensplaner that was widely used in California. Dr. Haley was inducted into the US Lawn Bowling Association Hall of Fame in 1999.

196x **The Maintenance of the Lawn Bowling Green.** First edition. Edgar R. Haley. Escondido, California, USA. 2 tables, no index, semisoft bound in dark green with black lettering, 8.5 by 11.0 inches (216 × 279 mm).

A simplified manual and personal account on the culture of lawn bowling greens, including the turfgrasses used, smoothing, turfgrass cultivation, and equipment needs. It is practically oriented to United States conditions, based on the experiences of the author in southern California.
Revised in 1975, 1980, and 1984.

1972 **The Building of a Lawn Bowling Green.** First edition. Edgar R. Haley. Escondido, California, USA. 51 pp., 5 illus., 2 tables, no index, textured semisoft bound in mustard yellow with black lettering, 8.5 by 11.0 inches (216 × 279 mm).

A simplified manual and personal account on the design, specifications, site development, construction, and planting of a lawn bowling green. It is practically oriented to United States conditions, based on the experiences of the author in southern California.
Revised in 1976, 1981, and 1990.

1975† **The Maintenance of the Lawn Bowling Green.** Second revised edition. Edgar R. Haley. Escondido, California, USA. 116 pp., 7 illus., 2 tables, no index, textured semisoft covers in green with black lettering in a comb binder, 8.4 by 10.9 inches (214 × 278 mm).

This edition is a major technical update and expansion of the subject matter. It contains a list of selected references in the back.
Revised in 1980 and 1984.

1976 **The Construction of the Lawn Bowling Green.** Second revised edition. Edgar R. Haley. Escondido, California, USA. 61 pp., 10 illus., 2 tables, no index, textured semisoft covers in mustard yellow with black lettering in a comb binder, 8.4 by 10.9 inches (214 × 278 mm).

This edition is a major technical update of the subject. It contains a list of selected references in the back.
Revised in 1981 and 1990.

1979† **Better Greens.** Edgar R. Haley. Escondido, California, USA. 88 pp., 8 illus., no tables, no index, textured semisoft covers in orange with black lettering in a comb binder, 8.4 by 11.0 inches (214 × 279 mm).

A small book on the turfgrass culture of bowling greens in the United States. It consists of reprints of selected articles that had appeared in **B.O.W.L.S.** of the American Lawn Bowls Association from 1974 to 1979.

1980† **Maintenance of the Lawn Bowling Green.** Third revised edition. Edgar R. Haley. Escondido, California, USA. 187 pp., 42 illus., 2 tables, no index, textured semisoft covers in dark green with black lettering in a comb binder, 8.4 by 10.9 inches (214 × 278 mm).

This is a major revision in both the organization and technical subject matter. It contains a list of references in the back.
Revised in 1984.

1981† **The Construction of the Lawn Bowling Green.** Third revised edition. Edgar R. Haley. Escondido, California, USA. 71 pp., 15 illus., 2 tables, no index, semisoft covers in mustard yellow with black lettering in a comb binder, 8.4 by 10.9 inches (214 × 278 mm).

This is a technical update on the practical procedures for site development, construction, and turfgrass establishment of lawn bowling greens in California.
Revised in 1990.

1984 **Maintenance of the Lawn Bowling Green.** Fourth revised edition. Edgar R. Haley, Prince George, British Columbia, Canada. 198 pp., 43 illus., 1 table, no index, textured semisoft bound in dark green with black lettering, 8.5 by 10.6 inches (215 × 269 mm).

An update and minor revision of the third edition.

1990 **The Construction of the Lawn Bowling Green.** Fourth revised edition. Edgar R. Haley. Escondido, California, USA. 91 pp., 25 illus., 1 table, no index, textured semisoft bound in mustard yellow with black lettering, 8.5 by 11.0 inches (216 × 279 mm).

An update and revision of the third edition.

Hall, David Walter (Dr.)

David Hall was born in New Orleans, Louisiana, USA, and earned a BSc in 1965 and an MSc in Systematic Botany in 1967, both from Georgia Southern University; he also earned a PhD in Systematic Botany from the University of Florida in 1978. He served at the University of Florida as a Research Associate from 1971 to 1973, as an Assistant in Botany from 1973 to 1981, and as Director of the Plant Identification and Information Service from 1981 to 1990. He then was Senior Scientist with KBN Engineering and Applied Services from 1991 to 1996 and with Golder Associates from 1996 to 1997. Since 1998, he has headed David W. Hall Consultant Inc. His expertise is plant identification and biology, including expertise as a Board Certified Forensic Examiner. Dr. Hall received the Florida Native Plant Society Green Palmetto Service Award in 1987, the Florida Department of Environmental Regulation Service Award in 1988, the University of Florida Outstanding Faculty Achievement and Performance Award in 1990, the Florida Association of County Agricultural Agents Distinguished Service Award in 1990, the National Association of County Agricultural Agents Distinguished Service Award in 1990, the Georgia Southern University Department of Biology Distinguished Alumni Award in 1991, Fellow of the American Academy of Forensic Sciences in 1993, a certification as a Diplomate of the American Board of Forensic Examiners in 1996, Fellow of the American College of Forensic Examiners in 1996, and the Outstanding Weed Scientist from the Florida Weed Science Society in 1999.

2001† **Color Atlas of Turfgrass Weeds.** Coauthor.
   See *McCarty, Lambert Blanchard—2001.*

2008 **Color Atlas of Turfgrass Weeds.** Revised second edition. Coauthor.
   See *McCarty, Lambert Blanchard—2008.*

Handreck, Kevin Arthur

Kevin Handreck was born in Hamilton, Victoria, Australia, and earned a BSc degree in Chemistry and Agricultural Science in 1959 and a Master of Agricultural Science (MAgSc) degree in Soil Science in 1967, both from the University of Melbourne. He became Scientific Liaison Officer in the Division of Soils at the Commonwealth Scientific and Industrial Research Organization (CSIRO) in Adelaide, South Australia, and, in 1970, at the Rural and Mining Industry Training Division of the New South Wales Technical and Further Education Commission from 1981 to 1996. He conducted research and adult education in growing media-soils and potting mixes. He is currently Managing Director of Netherwood Horticultural Consultants. Kevin Handreck received a Churchill Fellowship in 1982 and Fellow of the Australian Institute of Horticulture in 1990.

1979† **When Should I Water?** Kevin A. Handreck. Discovering Soils No. 8, CSIRO Division of Soils and Rellin Technical Publications, Adelaide, South Australia, Australia. 76 pp., 117 illus., 14 tables, no index, fold-stapled glossy semisoft covers in multicolor with white and black lettering, 6.5 by 9.2 inches (166 × 233 mm).

   A booklet on cultural approaches to water conservation for lawns and landscape gardens. It is oriented to the gardening enthusiasts and conditions in Australia.

1984† **Growing Media—for ornamental plants and turf.** First edition. Kevin A. Handreck and Neil D. Black. University of New South Wales Press, Randwick, New South Wales, Australia. 401 pp.; 393 illus.; 141 tables; indexed; glossy hardbound in multicolor with white, lavender, and brown lettering; 7.2 by 9.5 inches (182 × 241 mm).

   A text on growing media for plants, with emphasis on desired physical and chemical characteristics, environment for root growth, pH, nutrients, water relationships, and drainage. It includes three chapters specifically on turfgrasses and their soil management under field conditions.

   † *Reprinted in 1986, 1989, and 1991, with minor amendments.*
      Revised in 1994, 2002, and 2010.

1994† **Growing Media—for ornamental plants and turf.** Second revised edition. Kevin A. Handreck and Neil D. Black. University of New South Wales Press, Randwick, New South Wales, Australia. 448 pp., 489 illus., 196 tables, detailed index, glossy hardbound in multicolor with yellow, white. and blue lettering, 7.2 by 9.5 inches (182 × 241 mm).

   A major revision of this book on modified root zones, with key updates on such subjects as wetting agents, soil reaction adjustment, composting, root-zone mixes, controlled-release fertilizers, and water conservation. It is a unique book subject-wise.

   † *Reprinted in 1999.*
      Revised in 2002 and 2010.

2002 **Growing Media—for ornamental plants and turf.** Third revised edition. Kevin Handreck and Neil Black. University of New South Wales Press Limited, University of New South Wales, Sydney, New South Wales, Australia. 544 pp., 578 illus., 242 tables, extensive index, semisoft bound in multicolor with black and purple lettering, 6.9 by 9.6 inches (175 × 243 mm).

   This revision contains updates in technology such as minimizing pollution, water conservation, and efficient plant propagation.

Reprinted with some revisions in 2005.
Revised in 2010.

2010 **Growing Media—for ornamental plants and turf.*** Fourth revised edition. Kevin Handreck. University of New South Wales Press Limited, Sydney, New South Wales, Australia. 559 pp., 467 illus. in color, 246 tables, extensive index, semi-soft bound in multicolor with black and green lettering, 7.0 by 9.5 inches (178 × 190 mm).

An updated revision with minor revisions.

Hanson, Angus Alexander (Dr.)

Gus Hanson (1922–2005) was born in Chilliwack, British Columbia, Canada, and earned a BSA degree in Agronomy from the University of British Columbia in 1944, an MSc degree in Agronomy from McGill University in 1946, and a PhD degree in Agronomy from Pennsylvania State University in 1951. He served as Lecturer and Assistant Professor at McGill University from 1946 to 1949 and as Geneticist and Agronomist at the United States Department of Agriculture (USDA) Regional Pasture Lab from 1949 to 1952; he then moved to the USDA Center in Beltsville, Maryland, where he served as Investigation Leader of Grass & Turf from 1953 to 1965, as Chief of the Forage & Range Branch from 1965 to 1972, and as Director of Beltsville Agricultural Research Center from 1972 to 1979. He was elected President of the Crop Science Society of America from 1966 to 1967 and to the Board of Directors of the American Forage & Grassland Council (AF&GC) from 1966 to 1969 and from 1982 to 1986. Dr. Hanson was awarded the USDA Superior Service Award in 1961; Fellow of American Society of Agronomy and Crop Science Society of America, both in 1963; the Medallion Award of the AF&GC in 1972; Fellow of American Association for the Advancement of Science in 1975; the USDA Distinguished Service Award in 1979; and the Distinguished Grasslander Award of AF&GC in 1988.

1959† **Grass Varieties in the United States.*** First edition. A.A. Hanson. Agricultural Handbook No. 170, Agricultural Research Service, United States Department of Agriculture, Washington, D.C., USA. 72 pp., no illus., no tables, indexed, fold-stapled semisoft covers in green with black lettering, 7.7 by 10.2 inches (196 × 259 mm).

A small handbook on the origin, characteristics, and current status of named cultivars and experimental grass selections in the United States. The format includes source material, previous names or experimental numbers, characteristics ascribed to the cultivar by the originating breeder, and seed supply status. The handbook is organized alphabetically by genus. It is oriented to technical grass workers and researchers.

Revised in 1965, 1972, and 1994.

1965† **Grass Varieties in the United States.*** Second revised edition. A.A. Hanson. Agricultural Handbook No. 170, Agricultural Research Service, United States Department of Agriculture, Washington, D.C., USA. 102 pp., no illus., no tables, indexed, semisoft bound in light green with dark green lettering, 7.8 by 10.2 inches (199 × 259 mm).

The revision consists of those new cultivars and advanced experimental grass selections developed in the United States since 1959.

Revised in 1972 and 1994.

Figure 12.38. Toro Triad horse-drawn, three-gang, side-wheel-drive, reel mower, circa 1923. [5]

1969† **Turfgrass Science.** A.A. Hanson and F.V. Juska (editors) with 38 contributing authors. ASA Monograph Series No. 14, American Society of Agronomy, Madison, Wisconsin, USA. 715 pp., 173 illus., 76 tables, indexed, cloth hardbound in green with gold lettering, 5.9 by 9.0 inches (150 × 228 mm).

  ✳ A compilation of 28 chapters that address the basics of turfgrasses and their culture. There are discussions of five major types of cultural systems: athletic fields, putting greens, roadsides, golf fairways, turfed tees and roughs, and sod production. Also included are chapters on turfgrass equipment and on ground covers. It is oriented to United States conditions. A set of references is included at the back of each chapter. It is a must book for collectors of historical turfgrass books.

    Revised in 1992.
    See *Waddington, Donald VanPelt—1992.*

1972† **Grass Varieties in the United States.** Third revised edition. A.A. Hanson. Agricultural Handbook No. 170, Agricultural Research Service, United States Department of Agriculture, Washington, D.C., USA. 124 pp., no illus., no tables, indexed, semisoft bound in light green with dark green lettering, 7.8 by 10.2 inches (199 × 260 mm).

    The revision consists of those new cultivars and advanced experimental grass selections developed in the United States since 1965.

    Revised in 1994.
    See *Alderson, James Stillman—1994.*

Harradine, Donald Leslie

Donald Harradine (1911–1996) was born in Enfield, England, UK, and studied at Woolwich Polytechnic (London). His early training was as a golf professional, greenkeeper, and club maker under his stepfather, J.A. Hockey, who also did some golf course design. Harradine was active in golf course architecture from 1929 to 1989, primarily in Europe. He was a founding member of the British Association of Golf Course Architects (BAGCA) and founder of the International Greenkeepers Association (IGA) in 1968. He was elected Vice Chairman for 1977 to 1980, Chairman for 1980 to 1985, and President for 1985 to 1988 of the BAGCA; he served as President of the IGA from 1969 to 1992.

1955 **Der Rasen-Sportplatz (The Grass Sports Playing Field).** Second revised edition. Donald Harradine and Ralph F. Handloser, Verlag Paul Haupt, Bern, Switzerland. 68 pp. plus 6 pp. of advertising, 27 illus., 3 tables, no index, semisoft bound in greens with black lettering, 4.8 by 7.2 inches (121 × 184 mm). Printed entirely in the German language.

    A booklet concerning the site preparation, turfgrass selection, establishment, and culture of sports fields under the conditions in Germany.

    Revised in 1961.

1961† **Der Rasen-Sportplatz (The Grass Sports Playing Field).** Third revised edition. Donald Harradine, with translation to German by Ralph F. Handloser. Verlag Paul Haupt, Bern, Switzerland. 74 pp. plus 6 pp. of advertising, 27 illus., no tables, no index, semisoft bound in greens with black lettering, 5.3 by 8.0 inches (135 × 202 mm). Printed entirely in the German language.

    Contains minor revisions and updates.

Hardiman, W.[3]

  n.d. **Glimpses of Groundsmanship.** Coauthor.
  late See *Phillips, P.M.—late 195x.*
  195x

Harper II, John Clinton (Dr. and Professor Emeritus of Agronomy)

Jack Harper (1923–2005) was born in Harrisburg, Pennsylvania, USA. During WWII, he served in a US Army tank group in Europe, receiving a Purple Heart for his service. He then earned an Associate degree from North Georgia College in 1944 and a BSc degree in Agronomy from Pennsylvania State University in 1948; then he studied Agronomy at Rutgers University and received a PhD degree in Agronomy from Pennsylvania State University in 1952. Initially, Dr. Harper worked for the United States Department of Agriculture (USDA) as a Turfgrass Research Agronomist from 1952 to 1955, as an Agronomist and General Manager of the Lawn Grass Development Corporation from 1955 to 1956, and as Senior Agronomist for North America for the Toro Manufacturing Corporation from 1956 to 1958. In 1958, he became a Turfgrass Extension Specialist headquartered at Pennsylvania State University. He retired in 1988 after 30 years of educational presentations and writings. Dr. Harper received the Distinguished Service Award of the Pennsylvania Turfgrass Council in 1977 and the GCSAA Distinguished Service Award in 1978.

1967† **Athletic Fields—specification outline, construction, and maintenance.** Second revised edition. John C. Harper II. The Pennsylvania State University, College of Agriculture Extension Service, University Park, Pennsylvania, USA. 21 pp., 4 illus., 4 tables, no index, fold-stapled covers in white and blacks with black lettering, 8.4 by 10.9 inches (214 × 277 mm).

    A booklet on the layout, root zone, construction, turfgrass establishment, and maintenance of sports fields. It includes a representative set of contract specifications for the construction-establishment phase and a seasonal maintenance schedule. It was unique booklet for that time.

    Revised in 1986.

1986† **Athletic Fields—specification outline, construction, and maintenance.** Third revised edition. John C. Harper II. The Pennsylvania State University, College of Agriculture Extension Service, University Park, Pennsylvania, USA. 29 pp., 7 illus., 6 tables, no index, fold-stapled soft covers in cream and black with black lettering, 8.5 by 10.9 inches (216 × 277 mm).

    A revision and expansion of the second edition.

Hatsukade, Masayoshi (Dr. and *Professor* Emeritus)

Masayoshi Hatsukade was born in Hiroshima Prefecture, Japan, and earned a BSc degree in Entomology from Shizuoka University in 1964 and a PhD in Entomology from Tokyo Agricultural University in 1990. He served on the Entomology Faculty at Shizuoka University from 1975 to 2005, rising to Full Professor in 1990. Dr. Hatsukade conducted research on insects and soil microbes, with emphasis on turfgrasses.

1995† **Color Atlas: Major Insects of Turfgrasses and Trees on Golf Courses.** Masayoshi Hatsukade. Soft Science Inc., Tokyo, Japan. 128 pp., 323 illus. with 261 color, 46 tables, small index, cloth hardbound in green with gold lettering and a jacket in multicolor with green and white lettering, 10.0 by 7.1 inches (255 × 181 mm). Printed in the Japanese language.

    A book concerning the life cycle, damage symptoms, and control of the major insects occurring on turfgrasses and trees of golf courses under the conditions in Japan. Large color photographs provide key visual characteristics.

Haut-Commissariat à la Jeunesse, aux Loisirs et aux Sports (Office of the High Commission for Youth, Leisure, and Sports; Quebec, Canada)

1976 **L'Aménagement du Terrain de Baseball (The Creation of Baseball Fields).** Haut-Commissariat à la Jeunesse, aux Loisirs et aux Sports. Editeur Officiel du Québec, Quebec Province, Canada. 43 pp., 18 illus., 5 tables, no index, semi-soft covers in beige with black lettering in a plastic comb binder, 8.3 by 11.0 inches (210 × 280 mm). Printed in the French language.

    A booklet concerning the design, layout, drainage, and construction of baseball fields under the conditions in Quebec Province, Canada. It includes pitchers mound, home plate, and dual-use fields.

Hawes, Douglas Tilton

Doug Hawes was born in Dartmouth, Massachusetts, USA, and earned a BSc degree in Agronomy from the University of Massachusetts in 1962, an MSc degree in Floriculture and Ornamental Horticulture from Cornell University in 1965, and a PhD degree in Agronomy from the University of Maryland in 1972. He served as Instructor, Assistant Professor, and Associate Professor at the University of Maryland from 1966 to 1978; as Midcontinent Director of the United States Golf Association Green Section from 1978 to 1984; and as Turfgrass Consultant from 1984 to 2000. He wrote and published **Turf Com** from 1985 to 2005. In 1977, Dr. Hawes received an Excellence in Teaching Award from the Agricultural Alumni Chapter of the University of Maryland.

1973† **Limited Bibliography of Turf Literature.** Douglas T. Hawes. Institute of Applied Agriculture, University of Maryland, College Park, Maryland, USA. 122 pp., no illus., no tables, no index, textured semisoft covers with clear coverlays in greens with black lettering in a comb binder, 8.5 by 11.0 inches (215 × 279 mm).

    This book contains a listing of popular and semitechnical articles organized under multiple topics for the years 1965 to early 1973.

1994† **Turfgrass Lab Manual.** Coauthor.

    See *Mathias, Kevin—1994.*

Hawkes, George Rogers (Dr.)

George Hawkes was born in Preston, Idaho, USA, and earned a BA degree in Agronomy from Brigham Young University in 1949 and a PhD, majoring in Soil Science, from Ohio State University in 1952. He was a Soil Scientist with the United States Department of Agriculture at Pennsylvania State University and Beltsville, Maryland, from 1952 to 1957. He then joined Cal. Spray, Chevron Chemical Company, serving as Research Agronomist in the San

Figure 12.39. Lightweight Centaur power unit of the Central Tractor Company, Greenwich, Ohio, USA, with universal sulky, such as for center mounting of three-gang mowers, circa 1926. [2]

Joaquin Valley of California from 1957 to 1961, as Assistant National Manager of Agronomy from 1961 to 1973, as Manager-Fertilizer Research from 1973 to 1977, and as Environmental Science Specialist from 1977 to 1986. He was elected President of the California Chapter of the American Society of Agronomy (ASA) in 1980. Dr. Hawkes has been honored as Man of the Year of the Western Agricultural Chemical Association in 1980, Honoree of the California Chapter of the ASA in 1984, and Fellow of the American Association for the Advancement of Science in 1986.

1969†　**Turfgrass Fertilization.** George R. Hawkes (editor). California Fertilization Association, Sacramento, California, USA. 32 pp., 7 illus., 7 tables, no index, fold-stapled glossy semisoft covers in multicolor with white and black lettering, 5.5 by 8.5 inches (139 × 215 mm).

　　　　A booklet on turfgrass nutrition, nutrient carriers, fertilization practices, and soil amendments plus a small section on the culture of turfgrasses. There is a list of selected references in the back.

Hawthorn, Robert

Bob Hawthorn (1920–2001) was born in Newcastle-on-Tyne, Northumberland, England, UK. During WWII, he served in the British Army from 1939 to 1943, losing a leg. He was the sixth person to earn a National Diploma in Turfculture from the National Association of Groundsman in 1961. His first experience as a trainee groundsman was at Vickers-Armstrong from 1934 to 1937 and from 1943 to 1944, then he became Groundsman for Newcastle Church High School from 1944 to 1953, Head Groundsman at George Kents at Luton from 1953 to 1962, Head Groundsman at London Colney University sports ground from 1962 to 1967, and Athletics Ground Playing Field Officer for the Norfolk County Council Education Department from 1967 to 1985, when he retired. Starting in the 1960s, Bob Hawthorn became Secretary for the Education Committee of the National Association of Groundsman (NAG). In this capacity, he pioneered the development and administration of the national practical training and examinations of NAG for fifteen years. Bob Hawthorn was elected to the executive committee of NAG.

1972†　**Association Football Club Groundsman.** Second revised edition. R. Hawthorn. The Football Association, London, England, UK. 97 pp.; 8 illus.; 2 tables; no index; glossy semisoft bound in green, black, and white with light blue and white lettering; 4.9 by 7.2 inches (124 × 182 mm).

　　　　A small book on the construction, establishment, and culture of turfgrasses for soccer fields. It is oriented to cool-season turfgrass conditions of the United Kingdom.

　　　　A revision of the 1951 book.

　　　　See *The Football Association—1951.*

1977†　**Dawson's Practical Lawncraft.** Seventh revised edition. R. Hawthorn. Crosby Lockwood Staples, London, England, UK. 312 pp., 70 illus., 16 tables, indexed, cloth hardbound in brown with silver lettering and a glossy jacket in multicolor with white lettering, 6.0 by 9.2 inches (152 × 233 mm).

　　　　A book on the basics of turfgrass culture as related to conditions in England. Bob Hawthorn revised and updated R.B. Dawson's earlier work while retaining the general format.

　　　　For earlier editions, see *Dawson, Robert Brian—1968.*

Hawtree, Frederick William

"FW" Hawtree (1916–2000) was born in Bromley, Kent, England, UK, and earned a BA degree in French and German from Oxford University in 1938 and also received an MA degree in 1987. He was the second in a lineage of three golf course architects headquartered in England. F.W. Hawtree was a founding member of the British Association of Golf Course Architects in 1970 and was elected President from 1978 to 1980. He designed golf courses throughout Europe and in Africa during a 30-year period. He served as Vice President of the British and International Golf Greenkeepers Association (BIGGA) for many years. F.W. Hawtree was recipient of a BIGGA Honorary Life Membership in 1999.

1983† **The Golf Course—Planning, Design, Construction and Maintenance.** F.W. Hawtree. E. & F.N. Spon Limited, London, England, UK. 212 pp., 53 illus., 4 tables, indexed, glossy hardbound in multicolor with white lettering, 9.2 by 6.0 inches (235 × 152 mm).

One of the few books on golf course architecture and construction written in the 1980s. It is oriented to the cool-season turfgrass conditions in the United Kingdom.

† *Numerous reprintings.*

Hayes, Peter (Dr.)

Peter Hayes was born in Douglas, Isle of Man, UK, and earned a BSc degree majoring in Botany in 1957, a Bachelor of Agriculture (BAgr) degree with *first class honours*, majoring in Agricultural Botany in 1960, and a PhD in 1970, all from Queens University of Belfast, Ireland. He progressed from Lecturer to Senior Lecturer in Agricultural Botany at Queens University, Belfast, from 1960 to 1981, emphasizing agricultural seed technology and grassland agronomy. He also held a joint appointment as Principal Scientific Officer for the Department of Agriculture for Northern Ireland. In 1971, he was awarded the Stapledon Memorial Fellowship for research study in Australia. He then served as the third Director of the Sports Turf Research Institute in Bingley, England, from 1981 to 1995. Dr. Hayes received the National Turfgrass Council Award in 1994.

1984† **Technical Terms in Turf Culture.** Dr. Peter Hayes. The Sports Turf Research Institute, Bingley, West Yorkshire, England, UK. 76 pp., no illus., no tables, no index, fold-stapled glossy semisoft covers both in light brown and in yellow with brown lettering, 5.8 by 8.5 inches (148 × 217 mm).

This booklet contains a summary of selected turfgrass technical terms and their definitions. It focuses on terminology and some abbreviations as used in the United Kingdom.

1992† **The Care of the Golf Course.** First edition. P. Hayes, R.D.C. Evans, and S.P. Isaac (editors) with 23 contributing authors. The Sports Turf Research Institute, Bingley, West Yorkshire, England, UK. 270 pp., 22 illus., 15 tables, no index, glossy hardbound in multicolor with white lettering, 6.6 by 9.1 inches (167 × 231 mm).

This practical reference book is an assemblage of 102 short articles on turfgrass culture for golf courses by 23 STRI authors that were previously published in the quarterly **Bulletin of the Sports Turf Research Institute.** There is a bibliography in the back. It is oriented to cool-season turfgrass conditions in the United Kingdom.

Revised in 1996.

See *Perris, Jeffrey—1996.*

Hayashi, Shigeto

Shigeto Hayashi was born in Tokyo, Japan, and earned a BSc degree in Horticulture from Chiba University in 1981. He has worked as greenkeeper at Caledonian Golf Club from 1990 to 1992 and at The Greenbrier Westvillage Country Club from 1992 to 2003 and as Superintendent at Grandee Nasu-Shirakawa Golf Course since 2003. Shigeto Hayashi was elected to the Board of Directors of the Japanese Society of Turfgrass Science in 2004.

2007† **Cultivation for Turfgrass Management—Its Effects and Method Using Machinery.** Coeditor.

See *Inamori, Makoto—2007.*

Heller, Paul Robert (Dr. and Professor)

Paul Heller was born in Wooster, Ohio, USA, and earned a BA degree in Biology from Malone College, Canton, Ohio, in 1970; an MSc degree in 1972; and a PhD degree in 1976, with the latter two from Ohio State University. He then joined the faculty in the Entomology Department at Pennsylvania State University in 1976, where he rose to Full Professor in 2004.

1983† **Turfgrass Insect and Mite Manual.** First edition. Coauthor.

See *Shetlar, David John—1983.*

1988† **Turfgrass Insect and Mite Manual.** Second revised edition. Coauthor.

See *Shetlar, David John—1988.*

1990† **Turfgrass Insect and Mite Manual.** Third revised edition. Coauthor.

See *Shetlar, David John—2000.*

Hill, Steven Richard

Steve Hill was born in Hartford, Connecticut, USA, and earned a BSc degree in Biology from Bates College, Maine, in 1972; an MA in Biology from City University of New York in 1975; and a PhD in Botany from Texas A&M University in 1979. He served as Curator of the Herbarium and Instructor in the Department of Botany at the University of Maryland from 1979 to 1984, as Curator of the Herbarium in the Department of Biological Sciences at Clemson University from 1987 to 1994, as Associate Research Scientist in the Center for Biodiversity, Illinois Natural History Survey, from 1994 to 2003, and, since, as Botanist and Associate Technical Scientist in the Center for Wildlife and Plant Ecology, Illinois Natural History Survey, Champaign. He is involved in systematic botany, botanical surveys, herbarium development, and teaching.

1986† **A Checklist of Names for 3,000 Vascular Plants of Economic Importance.*** Coauthor.
　　　See *Terrell, Edward Everett—1986.*

Hilu, Khidir Wanni (Dr. and Professor)

Khidir Hilu was born in Baghdad, Iraq, and earned a BSc degree in 1966 and MSc in 1971, both from the University of Baghdad, and a PhD degree in Botany from the University of Illinois in 1976. He became Associate Professor in the Department of Botany at Virginia Polytechnic Institute and State University in 1981 and rose to Full Professor in 2004. He specializes in the biosystematics and evolution of grasses.

1986† **Grass Systematics and Evolution.*** Coeditor.
　　　See *Soderstrom, Thomas Robert—1985.*

Hitchcock, Albert Spear (Dr.)

Albert Hitchcock (1865–1935) was born in Owosso, Michigan, USA, with his early years spent in Nebraska and Kansas. He earned a BSA in Agriculture and an MSc from Iowa State Agriculture College in 1884 and 1886, respectively. He served as Assistant Chemist at Iowa State University in 1885; as Instructor of Chemistry at the University of Iowa from 1886 to 1889; as Botanical Assistant and Curator of the Herbarium at the Missouri Botanical Garden, Washington University, from 1889 to 1891; as Professor of Botany at Kansas State Agricultural College from 1892 to 1901; and as Assistant Chief of the Division of Agrostology of the United States Department of Agriculture (USDA) in Washington, DC, starting in 1901 under Dr. Frank Lamson-Scribner. In 1905, he was appointed Chief Botanist in charge of systematic agrostology for the USDA and Custodian of the Grass Section at the US National Herbarium, both positions formerly held by Dr. Charles V. Piper. He was a plant explorer, economic agrostologist, and taxonomist, who wrote over 250 publications on the taxonomy and description of grasses in 34 years, based on his collection of 25,000 specimens from around the world. Dr. Hitchcock was elected President of the Botanical Society in 1914 and President of the Washington Botanical Society in 1916. He was awarded an Honorary DSc degree from Iowa State College in 1920 and from Kansas State College in 1934.

Figure 12.40. Staude General Utility Golf Course Tractor with midundercarriage designed to attach any make of reel gang mower, St. Paul, Minnesota, USA, circa 1926. [2]

1914† **A Text-Book Of Grasses With Especial Reference To The Economic Species Of United States.**[*] A.S. Hitchcock. Rural Test-Book Series, The Macmillan Company, New York City, New York, USA. 276 pp. plus 4 pp. of advertising, 63 illus., 17 tables, indexed, cloth hardbound in green with black and gold lettering, 4.9 by 7.3 inches (125 × 185 mm).

This early textbook is organized in two parts. Part I addresses economic agrostology, while part II places emphasis on systematic agrostology and taxonomy. There are chapters on lawns, on grasses for miscellaneous purposes, and on grassy weeds. Keys to each genera are presented.

† *Reprinted in 1922.*

1920† **The Genera of Grasses of the United States with Special Reference to the Economic Species.**[*] First edition. A.S. Hitchcock. Bulletin 772, Bureau of Plant Industry, United States Department of Agriculture, Washington, D.C., USA. 307 pp., 189 illus., no tables, detailed index, semisoft bound in white with black lettering, 5.6 by 8.6 inches (143 × 219 mm).

A major taxonomy book consisting of botanical descriptions of the subfamilies, tribes, genera, and species within the grass family that are of economic importance either positive or negative in the United States. Included are turfgrasses and ornamental grasses. There are detailed drawings. Keys to the tribes and genera are presented. In 1920, Dr. Hitchcock stated the following:

> The most important lawn grasses are (1) in the North, bluegrass, Rhode Island bent, and creeping bent; (2) in the South, Bermuda grass, carpet grass, and St. Augustine grass.

Revised in 1936.

1935† **Manual of the Grasses of the United States.**[*] First edition. A.S. Hitchcock. Miscellaneous Publication No. 200, United States Department of Agriculture, Washington, D.C., USA. 1,040 pp., 1,696 illus., no tables, indexed, cloth hardbound in green with gold lettering, 5.8 × 8.9 inches (147 × 226 mm).

This monumental book contains descriptions of all grasses known to grow in the continental United States, excluding Alaska. There are 159 numbered genera and 1,100 numbered species, with most of them being illustrated. Dr. Hitchcock presented the following overview of grasses:

> One of the most widely distributed of the families of flowering plants, the grasses are found over the land surface of the globe, in marshes and in deserts, on prairies and in woodland, on sand, rocks, and fertile soil, from the Tropics to the polar region and from sea level to perpetual snow on the mountains.

There were eight reprintings.
Revised in 1950.

1936† **The Genera of Grasses of the United States with Special Reference to the Economic Species.**[*] Second revised edition. A.S. Hitchcock, as revised by A. Chase. Bulletin No. 772, Bureau of Plant Industry, United States Department of Agriculture, Washington, D.C., USA. 302 pp., 201 illus., no tables, indexed, semisoft bound in tan with black lettering, 5.9 by 9.2 inches (149 × 232 mm).

An updated revision of the 1920 first edition.

1950† **Manual of the Grasses of the United States.**[*] Second revised edition. A.S. Hitchcock, as revised by Agnes Chase. Miscellaneous Publication No. 200, United States Department of Agriculture, Washington, D.C., USA. 1051 pp., 1199 illus., no tables, indexed, cloth hardbound in teal with gold lettering, 5.6 by 8.8 inches (144 × 223 mm).

This revision of the second edition involves primarily an updating of the known grasses found in the continental United States, excluding Alaska. Ten genera and 298 species were added, for a total of 1,398 species in 169 genera. It is extensively referenced and contains a glossary in the back. The manual compilation was based on the United States National Herbarium Collection and was accomplished primarily by Mary Agnes Chase as Dr. Hitchcock died in 1935.

See also *Barkworth, Mary Elizabeth—2003 and 2007.*

1971† **Manual of the Grasses of the United States.**[*] Volumes I and II. A.S. Hitchcock and Agnes Chase. Dover Publications Inc., New York City, New York, USA. 569 pp. in vol. I and 482 pp. in vol. II, glossy semisoft bound in green and white for vol. I and in light brown and white for vol. II, both with white and black lettering, 6.0 by 9.0 inches (154 × 230 mm).

A reprinting of the 1950 second revised edition as a two-volume set.

Holborn, John Charles

John Holborn (1941–2008) was born in Slough, Berkshire, England, UK. He worked as a Health and Safety Inspector for the Royal Society for the Prevention of Accidents, as Director of Cottismore Landscaping, as Technical Director for the

Figure 12.41. Toro tractor with front-mounted, three-gang, side-wheel-drive, reel mower, circa 1927. [5]

National Playing Fields Association, and as Director for J.S. Bishop & Company Ltd. John Holborn also served as a member of the UK delegation to the EU Sports Turf Standards Committee in 1986 and as the Deputy Chairman of the National Turfgrass Committee in 1986.

1989†   **Sports Grounds Maintenance—An Elementary Guide.** Third revised edition. John C. Holborn (editor). National Playing Fields Association, London, England, UK. 40 pp.; 22 illus.; no tables; no index; fold-stapled glossy semisoft covers in greens, black, and white with black and green lettering; 5.7 by 8.2 inches (146 × 208 mm).

This revision is an update of a booklet organized as an elementary guide for club sports ground committees and agronomic staff. It encompasses the culture of turfgrasses for cricket grounds, bowling greens, lawn tennis courts, and sports fields under the conditions in England.

First and second editions published in 1967 and 1978.

See Bowles, William Henry—1967 and 1978.

Hoogerkamp, Meindert

Meindert Hoogerkamp was born in Veendam, the Netherlands, and earned a H BS-B diploma from the Winkler Prinslycaum at Veendam. He then fulfilled his military obligation. He earned an Engineering (Eng) diploma in Agronomy from Lanabouwhogeschool in Wageningen University in 1961. His employment history includes Proefstation voor Akker-en Weidebouw (Test Station for Land and Meadow Research; PAW) starting in 1961, Instituut voor Biologisch en Schei-kundig Onderzoek van Landbouwgewassen (Institute for Biological and Chemical Research of Agricultural Crops; IBS) in 1969, and Centrum voor Agrobiologisch Onderzoek (Center for Plant Physiological Research; CABO) in 1976. His research focused on the establishment and culture of pastures and turfgrass. He then became Head of the Department of Weed Science at CABO.

1975†   **Grasveldkunde (Grassfield Science).** M. Hoogerkamp and J.W. Minderhoud (editors) with 15 contributing authors: M.L. Hart, M. Hoogerkamp, P. Boekel, K.S. Smilde, J.H. Neuteboom, K. Wind, E. Dwarshuis, P. Zonderwijk, J.W. Minderhoud, R. den Engeles, P.J. Huisman, L.E.M. Klaar, J.Th. Moormans, J.P. van der Horst, and H.A. Kamp. Centrum voor Landbouwpublikaties en Landbouwdocumentatie, Wageningen, the Netherlands. 266 pp., 13 illus., 40 tables, no index, semisoft bound in multicolor with black lettering, 6.6 by 9.4 inches (169 × 239 mm). Printed entirely in the Dutch language.

A basic book on the establishment and culture of turfgrasses for sport fields and lawns in the Netherlands. It includes chapters on the ecophysiological background, the physical and chemical environment of grass fields, and the situation regarding roadside vegetation.

Hope, Frank (Dr.)

Frank Hope was born in 1949 in Manchester, England, UK. He earned a National Diploma in Horticulture and in Arborculture; he also earned a Master of Philosophy (MPhil) degree in 1987 and PhD in 1990, both in Biological Sciences from the University of Bath. He initially worked as a groundsman on municipal sports fields and lawns. He then lectured in Horticulture at the Cheshire College of Agriculture from 1974 to 1978 and served as the Education and Training Officer for the Institute of Groundsmanship from 1989 to 1990. He devotes most of his efforts as a tree consultant.

1978†   **Turf Culture—A Complete Manual for the Groundsman.** First edition. Frank Hope. Blandford Press Limited, Poole, England, UK. 293 pp., 113 illus. with 30 in color, 32 tables, indexed, cloth hardbound in light green with gold lettering and a glossy jacket in multicolor with brown and black lettering, 5.7 by 8.4 inches (145 × 213 mm).

   This book addresses the basics of site construction, establishment, and culture of turfgrasses as related to cool-season turfgrass conditions in England. Emphasis is placed on information used by turfgrass managers in charge of sports grounds, park lands, and garden lawns. It contains chapters on machinery and management aspects. There is a glossary of botanical terms in the back.

   Revised in 1990.

1983†   **Rasen—Anlage und Pflege von Zier-, Gebrauchs-, Sport- und Landschaftsrasen (Turfgrass—Construction and Maintenance for Aesthetic, Functional, Sport, and Landscaped Lawns).** Frank Hope and Heinz Schulz. Verlag Eugen Ulmer, Stuttgart, West Germany. 216 pp. plus 4 pp. of advertising; 60 illus. with 24 in color; 35 tables; indexed; cloth hardbound in beige, black, and white with black lettering; 5.9 by 8.9 inches (150 × 226 mm). Printed entirely in the German language.

   A book on soils and turfgrasses plus the establishment and culture of turfgrasses. It includes sections on turfgrasses for soccer fields, golf courses, and lawn tennis courts.

1990†   **Turf Culture—A Manual for the Groundsman.** Second revised edition. Frank Hope. Cassell Publishers Limited, London, England, UK. 293 pp., 136 illus. with 30 in color, 33 tables, indexed, cloth hardbound in dark green with gold lettering and a glossy jacket in multicolor with brown lettering, 5.6 by 8.4 inches (142 × 213 mm).

   There are minor revisions of this second edition.

Hopper, Henry Thomas[3]

n.d.   **The Provision and Maintenance of Playing Fields and Churchyards.** * H.T. Hopper. Trade & Technical Press
circa   Limited, Morden, Surrey, England, UK. 256 pp.; 74 illus.; 27 tables; indexed; cloth hardbound in green with gold
1967†   lettering and a glossy jacket in green, black, and white with white and green lettering; 5.3 by 8.4 inches (136 × 214 mm).

   This unique book addresses two distinct subjects. First is the layout, construction, establishment, and culture of sports fields. Included are soccer fields, rugby fields, lacrosse fields, baseball fields, hockey fields, netball fields, cricket grounds, and tennis courts. It also contains chapters on hard-surface tennis courts, artificial cricket grounds, and cinder running tracks. The second major subject covers the maintenance of church yards. The church yard–cemetery subject is seldom addressed in turfgrass books. Emphasis is placed on English climatic and soil conditions. Chapter subject titles for section II on churchyards include the following:

- Maintenance of Coarse and Rough Grass
- Laying and Maintenance of Lawns
- Trees and Shrubs
- Memorials
- Walls and Fences
- Paths
- Seats
- Conversion of Burial Areas to the Lawn Principle
- Disposal of Cremated Remains
- Closed Churchyards

Horton, William M.

Bill Horton was born in New Zealand. He was active as a greenkeeper during his early years. He was the first Greens Field Advisory Officer and also was Supervisor of Research Areas for the New Zealand Institute for Turf Culture from 1936 to 1952, when he died. Bill Horton was honored as a Life Member of the New Zealand Greenkeepers Association in 1953.

1950†   **Construction, Renovation, and Care of the Golf Course.** Coauthor.
   *   See *Levy, Enoch Bruce—1950.*

Hosotsuji, Toyoji (Dr. and Professor)

Toyoji Hosotsuji (1924–2005) was born in Kyoto, Kyoto Prefecture, Japan, and earned a BSc degree in 1949 and a DrAgr in 1979, both in Plant Pathology from Kyoto University. He joined the Research Division of Nihon Nohyaku Company Limited in 1949, serving through 1963; then he became Vice President of the Muromochi Chemical Industry through 1968. He served as Chief of the Plant Pathology Section at the Institute of Physical and Chemical Research Foundation in Tokyo from 1970 to 1994. He was a Visiting Professor at the Research Institute of Tokyo Agricultural University and part-time Instructor at Ehime, Shinshu, and Tottori Universities. Dr. Hosotsuji received the Achievement Award from the Pesticide Science Society of Japan in 1979.

Figure 12.42. Tractor-drawn, 11.5-foot-cutting-width (3.5 m), 5-gang, 6-blade, rear-wheel-drive, reel Jacobsen mower, 1939. [41]

1974† **Twelve Months of Turf Maintenance.** Toyoji Hosotsuji and Masayoshi Yoshida. Bunka Shuppankyoku, Tokyo, Japan. 176 pp., 232 illus. with 6 in color, 50 tables, indexed, glossy semisoft bound in white and blacks with black lettering and a glossy jacket in multicolor with blue and black lettering, 7.1 by 9.2 inches (181 × 233 mm). Printed entirely in the Japanese language.

A book on turfgrasses, establishment, and culture of lawns under conditions in Japan. There is a monthly maintenance calendar plus a glossary and a question & answer section in the back.

1979† **Color Atlas: Diseases, Insects and Weeds in Turfgrass.** First edition. Toyoji Hosotsuji and Masayoshi Yoshida. Zenkoku Noson Kyoiku Kyokai (Society of National Farm Village Education), Tokyo, Japan. 298 pp., 327 illus. with 251 in color, 30 tables, indexed in Japanese and English, glossy semisoft bound in multicolor with black lettering, 5.0 by 7.2 inches (128 × 182 mm). Printed in the Japanese language.

A book on the turfgrass pests found in Japan. Emphasis is placed on symptoms and identification of turfgrass diseases, insect damage, and weeds plus their control by cultural and chemical methods. It includes chapters on fungus and insect identification and on forecasting disease and insect pest occurrence. It also contains sections on the comparative descriptions of turfgrasses and weeds. The color photographs of pests are excellent.

Revised in 1989 and 1998.

1989† **Color Atlas: Diseases, Insects and Weeds in Turfgrass.** Second revised edition. Toyoji Hosotsuji and Masayoshi Yoshida. Zenkoku Noson Kyoiku Kyokai (Society of National Farm Village Education), Tokyo, Japan. 331 pp., 415 illus., 28 tables, indexed in Japanese and English, glossy semisoft bound in multicolor with black lettering, 5.1 by 7.2 inches (129 × 182 mm). Printed in the Japanese language.

There are small revisions of the first edition, with numerous illustrations added.

Revised in 1998.

1998† **Color Atlas: Diseases, Insects and Weeds in Turf.** Third revised edition. Toyoji Hosotsuji, Masayoshi Yoshida, Katsuyoshi Yoneyama, and Yoshiyuki Yamashita. Zenkoku Noson Kyoiku Kyokai (Society of National Farm Village Education), Tokyo, Japan. 319 pp.; 235 illus. with 222 in color; 27 tables; small index in Japanese and English; semisoft bound in tan with black lettering and a jacket in light green, white, and tan with black lettering; 5.0 by 7.2 inches (128 × 182 mm). Printed in the Japanese language.

There are small revisions and updates of the second edition.

Hotchkin, Stafford Vere (Colonel)

Colonel Hotchkin (1876–1953) was born in Woodhall Spa, Lincolnshire, England, UK. He served in the military during WWI and, subsequently, was a Conservative Member of Parliament from 1922 to 1923. He was active in the remodeling of Woodhall Spa Golf Club, which he purchased in 1920. He then formed Ferigna Limited, a firm involved in design, construction, maintenance, equipment, and seed supplies for golf courses. Colonel Hotchkin was active in the design of golf courses in South Africa starting in the mid-1920s and continuing up to WWII.

n.d. circa 1938 **The Principles of Golf Architecture, etc.** Col. S.V. Hotchkin. The Premier Press Limited, London, England, UK. 26 pp., no illus., no tables, no index, semisoft bound in yellow with greenlettering, 5.4 by 8.5 inches (137 × 216 mm).

   ✳   This booklet is a collection of articles that had been published in the **Club Sportsman**, as described by E.R.V. Knox in the foreword. The subjects addressed are site, soil, drainage, and equipment selection for golf courses plus a section on the culture of sports grounds.

Howard, David Russell

Dave Howard was born in Hastings, Hawkes Bay, New Zealand, and earned a BSc degree in Agriculture Science from Massey University in 1980. He has served as a Sports Turf Advisor for the New Zealand Sports Turf Institute since 1981. Originally, he worked as Regional Agronomist in the Southern North Island region, and in 1993, he moved to Dunedin where he has served the Otago region.

1993†   **Establishment and Management of Cotula Greens.** First edition. D.R. Howard. New Zealand Turf Culture Institute, Palmerston North, New Zealand. 218 pp., 133 illus. with 48 in color, 23 tables, small index, glossy semisoft covers in multicolor with green and white lettering in a wire ring binder, 8.1 by 11.5 inches (207 × 293 mm).

     A manual on the characteristics, establishment, culture, and cultural systems of cotula (*Leptinella* species) as used on lawn bowling greens. This is the first extended writing concerning a unique dicotyledon that is widely used on the South Island of New Zealand. The manual is oriented to conditions found in New Zealand.

     Revised in 2008.

     See *New Zealand Sports Turf Institute—2008.*

Howard, Frank Leslie (Dr. and *Professor* Emeritus)

Frank Howard (1903–1997) was born in Los Angeles, California, USA, and earned a BSc degree from Oregon State University in 1925 and a PhD degree in Mycology from the University of Iowa in 1930. Dr. Howard was a National Research Council Fellow at Harvard University from 1930 to 1932, pursuing studies of slime molds. He joined the Botany Faculty at Rhode Island State College in 1932, rose to Full Professor of Plant Pathology, and then was Head of the Department of Plant Pathology and Entomology at the University of Rhode Island from 1950 to 1969. Dr. Howard was active in teaching plant pathology and turfgrass disease research. His research in turfgrass disease control led to the development of some early organic fungicides. He retired in 1971. Dr. Howard was awarded Fellow of the American Phytopathological Society in 1967 and a charter member of the College of Resource Development Hall of Fame of the University of Rhode Island in 1992.

1951†   **Fungus Diseases of Turf Grasses.** F.L. Howard, J.B. Rowell, and H.L. Keil. Rhode Island Agricultural Experiment Station Bulletin 308, University of Rhode Island, Kingston, Rhode Island, USA. 56 pp., 8 illus. in color, 1 table, indexed by grass host and by pathogen, fold-stapled glossy semisoft bound in black and white with black lettering, 6.0 by 9.0 inches (152 × 228 mm).

   ✳   One of the early technical bulletins specifically on the diseases of cool-season turfgrasses and the first with color photographs of the injury symptoms. There are 46 diseases addressed, with reference citations in the back. It is a **very rare** booklet that is a must for collectors of historical turfgrass publications. The status of fungicides available for control of turfgrass diseases in 1951 was summarized by the authors as follows:

> The grasses that make up lawns, fairways and greens are attacked by 100 or more diseases. Just as it is unreasonable to expect any one grass to be immune to all of them, so is it to expect one fungicide to control them all. Mercurials were the only effective chemicals available a few years ago for the control of turf diseases. Mercury chloride mixtures (i.e., calomel and corrosive sublimate) are still used most and are very effective when properly applied for brown patch control. Thiuram has been reported to control dollar spot, turf diseases. Mercury chloride mixtures (i.e., calomel but in our trials it has not prevented copper spot or dollar spot. Research on turf fungicides has been a major project at the Rhode Island Station in recent years and improved products, including the organic cadmium- and phenyl mercury-containing materials, have been developed. These cadmium fungicides will both prevent and cure cases of copper spot and dollar spot, but they are not effective against brown patch. Heretofore *Helminthosporium* blights have gone uncontrolled, but indications are that the same water-soluble phenyl mercury compounds that are specific for crabgrass will also prevent some of these diseases.

Howell, William[3]

1968†   **Bowling Greens—Construction and Maintenance.** William Howell. Journal of Park Administration Limited, London, England, UK. 172 pp. with 12 pp. of advertising; 44 illus.; 4 tables; small index; hardbound in green with gold lettering and a glossy jacket in green, white, and black with green, white, black lettering and also as a semisoft bound version in tan with green lettering; 5.4 by 8.5 inches (138 × 215 mm).

A book devoted solely to the construction and culture of turfgrass bowling greens, both flat and crown. It is oriented to the soil and cool-season turfgrass conditions in the United Kingdom. There is a short chapter on indoor bowling greens in the back.

Hrabě, František (*Dr.* and *Professor*)

František Hrabě˙ was born in Hrušky, Czechoslovakia, and earned an MSc degree in 1966 and a PhD degree in 1978, both from Mendel University of Agriculture and Forestry in Brno. He became a Research Assistant in 1967, an Associate Professor in 1994, and a Full Professor in 1996 in the Forage Production Department of the Faculty of Agriculture at Mendel University of Agriculture and Forestry in Brno. His professional efforts range across fodder crops, pastures, grasslands, and turfgrass culture.

2004† **Trávy a Trávniky—Co o nich ještě nevíte (Grasses and Turfgrasses—What you do not know about them).** F. Hrabě (editor), with 13 contributing authors. Petr Baštan-Hanácká reklamni, Olomouc, the Czech Republic. 158 pp., 227 illus., 32 tables, no index, glossy hardbound in multicolor with green lettering, 8.3 by 11.6 inches (210 × 295 mm). Printed in the Czechoslovak language.

A book concerning turfgrasses for use under conditions in the Czech Republic, including their establishment and culture. Specific emphasis is on nutrition, water, and soil management plus weed, disease, and insect pests.

2009† **Trávniky: Pro zahradu, kraginu a sport (Turf for Garden, Landscape and Sports).** F. Hrabě (editor) with 19 contributing authors. P. Baštan, Olomouc, the Czech Republic. 335 pp., 417 illus., 39 tables, no index, glossy hardbound in multicolor with maroon and white lettering, 8.3 by 8.7 inches (210 × 220 mm). Printed in the Czechoslovak language.

A book concerning the establishment and culture of turfgrasses under conditions in the Czech Republic. Specific subjects addressed include soils, root-zone construction, cultural practices, irrigation, equipment, weeds, diseases, and insects. Special emphasis is placed on soccer fields and golf courses.

Huang, Bingru (*Dr.* and *Professor*)

Bingru Huang was born in Hebei, China, and earned a BSc degree in Agronomy from Hebei Agricultural University, China, in 1984; an MSc degree in Agronomy from Shandong Agricultural University, China, in 1987; and a PhD degree in Crop Science from Texas Tech University in 1991. She then pursued Postdoctoral study of drought in desert plants at the University of California, Los Angeles, from 1991 to 1993 and of waterlogging in wheat and turfgrass at the University of Georgia from 1994 to 1996. She was Assistant Professor at Kansas State University from 1996 to 2000 and then joined the faculty as Associate Professor in the Department of Plant Biology and Pathology at Rutgers University. She became a Full Professor in 2005. Her research has focused on the environmental stresses of plants with emphasis on drought and heat stress mechanisms of turfgrasses. Dr. Huang received the Young Crop Scientist Award of the Crop Science Society of America (CSSA) in 1997, Alfred Sloan Fellow in 1976, Fellow of the American Society of Agronomy in 2003, Fellow of the CSSA in 2004, and a Chang Jiang (Yangtze) Scholar by the Chinese Ministry of Education in 2006.

2004† **Applied Turfgrass Science and Physiology.** Coauthor.

See *Fry, Jack Douglas—2004.*

Hubbard, Charles Edward (*Dr.; OBE* and *CBE*)

Charles Hubbard (1900–1980) was born in Appleton, Norfolk, England, UK, and rose to become a world renowned grass taxonomist. Both his father and grandfather were estate groundsman. His grandfather was head gardener at the Leicestershire estate of Lord Berners, Keythorpe, and his father was head gardener for Princess Maud at the Sandringham Estate from 1892 to 1937. Charles E. Hubbard became acquainted with grasses at an early age while observing his father's activities at the Sandringham Estate in Norfolk. He then apprenticed as a Gardener at the Royal Gardens of Sandringham in Norfolk from 1916 to 1920, except for an eight-month service in the Royal Air Force during WWI. He entered Kew Botanic Gardens in 1920 as an Improver Gardener and then served as Technical Assistant in the Herbarium from 1922 to 1927, as Assistant Botanist from 1927 to 1930, and as Botanist starting in 1930. In 1935, he was placed in charge of the Graminae in the Herbarium. Subsequently, he became Keeper of the Herbarium and Library in 1957 and served as Deputy Director of the Kew Botanic Gardens from 1959 to 1965, when he retired. Dr. Hubbard became the accepted world authority on the classification and recognition of grasses. He was elected President of the Kew Guild from 1956 to 1957. He served on the Councils of the Linnean Society of London, the British Ecological Society, the Botanical Society of the British Isles, and the Systematics Association. Dr. Hubbard was inducted as an OBE in 1954 and was advanced to Commander in 1965 plus the Linnean Gold Medal in 1967 and the Royal Horticultural Society Veitch Memorial Gold Medal in 1970. The University of Reading conferred on Charles Hubbard an Honorary Doctorate of Science *honoris causa* in 1960.

1954† **Grasses.**˙ First edition. C.E. Hubbard. Penguin Books Limited, Harmondsworth, Middlesex, England, UK. 440 pp.; 350 illus.; no tables; indexed by common and by botanical names; cloth hardbound in gray with gold lettering and a

Figure 12.43. Roseman tractor with dump body, Evanston, Illinois, USA, circa 1926. [2]

jacket in multicolor with black lettering and also semisoft bound in blue, white, and black with black and white lettering; 4.3 by 7.0 inches (110 × 179 mm).

* This classic manual is a guide to the structure, identification, uses, and distribution of grasses in the British Isles. Included are 152 different species of grasses contained in the collections at the Royal Botanic Gardens at Kew. There is a list of selected references in the back.

† *Reprinted in 1959.*
  Revised in 1968 and 1984.

1968† **Grasses.** Second revised edition. C.E. Hubbard. Penguin Books Limited, Harmondsworth, Middlesex, England, UK. 463 pp.; 316 illus.; no tables; indexed by both common and by botanical names; glossy semisoft bound in blue, black, and green with white lettering; 4.3 by 7.1 inches (109 × 180 mm).

The revisions consist of updating the grass distribution data and the nomenclature plus the addition of a few species. There is a small glossary and a bibliography in the back.

† *Reprinted in 1972, 1974, 1976, 1978, and 1980.*
  Revised in 1984.

1984† **Grasses—A guide to their Structure, Identification, Uses, and Distribution in the British Isles.** Third revised edition. C.E. Hubbard and J.C.E. Hubbard. Penguin Books Limited, Harmondsworth, Middlesex, England, UK. 480 pp., 163 multiple illus., 1 table, indexed, glossy semisoft bound in multicolor with brown and black lettering, 4.3 by 7.1 inches (109 × 180 mm).

This third revised version includes a section on the subspecies *Festuca rubra* and an expansion of information on amenity turfgrass uses in the British Isles.

Huck, Michael Thomas

Mike Huck was born in Racine, Wisconsin, USA, and earned a Diploma in Automotive Mechanics from Gateway Technical Institute of Kenosha, Wisconsin, in 1976 and a BSc degree in Ornamental Horticulture and Turfgrass from California State University, Pomona, in 1982. He has been Assistant Golf Course Superintendent and Equipment Mechanic at Maplecrest Country Club in Somers, Wisconsin from 1974 to 1979; Spray and Application Technician at the Babe Zaharias Course, Industry Hills Resort, City of Industry, California, from 1980 to 1982; Assistant Golf Course Superintendent at the Dwight D. Eisenhower Course, Industry Hills Resort, from 1982 to 1983; Golf Course Superintendent at the Industry Hills Resort from 1983 to 1986; Golf Course Superintendent at Mission Viejo Country Club in Mission Viejo, California, from 1988 to 1994; Golf Course Superintendent at the SCGA Members Course in Murrieta, California, from 1994 to 1995; Southwest Regional Agronomist for the United States Golf Association Green Section from 1995 to 2001; and Independent Consultant for Irrigation and Turfgrass Services of Dana Point, California, since 2001. Mike Huck received the Affiliate Merit Award from the Southern California Golf Course Superintendents Association in 2006.

2009† **Turfgrass and Landscape Irrigation Water Quality: Assessment and Management).** Coauthor.
See *Duncan, Ronny Rush—2009.*

Hughes, Stephen Michael

Stephen Hughes was born in Sydney, New South Wales, Australia, and earned a BSc degree in Forestry from the Australian National University in 1975 plus a Postgraduate Diploma in Horticultural Science from Sydney University in 1983. He has served as Senior Technical Officer in charge of Turf and Irrigation for the Horticultural Services Unit of the Australian Capital Territory (ACT) Parks, starting in 1982; as Manager of City Parks from 1995 to 1996; and as Manager of ACT Parks and Conservation Service from 1997 to 2007. Since 2006, Stephen Hughes has been Manager of the ACT Parks and Reserves.

1988† **Turf Irrigation.** Coauthor.
> See *McIntyre, David Keith—1988.*

Hunt, Roderick (Dr.)

Rod Hunt was born in Poole, Dorset County, UK, and earned a BSc degree in Botany and Zoology in 1967, a PhD degree in Plant Physiology and Ecology in 1971, and a DSc degree in 2001, all from the University of Sheffield. He served as Junior Fellow in Botany at the University of Bristol from 1970 to 1973 and as Liaison Officer at the Unit of Comparative Plant Ecology, Department of Animal and Plant Sciences, from 1973 to 1975. He also served as Honorary Lecturer in 1975, Honorary Senior Lecturer in 1990, and Honorary Reader in 1998, all at the University of Sheffield. Since 2003, he has been Visiting Professor in Biological Sciences at the University of Exeter. He also has been associated with the **Annals of Botany**, as Editor from 1985 to 1990, as Chief Editor from 1990 to 1996, and as Company Secretary and Treasurer since 1997. Dr. Hunt was named Fellow in the Institute of Biology, UK, in 1996. He also is an accomplished opera singer.

1980† **Amenity Grassland—An Ecological Perspective.** Coeditor.
> See *Rorison, Ian Henderson—1980.*

Hurdzan, Michael John (Dr. and *Colonel*)

Mike Hurdzan was born in Wheeling, West Virginia, USA, and earned a BSc degree in Turfgrass Management from Ohio State University in 1965 and both an MSc degree in Turfgrass and Soil Science in 1968 and a PhD degree in Environmental Plant Physiology from the University of Vermont in 1974. Colonel Hurdzan devoted 23 years to the US Army Special Forces Reserve Command of Green Beret, Infantry, and Psychological Operations Units. He became a full-time golf course architect in the early 1970s and was associated with architect Jack Kidwell of Columbus, Ohio. He currently is President and owner of Hurdzan and Fry Golf Course Design. He was elected to the Board of Directors of the American Society of Landscape Architects in 1977 and President of the American Society of Golf Course Architects in 1985. Dr. Hurdzan received the Ohio Turfgrass Foundation Award of Professional Excellence in 1989, the Ohio Turfgrass Foundation Man of the Year Award in 1991, the GCSAA Distinguished Service Award in 1995, the GCSAA President's Award for Environmental Leadership in 1997, the Ohio State University Distinguished Alumni Award in 1997, the Distinguished Alumni Award from the University of Vermont in 2000 and 2002, the Donald Rossi Award from the Golf Course Builders Association in 2002, and the Donald Ross Award from the American Society of Golf Course Architects in 2007.

1985† **Evolution of the Modern Green.** Michael J. Hurdzan. PGA Magazine, Professional Golfers Association of America, Palm Beach Gardens, Florida, USA. 24 pp., 35 illus., no tables, no index, fold-stapled glossy soft covers in dark green with white lettering, 8.4 by 10.7 inches (213 × 272 mm).
> A reprint of a four-part series of articles previously published in the **PGA Magazine**. It represents the authors thoughts on the historical development, construction, turfgrass establishment, and culture of putting greens.

1996† **Golf Course Architecture—Design, Construction, & Restoration.** First edition. Dr. Michael J. Hurdzan. Sleeping Bear Press, Chelsea, Michigan, USA. 424 pp., 99 illus. with 315 in color, 9 tables, indexed, glossy hardbound in multicolor with white and orange lettering, 8.4 by 11.0 inches (213 × 279 mm).
> A large book on the theory and practice of golf course architecture, construction, and turfgrass establishment. There are extensive, quality photographs throughout. Modern methods of construction are addressed within an interesting historical perspective.
> *Revised in 2006.*

1998 **Golf Course Architecture—Design, Construction, & Restoration.** Author.
> A Korean version of the 1996 book translated and published by Orange Engineering, Seoul, South Korea.

1999 **Golfplatz Architektur—Design, Konstruktion & Platzerneuerung (Golf Course Architecture—Design, Construction, & Restoration).** Dr. Michael J. Hurdzan.
> A German version of the 1996 book translated by Georg Boehm and published by E. Albrecht Verlags-KG, München, Germany.

2004† **Golf Greens—History, Design and Construction.** Dr. Michael J. Hurdzan. John Wiley & Sons Inc., Hoboken, New Jersey, USA. 334 pp.; 295 illus. with 31 in color; 2 tables; large index; glossy hardbound in multicolor with white, green, blue, and black lettering; 7.8 by 9.1 inches (198 × 231 mm).
> A book on putting green evolution, surface design, root-zone profile approaches, construction methods, turfgrasses versus artificial turf, legal liabilities, and current trends. The author has been willing to present his personal opinions on a number of aspects in the selection and installation of various construction materials and techniques.

2006† **Golf Course Architecture—Evolution of Design, Construction, and Restoration Technology.** Second revised edition. Michael J. Hurdzan. John Wiley & Sons Inc., Hoboken, New Jersey, USA. 464 pp., 521 illus. with 335 in color, 16 tables, large index, glossy hardbound in multicolor with green lettering, 8.5 by 10.8 inches (215 × 274 mm).

An update and revision of the 1996 first edition. Changes influencing the revision include increased site use restrictions, environmental protection issues, new closely-cut turfgrass cultivars, nonsteel shoe spikes, and wildlife habitat needs.

Hutchinson, Horatio Gordon

Horace Hutchinson (1859–1932) was born in London, England, UK, and earned a BA degree from Corpus Christi College, Oxford University, in 1881. Through his 16 books on golf, he became the game's first widely read author. He evolved his playing expertise at Royal North Devon Golf Club, known as Westward Ho! Horace Hutchinson was a successful golfer, winning the first two official British Amateur Championships in 1886 and 1887. He was elected the first English Captain of the Royal and Ancient Golf Club of St Andrews in 1908. In summary, Horace Hutchinson was a player, writer, and historian of golf, including providing leadership as editor of the earliest full-sized book on the design, construction, and turfgrass culture of golf courses.

1906† **Golf Greens and Green-Keeping.** Horace G. Hutchinson (editor) with twelve contributing authors. The Country Life Library of Sport, Country Life Limited and George Newnes Limited, London, England, UK, and Charles Scribner's Sons, New York City, New York, USA. 219 pp.; 38 illus. with a folding plate; 2 tables; no index; textured cloth hardbound in maroon with black, gold, and gilt lettering and also dark red with gold lettering; 5.7 by 8.9 inches (145 × 227 mm).

　＊　 One of the oldest bound books published on the subject of turfgrasses. It is a **very rare** book that is a must for collectors of historical turfgrass and/or golf books. It is devoted solely to the design, construction, and maintenance of golf course turfgrasses. An assemblage of 16 papers written by 12 authors with five of these contributing authors eventually writing their own separate golf-turfgrass books. The chapter titles and contributing authors include the following:

1. Introduction—Horace G. Hutchinson
2. The Formation of Turf—Gilbert Beale (Messrs. Carter-Seedsman)
3. Treatment and Upkeep of Seaside Links—Hugh Hamilton
4. Typical East Coast Links-Deal—Henry Hunter (golf professional and greenkeeper)
5. Treatment and Upkeep of a Golf Course on Light Inland Soil—H.S. Colt (golf course architect)
6. Construction and Upkeep of Heath Land Courses—William Herbert Fowler (golf course architect)
7. Formation and Upkeep of Courses Made Out of Pine Forests—S. Mure Ferguson
8. Treatment of an Inland Green on Medium Soil—Peter W. Lees (greenkeeper and both golf course constructor and architect)
9. Upkeep of a Golf Course on Chalk Downs—Leonard Keyser
10. Golf Course on Heavy Soil—James Braid (golf professional and golf course architect)
11. Making a Park Course—E. Mepham
12. Formation and Placing of Hazards—Cecil Key Hutchison (amateur golfer and golf course architect)
13. Remarks on the Laying Out of Courses—H.H. Hilton
14. The Championship Courses—H.H. Hilton
15. General Deductions on Greenkeeping—Horace G. Hutchinson
16. A Few Leading Principles in Laying Out Links—Horace G. Hutchinson

Newly evolving areas of emphasis in this classic, pioneering book were less frequent rolling, switching from heavy iron rollers to light wood rollers; maintenance of greens varying with the soil type; mowing of fairways preferred to the use of sheep; the need for rabbit control; and an appreciation of the greenkeeper's contributions. It contains numerous early golfing photographs.

2001† **Golf Greens and Green-Keeping.** Horace G. Hutchinson (editor) with twelve contributing authors. Reprinted by Ann Arbor Press, Chelsea, Michigan, USA. In a similar format.

Horace Hutchinson gave a perspective of early turfgrass technical developments on golf courses in Scotland as the game spread from the seaside to inland areas in the late 1800s:

If we had never departed from the view that golf was good only by the sea, it would not have been any use to collect all this book of wisdom. But, happily, we have ceased to be so exclusive, and almost of necessity, for if all the golfers of today were to insist on playing by the seaside, they would jostle each other into the waves. We have learned that golf can be played on inland soils, and in course of learning that lesson we have, in the passage of the last fifteen years, been learning a great deal about the best manner of preparing these inland soils to make the golf that we play on them as good, and as like the real seaside thing, as possible. The seaside did not want

Figure 12.44. Worthington turf tractor with flat metal wheels, studded on the back, and weighing ~1,000 pounds, circa 1924. [34]

so much help from art in order to be useful to the golfer as the inland soil. On some links he had but to cut a hole, and he could start and play forthwith. Westward Ho at its commencement was of this happy constitution. Heavier soils required much more preparation, and by the time the golfer of the inland greens had succeeded in bringing them up to a certain state of imperfection, his views as to what was wanted for a golf course had been educated. He began to demand more, even of his seaside links, so that tees had to be levelled, putting greens had to be extended, and so on, to meet his cultured tastes. However, the lesson began to be learned that putting greens, even on seaside links, do not go on recuperating their vigour for ever, so that the golfer may go on putting on them and never give them back any care in return. Some of our most splendid links—St. Andrews and Hoylake are examples—had a moment in their history when the putting greens were parlous bad. I fear, from what I hear, that others of the best are going through that bad time even now. But in the process of these years we have learnt not only that greens require care, but also the kind of care that they require. These have been years of empiricism, of experiment, of failure leading to success. Lessons have been learnt in these years. One of the lessons is that different courses require very different treatment—different because their soil is different, their climate is different, they differ in the circumstances of the amount of play on them: it makes a difference whether they are surrounded with lands that are richly tilled, from which the soil and the dressing may blow over them, or whether barren lands or more barren sea are about them; they differ as to their natural grasses.

E. Mepham commented the following:

Sweeping of greens is necessary every morning where sheep and rabbits feed on them, but they should be swept as lightly as possible so as not to injure the grass.

I

Imaizumi, Hisao

Hisao Imaizumi was born in Tokyo, Japan, and earned a Law degree from Dokkyo University in 1978. He has served as Technical and Sales Manager of Clinton (Japan) Inc. since 1998. Hisao Imaizumi was elected a Director of the Japanese Society of Turfgrass Science in 2008.

2007† **Cultivation for Turfgrass Management—Its Effects and Method Using Machinery.** Coeditor.
See *Inamori, Makoto—2007.*

Inamori, Makoto

Makoto Inamori was born in Yokosuka City, Kanagawa Prefecture, Japan, and earned a BSc degree in Agriculture from Shizuoka University in 1977. He has served as President and on the Board of Directors of the East-Japan Turfgrass Research

Institution since 1987. Makoto Inamori was elected to the Board of Directors of the Japanese Society of Turfgrass Science for 1994.

2007†   **Cultivation for Turfgrass Management—Its Effects and Method Using Machinery.** Makoto Inamori, Hisao Imaizumi, Yuichiro Ushiki, Katsuhiro Owata, Shoichi Kimura, Hideaki Tonogi, and Shigeto Hayashi (editors). Soft Science Inc., Tokyo, Japan. 195 pp.; 399 illus.; 8 tables; small index; cloth hardbound in green, orange, and white with gold lettering and a glossy jacket in multicolor with blue and black lettering; 7.2 by 10.1 inches (182 × 257 mm). Printed in the Japanese language.

    A speciality book concerning turfgrass cultivation objectives, types, research summary, specific models, and associated cultural practices. Numerous root-zone profiles are presented relative to the type and depth of turfgrass cultivation tines.

The Irrigation Association (IA)

The Irrigation Association was formed in the United States in 1949. It is a nonprofit organization that promotes and supports the irrigation industry. The membership encompasses equipment manufacturers, distributors and dealers, and irrigation system designers, contractors, educators, researcher, and technicians. Activities include standards, education courses, and professional certifications. The IA sponsors the International Trade Show annually and publishes the semiannual **Member Update**.

2000†   **Landscape Drainage Design.** The Irrigation Association, Falls Church, Virginia, USA. 144 pp., 100 illus., 19 tables, no index, glossy semisoft bound in multicolor with green lettering, 8.3 by 10.7 inches (212 × 273 mm).

    A manual on soil water relationships, drainage design, and drainage installation. There is a glossary of selected terms in the back.

Isaac, Steven Paul

Steve Isaac was born in Wallasey, Cheshire, England, UK, and earned a BSc degree with *honours* in Applied Biology from Liverpool Polytechnic in 1982. He served as a Turfgrass Advisory Agronomist for the Sports Turf Research Institute (STRI) starting in 1985 and as STRI Area Manager for Scotland and Ireland until 2000. He became Assistant Director of Golf Course Management for the Royal and Ancient (R&A) Golf Course Committee at St Andrews in 2003. Steven Isaac has led the Best Practices for Golf Courses Program of the R&A.

1992†   **The Care of the Golf Course.** Coeditor.

    See *Hayes, Peter—1992.*

Ito, Hidemasa[3]

1994†   **Management of Turfgrass for Public Space.** Coauthor.

    See *Kitamura, Fumio—1994.*

J

Jackson, Noel (*Dr.* and *Professor* Emeritus)

Noel Jackson was born in Northallerton, Yorkshire, England, UK, and earned a BSc degree with *honours* in Agricultural Botany from Kings College, Newcastle-upon-Tyne, in 1954 and a PhD degree in Agronomy and Forages from Durham University in 1960. He served at the Sports Turf Research Institute as a Biologist working with turfgrass disease research from 1958 to 1965. He then immigrated to the United States and joined the faculty in the Department of Plant Sciences at the University of Rhode Island (URI) where he progressed to Full Professor of Plant Pathology involved in turfgrass pathology teaching, adult extension, and research. He advanced the knowledge of take-all patch and other rhizosphere diseases, such as anthracnose and yellow tuft. Dr. Jackson served as Acting Chairman of the Plant Science Department of URI in 1978, 1991, 1993, and 1994, and he retired in 2003. He was active in the founding of the Rhode Island Turfgrass Conference and the merger to form the New England Regional Turfgrass Conference. Dr. Jackson received the GCSAA Distinguished Service Award in 1991 and the USGA Green Section Award in 1999.

1965†   **Fungal Diseases of Turf Grasses.** Second revised edition. N. Jackson and J. Drew Smith. Sports Turf Research Institute, Bingley, West Yorkshire, England, UK. 97 pp., 32 illus., 8 illus. in color, 3 tables, indexed, cloth hardbound in green with gold lettering, 5.9 × 9.0 inches (150 × 228 mm).

    A small book devoted to the basic pathology of turfgrass diseases, especially as related to conditions in England. Each disease is discussed in terms of host range, symptoms, causal fungus, conditions influencing occurrence, cultural controls, and suggested fungicide controls. The text is documented by extensive references.

The first edition under the same title was authored by Smith alone (see Smith, Jeffrey Drew—1959), while the third edition involves three authors (see Smith, Jeffrey Drew—1989).

1989†   **Fungal Diseases of Amenity Turf Grasses.** Coauthor.

See *Smith, Jeffrey Drew—1989.*

Jakobsen, Bent Leo-Frank (Dr.)

Bent Jakobsen was born in Monsted, Denmark, and earned a BAgr in Agriculture in 1964 and a PhD in Soil Science in 1968 both from the Royal Veterinary and Agricultural University in Copenhagen, Denmark. He served as a Lecturer in irrigation, drainage, and soil management at the Agricultural University, Copenhagen, from 1968 to 1979, as a Research Scientist in soil compaction by forest operations for the Commonwealth Scientific and Industrial Research Organization (CSIRO) Division of Forest Research, Canberra, Australia, from 1979 to 1984, as Research Scientist in wheat for University of Adelaide, South Australia, from 1986 to 1988, and as Technical Officer with the City Park Technical Services Unit in Canberra, Australian Capital Territory. In the City Park Unit, he was involved in soil testing and sportsfield management from 1988 to 1999. Since then, he has been Chief Soil Scientist for Rootzone Laboratories International in Canberra.

2000†   **Practical Drainage for Golf, Sportsturf, and Horticulture.** Coauthor.

See *McIntyre, David Keith—2000.*

Japanese Society of Turfgrass Science (JSTS)

Formed in 1983, the JSTS or Nihon-Shibakusa-gakai is an organization of Japanese turfgrass scientists headquartered in Tokyo. It sponsors two meetings per year and publishes the **Journal of the Japanese Society of Turfgrass Science** two times per year.

1977†   **The Elements of Turf and Turfgrass.** Japanese Society for Turfgrass Research with 45 contributing authors. Soft Science Inc., Tokyo, Japan. 480 pp.; 317 illus. with 64 in color; 258 tables; index by author, subject, and plant and animal names; cloth hardbound in green with gold lettering; 8.5 by 12.0 inches (215 × 304 mm). Printed entirely in the Japanese language.

A basic text on turfgrass science as it applies to conditions in Japan. Included are turfgrass characteristics, establishments, and cultures plus soils aspects, pests and their control, equipment, and cultural systems for specific uses. There is a list of selected references at the back of each chapter. It was published to celebrate the fifth anniversary of the Japan Society of Turfgrass Science.

1988†   **Turf: Science and Culture For Green Vegetation.** Japanese Society of Turfgrass Science with 41 contributing authors. Soft Science Inc., Tokyo, Japan. 564 pp., 285 illus. with 64 in color, 194 tables, indexed, cloth hardbound in dark green with gold imprinted lettering and a glossy jacket in greens with dark green and white lettering, 7.2 by 10.1 inches (183 × 256 mm). Printed entirely in the Japanese language.

An update of a basic text on turfgrass science and culture as it applies to conditions in Japan. Included are turfgrass characteristics, establishments, and cultures plus soils aspects, pest management, equipment, and cultural systems for specific uses. There is a list of selected references at the back of each chapter. It was published to celebrate the tenth anniversary of the Japan Society of Turfgrass Science.

†   *Reprinted in 1989, 1991, 1993, and 1995.*

2001†   **Handbook: Management of Turf and Turfgrass Research.** Japanese Society of Turfgrass Science. Soft Science Inc., Tokyo, Japan. 423 pp., 457 illus. with 95 in color, 123 tables, indexed, leatherette bound in dark blue with gold lettering and a jacket in yellow and green with dark blue lettering, 7.1 by 10.0 inches (180 × 255 mm). Printed in the Japanese language.

Lists of selected references are included at the ends of certain chapters.

2003†   **Dictionary of Turfgrass Science.** Japanese Society of Turfgrass Science. Shokucho-kaikan, Tokyo, Japan. 258 pp., 9 illus., 5 tables, extensive index in English, semisoft bound in green with gold lettering and a jacket in blues and white with dark blue lettering, 5.8 by 8.3 inches (147 × 210 mm). Printed in the Japanese language.

A dictionary encompassing 1,500 turfgrass related terms and their respective definitions.

Jensen, Ellis Ray

Ray Jensen was born in Fountain Green, Utah, USA. He served in the Counter Intelligence Corp of the US Army. He was a Soil Scientist with the United States Department of Agriculture (USDA) Soil Conservation Service in Tifton, Georgia, from 1946 to 1955 and then formed Southern Turf Nurseries. Ray Jensen pioneered the production and marketing of centipedegrass seed starting in 1950 and, in 1956, also initiated pioneering efforts to vegetatively plant hybrid bermudagrass by mechanical means that he developed. Ray Jensen has been awarded Honorary Alumnus of Abraham Baldwin Agricultural College, Honorary Member of the Future Farmers of America, and inductee to the Georgia Agriculture Hall of Fame in 2006.

n.d.† **Tifton Dwarf-Bermudagrass Manual For Greens & Fairways.** E. Ray Jensen. Southern Turf Nurseries, Tifton, Geor-
circa gia, USA. 26 pp., 17 illus. with 8 in color, 3 tables, no index, fold-stapled glossy semisoft covers in multicolor with black
1965 lettering, 5.5 by 8.4 inches (139 × 213 mm).

A small booklet devoted to the description of the Tifton Dwarf cultivar of hybrid bermudagrass (*Cynodon dactylon* × *C. transvaalensis*) plus methods for the construction, planting, and establishment on greens. It also includes sections on turfgrass culture, winter overseeding, and pest control. The background source of this genotype was not the same as Tifdwarf hybrid bermudagrass.

John George Cunningham Limited

John G. Cunningham Limited of Burntisland, Fife, Scotland, was formed circa 1934. It was an independent agricultural business that also supplied golf courses. The company ceased operations circa the 1970s.

n.d. **The Establishment and Care of Fine Turf for Lawns and Sports Grounds.** John G. Cunningham Limited, Burntis-
circa land, Scotland, UK. 126 pp., 51 illus., 4 tables, no index, semisoft bound in green with black lettering, 5.4 by 8.5 inches
193x (137 × 216 mm).

⁎ A **very rare** book that is a must for collectors of historical turfgrass books. It covers the establishment, renovation, culture, and pests of turfgrasses in lawns, bowling greens, tennis courts, and sports fields. A very similar book with the same title, but under the authorship of Dr. David Clouston, was subsequently published in 1937.

Johns, Donald, Jr. (Dr.)

Don Johns was born in Corsicana, Texas, USA, and earned a BA degree from the University of Texas in 1970, an MSc degree from Texas A&M University in Soil Science and Turfgrass in 1976, and a PhD degree in Turfgrass Physiology in 1980 from Texas A&M University, studying with Dr. Jim Beard, plus an MSc in Statistics from Purdue University at Indianapolis in 1994. He served the Eli Lilly Company as a Science Research Representative from 1981 to 1987, as Senior Scientific Information Analyst from 1987 to 1988, and as Statistician since 1989. Dr. Johns has specialized in turfgrass water relations, experimental design, and categorical data analysis.

1979† **Introduction To Turfgrass Science And Culture Laboratory Exercises.** Coauthor.
See *Beard, James B—1979.*

Johns, Rodney Dale

Rodney Johns was born in Davenport, Iowa, USA, and earned a 1-Year Turf and Greenhouse Management Certificate from Indian Hills Community College of Ottumwa, Iowa, in 1993, an Associate of Science Degree from John Wood Community College of Quincy, Illinois, in 1995, and a BSc degree in Plant Science and Horticulture from the University of Missouri-Columbia in 1997. He has worked at River Valley Country Club as Assistant Superintendent from 1988 to 1992 and as Superintendent from 1992 to 1997. He currently owns Arki-Tec Landscaping and Sales LLC in Canton, Missouri. Rodney Johns serves as Chair of the Tree Board in Canton, Missouri, and received the Green Award from Culver Stockton College.

2004 **Turfgrass Installation—Management and Maintenance.** Rodney Johns. McGraw-Hill Companies Inc., New York City, New York, USA. 599 pp., 194 illus., 25 tables, small index, glossy hardbound in multicolor with white and green lettering, 5.9 by 8.9 inches (150 × 226 mm).

A book on a diverse array of turfgrass subjects totaling 21 chapters, including the lawn care business and golf course management. Five of the chapters on golf course management and on soil and water testing were authored and also are similar to the ones found in the 2004 book by Danny H. Quast and Wayne Otto. Furthermore, three of the chapters on weeds, diseases, and pesticides were authored and also are similar to the ones found in the 2004 book by Matthew James Fagerness. The irrigation chapter was written by Bruce Carleton.

2004 **Turfgrass Chemicals and Pesticides—A Practitioner's Guide.** Coauthor.
See *Fagerness, Matthew James—2004.*

2006 **Cao ping quan li: Jingying yu guanli.** Chinese translation of **Turfgrass Installation: Management and Maintenance** (2004), as translated by Jing Sun and Xiaospei Wang. McGraw-Hill Education (Asia) Co. and China Electric Power Press, Beijing, China. 365 pp. Printed in the Chinese language.

2006 **Tips & Traps for Growing and Maintaining the Perfect Lawn.** Rodney Johns. The McGraw-Hill Companies, New York City, New York., USA. 204 pp.; 66 illus.; 3 tables; small index; glossy semisoft bound in multicolor with yellow, black, blue, and white lettering; 5.9 by 9.0 inches (150 × 228 mm).

A book on the planting, care, and pest management of lawns. It is oriented to home lawn enthusiasts under conditions in the United States.

Joo, Young Kyoo (Dr. and *Professor*)

1991   **Characteristics, Species, and Establishment of Turfgrass.** In Home Horticulture. Young Kyoo Joo with four contributing authors. Korean Open University Press, Seoul, South Korea. 345 pp., 221 illus., 45 tables, indexed, glossy hardbound in multicolor with black lettering, 7.5 by 10.0 inches (190 × 255 mm). Printed in the Korean language.

   A book on the fundamentals of turfgrass science and culture. Included are the characteristics, species, and establishment of turfgrass for home horticulture. This book was written for a college course.

   Reprinted in 2002.

1992   **Golf Course Planning in Korea: Theory and Practice.** Young Kyoo Joo with six contributing authors: Gwi Gon Kim, Myeong Kil Kim, Ji Deok Kim, Hwi Young Oh, Dong Geum Lee, and Sang Ha Lim. Chogyeong Publishing Company, Seoul, South Korea. 567 pp., 206 illus., 46 tables, large index, cloth hardbound in blue with gold lettering, 7.5 by 10.2 inches (190 × 260 mm). Printed entirely in the Korean language.

   A book on golf course development and operation. Included are the history, design, planning, construction, turfgrass maintenance, pesticides, equipment, and personnel management for golf courses.

1999   **Fundamental Turfgrass Science and Establishment of Turfgrass.** In Home Practical Horticulture. Young Kyoo Joo with eight contributing authors. Hyangmoonsa Press, Seoul, South Korea, for the Korean Society for Horticulture. 498 pp., 200 illus., 40 tables, indexed, glossy hardbound in multicolor with black lettering, 7.5 by 10.0 inches (190 × 255 mm). Printed in the Korean language.

   A book concerning turfgrasses and their establishment and culture, including the soils aspects. It is oriented to the climatic and soil conditions in Korea.

2002   **KPGA Golf Course Manual.** Young Kyoo Joo with three contributing authors. Doosan Dong—A Press for the Korean Professional Golf Association, Seoul, South Korea. 386 pp., 178 illus. in color, 60 tables, indexed, hardbound in green with blue lettering, 7.7 by 10.4 inches (195 × 265 mm). Printed in the Korean language.

   A book concerning the fundamentals of turfgrass culture and maintenance for golf course superintendents and ground workers, including golf course construction, personnel management, and laws and regulations concerning design, construction, and operation.

Joyner, Bobby Gerald (Dr.)

Bobby Joyner was born in Plant City, Florida, USA, and earned both BSc and MSc degrees in Plant Pathology from the University of Florida in 1970 and 1972, respectively, and a PhD degree in Plant Pathology from Virginia Polytechnic Institute & State University in 1976, studying with Dr. Houston Couch. He has held positions as Director of the Diagnostic Lab for ChemLawn from 1976 to 1988, as Technical Services Manager for ChemLawn and TruGreen/ChemLawn from 1988 to 1998, as Director of Research for TruGreen/ChemLawn from 1998 to 2003, and as Director of Technical Services for TruGreen/ChemLawn since 2003.

1984†   **ChemLawn Turf Field Guide.** Coauthor.

   See *Kick-Raack, Joanne Marie—1984.*

Juska, Felix Victor (Dr.)

Felix Juska (1914–1973) was born in Chicago, Illinois, USA, and earned a BSc in 1946, an MSc in 1952, and a PhD in 1955, all from Michigan State University specializing in Soil Science, studying with Dr. Jim Tyson. He served as a Vocational Agricultural Instructor in Michigan from 1946 to 1950, as County Supervisor for the Farmer's Home Administration from 1950 to 1953, and as Research Agronomist for the Forage and Range Research Branch, Plant Science Research Division, United States Department of Agriculture, in Beltsville, Maryland, from 1955 to 1971. He conducted research in turfgrass culture and cultivar evaluation. He was elected Turfgrass Management Division C-5 Chair for 1963. Dr. Juska was awarded Fellow of the American Society of Agronomy in 1966.

1969†   **Turfgrass Science.** Coeditor.

   See *Hanson, Angus Alexander—1969.*

K

Kansai Golf Union, Green Section

The Kansai Golf Union, Green Section, was formed in 1960, with the objective to disseminate scientific knowledge and methods concerning turfgrass and golf course management. The Green Section maintains a Research Center in Takarazuka, Hyogo-ken, Japan, that includes a main research building with four laboratories plus offices and seminar rooms. There is a field research facility nearby. The Green Section has published the **Turf Research Bulletin** annually since 1961 the **Turf News** three times per year since 1967, and the **Green Letter** monthly since 2003.

1983† **Handbook of Turf and Golf Course Management.** First edition. Kansai Golf Union Green Section, Takarazuka City, Hyogo Prefecture, Japan. 288 pp., 154 illus. with 108 in color, 44 tables, indexed, cloth hardbound in yellowish green with silver lettering, 5.8 by 8.2 inches (147 × 209 mm). Printed in the Japanese language.

A book concerning the turfgrasses used in Japan plus their establishment and culture. Emphasis is placed on pest, environmental, and soil problems. There is a glossary in the back.

Revised in 1991.

1991† **Handbook of Turf and Golf Course Management.** Revised second edition. Kansai Golf Union Green Section Research Center, Takarazuka City, Hyogo, Japan. 374 pp., 354 illus. with 110 in color, 46 tables, indexed, cloth hardbound in dark blue with silver lettering, 5.8 by 8.2 inches (147 × 209 mm). Printed in the Japanese language.

A revision and update of the 1983 first edition. Chapters concerning turfgrass physiology and overseeding techniques have been added. There is a glossary in the back.

Karnok, Keith J (Dr. and Professor)

Keith Karnok was born in Cleveland, Ohio, USA, and earned a BSA degree in Agronomy and Plant Pathology in 1973 and an MSc degree in Agronomy and Turfgrass in 1974, both from the University of Arizona, studying with Dr. Bob Kneebone; he also earned a PhD degree in Turfgrass Physiology from Texas A&M University in 1977, studying with Dr. Jim Beard. He has been active in turfgrass teaching and research at Ohio State University from 1977 to 1982 and since then at the University of Georgia (UGA), becoming a Full Professor in 1993. He was elected Turfgrass Science Division C-5 Chair for 1994, to the Crop Science Society of America (CSSA) Board of Directors from 1993 to 1995; he was also elected President of the North American College Teachers of Agriculture (NACTA) for 2005 to 2006. Dr. Karnok received the D.W. Brooks Award for Excellence in Teaching from the College of Agriculture and Environmental Science, University of Georgia (UGA), in 1992; the Teaching Award of Merit of NACTA in 1993; Fellow of NACTA in 1993; the Agronomic Resident Education Award of the American Society of Agronomy (ASA) in 1994; the Excellence in Teaching Award of the UGA Chapter of Gamma Sigma Delta in 1995; the Southern Regional Outstanding Teaching Award of NACTA in 1995; a Fellow of ASA in 1995; Fellow of CSSA in 1995; the Outstanding Professor Award of the UGA Chapter of Alpha Gamma Rho in 1995; the Media Award of Excellence of NACTA in 1996; the Josiah Meigs Award for Excellence in Teaching from UGA in 1996; the Teaching Award of Excellence of NACTA in 1997; the Turfgrass Science Award of the CSSA in 1997; the Crop Science Teaching Award of the CSSA in 1997; the Southern Region Excellence in Teaching Award of the United States Department of Agriculture in 1997; Fellow of the American Association for the Advancement of Science in 1997; UGA Teaching Fellow in 1999; the Outstanding Education Award from the Georgia Green Industry Association in 2002; the John Deere Award of NACTA in 2004; and the Certificate of Appreciation for Excellence in Teaching from the UGA Student Government Association in 2007.

Figure 12.45. Scything tall grass in rough of golf course, circa 1940s. [30]

1979† **Introduction To Turfgrass Science And Culture Laboratory Exercises.** Coauthor.
See *Beard, James B—1979.*

1997† **Turfgrass Management Information Directory.** First edition. Keith J. Karnok (editor). Ann Arbor Press, Chelsea, Michigan, USA. 105 pp., no illus., no tables, no index, glossy semisoft bound in maroon with white lettering, 8.5 by 10.9 inches (217 × 278 mm).

A compilation of teaching resources, service labs, green industry organizations, university and industry personnel, and publications allied with turfgrass. The book was developed by the Turfgrass Science Division C-5 of the Crop Science Society of America.
Revised in 1998 and 2000.

1998† **Turfgrass Management Information Directory.** Second revised edition. Keith J. Karnok (editor). Ann Arbor Press, Chelsea, Michigan, USA. 188 pp., 25 illus., 5 tables, no index, glossy semisoft bound in green with white lettering, 8.5 by 10.9 inches (217 × 278 mm).

This second revision has been updated and expanded by the addition of sections on turfgrass chemicals, scientific names, US climatic maps, mathematical conversions and calculations, websites, and a glossary.
Revised in 2000.

2000† **Turfgrass Management Information Directory.** Third revised edition. Keith J. Karnok (editor). Ann Arbor Press, Chelsea, Michigan, USA. 256 pp., 25 illus., 5 tables, no index, glossy semisoft covers in dark green with white lettering in a ring binder, 8.5 by 10.9 inches (216 × 278 mm).

This third edition is an updated version.

Keil, Harry Louis (Dr.)

Harry Keil (1915–1978) was born in Hudson, Pennsylvania, USA, and earned a BSc degree in 1937, an MSc degree in 1939, and a PhD degree in Plant Pathology in 1946, all from Pennsylvania State University. He was Assistant Research Professor and Agricultural Agent of Plant Pathology at the University of Rhode Island from 1942 to 1946. During this time he codeveloped, with Dr. Frank Howard, early fungi toxicity information of chemicals, including the organic mercuries and cadmium compounds. From 1946 to 1958, he continued fungi toxicity research with the Rohm and Hass Chemical Company, developing the dithiocarbamates. In 1958, he accepted a position as Research Plant Pathologist with the United States Department of Agriculture in Beltsville, Maryland, and continued this work until his death in 1978. Dr. Keil was elected President of the Potomac Division of the American Phytopathological Society for 1973.

1951† **Fungus Diseases of Turf Grasses.** Coauthor.
See *Howard, Frank Leslie—1951.*

Kenna, Michael Patrick (Dr.)

Mike Kenna was born in La Mesa, California, USA, and earned a BSc in Ornamental Horticulture from the California State Polytechnic University in Pomona in 1979 and both an MSc degree in Agronomy in 1981 and a PhD degree in Crop Science in 1984 from Oklahoma State University. He joined the faculty at Oklahoma State University as an Assistant Professor in 1985 and was involved in turfgrass research and extension. In 1990, he became Director of Research for the Green Section of United States Golf Association and became involved in administration of a major turfgrass research fund through the Turfgrass and Environmental Research Program. Dr. Kenna was honored as a Distinguished Alumnus by the Alumni Association of California State Polytechnic University in 2003.

1998† **Turfgrass Biotechnology—Cell and Molecular Genetic Approaches to Turfgrass Improvement.** Coeditor.
See *Stricklen, Mariam B.—1998.*

2008† **Water Quality and Quantity Issues for Turfgrasses in Urban Landscapes.** Coeditor.
See *Beard, James B—2008.*

Key, Arthur

Arthur Key was born in Hempshill, Nottinghamshire, England, UK, and earned a BA degree with *honours* in Natural Science (Chemistry) from Wadham College, Oxford University, in 1895 and an MA from the same college in 1898. He later became associated with the Key Fertilizer Company.

n.d.
circa
1924
Hints on Lawns. Arthur Key. The Key Fertilizer Company, London, England, UK. 21 pp., no illus., 1 table, no index, fold-stapled soft covers in tan with blue lettering, 7.3 by 4.7 inches (186 × 120 mm).

* A **rare**, small booklet on fertilization and weed control for lawns. It is oriented to lawn enthusiasts and cool-season turfgrass conditions in the United Kingdom.

Kick-Raack, Joanne Marie

Joanne Kick-Raack was born in Dalton, Ohio, USA, and earned an Associate of Applied Science (AAS) degree from the Agricultural Technical Institute, Ohio State University, in 1975, a BSc degree in Plant Pathology from Cornell University in 1978, and an MSc degree in Agricultural Education from Ohio State University (OSU) in 1995. She was Plant Pathologist and Assistant Manager of the ChemLawn Plant Diagnostic Laboratory from 1981 to 1987 and has been Coordinator of Pesticide Education Programs at OSU since 1995.

1984† **ChemLawn Turf Field Guide.** J.M. Kick-Raack and B.G. Joyner. ChemLawn Diagnostic Laboratories, ChemLawn Corporation, Delaware, Ohio, USA. 188 pp., 471 illus. in color, 1 table, no index, side tabs, stiff plastic covers in white with green lettering in a 6-hole binder, 5.5 by 8.4 inches (139 × 214 mm).

A practical field guide to the identification of pests of both cool- and warm-season turfgrasses in North America. It contains quality color photographs.

Kiely, William Avanton

William Kiely was born in 1870 in Westport, New Zealand, and earned a degree from Auckland College. He served as Vice Chairman and Chairman of the New Zealand Golf Council (NZGC) located in Wellington. He was a founding member of the New Zealand Greenkeeping Research Committee in 1932.

1950† **Construction, Renovation, and Care of the Golf Course.** Coauthor.
See *Levy, Enoch Bruce—1950.*
The authors made the following comments regarding fairway irrigation in 1950:

> We very much doubt if time and expense in watering fairways is justified. If the same money were spent in getting in appropriate grasses for the fairways and in weed control so that these grasses may grow unencumbered by rank competitors a greater satisfaction would accrue to players than where the fairways were watered and still a mass of weeds and the sward a mosaic of brown and green patches. Whatsoever watering does it will increase the need of fertiliser treatment and the necessity to pay greater attention to weed and clover control by toxic fertilisers or chemical sprays.

Kim, Doo Hwan (Dr. and Professor)

Doo Hwan Kim was born in Seoul, South Korea, and earned a BSc degree in Horticulture from Seoul National University in 1982; he also earned an MSc degree in Horticulture in 1986 and a PhD degree in Crop Breeding in 1990, both from the University of Illinois. He serves as a Professor in the Department of Molecular and Biotechnology at Konkuk University and as President of KV Bio Inc., working on the breeding of turfgrasses—zoysiagrass (*Zoysia* spp.) and bermudagrass (*Cynodon* spp.). He was elected Vice President of the Turfgrass Society of Korea. Dr. Kim received the Innovation Award from the Ministry of Agriculture and Forestry, Republic of Korea, in 2004.

2005 **Design, Construction, Maintenance, and Management of Golf Course.** Doo Hwan Kim with 21 contributing authors. Korean Golf Course Management Association, Seoul, South Korea. 466 pp., 292 illus. in color, 122 tables, no index, soft bound in silver with black lettering, 8.3 by 10.6 inches (210 × 270 mm). Printed entirely in the Korean language.

A book concerning the development, design, construction, and management of golf courses plus the equipment personnel and business administration requirements under conditions in South Korea.

2006 **Turfgrass Dictionary.** Doo Hwan Kim with 16 contributing authors. Korea Golf Course Management Association, Seoul, South Korea. 252 pp., 184 illus., 8 tables, indexed, glossy semisoft bound in blue with gold lettering, 6.2 by 8.5 inches (158 × 216 mm). Printed in the Korean language.

A book concerning the detailed description of the terminologies used in turfgrass science and industry. Written by members of the Korean Turfgrass Society.

2009 **Ways to Establish and Maintain Turf Ground.** Doo Kwan Kim with 34 contributing authors. Seoul Olympic Sports Promotion Foundation, Seoul, South Korea. 150 pp., 73 illus., 33 tables, no index, soft bound in green with black lettering, 6.0 by 8.9 inches (152 × 225 mm). Printed in the Korean language.

A book concerning (a) general outlines of turfgrass use and its botany, (b) turfgrass use in school sports grounds in Korea, (c) establishment of turfgrass sports grounds, (d) maintenance of turfgrass sports grounds, and (e) examples of turfgrass sports grounds in Korea and abroad.

Kim, In Seob (Dr.)

In Seob Kim was born in Daegu, South Korea, and earned a BSc degree in Agronomy, an MSc degree in Turfgrass Science, and a PhD in Turfgrass Management, all from Kyungpook National University in 1980, 1984, and 1988, respectively. He served as an Assistant Professor at Kyungpook National University from 1984 to 1986 and is currently Adjunct Professor at Kyungpook National University. He was Senior Researcher at the Korean Turfgrass Research

Figure 12.46. Horse- or tractor-drawn Massey-Harris manure spreader with metal wheels and two discharge beaters, circa 1924. [34]

Institute from 1989 to 1996. Then he was Management Director at Suwon Country Club from 1996 to 1999. Subsequently, he founded and is CEO of the TechnoGreen Company Limited, a turfgrass management firm. Dr. Kim was elected a Director of the Turfgrass Society of Korea in 1990 and a Director of the Korean Society of Weed Science in 1990.

1992　**Basic and Practice in Golf Course Management.** In Seob Kim with seven contributing authors. The Korea Turfgrass Research Institute Press, Seoul, South Korea. 772 pp., 421 illus., 248 tables, no index, hardbound in green with blue lettering, 7.5 by 10.2 inches (190 × 260 mm). Printed in the Korean language.

　　An early Korean textbook on the fundamentals of turfgrass establishment and culture for golf courses under the conditions in Korea.

1999　**System and Leadership for Golf Courses.** In Seob Kim and Yong Tae Ahn. GMI Media Press, Seoul, South Korea. 154 pp., 22 illus., 20 tables, no index, glossy semisoft bound in green with black lettering, 7.5 by 10.2 inches (190 × 260 mm). Printed in the Korean language.

　　A book concerning the fundamentals of leadership and personnel management involved in golf course operations under the conditions in Korea, with some case studies. Included are superintendent qualifications, communication skills, and governance system.

Kim, Kyeung Nam (Dr. and Assistant Professor)

Kyeung Nam Kim was born in Chuncheon, Kangweondo, South Korea, and earned BSc and MSc degrees in Horticultural Science from Seoul National University in 1984 and 1986, respectively, and a PhD in Horticulture and Forestry from the University of Nebraska in 1992, studying with Drs. Garald Horst and Robert Shearman. He worked as Senior Researcher at the Turfgrass & Environmental Research Institute from 1993 to 1998 and as Senior Technical Advisor at Samsung Everland Inc. from 1998 to 2004 and, since, has been Assistant Professor in the Department of Horticultural Science at Samyook University and Director of the Asia Sports Turf Research Institute. Dr. Kim received an Honorary Award for Contribution to the 2002 FIFA World Cup Soccer Competition Korea/Japan from the Korean Ministry of Culture and Tourism.

2006　**Introductory Turfgrass Science.** Kyeung Nam Kim. Samyook University Press, Seoul, South Korea. 359 pp., 243 illus., 76 tables, indexed, glossy semisoft bound in white with black lettering, 7.5 by 10.2 inches (190 × 260 mm). Printed entirely in the Korean language.

　　An introductory book concerning cool- and warm-season turfgrasses, including characteristics, adaptation, and culture under Korean conditions. It is the first of a three volume series.

2006　**Turfgrass Management.** Kyeung Nam Kim. Samyook University Press, Seoul, South Korea. 433 pp., 101 illus., 97 tables, indexed, glossy semisoft bound in white with black lettering, 7.5 by 10.2 inches (190 × 260 mm). Printed entirely in the Korean language.

　　A book concerning the cultural practices for turfgrasses under conditions in Korea, with emphasis on irrigation, mowing, and fertilization. It also includes cultural systems for golf courses. The second in a three volume series.

2007   **Turfgrass Establishment.** Kyeung Nam Kim. Samyook University Press, Seoul, South Korea. 417 pp., 740 illus., 297 tables, indexed, glossy semisoft bound in white with black lettering, 7.5 by 10.2 inches (190 × 260 mm). Printed entirely in the Korean language.

     A book concerning turfgrass establishment including soil properties, soil preparation, and planting procedures—both seeding and vegetative propagation—under conditions in Korea. The third in a three volume series.

Kim, Myeong Kil

Myeong Kim was born in Ulsan, Gyueongsangnamdo Province, South Korea, and earned a BSc degree from the Korean Air Force Academy in 1961 and a BSc degree from the Department of Civil Engineering at Seoul National University in 1965. He is a golf course architect and President of Field Consultant Company Limited. Myeong Kim received the Engineering Prize from the Korean Society of Civil Engineers in 1986.

1993   **How to Make Golf Course.** Myeong Kil Kim. Jacobsen Textron Publishing Company, Seoul, South Korea. 343 pp., 63 illus. with 35 in color, 8 tables, no index, cloth hardbound in green with gold lettering, 7.5 by 10.2 inches (190 × 260 mm). Printed entirely in the Korean language.

     A book concerning the design architecture, planning, construction, and landscaping of golf courses under conditions in Korea.

Kimura, Shoichi (Dr.)

Shoichi Kimura was born in Chiba City, Chiba Prefecture, Japan, and earned a BSc degree in 1985 and a PhD degree in 1993, both in Agronomy from the University of Tokyo. He has been Director of Research and Development for the Toyo Green Company Limited since 1993. Dr. Kimura was elected Director of the Japanese Society of Turfgrass Science since 1996.

2007†   **Cultivation for Turfgrass Management—Its Effects and Method Using Machinery.** Coeditor.
     See *Inamori, Makoto—2007.*

Kitamura, Fumio (Dr. and Professor)

Fumio Kitamura was born in Asahikawa City, Hokkaido, Japan, and earned a BAg degree in 1951 and a DrAgr in Horticulture in 1970, both from the University of Tokyo. He became Professor of Agriculture at the University of Tokyo, serving from 1976 to 1984. He also served as Director of the University's Kemigawa Arboretum from 1978 to 1982, then as Professor of Agriculture at Shinshu University National University Corporation from 1984 to 1989, and, later, as Professor at Osaka University of Arts from 1989 to 1995. He was elected President of the Japanese Institute of Landscape Architecture for 1983 to 1985, President of the Japanese Society of Turfgrass Science for 1980 to 1984 and 1990 to 1994, and President of the Japanese Society of Revegetation Technology for 1991 to 1993. Dr. Kitamura received the Scientific Award from the Japanese Institute of Landscape Architecture in 1971, the Scientific Award from the Japanese Society of Turfgrass Science

Figure 12.47. Centrifugal, dry-particle Vesey's universal spreader, mid-1910s. [39]

in 1995, the Scientific Award and Service Award from the Japanese Society of Revegetation in 1996, and the Award of Uehara Price from the Japanese Institute of Landscape Architecture in 1996. Dr. Kitamura was named Honorary Member of the Japanese Association of Society of Botanical Gardens in 1984, the Japanese Society of Revegetation Technology in 1993, the Japanese Society of Turfgrass Science in 1994, and the Japanese Institute of Landscape Architecture in 1995.

1966† **Lawn and Lawn Grasses for Use as Turf.** Fumio Kitamura. Kashima-Shoten Book Company, Toshima-ku, Tokyo, Japan. 232 pp., 100 illus., 21 tables, extensive index including one in English, textured cloth hardbound in green with gold lettering and a jacket in multicolor with red and white lettering plus a slip case in tan with black lettering, 6.0 by 8.2 inches (152 × 209 mm). Printed entirely in the Japanese language.

    A book on the basics of turfgrasses and their culture under conditions found in Japan. There is a section on the earlier history of turfgrass use in Japan from the Edo era & from before (720 BC). Considerable emphasis is placed on descriptions of the turfgrasses. It contains chapters on cultural systems for lawns, golf courses, sports fields, and public grounds. There is a list of selected references included at the back of many chapters.

1977† **General Views on Turf and Turfgrasses.** Coeditor.
    See *Ehara, Kaoru—1977.*

1994† **Management of Turfgrass for Public Space.** Fumio Kitamura, Mitsuo Kondo, Hidemasa Ito, and Hiroshi Takato (editors) with twelve contributing authors. Soft Science Inc., Tokyo, Japan. 264 pp.; 255 illus. with 41 in color; 66 tables; indexed; cloth hardbound in dark blue with gold lettering and a jacket in multicolor with dark blue, green, and white lettering; 7.1 by 10.0 inches (181 × 256 mm). Printed in the Japanese language.

    A guide book for the maintenance of public turfgrass areas, especially parks. The design and management of parks are addressed. Emphasis is placed on the various functional turfgrass uses and characteristics of spaces within parks, such as for leisure, recreation, sports, golf courses, roadsides, unmowed areas of grass and wild flowers, and grassed roof tops. Lists of selected references are included at the ends of certain chapters.

1997† **A Handbook of Turfgrass and Turf.** Fumio Kitamura and Yoshisuke Maki (editors) with twelve contributing authors. Hakuyu-sha, Tokyo, Japan. 348 pp., 148 illus., 117 tables, indexed, cloth hardbound in dark blue with gold lettering plus a slip cover in multicolor with black lettering, 7.5 by 10.4 inches (190 × 263 mm). Printed entirely in the Japanese language.

    A text book on the science and culture of turfgrasses as applied to conditions in Japan. Included are the history of turfgrass use and research, germplasm introduction, pest management, and environmental aspects. There is a list of selected references at the back of each chapter.

2001† **The Tale of Turfgrass.** Fumio Kitamura. Soft Science Inc., Tokyo, Japan. 273 pp., 95 illus. with 32 in color, no tables, index, cloth hardbound in green with gold lettering and a jacket in multicolor with black lettering, 5.8 by 8.3 inches (147 × 210 mm). Printed in the Japanese language.

    A unique book on the history of native turfgrasses and lawns in Japan up to when landscape gardening technology of modern times was introduced from Western culture. Examples include the imperial courts, Buddhist temples, Shinto shrines, and samurai mansions. It also includes the early phases of turfgrass culture on golf courses and of turfgrass science.

Klug, John Robert

John Klug was born in Bulawago, Rhodesia, and earned a BSc degree in 1973 and an MSc degree in 1989, both in Agriculture from the University of Natal, South Africa. He was an Extension Officer in livestock and range management and in farm planning in Rhodesia from 1974 to 1977. Then in 1979, he accepted a position at the University of Natal, serving as a Lecturer and becoming Senior Lecturer in 1990. John Klug has taught courses in land use planning and in range management and, since 1994, has been emphasizing instruction in sports turf management.

2002† **The Cricket Pitch and its Outfield.** Coauthor.
    See *Tainton, John Melbourne—2002.*

Knoop, William Edward (Dr. and *Professor and Turfgrass Specialist* Emeritus)

Bill Knoop was born in Norwalk, Ohio, USA, and earned a BSc degree in Horticulture from Iowa State University in 1965, an MScA in Ornamental Horticulture from the University of Florida in 1969, and a PhD in Plant Science from the University of New Hampshire in 1976. He was Turfgrass Specialist at the University of New Hampshire from 1970 to 1974, Director of Education for the Golf Course Superintendents Association of America from 1974 to 1976, Assistant Professor in adult education at Iowa State University from 1976 to 1978, and Professor and Turfgrass Specialist at Texas A&M University, Dallas Center, from 1978 to 1995, when he retired. Dr. Knoop received the EPA Environmental Excellence Award in 1990, United States Department of Agriculture Award for Superior Service in 1991, and the Texas Governor's Award for Environmental Excellence in 1995.

1974† **Pesticide Usage—Reference Manual.** William E. Knoop. Golf Course Superintendents Association of America, Lawrence, Kansas, USA. 135 pp., 69 illus., 16 tables, no index, fold-stapled soft covers in light blue with black lettering, 8.5 by 11.0 inches (215 × 279 mm).

A reference booklet on pest diagnosis and the proper selection, handling, and application of pesticides for golf courses. Considerable emphasis is placed on the safety aspects.

n.d.† **The Mathematics of Turfgrass Maintenance.** Coauthor.
circa
1977

    See *Beard, James B—circa 1977.*

1977† **The Golf Professional's Guide to Turfgrass Maintenance.** First edition. William E. Knoop. The Professional Golfer's Association of America, Lake Park, Florida, USA. 20 pp., 59 illus., 1 table, no index, fold-stapled glossy semisoft covers in browns with black lettering, 8.4 by 11.0 inches (214 × 279 mm).

    An abbreviated booklet in which the turfgrass cultural practices used on golf courses are outlined. It was used in training for PGA certification.

    Revised in 1980.

1980† **The Golf Professional's Guide to Turfgrass Maintenance.** Second revised edition. William E. Knoop. The Professional Golfer's Association of America, Lake Park, Florida, USA. 21 pp., 60 illus., 1 table, no index, fold-stapled glossy semisoft covers in greens with black lettering, 8.4 by 10.9 inches (214 × 278 mm).

    There are only small revisions of the first edition.

Kondoh, Mitsuo (Dr. and Professor)

Mitsuo Kondoh was born in Yokohama City, Kanagawa, Japan, and earned a BSc degree in Agriculture from Tokyo Agricultural University in 1971 and a DrAgr degree in Landscape Architecture from Tokyo Agricultural University in 1987. He has served on the faculty at the Department of Landscape Agriculture Science in regional environmental science at Tokyo Agricultural University since 1989. He was elected President of the Japanese Society of Turfgrass Science for 2004 through 2008 and to the Board of Directors of the Japanese Institute of Landscape Architecture and of the Japanese Society of Revegetation Technology. Dr. Kondoh received the Scientific Award from the Japanese Society of Landscape Architecture in 1989.

1994† **Management of Turfgrass for Public Space.** Coauthor.

    See *Kitamura, Fumio—1994.*

2003† **Turf Science and Maintenance of School Grounds.** Mitsuo Kondoh (editor) with twelve contributing authors. Soft Science Inc., Tokyo, Japan. 147 pp. plus 18 pp. of advertising, 25 illus. with 71 in color, 37 tables, no index, glossy semisoft bound in multicolor with green and black lettering, 8.2 by 11.7 inches (208 × 297 mm). Printed in the Japanese language.

    A guide book concerning the benefits, construction, establishment, and culture of turfgrasses on school grounds under the conditions in Japan. There also is a section on costs. Lists of selected references are included in the ends of certain chapters.

Figure 12.48. Manual-push, 3-foot-wide, gravity-drop spreader with finger-touch control, 3-cubic-foot hopper, and triple roll agitation, circa 1934. [41]

Krans, Jeffrey Venon (Dr. and *Professor*)

Jeff Krans was born in Iron Mountain, Michigan, USA, and earned a BSc degree in Resource Management from the University of Wisconsin, Stevens Point, in 1970; an MSc degree in Soil Science from the University of Arizona in 1973, studying with Dr. Bob Kneebone; and a PhD in Crop Physiology and Turfgrass from Michigan State University in 1975, studying with Dr. Jim Beard. He then joined the faculty in the Agronomy Department at Mississippi State University (MSU) with responsibilities in turfgrass physiology research and teaching. Dr. Krans conducted pioneering research in tissue culture of turfgrasses and released three bermudagrass (*Cynodon*) cultivars. He rose to Associate Professor in 1982 and to Full Professor in 1985 and was honored in 1989 with the special designation of Distinguished Professor. In 1988, he was a Visiting Scientist at the New Zealand Department of Scientific and Industrial Research, Grasslands Division, conducting tissue culture research. In 2002, he retired from MSU and became active in consulting for several national turfgrass organizations. He was elected Turfgrass Science Division C-5 Chair for 1995 and was elected to the Crop Science Society of America Board of Directors from 1994 to 1996. Dr. Krans received the Teaching Award of Merit from the National Association of Colleges and Teachers of Agriculture in 1986; the University Faculty Award for Research and Teaching in 1987, which is given to one MSU faculty member per year; the Distinguished Service Award of the Mississippi Turfgrass Association in 1987; and Fellow of the American Society of Agronomy in 2000.

1986† **Basic Turfgrass Botany and Physiology.** First edition. Coauthor.
    See *Beard, James B—1986.*

1993† **Basic Turfgrass Botany and Physiology.** Second revised edition. Coauthor.
    See *Beard, James B—1993.*

1999† **Sports Fields—A Manual for Design, Construction and Maintenance.** Coauthor.
    See *Puhalla, James—1999.*

2003† **Baseball and Softball Fields.** Coauthor.
    See *Puhalla, James—2003.*

L

Lamson-Scribner, Frank (Dr.)

Frank Lamson-Scribner (1851–1938) was born in Cambridgeport, Massachusetts, USA, and earned a BSc degree from Maine State College of Agriculture and Mechanical Arts in 1870. He served as secretary to the Maine State Board of Agriculture from 1873 to 1875 and then taught botany at high schools in Bangor, Maine, and Philadelphia, Pennsylvania, from 1875 to 1885. Scribner was then appointed to the Division of Botany, United States Department of Agriculture (USDA), as the first scientist commissioned to study plant diseases. In 1888, he became Professor of Botany and Horticulture at the University of Tennessee, and in 1890, he also was named Director of the Agricultural Experiment Station. In 1894, he returned to Washington to lead the newly formed Division of Agrostology, United States Department of Agriculture (USDA), and served in several other USDA positions until retiring in 1922. He was especially active in plant exploration and taxonomy of grasses. Dr. Lamson-Scribner was a charter member of the Botanical Society of America and received the Chevaliers' Cross from France in 1890 and an Honorary Doctor of Laws (LLD) degree from the University of Maine in 1922.

1896† **Useful and Ornamental Grasses.**˙ F. Lamson-Scribner. Bulletin No. 3, Division of Agrostology, United States Department of Agriculture, Washington, D.C., USA. 119 pp., 89 illus., no tables, no index, fold-stapled soft covers in white with black lettering, 5.2 by 8.5 inches (131 × 216 mm).

    ∗ A **rare**, pioneering bulletin concerning the characteristics, adaptations, and uses, including turfgrass lawn and unmowed ornamental applications, for 88 grass species in the United States.

1898† **Economic Grasses.**˙ F. Lamson-Scribner. Bulletin No. 14, Division of Agrostology, United States Department of Agriculture, Washington, D.C., USA. 85 pp., 96 illus., no tables, detailed index, fold-stapled soft covers in tan with black lettering, 6.0 by 9.4 inches (152 × 240 mm).

    ∗ A **rare**, somewhat condensed version of the 1896 Bulletin No. 3 by the same author. There are 91 grasses described, all of which are found in the United States.

Langvad, Bjarne Johan

Bjarne Langvad (1925–1969) was born in Alesund, Norway, and earned a degree from Ås College in Oslo, Norway, in 1949, specializing in Horticulture. Subsequently, he earned an MSc degree in Horticulture from Michigan State University in 1956. His professional career focused on turfgrass research and advising while employed at the Weibullsholm Plant Breeding Institute in Landskrona, Sweden. These efforts resulted in the Weigrass Method of hydroseeding and the Weigrass sandbed root-zone construction system; in pioneering work in turfgrass soil warming via hot air, hot water, and electric heating; and in the turfgrass cultivars, Evergreen turf timothy, Sydsport Kentucky bluegrass, and Erika red fescue. Bjarne tragically died in an automobile accident.

1971† **Våra Grönytor—Anläggning och skötsel (Our Green Areas—Establishment and Maintenance).** Bjarne Langvad. LTs förlag LTK, Borås, Sweden. 184 pp., 95 illus. with 12 in color, 11 tables, small index, glossy hardbound in multi-color with black lettering, 5.3 by 7.9 inches (134 × 201 mm). Printed entirely in the Swedish language.

A book on the basics of turfgrass establishment and culture in relation to Swedish conditions. It includes sections on lawns, parks, soccer fields, golf courses, airfields, roadsides, and sod production. There is a unique section on sod roofs.

Latham, James Maston, Jr.

Jim Latham (1928–2008) was born in Hillsboro, Texas, USA, and earned both a BSc degree in Soil Science and an MSc degree in Agronomy from Texas A&M University in 1952 and 1954, respectively. He served as Assistant Turf Specialist at Georgia Coastal Plain Experiment Station from 1954 to 1956, as Regional Agronomist for the USGA Green Section from 1956 to 1960, as Turf Agronomist for the Milorganite Division of the Milwaukee Metro Sewerage District from 1960 to 1984, and as Director of the Great Lakes Region, USGA Green Section, from 1985 to 1995. Jim Latham received the Distinguished Service Award of the O.J. Noer Research Foundation in 1985 and the Distinguished Service Award of the Wisconsin GCSA in 1990 and 1994.

1966† **Winter Injury.** Coauthor.
See *Wilson, Charles Granville—1966.*

1966† **Better Fairways: Northern Golf Courses—Theory and Practice.** Coauthor.
See *Wilson, Charles Granville—1966.*

1970† **Better Bentgrass Greens—Fertilization & Management.** Coauthor.
See *Wilson, Charles Granville—1970.*

1971† **Better Fairways: Northern Golf Courses—Theory and Practice.** Coauthor.
See *w—1971.*

1972† **Permanent Fertilization Record and Handbook.** Coauthor.
See *Wilson, Charles Granville—1972.*

Lawson, David Mutrie (Dr.)

David Lawson was born in Paisley, Renfrewshire, Scotland, UK, and earned a BSc degree with *honours* in Agricultural Chemistry from Glasgow University in 1978 and a PhD in Soil Chemistry from Edinburgh University in 1982. Since 1983, he has served as a Research Chemist, and, in 1999, he became Senior Research Officer with the Sports Turf Research Institute.

1991† **Fertilisers for Turf.** Second revised edition. David M. Lawson. The Sports Turf Research Institute, Bingley, West Yorkshire, England, UK. 41 pp., 11 illus., 2 tables, no index, fold-stapled glossy semisoft covers in beige with brown lettering, 5.8 by 8.3 (147 × 209 mm).

Figure 12.49. Fertilizing a fairway in South Africa with a gravity-drop spreader drawn by a team of oxen, circa 1930s. [24]

Figure 12.50. Tractor-drawn fertilizer spreader with attached drag mat to enhance uniform fertilizer distribution, circa 1931. [2]

An update of a simplified booklet on turfgrass nutrition, soil chemical relationships, fertilizer carriers, and fertilization programs. It is oriented to climatic soil conditions in the United Kingdom.

For first edition, see *Sports Turf Research Institute—1978.*

Revised in 1996.

1996† **Fertilisers for Turf.** Third revised edition. D.M. Lawson. The Sports Turf Research Institute, Bingley, West Yorkshire, England, UK. 47 pp.; 8 illus.; 9 tables; no index; fold-stapled glossy semisoft covers in beige, black, and green with black and white lettering; 5.7 by 7.8 inches (145 × 199 mm).

An update of the 1991 second revised edition.

Leavens, George D.

George Leavens (1875–1915) was born in Rhode Island, USA. In the 1900 census, he was living in Grafton, Massachusetts, and was listed as being a farmer. He subsequently served as a Chemist in the Department of Fertilizers at the Massachusetts Agricultural Experiment Station from 1903 to 1910. He specialized in grass fertilization and eventually became a Manager, possibly of the Coe-Mortimer Company in New York City as he resided in Brooklyn, New York in 1910.

1914 **Lawns, Golf Courses, Polo Fields, Turf Courts, and How to Treat Them.** Coauthor.

See *Cunningham, Samuel Alfred—1914.*

Lee, Jae Pil (Dr. and Professor)

Jae Lee was born in Yir-rong Gun, South Korea, and earned a BSc degree in Horticulture in 1993, an MSc degree in Turfgrass Science in 1995, and a PhD in Turfgrass Management and Construction in 2003, all from Konkuk University. Since 2006, he has served as Plural Professor of the Long Life Education Center at Konkuk University and also has been Director of KV Bio Cooperation. His area of emphasis is sports field construction and maintenance.

2005 **Design, Construction, Maintenance and Management of Golf Course.** Jae Pil Lee, with twenty contributing authors. Chungyoun Publishing Company, Seoul, South Korea. 460 pp., 292 illus., 122 tables, no index, semisoft bound in silver with black lettering, 8.3 by 10.6 inches (210 × 270 mm). Printed in the Korean language.

A book concerning the design, planning, construction, turfgrass establishment, and culture of golf courses under the conditions in South Korea. It includes both the culture and managment aspects. There is a list of selected references in the back.

2007 **Question and Answer for Turfgrass, Construction, Maintenance and Management of Golf Course.** Jae Pil Lee. Hackness Press, Seoul, South Korea. 378 pp., 150 illus., 100 tables, no index, semisoft bound in blue with black lettering, 8.3 by 11.6 inches (210 × 294 mm). Printed in the Korean language.

A book oriented to the preparation of individuals who wish to take a test to become a member of a golf course maintenance and/or management staff in South Korea.

Lee, Sang Jae (Dr.)

Sang Jae Lee was born in Goseong, Gyeongnam Province, South Korea, and earned a BSc degree in 1982 and an MSc degree in 1992, both in Landscape Architecture from Cheongju University, and a PhD degree in Environmental Horticulture from Sungkyunkwan University in 1999. He has apprenticed with the New Zealand Greenkeepers Association, Robert Trent Jones II Company in the United States and Kaki and SUNA Golf Enterprise Com in Japan. He was superintendent at Gonjiam Golf Course (LG Group) in South Korea from 1985 to 1999. Since then, he has been a consultant in golf course construction in South Korea and other Pacific rim countries and the President of LSJ Golf Engineering. He also served as Adjunct Professor at Konkuk University from 2005 to 2008. Sang Lee was elected President of the Turfgrass Society of Korea for 2008 to 2009.

1994   **Golf Course Construction and Turfgrass Management.** Sang Jae Lee. Seowon Yanghang, Seoul, South Korea. 552 pp., 72 illus., 37 tables, indexed, semisoft bound in green with light green lettering, 7.5 by 10.2 inches (190 × 260 mm). Printed entirely in the Korean language.

   A book on the establishment and culture of turfgrasses on golf courses under conditions in South Korea. There is a section on landscaping golf courses with trees.

Lees, Peter Whitecross

Peter Lees (1868–1923) was born in Gullane in the parish of Dirleton in Haddingtonshire (now known as East Lothian), Scotland, UK. While not in school, he worked as a golf caddy. His first greenkeeper experience was at Archerfield Golf Club, then at Mortonhall Golf Club (Edinburg) for two years, at Braid Hills Course (Edinburg Municipal) for one year, and at Edinburg Burgess Golfing Society at Barnton for nine years. By 1904, he was Head Greenkeeper for the Royal Burgess Golfing Society, Edinburgh, Scotland, and, in 1907, became Head Greenkeeper at Mid-Surrey Golf Club, Richmond, England, UK, where he was responsible for course revisions in 1911. He is credited with developing the original earthworm irritant control, mowrah meal, for greens in the United Kingdom in the late 1890s while working as a golf greenkeeper (see Reginald Beale). Peter Lees was a founding member of the British Greenkeepers Association in 1906. He immigrated to the United States in 1914 via Pennsylvania. Peter Lees agreed to construct and then serve as greenkeeper at the Lido Golf Club on Long Island, New York, and lived from 1914 to 1923 in nearby Lynbrook. He was active as a golf construction contractor for architects such as Charles Blair Macdonald and A.W. Tillinghast and also did some golf course architecture, including Hempstead Golf and Country Club (9 holes) in 1920 to 1921; he revised Ives Hill Country Club in 1920, Garden City, and Green Meadows.

1918   **Care of the Green.** Peter W. Lees. C.B. Wilcox, New York City, New York, USA. 91 pp. plus 2 pp. of advertising, no illus., no tables, no index, cloth hardbound in tan with black lettering, 6.8 by 4.9 inches (171 × 125 mm).

   *   A **truly rare** book that is a must for collectors of historical turfgrass books. It is a small book on the construction, establishment, and culture of turfgrasses on putting greens, turfed tees, and fairways of golf courses in the United States. The last six chapters encompass a seasonal calendar of turfgrass maintenance practices. The chapter subject titles include the following:

• Course Construction	• Spring Work on the Golf Course
• Building the Green	• Early Work on the Golf Course
• Hillside Greens	• Early Summer Work on the Golf Course
• Rolling the Green	• Summer Work on the Golf Course
• Worms on the Green	• Mid-Summer Work on the Golf Course
• Care of the Green	• General Review of the Greenkeeper's Work

2006†   **Care of the Green.** Peter W. Lees. An archival reprint by Ventura Pacific Limited in a 8.1 by 10.6 inch (205 × 270 mm) format.

Lefton, Jeff Lynn

Jeff Lefton was born in Lafayette, Indiana, USA, and earned a BSc degree with *distinction* in 1967 and an MSc degree in 1973, both in Agronomy from Purdue University, studying with Dr. Bill Daniel. He served as Assistant Professor at Ohio State University from 1973 to 1977, as Turfgrass Extension Specialist at Purdue University from 1987 to 1991, and as Vice President Marketing and Technical Services of Mainscape Inc. since 1993. Jeff Lefton was elected President of the Midwest Regional Turf Foundation in 1978.

1989†   **Turfgrass Management Field Guide.** Jeff Lefton. Turfgrass Technology Center, West Lafayette, Indiana, USA. 167 pp., 9 illus., 59 tables, indexed, glossy semisoft bound in green with black lettering, 3.4 by 6.1 inches (86 × 155 mm).

   A small booklet or field guide on turfgrass culture, with emphasis on fertilization and pest management. It is oriented to professional turfgrass managers and conditions in the United States.

   Reprinted in 1990 in a green cover.

Figure 12.51. Peter Lees, circa 1914. [7]   **Figure 12.52.** Sir Bruce Levy, 1957. [23]

Levy, Enoch Bruce (Dr.; Sir and OBE)

Sir Bruce Levy (1892–1985) was born in the rural Auckland area of New Zealand. Upon completion of the civil service junior
examination, he became a clerical cadet in the Department of Agriculture, Commerce and Tourists in Wellington in
1911. He earned a BSc degree from Victoria University College in 1928. In 1928, he was appointed Head of the Plant
Research Station of the Department of Agriculture and, in 1936, became Director of the Grasslands Division, Plant
Research Bureau of the Department of Scientific and Industrial Research in Palmerston North, serving until his retirement in 1951. He was founder and Chair of the New Zealand Greens Research Committee from 1932 to 1951 and
Director of the New Zealand Institute of Turf Culture from 1951 to 1958. Sir Levy was awarded the first Life Member
of the New Zealand Institute for Turf Culture in 1974, an OBE initiation in 1950, the R B. Bennett Empire Prize of
the Royal Society of Arts in 1951, and an Honorary Doctor of Science degree from the University of New Zealand in
1953; he was Knighted by the Queen in 1953.

n.d.   **Construction, Renovation and Care of the Bowling Green.** E. Bruce Levy. New Zealand Greenkeeping Research
circa   Committee, Palmerston North, New Zealand. 84 pp., 44 illus., 4 tables, no index, cloth hard board bound in blue with
1949   gold imprint lettering, 5.5 by 8.4 inches (140 × 213 mm).
  ∗   An important, **rare** book for collectors of historical turfgrass books. This book represents two firsts: the first book
         devoted solely to the construction, establishment, and culture of lawn bowling greens and the first turfgrass book
         authored and published in New Zealand. It is written specifically for New Zealand conditions within a background of
         soundly based principles known at that time. It contains a good series of photographs. The use of the *Leptinella* (syn.
         Cotula) species on bowling greens is introduced. Dr. Levy outlined the steps in renovation of bowling greens, including
         the mud bath, as follows:

> TENTH OPERATION—THE MUD-BATH.
> For this treatment some thirty-six small barrow loads of top-soil are required for the full-sized green. This
> is barrowed on to the green on planks and is roughly spread on the green and brought to a thin, gruel-like
> consistency by application of water so that it may be readily swilled uniformly over the green with the light
> six-foot straight-edge.

n.d.†   **Construction, Renovation and Care of the Golf Course.** E. Bruce Levy, W.A. Kiely, and W.M. Horton. New Zealand
circa   Institute for Turf Culture, Palmerston North, New Zealand. 101 pp., 43 illus., 4 tables, no index, textured semisoft
1950   bound in gray with black lettering, 5.1 by 8.2 inches (130 × 210 mm).

* An important, **rare** book for collectors of historical turfgrass books. It is a soundly based book written in a how-to format. Included are design, construction, establishment, and culture of golf course turfgrasses under New Zealand conditions. It contains a chapter on golf course architecture and a section on equipment. Also presented are summaries of turfgrass research findings conducted in New Zealand. This book is the second in a series by the New Zealand Institute for Turf Culture. The status of animal grazing versus the rotary mower for golf turfgrass culture in New Zealand in 1950 is summarized by the authors:

> The standard Golf course is a vast area of land over which to establish and maintain appropriate swards of turf grasses. The elimination of the animal and its replacement by rotary mowers has to a marked degree added to the problem and cost of upkeep. Howbeit golf in its fullest attainment and realisation as a game just cannot tolerate the filth of the animal nor the irregular turf surface produced under the influence of its treading and grazing. The rotary mower with all its added deficiencies and problems is here to stay and it is necessary to fashion a system of golf course upkeep with the rotary mower as the central and main ecological factor bearing on the establishment, maintenance and care of the sward.

Lewis, Idris Giles

Idris Lewis was born in Merthyr Tydfil, Breconshire, Mid-Glamorgan, Wales, UK, and earned a National Diploma in Agriculture from University College of Wales, Abersystwyth, in 1927. He served the Board of Greenkeeping Research (BGR) as the first Advisory Assistant from 1931 to 1936. Lewis then joined with Dr. T.W. Evans, also formerly with BGR, to become Editors and Publisher of **The Agrostologist** first published in the autumn of 1936 from their offices in London. They also were active as Consulting Agrostologists to turfgrass facilities. WWII brought the cessation of these organizations in the summer of 1940. Idris Lewis subsequently formed I.G. Lewis and Company Limited of Penn, Buckshire.

1948† **Turf.** Idris Giles Lewis. Faber and Faber Limited, London, England, UK. 141 pp., 17 illus., 7 tables, small index, cloth hardbound in dark green with gold lettering and a glossy jacket in light green and white with brown lettering, 5.5 by 8.6 inches (140 × 218 mm).

* A book on the establishment and culture of turfgrasses on lawns, bowling greens, soccer fields, rugby fields, tennis courts, cricket grounds, and golf courses of England. The renovation of turfgrasses following WWII is addressed. In 1948, Idris Lewis supported the acid theory for turfgrass culture:

> One outcome of this work was the famous Acid Theory; a heated controversy which swept this country and America for several years claiming thousands of converts to the treatment which left in its wake hundreds of lawns and golf greens infinitely better than before while causing severe damage to others.
>
> In attempting to modify the soil and achieve that state which, under natural conditions, supports, a fine type of turf, it was overlooked that the agent employed, being a nitrogenous fertilizer, influenced greatly the growth of grass; an influence which in some cases proved to be not of the best. Today, the Acid Theory takes its proper place in turf maintenance as one of the factors concerned, not as the panacea for all ills. But the principle on which the theory was founded still holds good and we would do well to remember this. All available evidence supports the view that an acid medium is desirable for the development of the finest turf. The dwarf grasses thrive in acid soils which are inimical to the growth of most broad-leaved grasses and many weeds. Further, the formation of the desired mat of fibre is also encouraged in such soils.

Liffman, Karl

Karl Liffman was born in Melbourne, Victoria, Australia, and earned a Diploma in Horticultural Science from Burnley Horticultural College in 1974, a BEd degree from the Hawthorn Institute in 1989, and a Diploma in Forestry from the School of Forestry, Creswick, in 2001. His activities have emphasized education with a focus on technical training specialities. He served as Instructor of horticulture at Bendigo College of Technical and Further Education (TAFE) from 1983 to 1994, as an Instructor in Horticulture at the North Melbourne Institute of Technology from 1995 to 1998, as Field Skills Instructor in Forestry at the University of Melbourne, Creswick, from 1998 to 2006, and as Trainer and Assessor in Forestry Field Skills for Timber Training Creswick Limited since 2007.

1984† **Bowling Greens: A Practical Guide.** Karl Liffman with the assistance of Malcolm Bartlett and seven other contributors. TAFE Publications Unit, Collingwood, Victoria, Australia. 130 pp., 92 illus. with 12 in color, 13 tables, no index, stapled glossy semisoft covers in white and green with white lettering, 8.2 by 11.6 inches (209 × 294 mm).

A practical guide to the culture of lawn bowling greens, with one chapter on construction. Much of the guide is devoted to the pests of turfgrasses. It is oriented to Australian conditions.

1984† **Sports Grounds & Turf Wickets—A Practical Guide.** Karl Liffman. TAFE Publications Unit, Collingwood, Victoria, Australia. 110 pp.; 94 illus.; 8 tables; no index; glossy semisoft bound in white, green, and black with black and white lettering; 8.2 by 10.9 inches (209 × 278 mm).

A small, practical guide on the construction, establishment, culture, and pest management of turfgrasses on sports fields, including such sports as Australian-rules football, cricket, rugby league, rugby union, and soccer. It is oriented to Australian conditions. There is a small glossary of terms and a list of selected references in the back.

Reprinted in 1986.

Lock, Herbert Christmas

Bert Lock (1903–1978) was born in East Molesey, Surrey, England, UK. He was a professional cricketer and groundsman for the Surrey County Cricket Club from 1922 to 1932 and for the Devon County Cricket Club from 1934 to 1939. He competed as a medium-paced, right-arm bowler at the first XI level in 1926 and from 1928 to 1932 plus at the second XI level in 1924 to 1932. During WWII, he served in the Royal Air Force from 1939 to 1945. He subsequently was Head Groundsman for the Surrey Cricket Club at the Kennington Oval from 1945 to 1964. He served as Advisor on Pitches for the Test and County Cricket Board of England, UK. He was elected Chairman of the National Association of Groundsman from 1959 to 1962. Bert Lock was awarded the Queen's Jubilee Medal in 1977.

1957† **Cricket Ground Maintenance.** First edition. H.C. Lock. Surrey Junior Cricket Committee, Kennington, Surrey, England, UK. 33 pp. plus 10 pp. of advertising, 15 illus., 3 tables, small index, fold-stapled semisoft covers in green with black lettering, 5.6 by 8.0 inches (141 × 203 mm).

A small book concerning the construction and culture of cricket grounds. It includes construction and culture of cricket squares and preparation of wickets for competition. There is a calendar for groundsmen in the back.

Revised in 1963 as Lawn and Turf Culture for Gardens and Grounds.

1963† **Lawn and Turf Culture for Gardens and Grounds.** Second revised edition. H.C. Lock. Surrey County Cricket Club, Wimbledon, Surrey, England, UK. 56 pp. with 18 pp. of advertising, 15 illus., no tables, small index, fold-stapled semisoft covers in light green with brown lettering, 4.8 by 7.2 inches (123 × 183 mm).

A small booklet on the construction, establishment, and culture of turfgrasses for lawns and cricket pitches or squares. It is oriented to cool-season turfgrass conditions in the United Kingdom.

1972† **Cricket—Take Care of Your Square.** First edition. H.C. Lock. EP Publishing Limited and National Cricket Association, Wakefield, Yorkshire, England, UK. 33 pp.; 15 illus.; 3 tables; no index; fold-stapled glossy semisoft covers in white, green, and black with green, orange, and black lettering; 5.2 by 8.0 inches (133 × 203 mm).

A small guide to the construction, establishment, and culture of the cricket wicket that is oriented to the cool-season turfgrass conditions in England. It includes a maintenance calendar. The foreword is by A.E.R. Gilligan.

Revised in 1987.

1987† **Cricket—Take Care of Your Square.** Second revised edition. H.C. Lock and P.L.K. Dury. National Cricket Association, England, UK. 32 pp., 29 illus., 2 tables, no index, fold-stapled semisoft covers in multicolor with green and black lettering, 5.8 by 8.2 inches (148 × 208 mm).

It contains some revisions, especially to the illustrations.

Luff, Richard Ten Eyck

Richard Luff was born in Exeter, New Hampshire, USA, and earned a BA degree in History and Geography from the University of Vermont in 1990. He has worked as Superintendent from 1990 to 1994 and General Manager from 1995 to 2001 at Sagamore-Hampton Golf Club and then as President of Sagamore Golf Inc. since 2002. Richard Luff received the Paul Harris Fellowship Award from the Rotary International Club, Portsmouth, New Hampshire, in 2003 and the Leadership Award from the First Tee of New Hampshire in 2007.

2002† **Ecological Golf Course Management.** Coauthor.

See *Sachs, Paul David—2002.*

The Luther Burbank Society

1915 **Beautify the Lawn to Reflect Your Personality.** Plant Life Series No. 9, The Luther Burbank Society, Luther Burbank Press, Santa Rosa, California, USA. 29 pp., no illus., no tables, no index, 5.2 by 7.7 inches (132 × 195 mm).

＊ An early, general booklet concerning the planting and care of lawns. It is oriented to lawn enthusiasts in the United States.

M

Mabee, Carleton Hart (DDS)

Carl Mabee was born in Winchester, Massachusetts, USA, and earned a BSc degree *cum laude* from Northeastern University in 1967 and a Doctor of Dental Surgery (DDS) degree from Fairleigh-Dickinson University in 1971. He is a dentist by profession. An avid croquet player, he was named United States Croquet Association (USCA) Rookie of the Year in 1988. He won the Croquet Singles National Championship in 1994 and the Croquet Doubles National Championship in 1994 and 1996. Dr. Mabee has served as Chairman of the USCA Courts and Greens Committee.

1991† **A Guide to Croquet Court Planning, Building, & Maintenance.** Carleton H. Mabee. Bass Cover Books, Kennebunkport, Maine, USA. 120 pp., 120 illus., 15 tables, indexed, glossy semisoft bound in greens and white with green and black lettering, 10.8 by 8.3 inches (275 × 211 mm).

The first book published that is devoted solely to turfgrass culture of lawn croquet courts. Site selection, construction, planting, culture, pest, and equipment aspects are addressed. It includes a small glossary and a selected set of references in the back. The book is oriented to climatic conditions in the United States. There is a preface by Jack R. Osborn.

MacDonald, I. James C.

James MacDonald (1855–1930) was born in Huntingtower, Perthshire, Scotland, UK, and studied at Perth Academy, where he embarked on a study and collection of the Graminae in the Perthshire area. He moved south to work at the nurseries of Messrs. Dicksons of Chester and later at Messrs. James Veitch and Sons at Chelsea. James MacDonald then decided to specialize in turfgrasses. He founded a firm of grass specialists located at Harpenden, Hertfordshire, England, UK, and developed a turfgrass research nursery at Harpenden. He is acknowledged in the foreword by R.V. Giffard Woolley, editor of **The Garden**, as having "the distinction of being the first Englishman to devote the whole of his energies to the cultivation of lawn grasses and the production of beautiful greensward."

1923† **Lawns, Links, & Sportsfields.** James MacDonald. Country Life Library Series, Country Life Limited and George Newnes Limited, London, England, UK, and Charles Scribner's Sons, New York City, New York, USA. 78 pp., 25 illus., no tables, indexed, cloth hardbound in green with white and dark green lettering, 4.7 by 7.2 inches (121 × 184 mm).

* A **very rare** book that is a must for collectors of historical turfgrasses books. An interesting early book oriented to English systems of turfgrass establishment and culture on lawns, golf courses, tennis courts, croquet courts, cricket grounds, football fields, hockey fields, bowling greens, and polo grounds. There is a foreword by R.V. Giffard Woolley. The chapter subject titles include the following:

I.	Drainage	X.	Golf Courses (Teeing Grounds)
II.	Cultivation	XI.	Tennis Courts and Croquet Lawns
III.	Leveling	XII.	Cricket Grounds
IV.	Seeding a Lawn	XIII.	Football Grounds and Hockey Fields
V.	The Selection of Seeds	XIV.	Bowling Greens
VI.	Turfing Lawns	XV.	Polo Grounds
VIII.	Mowing and Rolling	XVI.	Lawn Pests and Troubles
VIII.	Golf Courses (General)	XVII.	Manures and Manuring
IX.	Golf Courses (Putting Greens)		

Regarding mowing turfgrasses in 1923 by the manual scythe versus the mechanical reel mower, MacDonald stated the following:

> With the advent of the lawn mower, a great change occurred in the treatment and condition of lawns. The scythe was an implement which required a very considerable amount of skill and concentration of energy to keep the turf to the required smoothness. With the lawn mower the work is of a more mechanical nature, and the lawns have suffered accordingly. About the middle of the last century it was one of the most exhilarating sights possible to see in the early morning a band of mowers swinging their scythes in unison as they cut the grass on the lawns whilst it was yet wet with dew.

n.d.† **Lawns—Sports Grounds—Meadow Lands.** James MacDonald. Harpenden, Hertfordshire, England, UK. 20 pp.,
circa 12 illus., no tables, no index, fold-stapled semisoft covers in cream with black lettering, 4.7 by 6.4 inches (120 ×
192x 163 mm).

Figure 12.53. James MacDonald, circa 1920s. [28]

Figure 12.54. MacDonald's turf nursery in Harpenden, England, UK, circa 1922. [28]

*   A **very rare**, small booklet on the establishment and culture of turfgrasses for lawns, tennis courts, bowling greens, and golf courses under the cool-season turfgrass conditions in England. This booklet consists of a series with more than five editions.

Mackenzie, Alister (MD)

Alister Mackenzie (1870–1934) was born in Normanton, Yorkshire, England, UK, and earned a BA degree in Natural Science in 1891 and Bachelor of Medicine, Bachelor of Surgery, and MA degrees in 1897, all from Cambridge University. He enrolled in the university under the name Alexander MacKenzie. Also, he became a member of the Royal College of Surgeons and a licentiate of the Royal College of Physicians. From 1899 to 1902, he served as a Civil Surgeon with the Somerset Light Infantry during the Boer War in South Africa. Dr. Mackenzie practiced medicine as House Surgeon at the General Infirmary, Leeds, England. Between 1900 and 1910, he became gradually more interested in the design of golf courses. During WWI, he returned to medicine as an Army Surgeon but soon transferred to the Royal Engineers where Major Mackenzie pioneered innovative camouflage techniques and was awarded the Order of St Stanisloff. Then in 1918, Dr. Mackenzie joined in a partnership with H.S. Colt and C.H. Alison that was gradually dissolved. He was involved in the design of more than 400 golf courses and also did some advising on turfgrass culture for golf courses. Dr. Mackenzie settled in the United States in 1925 at Pasatiempo in Santa Cruz, California, USA. During 1926 and 1927, he embarked on a world tour involved in golf course design in Australia, New Zealand, and South America.

1920†   **Golf Architecture—Economy In Course Construction And Green-Keeping.** Dr. A. Mackenzie. Simpkin, Marshall, Hamilton, Kent, and Company Limited, London, England, UK. 135 pp., 21 illus., no tables, no index, cloth hardbound in dark green with black lettering, 4.0 by 6.3 inches (102 × 161 mm).

* A **very rare**, important book on golf course architecture and construction written by one of the great architects. The introduction is by his "senior partner" Harry S. Colt. It encompasses four subject chapters titled as follows:

> I.   General Principles of Economy in Course Construction and Green-Keeping
> II. Some Further Suggestions
> III. Ideal Holes
> IV. The Future of Golf Architecture

Dr. Alister Mackenzie emphasized the merits of designing a golf course by taking advantage of the natural features of a site rather than using extensive soil excavations. He practiced his art worldwide. The book emphasizes English golf courses. There also is a small section on golf turfgrass maintenance. An early classic book that is a must for collectors of historical golf books. Dr. Mackenzie gave the following thoughts concerning earthworm problems and their control on golf courses:

> It is now an absolutely exploded fallacy that worms are of any use on a golf course; they should be got rid of by the use of charcoal obtained from steel furnaces: ordinary wood charcoal is almost useless. Charcoal in this form acts mechanically, owning to the small sharp pieces of steel attached to it: it scratches the worms and prevents them getting through.

† *Reprint by Grant Books Limited—1982.*

1982† **Dr. Mackenzie's Golf Architecture.** Alister Mackenzie. Grant Books Limited, Droitwich, Worcestershire, England, UK. 65 pp., 18 illus. with 12 in color, no tables, no index, cloth hardbound in green with gold and gilt lettering, 6.7 by 9.4 inches (169 × 238 mm).

A reprinting of Dr. Alister Mackenzie's classic 1920 book **Golf Architecture—Economy In Course Construction And Green-Keeping.** Included is an introduction by Robert Trent Jones and a commentary by Peter Thomson and Michael Wolveridge. It also contains a short 1932 article by Dr. Mackenzie titled "Plans for the Ideal Golf Course" and another article titled "Hints on Green-keeping." All the photographs have been replaced with different ones. It was a limited reprinting of 700 copies. Dr. Mackenzie stated some of his philosophy of golf architecture in 1920:

> Never follow the advice of a golfer, however good a player he may be, unless he is broad minded enough to disregard his own game and recognise that not only has the beginner to be considered, but also that a very high standard of golf architecture improves everyone's play.
>   Golf Course construction is a difficult art (like sculpture) and still in its infancy. Endeavor to make every feature indistinguishable from a natural one.
>   Most courses have too many bunkers. They should be constructed mainly from a strategical and not from a penal point of view.
>   Fiercely criticised holes often improve the standard of play and ultimately become most popular.

See original printing, *Mackenzie, Alister—1920.*

1987† **Golf Architecture.** Reprint. A. Mackenzie. The Classics of Golf, Ailsa Inc., Stamford, Connecticut, USA. 206 pp., 35 illus., 2 tables, no index, cloth hardbound in dark green with black lettering, 5.0 by 6.9 inches (126 × 176 mm).

A reprinting of the early classic of Dr. Alister Mackenzie titled **Golf Architecture.** Added are a foreword by Herbert Warren Wind and an afterword by Lewis A. Lapham. It also has an appendix containing a 1928 prospectus of the Cypress Point Golf Club, including the original Mackenzie layout and 11 photographs of the sites for various holes and greens. There is a photograph of Dr. Alister Mackenzie in the front.

1995† **The Spirit of St. Andrews.** Alister Mackenzie. Sleeping Bear Press, Chelsea, Michigan, USA. 269 pp., 67 illus., 3 tables, no index, cloth hardbound in dark green with gold lettering and a glossy jacket in multicolor with black lettering, 6.2 by 9.0 inches (157 × 228 mm).

A book written by Dr. Alister Mackenzie prior to his death in 1934 but not published until discovered 60-plus years later in 1995. He discusses the evolution of golf, general principles of golf course architecture, economy in construction, ideal golf holes, and greenkeeping. Photographs from the 1930s have been added. There is a foreword by Robert T. Jones Jr. Dr. Mackenzie describes the character of golf courses in the 1890s as follows:

> The more money these clubs have had to spend, the more their courses have deteriorated. Not only has the turf been ruined, but there has been a wanton destruction of many of the natural features. The greens have been flattened out, sand hazards which created the interest and strategy of the holes have been

filled up, and in too many cases even the undulations of the fairways have been destroyed. The features of the links land, which the modern golf course architect attempts to imitate, sometimes with indifferent success, have disappeared, so that these glorious natural courses have only too frequently become as dull and insipid as a second rate inland course.

There was a course in the wilds of the west coast of Scotland among most spectacular sand dunes named Machrihanish. It was scores of miles from any railway station and the only way to get to it was by steamboat and a long carriage drive. Notwithstanding its inaccessibility, the course was so good, and the climate so bracing, that it became very popular and attracted an increasing number of men who had reputations as players. Some of these players were open or amateur champions, and the natives, dazzled by their reputations, avidly agreed to their suggestions for "improving the course." I have been visiting Machrihanish at intervals for over thirty years and each time I have found that the course has appreciably deteriorated.

When I made my first visit, the course was kept by one greenkeeper, it was not mown except by the rabbits, nor was it rolled. The greenkeeper's duties consisted in simply cutting the holes and filling up any scrapes the rabbits might have made on the fairways and greens. This was rarely necessary, as rabbits very seldom scrape well-trodden turf, but confine their attentions to bare places on the outskirts of the course. Even the depredations of the rabbits therefore involved remarkably little labour, and in these early days the turf on the greens and fairways was superior to any I have seen before or since. There was a complete freedom from weeds, daisies, or worm casts, and one's ball sat up on the closely cropped turf in a remarkable manner. Each hole was an adventure. There were no guiding flags and no fixed routes. One could frequently beat an opponent who had greater length and skill by superior strategy.

The course was planned by Old Tom Morris. The annual subscription was ten shillings and there was no initiation fee.

1997 Published in Japanese by Quatre-Quart Company Limited, Japan, as translated by Ko Sakota. 162 pp.
2001 Published in Chinese by Shang Zhou Chu Ban, Taipei, Taiwan, as translated by Chunyue Dai. 302 pp.

Macself, Albert James

Albert Macself (1878–1952) was born in West Derby, Lancashire, England, UK. He began horticultural apprenticeship by working in various nurseries in England and Scotland. In 1907, he embarked on his journalistic career with the **Gardeners Magazine** by joining the advertising staff; he then moved to the **Horticultural Trade Journal.** He subsequently served as Editor of **Amateur Gardening** for 20 years, from 1926 to 1946. Albert Macself received the Royal Horticultural Society Victoria Medal of Honour in Horticulture in 1950.

1924† **Grass—A new and thoroughly practical book on Grass for ornamental Lawns and all purpose of Sports and Games.** A.J. Macself. Cecil Palmer, London, England, UK. 204 pp., 64 illus. with 1 in color, no tables, small index, cloth hardbound in light green with gold lettering and a jacket in multicolor with brown lettering, 5.5 by 8.6 inches (140 × 218 mm).

  * A **rare**, early book that is a must for collectors of historical turfgrass books. It is an elementary book devoted primarily to lawn and sports field establishment and culture under the cool-season turfgrass conditions in England. The chapter subject titles include the following:

I.	Grass	XI.	Weeds and their Eradication
II.	On the Preliminary Treatment of Soil	XII.	More Concerning Weeds
III.	Further Details Regarding Soil Preparation	XIII.	A Few Other Unwelcome Intruders
IV.	Turf	XIV.	Insect Enemies of Grass
V.	Lawn Making from Seeds	XV.	Tennis and Croquet Lawns
VI.	The Care of Young Seedling Grass	XVI.	Bowling Greens
VII.	The Finer Grasses	XVII.	Cricket Grounds
VIII.	A Few Grasses of Lesser Importance	XVIII.	Football and Hockey Pitches
IX.	Concerning the Nourishment of Established Grass	XIX.	Golf Links and Putting Greens
X.	Mowing, Rolling, and Other Details of Routine Work	XX.	A Chapter of Miscellanies

It contains numerous botanical drawings of turfgrasses and weeds by G.E. Lee. In 1924, Albert Macself advised on the use of horse boots as follows:

> Horse-boots are an essential wherever animals are used on grass, and it is imperative that these shall fit well and be kept in good repair. A badly-fitting or loosely fastened boot may easily cause as much damage as an unshod hoof.

n.d.†   **Lawns and Sports Greens.** First edition. A.J. Macself. W.H. & L. Collingridge Limited, London, England, UK. 144
circa   pp., 40 illus., no tables, no index, cloth hard stitch-bound in light green with burgundy lettering and a jacket in greens
1930   and white with black lettering, 4.5 by 7.0 inches (114 × 178 mm).

*   A **relatively rare**, small book covering turfgrass establishment and culture, with emphasis on lawns. It is oriented to the amateur home lawn enthusiasts in England. One chapter briefly discusses tennis courts, bowling greens, croquet courts, putting greens and clocks, cricket grounds, and football grounds. In 1930, Albert Macself recommended sweeping English lawns:

> Sweeping a lawn is always beneficial and will free the grass from many impurities which choke the breathing pores of the blades and smother the crowns, thereby hampering the development of new growth.

Revised in 1947.

1947†   **Lawns and Sports Greens.** Second revised edition. A.J. Macself. W.H. & L. Collingridge Limited, London, England, UK. 103 pp., 45 illus., no tables, small index, cloth hardbound in green with dark green lettering and a jacket in greens and white with black and brown lettering, 4.8 by 7.2 inches (122 × 184 mm).

*   This small book on turfgrass culture has been significantly revised after the effects of WWII, including photographs and the addition of an index. The status of the hand scythe versus the mechanical mower in 1947 was summarized by Albert Macself:

> A century ago every gardener had to make himself competent in the art of wielding and honing a scythe, and even now those of us who write continue to advise that young seedling grass should be scythed once or twice before being subjected to mowing by machine. That advice is theoretically good, but from the severely practical standpoint it stands sadly at a discount, for the simple reason that men who are capable of deftly severing the tops of soft young grass with a scythe are becoming almost as scarce as are ladies who sit at the spinning wheel.

Madden, Ernest Alexander

Ernie Madden (1903–1967) was born in Aukland, New Zealand, and earned a Diploma in Agriculture (HDA) degree from Hawkesbury Agricultural College of Richmond, New South Wales, Australia, in 1923. Initially, he was involved in practical dairy farming in New Zealand. He became Assistant Instructor in Agriculture at Whangamomona, Taranaki, in 1928; became Assistant Agrostologist in 1929; was appointed Agrostologist in 1937; and rose to Assistant Director of the Grasslands Division in 1940, all within the Plant Research Bureau for the New Zealand Department of Scientific and Industrial Research. He served until 1951 and, subsequently, became Agrostologist for the Department of Agriculture in Palmerston North from 1951 to 1962. He pioneered aerial grass seeding and fertilizing of pastures. During WWII he served in the Works Department and was involved in aerial surveillance, grass runway construction, and improvement for aerodromes in the South Pacific. He originated the idea of camouflaging grass airfields overseas during the war by dyeing and treating them with fertilizers so they looked like paddocks. For his pioneering contributions, Ernie Madden was awarded Companion of the Royal Aeronautical Society in 1955. He also served as Deputy Director of the New Zealand Greenkeeping Research Committee from 1936 to 1964. He supervised the first New Zealand turfgrass research at Manawatu Golf Club starting in 1932, and, subsequently, at Hokowhitu Golf Coarse in Palmerston North. He also served as examiner for the Diploma in Turf Culture. Ernie Madden was awarded Life Membership in both the New Zealand Golf Greenkeepers Association and the New Zealand Institute for Turf Culture.

1952†   **Garden Lawns and Playing Greens.** E.A. Madden. Whitcombe and Tombs Limited, Christchurch, New Zealand. 93 pp.; 29 illus.; 5 tables; indexed; cloth hardbound in green with black lettering and a jacket in green, black, and white with black and green lettering; 5.5 by 8.5 inches (140 × 216 mm).

*   A **rare** book on the establishment and culture of lawns. This practical book is oriented to the lawn enthusiasts and conditions in New Zealand. In 1952, Ernie Madden took the following position on clippings removal versus return to the turfgrass:

> There is much controversy whether or not the clippings should be left on the lawn. They may decompose and provide plant food, but this can be better supplied by artificial fertilizers.

Figure 12.55. Tractor-drawn West Point core cultivator being utilized on fairway turfgrass, circa 1960. [29]

The clippings may have a mulching effect and prevent loss of soil moisture by evaporation, but ultimately they may cause an opening of the turf by smothering some of the grasses. Another factor is that during the decomposing or rotting process favourable conditions are created for fungous diseases. The removal of the clippings will ensure the removal of many seeds of such weeds as *Poa annua* and suckling clover. The removal of clippings is definitely recommended.

Madison, John Herbert, Jr. (Dr.)

John Madison (1918–2005) was born in Burlington, Iowa, USA. He shared his childhood between Burlington, Iowa, and
    Hartford, Connecticut. In Hartford, he attended vocational school and was trained as a machinist before going on to
    Oberlin College in 1939. He left Oberlin in 1941 to work as a tool and die maker for Colt Arms in Connecticut; he
    then entered the US Army Corps in 1944 and served the remainder of the war as an airplane mechanic. After completing an AB degree in Botany from Oberlin College in 1946, he started a career in farming in Ohio and later in New
    York state. He subsequently earned a PhD degree in Plant Physiology from Cornell University in 1953. Dr. Madison
    joined the faculty at the University of California at Davis in 1953, serving as an Assistant and Associate Professor of
    Landscape Horticulture through 1982; he was involved in turfgrass research and teaching activities. He was passionate
    about growing plants, was instrumental in founding the UC Davis Arboretum, and was a frequent speaker and adviser
    on horticulture in California. Dr. Madison received the GCSAA Distinguished Award in 1980. Following retirement
    from UC Davis, he lived in Mendocino County where he was active in the California Native Plant Society, editing its
    newsletter, leading nature walks, and writing botanical field guides to interesting sites. At the age of 80, he returned to
    Ithaca, New York, and bought a farm where he planted a wide selection of heirloom apple cultivars.

1971†   **Principles of Turfgrass Culture.** John H. Madison. Van Nostrand Reinhold Company, New York City, New York,
    USA. 420 pp., 121 illus., 49 tables, indexed by author and by subject, cloth hardbound in green with gold lettering and
    a glossy jacket in multicolor with white and black lettering, 6.0 by 8.9 inches (152 × 226 mm).

    A basic text on turfgrass science covering turfgrass botany, soils, fertility, irrigation, and drainage. The discussions are
    supported by references. A list of references is presented at the back of each chapter. A glossary is included in the back.
    This book is essentially the first of a two-volume set, although not identified as such.

1971†   **Practical Turfgrass Management.** John H. Madison. Van Nostrand Reinhold Company, New York City, New York,
    USA. 466 pp., 138 illus., 49 tables, indexed by author and by subject, cloth hardbound in green with gold lettering and
    a glossy jacket in multicolor with white and green lettering, 6.0 by 8.9 inches (152 × 226 mm).

    A basic text on turfgrass science covering turfgrass species, cultivars, mowing, auxiliary practices, insects, diseases,
    and weeds. The discussions are supported by references. Dr. Madison often introduces his personal philosophy. There
    is a list of references at the back of each chapter. A small glossary is included in the back. This book is essentially the
    second of a two-volume set, although not identified as such.

    Reprinted in 1982 by R.E. Krieger Publishing Company, Malabar, Florida, USA.

Maki, Yoshisuke (Dr. and Professor)

Yoshisuke Maki (1924–2007) was born in Yamagata Prefecture, Japan, and earned a BSc degree in Agronomy from the Tokyo University of Education in 1954; an MSc degree in Plant Breeding from the University of Wisconsin, Madison, in 1961; and a PhD degree in Grass Breeding from Kyushu University in 1965. He served as Technical Officer of Agronomy for the Japanese Ministry of Agriculture, Forestry and Fisheries from 1955 to 1981; as Professor of Agronomy at Akita Prefectural College of Agriculture from 1981 to 1990, which included responsibilities such as Director of the Library from 1983 to 1984 and Dean of Students from 1986 to 1989; and, since then, as President of Maki Turfgrass Doctors International Limited. He was elected East Asia Director of the International Grassland Congress from 1975 to 1985, President of the International Turfgrass Society (ITS) from 1985 to 1989 while serving on the ITS Board of Directors from 1981 to 1993, President of the Japanese Society of Turfgrass Science from 1986 to 1990, Vice President of the Japanese Society of Golf Science from 1989 to 1995, Vice President of the Turfgrass Research & Development Organization of Japan in 1990, and Vice President of the Japanese Society of Environmental Research on Green Tech for 1990 to 1991. Dr. Maki received the Scientific Award from the Japanese Society of Grassland Science in 1974 and the Distinguished Service Award from the Japanese Society of Turfgrass Science in 1996.

1991† **Bentgrass—Characteristics and Newest Methods of Golf Green Construction and Maintenance.** Yoshisuke Maki, Hisashi Yanagi, and Sho Okubo (editors) with 27 contributing authors. Soft Science Inc., Tokyo, Japan. 345 pp.; 327 illus. with 140 in color; 88 tables; indexed by author and by subject, cloth hardbound in green with gold lettering and a glossy jacket in multicolor with yellow, black, and white lettering; 7.0 by 10.1 inches (180 × 256 mm). Printed entirely in the Japanese language.

An extensive book on bentgrass (*Agrostis* species) putting greens for golf courses in Japan. It encompasses cultivar selection, root-zone construction, turfgrass establishment, culture, and specific cultural systems. There is an extensive review of the literature.

1992† **Turf Development and Management for Turfgrass Managers.** Yoshisuke Maki (editor) with 19 contributing authors. The Association of the National Farm-Village Education, Tokyo, Japan. 292 pp., 200 illus. with 110 in color, 61 tables, no index, textured hardbound in dark green with silver lettering and a jacket in multicolor with black and white lettering plus a slip case in beige with black lettering, 7.0 by 10.4 inches (188 × 265 mm). Printed in the Japanese language.

A book on the fundamentals of turfgrass science and culture. Included are the planning, construction, establishment, and culture of turfgrasses on golf courses and sports fields under the conditions in Japan.

Reprinted in 1993.

1997† **Terminology of Turfgrass Maintenance.** First edition. Yoshisuke Maki. IKKI-Shuppan, Tokyo, Japan. 394 pp., 31 illus., 2 tables, English index, textured semisoft bound in brown with gold lettering plus a slip case in tan with black lettering, 4.3 by 6.9 inches (110 × 176 mm). Printed in the Japanese language, with each term also listed in English.

A dictionary of terms employed in turfgrass science and culture as used in Japan. There is a list of selected references in the back.

Revised in 2001.

1997† **A Handbook of Turfgrass and Turf.** Coeditor.

See *Kitamura, Fumio—1997.*

2001† **The Terminology of Turfgrass Management.** Second revised edition. Yoshisuke Maki. IKKI Publishing Company, Tokyo, Japan. 400 pp., 83 illus., 2 tables, extensive index in English, glossy semisoft bound in brown with gold lettering plus a slip case in tan with black lettering, 4.3 by 6.9 inches (110 × 175 mm). Printed in the Japanese language.

An update of the 1997 first edition.

Malaysian Golf Association (MGA)

The MGA is an organization of golf clubs in Malaysia. It was formed in 1962, with the first president being Henry H.S. Lee. It is headquartered in Kuala Lumpur, Malaysia. The MGA sponsors an annual conference and publishes **FORE** quarterly.

1984† **Golf Course and Turf Grass Management.** James B Beard, Ronald W. Fream, and Chan Heun Yin. Malaysian Golf Association, Kuala Lumpur, Malaysia. 109 pp., 18 illus., 21 tables, no index, glossy semisoft bound in multicolor with red and black lettering, 7.1 by 9.9 inches (181 × 251 mm).

A manual summarizing the lectures presented during the first turfgrass seminar held in Malaysia. This seminar represented the first formal effort at adult education for golf course greenkeepers in southeast Asia.

Mansfield, Peter John (BEM)

Peter Mansfield was born in Bethnal Green, Middlesex, England, UK. After service in the Royal Marines during WWII, he followed in his father's horticultural footsteps. His apprentice training was as a gardener for various park municipalities plus gaining the National Diploma in Turfculture from the Institute of Groundsmanship in 1969. He was Ground's Supervisor for Marlborough College from 1968 to 1997 and served as Chairman of Education for the Institute of

Groundsmanship and as an Advisor for the National Cricket Advisory. Peter Mansfield received the British Empire Medal in 1990.

n.d.†
circa
1980
Maintenance of the Club Cricket Ground. Peter Mansfield. The Cricketer International, Tunbridge Wells, Kent, England, UK. 26 pp. with 3 pp. of advertising; 21 illus.; no tables; no index; fold-stapled glossy semisoft covers in green, white, and black with black lettering; 5.9 by 8.3 inches (150 × 211 mm).

    A small booklet on the layout and maintenance of a cricket square or table and of the individual wickets. There is a small section on the cricket outfield. It is oriented to cool-season turfgrass conditions in the United Kingdom.

Markuson, Miner John (Associate Professor)

Miner Markuson (1896–1959) was born in Faribault, Minnesota, USA. During WWI, he served with the US Army Engineers from 1918 to 1919. He then earned a BSc degree in Architecture from the University of Minnesota in 1923. He taught architecture structure and surveying at Virginia Polytechnic Institute from 1923 to 1925, then became an Assistant Professor of Agricultural Engineering at Massachusetts State College, and subsequently advanced to Associate Professor. He retired in 1957, after 32 years teaching structure drawing, drainage, and house planning courses. Miner Markuson was elected President of the Western Massachusetts Architectural Society and was on special assignment with the War Department in 1941.

1933† **Soil Management For Greenkeepers.** Coauthor.
   ✳ See *Cubbon, Miles Hugo—1933.*

> If water is present in large amounts, it fills the air spaces and stops plant roots from seeking food at greater depths. Then in a dry year the plants are not able to adjust themselves quickly enough to get necessary moisture for normal growth. Continuous tramping over a wet area produces a hard imervious condition which discourages plant growth. This action is called puddling. If air is present, the pudding process is not serious.

Martin, David Paul (Dr.)

Dave Martin was born in New Holland, Pennsylvania, USA, and earned a BA degree in Biology from Goshen College in 1966, an MSc degree in Crop Ecology and Turfgrass from Michigan State University in 1970, and a PhD degree in Crop Physiology and Turfgrass from Michigan State University in 1972, the latter two while studying with Dr. Jim Beard. He served on the Agronomy Faculty at Ohio State University from 1973 to 1979 and then joined TruGreen/ChemLawn where he served as Director of Research from 1980 to 1998. He then joined Grace Brethren International Missions as Administrator for Argentina living in Don Bosco. He was elected Turfgrass Science Division C-5 Chair in 1984 and to the Crop Science Society of America Board of Directors from 1983 to 1985. Dr. Martin was

Figure 12.56. Early motor-powered Verti-cut, Pennsylvania, USA, 1953. [4]

awarded the Ohio Turfgrass Foundation Man of the Year Award in 1979 and the American Library Association Oberly Award in 1979.

1977† **Turfgrass Bibliography—From 1672 to 1972.** Coauthor.
See *Beard, James B—1977.*

Mascaro, Thomas Charles

Tom Mascaro (1915–1997) was born in Philadelphia, Pennsylvania, USA, and earned an Associate degree in Specialized Business from the Lansdale School of Business in 1936. He also gained experience as an apprentice engineer at the George Clouser Tool and Die Company in West Point, Pennsylvania. In 1936, he formed the West Point Products Corporation in West Point, Pennsylvania, and served as its President for 33 years. He pioneered the production and marketing of potting soil mixes for plants and subsequently was active in the invention of specialized mechanical equipment for turfgrass culture. Tom Mascaro was awarded 12 US patents, including the first practical Aerifier for turfgrass core cultivation (1946), the Verti-cut for vertical cutting (1952), the power drag mat, and a battery powered greensmower. He lectured extensively at turfgrass conferences across North America, especially in their formative years, and made a further contribution by recording, publishing, and donating to attendees the proceedings of numerous early turfgrass conferences in the 1950s. He also published **West Pointers.** In 1969, West Point Products Corporation merged with Hahn Inc. with Tom serving as Head of Product and Market Development of the Turf Division. Tom moved to Florida in 1974 and formed the Turfgrass Products Company in 1976. Tom Mascaro was cofounder and was subsequently elected President of the Pennsylvania Turfgrass Council for 1961 through 1963. He received the USGA Green Section Award in 1971 and the GCSAA Distinguished Service Award in 1976.

1955† **Improving Athletic Field Turfgrass.** First edition. Dr. Fred V. Grau. Product Use Bulletin No. 2, West Point Products Corporation, West Point, Pennsylvania, USA. 27 pp., 21 illus., no tables, no index, fold-stapled soft covers in yellow with black lettering, 6.0 by 9.0 inches (152 × 229 mm).

A **rare**, small handbook on the establishment, culture, and renovation of turfgrasses on sports fields, with emphasis on the maintenance aspects. Authored by Tom Mascaro, although not acknowledged as such in the text.
Revised circa 1959.

1958† **Fall Renovation of Greens and Fairways.** First edition. Tom Mascaro. Product Use Bulletin No. 3, West Point Products Corporation, West Point, Pennsylvania, USA. 62 pp., no illus., no tables, no index, stapled glossy semisoft covers in yellow with black lettering and a stapled side, 5.9 by 10.9 inches (214 × 278 mm).

A **rare**, early booklet on the methods of renovating established turfgrasses by coring and vertical cutting. The first section addresses approaches for the analysis of turfgrass problems. The author was the inventor and manufacturer of the first truly successful, mechanically powered coring and vertical cutting machines.
Revised in 1960.

n.d.† **Improving Athletic Field Turfgrass.** Second revised edition. Tom Mascaro. Product Use Bulletin No. 2, West Point
circa Products Corporation, West Point, Pennsylvania, USA. 24 pp.; 21 illus.; 1 table; no index; fold-stapled glossy soft covers
1959 in yellow, black, and white with black lettering; 6.0 by 9.1 inches (152 × 232 mm).
There are minor revisions.

1960† **Fall Renovation of Greens and Fairways.** Second revised edition. Tom Mascaro. Product Use Bulletin No. 3, West Point Products Corporation, West Point, Pennsylvania, USA. 48 pp., 16 illus., no tables, no index, fold-stapled semisoft covers in tan with green lettering, 6.0 by 9.1 inches (152 × 231 mm).
The revision consists primarily of the addition of black and white photographs.

1964† **Turfgrass Course.** First edition. Tom Mascaro and H.B. Musser. The Professional Golfer's Association of America, Lake Park, Florida, USA. 105 pp., 20 illus., 5 tables, no index, glossy semisoft bound in yellow with gold lettering, 8.0 by 11.0 inches (203 × 279 mm).

A simplified, small book designed as part of a training program for golf course superintendents. There are sections on turfgrass selection, establishment, culture, pest control, and problem diagnosis. This book began as a correspondence course for salesman authored by Tom Mascaro.
Revised in 1974.

1974† **Turfgrass Course.** Second revised edition. Tom Mascaro, H.B. Musser, and Warren Bidwell. The Professional Golfer's Association of America, Lake Park, Florida, USA. 118 pp., 20 illus., 5 tables, no index, glossy semisoft bound in yellow with black lettering, 8.0 by 9.9 inches (202 × 252 mm).

A simplified, small book designed as part of a training program for the golf professional. Sections encompass turfgrass selection, establishment, culture, pest control, and analysis of turfgrass problems. This book began as a correspondence course for salesmen authored by Tom Mascaro and was revised by Professor Burt Musser and Tom Mascaro for the golf course superintendents. Subsequently, it was revised by Professor Musser for the Professional Golfer's Association in 1964 and updated by Warren Bidwell in 1974, including photographs.

Figure 12.57. A vertical-cutting renovation attachment for a motor-powered Toro Park mower, 1930s. [5]

1992†   **Diagnostic Management for Golf Greens.** Tom Mascaro. Oakland Park, Florida, USA. 387 pp., 228 illus. with 105 in color, 8 tables, no index, textured hardbound in green with gold lettering, 5.4 by 8.3 inches (138 × 212 mm).

      A book that presents guidelines and techniques for diagnosis of impending problems on turfed golf greens. Also available in coordinated software for use with a computer.

Mathias, Kevin[2]

1994†   **Turfgrass Lab Manual.** Kevin Mathias and Doug Hawes. Institute of Applied Agriculture, University of Maryland, College Park, Maryland, USA. 106 pp., 71 illus., 10 tables, no index, glossy semisoft covers in multicolor with red lettering, 8.5 by 11.0 inches (216 × 279) mm.

      A manual of twelve laboratory exercises for use in conjunction with a basic turfgrass course. It is oriented to Maryland usage.

Matthias, Claus G.[2]

2002   **Praxis des Sportplatzbaus (Practices in Sport Field Construction).** First edition. Claus G. Matthias. Kontakt & Studium Band 634, Expert Verlag, Renningen-Malmsheim, Germany. 109 pp., no illus., 39 tables, no index, glossy semisoft covers in blues and oranges with white lettering, 5.8 by 8.2 inches (147 × 209 mm). Printed entirely in the German language.

      A small book regarding the prevention or diagnosis of errors in the design, specifications, and construction of sports fields. There is a construction project checklist in the back.

2008   **Praxis des Sportplatzbaus (Practices in Sport Field Construction).** Second revised edition. Claus G. Matthias and Wilfred J. Bartz. Kontakt & Studium Band 634, Expert Verlag, Renningen, Germany. 118 pp., no illus., 43 tables, no index, glossy semisoft covers in blues and oranges with white lettering, 5.8 by 8.2 inches (147 × 209 mm). Printed entirely in the German language.

      Contains small revisions to the 2002 edition.

McCarty, Lambert Blanchard (Dr. and Professor)

Bert McCarty was born in Columbia, South Carolina, USA, and earned a BSc degree in Agronomy and Soils from Clemson University in 1981, an MSc degree in Crop Science from North Carolina State University in 1983, and a PhD degree in Plant Pathology and Plant Physiology from Clemson University in 1986. He served as Assistant and Associate Professor at the University of Florida from 1987 to 1996 and as Full Professor in the Horticulture Department at Clemson University since 1996. He is active in adult education and research in turfgrass culture, with emphasis on weed management and plant growth regulators. He was elected President of the Florida Weed Science Society in 1996. Dr. McCarty was honored as Outstanding Young Weed Scientist of the Southern Weed Science Society in 1997 and Outstanding Young Weed Scientist of the Weed Science Society of America in 1999.

1990† **Florida Lawn Handbook.** First edition. Lambert B. McCarty, Robert J. Black, and Kathleen C. Ruppert (editors) with nine contributing authors. Florida Cooperative Extension Service, University of Florida, Gainesville, Florida, USA. 216 pp., 115 illus. with 88 in color, 36 tables, no index, glossy semisoft bound in multicolor with yellow lettering, 8.1 by 10.7 inches (205 × 272 mm).

    A handbook regarding the selection, establishment, and maintenance of Florida lawngrasses. The book was developed as a training guide for Florida Master Gardeners, who in turn work to inform lawn gardeners.

    Revised in 1997 and 2005.

1993† **Best Management Practices for Florida Golf Courses.** First edition. L.B. McCarty and Monica L. Elliott. Department of Environmental Horticulture, University of Florida, Gainesville, Florida, USA. 173 pp.; 112 illus. with 101 in color; 72 tables; no index; glossy semisoft bound in multicolor with blue, white, and black lettering; 8.3 by 10.9 inches (212 × 278 mm).

    A reference guide on turfgrass culture for golf courses in Florida. It includes nine sections on putting green construction, irrigation, fertilization, other cultural practices, and pest management. There are selected references at the end of each section.

    Revised in 1999.

    See *Unruh, Joseph Bryan—1999.*

2001† **Color Atlas of Turfgrass Weeds.** First edition. L.B. McCarty, John W. Everest, David W. Hall, Tim R. Murphy, and Fred Yelverton. Ann Arbor Press, Chelsea, Michigan, USA. 287 pp., 578 illus. in color, 7 tables, indexed, glossy hardbound in multicolor with white and yellow lettering, 7.0 by 9.9 inches (178 × 251 mm).

    A reference manual concerning the diagnosis of problem weeds occurring in turfgrass in the United States. For each weed species there is a description, means of propagation, occurrence distribution globally, and control strategies plus color pictorials. There is a glossary of taxonomic terms in the back.

2001† **Best Golf Course Management Practices.** First edition. L.B. (Bert) McCarty with 27 contributing authors. Prentice-Hall Inc., Upper Saddle River, New Jersey, USA. 700 pp., 357 illus. with 102 in color, 171 tables, indexed, glossy hardbound in multicolor with green and white lettering, 8.3 by 10.9 inches (210 × 277 mm).

    An extensive reference guide on turfgrass culture of golf courses. The eight sections address turfgrasses, soils, construction and establishment, fertilization, irrigation, management practices, pest management, and pesticide and nutrient management. There is a glossary and a small list of selected references in the back.

    Revised in 2005.

2002 **Managing Bermudagrass Turf.** L.B. (Bert) McCarty and Grady Miller. Ann Arbor Press, Chelsea, Michigan. USA. 235 pp., 175 illus. with 67 in color, 35 tables, small index, glossy hardbound in beige with black lettering, 7.0 by 10.0 inches (177 × 254 mm).

    A speciality book on bermudagrass as utilized for turfgrass purposes. There are four parts: characteristics, establishment, culture for golf putting greens, and pests. There is a short, selected bibliography in the back.

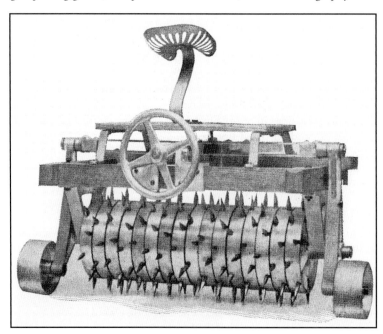

Figure 12.58. Horse-drawn spike roller of 3-foot (91-cm) effective width and 1,600 pounds (725.7 kg), 1920s. [41]

2003† **Fundamentals of Turfgrass and Agricultural Chemistry.** L.B. McCarty, Ian R. Rodriquez, B. Todd Bunnell, and F. Clint Waltz. John Wiley & Sons Inc., Hoboken, New Jersey, USA. 384 pp.; 100 illus.; 90 tables; indexed; glossy hardbound in multicolor with orange, gray, and black lettering; 7.0 by 10.0 inches (177 × 254 mm).

A book on chemistry as applied to plants, water, soils, fertilizers, and pesticides utilized in turfgrass culture. Practice problems are included at the ends of chapters. There is a fairly large glossary in the back.

2005 **Best Golf Course Management Practices.** Second revised edition. L.B. McCarty (editor) with 23 contributing authors. Pearson-Prentice Hall Inc., Upper Saddle River, New Jersey, USA. 896 pp.; 560 illus. with 123 in color; 106 tables; large index; glossy hardbound in multicolor with white, yellow, and black lettering; 8.25 by 10.8 inches (210 × 274 mm).

An update of a reference guide concerning turfgrass culture of golf course facilities.

2005† **Designing, Constructing, and Maintaining Bermudagrass Sports Fields.** Second revised edition. Bert McCarty, Grady Miller, Clint Waltz, and Trent Hale. Ec 698/SP 361/Bulletin 1292 Clemson University Public Service Publishing, Clemson, South Carolina, USA. 104 pp., 146 illus. with 122 in color, 20 tables, no index, glossy semisoft covers in multicolor with black lettering, 8.3 by 10.7 inches (212 × 276 mm).

A manual concerning the design, root-zone construction, bermudagrass cultivars, site preparation, planting, establishment, and culture of team sports fields. It is oriented to warm-humid conditions in the United States.

2008† **Color Atlas of Turfgrass Weeds.** Second revised edition. L.B. (Bert) McCarty, John W. Everest, David W. Hall, Tim R. Murphy, and Fred Yelverton. John Wiley & Sons Inc., Hoboken, New Jersey, USA. 442 pp., 1,006 illus. in color, 15 tables, indexed, glossy hardbound in multicolor with yellow and white lettering, 7.5 by 9.2 inches (190 × 233 mm).

This revised second edition represents a major expansion to 350 turfgrass weeds and their associated color photographs at various growth stages designed for use in identification. There is a CD-ROM included.

McIntyre, David Keith

Keith McIntyre was born in Armidale, New South Wales, Australia, and earned a BSc degree with *honours* from the University of New England in 1964 and an MSc degree in Botany from the Australian National University in 1967. From 1969 to 1977, he worked for the Australian National Botanic Gardens. He then served as the Science Officer at the Horticultural Technical Services Unit (TSU) of the ACT Parks and Conservation Service in Canberra, Australia Capital Territory, Australia, from 1977 to 1995. He subsequently formed the Horticultural Engineering Consultancy in Canberra. Keith McIntyre received the Australian Award in Park Administration from the Royal Australian Institute of Parks and Recreation in 1992; he was also member of a team that was the recipient of the ACT Government Achievement Award in 1993, team leader of a team that was given the ACT Electricity and Water Environmental Award in 1995, and Life Member of the Royal Australian Institute of Parks and Recreation in 1997.

1988† **Turf Irrigation.** Second revised edition. D.K. McIntyre and S.M. Hughes. Management Aid No. 2. Royal Australian Institute of Parks and Recreation, Dickson, Australia Capital Territory, Australia. 60 pp. plus 3 pp. of advertising in color, 53 illus., 8 tables, no index, fold-stapled semisoft covers in green with black lettering, 7.1 by 9.7 inches (180 × 246 mm).

A manual on water management in turfgrass irrigation. Included are water movement in soils, plant-water relations, types of irrigation, irrigation system design and installation, and turfgrass irrigation practices. The first edition published in 1978 was authored by W.A. Martin and M.J. Mortimer (see chapter 19).

2000† **Practical Drainage for Golf, Sportsturf, and Horticulture.** Keith McIntyre and Bent Jakobsen. Ann Arbor Press, Chelsea, Michigan, USA. 220 pp., 103 illus., 18 tables, small index, glossy hardbound in greens with white lettering, 6.0 by 8.9 inches (150 × 226 mm).

A specialty book on the basics of water movement into and through soils, with emphasis on turfgrass areas subjected to intense traffic. Practical, effective means of facilitating removal of excess water by surface and subsurface drainage techniques are discussed.

2001† **Cricket Wickets—Science v Fiction.** Keith McIntyre and Don McIntyre. Horticultural Engineering Consultancy, Kambah, Australian Capital Territory, Australia. 292 pp. plus 1 p. advertising, 113 illus., 6 tables, no index, glossy semisoft bound in multicolor with black and blue lettering, 6.25 by 9.0 inches (158 × 228 mm).

A specialty book on cricket wickets—the incongruous situation of growing grass on an intentionally compacted, dense, almost pure clay. It includes the clay properties, testing methodology, and physics of ball bounce, turn, cut and lift, plus wicket construction, maintenance, and competition preparation. There is a list of selected references in the back.

2004† **Problem Solving for Golf Courses, The Landscape, Sportsgrounds & Racecourses.** Keith McIntyre. Horticultural Engineering Consultancy, Kambah, Australian Capital Territory, Australia. 440 pp., 148 illus., 16 tables, no index, glossy semisoft bound in multicolor with black lettering, 6.3 by 9.1 inches (160 × 230 mm).

A speciality book concerning soils and drainage as related to turfed sports fields, golf courses, and racecourses. It is organized in two parts, with part I addressing the science and technology of soils and part II focused on the major problems and solutions encountered on golf courses, landscapes, sports fields, and racecourses. There is a chapter on soil testing methodology and a list of selected references in the back.

2007† **How To Build A Sand Based Golf Green.** Keith McIntyre, Mark Parker, and David Warwick. Horticultural Engineering Consultancy, Kambah, Australian Capital Territory, Australia. 250 pp. plus 1 p. advertising, 111 illus. with 85 in color, 7 tables, no index, semisoft bound in multicolor with black lettering, 6.3 by 9.0 inches (159 × 229 mm).

A specialty book concerning planning, construction, and turfgrass establishment of putting greens, surrounds, and bunkers for golf courses. There are in-depth discussions of the profile, particle size specifications, and drains testing for the root zone and base plus irrigation design, specifications, and installation.

McIntyre, Donald Sidney (Dr.)

Don McIntyre was born in Darling Downs, Queensland, Australia, and earned a BSc degree in Physics from the University of Queensland in 1946, an MSc degree in Physics from the University of Queensland in 1951, and a PhD in Soil Physics from the University of Wisconsin in 1957. From 1958 to 1985, he served as a Soil Physicist with the Commonwealth Scientific and Industrial Research Organization (CSIRO) Soils, specializing in research concerning cracking clay soils including those used in cricket wickets.

2001† **Cricket Wickets—Science v Fiction.** Coauthor.
See *McIntyre, David Keith—2001.*

McKell, Cyrus Milo (Dr. and Professor)

Cyrus McKell was born in Payson, Utah, USA, and earned both BSc and MSc degrees from the University of Utah in 1949 and 1950, respectively, and a PhD in Botany from Oregon State University in 1956. He served as Range Science Plant Physiologist for ARS-USDA at UC Davis from 1956 to 1961, as Associate Professor of Agronomy at the University of California, Riverside, from 1961 to 1969, as Professor of Range Science and Director of the Institute of Land Rehabilitation at Utah State University from 1969 to 1981, as Vice President and Director of Research at Native Plants Inc. from 1981 to 1988, and as Professor of Botany and Dean of Science at Weber State University from 1988 to 1994. Dr. McKell has received honors as a Fulbright Scholar in 1967 and as a Fellow of the American Association for the Advancement of Science in 1989; he has also received the Utah Governor's Medal in Science and Technology in 1991 and the Willard Gardner Prize in Science from the Utah Academy of Science, Arts and Letters in 1999.

1972† **The Biology and Utilization of Grasses.** Coeditor.
See *Youngner, Victor Bernarr—1972.*

McKernan, Dennis

Dennis McKernan was born in Canada and earned a BA degree in Psychology and Religion at Carleton University in Ottawa, Ontario, and an Associate diploma in Horticulture and Turfgrass Management from the University of Guelph in Ontario. He worked as a golf course superintendent from 1976 to 1979. He then became Grounds Supervisor at Olds College in Olds, Alberta, Canada, and, in 1986, became an Instructor at Olds College, teaching turfgrass and irrigation management subjects. Then in 2003, Dennis McKernan became Education Director of The Irrigation Association (IA) headquartered in Falls Church, Virginia, USA. He was responsible for the industry's irrigation-specific education program designed to assist professionals advance within the industry. Previously, he was active in the IA as a Volunteer Regional Authorized Instructor and had earned both Certified Landscape Irrigation Auditor (CLIA) and Certified Irrigation Designer (CID), Golf Course, certifications. In 2008, he formed Lifeworks Design and Consulting involved in consultant activities.

1994† **Great Plains Turfgrass Manual.** Second revised edition. Dennis McKernan. Patterson Productions, Didsbury, Alberta, Canada. 206 pp., 46 illus., 18 tables, no index, glossy semisoft bound in green and yellow with orange and green lettering, 8.25 by 10.9 inches (210 × 275 mm).

An instruction manual on the establishment and culture of turfgrasses in the Great Plains region of Canada and northern United States. There are specific chapters on construction of sports fields and on golf and bowling greens. A set of study questions for each chapter is included in the back.

1997† **Great Plains Turfgrass Manual.** Third revised edition. Dennis McKernan. Life Works Design & Consulting, Olds, Alberta, Canada. 218 pp., 56 illus., 14 tables, no index, glossy semisoft bound in multicolor with orange and green lettering, 8.5 by 10.9 inches (216 × 277 mm).

A significant revision and update of the 1994 second edition. The study questions have been eliminated.

McMaugh, Peter Edwin

Peter McMaugh was born in Sydney, New South Wales, Australia, and earned a BSc degree in Agriculture from the University of Sydney in 1964. He served as Research Officer-in-Charge and then as Foundation Director at the Australian Turfgrass Research Institute in Sydney, New South Wales, for 14 years, from 1964 to 1978; as Managing Director of Turfgrass Scientific Services for 22 years; and as Managing Director of Qualturf P/L for 31 years. He was elected Foundation President of the NSW Growers Association in 1968 and President of the International Turfgrass Society (ITS) for 1993 to 1997 while serving on the ITS Board of Directors from 1985 to 2001. Peter McMaugh was a recipient of the Australian Sports Medal from the Australian Government in 2000, the Distinguished Service Award from the Australian Golf Course Superintendents Association in 2000, and the Graham Gregory Award from Horticultural Australia in 2009.

1968† **Bowling Green Construction.** P. McMaugh. Grass Research Bureau (NSW) Limited, Sydney, New South Wales, Australia. 33 pp., 11 illus., 5 tables, no index, fold-stapled textured semisoft covers in greens with black lettering, 5.0 by 8.0 inches (128 × 203 mm).

   A small guide to the planning, design, contracts, construction, establishment, and equipment procurement for turfgrass bowling greens under the conditions in Australia.

   † *Reprinted in 1970.*

1970† **Golf Green Construction.** P. McMaugh. Australian Turf Grass Research Institute, Sydney, New South Wales, Australia. 36 pp., 52 illus., 2 tables, no index, fold-stapled textured semisoft covers in tan with blue lettering, 6.8 by 8.7 inches (172 × 222 mm).

   A small booklet on the construction of putting greens for golf courses in Australia. Emphasis is placed on those aspects that influence subsequent efficient maintenance of a high quality turfgrass surface.

McNaughton, E.J.[3]

n.d. **Turf Production and Maintenance.** E.J. McNaughton. Imperial Chemical Industries (Malaya) Limited, Kuala Lumpur, Malaysia. 44 pp., 9 illus., 5 tables, no index, semisoft bound in yellow and green with black lettering, 5.2 by 8.2 inches (133 × 208 mm).

circa

1934

 * A **truly rare**, pioneering booklet on the construction, establishment, and culture of turfgrasses for golf courses, lawns, tennis courts, and sports fields. It is the first turfgrass book published in southeast Asia and the second in the southern hemisphere. The book is oriented to warm-season turfgrasses and tropical conditions in the Malaysia and Singapore region. In 1934, E.J. McNaughton advocated the use of explosives for the following purpose:

> Hard Pan
> Where a "hard pan" underlies an area selected for a recreation ground, as sometimes occurs under Malayan conditions, it must be broken by means of a crowbar or pickaxe, or by explosives. The latter method is inexpensive besides being expeditious and effective. Gelignite, which has a horizontal rending effect is the best explosive for the purpose. The size of charge and depth and spacing to adopt are matters for trial, since the effect varies so much in each soil. It is not desired to blow out a hole, but merely to "heave" the soil surface, breaking the "pan." Preliminary trial could be made with half-sticks of ⅞ inch Gelignite cartridges with No. 6 detonators and black safety fuse. A suitable depth would be 2 ft. to 2 ½ ft., but much would depend on the depth of the pan. The hole prepared to receive the charge should be made with a crowbar and be as close a fit as possible. The half cartridge with detonator and 3 ft. of fuse should then be set into the hole and carefully tamped with a wooden rod. Should the first explosion blow out a hole, increase the depth to 3 ft. or alternatively use one-third of a cartridge. Charges could be spread about 10 ft. to 15 ft. apart; but this distance is best determined after observation of the effect of the first explosion. Cost would be about 7 to 10 cents per charge allowing for 3 ft. of fuse.
>
> White ants cause considerable damage to fine grass by eating the young shoots through at the bases, producing scattered bare patches in "stepping-stone" arrangement over the greens. The upper tunnels of these grass-eating termites are not continuous, but come to the surface at intervals where the pests have feeding grounds, with the result that ordinary fumigation methods have proved unsatisfactory. Until recently all control measures tried have met with failure, or at most very indifferent success; but Mr. Robertson of Singapore Golf Club, after prolonged trials, has at last discovered a method which, although laborious, appears to yield excellent results.
>
> The entire surface of the infected green is probed with a crowbar at 18 inch centres and to a depth of about 2 ft. Each hole is eased open to a width of about 2 inches by moving the crowbar and the mouths of the holes are then plugged with pads of loose grass. The idea of this procedure is to discover deep-seated termite galleries. If a white ant tunnel has been broken by a crowbar hole, the termites soon commence to repair the damage, under cover of darkness, and removal of the grass plugs and examination of holes several hours later will immediately reveal whether a gallery has been struck or not. Into each infected hole, about a cigarette tinful of carbon bisulphide is then poured and ignited. The explosive action propels flames and fumes along the tunnels destroying all life. Where termite galleries have been discovered and

Figure 12.59. Manual-push perforating roller with multiple sections, weighing ~200 pounds (90.7 kg), 1920s. [34]

treated in this way, extermination of the colony is certain. The central colonies are almost invariably situated within the confines of the green, usually about 2 ft. below the green surface or in the surrounding bunkers and banks, a few yards only away from the scene of activities, with the result that the fumigant readily reaches the main nest, penetrating every gallery and destroying royalties, brood and adult termites.

Grass-eating termites, once established play hovac with a green, and no effort should be spared to eradicate the pest. As many as 2000 separate colonies have been discovered in one season on the greens of a single badly-infested golf course.

Mellor, David Reynold

David Mellor was born in Piqua, Ohio, USA, and earned a BSc degree in Agronomy and Turfgrass Management and Landscape Horticulture from Ohio State University in 1987. He was Assistant Director of Grounds for the Milwaukee Brewers Baseball Club and Milwaukee County Stadium from 1987 to 1995 and has been Director of Grounds for the Boston Red Sox Baseball Club since 2000. He has been active nationally in speaking and writing. David Mellor received the Outstanding Young Professional Achievement Award from Ohio State University in 1999.

2001† **Picture Perfect—Mowing Techniques for Lawns, Landscapes, and Sports.** David R. Mellor. Ann Arbor Press, Chelsea, Michigan, USA. 192 pp., 143 illus. with 73 in color, 4 tables, no index, glossy hardbound in multicolor with white lettering, 6.0 by 9.9 inches (152 × 252 mm).

A unique, specialty book on the use of mowing techniques to provide distinct ornamental patterns. Chapters include types of mowers used, basic patterns, and proper operation of mowers in pattern development. Well illustrated with color photographs.

Merino Merino, Domingo

Domingo Merino was born in Pozuelo de Calatrav, Ciudad Real, Spain, and earned a BSc degree in Agriculture Technical Engineering from Madrid Polytechnic University in 1977. He worked as Agricultural Consultant from 1982 to 1999. He has worked as Laboratory Director of the Agriculture and Environment Department for the Diputacion Foral de Guipúzcoa. Domingo Merino is Secretary of ANORCADE (Association for Normalization and Quality Standards for Turfgrass Sportfields Construction in Spain) and also is active as a consultant for sport fields.

1998† **Césped Deportivo: Construcción y Mantenimiento (Sport Turfgrass: Construction and Maintenance).** Domingo Merino Merino and Javier Ansorena Miner. Ediciones Mundi-Prensa, Madrid, Spain. 386 pp.; 177 illus. with 148 in color; 75 tables; 12 appendixes; no index; glossy hardbound in multicolor with yellow, green, and white lettering; 6.9 by 9.9 inches (175 × 250 mm). Printed in the Spanish language.

A book focused on sport fields and normalization standards. Included are climate, soils, construction, drainage systems, root-zone materials selection, and turfgrass culture. It is oriented to conditions in Spain. There is a small list of selected references in the back.

Merritt, Mortimer G.

Mortimer Merritt (1859–194x) was born in Springfield, Massachusetts, USA. He was issued a US patent in 1888 for a typewriter that was originally made by Merritt & Manufacturing Company in Springfield and, subsequently, produced at Lyon & Manufacturing Company in New York. He had moved to Rome, New York, by 1900. He was a landscape gardener in the 1920s and, by the 1930s, was listed in the US census as a landscape architect active in the Oneida County area of New York state. Later, he ran a question and answer department concerning homes and lawns for a Springfield, Massachusetts, newspaper.

1939† **Practical Lawn Care.** Mortimer G. Merritt. A.T. De La Mare Company Inc., New York City, New York, USA. 32 pp., 14 illus., no tables, no index, cloth hardbound in light green with dark green lettering, 4.8 by 7.0 inches (122 × 178 mm).

   * A **relatively rare**, small book on lawn establishment and culture. It is practically oriented to the home lawn enthusiasts and conditions in North America. Mortimer Merrit proposed the following approaches for the renovation of lawns in 1939:

> Few gardeners realize what an enormous crop is harvested every year from a lawn mowed every week, especially if the grass is dense and vigorous. Even though the soil has been well prepared with humus and fertilizers, occasional dressings of sandy loam, or better still, of sharp sand and loam, are essential to enable the roots to support this frequent cropping of the top growth.
>
> The old-fashioned custom of spreading barnyard manure on the lawn in the Fall and letting it lie all Winter does provide humus and fertilizer, but, unfortunately, it is impossible to obtain such manure free of large numbers of weed seeds, and especially those of coarse field and meadow grasses which outgrow and smother the finer grasses. This fact has caused the almost universal abandonment of barnyard manure by expert lawn keepers.
>
> The use of sand on an established lawn is highly important. Sand is silica, a material that gives all grasses their stiffness. Without it, the blades of grass would be soft and flabby. The roots literally dissolve the silica, and it is this silica that gives grass the knife-like edges, noticeably in the coarser blades. Sand also keeps the soil from baking hard, making it soft and porous. Sand, therefore, should be applied to a lawn at intervals, certainly every three or four years.

Metsker, Stanley Eugene

Stan Metsker was born in Wichita, Kansas, USA, and earned a BSc degree in Agriculture from Colorado State University in 1958. He served as Golf Course Superintendent at Lakewood Country Club from 1961 to 1963, at Boulder Country Club from 1963 to 1971, and at Country Club of Colorado from 1972 to 2001, when he retired. He was elected President of the Rocky Mountain GCSA in 1968 and founded and edited the **RMGCSA Reporter** starting in 1966. Stan Metsker was inducted into the Colorado Golf Hall of Fame in 1999 and received the GCSAA Distinguished Service Award in 2007.

1996† **On The Course—The Life and Times of a Golf Superintendent.** Stan Metsker. Metsker Publishing, Colorado Springs, Colorado, USA. 116 pp., 6 illus., 4 tables, indexed, semisoft bound in greens and white with black lettering, 5.5 by 8.5 inches (139 × 216 mm).

    A small book on the life experiences of the author, with the primary emphasis being on the golf course superintendent aspects. His studies were interrupted by military service during WWII.

Middlesex County Cricket Club

n.d.† **Cricket Pitches—Preparation and Maintenance.** Junior Cricket Committee, Middlesex County Cricket Club and
circa Sutton Seeds Limited, Reading, England, UK. 25 pp., 27 illus., 2 tables, 1956, no index, fold-stapled textured semisoft
1956 covers in blue with black lettering, 5.8 by 8.7 inches (148 × 221 mm).

    A small booklet on the culture of cricket pitches or squares under the cool-season turfgrass conditions in England. There are short sections on maintenance of the turfed outfield and on spring renovation of outfields where winter soccer play has occurred plus a seasonal maintenance calendar.

Miller, Grady Lance (Dr. and *Professor*)

Grady Miller was born in Shreveport, Louisiana, USA, and earned a BSc degree *magna cum laude* in Agriculture from Louisiana Tech University in 1988, an MSc degree in Agronomy and Forages from Louisiana State University in 1990, and a PhD degree in Agronomy and Turfgrass from Auburn University in 1995, studying with Dr. Ray Dickens. He was a Research Technician for Rohm & Hass Chemical Company in 1988. He joined the faculty at the University of Florida as an Assistant Professor in 1995 and rose to Full Professor in 2006. In 2006, he became a Full Professor and Extension Specialist at North Carolina State University and was involved in extension, research, and teaching responsibilities in the turfgrass area, with a focus on water and nutrient influences and cultivar assessments. Dr. Miller received a Musser International Turfgrass

Foundation Award of Excellence in 1993, the Potash & Phosphate Institute Fellowship in 1994, the Outstanding Education Materials Award from the American Society of Agronomy in 1998, and the Outstanding Extension Publication Award from the American Society for Horticultural Science in 2006.

2002†   **Managing Bermudagrass Turf.** Coauthor.
　　　　　　See *McCarty, Lambert Blanchard—2002.*

2005†   **Designing, Constructing, and Maintaining Bermudagrass Sports Fields.** Coauthor.
　　　　　　See *McCarty, Lambert Blanchard-2005.*

Mills, Charles Bright

"Chid" Mills (1893–1980) was born in Marysville, Ohio, USA, and earned a BA degree in Economics from Ohio Wesleyan College in 1920. He served in the 309th Engineer Corps in France during WWI. He joined Scotts when he was an eighth grader at the age of 14. He was the fifth employee and worked a long day for $0.10 per hour. He rose to President of the O.M. Scott and Sons Company, serving from 1948 to 1956, and to Chairman of the Board from 1956 to 1974. He also served as Trustee of Lakeside Associates, the Dana Morey Foundation, and the Methodist Theological School in Ohio. "Chid" Mills was elected President of the American Seed Trade Association in 1958 and 1959, President of the Board of Trustees at Ohio Wesleyan University from 1956 to 1960, Director of the Ohio Chamber of Commerce, and director of the First National Bank of Marysville.

1961†   **First in Lawns: O.M. Scott & Sons.** Charles B. Mills. Newcomen Society in North America, New York City, New York, USA. 24 pp., 5 illus., no tables, no index, fold-stapled semisoft covers in white and red with black lettering, 6.0 by 9.0 inches (152 × 229 mm).
　　　　　　A small booklet on the successful history of the O.M. Scott and Sons Company as a researcher, manufacturer, and supplier of materials for the establishment and care of grass lawns.

　　†   *Reprinted in 1964.*

Miln Marsters Group Limited

n.d.   **Grasses for Lawns, Sports Fields and other Amenity Areas.** Miln Marsters Group Limited, King's Lynn, Norfolk, circa England, UK. 32 pp., 29 illus., 3 tables, no index, soft bound in multicolor with white lettering, 5.8 by 8.2 inches (148 19xx   × 208 mm).
　　＊   A small booklet on the establishment of turfgrasses, with emphasis on sports fields. The characteristics of the cool-season turfgrass species used in the United Kingdom are described. There is a small section on the culture of turfgrasses.

Minderhoud, Jan Willem (Dr.)

Jan Minderhoud (1924–2001) was born in the Netherlands and earned a PhD degree in Agricultural Engineering from Wageningen Agricultural University, Wageningen, in 1960. He worked for the Centraal Instituut voor Landbourvkundig Onderzoek at the Proefstation voor de Akker-en Weidebouw (Test Station for Land and Meadow Research), specializing in land and water management.

1975†   **Grasveldkunde (Grassfield Science).** Coeditor.
　　　　　　See *Hoogerkamp, Meindert—1975.*

Ministére de l'Éducation Nationale (Ministry of National Education—France)

1960   **Équipement Sportif: Établissement et Entretien des Sols (Sports Facilities: Establishment and Maintenance of Soils).** Ministére de l'Éducation Nationale, Paris, France. 90 pp., 16 illus., 10 tables, no index, semisoft bound in green and white with black lettering, 8.3 by 10.6 inches (210 × 270 mm). Printed in the French language.
　　　　　　A small book on the design, construction, and turfgrass establishment of sports fields under conditions in France. It includes soil amendments and drainage plus a section on trees and shrubs in the back.

Ministère de la Jeunesse et des Sports (Ministry of Youth and Sports—France)

1992   **Les Sols Sportifs de Plein Air (Outdoor Sports Soils).** Ministère de la Jeunesse et des Sports. Editions du Moniteur, Paris, France. 248 pp., 62 illus., 35 tables, indexed, glossy semisoft bound in multicolor with black and white lettering, 8.3 by 10.6 inches (210 × 270 mm). Printed entirely in the French language.
　　　　　　A book on the construction of sports fields. Included are such sport facilities as soccer, rugby, and golf courses. It is oriented to the climatic conditions in France.

Ministry of Education, Great Britain

1955†   **New School Playing Fields.** First edition. Building Bulletin No. 10, Ministry of Education, London, England, UK. 87 pp. plus 2 pp. of advertising, 23 illus., 10 tables, no index, fold-stapled glossy semisoft covers in blacks and white with black lettering, 6.6 by 8.2 inches (168 × 208 mm).

Figure 12.60. Manual-push, walk-behind, 30-inch-wide, Wilder-Strong spike discer
with 9½-inch-diameter discs and flip carrier, circa 1936. [9]

A bulletin concerning development of sports field facilities for primary and secondary schools. It includes site selection, layout, specifications, construction, seedbed preparation, planting, and maintenance of turfgrasses for playing fields for Great Britain. There are drawings of field layouts for a wide range of sports. There are sections on nonturf surfaces. A small list of selected references is in the back.
Revised in 1966.
See *Department of Education and Science—1966.*

Mocioni, Massimo (Dr.)

Massimo Mocioni was born in Moncalieri, Italy, and earned BSc and PhD degrees in Agricultural Science and Plant Pathology, both from the University of Torino in 1990 and 2001, respectively. Dr. Mocioni is Vice Coordinator and Agronomist of the Italian Golf Federation Green Section and Research Collaborator of Agroinnova, University of Torino.

2000† **La difesa dei tappeti erbosi: Malattie fungine, nemici animali e infestanti (Pest control on turfgrasses: Diseases, animal enemies and weeds).** Coauthor.
See *Gullino, Maria Lodovica—2000.*

2003† **La Pianificazione della Costruzione di un Percorso di Golf (Planning the Construction of a Golf Course).** Coauthor.
See *Beard, James B—2003.*

2004† **Il campo pratica per il gioco del golf—Costruzione, manutenzione e gestione (The Practice golf range—Construction, maintenance and management).** Coauthor.
See *Casati, Maria Paolo—2004.*

Modestini, Francesco Saverio

Francesco Modestini was born in Rome, Italy, and earned a BSc degree in Agronomy from the University of Perugia in 1987 and studied Turfgrass Science at Texas A&M University in 1988 with Dr. Jim Beard. He served as an Instructor at the Turfgrass Technical School of the Italian Golf Federation from 1988 to 1995, as Superintendent of Olgiata Golf Club from 1995 to 2000, and, since, in consultant activities for golf courses and sports fields in central Italy.

2000† **Tappeti Erbosi (Turfgrasses).** Coauthor.
See *Panella, Adelmo—2000.*

2006† **Tappeti Erbosi—Cura, gestione e manutenzione, delle aree verdi pubbliche e private (Turfgrasses—Care, management and maintenance of public and private green areas).** Coauthor.
See *Croce, Paolo—2006.*

Monje-Jiménez, Rafael Jesús

Rafael Monje-Jiménez was born in Dos Hermanas, Seville, Spain, and earned a Technical Agricultural Engineer degree from the University of Seville in 1984. He worked as Head Greenkeeper for more than 11 years and was a turfgrass consultant for several golf courses and athletic fields for the CCAA of Andalucia. Since 1996, he has been an executive member of the Rules and Golf Course Committee of the Real Federación Andaluza de Golf. Currently he works for the Consejería de Agricultura y Pesca de la Junta de Andalucia. He also is Professor in the Agroforestry Department at the University of Seville, serving as Coordinator of the Turfgrass Studies.

1997   **Céspedes en Compos de Golf: Su Mantenimiento y Otras Consideraciones (Golf Course Turf: Maintenance and Other Considerations).** Rafael Monje-Jiménez. Andalucía-Consejeria de Agricultura y Pesca-Federación Andaluza de Golf, Seville, Spain. 121 pp., 50 illus. in color, 12 tables, indexed, glossy semisoft bound in greens and white with white and black lettering, 7.0 by 9.5 inches (170 × 240 mm). Printed in the Spanish language.

A small book concerning the agronomic practices involved in golf course culture under the conditions in Spain. There is a list of selected references in the back.

2000   **Manejo de Céspedes Con Bajo Consumo de Agua (Management of Turfgrass with Low Water Consumption).** First edition. Rafael J. Monje-Jiménez. Junta de Andalucia, Consejeria de Agricultura y Pesca, Andalucia, Spain. 101 pp., 15 illus., 27 tables, small index, glossy semisoft bound in greens and white with white and black lettering, 7.0 by 9.5 inches (170 × 240 mm). Printed in the Spanish language.

A small book concerning the evapotranspiration rate and drought resistance of various turfgrass species and the turfgrass cultural practices that conserve water. There is a list of selected references in the back.

2002   **Mantenimiento de Campos de Golf (Golf Course Maintenance).** Rafael Monje-Jiménez. Ediciones Mundi-Prensa, Madrid and Andalucía-Consejería de Agricultura y Pesca, Seville, Spain. 264 pp., 92 illus. in color, 46 tables, indexed, glossy semisoft bound in greens and white with white and black lettering, 7.0 by 9.5 inches (170 × 240 mm). Printed in the Spanish language.

A book concerning golf as an economic and sports activity. Golf course maintenance is addressed, including turfgrass morphology and physiology, species and cultivars, equipment, and greens staff management plus a chapter on environmental issues. There is a list of selected references in the back.

2004   **Grama basta** *Stenotaphrum secundatum* **en parques y jardines (St. Augustine Grass** *(Stenotaphrum secundatum)* **in Parks and Lawns).** Rafael Monje-Jiménez. Andalucia-Consejería de Agricultura y Pesca, Seville, Spain. 67 pp., 1 illus., 8 tables, indexed, glossy semisoft bound in greens and white with white and black lettering, 6.6 by 9.4 inches (170 × 240 mm). Printed entirely in the Spanish languages.

A small book concerning the culture of St. Augustinegrass for turfgrass use in parks, gardens, and lawns under the warm-season conditions in Spain.

2006†   **Manejo de Céspedes Con Bajo Consumo de Agua (Management of Turfgrass with Low Water Consumption).** Second revised edition. Rafael J. Monje-Jiménez. Junta de Andalucia-Consejerío de Agricultura y Pesca, Seville, Spain. 112 pp.; 40 illus. with 30 in color; 28 tables; no index; glossy semisoft bound in multicolor with white, green, and black lettering; 6.6 by 9.4 inches (170 × 240 mm). Printed entirely in the Spanish language.

An updated revision of the 2000 edition. It includes an irrigation model, called the Seasonal Quality System, that addresses the relationship between irrigation and rainfall and the grade of turfgrass utilization. There is a selected list of references in the back.

2008†   **Céspedes Ornamentales y Deportiva (Ornamentals and Sport Turf).** Rafael Monje-Jiménez. Andalucía-Consejería de Agricultura y Pesca, Seville, Sprain. 527 pp., 297 illus. in color, 160 tables, indexed, glossy semisoft bound in greens and white with white and black lettering, 6.6 by 9.8 inches (170 × 250 mm). Printed entirely in the Spanish language.

A book concerning the turfgrass species descriptions plus their establishment and culture. The diagnoses and control of turfgrass pests are addressed. There are turfgrass cultural systems for golf courses, sports fields, tennis courts, and ornamental lawns under the conditions in Spain.

Monteith, John L., Jr. (Dr.)

John Monteith (1893–1980) was born in Chatham, New Jersey, USA, and earned a BSc degree in Plant Pathology and Plant Breeding in 1916 plus *honors* in the Military Department as Cadet First Lieutenant from Rutgers College and a PhD in Plant Pathology from the University of Wisconsin in 1923. He served as a line officer in the US Army Infantry during WWI from 1917 to 1919. He was employed in 1924 as a Plant Pathologist by the United States Department of Agriculture in Washington, DC, conducting research on vegetable diseases. Then from 1928 to 1942, he served as a Plant Pathologist and Director of the United States Golf Association (USGA) Green Section, where he provided key leadership in golf course turfgrass research and advisory services. Dr. Monteith developed the first effective controls for two major turfgrass diseases—Rhizoctonia brown patch and dollar spot. During WWII, he became Chief of the Turf Unit in the Construction Branch of the United States Army Engineer Corps. Dr. Monteith received the GCSAA Distinguished Service Award in 1959 and the first USGA Green Section Award in 1961.

1932† **Turf Diseases and Their Control.** John Monteith Jr. and Arnold S. Dahl. Volume 12, Number 4, The Bulletin of the United States Golf Association Green Section, New York City, New York, USA. 102 pp., 60 illus., 2 tables, no index, semisoft bound in gray with black lettering, 6.9 by 10.0 inches (175 × 255 mm).

   ∗   This is a **relatively rare** publication that is a must for collectors of historical turfgrass books. It is the first publication devoted to the diseases of turfgrasses and the authoritative source of information concerning turfgrass diseases during the first half of the twentieth century. Included are symptom descriptions, cause, cultural influences, and chemical controls for more than 20 diseases. It was written by two pioneers in turfgrass disease research. Both studied plant pathology for their PhD degrees at the University of Wisconsin with Professor L.R. Jones. It also includes discussions of turfgrass injury caused by acidity, chemicals, frost injury, lightning, poor drainage, salinity, scalping, scums, shading, thatch, water stress, and winterkill. It was first published in **The Bulletin** of the USGA Green Section and subsequently as a separate bound book in semisoft covers. Coauthors Monteith and Arnold Dahl provided a historical perspective on the initial identification of fungus-caused disease problem on turfgrasses termed brown patch:

   > The modern interest in turf diseases seems to trace back to the definite recognition of a disease in the turf garden of Fred W. Taylor at his home in Philadelphia in 1914. In 1915, from browned patches of turf, a fungus was isolated which later was proved to be the cause of this injury. From that time there has been a constantly increasing interest in turf ailments of all kinds.

Moodie, Andrew William Stuart

Andrew Moodie (1900–1947) was born in Leichardt, New South Wales (NSW), Australia, and earned a Diploma in Agriculture (HDA) degree in 1918 and an HDD degree with a Gold Medal in Dairy in 1919, both from Hawkesbury Agricultural College at Richmond, New South Wales. In 1920, he was appointed an Experimentalist of the NSW Department of Agricultural at the College and Grafton Experiment Farm. In 1922, he was named Assistant Agrostologist with the NSW Department of Agriculture; he was promoted to Senior Assistant Agrostologist in 1937 and to Senior Agrostologist in 1939. In 1944, he was appointed Chief Instructor in Agriculture for NSW with the NSW Department of Agriculture. His initial focus was on grassland research and advisory work. During the 1930s and early 1940s, he was an active pioneer in organizing turfgrass training courses for golf course staff and in advising on turfgrass culture for golf, bowling, and sports grounds in NSW. Andrew Moodie was elected President of the NSW Branch of the Australian Institute of Agricultural Sciences in 1920. He also served as a member of the NSW Golf Council and as President of the Greenkeepers Association of NSW.

n.d.† **The Lawn Beautiful: Its Care and Management.** A.W.S. Moodie. Department of Agriculture, New South Wales, circa Australia. 32 pp. with 10 p. advertising, 1 illus., no tables, no index, fold-stapled semisoft covers in light green with dark 193x green lettering, 5.5 by 8.6 inches (140 × 218 mm).

   ∗   A booklet about the characteristics, planting, and care of turfgrasses for lawns in the Sydney, New South Wales, area. It is the earliest turfgrass book published in Australia.

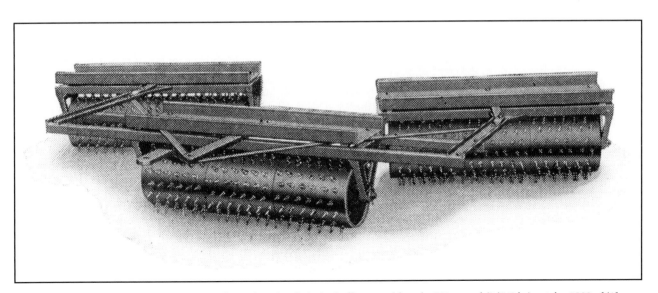

Figure 12.61. Tractor-drawn, 3-gang spike roller with 104-inch (2.6-m) effective width and a 550-pound (249.5-kg) weight, 1920s. [41]

Mulford, Furman Lloyd

Lloyd Mulford (1869–1953) was born in Millville, New Jersey, USA, and earned a BSc degree in Agriculture from Cornell University in 1892. Initially he was Superintendent of Parks in Harrisburg, Pennsylvania, and, subsequently, was a Landscape Gardener in the Bureau of Plant Industry, starting in 1911. He later became an Associate Horticulturist in the Division of Fruit and Vegetable Crops and Diseases in the Bureau of Plant Industry, retiring in 1939. Lloyd Mulford was active in horticulture research and was widely known for his practical bulletins on ornamental horticulture.

1912† **Lawn Soils and Lawns.** Coauthor.

   *   See *Schreiner, Oswald—1912.*

Murdoch, Joseph Simpson Ferguson

Joe Murdoch (1919–2000) was born in Philadelphia, Pennsylvania, USA, and served as a Master Sergeant in the US Army during WWII. He completed the course of study from the Charles Morris Price School of Advertising and Journalism in 1947. He was a copy editor for the **Philadelphia Inquirer** from 1947 to 1951 and then joined the Sun Oil Company as advertising manager and later as industrial products supervisor, retiring in 1981. Joe Murdoch was an early collector of golf books in the 1950s and 1960s and was a cofounder of The Golf Collectors Society (GCB) in 1970.

1968† **The Library of Golf—1743–1966.** Joseph S.F. Murdoch. Gale Research Company, Detroit, Michigan, USA. 314 pp.; 44 illus.; no tables; indexed by author, by year published, and by book title; cloth hardbound in green and black with gold lettering and a slip case in black and cream with green and black lettering; 5.9 by 8.9 inches (151 × 225 mm).

    This book is a compilation of golf books that have been published. Included is a complete citation for each book, followed by a short description. It is a pioneering reference book that is a must for collectors of historical golf books. A supplement was published in 1978.

1978† **Golf—A Guide to Information Sources.** Joseph S.F. Murdoch and Janet Seagle. Volume 7 in the Sports, Games, and Pasttimes Information Guide Series, Gale Research Company, Detroit, Michigan, USA. 232 pp.; no illus.; no tables; indexed by author, title, and subject; cloth hardbound in red and black with black and gold lettering; 5.4 by 8.4 inches (136 × 214 mm).

    This book contains selected, annotated listings of books and periodicals on golf, including golf turfgrass. Also included are listings of governing bodies, libraries, museums, equipment manufacturers, golf course architects and contractors, and educational programs.

1991† **The Murdoch Golf Library.** Joseph S.F. Murdoch. Grant Books, Droitwich, Worcestershire, England, UK. 233 pp., 49 illus. with 1 in color, no tables, no index, cloth hardbound in dark blue with gold lettering and a glossy jacket in navy blue with gold lettering, 7.2 by 9.5 inches (182 × 240 mm).

    A compilation of 3,000-plus books in the personal collection of Joe Murdoch. Included are acquisition notations, a short description, and random personal notes. It was a limited printing of 950 copies.

Murphy, Timothy Richard (Dr. and Professor)

Tim Murphy was born in Knoxville, Tennessee, USA, and earned a BSc degree in Agriculture from Berea College in 1975, and he earned an MSc degree in 1979 and a PhD degree in 1985, both in Weed Science from Clemson University. He served as a Weed Research Technician at Clemson University for seven years while studying for the latter two degrees. In 1985, he joined the faculty at the University of Georgia, rising to Full Professor in 1995. He has been involved in extension, research, and teaching activities, with emphasis on weed control of turfgrasses. He retired in 2007. Dr. Murphy received the Outstanding Extension Worker Award of the Weed Science Society of America in 2000.

2001† **Color Atlas of Turfgrass Weeds.** Coauthor.

    See *McCarty, Lambert Blanchard—2001.*

2008† **Color Atlas of Turfgrass Weeds.** Second revised edition. Coauthor.

    See *McCarty, Lambert Blanchard—2008.*

Murray, Charles Molteno (DSO)

Charles Murray (1877–1950) was born at Claremont, Cape Colony, South Africa. He earned a BA degree in 1898, an MA degree in 1903, and a Bachelor of Medicine and Bachelor of Surgery degree in 1904, all from Pembroke College, Cambridge University, UK. He interned at Guy's Hospital, London. He was a member of the Royal College of Surgeons and a Licentiate of the Royal College of Physicians. He practiced medicine at Kenelworth, South Africa. During WWI, he served in the South African Medical Corps as a Lieutenant Colonel from 1914 to 1919 and received the Victoria Cross and Distinguished Service Order. Subsequently he was a surgeon at Victoria Cottage Hospital in Wynberg, South

Africa, retiring in 1933. He was active in promoting research and turfgrass development for golf courses and also as a golf course architect and construction supervisor. He was elected President of the Cape Western Branch, British Medical Association, in 1923. Dr. Murray was a pioneer in turfgrass improvements for South Africa whose assessments date back to 1907. He conducted numerous turfgrass tests at his home grounds and on the links of the Royal Cape Golf Club at Wynberg.

1932 **Greenkeeping in South Africa.** C.M. Murray. South African Golf, Cape Town, South Africa. 104 pp., 5 illus., 1 table, no index, cloth semisoft bound in dark green with black lettering, 4.6 by 7.2 inches (118 × 183 mm).

✳ This is a **truly rare** book that is a must for collectors of historical turfgrass publications. It is the first turfgrass book published in South Africa and in the southern hemisphere. It covers the establishment and culture of turfgrasses for golf courses under the climatic conditions in South Africa. This book represents the observations of Dr. Murray over a 25-year period. It is a reprint in book format of a series of articles that had been published in **South African Golf.** The foreword is by Robert Craig. Dr. Murray summarized his early observations on turfgrasses with a creeping growth habit as follows:

> When I commenced to take an interest in turf, I felt we laboured under a disadvantage in being unable to make use of the seed-sown grasses as used in England. My first consoling discovery was made at North Berwick in 1907. There were some particularly pleasing greens close to the seashore. I dug up a small slip of turf from the edge of one of these greens, and found to my surprise that it was composed of a branching creeping grass (creeping bent). The turf was so much superior to that produced from the various mixtures sown for turf that it has always been a puzzle why more use was not made of this type of grass. The interest which has been aroused in turf research in recent years in England enabled me in 1930 to see a very wide range of turf grasses under observation in test plots. All I can say is, what I saw more firmly endorses what I thought in 1907. Turf has been defined as a mat of grass in which the old are constantly being replaced by a growth of new shoots. Creeping grasses fulfill this better than any other, in that properly cared for they go on forever. From our South African Cynodon grasses it is possible to form a better sporting turf than I believe is possible elsewhere than where the same type of grass will grow.

Musser, Howard Burton (Professor Emeritus of Agronomy)

Burt Musser (1893–1968) was born in South Williamsport, Pennsylvania, USA, and earned an AB degree in Prelaw from Bucknell University in 1914 and a BSc degree in Agronomy from Pennsylvania College in 1917. From 1918 to 1919, he was an Assistant Scientist with the United States Department of Agriculture (USDA) and, from 1920 to 1922, was a USDA Grain Grading Specialist. He subsequently served on the Agronomy Faculty at Pennsylvania State University from 1922 to 1959, except from 1942 to 1945 when he served as Lieutenant Colonel in the US Air Corps, serving as Head Liaison with the Army Corps of Engineers for airfield dust and erosion control. Initially, he was active in red clover breeding and seed production. Starting in 1929, he became active in turfgrass research and teaching. These efforts resulted in the development of Penncross creeping bentgrass (*Agrostis stolonifera*) and Pennlawn red fescue (*Festuca rubra*). He was elected Turf Management Division XI Chair of the American Society of Agronomy in 1959. Professor Musser received the GCSAA Distinguished Service Award in 1958 and the USGA Green Section Award in 1966. The Musser International Turfgrass Foundation Inc. was formed in honor of Professor Musser.

1950† **Turf Management.** First edition. H. Burton Musser. McGraw-Hill Book Company Inc., New York City, New York, USA. 354 pp., 66 illus., 35 tables, indexed, cloth hardbound in blue with gold lettering and a jacket in green and black with white lettering, 6.0 by 8.9 inches (152 × 226 mm).

A book that addresses the fundamentals of turfgrass establishment and culture within the framework of a how-to approach. Although oriented to golf courses, it served as the standard reference source on turfgrass in North America for two decades. The foreword is by James D. Standish Jr. There is a chapter on golf course architecture written by Robert Bruce Harris and Robert Trent Jones. It also contains a chapter on golf course operations written by Marshall E. Farnham, a golf course superintendent, in collaboration with superintendents T.M. Baumgartner, Ray Gerber, W.H. Glover, and E.W. van Gorder. Members of the Editorial Board were Fred V. Grau, Marshall E. Farnham, Herb Graffis, and O.J. Noer. The book was sponsored by the United States Golf Association, as was an earlier book by Drs. C.V. Piper and Russell Oakley (1917) and later books by Dr. Jim Beard (1982 and 2002). There is an extensive appendix. This book is a must for collectors of historical turfgrass books. The advances in equipment for turfgrass maintenance were reviewed in 1950 by Professor Musser:

> Both the art and science of turf management have developed rapidly during the last three decades. Among the factors primarily responsible for this development are an increased interest in the broad uses

of special-purpose turf and the beginnings of a realization of the magnitude of the services which it involves. The art of growing grass has made particularly rapid strides. This is perhaps best evidenced in the field of laborsaving equipment. The power shovel and bulldozer have replaced the horse-drawn scoop and pick and shovel in construction operations. The nine-gang tractor-drawn mower, cutting a swath of approximately 20 ft. at a speed of 6 to 8 miles per hour, has solved the clipping problem. Pop-up irrigation systems which will water large areas of the course by the turn of a single valve are becoming increasingly common, and high-pressure sprayers have completely replaced the sprinkling can and gravity-flow barrel. These, and similar developments in many other fields, emphasize the progress which has been made.

Revised in 1962.

1962† **Turf Management.** Second revised edition. H. Burton Musser. McGraw-Hill Book Company Inc., New York City, New York, USA. 356 pp., 57 illus., 31 tables, indexed, cloth hardbound in green with gold lettering and a jacket in green and black with white lettering, 5.9 by 9.0 inches (150 × 228 mm).

This revision contains an update on turfgrasses and cultivars plus the advances in weed, disease, and insect control in North America. The golf course architecture chapter was deleted. Members of the Editorial Board were William H. Bengeyfield, William C. Chapin, Joseph C. Dey Jr., Marvin H Ferguson, Charles K. Hallowell, and Alexander M. Radko. The book was sponsored by the United States Golf Association, as were an earlier book by Drs. C.V. Piper and Russell Oakley (1917) and later books by Dr. Jim Beard (1982 and 2002).

1964† **Turf Grass Course.** H.B. Musser. PGA Home Study Program II, Professional Golfers' Association of America, Dunedin, Florida, USA. 69 pp., no illus., no tables, no index, fold-stapled textured glossy semisoft covers in green with black lettering, 8.0 by 10.0 inches (203 × 253 mm).

This is a reprinting of material contained in the West Point Turfgrass Correspondence Course. It is a simplified overview of turfgrass establishment and culture of both cool- and warm-season turfgrasses on golf courses.

See also *Mascaro, Thomas Charles—1964.*

1974† **Turfgrass Course.** Coauthor.

See *Mascaro, Thomas Charles—1974.*

Figure 12.62. Burt Musser, late 1950s. [6]

N

Nakahara, Hisakazu[3]

1994† **Turf Management in Soccer Fields: The Construction and Maintenance.** Hisakazu Nakahara and Hisashi Yanagi (editors). Soft Science Inc., Tokyo, Japan. 236 pp., 341 illus. with 168 in color, 73 tables, indexed, cloth hardbound in maroon with gold lettering and a jacket in multicolor with white and red lettering, 10.1 by 7.1 inches (257 × 181 mm). Printed in the Japanese language.

A specialty book concerning the construction, planting, and culture of turfgrass soccer fields under conditions in Japan. Lists of selected references are included at the backs of certain chapters.

Nakamura, Naohiko (Dr. and *Professor* Emeritus*)*

Naohiko Nakamura was born in Tokyo, Japan, and received a BSc degree in Agronomy from Hokkaido University in 1941 and a DrAgri in Crop Science from Hokkaido University in 1959. He became Professor in the Department of Agriculture at Kobe University in 1969, specializing in the plant physiology of zoysiagrasses (*Zoysia* spp.). He became Professor *Emeritus* at Kobe University in 1981 and then joined the Kansai Golf Union Green Section, serving until 1999 as Manager of Turfgrass Physiology Research. Dr. Nakamura received the Society Award from the Japanese Society of Turfgrass Science in 1994.

1993† **Zoysiagrass: Science and Management for Golf Course.** Naohiko Nakamura (editor). Soft Science Inc., Tokyo, Japan. 377 pp.; 363 illus. with 70 in color; 103 tables; small index; cloth hardbound in green with gold lettering with a jacket in multicolor with dark blue, yellow, green, and white lettering; 7.1 by 10.0 inches (181 × 256 mm). Printed in the Japanese language.

A book concerning the characteristics and culture of *Zoysia japonica* and other *Zoysia* species. It includes soil adaptation, diseases, insects, and their management. It also encompasses the culture of zoysiagrasses for golf course uses and sod production.

National Golf Foundation (NGF)

The National Golf Foundation is a nonprofit organization of golf equipment manufacturers founded in 1936. The primary function is to provide information on planning, construction, maintenance, and operation of golf courses and thereby stimulate greater interest and participation in golf. This is accomplished via printed materials, actual visitations, and lectures.

1947† **Golf Range Operator's Handbook.** National Golf Foundation Inc., Chicago, Illinois, USA. 34 pp., 66 illus., no tables, no index, semisoft bound in browns and white with white and brown lettering, 8.4 by 11.0 inches (214 × 279 mm).

* The **rare** first book on the planning, development, and operation of golf practice driving ranges, including target greens and sand bunkers.

1949† **Golf Facilities: Organization, Construction, Management, Maintenance.** Herb Graffis (editor). National Golf Foundation Inc., Chicago, Illinois, USA. 80 pp., 80 illus., 5 tables, no index, semisoft bound in green, black, and tans with black lettering, 8.6 by 10.9 inches (219 × 276 mm).

* A **rare** handbook on the organization, planning, construction, maintenance, and management of the golf course and golf club in the United States.

1956† **Golf Operators Handbook.** First edition. Ben Chlevin (editor). National Golf Foundation Inc., Chicago, Illinois, USA. 104 pp.; 228 illus.; no tables; no index; semisoft bound in green, white, and black with white and black lettering; 8.2 by 10.9 inches (209 × 277 mm).

* A practical book on the planning, construction, maintenance and operation of a miniature golf putting course, golf driving range, and par-3 golf course, including turfgrass aspects.
   Revised in 1961.

n.d.† **Planning and Building the Golf Course.** First edition. National Golf Foundation Inc., Chicago, Illinois, USA. 30
circa   pp.; 42 illus.; 2 tables; no index; semisoft bound in white, black, and green with black lettering; 8.4 by 10.9 inches (214
196x   × 278 mm).

* This small booklet outlines the planning, design, and construction of a golf course in North America plus a section in the back on water and irrigation systems.
   See also *Golf Facilities—1949.*
   Revised in 1971.

1961† **Golf Operators Handbook.** Second revised edition. Ben Chlevin (editor). National Golf Foundation Inc., Chicago, Illinois, USA. 104 pp., 220 illus., 3 tables, no index, semisoft bound in green with white and black lettering, 8.4 by 11.0 inches (214 × 279 mm).

A slight revision of a practical book on the planning, construction, maintenance, and operation of a golf driving range, miniature golf putting course, and a par-3 golf course.

1971† **Planning and Building the Golf Course.** Third revised edition. National Golf Foundation Inc., Chicago, Illinois, USA. 27 pp., 37 illus., no tables, no index, semisoft bound in light green with dark green lettering, 8.4 by 10.9 inches (214 × 278 mm).

    This third revision contains some changes.
    Revised in 1975 and 1981.

1975† **Planning and Building the Golf Course.** Fourth revised edition. National Golf Foundation Inc., Chicago, Illinois, USA. 28 pp., 36 illus., no tables, no index, semisoft bound in light green with dark green lettering, 8.5 by 11.1 inches (216 × 282 mm).

    This fourth edition contains minor changes.
    Revised in 1981.

1981† **Planning and Building the Golf Course.** Fifth revised edition. National Golf Foundation Inc., North Palm Beach, Florida, USA. 48 pp., 23 illus., 2 tables, no index, fold-stapled textured semisoft covers in yellow with black lettering, 8.3 by 10.9 inches (212 × 278 mm).

    This fifth edition is a major revision that is authored by golf course architects Ken Killian, Dave Nugent, and Jeff Brauer.

1995† **Guidelines for Planning and Developing a Public Golf Course.** National Golf Foundation Inc., Jupiter, Florida, USA. 83 pp., 6 illus., 10 tables, no index, semisoft bound in white and red with red and black lettering, 8.3 by 10.7 inches (211 × 272 mm).

    A guide concerning the feasibility assessment, financing, planning, construction, and operation of a public golf course.

New South Wales Cricket Association

1981 **Turf Cricket Wickets.** New South Wales Cricket Association, Sydney, New South Wales, Australia. 32 pp., 16 illus., no tables, no index, semisoft stapled covers in light green with black lettering, 5.5 by 8.5 inches (140 × 215 mm).

    A booklet concerning the design, construction, culture, match preparation, and soil selection and processing for cricket wickets as practiced in New South Wales, Australia, in the early 1980s.

New Zealand Institute for Turf Culture

The New Zealand Institute for Turf Culture was formed in 1949, as reconstituted from the New Zealand Greenkeeping Research Committee established in 1932 by the New Zealand Golf Association, and under the leadership of Bruce Levy. The membership is composed of sports organizations and professional turfgrass managers. It was renamed the New Zealand Turf Culture Institute in 1949 and the New Zealand Sports Turf Institute (NZSTI) in 1995. Turfgrass research was the main emphasis of the institute from 1932 to the early 1960s, when a formal advisory service was organized and expanded. Periodicals published include the **New Zealand Turf Culture Institute Newsletter** from 1959 to 1976, **Sports Turf Review** from 1976 to 1986, and the quarterly **New Zealand Turf Management Journal** since 1986.

1961† **Turf Culture.** First edition. C. Walker, E.H. Arnold, E.A. Madden, G.S. Harris, L.J. Matthews, R.M. Brien, S.M.J. Stockdill, J.P. Salinger, and J.L. Doutre (contributing authors). The New Zealand Institute for Turf Culture, Palmerston North, New Zealand. 179 pp., 40 illus., 11 tables, no index, textured hard board bound in greens with black lettering, 5.4 by 8.3 inches (138 × 212 mm).

    A book on the establishment and culture of turfgrasses under New Zealand conditions, except for the construction aspects. Emphasis is placed on sports turfgrass surfaces such as golf courses, lawn bowling greens, lawn tennis courts, cricket fields, and croquet courts. It contains substantial sections on turfgrasses, weeds, and turfgrass equipment.

  †   *Revised in 1971.*
    See *Walker, Cyril—1971.*

1995† **A Practical Handbook for Sports Turf Managers.** Compiled by T.J. McLeod and edited by R.J. Gibbs, B.A. Way, and K.W. McAuliffe. New Zealand Turf Culture Institute, Palmerston North, New Zealand. 331 pp., 51 illus. with 7 in color, 22 tables, small index, hard covers in white and black with white lettering in a 3-hole binder, 8.2 by 11.7 inches (208 × 297 mm).

    A reference book for sports turfgrass managers concerning New Zealand legislative and environmental issues. It addresses employee contracts and training, pesticide safety, erosion control, wetland management, water conservation and quality, and machinery safety.

New Zealand Sports Turf Institute (NZSTI)

The New Zealand Sports Turf Institute was organized in 1995 from the New Zealand Turf Culture Institute. NZSTI now has five permanent offices in New Zealand and an office in both Brisbane, Australia, and Kuala Lumpur, Malaysia.

2000† **School Sportsfields Handbook.** Edward Hall and Keith McAuliffe (editors). New Zealand Sports Turf Institute, Palmerston North, New Zealand. 55 p.; no illus.; 12 tables; no index; semisoft bound in light greens, grays, and black with black lettering; 8.2 by 11.4 inches (207 × 290 mm).

    A small booklet concerning the establishment and culture of turfgrasses for school sports and playing fields under the conditions in New Zealand.

2008† **Establishment and Management of Natural Bowling Greens in New Zealand.** Second revised edition. Compiled by David Ormsby. New Zealand Sports Turf Institute, Palmerston North, New Zealand. 234 pp.; 184 illus. with 177 in color; 33 tables; small index; glossy semisoft covers in multicolor with lavender, green, black, and white lettering in a wire ring binder; 8.2 by 11.7 inches (209 × 297 mm).

    A major revision and update of a detailed manual concerning the establishment and culture of cotula and starweed bowling greens under the conditions in New Zealand. This is a one-of-a-kind manual about the *Leptinella* and *Plantago* species as used only in New Zealand.

2009† **Establishment and Management of Croquet Lawns in New Zealand.** Compiled by David Ormsby. New Zealand Sports Turf Institute, Palmerston North, New Zealand. 236 pp., 152 illus. in color, 38 tables, small index, glossy semisoft covers in multicolor with white lettering in a ring binder, 5.8 by 8.2 inches (148 × 209 mm).

    A unique, specialized book concerning the root-zone construction, drainage, planting, and culture of turfgrass for croquet lawns under the conditions in New Zealand. Renovation and leveling are addressed, along with suggested equipment and servicing. It also includes sections on record keeping and budgeting.

Niemczyk, Harry Donald (Dr. and *Professor* Emeritus)

Harry Niemczyk was born in Grand Rapids, Michigan, USA, and earned a BSc degree in 1957, an MSc degree in 1958, and a PhD degree in 1961, all from Michigan State University. He served as a Research Officer for Canada Agriculture from 1961 to 1964 and as an Assistant Professor through Full Professor of Entomology at Ohio State University in Wooster from 1964 to 1992. He has conducted research on turfgrass insects, control of their damage to turfgrasses, and turfgrass pesticide behavior, fate, and mobility. Dr. Niemczyk received the Professional Excellence Award of the Ohio Turfgrass Foundation in 1982, the Wilkie Lectureship at Michigan State University in 1998, and the Honorary Member Award from Turfgrass Producers International in 2008.

1981† **Destructive Turf Insects.** First edition. Harry D. Niemczyk, Wooster, Ohio, USA. 48 pp., 162 illus. with 131 in color, 2 tables, no index, textured semisoft covers in green and white with white lettering in a comb binder, 8.2 by 11.0 inches (208 × 279 mm).

    A small handbook on the identification, turfgrass injury symptoms, occurrence, and control of the major insects that damage turfgrasses. Emphasis is on the pests of C_3 cool-season turfgrasses in North America. It contains a section on the more sophisticated methods of insect detection. There are extensive color photographs.

    Revised in 2000.

2000† **Destructive Turf Insects.** Second revised edition. Harry D. Niemczyk and David J. Shetlar. H.D.N. Books, Wooster, Ohio, USA. 148 pp., 481 illus. with 432 in color, 9 tables, no index, textured semisoft covers in scarlet with gray lettering in a ring binder, 8.5 by 11.0 inches (216 × 280 mm).

    A major revision of a practical book that addresses the diagnosis, life cycle, and control of insect pest problems in turfgrasses. There are new sections on insect pests of warm-season turfgrasses and a major expansion of the color photographs. There is a list of references in the back.

Nikolai, Thomas Anthony (Dr.)

Thom Nikolai was born in Dearborn, Michigan, USA, and earned a Two-Year Turfgrass Certificate in 1986, a BSc degree in Soils and Turfgrass in 1993, and a PhD degree in Soils and Turfgrass in 2003, studying with Dr. Paul Rieke, all from Michigan State University (MSU). He served as a Research Technician for Dr. Paul Rieke from 1993 to 2000 and as Turfgrass Academic Specialist since 2001, both at MSU. Dr. Nikolai was named MSU Outstanding Faculty Member by the Ag-Tech students in 2003.

2005† **The Superintendent's Guide to Controlling Putting Green Speed.** Thomas A. Nikolai. John Wiley & Sons Inc., Hoboken, New Jersey, USA. 160 pp., 41 illus., 12 tables, indexed, glossy hardbound in greens with white and orange lettering, 6.0 by 8.9 inches (152 × 226 mm).

    A speciality book on the ball roll distance of golf putting greens. The influence of weather, root zones, turfgrass species, and cultural practices, including a focus on rolling, are discussed.

2005   Also published in Canada at same time.

Figure 12.63. A tractor-drawn, three-gang Aerifier coring unit with hydraulic lift for transport, circa 1949. [4]

Niwa, Teizo (Dr. and *Professor* Emeritus)

Teizo Niwa (1891–1967) was born in Yokohama, Kanagawa Prefecture, Japan, and earned a BSc degree from Tokyo Imperial University and a DAgri from Tokyo Imperial University in 1917. He served on the faculties at Mie College of Agriculture of Tokyo University and at Meiji University.

1943   **Japanese Lawn Grasses.** First edition. Teizo Niwa. Meibundo Inc., Tokyo, Japan. 150 pp., 10 illus., small index, cloth hardbound, 5.2 by 7.5 inches (130 × 190 mm). Printed entirely in the Japanese language.

   ∗   An introductory book on turfgrass establishment and culture, emphasizing *Zoysia* species. It includes sections on the history of turfgrass utilization in Japan and the introduction of cool-season turfgrass from overseas in the back. It is a key book for collectors of historical turfgrass books.

1958   **Japanese Lawn Grasses.** Second revised edition. Teizo Niwa. Meibundo Inc., Tokyo, Japan. 174 pp. 86 illus., 41 tables, indexed, cloth hardbound, 5.2 by 7.5 inches (130 × 190 mm). Printed entirely in the Japanese language.

     A revision of an early Japanese turfgrass book. Included are the history; effectiveness of the lawn; construction; culture; renovation; sod production; flowering, seed setting, and seed characteristics of Japanese lawngrasses (*Zoysia* species); establishment of the lawn and lawn garden; and introduction, culture, and pest control of Western turfgrasses in Japan.

Nixon, Philip Leon (Dr.)

Phil Nixon was born in Hutchinson, Kansas, USA, and earned an Associate (AS) degree in Biology from Lincoln Land Community College in 1972, both a BSc degree in 1974 and an MSc degree in 1976 in Botany from Southern Illinois University, and a PhD degree in Entomology from Kansas State University in 1979. He served as Area Extension Advisor of Entomology for the Cooperative Extension Service, University of Illinois (UL), from 1980 to 1987 and, since, has been Extension Entomologist in the Department of Natural Resources and Environmental Sciences, University of Illinois, in charge of coordinating Pesticide Applicator Training in Illinois. He has been elected to the Board of Directors of the American Association of Pesticide Safety Educators for 2005 to 2006 and the North Central American Association of Pesticide Safety Educators for 2005 to 2006. Dr. Nixon received the Academic Professional Award of Excellence from the UL Department of Natural Resources and Environmental Sciences, the UL Team Extension Award of Excellence for Outstanding or Innovative Program for both 2004 and 2005, and both the UL College of Agricultural, Consumer and Environment Sciences Professional Staff Award for Sustained Excellence for 2005 and the Team Award for Excellence for 2005.

1997†   **Controlling Turfgrass Pests.** Second revised edition. Coauthor.
     See *Fermanian, Thomas Walter—1997.*

2003†   **Controlling Turfgrass Pests.** Third revised edition. Coauthor.
     See *Fermanian, Thomas Walter—2003.*

Noer, Oyvind Juul

"O.J." Noer (1890–1966) was born in Stoughton, Wisconsin, USA, and earned a BSc degree in Soils from the University of Wisconsin in 1912. He conducted some of the early soil surveys in Wisconsin. O.J. served as a Captain in the Chemical Warfare Service of the US Army during WWI. Subsequently, he worked as Sales Manager for the Stoughton Wagon Company. From 1924 to 1960, he served as a Turfgrass Agronomist for the Turf Service Bureau of the Milwaukee

Figure 12.64. O.J. Noer, circa 1950s. [6]

Sewerage Commission. He was one of the pioneer agronomists whose most significant contributions were through his innumerable talks and golf course technical visitations throughout North America at a time when very few knowledgeable specialists were available. O.J. Noer received the GCSAA Distinguished Service Award in 1952, 1959, and 1960, and the third United States Golf Association Green Section Award in 1963. The O.J. Noer Research Foundation was established in his memory in 1959 and the O.J. Noer Memorial Turfgrass Collection was established at the Michigan State University Library in 1968.

1928†   **The ABC of Turf Culture.** O.J. Noer. The National Greenkeeper Inc., Cleveland, Ohio, USA. 95 pp., 8 illus., 13 tables, no index, cloth semihardbound in brown with gold lettering, 5.9 by 9.0 inches (151 × 228 mm).

   ✳   This is a **very rare** book that is a must for collectors of historical turfgrass books. The book is devoted specifically to the physical and chemical dimensions of soils and the turfgrass nutritional aspects of golf course turfgrass maintenance in the United States. It was a major contribution in its time. The introduction is by John Morley. There is a photograph of a young O.J. Noer in the front. The chapter subject titles include the following:

     I.   Factors Affecting Turf Growth
     II.   Effect of Size and Arrangement of Soil Particles on Turf Growth
     III.   The Part Water Plays in the Growth of Turf
     IV.   The Functions of Organic Matter in the Soil
     V.   Soil Composition and How Plant Food Becomes Available
     VI.   The Nature of Soil Acidity and Effect on Fertilizer Materials on Soil Reaction
     VII.   Lime in Sand, Soil, or Water Often Overcomes Acidic Properties of Sulphate in Ammonia
     VIII.   Essential Plant Food Elements and How Plants Feed
     IX.   Elements and Characteristics of Various Groups of Fertilizer Materials
     X.   Composition and Properties of Individual Fertilizer Materials
     XI.   Principles Underlying the Practical Use of Fertilizers on Greens and Fairways

O.J. articulated his philosophy of nitrogen fertilization of turfgrass in 1928 as follows:

In order to maintain uniform growth the turf must obtain a uniform and continuous supply of nitrogen. It is not feasible to build up large reserves of nitrogen in the soil because of unavoidable losses from leaching and denitrification. This danger exists even with insoluble organic nitrogen, because it is converted into soluble forms by bacteria in the soil, and if the amount formed is larger than the turf roots can take up and utilize, loss occurs.

     Too much nitrogen tends to produce coarse broad leaves, and a weak succulent turf, particularly if readily available nitrogen is used. Such turf is probably more susceptible to diseases such as brown-patch.

> All things considered best results are obtained from moderate applications, at frequent intervals, rather than occasional heavy applications.
>
> Where good top dressing containing well rotted manure is used very little response is obtained from additional applications of phosphoric acid and potash. Both tend to encourage clover so their use should be based on trials which demonstrate the need for larger amounts than are contained in top dressing mixtures.
>
> All carefully conducted tests indicate that sulphate of ammonia encourages the growth of the finer textured grasses, and discourages clover, and nitrate of soda has the opposite effect. Consequently sulphate of ammonia should be chosen as the source of quickly available nitrogen, and used to produce these effects. Ammo-phos is an excellent material where additional phosphoric acid is required.
>
> There is also a need for more slowly available nitrogen, to insure a uniform supply. In the past this was supplied by the manure used in compost piles. Near large cities manure is difficult to obtain and many clubs are substituting such materials as cottonseed meal, poultry manure and Milorganite. None of these require long composting and should be mixed with the top dressing just previous to top dressing the green, or they can be spread broadcast over the green and top dressing mixture applied over them.
>
> The amount and character of turf growth must be used as a guide in determining the amount of nitrogenous fertilizer to apply. Because of the danger of burning, sulphate of ammonia applications should not exceed three to five pounds per 1000 square feet in the spring and fall, and one to three pounds in the hot summer months. The organic materials can be applied at rates of 15 to 30 pounds per 1000 square feet. The heavier rates are safe during cool weather, and the lighter amounts during the hot summer months. Naturally heavier applications should be made where the turf is poor.

1937† **The ABC of Turf Culture.** O.J. Noer. The Service Bureau, Milwaukee Sewerage Commission, Milwaukee, Wisconsin, USA. 28 pp., 9 illus., 2 tables, no index, textured semisoft covers in tan with a green imprint and white and green lettering in a 3-hole binder; 6.0 by 8.9 inches (152 × 226 mm).

∗ A **rare**, completely different booklet than the author's 1928 version. It is a short publication originally published serially in 1937 in **Golfdom** and **Greenkeepers' Reporter**. It is oriented primarily to the culture of turfgrasses on golf courses in the United States, with emphasis on soils and nutrition.

1947† **The Role of Lime in Turf Management.** First edition. O.J. Noer. Bulletin No. 1, Turf Service Bureau, Sewerage Commission, Milwaukee, Wisconsin, USA. 23 pp., 1 illus., 2 tables, no index, top-stapled semisoft covers in green with black lettering, 8.4 by 10.9 inches (213 × 278 mm).

∗ A **very rare** booklet that addresses the theory, justifications, and practice of liming to correct acidic soils and provide more optimum growing conditions for turfgrasses under conditions in North America.
    Revised in 1955, 1958 and 1970.

1946† **Better Bent Greens—Fertilization and Management.** First edition. O.J. Noer. Bulletin No. 2, Turf Service Bureau, Sewerage Commission, Milwaukee, Wisconsin, USA. 25 pp., no illus., no tables, no index, side-stapled semisoft covers in green with black lettering, 8.4 by 10.9 inches (213 × 278 mm).

∗ A **very rare** booklet on the cultural practices for bentgrass (*Agrostis* species) putting greens under the conditions in North America.
    Revised in 1947 and 1959.

1946† **Better Fairway Turf—Theory & Practice.** First edition. O.J. Noer. Bulletin No. 3, Turf Service Bureau, Sewerage Commission, Milwaukee, Wisconsin, USA. 34 pp., 2 illus., no tables, no index, side-stapled semisoft covers in dark green with black lettering, 8.4 by 10.9 inches (213 × 278 mm).

∗ A **very rare** booklet organized in two parts: the theory and the practices of fairway turfgrass culture. Liming, fertilization, and weed control practices in North America are emphasized. The following statement represents "O.J.'s" views on potassium fertilization in the 1940s:

> Fairways seldom need potash fertilizer because most soils contain an abundance of potash. The soil supply is constantly augmented by the decay of clippings. The use of potash may aggravate clover. Poor sandy soils and peats are the only ones that may need occasional applications of potash.

    Revised in 1950 and 1955.

1947† **A Sensible Fertilizer Program for Greens and Turf Nurseries.** Second revised edition. O.J. Noer. Bulletin No. 2, Turf Service Bureau, Sewerage Commission, Milwaukee, Wisconsin, USA. 29 pp., no illus., no tables, no index, top-stapled semisoft covers in green with black lettering, 8.4 by 10.9 inches (213 × 278 mm).

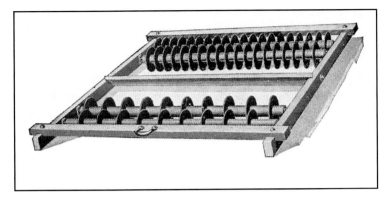

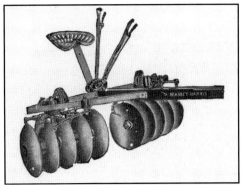

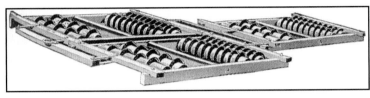

Figure 12.65. A Meeker "disking" turfgrass renovation apparatus: a single unit on the right and a three-gang, horse-drawn unit on the left, 1920s. [41]

⁎ A **very rare** booklet on the cultural practices for bentgrass (*Agrostis* spp.) putting greens and the nursery. Emphasis is placed on the turfgrass fertilization practices for North America.

Revised in 1959 and 1970.

1950† **Better Bent Greens—Fertilization and Management.** Second revised edition. O.J. Noer. Bulletin No. 2, Turf Service Bureau, Sewerage Commission, Milwaukee, Wisconsin, USA. 34 pp., 1 illus., no tables, no index, top-stapled semisoft covers in green with black lettering, 8.4 by 10.9 inches (213 × 278 mm).

⁎ A **rare** booklet concerning the cultural practices for bentgrass (*Agrostisi* species) putting greens under the conditions in North America.

Revised in 1955, 1966, and 1971.

1955† **The Role of Lime in Turf Management.** Second revised edition. O.J. Noer. Bulletin No. 1, Turf Service Bureau, Sewerage Commission, Milwaukee, Wisconsin, USA. 25 pp., 1 illus., 2 tables, no index, side-stapled semisoft covers in green with black lettering, 8.4 by 10.9 inches (213 × 278 mm).

⁎ Only slight revisions of this **rare** booklet.

Revised in 1958.

1955† **Better Fairways—Theory & Practice.** Third revised edition. O.J. Noer. Bulletin No. 3, Turf Service Bureau, Sewerage Commission, Milwaukee, Wisconsin, USA. 48 pp., 1 illus., 1 table, no index, side-stapled semisoft covers in green with black lettering, 8.4 by 10.9 inches (213 × 278 mm).

⁎ There are some revisions to this **rare** booklet.

Revised in 1966 and 1971.

See Wilson, Charles Granville—1966 and 1971.

1958† **The Role of Lime in Turf Management.** Third revised edition. O.J. Noer. Bulletin No. 1, Turf Service Bureau, Sewerage Commission, Milwaukee, Wisconsin, USA. 21 pp., 1 illus., 2 tables, indexed, side-stapled semisoft covers in green with black lettering, 8.4 by 10.9 inches (213 × 278 mm).

⁎ This edition of the booklet contains major revisions and updates of liming practices for turfgrasses under the conditions in North America.

1959† **Better Bent Greens—Fertilization and Management.** Third revised edition. O.J. Noer. Bulletin No. 2, Turf Service Bureau, Sewerage Commission, Milwaukee, Wisconsin, USA. 33 pp., 1 illus., no tables, no index, side-stapled semisoft covers in green with black lettering, 8.4 by 10.9 inches (213 × 278 mm).

⁎ An updated booklet that briefly discusses a diverse array of cultural practices used on bentgrass (*Agrostis* species) greens in North America. Emphasis is placed on turfgrass fertilization practices.

Revised in 1970.

See Wilson, Charles Granville—1970.

Noma, Yutaka[2]

1997† **Handbook of Turfgrass and Turf.** Yutaka Noma. Hakuyusha Ltd., Tokyo, Japan. 353 pp.; 102 illus. with 12 in color; 114 tables; no index; cloth hardbound in green with gold lettering and a jacket in multicolor with black, green, and blue lettering; 7.1 by 10.1 inches (181 × 256 mm). Printed in the Japanese language.

Figure 12.66. Dr. Russell Oakley, circa 1920s. [7]

A book concerning turfgrasses and their culture under the conditions in Japan. Lists of selected references are included at the ends of certain chapters.

Norfolk City

1918   **Norfolk Lawns and How to Grow Them.** The Commission for Beautifying the City, Norfolk, Virginia, USA. 30 pp., 18 illus., 1 table, no index, semisoft bound in tan with black lettering, 6.1 by 8.9 inches (156 × 227 mm).

  ✳   A **rare**, unique, pioneering booklet developed by the homeowners of Norfolk, Virginia. It focused on strategies for correcting the ongoing problems of growing an acceptable lawn under the local soil and climatic conditions.

O

Oakley, Russell Arthur (Dr.)

Russell Oakley (1880–1931) was born in Marysville, Kansas, USA, and earned a BSc degree from Kansas Agricultural College in 1903 and a DSc degree from Iowa State College in 1920. He was appointed Agrostologist for the United States Department of Agriculture (USDA) in 1903 and Agronomist in charge of seed distribution in 1913. In 1926, he became Senior Agronomist in charge of the USDA Division of Forage Crops and Diseases and Chairman of the Research Committee of the United States Golf Association Green Section. He continued to serve in these capacities until his death. Dr. Oakley and Dr. C.V. Piper were responsible for establishment of the turfgrass research program within the USDA in Washington, DC, and also for creation of **The Bulletin** of the USGA. He served as Chairman of the Seed Stocks Committee, Vice Chairman of the Federal Horticultural Board, and member of the Advisory Federal Plant Quarantine Board. Dr. Oakley was awarded Fellow of the American Association for the Advancement of Science in 1922 and Fellow of the American Society of Agronomy in 1927.

1917†   **Turf for Golf Courses.** Coauthor.
        See *Piper, Charles Vancouver—1917.*
        In 1917, coauthors C.V. Piper and Russell Oakley commented on the variability in conditions both between and within golf courses:

> Scarcely any two golf courses are sufficiently similar so that one may confidently affirm that the methods found satisfactory on one will apply to the other. Nearly every golf course is confronted with turf-growing problems that it must solve for itself.

> The experience in golf course management has not resulted in much increase of accurate knowledge so far as turf-growing is concerned. Every putting-green on a golf course has in most cases been subjected to so many kinds of treatment that it is impossible for any one to determine just which factors were good and which were bad. Add to this that the exact record of the history and treatment of a particular piece of turf is rarely preserved by any golf club, and it is not difficult to understand why there is so great difference of opinion as to what is desirable and what undesirable in growing turf in a particular region.

Odorless Plant Food Company

The Odorless Plant Food Company operated in Takoma Park, Washington, DC, USA. The company name is in reference to nonmanure based fertilizers, as, previously, manure had been the main type of fertilizer available.

1916   **The Maintenance of Lawns—Applicable also to Golf Courses.** Odorless Plant Food Company, Washington, D.C., USA. 28 pp., no illus., no tables, no index, fold-stapled textured covers in greens with red lettering, 5.3 by 7.6 inches (135 × 192 mm).

    ✳   An early booklet on the establishment and care of lawns, with emphasis on fertilizing. It is oriented to the lawn enthusiasts in the East Coast, mid-Atlantic region of the United States.

Okubo, Sho

Sho Okubo was born in Jousou-Shi, Ibaraki Prefecture, Japan, and received a BSc degree in Landscape Architecture from Chiba Horticulture College in 1950. He worked at Tokyo Drake Golf Course from 1954 through 1957. He then worked as Golf Course Superintendent at Ryugasaki Country Club from 1958 through 1971 and as Vice President and also Chairman for Toyo Green Sangyo Company Limited from 1971 to 1995. Since July 1995, he has been President of Okubo G Design Company Limited and has been active as a golf course architect. He was elected Vice President of The Japan Society of Turfgrass Science for 1991 to 1994. Sho Okubo received the Society Award of the Japanese Society of Turfgrass Science in 1993.

1975   **Turf and Lawn Garden Construction.** Sho Okubo and Kunio Someno. Seibundo Shinkosha, Tokyo, Japan. 200 pp., 42 illus. with 16 in color, 27 tables, indexed, glossy semisoft bound in multicolor, 7.5 by 9.1 inches (190 × 230 mm). Printed in the Japanese language.

    A guide book on turfgrasses plus their establishment and care for lawns under conditions in Japan.

1991†   **Bentgrass: Characteristics and Newest Methods of Golf Green Construction and Maintenance.** Coeditor. See *Maki, Yoshisuke—1991.*

Olcott, Luther James Bradford

James Olcott (1830–1910) was born in Manchester, Connecticut, USA. From 1849 to 1853, he worked in California where he became active in booking ships to import exotic produce and also started a restaurant and boarding house in Sacramento. He then returned to Connecticut, eventually settling in South Manchester circa 1865, working in farming,

Figure 12.67. View of Olcott Grass Garden in South Manchester, Connecticut, USA, circa 1910. [7]

landscaping, and road construction for varying periods of time. In 1868, Olcott started his own farm operation. He was a key member of the first Board of Trustees of Storrs Agricultural School in 1881, which eventually became the University of Connecticut. The Connecticut State Agricultural Experiment Station initiated a request in 1890 to James Olcott to begin collecting grasses that would be beneficial to Connecticut. During travels and through contacts around the world, he collected different types of grasses that he planted near his Connecticut home. Each was increased and transplanted into 4 by 4 foot (1.2 × 1.2 m) plots for observations of adaptation to Connecticut and seed yields.

1892† **Talk About A Grass Garden.** James Bradford Olcott. The Case, Lockwood & Brainard Company, Hartford, Connecticut, USA. 26 pp., 3 illus. plus a 4-unit foldout of the Grass-Sward Garden plot plan, no tables, no index, fold-stitched soft covers in tan with black lettering, 5.9 by 9.0 inches (150 × 230 mm).

* A **very rare,** small booklet describing the initiation of an early Grass-Sward Garden at a Research Station of the Connecticut State Agriculture Agency located at South Manchester. He proposed that Connecticut should develop a major perennial grass seed and sod production business for marketing nationwide. His tests were focused on this objective. Unfortunately, there was no published data concerning the research findings therein or in other subsequent publications. Numerous authors have attributed the first turfgrass research to Olcott, but this is clearly untrue as Dr. William Beal published his research much earlier. This booklet is a reprinting from the 1892 Connecticut Agricultural Report.

O.M. Scott and Sons Company

Orlando McLean "O.M." Scott (1837–1923) was born in Licking County, Ohio, USA. O.M. served in Company H of the 121st Regiment of the Ohio Volunteer Infantry. He was discharged from the Union Army as a First Lieutenant in 1865 after being wounded in the battle of Bentonville, North Carolina. By 1866, he was operating a country elevator in Raymond, Ohio, eight miles south of Marysville. The O.M. Scott and Brother Company was founded in 1868 in Marysville, Ohio, by O.M. Scott. The company initially marketed hardware and farm items, including buggies, bicycles, motorcycles, and electric cars plus farm seeds. O.M. was an energetic and very successful businessman. The major problem in growing crops and grasses at that time was the high weed content in the seed used for planting. O.M. Scott identified this major problem and soon purchased a wooden, hand-cranked seed cleaner, a Clipper Mill, and embarked on techniques to remove weed seeds. This weed-free seed was in great demand. Subsequently, O.M. Scott pioneered the use of seed testing to monitor seed quality long before it was mandated by state laws.

Charles B. Mills described O.M. Scott as follows:

> O.M. Scott was a unique character. You may get the idea that O.M. Scott was a gentleman of opinions—and you're right. His views on all subjects were definite. By his measurements a man had better than the average chance of getting to Heaven if he abstained from smoking and drinking—belong to and attended, the Congregational Church regularly—and never swerved from voting the Republican ticket. O M was a crank on Sabbath observance. Food preparations at his home for Sunday were made on Saturday and they were important preparations, because the Scotts liked to eat. A favorite was Ice Cream, and on any Saturday afternoon, you'd see the Scott freezer hard at work on the back porch.

The company name was changed to O.M. Scott and Sons in 1903. The hardware store was sold in 1904 as O.M. foresaw a major threat from the emerging national mail order firms, i.e. Sears & Roebuck and Montgomery Ward. The company emphasis changed to farm seeds. O.M.'s first son, Dwight Guthrie Scott (1875–1966), joined his father's firm as a merchant and shortly thereafter initiated a mail order business for grass seeds in 1906. In 1916, a major order from Brentwood in the Pines of Long Island, New York, was received for 5,000 pounds (2,268 kg) of Kentucky bluegrass (*Poa pratensis*) seed for use as a fairway planting. This major event resulted in the O.M. Scott and Sons Company changing to a major emphasis on marketing to the emerging turfgrass industry. The company established the first nationwide turfgrass marketing program in 1919; opened their first retail outlet through the J.L. Hudson Company in Detroit in 1924; initiated the famous **Lawn Care** magazine in 1928; introduced the first branded, organic, slow-release lawn fertilizer—Turf Builder*—in 1928, with soybean meal being the main constituent of Turf Builder up to WWII; and developed the homeowner oriented Scotts Spreader* in 1930. For the first time, grass seed as well as fertilizer could be applied uniformly to lawns. After WWII, there was a major expansion in the Scotts research staff to 23 turfgrass researchers and more than 10 acres (4 ha) of experimental plots in 1950. Growth continued to 75 full-time research employees by 1973 and encompassed more than 800 acres (324 ha) of experimental field plots. Manufacturing plants were constructed for fertilizers in 1939, seed processing in 1953, and chemical products in 1956. The Pro-Turf Division was formed in 1970.

n.d.† **Weedless Lawns and Golf Courses.** O.M. Scott and Sons Company, Marysville, Ohio, USA. 31 pp., 7 illus., no tables, circa no index, fold-stapled soft covers in white with green and black lettering, 3.5 by 6.0 inches (80 × 152 mm).
1921

Figure 12.68. Orlando M. Scott. [6]

* A tr**uly rare,** small booklet on establishing quality lawns by the selection of quality grass seed, especially freedom from weeds. There are small sections on golf course putting greens and fairways. The predecessor to this book was a pioneering little booklet called **Weeds and Seeds and How to Know Them.**

1922† **The Seeding and Care of Golf Courses.** First edition. O.M. Scott and Sons Company, Marysville, Ohio, USA. 54 pp.; many sketches in green; no tables; no index; cloth hardbound in tan, green, and orange with orange lettering; 5.5 by 8.0 inches (140 × 203 mm).

* A **very rare,** small book that is a must for collectors of historical turfgrass books. It addresses site preparation, establishment, and culture of turfgrasses for golf courses. The subjects are covered in a fairly detailed manner, including an early mention of the Rhizoctonia brown patch disease. The chapter subject topics addressed include the following:

· Turf Problems	· The Care of Turf
· The Soil	· Fertilizers
· Making the Fairway	· Weeds
· Building the Putting Green	· Turf Grasses

O.M.'s brother-in-law, Reverend Harvey Colburn, was the primary author of this book and other early Scotts publications. Within three years after this 1922 publication, O.M. Scott and Sons had sold seed to one out of every five golf courses in the United States, most of which was sold by mail order. Scotts imported and sold a blend of bentgrass (*Agrostis* spp.) seeds known as South German bentgrass, with the center of seed production being around Darmstadt, Germany, just southeast of Frankfurt. Five tons of seed was imported the first year, circa 1921.

The historical approach of using a large number of grass species in a seed mixture versus a selected few species specifically adapted to the climatic, soil, and use conditions of a site was addressed in 1922:

> In selecting seed, sacrificing quality is never a saving. The best seed proves the least expensive, for being free from impurities, it goes farther than cheap seed, produces a thicker, cleaner stand of grass and makes necessary a smaller expenditure for weed eradication. Only when one has several superior lots of seed to select from should price be made the basis of decision.
>
> Messrs. Piper and Oakley, to whom reference is made in the introduction, have added much to popular knowledge of the best varieties and their special adaptations, and so have greatly aided efficient buying.

> Through scientific study the fallacy of mysterious ready-made mixtures is now exposed. They serve no logical purpose, but instead, introduce undesirable grasses and add greatly to expense. In purchasing different varieties separately, one is better able to judge quality, to secure the desired proportions and to reduce the cost.

Revised in 1923.

1923† **The Seeding and Care of Golf Courses.** Second revised edition. O.M. Scott and Sons Company, Marysville, Ohio, USA. 55 pp.; many illus. as vignettes in green; no tables; no index; cloth hardbound in tan, green, and orange with orange lettering; 5.5 by 8.0 inches (140 × 203 mm).

* There are minor revisions of the first edition, with the addition of a section on rough bluegrass (*Poa trivialis*). It is a **very rare** book.

1928† **Bent Lawns.** First edition. O.M. Scott and Sons Company, Marysville, Ohio, USA. 23 pp.; 8 illus. with 2 in color; no tables; no index; fold-stapled soft covers in light green, black, and white with white lettering; 3.5 by 6.7 inches (88 × 170 mm).

* A **very rare**, small booklet on the advantages of creeping bentgrass (*Agrostis stolonifera*) for lawns and its establishment by both vegetative stolons and by seed. Scotts initiated the production of creeping bentgrass stolons for marketing and vegetative planting in 1918. A memorable day was when a fully-iced, express railroad car loaded with shredded creeping bentgrass stolons was shipped on the afternoon passenger train from the Marysville station for the Polo Field, Boston. It is a must for collectors of historical turfgrass books.

† *Revised in 1931.*

1931† **Bent Lawns.** Second revised edition. O.M. Scott and Sons Company, Marysville, Ohio, USA. 20 pp., 19 illus., no tables, no index, fold-stapled glossy semisoft covers in multicolor with black lettering, 3.4 by 6.2 inches (86 × 158 mm).

* There are small revisions of this **very rare** booklet.

† *Revised circa 1935 as a 16 page booklet.*

n.d. **The Seeding and Care of Lawns.** O.M. Scott and Sons Company, Marysville, Ohio, USA. 24 pp., no illus., no tables,
circa no index, fold-stapled soft covers in multicolor with black lettering, 3.4 by 5.7 inches (86 × 146 mm).
193x

* A **very rare** small booklet concerning the planting and care of cool-season turfgrass lawns in the United States.

1931† **The Putting Green—Its Planting and Care.** O.M. Scott and Sons Company, Marysville, Ohio, USA. 39 pp., 14 illus. mostly in color, no tables, small index, fold-stapled semisoft covers in multicolor with black lettering, 5.9 by 8.6 inches (150 × 218 mm).

* A **very rare** manual that is a must for collectors of historical turfgrass books. It is a well-prepared, small manual on the construction, planting, and culture of turfgrasses on putting greens in North America. New dimensions introduced include poling greens and winter overseeding of bermudagrass greens plus discussions of such problems as Pythium blight, snow molds, and winterkill. The 1931 status of bentgrass (*Agrostis* species) and fine-leaf fescue (*Festuca* species) species and cultivars for greens was summarized as follows:

> *Colonial Bent*, also called Rhode Island and Browntop. This is a generally non-creeping variety that does not form as matted and compact a sod as true creeping bent. It is less exacting in its demand for moisture than creeping bent and will even thrive on dry, sandy or clay soil. It has a wide range of usefulness.
>
> *Velvet Bent*. As its name implies Velvet produces a fine stem and leaf growth of velvet-like character. It spreads slowly from overground creeping stems. Pure Velvet Bent seed is almost unknown, as it is usually mixed with other varieties such as South German Mixed Bent. In some instances seed of 50% pure Velvet Bent may be obtained but the cost is almost prohibitive.
>
> *South German Mixed Bent*. This grass is a mixture of Colonial, Velvet and Creeping Bent with some Redtop. It is largely used for putting greens and a good grade of it produces excellent turf. The turf, however, will be variegated in appearance and texture because of the different kinds of bent in the mixture.
>
> *Seaside Bent*, also known as Coos County Bent. Seaside is produced chiefly in Oregon and Washington. It is distinctly a creeping grass spreading by trailing and rooting overground stems. Seaside has been commercially available for only a few years, but in that time has become quite popular. It is being used in increasingly larger amounts each year as its good features are appreciated.
>
> Particularly encouraging have been the results with Seaside in Oklahoma. Other varieties of bent will not survive so far south, but since its introduction into Oklahoma in 1928, Seaside has proven very satisfactory even in the extreme southern parts of the state.
>
> *Creeping Bent*. Creeping Bent is a widely used expression which has lost its value as referring to any particular kind of bent. Formerly it was used to describe South German Mixed Bent but that was misleading as South Germany Mixed Bent contains a very small percentage of true Creeping Bent. There is practically no

> true Creeping Bent seed available except as it may be a part of South German Mixed Bent. Seaside comes nearer to the classification of Creeping Bent than any of the others but it has not been recognized and named as such by the U.S. Department of Agriculture.
>
> Creeping Bent has many different strains and some of them are entirely unsatisfactory for putting greens because they develop a decided grain or they are comparatively coarse. Of the many strains the Washington has emerged as the best, and is now used almost exclusively for the vegetative planting of greens. There is one other strain which compares with it, that being the Metropolitan. It is quite popular in the East.
>
> Other Grasses for Greens
> While bent grasses are used mostly on putting greens, other varieties may be planted because of the necessity of using something cheaper, or because bent is not adapted, or because watering facilities are not available. In general, bent grasses are not suitable for use on greens which are not to be watered regularly during dry seasons.
>
> *Fescue* is a grass adapted for putting greens as it stands fairly close cutting and holds up well under heavy wear. Chewings New Zealand Fescue is the only safe variety of fescue to seed on greens. European Red Fescue, so highly recommended, is often badly mixed with either Hard or Sheep's Fescue. These are coarse and inclined to bunch, and are unsatisfactory on a golf course except in the rough.

1932† **Lawn Making and Maintenance.** O.M. Scott and Sons Company, Marysville, Ohio, USA. 59 pp., 6 illus., no tables, indexed, fold-stapled textured semisoft covers in blues with dark blue lettering, 6.0 by 9.0 inches (152 × 228 mm).

* A **truly rare** publication that is a must for collectors of historical turfgrass books. It is a small manual on the establishment and culture of lawns. It is oriented to the amateur home lawn enthusiasts and turfgrass soil–climatic conditions in North America. This is a practical, well-written guidebook of its time. A total of 6,000 copies were printed.

1932† **Campus and Athletic Field.** O.M. Scott and Sons Company, Marysville, Ohio, USA. 66 pp., 7 illus., 1 table, small index, fold-stapled semisoft textured covers in tans with dark brown lettering, 6.0 by 8.9 inches (150 × 227 mm).

* A **truly rare** publication that is a must for collectors of historical turfgrass books. It is a handbook on the establishment and culture of lawns and sports fields for educational facilities in the United States. The first book published on this subject in North America. It contains two tables in the back that summarize these practices on football, soccer, and baseball fields plus tennis courts, polo fields, bowling greens, and croquet courts. A total of 6,000 copies were printed. A suggested 1932 approach to turfgrass rolling is discussed as follows:

> **Rolling.** Early spring rolling presses into the ground roots and grass crowns which have been heaved by the alternate freezing and thawing of winter. An occasional compacting of most soils insures proper capillarity, and presses the soil around the grass roots, so that they will be able to draw on the supply of moisture and plant food.
>
> Proper moisture conditions for rolling occur for just a few days every spring. One must watch carefully, as lawns are drying out, to be sure that the soil is in the right condition to roll. Light or sandy soils may be rolled when wet without injury. If a heavy clay soil is rolled when wet it becomes puddled and packed, with consequent injury to the turf. A well drained soil is in condition to roll earlier in the spring than one poorly drained.

1932† **Turf for Parks.** O.M. Scott and Sons Company, Marysville, Ohio, USA. 67 pp., 7 illus., 2 tables, indexed, fold-stapled textured semisoft covers in burnt orange with black lettering, 5.9 by 8.9 inches (150 × 226 mm).

* A **truly rare** publication that is a must for collectors of historical turfgrass books. This is a pioneering handbook on the establishment and culture of turfgrasses for lawns and sports fields, including greens. It is oriented to turfgrass soil conditions in the United States. A total of 1,500 copies were printed.

1932† **Cemetery Lawns—Makingand Maintenance.** O.M. Scott and Sons Company, Marysville, Ohio, USA. 66 pp., 9 illus., 2 tables, indexed, both a textured hardbound in green with gold lettering and fold-stapled semisoft covers in teal with dark blue lettering, 6.0 by 8.9 inches (151 × 227 mm).

* A **truly rare** publication that is a must for collectors of historical turfgrass books. It is the first book on the construction, establishment, and culture of cemetery turfgrasses. It is practically oriented to turfgrass soil conditions in the United States. A total of 1,500 copies were printed.

n.d.† **Lawns.** O.M. Scott and Sons Company, Marysville, Ohio, USA. 24 pp., no illus., no tables, no index, fold-stapled
circa semisoft covers in multicolor with black lettering, 3.3 by 5.6 inches (84 × 143 mm).
1933

✳   A **rare**, small booklet on the construction, culture, and renovation of lawns. It is oriented to cool-season turfgrass conditions in the United States. A total of 48,000 copies were printed.

n.d.†   **Lawn Care.** O.M. Scott and Sons Company, Marysville, Ohio, USA. 56 pp., 24 illus., 1 table, no index, fold-stapled
circa   soft covers in white with black lettering, 5.9 by 8.9 inches (151 × 226 mm).
1934

✳   A **relatively rare** bound set of the first 27 issues of Nos. 1 through 27 of the **Lawn Care** bulletin published from 1928 through 1933. Emphasis is on the pest management of individual weed and insect problems of turfgrasses. A must book for the collectors of historical turfgrass books. In 1928, the direct mail circulation of **Lawn Care** was around 5,000 and, by 1961, it was around 4 million. The status of broadleaf plantain (*Plantago major*) weed control in lawns in 1928 was summarized as follows:

> Control Measures
> Any attack should take place before the weed starts to seed. Scattered plants may be destroyed by individual treatments with carbolic acid without much defacement of the turf. Squirt a few drops of carbolic acid into the crown of each plant with an ordinary machine oil can.
>
>   Of course, the separate plants may be cut out but care should be taken to remove most of the taproot.
>
>   Where Plantain are too thick to make individual treatment practical and yet the area is not the proper size for cultivation, it is worthwhile to destroy them with iron sulfate. Dissolve 3 pounds of iron sulfate in 5 to 10 gallons of water and sprinkle or spray this solution on 1000 square feet of lawn area. Several treatments may be necessary for best results. Repeat applications should be carried out at the first signs of recovery shown by the weeds. This chemical in solution leaves an ugly rust stain on clothing, metal and stonework. Care should be taken in handling it to avoid spilling and splashing.
>
>   Badly infested lawns should be plowed under, cultivated, fertilized and replanted with clean seed. Regularly feeding an established lawn is the most practical weed control measure. A partially inorganic grass food used frequently makes possible a weed free lawn. Usually weeds, even Plantain, invade only that turf which is not strong enough to resist their presence.

    See *Lawn Care—1938.*

1938†   **Lawn Care.** O.M. Scott and Sons Company, Marysville, Ohio, USA. 54 pp., 26 illus., no tables, no index, fold-stapled semisoft covers in white with black lettering, 5.4 by 8.2 inches (138 × 208 mm).

    This small book is a condensation of **Lawn Care** issue Nos. 1 through 47 from 1928 to 1938, with a numbering system of 40 through 55.

n.d.†   **Good Lawns—Planting and Maintaining.** First edition. O.M. Scott and Sons Company, Marysville, Ohio, USA. 32
circa   pp., no illus., no tables, no index, fold-stapled semisoft covers in yellow and greens with yellow lettering, 3.5 by 6.2
1938   inches (88 × 158 mm).

✳   A **rare**, small guide to lawn establishment from seed, with a smaller section on lawn care in the back. It is practically oriented to conditions in North America.

    Revised circa 1939.

n.d.†   **Lawn Care.** O.M. Scott and Sons Company, Marysville, Ohio, USA. 48 pp., 21 illus., no tables, no index, fold-stapled
circa   soft covers in white with black lettering, 5.9 by 8.9 inches (151 × 227 mm).
1938

✳   A **relatively rare** bound set of issue Nos. 28 through 47 of the **Lawn Care** bulletin from 1934 through 1937. Control of individual weeds and turfgrass soil cultural practices for lawns is emphasized. It is a must book for collectors of historical turfgrass books.

    See *Lawn Care—1950.*

1939†   **Good Lawns—planting and maintaining.** Second revised edition. O.M. Scott and Sons Company, Marysville, Ohio, USA. 37 pp.; 15 illus. as vignettes; no tables; no index; fold-stapled semisoft covers in greens, yellow, and white with yellow lettering; 3.5 by 6.2 inches (88 × 157 mm).

✳   An expansion of a **very rare**, small guide to lawn establishment, renovation, and culture under the climatic conditions in North America.

    Revised circa 1947.

1944†   **Lawn Care.** O.M. Scott and Sons Company, Marysville, Ohio, USA. 228 pp.; 93 illus.; 2 tables; indexed; semisoft covers in green, black, and yellow with yellow and black lettering in a 2-hole binder; 6.0 by 9.0 inches (153 × 229 mm).

✳   A **relatively rare** composite of the 77 issues of **Lawn Care** published between 1928 and 1944. It represents a chronological history of lawn turfgrass developments in North America from 1928 to 1944.

1944† **Scotts Lawn Care.** O.M. Scott and Sons Company, Marysville, Ohio, USA. 48 pp., 39 illus., 1 table, no index, fold-stapled semisoft covers in white with black lettering, 5.4 by 8.3 inches (138 × 210 mm).

   ＊   This **rare,** small book is a condensation of **Lawn Care** issue Nos. 47 through 77 from 1938 to 1944, with a numbering system of 56 through 78.

n.d.† **Good Lawns—Planting and Maintaining.** Third revised edition. O.M. Scott and Sons Company, Marysville, Ohio,
circa   USA. 44 pp.; 16 illus.; no tables; no index; fold-stapled semisoft covers in greens, black, and white with black lettering;
1947   3.5 by 6.3 inches (88 × 159 mm).

   ＊   There are only minor revisions of a **very rare** guide booklet. Multigraphed from rubber plates.

1950† **Lawn Care.** O.M. Scott and Sons Company, Marysville, Ohio, USA. 72 pp.; 55 illus. with 8 in color; no tables; indexed; soft bound in greens, black, and white with black lettering; 5.3 by 8.2 inches (134 × 208 mm).

   ＊   A **rare** digest of the first 100 issues of **Lawn Care** from 1928 to 1950. It is organized in 13 chapters covering the establishment and care of lawn turfgrasses in the United States.
          Reprinted in 1952.

1950† **Lawn Care.** Region editions. O.M. Scott and Sons Company, Marysville, Ohio, USA. 140 pp., 98 illus., no tables, indexed, semisoft covers in multicolor with black lettering in a 2-hole binder, 5.5 by 8.3 inches (140 × 210 mm).

   ＊   A book composed of two bound parts. Part one is a condensed bound set of the current issue Nos. 98 through 115 of the **Lawn Care** bulletin that was published five times per year. Part two is a digest of 13 key back issues published prior to No. 98 that are updated. The major adjustments necessitated by the war effort during the 1940s are illustrated by the following statement:

> ### War Conservation
> As Lawn Care goes to press, word comes from the War Production Board asking for a voluntary curtailment in the use of mineral Nitrogen. This was one of the several ingredients of Turf Builder (Scotts special grass food) but no more of it will be used until there is again a plentiful supply. This ruling had been anticipated for some time so we are ready with a new formula for the duration.
>
>   In developing Turf Builder from the first ton produced in 1928 to the present time we have had only one thought—to make it the best grass food that could be produced. Now, though we must make a change in the ingredients, the goal is the same—to make it the best that can be made from those materials available in war times.

         Revised in 1952 and 1953.

1956† **Turf Talks.** O.M. Scott and Sons Company, Marysville, Ohio, USA. 92 pp., 34 illus., 4 tables, no index, semisoft bound in yellow with black lettering, 6.0 by 9.0 inches (151 × 228 mm).

   ＊   A compilation of 20 issues of **Turf Talks** published from 1936 to 1956. The focus is on the establishment and culture of turfgrasses on golf courses. The expertise of turfgrass scientists was in great demand during the WWII effort as is shown in the following:

> First off is Dr. John Monteith, Jr., former head of the Green Section, who is attached to the office of the Chief of Engineers, U.S. Army, at Washington. He has developed a turf unit in the construction branch and is interested in getting grass on new airfields. He is assisted in Washington and in the field by Dr. Fred Grau, formerly of Penn State.
>
>   Dr. H.B. Sprague, formerly at New Brunswick, N.J., is a Captain attached to the Army Air Corps. Dr. H.B. Musser, formerly of Penn State College, has been commissioned a Major in the Engineers to co-operate in turf work. His headquarters are in Atlanta, Georgia. Captain George E. Harrington is Army Air Forces Liaison Officer in Atlanta, Georgia; John W. Bengston is in the U.S. District Engineer's Office at Mobile, Alabama, and Alton E. Rabbit, of Capitol City Parks fame, is with the Bureau of Aeronautics in the Navy Department and still in Washington. Gordon H. Jones is with the U.S. Engineers at Dallas, Texas.

n.d.† **What's that weed?** O.M. Scott and Sons Company, Marysville, Ohio, USA. 23 pp., 34 illus. in color, no tables, no
circa   index, fold-stapled soft covers in dark green and white with light green and white lettering, 8.9 by 7.0 inches (227 ×
1965   178 mm).

      A small booklet on weed identification and controls in North America. Excellent close-up color photographs.

1971† **Scotts Professional Turf Seminar.** The O.M. Scott & Sons Company, Marysville, Ohio, USA. 230 pp., 261 illus. with 1 in color, 8 tables, no index, glossy semisoft covers in black with gold lettering in a comb binder, 8.5 by 11.0 inches (215 × 278 mm).

Figure 12.69. The walking Little Wonder centrifugal grass seeder, circa 1911. [45]

A manual concerning seed labeling, blending, cultivars, planting, cultural practices, and pests. There is a glossary of selected terms in the back.

1972† **Scotts Professional Turf Manual.** The O.M. Scott & Sons Company, Marysville, Ohio, USA. 179 pp., 267 illus., 4 tables, no index, glossy semisoft covers in black with gold lettering in a comb binder, 8.3 by 11.0 inches (210 × 278 mm).

A manual concerning cultivars, seed labels, planting, cultural practices, and pests of turfgrasses. A glossary of selected terms is included in the back.

1972† **Scotts Professional Turf Manual.** O.M. Scott and Sons Company, Marysville, Ohio, USA. 252 pp., 199 illus., 2 tables, no index, board covers in red with white lettering in a 3-hole binder, 8.5 by 11.0 inches (217 × 279 mm).

A manual on turfgrass species and cultivar characteristics; seeding and sodding; turfgrass culture; and diseases, insects, and weeds. It is practically oriented to the turfgrasses and conditions in the United States. There is a glossary and summary of products in the back.

1973† **Professional Turf Seminar.** O.M. Scott and Sons Company, Marysville, Ohio, USA. 341 pp., 286 illus. with 1 in color, 2 tables, no index, glossy semisoft covers in reds with gold lettering in a comb binder, 8.5 by 10.9 inches (215 × 278 mm).

A manual concerning turfgrass species and cultivars, seeds and seeding, turfgrass culture, diseases, insects, and weeds. It is oriented to turfgrass practitioners and conditions in the United States.

1973† **Scotts Professional Turf Manual.** ProTurf Division, O.M. Scott and Sons Company, Marysville, Ohio, USA. 265 pp., 170 illus., 14 tables, no index, board covers in white with red lettering in a 3-hole binder, 8.4 by 10.9 inches (213 × 278 mm).

A manual on turfgrass species and cultivar characteristics, seeding, turfgrass culture, diseases, insects, and weeds. It is practically oriented to the turfgrasses and conditions found in the United States. There is a glossary in the back.

1974† **Lawn and Product Information Manual.** Regional editions. Scotts Training Institute, O.M. Scott & Sons Company, Marysville, Ohio, USA. 228 pp., 214 illus., 15 tables, no index, semisoft covers in green with gold lettering in a comb binder, 8.4 by 10.9 inches (214 × 277 mm).

A manual on turfgrass species and cultivars, establishment, culture, pest management, and Scott products for lawns in North America. It is oriented to retailers of products to home lawn owners. The manual contains a glossary in the back.

1974† **Professional Turf Manual.** ProTurf Division, O.M. Scott and Sons Company, Marysville, Ohio, USA. 125 pp., 22 illus., 2 tables, no index, semisoft covers in black and orange with white and gold lettering in a comb binder, 8.4 by 10.9 inches (214 × 277 mm).

A manual on turfgrass growth, diseases, fertilization, seeds, insects, and nematodes. It is practically oriented to the turfgrasses and conditions in the United States. There is a glossary in the back.

1975† **Professional Turf Seminar Manual.** ProTurf Division, O.M. Scott and Sons Company, Marysville, Ohio, USA. 145 pp., 53 illus., 17 tables, no index, glossy semisoft covers in cream and red with white and gold lettering in a comb binder, 8.4 by 10.9 inches (214 × 277 mm).

A manual on turfgrass growth, seed selection, soil management, nutrition and fertilizers, diseases, and insects. It is practically oriented to the turfgrasses and conditions in the United States.

1975† **Lawn and Product Information Manual—Southern Edition.** O.M. Scott and Sons Company, Marysville, Ohio, USA. 163 pp., 156 illus., 9 tables, no index, glossy semisoft covers in dark green with gold lettering in a comb binder, 8.0 by 11.0 inches (204 × 279 mm).

    A manual on turfgrass species for lawns used in warm climates plus their planting and culture, including weeds, insects, and diseases. There is a glossary in the back.

1976† **Professional Turf Seminar Manual.** ProTurf Technical Services Department, The O.M. Scott & Sons Company, Marysville, Ohio, USA. 167 pp., 76 illus., 26 tables, no index, glossy semisoft covers in multicolor with white and blue lettering in a comb binder, 8.5 by 11.0 inches (216 × 278 mm).

    A manual concerning soil chemistry, soil testing, fertilizers, turfgrass seeds, water quality, and water testing.

1976† **Scotts Professional Turf Manual.** ProTurf Division, O.M. Scott and Sons Company, Marysville, Ohio, USA. 261 pp., 222 illus., 15 tables, no index, board covers in white with green lettering in a 3-hole binder, 8.4 by 11.0 inches (213 × 279 mm).

    A manual on turfgrass species and cultivar characteristics, seeding, turfgrass culture, diseases, insects, nematodes, and weeds. It is practically oriented to the turfgrasses and conditions in the United States. There is a glossary in the back.

1977† **Professional Turf Seminar Manual.** Scotts ProTurf, The O.M. Scott & Sons Company, Marysville, Ohio, USA. 149 pp., 116 illus., 11 tables, no index, glossy semisoft covers in white and orange with green lettering in a comb binder, 8.5 by 11.0 inches (215 × 278 mm).

    A manual concerning soils, pests, and seeds as related to turfgrasses.

1979† **Professional Turf Seminar—Presentation Notes.** Scotts Proturf, O.M. Scott and Sons Company, Marysville, Ohio, USA. 68 pp., 14 illus., 2 tables, no index, semisoft covers in white and black with white and black lettering in a comb binder, 8.5 by 11.0 inches (215 × 278 mm).

    A booklet on the culture of extensive turfgrass areas in the United States. Emphasis is placed on turfgrass nutrition and fertilization practices plus a section on disease and insect control.

1979† **Information Manual for Lawns.** O.M. Scott and Sons Company, Marysville, Ohio, USA. 96 pp., 177 illus. with 133 in color, 4 tables, no index, both glossy hardbound and semisoft bound versions in multicolor with green and white lettering, 7.5 by 10.0 inches (190 × 254 mm).

    A quality book on the establishment and culture of lawn turfgrasses. It is practically oriented to the home lawn enthusiasts and conditions in the United States. It contains extensive chapters on weeds, diseases, and insects. There are excellent color photographs throughout.

1980† **Professional Turf Seminar—Presentation Notes.** Scotts ProTurf, The O.M. Scott & Sons Company, Marysville, Ohio, USA. 39 pp., 13 illus., 2 tables, no index, fold-stapled glossy semisoft covers in multicolor with white lettering, 8.4 by 11.0 inches (214 × 278 mm).

    A small manual concerning fertilizers, soil testing, and pests.

1982† **Scotts Guide to the Identification of Grass Diseases and Insects.**

    See *Converse, James Thayer—1982.*

1985† **Scotts Guide to the Identification of Grasses.** The O.M. Scott and Sons Company, Marysville, Ohio, USA. 84 pp., 73 illus., no tables, index in front, textured semisoft bound in green with gold lettering, 6.3 by 3.3 inches (159 × 83 mm).

    This is essentially a reprinting.

    See *Converse, James Thayer—1979.*

    † *Reprinted in 1995* with fold-stapled textured glossy covers in green with white lettering.

1985† **Scotts Guide to the Identification of Dicot Turf Weeds.** The O.M. Scott and Sons Company, Marysville, Ohio, USA. 119 pp., 100 illus., no tables, index in front, textured semisoft covers in black with white lettering in a ring binder, 6.3 by 3.3 inches (159 × 83 mm).

    This is essentially a reprinting.

    See *Converse, James Thayer—1979.*

    † *Reprinted in 1995* with fold-stapled textured glossy covers in black with white lettering.

Organic Gardening and Farming

1970† **Lawn Beauty—The Organic Way.** Compiled by Glenn F. Johns and the editors of Organic Gardening and Farming. Rodale Books Inc., Emmaus, Pennsylvania, USA. 411 pp.; 34 illus.; 12 tables; indexed; cloth hardbound in light green with black lettering and a glossy jacket in multicolor with black, white, and green lettering; 5.7 by 8.7 inches (145 × 221 mm).

    A book of the establishment and culture of lawns using organic means. Included are organic nutritional sources and proposed organic methods of weed, disease, and insect control.

    Reprinted in 1971.

Figure 12.70. Soil being prepared for seeding a fairway in central Ohio, USA, circa 1937. [32]

Otto, Wayne Dennis

Dennis Otto (1939–2004) was born in Mequon, Wisconsin, USA, and completed the 2-Year Turf Management program at Pennsylvania State University in 1960. He served as Golf Course Superintendent at Ozaukee Country Club, Wisconsin, from 1969 to 2002. He then became a spokesperson for the Milwaukee Metropolitan Sewerage District and formed his own business, Turfgrass Support Services. He was elected President of the Wisconsin Golf Course Superintendents Association (WGCSA) for 1977 to 1978. Dennis Otto received the WGCSA Distinguished Service Award in 2002, and the Wee One Foundation was formed as a tribute to him.

2004†   **Turf Management—Tools and Techniques.** Coauthor.
       See *Quast, Danny Herman—2004.*

Owada, Katsuhiro

Katsuhiro Owada was born in Chiba City, Chiba Prefecture, Japan. He has worked for the Soubu City Development Company Group responsible as General Manager of superintendents for ten golf courses. Katsuhiro Owada was elected President of the Japan Turfgrass Superintendents Society from 1996 to 2008.

2007†   **Cultivation for Turfgrass Management—Its Effects and Method Using Machinery.** Coeditor.
       See *Inamori, Makoto—2007.*

P

Palmer, Arthur E.[3]

n.d.   **Garden Lawns and Greens: Their Maintenance, Improvement and Renovation.** Arthur E. Palmer. The Country
circa   Gentleman's Association Limited, Letchworth, Hertfordshire, England, UK. 44 pp., 12 illus., 2 tables, no index, fold-
193x   stapled semisoft covers in green with gold lettering, 4.7 by 7.2 inches (120 × 183 mm).
  *   A **very rare**, small booklet on the culture, renovation, and pest control of turfgrasses, with emphasis on lawns in En-
       gland. It includes small sections on lawn bowling greens, putting greens, cricket pitches, and sports fields. Arthur Palmer
       discussed the characteristics of various fertilizers and topdressings available in England in the 1930s:

> MANURES and DRESSINGS
> Basic Slag.—While this is a splendid fertiliser for grass land, it should not be used on lawns, bowling or putting greens, for it encourages clover.
>   Bone Meal.—is a very suitable dressing for lawns on heavy soils. It must be finely ground, or birds will pick up and carry off any coarse particles. It may be used at the rate of 3 or 4 ounces per square yard, or

say, 6 to 8 lbs. to each pole (30¼ square yards). It should be applied in autumn or early spring, being well brushed in, the lawn then being rolled.

Charcoal is sometimes employed as a top-dressing to lawns and greens. It keeps the surface soil sweet, purifies and aerates it, absorbs Ammonia and prevents it being washed away, giving it up as the roots of the grass require it. It improves the colour of the grass and encourages strong growth. It should be used in a ground or finely granulated state and is specially suitable for clay and peaty soils. It may be applied at the rate of ½ to 1 lb. per square yard, being brushed into the turf and then rolled.

Kainit supplies Potash and is useful on light soils and may be employed in the proportion of about 2 lbs. per square rod. It is better, however, when Potash is needed, to apply Sulphate of Potash.

Lawn Fertiliser.—For most lawns a complete fertiliser, that is, one containing the correct proportions of Nitrogen, Phosphoric Acid and Potash, is the best. It is used in various proportions, according to its strength and composition. It can be prepared to suit any kind or condition of soil.

Lawn Sand.—There are a variety of preparations sold by different firms under the name of "Lawn Sand," and while some are very efficacious and satisfactory, others are not. A first class make of Lawn Sand is a valuable adjunct to the gardener's remedies. By its use many of the weeds growing on a lawn can be destroyed without damage to the grass, thus saving an enormous amount of labour.

Some people wonder how it is that Lawn Sand will destroy weeds and not grass. This is due to the fact that most weeds are covered with hairs to which the Lawn Sand adheres, extracting the moisture and drying them up; but in the case of grass, the blades are smooth and erect so that the powder does not adhere to them, but falls on the ground near the roots. Some strong growing weeds will require more than one application, and those with tap roots should have the crowns cut off, the Lawn Sand being applied immediately to the cut surfaces.

Panella, Adelmo (Professor Emeritus*)*

Adelmo Panella (1912–1992) was born in Spoleto, Italy, and earned a LAUREA degree from the University of Perugia in 1937. He joined the faculty at the University of Perugia in 1946 and was promoted to Lecturer in 1951, to Assistant Professor in 1953, and to Full Professor in 1959. He also served as Director of the Plant Breeding Institute at the University of Perugia from 1959 to 1981. He was active in grass breeding and turfgrass cultivar evaluation. Professor Panella was elected President of the Italian Society of Agronomy for 1972 through 1978 and President of the Italian Society of Plant Genetics from 1978 to 1981.

1972† **Tappeti Erbosi: Impianto, manutenzione, impieghi (Turfgrasses: Establishment, maintenance, and use).** First edition. Adelmo Panella. Edagricole, Bologna, Italy. 199 pp., 181 illus. with 30 in color, 6 tables, indexed, cloth hardbound in multicolor with white lettering, 8.0 by 10.3 inches (205 × 262 mm). Printed entirely in the Italian language.

Figure 12.71. Manual-push wheelbarrow seeder, as used to plant grass seed in central Ohio, USA, circa 1927. Eighty acres were seeded in three days, working from daybreak to darkness. [32]

This is the first turfgrass book published in Italian. It is organized in three parts: soils and turfgrass selection; construction and turfgrass culture; and ornamental, recreational, and functional turfs—soccer and golf course turfgrasses. It is oriented to the diversity of environmental conditions in Italy.

Revised in 1981 and 2000.

1981† **Tappeti Erbosi: Impianto, manutenzione e impieghi (Turfgrasses: Establishment, maintenance, and use).** Second revised edition. Adelmo Panella. Edagricole-Edizioni Agricole, Bologna, Italy. 199 pp., 183 illus. with 41 in color, 6 tables, indexed, cloth hardbound in multicolor with white lettering, 8.0 by 10.4 inches (204 × 264 mm). It is printed in the Italian language.

Essentially a reprinting of the first edition.

Revised in 2000.

2000† **Tappeti Erbosi (Turfgrasses).** First edition. A. Panella, P. Croce, A. De Luca, M. Falcinelli, F.S. Modestini, and F. Veronesi. Calderini Edagricole, Bologna, Italy. 475 pp., 242 illus. with 234 in color, 50 tables, indexed, textured cloth hardbound in multicolor with white lettering, 8.0 by 10.4 inches (203 × 265 mm). It is printed in the Italian language.

A science-based book on the establishment and culture of both cool- and warm-season turfgrasses for use on golf courses, lawns, and sports fields. It is oriented to the diversity of environmental conditions in Italy. There is a list of selected references in the back.

Park, Edward Walker (LDS and RSC)

Eddie Park (1926–1989) was born in Silecroft, Cumberland, England, UK. His school, St Bees, possessed a 9-hole golf course which he maintained minimally with a horse during WWII. He then qualified as a Dental Surgeon at Edinburgh Dental School, Scotland, in 1947. He moved to Sheffield, England, in 1950, where he was professionally active as a dental surgeon. Dr. Park developed a lifelong interest in golf courses. He was a regular contributor to **Golf Monthly,** writing golf course maintenance articles. Dr. Park was elected Chairman of the Green Committee from 1966 to 1974 and Captain in 1975 of Lindrick Golf Club and also President of the Sheffield Golf Union in 1977.

1990† **Real Golf.** Eddie Park. A. Quick & Company, Dovercourt, Essex, England, UK. 179 pp., 184 illus. with 62 in color, 1 table, no index, cloth hardbound in green with gold lettering, 7.9 by 11.2 inches (200 × 285 mm).

This book consists of an interesting collection of golf course–maintenance and golf turfgrass–oriented articles published in **Golf Monthly** and authored by Eddie Park that was assembled and published after his death by Nancie Park. It also includes a few selected articles by their son, Nicholas Park, plus an introductory set of historical articles by "Historicus," the pseudonym used by Eddie Park. The focus is on golf turfgrass issues and developments in the United Kingdom.

Parker, Charles Wilson

Charles Parker (1896–1971) was born in Bath, Maine, USA. By the 1930s, he was involved in greenkeeping in Middlesex County, Massachusetts. He was Head Greenkeeper at Belmont Spring Golf Club in Massachusetts and then at Agawam Hunt Club in Rhode Island from 1953 to 1966, when he retired.

1939† **The Lawn—How to Make It, and How to Maintain It.** Charles W. Parker. The Gardener's Library, Hale, Cushman & Flint, Lexington, Massachusetts, USA. 118 pp., 33 illus., 8 tables, small index, cloth hardbound in light green with dark green lettering and a jacket in greens with black and white lettering, 5.4 by 8.5 inches (137 × 216 mm).

* A **relatively rare**, small book on lawn establishment and culture. It evolved from a bulletin written for the Massachusetts Horticultural Society. There is an attempt to make the content more science-based than many lawn books published in the 1930s. In a section titled "Western Lawn Practice," Charles Parker stated the following:

> Mowing grass too low and too often is a practice that is doing a great deal of injury to lawns. This is the last word in the care of lawns and is based on long continued experiments by the Kansas State College. Three inches is the best height to set the lawn mower for mowing a Blue Grass lawn, according to the newest theory. Cutting the grass any shorter reduces its vigor. The Kansas authorities go so far as to say that mowing need only be done when the grass gets so tall that it is ready to fall over. All of these revolutionary statements are made from the viewpoint of maintaining a dense turf.

Parker, Mark

Mark Parker was born in Sydney, New South Wales, Australia, and earned Certificates in Greenkeeping from Wollongoing Training & Further Education in 1987 and in Horticulture from Ryde Technical & Further Education (TAFE) in 1990 and a Masters of Agriculture degree in Turf Management from Sydney University in 1996. He served as Assistant Golf Course Superintendent from 1987 to 1988 and as Superintendent at Concord Golf Club since 1988. He also has been Director of Eclectic Golf Construction P/L from 1999 to 2007 and of Mark Parker Golf P/L since 2007. Mark Parker was elected President of the New South Wales Superintendent's Association in 1988.

2007† **How To Build A Sand Based Golf Green.** Coauthor.
See *McIntyre, David Keith—2007.*

Parsons, Samuel Howland, Jr.

Samuel Parsons (1844–1923) was born in New Bedford, Massachusetts, USA, the son of noted horticulturalist Samuel Parson Sr., who operated a nursery in Flushing, Queens, New York. He earned a Bachelor in Philosophy degree from Yale University in 1862. After apprenticing in farming, he returned to New York City to become an apprentice with landscape architect Calvert Vaux from 1879 to 1884 and then became his partner, serving until 1895. He next was the Head Landscape Architect of New York City, serving from 1895 to 1911. He became a well-known American landscape architect. His key designs were the Beaux-Arts park effects in New York City, the development of Central Park, and the San Diego Balboa Park. Samuel Parsons was elected President of the American Society of Landscape Architects from 1905 to 1907 and was one of 10 founding members.

1891 **Landscape Gardening.** Samuel Parsons Jr. G.P Putnam's Sons, New York City, New York, USA. 341 pp., 230 illus., no tables, large index, cloth hardbound in green with gold lettering, 6.7 by 9.6 inches (170 × 245 mm).

A late 1800s landscape gardening book that has an unusually significant discussion for that time concerning lawns in American gardening. The author emphasizes the low cost of lawns compared to that of other landscape plants available at that time. There are sections on lawns and landscapes for parks, railway yards, church yards, and cemeteries. Samuel Parsons laments the lack of quality grass seed as follows:

> But I will say now and here, that sad experience has proved long ago that want of pure grass seed, and the right variety of grass seed, is one of the chief causes of the failure and uncertainty of lawns. Seedsmen cannot furnish pure grass seed, because no one grows pure grass seed, and certainly not the best sorts of seed for making good greensward.

Perris, Jeffrey

Jeff Perris was born in Rhondda, Mid-Glamorgan, Wales, UK, and earned a BSc degree from the University College of North Wales, Bangor, in 1966. He then joined the Sports Turf Research Institute (STRI) as an Agronomist involved in turfgrass advisory and lecture activities. He became a Senior Advisory Agronomist in 1987, the STRI Director

Figure 12.72. Comparative control of English daisy with lawn sand on left side of a green, circa 1911. Lawn sand contains ammonium sulfate and iron sulfate that kills the aboveground foliage by foliar burn. [46]

of Consultancy Services in 1988, and an STRI Senior Consultant in 2006. He was named a Director of STRI in 1998.

1996†   **The Care of the Golf Course.** Second revised edition. J. Perris and R.D.C. Evans (editors) with 22 contributing authors. The Sports Turf Research Institute, Bingley, West Yorkshire, England, UK. 340 pp., 36 illus., 14 tables, no index, glossy hardbound in multicolor with white lettering, 6.7 by 9.0 inches (170 × 231 mm).

A revision and update of a practical reference book that is an assemblage of 132 short articles on golf course turfgrass culture by 22 STRI authors that were previously published in the quarterly **Bulletin of the Sports Turf Research Institute**. It is oriented to the cool, moist environmental conditions of the United Kingdom. There is a bibliography in the back.

First edition published in 1992.
See Hayes, Peter—1992.

2000†   **Grass Tennis Courts—How To Construct And Maintain Them.** J. Perris (editor) with 13 contributing authors. The Sports Turf Research Institute, Bingley, West Yorkshire, England, UK, in partnership with The All England Lawn Tennis Club, Wimbledon, England, UK. 160 pp. including 1 p. advertising, 108 illus. with 58 in color, 2 tables, no index, glossy hardbound in multicolor with white lettering, 6.8 by 9.6 inches (172 × 243 mm).

A speciality book on the history, playing characteristics, construction, planting, culture, renovation, and line marking of turfgrass courts for tennis. It is oriented to conditions in the United Kingdom. One of the few books published on this subject. There is an extensive bibliography in the back.

2008†   **All About Bowls.** Third revised edition. Jeff Perris (editor). Sports Turf Research Institute, Bingley, West Yorkshire, England, UK. 224 pp.; 211 illus. with 149 in color; 23 tables; no index; glossy semisoft bound in greens and white with brown, black, and white lettering; 6.3 by 9.1 inches (160 × 230 mm).

A book concerning the construction, establishment, and culture of turfgrasses for bowling greens under the conditions in the United Kingdom. It also includes an interesting historical review of bowls starting in the late 1200s. There is a foreword by Tony Allcock. Reprints of three research papers from the **Journal of the Sports Turf Research Institute** are included in the back.

First and second editions published in 1988 and 1992.
See Evans, Roger David Calder—1988 and 1992.

Perry, Floyd D.[2]

1994†   **Pictorial Guide To Quality Groundskeeping "Covering All The Bases."** Floyd Perry. Holland Communications Company, Orlando, Florida, USA. 103 pp., 359 illus. with 91 in color, no tables, no index, glossy semisoft covers in multicolor with black lettering in a plastic comb binder, 11.0 by 8.5 inches (279 × 216 mm).

This guide is primarily a visual representation of techniques and facilities used for lower-budget baseball and softball fields. It includes dugouts, batting cages, field tarps, and signage. Outlines of a seasonal turfgrass maintenance schedule, field preparation for planting, and game-day maintenance schedule are presented in the book.

1995†   **Pictorial Guide To Quality Groundskeeping "There Ain't No Rules."** Floyd Perry. Grounds Maintenance Services, Orlando, Florida, USA. 109 pp. plus 1 p. advertising, 470 illus. with 203 in color, 2 tables, no index, glossy semisoft covers in multicolor with black lettering in a plastic comb binder, 11.0 by 8.5 inches (279 × 216 mm).

This guide is primarily a visual representation of techniques and facilities used for soccer, football, Little League fields, and multiple-use facilities. In the back, there are short articles on managing sports turfgrass by Dr. Tom Samples, warm-season turfgrass weeds by Dr. L.B. McCarty, sports turfgrass cultivation by Dr. Gil Landry, fertilization by Eugene Mayer, and growth regulators by Drs. Dennis P. Shepard and Joseph M. DiPaola.

1997†   **Pictorial Guide To Quality Groundskeeping "Maintain It Easy—Keep It Safe."** Floyd Perry. Grounds Maintenance Services, Orlando, Florida, USA. 125 pp., 695 illus. with 158 in color, no tables, no index, glossy semisoft covers in multicolor with black lettering in a comb binder, 11.0 by 8.5 inches (279 × 215 mm).

This guide is primarily a visual representation of techniques used on low-budget softball fields and facilities.

Pessarakli, Mohammad (Dr.)

Mohammad Pessarakli was born in Gorgan, Golestan Province, Iran, and earned an Associate of Arts (AA) degree in Forestry from the University of Agriculture and Natural Resources of Gorgan in 1973, a BSc degree and an MSc degrees from Arizona State University in 1977 and 1978, respectively, and a PhD degree from the University of Arizona in 1981. He then joined the University of Arizona as an Instructor from 1981 to 1983, as Research Associate from 1983 to 1986, as Senior Research Specialist from 1989 to 2005, and as Associate Research Professor since 2005. He has been involved in teaching, research, and adult education related to turfgrass culture and management. Dr. Pessarakli received the Five Star Faculty Teaching Award from the University of Arizona in 1984, 1989, and 2001.

2008† **Handbook of Turfgrass Management and Physiology.** Mohammad Pessarakli (editor). CRC Press, Boca Raton, Florida, USA. 716 pp., 109 illus. with 28 in color, 88 tables, large index, glossy hardbound in multicolor with green and black lettering, 6.9 by 10.0 inches (175 × 253).

A compilation of papers on a diverse range of subjects related to turfgrasses. There are a total of 47 contributing authors from Argentina, Guam, Iran, Italy, and the United States.

Petersen, Martin

Martin Petersen was born in Aarhins, Denmark, and earned an MSc degree in Agricultural Science from the Royal Veterinary and Agricultural University, Copenhagen, Denmark, in 1951. He served as Research Scientist at the Stole Experiment Station from 1951 to 1953, as Research Scientist at the Royal Veterinary and Agricultural University for 1954, and as Cytologist at the Weibull Plant Breeding Institute from 1955 to 1956. Then from 1956 to 1964, he was Head of Plant Breeding for Daehnfeldt Seed Company, with emphasis on turfgrasses. Martin Petersen was elected to the International Turfgrass Society Board of Directors from 1993 to 2001.

1975 **Anlæg af sportspladser: Jord/Vand/Luft (Construction of Sportsgrounds: Soil-Water-Air).** Martin Petersen. Daehnfeldt Markfrø, Odense, Denmark. 163 pp., 200 illus., 54 tables, indexed, textured hardbound in brown with silver lettering, 6.0 by 8.2 inches (152 × 209 mm). Printed entirely in the Danish language.

A book on the construction of turfgrass sports fields under conditions in Denmark. It includes sections on soils, sands, organic matter, soil amendments, root growth, water movement, and drainage systems. There is a list of selected references in the back.

1978 **Slid på Sportspladser: Komprimering og Renovering (Wear on Sportsgrounds: Compaction and Renovation).** Martin Petersen. Daehnfeldt Markfrø, Odense, Denmark. 128 pp., 194 illus., 17 tables, no index, glossy semisoft bound in multicolor with black lettering, 6.0 by 8.4 inches (153 × 213 mm). Printed entirely in the Danish language.

A book addressing turfgrass traffic stress on sports fields in Denmark. It includes sections on wear tolerance of turfgrass species, soil compaction, soil aeration porosity, and earthworm activity plus sections on turfgrass renovation, reconstruction, and machinery for sports fields.

1981† **Græsplæner—Principper and Funktioner (Turfgrass Areas—Principles and Functions).** Martin Petersen. A/SL. Daehnfeldt, Odense, Denmark. 362 pp., 207 illus., 86 tables, no index, glossy semisoft bound in greens and white with yellow and white lettering, 8.1 by 11.7 inches (207 × 297 mm). Printed entirely in the Danish language.

A book on the establishment and culture of cool-season turfgrasses under the conditions in Denmark. Included are sections on seed mixtures, polystand compatibility, root growth, mowing, fertilization, environmental stresses, wear tolerance, and diseases.

Pettigrew, William Wallace

William Pettigrew (1867–1947) was born in Auchinleck, Ayrshire, Scotland, UK, and obtained his early apprentice training in horticulture as a Kew gardener from 1888 until 1890. In 1891, he became Head Gardener to the Cardiff Corporation, where he was active in landscape design and construction. Subsequently in 1915, he became General Superintendent of the Manchester Parks Department, serving for 32 years. Formation of the Institute of Park Administration was largely due to his efforts and he was elected President of the Institute for three years. William Pettigrew was awarded the Victoria Medal of Honour in Horticulture from the Royal Horticultural Society for 1926.

1937† **Municipal Parks—Layout, Management & Administration.** * W.W. Pettigrew. The Journal of Park Administration Limited, London, England, UK. 279 pp. plus 16 pp. of advertising, 136 illus., 13 tables, indexed, cloth hardbound in green with gold lettering, 6.1 by 9.6 inches (155 × 244 mm).

An early book on the management and administration of public parks. The origin of parks is discussed in the first chapter. There are chapters on the design and construction of turfed parks, sports fields, bowling greens, tennis courts, putting greens, and recreation grounds. Forty-three sample forms and record formats are included in the appendix. It is oriented to conditions in the United Kingdom.

Phillips, P.M.³

n.d.
circa
late
195x
Glimpses of Groundsmanship. Compiled by P.M. Phillips and W. Hardiman from the articles of 16 authors. National Association of Groundsmen, Wimbledon, England, UK. 146 pp. plus 34 pp. of advertising, 37 illus., 1 table, no index, semisoft bound in light green with dark green lettering, 5.5 by 8.4 inches (140 × 213 mm).

A book oriented to sports ground construction, establishment, and culture under conditions in the United Kingdom. Subjects addressed include soils, fertilizers, turfgrass characteristics, weeds, pests, diseases, snow removal, and equipment. There is a glossary of selected terms in the back.

Figure 12.73. Horse-drawn, 10-foot-wide, gravity-drop, mechanical grass seed planter, circa 1910. [12]

Pierce, S.J.

1951 **Bowling Greens: Their Construction and Maintenance.** Revised second edition. S.J. Pierce and C.B. Rigney. New South Wales Bowling Association, Sydney, New South Wales, Australia. 31 pp., 6 illus., no tables, no index, semisoft stapled covers in fawn with black lettering, 5.5 by 8.7 inches (140 × 220 mm).

An early booklet concerning the construction and culture of bowling greens as practiced in New South Wales, Australia, in the 1950s.

Piper, Charles Vancouver (Dr.)

Charles Piper (1867–1926) was born in Victoria, British Columbia, Canada, and grew up in Seattle, Washington, USA. He earned a BSc degree in 1885 and a MSc degree in 1892, both from the Territorial University of Washington. He served as Professor of Botany and Zoology at the Washington Agricultural College and School of Science from 1892 to 1903, where he documented the early flora of Washington and Idaho and conducted pioneering research on the crop diseases of the region. He also devoted one year as a Fellow at the Gray Herbarium, Harvard University, where he earned an MSc degree in 1900. He then joined the United States Department of Agriculture in Washington, DC, as a Systematic Taxonomist and Agrostologist in 1903 to 1905 and then was in charge of forage crop investigations until his death in 1926. During this period, he discovered sudangrass (*Sorghum × drummondii*) in Africa and introduced it into the US as a forage grass. He was advising on turfgrass problems, such as establishing grass on sand dunes at the National Golf Links, by 1908. He was a botanist who began his golf course turfgrass investigations in 1912 and developed an increasing interest in turfgrasses, although originally responsible for forage research. When the Green Section of the United States Golf Association (USGA) was established in 1921, Dr. Piper became Chairman of the USGA Research Committee. He continued in this capacity until his death. He traveled to Europe to study the turfgrasses used on golf courses in 1924. He is attributed with developing methods for vegetatively propagating creeping bentgrass (*Agrostis stolonifera*) for golf putting greens. Dr. Piper was elected President of the Washington Botanical Society in 1908 and president of the American Society of Agronomy (ASA) in 1914. He was awarded an American Association for the Advancement of Science Fellow in 1905, ASA Fellow in 1925, and an Honorary DSc degree from Kansas Agricultural College in 1921.

1917† **Turf for Golf Courses.** Charles V. Piper and Russell A. Oakley. The Macmillan Company, New York City, New York, USA. 262 pp. plus 5 pp. of advertising, 73 illus., 2 tables, indexed, cloth hardbound in dark green with gold lettering and a jacket in yellow and brown with brown lettering, 5.5 by 8.5 inches (140 × 216 mm).

＊ This book is a must for collectors of historical turfgrass books. It is a **relatively rare**, pioneering treatise. This important early book on golf course construction, establishment, and maintenance was the first authored by formally educated scientists with doctor of philosophy degrees and specializing as agrostologist of grasses and agronomist of soils. The authors' approach was based on known fundamental science rather than only the observations and experiences of golf enthusiasts. It contained the most comprehensive characterization of turfgrasses that had been assembled up to that time. The chapter subject titles include the following:

Figure 12.74. Dr. Charles Piper, circa 1920s. [7]

I.	General View	VIII.	The Turf Grasses for Different Purposes
II.	Soils for Turf Grasses	IX.	The Making of the Turf
III.	Fertilizers	X.	Subsequent Care
IV.	Manures, Composts, and Other Humous Materials	XI.	Weeds and Their Control
		XII.	Animal Pests
V.	Lime and Its Use	XIII.	Turf Machinery
VI.	The Important Turf Plants	XIV.	Experimental Work on Golf Courses
VII.	How to Distinguish Different Kinds of Turf	XV.	Personal Experiences

Note that diseases of turfgrasses are not discussed, indicating that pathogens were not thought to be a significant problem of turfgrasses in 1917. Both authors were employees of the United States Department of Agriculture. The United States Golf Association assisted in publication of the book. Chapter 14 contains descriptions of the experimental work of James B. Olcott in Connecticut and Fred W. Taylor near Philadelphia. Presentations of personal experiences in growing turfgrasses on golf courses near Philadelphia at the Merion Golf Club by Hugh I. Wilson, on Long Island at the National Golf Links of America at Shinnecock Hills by Charles B. Macdonald, and near Washington, DC, at the Columbia Golf Club by Dr. Walter S. Harban are also of interest. The approach to root-zone construction of putting greens suggested by Drs. C.V. Piper and Russell Oakley in 1917 was as follows:

Sandy soils are bettered by the addition of silt or clay, or both, so as to obtain in the surface foot about one-third of these materials. Where clay is used, it should be dry and pulverized, as otherwise a good mixture is not secured. Humus-forming materials should be added in large quantity, preferably enough to cover the green to a depth of three to four inches. The thorough mixing of these different elements will form a satisfactory sandy loam soil. In some seaside courses the error has been made of attempting to build a green by making a layer of peat eight to twelve inches thick and covering with a few inches of soil, but good results are not to be expected from any such method. Peat remains practically inert unless well mixed with soil. Preferably, it should be composted for a full year before using.

Stiff clay soils are best improved by the admixture of sand and humous materials. Three to four inches of sand may advantageously be incorporated in the top twelve inches, together with a liberal amount of humus-forming materials. Enough of the latter to cover the ground to a depth of four inches is not excessive. When the clay, sand, and humous materials are thoroughly mixed by plowing and cultivating, a very fair substitute for a clay loam is secured.

†   *Reprinted in 1923 and in 1929 in the Rural Science Series with jacket.*

Pohl, Richard Walter (Dr. and Emeritus *Distinguished Professor)*

Richard Pohl (1916–1993) was born in Milwaukee, Wisconsin, USA, and earned a BSc degree *summa cum laude* from Marquette University in 1939 and a PhD degree in Botany from the University of Pennsylvania in 1947. He then joined the faculty in the Department of Botany and Plant Pathology at Iowa State College in 1947 as an Assistant Professor involved in plant taxonomy teaching and research. He was named an Associate Professor in 1951 and a Full Professor in 1956. In 1950, he became Director of the Department Herbarium, serving until his retirement in 1986. He was elected President of the American Society of Plant Taxonomists for 1973. Dr. Pohl was named Distinguished Professor of Sciences and Humanities at Iowa State University in 1975 and Distinguished Fellow of the Iowa Academy of Science in 1982.

1996   **Agnes Chase's First Book of Grasses: The Structure of Grasses Explained for Beginners.** Coauthor.
　　　　See *Clark, Lynn Gail—1996.*

Potter, Daniel Andrew (Dr. and *Professor)*

Dan Potter was born in Ames, Iowa, USA, and earned a BSc degree from Cornell University in 1974 and a PhD degree in Entomology from Ohio State University in 1978. He joined the Entomology faculty at the University of Kentucky in 1979 and rose to Full Professor in 1989. He is active in entomology research and teaching involving the biology and management of turfgrass insects. Dr. Potter received the Distinguished Achievement Award in Urban Entomology from the Entomological Society of America in 1995, the Master Teacher Award from the University of Kentucky in 1998, the Distinguished Research Professorship at the University of Kentucky in 1998, and the Distinguished Achievement Award in Teaching from the Entomological Society of America in 1999.

1998†   **Destructive Turfgrass Insects—Biology, Diagnosis and Control.** Daniel A. Potter. Turfgrass Science and Practice Series, Ann Arbor Press, Chelsea, Michigan, USA. 344 pp., 386 illus. with 235 in color, 9 tables, large index, glossy hardbound in multicolor with white lettering, 7.0 by 10.0 inches (178 × 254 mm).

　　　　A practical book on the insect, mite, and vertebrate pests of both cool- and warm-season turfgrasses. Selected references are presented at the end of each chapter. An extensive glossary of terms is included in the back, along with a list of selected reference sources. There is a preface by Dr. James B Beard.

Figure 12.75. Turfgrass experimental plots at Arlington Turf Garden, Washington, DC, USA, circa 1930. [21]

Powell, Will D.[3]

1947   **"The Green Manual"—The Construction of Bowling Greens and Their Maintenance.** W.D. Powell (editor) with 12 contributing authors: W.D. Powell, Billy Powell, Peter Fraser, F.A. Sharman, Dr. Thos. D. Hall, Harry W. Johns, A.R.P. Walker, A.B.M. Whitnall, D. Meredith, S.B. Walters, A.F. Furniss, and A.A. Langford. Official Handbook of the Johannesburg and Southern Transvaal S.D.B.A. Greens Advisory Committee, Johannesburg, South Africa. 122 pp., 37 illus., 1 table, no index, semisoft bound in light green with black lettering, 5.5 by 8.6 inches (140 × 218 mm).

　*　　This is a **truly rare** publication that is a must for collectors of historical turfgrass books. It is a small, early handbook on the construction, establishment, and culture of turfgrasses on bowling greens. It is practically oriented to conditions in the Transvaal region of South Africa. Peter Fraser made the following comments about turfgrasses for bowling greens in South Africa:

> Opinions differ somewhat regarding the variety of grass best suited for bowling greens. Especially in the Transvaal, where there are several varieties to choose from. Although there are several, they are all Cynoden, or sub-types, the difference (although very little) having been caused by climate, soils, and also by the effluxion of time. Personally, I recommend Cynoden Transvaalensis, or better known as "Florida," as it has proved the best in many respects—more especially in the high veld. Its chief advantages are: it can be treated during any period of its existence with impunity, more especially as regards top-dressing. It can also be "shaved off" to earth level should a matted surface accumulate. Weakness in growth, especially at rink-ends, can be treated either by puncturing and top-dressing, or be returfed during active growth, whereas Bradley must be treated while dormant, neither can it stand up to "shaving off." Of recent years it has become most susceptible to attacks from eelworm, therefore Bradley cannot be recommended as a grass suitable for bowling greens.
>
> Maginnes has proved good in less fertile soils, such as Kimberley and districts, but in the Johannesburg area it is much too rampant in growth even when grown on the poorer soils, and consequently "Stubbles" with close mowing. This renders it an inferior bowling green grass. In some localities there are indigenous grasses which do as well under cultivation as the best varieties of Cynoden. In some parts of the Orange Free State the variety, Harrismith, which is also a Cynoden, has proved an ideal grass.
>
> Coastal conditions for grass growing are entirely different from those inland. Therefore different types have to be employed. I do know different types of the fine Cynodens have been tried to no avail, therefore, the coarser local types have still to be adhered to. A fine textured sort, known as Umgeni, was tried out in Durban, and although a success on golf putting greens, failed to withstand the "wear and tear" associated with the game of bowls. A grass in Durban which has been introduced and does show signs of promise, is "Campbell" grass, being finer in texture than many of the kweeks at present in use. The best of the coastal kweeks is that known as "Cynoden Dactylon," although not to be compared in texture refinement with Cynoden Transvaalensis, it does provide a good keen bowling surface under intelligent supervision.

Price, Robert John (Dr.)

Robert Price was born in Cardiff, Wales, UK, and earned a BSc degree in Geography and Geology from the University of Wales in 1957, a PhD in Geology and Geomorphology from the University of Edinburgh in 1961, and a DSc from the University of Wales in 1984. He served as a Reader in Geography at the University of Glasgow from 1962 to 1990, conducting research and teaching in quaternary environments and glacial geomorphology. Since then, he has been a private golf consultant in the management, marketing, and design of golf courses in Scotland. Dr. Price was elected President of the British Geomorphological Research Society in 1979 and President of Geography (Section E) of the British Association for the Advancement of Science in 1983.

1989†   **Scotland's Golf Courses.** First edition. Robert J. Price. Aberdeen University Press, Aberdeen, Scotland, UK. 235 pp., 137 illus. with 8 in color, 10 tables, indexed, hardbound in multicolor with white lettering, 5.9 by 9.0 inches (150 × 228 mm).

An interesting, unique book on 425 golf courses in Scotland. Each course location is assessed relative to the geological history, landforms, and native vegetation. The golf courses are presented in four regions via extensive maps. A selected bibliography is included in the back.

2002†   **Scotland's Golf Courses.** Second revised edition. Robert J. Price. The Mercat Press, Edinburgh, Scotland, UK. 243 pp., 145 illus. in color, 8 tables, indexed, glossy semisoft bound in multicolor with white lettering, 6.2 by 9.2 inches (158 × 234 mm).

This revision and update of a unique book now encompasses 538 golf courses in Scotland.

Puhalla, James

Jim Puhalla was born in Youngstown, Ohio, USA. He was owner of Valley Landscaping for 18 years and President and owner of Sportscape International Inc. since 1978. Jim Puhalla is active in the design, construction, and renovation of sports fields.

1999† **Sports Fields—A Manual for Design, Construction and Maintenance.** Jim Puhalla, Jeff Krans, and Mike Goatley. Ann Arbor Press, Chelsea, Michigan, USA. 478 pp., 222 illus., 62 tables, large index, glossy hardbound in blues and white with white lettering, 7.0 by 10.0 inches (177 × 253 mm).

A technical book concerning the design, construction, culture, and renovation of sports fields under the conditions in North America. It is organized in five parts: "I. Principles of Sports Turf Culture"; "II. Sports Fields," encompassing baseball, softball, football, rugby, soccer, lacrosse, field hockey, lawn bowling, and croquet; "III. Other Sports Surfaces," such as tennis, track and field, volleyball, playgrounds, and bocce; "IV, Quality, Evaluation and Safety of Sports Facilities"; and "V. Ancillary Information." There is a glossary in the back.

2003† **Baseball and Softball Fields: Design, Construction, Renovation, and Maintenance.** Jim Puhalla, Jeff Krans, and Mike Goatley. John Wiley & Sons Inc., Hoboken, New Jersey, USA. 235 pp.; 133 illus.; 22 tables; indexed; glossy hardbound in multicolor with yellow, crème, green, and white lettering; 7.0 by 10.0 inches (177 × 253 mm).

A specialty book concerning the practical aspects involved in the layout, construction, and maintenance of baseball and softball facilities under conditions in the United States. Included are skinned areas, pitcher's mound, warning track, batter's box, marking and lines, backstop, dugout, bullpen, and fencing.

Q

Quast, Danny Herman

Danny Quast was born in Dayton, Ohio, USA, and earned an Associate's degree in Turf Management from the Stockbridge School of Agriculture, University of Massachusetts, in 1963. He was Superintendent of Grounds at the W.A. Cleary Corporation from 1963 to 1965; at Troy Country Club, Ohio, from 1965 to 1968; at Springfield Country Club, Ohio, from 1968 to 1973; and at Milwaukee Country Club, Wisconsin, from 1973 to 1989; he then became Grounds Manager at Medinah Country Club, Illinois, from 1989 to 2000. He has since been owner of DHD Tree Products Inc. in Juneau, Wisconsin. He has been elected Director of Wisconsin Golf Course Superintendents Association in 1974, President of Midwest Regional Turf Foundation in 1976, and President of the Wee One Foundation from 2005 to 2007. Danny Quast received the GCSAA Distinguished Service Award in 2004.

2004† **Golf Course Turf Management—Tools and Techniques.** Danny H. Quast and Wayne Otto. McGraw Hill Companies Inc., New York City, New York, USA. 538 pp., 154 illus., 41 tables, small index, glossy hardbound in multicolor with white and green lettering, 5.9 by 9.0 inches (150 × 226 mm).

A book on management of a golf course from the perspective of two golf course superintendents. It is focused on cool-season turfgrasses for the cool climatic regions of the United States. The chapters on weeds, insects, pesticides, and best management practices are written by Dr. Matthew Fagerness and are basically similar to chapters in the 2004 book by Fagerness and Donald Johns.

R

Randell, Roscoe (Dr. and Professor Emeritus)

Roscoe Randell was born in Tuscola, Illinois, USA, and earned a BSc degree in Agronomy, an MSc degree in Education, and a PhD degree in Entomology, all from the University of Illinois in 1951, 1959, and 1970, respectively. He served as County Extension Agent and then Area Extension Agent in Illinois from 1956 to 1964 and as Extension Entomologist at the University of Illinois from 1965 to 1991. He conducted adult educational and research programs for the management of horticultural insects. Dr. Randell received the Distinguished Service Award of the Illinois Turfgrass Foundation in 1985, the Sustained Excellence Award of the University of Illinois Cooperative Extension Service in 1990, and the Certificate of Meritorious Service from Epsilon Sigma Phi, University of Illinois in 1990.

1979† **How to control Lawn Diseases and Pests.** Coauthor.
See *Shurtleff, Malcolm C.—1979.*

1987†,   **Controlling Turfgrass Pests.** Coauthor.
1997†,
   and
2003†

      See *Shurtleff, Malcolm C.—1987, 1997, and 2003.*

Reed, Frederick John

Frederick Reed (1904–1970) was born in England, UK. He became Director of Ryder's Lawn Turf Advisory Department in St
    Albans, Hertfordshire, England, UK.

1950†   **Lawns and Playing Fields.** F.J. Reed. Faber and Faber Limited, London, England, UK. 212 pp., 31 illus., 35 tables,
    indexed, cloth hardbound in green with gold lettering and a jacket in green and white with black lettering, 5.5 by 8.5
    inches (140 × 216 mm).

       The basics of turfgrass establishment and culture are addressed in this book. Considerable emphasis is placed
on descriptions of the turfgrasses and associated weeds, insects, and diseases. In addition to coverage of lawns,
there are sections on turfgrass establishment and culture for cricket, lawn bowls, lawn tennis, golf, parks, soccer,
rugby, and field hockey. It contains a glossary in the back. Reed made the following comments about herb lawns
in 1950:

> Uniformity of texture required for many outdoor games has resulted in the indigenous herbs disappearing
> from lawns and sports greens, in fact many delightful species are regarded as weeds and treated as such.
>   While the superiority of lawn grasses for field games is undoubted the herb lawn should be cultivated
> far more than it is today. It is true that all dwarf herbs may not wear well but it is possible to form and to
> maintain a golf green from chamomile as is evidenced by the chamomile greens which at one time were a
> feature of the course of the Royal Guernsey Golf Club.
>   The celebrated John Evelyn knew chamomile well, for in his famous *Kalendarium Hortense,* writing
> under October he states: "It will now be good to beat, roll and mow carpet walks and chamomile."
>   Charm and variety could be added to garden lawns and increased interest brought to ornamental turf
> in public parks by the establishment of herb lawns.
>   The time is ripe for a revival of interest in this link with Elizabethan gardens.

n.d.   **Lawn Grasses.** First edition. F.J. Reed. Ryder and Son, St Albans, Hertfordshire, England, UK. 28 pp., 25 illus., 2
circa   tables, no index, fold-stapled semisoft covers in green with green lettering, 5.0 by 8.0 inches (127 × 205 mm).
1954

      A small guide to the identification and uses of turfgrasses under conditions in the United Kingdom.

1963†   **Lawn Grasses.** Second revised edition.
    Contains minor revisions.

Rees, John Leslie

John Rees (1890–1955) was born in Port Talbot County, Wales, UK. He studied at Swansea Technical College, Swansea, in 1915
    and then was Assistant Lecturer in Agriculture at the University College, Bangor, and a teacher at Caernarvon, Mer-
ioneth. He earned a BSc degree in Agriculture in 1927 and an MSc degree in 1928, both from the University College
of Wales, Aberystwyth. He accepted a grassland research position in 1928 at the Waite Agricultural Research Institute,
University of Adelaide, Australia. During WWII, John Rees was a Grassland Officer in the Ministry of Agriculture,
United Kingdom. He then was a turfgrass consulting agronomist in Sydney, New South Wales, Australia.

1962†   **Lawns, Greens, and Playing Fields—Their Making and Maintenance.** J.L. Rees. Angus and Robertson Limited,
    Sydney, Australia. 290 pp., 60 illus., 20 tables, indexed, cloth hardbound in green with gold lettering and a jacket in
    multicolor with white lettering, 5.3 by 8.2 inches (134 × 210 mm).

      This book addresses the basics of turfgrass establishment and culture, with emphasis on Australian conditions. In
the back, there are chapters on turfgrasses for home lawns, cricket grounds, lawn bowls, golf courses, and sports fields.

Rice, William E.

Bill Rice (1926–1987) was born in Philadelphia, Pennsylvania , USA. He was a Research Associate in the Department of Botany
    at the University of Maryland. Later he conducted contract work for the Botany Department, National Museum of
Natural History, Smithsonian Institution, on the endangered flora project from 1976 to 1983 and for the United States
Department of Agriculture on a check list of introduced plants from 1983 to 1985.

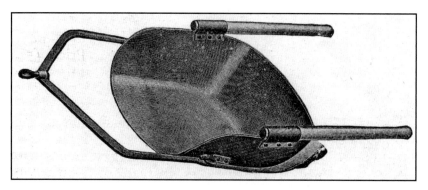

Figure 12.76. Earth scoop for volume movement of soil in which one person drove the horse and one walked behind the scoop to guide the depth of soil collection and location of dumping, 1939. [16]

1986† **A Checklist of Names for 3,000 Vascular Plants of Economic Importance.** Coauthor.
　　See *Terrell, Edward Everett—1986.*

Rieke, Paul Eugene (Dr. and *Professor* Emeritus)

Paul Rieke was born in Kankakee, Illinois, USA, and earned a BSc degree in 1956 and an MSc degree in 1958, both in Agronomy from the University of Illinois, and a PhD degree in Soil Chemistry from Michigan State University (MSU) in 1963. He then joined the Soil Science faculty at Michigan State University in 1963 as an Assistant Professor. He became Associate Professor in 1969, Professor in 1976, and retired in 1999. He served as an Extension Specialist in Greenhouse Soil Management from 1963 to 1965 and in Turfgrass Management from 1980 to 1999; he also has been active in turfgrass teaching and research, with emphasis on soil management, turfgrass cultivation, topdressing, fertilization, and soil testing. He was elected Turfgrass Science Division C-5 Chair in 1973, to the Crop Science Society of America (CSSA) Board of Directors for 1972 through 1974, and to the American Society of Agronomy Board of Directors from 1978 through 1981. Dr. Rieke was awarded CSSA Fellow in 1986, the Outstanding Faculty Award from MSU Institute of Agricultural Technology in 1989 and 1999, the Distinguished Faculty Award of MSU College of Agriculture and Natural Resources Alumni Association in 1990, the Meritorious Service Award of Michigan Turfgrass Foundation in 1994, the MSU Outstanding Extension Specialist Award in 1995, the GSAA Distinguished Service Award in 1996, the USGA Green Section Award in 1997, the USGA Piper and Oakley Award in 1999, the Turfgrass Producers International Honorary Membership in 1999, the Golf Association of Michigan Distinguished Service Award in 2000, and the CSSA Turfgrass Science Award in 2003.

1977† **The Mathematics of Turfgrass Maintenance.** Coauthor.
　　See *Beard, James B—circa 1977.*

2001† **Turfgrass Soil Fertility and Chemical Problems—Assessment and Management.** Coauthor.
　　See *Carrow, Robert Norris—2001.*

Roberts, Beverly Cruickshank

Beverly Cruickshank was born in Westerly, Rhode Island, USA, and earned a BSc degree in Biology from New Jersey College for Women (Rutgers) in 1953 and an MA degree in Education from the University of Rhode Island in 1978. She was founding President of Focus Associates of Rhode Island from 1975 to 1982; Director of Christian Education at Kingston Congregational Church from 1976 to 1978 and of First Baptist Church from 1979 to 1980; and Office Manager, Graphic Artist, and Writer for the Lawn Institute from 1982 to 1992. She was elected President of Sigma Kappa Sorority for 1967 to 1968.

1991† **Lawn Institute Special Topics Sheets.** Coauthor.

See *Roberts, Eliot Collins—1991.*

Roberts, Eliot Collins (Dr.)

Eliot Roberts was born in Camden, New Jersey, USA, and earned a BSc degree in Agricultural Chemistry from Rhode Island State University in 1950 and both MSc and PhD degrees in Soil Science and Plant Physiology from Rutgers University in 1952 and 1955, respectively. He became Assistant Professor from 1954 to 1957 and Associate Professor from 1957 to 1959, both in Agronomy at the University of Massachusetts, and, subsequently, Associate Professor from 1959 to 1964 and Professor from 1964 to 1967, both in Agronomy and Horticulture at Iowa State University. He was very involved in teaching turfgrass science and management and also conducted turfgrass research with emphasis on nutrition and soils. He became Chairman of the Ornamental Horticulture Department at the University of Florida from 1967 to 1970 and then Chairman of the Plant and Soil Science Department at the University of Rhode Island (URI) from 1970 to 1973. He then became Professor of Plant and Soil Science at URI from 1973 to 1982, where he was Rhode Island State Research and Extension Coordinator for Rural Development. From 1982 to 1992, he served as Executive Director of the Lawn Institute and, from 1992 to 2006, as Principal of Rosehall Associates in Pleasant Hill, Tennessee. He was elected to the Board of Directors of the American Society of Agronomy (ASA) for 1961 to 1963, the Chair of the Turf Division 7 of the Weed Science Society of America in 1962, the Chair of the Turfgrass Management Division C-5 for 1965, and the Board of Directors of the Crop Science Society of America (CSSA) from 1964 to 1966. Dr. Roberts received ASA Fellow in 1971, CSSA Fellow in 1985, the Meritorious Service Award from the Iowa Turfgrass Institute in 1988, the Long Standing Support Award of the Missouri Valley Turfgrass Association in 1990, the Hall of Fame Award of the Oklahoma Turfgrass Research Foundation in 1992, and the Presidential Award of the Nebraska Turfgrass Association in 1994.

1991† **Lawn Institute Special Topics Sheets.** Eliot C. Roberts and Beverly C. Roberts. The Lawn Institute, Pleasant Hill, Tennessee, USA. 385 pp., 99 illus., 17 tables, no index, glossy hard covers in white with green lettering in a 3-hole binder, 8.5 by 11.0 inches (217 × 279 mm).

A manual on turfgrass trends, seed strategies, turfgrass cultivars, cultural practices, pest control, and turfgrass establishment.

Rockwell, Frederick Frye

Fred Rockwell (1884–1976) was born in Brooklyn, New York, USA. Rockwell was a gardening writer and was the author of 25 books. He also was active as a gardening lecturer. He began his career in New York City, first as Circulation Manager of **Garden Magazine** from 1906 to 1907, as Manager of the Wiltshire Book Company from 1907 to 1908, and as a publicity man for the **New York Call** in 1908 to 1909. From 1917 to 1919, he was Director of the Service Department and Manager of Burpee's Seed Farms, Doylestown, Pennsylvania, returning in the latter year to New York City where he was Advertising Manager with the Tuthill Advertising Agency from 1919 to 1921. He then was Manager of the nursery and bulb department and of publicity and advertising at Seabrook Company (Seabrook Farms), Bridgeton, New Jersey, from 1921 to 1925. In 1933, he became the first Garden Editor of the **New York Sunday Times**, continuing until 1943, when he founded and began to edit his own magazine, **The Home Garden**. This was combined with **Flower Grower Magazine** in 1953, and Fred Rockwell remained Senior Editor of **Flower Grower—Home Garden** until 1965. He was a founder of the Men's Garden Club of New York and was elected President of the Men's Garden Clubs of America from 1941 to 1946.

1929† **Lawns.** F.F. Rockwell. The Home Garden Handbooks, The Macmillan Company, New York City, New York, USA. 87 pp., 17 illus., 2 tables, small index, cloth hardbound in light green with dark green lettering and a jacket in multicolor with black lettering, 4.7 by 7.3 inches (120 × 185 mm).

* A **rare**, practical book on lawn establishment and culture. It is designed for the lawn enthusiasts and conditions in North America. Fred Rockwell emphasized the distinct differences in cultural philosophies between lawns and crops:

> Grass, as has already been pointed out, must be looked upon as a growing crop, and like any other crop requires methods of culture adapted to it. Speaking of the lawn as a "carpet" of green, we are prone to

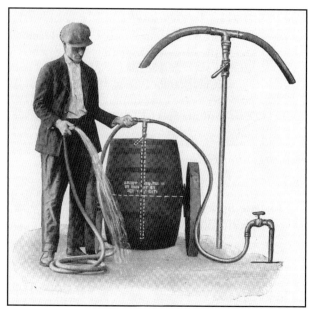

Figure 12.77. Early suggested methods for applying water-soluble fertilizer, such with a manual water can (left) and a hose proportioning device plus water barrow and truck (right), 1920s. [18, 41]

forget that this particular carpet has roots, and must be fed, cultivated, and protected from its enemies like any other plants about the place. The fact that the culture required for grass plants differs from that accorded others makes it none the less essential.

In fact, if one stops to consider the situation a moment, it is evident that the grass plants must continue their existence under peculiarly difficult conditions. The digging up and enriching of the soil from season to season; the cultivation of the soil during spring and summer to destroy weeds and to conserve moisture; the application of a mulch in summer for the same purpose, or in winter to afford protection, is out of the question. Moreover, the plants must withstand wear and tear to which nothing else is subjected. Altogether we ask a great deal of the tiny plants which go to make up our lawns.

†   *Reprinted in 1930 with a different jacket and in 1935 and 1940.*

Rodriguez, Ian Rex (*Dr.* and *Associate Professor*)

Ian Rodriguez was born in Tampa, Florida, USA, and earned a BSc degree in 1994 and an MSc degree in 1998, both in Environmental Horticulture from the University of Florida, and a PhD degree in Plant Physiology from Clemson University in 2003. Since 2002, he has been on the faculty at Lake City Community College, Lake City, Florida. Dr. Rodriguez instructs and advises two-year students in golf course management, landscaping, pest control, and irrigation.

2003†   **Fundamentals of Turfgrass and Agricultural Chemistry.** Coauthor.

       See *McCarty, Lambert Blanchard—2003.*

Rorison, Ian Henderson (*Dr.* and *Professor* Emeritus)

Ian Rorison was born in Dartford, Kent, UK, and earned BA degree in Botany in 1952 and both MA and DPhil degrees in Biological Sciences in 1956, all from Oxford University. He served as Head of the Grassland Research Unit of the Nature Conservancy Council from 1969 and 1973, as Director of the Plant Ecology Unit of the Natural Environmental Research Council from 1974 to 1989, as Honorary Lecturer in the Department of Animal and Plant Sciences at the University of Sheffield. Dr. Rorison has studied mineral nutrition and climatic interactions relating to environmental change. He was elected Vice President of the British Ecological Society from 1983 to 1985.

1980†   **Amenity Grassland—An Ecological Perspective.** I.H. Rorison and Roderick Hunt (editors). John Wiley and Sons Limited, Chichester, England, UK. 261 pp., 72 illus. with 2 in color, 49 tables, indexed, textured cloth hardbound in burgundy with gold lettering and a glossy jacket in multicolor with green and burgundy lettering, 5.9 by 9.0 inches (150 × 228 mm).

A compilation of chapters by 14 individual authors covering specific subjects addressed during a meeting held at the University of Sheffield in 1978. The discussions encompass both intensively managed turfgrasses and the extensive seminatural grasslands that are not used primarily for agricultural production. It is oriented primarily to cool-season conditions in the United Kingdom.

Rowell, John Bartlett (Dr. and Professor)

John Rowell was born in Pawtucket, Rhode Island, USA, and earned a BSc degree from Rhode Island State College in 1941 and an MSc degree from the University of Minnesota in 1942. He then served in the US Coast Guard from 1942 to 1945. Subsequently, he earned a PhD degree in Plant Pathology from the University of Minnesota in 1949. He was Assistant Research Professor of Plant Pathology at the University of Rhode Island from 1948 to 1949. Then he accepted a position with the Agricultural Research Service, United States Department of Agriculture, at the University of Minnesota in 1949. His pathology research focused on rust diseases and *Puccinia* culture techniques. Dr. Rowell retired in 1980.

1951† **Fungus Diseases of Turf Grass.** Coauthor.
See *Howard, Frank Leslie—1951.*

Ruhr-Stickstoff Aktiengesellschaft

n.d. **Rasenpflege (Lawncare).** Ruhr-Stickstoff-Bochum, Fernruf, Germany. 33 pp., 21 illus., 1 table, no index, fold-stapled
circa glossy semisoft covers in multicolor with white and black lettering, 5.0 by 8.0 inches (146 × 210 mm).
19xx

A guide for the care of large turfgrass surfaces such as parks, green space, sports fields, equestrian fields, and airfields.

Ruppert, Kathleen Carlton (Dr.)

Kathleen Ruppert was born in DeLand, Florida, USA, and earned BSc degrees in Ornamental Horticulture and Environment Studies in 1977 and in Fruit Crops in 1986, an MSc degree in Agricultural & Extension Education in 1981, and a Doctor of Education (EdD) degree in Education Administration in 1991, all from the University of Florida. She served as Assistant Professor in the Department of Environmental Horticulture, University of Florida, from 1993 to 1997 and was involved in teaching and adult education activities. She then served as Assistant Extension Scientist from 1997 to 2004 and rose to Associate Extension Scientist in 2004 and Extension Scientist since 2009. She was elected a Board Member for 1994 to 1995 and Educational Advisor for 1996 to 1997 of the Florida Chapter of the American Horticultural Therapy Association and also Vice President of the Garden and Landscape Section of the Florida State Horticultural Society for 1993. Dr. Ruppert has received numerous awards as author of various publications and was honored as Outstanding State Specialist by the Florida Association of County Agricultural Agents in 1996; she was awarded the Wachovia Professional Improvement and Enhancement Award by the Florida Cooperative Extension Service in 2003 and 2006; and she was honored as the Outstanding State Specialist by the Florida Extension Association of Family and Consumer Sciences in 2004.

1990† **Florida Lawn Handbook.** First edition. Coeditor.
See *McCarty, Lambert Blanchard—1990.*

1997† **Florida Lawn Handbook.** Second revised edition. Kathleen C. Ruppert and Robert J. Black (editors) with 14 contributing authors. Department of Ornamental Horticulture, University of Florida, Gainesville, Florida, USA. 223 pp.; 135 illus. with 103 in color; 54 tables; no index; glossy semisoft bound in multicolor with white, blue, and black lettering; 8.4 by 10.1 inches (213 × 257 mm).

A major update of the 1990 first edition. Additions include certain turfgrass species, insects, and weeds that occur in Florida.
Revised in 2005.
See *Trenholm, Laurie Elizabeth—2005.*

1997† **Mowing for Money: A Dollar and Sense Guide to Lawn Care Workbook/Business Recordbook.** Kathleen C. Ruppert, Arlene Zavocki Stewart, Brian Owens, and Robert Degner. 4HPSM3O/4HPSM31, Department of Environmental Horticulture, University of Florida, Gainesville, Florida, USA. 37/21 pp., 13/0 illus., 7/15 tables, no index, two sets soft bound in white with black lettering, 8.5 by 11.4 inches (216 × 289 mm).

A simplified two set guide to lawn mowing for pay for youths 14 years of age and older under the laws in Florida. Included are mower operation, maintenance and safety, proper mowing practices, and business aspects plus recordkeeping procedures with representative forms.

S

Sachs, Paul David

Paul Sachs was born in Boston, Massachusetts, USA, and earned a BA degree in Psychology from Windham College, Putney, Vermont, in 1974. He was founder and owner of Newbury Construction Company from 1975 to 1983 and then was associated with Vermont Organic Fertilizer Company from 1988 to 1991 and North Country Organics from 1983 to the present. He served on the Technical Advisory Panel for the National Organic Standards Board of the United States Department of Agriculture from 1988 to 1991. Paul Sachs received the Certificate of Merit from the New York State Turf and Landscape Conference in 1994, the Certificate of Merit from the New York State Turfgrass Convention in 1995, and the Guardian of Nature Award from the World Wildlife Fund in 1997.

2002† **Ecological Golf Course Management.** Paul D. Sachs and Richard T. Luff. Ann Arbor Press, Chelsea, Michigan, USA. 213 pp., 43 illus., 19 tables, indexed, glossy hardbound in multicolor with green and white lettering, 7.0 by 10.0 inches (177 × 254 mm).

A book focused mainly on nonchemical methods for maintenance of turfgrasses on golf courses. Six chapters address the following: soil systems, fertility, compost, analyses, pests, and cultural practices. There is a small glossary and list of selected references in the back.

2004† **Managing Healthy Sports Fields.** Paul D. Sachs. John Wiley & Sons Inc., Hoboken, New Jersey, USA. 244 pp.; 99 illus.; 13 tables; small index; glossy hardbound in green and white with yellow, white, and black lettering; 6.2 by 9.2 inches (156 × 232 mm).

A book focused on using organic materials to maintain low-impact and chemical-free playing fields, with emphasis on soil organisms, composting, chemical analyses, pest management, and mechanical culture techniques.

Samardžija, Nikola Mile

Nikola Samardžija was born in Zagreb, Hrvatska, Yugoslavia, and earned a Mr degree from the University of Zagreb in 1998. Nikola Samardžija has served as a landscape specialist for Zrinjevac in Zagreb, Croatia, with emphasis on parks, lawns, and sports fields.

1998† **Travnjaci—Sportski, parkovni, ukrasni (Turfgrasses—Sports, parks, decorative lawns).** Nikola Samardžija. Zrinjevac d.d., Zagreb, Hrvatska. 144 pp., 106 illus. in color, 23 tables, indexed, hardbound in green with dark green lettering, 7.0 by 9.4 inches (170 × 2240 mm). Printed entirely in the Croatian language.

A small book concerning the establishment and culture of sport fields, parks, and lawns under conditions in Croatia. Total of 1,000 copies printed.

2007† **Ukrasni Travnjak (Decorative Turfgrass).** Coauthor.

See *Čižek, Jan Josip—2007.*

Sampson, Kathleen

Kathleen Sampson (1892–1980) was born in Chesterfield, Derbyshire, England, UK. She earned a BSc degree from the University of London and an MSc degree from Royal Holloway College in 1917. She was Assistant Lecturer in Agricultural Botany at Leeds University from 1915 to 1917 and then became a Lecturer at the University College, Aberystwyth, in 1919; she also served as a Plant Pathologist at the Welsh Plant Breeding Station, Aberystwyth, until 1945. She was active in teaching and research of plant pathogens and the resultant diseases. Kathleen Sampson was elected President of the British Mycological Society in 1938.

Figure 12.78. Horse-drawn wagon with motor-powered apparatus for the shredding and mixing of composted material for topdressing, circa 1916. [35]

1941† **Diseases of British Grasses and Herbage Legumes.*** First edition. Kathleen Sampson and J.H. Western. Cambridge University Press, London, England, UK. 85 pp., 50 illus., 3 tables, indexed, semisoft bound in green with black lettering, 6.0 by 9.7 inches (152 × 246 mm).

* This **rare** treatise discusses the diseases of forage grasses, turfgrasses, and legumes of England in considerable detail. An authoritative work of its time that is supported by an extensive set of references. The authors characterized fairy rings in England as follows:

> Fairy rings occur in all parts of the world on closely grazed or cut turf. *Marasmius oreades* (Bolt.) Fr., several species of *Tricholoma* and other higher fungi are responsible, producing by their centrifugal growth ever-widening circles of a darker green. Rings on English downland, clearly visible in aerial photographs, are many hundreds of feet in diameter and must be centuries old.

† *Reprinted in 1942.*
Revised in 1954.

1954† **Diseases of British Grasses and Herbage Legumes.*** Second revised edition. Kathleen Sampson and J.H. Western. Cambridge University Press, London, England, UK. 118 pp., 58 illus., 5 tables, indexed, cloth hardbound in green with gold lettering, 6.0 by 9.6 inches (153 × 244 mm).

A major revision and update of the first edition. There is an expansion of the diseases discussed and references cited. This edition was published with assistance of the British Mycological Society.

Sanders, Patricia Louise (Associate Professor)

Pat Sanders was born in Cumberland, Maryland, USA, and earned a BSc degree in Medical Technology from West Virginia University in 1952 and a Master of Education (Med) degree in Biological Science from Pennsylvania State University in 1972. In 1969, she became a Research Assistant in the Department of Plant Pathology at Pennsylvania State University and rose to Associate Professor in 1984. She was active in research, adult extension, and teaching in Turfgrass Pathology for 30 years, retiring in 1996.

1977† **A Turf Manager's Guide—Microscopic Identification of Common Turfgrass Pathogens.** Patricia O'Connor Sanders. Pennsylvania Turfgrass Council Inc., University Park, Pennsylvania, USA. 28 pp., 52 illus. with 30 in color, no tables, no index, soft covers in white with green lettering in a comb binder, 6.1 by 8.9 inches (156 × 227 mm).

A small, specialty booklet on the laboratory procedures for microscopic examination of turfgrass fungal pathogens as found in the field. It contains photographs of fungal structures as they appear under a microscope and the common symptoms of nine diseases as they occur on cool-season turfgrasses. There is a short section on nematodes plus a small glossary in the back.

1993† **The Microscope in Turfgrass Disease Diagnosis.** Patricia L. Sanders. Pennsylvania Turfgrass Council Inc., University Park, Pennsylvania, USA. 32 pp., 84 illus. with 29 in color, no tables, no index, glossy semisoft covers in multicolor with black lettering in a comb binder, 6 by 9 inches (152 × 228 mm).

A specialty booklet on the selection and use of a microscope in the identification of fungal pathogens that cause turfgrass diseases. Also it addresses sample preparation and includes keys to pathogen and disease identification. There is a small glossary of selected terms in the back.

Sanders, Thomas William

Thomas Sanders (1855–1926) was born in Martley, Worcestershire, England, UK. He was initially apprenticed to a structure builder but chose to become a gardener. Accordingly, he apprenticed as a gardener at several large English gardens then at Versailles, France, and at several famous nurseries in England. In 1884, he took charge of the gardens at Firs owned by John Wingfield Larking at Lee in Kent, England. He was self-educated in practical horticulture and became a noted, early horticultural journalist in England, being author of over 37 gardening books including **The Encyclopaedia of Gardening**. From 1887 to 1926, he was editor of **Amateur Gardening** published by W.H. and L. Collingridge. He founded the National Amateur Gardeners' Association in 1890 and served as President for 21 years. Thomas Sanders was awarded Fellow of the Royal Horticultural Society in 1891 and Knighthood of the Royal Order of Vasa of Sweden in 1906.

1910† **Lawns and Greens: Their Formation and Management.** First edition. T.W. Sanders. W.H. & L. Collingridge, London, England, UK. 137 pp. plus 10 pp. of advertising, 49 illus., 3 tables, small index, cloth hard boards bound in green with red trim and black lettering, 4.9 by 7.0 inches (125 × 179 mm).

* This **truly rare** book is a must for collectors of historical turfgrass books. This handbook on turfgrass culture is oriented to the practitioner, with emphasis on lawns. One of the early books to encompass cricket grounds, bowling greens, lawn tennis, and croquet courts as well as lawns and golf courses. Subjects addressed included grass steps, horse boots, manures, rolling, scything, and worming. The chapters on cricket grounds and golf greens were authored by J.C.

Newsham, Headmaster of the Farm School, Old Basing, Basingstoke. The book is organized in four parts with chapter subject titles as follows:

Part I. Formation of Lawns	VI. Lawn Pests
I. Soils and Their Treatment	VII. Lawn Weeds
II. Drainage	
III. Preparing and Leveling Sites	Part III. Tennis Lawns, Greens, etc.
IV. Turf versus Seeds	I. Tennis and Croquet Lawns
V. Seeds for Lawns	II. Bowling Greens
VI. Seed Sowing	III. Cricket Grounds
VII. Turf Laying	IV. Golf Greens
	V. Grass Paths and Steps
Part II. Management of Lawns	
I. Mowing Lawns	Part IV. Appliances
II. Rolling Lawns	I. Lawn Mowers
III. Watering Lawns	II. Garden Rollers
IV. Renovating Lawns	III. Tools and Appliances
V. Manures for Lawns	IV. Miscellaneous Data

In 1910 Thomas Sanders made the following observations about the contributions of lawn mowers in England:

> For upwards of seventy years, at least, the mowing machine has played an important part in the maintenance and success of the lawn, the lovely, velvety turf of which has long been one of the chief charms of a well-ordered British garden. Prior to its introduction, our forebears had to rely upon that useful, but nevertheless by no means easy, tool to handle—the scythe. In those days to be able to use and sharpen a scythe successfully was regarded as no mean accomplishment. Moreover, its use meant early rising, in order to cut the grass whilst its blades were succulent and covered with dew, the only period in the day when it was possible to mow evenly and quickly. Mowing by scythe was therefore a laborious and costly business, and only those of ample means could indulge in the luxury of a lawn. When the mowing machine superseded the scythe, and rival manufacturers had succeeded in improving its mechanism and reducing its cost, then the desire for adding the charming feature of a lawn to the garden became more general, with the result that to-day even the smallest garden is not considered perfect without its patch of turf. The mowing machine may, indeed, claim to have accomplished more for the beauty and charm of a garden than any other appliance in existence.
>
> There is no lack of diversity in form, in size, and in cost among modern mowing machines. They can be obtained to suit the strength of a youth or a lady, the more powerful muscles of a man or two men, or a donkey, pony, or horse, or the superior force of a motor engine; and the range of prices is equally variable, the cheapest being available at a guinea, and the most expensive—the motor type—costing upward of £150. The cheapest type is mostly of American manufacture, and not so durable or easily repairable as those of British make at a slightly enhanced cost.

Revised circa 1911 and in 1920.

n.d.†
circa
1911
  *
Lawns and Greens: Their Formation and Management. Second revised edition. T.W. Sanders. W.H. & L. Collingridge, London, England, UK. 142 pp. plus 11 pp. of advertising, 55 illus., 3 tables, small index, cloth hardbound in green with red trim and black lettering, 4.7 by 7.0 inches (119 × 177 mm).
There are small revisions of this **truly rare** book. The book is a must for collectors of historical turfgrass books. In 1911, Thomas Sanders advised on the topdressing of turfgrasses:

> Top-dressings.—In addition to the artificials mentioned in the preceding notes, it is advisable to give lawns, tennis courts, and bowling greens an occasional, if not an annual, top-dressing with a good mixed compost. This should consist primarily of good loamy soil, such as has not been taken from a weedy spot. In connection with the latter fact, it should clearly be understood that very great care must be selected in the choice of soil, as there is a great risk of daisy, dandelion, and other weed seeds being introduced through its agency to the turf. Another important ingredient is well-decayed manure, such as has been used for a hot-bed or a mushroom-bed. Mix two parts of soil with one of the manure, and pass the whole through a half-inch screen or sieve; in fact, it would answer still better if a quarter-inch mesh were used. To each cubic yard or cartload of this mixture add one

> hundredweight of lime (ground lime, if possible). Mix thoroughly, and then apply at the rate of a good barrow-load to every six or eight square yards. November is the best time to apply it. Spread evenly over the surface, and then sweep with a birch broom. This is a simple, plain top-dressing for an average lawn.
>
> Where the grass is thin and poor, or is much used and worn, as in the case of a bowling green or a tennis court, then a much richer compost should be applied. For example, in addition to the ingredients just mentioned, there should be added to each cartload about 3cwt. of wood-ashes, 1cwt. of fine charcoal, and ½cwt. of bone-meal. Mix all thoroughly together, and apply at the rate of one barrow-load to eight or ten square yards, in November. The addition of charcoal is most beneficial. It encourages a sturdy, green growth, and helps to keep the surface soil sweet. The surface roots, moreover, like it, and cluster around each particle in search of the food stored within its pores. The above compost is a capital top-dressing for moss-infested lawns.
>
> For a spring top-dressing we have found 4lb. of dissolved bones and one peck of fine charcoal per square rod a splendid mixture for improving the turf of bowling greens and tennis courts. This can be tried in place of the mixed composts previously mentioned.

Revised in 1920.

1920† **Lawns and Greens—Their Formation and Management.** Third revised edition. T.W. Sanders. W.H.L. Collingridge, London, England, UK. 113 pp. plus 3 pp. of advertising, 49 illus., 2 tables, small index, cloth hardbound in green with black lettering and a jacket with a multicolor plate on a light brown cover with black lettering, 4.7 by 7.1 inches (120 × 181 mm).

    ∗   There are small revisions in the third edition of this **rare** book, including the illustrations. The costs for materials and labor to accomplish various turfgrass establishment and culture activities in 1920 are presented throughout the book. This book is a must for collectors of historical turfgrass books.

Schery, Robert Walter (Dr.)

Bob Schery (1917–1987) was born in St. Louis, Missouri, USA, and earned an AB degree in 1938, an MSc degree in Botany in 1940, and a PhD degree in Botany in 1942, all from Washington University of St. Louis, where he also was on the research staff of the Missouri Botanical Garden. He worked as a Botanist for the Monsanto Chemical Company and the O.M. Scott & Sons Company. He was the founding Executive Director of the Lawn Institute, serving from 1957 to 1982, during which he wrote numerous books and feature articles on lawns.

1961† **The Lawn Book.** Robert W. Schery. The Macmillan Company, New York City, New York, USA. 207 pp.; 109 illus.; 17 tables; good index; cloth hardbound in green with yellow and black lettering and a jacket in white, black, and green with white, black, and green lettering; 5.4 by 8.2 inches (138 × 208 mm).

A small book on the basics of turfgrass establishment and culture within a how-to approach. It is practically oriented to the home lawn enthusiasts and conditions in the United States. The drawings of turfgrasses and weeds are well done.
Revised in 1973 with a new title.

1973† **A Perfect Lawn—The Easy Way.** Revised edition of a book originally titled **The Lawn Book—1961.** Robert W. Schery. Macmillan Publishing Company Inc., New York City, New York, USA. 304 pp.; 162 illus.; 28 tables; indexed; cloth hardbound in green with silver lettering and a glossy jacket in green, white, and black with white, black, and green lettering; 6.3 by 9.2 inches (159 × 233 mm).

An update and expansion on the basics of turfgrass establishment and culture within a how-to approach. It is practically oriented to the home lawn enthusiasts and conditions in the United States and Canada. The drawings of turfgrasses and weeds are well done. It contains a glossary in the back.

1976† **Lawn Keeping.** Robert W. Schery. Prentice-Hall Inc., Englewood Cliffs, New Jersey, USA. 232 pp.; 156 illus.; 27 tables; indexed; semisoft bound in multicolor with white, green, and black lettering and also cloth hardbound in green with white lettering and a jacket in multicolor with white, green, and black lettering; 6.8 by 9.2 inches (172 × 234 mm).

A well-written book on the establishment and culture of lawn turfgrasses. The book is practically oriented to lawn enthusiasts and conditions in North America. It contains a glossary in the back.

Schnabel, Astulf (Dr.)

Astulf Schnabel (1933–2002) was born in Bautzen, Germany. After apprenticeship as a gardener, he earned a DrRerMat (PhD) degree in Landscape Architecture from Humboldt University of Berlin. Dr. Schnabel served at Wissenschaftlich-Technisches Zentrum (WTZ) Sportbauten des Staatssekretariats für Körperkultur und Sport, Leipzig, from 1959 to 1989 and then was a landscape architect with official appointed authority for sports grounds in Leipzig from 1989 to 2002.

1976† **Rasen für Sport und Spiel (Turf for Sports and Games).** Coauthor.
See *Gandert, Klaus-Dietrich—1976.*

Schreiner, Oswald (Dr.)

Oswald Schreiner (1875–1965) was born in Nassau, Germany. His family immigrated to the United States in 1883 and settled in Baltimore, Maryland. He graduated from Baltimore Polytechnic in 1892, earned a Graduate in Pharmacy (PhG) from the University of Maryland in 1894, and attended Johns Hopkins University from 1894 to 1895. He then became a US Pharmacopoeia Fellow at the University of Wisconsin where he earned a BSc degree in 1897, an MSc in 1898, and a PhD degree in 1902, studying sesquiterpenes. He also was awarded the Ebert Prize by the American Pharmaceutical Association in 1900. He joined the United States Department of Agriculture (USDA), serving as Chemist in the Bureau of Soils from 1903 to 1906, as Chief of the Division of Soil Fertility Investigations in the Bureau of Plant Industry from 1906 to 1923, as Biochemist in charge of soil fertility investigations in the Bureau of Plant Industry from 1923 to 1940, and as Assistant to the Chief of the Bureau of Plant Industry from 1940 to 1944, when he retired. He conducted research on the chemistry, fertility, and organic matter aspects of soils. Oswald Schreiner was a pioneer in documenting plant-soil allelopathy, being the first researcher to isolate and identify organic phytotoxins associated with unproductive soils. He also demonstrated that these organic substances in soils not only affected plant growth but also affected nutrient uptake by roots. He was elected President of the Association of Agricultural Chemists in 1928. Dr. Schreiner received the Longstreth Medal of the Franklin Institute in 1912 and Fellow of the American Association for the Advancement of Science.

1911† **Lawn Soils.** Oswald Schreiner and J.J. Skinner. Bulletin No. 75, Bureau of Soils, United States Department of Agriculture, Washington, D.C., USA. 55 pp., 26 illus., 12 tables, no index, fold-stapled semisoft covers in white with black lettering, 5.7 by 8.9 inches (146 × 225 mm).

  ✻ An early, pioneering bulletin by two leading soil scientists that addresses in some detail the soil factors involved in the establishment and culture of lawns in United States, with special emphasis on parklands and similar large tracts of land for public use. It is a **very rare** pioneering booklet on the soils dimensions of lawn construction and culture written from a reasonable base in soil science. The role of the civic art movement in expanded use of lawns in public areas is noted as follows:

> The widespread movement of civic art to improve and beautify cities and towns by park and art commissions, civic associations, and individuals naturally creates a demand for information concerning lawns, their improvement, soil requirements, fertilization, maintenance, soil suited for filling-in or top dressing, and similar questions. It is to supply this general demand for soil information regarding lawns that the present bulletin is designed.

1912† **Lawn Soils and Lawns.** Oswald Schreiner, J.J. Skinner, L.C. Corbett, and F.L. Mulford. Farmers Bulletin 494, United States Department of Agriculture, Washington, D.C., USA. 48 pp., 18 illus., 2 tables, no index, fold-stitched soft covers in white with black lettering, 5.7 by 9.0 inches (144 × 228 mm).

Figure 12.79. Portable Royer compost mixer and soil shredder with belt drive to power source and capacity of 5–8 cubic yards per hour, circa 1934. [41]

* An early, small bulletin on lawn-soil construction and on turfgrass establishment and culture of lawns under conditions in the United States. It is a **very rare** publication.

Schroeder, Charles Bernard (Professor)

Chuck Schroeder was born in Alton, Illinois, USA, and earned a BSc degree in Ornamental Horticulture in 1965 and an MSc degree in Agriculture Education in 1970, both from the University of Illinois. He then joined the faculty at Danville Area Community College teaching ornamental horticulture, with emphasis on turfgrass and golf course management subjects. He rose to Full Professor in 1990. After 41 years of educational activities, he retired in 2007. Professor Schroeder received the Those Who Excel Award from the State of Illinois in 1993.

1994† **Turf Management Handbook.** Fourth revised edition. Charles B. Schroeder and Howard B. Sprague. Interstate Publishers Inc., Danville, Illinois, USA. 198 pp., 170 illus. with 24 in color, 6 tables, indexed, glossy hardbound in multicolor with white lettering, 7.0 by 9.6 inches (177 × 243 mm).

This fourth revision is a practically oriented, introductory guidebook on the establishment and culture of turfgrasses for lawns, playing fields, and parks under conditions in the United States.

Previous revision in 1982.
See *Sprague, Howard Bennett—1982.*
Revised in 1996.

1996† **Turf Management Handbook.** Fifth revised edition. Charles B. Schroeder and Howard B. Sprague. Interstate Publishers Inc., Danville, Illinois, USA. 206 pp., 185 illus. with 24 in color, 6 tables, indexed, glossy hardbound in multicolor with white lettering, 7.0 by 10.0 inches (177 × 255 mm).

A major revision involving primarily the addition of a small chapter on irrigation systems.

Schulz, Heinz (Dr.)

Heinz Schulz (1931–2009) was born in Chladowa, Posen, Germany. He studied Agriculture at Humboldt University of Berlin, finishing in 1962, and earned a DrAgr from the University of Hohenheim in 1967. He then joined the Institut für Pflanzenbau und Grünland at the University of Hohenheim, serving as Scientific Assistant from 1967 to 1975 and as Academic Director from 1975 to 1996, when he retired. He was a leading educator of turfgrass science in Germany and was a 1989 founding member of the Greenkeeping Qualification Program. Dr. Schulz was elected President of the German Turfgrass Society from 1996 to 2000 and received the award of Honorary Member of the German Greenkeeper Association in 2000.

1983† **Rasen—Anlage und Pflege von Zier-, Gebrauchs-, Sport- und Landschaftsrasen (Turfgrass—Construction and Maintenance for Aesthetic, Functional, Sport and Landscaped Lawns).** Coauthor.
See *Hope, Frank—1983.*

Schumann, Gail Lynn (Dr.)

Gail Schumann was born in Cincinnati, Ohio, USA, and earned a BSc degree in Botany from the University of Michigan in 1972 and both MSc and PhD degrees in Plant Pathology from Cornell University in 1975 and 1978, respectively. She joined the faculty of the University of Massachusetts in 1989 as an Assistant Professor and became an Associate Professor in 1995. She was active in teaching, research, and extension, including operation of a turfgrass diagnostic laboratory. She was elected President of the Northeast American Phytopathology Society in 1995. Dr. Schumann received the Excellence in Teaching Award from the American Phytopathology Society (APS) in 1993, the Distinguished Teaching Award from the University of Massachusetts in 1995, and the Award of Merit from the Northeast APS in 1996. Then in 2003, she became Adjunct Professor in the Department of Biological Sciences at Marquette University, Milwaukee, teaching biology.

1998† **IPM Handbook for Golf Courses.** Gail L. Schumann, Patricia J. Vittum, Monica L. Elliott, and Patricia P. Cobb. Ann Arbor Press, Chelsea, Michigan, USA. 272 pp., 188 illus. with 32 in color, 9 tables, indexed, glossy hardbound in multicolor with white lettering, 6.9 by 9.9 inches (175 × 251 mm).

A book devoted to the basics and practical applications of integrated pest management to golf courses. There are chapters that specifically address insects, diseases, nematodes, and weeds. A glossary is included in the book.

Seagle, Janet Marilyn

Janet Seagle (1924–2003) was born in Spokane, Washington, USA, and earned a BA degree in Art from New York University in 1945. She served as Art Director for both the **United States Golf Association (USGA) Journal** from 1963 to 1986 and the **USGA Green Section Record** from 1963 to 1986. She was the Museum Curator and Librarian for the United States Golf Association from 1964 to 1989. Miss Seagle specialized in golf history and memorabilia.

1978† **Golf—A Guide To Information Sources.*** Coauthor.
See *Murdoch, Joseph Simpson Ferguson—1979.*

Shaffer, Randolph[3]

1932 **Grass—Golf Clubs Can Do It, So Can You.** Coauthor.
  * See *Goit, Whitney—1932.*

Sharp, William Curtis

William Sharp was born in Frost, West Virginia, USA, and earned a BSc degree from West Virginia University in 1953 and an MSc degree from Pennsylvania State University in 1955. He served the United States Department of Agriculture (USDA) Soil Conservation Service, now the Natural Resource Conservation Service, as Plant Materials Manager from 1970 to 1974, as Regional Plant Materials Technical Leader from 1974 to 1985, and as National Plant Materials Program Leader from 1985 to 1993. His emphasis was on plant conservation development. William Sharp received the USDA Superior Service Award in 1984 and 1992.

1994† **Grass Varieties in the United States.**[*] Coauthor.
  See *Alderson, James Stillman—1994.*

Shaw, Robert Blaine (Dr. and Professor)

Bob Shaw was born in Commerce, Texas, USA, and earned a BSc degree with *honors* from Southwest Texas State University in 1973 and both MSc and PhD degrees from Texas A&M University in 1976 and 1979, respectively, all majoring in Range Science. He served on the faculty at the University of Florida for two years and then at Colorado State University from 1981 to 2005, rising from Assistant Professor to Full Professor. In 2007, he moved to Texas A&M University as a Full Professor in the Department of Ecosystem Science and Management and as Associate Director of the Institute of Renewable Natural Resources conducting research and extension activities.

1983† **Grass Systematics.** Second revised edition. Coauthor.
  See *Gould, Frank Walton—1983.*

Sheard, Robert Wesley (Dr. and Professor)

Bob Sheard was born in Maryfield, Saskatchewan, Canada, and earned a BSA degree majoring in Soil Science from Saskatchewan University in 1951, an MSA degree in Soil Science from the University of Toronto in 1954, and a PhD degree in Agronomy from Cornell University in 1961. He joined the Soil Science Faculty at the Ontario Agricultural College, University of Guelph, in 1954 and was appointed Assistant Professor in 1956, Associate Professor in 1964, and Full Professor in 1972. He was active in soils teaching, research, and extension, with emphasis on turfgrasses. He retired in 1990. He served as Secretary of the Board of Management of the Sports Turf Association (STA) of Ontario from 1989 to 2008. He was elected to the Board of Directors of the International Turfgrass Society for 1981 through 1989. Dr. Sheard was awarded an Honorary Life Membership to the STA in 1998 and also was honored by the STA in 2007 through the establishment of the R.W. Sheard Scholarship.

2000† **Understanding Turf Management.** First edition. R.W. Sheard. Sports Turf Association of Ontario, Guelph, Ontario, Canada. 168 pp., 47 illus., 31 tables, indexed, glossy semisoft bound in multicolor with black lettering, 5.9 by 8.9 inches (150 × 226 mm).

  A book based on an assemblage of 24 articles that appeared in **Sports Turf Manager**. The focus is on turfgrass soils and their culture for sports fields from both a physical and chemical standpoint under the conditions in Canada. There also are chapters on the four major cool-season turfgrass species.
  Revised in 2005.

2005† **Understanding Turf Management.** Second revised edition. R.W. Sheard. The Sports Turf Association, Guelph, Ontario, Canada. 170 pp., 41 illus., 33 tables, small index, glossy semisoft bound in multicolor with black lettering, 5.9 by 8.9 inches (150 × 226 mm).

  An update with a few revisions of the first edition.

2008† **Athletic Field Construction Manual.** R.W. Sheard (editor) with six contributors. Sports Turf Association Inc., Guelph, Ontario, Canada. 73 pp., 34 illus., 5 tables, no index, glossy semisoft covers in multicolor with white and black lettering in a ring binder, 8.5 by 11.0 inches (215 × 278 mm).

  A small manual concerning a field classification system of five categories, user's expectations in hours per season, existing field evaluation check list, and field dimensions for most field sports played on turfgrass surfaces.

Shearman, Robert Curtice (Dr. and Professor)

Bob Shearman was born in Dallas, Oregon, USA, and earned a BSc degree in Agronomy from Oregon State University in 1967 and both an MSc degree in 1971 and PhD degree in 1973 in Crop Physiology and Ecology and Turfgrass from Michigan State University with Dr. Jim Beard. He served in the US Peace Corps as an Agronomist in Honduras from 1967 to 1968. From 1973 to 1975, he worked as a Turfgrass Agronomist for the O.M. Scott and Sons Company at their

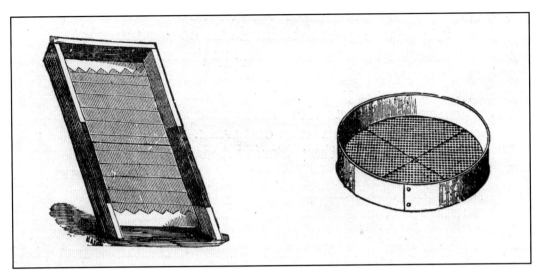

Figure 12.80. Soil screen (left) and sieve (right) used in preparation of compost and materials for topdressing via manual shoveling or shaking, respectively. Mesh sizes available varied from ⅛ to 1.0 inches, 1924. [14]

Gervis, Oregon, Research Center. In 1975, he joined the faculty at the University of Nebraska working in turfgrass extension, research, and teaching, with emphasis on turfgrass physiology and integrated turfgrass management. From 1989 to 1993, he served as Head of the Agronomy Department and then returned to turfgrass research activities and to being Executive Director of the National Turfgrass Evaluation Program (NTEP) from 1993 to 1998 and Special Project Coordinator from 1998 to 2001. In 2006, he was named the Sunkist Fiesta Bowl Professor of Agronomy. He was elected Turfgrass Science Division C-5 Chair in 1983, Board Member of the Crop Science Society of America (CSSA) Board of Directors from 1982 through 1984 and 1994 through 1996, Board Member of the American Society of Agronomy (ASA) Board of Directors from 1988 through 1990, and President of CSSA in 1995. Dr. Shearman was awarded Fellow of the ASA in 1988, Fellow of CSSA in 1989, the CSSA Turfgrass Science Award in 1992, Fellow of the American Association for the Advancement of Science in 1998, and the USGA Green Section Award in 2006.

1992† **Turfgrass.** Coeditor.
　　　See *Waddington, Donald VanPelt—1992.*

Shetlar, David John (*Dr.* and *Associate Professor*)

Dave Shetlar was born in Columbus, Ohio, USA, and earned a BSc degree in 1969 and an MSc degree in 1975, both in Zoology from the University of Oklahoma, and a PhD degree in Entomology from Pennsylvania State University in 1977. He served as Assistant Professor of Entomology at Pennsylvania State University from 1977 to 1983, as Research Scientist at ChemLawn Services Corporation from 1984 to 1990, and as Associate Professor of Landscape Entomology at Ohio State University since 1990. Dr. Shetlar is active in extension, teaching, and research of turfgrass entomology, including biological controls, insect development models, and insect control evaluation.

1983† **Turfgrass Insect and Mite Manual.** First edition. David J. Shetlar, Paul R. Heller, and Peter D. Irish. The Pennsylvania Turfgrass Council Inc., Bellefonte, Pennsylvania, USA. 63 pp., 60 illus. with 44 in color, 1 table, no index, fold-stapled glossy semisoft covers in multicolor with white lettering, 6.0 by 9.0 inches (152 × 228 mm).
　　　A small handbook on the diagnosis of turfgrass injury symptoms and identification of the associated insect or mite. The description, life cycle, damage symptoms, and control of insects and mites are covered, along with extensive color illustrations. It is oriented to conditions in the northeastern United States. It also contains an introductory section on pesticide usage. A glossary of selected terms is presented in the back.
　　　Revised in 1988 and 1990.

1988† **Turfgrass Insect and Mite Manual.** Second revised edition. David J. Shetlar, Paul R. Heller, and Peter D. Irish. The Pennsylvania Turfgrass Council Inc., Bellefonte, Pennsylvania, USA. 67 pp., 67 illus. with 46 in color, 2 tables, no index, fold-stapled glossy semisoft covers in greens with white lettering, 5.7 by 8.8 inches (145 × 223 mm).
　　　A revision, with the green June beetle and oriental beetle added.
　　　Revised in 1990.

1990† **Turfgrass Insect and Mite Manual.** Third revised edition. David J. Shetlar, Paul R. Heller, and Peter D. Irish. The Pennsylvania Turfgrass Council Inc., Bellefonte, Pennsylvania, USA. 67 pp., 70 illus. with 49 in color, no index, fold-stapled glossy semisoft covers in greens with white lettering, 5.9 by 9.0 inches (149 × 230 mm).
　　　Contains updates and minor revisions.

1995† **Managing Turfgrass Pests.** Coauthor.
    See *Watschke, Thomas Lee—1995.*

2000† **Destructive Turf Insects.** Coauthor.
    See *Niemczyk, Harry Donald—2000.*

Shildrick, John Patrick

John Shildrick (1928–1994) was born in Dublin, Ireland. He served in the British Army from 1946 to 1948 and then earned a BA degree in Agriculture from Emmanuel College, Cambridge University, in 1951. He served as Head of the Herbage Section, Seed Production Branch of the National Institute of Agricultural Botany at Cambridge, from 1952 to 1964. Then from 1968 to 1988, John Shildrick worked as a researcher at the Sports Turf Research Institute (STRI) and also served as STRI Assistant Director. He served as Secretary and Administrator of the National Turfgrass Council from 1980 to 1994 and Permanent Secretary of the British Turf & Landscape Irrigation Association (BTLIA) from 1986 to 1994. John Shildrick was honored by the BTLIA with establishment of the John P. Shildrick Award in 1995.

1984† **Turfgrass Manual.** John Shildrick. The Sports Turf Research Institute, Bingley, West Yorkshire, England, UK. 60 pp., 6 illus., 25 tables, no index, fold-stapled glossy semisoft covers in light blue with dark blue lettering, 5.8 by 8.1 inches (147 × 207 mm).

    A booklet concerning the characteristics of seeds and cool-season turfgrasses for use on lawns, sport fields, and golf courses. Both monostand and polystand turfgrass performances under the conditions in England are discussed.

1986† **Grass Seed Mixtures for Children's Play Areas.** John Shildrick. National Playing Fields Association, London, England, UK. 42 pp.; 27 illus.; 2 tables; no index; soft bound in white, greens, and black with green, white, and black lettering; 5.8 by 8.2 inches (148 × 208 mm).

    A booklet on the suggested turfgrass seed mixtures for various types of play areas in England. It also contains a section on turfgrass-soil stabilization materials.

Shim, Gyu Yul (Dr.)

Gyu Shim was born in Hadong-Gun, South Korea, and earned BSc, MSc, and PhD degrees in Plant Pathology from Gyeongsang National University in 1986, 1988, and 1995, respectively. He has worked as an administrator and lecturer for the Green Keeper's Academy and the Golf Course Manager's Certification Program. He also has been active in education and research concerning disease control for turfgrasses on golf courses. Since 1989, he has been General Manager of the Turfgrass Society of Korea. Dr. Shim was elected as Vice President of the Korea Turfgrass Research Institute for 1989 to 2008 and as a Director of the Korean Society of Plant Pathology from 1995 to 2008.

1992 **Basics and Application in Golf Course Management.** Gyu Yul Shim with nine contributing authors. Korean Turfgrass Research Institute, Seongnam, South Korea. 772 pp., 120 illus. in color, 213 tables, no index, cloth hardbound in light green with blue lettering, 7.5 by 10.2 inches (190 × 260 mm). Printed entirely in the Korean language.

    A scholarly book concerning the layout, construction, turfgrass establishment, and culture for golf courses under conditions in Korea. There are individual sections on the cultural systems for putting greens, turfgrass tees, fairways, and roughs. Appropriate references are included.

2006 **Explanation of Turfgrass Terminology.** Gyu Yul Shim with 16 contributing authors. Korean Turfgrass Research Institute, Seongnam, South Korea. 252 pp., 134 illus. with 129 in color, 8 tables, large index, semisoft covers in blue with gold lettering, 6.1 by 8.5 inches (155 × 215 mm). Printed entirely in the Korean language, with comparable Chinese and English terminologies.

    A book concerning the Korean turfgrass terminology and their respective definitions and explanations. It includes lists of weeds, diseases, and insects.

Shoor, Jakob

Jakob Shoor (1905–1977) was born in Russia and immigrated to Israel in 1930. He was Chief Gardener at the Kibbutz Ashdot Yakov from 1935 to 1977. He pioneered the development and culture of gardens and public parks in Israel and served on the Editorial Board of **Garden Landscape.** Jakob Shoor was elected to the Board of Directors of the Israel Gardeners Association from 1944 to 1975.

1963† **The Lawn and Its Care.** Coauthor.
    See *Bahir, Zvi—1963.*

Shurtleff, Malcolm C., Jr. (Dr. and Professor Emeritus)

"Mal" Shurtleff was born in Fall River, Massachusetts, USA, and earned a BSc degree in Biology from the University of Rhode Island in 1943. He served in the US Navy during WWII. Subsequently, he earned MSc and PhD degrees in Plant

Figure 12.81. Stationary belt-drive, grain thrasher that has been extensively modified for compost and soil shredding, mixing, and shaker screening, circa 1921. [7]

Pathology from the University of Minnesota in 1950 and 1953, respectively. He served as Assistant Professor at the University of Rhode Island from 1950 to 1954, as Associate Professor at Iowa State University from 1954 to 1958, and then as Full Professor at the University of Illinois until his retirement in 1992. Mal was active in adult education and teaching programs to manage turfgrass diseases through cultural and chemical practices. Dr. Shurtleff was awarded Fellow of the American Phytopathological Society in 1971, and the United States Department of Agriculture Distinguished Service Award in 1986.

1974† **How to control Lawn Diseases and Pests.** Malcolm C. Shurtleff and Roscoe Randell. Intertec Publishing Corp., Kansas City, Missouri, USA. 97 pp., 150 illus. with 66 in color, 14 tables, indexed, glossy semisoft bound in multicolor with white and black lettering, 8.0 by 11.0 inches (204 × 280 mm).

A small handbook on the disease, insect, and pest problems of turfgrasses in North America. The format includes injury symptom diagnosis, pest identification, and appropriate controls. At the back, there is a glossary and an appendix of tables on conversions, calibrations, and calculations.

1987† **Controlling Turfgrass Pests.** First edition. Malcolm C. Shurtleff, Thomas W. Fermanian, and Roscoe Randell. A Reston Book, Prentice-Hall Inc., Englewood Cliffs, New Jersey, USA. 449 pp., 368 illus. with 60 as color plates, 46 tables, detailed index, glossy hardbound in multicolor with white and orange lettering, 6.9 by 9.2 inches (175 × 233 mm).

This book addresses the biology, management, and control of the weeds, insects, and diseases of turfgrasses, both cool- and warm-season. Integrated pest management approaches are discussed, along with chemical controls. It contains a glossary in the back plus a photograph of the three authors in the front.

Revised in 1997 and 2003.

1997† **Controlling Turfgrass Pests.** Second revised edition. Coauthor.

See *Fermanian, Thomas Walter—1997.*

2003† **Controlling Turfgrass Pests.** Third revised edition. Coauthor.

See *Fermanian, Thomas Walter—2003.*

Singapore Parks and Recreation Department

1977† **A Handbook on the Maintenance of School Fields.** Chan Kai Yau. Parks and Recreation Department, Ministry of National Development and Ministry of Education, Singapore. 32 pp., 17 illus. in color, 1 table, no index, fold-stapled semisoft covers in multicolor with black lettering, 5.7 by 8.2 inches (145 × 207 mm).

A handbook on the culture of turfgrasses for school sports fields. It includes descriptions and color photographs of the turfgrass species used and major weeds found in Singapore.

Skinner, Joshua John (Dr.)

Joshua Skinner (1882–1969) was born in Hertford, North Carolina, USA, and earned a BSc degree in 1903 and an MSc degree in 1904, both in Chemistry from the University of North Carolina, and a PhD in Agriculture from American University in 1917. He served as Soil Scientist-Chemist with the United States Department of Agriculture (USDA), Bureau of Soils, Washington, DC, from 1904 to 1914 and as Biochemist with the USDA Bureau of Plant Industry from 1914 to

1945, when he retired. He conducted research on the biochemistry of plants, soils, and fertilizers. Dr. Skinner received the Longstreth Medal from the Franklin Institute in 1919.

1911† **Lawn Soils.** Coauthor.
   * See *Schreiner, Oswald—1911.*
1912† **Lawn Soils and Lawns.** Coauthor.
   * See *Schreiner, Oswald—1912.*

Skirde, Werner-Konrad (Dr.)

Werner Skirde was born in Komienen, Ostpreussen, Germany, and earned a degree in Agriculture in 1952 and a DrAgr in 1958, both from University of Rostock. He subsequently was employed at Justus Liebig University, Giessen, serving as Scientific Assistant from 1957 to 1967, as Academic *Rat* and Academic *Oberrat* from 1967 to 1980, and as Professor of Landscape Development and Landscape Construction from 1980 to 1994, when he retired. His research specialty was soil and vegetation technology, with emphasis on turfgrasses. Leadership efforts included such working groups as Deutsches Institut für Normung (DIN) and Deutscher Fussballbund DFB. Dr. Skirde was elected to the Executive Committee of the International Turfgrass Society for 1973 to 1977 and received the BGL Award "Silberne Landschaft."

1978† **Vegetationstechnik Rasen und Begrünungen (Techniques of Growing Lawns and Green Spaces).** Werner Skirde. Series No. 1 Landscape and Sports Grounds, Patzer Verlag GmbH u. Company KG, Berlin and Hanover, West Germany. 240 pp., 81 illus., 38 tables, indexed, glossy semisoft bound in greens and white with black lettering, 6.3 by 9.3 inches (160 × 236 mm). Printed entirely in the German language.

   A book on turfgrasses and their use in the construction, establishment, and culture of lawns and sports fields under the conditions found in the middle European region. It is oriented to the field practitioner. Reference is made to the principles of turfgrass maintenance for sports fields, golf courses, tennis courts, horse racetracks, and the revegetation of disturbed areas. The text is supported by selected reference citations.

1980† **Erhaltung von Sportplätzen (Maintaining Sports Grounds).** Werner W. Skirde, Walter Büring, Heiner Pätzold, Helmut Tietz, Klaus Trojahn, Franz Müller, and Alfred Niesel. Series No. 2 in Landscape and Sports Field Construction, Patzer Verlag GmbH u. Company KG, Berlin, West Germany. 208 pp., 122 illus., 61 tables, indexed, glossy semisoft bound in greens and white with black lettering, 6.3 by 9.3 inches (161 × 236 mm). Printed entirely in the German language.

   A book on the construction, culture, and renovation of turfgrass sports fields plus chapters on gravel and aggregate and artificial sport surfaces, on maintenance equipment, and on comparative costs.

1983† **Untersuchungen zur Be- und Entwässerung von Rasensportflächen (Analysis of Irrigation and Drainage of Sports Turf Surfaces).** Werner Skirde. Schriftenreihe Sport- und Freizeitanlagen. Berichte, B2/82, Bundesinstitut für Sportwisssenschaft, Köln, Germany. 82 pp., 67 illus., 45 tables, no index, glossy semisoft bound in blues with white lettering, 8.1 by 7.9 inches (206 × 200 mm). Printed in the German language.

   A small book on root-zone construction techniques for sports fields including materials selection and particle size specifications as used in Germany. There is a list of selected references in the back.

Sleper, David Allen (Dr. and Professor)

Dave Sleper was born in Buffalo Center, Iowa, USA, and earned BSc and MSc degrees in Agronomy from Iowa State University in 1967 and 1969, respectively, and a PhD in Plant Breeding and Genetics from the University of Wisconsin, Madison in 1973. He served on the faculty at the University of Florida from 1973 to 1974 and then joined the faculty at the University of Missouri, becoming an Associate Professor in 1978 and a Full Professor in 1984. He focused on the breeding and genetics of cool-season forage and turfgrasses from 1973 to 1994 and of soybeans since 1994. Dr. Sleper was elected President of the Crop Science Society of America (CSSA) in 2001 and President of the American Society of Agronomy (ASA) in 2006. Dr. Sleper was awarded Fellow of CSSA in 1986 and the ASA in 1985.

1989† **Contributions from Breeding Forage and Turf Grasses.** D.A. Sleper, K.H. Asay, and J.F. Pedersen (editors) with contributing authors of the turfgrass chapters being Drs. K.H. Asay, D.A. Sleper, William A. Meyer, C. Reed Funk, and Philip Busey. CSSA Special Publication Number 15, Crop Science Society of America, Madison, Wisconsin, USA. 140 pp.; 10 illus.; 15 tables; no index; glossy semisoft bound in black, gray, and white with white and black lettering; 6.0 by 8.9 inches (152 × 227 mm).

   A small book updating the research advances that have been made in the breeding and genetic improvement of perennial grasses, including turfgrasses.

Smiley, Richard Wayne (Dr. and Professor)

Richard Smiley was born in Paso Robles, California, USA, and earned a BSc degree in Soil Science from California State Polytechnic University, San Luis Obispo, in 1965; an MSc degree in Soil Science in 1969; and a PhD degree in Plant Pathology in 1972, the latter two from Washington State University. He was a NATO Postdoctoral Fellow at the Commonwealth Scientific and Industrial Research Organization (CSIRO) in Adelaide, Australia. He then served on the faculty at Cornell University from 1973 through 1985, becoming an Associate Professor. He conducted research on the etiology and control of diseases of turfgrasses and wheat, with emphasis on soilborne root and crown rot diseases. Since 1985, he has served as Professor and Superintendent of the Columbia Basin Agricultural Research Center of Oregon State University in Pendelton, Oregon. Dr. Smiley was awarded an Honorary Membership to the American Sod Producers Association in 1986, Fellow of the American Phytopathological Society in 1994, and the Biskey Award for Faculty Excellence from Oregon State University in 1995.

1983† **Compendium of Turfgrass Diseases.** First edition. Richard W. Smiley. The American Phytopathological Society, St. Paul, Minnesota, USA. 102 pp., 303 illus. with 185 in color, 24 tables, indexed, fold-stapled glossy semisoft covers in multicolor with white lettering, 8.5 by 11.0 inches (216 × 280 mm).

  A compilation of the common diseases occurring on turfgrasses throughout the world. It is oriented to turfgrass managers and pathologists, including both technical and practical dimensions. It includes illustrations of both fungal life cycles and spores. A selected set of references is included at the end of each section. There is a glossary in the back. The color photographs of diseases are very extensive.

  Revised in 1992 and 2005.

1992† **Compendium of Turfgrass Diseases.** Second revised edition. Richard W. Smiley, Peter H. Dernoeden, and Bruce B. Clarke. APS Press, American Phytopathological Society, St. Paul, Minnesota, USA. 98 pp., 304 illus. with 172 in color, 22 tables, indexed, fold-stapled glossy semisoft covers in multicolor with white lettering, 8.4 by 10.9 inches (213 × 277 mm).

  A substantial revision of the first edition. New diseases that have been described since 1983 were added and one-third of the photographs were replaced. It was published in English and translated into the Chinese, Japanese, and Spanish languages.

  Revised in 2005.

1996† **Plagas y Enfermedades de los Cespedes (Compendium of Turfgrass Diseases).** Richard W. Smiley, Peter H. Dernoeden, and Bruce B. Clarke, with translation by Begoña de Sebastian Celaya. Ediciones Mundi-Prensa, Madrid, Spain. 103 pp., 236 illus. with 86 in color, 13 tables, indexed, glossy semisoft bound in multicolor with white lettering, 8.4 by 11.0 inches (218 × 279 mm). Printed entirely in the Spanish language.

  A Spanish translation of the 1992 second revised edition.

2005† **Compendium of Turfgrass Diseases.** Third revised edition. Richard W. Smiley, Peter H. Dernoeden, and Bruce B. Clarke. The American Phytopathological Society, St. Paul, Minnesota, USA. 175 pp., 430 illus. with 120 in color, 10 tables, indexed, glossy semisoft bound in multicolor with white and green lettering, 8.4 by 10.8 inches (213 × 2274 mm).

  A major revision of the 1992 edition, including updated illustrations.

Smith, Jeffrey Drew

Drew Smith (1922–2004) was born in Ireshopeburn, County Durham, England, UK. He earned a BSc degree with *distinction* in Botany from the University of Durham in 1946 and an MSc degree in Agriculture from the University of Durham King's College, Newcastle-upon-Tyne, in 1957. He served as Plant Pathologist for the Sports Turf Research Institute from 1954 to 1958, as Plant Pathologist at the North of Scotland College of Agriculture from 1958 to 1960, as Senior Research Scientist at the Saskatoon Research Station of the Research Branch of Agriculture, Canada, from 1965 to 1984, and as Visiting Professor in the Department of Horticultural Science at the University of Saskatchewan, Saskatoon, Canada, from 1989 to 2004 plus as Associate Director (Technical) of the Saskatchewan Golf Association since 1990. His research emphasis has been on forage and turfgrass diseases and grass breeding for disease resistance. Drew Smith received the D.L. Bailey Award of the Canadian Phytopathological Society in 1990.

1959† **Fungal Diseases of Turf Grasses.** First edition. J. Drew Smith. Sports Turf Research Institute, Bingley, West Yorkshire, England, UK. 90 pp., 16 illus., 3 tables, indexed, cloth hardbound and also semisoft bound in white with green lettering, 5.7 by 8.7 inches (146 × 220 mm).

  A **relatively rare** book devoted to the basics of 20 turfgrass diseases, especially as related to conditions in England. Some rare invaders, such as lichens, are discussed. It is supported by selected references to scientific papers.

  Revised under the authorship of Noel Jackson and Jeffrey Drew Smith.

  See *Jackson, Noel—1965.*

1965† **Fungal Diseases of Turf Grasses.** Coauthor.

  See *Jackson, Noel—1965.*

1989† **Fungal Diseases of Amenity Turf Grasses.** Third revised edition. J.D. Smith, N. Jackson, and A.R. Woolhouse. E. & F.N. Spon Limited, London, England, UK. 401 pp., 182 illus. with 65 color plates, 17 tables, indexed, cloth hardbound in gray with gold lettering and a jacket in light green with white lettering, 7.3 by 9.7 inches (185 × 246 mm).

  A basic text on the pathology of turfgrass diseases. It is a major revision that now encompasses diseases of warm-season turfgrasses as well as cool-season species. Numerous fungal structures and spores of various pathogens are illustrated. The book is extensively referenced.

Figure 12.82. Stationary belt-drive, compost and soil shredder, and rotating screen, absent shields over the shredder, circa 1922. [7]

Smith, Percy White[3]

1948   **Specification of Playing Facilities.** First edition. P.W. Smith. The National Playing Fields Association, London, England, UK. 60 pp., 16 illus., no tables, small index, fold-stapled semisoft covers in green and white with black lettering, 5.5 by 8.4 inches (139 × 213 mm).

   ✳   A **rare** manual of the specifications for the construction of various types of playing facilities in the United Kingdom. The emphasis is on turfgrass surfaces, with small sections on the cultural aspects. Included are cricket grounds, lawn tennis courts, putting greens, and team sports fields. In 1948, Percy Smith proposed the following construction profile for bowling greens:

> The foundation of the green to consist of a layer of clean boiler clinker (3" to 3/4" gauge) spread over area of green to a depth of 4 1/2" consolidated. Carefully levelled and consolidated and covered with a layer of clean clinker ash (1/2" to fine) well mixed with an equal quantity of good vegetable top soil (through 1/2" screen) and spread and levelled to a consolidated depth of 3.

> *Revised in 1965.*
> See *Gooch, Robert Butler—1965.*

1950†   **The Planning, Construction and Maintenance of Playing Fields.** Percy White Smith. Oxford University Press, London, England, UK. 224 pp.; 103 illus.; 23 tables; indexed; textured hardbound in green with gold lettering and a jacket in greens and white with black, white, and green lettering; 7.2 by 9.7 inches (183 × 246 mm).

     This major work is devoted to site selection, layout, design, specifications, construction, and planting of sports fields in England. It is the first book of its kind and is **relatively rare**. The chapter subject titles include the following:

> I.   Choosing a Site
> II.   Space Requirements for Games and Athletics
> III.   Planning to Playing Field
> IV.   The Construction of Playing Fields
> V.   Management and Maintenance of Playing Fields
> VI.   The Floodlighting of Playing Fields
> VII.   Specification of Playing Facilities

     It also contains a small chapter on maintenance of sports field turfgrasses. Included are plans for soccer, rugby, hockey, lacrosse, cricket, netball, lawn tennis, lawn bowling, baseball, softball, rounders, hazena, field handball, longball, stoolball, volley ball, and running track facilities. The development of this book was sponsored by the National

Playing Fields Association. The foreword is by Sir George L. Pepler. In 1950, Percy Smith outlined the following procedures for installation of irrigation systems in sports fields in the United Kingdom:

> There are various methods of overhead irrigation in common use on sports fields, and in the writer's opinion they are much to be preferred to underground arrangements for flooding playing surfaces which it is claimed reduces loss by evaporation. Any faults in watering equipment above ground can be instantly detected and easily rectified. Underground defects, however, may pass undetected for some time and correction may entail much trouble and suspension of play.
>
> Overhead sprinklers and spray lines have reached a high degree of efficiency where properly adjusted to the working pressures available and, used in the proper circumstances, are as near an approach to natural rainfall as can be obtained. There is always a possibility of soil washing or scouring from badly adjusted or otherwise defective underground distributors which may sooner or later upset the stability if not the growth of the playing surface.
>
> It must be realized also that rainfall not only wets the soil but cleanses the foliage as well and any artificial watering system which ignores this fact only partially fulfills its functions. An underground system would be no use whatever in assisting the penetration of fertilizer and other dressings.
>
> While there are sprinklers designed to operate on very low pressures (6 lb. per square inch), the most serviceable types require water-pressures of from 20 lb. to 50 lb. per square inch, and the normal size hose for the average revolving sprinkler is ¾-inch diameter. There should be, therefore, a ¾-inch hydrant fixed at a convenient point, for each facility to be fed by a sprinkler, to which the supply hose can be attached. Where a movable sprinkler is employed about 120 feet of hose would be ample to cover a single tennis court, bowling green, or cricket square of average dimensions, but where the area is large, as in the case of a range of tennis courts in one enclosure, the number of hydrants will have to be increased to suit the number of sprinklers and lengths of hose available.

Snook, Gerald Frederick

Gerald Snook was born in Preston, Lancashire, England, UK, and earned a Royal Horticulture Society General Certificate in Horticulture. He was employed as Propagating Gardener at Preston and Chorley Hospitals in 1965, as Head Gardener for Manchester City Council in 1969, as Assistant in 1971 and as Superintendent in 1973 of Parks and Cemeteries for Benfleet Urban District Council, and as Horticultural Officer in 1990 and Parks Development Officer in 2003 for the Castle Point Borough Council. He was named Technical Officer in 1973, Vice Chairman in 1988, and Vice President in 2004 for the Essex Playing Fields Association. Gerald Snook received the Duke of Edinburgh Award for work on playing fields in the County of Essex.

1978   **Concise Guide to the Maintenance of Sports Grounds.** Gerald Snook. Essex County Playing Fields Association, Chelmsford, Essex, England, UK. 20 pp., 6 illus., no tables, no index, fold-stapled semisoft covers in white with black lettering, 4.5 by 6.3 inches (114 × 160 mm).

A small reference guide in a brief format concerning the culture of sports fields in England. It includes rugby, cricket, bowling green, field hockey, and lacrosse.

Snow, James Taft

Jim Snow was born in Ithaca, New York, USA, and earned a BSc in 1974 and an MSc in 1976, both from Cornell University. He joined the United States Golf Association (USGA) Green Section as an Agronomist in the Northeast in 1976, became Northeastern Regional Director in 1982, and National Director in 1990. As National Director, he administered the programs of the department, where he was involved with the USGA Turf Advisory Service, the Turfgrass & Environmental Research Program, the Internship program, the regional and national annual conferences, the editorial responsibilities of the **Green Section Record** magazine, the USGA championship involvement; the publishing responsibilities involving books and other publications; the oversight and involvement with the Audubon Cooperative Sanctuary Program For Golf Courses (administered by Audubon International); and the oversight and involvement with the Wildlife Links Program (administered by the National Fish and Wildlife Foundation). Jim Snow received the Hall of Fame Award of the New Jersey Turfgrass Association in 2003 and the John Reid Lifetime Achievement Award of the Metropolitan GCSA in 2003.

1993†   **USGA Recommendations for Putting Green Construction.** James T. Snow (editor). United States Golf Association, Far Hills, New Jersey, USA. 36 pp., 19 illus. with 16 in color, 12 tables, no index, fold-stapled glossy semisoft covers in multicolor with green and black lettering, 8.5 by 10.8 inches (215 × 275 mm).

A major revision of the recommendations for the USGA method of root-zone construction for putting greens. It includes the laboratory standards for testing of root-zone materials and mixes. There is an extensive list of references.

Société Francaise des Gazons (French Turfgrass Society)

1990† **L'Encyclopédie des Gazons (Turf Encyclopedia).** Edited by a 16 member committee of La Société Francaise des Gazons, with 42 contributing authors. Editions S.E.P.S., Boulogne, France. 360 pp., 461 illus. with 297 in color, 125 tables, no index, glossy hardbound in multicolor with white and black lettering, 8.2 by 11.7 inches (208 × 297 mm). Printed entirely in the French language.

A major book on the science and culture of turfgrasses in France by members of La Société Francaise Des Gazons. It is organized in two parts: theory and application, which include ornamental, sport, and functional turfgrasses. The illustrations are well done. There is a small glossary and a selected bibliography in the back, including ornamental, sport, and functional turfgrasses.

Soderstrom, Thomas Robert (Dr.)

Thomas Soderstrom (1936–1987) was born in Chicago, Illinois, USA, and earned a BSc degree from the University of Illinois in 1957 and MSc and PhD degrees in Botany from Yale University in 1958 and 1961, respectively. He joined the National Museum of Natural History at the Smithsonian Institute as Assistant Curator in 1960, became Associate Curator in 1962, and became Curator of the Department of Botany in 1964. His expertise was the taxonomy and biology of bamboos. Dr. Soderstrom was a recipient of the Gold Seal Award from the National Council of State Garden Clubs, an Honorary Associate of the Botanical Society of Brazil in 1984, and a Fellow of the American Association for the Advancement of Science in 1986, and received the Special Achievement Award from the Smithsonian Institution.

1986† **Grass Systematics and Evolution.** Thomas R. Soderstrom, Khidir W. Hilu, Christopher S. Campbell, and Mary E. Barkworth (editors). Smithsonian Institution Press, Washington, D.C., USA. 474 pp., 217 illus., 68 tables, detailed taxonomic index, cloth hardbound in green with gold lettering and a jacket in light gold and light green with black lettering, 8.4 by 11.0 inches (214 × 279 mm).

A scholarly book composed of 33 papers that summarize the current knowledge on systematics and evolution of grasses. It contains an extensive bibliography of over 2,300 references in the back.

Souma, Taketane

Taketane Souma (1889–1936) was born in Tokyo, Japan. He was a relative of the Japanese Emperor. His father conducted studies concerning winter color enhancement of turfgrass, which Taketane Souma discusses in the book.

1937 **Evergreen Turfgrass.** Taketane Souma. Meibundo Inc., Tokyo, Japan. 91 pp., 64 illus., 27 tables, no index, bound in green with black lettering, 7.0 by 10.0 inches (180 × 255 mm). Printed in the Japanese language.

∗ A **very rare** handbook concerning the construction and maintenance of turfgrasses at the Tokyo Golf Club. Emphasis is placed on techniques to sustain year-round green color.

1991† **Evergreen Turfgrass.** Reprinting. Taketane Souma. Soft Science Inc., Tokyo, Japan. 106 pp., 64 illus., 27 tables, no index, textured hardbound in dark green with gold lettering, 7.1 by 10.0 inches (180 × 255 mm). Printed entirely in the Japanese language.

A reprint of the historical 1937 edition, including the original photographs.

Spencer, Jerry[2]

2008† **Nutrition of Sports Turf in Australia.** Jerry Spencer. Landlinks, Collingwood, Victoria, Australia. 206 pp., 70 illus., 66 tables, small index, glossy semisoft bound in multicolor with white lettering, 6.7 by 9.6 inches (170 × 244 mm).

A book concerning aspects of plant nutrition, fertilizer carriers, and fertilization techniques for turfgrasses under conditions in Australia. There is a list of selected references in the back.

Sport England

Formerly known as the English Sports Council, Sport England is focused on the creation of a world-leading community sport system. Funding is from the National Lottery and Exchequer; it is allocated to organizations and projects that will grow and sustain participation in grass roots sport.

2000 **Natural Turf for Sport.** Sport England. Sport England Publications, Wetherby, England, UK. 35 pp.; 42 illus. in color; 7 tables; no index; glossy soft bound in multicolor with black, red, and blue lettering; 8.3 by 11.6 inches (210 × 296 mm). A booklet concerning the layout, construction, and planting of turfgrasses for sports fields in England, UK.

Sports Turf Association of Ontario (STA)

The Sports Turf Association of Ontario was formed in 1987 with the objective to promote safe sports turfgrass surfaces through education and professional programs. Membership encompasses persons engaged in the management of athletic fields, golf course superintendents, suppliers of materials, and landscape architects. The STA sponsors an annual conference and a summer field day plus publication of the quarterly **Sports Turf Manager**.

Figure 12.83. Belt-drive or manual-powered, 7-foot-long (2.1 m), 30-inch-diameter (76 cm) rotary soil screen with three replaceable sections and 1.5 to 5 cubic yards (1.1 to 3.8 m³) per hour capacity, depending on condition of the material, circa 1928. [41]

1991† **An Athletic Field Managers Guide For Construction and Maintenance.** First edition. Sports Turf Association, Guelph, Ontario, Canada. 39 pp.; 36 illus. with 15 in color; 3 tables; no index; glossy semisoft bound in cream, green, and brown with green lettering; 6.3 by 9.4 inches (159 × 238 mm).

A booklet on the construction, turfgrass establishment, and culture of sports fields in Canada. It contains a list of references and drawings of field dimensions in the back.

Revised in 2005.

See *Sheard, Robert Wesley—2005.*

Sports Turf Research Institute (STRI)

STRI is an organization of golf clubs and turfgrass organizations in the United Kingdom, formed to support turfgrass research and provide advisory services. It is headquartered in Bingley, West Yorkshire, England, UK. Originally formed in 1929 under the name Board of Greenkeeping Research (BGR), it was changed to the STRI in 1951. The BGR evolved from golf interests that initiated an Expert Advisory Committee in 1924, led by Norman Boase of the Royal and Ancient Golf Club of St Andrews. The first Director was Robert B. Dawson. The name change represented a broadening of activities to include all types of sport and recreational areas.

1978† **Fertilisers In Turf Culture.** First edition. Sports Turf Research Institute, Bingley, West Yorkshire, England, UK. 43 pp., 9 illus., 5 tables, no index, fold-stapled semisoft covers in light green with dark green lettering, 5.7 by 8.2 inches (146 × 208 mm).

A practitioner-oriented booklet on chemical soil testing; turfgrass nutrition; nutrient carrier sources for N, P, K, Mg, and Fe; fertilizer use; and turfgrass fertilization programs. It is oriented to cool-season turfgrasses and conditions in the United Kingdom.

Revised in 1991.

See *Lawson, David Mutrie—1991.*

1979† **Turfgrass Diseases.** First edition. Sports Turf Research Institute, Bingley, West Yorkshire, England, UK. 36 pp., 8 illus. in color, no tables, no index, semisoft bound in blue with dark blue lettering, 5.9 by 8.2 inches (149 × 208 mm).

A practitioner-oriented booklet in which 16 common diseases of turfgrasses found in the British Isles are described, along with cultural and chemical methods for their control.

Revised in 1987.

1979† **Golden Jubilee Celebrations.** The Sports Turf Research Institute, Bingley, West Yorkshire, England, UK. 26 pp. plus 42 pp. of advertising, 3 illus., no tables, no index, fold-stapled semisoft covers in light green with gold and green lettering, 6.0 by 9.0 inches (152 × 229 mm).

A booklet consisting of the day's program of events, three papers titled (1) Fifty Years Young, (2) Grasses for Sports Fields, and (3) Modern Sports Turfgrass Drainage plus a large section of advertisements in the back portion.

Sprague, Howard Bennett (Dr. and Professor Emeritus)

Howard Sprague (1898–1988) was born in Cortland, Nebraska, USA, and earned BSc and MSc degrees, both from the University of Nebraska in 1921 and 1923, respectively, and a PhD degree from Rutgers University in 1926. He joined the faculty

Figure 12.84. Aerial view of the 10-acre experimental ground at the STRI St Ives Research Station, Yorkshire, England, UK, circa 1954. [3]

at the University of Minnesota for one year and then returned to Rutgers University in 1927 through 1942 and was Head of the Agronomy Department. During the 1930s, Dr. Sprague pioneered the establishment of a turfgrass research program at Rutgers University and was the senior author on several early Agricultural Experiment Station Bulletins on turfgrass nutrient and lime requirements (1934) and on *Poa annua* (1937). He served his country in three wars—WWI, WWII, and Korea. Following WWII, he joined the Renner Research Foundation in Texas as Director of the Agriculture Division. He then served as Head of the Agronomy Department at Pennsylvania State University from 1953 to 1964. He was with the National Academy of Sciences as Executive Secretary to its Agricultural Board from 1964 to 1969. Then from 1969 to 1987, he was an agricultural consultant for the US Agency for International Department (USAID). Dr. Sprague was elected President of the American Forage and Grassland Council from 1953 to 1956, President of the Crop Science Society of America (CSSA) in 1960, and President of the American Society of Agronomy (ASA) in 1964. He was recipient of the ASA Fellow in 1941, CSSA Fellow in 1941, and USGA Green Section Award in 1974.

1940†   **Better Lawns For Homes and Parks.** Howard B. Sprague. A Whittlesey House Garden Series Book, Whittlesey House, McGraw-Hill Book Company Inc., New York City, New York, USA. 205 pp., 36 illus., 6 tables, indexed, cloth hardbound in light green with black lettering and a jacket in black and light green with light green and black lettering, 5.4 by 8.0 inches (138 × 202 mm).

     This book addresses the basics of turfgrass establishment and culture in a practical manor. Emphasis is placed on the lawns of North America. Dr. Sprague proposed the following procedure for sodding lawns in 1940:

> The sod should be cut to uniform thickness, not less than 1½ nor more than 2½ inches, and to uniform width to facilitate making a smooth surface with the newly laid sod. The joints in the sod are then filled with rich screened loam soil and the sod rolled to firm it in place. On slopes, it is desirable to stake the turf down until it is rooted, to prevent slippage during heavy rains. In handling sod in cool weather, there is little danger of killing the grass while it is being hauled or piled preparatory to laying. In warm weather, to avoid injury sod must not be piled for more than 24 hours.

     †   *Second printing in 1940.*

1945†   **Better Lawns.** Howard B. Sprague. The American Garden Guild Inc. and Doubleday, Doran, and Company Inc., New York City, New York, USA. 205 pp., 36 illus., 6 tables, indexed, cloth hardbound in slate with black and gold lettering and a jacket in multicolor with white and black lettering, 5.3 by 7.9 inches (135 × 201 mm).

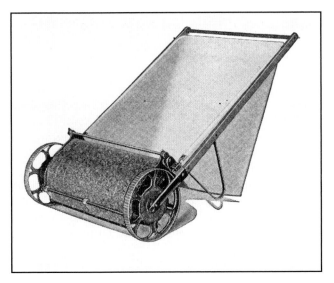

Figure 12.85. Springfield sweepers, both manual push (left) and horse drawn (right), 1920s. [41]

Essentially a reprinting of the 1940 book.

1970† **Turf Management Handbook.** First edition. Howard B. Sprague. The Interstate Printers and Publishers Inc., Danville, Illinois, USA. 253 pp., 147 illus., 10 tables, indexed, cloth bound in green with gold and black lettering, 5.9 by 8.9 inches (149 × 226 mm).

An introductory book on the basics of turfgrasses plus their establishment and culture. It addresses both cool- and warm-season turfgrasses for various types of use.

Revised in 1976, 1982, 1994, and 1996.

1976† **Turf Management Handbook.** Second revised edition. Howard B. Sprague. The Interstate Printers and Publishers Inc., Danville, Illinois, USA. 255 pp., 131 illus., 6 tables, indexed, cloth hardbound in greens and white with white and black lettering, 5.9 by 9.0 inches (149 × 228 mm).

This second revision contains minor changes.

Revised in 1982, 1994, and 1996.

1982† **Turf Management Handbook.** Third revised edition. Howard B. Sprague. The Interstate Printers and Publishers Inc., Danville, Illinois, USA. 255 pp., 132 illus., 7 tables, indexed, cloth hardbound in greens and white with white and black lettering, 5.9 by 8.9 inches (150 × 226 mm).

This third revision contains only minor changes.

Revised in 1994 and 1996.

1994† **Turf Management Handbook.** Fourth revised edition. Coauthor.

See *Schroeder, Charles Bernard—1994.*

1996† **Turf Management Handbook.** Fifth revised edition. Coauthor.

See *Schroeder, Charles Bernard—1996.*

Stanley Thompson & Company

A golf course architecture firm founded by Stanley Thompson (1874–1952) in Toronto, Ontario, Canada. Born in Scotland, he immigrated with his family to Toronto before WWI. He trained many assistants who also became successful golf course architects, including Robert Trent Jones Sr. and Geoffrey S. Cornish.

n.d. **About Golf Courses: Their Construction and Up-Keep.** Stanley Thompson & Company Limited, Toronto, Ontario,
circa Canada. 32 pp., 13 illus., no tables, no index, fold-stapled semisoft covers in gray with gold embossed and red designs,
193x 6.8 by 9.5 inches (169 × 241 mm).

A **very rare**, small early booklet concerning the design, construction, and planting of golf courses under Canadian conditions plus the culture of putting greens and fairways.

Printed as a facsimile by the Stanley Thompson Society of Etobicoke, Ontario, Canada, in 2003.

Stevens, W.J.[3]

1909 **Lawns.** W.J. Stevens. One & All Garden Books II No. 19, Agricultural & Horticultural Association Ltd., London, England, UK. 20 pp., 26 illus., no tables, no index, stitched semisoft covers in white with black lettering, 5.3 by 8.3 inches (135 × 210 mm).

*   A **very rare**, small booklet concerning cool-season turfgrasses and their planting and care under conditions in the United Kingdom. There also is a small section in the back on cricket and tennis grounds by Tom Hearne of Lord's. Also, the origins of turfgrass seed production in Germany are described.

Stewart and Company

In December 1894, James Cumming Stewart (1864–1940) started the business as Stewart and Low, Seedsmen in St Giles Street, Edinburgh, Scotland, UK. The proximity to the ancient fortress of Edinburgh Castle inspired him to utilize the image from the southern profile combined with the name "Stronghold" to become the company's trademark and brand name, respectively. At the outset the company's customers were largely professional gardeners and farmers. James Stewart also became active in turfgrass advisory visits to golf courses and sport fields. In 1900, John Lockhart Forbes (1883–1967) joined the business and focused on expanding the turfgrass sports market. Working closely with many of the leading greenkeepers of the day such as William Gault, Peter Lees, and Hugh Hamilton, the company pioneered new grass seed mixtures and turfgrass management products. The business name was changed to Stewart and Company, Seed Merchants in 1914. In 1936, Stewart's son, James Ian Stewart (1905–1988), became a Director, and the business was registered as a limited company, Stewart & Co. Seedsmen Ltd., based in South St Andrew Street, Edinburgh. James Cumming Stewart died in 1940 during WWII. At the cessation of hostilities, Ian Forbes (1913–1998), son of John L. Forbes, returned from active service to rejoin the business as Managing Director. Domestic gardening was very popular during those years of austerity and the establishment of a vigorous retail business was undertaken. The sports turf business was still there but was much depleted due to the ravages of wartime and the postwar shortages. During this period, Ian Stewart oversaw the agricultural business. In 1961, John Forbes retired from 60 years active participation in the business. The mid-1960s saw further decline in the agricultural seed business, and Ian Stewart retired, selling his shares to the Forbes family. Graeme David Forbes joined the business in 1967. This was a time of great activity in the sports turf industry due largely to the increasing use of automatic irrigation and minicrumb fertilizers. The company moved to Dalkeith, Midlothian, Scotland, in 1975, and Graeme became Managing Director. In doing so, they ended the long seed-trade connection with agriculture, professional horticulture, and domestic gardening to concentrate on the rapidly expanding sports market. Following a visit to St. Louis, Michigan, USA, Graeme Forbes joined Derek L. Edwards in a turfgrass farming business forming Turfgrass Services International Ltd. in Pocklington, York, England, UK, trading as Inturf. In 1986, Stewarts commenced turfgrass growing on 50 acres of rented land in East Lothian, Scotland. Ian Forbes retired from the business in 1988. In 1996, Grahamstone Farm, Kinnesswood, Kinross, was purchased to expand the turfgrass growing business alongside the land rented in East Lothian. In addition to being Scotland's largest turfgrass growers, the sports business continued selling grass seed, pesticides, and fertilizers accompanied by a manufacturing operation producing topdressings and golf course accessories. The company had been the first in the UK to produce nylon golf flags, fiber glass flag pins, and had invented the telescopic turf switch. In addition, there was a thriving contracts department providing specialist services to sports turfgrass businesses. In 2000, the company employed 52 people and, in 2005, Duncan John Forbes also became a Director. In 2006, a management buyout was agreed upon for the contracting business, leading to the formation of Sportsmasters (UK) Ltd. In the same year, the sport turfgrass supply business of Stewart & Co. Seedsmen Ltd. was sold to Richard Aitken (Seedsmen) Ltd., Glasgow. The company had decided to concentrate on turfgrass farming and moved to offices created at Grahamstone Farm and the name was changed to Stewartsturf Ltd. In 2009, Duncan was appointed Managing Director. Stewart and Company published an early series of booklets starting circa 1900 through the late 191x. Examples included herein are circa 1908, 1910, 1914, and 1916.

1908†   **Links and Lawn Grasses.** Stewart and Company, Edinburgh, Scotland, UK. 26 pp., 12 illus., no tables, no index, fold-stapled semisoft covers in green with black lettering, 8.6 by 5.5 inches (218 × 140 mm).

*   A **very rare** booklet outlining site preparation and seeding procedures for turfgrasses plus renovation, weed control, and worm irritant use.

1909†   **Descriptive List of Links and Lawn Grasses.** Stewart and Company, Edinburgh, Scotland, UK. 38 pp., 9 illus., no tables, no index, fold-stapled semisoft covers in gray with black lettering, 8.6 by 5.5 inches (218 × 140 mm).

*   A **very rare**, revised, expanded booklet of the 1908 edition.

1911†   **Links and Lawn Grasses.** Stewart and Company, Edinburgh, Scotland, UK. 50 pp., 15 illus., no tables, no index, fold-stapled semisoft covers in brown and green with black lettering, 8.6 by 5.5 inches (218 × 140 mm).

*   A **very rare** booklet that contains brief sections on the establishment and culture of turfgrasses for golf courses, sports fields, and lawns in Scotland, along with presentations of their products.

1913†   **Formation and Treatment of Links and Lawn Turf.** Stewart and Company, Edinburgh, Scotland, UK. 70 pp., 26 illus.,
circa   1 table, no index, fold-stapled semisoft covers in white and gold with gold lettering, 5.5 by 8.4 inches (140 × 210 mm).

*   An updated expanded revision, with less promotion of the turfgrass products. It is a **very rare** booklet.

1916†   **Links and Lawn Turf—Formation and Upkeep.** Stewart and Company, Edinburgh, Scotland, UK. 42 pp., 9 illus., 1
circa   table, no index, fold-stapled semisoft covers in white and gold with gold lettering, 5.5 by 8.4 inches (140 × 210 mm).

*   An updated revision of reduced length. A **very rare** booklet.

Stewart, Victor Innes (Dr.)

Vic Stewart was born in London, England, UK. He served in the British Military as an Artillery Officer from 1942 to 1947. He earned a BSc degree with *first class honours* in Botany in 1954 and a PhD degree in Soil Science in 1958, both from the University of Aberdeen, Scotland. He then became a Soil Science Lecturer in the Department of Agricultural Chemistry at the University College of Wales (UCW), Aberystwyth, and was promoted to Senior Lecturer in charge of the Soil Science Unit in 1971. His emphasis was on research and teaching concerned with land reclamation back to agriculture and later the construction, drainage, and culture of sports fields. Drs. Stewart and Bill Adams conducted the original research on a slit drainage system involving vertical minislits at right angles and connected with main slits with permeable fill that are oriented over drain lines. He retired in 1982 to focus on consulting activities while continuing as an Honorary Lecturer in Applied Soil Science at UCW, Aberystwyth. Dr. Stewart received the Bob Mackenzie Award from the Association of Playing Fields Officers and Landscape Managers in 1981 and the President's Certificate Award from the National Playing Fields Association in 1985.

1994† **Sports Turf—Science, construction and maintenance.** V.I. Stewart. Published in association with the National Playing Fields Association. E. & F.N. Spon Limited, London, England, UK. 260 pp., 56 illus., 22 tables, small index, glossy hardbound in green with red and white lettering, 7.3 by 9.6 inches (185 × 245 mm).

A specialty book on design specifications and construction of root zones for sports fields under climatic conditions similar to those in the United Kingdom. A scientific appraisal based on soil physics is given for the various practices in use. The turfgrass root zone aspects are addressed in depth, while the turfgrass plant aspects are nominally mentioned. There is a short list of selected references in the back.

Sticklen, Mariam Behrouz (Dr. and Professor)

Mariam Sticklen was born in Ramhormuz, Iran, and earned a BSc degree in weed physiology from Palavi University, Shirzah, in 1967; an MSc degree in Herbicide Physiology from American University in 1970; and a PhD degree specializing in Biotechnology from Ohio State University in 1981. She served as a nontenured Assistant Professor at Clemson University from 1981 to 1982 and as an Instructor in the Agronomy Department at Ohio State University from 1983 to 1987 and then, in 1987, joined the faculty in the Department of Crop and Soil Science and Department of Entomology at Michigan State University (MSU). She rose from Assistant Professor in 1987, to Associate Professor in 1995, and to Full Professor in 2000. She is actively involved in research emphasizing cell and molecular genetic improvement of grasses and in teaching of biotechnology courses. From 1991 to 1994, she was Research Director of the Agricultural Biotechnology for Sustainable Productivity Project at MSU.

1998† **Turfgrass Biotechnology—Cell and Molecular Genetic Approaches to Turfgrass Improvement.** Mariam B. Sticklen and Michael P. Kenna with 67 contributing authors. Ann Arbor Press, Chelsea, Michigan, USA. 278 pp., 28 illus., 33 tables, indexed, glossy hardbound in multicolor with white lettering, 6.2 by 9.2 inches (157 × 234 mm).

A specialty book reviewing the techniques and applications of biotechnology to improve turfgrasses. Four major aspects to enhance turfgrass breeding efforts are addressed: molecular marker analysis, biological control, genes for improvement, and tissue culture and genetic engineering.

Stinson, Richard Floyd (Dr. and Professor Emeritus)

Richard Stinson (1921–2006) was born in Euclid, Ohio, USA, and earned a BSc degree in Floriculture, an MSc degree in Horticulture, and a PhD degree in Horticulture in 1943, 1947, and 1952, respectively, all from Ohio State University. He served as Assistant Professor of Floriculture at the University of Connecticut from 1948 to 1955, as an Assistant and then Associate Professor of Floriculture from 1955 to 1967 at Michigan State University, and as Associate Professor of Agricultural Education and Horticulture at Pennsylvania State University from 1967 to 1989, when he retired. Dr. Stinson was active in horticulture education.

1968† **Turfgrass Maintenance and Establishment.** Richard F. Stinson (director-editor). Teacher Education Series Vol. 9, No. 4, Department of Agricultural Education, Pennsylvania State University, University Park, Pennsylvania, USA. 154 pp., 104 illus., 2 tables, no index, semisoft bound in greens and black with black lettering, 8.4 by 10.8 inches (214 × 274 mm).

A basic turfgrass handbook for vocational students organized in six instructional units, with a set of test questions at the end of each unit. There is a list of selected references in the back.

1981† **Greenkeeping.** Richard F. Stinson (editor). Teacher Education Series Vol. 22, No. 6, AGDEX 273, Instructional Materials Services, College of Agriculture, Pennsylvania State University, University Park, Pennsylvania, USA. 101 pp., 130 illus., 3 tables, no index, semisoft bound in tan and greens with white and green lettering, 8.4 by 10.7 inches (213 × 273 mm).

Figure 12.86. Apparatus for steam sterilization at ~200 °F of soil and compost for topdressing, early 1930s. [43]

An introductory instructional manual designed for novice individuals preparing for or just starting employment as a golf course agronomic staff member. It includes 18 instructional units plus a competency check list in the back. It was a unique publication for that time.

Sudell, Richard

Richard Sudell (1900–1968) was born on a farm in Lancashire, England, UK, and was a student gardener at the Kew Botanical Garden. He then joined a firm of landscape contractors and became a landscape architect specializing in recreation grounds. In 1927, he was one of the founders of the British Institute of Landscape Architects. Richard Sudell was elected President of the British Institute of Landscape Architects in 1955.

1957†   **Sports Buildings and Playing Fields.*** Richard Sudell and D. Tennyson Waters. B.T. Batsford Limited, London, England, UK. 240 pp.; 210 illus.; 6 tables; indexed; cloth hardbound in light green with gold lettering and a jacket in light green, black, and white with white and black lettering; 7.2 by 9.6 inches (183 × 245 mm).

A book on the design and construction of sports playing fields and building. It includes turfgrass surfaces for team sports fields, cricket grounds, tennis courts, bowling greens, golf courses, and track & field facilities.

Sutton & Sons Limited

House of Sutton was a seed merchant firm founded in 1806 in Reading, Berkshire, England, UK, by John Sutton (1777–1863). It was formed one year after the formation of the Horticultural Society of London, forerunner of the Royal Horticultural Society. The firm initially marketed flour and agricultural seeds—chiefly corn and pasture seeds. In 1836, the name of the firm was changed to John Sutton and Son. For over 170 years, the company was run by five generations of the Sutton family up to 1979.

By 1904, Suttons was marketing its own artificial manure and the Sutton lawn mower. The company established a seed-testing laboratory for germination and purity in 1840. A Sports Ground Construction Department had been formed by Sutton & Sons by 1925. In 1985, Sutton & Sons bought Carters Tested Seeds, then owned by the Horticultural Botanical Association, and also purchased both R & G Cuthberts and Samuel Dobie and Son.

A second son, **Martin Hope Sutton** (1815–1901), was actively involved in horticultural plants and grasses, even as a young boy. He joined the firm in 1828 and became a Full Partner in his father's firm in 1836, where he continued to work until retirement in 1888. He led an increasing emphasis on garden seeds and grasses during the 1830s. A grass garden was formed at Reading in 1864 under his direction. He was author of **Laying Down of Land to Permanent Pasture** in 1886. M.H. Sutton of Sutton & Sons started a test grass garden at Reading in 1863 and subsequently formed more formal turfgrass plots near Reading, England, in the 1880s. The history of Sutton & Sons was published in 2006.

~

2006†   **Sutton Seeds—A History 1806–2006.** Early Local History Group with six principal contributing authors. The University of Reading, Reading, England, UK. 256 pp., 172 illus., 8 tables, small index, semisoft bound in multicolor with white and black lettering, 8.3 by 9.4 inches (209 × 239 mm).

A unique, interesting book concerning the 200-year history of Sutton & Sons Limited. It was researched and written primarily by long-term employees of the company led by Gerry Westall and Ron Butler.

~

A full page listing of turfgrass seeds for lawns appeared in Sutton's 1857 **Spring Catalogue and Amateur's Guide**. This listing was expanded to multiple pages from 1867 onward. In 1868, this same publication also included information on turfgrass culture titled **"Hints on the Formation and Improvement of Garden, Croquet Grounds, Cricket Grounds, etc."** Subsequently, **"The Formation of Lawns From Seeds"** of eight pages was published as a separate in the 1890s. From these beginnings, a series of seventeen pioneering turfgrass publications of twenty or more pages evolved. Fifteen are described herein as published in 1892, 1896, 1899, 1902, 1904, 1906, 1908, 1909, 1912 (eleventh), 1914 (twelfth), 1920 (thirteenth), 1931 (fourteenth), 1937 (fifteenth), 1948 (sixteenth), and 1962 (seventeenth).

1892† **Formation of Lawns, Tennis & Cricket Grounds from Seed.** Sutton & Sons of Reading. Simpkin, Marshall, Hamilton, Kent and Company Limited, London, England, UK. 25 pp. plus 3 pp. of advertising, no tables, no index, fold-stitched textured semisoft covers in maroon with gold trim and gold lettering, 5.6 by 8.5 inches (142 × 216 mm).

* This is a special, **truly rare** publication that is a must for collectors of historical turfgrass books, as is the entire series of revised editions. It is the first known publication of more than 20 pages that is devoted entirely to turfgrass concerns. Section titles for the 1892 booklet include the following:

- Grass Slopes
- Drainage
- Preparatory Work
- Weed Seeds in Soil
- Enriching the Soil
- Surface Preparation
- Selection of Seeds
- Fine Grasses and Clovers
- Quantity of Seed
- Time and Mode of Sowing
- Worm Casts
- Water and Shade
- Bird Scares
- Mowing
- Rolling
- Destruction of Weeds
- Improving Old Lawns
- Observations Relating Exclusively to Cricket Grounds

Figure 12.87. John Sutton, circa 1850s (left), and Martin Audley Sutton, circa 1950s (right). [20]

This small pioneering booklet is devoted to information regarding turfgrass establishment and culture of lawns and tennis courts, with a short section on cricket grounds. Note the emphasis placed on worm casts and bird scars. It also is interesting to note the early focus on turfgrass surfaces for tennis and cricket. The introduction to this pioneering publication was as follows:

> A CLOSE AND VERDANT LAWN is a beautiful object in itself, and it enhances the charm of every flower, shrub, and tree in the garden. The popularity of tennis and bowls has made a piece of turf almost indispensable to every residence where a court or bowling green can be secured. It is not our business, however, to dwell on the pleasure to be derived from lawns, nor on their influence in the promotion of healthy enjoyment. We have merely to consider the best and least expensive means of creating them.

1896†   **Lawns, Tennis And Cricket Grounds from Seed.** Revised edition. Sutton & Sons Limited, Simpkin, Marshall, Hamilton, Kent & Company Limited, London, England, UK. 32 pp. with 4 pp. of advertising, 10 illus., no tables, no index, fold-stitched semisoft covers in black with gold lettering, 7.0 by 11.0 inches (177 × 281 mm).

＊   A very **truly rare**, early booklet on the seed establishment and subsequent culture of turfgrasses on estate lawns and tennis courts, with small sections on cricket grounds and putting greens. It contains interesting large photographs of typical turfgrass uses in the 1890s. The suggested enrichment of soil prior to planting turfgrasses in England involved the following procedures in the 1890s:

> Grass is a fixed crop, chiefly deriving its nourishment from a few inches near the surface, and the only way of refreshing it is by raking or harrowing and top-dressing. Hence there are obvious reasons for putting the land into good heart before sowing. Well-rotted stable manure is always beneficial, but that which is fresh should be avoided because of its tendency to make the soil hollow. Where artificial manure is more convenient, two cwt. of superphosphate of lime, one cwt. of Peruvian guano, and one cwt. of bone dust, mixed together, constitute an excellent dressing. The quantities named are sufficient for an acre, and the mixture can be sown when the processes of raking and rolling are in progress. Sutton's A1 Lawn Manure can also be recommended with confidence, as it contains all the constituents essential to the growth of fine grasses and clovers. Three cwt. required per acre. After the application of artificial manure not less than ten days should elapse before sowing the grasses.

1899†   **Lawns: Garden Lawns, Tennis Lawns, Putting Greens, Cricket Grounds.** Revised edition. Sutton & Sons of Reading Limited. Simpkin, Marshall, Hamilton, Kent & Company Limited, London, England, UK. 34 pp. with 6 pp. of advertising, 6 plates with coloration and tissue protectants, no tables, no index, thick cloth hardbound with distinct beveled edges in green with gold lettering and also fold-stitched semisoft covers in green with orange lettering, 5.4 by 8.4 inches (138 × 213 mm).

＊   A **very rare**, small book on the seed establishment and culture of lawns and tennis courts, with a small section on putting greens and cricket grounds.

1902   **Lawns: Garden Lawns, Tennis Lawns, Bowling Greens, Croquet Grounds, Putting Greens, Cricket Grounds.** Revised edition. Sutton & Sons of Reading Limited, Simpkin, Marshall, Hamilton, Kent & Company Limited, London, England, UK. 36 pp. plus 2 pp. of advertising, no illus., no tables, no index, fold-stitched semisoft covers in green with embossed red lettering, 7.2 by 8.5 inches (138 × 215 mm).

＊   A **very rare**, small book concerning the seed establishment and early culture of lawns plus short sections on putting greens and cricket grounds.

1904†   **Garden Lawns, Tennis Lawns, Bowling Greens, Croquet Grounds, Putting Greens, Cricket Grounds.** Revised edition. Sutton & Sons Limited. Simpkin, Marshall, Hamilton, Kent & Company Limited, London, England, UK. 68 pp. including 3 pp. of advertising, 20 illus., no tables, no index, fold-stitched semisoft covers in green with embossed gold lettering and imprint, 5.5 by 8.6 inches (139 × 218 mm).

＊   This is an expanded revision of a **very rare** booklet, with photographs added.

1906†   **Lawns: Garden Lawns, Tennis Lawns, Bowling Greens, Croquet Grounds, Putting Greens, Cricket Grounds.** Revised edition. Sutton & Sons of Reading. Simpkin, Marshall, Hamilton, Kent & Company Limited, London, England, UK. 49 pp. plus 7 pp. of advertising, 18 illus., no tables, no index, fold-stitched semisoft covers in greens with embossed gold lettering, 5.5 by 8.7 inches (139 × 221 mm).

＊   A **very rare**, shortened version of the earlier edition. The issue of seed selection involving turfgrasses plus clover versus turfgrasses alone was addressed in 1906:

> The selection of grasses and clovers which are to form a fine dense sward should be regarded as in the highest degree important. They must be permanent in character, adapted to the soil, and free from

Figure 12.88. A road scraper being used to eliminate small undulations in the fairway of a golf course in the late 1920s. [41]

coarse-growing varieties. On land which is liable to burn, clovers maintain their verdure under a hot sun after grasses have become brown. Still, the two classes of plants must be carefully proportioned, for although clovers show conspicuously in summer, they partially disappear in winter, when the grasses are predominant. Clovers are also very slippery, they show signs of wear earlier than grasses, hold moisture longer after a shower, and quickly discolour tennis balls. It is therefore advisable to sow grasses only for tennis lawns, putting and bowling greens, unless the turf is peculiarly liable to scorch in summer. Then it is an open question whether an admixture of clovers may be regarded as the lesser of two evils. For the reasons stated it is our practice to omit clover seeds from mixtures to be sown for the purposes named, unless a wish to the contrary is expressed.

1908   **Lawns: Garden Lawns, Tennis Lawns, Bowling Green, Croquet Grounds, Putting Greens, Cricket Grounds.** Revised edition. Sutton & Sons of Reading Limited. Simpkin, Marshall, Hamilton, Kent & Company Limited, London, England, UK. 54 pp. with 6 pp. of advertising, 16 illus., no tables, no index, fold-stitched semisoft covers in green with gold lettering, 5.5 × 8.7 inches (139 × 220 mm).

   ✳   This **very rare** booklet contains small revisions.

1909   **Lawns: Garden Lawns, Tennis Lawns, Bowling Greens, Croquet Grounds, Putting Greens, Cricket Grounds.** Revised edition. Sutton & Sons of Reading. Simpkin, Marshall, Hamilton, Kent & Company Limited, London, England, UK. 63 pp. plus 9 pp. of advertising, 14 illus., no tables, no index, fold-stitched semisoft covers in green with gold lettering, 5.4 by 8.4 inches (138 × 214 mm).

   ✳   A **very rare** booklet containing some revisions.

1912†   **Lawns: Garden Lawns, Tennis Lawns, Croquet Grounds, Bowling Greens, Putting Greens, Cricket Grounds.** Eleventh revised edition. Sutton & Sons of Reading Limited. Simpkin, Marshall, Hamilton, Kent & Company Limited, London, England, UK. 75 pp. plus 5 pp. of advertising, 21 illus., no tables, no index, fold-stitched semisoft covers in green with embossed silver lettering, 5.5 by 8.5 inches (139 × 215 mm).

   ✳   This booklet contains some revisions, including new photographs. It is a **very rare** booklet.
      Revised in 1914.

1914   **Lawns: Garden Lawns, Tennis Lawns, Croquet Grounds, Bowling Greens, Putting Greens, Cricket Grounds.** Twelfth revised edition. Sutton & Sons of Reading Limited. Simpkin, Marshall, Hamilton, Kent & Company Limited, London, England, UK. 75 pp. plus 5 pp. of advertising, 17 illus., no tables, no index, fold-stitched semisoft covers in green with embossed silver lettering, 5.5 by 8.5 inches (140 × 216 mm).

   ✳   This booklet contains small revisions. It is a **very rare** booklet.
      Revised in 1920.

1920† **Lawns: Garden Lawns, Tennis Lawns, Croquet Grounds, Putting Greens, Cricket Grounds, Bowling Greens.** Thirteenth revised edition. Sutton & Sons Limited of Reading. Simpkin, Marshall, Hamilton, Kent & Company Limited, London, England, UK. 80 pp. plus 4 pp. of advertising, 21 illus., no tables, no index, fold-stapled semisoft covers in green with embossed silver lettering, 5.4 by 8.2 inches (137 × 209 mm).

* There are only small revisions. It is a **very rare** booklet. The issue of worker safety in the use of nonselective weed killers in 1920 in England was expressed:

> Another mode of killing weeds is to dip a wooden skewer into sulphuric acid, strong carbolic acid, or one of the liquid weed destroyers, and then plunge the skewer into the heart of the plant. The result is deadly and instantaneous. In using these destructive fluids, however, great caution must be exercised to avoid personal injury or the burning of clothing. But when every care is taken to prevent an accident, the fact remains that these corrosive fluids may prove to be a source of grave danger, even when the bottle containing the liquid is kept in a place of supposed security. For this reason poisons are better avoided, especially as the object can be attained by other means which are free from danger.

Revised in 1931.

1931† **Lawns: Garden Lawns, Tennis Courts, Croquet Grounds, etc.** Fourteenth revised edition. Sutton & Sons of Reading. Simpkin Marshall Limited, London, England, UK. 60 pp. plus 4 pp. of advertising, 20 illus., no tables, no index, fold-stapled textured semisoft covers in green with dark green raised lettering, 7.2 by 9.7 inches (183 × 247 mm).

* An interesting, **very rare** handbook on the establishment and culture of turfgrasses in England. The status of annual bluegrass (*Poa annua*) in English turfgrasses in 1931 was summarized as follows:

> It is not uncommon to find an old lawn largely, if not entirely, composed of Annual Meadow Grass. With considerable care and attention it is possible to keep such turf satisfactory, although it may be rather soft, and possess a peculiar tinge at certain seasons. Turf composed of Annual Meadow Grass, however, is extremely sensitive to the weather, and is very liable to die off in patches following periods of drought or severe cold. While recovery may be comparatively rapid, the bare spots afford an opportunity for weeds to become established.
>   Close cutting will not destroy Annual Meadow Grass; in point of fact, it flourishes under such treatment. The only remedy we know is hand weeding, and those who value a verdant lawn will do well in taking some trouble to remove by hand any specimens which may become established. Unfortunately, the presence of this grass is not always noticed, and owing to the freedom with which seed is produced it quickly becomes intermixed with the turf, and is then very difficult to deal with.

Revised in 1937.

n.d.† **Useful Notes for Green Committees.** Sutton & Sons Limited, Reading, England, UK. 28 pp., no illus., 4 tables, no
circa index, fold-stapled semisoft covers in green with dark blue lettering, 3.6 by 6.0 inches (92 × 153 mm).
193x
* A **very rare**, small booklet on the establishment and culture of turfgrasses for golf courses under the conditions in England. There is a monthly calendar of cultural practices.

1937† **Lawns: Garden Lawns, Tennis Courts, Croquet Grounds, etc.** Fifteenth revised edition. Sutton & Sons Limited of Reading. Simpkin Marshall Limited, London, England, UK. 60 pp. plus 4 pp. of advertising, 31 illus., no tables, no index, fold-stapled textured semisoft covers in green with raised black lettering, 7.2 by 9.6 inches (182 × 245 mm).

* A slight revision of a **very rare** booklet, primarily the photographs.
Revised in 1948.

1948† **Lawns and Sports Grounds.** Sixteenth revised edition. Sutton's of Reading, Sutton & Sons Limited, Reading, England, UK. 76 pp. plus 4 pp. of advertising, 34 illus., no tables, no index, fold-stapled semisoft covers in dark green with black embossed lettering, 5.4 by 8.5 inches (138 × 215 mm).

* An interesting, well-prepared handbook on the establishment and culture of turfgrasses in England. It is a major revision of the 1937 edition. It contains new chapters on the culture of turfgrasses for shaded lawns, cricket grounds, rugby fields, soccer fields, hockey fields, and bowling greens. The photographs are good for the 1940s. There are drawings for six sports layouts in the back. A procedure for correcting soil compaction in 1948 in England is presented in the following:

> Not only is the surface of turf liable to become hidebound, but the underlying soil also may be so consolidated that beneficial elements cannot penetrate: this is especially the case with areas used for games. It

is therefore desirable that lawns should be deeply pricked from time to time. This process allows air and water to penetrate evenly and improves surface drainage; plant food is carried to a greater depth and deeper root formation is encouraged. These factors naturally increase drought resistance.

This aeration can be most conveniently effected by the use of a fork with either hollow or solid tines, or by one of the special machines made for the purpose. On most soils it should be sufficient if the operation is carried out with hollow tines once every couple of years, alternating with solid tining. A fork with solid tines is sometimes very beneficial in cases where specially deep piercing is needed. The use of a solid tine machine may with advantage be repeated several times in the course of a year, when there is sufficient moisture in the soil.

Revised in 1962.
See *Sutton, Martin Audley Foquet—1962.*

1952   **Turf Weeds: Identification & Control.** First edition. Sutton's Grass Advisory Service, Sutton & Sons of Reading Limited, Reading, England, UK. 20 pp., 42 illus., no tables, small index, soft bound in light green with white and black or green and black lettering, 5.5 by 8.4 inches (140 × 214 mm).

   ✱   A small booklet on the botanical descriptions and controls for weeds of turfgrasses commonly found in the United Kingdom.

Slight revised controls published in 1956, 1965, 1971, and 1975.

Sutton, Martin Audley Fouquet

Audley Sutton (1900–1963) was born in Reading, Berkshire, England, UK. He was the son of Martin Hubert Fouquet Sutton and great-grandson and fourth in the line of descent of John Sutton, the founder of seed merchandiser Sutton & Sons of Reading, Berkshire, England, UK. He joined the company in 1923 and was made a partner in 1926. He eventually directed the Grass Research and Advisory Service of the company for many years. This included presenting some of the earliest educational courses concerning greenkeeping offered in the United Kingdom. Martin Sutton was awarded Fellow of the Royal Horticultural Society in 1925.

1933†   **Golf Courses—Design, Construction and Upkeep.** First edition. Martin A.F. Sutton (editor) with seven contributing authors: Bernard Darwin, T. Simpson, C.H. Alison, Martin A.F. Sutton, H.O. Hobson, P. Mackenzie Ross, and H.S. Colt. Simpkin Marshall Limited, London, England, UK. 152 pp., 75 illus., 21 tables, indexed, cloth hardbound in green with gold lettering and a jacket in green with black lettering, 7.3 by 9.7 inches (185 × 247 mm).

   ✱   A **rare** book that is a must for collectors of historical turfgrass and/or golf books. It is a compendium on golf course architecture and construction and on the associated establishment and culture of turfgrasses. The introduction is by Bernard Darwin. It is a very interesting book that is oriented to English conditions. The chapter subject titles and contributing authors include the following:

- The Design and Construction of a Golf Course—T. Simpson
- A Precis of Golf Architecture—C.H. Alison
- The Formation and Upkeep of Golf Courses and Putting Greens—Martin A.F. Sutton
- Golf Club Management—H.O. Hobson
- The Concentrated Golf Course—P. Mackenzie Ross
- Reminiscences and Reflections—H.S. Colt
- Tables and Other Useful Information

In his introduction Bernard Darwin summarized the technical status of greenkeeping in England in 1933 as follows:

Once upon a time the whole art and mystery of greenkeeping might almost have been summed up in Tom Morris's classic prescription "Mair saund, Honeyman," and how complex a business it has now become with men of science and laboratories devoted to it! Leaving all the rest of the course on one side, the putting greens have reached so high a standard that we have become sadly spoilt and pampered, and I am convinced that even when greens are just as good as they can be there will always be some members in the club to complain of them simply because life is "extremely flat with nothing whatever to grumble at." Both Mr. Sutton and his architectural collaborators lay great stress on the tremendous importance of good greens, and no doubt a course is to a large extent judged by its greens; if they are poor no other splendours make amends, and if they are good they cover a multitude of sins.

It is interesting, as showing how elaborate and technical a business course construction has become, to find Mr. Sutton saying that "where all the constructive work is correct an artificial green is far easier to

> maintain than a natural one." How far we have travelled since the old days when "natural" golf seemed the one thing to be desired! I suppose the fact is that golf has become so popular that we cannot eat our cake and have it; we cannot have purely natural golf any more when so many of us want to play it. Even St. Andrews, wonderful though it is, has perforce had to take on many of the qualities of an inland course, or it would have been trodden into nothingness by its admirers.

Revised in 1950.

1950† **Golf Courses—Design, Construction and Upkeep.** Second revised edition. Martin A.F. Sutton (editor) with seven contributing authors: Bernard Darwin, P. Mackenzie Ross, T. Simpson, Robert Trent Jones, Martin A.F. Sutton, C.H. Alison, and C.K. Cotton. Sutton & Sons, Reading, England, UK. 192 pp. with 8 pp. of advertising, 86 illus. with a color frontis page and tissue protection, 30 tables, detailed index, cloth hardbound in light green with gold lettering and a jacket in light green with black lettering, 7.4 by 9.7 inches (188 × 247 mm).

   ✳   A **relatively rare** compendium on golf course architecture and construction plus the associated establishment and culture of turfgrasses. An interesting book oriented primarily to conditions in the United Kingdom. The chapter subject titles and contributing authors include the following:

> - The Scottish Golf Links: Their Influence on World Golf Course Architecture—Philip Mackenzie Ross
> - The Design and Construction of a Golf Course—T. Simpson
> - Golf Course Architecture in the United States—Robert Trent Jones
> - The Formation and Upkeep of Golf Courses and Putting Greens—Martin A.F. Sutton
> - Golf Courses Overseas—C.H. Alison
> - Golf Club Management—C.K. Cotton
> - Bibliography
> - Appendix of Tables and Other Useful Information

There is an introduction by Bernard Darwin. Robert Trent Jones and Charles H. Alison briefly address golf courses in the United States, Europe, Japan, Australia, and South Africa. The main text, which is authored by Martin A.F. Sutton, has only small revisions of the first edition. An approach to root zone construction of putting greens in England in 1950 is discussed by Martin Sutton as follows:

> When filling in the trenches, the pipes should first be covered with a layer of fair-sized stones or rough clinker, and then with ashes to the level of the subsoil. The material should be well packed round the pipes and the ash carefully rammed from the top, before replacing the top soil.
>
> Where the soil is of an exceptionally heavy character the most thorough procedure is to spread a complete layer of rough ashes or clinker over the subsoil, in order to give full working effect to the drains. Approximately 100 cubic yards would be needed to cover an average green to a consolidated depth of 4 inches. If funds permit, the ash should be covered with 6 inches of lighter soil imported for the purpose, but should this be out of the question, the original surface soil must be replaced and improved in texture by incorporating as much sand as possible.

1962† **Lawns and Sports Grounds.** Seventeenth revised edition. Martin A.F. Sutton. Sutton & Sons Limited, Reading, England, UK. 248 pp., 127 illus., 22 tables, no index, cloth hardbound in dark green with silver lettering and a jacket both in green with dark green lettering and in multicolor with red and black lettering, 5.4 by 8.3 inches (136 × 211 mm).

An interesting handbook on the construction, establishment, and culture of turfgrass areas, with emphasis on conditions in England. The basics are addressed within a how-to approach. Presented in the back are chapters on turfgrass culture for sports fields, cricket grounds, bowling greens, tennis courts, and lawns. This revised edition is the seventeenth and last in a long series published by Sutton & Sons of Reading that was first published in 1892. It is quite an accomplishment.

Sutton, Martin Hubert Fouquet

Martin Sutton (1875–1930) was born in Reading, Berkshire, England, UK. He was the eldest son of Martin John Sutton and the eldest grandson of Martin Hope Sutton of Reading, England, UK. He was educated at Harrow and received a BA degree from Christ Church-Oxford University in 1895. He became a partner in the seed firm of Sutton & Sons, Reading, Berkshire, England, UK, in 1897 and was the Second Senior Partner of the firm at the time of his death. He was active in selection and breeding experiments for the improvement of sport turfgrasses and pastures, as well as clovers

Figure 12.89. A set of four manual-push wheelbarrow grass seeders in operation on an extensive turfgrass construction area, in late 1930s. [41]

and potatoes. He was selected as Vice President of the Golf Greenkeeper Association at the time it was formed in 1912. He also was a Justice of the Peace (JP) and served as a Governor of the Royal Agricultural Society and of the University College, Reading.

n.d.† **The Book Of The Links—A Symposium on Golf.** Martin H.F. Sutton (editor) with six contributing authors: Mar-
circa tin H.F. Sutton, H.S. Colt, A.D. Hall, Bernard Darwin, Sir George Riddell, and W. Kirkpatrick. W.H. Smith & Son,
1912 London, England, UK. 212 pp. with 12 pp. of advertising, 58 illus. and cartoons, 40 pp. of tables, detailed index, cloth hardbound in blue with gold lettering on a red imprint, 5.8 by 9.4 inches (148 × 239 mm).

* A **very rare** book that is a must for collectors of historical turfgrass and/or golf books. It is an important early com-
pendium encompassing golf course architecture, construction, turfgrass establishment, and culture under the conditions in the United Kingdom. It includes 40 pages of general information and descriptions of manures in the back. There are unique cartoons by Tom Wilkinson. The section subject titles and contributing authors include the following:

- The Construction of New Courses—H.S. Colt, golf course architect
- The Formation and Maintenance of Putting Greens and Teeing Grounds—Martin H.F. Sutton.
- The Manuring of Golf Greens and Courses—A.D. Hall, former director, Rothamsted Experimental Station
- Grasses and Grass Seeds—Martin H.F. Sutton.
- Golf Architecture—H.S. Colt
- The Influence of Courses Upon Player's Style—Bernard Darwin
- Caddies—Sir George Riddell
- Essay on Some Questions in Greenkeeping—W. Kirkpatrick, greenkeeper, Rye Golf Club
- The Vegetation of Golf Links—By a golfing botanist
- Finance
- Notes on Organic and Artificial Manures
- Tables and General Information

The section titled "Essay on Some Questions in Greenkeeping" was authored by the greenkeeper at Rye Golf Course by W. Kirkpatrick. It represents the first published paper written by a practicing golf greenkeeper in the United Kingdom. The costs of individual cultural practices and implements are presented as well as the total budget for course maintenance and a table of wages. The cartoons by Tom Wilkinson are unique. One set is a tongue-in-cheek depiction of sand bunkers on wheels and of golfers riding across the fairways in four-wheeled motorized

Figure 12.90. Manual pull and push, mechanical Toro topdresser, circa 1926. [2]

vehicles. Impossible? Manuring was a term used during the first part of the twentieth century to encompass fertilizers ranging from natural organics to synthetics, with the latter termed artificial fertilizers. There is a presentation concerning various types of natural manures used in English in 1912 and also the use of salt to obtain a fertilizer response:

Natural (Organic) Manures

In addition to farmyard manure other excreta are sometimes used, such as:

Horse Dung, which rapidly decomposes, and therefore its fermentation gives rise to considerable heat, being a "hot" manure.

Pig Dung slowly decomposes, and is a very rich manure. It is best mixed with other animal manures.

Sheep Dung requires a little longer time than horse dung for its decomposition. It is rich in solid matter.

Cow Dung slowly decomposes, and contains the smallest percentage of solid matter of the four animal manures. It is of far less value as a manurial agent than horse, pig, or sheep manures.

The following is given as the composition of the above-named manures:

	Horse	Pig	Sheep	Cow
Water	77.25	77.13	56.47	82.47
Solid matter	22.75	22.87	43.53	17.53
	100.00	100.00	100.00	100.00

Salt.—The general properties of salt from an agricultural point of view are:

(a) In small quantities it promotes the decomposition of animal and vegetable matters contained in all cultivated soils.

(b) It acts as a direct plant food when used in small proportions.

(c) It has the power of destroying noxious insects, slugs, and weeds when applied to fallow land.

(d) It possesses stimulating powers on growing plants.

(e) It increases the power of certain soils of absorbing moisture from the atmosphere.

(f) It has the power of preserving the juices of plants and the soils on which they grow from the effects of sudden fluctuations in the temperature of the atmosphere.

Also described are the natural organic and synthetic fertilizers used on golf course turfgrasses in England in 1912. As in many of the writings of that time, the use of potassium fertilizers was discouraged because they tended to encourage legume growth.

NOTES ON ORGANIC AND ARTIFICIAL MANURES

The following notes on manures (organic and artificial), which have been of use in certain exceptional circumstances, may be of interest. It should, however, be remembered that all *artificial* fertilisers need applying with discretion and after due consideration of local conditions, and expert advice on the subject should be sought. A feature of artificial manure is that no danger of weed seeds accompanies its use, and, generally speaking, it is best applied in moist weather or watered in.

Sutton's Grass Manure is a valuable "complete" manure, which has been prepared in consultation with the highest scientific authorities of the day. When ground is being got ready for seeds it should be harrowed or raked in (at the rate of from four to five pounds per rod or pole of ground) a few days before sowing the seed. For an existing turf it may be used throughout the year either alone (at the rate of two to four pounds per rod or pole) or mixed with fine, dry earth in equal quantity. During the growing period the dressing need not be confined to one application, but may be used as occasion requires to stimulate growth through the spring and summer.

Dried Blood is a valuable nitrogenous artifical fertiliser for late winter or early spring use. Its action is gradual. It benefits sandy loam soils, and should be used at the rate of 3 cwt. per acre.

Rape Meal is highly esteemed, not only as a top dressing for young grass and newly-sown greens, but also on old turf in early spring, autumn, or winter. It is somewhat quick in action, and the effect of applying it is only temporary. It may be used on all soils at the rate of 8 cwt. per acre. It is nitrogenous in character.

Soot is a rapidly-acting nitrogenous manure, which especially encourages the growth of grasses. Its great drawback is that it remains on the grass for a considerable period and stains golf balls. Forty bushels per acre is approximately the correct dressing.

Peat Moss Manure of high grade is useful for digging in on light sandy soils to assist in the formation of humus, and may be employed also as a winter top-dressing, but should occasionally be raked about. Its action is lasting.

Bone Meal, Dissolved Bones, and Bones (½ in.) are generally looked upon as phosphatic rather than nitrogenous. The first and last named act slowly. Unfortunately there is great risk in their use, as they invariably encourage a growth of clover. Quantity varies from 2 to 6 cwt. per acre.

Muriate of Potash is, as its name indicates, a potassic fertiliser, and when applied, especially on light soils, where it is generally most needed, its action is lasting and somewhat slow; 1 cwt. per acre is the utmost that should be used.

Sulphate of Potash of high grade contains about 54 percent. of potash, and is a useful manure for applying to land deficient in that constituent. It is, however, very apt to encourage a growth of clover, besides which it is generally too expensive to admit of universal use.

Kainit, although frequently employed alone, is, as a rule, best used in conjunction with phosphatic manures. It contains from 12 to 14 per cent. of potash, hence its great tendency to promote a leguminous growth; 2 cwt. per acre is a suitable dressing.

191x†   **The Laying Out and Upkeep of Golf Courses and Putting Greens.** Martin H.F. Sutton. Simpkin, Marshall, Hamilton & Kent, London, England, UK. 40 pp. plus 7 pp. of advertising, 53 illus., no tables, no index, semisoft bound in red with a circular window showing a black and white photograph of James Braid driving a golf ball and with black lettering, 9.4 by 7.2 inches (239 × 183 mm).

*   A **truly rare** booklet that is a must for collectors of historical turfgrass and/or golf books. It is an early booklet on the construction, turfgrass establishment, and culture of golf courses in England, with emphasis on putting greens. The text includes part of a lecture presented by Martin Sutton before the Royal Horticultural Society. It is well illustrated. Martin H.F. Sutton provided the following hints on turfgrass establishment of putting greens:

Preparing land for greens needs greater care than for courses.
The finished surface must be smooth and true.
A loose soil will not grow fine turf.
Roll until the foot scarcely makes an impression.
Then crumble the surface finely with a rake, or
As an alternative sow on the flat surface and cover thinly with sifted soil or fine sand.
Roll down twice, first North and South, then East and West.
Be liberal with seed to ensure a dense plant.
Sow on a calm day.
The mixture to consist of the very finest grasses only.
Prevent birds from disturbing the surface.
Should weeds appear remove while quite small.
Early mowing is vital to success.

> Use a scythe at first—say three or four times.
> Follow with mower set rather high.
> See that the machine is in perfect order.
> The green-keeper to wear heelless shoes.
> In mowing vary the direction.
> Roll frequently, but with discretion.
> Water when necessary, and invariably in the evening.
> Charcoal of fine texture is useful on wet surfaces.

T

Tainton, Neil Melbourne (Dr. and *Professor* Emeritus)

Neil Tainton was born in King Williams Town, Cape Province, South Africa, and earned a BSc degree *cum laude* in Agriculture in 1956 and an MSc degree *cum laude* in Agriculture in 1958, both from the University of Natal, and a PhD from the University of Wales, Aberystwyth, in 1967. He served the University of Natal as Lecturer from 1959 to 1969, as Senior Lecturer from 1969 to 1975, as Associate Professor from 1975 to 1978, and as Professor from 1979 to 1994, when he retired. He served as Dean of the Faculty of Agriculture from 1982 to 1984. He specialized in the management of natural grasslands, forages, and sports turfgrasses. Dr. Tainton has received the Agricultural Leadership Award from the Witwatersrand Agricultural Society in 1985, the Research Fellow from the University of Natal in 1989, the Director's Award from the Agricultural Research Council in 1995, and an Honorary Life Membership in the Grassland Society of Southern Africa in 1998.

2002† **The Cricket Pitch and its Outfield.** Neil Tainton and John Klug. University of Natal Press, Scottsville, South Africa. 189 pp., 72 illus. with 33 in color, 8 tables, indexed, glossy hardbound in multicolor with white lettering, 5.9 by 9.0 inches (150 × 228 mm).

    A practically oriented book on the construction, establishment, and culture of cricket grounds under the conditions in South Africa. There are chapters on turfgrasses, soils, cricket pitch or wicket, cricket outfield, and pitch testing. There is a small glossary and list of selected references in the back.

Takematsu, Tetsuo (Dr. and *Professor* Emeritus)

Tetsuo Takematsu (1921–2006) was born in Nagano Prefecture, Japan, and earned a BSc degree from Utsunomiya Agricultural College in 1942 and a PhD degree from Hokkaido University in 1961. He served in the army from 1942 to 1945. He then was Lecturer at Utsunomiya Agricultural College from 1946 to 1951, Associate Professor in the College of Agriculture at Utsunomiya University from 1951 to 1967, and Professor and Director of the Weed Control Research Institute at Utsunomiya University from 1967 to 1987, when he retired. He conducted weed science teaching and research. Dr. Takematsu was elected President of both the Weed Science Society of Japan in 1980 and the Japanese Society of Plant Growth Regulation in 1984.

1971† **Easy Weed Control of Garden, Lawn, and Housing Sites.** Tetsuo Takematsu. Hakuyu-sha, Tokyo, Japan. 245 pp., 110 illus., 10 tables, indexed, cloth hardbound, 5.1 by 7.3 inches (130 × 185 mm). Printed entirely in the Japanese language.

    A practical book concerning the control of weeds in lawns and landscapes under the conditions in Japan.

1985† **Theory and Practice of Weed Control in Turf.** First edition. Tetsuo Takematsu and Yasutomo Takeucki. Hakuyu-sha, Tokyo, Japan. 330 pp., 132 illus., 52 tables, indexed, cloth hardbound in dark blue with gold lettering, 7.1 by 10.1 inches (181 × 257 mm). Printed entirely in the Japanese language.

    A book concerning weed biology, physiology of herbicides, and weed control practices in turfgrasses under conditions in Japan.

    Revised in 1991.

1991† **Theory and Practices of Weed Control in Turf.** Second revised edition. Tetsuo Takematsu and Yasutomo Takeucki. Hakuyu-sha, Tokyo, Japan. 334 pp., 140 illus., 140 tables, cloth hardbound, 7.3 by 10.4 inches (185 × 265 mm). Printed entirely in the Japanese language.

    A minor updated revision.

Takeuchi, Yasutomo (Dr. and *Professor* Emeritus)

Yasutomo Takeuchi was born in Niigata Prefecture, Japan, and earned a BSc degree in 1965 and an MSc degree in 1966, both in Agriculture from Utsunomiya University, and a PhD degree in Agriculture from the University of Tokyo in 1975.

He joined the faculty of Utsunomiya University as an Assistant Professor in 1967, rose to Associate Professor in 1977 and Full Professor in 1987, and retired in 2007. He conducted research and teaching in weed control and plant growth regulation. Dr. Takeuchi was elected President of the Japanese Society of Turfgrass Science for 2002 through 2004 and the Weed Science Society of Japan for 2004 through 2006.

1985† **Theory and Practice of Weed Control in Turf.** Coauthor.
See *Tetsuo, Takematsu—1985.*

1991† **Theory and Practice of Weed Control in Turf.** Coauthor.
See *Tetsuo, Takematsu—1991.*

Tanaka, Akemi (Dr.)

Akemi Tanaka was born in Osaka City, Osaka, Japan, and earned a BSc degree in Plant Pathology from Kagawa University in 1985, an MSc degree in Plant Pathology from Kagawa University in 1987, and a Doctor of Arts (DA) degree in Plant Pathology from Ehime University in 1993. She served as Chief Scientist of the Institute for Green Science from 1993 to 2001. She then became President of the Institute for Green Science in 2001. Dr. Tanaka received the Young Scientist Award from the Japanese Society of Turfgrass Science in 2001.

2003† **Color Atlas of Turfgrass Diseases.** Second revised edition. Akemi Tanaka. Soft Science Inc., Tokyo, Japan. 151 pp.; 375 illus. in color; 5 tables; no index; cloth hardbound in green with gold lettering and a jacket in multicolor with white, green, pink, and orange lettering; 10.1 by 7.2 inches (217 × 182 mm). Printed in the Japanese language.

A revision and update of the 1991 edition by Toshikazu Tani. The petri plate photographs of fungal symptoms development have been removed, but close-up photographs of fungal mycelium and spores have been added for certain causal pathogens.

Tani, Toshikazu (Dr. and *Professor* Emeritus)

Toshikazu Tani (1929–2001) was born in Okawa, Kagawa Prefecture, Japan, and earned a BSc degree from Kagawa University in 1954 and a PhD degree from Kyoto University in 1964. He served on the Plant Pathology Faculty at Kagawa University from 1954 to 1993, as Dean of Student Affairs from 1986 to 1989, and as Dean of the Faculty of Agriculture from 1989 to 1993. Dr. Tani was CEO of the Institute of Environmental Green Science from 1994 till his death. He was active in teaching and research concerning the diseases of grasses and published ten books and more than 100 scientific articles. He received the Society Award of the Phytopathological Society of Japan in 1980 and the Society Award of the Japanese Society of Turfgrass Science in 1993.

1991† **Color Atlas: Turfgrass Diseases.** Toshikazu Tani. Soft Science Inc., Tokyo, Japan. 122 pp., 243 illus. in color, 5 tables, no index, cloth hardbound in black with silver lettering and a jacket in multicolor with light green and white lettering, 10.0 by 7.1 inches (255 × 180 mm). Printed entirely in the Japanese language.

A quality book on diseases of turfgrasses and their management, with emphasis on the diseases that occur on golf courses in Japan. There are numerous, high-quality, color photographs of disease symptoms. It contains a section on diagnostic techniques for turfgrass fungal pathogens, with a series of quality color diagnostic photographs in the back. It is a unique disease diagnostic book.

† *Reprinted in 1992 and 1994.*
Translated into a revised and expanded English edition in 1997.
Revised in 2003.
See *Tanaka, Akemi—2003.*

1997† **Color Atlas of Turfgrass Diseases.** Toshikazu Tani and James B Beard. Ann Arbor Press, Chelsea, Michigan, USA. 245 pp., 431 illus. in color, 33 tables, small index, glossy hardbound in multicolor with white lettering, 9.8 by 6.9 inches (250 × 176 mm).

An expanded edition of a quality book on the diseases of turfgrasses and their management. Fifty-five diseases of both warm- and cool-season turfgrasses are addressed worldwide. Emphasis is placed on the diagnostic aspects via color photographs of the multiple symptom phases in development of each disease. It contains a section on diagnostic techniques for turfgrass fungal pathogens, with a series of quality color photographs. A glossary of selected terms is included in the back. It is a unique disease diagnostic book.

1997† **Strategy for Integrated Pest Management.** Toshikazu Tani. Soft Science Inc., Tokyo, Japan. 153 pp., 125 illus. with 24 in color, 96 tables, indexed, textured semisoft bound in tan with dark blue lettering and a jacket in multicolor with dark blue and white lettering, 7.1 by 10.1 inches (180 × 257 mm). Printed in the Japanese language.

A specialty book concerning the use of integrated pest management (IPM) on golf courses. It includes the life cycles of turfgrass diseases, insects, and weeds under the conditions in Japan. It also includes IPM strategies where chemical controls can and cannot be employed.

Figure 12.91. Manual-push Macgregor compost distributor, Wheaton, Illinois, USA, circa 1926. [2]

Tashiro, Haruo (Dr. and Professor Emeritus of Entomology)

"Tash" Tashiro (1917–2009) was born in Selma, California, USA, and earned a BSc degree in Botany and Zoology from Wheaton College in 1945 and both MSc and PhD degrees in Entomology from Cornell University in 1946 and 1950, respectively. He served as Research Entomologist in the United States Department of Agriculture (USDA) at Geneva, New York, from 1950 to 1963 and as USDA Investigations Leader and Citrus Research Entomologist at Riverside, California, from 1963 to 1967. He then returned as Professor of Entomology at the Cornell University New York State Agricultural Experiment Station, Geneva, New York, in 1967, from where he retired in 1983. Dr. Tashiro conducted bionomics research on the biology and management of turfgrass insects, including European chafer, black turfgrass ataenius, and hyderodes weevil. He was Visiting Fellow at the College of Tropical Agriculture, University of Hawaii, in 1975.

1987†   **Turfgrass Insects of the United States and Canada.** First edition. Haruo Tashiro, Comstock Publishing Associates, Cornell University Press, Ithaca, New York, USA. 391 pp., 521 illus. with 381 in color, 34 tables, detailed index, cloth hardbound in red with gold lettering and a jacket in multicolor in white and yellow lettering, 6.9 by 9.4 inches (175 × 238 mm).

This comprehensive, scholarly text addresses the biology of insects, mites, and vertebrate pests of turfgrasses. It also includes sections on detection, diagnosis, population surveys, and control. It contains a reference list and a glossary of selected terms in the back. There are excellent, extensive color photographs.

Revised in 1999.

1999†   **Turfgrass Insects of the United States and Canada.** Second revised edition. Coauthor.

See *Vittum, Patricia Joan—1999.*

Tasmanian Department of Agriculture

1978†   **Turf Manual.** Tasmanian Department of Agriculture Bulletin 53, Hobart, Tasmania, Australia. 46 pp., 2 illus., 10 tables, no index, stapled covers in light green with black lettering, 6.9 by 9.6 inches (179 × 243 mm).

A booklet concerning the establishment and culture of turfgrasses, with emphasis on sports fields. There are sections on cricket wickets and bowling greens.

Terrell, Edward Everett (Dr.)

Ed Terrell was born in Wilmington, Ohio, USA, and earned an AB degree from Wilmington College in 1947, an MSc degree from Cornell University in 1949, and a PhD degree from the University of Wisconsin in 1952. He served as Muellhaupt Scholar in Botany at Ohio State University from 1952 to 1953; as Associate Professor and Head of the Department of Science at Pembroke State College from 1954 to 1956; as Associate Professor of Biology at Guilford College from 1956 to 1960; as Botanist in the Science and Education Administration, Agricultural Research Service of the United States Department of Agriculture, from 1960 to 1985; and as Research Associate in Botany at the University of Maryland from 1986 to 1991 and from 1995 to 1997. Dr. Terrell conducted research on plant taxonomy of grasses.

1977†   **A Checklist of Names for 3,000 Vascular Plants of Economic Importance.*** First edition. Edward E. Terrell. Agriculture Handbook No. 505, Agricultural Research Service, United States Department of Agriculture, Washington, D.C.,

USA. 201 pp., no illus., no tables, index by scientific and common names, semisoft bound in light green with black lettering, 7.9 by 10.2 inches (200 × 260 mm).

This book presents a compilation of taxa for more than 3,000 vascular plants plus representative common names as used in the United States. Approximately 700 synonyms also are included. There is a short list of selected references in the back. *Revised in 1986.*

1986† **A Checklist of Names for 3,000 Vascular Plants of Economic Importance.**˙ Second revised edition. Edward E. Terrell, Steven R. Hill, John H. Wiersema, and William E. Rice. Agriculture Handbook No. 505, U.S. Department of Agriculture, Washington, D.C., USA. 241 pp., no illus., no tables, indexed by scientific and common names, semisoft bound in cream with green lettering, 8.3 by 10.8 inches (210 × 275 mm).

An updating of the nomenclature for the economic vascular plants of the United States, encompassing 1,241 genera and 3,296 species plus 983 synonyms. There is a list of selected references in the back.

Thomas, Harry Higgott

Harry Thomas (1876–1956) was born in Drayton Manor, Staffordshire, England, UK, and trained in the Royal Gardens at Windsor where his father was the Head Gardener. He later worked at the gardens of Baron de Rothschild at Ferrières-en-Brie near Paris and at Cannes on the French Riviera. He gained admittance as a Kew gardener in 1897. He joined the staff of **The Gardener** in 1899 as editor and also was editor of **Popular Gardening** for 40 years, from 1907 to 1947. Harry Thomas was a prolific horticultural journalist. In 1913, he also became gardening correspondent of the newspaper the **Morning Post Daily Telegraph** in which he continued until retirement in 1953. He was elected President of the Kew Guild for 1953 to 1954. He was awarded the Victoria Medal of Honour in Horticulture of the Royal Horticultural Society in 1948.

n.d.†
circa
1929
*

Fertilisers for Gardens and Lawns: How To Use Them To Advantage.˙ H.H. Thomas. Charles Norrington & Company Limited, Plymouth, England, UK. 48 pp., 1 illus., no tables, no index, cloth hardbound with one-fifth tan and two-fifths velvet boards in green with gold lettering, 4.7 by 7.3 inches (120 × 85 mm).

This **very rare**, small manual emphasizes the use of fertilizers and liming materials in lawn care. The simplicity of presentation is oriented to the amateur lawn enthusiasts of England. There are short sections on lawn establishment and care. Descriptions and prices of lawn fertilizers and soil conditioners are presented in the back. It was printed for private circulation. In 1929, Harry Thomas described his approach for the use of manures in feeding lawns as follows:

> Natural manures from the stable or farmyard are not so suitable as chemical manures for use on lawns because of their bulkiness. If used too freely or if not sufficiently decayed they will do more harm than good and may indeed cause many of the finer grasses to perish. If the lawn is on light land that dries out quickly in Summer, a dressing of thoroughly decayed manure may be put on in January; much of it will have been washed in by the end of March and the residue can then be swept off: much of it will have been washed in by end of March and the residue can then be swept off: such a dressing will help in some measure to provide humus, will supply certain manurial elements and, if the grass is not smothered, it will be beneficial. It is, however, unwise to apply stable manure or farmyard manure to a lawn on heavy land; the result will almost certainly be to kill or weaken some of the grasses: a far better procedure is to prepare a compost of sifted soil, sand and thoroughly decayed manure, *e.g.*, that

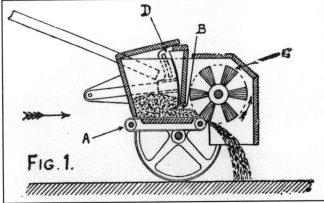

Figure 12.92. Overall view and patent diagram showing cross-section of the manual-push, 29-inch-wide Coultas Fertilizer Distributor, Grantham, England, UK, 1934. [26]

> from an old mushroom bed, and use it only if such cannot be kept in good condition for an indefinite period unless deficiencies in the supply of plant foods are made good. As the grass grows it extracts certain elements from the soil just as flowering plants do and, as a consequence, the soil in time becomes impoverished and the grass becomes thin, light green instead of deep green, and weeds begin to spread. Now the only way in which the loss of plant food can be made good is by "topdressing," or by applying suitable manures.

Thomas, Robert[3]

1974† **Les Gazons (Turfs).** First edition. Robert Thomas. Collection des Techniques Horticoles Spécialisées, Éditions J.-B. Bail-liére, Paris, France. 180 pp., 128 illus., 21 tables, no index, glossy semisoft bound in multicolor with green lettering, 7.1 by 8.8 inches (180 × 224 mm). Printed entirely in the French language.

   A book on the basics of turfgrass establishment and culture within a how-to approach. It encompasses both lawns and sports fields, as related to conditions in France. There is a small bibliography in the back.
   Revised in 1985 and 1988.

1976† **Los Cespedes (Turfs).** First edition translation. Robert Thomas. Ediciones Mundi-Prensa, Madrid, Spain. 274 pp., 130 illus., 24 tables, no index, glossy semisoft bound in multicolor with white and black lettering, 5.4 by 8.4 inches (137 × 213 mm). Printed entirely in the Spanish language.

   A Spanish translation of the original 1974 French first edition of **Les Gazons (Turfs)**, as translated by Jose M. Mateo Box.
   Revised in 1985 and 1988.

1985† **Los Cespedes (Turfs).** Second edition translation. Robert Thomas and Jean-Paul Guérin. Ediciones Mundi-Prensa, Madrid, Spain. 317 pp.; 136 illus.; 28 tables; no index; glossy semisoft bound in multicolor with green, white, and black lettering; 5.4 by 8.4 inches (137 × 213 mm). Printed entirely in the Spanish language.

   A Spanish translation of the original French second edition of **Les Gazons (Turfs)**, as translated by Jose M. Mateo Box and P. Ubano Terron. It contains a list of selected references in the back.

J.M. Thorburn and Company

The seedhouse of J.M. Thorburn and Company was originally formed in 1802 in New York City as G. Thorburn and changed in 1821 to G. Thorburn & Sons. The company was an early US marketer of manures and seeds for gardens. The firm ceased operations in 1921 after 119 years in business. Grant Thorburn (1773–1863) was born in Dalkeith, Scotland, and immigrated to New York in 1794. He became a florist, seedsman, and prolific horticultural author. The company claims to have been the first marketer of grass seeds in the United States:

> For nearly a hundred years our house has made a specialty of grasses. We were probably the first to introduce to this country, for commercial distribution, the many valuable natural and cultivated foreign species that now take such an important part in our agricultural economy. The formation of permanent pastures and meadows has for many years been a subject of careful study with us, and our celebrated formulae for seeding grasslands have received universal recognition.
>
> We have given similar attention to lawngrasses, and our thorough knowledge of the subject enables us to prescribe the most suitable kind for all soils, situations and climatic conditions. Most of the very finest lawns from Florida to Canada were produced by seed prescribed and furnished by us.

1899 **The Seeding and Preservation of Golf Links.** First edition. J.M. Thorburn & Co., New York City, New York, USA. 24 pp. with 9 pp. of advertising, 2 illus., no tables, no index, sown-textured semisoft bound in greens with black lettering in a jacket, 5.5 by 8.0 inches (140 × 203 mm).

*   A **very rare** booklet concerning turfgrasses and their establishment and culture for golf courses. It is oriented to conditions in the United States and is one of the earliest turfgrass publications in the United States. The company's opinion on the use of manure is discussed as follows:

> STABLE OR BARNYARD MANURE. This is decidedly the best and most economical form of fertilizer to use in the initial preparation of the ground for permanent grass lands. A too liberal quantity can hardly be applied. A very heavy dressing, plowed in in the fall, and another dressing cross-plowed in the following spring are recommended. This manure contains all the "essential manurial elements" but in a form more or less insoluble. They become available slowly and continuously, as the manure decays. Besides being one of the best direct fertilizers, this form of manure has a most valuable mechanical effect in improving the condition of the soil—the vegetable matter in it making stiff, clayey lands more porous and a too light soil more compact. But, although there is nothing better for plowing under, it is by no

> means as economical when used as a top-dressing. For when so applied the mechanical effect above mentioned is only slightly exercised, and much of the fertilizing properties is lost by the long exposure to the atmosphere. These considerations, and the extra labor involved in spreading in the fall and removing the coarse vegetable parts remaining in the spring, make the natural manure less desirable for surface application than some of the other forms of fertilizer. Its liability to contain weed seeds is still another objection to stable manure for top-dressing.

1908   **The Seeding and Preservation of Golf Links.** Second revised edition. J.M. Thorburn & Co., New York City, New York, USA. 24 pp. plus 9 pp. of advertising, 8 illus., no tables, no index, fold-stapled semisoft covers in greens with black lettering, 5.5 by 8.0 inches (140 × 203 mm).

   *   A revised, **very rare,** early booklet concerning turfgrasses and their establishment and culture for golf courses. Main revision is emphasis on turfgrass seed mixtures.

Thorn, James Frederick (OBE)

Jim Thorn (1926–2005) was born in Wales, UK. He served as tennis groundskeeper at St George Hill Lawn Tennis Club at Weybridge and was Head Groundsman at the All England Lawn Tennis and Croquet Club at Wimbledon from 1982 to 1991, when he retired. He was inducted as an OBE in 1990 and received the Annual Award from the Tennis Sports Writers Association in 1991.

1991†   **The Lawns of Wimbledon (The Construction and Maintenance of Grass Courts).** Jim Thorn. The All England Lawn Tennis and Croquet Club, Wimbledon, England, UK. 55 pp., 25 illus., no tables, no index, glossy semisoft bound in multicolor with green and white lettering, 5.8 by 8.2 inches (147 × 208 mm).

     A booklet on the construction, culture, and repair of lawn tennis courts. It is oriented to cool-season turfgrass conditions in the United Kingdom. The author relates his experience based on 40 years in caring for grass tennis facilities.

Tonogi, Hideaki

Hidea Tonogi was born in Fuji-City, Shizuoka Prefecture, Japan, and earned a BSc degree in Horticulture from Chiba University in 1985 and studied Turfgrass Science at Texas A&M University in 1988 with Dr. Jim Beard. He worked on the Green Staff at Shin-Numazu Country Club from 1985 through 1987, as an Agronomist with Nichino Ryokka Company Limited from 1988 to 2001, as an Agronomist and in sales with Scotts Japan Company Limited from 2001 to 2005, and as Manager of the Sales Department of Certis Japan Company Limited since 2005. Hidea Tonogi has been elected to the Board of Directors of the Japanese Society of Turfgrass Science since 1998.

2007†   **Cultivation for Turfgrass Management—Its Effects and Method Using Machinery.** Coeditor.

     See *Inamori, Makoto—2007.*

Toogood, Edward Kemp

Edward Toogood was born in Southampton, Hampshire, England, UK. He was a seed merchant, continuing a family lineage, including being Director of the company. The company of Toogood & Sons was established in 1815 and for over 50 years was the Royal Seedsman.

1900†   **Toogood's Treatise on Lawns, Cricket Grounds, &c.** E. Kemp Toogood. Toogood & Sons, Southampton, England, UK. 20 pp., 5 illus., 1 table, no index, fold-stapled semisoft covers in green with black lettering, 4.9 by 7.2 inches (125 × 182 mm).

   *   A **very rare,** small booklet on the site preparation, establishment, and culture of lawn turfgrasses under the cool-season conditions in England. Edward Toogood had the following advice for mowing turfgrasses in 1900:

> The young plant should be topped with a scythe when it is some three inches or so high, to encourage the grasses to tiller out as much as possible; and rolling must follow this and every future mowing. By continuing to mow and roll frequently during summer and autumn, the grass is much strengthened, and annual weeds are kept in check. The scythe is to be preferred to the mowing-machine for the first few cuttings.

Torello, William Anthony (Dr. and *Associate Professor*)

Bill Torello was born in New York City, New York , USA, and served in the US Navy from 1970 to 1974. He earned a BSc degree in Plant and Soil Science and Chemistry in 1977 and an MSc degree in Turfgrass Science and Management in 1979, both from the University of Rhode Island, and a PhD in Horticulture and Turfgrass from the University of Illinois in 1981, studying with Dr. Al Turgeon. In 1981, he joined the Plant and Soil Sciences faculty at the University of Massachusetts, Amherst, as an Assistant Professor and became Associate Professor. His main activity was turfgrass teaching,

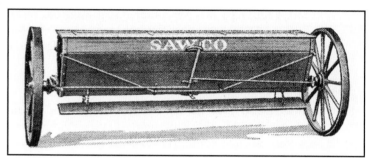

Figure 12.93. Two early fertilizer spreaders used on turfgrass areas in the 1920s: manual-push, 6-foot-wide (1.8 m) gravity distributor (left) and a horse or tractor drawn, 8-foot-wide (2.4 m), spreader with revolving agitator (right). [41]

and he served as Director of the Turfgrass Science and Management Program from 1994 to 1998. He left the University of Massachusetts in 2004 and formed Professional Turfgrass Consulting. He then became Director of Research and Product Development for Converted Organics Inc. of Boston.

1982† **Lawns—Basic Factors, Construction, and Maintenance of Fine Turf Areas.** Coauthor.
See *Vengris, Jonas—1982.*

Trenholm, Laurie Elizabeth (Dr. and *Associate Professor*)

Laurie Trenholm was born in Rochester, New York, USA, and earned BSc and MSc degrees both in Environmental Horticulture and Turfgrass from the University of Florida in 1994 and 1996, respectively, and a PhD degree in Crop Science and Turfgrass Physiology from the University of Georgia in 1999. She joined the University of Florida as Urban Turfgrass Specialist in the Department of Environmental Horticulture, serving as an Assistant Professor from 1999 to 2005, and then became an Associate Professor. Her main activities involve working with the lawn care industry and education concerning best management practices. Dr. Trenholm was awarded the Turfgrass Educators Award from Turfgrass Producers International in 2008.

2005† **The Florida Lawn Handbook.** Third revised edition. Laurie E. Trenholm and J. Bryan Unruh (editors) with eight contributing authors. University Press of Florida, Gainesville, Florida, USA. 182 pp., 154 illus. with 142 in color, 30 tables, no index, glossy hardbound in multicolor with white lettering, 6.7 by 9.8 inches (171 × 249 mm).

This third edition is a major revision focused on the best management practices for lawns under Florida conditions. It includes turfgrass selection, planting, fertilization, care, pest management, environmental stresses, and organic lawn care.

2005† Also printed as a glossy semisoft bound version in multicolor with yellow and white lettering, 6.9 by 10.0 inches (176 × 253 mm).
See also *McCarty, Lambert Blanchard—1990* and *Ruppert, Kathleen Carlton.—1991.*

Troughton, Arthur (Dr. and *Dr.Sc.*)

Arthur Troughton (1927–1982) was born in Middlesborough, Yorkshire, England, UK, and earned a BSc degree in Agriculture with *first class honours* from Kings College of Durham University in 1948 and a PhD degree from the University of Wales, Aberystwyth, in 1958. He served as Scientific Officer and Grassland Agronomist at the Welsh Plant Breeding Station, University College of Wales, Aberystwyth, Wales, starting in 1948, became Senior Scientific Officer in 1958, and Principal Scientific Officer in 1964. He conducted research on the rooting and root-shoot relationships of grasses, until his early death in 1982. He was elected President of the University College, Aberystwyth Agricultural Society, for 1974 to 1975. Dr. Troughton received a National Research Fellowship to New Zealand for 1958 to 1959 and an Honorary DSc degree from the University of Newcastle-upon-Tyne in 1981.

1957† **The Underground Organs of Herbage Grasses.*** Arthur Troughton. Bulletin No. 44, Commonwealth Bureau of Pastures and Field Crops, Hurley, Berkshire, England, UK. 163 pp., 17 illus., 56 tables, indexed, cloth hardbound in green with dark green lettering and a jacket in green with black lettering, 7.2 by 9.6 inches (184 × 245 mm).

This book provides an important comprehensive review of the roots, rhizomes, corms, and leaf bases of grasses, including their morphology, anatomy, life history, and chemical composition. The environmental and cultural influences also are reviewed. It contains a detailed bibliography in the back.

Tucker, William Henry

William Tucker (1871–1954) was born in Redhill, Surrey, England, UK. His apprenticeship was in golf course construction for both Tom Dunn and Willie Dunn in England, France, and Switzerland. He immigrated to the United States in 1895, where he joined his brother Samuel in the Tucker Brothers firm, which was involved in handmade Defiance golf clubs. In 1898, he worked on the design of the new St Andrews Golf Club with Harry Tallmadge and, subsequently, served as the greenkeeper. He also continued to be active in golf course design. In the early 1900s he was the professional and greenkeeper at the Chevy Chase Club near Washington, DC. During the 1920s, he formed full-time golf course architecture and construction firms with his son William H. Tucker Jr. William Tucker also was involved in turfgrass culture of sports fields and lawn tennis courts, including Yankee Stadium and Forest Hills.

1915   **Practical Illustrations of Turf Production.** William Tucker. William Tucker, New York City, New York, USA. 36 pp. including 3 pp. of advertising, 34 illus., no tables, no index, fold-stapled soft covers in tan with green lettering, 7.8 by 10.1 inches (197 × 257 mm).

   *   A **truly rare** publication that is a must for collectors of historical turfgrass books. It is a small booklet on the Tucker Method of construction, establishment, and culture of turfgrasses for golf courses, tennis courts, bowling greens, and estate lawns. It is oriented to North American conditions and extensively illustrated by photographs.

Tulloch, William Weir (DD)

William Tulloch (1846–1920) was born in Dundee, Forfarshire, Scotland, UK, and earned an MA from United College, University of St Andrews, in 1868 and a Bachelor of Divinity (BD) degree in 1870 and Doctor of Divinity (DD) degree in 1891, both from St Mary's College, University of St Andrews. He subsequently served the church ministry. He was licensed by the St Andrews Presbytery in 1870 and subsequently was minister of North Parish, Greenock, from 1871 to 1874, Kelso from 1874 to 1877, and Maxwell, Glasgow, 1877 to 1901. Reverend Tulloch also was an author and a member of the Royal & Ancient Golf Club of St Andrews.

1908   **The Life of Tom Morris—With Glimpses of St Andrews and Its Golfing Celebrities.** W.W. Tulloch. T. Werner Laurie, London, England, UK. 346 pp., 27 illus., no tables, indexed, cloth hardbound in multicolor with black lettering, 5.6 by 8.6 inches (143 × 219 mm).

   *   Tom Morris (1821–1908) was born in St Andrews, Scotland, and spent his early days there. He was Keeper of the Green at Prestwick Golf Club from 1851 to 1864. He then returned to St Andrews also as professional golfer and greenkeeper known then as a Keeper of the Green, a position he held until his retirement in 1904. He won four British Opens—1861, 1862, 1864, and 1867. "Old Tom" also designed over 30 golf courses including Castletown (1892), Dunblane (1891), Glasgow (1904), Ladybank (1879), Lahinch (1893), Muirfield (1891), Perth (1864), Prestwick (1851 and 1864), Royal County Down (1891), Royal Dornoch (1887), Royal North Devon (1874), St Andrews New Course (1894), and Sterling (1892), plus remodeling of Carnoustie (1867), Cullen (1892), and Machrihanish (1879). Rev. Tulloch described the transition of Tom Morris to the Royal and Ancient Club at St Andrews as follows:

> The Minutes of the meeting of the Royal and Ancient Club for September 1863 state that Alexander Herd had resigned the custody of the links, and the Committee accordingly were authorised to appoint a custodian at a salary considerably larger than that hitherto given. Thereupon Major Boothby gave notice of a motion that a professional golfer should be employed as a servant of the Club, and that the entire charge of the course should be entrusted to him. Subsequently a large majority were found to be in favour of the proposal, as a result of which Tom Morris was introduced from Prestwick. His duties were explained to him: to keep the putting-greens in good order, to repair when necessary, and to make the holes. For heavy work, carting, etc., he was to be allowed assistance at the rate of one man's labour for two days in the week, and it was understood that he was to work under the Green Committee. Emblems of office were then handed over to him, to wit, a barrow, a spade and a shovel in prophetic instinct, belike that "saund" and even "mair saund, Honeyman" would be in future ages the watchword of the newly-appointed Chief of the Links. The sum of £50 per annum was voted by the Union Club for payment of the custodian's salary, and £20 for the upkeep of the links.

1982   **The Life of Tom Morris—With Glimpses of St Andrews and Its Golfing Celebrities.** Published by Ellesborough Press Limited for the Golf Course Superintendents Association of America, Lawrence, Kansas, USA. 362 pp., 27 illus., 8 tables, indexed, cloth hardbound in multicolor with black and gold lettering, 5.5 by 8.5 inches (140 × 216 mm).

A reprinting of the 1908 edition about the golfer, architect, and greenkeeper, Tom Morris of St Andrews, Scotland, UK. It was a limited edition printing of 300 copies.

1982   **The Life of Tom Morris.** A facsimile edition. Ellesborough Press Limited, London, England, UK. 349 pp., 27 illus., no tables, indexed, textured hardbound in dark green with gold lettering and a cloth board slip case in light green, 5.6 by 8.5 inches (142 × 216 mm).

A facsimile of the 1908 edition, with a limited printing of 200 copies.

1992†   **The Life of Tom Morris—With Glimpses of St Andrews and Its Golfing Celebrities.** A facsimile edition. Rare Book Library, United States Golf Association, Far Hills, New Jersey, USA. 353 pp., 27 illus., no tables, indexed, cloth hardbound in light brown with gold lettering and a tan slip case, 5.5 by 8.5 inches (140 × 216 mm).

A facsimile of the 1908 edition, with a limited printing of 1,500 copies.

Turfgrass Science Division C-5, Crop Science Society of America (CSSA)

Turfgrass Science Division C-5 is a division of the Crop Science Society of America (CSSA). It was officially formed in 1955 as a section of the American Society of Agronomy (ASA). An annual meeting is held in the United States, where more than 200 turfgrass research papers are presented and information exchanged.

1976†   **Post 1976 Turfgrass Industry Challenges In Research, Teaching And Continuing Education.** The seven contributing authors include James B Beard, Elwyn E. Deal, Wayne W. Huffine, John R Kelsheimer, Jerrell B. Powell, A.M. Radko, and James R. Watson. Turfgrass Science Division C-5, Crop Science Society of America, Madison, Wisconsin, USA. 33 pp., no illus., 4 tables, no index, semisoft bound in tan with black lettering, 8.0 by 10.4 inches (202 × 265 mm).

An assemblage of papers derived from a Bicentennial Symposium of the American Society of Agronomy (ASA). The seven papers provided a historical perspective, current assessments, and future projections concerning turfgrass research, teaching, and adult education.

Turfgrass Information Center/File (TGIF), Michigan State University

The Turfgrass Information Center encompasses the Turfgrass Information File which is a computer-based automated-retrieval system for the published research and technical literature on turfgrasses. It is headquartered at the Michigan State University Library in East Lansing, Michigan, USA. It was originally based on the O.J. Noer Memorial Collection. Then the James B Beard Turfgrass Collection and Room were added in 2003, which represents by far the largest turfgrass literature collection in the world.

1992†   **Cumulative Index—1972–1992 Proceedings of the Michigan Turfgrass Conference.** Turfgrass Information Center, Michigan State Libraries and Michigan Turfgrass Foundation, East Lansing, Michigan, USA. 501 pp., 2 illus., no tables, indexed, textured semisoft covers in cream and greens with black lettering in a comb binder, 8.5 by 10.9 inches (216 × 278 mm).

A listing of citations by author, subject, and chronological order for the Michigan Turfgrass Conference Proceedings from 1972 to 1992, as produced by Peter Cookingham. It contains references for 1,020 turfgrass papers and research reports.

1994†   **The Turfgrass Index.** Volume 1, Nos. 1 & 2 (January–June) and 3 & 4 (July–December). Turfgrass Information Cen-
&   ter, Michigan State University, East Lansing, Michigan, USA. 270/408 pp., no illus., no tables, fully indexed by author
1995†   and subject, semisoft covers in greens with black lettering in a comb binder, 8.5 by 10.9 inches (216 × 278 mm).

This index provides both author and subject index citations encompassing the full range of turfgrass-related literature published in 1993. It was produced from the Michigan State University Turfgrass Information File by Peter Cookingham.

Turfgrass Resource Center

2000†   **Water Right: Conserving Our Water—Preserving Our Environment.** Turfgrass Resource Center, Rolling Meadows, Illinois, USA. 67 pp.; 107 illus. in color; 1 table; no index; glossy semisoft bound in multicolor with blue, green, white, and black lettering; 8.5 by 10.9 inches (216 × 278 mm).

A small book on the need for water conservation along with the need for functional benefits derived from landscaped lawns, especially in populated urban areas. There are 10 case histories presented in the book.

n.d.†   **Facts About Artificial Turf and Natural Grass.** Turfgrass Resource Center, East Dundee, Illinois, USA. 28 pp., 2 illus.,
circa   no tables, no index, glossy soft bound in multicolor with white lettering, 5.4 by 8.5 inches (137 × 215 mm).
2000x

A booklet comparing artificial surfaces to natural turfgrass in terms of cost, durability, health, and mental and environmental effects. There is a list of selected references in the back.

†   *Reprinted in 2007.*

2006†   **Natural Grass and Artificial Turf: Separating Myths and Facts.** Turfgrass Resource Center, East Dundee, Illinois, USA. 31 pp., 45 illus. with 42 in color, 6 tables, no index, fold-stapled glossy soft covers in multicolor with white lettering, 5.8 by 8.5 inches (148 × 215 mm).

Figure 12.94. Horse-drawn lawn-sweeping machine advertised to collect and deposit the sweepings into heaps while the machine was in motion, circa 1911. [37]

A booklet that compares the playing and environmental characteristics of a turfgrass field surface with an artificial field surface.

Turfgrass Society of Korea

The Turfgrass Society of Korea was formed in 1987. The society objectives are to exchange information of scientific research and technical development of turfgrasses and groundcovers, to distribute this information, and, finally, to contribute to land development and greening. Membership is open to professors, researchers, superintendents and managers, contractors, designers, students, sod producers, and material manufacturers. The headquarters is officed in Seoul, South Korea. The society sponsors an annual conference and the **Korean Journal of Turfgrass Science**, which is published two times annually.

2002　**Ways to Establish and Maintain Turf Ground.** Turfgrass Society of Korea with 34 contributing authors. Seoul Olympic Sports Promotion Foundation, Seoul, South Korea. 165 pp., 74 illus. with 66 in color, 34 tables, no index, semisoft covers in light green with black lettering 5.9 by 8.9 (150 × 225 mm). Printed entirely in the Korean language.

A book concerning the establishment and culture of turfgrasses for sports fields. It is oriented primarily to the conditions in Korea. There are a list of selected references and a glossary in the back.

Turgeon, Alfred Joseph (Dr. and Professor)

Al Turgeon was born in White Plains, New York, USA, and earned a BSc degree in Plant Science and Turfgrass from Rutgers University in 1965. He served three years on active duty with the US Army from 1965 to 1968, including one year in Vietnam as a combat helicopter pilot. He rose to the rank of Captain, with military decorations including the Distinguished Flying Cross, Bronze Star, Air Medal, Army Commendation Medal, and Purple Heart. He resumed his education, earning both an MSc degree in 1970 and a PhD degree in 1971 in Crop Science and Turfgrass from Michigan State University. Dr. Turgeon joined the faculty at the University of Illinois as an Assistant Professor working in turfgrass extension, research, and teaching, becoming an Associate Professor in 1976. In 1980, he became Full Professor and Director of Research for the Texas Agricultural Experiment Station Center at Dallas. He then served as Vice President for Research and Technical Services for the TruGreen Corporation from 1983 to 1986. Next, he was Head of the Department of Agronomy at Pennsylvania State University from 1986 to 1994, when he became Director of Educational Technologies in the College of Agricultural Sciences and Professor-in-Charge of the Turfgrass Management Program at the Penn State World Campus since 1998. He was elected Turfgrass Science Division C-5 Chair in 1981 and to the Crop Science Society of America (CSSA) Board of Directors from 1980 through 1982. Dr. Turgeon received the Illinois Turfgrass Foundation Distinguished Service Award in 1988, a Fellow of the American Society of Agronomy in 1988, and a Fellow of CSSA in 1989.

1980†　**Turfgrass Management.** First edition. A.J. Turgeon. Reston Publishing Company, Reston, Virginia, USA. 391 pp., 125 illus. with 28 in color, 30 tables, indexed, cloth hardbound in dark green with gold lettering and a glossy jacket in white with brown and green lettering, 6.9 by 9.1 inches (176 × 232 mm).

An introductory book on the construction, establishment, and culture of turfgrass areas. Emphasis is placed on basic principles for beginning students. Illustrations are mostly in the form of drawings by Floyd Giles; there are also color photographs of diseases. There is a glossary of selected terms in the back.

Revised in 1985, 1991, 1996, 1999, 2002, 2005, and 2008.

1985† **Turfgrass Management.** Second revised edition. A.J. Turgeon. Reston Publishing Company Inc., Reston, Virginia, USA. 416 pp., 240 illus. plus 28 in color, 37 tables, indexed, glossy hardbound in multicolor with white lettering, 6.9 by 9.1 inches (176 × 232 mm).

This revision has some changes, especially in the drawings.
Revised in 1991, 1996, 1999, 2002, 2005 and 2008.

1991† **Turfgrass Management.** Third revised edition. A.J. Turgeon. Prentice-Hall Inc., Englewood Cliffs, New Jersey, USA. 418 pp., 251 illus. with 28 in color, 37 tables, indexed, glossy hardbound in multicolor with black lettering, 6.9 by 9.1 inches (176 × 232 mm).

This revision contains minor changes, such as a pesticide update.
Revised in 1996, 1999, 2002, 2005, and 2008.

1994† **Turf Weeds and Their Control.** Alfred J. Turgeon (editor) with 12 contributing authors: George B. Beestman, B.E. Branham, Lloyd M. Callahan, Ralph E. Engel, David W. Hall, William J. Kowite, Lambert B. McCarty, Tim R. Murphy, Donald Penner, Thomas F. Reed, Steve Tilko, and Thomas L. Watschke. American Society of Agronomy and Crop Science Society of America, Madison, Wisconsin, USA. 259 pp., 254 illus. with 133 in color, 25 tables, indexed, glossy hardbound in greens, black and white with white and green lettering, 6.0 by 9.0 inches (152 × 228 mm).

A technical book on turfgrass weeds and their cultural and chemical control. The herbicide aspects encompass mode of action, phytotoxcity, fate, formulation, and application dimensions. Each of the eight chapters has an extensive list of references at the back. Most of the 133 color plates are close-ups of weeds.

1996† **Turfgrass Management.** Fourth revised edition. A.J. Turgeon. Prentice-Hall Inc., Upper Saddle River, New Jersey, USA. 406 pp., 255 illus. with 28 in color, 31 tables, indexed, glossy hardbound in multicolor with orange and green lettering, 7.0 by 9.1 inches (179 × 232 mm).

This revision has some changes.
Revised in 1999, 2002, 2005, and 2008.

1999† **Turfgrass Management.** Fifth revised edition. A.J. Turgeon. Prentice-Hall Inc., Upper Saddle River, New Jersey, USA. 392 pp.; 310 illus. with 48 in color; 15 tables; indexed; glossy hardbound in multicolor with black, white, and green lettering; 7.0 by 9.2 inches (177 × 234 mm).

This revision has some changes and updates, with significantly more color photographs.
Revised in 2002, 2005, and 2008.

2002† **Turfgrass Management.** Sixth revised edition. A.J. Turgeon. Prentice-Hall Inc., Upper Saddle River, New Jersey, USA. 448 pp.; 216 illus. with 58 in color; 33 tables; indexed; glossy hardbound in multicolor with green, white, and blue lettering; 7.0 by 9.2 inches (177 × 234 mm).

An updated revision
Revised in 2005 and 2008.

2005† **Turfgrass Management.** Seventh revised edition. A.J. Turgeon. Pearson Prentice-Hall, Upper Saddle River, New Jersey, USA. 427 pp., 530 illus. with 85 in color, 40 tables, indexed, glossy hardbound in multicolor with black lettering, 7.0 by 9.2 inches (177 × 234 mm).

An updated revision.
Revised in 2008.

2006† **The Turf Problem Solver—Case Studies and Solutions for Environmental, Cultural, and Pest Problems.** A.J. Turgeon and J.M. Vargas Jr., John Wiley & Sons Inc., Hoboken, New Jersey, USA. 252 pp.; 91 illus. with 26 in color; 4 tables; indexed; glossy hardbound in multicolor with white, green, and orange lettering; 6.1 by 9.2 inches (154 × 234 mm).

A book of turfgrass case studies representing a range of typical problems, coupled with an analysis and solution for each. The case studies are presented in five topic groups: turfgrass, environmental, cultural, pest, and people.

2008† **Turfgrass Management.** Eighth revised edition. A.J. Turgeon. Pearson Prentice Hall, Upper Saddle River, New Jersey, USA. 448 pp.; 257 illus. with 68 in color; 35 tables; indexed; glossy hardbound in multicolor with green, white, and blue lettering; 7.0 by 9.2 inches (177 × 234 mm).

An updated revision.

Turner, A. Leslie[3]

1962† **Sports Field Drainage—An Outline of Field Drainage with Particular Reference to the Drainage of Sports Fields.** A.L. Turner. Educational Book Series No. 1, National Association of Groundsmen, West Ewell, Surrey, England, UK. 63 pp. plus 11 pp. of advertising, 30 illus., 13 tables, no index, fold-stapled textured semisoft covers in light blue with blue lettering, 5.5 by 8.4 inches (139 × 214 mm).

A small book on field drainage, with emphasis on sports fields. The design and installation of drainage systems are discussed. The foreword is by A.E.R. Gilligan. There is a bibliography of selected references in the back.

U

Uehara, Keiji (Dr. and *Professor*)

Keiji Uehara (1889–1981) was born in Tokyo, Japan, and earned a BSc degree in Forestry in 1914 and a Dr. of Forestry degree in 1920, both from the Imperial University of Tokyo. He then became associated with the Ministry of Education in 1920. In 1924, he founded and was President of the Professional High School for Landscape Architecture, which was merged with Tokyo Agricultural University in 1942. He served as Professor at the Tokyo Agriculture University from 1953 to 1975.

1934† **Turfgrass, Turf, and Home Lawn.** First edition. Keiji Uehara. Kokonshoin Publishing Company, Tokyo, Japan. 267 pp., 46 illus., 28 tables, indexed, textured cloth hardbound in black with gold lettering, and a tan slip case with dark blue lettering, 5.0 by 7.1 inches (127 × 181 mm). Printed entirely in the Japanese language.

    *   A **very rare** guidebook on turfgrass characteristics and use in landscapes under conditions in Japan. This is the oldest turfgrass book published in Japan and a must for collectors of historical turfgrass books.

        Revised in 1959.

1959† **Turfgrass and Turf.** Second revised edition. Keiji Uehara. Kajima-shoten Book Company, Tokyo, Japan. 244 pp., 95 illus., 18 tables, indexed, textured cloth hardbound in black with gold lettering and a slip case in tan with black lettering, 5.0 by 7.2 inches (129 × 183 mm). Printed entirely in the Japanese language.

    An early Japanese book on the basics of turfgrasses and their culture in the lawn garden. Included are sections on Japanese lawn grasses, Western turfgrasses, and lawn garden construction. Emphasis is placed on the morphology, characteristics, culture, sodding, and renovation of Japanese lawn grasses (*Zoysia* species) and the evergreen (cool-season) turfgrass species.

United States Baseball Federation (USBF)

Via an act of Congress in 1978, the United States Baseball Federation (USBF) became the national governing body for the sport, representing all of amateur baseball in America on the United States Olympic Committee and on the International Baseball Association. The USBF governs over 19 million amateur athletes in the United States. Membership is open to all amateur baseball groups and supporters who wish to join.

1980 **A Baseball Facility: Its Construction and Care.** First edition. United States Baseball Federation, Greenville, Illinois, USA. 53 pp., 17 illus., no tables, no index, fold-stapled glossy semisoft covers in multicolor with green lettering, 6.0 by 9.0 inches (151 × 227 mm).

    A guide book on the building and care of a baseball field, including site selection, soils, construction, turfgrass establishment, and maintenance. Associated structures and facilities also are discussed.

    Minor revisions in 1984 and 1987 and significantly revised in 1992.

1992 **A Baseball Facility: Its Construction and Care.** Fourth revised edition. United States Baseball Federation, Trenton, New Jersey, USA. 87 pp., 24 illus., no tables, no index, fold-stapled glossy semisoft covers in multicolor with green and black lettering, 6.0 by 9.0 inches (151 × 227 mm).

    A significant revision of the 1987 edition. The booklet addresses the siting, soils, layout, construction, and maintenance of turfgrass on baseball fields plus the infield, bases, pitcher's mound, bullpen, and associated physical facilities.

    Earlier editions printed in 1980, 1984, and 1987.

United States Golf Association (USGA)

The USGA is a large organization of mostly amateur golfers that was formed in 1894. It is headquartered in Far Hills, New Jersey, USA. The *Rules of Golf* are developed jointly by the United States Golf Association and the Royal and Ancient Golf Club of St Andrews, Scotland, and are published annually. The USGA is the sponsor of the US Open and several US amateur championship competitions. It also supports the USGA Green Section, turfgrass research, and a major Golf Museum and Library.

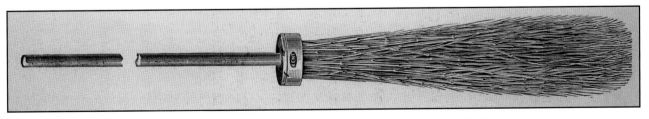

Figure 12.95. Bamboo broom used for sweeping greens, circa 1913. [40]

1968† **USGA Golf Championship Manual.**˙ United States Golf Association, Far Hills, New Jersey, USA. 49 pp., no illus., 1 table, no index, soft bound in cream with black lettering, 5.9 by 8.9 inches (151 × 227 mm).

This booklet provides guidelines to follow in preparation for a major golf championship in the United States. The first portion addresses grounds preparation, including the turfgrass areas.

1994† **Wastewater Reuse for Golf Course Irrigation.**˙ United States Golf Association. Lewis Publishers, Chelsea, Michigan, USA. 302 pp., 31 illus., 29 tables, indexed, glossy hardbound cover in multicolor with white and green lettering, 6.1 by 9.3 inches (156 × 235).

A compilation of papers presented at a symposium devoted to updating the information available on the use of effluent water for irrigation of golf course turfgrasses. Nineteen authors made contributions. A list of US golf courses utilizing effluent water is presented in the back.

1994† **Golf and the Environment.** United States Golf Association, Far Hills, New Jersey, USA. 20 pp., 10 illus., 1 table, no index, glossy semisoft bound in multicolor with green lettering, 8.5 by 11.0 inches (216 × 279 mm).

A booklet summarizing the benefits of turfgrasses to the environment and human activities. It is based on the scholarly research review of Drs. James B Beard and Robert L. Green as published in the **Journal of Environmental Quality** in 1994.

2002† **Building the USGA Green: Tips for Success.**˙ James F. Moore (editor) with the Green Section Staff being contributing authors. The United States Golf Association, Far Hills, New Jersey, USA. 32 pp., 25 illus. in color, 2 tables, no index, glossy semisoft bound in multicolor with white and black lettering, 8.5 by 10.9 inches (215 × 276 mm).

A guide concerning construction of a putting green by the USGA method. Sections include planning, subgrade, wicking barrier, drainage, gravel drain blanket, intermediate layer, root-zone mix, fumigation, planting, and grown-in.

Unruh, Joseph Bryan (Dr. and Associate Professor)

Bryan Unruh was born in Las Cruces, New Mexico, USA, and earned a BSc degree in Horticulture from Kansas State University in 1989, studying with Dr. Jeff Nus; an MSc degree in Horticulture from Kansas State University in 1992, studying with Dr. Roch Gaussoin; and a PhD in Horticulture from Iowa State University in 1995, studying with Dr. Nick Christians. He has served as an Extension Specialist at the West Florida Research and Education Center at Jay, Florida, and as Assistant Professor from 1996 to 2002 and Associate Professor since then. Dr. Unruh received the John & Martha Woeste Professional Development Award from the University of Florida in 2004 and the Outstanding Extension Specialist Award from the Florida Association of County Agriculture Agents in 2004.

1999† **Best Management Practices for Florida Golf Courses.** Second revised edition. J. Bryan Unruh and Monica L. Elliott with nine contributing authors: G.L. Miller, J.L. Cisar, A.E. Dudeck, J.B. Sartain, G.H. Snyder, P. Busey, J. Howard Frank, R.A. Dunn, and B.J. Brecke. University of Florida, Institute of Food and Agricultural Services, Gainesville, Florida, USA. 232 pp.; 126 illus. with 107 in color; 77 tables; no index; glossy semisoft bound in multicolor with blue, yellow, and black lettering; 8.4 by 10.9 inches (213 × 277 mm).

An updated reference guide on turfgrass culture of golf courses in Florida. There are five sections that address putting green construction, irrigation water management, fertilizer and fertilization practices, cultural practices, and pest management.

For first edition, see *McCarty, Lambert Blanchard—1993*.

2005† **The Florida Lawn Handbook.** Third revised edition. Coauthor.

See *Trenholm, Laurie Elizabeth—2005*.

Ushiki, Yuichiro

Yuichiro Ushiki was born in Saitama-City, Saitama Prefecture, Japan, and earned a degree in Agriculture and Life Science from Hirosaki University in 1973. He has been Golf Course Superintendent at Taheiyo Club Inc. since 1999. Yuichiro Ushiki was elected to the Board of Directors of the Japanese Society of Turfgrass Science in 2004.

2007† **Cultivation for Turfgrass Management—Its Effects and Method Using Machinery.** Coeditor.

See *Inamori, Makoto—2007*.

V

van de Walle, Henri[3]

1942 **Engazonnement des Aérodromes et Terrains de Sports (Turf Establishment of Airfields and Sports Grounds).** Henri van de Walle. Editions Imprimeries Réunies de Senlis, Senlis, France. 203 pp., 46 illus., 16 tables, small index, soft bound in light brown with brown lettering, 6.0 by 9.2 inches (152 × 234 mm). Printed entirely in the French language.

＊   A **relatively rare**, unique book on the construction, establishment, and culture of turfgrasses on airfields under conditions in France. A small section in the back addresses sports field turfgrasses. The preface is by R. Lemaire.

Vargas, Joseph Martin, Jr. (Dr. and Professor)

Joe Vargas was born in Fall River, Massachusetts, USA, and earned a BSc degree in Agricultural Science from the University of Rhode Island in 1963, an MSc degree in Botany and Plant Pathology from Oklahoma State University in 1965, and a PhD degree in Plant Pathology from the University of Minnesota in 1968. Since 1968, Dr. Vargas has served on the Plant Pathology Faculty at Michigan State University and has been involved in the research and teaching of pathology and diseases of turfgrasses, with emphasis on fungicide resistance and biological control. Dr. Vargas received the Michigan Turfgrass Foundation Distinguished Service Award in 1996, the GCSAA Distinguished Service Award in 1997, and the USGA Green Section Award in 2007.

1981†   **Management of Turfgrass Diseases.** First edition. J.M. Vargas Jr. Burgess Publishing Company, Minneapolis, Minnesota, USA. 204 pp., 115 illus. with 23 in color, 32 tables, indexed, cloth hardbound in blue and orange with blue lettering, 7.0 by 10.0 inches (178 × 254 mm).

   A text on the basics of turfgrass diseases, with special emphasis on the turfgrass cultural dimensions of disease prevention. Each disease is discussed in terms of the causal pathogen, hosts, symptoms, occurrence, cultural influences, resistant cultivars, and chemical controls. In addition to the fungal diseases, the nematodes, bacterial, viral, and mycoplasma-like diseases are addressed. The author has a personalized writing style.
   Revised in 1994 and 2005.

1994†   **Management of Turfgrass Diseases.** Second revised edition. J.M. Vargas Jr. Advances in Turfgrass Science Series. Lewis Publishers, Ann Arbor, Michigan, USA. 294 pp., 100 illus. with 72 in color, 23 tables, indexed, hardbound in multicolor with white lettering, 6.0 by 9.2 inches (152 × 233 mm).

   This revision updates the advances made in the science and management of turfgrass diseases. It contains a large increase in color photographs. It includes both a glossary of selected terms and a bibliography in the back. The preface is by Dr. James B Beard.
   Revised in 2005.

2004†   ***Poa annua*—Physiology, Culture, and Control of Annual Bluegrass.** J.M. Vargas Jr. and A.J. Turgeon. John Wiley & Sons Inc., Hoboken, New Jersey, USA. 175 pp., 52 illus. with 35 in color, 12 tables, indexed, glossy hardbound in greens and black with white and black lettering, 7.5 by 9.2 inches (190 × 233 mm).

   A book specifically focused on annual bluegrass (*Poa annua*) in turfgrass situations. The biology of this species is addressed in detail. There are chapters on the culture of annual bluegrass for putting greens, tees, and fairways as well as annual bluegrass control approaches in other turfgrass species. There is a bibliography in the back.

2005†   **Management of Turfgrass Diseases.** Third revised edition. J.M. Vargas Jr. John Wiley & Sons Inc., Hoboken, New Jersey, USA. 337 pp.; 122 illus. with 72 in color; 23 tables; indexed; glossy hardbound in multicolor with white, yellow, and black lettering; 6.0 by 9.2 inches (152 × 233 mm).

   An update of the cultural practices that can be used in the management of both turfgrasses and potential pathogens in order to minimize disease problems.

Figure 12.96. A 50-gallon (189.3-l) water barrow mounted on a 2-wheel, manual-push truck, with a 50-inch-wide (1.3 m), perforated sprinkler boom, 1920s. [41]

2006† **The Turf Problem Solver—Case Studies and Solutions for Environmental, Cultural, and Pest Problems.** Coauthor.
  See *Turgeon, Alfred Joseph—2006.*

Vengris, Jonas (Dr. and *Professor* Emeritus*)*

Jonas Vengris (1909–2004) was born in Daglienai, Lithuania, and earned BSc and MSc degrees in Agronomy from the Agricultural College at Dotnuva, Lithuania, in 1934 and 1936, respectively, and a Doctor of Agricultural Science (DrAgrSc) degree in Plant Physiology and Biochemistry from the University of Wisconsin in 1958. He was Lecturer from 1941 to 1943 and Assistant Professor from 1943 to 1944 at the Agricultural College at Dotnuva, Lithuania, and was Associate Professor at Baltic University in Hamburg-Pinneberg, West Germany from 1946 to 1949. He then immigrated to the United States to join the faculty at the University of Massachusetts as Assistant Professor from 1949 to 1963, as Associate Professor from 1963 to 1970, and as Full Professor from 1970 to 1981, when he retired. He was active in weed control and also taught agronomy courses. Dr. Vengris received the Award of Merit from the Northeastern Weed Science Society in 1979.

1969† **Lawns—Basic Factors, Construction, and Maintenance of Fine Turf Areas.** First edition. Jonas Vengris. Thomson Publications, Davis, California, USA. 229 pp., 107 illus., 27 tables, indexed, semisoft covers in white with green and black lettering in a comb binder, 8.4 by 10.8 inches (213 × 275 mm).
  An introductory manual on the basics of turfgrass establishment and culture in the United States. References are cited at the end of each chapter, with problem sets found at the ends of some chapters.
  Revised in 1973 and 1982.

1973† **Lawns—Basic Factors, Construction, and Maintenance of Fine Turf Areas.** Second revised edition. Jonas Vengris. Thomson Publications, Indianapolis, Indiana, USA. 247 pp.; 110 illus.; 27 tables; indexed; semisoft bound in green and white with white, green, and black lettering; 8.4 by 10.9 inches (214 × 278 mm).
  A substantial revision.
  Revised in 1982.

1982† **Lawns—Basic Factors, Construction, and Maintenance of Fine Turf Areas.** Third revised edition. Jonas Vengris and William A. Torello. Thomson Publications, Fresno, California, USA. 195 pp.; 105 illus.; 27 tables; indexed; glossy semisoft bound in green and white with white, green, and black lettering; 8.4 by 10.9 inches (214 × 277 mm).
  The manual contains some revisions and an improved format.

Veronesi, Fabio

Fabio Veronesi was born in Rome, Italy, and earned a BSc degree from the University of Perugia in 1975. He was a Visiting Scientist to the Agronomy Department at the University of Wisconsin 1982. He served as a Researcher at the University of Perugia from 1980 to 1986, as Associate Professor of Plant Breeding at the University of Sassri from 1986 to 1990, as Professor of Plant Genetics at Universities of Udine and Ancona from 1990 to 1995, and as Professor of Genetic Biotechnologies at the University of Perugia since 1995. He is actively involved in cytology, biotechnology, and breeding of turf and forage grasses. Fabio Veronesi was elected President of the Eucarpia Fodder Crops and Amenity Grasses Section from 1994 to 1999.

2000† **Tappeti Erbosi (Turfgrasses).** Coauthor.
  See *Panella, Adelmo—2000.*

2006† **Tappeti Erbosi—Cura, gestione e manutenzione, delle aree verdi pubbliche e private (Turfgrasses—Care, management and maintenance of public and private green areas).** Coauthor.
  See *Croce, Paolo—2006.*

Victoria Department of Agriculture, Australia

n.d. **Lawns and Playing Fields.** F.R. Drake, C.J.R. Johnston, L.C. Jones, R.J. Newman, and W.I. Walbran. Department of
circa Agriculture, Melbourne, Victoria, Australia. 32 pp., 30 illus., 2 tables, no index, fold-stapled glossy semisoft covers in
19xx black and white with red lettering, 7.2 by 9.8 inches (184 × 248 mm).
  A practical guide to the planting and care of lawns and bowling and golf greens under the conditions in southeastern Australia.

1963 **Turf Manual.** R.J. Newman, W.I. Walbran, E.B. Littlejohn, L.C. Jones, and H.B. Wilson (contributing authors). Victoria Department of Agriculture, Melbourne, Victoria, Australia. 87 pp., 25 illus. with 6 in color, 8 tables, no index, fold-stapled semisoft stapled covers in multicolor with black lettering, 7.0 by 9.0 inches (184 × 240 mm).
  A general guide book concerning turfgrass characteristics, soils and site preparation, culture, and pests. It is oriented to conditions in southwest Australia.
  Revised in 1963.

1971† **Turf Manual.** R.J. Newman, W.I. Walbran, E.B. Littlejohn, I.C. Jones, and H.B. Wilson (contributing authors). Victoria Department of Agriculture, Victoria, Australia. 89 pp., 25 illus. with 6 in color, 7 tables, no index, textured semisoft bound in multicolor with black lettering, 7.2 by 9.5 inches (183 × 240 mm).

An Australian compendium of subjects concerning the establishment and culture of turfgrasses for golf courses, bowling greens, and sports grounds. Included are sections on soil physical and chemical properties as well as problem diagnosis.

Viergever, Richard Leonard

Dick Viergever (1913–1994) was born in Mankato, Minnesota , USA. He started working on golf courses in various capacities in 1926 and also as an athletic instructor and director. He was Pro-Superintendent of the Legion Golf Course, Manhattan, Kansas, from 1935 to 1937, of the Hutchinson Country Club, Hutchinson, Kansas, from 1940 to 1941, and of the Owensboro Country Club, Owensboro, Kentucky, in 1946. During WWII, he served in the US Army in the Philippines. He was the Golf Course Superintendent of Elks Country Club, DuQuion, Illinois, from 1950 to 1954; of Woodbridge Golf & Country Club, Lodi, California, from 1955 to 1959; of the Olympic Club, San Francisco, California, from 1959 to 1965; of Valley Hi Country Club, Sacramento, California, in 1965; and of the City of Sunnyvale from 1966 to 1978, when he retired. Dick Viergever was elected a director in 1956, Secretary-Treasurer for 1958, Vice President for 1959 and 1965, and President for 1960 and 1966 of the Golf Course Superintendents Association of Northern California and as President of the Northern California Turf Grass Council from 1960 to 1966.

1970† **The Modern Golf Course Superintendent.** Richard Viergever. Thomson Publications, Fresno, California, USA. 107 pp., 14 illus., 12 sample forms, no tables, no index, semisoft covers in green and white with white and black lettering in a comb binder, 8.3 by 10.8 inches (212 × 275 mm).

A small book on the management aspects of a golf course operation presented from a superintendent's view point. Included are discussions of job descriptions, staff training, purchasing, and communications. It contains some early photographs of turfgrass maintenance activities. Dick Viergever describes an interesting historical perspective concerning golf course turfgrass practices in the United States between 1920 and 1950:

> Most of the work, in those days, was done by hand. Even in the early twenties, many fairways were still cut with horse drawn mowers. Then came the tractor with wide steel wheels equipped with short studs for traction. The power mower was not yet in use. In some ways these methods were better than todays. High costs nowadays make mass production and automation necessities, and the cost of labor precludes much of the detailed and careful handwork that could be done in the past.
>
> Chemicals, with exception of calomel, corrosive sublimate and one or two others for the control of fungus diseases, were virtually unknown. Weeds were killed by plunging a sliver of wood, which had been dipped in acid, into the crown of the weed, or by hand pulling. It was not unusual, at the larger Clubs, to see fifty or sixty people going down a fairway on hands and knees, pulling weeds or, at Clubs with limited budgets, even the Members digging weeds out of the grass.
>
> Fertilizer was usually applied in top-dressing, and was mostly in the form of manure. It was the usual custom to topdress greens several times a year.
>
> Very few courses had irrigated fairways; in fact, except on large courses and those in the arid West, watered fairways were not common until after World War II.
>
> Mowing was not as frequent as today, nor as close. Height of cut, on a normal first class course, might have been: fairways, one and one half to two inches, greens, three eights inch, and roughs not at all or possibly once a year after the grass had been allowed to go to seed. Tees were often small leveled mounds of oiled clay.
>
> Construction, when done by the real experienced greenkeeper, was comparable to that of today, although the old "chocolate drop" mounds, steep grass bunkers and other steep banks and hard to maintain features have been nearly eliminated. Drainage, as today, was considered of utmost importance and, the old cinder base green, even without tile, could well replace some of the shoddily constructed greens being built today. These greens consisted of a heavy layer of coarse cinders or other large material covered with pea gravel or fine crushed rock on which was added about twenty inches of fine topsoil. This type of construction will make a satisfactory putting green even by present standards.

Vietti, Mario

Mario Vietti was born in Turin, Italy, and earned a BSc degree from the University of Pigcenda in 1973. Since then, he has worked as a landscape architect involved in the planning and restoration of gardens and parks in Italy.

Figure 12.97. Pressurized popup irrigation heads as installed in-ground and controlled by an automatic zone valve, circa 1930. [21]

1991†   **Il Prato Ornamentale (Ornamental Turf).** Mario Vietti. Instituto Geografico De Agostini, Novara, Italy. 192 pp.; 485 illus. with 147 in color; 25 tables; no index; textured cloth hardbound in green with gold lettering and a glossy jacket in multicolor with yellow, white, and black lettering; 7.3 by 9.6 inches (185 × 244 mm). Printed entirely in the Italian language.

A practical book on the construction, planting, and culture of turfgrass, including lawns and sports facilities. It is oriented to climatic and soil conditions in Italy. There is a chapter on golf course and sports field turfgrasses in the back plus a small glossary of selected terms and a list of selected references.

Villani, Michael Gerard (Dr. and Professor)

Mike Villani (1953–2001) was born in San Antonio, Texas, USA, and earned a BA degree *magna cum laude* in Liberal Arts from the State University of New York at Stony Brook in 1979 and a PhD degree in Entomology from North Carolina State University in 1984. In 1985, he joined the faculty as Assistant Professor in the Department of Entomology at the New York State Agricultural Experiment Station, Cornell University, and rose to Full Professor of Soil Insect Ecology. His activities encompass research, teaching, and extension activities, specializing in the ecology and management of soil insect pests of turfgrasses and horticultural crops. He was coeditor of Environmental Entomology. Dr. Villani received the National Recognition Award in Urban Entomology from the Entomological Society of America in 1997, the Citation of Merit from the New York State Turfgrass Association in 1999, and the Outstanding Service Award of the Turfgrass Council of North Carolina in 2001.

1995†   **Handbook of Turfgrass Insect Pests.** Coeditor.
See *Brandenburg, Rick Lynn—1995.*

1999†   **Turfgrass Insects of the United States and Canada.** Second revised edition. Coauthor.
See *Vittum, Patricia Joan—1999.*

Vittum, Patricia Joan (Dr. and Professor)

Pat Vittum was born in Geneva, New York, USA, and earned a BA degree in Chemistry from the College of Wooster, Ohio, in 1974; an MSc degree in 1979; and a PhD degree in Entomology from Cornell University in 1980, studying with Dr. Haruo Tashiro. She then joined the faculty in the Entomology Department at the University of Massachusetts as an Assistant Professor in 1980 and rose to Associate Professor in 1987 and to Full Professor in 2000. She is active in turfgrass teaching and adult extension and serves as the Pesticide Coordinator for the state of Massachusetts.

1998†   **IPM Handbook for Golf Courses.** Coauthor.
See *Schumann, Gail Lynn—1998.*

1999†   **Turfgrass Insects of the United States and Canada.** Second revised edition. Patricia J. Vittum, Michael G. Villani, and Haruo Tashiro. Comstock Publishing Associates of Cornell University Press, Ithaca, New York, USA. 422 pp.; 717 illus. with 576 in color; 14 tables; well indexed; cloth hardbound in blue with black lettering and a jacket in multicolor with white, yellow, and black lettering; 7.0 by 10.0 inches (178 × 254 mm).

This is an update of a comprehensive, scholarly text that addresses the biology of insects, mites, and vertebrate pests of turfgrasses. There are new chapters on biological control, sampling techniques, setting thresholds, and integrated pest management. The chapters on chemical control and insect habitats have been expanded. A glossary of selected terms and list of references are included in the back.

Voykin, Paul Nicholas

Paul Voykin was born in Saskatoon, Saskatchewan, Canada. His childhood experiences were at the family homestead in Red Pheasant, and he apprenticed at several golf courses in the Alberta region. He became a United States citizen in 1963. He has been Golf Course Superintendent at Briarwood Country Club in Deerfield, Illinois, since 1961 and, previously, was Golf Course Superintendent at Calumet Country Club for three years and assistant superintendent at Olympia Fields Country Club for two years. He was a pioneer and strong proponent of introducing unmowed nature areas and wild flowers on golf courses. Paul Voykin was elected President of the Midwest Association of Golf Course Superintendents in 1972. He received the GCSAA Leo Feser Award in 1981 and Golf Super News Superintendent of the Year Award in 2004.

1969† **A Perfect Lawn the Easy Way.** First edition. Paul N. Voykin. Rand McNally & Company, New York City, New York, USA. 155 pp.; no illus.; no tables; indexed; cloth hardbound in light green with black lettering and a jacket in multicolor with blue, green, and black lettering; 5.4 by 8.2 inches (137 × 208 mm).

A small book on the establishment and culture of lawns organized on a calendar basis. It is practically oriented to the home lawn enthusiasts and conditions in the United States. It includes a small chapter on building a home putting green. It has a different writing style.

Revised in 1971.

1969† **A Perfect Lawn the Easy Way.** Paul N. Voykin. Rand McNally & Company, New York City, New York, USA. 125 pp.; no illus.; no tables; indexed; glossy semisoft bound in multicolor with blue, green, and black lettering and a jacket in glossy multicolor with blue, green, and black lettering plus a clear overlay; 6.1 by 9.1 inches (155 × 231 mm).

This is a soft bound version of the previous book.

1969† **A Perfect Lawn the Easy Way.** Paul N. Voykin. Documentary Books, New York City, New York, USA. 155 pp.; no illus.; no tables; indexed; cloth hardbound in mustard and light green with gold lettering and a jacket in cream, green, and black with green and black lettering; 5.1 by 8.3 inches (130 × 210 mm).

A reprinting of the 1969 first edition.

A 1971 reprinting of the 1969 book was named Book of the Month by Doubleday.

1976† **Ask the Lawn Expert.** Paul N. Voykin. Macmillan Publishing Company Inc., New York City, New York, USA. 249 pp.; no illus.; 5 tables; indexed; cloth hardbound in green with gold lettering and a jacket in multicolor with orange, green, gray, white, blue, and black lettering; 5.4 by 8.2 inches (138 × 208 mm).

A book on lawn establishment and culture organized in a question-and-answer format. It is practically oriented to home lawn enthusiasts and conditions in North America.

W

Waddington, Donald VanPelt (Dr. and *Professor* Emeritus *of Soil Science)*

Don Waddington was born in Norristown, Pennsylvania, USA, and earned a BSc degree from Pennsylvania State University in 1953, an MSc degree in Soils from Rutgers University in 1960, and a PhD degree in Agronomy from the University of Massachusetts in 1964. He served as Instructor of Turfgrass at the University of Massachusetts from 1960 to 1965. He then joined the faculty at Pennsylvania State University in 1965, rising to Full Professor of Turfgrass and Soil Science. He served in teaching and research, with emphasis on soil amendments, turfgrass nutrition and fertilizers, and turfgrass sports fields. He retired in 1991. He was elected to the Turgrass Science Division C-5 Chair for 1986 and to the Crop Science Society of America (CSSA) Board of Directors for 1985 through 1987. Dr. Waddington was awarded Fellow of the American Society of Agronomy in 1986 and the CSSA Turfgrass Science Award in 1993.

1992† **Turfgrass.** D.V. Waddington, R.N. Carrow, and R.C. Shearman (editors). Agronomy Series No. 32, American Society of Agronomy, Crop Science Society of America, and Soil Science Society of America, Madison, Wisconsin, USA. 805 pp.; 77 illus.; 86 tables; indexed; cloth hardbound in green with gold lettering; 6.0 by 9.0 inches (152 × 228 mm).

An extensive monograph updating the research findings and methodologies generated by 43 turfgrass scientists. It is organized in 22 chapters divided into five sections: the turfgrass industry, turfgrass physiology, soils and water, management, and research methods. It is extensively referenced. The monograph is the result of the CSSA Turfgrass Science Division C-5 member activities. This monograph is an update and major revision of an earlier edition.

See the earlier version: *Hanson, Angus Alexander—1969.*

2001† **Turfgrass Soil Fertility and Chemical Problems—Assessment and Management.** Coauthor.

See *Carrow, Robert Norris—2001.*

Walker, Cyril

Cyril Walker (1899–1981) was born in Auckland, New Zealand, and earned a Diploma in Agriculture (HDA) degree from
Hawkesbary Agricultural College of Richmond, New South Wales, Australia, in 1924. From the mid-1920s, he was
Assistant Instructor in Agriculture for the New Zealand Department of Agriculture at TePuke, Bay of Plenty, work-
ing in the breeding of maize. In the late 1940s, he served as Officer-in-Charge of Turf Research for the New Zealand
Department of Agriculture in Palmerston North. He joined the Executive of the New Zealand Turf Culture Institute
(NZTCI) in 1954. He was a Senior Advisory Officer for the New Zealand Institute for Turf Culture from 1956 to 1964
and Director of Green Services from 1964 to 1977. In 1959, he originated and was editor of the **New Zealand Turf
Culture Institute Newsletter/Sports Turf Review** for eighteen years. Cyril Walker retired from the NZTCI in 1977.

1971† **Turf Culture.** Second revised edition. C. Walker (editor) with thirteen contributing authors: H.S. Gibbs, D.G. Bowler,
G.S. Harris, G.A.H. Helson, G.C.M. Latch, F.W. Gibbens, K.H. Marcussen, R. Beveridge, A.C. Burgess, J.P. Salinger,
J.W. Goodwin, C. Walker, and G.S. Robinson. New Zealand Institute for Turf Culture, Palmerston North, New Zealand. 362 pp., 186 illus. with 8 in color, 25 tables, indexed, textured cloth hardbound in green with gold lettering, 5.9
by 9.4 inches (150 × 239 mm).

An update of a book on the construction, establishment, and culture of turfgrasses under the conditions in New
Zealand. Turfgrass cultural systems addressed include golf putting greens, turfed tees, fairways, bunkers, croquet greens,
tennis courts, cricket grounds, horse racecourses, lawns, and parklands. The foreword is by G.S. (Robbie) Robinson.

See earlier version, *New Zealand Institute for Turf Culture—1961.*

Walmsley, William Hugh

Bill Walmsley was born in Auckland, New Zealand, and earned a BSc degree in Horticultural Science from Massey University in
1977. He served as an Agronomist with the New Zealand Turf Culture Institute from 1977 to 2003. Since then he has
been a Turf Agronomist for PGG Wrightson Turf, Christchurch.

1990† **Fungicides for Turfgrass Disease Control.** W.H. Walmsley (editor). New Zealand Turf Culture Institute, Palmerston
North, New Zealand. 26 pp., no illus., 6 tables, no index, soft bound in black and white with green lettering, 8.1 by
11.5 inches (206 × 292 mm).

A small booklet on fungicides for the control of turfgrass diseases in New Zealand. It includes the characteristics,
selection, application, and usage programs for fungicides.

Waltz, F. Clinton, Jr.[2]

2003† **Fundamentals of Turfgrass and Agricultural Chemistry.** Coauthor.
See *McCarty, Lambert Blanchard—2003.*

2005† **Designing, Constructing, and Maintaining Bermudagrass Sports Fields.** Coauthor.
See *McCarty, Lambert Blanchard—2005.*

Figure 12.98. Portable, manual-push, 50-gallon water barrow with attachments, such as a gravity sprinkler, circa 1924. [34]

Warren, Benedict Orlando, Jr.

Ben Warren (1909–1992) was born in Virdin, Illinois, USA, and apprenticed for nursery and greenhouse businesses for 12 years. He founded Warren's Turf Nursery in 1938 on ten acres of land in Palos Township, Illinois. He continued as President for 42 years, with the company growing to over 4,000 production acres in six states. From 1942 into 1945, he served in the US Navy during WWII, receiving the Bronze Star for his actions at Bougainville. He was one of the pioneers and leaders in the development of the commercial sod production industry. Ben was active in a broad range of technologies in addition to managing the largest sod production company in the United States. He held a number of patents for sod harvesting equipment, for vacuum sod cooling, and for the original sod washing technique. His efforts encompassed turfgrass research and breeding, with three improved Kentucky bluegrass (*Poa pratensis*) cultivars released—A-10, A-20, and A-34 (BenSun). He was the key leader in organizing the American Sod Producers Association (ASPA) from 1967 to 1968. Ben Warren was elected the first President of the American Sod Producers Association in 1967 and Founding President of the Illinois Turfgrass Foundation from 1959 to 1961. He was awarded the Distinguished Service Award of the Illinois Turfgrass Foundation in 1963, the American Sod Producers Association Honorary Member in 1979, and the GCSAA Distinguished Service Award in 1982.

1959† **A Survey in Turf Transplanting—How Thick Should Sod Be Cut?** Ben O. Warren. Warren's Turf Nursery, Palos Park, Illinois, USA. 21 pp., 2 illus., 1 table, no index, fold-stapled textured semisoft covers in tan with brown lettering, 8.4 by 10.9 inches (214 × 278 mm).

　　A booklet documenting the research available concerning the preferred depth for harvesting transplanting sod. The issue was that the state highway specification required sod be cut with excessive amounts of thick soil.

Warwick, David Brian

David Warwick was born in Sydney, New South Wales, Australia, and earned a Certificate of Greenkeeping in 1987 and a Certificate of Horticulture in 1990, both from Ryde Technical & Further Education (TAFE). He apprenticed as a greenkeeper at Concord Golf Club from 1983 to 1987 and as greenkeeper at the Royal Sydney Golf Club in 1988. He was Assistant Superintendent at Bonnie Doon Golf Club, Sydney, from 1989 to 1990. Next, he was Assistant Superintendent and the Superintendent at Arundel Hill Country Club in Queensland from 1990 to 1996, and since 1996, he has been Superintendent at Avondale Golf Club in Sydney. He was elected to the Board of Directors of the New South Wales Superintendent's Association from 1998 to 2001 and the Australian Golf Course Superintendents Association (AGCSA) from 2002 to 2003. David Warwick received the AGCSA Claude Crockford Environmental Management Award in 2004 and the AGCSA Excellence in Golf Course Management Award in 2010.

2007† **How To Build A Sand Based Golf Green.** Coauthor.
　　See *McIntyre, David Keith—2007.*

Watschke, Thomas Lee (Dr. and *Professor*)

Tom Watschke was born in Charles City, Iowa, USA, and earned a BSc degree in Agronomy from Iowa State University in 1967 and both an MSc degree in Agronomy and Turfgrass in 1969 and a PhD degree in Agronomy and Turfgrass in 1971 from Virginia Polytechnic Institute and State University, studying with Dr. Dick Schmidt. He joined the faculty at Pennsylvania State University (PSU) in 1970 and was promoted to Full Professor in 1981. He has worked in the turfgrass research and teaching areas, with emphasis on weed control, plant growth regulators, fertilizers, and related water quality effects. He was elected Turfgrass Science Division C-5 Chair in 1990 and to the Crop Science Society of American (CSSA) Board of Directors for 1989 through 1991. Dr. Watschke received the PSU Gamma Sigma Delta Outstanding Teaching Award in 1982, the Distinguished Service Award of the Pennsylvania Turfgrass Council in 1989, and the GCSAA Distinguished Service Award in 1990; he was also recognized as a Fellow of the American Society of Agronomy in 1991, a Fellow of the CSSA in 1992, and an American Sod Producers Association Honorary Member in 1992.

1995† **Managing Turfgrass Pests.** Thomas L. Watschke, Peter H. Dernoeden, and David J. Shetlar. Advances in Turfgrass Science Series, Lewis Publishers, Ann Arbor, Michigan, USA. 361 pp., 199 illus. with 92 in color, 18 tables, indexed, glossy hardbound in multicolor with white lettering, 6.0 by 9.2 inches (152 × 233 mm).

　　This practical book encompasses the integrated management of the pests of turfgrass, which involves IPM strategies. It is organized in three parts: weeds, disease, and insects and mites. There are 92 color photographs of the major diseases and insects. There is a preface by Dr. James B Beard.

Watson, Leslie (Dr.)

Les Watson was born in Leek, Staffordshire, England, UK, and earned a BSc degree with *honours* in Plant Biology in 1959 and a MSc degree in Science in 1960, both from Manchester University in England, and a DSc degree from the Australian University in 1990. He taught basic systemic botany at Southampton University, England, from 1962 to 1969. He then became a Fellow in the Research School of Biological Sciences at the Australian National University, Canberra, Australian Capital Territory, Australia, from 1969 to 1977 and was Senior Fellow from 1977 to 1993, conducting plant taxonomic research including grasses. While formally retiring in 1994, Dr. Watson has continued to maintain web data bases with interactive keys and descriptions for grass genera.

1977† **Identifying Grasses: data, methods, and illustrations.** Coauthor.
> See *Clifford, Harold Trevor—1977.*

1992 **The Grass Genera of the World.** First edition. Leslie Watson and Michael J. Dallwitz. CAB International, Wallinford, Oxon, UK. 1042 pp., no illus., 1 table, indexed, glossy hardbound in multicolor with black and white lettering, 6.0 by 9.2 inches (152 × 233 mm).

> A major taxonomic reference text detailing the descriptions of 780 genera in the grass (*Poaceae*) family found in the world. The descriptions consist of 496 characters including morphological, anatomical, physiological, biochemical, and cytological dimensions.
> *Revised in 1994.*

1994 **The Grass Genera of the World.** Second revised edition. Leslie Watson and Michael J. Dallwitz. CAB International, Wallinford, Oxon, UK. 1081 pp., no illus., 1 table, indexed, glossy hardbound in multicolor with black lettering, 6.0 by 9.2 inches (152 × 233 mm).

> This revision of the original text involves the addition of eight genera of *Poaceae*, an expansion of anatomical descriptions for the *Chloridoideae*, and other minor corrections and improvements.

Welch, Robert J.

Bob Welch (1928–2003) was born in Milwaukee, Wisconsin, USA. He was a long time employee of the Milwaukee Metropolitan District, including the Milorganite Division. His activities included interaction with distributors, organizing trade show exhibits, and managing fertilizer test plots. He retired in 1988. Bob Welch served on the USGA Public Links Committee as the Wisconsin representative and as Secretary of the O.J. Noer Research Foundation.

1966† **Winter Injury.** Coauthor.
1966† **Better Fairways: Northern Golf Courses—Theory and Practice.** Coauthor.
1970† **Better Bent Greens—Fertilization & Management.** Coauthor.
1971† **Better Fairways: Northern Golf Courses—Theory and Practice.** Coauthor
1972† **Permanent Fertilization Record and Handbook.** Coauthor.
> See *Wilson, Charles Granville—1966, 1966, 1970, 1971, and 1973, respectively.*

Welterlen, Mark Stephen (Dr.)

Mark Welterlen was born in Fitchburg, Massachusetts, USA, and earned an Associate of Science certificate in Turf Management from Essex Agricultural and Technical Institute in 1975, a BSc degree in Plant Science and Turfgrass from the

Figure 12.99. Laterally mobile, 800-foot-long sod conveyor for loading onto trucks, circa 1959. [48]

University of Rhode Island in 1977, and MSc and PhD degrees in Agronomy and Turfgrass from Pennsylvania State University in 1980 and 1982, respectively, studying with Dr. Tom Watschke. He served as Assistant Professor in turfgrass teaching and research at the University of Maryland from 1982 to 1988. Dr. Welterlen then joined Intertec Publishing Inc. as Editor and as Editor-in-Chief of **Grounds Maintenance** magazine from 1988 to 1999 and then held the positions of Associate Publisher, Publisher, and Group Publisher until 2006 when this publisher ceased operations. Then in 2007, he joined the PBI/Gordon Corporation as Manager of Product Development. Dr. Welterlen received the President's Award for Academic Excellence in Plant and Soil Science from the University of Rhode Island in 1977 and the Turf and Ornamental Communicators Association and Terra Industries Environmental Communicator of the Year Award in 1999.

1987† **Laboratory Manual for Turfgrass Management.** Mark S. Welterlen. A Reston Book, Prentice-Hall Inc., Englewood Cliffs, New Jersey, USA. 202 pp., 118 illus., 22 tables, no index, glossy semisoft bound in green and yellow with green and yellow lettering, 8.3 by 10.9 inches (211 × 278 mm).

This turfgrass laboratory manual contains 17 instructional units. A glossary of terms is presented in the back.

Western, John Henry (Dr. and *Professor*)

John Western (1908–1982) was born in the West County, England, UK. He spent a year at Avoncroft (Midland) Agricultural College and then earned a BSc degree in Agricultural Botany with *first class honours* in 1932 at the University College, Aberystwyth, Wales. He pursued further advanced study while working in agricultural botany research from 1933 to 1935 at Aberystwyth and also studied at the University of Minnesota from 1935 to 1936 with Dr. E.C. Stakman. He earned a PhD degree in Plant Pathology in 1936, studying with Kathleen Sampson. He taught agricultural botany at Manchester along with conducting advisory work for the Ministry of Agriculture and Fisheries. In 1946, he became Regional Plant Pathologist for the National Agricultural Advisory Service for the Northern Region at Newcastle-upon-Tyne. Then he became Senior Lecturer in Agricultural Botany at the University of Leeds in 1951 and assumed the position of Chair and Professor of Agricultural Botany in 1959. He retired in 1971. Dr. Western served as President of the Association of Applied Biologists from 1964 to 1965.

1941† **Diseases of British Grasses and Herbage Legumes.** Coauthor.
&
1954†

See *Sampson, Kathleen—1941 and 1954.*

White, Charles Benjamin

Bud White was born in Greensboro, North Carolina, USA, and earned an Associate degree in Golf Course and Turfgrass Management from Catawba Valley Community College in 1975, a BSc degree in Plant and Soil Science from Tennessee Technical University in 1977, and an MSc degree in Horticultural Science and Turfgrass from Clemson University in 1979. He then was Southeastern Agronomist and Director of the USGA Green Section from 1978 to 1987. After a period supervising golf course construction, he formed Total Turf Service and was involved in consultancy for golf courses. Bud White rejoined the USGA Green Section in 2002 as Senior Agronomist for the Midcontinent Region.

2000† **Turf Managers' Handbook for Golf Course Construction, Renovation, and Grow-In.** Charles B. White. John Wiley & Sons Inc., Hoboken, New Jersey, USA. 328 pp., 132 illus. with 16 in color, 19 tables, indexed, glossy hardbound in multicolor with white lettering, 6.0 by 9.0 inches (152 × 228 mm).

A specialty handbook with emphasis on the turfgrass planting and establishment phases of golf course construction. There is an extensive appendix of tables and illustrations plus a small glossary and selected list of references in the back.

White, Laurence Walter

Laurie White was born in 1904 in Watford, Hertfordshire, England, UK, as the son of the Head Groundsman at Lords and at Luton Cricket Grounds, Harry White. In 1926, he worked as a Groundsman at Merchant Taylor's School and then served as Head Groundsman for the National Association of Boys Clubs Ground from 1932 to 1935. He then served as Head Groundsman at St Bartholomew's Hospital Athletic Ground from 1937 to 1952, except for Royal Air Force military service as a flying instructor from 1940 to 1945 during WWII. He retired in 1969. Laurie White was a founding member of the National Association of Groundsman and served on the Executive Committee as Treasurer and as Vice President.

1952† **Practical Groundsmanship.** L.W. White and W.H. Bowles. English Universities Press Limited, London, England, UK. 258 pp., 15 illus., 9 tables, indexed, cloth hardbound in light green with black lettering and a jacket in green and white with white and black lettering, 4.8 by 7.2 inches (122 × 183 mm).

＊   A book on the construction, establishment, and culture of sports fields under the cool-season turfgrass conditions in England. There are specific chapters on turfgrasses for cricket, tennis, lawn bowling, field hockey, soccer, and running tracks plus garden lawns and paths. The foreword is by the Rt. Hon. Lord Luke.

White, Susan Kay

Susan White was born in Flint, Michigan, USA, and earned a BSc degree with *distinction* in Agricultural Education in 1980 and an MSc in Agronomy in 1986, both from the University of Minnesota. She served as Extension Associate in the Department of Agronomy and Turfgrass at Ohio State University from 1986 to 1991 and, since 1992, has been a Research Fellow at the University of Minnesota serving in the Department of Soil, Water and Climate from 1992 to 1994 and then in the Department of Agronomy and Plant Genetics.

1989†   **Turf Management.** Revised edition. Susan K. White. Ohio Agricultural Education Curriculum Materials Service, AGDEX 273, Ohio State University, Columbus, Ohio, USA. 109 pp., 103 illus., 11 tables, no index, semisoft bound in green and black with black lettering, 8.25 by 11.0 inches (210 × 280 mm).

   A basic instructional manual on turfgrasses including soils, drainage, species and cultivars, propagation, and culture. It is oriented for use in vocational schools. The original **Turf Management** student manual was written by John C. Billick, Vocational Agriculture Teacher, at Perkins High School, Sandusky, Ohio.

Wiecko, Greg (Dr.)

Greg Wiecko was born in Bydgoszcz, Poland, and earned an MSc degree in Agriculture from the Agricultural University of Bydgoszcz, Poland, in 1981 and a PhD degree in Agronomy from the University of Georgia in 1990. He served on the faculty at the Agricultural University of Bydgoszcz from 1981 to 1985 and as Product Development Manager for the Monsanto-Europe Company from 1990 to 1991. He was Professor at the University of Guam from 1994 to 2005 and was involved in turfgrass adult education, teaching, and research. Since 2005, he has been Associate Director of the Agricultural Experiment Station at the University of Guam.

2006†   **Fundamentals of Tropical Turf Management.** Greg Wiecko. CAB International, Cambridge, Massachusetts, USA. 213 pp.; 184 illus.; 11 tables; indexed; glossy semisoft bound in multicolor with white, yellow, and green lettering; 6.1 by 9.1 inches (156 × 232 mm).

   An introductory book on establishment and culture of certain warm-season turfgrasses under tropical conditions of the Pacific region. There is a glossary of selected terms in the back.

Wiersema, John Harry (Dr.)

John Wiersema was born in Muskegon, Michigan, USA, and earned a BSc degree *magna cum laude* from Western Michigan University in 1974 and both MSc and PhD degrees from the University of Alabama in 1979 and 1984, respectively. He served as a Research Associate at the University of Maryland from 1984 to 1989 and, since, as a Botanist with the Agricultural Research Service of the United States Department of Agriculture at Beltsville, Maryland, conducting research concerning the taxonomy and nomenclature of economic plants. Dr. Wiersema received the Jesse Greenman Award from the Missouri Botanical Garden in 1988.

1986†   **A Checklist of Names for 3,000 Vascular Plants of Economic Importance.**＊ Coauthor.
   See *Terrell, Edward Everett—1986.*

Wilkinson, Henry Thomas (Dr. and *Professor*)

Hank Wilkinson was born in Princeton, New Jersey, USA, and earned a BSc degree in Crop Protection from Purdue University in 1973 and both MSc and PhD degrees in Plant Pathology from Cornell University in 1977 and 1980, respectively. He was a Postdoctoral Fellow in Plant Pathology at Washington State University from 1980 to 1982. He joined the Faculty at the University of Illinois in 1982, conducting research and teaching in turfgrass pathology. He rose to Associate Professor in 1986 and Full Professor in 1997.

1997†   **Controlling Turfgrass Pests.** Second revised edition. Coauthor.
   See *Fermanian, Thomas Walter—1997.*
2003†   **Controlling Turfgrass Pests.** Third revised edition. Coauthor.
   See *Fermanian, Thomas Walter—2002.*

Wilson, Charles Granville

Charlie Wilson was born in Port Jervis, New York, USA. He served four years in the US Navy on a motor torpedo boat during WWII. In 1946, he enrolled at the University of Maryland while also working part time for the United States Golf Association (USGA) Green Section at the research plots in Beltsville. He earned a BSc degree in Agronomy from the University of Maryland in 1950 and became a full time employee of the USGA Green Section at Beltsville. He then was appointed to establish the first Field Office of the USGA Green Section in Davis, California, in 1952. In 1955, he accepted an Agronomist position with the Sewerage Commission in Milwaukee, Wisconsin, became Director of Agronomy and Marketing in 1960, and retired in 1974. He conducted golf course advisory visitations and lectured frequently at turfgrass conferences throughout North America. He was elected Chair of the Turf Management Division XI of the Crop Science Society of America in 1956. Charlie Wilson was instrumental in gifting the O.J. Noer Collection to the Michigan State University Library. In 1982, he was recipient of the USGA Green Section Award.

1966† **Winter Injury.** C.G. Wilson, J.M. Latham, and R.J. Welch. Bulletin No. 5, Turf Service Bureau, Sewerage Commission, Milwaukee, Wisconsin, USA. 34 pp., no illus., no tables, no index, fold-stapled semisoft covers in green with black lettering, 8.4 by 10.9 inches (213 × 278 mm).

A pioneering synopsis of the key aspects of winter injury of turfgrasses in North America. It was based on the presentations of 16 symposium speakers. It also includes a list of references in the back. This was the first Golf Turf Symposium that became an annual event.

Also printed in the French language as translated by Robert Paris of Quebec, Canada, and titled **Dommages Causes Par l'Hiver (Damage Caused by Winter).**

1966† **Better Fairways: Northern Golf Courses—Theory and Practice.** Third revised edition. C.G. Wilson, J.M. Latham Jr., and R.J. Welch. Bulletin No. 3, Turf Service Bureau, Sewerage Commission, Milwaukee, Wisconsin, USA. 27 pp., 2 illus., no tables, no index, fold-stapled semisoft covers in green with black lettering, 8.4 by 10.9 inches (214 × 278 mm).

An update and revision of a booklet concerning the cultural practices for golf course fairways in the cool climatic regions of North America.

Revised in 1971.

1970† **Better Bent Greens—Fertilization & Management.** Fourth revised edition. C.G. Wilson, J.M. Latham Jr., and R.J. Welch. Bulletin No. 2, Turf Service Bureau, Sewerage Commission, Milwaukee, Wisconsin, USA. 20 pp., 1 illus., 1 table, no index, fold-stapled semisoft covers in green with black lettering, 8.5 by 10.9 inches (217 × 278 mm).

A major updated version of the 1959 edition, especially in terms of bentgrass cultivars, root-zone construction, and control of turfgrass pests.

Figure 12.100. Slanted flat wire screen setup for manually processing compost, sand, and soil to the desired particle size, 1920s (top), and a compost mix being distributed onto turfgrass by manually slinging or casting with a shovel from small piles at a convenient spacing dumped from a wheelbarrow (bottom). [41]

1971† **Better Fairways: Northern Golf Courses—Theory and Practice.** Fourth revised edition. C.G. Wilson, J.M. Latham Jr., and R.J. Welch. Bulletin No. 3, Turf Service Bureau, Sewerage Commission, Milwaukee, Wisconsin, USA. 21 pp., 2 illus., no tables, no index, fold-stapled semisoft covers in green with black and green lettering, 8.4 by 10.9 inches (214 × 278 mm).

   A major revision of a booklet that addresses the cultural practices employed on golf course fairways in the cool climatic regions of North America.

1972† **Permanent Fertilization Record and Handbook.** C.G. Wilson, J.M. Latham Jr., and R.J. Welch. Bulletin No. 4, Turf Service Bureau, Sewerage Commission, Oak Creek, Wisconsin, USA. 43 pp., 2 illus., 6 tables and forms, no index, fold-stapled semisoft covers in green with black and green lettering, 10.9 by 8.4 inches (278 × 213 mm).

   A handbook that provides the format for organization of soil test records, monthly fertilizer balance sheets, and annual fertilizer application summaries. These forms were originally devised by golf course superintendent Robert M. Williams.

Windows, Richard James

Richard Windows was born in Newcastle-upon-Tyne, Tyne & Wear, UK, and earned a BSc with *honours* in Plant Sciences from the University of Sheffield in 1999. He has been a Turfgrass Agronomist with the Sports Turf Research Institute since 2000.

2007† **STRI Disturbance Theory.** Coauthor.

   See *Bechelet, Henry Charles—2007.*

Wise, Louis Neal (*Dr.* and *Professor*)

Louis Wise (1921–1990) was born in Slagle, Louisiana, USA, and earned a BSc in Education Science from Northwestern State University (Louisiana) in 1942. During WWII. he served in the military as a pilot in the US Army Air Corps from 1943 to 1946. He then earned a BSc in General Agriculture from Louisiana State University in 1946, an MSc in Agronomy from Louisiana State University in 1947, and a PhD in Agronomy and Turfgrass from Purdue University in 1950 studying with Dr. G.O. Mott. He then joined the Agronomy Faculty at Mississippi State University (MSU), was named Director of the Mississippi Seed Technology Lab in 1952, and became a Full Professor in 1954. He served as Dean of the MSU College of Agriculture from 1961 to 1966 and as Vice President of MSU Agriculture, Forestry and Veterinary Medicine from 1966 to 1986, when he retired. He was elected President of the Mississippi Section of the American Society of Agronomy in 1959 and President of the Southern Association of Agricultural Scientists in 1974. Dr. Wise was recipient of the Man of the Year Award of the Southern Seedsmen's Association in 1958, the Man of the Year in Service to Mississippi Agriculture Award from Progressive Farmer in 1972, the Man of the Year in Service to Agriculture Award from Progressive Farmer in 1974, a Fellow of the American Society of Agronomy in 1979. He was Knighted by the Brotherhood of the Vines in 1985.

1961† **The Lawn Book.** L.N. Wise. W.R. Thompson, State College, Mississippi, USA. 250 pp., 102 illus., 27 tables, indexed, cloth hardbound in green with gold lettering and a jacket in multicolor with black lettering, 6.0 by 9.0 inches (152 × 228 mm).

   A book on the basics of turfgrass establishment and culture within a how-to approach. It is oriented to the home lawn enthusiasts and warm-season turfgrass conditions in the southern United States.

Witteveen, Gordon Cornelis

Gordon Witteveen (1934–2010) was born in Harlingen, the Netherlands, and immigrated to Canada after completing high school in 1954. He earned a BSc degree from Ontario Agricultural College in Guelph, Ontario, Canada, in 1958. He served as superintendent at the Highland Country Club near London, Ontario, from 1958 to 1960, Northwood Golf Course in Toronto from 1961 to 1972, and the 45-hole Board of Trade Country Club in Woodbridge, Ontario, from 1973 to 1999. He was then owner and operator of Pleasant View Golf Club near Brantford, Ontario, Canada. He was elected President of the Canadian Golf Course Superintendents Association (CGCSA) in 1970, to the Board of Directors of the Golf Course Superintendents Association of America (GCSAA) from 1973 to 1976, and the O.J. Noer Turfgrass Research Foundation Board from 1980 to 2010. He initiated the CGSA **Green Master** publication in 1966. Gordon Witteveen received the GCSAA Leo Feser Award in 1983, the CGCSA Superintendent of Year Award in 1983, and the John B. Steel Distinguished Service Award of the CGCSA in 1998.

1998† **Practical Golf Course Maintenance—The Magic of Greenkeeping.** First edition. Gordon Witteveen and Michael Bavier. Ann Arbor Press, Chelsea, Michigan, USA. 262 pp., 94 illus., 1 table, indexed, glossy hardbound in multicolor with white and black lettering, 7.0 by 10.0 inches (177 × 254 mm).

   A book concerning hands-on, day-to-day operations involved in golf course maintenance. The authors describe personal experiences, both good and bad. They have employed an interesting approach to the topics.

   Revised in 2005.

2003† **Guía Práctica para Manejo de Pastos en Campos de Golf (Practical Golf Course Maintenance).** Gordon Witteveen and Michael Bavier. John Wiley & Sons Inc., Hoboken, New Jersey, USA. 237 pp., 79 illus., 1 table, indexed, glossy semisoft bound in multicolor with white and black lettering, 6.9 by 9.9 inches (175 × 250 mm). Printed entirely in the Spanish language.

A Spanish translation of the 1998 first edition of **Practical Golf Course Management**.

2005† **Practical Golf Course Maintenance—The Magic of Greenkeeping.** Second revised edition. Gordon Witteveen and Michael Bavier. John Wiley & Sons Inc., Hoboken, New Jersey, USA. 254 pp.; 59 illus.; 2 tables; small index; glossy hardbound in multicolor with green, white, orange, and yellow lettering; 7.0 by 9.9 inches (179 × 251 mm).

Additions to the second revised edition include pesticide issues, putting green ball roll, and warm-season turfgrass culture, including winter overseeding and sprigging.

Woolhouse, Alan Roy

Roy Woolhouse was born in Sheffield, England, UK, and earned a BSc degree in Botany from Sheffield University in 1964. He was employed as a Biologist at the Sports Turf Research Institute from 1968 to 1986. He then left the turfgrass field to set up his own computer business.

1989† **Fungal Diseases of Amenity Turf Grasses.** Coauthor.
See *Smith, Jeffrey Drew—1989.*

Y

Yanagi, Hisashi

Hisashi Yanagi was born in Ashikaga, Tochigi Prefecture, Japan, and earned a BSc degree in Plant Pathology from Chiba University in 1956. He served with Toyo Green Company Limited, including as Chairman from 1995 to 1999. He was elected Director of the International Turfgrass Society from 1989 to 1997 and Vice President of the Japan Society of Turfgrass Science from 1994 to 1997. Hisashi Yanagi received the Academy Award of the Japanese Society of Turfgrass Science in 1992.

1991† **Bentgrass: Characteristics and Newest Methods of Golf Green Construction and Maintenance.** Coeditor.
See *Maki, Yoshisuke—1991.*

1994† **Turf Management in Soccer Fields: The Construction and Maintenance.** Coeditor.
See *Nakahara, Hisakazu—1994.*

2002† **New Creeping Bentgrass: Its Characteristics and Management for Putting Green.** Hisashi Yanagi (editor) with nine contributing authors. Soft Science Inc., Tokyo, Japan. 223 pp., 196 illus. with 149 in color, 39 tables, no index, cloth hardbound in dark blue with gold lettering and a jacket in multicolor with dark blue lettering, 7.1 by 10.1 inches (181 × 247 mm). Printed in the Japanese language.

A specialty book concerning the newer high-density, low-growing creeping bentgrasses (*Agrostis stolonifera*) that tolerate very close mowing. It includes history, characteristics, cultural requirements, pest management, and introductory conversion. Lists of selected references are included at the ends of certain chapters.

Yelverton, Fred Hinnant (Dr. and *Professor & Extension Specialist*)

Fred Yelverton was born in Wilson, North Carolina, USA, and earned a BSc degree in 1981, an MSc degree in 1983, and a PhD degree in 1990, all from North Carolina State University. He joined the faculty at North Carolina State University (NCSU) as an Assistant Professor and rose to Full Professor. He is active in research, adult education, and teaching of weed management and plant growth regulators in turfgrasses. He also serves as Co-Director for the NCSU Center for Turfgrass Environmental Research and Education. Dr. Yelverton received the Outstanding Young Weed Scientist Award from the Southern Weed Science Society in 1999, the Certificate of Excellence in Educational Materials from the American Society of Agronomy in 1999, the State Midcareer Award from Epsilon Sigma Phi in 2002, the Outstanding Service Award from the Turfgrass Council of North Carolina in 2003, the Outstanding Extension Service Award from NCSU in 2005, and the Alumni Outstanding Extension and Outreach Award from the NCSU Alumni Association in 2005.

2001† **Color Atlas of Turfgrass Weeds.** First edition. Coauthor.
See *McCarty, Lambert Blanchard—2001.*

2008† **Color Atlas of Turfgrass Weeds.** Second revised edition. Coauthor.
See *McCarty, Lambert Blanchard—2008.*

Yoshida, Masayoshi (Dr. and *Professor* Emeritus)

Masayoshi Yoshida (1917–1990) was born in Oita Prefecture, Japan, and earned a BSc degree in 1948 and a PhD degree in Entomology in 1951, both from Kyoto University. He was Professor of Entomology at Shizuoka University in Shizuoka, Japan, from 1948 to 1981.

1974† **Twelve Months of Turf Maintenance.** Coauthor.
  See *Hosotsuji, Toyoji—1974.*
1979† **Color Atlas: Diseases, Insects and Weeds of Turfgrass.** First edition. Coauthor.
  See *Hosotsuji, Toyoji—1979.*
1989† **Color Atlas: Diseases, Insects and Weeds of Turfgrass.** Second revised edition. Coauthor.
  See *Hosotsuji, Toyoji—1989.*
1998† **Color Atlas: Diseases, Insects and Weeds of Turfgrass.** Third revised edition. Coauthor.
  See *Hosotsuji, Toyoji—1998.*

Yoshida, Shigeharu[3]

1976 **Lawn: Ecology and Production Technique.** Shigeharu Yoshida. Yokendo Limited, Tokyo, Japan. 269 pp., 100 illus. with 5 in color, 59 tables, small index, cloth hardbound in green with gold lettering, 5.8 by 8.2 inches (147 × 208 mm). Printed in the Japanese language.

  A book concerning the construction, establishment, and culture of turfgrasses under the conditions in Japan. There is a list of selected references in the back.

Young, Orville Willard

Orville Young (1906–1997) was born in Brookville, Ohio, USA, and graduated from Brookville High School in 1925. Starting in 1928, he apprenticed with his father, Clarence L. Young, who was Head Greenkeeper at Springfield Country Club, Springfield, Ohio. Orville Young served as Golf Course Superintendent at the Moraine Country Club in Dayton, Ohio, from 1932 to 1957 and then as Superintendent at Brown's Run Country Club, Middletown, Ohio, from 1957 to 1963, when he retired.

n.d. **Better Lawns and Turf.** Orville W. Young. Orville Young Pub., Dayton, Ohio, USA. 62 pp., 22 illus., no tables, small
circa index, textured semisoft bound in greens with dark green lettering, 4.5 by 5.8 inches (114 × 147 mm).
1951
  ✳ A **truly rare**, small book on the establishment and culture of lawns. It is practically oriented to home lawn enthusiasts with emphasis on the cool-humid climatic regions of the United States.

Youngner, Victor Bernarr (Dr. and Professor)

Vic Youngner (1922–1984) was born in Nelson, Minnesota, USA. He served from 1942 to 1945 in the Army Air Corps during WWII, and, subsequently, earned BSc and PhD degrees both in Horticulture from the University of Minnesota in 1948 and 1952, respectively. Dr. Youngner then worked as a plant breeder for several commercial seed companies. He joined the Agronomy Faculty at the University of California, Los Angeles, in 1955 and at the University of California, Riverside, from 1966 to 1984. He was active in teaching and research with grasses and was involved in the breeding of turfgrasses, with emphasis on zoysiagrasses (*Zoysia* spp.) and bermudagrasses (*Cynodon* spp.). He served as editor of the quarterly **California Turfgrass Culture** from 1955 to 1984. He was elected Turf Management Division XI Chair

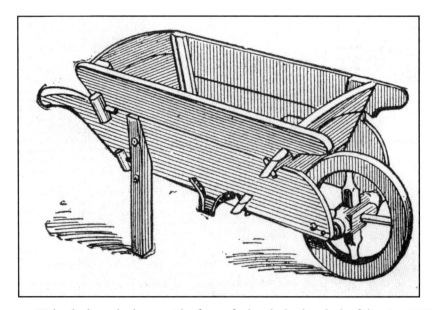

Figure 12.101. Wide-wheel, wooden barrow with a frame of oak and ash, plus a body of elm, circa 1920s. [14]

Figure 12.102. Using a barn broom to sweep earthworms into piles for physical removal, circa 1926. [2]

in 1961. Dr. Youngner received a Fellow of the American Society of Agronomy in 1968 and the United States Golf Association Green Section Award in 1985.

1972† **The Biology and Utilization of Grasses.** V.B. Youngner and C.M. McKell (editors) with 30 contributing authors. Academic Press Inc., New York City, New York, USA. 426 pp., 84 illus., 34 tables, indexed, cloth bound in light brown and green with gold and black lettering and a jacket in orange and white with white and black lettering, 6.0 by 9.0 inches (152 × 228 mm).

This generic book is a compilation of 28 chapters that evolved from a grass biology symposium held in Riverside, California, in 1969. Emphasis is placed on the evolution, breeding, growth, development, physiology, nutrition, and community dynamics of perennial grasses. Selected references are included at the back of each chapter.

Z

Zanin, Guiseppe (Professor)

Guiseppe Zanin was born in Bassano del Grappa, Vicenza, Italy, and earned a BSc degree in Agricultural Sciences from the University of Padua in 1975. He then accepted a position in weed science with the National Research Council (NRC) and, since, 1988 has been a Full Professor on the Agriculture Faculty at Padua University involved in teaching and research in weed science.

2002† **La difesa dei tappeti erbosi: Malattie fungine, menici animali e infestanti (Pest control on turfgrasses: Disease, animal enemies and weeds).** Coauthor.

See *Gullino, Maria Lodovica—2000.*

B. CHRONOLOGICAL LISTING OF TURFGRASS BOOKS (1892 TO 2010)

The turfgrass books listed in this section are presented chronologically by year. The criteria for books included in this list are (1) those devoted, for the most part, to the subject of turfgrass science, establishment, and culture, (2) those 20 or more pages in length, both cloth and soft bound, and (3) a representation portion of the reviewed editions.

A total of 806 turfgrass books and revisions have been published between 1892 and 2010. A summary of the number of turfgrass books published at five-year intervals for six areas of the world is presented in table 12.2.

Prior to 1945 a total of 169 books had been authored throughout the world. Of this number, by far the majority had been written and published in either the United Kingdom or the United States, with five books published in Europe, three in Japan, and one each in Malaysia and in South Africa. Since 1945, a

TABLE 12.2. SUMMARY OF THE NUMBER OF TURFGRASS BOOKS OF 20 OR MORE PAGES AND ANY REVISIONS PUBLISHED SINCE 1892, AT FIVE-YEAR INTERVALS, PLUS SUMMARIES OF THE NUMBER OF BOOKS PUBLISHED IN SIX MAJOR REGIONS OF THE WORLD.

Time span	Total number of books	Number of books published for six areas of the world					
		Africa and Near East	Japan and Asia	Australia and New Zealand	Europe	North America	United Kingdom
1890–94	2	—	—	—	—	1	1
1895–99	7	—	—	—	—	4	3
1900–1904	3	—	—	—	—	—	3
1905–09	18	—	—	—	2	3	13
1910–14	19	—	—	—	—	7	12
1915–19	8	—	—	—	—	5	3
1920–24	24	—	—	—	—	10	14
1925–29	17	—	—	—	1	5	11
1930–34	31	1	2	—	—	14	14
1935–39	32	—	1	1	—	13	17
1940–44	8	—	1	—	2	3	2
1945–49	24	1	—	1	1	14	7
1950–54	27	—	—	2	1	9	15
1955–59	30	—	3	1	1	15	10
1960–64	3	1	—	3	4	12	10
1965–69	32	—	3	1	1	17	10
1970–74	44	—	3	4	6	27	4
1975–79	58	—	8	3	9	29	9
1980–84	51	—	5	3	8	27	8
1985–89	46	—	3	4	5	26	8
1990–94	100	—	17	15	12	40	16
1995–99	66	—	8	6	4	40	8
2000–2004	71	1	9	5	14	38	4
2005–9	58	—	8	5	8	30	7
Total	806	4	71	54	79	389	209

major portion of the turfgrass books have been published in the United States, followed by the United Kingdom and Europe, Japan and Asia, and Australia and New Zealand.

FORMAT

The citation style used in this chapter is as follows:
- Year published.
- Title of book.
- Edition number, if more than one.
- Author last name(s) or sponsor if author is anonymous.
- Number of pages.
- Country of publication.
- List of reprinting(s) and revisions with dates.
* Indicates the book contains a chapter or section on turfgrass but addresses a broader array of subjects.

1890 through 1899

1892   **The Formation of Lawns, Tennis & Cricket Grounds from Seed.** First ed. Sutton & Sons. 25 pp. UK. There are 17 revised editions in this series, with the 1892, 1896, 1899, 1902, 1904, 1906, 1908, 1909, 1912, 1914, 1920, 1931, 1937, and 1962 editions listed herein.

1892   **Talk about a Grass Garden.** J.B. Olcott. 26 pp. USA.

1895   **The Improvement of Cricket Grounds on Economical Principles.** J.A. Gibbs. 42 pp. UK.

1896   **Grasses of North America.** W. Beal. 715 pp. USA.

1896   **Lawns, Tennis and Cricket Grounds from Seed.** Second rev, ed. Sutton & Sons. 32 pp. UK.

1896   **Useful and Ornamental Grasses.** F. Lamson-Scribner. 119 pp. USA.

1898   **Economic Grasses.** F. Lamson-Scribner. 85 pp. USA.

1899   **Lawns: Garden Lawns, Tennis Lawns, Putting Greens, Croquet Grounds, Cricket Grounds.** Third rev. ed. Sutton & Sons. 34 pp. UK.

1899   **The Seeding and Preservation of Golf Links.** First ed. J.M. Thorburn and Co. 24 pp. USA.

1900 through 1904

1900   **Toogood's Treatise on Lawns, Cricket Grounds; &c.** E.K. Toogood. 20 pp. UK.

1902   **Lawns: Garden Lawns, Tennis Lawns, Bowling Greens, Croquet Grounds, Putting Greens, Cricket Grounds.** Fourth rev. ed. Sutton & Sons. 36 pp. UK.

1904   **Garden Lawns, Tennis Lawns, Bowling Greens, Croquet Grounds, Putting Greens, Cricket Grounds.** Fifth rev. ed. Sutton & Sons. 65 pp. UK.

1905 through 1909

1906   **Golf.** R. Beale. 32 pp. UK.

1906   **Golf Greens and Green-Keeping.** H.G. Hutchinson (ed). 219 pp. UK.

1906   **The Lawn.** L.C. Corbett. 20 pp. USA.

1906   **Lawns and How to Make Them.** First ed. Leonard Barron. 174 pp. USA. Revised in 1923.

1906   **Lawns: Garden Lawns, Tennis Lawns, Bowling Greens, Cricket Grounds.** Sixth rev. ed. Sutton & Sons. 49 pp. UK.

1906   **The Laying Out and Upkeep of Golf Courses and Putting Greens.** M.H.F. Sutton. 47 pp. UK.

1908   **Les Gazons (Turfs).** J.C.N. Forestier. 136 pp. France.

1908   **Lawns: Garden Lawns, Tennis Lawns, Bowling Greens, Croquet Grounds.** Rev. ed. Sutton & Sons. 54 pp. UK.

1908   **The Life of Tom Morris—With Glimpses of St Andrews and Its Golfing Celebrities.** W.W. Tulloch. 346 pp. UK.

1908   **Links and Lawn Grasses.** Stewart and Co. 26 pp. UK.

1908   **Manures and Composts for Golf Greens, &c.** R. Beale. 20 pp. UK.

1908   **The Practical Greenkeeper.** First ed. R. Beale. 69 pp. UK. There are 31 revised editions in this series, with the 1908, 1910, 1911, 1913, 1914, 1920, 1921, 1922, and 1924 through 1939 listed herein plus three American editions in 1914, 1916, & 1921.

1908   **The Seeding and Preservation of Golf Links.** Second rev. ed. J.M. Thorburn & Co. 24 pp. USA.

1909   **Descriptive List of Links and Lawn Grasses.** Stewart and Co. 50 pp. UK.

1909   **Lawns.** W.J. Stevens. 20 pp. UK.

1909   **Lawns: Garden Lawns, Tennis Lawns, Bowling Greens, Croquet Grounds, Putting Greens, Cricket Grounds.** Rev. ed. Sutton & Sons. 75 pp. UK.

19xx   **Grasses for Lawns, Sports Fields and Other Amenity Areas.** Miln Marsters Group. 32 pp. UK.

19xx   **Rasenpflege (Lawncare).** Ruhr-Stickstoff Aktiengesellschaft. 33 pp. Germany.

1910 through 1914

1910   **Lawns and Greens: Their Formation and Management.** First ed. T.W. Sanders. 137 pp. UK. Revised in 1911 and 1920.

1910   **The Practical Greenkeeper.** Third rev. ed. R. Beale. 103 pp. UK. Revised in 1911.

1911   **Lawns and Greens: Their Formation and Management.** Second rev. ed. T.W. Sanders. 142 pp. UK. Revised in 1920.

1911   **Lawn Soils.** O. Schreiner and J.J. Skinner. 55 pp. USA.

1911   **Links and Lawn Grasses.** Stewart and Co. 50 pp. UK.

1911   **The Practical Greenkeeper.** Fourth rev. ed. R. Beale. 104 pp. UK. Revised in 1913.

1912   **The Book Of The Links—A Symposium on Golf.** Sutton & Sons. 212 pp. UK.

1912   **Lawns: Garden Lawns, Tennis Lawns, Croquet Grounds, Bowling Greens, Putting Greens, Cricket Grounds.** Eleventh rev. ed. Sutton & Sons. 75 pp. UK. Revised in 1914.

1912   **Lawn Soils and Lawns.** O. Schreiner, J.J. Skinner, L.C. Corbett, and F.L. Mulford. 48 pp. USA.

1912   **Making A Lawn.** First ed. L.J. Doogue. 51 pp. USA.

1913   **Formation and Treatment of Links and Lawn Turf.** Stewart and Co. 70 pp. UK.

1913   **Practical Golf Greenkeeping.** W.K. Gault. 101 pp. UK.

1913   **The Practical Greenkeeper.** Fifth rev. ed. R. Beale. 103 pp. UK. Revised in 1914.

1914   **Carters Practical Greenkeeper.** Second rev. American ed. R. Beale. 100 pp. USA. Revised in 1916.

1914   **Lawns: Garden Lawns, Tennis Lawns, Croquet Grounds, Bowling Greens, Putting Greens, Cricket Grounds.** Twelfth rev. ed. Sutton & Sons. 75 pp. UK. Revised in 1920.

1914   **Lawns, Golf Courses, Polo Fields, Turf Courts, and How to Treat Them.** S.A. Cunningham and G.D. Leavens. 40 pp. USA.

1914   **The Practical Greenkeeper.** Sixth rev. ed. R. Beale. 103 pp. UK.

1914   **The Proper Care of Lawns.** L.J. Doogue. 22 pp. USA.

1914   **A Text-Book of Grasses.** A.S. Hitchcock. 276 pp. USA.

1915 through 1919

1915   **Practical Illustrations of Turf Production.** W.K. Tucker. 36 pp. USA.

1916   **Carters Practical Greenkeeper.** Third rev. American ed. R. Beale. 64 pp. USA. Revised in 1921.

1916   **Links and Lawn Turf—Formation and Upkeep.** Stewart and Co. 42 pp. UK.

1916   **The Maintenance of Lawns—Applicable also to Golf Courses.** Odorless Plant Food Co. 28 pp. USA.

1917   **Turf for Golf Courses.** C.V. Piper and R.A. Oakley. 267 pp. USA.

1918   **Care of the Greens.** P.W. Lees. 91 pp. UK.

1918   **Norfolk Lawns and How to Grow Them.** Commission for Beautifying Norfolk. 30 pp. USA.

191x   **The Laying Out and Upkeep of Golf Courses and Putting Greens.** M.H.F. Sutton. 40 pp. UK.

1920 through 1924

1920   **The Genera of Grasses of the United States with Special Reference to the Economic Species.** First ed. A.S. Hitchcock. 307 pp. USA. Revised in 1936.

1920   **Golf Architecture—Economy In Course Construction And Green-Keeping.** A. Mackenzie. 135 pp. UK.

1920   **Lawns and Greens—Their Formation and Management.** Third rev. ed. T.W. Sanders. 113 pp. UK.

1920   **Lawns and Sports Grounds.** R. Beale. 40 pp. USA.

1920 **Lawns: Garden Lawns, Tennis Lawns, Croquet Grounds, Putting Greens, Cricket Grounds, Bowling Greens.** Thirteenth rev. ed. Sutton & Sons. 80 pp. UK. Revised in 1931.

1920 **The Practical Greenkeeper.** Seventh rev. ed. R. Beale. 48 pp. USA.

1920 **Some Essays on Golf-Course Architecture.** H.S. Colt and C.H. Alison. 69 pp. UK.

1920 **Weedless Lawns and Golf Courses.** O.M. Scott & Sons Co. 31 pp. USA.

1921 **Carters Practical Greenkeeper.** Fourth rev. ed. R. Beale. 48 pp. USA.

1921 **The Practical Greenkeeper.** Fourth rev. American ed. R.E.C. Beale. 48 pp. UK.

1921 **Weedless Lawns and Golf Courses.** O.M. Scott & Sons Co. 31 pp. USA.

1922 **First Book of Grasses: The Structure of Grasses Explained for Beginners.** First ed. A. Chase. 129 pp. USA. Revised in 1937 and 1977.

1922 **The Practical Greenkeeper.** Tenth rev. ed. R. Beale. 50 pp. UK.

1922 **The Seeding and Care of Golf Courses.** First ed. O.M. Scott & Sons Co. 54 pp. USA. Revised in 1923.

1923 **Lawn Making.** Second rev. ed. Leonard Barron. 176 pp. USA.

1923 **Lawns and Sports Grounds.** R. Beale. 40 pp. UK.

1923 **Lawns, Links, and Sportsfields.** I.J.C. MacDonald. 78 pp. UK.

1923 **The Seeding and Care of Golf Courses.** Second rev. ed. O.M. Scott & Sons Co. 55 pp. USA.

1924 **Grass—A new and thoroughly practical book on Grass for ornamental Lawns and all purposes of Sports and Games.** A.J. Macself. 204 pp. UK.

1924 **Hints on Lawns.** A. Key. 21 pp. UK.

1924 **Lawns and Sports Grounds.** Spring ed. R. Beale. 40 pp., UK.

1924 **Lawns for Sports—Their Construction and Upkeep.** R. Beale. 276 pp. UK.

1924 **The Practical Greenkeeper.** Fourteenth rev. ed. R. Beale. 52 pp. UK.

1924 **Supplement Issued With Lawns for Sports—Materials, Tools And Fittings.** R. Beale. 67 pp. UK.

1925 through 1929

1925 **The Practical Greenkeeper.** Sixteenth rev. ed. R. Beale. 52 pp. UK.

1926 **The Practical Greenkeeper.** Seventeenth rev. ed. R. Beale. 52 pp. UK.

1927 **Lawns and Weeds.** R. Beale. 32 pp. UK.

1927 **The Practical Greenkeeper.** Eighteenth rev. ed. R. Beale. 52 pp. UK.

1927N **The Practical Greenkeeper.** Rev. new ed. R. Beale. 56 pp. UK.

1928 **The ABC of Turf Culture.** O.J. Noer. 95 pp. USA.

1928 **Bent Lawns.** First ed. O.M. Scott & Sons Co. 23 pp. USA. Revised in 1931.

1928 **Lawns and Weeds.** R. Beale. 32 pp. UK.

1928 **Le Parcours de Golf—Sa Construction—son Entretien—les Gazons en General (The Golf Course—Construction, Maintenance, Turfs in General).** M.L. Bandeville. 147 pp. France.

1928N **The Practical Greenkeeper.** Rev. new ed. R. Beale. 28 pp. UK.

1928 **Soil Acidity: The Vital Importance of Top-Dressing And Other Notes.** N. Hackett. 37 pp. UK.

1929 **Fertilizers for Gardens and Lawns: How To Use Them To Advantage.** H.H. Thomas. 48 pp. UK.

1929 **Lawns.** F.F. Rockwell. 87 pp. USA.

1929N **The Practical Greenkeeper.** R. Beale. 56 pp. UK.

1929 **Roadside Development.** J.M. Bennett. 265 pp. USA.

192x **Lawns, Golf and Sport Turf.** W.J. Collins. 29 pp. USA.

192x **Lawns—Sports Grounds—Meadow Lands.** First ed. I.J.C. MacDonald. 20 pp. UK.

1930 through 1934

1930 **The Lawn—The Culture of Turf in Park, Golfing and Home Areas.** First ed. L.S. Dickinson. 128 pp. USA.

1930 **Lawns and Sports Greens.** First ed. A.J. Macself. 144 pp. UK.

1930N **The Practical Greenkeeper.** R. Beale. 56 pp. UK.

1931 **Bent Lawns.** Second rev. ed. O.M. Scott & Sons Co. 20 pp. USA.

1931 **The Book of The Lawn—A Complete Guide to the Making and Maintenance of Lawns and Greens for all purposes.** First ed. R. Beale. 151 pp. UK. Revised in 1952.

1931 **Golf Course Commonsense.** G.A. Farley. 256 pp. USA.

1931 **Guide To The Experiments of the St Ives Research Station.** Board of Greenkeeping Research. 46 pp. UK.

1931   **Lawns and Weeds.** R. Beale. 32 pp. UK.
1931   **Lawns: Garden Lawns, Tennis Courts, Croquet Courts, etc.** Fourteenth rev. ed. Sutton & Sons. 60 pp. UK. Revised in 1937.
1931   **The Lawn—The Culture of Turf in Park, Golfing, and Home Areas.** Second rev. ed. L.S. Dickinson. 128 pp. USA.
1931N   **The Practical Greenkeeper.** Rev. new ed. R. Beale. 56 pp. UK.
1931   **The Putting Green—Its Planting and Care.** O.M. Scott & Sons Co. 39 pp. USA.
1932   **Campus and Athletic Fields.** O.M. Scott & Sons Co. 66 pp. USA.
1932   **Cemetery Lawns—Making and Maintenance.** O.M. Scott & Sons Co. 66 pp. USA.
1932   **Grass—Golf Clubs Can Do It So Can You.** W. Goit and R. Shaffer. 30 pp. USA.
1932   **Greenkeeping for Golf Courses.** R.B. Dawson and T.W. Evans. 60 pp. UK.
1932   **Greenkeeping in South Africa.** C.M. Murray. 104 pp. South Africa.
1932   **The Improvement of Lawns, Golf Courses and Tennis Courts.** First ed. Imperial Chemical Industries Ltd. 20 pp. UK. Revised in 1933.
1932   **Lawn Making and Maintenance.** O.M. Scott & Sons Co. 59 pp. USA.
1932   **Lawns and Sports Greens.** First ed. A.J. Macself. 144 pp. UK. Revised in 1947.
1932N   **The Practical Greenkeeper.** Rev. new ed. R. Beale. 56 pp. UK.
1932   **Turf Diseases and Their Control.** J.L. Monteith Jr. and K.A.S. Dahl. 102 pp. USA.
1932   **Turf for Parks.** O.M. Scott & Sons Co. 67 pp. USA.
1933   **Golf Courses—Design, Construction and Upkeep.** First ed. M.A.F. Sutton. 152 pp. UK. Revised in 1950.
1933   **Improvement of Lawns, Golf Greens and Fairways.** Second rev. ed. Imperial Chemical Industries Ltd. 23 pp. UK.
1933   **Lawns.** O.M. Scott & Sons Co. 24 pp. USA.
1933   **Soil Management For Greenkeepers.** M.H. Cubbon and M.J. Markuson. 152 pp. USA.
1934   **Lawn Care.** O.M. Scott & Sons Co. 56 pp. USA.
1934N   **The Practical Greenkeeper.** R. Beale. 56 pp. UK.
1934   **Turfgrass, Lawn and Lawn Garden.** First ed. K. Uehara. 267 pp. Japan.
1934.   **Turf Production and Maintenance.** E.J. McNaughton. 44 pp. Malaysia.

1935 through 1939

1935   **Lawns & Weeds.** R. Beale. 32 pp. UK.
1935   **Manual of the Grasses of the United States.** First ed. A.S. Hitchcock. 1040 pp. USA. Revised in 1950.
1935N   **The Practical Greenkeeper.** Rev. new ed. R. Beale. 56 pp. UK.
1936   **The Genera of Grasses of the United States with Special Reference to the Economic Species.** Second rev. ed. A.S. Hitchcock as rev. by A. Chase. 302 pp. USA.
1936N   **The Practical Greenkeeper.** Rev. new ed. R. Beale. UK.
1936   **Roadsides—The Front Yard of the Nation.** J.M. Bennett. 250 pp. USA.
1937   **The ABC of Turf Culture.** O.J. Noer. 28 pp. USA.
1937   **The Establishment and Care of Fine Turf for Lawns and Sports Grounds.** First ed. D. Clouston. 121 pp. UK. Revised in 1939.
1937   **Evergreen Turfgrass.** T. Souma. 91 pp. Japan.
1937   **First Book of Grasses: The Structure of Grasses Explained for Beginners.** Second rev. ed. A. Chase. 140 pp. USA. Revised in 1959.
1937   **The Lawn—How to Make and Maintain It.** W.P. Brett. 144 pp. UK.
1937   **Lawns: Garden Lawns, Tennis Courts, Croquet Grounds, etc.** Fifteenth rev. ed. Sutton & Sons. 60 pp. UK. Revised in 1948.
1937   **Municipal Parks—Layout, Management and Administration.** W.W. Pettigrew. 279 pp. UK.
1937N   **The Practical Greenkeeper.** Rev. new ed. R. Beale. 56 pp. UK.
1938   **Good Lawns—Planting and Maintaining.** First ed. O.M. Scott & Sons Co. 32 pp. USA. Revised in 1939.
1938   **Lawn Care.** O.M. Scott & Sons Co. 54 pp. USA.
1938   **Lawn Care.** O.M. Scott & Sons Co. 48 pp. USA.
1938N   **The Practical Greenkeeper.** Rev. new ed. R. Beale. 56 pp. UK.
1938   **The Principles of Golf Architecture, etc.** S.V. Hotchkin. 26 pp. UK.
1939   **The Establishment and Care of Fine Turf for Lawns and Sports Grounds.** Second rev. ed. D. Clouston. 126 pp. UK.
1939   **Good Lawns—Planting and Maintaining.** Second rev. ed. O.M. Scott & Sons Co. 37 pp. USA. Revised in 1947.
1939   **How To Make and Keep A Lawn.** Plant Protection Ltd. 21 pp. UK.
1939   **The Lawn—How to Make It, and How to Maintain It.** C.W. Parker. 118 pp. USA.

1939N **The Practical Greenkeeper.** Rev. new ed. R. Beale. 56 pp. UK.

1939 **Practical Lawn Care.** M.G. Merritt. 32 pp. USA.

1939 **Practical Lawn Craft.** First ed. R.B. Dawson. 300 pp. UK. Revised in 1945.

193x **About Golf Courses: Their Construction and Up-Keep.** Stanley Thompson & Co. 32 pp. Canada.

193x **The Establishment and Care of Fine Turf for Lawns and Sports Grounds.** John George Cunningham Ltd. 126 pp. UK.

193x. **Garden Lawns and Greens: Their Maintenance, Improvement and Renovation.** A.E. Palmer. 44 pp. UK.

193x **The Lawn.** W.P. Brett. 144 pp. UK.

193x **The Lawn Beautiful: Its Care and Management.** A.W.S. Moodie. 32 pp. Australia.

193x **The Seeding and Care of Lawns.** O.M. Scott & Sons Co. 24 pp. USA.

1940 through 1944

1940 **Better Lawns For Homes and Parks.** First ed. H.B. Sprague. 205 pp. USA. Revised in 1945.

1940 **Recreation Grounds.** Army Sport Control Board. 44 pp. UK.

1941 **Diseases of British Grasses and Herbage Legumes.** First ed. K. Sampson and J.H. Western. 85 pp. UK. Revised in 1954.

1942 **Engazonnement des Aérodromes et Terrains de Sports (Turf Establishment of Airfields and Sports Grounds).** H. Van de Walle. 203 pp. France.

1942 **Les Pelouses et Terrains de Sport—Construction and Entretien: Les Gazons en Général (Lawns and Sports Grounds—Construction and Maintenance: Turfs in General).** M.L. Bandeville. First ed. 159 pp. France. Revised in 1952.

1943 **Japanese Lawn Grasses.** First ed. T. Niwa. 150 pp. Japan. Revised in 1953.

1944 **Scotts Lawn Care.** O.M. Scott & Sons Co. 48 pp. USA.

1944 **Lawn Care.** O.M. Scott & Sons Co. 228 pp. USA.

1945 through 1949

1945 **Better Lawns.** H.B. Sprague. 205 pp. USA.

1945 **Practical Lawn Craft.** Second rev. ed. R.B. Dawson. 306 pp. UK. Revised in 1947.

1946 **Better Bent Greens—Fertilization and Management.** First ed. O.J. Noer. 25 pp. USA. Revised in 1947 and 1959.

1946 **Better Fairway Turf—Theory & Practice.** First ed. O.J. Noer. 34 pp. USA. Revised in 1950 and 1955.

1946 **Terrains de Sports—Stades, gazons, drainage, irrigation (Sports Grounds—Stadiums, turfs, drainage, irrigation).** Charles Bouhana. 517 pp. France.

1947 **Golf Range Operator's Handbook.** National Golf Foundation. 34 pp. USA.

1947 **Good Lawns—Planting and Maintaining.** Third rev. ed. O.M. Scott & Sons Co. 44 pp. USA.

1947 **"The Green Manual"—On the Construction of Bowling Greens and Their Maintenance.** W.D. Powell (ed). 122 pp. South Africa.

1947 **Good Lawns—Planting and Maintaining.** Third rev. ed. O.M. Scott & Sons Co. 42 pp. USA.

1947 **Lawns and Sports Greens.** Second rev. ed. A.J. Macself. 103 pp. UK.

1947 **Planting Turf.** Army, United States Corps of Engineers. 25 pp. USA.

1947 **Practical Lawn Craft.** Rev. ed. R.B. Dawson. 306 pp. UK. Revised in 1949, 1954, 1959, and 1968.

1947 **The Role of Lime in Turf Management.** First ed. O.J. Noer. 23 pp. USA. Revised in 1955.

1947 **A Sensible Fertilizer Program for Greens and Turf Nurseries.** Second rev. ed. O.J. Noer. 29 pp. USA.

1947 **Turf Weed Control With 2,4-D.** F.F.G. Davis. 59 pp. USA.

1948 **Lawns and Sports Grounds.** Sixteenth rev. ed. Sutton & Sons. 76 pp. UK. Revised in 1962.

1948 **Specification of Playing Facilities.** First ed. P.W. Smith. 60 pp. UK. Revised in 1965 by R.B. Gooch.

1948 **Turf.** I.G. Lewis. 141 pp. UK.

1949 **Airport Turfing.** Civil Aeronautics Administration. 40 pp. USA.

1949 **Construction, Renovation and Care of the Bowling Green.** E.B. Levy. 84 pp. New Zealand.

1949 **Golf Facilities: Organization, Construction, Management, Maintenance.** H. Graffis. 80 pp. USA.

1949 **Grounds Construction and Maintenance.** US Department of the Air Force. 77 pp. USA.

1949 **Practical Lawn Craft.** Third rev. ed. R.B. Dawson. 315 pp. UK. Revised in 1954, 1959, and 1968.

194x **The Establishment and Maintenance of Turf on Putting and Bowling Greens.** J.H. Boyce. 21 pp. USA.

1950 through 1954

1950   **Better Fairways—Theory and Practice.** Second rev. ed. O.J. Noer. 45 pp. UK. Revised in 1955, 1966, and 1971.

1950   **The Construction and Care of Lawns.** J.H. Boyce. 24 pp. Canada.

1950   **Construction, Renovation and Care of the Golf Course.** E.B. Levy, W.A. Kiely, and W.M. Horton. 101 pp. New Zealand.

1950   **Golf Courses—Design, Construction and Upkeep.** Second rev. ed. M.A.F. Sutton. 192 pp. UK.

1950   **Lawn Care.** O.M. Scott & Sons Co. 72 pp. USA.

1950   **Lawn Care.** Region ed. O.M. Scott & Sons Co. 140 pp. USA.

1950   **Lawns and Playing Fields.** F.J. Reed. 212 pp. UK.

1950   **Lawn Problems of the Southwest.** H.J. Dittmer. 76 pp. USA.

1950   **Manual of the Grasses of the United States.** Second rev. ed. A.S. Hitchcock and A. Chase. 1051 pp. USA.

1950   **The Planning, Construction, and Maintenance of Playing Fields.** P.W. Smith. 224 pp. UK.

1950   **The Science of Turf Cultivation.** R.P. Faulkner. 64 pp. UK.

1950   **Turf Management.** First ed. H.B. Musser. 354 pp. USA. Revised in 1962.

1951   **Better Lawns and Turf.** O.W. Young. 62 pp. USA.

1951   **Fungus Diseases of Turf Grasses.** F.L. Howard, J.B. Rowell, and H.L. Keil. 56 pp. USA.

1951   **The Soccer Club Groundsman.** First ed. Football Association. 108 pp. UK. Revised in 1972 by R. Hawthorn.

1952   **The Book Of The Lawn—A Complete Guide to the Making and Maintenance of Lawns and Greens for all purposes.** Second rev. ed. R. Beale. 151 pp. UK.

1952   **Garden Lawns and Playing Grounds.** E.A. Madden. 93 pp. New Zealand.

1952   **Pelouses et Terrains de Sports (Lawns and Sports Grounds).** M.L. Bandeville. Second rev. ed. 174 pp. France.

1952   **Practical Groundsmanship.** L.W. White and W.H. Bowles. 258 pp. UK.

1952   **Turf Weeds: Identification & Control.** First edition. Sutton & Sons. 20 pp. UK.

1953   **Public Grounds Maintenance Handbook.** First ed. H.S. Conover. 495 pp. USA. Revised in 1958 and 1977.

1954   **Diseases of British Grasses and Herbage Legumes.** Second rev. ed. K. Sampson and J.H. Western. 118 pp. UK.

1954   **Grasses.** First ed. C.E. Hubbard. 440 pp. UK. Revised in 1968 and in 1984 by J.C.E. Hubbard.

1954   **Handbook of Organic Fertilisers—Their Properties and Uses.** Fertiliser Journal. 84 pp. UK.

1954   **Lawn Grasses.** First ed. F.J. Reed. 28 pp. UK.

1954   **Practical Lawn Craft and Management of Sports Turf.** Fourth rev. ed. R.B. Dawson. 320 pp. UK. Revised in 1959.

1954   **Treatment of Golf Courses and Sports Grounds.** Carters Tested Seeds Ltd. 48 pp. UK.

1955 through 1959

1955   **Better Fairways—Theory and Practice.** Third rev. ed. O.J. Noer. 48 pp. USA.

1955   **Cricket Pitches—Preparation and Maintenance.** Middlesex County Cricket Club. 21 pp. UK.

1955   **Der Rasen-Sportplatz (The Grass Sports Playing Field).** Second rev. ed. D. Harradine and R.F. Handloser. 68 pp. Switzerland.

1955   **Improving Athletic Field Turfgrass.** First ed. F.V. Grau. 27 pp. USA.

1955   **Improving Athletic Field Turfgrass.** First ed. T.C. Mascaro. 25 pp. USA. Revised in 1959.

1955   **New School Playing Fields.** First ed. Ministry of Education, Great Britain. 87 pp. UK.

1955   **The Role of Lime in Turf Management.** Second rev. ed. O.J. Noer. 25 pp. USA. Revised in 1958 and 1970.

1956   **Cricket Pitches—Preparation and Maintenance.** Middlesex County Cricket Club. 25 pp. UK.

1956   **Golf Operators Handbook.** First ed. B. Chlevin (ed). 104 pp. USA. Revised in 1961.

1956   **Turf Talks.** O.M. Scott & Sons Co. 92 pp. USA.

1957   **Cricket Ground Maintenance.** First ed. H.C. Lock. 33 pp. UK.

1957   **Practical Lawn Craft and Management of Sports Turf.** Fourth rev. ed. R.B. Dawson. 221 pp. Russia.

1957   **Sports Buildings and Playing Fields.** R. Sudell and D.T. Waters. 240 pp. UK.

1957   **The Underground Organs of Herbage Grasses.** A. Troughton. 163 pp. UK.

1958   **Fall Renovation of Greens and Fairways.** First ed. T.C. Mascaro. 62 pp. USA. Revised in 1960.

1958   **Grounds Maintenance Handbook.** Second rev. ed. H.S. Conover. 501 pp. USA. Revised in 1977.

1958   **Japanese Lawn Grasses.** Second rev. ed. T. Niwa. 174 pp. Japan.

1958   **Repairs and Utilities—Grounds Maintenance and Land Management.** US Army. 126 pp. USA.

1958   **The Role of Lime in Turf Management.** Third rev. ed. O.J. Noer. 21 pp. USA.

1959   **Better Bent Greens—Fertilization and Management.** Third rev. ed. O.J. Noer. 33 pp. USA. Revised in 1970.

1959   **First Book of Grasses: The Structure of Grasses Explained for Beginners.** Third rev. ed. A. Chase. 149 pp. USA.

1959   **Fungal Diseases of Turf Grasses.** First ed. J.D. Smith. 90 pp. UK. Revised in 1965.

1959 **Grass Varieties in the United States.** First ed. A.A. Hanson. 72 pp. USA. Revised in 1965, 1972, and 1994.

1959 **Improving Athletic Field Turfgrass.** Second rev. ed. T.C. Mascaro. 24 pp. USA.

1959 **Lawn and Lawn Garden.** K. Uehara. 244 pp. Japan.

1959 **Practical Lawn Craft and Management of Sports Turf.** Fifth rev. ed. R.B. Dawson. 320 pp. UK. Revised in 1968.

1959 **Selection and Layout of Land for Playing Fields and Playgrounds.** Second rev. ed. R.B. Gooch. 92 pp. UK.

1959 **A Survey in Turf Transplanting—How Thick Should Sod Be Cut?** B.O. Warren. 21 pp. USA.

1959 **Turfgrass and Turf.** Second rev. ed. K. Uehara. 244 pp. Japan.

195x **Glimpses of Groundsmanship.** P.M. Phillips and W. Hardiman. 34 pp. UK.

195x **Lawns and Playing Fields.** F.R. Drake, C.J.R. Johnston, L.C. Jones, R.J. Newman, and W.I. Walbran. 32 pp. Australia.

1960 through 1964

1960 **Équipement Sportif: Éstablissement et Entretien des Sols (Sports Facilities: Establishment and Maintenance of Soils).** Ministère de l'Education Nationale. 90 pp. France.

1960 **Fall Renovation of Greens and Fairways.** Second rev. ed. T.C. Mascaro. 48 pp. USA.

1960 **Lawns—For Garden and Playing Field.** First ed. R.B. Dawson. 176 pp. UK. Revised in 1963.

1960 **Municipal Handbook for Golf Course Committees.** First ed. T.C. Mascaro. 27 pp. USA. Revised in 1962.

1960 **Rasen: Bedeutung, Anlage, Pflege (Turf: Importance, Installation, Care).** K.D. Gandert. 156 pp. East Germany.

1961 **Der Rasen-Sportplatz (The Grass Sports Playing Field).** Third rev. ed. D. Harradine with translation to German by R.F. Handloser. 74 pp. Switzerland.

1961 **First in Lawns: O.M. Scott & Sons.** C.B. Mills. 24 pp. USA.

1961 **Golf Operators Handbook.** Second rev. ed. B. Chlevin. 104 pp. USA.

1961 **The Lawn Book.** First ed. R.W. Schery. 207 pp. USA.

1961 **The Lawn Book.** L.N. Wise. 250 pp. USA.

1961 **Planting Turf—Engineering and Design Manual.** Corps of Engineers. 45 pp. USA.

1961 **Turf Culture.** First ed. New Zealand Institute for Turf Culture. 179 pp. New Zealand.

1961 **The Use and Control of Vegetation on Roads and Airfields Overseas.** K.E. Clare. 48 pp. USA.

1962 **Diseases of Turfgrasses.** First ed. H.B. Couch. 289 pp. USA. Revised in 1973.

1962 **Lawns and Sports Grounds.** Seventeenth rev. ed. M.A.F. Sutton. 248 pp. UK.

1962 **Lawns, Greens and Playing Fields—Their Making and Maintenance.** J.L. Rees. 290 pp. Australia.

1962 **Rasen, Gras und Grünflächen (Turf, Grasses, and Green Spaces).** First ed. C. Eisele. 135 pp. West Germany. Revised in 1973.

1962 **Sports Field Drainage—An Outline of Field Drainage with Particular Reference to the Drainage of Sports Fields.** A.L. Turner. 63 pp. UK.

1962 **Turf Culture.** I.R. Greenfield. 364 pp. UK.

1962 **Turf Management.** Second rev. ed. H.B. Musser. 356 pp. USA.

1963 **The Lawn and Its Care.** Z. Bahir and J. Shoor. 64 pp. Israel.

1963 **Lawn and Turf Culture for Gardens and Grounds.** Second rev. ed. H.C. Lock. 56 pp. UK.

1963 **Lawn Grasses.** Second rev. ed. F.J. Reed. UK.

1963 **Lawns—For Garden and Playing Field.** Second rev. ed. R.B. Dawson. 176 pp. UK.

1963 **The Management of Vegetation on Airfields Overseas.** K.E. Clare. 35 pp. UK.

1963 **Selection and Layout of Land for Playing Fields and Playgrounds.** Third rev. ed. R.B. Gooch. 107 pp. UK.

1963 **Turf Manual.** R.J Newman, W.I. Walbran, E.B. Littlejohn, L.C. Jones, and H.B. Wilson. 87 pp. Australia.

1964 **Grasses and Grasslands.** C. Barnard (ed). 269 pp. UK.

1964 **Turf Grass Course.** H.B. Musser. 69 pp. USA.

1964 **Turfgrass Course.** First ed. T.C. Mascaro. 105 pp. USA.

1965 through 1969

1965 **Fungal Diseases of Turf Grasses.** Second rev. ed. N. Jackson and J.D. Smith. 97 pp. UK. Revised in 1989.

1965 **Gazons en Sportvelden (Lawns and Sportsfields).** R.B. Dawson. 168 pp. The Netherlands.

1965 **Grass Varieties in the United States.** Second rev. ed. A.A. Hanson. 102 pp. USA. Revised in 1972 and 1994.

1965 **Sports Ground Construction—Specifications For Playing Facilities.** First ed. R.B. Gooch and J.R. Escritt. 104 pp. UK. Revised in 1975.

1965 **Tifton Dwarf-Bermudagrass Manual For Greens & Fairways.** E.R. Jensen. 26 pp. USA.

1965 **What's that weed?** O.M. Scott & Sons Co. 23 pp. USA.

1966 **Better Fairways: Northern Golf Courses—Theory and Practice.** Third rev. ed. C.G. Wilson, J.M. Latham Jr., and. R.J. Welch. 27 pp. USA.

1966 **Lawn and Lawn Grasses for Use as Turf.** F. Kitamura. 232 pp. Japan.

1966 **Playing Fields and Hard Surface Areas.** Second rev. ed. Department of Education and Science. 92 pp. UK.

1966 **Treatment of Golf Courses and Sports Grounds.** Carters Tested Seeds Ltd. 31 pp. UK.

1966 **Winter Injury.** C.G. Wilson, J.M. Latham Jr., and R.J. Welch. 34 pp. USA.

1967 **Athletic Fields—Specification Outline, Construction, and Maintenance.** Second rev. ed. J.C. Harper II. 21 pp. USA.

1967 **Cave's Guide to Turf Culture.** L.W. Cave. 188 pp. UK.

1967 **The Provision and Maintenance of Playing Fields and Churchyards.** H.T. Hopper. 256 pp. UK.

1967 **Sports Ground Maintenance—an elementary guide to club committees.** First ed. W.H. Bowles. 40 pp. UK. Revised in 1978.

1967 **Turfgrass Establishment and Care of the Home Lawn—Including Home Garden Design.** K. Ehara, Y. Maki, and H. Chono (eds). 237 pp. Japan.

1968 **Bowling Green Construction.** P.E. McMaugh. 33 pp. Australia.

1968 **Bowling Greens—Construction and Maintenance.** W. Howell. 172 pp. UK.

1968 **Building Golf Holes for Good Turf Management.** M.H. Ferguson (ed). 55 pp. USA.

1968 **Grasses.** Second rev. ed. C.E. Hubbard. 463 pp. UK. Revised in 1984 by J.C.E. Hubbard.

1968 **Grass Systematics.** First ed. F.W. Gould. 382 pp. USA.

1968 **The Library of Golf—1743–1966.** J.S.F. Murdock. 314 pp. USA. Supplement in 1978.

1968 **Practical Lawn Craft and Management of Sports Turf.** Sixth rev. ed. R.B. Dawson. 320 pp. UK. Revised in 1977.

1968 **Turfgrass and Turf—Establishment and Maintenance.** First ed. K. Ehara. 500 pp. Japan. Revised in 1971.

1968 **Turfgrass Maintenance and Establishment.** R.F. Stinson. 154 pp. USA.

1968 **USGA Golf Championship Manual.** United States Golf Association. 49 pp. USA.

1969 **Lawns—Basic Factors, Construction, and Maintenance of Fine Turf Areas.** First ed. J. Vengris. 229 pp. USA. Revised in 1973 and 1982.

1969 **A Perfect Lawn the Easy Way.** First ed. P.N. Voykin. 155 pp. USA. Revised in 1971.

1969 **Turfgrass Fertilization.** G.R. Hawkes. 32 pp. USA.

1969 **Turfgrass Science.** A.A. Hanson and F.V. Juska (eds). 715 pp. USA. Revised in 1992 with D.V. Waddington and R.C. Shearman (eds).

196x **The Maintenance of the Lawn Bowling Green.** First ed. E.R. Haley. USA. Revised in 1975.

196x **Planning and Building the Golf Course.** First ed. National Golf Foundation. 30 pp. USA. Revised in 1971.

1970 through 1974

1970 **Better Bentgrass Greens—Fertilization Management.** Fourth rev. ed. C.G. Wilson, J.M. Latham Jr., and R.J. Welch. 20 pp. USA.

1970 **Golf Green Construction.** P.E. McMaugh. 36 pp. Australia.

1970 **Long-Lasting Turfed Areas for Sport and Leisure.** G.G. Abramashvili. 104 pp. Russia.

1970 **The Modern Golf Course Superintendent.** R. Viergever. 107 pp. USA.

1970 **Turf Management Handbook.** First ed. H.B. Sprague. 253 pp. USA. Revised in 1976, 1982, 1994, and 1996.

1971 **Better Fairways: Northern Golf Courses—Theory and Practice.** Fourth rev. ed. C.G. Wilson, J.M. Latham Jr., and R.J. Welch. 21 pp. USA.

1971 **Easy Weed Control of Garden, Lawn, and Housing Sites.** T. Takematsu. 245 pp. Japan.

1971 **Manual of the Grasses of the United States.** A.S. Hitchcock and A. Chase. 569 pp. USA.

1971 **Planning and Building the Golf Course.** Third rev. ed. National Golf Foundation. 27 pp. USA. Revised in 1975 and 1981.

1971 **Podizanje I Održavanje Travnatih Terena (Construction and Culture of Turfgrass Fields).** P.M. Bošković and F. Bureš. 84 pp. Yugoslavia.

1971 **Practical Turfgrass Management.** J.H. Madison. 466 pp. USA.

1971 **Principles of Turfgrass Culture.** J.H. Madison. 420 pp. USA.

1971 **Scotts Professional Turf Seminar.** O.M. Scott & Sons Co. 230 pp. USA.

1971 **Turf and Turfgrasses—Establishment and Maintenance.** Second rev. ed. K. Ehara. 550 pp. Japan. Revised in 1976.

1971 **Turf Culture.** Second rev. ed. C. Walker (ed). 362 pp. New Zealand.

1971 **Turf Manual.** W.I. Walbran, E.B. Littlejohn, R.J. Newman, A.W. Kelloch, and H.B. Wilson. 89 pp. Australia.

1971 **Våra Grönytor—Anläggning och skötsel (Our Green Areas—Establishment and Maintenance).** B.J. Langvad. 184 pp. Sweden.

1972 **Association Football Club Groundsman.** Second rev. ed. R. Hawthorn. 97 pp. UK.

1972 **The Biology and Utilization of Grasses.** V.B. Youngner and C.M. McKell (eds). 426 pp. USA.

1972 **The Building of a Lawn Bowling Green.** First ed. E.R. Haley. 51 pp. USA.

1972 **Cricket—Take Care of Your Square.** First ed. H.C. Lock. 33 pp. UK.

1972 **Grass Varieties of the United States.** Third rev. ed. A.A. Hanson. 124 pp. USA. Revised in 1994 by J.S. Alderson.

1972 **Lawns—A manual of Lawn care.** W.V. DeThabrew. 184 pp. UK.

1972 **Permanent Fertilization Record and Handbook.** C.G. Wilson, J.M. Latham Jr., and R.J. Welch. 43 pp. USA.

1972 **Scotts Professional Turf Manual.** O.M. Scott & Sons Co. 179 pp. USA.

1972 **Scotts Professional Turf Manual.** O.M. Scott & Sons Co. 252 pp. USA.

1972 **Tappeti Erbosi: Impianto—manutenzione—impieghi (Turfgrasses: Establishment, maintenance, use).** First ed. A. Panella. 199 pp. Italy.

1973 **Diseases of Turfgrasses.** Second rev. ed. H.B. Couch. 348 pp. USA. Revised in 1995.

1973 **Lawns—Basic Factors, Construction, and Maintenance of Fine Turf Areas.** Second rev. ed. J. Vengris. 247 pp. USA. Revised in 1982.

1973 **Lawns, Sportsgrounds and Playing Areas—Including Non-Grass Surfaces.** W.V. De Thabrew. 279 pp. UK.

1973 **Limited Bibliography of Turf Literature.** D.T. Hawes. 122 pp. USA.

1973 **A Perfect Lawn—The Easy Way.** R.W. Schery. 304 pp. USA.

1973 **Professional Turf Seminar.** O.M. Scott & Sons Co. 341 pp. USA.

1973 **Rasen, Gras und Grünflächen (Turf, Grasses and Green Spaces).** Second rev. ed. C. Eisele. 144 pp. West Germany.

1973 **Scotts Professional Turf Manual.** ProTurf Division, O.M. Scotts & Sons Co. 265 pp. USA.

1973 **Turfgrass: Science and Culture.** J. B Beard. 658 pp. USA.

1974 **Les Gazons (Turfs).** First ed. R. Thomas. 180 pp. France. Revised in 1985.

1974 **How to control Lawn Diseases and Pests.** M.C. Shurtleff and R. Randell. 97 pp. USA.

1974 **Lawn and Product Information Manual.** Regional ed. O.M. Scott & Sons Co. 218 pp. USA.

1974 **Pesticide Usage—Reference Manual.** W.E. Knoop. 135 pp. USA.

1974 **Practical Golf Course Design and Construction.** G.C. Riddell. 45 pp. Australia.

1974 **Professional Turf Manual.** ProTurf Division. O.M. Scott & Sons Co. 125 pp. USA.

1974 **Turfgrass Course.** Second rev. ed. T.C. Mascaro, H.B. Musser, and W. Bidwell. 118 pp. USA.

1974 **Twelve Months of Turf Maintenance.** T. Hosotsuji and M. Yoshida. 176 pp. Japan.

1975 through 1979

1975 **Analeg af Sportsplader. Jord-Vand-Luft (Construction of Sportsgrounds: Soil, Water, Air).** M. Petersen. 161 pp. Denmark.

1975 **Biološki Problemi Travnjaka na Terenima Jugoslavije (Turf Biological Problems on Yugoslavian Playing Fields).** P.M. Bošković. 83 pp. Yugoslavia.

1975 **Grasveldkunde (Grassfield Science).** M. Hoogerkamp and J.W. Minderhoud (eds). 266 pp. The Netherlands.

1975 **How to have a Beautiful Lawn.** First ed. J. B Beard. 113 pp. USA. Revised in 1979, 1983, 1988, and 1993.

1975 **Lawn and Product Information Manual—Southern Edition.** O.M. Scott & Sons Co. 163 pp. USA.

1975 **The Maintenance of the Lawn Bowling Green.** Second rev. ed. E.R. Haley. 116 pp. USA. Revised in 1980.

1975 **Planning and Building the Golf Course.** Fourth rev. ed. National Golf Foundation. 28 pp. USA.

1975 **Professional Turf Seminar Manual.** ProTurf Division, O.M. Scott & Sons Co. 145 pp. USA.

1975 **Sports Ground Construction—Specifications.** Second rev. ed. R.B. Gooch and J.R. Escritt. 126 pp. UK.

1975 **Travnatá Hřiště (Turf Playing Fields).** F. Bureš and O. Blažek. 224 pp. Czechoslovakia.

1975 **Travnate Površine—Izgradnja i održavanje (Turfgrass Areas—Construction and Culture).** P. Bošković. 91 pp. Yugoslavia.

1975 **Turf and Lawn Garden Construction.** S. Okubo and K. Someno. 200 pp. Japan.

1976 **Ask the Lawn Expert.** P.N. Voykin. 249 pp. USA.

1976 **The Construction of the Lawn Bowling Green.** Second rev. ed. E.R. Haley. 61 pp. USA.

1976 **L'Aménagement du Terrain de Baseball (The Creation of Baseball Fields).** Haut-Commissariat à la Jeunesse, aux Loisirs et aux Sports. 43 pp. Canada.

1976 **Lawn: Ecology and Production Technique.** S. Yoshida. 269 pp. Japan.

1976 **Lawn Keeping.** R.W. Schery. 232 pp. USA.

1976   **Post 1976 Turfgrass Industry Challenges In Research, Teaching And Continuing Education.** Turfgrass Science Division C-5, Crop Science Society of America (CSSA). 33 pp. USA.

1976   **Professional Turf Seminar Manual.** ProTurf Division, O.M. Scott & Sons Co. 167 pp. USA.

1976   **Rasen für Sport und Spiel (Turf for Sports and Games).** K.D. Gandert and A. Schnabel. 171 pp. East Germany.

1976   **Scotts Professional Turf Manual.** ProTurf Division, O.M. Scott & Sons Co. 261 pp. USA.

1976   **Turf and Turfgrasses—Establishment and Maintenance.** Third rev. ed. K. Ehara. 550 pp. Japan.

1976   **Turf Management Handbook.** Second rev. ed. H.B. Sprague. 225 pp. USA. Revised in 1982, 1994, and 1996.

1977   **A Checklist of Names for 3,000 Vascular Plants of Economic Importance.** First ed. E.E. Terrell. 201 pp. USA. Revised in 1986.

1977   **Dawson's Practical Lawncraft.** Seventh rev. ed. R. Hawthorn. 312 pp. UK.

1977   **The Elements of Turf and Turfgrass.** Japanese Society for Turfgrass Research. 480 pp. Japan.

1977   **General Views on Turf and Turfgrasses.** K. Ehara and F. Kitamura (eds). 480 pp. Japan.

1977   **The Golf Professional's Guide to Turfgrass Maintenance.** First ed. W.E. Knoop. 20 pp. USA. Revised in 1980.

1977   **Grounds Maintenance Handbook.** Third rev. ed. H.S. Conover. 631 pp. USA.

1977   **A Handbook on the Maintenance of School Fields.** C.K. Yau. 32 pp. Singapore.

1977   **Identifying Grasses: data, methods, and illustrations.** H.T. Clifford and L. Watson. 146 pp. Australia.

1977   **The Mathematics of Turfgrass Maintenance.** J. B Beard, P.E. Rieke, W.E. Knoop, and T.J. Rogers. 58 pp. USA.

1977   **Monograph of Turf Culture and Turfgrasses.** K. Ehara and F. Kitamura. 480 pp. Japan.

1977   **Professional Turf Seminar Manual.** ProTurf Division, O.M. Scott & Sons Co. 149 pp. USA.

1977   **Turfgrass Bibliography—From 1672 to 1972.** J. B Beard, H.J. Beard, and D.P. Martin. 730 pp. USA.

1977   **A Turf Manager's Guide—Microscopic Identification of Common Turfgrass Pathogens.** P.L. Sanders. 52 pp. USA.

1978   **ABC of Turf Culture.** J.R. Escritt. 248 pp. UK.

1978   **Concise Guide to the Maintenance of Sports Grounds.** G. Snook. 20 pp. UK.

1978   **Fertilisers In Turf Culture.** First ed. Sports Turf Research Institute. 43 pp. UK. Revised in 1991 by D.M. Lawson.

1978   **Golf—A Guide To Information Sources.** J.S.F. Murdoch and J.M. Seagle. 232 pp. USA.

1978   **Slid på Sportsplanderk Komprimering og Renovering (Wear on Sportsgrounds, Compaction and Renovation).** M. Peterson. 128 pp. Denmark.

1978   **Sports Ground Maintenance—an elementary guide to club committees.** Second rev. ed. W.H. Bowles. 31 pp. UK.

1978   **Turf Culture—A Complete Manual for the Groundsman.** First ed. F. Hope. 293 pp. UK.

1978   **Vegetationstechnik Rasen und Begrünungen (Techniques of Growing Lawns and Green Spaces).** W.W. Skirde. 240 pp. West Germany.

1979   **Better Greens.** E.R. Haley. 88 pp. USA.

1979   **Color Atlas: Diseases, Insects and Weeds in Turfgrass.** First ed. T. Hosotsuji and M. Yoshida. 298 pp. Japan.

1979   **Golden Jubilee Celebrations.** Sports Turf Research Institute. 26 pp. UK.

1979   **Grounds Maintenance Handbook for Schools.** Education Department of South Australia. 37 pp. Australia.

1979   **How to have a Beautiful Lawn.** Second rev. ed. J. B Beard. 113 pp. USA. Revised in 1983, 1988, and 1993.

1979   **Information Manual for Lawns.** O.M. Scott & Sons Co. 96 pp. USA.

1979   **Introduction To Turfgrass Science And Culture Laboratory Exercises.** J. B Beard, J.M. DiPaola, D. Johns, K.J. Karnok, and S.M. Batten. 217 pp. USA.

1979   **Professional Turf Seminar—Presentation Notes.** ProTurf Division, O.M. Scott & Sons Co. 68 pp. USA.

1979   **Scotts Guide to the Identification of Dicot Turf Weeds.** J.T. Converse. 119 pp. USA.

1979   **Scotts Guide to the Identification of Grasses.** J.T. Converse. 82 pp. USA.

1979   **Town and Sporting Grass Fields.** G.G. Abramashvili. 104 pp. USSR.

1979   **Turfgrass Diseases.** First ed. Sports Turf Research Institute. 36 pp. UK. Revised in 1987 by N.A. Baldwin.

1979   **Turf Managers' Handbook.** First ed. W.H. Daniel and R.P. Freeborg. 424 pp. USA. Revised in 1987.

1979   **When Should I Water?** K.A. Handreck. 76 pp. Australia.

1980 through 1984

1980   **Amenity Grassland—An Ecological Perspective.** I.H. Rorison and R. Hunt. 261 pp. UK.

1980   **A Baseball Facility: Its Construction and Care.** First ed. US Baseball Federation. 53 pp. USA.

1980   **The Bossard Sportsfields Landscape Groundskeeping Pedology Handbook.** B.J. Bossard. 63 pp. USA.

1980   **Development of Beautiful Lawn Garden—Understandable at a Glance.** K. Ehara. 158 pp. Japan.

1980   **Erhaltung von Sportplätzen (Maintaining Sports Grounds).** W.W. Skirde, W. Büring, H. Pätgold, H. Tietz, K. Trojahn, F. Müller, and A. Niesel. 208 pp. West Germany.

1980   **The Golf Professional's Guide to Turfgrass Maintenance.** Second rev. ed. W.E. Knoop. 21 pp. USA.

1980   **Les Terrains de Sports (Sports Grounds).** Association Francaise pour le Développement des Équipements de Sportifs et de Loisirs. 284 pp. France.

1980   **Maintenance of the Club Cricket Ground.** P. Mansfield. 26 pp. UK.

1980   **Maintenance of the Lawn Bowling Green.** Third rev. ed. E.R. Haley. 187 pp. USA.

1980   **Professional Turf Seminar—Presentation Notes.** ProTurf Division, O.M. Scott & Sons Co. 39 pp. USA.

1980   **Turfgrass Maintenance, Foreign Experience.** G.G. Abramashvili. 120 pp. USSR.

1980   **Turfgrass Management.** First ed. A.J. Turgeon. 391 pp. USA. Revised in 1985, 1991, 1996, 1999, 2002, 2005, and 2008.

1981   **The Construction of the Lawn Bowling Green.** Third rev. ed. E.R. Haley. 71 pp. USA.

1981   **Destructive Turf Insects.** First ed. H.D. Niemczyk. 48 pp. USA. Revised in 2000.

1981   **Græsplæner—Principper and Funktioner (Turfgrass Areas—Principles and Functions).** M. Petersen. 362 pp. Denmark.

1981   **Greenkeeping.** R.F. Stinson (ed). 101 pp. USA.

1981   **Management of Turfgrass Diseases.** First ed. J.M. Vargas. 204 pp. USA. Revised in 1994 and 2005.

1981   **Planning and Building the Golf Course.** Fifth rev. ed. National Golf Foundation. 48 pp. USA.

1981   **Tappeti Erbosi: Impianto, manutenzione e impieghi (Turfgrasses: Establishment, maintenance and use).** Second rev. ed. A. Panella. 199 pp. Italy.

1982   **Greenkeeping.** J.K. Campbell. 88 pp. UK.

1982   **Knowledge of Course Management of Golf Club.** K. Ehara. 561 pp. Japan.

1982   **Lawns—Basic Factors, Construction, and Maintenance of Fine Turf Areas.** J. Vengris and W.A. Torello. 195 pp. USA.

1982   **Scotts Guide To The Identification of Turfgrass Diseases and Insects.** J.T. Converse. 105 pp. USA.

1982   **Turfgrass Manual.** Second rev. ed. R.J.C. Hawarth. 47 pp. UK.

1982   **Turf Management for Golf Courses.** J. B Beard. 641 pp. USA. Revised in 2002.

1982   **Turf Management Handbook.** Third rev. ed. H.B. Sprague. 255 pp. USA. Revised in 1994 and 1996.

1983   **Better Turfgrass Nutrition.** J. B Beard. 22 pp. USA.

1983   **Compendium of Turfgrass Diseases.** First ed. R.W. Smiley. 102 pp. USA. Revised in 1992 and 2005.

1983   **Construction and Maintenance of Natural Grass Athletic Fields.** R.L. Goss and T.W. Cook. 27 pp. USA.

1983   **Golf Course: Planning, Design, Construction, and Maintenance.** F.W. Hawtree. 212 pp. UK.

1983   **Grass Systematics.** Second rev. ed. F.W. Gould and R.B. Shaw. 397 pp. USA.

1983   **Handbook of Turf and Golf Course Management.** First ed. Kansai Golf Union. 288 pp. Japan.

1983   **How to have a Beautiful Lawn.** Third rev. ed. J. B Beard. 113 pp. USA. Revised in 1988 and 1993.

1983   **Rasen—Anlage und Pflege von Zier-, Gebrauchs-, Sport- und Landschaftsrasen (Turfgrass—Construction and Maintenance for Aesthetic, Functional, Sport, and Landscaped Lawns).** F. Hope and H. Schulz. 216 pp. West Germany.

1983   **Turfgrass Insect and Mite Manual.** First ed. D.J. Shetlar, P.R. Heller, and P.D. Irish. 63 pp. USA.

1983   **Untersuchungen zur Be- und Entwässerung von Rasensportflächen (Analysis of Irrigation and Drainage of Sports Turf Surfaces).** W. Skirde. 82 pp. Germany.

1984   **Bowling Greens: A Practical Guide.** K. Liffman. 130 pp. Australia.

1984   **ChemLawn Turf Field Guide.** J.M. Kick-Raack and B.G. Joyner. 188 pp. USA.

1984   **Football Grounds—Their Construction and Standards of Maintenance.** P.M. Bošković. 80 pp. Yugoslavia.

1984   **Golf Course and Turf Grass Management.** J.B. Beard, R.W. Fream, and C.H. Yin. 109 pp. Malaysia.

1984   **Golf Course Management.** J. B Beard. 71 pp. USA.

1984   **Grasses—A Guide To Their Structure, Identification, Uses, and Distribution in the British Isles.** Third rev. ed. C.E. Hubbard and J.C.E. Hubbard. 480 pp. UK.

1984   **Growing Media for Ornamental Plants and Turf.** First ed. K.A. Handreck and N.D. Black. 401 pp. Australia.

1984   **Maintenance of the Lawn Bowling Green.** Fourth rev. ed. E.R. Haley. 198 pp. Canada.

1984   **Sports Grounds & Turf Wickets—A Practical Guide.** K. Liffman. 110 pp. Australia.

1984   **Technical Terms in Turf Culture.** P. Hayes. 76 pp. UK.

1984   **Turf and Turfgrasses—Establishment and Maintenance.** Fourth rev. ed. 572 pp. Japan.

1984   **Turfgrass Insect and Mite Manual.** First ed. D.J. Shetlar, P.R. Heller, and P.D. Irish. 63 pp. USA.

1984   **Turfgrass Manual.** J. Shildrick. 60 pp. UK.

1984   **Turfgrass Plant Physiology and Botany.** J. B Beard. 20 pp. USA. Revised in 1986.

1984   **Turfgrass Science and Management.** First ed. R.D. Emmons. 451 pp. USA.

1985 through 1989

1985   **Evolution of the Modern Green.** M.J. Hurdzan. 24 pp. USA.

1985   **Fotbalová Travnatá Hřiště (Grassy Football Playgrounds).** F. Bureš. 104 pp. Czechoslovakia.

1985   **Scotts Guide to the Identifications of Dicot Turf Weeds.** O.M. Scott & Sons Co. 119 pp. USA.

1985   **Scotts Guide to the Identification of Grasses.** O.M. Scott & Sons Co. 84 pp. USA.

1985   **Theory and Practice of Weed Control in Turf.** First ed. T. Takematsu and Y. Takeuchi. 330 pp. Japan.

1985   **Turfgrass Management.** Second rev. ed. A.J. Turgeon. 416 pp. USA. Revised in 1991, 1996, 1999, 2002, 2005, and 2008.

1985   **Turfgrass Water Conservation.** V.A. Gibeault and S.T. Cockerham (edu).155 pp. USA.

1986   **Athletic Fields—Specification outline, construction, and maintenance.** Third rev. ed. J.C. Harper II. 29 pp. USA.

1986   **Basic Turfgrass Botany and Physiology.** First ed. J. B Beard and J.V. Krans. 44 pp. USA.

1986   **A Checklist of Names for 3,000 Vascular Plants of Economic Importance.** Second rev. ed. E.E. Terrell, S.R. Hill, J.H. Wiersema, and W.E. Rice. 241 pp. USA.

1986   **Grass Seed Mixtures for Children's Play Areas.** J. Shildrick. 42 pp. UK.

1986   **Grass Systematics and Evolution.** T.R. Soderstrom, K.W. Hilu, C.S. Campbell, and M.E. Barkworth. 474 pp. USA.

1987   **Controlling Turfgrass Pests.** First ed. M.C. Shurtleff, T.W. Fermanian, and R. Randell. 449 pp. USA.

1987   **Cricket—Take Care of Your Square.** Second rev. ed. H.C. Lock and P.L.K. Dury. 32 pp. UK.

1987   **Disease, Insect and Weed Control in Turf.** David Worrad and the Australian Turfgrass Research Institute Ltd. 34 pp. Australia.

1987   **Klippta Gräsytor (Mown Grass Areas).** S. Dahlsson. 165 pp. Sweden.

1987   **Laboratory Manual for Turfgrass Management.** M.S. Welterlen. 202 pp. USA.

1987   **Turfgrass Insects of the United States and Canada.** First ed. H. Tashiro. 391 pp. USA.

1987   **Turf Managers' Handbook.** Second rev. ed. W.H. Daniel and R.P. Freeborg. 437 pp. USA.

1988   **Bowling Greens—Their History, Construction and Maintenance.** First ed. R.D.C. Evans. 196 pp. UK.

1988   **Building a Healthy Lawn: A Safe and Natural Approach.** S. Franklin. 168 pp. USA.

1988   **How to have a Beautiful Lawn.** Fourth rev. ed. J. B Beard. 117 pp. USA.

1988   **Lawn Care—A Handbook for Professionals.** H.F. Decker and J.M. Decker. 270 pp. USA.

1988   **Manuale per la Conduzione dei Tappeti Erbosi dei Campi di Golf Italiani (Manual on Italian Golf Course Turf Management).** J. B Beard. 241 pp. Italy.

1988   **Turfgrass for Sport Fields.** G.G. Abramashvili. 160 pp. USSR.

1988   **Turfgrass: Identification, Description, Culture.** J. B Beard. 49 pp. USA.

1988   **Turfgrass Insect and Mite Manual.** Second rev. ed. D.J. Shetlar, P.R. Heller, and P.D. Irish. 67 pp. USA.

1988   **Turfgrass Manual.** Third rev. ed. British Seed Houses Ltd. 60 pp. UK.

1988   **Turfgrass Sod Production.** S.T Cockerham. 84 pp. USA.

1988   **Turf Irrigation.** Second rev. ed. D.K. McIntyre and S.M. Hughes. 60 pp. Australia.

1988   **Turf Market.** Australian Turfgrass Research Institute Ltd. 46 pp. Australia.

1988   **Turf: Science and Culture For Green Vegetation.** Japanese Society of Turfgrass Science. 564 pp. Japan.

1989   **Color Atlas: Diseases, Insects and Weeds in Turfgrass.** Second rev. ed. T. Hosotsuji and M. Yoshida. 331 pp. Japan.

1989   **Contributions from Breeding Forage and Turf Grasses.** D.A. Sleper, K.H. Assay, and J.F. Pedersen. 140 pp. USA.

1989   **Disease, Insect and Weed Control in Turf.** Australian Turfgrass Research Institute. 54 pp. Australia.

1989   **Fungal Diseases of Amenity Turf Grasses.** Third rev. ed. J.D. Smith, N. Jackson, and A.R. Woolhouse. 401 pp. UK.

1989   **Golf Courses: A Guide to the Construction and Maintenance of Low to Medium Budget Facilities.** R.W. Daniels. 66 pp. Canada.

1989   **Malattie Parassitarie, Attacchi degli Insetti e le Infestanti dei Tappeti Erbosi (Diseases, Insects and Weeds of Turf-grasses).** J. B Beard. 20 pp. Italy.

1989   **Southern Turfgrasses: Their Management and Use.** First ed. R.L. Duble. 351 pp. USA.

1989   **Specifications for a Method of Putting Green Construction.** W.H. Bengeyfield. 25 pp. USA.

1989   **Sports Grounds Maintenance—An Elementary Guide.** Third rev. ed. J.C. Holborn. 40 pp. UK.

1989   **Turf Management.** Rev. ed. S.K. White. 109 pp. USA.

198x   **Cricket Pitch Management.** P.L.K. Dury. 57 pp. UK.

198x   **Proturf Guide to the Identification of Dicot Weeds.** J.T. Converse. 119 pp. USA.

198x   **Proturf Guide to the Identification of Grasses.** J.T. Converse. 82 pp. USA.

198x   **Sportsground Construction & Management.** P.L.K. Dury. 84 pp. UK.

1990 through 1994

1990　**The Construction of the Lawn Bowling Green.** Fourth rev. ed. E.R. Haley. 91 pp. USA.

1990　**Florida Lawn Handbook.** First ed. L.B. McCarty, R.J. Black, and K.C. Ruppert. 216 pp. USA.

1990　**Fungicides for Turfgrass Disease Control.** W.H. Walmsley. 26 pp. New Zealand.

1990　**Guida per la Costruzione e la Manutenzione dei Greens nei Campi di Golf Italiani (Italian Guide to Putting Green Construction and Maintenance).** J. B Beard. 90 pp. Italy.

1990　**L'Encyclopédie des Gazons (Turf Encyclopedia).** Société Francaise des Gazons (French Turfgrass Society). 360 pp. France.

1990　**A Manual For Construction and Maintenance of Bentgrass Putting Greens.** J. B Beard. 44 pp. Japan.

1990　**Real Golf.** E. Park. 179 pp. UK.

1990　**Recreational Fields—A Guide to Site Selection, Construction and Maintenance.** R.W. Daniels. 72 pp. Canada.

1990　**Sands for Sports Turf Construction and Maintenance.** Australian Turfgrass Research Institute. 67 pp. Australia.

1990　**The Sand Putting Green—Construction and Management.** W.B. Davis, J.L. Paul, and D.C. Bowman. 22 pp. USA.

1990　**Scuola Nazionale di Golf (National School of Golf).** Federazione Italiana Golf. 32 pp. Italy.

1990　**Svenska Golfförbundets Skötselmanual för Golfgräs (A Manual on Swedish Golf Course Turf).** J. B Beard. 116 pp. Sweden.

1900　**Toogood's Treatise on Lawns, Cricket Grounds, &c.** E.K. Toogood. 20 pp. UK.

1990　**Turf Culture—A Manual for the Groundsman.** Second rev. ed. F. Hope. 293 pp. UK.

1990　**Turfgrass Insect and Mite Manual.** Third rev. ed. D.J. Shetlar, P.R. Heller, and P.D. Irish. 67 pp. USA.

1991　**An Athletic Field Managers Guide For Construction and Maintenance.** First ed. Sports Turf Association of Ontario. 39 pp. Canada. Revised in 2005 by R.W. Sheard.

1991　**Bentgrass—Characteristics and Newest Methods of Golf Green Construction and Maintenance.** Y. Maki and S. Okubo. 345 pp. Japan.

1991　**Characteristics, Species, and Establishment of Turfgrass.** J.Y. Joo. 345 pp. South Korea.

1991　**Color Atlas: Turfgrass Diseases.** T. Tani. 122 pp. Japan.

1991　**Cricket Grounds—The Evolution, Construction and Maintenance of Natural Turf Cricket Tables and Outfields.** R.D.C. Evans. 280 pp. UK.

1991　**Fertilisers for Turf.** Second rev. ed. D.M. Lawson. 41 pp. UK.

1991　**Guida per la Costruzione e la Manutenzione di Fairway e Tee nei Campi di Golf Italiani (Italian Guide to Fairway and Tee Construction and Maintenance).** J. B Beard. 107 pp. Italy.

1991　**A Guide to Croquet Court Planning, Building, & Maintenance.** C.H. Mabee. 120 pp. USA.

1991　**Handbook of Turf and Golf Course Management.** Second rev. ed. Kansai Golf Union. 374 pp. Japan.

1991　**Handbuch Rasen: Grundlagen—Anlage—Pflege (Handbook for Turfgrass: Basics, Installation, Maintenance).** K.D. Gandert and F. Bureš. 364 pp. Germany.

1991　**Il Prato Ornamentale (Ornamental Turf).** M. Vietti. 192 pp. Italy.

1991　**Lawn Institute Special Topics Sheets.** E.C. Roberts and B.C. Roberts. 385 pp. USA.

1991　**The Lawns of Wimbledon (The Construction and Maintenance of Grass Courts).** J. Thorn. 55 pp. UK.

1991　**A Manual on Golf Course Construction, Turf Establishment and Cultural Practices.** J. B Beard. 83 pp. UK.

1991　**A Manual on Malaysian Golf Course Turf Construction, Establishment and Maintenance.** J. B Beard. 125 pp. Malaysia.

1991　**The Murdoch Golf Library.** J.S.F. Murdoch. 233 pp. UK.

1991　**Sportsturf Water Management Manual.** Australian Turfgrass Research Institute. 53 pp. Australia.

1991　**Theory and Practice of Weed Control in Turf.** Second rev. ed. T. Takematsu and Y. Takeuchi. 334 pp. Japan.

1991　**Turfgrass Culture and Pest Management for the Southeastern United States.** J. B Beard. 96 pp. USA.

1991　**Turfgrasses and Soils for the Southeastern United States.** J. B Beard. 71 pp. USA.

1991　**Turfgrass Management.** Third rev. ed. A.J. Turgeon. 418 pp. USA. Revised in 1996, 1999, 2002, 2005, and 2008.

1992　**A Baseball Facility: Its Construction and Care.** Fourth rev. ed. United States Baseball Federation. 87 pp. USA.

1992　**Basic and Practice in Golf Course Management.** I.S. Kim. 772 pp. South Korea.

1992　**Basics and Application in Golf Course Management.** G.Y. Shim. 772 pp. South Korea.

1992　**Bowling Greens—Their History, Construction and Maintenance.** Second rev. ed. R.D.C. Evans. 211 pp. UK.

1992　**The Care of the Golf Course.** First ed. P. Hayes, R.D.C. Evans, and S.P. Isaac. 270 pp. UK.

1992　**Compendium of Turfgrass Diseases.** Second rev. ed. R.W. Smiley, P.H. Dernoeden, and B.B. Clarke. 98 pp. USA.

1992　**Criteri Generali per la Pianificazione della Costruzione di un Percorso di Golf (Golf Course Planning and Development Guidelines).** J. B Beard. 48 pp. Italy.

1992 **Cumulative Index—1972–1992, Proceedings of the Michigan Turfgrass Conference.** Turfgrass Information Center File (TGIF), Michigan State University. 501 pp. USA.

1992 **Diagnostic Management for Golf Greens.** T.C. Mascaro. 387 pp. USA.

1992 **Golfbanesseminarium—Handbook för Anlägguing och skötsel an Golfbanor I Sverige (A Manual on Swedish Golf Course Turf Construction, Establishment, and Maintenance).** J. B Beard. 118 pp. Sweden.

1992 **Golf Course Planning in Korea: Theory and Practice.** Y.K. Joo. 567 pp. South Korea.

1992 **Golf Course Presentation.** J. Hacker and G. Shiels. 45 pp. UK.

1992 **The Grass Genera of the World.** First ed. L. Watson and M.J. Dallwitz. 1042 pp. UK.

1992 **Les Maladies Cryptogamiques des Gazons: Biologie et Contrôle (Fungal Diseases of Turfgrasses: Biology and Control).** First ed. A. Dehaye. 128 pp. France.

1992 **Les Sols Sportifs de Plein Air (Outdoor Sports Soils).** Ministère de la Jeunesse et des Sports. 248 pp. France.

1992 **A Manual on Turfgrasses, Root Zone Construction, Turf Establishment and Cultural Practices for Japan.** J. B Beard. 85 pp. Japan.

1992 **The Mathematics of Turfgrass Maintenance.** First ed. M.L. Agnew and N.D. Christians. 77 pp. USA.

1992 **Soil Manual I.** M.G. Waillis. Australian Turfgrass Research Institute. 22 pp. Australia.

1992 **Sportsturf Water Management Manual.** Australian Turfgrass Research Institute. 66 pp. Australia.

1992 **Turf Development and Management for Turfgrass Managers.** Y. Maki. 292 pp. Japan.

1992 **Turfgrass.** D.V. Waddington, R.N. Carrow, and R.C. Shearman (eds). 805 pp. USA.

1992 **Turfgrass Culture and Pest Management for the Midwestern United States.** J. B Beard. 95 pp. USA.

1992 **Turfgrasses and Soils for the Midwestern United States.** J. B Beard. 72 pp. USA.

1992 **Turfgrass Variety Manual.** Australian Turfgrass Research Institute. 50 pp. Australia.

1993 **Basic Turfgrass Botany and Physiology.** Second rev. ed. J. B Beard and J. Krans. 77 pp. USA.

1993 **Best Management Practices for Florida Golf Courses.** First ed. L.B. McCarty and M.L. Elliott. 173 pp. USA.

1993 **The Complete Golf Course: Turf and Design.** C. Crockford. 150 pp. Australia.

1993 **Disease, Insect & Weed Control in Turf.** Australian Turfgrass Research Institute. 85 pp. Australia.

1993 **Establishment and Management of Cotula Greens.** First ed. D.R. Howard. 218 pp. New Zealand.

1993 **A Guide to Integrated Control of Turfgrass Diseases. Volume I: Cool Season Turfgrasses.** L.L. Burpee. 242 pp. USA.

1993 **How to have a Beautiful Lawn.** Fifth rev. ed. J. B Beard. 169 pp. USA.

1993 **How to Make Golf Course.** M.K. Kim. 343 pp. South Korea.

1993 **A Manual for Construction and Maintenance of Creeping Bentgrass Putting Greens.** J. B Beard. 56 pp. USA.

1993 **The Microscope in Turfgrass Disease Diagnosis.** P.L. Sanders. 32 pp. USA.

1993 **Turf and Turfgrass—Establishment and Maintenance.** Fifth rev. ed. K. Ehara. 572 pp. Japan.

1993 **Turf Disease Manual.** Australian Turfgrass Research Institute. 45 pp. Australia.

1993 **Turfgrass Ecology & Management.** T.K. Danneberger. 201 pp. USA.

1993 **Turfgrass Handbook for Southeast Asian Golf Course and Sport Facilities.** J. B Beard. 137 pp. USA.

1993 **Turfgrass Manual.** Fourth rev. ed. British Seed Houses Ltd. 67 pp. UK.

1993 **Turfgrass Patch Diseases Caused by Ectotrophic Root-Infecting Fungi.** B.B. Clarke and A.B. Gould. 161 pp. USA.

1993 **Turfgrass Weed Manual.** Australian Turfgrass Research Institute. 23 pp. Australia.

1993 **USGA Recommendations for Putting Green Construction.** J.T. Snow. 36 pp. USA.

1993 **Zoysiagrass: Science and Management for Golf Course.** N. Nakamura. 377 pp. Japan.

1994 **Campi di Golf e Ambiente: Localizzazione, Progettazione e Gestione (Golf Courses and Environment—Site, Design, and Management).** Federazione Italiana Golf. 83 pp. Italy.

1994 **Golf and the Environment.** United States Golf Association. 20 pp. USA.

1994 **The Grass Genera of the World.** Second rev. ed. L. Watson and M.J. Dallwitz. 1081 pp. UK.

1994 **Grass Varieties in the United States.** J.S. Alderson and W.C. Sharp. 296 pp. USA.

1994 **Great Plains Turfgrass Manual.** Second rev. ed. 206 pp. Canada.

1994 **Growing Media for Ornamental Plants and Turf.** Second rev. ed. K.A. Handreck and N.D. Black. 448 pp. Australia.

1994 **Management of Turfgrass Diseases.** Second rev. ed. J.M. Vargas Jr. 294 pp. USA.

1994 **Management of Turfgrass for Public Space.** F. Kitamura, M. Kondo, H. Ito, and H. Takato. 264 pp. Japan.

1994 **Natural Turf for Sport and Amenity: Science and Practice.** W.A. Adams and R.J. Gibbs. 404 pp. UK.

1994 **Soil Chemistry Manual.** Australian Turfgrass Research Institute. 38 pp. Australia.

1994 **Sports Turf—Science, construction and maintenance.** V.I. Stewart. 260 pp. UK.

1994 **The Turfgrass Index.** Turfgrass Information Center, Michigan State University. 408 pp. USA.

1994 **Turfgrass Lab Manual.** K. Mathias. 106 pp. USA.

1994 **Turf Management Handbook.** Fourth rev. ed. C.B. Schroeder and H.B. Sprague. 198 pp. USA. Revised in 1996.

1994   **Turf Management in Soccer Fields: The Construction and Maintenance.** H. Nakahara and H. Yanagi. 236 pp. Japan.

1994   **Turf Tips Compendium Volume I.** Australian Turfgrass Research Institute. 88 pp. Australia.

1994   **Turf Nutrition and Fertilizers.** Australian Turfgrass Research Institute. 48 pp. Australia.

1994   **Turf Weeds and Their Control.** A.J. Turgeon. 259 pp. USA.

1994   **Turfgrass Lab Manual.** K. Mathias and D. Hawes. 106 pp. USA.

1994   **Wastewater Reuse for Golf Course Irrigation.** United States Golf Association. 302 pp. USA.

1994   **Winter Games Pitches—The Construction and Maintenance of Natural Turf Pitches for Team Games.** R.D.C. Evans. 209 pp. UK.

1995 through 1999

1995   **Agnes Chase's First Book of Grasses: The Structure of Grasses Explained for Beginners.** Fourth rev. ed. L.G. Clark, R.W. Pohl, and A. Chase. 127 pp. USA.

1995   **Color Atlas: Major Insects of Turfgrasses and Trees on Golf Courses.** M. Hatsukade. 128 pp. Japan.

1995   **Diseases of Turfgrasses.** Third rev. ed. H.B. Couch. 421 pp. USA.

1995   **Guide d'Aménagement des Terrains Extérieurs: Baseball, Soccer, Softball (Guide for Installation of Outdoor Fields: Baseball, Soccer, Softball).** D. Brown. 40 pp. Canada.

1995   **Guidelines for Planning and Developing a Public Golf Course.** National Golf Foundation. 83 pp. USA.

1995   **A Guide to Integrated Control of Turfgrass Diseases. Volume II: Warm Season Turfgrasses.** L.L. Burpee. 175 pp. USA.

1995   **Handbook of Turfgrass Pests.** R.L. Brandenburg and M.G. Villani. 141 pp. USA.

1995   **Managing Turfgrass Pests.** T.L. Watschke, P.H. Dernoeden, and D.J. Shetlar. 361 pp. USA.

1995   **Manual on Diagnosing Turfgrass Problems and Corrective Actions.** J. B Beard. 78 pp. USA.

1995   **A Practical Handbook for Sports Turf Managers.** T.J. McLeod. 331 pp. New Zealand.

1995   **The Spirit of St. Andrews.** A. Mackenzie. 269 pp. USA.

1995   **Turfgrass Identification Manual.** Australian Turfgrass Research Institute. 32 pp. Australia.

1995   **The Turfgrass Index.** Turfgrass Information Center, Michigan State University. 408 pp. USA.

1995   **Turfgrass Science and Management.** Second rev. ed. R.D. Emmons. 512 pp. USA.

1995   **Turfgrass Science and Management Lab Manual.** Second rev. ed. R.D. Emmons and R.W. Boufford. 163 pp. USA.

1996   **The Biology of Grasses.** G.P. Chapman. 273 pp. UK.

1996   **The Care of the Golf Course.** Second rev. ed. J. Perris and R.D.C. Evans. 340 pp. UK.

1996   **The Cricket Groundsman's Companion—A Basic Guide to the Maintenance of Cricket Tables & Outfields.** R.D.C. Evans. 21 pp. UK.

1996   **Disease, Insect & Weed Control in Turf.** Australian Turfgrass Research Institute. 116 pp. Australia.

1996   **Fertilisers for Turf.** Third rev. ed. D.M. Lawson. 47 pp. UK.

1996   **Golf Course Architecture—Design, Construction, & Restoration.** First ed. M.J. Hurdzan. 424 pp. USA.

1996   **On The Course—The Life and Times of a Golf Superintendent.** S.E. Metsker. 116 pp. USA.

1996   **The STRI Golf Greenkeepers Training Course Craft Level Theory.** R.D.C. Evans. 198 pp. UK.

1996   **Turfgrasses: Their Management and Use in the Southern Zone.** Second rev. ed. R.L. Duble. 331 pp. USA.

1996   **Turfgrass Management.** Fourth rev. ed. A.J. Turgeon. 406 pp. USA. Revised in 1999, 2002, 2005, and 2008.

1996   **Turf Management Handbook.** Fifth rev. ed. C.B. Schroeder and H.B. Sprague. 206 pp. USA.

1997   **Céspedes en Campos de Golf: Su Mantenimiento y Otras Consideraciones (Golf Course Turf: Maintenance and Other Considerations).** R. Monje-Jiménez. 121 pp. Spain.

1997   **Color Atlas of Turfgrass Diseases.** First English ed. T. Tani and J. B Beard. 245 pp. USA.

1997   **Controlling Turfgrass Pests.** Second rev. ed. T.W. Fermanian, M.C. Shurtleff, R. Randall, H.T. Wilkinson, and P.L. Nixon. 655 pp. USA.

1997   **Florida Lawn Handbook.** Second rev. ed. K.C. Ruppert and R.J. Black. 223 pp. USA.

1997   **Great Plains Turfgrass Manual.** Third rev. ed. D. McKernan. 218 pp. Canada.

1997   **A Handbook of Turfgrass and Turf.** F. Kitamura and Y. Maki. 348 pp. Japan.

1997   **Handbook of Turfgrass Turf.** Y. Noma. 353 pp. Japan.

1997   **The Mathematics of Turfgrass Maintenance.** Second rev. ed. N.D. Christians and M.L. Agnew. 149 pp. USA.

1997   **Mowing for Money.** K.C. Ruppert, A.Z. Stewart, B. Owens, and R. Degner. 37/21 pp. USA.

1997   **On Course Field Guide to the Identification of Golf Course Grasses.** R.D.C. Evans. 24 pp. UK.

1997   **Practical Greenkeeping.** Jim Arthur. 271 pp. UK.

1997 **Référence Gazon: Avec applications sur tableur (Lawn References: With application on spreadsheets).** P. Collen. 198 pp. France.

1997 **Strategy for Integrated Pest Management.** T. Tani. 153 pp. Japan.

1997 **Terminology of Turfgrass Maintenance.** First ed. Y. Maki. 394 pp. Japan.

1997 **Turfgrass Management Information Directory.** First ed. K.J. Karnok. 105 pp. USA.

1997 **Turf Tips Compendium Volume I.** Australian Turfgrass Research Institute. 71 pp. Australia.

1997 **Turf Tips Compendium Volume II.** Australian Turfgrass Research Institute. 70 pp. Australia.

1998 **Césped Deportivo: Construcción y Mantenimiento (Sport Turfgrass: Construction and Maintenance).** D.M. Merino. 386 pp. Spain.

1998 **Color Atlas: Diseases, Insects and Weeds in Turf.** Third rev. ed. T. Hosotsuji and M. Yoshida. 319 pp. Japan.

1998 **Fundamentals of Turfgrass Management.** First ed. N. Christians. 301 pp. USA.

1998 **Golf Course Design.** R.M. Graves and G.S. Cornish. 446 pp. USA.

1998 **Grounds Maintenance: Managing Outdoor Sport and Landscape Facilities.** P.L.K. Dury. 150 pp. UK.

1998 **IPM Handbook for Golf Courses.** G.L. Schumann, P.J. Vittum, M.L. Elliott, and P.P. Cobb. 272 pp. USA.

1998 **Practical Golf Course Maintenance—The Magic of Greenkeeping.** First ed. G. Witteveen and M. Bavier. 262 pp. USA.

1998 **Salt-Affected Turfgrass Sites: Assessment and Management.** R.N. Carrow and R.R. Duncan. 185 pp. USA.

1998 **Travnjaci—Sportski, parkovni, ukrasni (Turfgrasses—Sports, parks, decorative lawns).** N.M. Samaradžija. 144 pp. Croatia.

1998 **Turfgrass Biotechnology—Cell and Molecular Genetic Approaches to Turfgrass Improvement.** M.B. Striklen and M.P. Kenna (eds). 278 pp. USA.

1998 **Turfgrasses and Their Cultivars.** Y. Asano and K. Aoki. 405 pp. Japan.

1998 **Turfgrass Management Information Directory.** Second rev. ed. K.J. Karnok. 188 pp. USA.

1998 **Turfgrass Trends.** J. B Beard. 55 pp. USA.

1998 **Turf Management: Principles and Practices—Study Guide.** J.L. Eggens. 244 pp. Canada.

1999 **Best Management Practices for Florida Golf Courses.** Second rev. ed. J.B. Unruh and M.L. Elliott. 232 pp. USA.

1999 **Fundamental Turfgrass Science and Establishment of Turfgrass.** Y.K. Joo. 498 pp. South Korea.

1999 **International Turf Management Handbook.** D. Aldous. 430 pp. Australia.

1999 **Linee guida generali per una manutenzione ecocompatibile dei percorsi di golf italiani (Guidelines for the environment-friendly maintenance of Italian golf courses).** Federazione Italiana Golf. 31 pp. Italy.

1999 **Sports Fields—A Manual for Design, Construction, and Maintenance.** J. Puhalla, J. Krans, and M. Goatley. 478 pp. USA.

1999 **Sports Turf Management Program.** Ashman and Associates. 130 pp. USA.

1999 **System and Leadership for Golf Courses.** I.S. Kim and Y.T. Ahn. 154 pp. South Korea.

1999 **Turfgrass Insects of the United States and Canada.** Second rev. ed. P.J. Vittum, M.G. Villani, and H. Tahiro. 422 pp. USA.

1999 **Turfgrass Management.** Fifth rev. ed. A.J. Turgeon. 392 pp. USA. Revised in 2002, 2005, and 2008.

2000 through 2004

2000 **Creeping Bentgrass Management: Summer Stresses, Weeds, and Selected Maladies.** P.H. Dernoeden. 144 pp. USA.

2000 **Destructive Turf Insects.** Second rev. ed. H.D. Niemczyk and D.J. Shetlar. 148 pp. USA.

2000 **Grass Tennis Courts—How To Construct and Maintain Them.** J. Perris. 160 pp. UK.

2000 **La difesa dei tappeti erbosi: Malattie fungine, nemici animali e infestanti (Pest control on turfgrasses: Diseases, animal enemies, and weeds).** M.L. Gullino, M. Mocioni, G. Zanin, and A. Alma. 196 pp. Italy.

2000 **Landscape Drainage Design.** Irrigation Association. 144 pp. USA.

2000 **Manejo de Céspedes Con Bajo Consumo de Agua (Management of Turfgrass with Low Water Consumption).** First ed. R.J. Monje-Jiménez. 101 pp. Spain.

2000 **The Mathematics of Turfgrass Maintenance.** Third rev. ed. N.E. Christians and M.L. Agnew. 176 pp. USA.

2000 **Natural Turf for Sport.** Sport England. 35 pp. UK.

2000 **Practical Drainage for Golf, Sportsturf, and Horticulture.** K. McIntyre and B.L.F. Jakobsen. 220 pp. USA.

2000 **School Sportsfields Handbook.** E. Hall and K. McAuliffe. 55 pp. New Zealand.

2000 **Tappeti Erbosi (Turfgrasses).** First ed. A. Panella, P. Croce, A. DeLuca, M. Falcinelli, F.S. Modestini, and F. Veronesi. 475 pp. Italy.

2000 **The Turfgrass Disease Handbook.** H.B. Couch. 209 pp. USA.

2000   **Turfgrass Management Information Directory.** Third rev. ed. K.J. Karnok. 256 pp. USA.
2000   **Turfgrass Maintenance Reduction Handbook.** A.D. Brede. 388 pp. USA.
2000   **Turfgrass Science and Management.** Third rev. ed. R.D. Emmons. 528 pp. USA.
2000   **Turf Managers' Handbook for Golf Course Construction, Renovation, and Grow-In.** C.B. White. 328 pp. USA.
2000   **Understanding Turf Management.** First ed. F.W. Sheard. 168 pp. Canada. Revised in 2005.
2000   **Water Right: Conserving Our Water, Preserving Our Environment.** Turfgrass Resource Center. 67 pp. USA.
2001   **Best Golf Course Management Practices.** First ed. L.B. McCarty. 700 pp. USA.
2001   **Color Atlas of Turfgrass Weeds.** First ed. L.B. McCarty, J.W. Everest, D.W. Hall, T.R. Murphy, and F. Yelverton. 287 pp. USA.
2001   **Cricket Wickets—Science v Fiction.** D.K. McIntyre and D.S. McIntyre. 292 pp. Australia.
2001   **Fudbalski Stadioni i Igrališta-Tehnički Kriteriji FSJ, UEFE i FIFE—Projektovanje, Izgradnja i Bezbednost-Evropska Igrališta i Održavanje (Football Stadium and Playground: Technical Requirement, Regulation, Safety on the FSJ, UEFA, FIFA—Design, Construction and Maintenance).** P.M. Bošković. 135 pp. Yugoslavia.
2001   **Handbook: Management of Turf and Turfgrass Research.** Japanese Society of Turfgrass Science. 423 pp. Japan.
2001   **Picture Perfect—Mowing Techniques for Lawns, Landscapes, and Sports.** D.R. Mellor. 192 pp. USA.
2001   **The Tale of Turfgrass.** F. Kitamura. 273 pp. Japan.
2001   **The Terminology of Turfgrass Management.** Second rev. ed. Y. Maki. 400 pp. Japan.
2001   **Turfgrass Soil Fertility and Chemical Problems—Assessment and Management.** R.N. Carrow, D.V. Waddington, and P.E. Rieke. 412 pp. USA.
2001   **Turf Management for Golf Courses.** Second rev. ed. J. B Beard. 793 pp. USA.
2002   **Building the USGA Green: Tips for Success.** United States Golf Association. 32 pp. USA.
2002   **The Cricket Pitch and its Outfield.** N.M. Tainton and J.R. Klug. 189 pp. South Africa.
2002   **Ecological Golf Course Management.** P.D. Sachs and R.T.E. Luff. 213 pp. USA.
2002   **Football Stadium and Playground For Hot Arid, Semiarid, and Wet Regions.** P.M. Bošković. 117 pp. Yugoslavia.
2002   **Growing Media for Ornamental Plants and Turf.** Third rev. ed. K.A. Handreck and N.D. Black. 544 pp. Australia.
2002   **KPGA Golf Course Manual.** Y.K. Joo. 386 pp. South Korea.
2002   **Managing Bermudagrass Turf.** L.B. McCarty and G.L. Miller. 235 pp. USA.
2002   **Mantenimiento de Campos de Golf (Golf Course Maintenance).** R. Monje-Jiménez. 264 pp. Spain.
2002   **New Creeping Bentgrass: Its Characteristics and Management for Putting Green.** H. Yanagi (ed). 223 pp. Japan.
2002   **Praxis des Sportplatzbaus (Practices in Sport Field Construction).** First ed. C.G. Matthias. 109 pp. Germany.
2002   **Scotts Lawns—Your Guide to a Beautiful Lawn.** N.E. Christians and A. Ritchie. 192 pp. USA.
2002   **Sports turf & amenity grasses: a manual for use and identification.** E I. Aldous and I.H. Chivers. 120 pp. Australia.
2002   **Turfgrass Management.** Sixth rev. ed. A.J. Turgeon. 448 pp. USA. Revised in 2005 and 2008.
2002   **Ways to Establish and Maintain Turf Ground.** Turfgrass Society of Korea. 165 pp. South Korea.
2003   **Baseball and Softball Fields: Design, Construction, Renovation, and Maintenance.** J. Puhalla, J. Krans, and M. Goatley. 235 pp. USA.
2003   **Color Atlas of Turfgrass Diseases.** Second rev. ed. A. Tanaka. 375 pp. Japan.
2003   **Controlling Turfgrass Pests.** Third rev. ed. T.W. Fermanian, M.C. Shurtleff, R. Randall, H.T. Wilkinson, and P.L. Nixon. 668 pp. USA.
2003   **Dictionary of Turfgrass Science.** Japanese Society of Turfgrass Science. 258 pp. Japan.
2003   **Flora of North America, Volume 25, Magnoliophyta: Commelinidae (in part): Poaceae, Part 2.** M.E. Barkworth (ed). 809 pp. USA.
2003   **Fundamentals of Turfgrass Agricultural Chemistry.** L.B. McCarty, I.R. Rodriquez, B.T. Bunnell, and F.C. Waltz. 384 pp. USA.
2003   **La Pianificazione della Costruzione di un Percorso di Golf (Planning the Construction of a Golf Course).** J. B Beard, P. Croce, A. DeLuca, and M. Mocioni. 52 pp. Italy.
2003   **Moj Travnjak (My Lawn).** P.M. Bošković. 72 pp. Serbia.
2003   **Practical Greenkeeping.** Jim Arthur. 213 pp. UK.
2003   **Turfgrass Biology, Genetics, and Breeding.** M.D. Casler and R.R. Duncan (eds). 377 pp. USA.
2003   **Turf Science and Maintenance of School Grounds.** M. Kondoh. 147 pp. Japan.
2004   **Applied Turfgrass Science and Physiology.** J.D. Fry and B. Huang. 320 pp. USA.
2004   **Design and Maintenance of Outdoor Sports Facilities.** P.L.K. Dury. 104 pp. UK.
2004   **Die Rasenfibel (The Turf Primer).** B. Billing. 190 pp. Switzerland.
2004   **Establishing and Maintaining the Natural Turf Athletic Field.** S.T. Cockerham, V.A. Gibeault, and D.B. Silva. 60 pp. USA.
2004   **Fundamentals of Turfgrass Management.** Second rev. ed. N. Christians. 365 pp. USA.

2004 **Golf Course Turf Management—Tools and Techniques.** D.H. Quast. 538 pp. USA.

2004 **Golf Greens—History, Design and Construction.** M.J. Hurdzan. 334 pp. USA.

2004 **Grama basta** *Stenotaphrum secundatum* **en parques y jardines (St. Augustine Grass [*Stenotaphrum secundatum*] in parks and lawns).** R. Monje-Jiménez. 67 pp. Spain.

2004 **Il campo pratica per il gioco del golf—Costruzione, manutenzione e gestione (The practice golf range—Construction, maintenance, and management).** M.P. Casati, P. Croce, A. DeLuca, and M. Mocioni. 82 pp. Italy.

2004 **Managing Healthy Sports Fields.** P.D. Sachs. 244 pp. USA.

2004 *Poa annua*—**Physiology, Culture, and Control of Annual Bluegrass.** J.M. Vargas Jr. and A.J. Turgeon. 175 pp. USA.

2004 **Problem Solving for Golf Courses, The Landscape, Sportsgrounds & Racecourses.** D.K. McIntyre. 440 pp. Australia.

2004 **Sve o Savremenim Fudbalskim i Golf Terenima (All about Football and Golf Fields).** P.M. Bošković and B.P. Bošković. 134 pp. Serbia.

2004 **Trávy a Trávniky—Co o nich ještĕ nevíte (Grasses and Turfgrasses—What you do not know about them).** F. Hrabĕ. 158 pp. The Czech Republic.

2004 **Turfgrass Chemicals and Pesticides—A Practitioner's Guide.** M.J. Fagerness and R. Johns. 52 pp. USA.

2004 **Turfgrass Installation—Management and Maintenance.** R.D. Johns. 599 pp. USA.

2004 **Turf Management In The Transition Zone.** J.H. Dunn and K.L. Diesburg. 288 pp. USA.

2004 **Turf Management—Tools and Techniques.** D.H. Quast and W.D. Otto. 538 pp. USA.

2005 through 2009

2005 **Beard's Turfgrass Encyclopedia—for Golf Courses, Grounds, Lawns, Sports Fields.** J. B Beard and H.J. Beard. 511 pp. USA.

2005 **Best Golf Course Management Practices.** Second rev. ed. L.B. McCarty. 896 pp. USA.

2005 **Compendium of Turfgrass Diseases.** Third rev. ed. R.W. Smiley, P.H. Dernoeden, and B.B. Clarke. 175 pp. USA.

2005 **Design, Construction, Maintenance, and Management of Golf Course.** D.H. Kim. 466 pp. South Korea.

2005 **Designing, Construction, and Maintaining Bermudagrass Sports Fields.** Second rev. ed. L.B. McCarty, G. Miller, C. Waltz, and T. Hale. 104 pp. USA.

2005 **The Florida Lawn Handbook.** Third rev. ed. L.E. Trenholm and J.B. Unruh. 182 pp. USA.

2005 **Guide d'Aménagement et d'Entretien desTerrains de Soccer Extérieurs (Guide for Installation and Maintenance of Outdoor Soccer Fields).** L. Gionet and D. Brown. 77 pp. Canada.

2005 **History Records of the Division C-5—Turfgrass Science 1946–2004. Volumes I through IV.** J. B Beard and H.J. Beard. 1,157 pp. USA.

2005 **Management of Turfgrass Diseases.** Third rev. ed. J.M. Vargas Jr. 337 pp. USA.

2005 **Practical Golf Course Maintenance—The Magic of Greenkeeping.** Second rev. ed. G. Witteveen and M. Bavier. 254 pp. USA.

2005 **Sports Turf & Amenity Grassland Management.** S. Brown. 192 pp. UK.

2005 **STRI Guidelines for Golf Green Construction in the United Kingdom.** S. Baker. 20 pp. UK.

2005 **The Superintendent's Guide to Controlling Putting Green Speed.** T.A. Nikolai. 160 pp. USA.

2005 **Turfgrass Management.** Seventh rev. ed. A.J. Turgeon. 427 pp. USA. Revised in 2008.

2005 **Understanding Turf Management.** Second rev. ed. R.W. Sheard. 170 pp. Canada.

2006 **Explanation of Turfgrass Terminology.** G.Y. Shim. 252 pp. South Korea.

2006 **Fundamentals of Tropical Turf Management.** G. Wiecko. 213 pp. USA.

2006 **Golf Course Architecture—Evolution of Design, Construction, and Restoration Technology.** Second rev. ed. M.J. Hurdzan. 464 pp. USA.

2006 **Handboek Greenonderhoud (Golf Green Maintenance Handbook).** S. Baker. 71 pp. UK.

2006 **Introductory Turfgrass Science.** K.N. Kim. 359 pp. South Korea.

2006 **Manejo de Céspedes Con Bajo Consumo de Agua (Management of Turfgrass with Low Water Consumption).** Second rev. ed. R.J. Monje-Jiménez. 112 pp. Spain.

2006 **Natural Grass and Artificial Turf: Separating Myths and Facts.** Turfgrass Resource Center. 31 pp. USA.

2006 **Rootzones, Sands and Top Dressing Materials for Sports Turf.** S. Baker. 104 pp. UK.

2006 **Sutton Seeds—A History 1806–2006.** Sutton & Sons Ltd. 256 pp. UK.

2006 **Tappeti Erbosi—Cura, gestione e manutenzione, delle aree verdi pubbliche e private (Turfgrasses—Care, management and maintenance of public and private green areas).** Second rev. ed. P. Croce, A. DeLuca, M. Falcinelli, F.S. Modestini, and F. Veronesi. 231 pp. Italy.

2006 **Tips & Traps for Growing and Maintaining the Perfect Lawn.** R. Johns. 204 pp. USA.

2006 **Turfgrass Dictionary.** D.H. Kim. 252 pp. South Korea.

2006 **Turfgrass Management.** K.N. Kim. 433 pp. South Korea.

2006 **The Turf Problem Solver—Case Studies and Solution for Environmental, Cultural, and Pest Problems.** A.J. Turgeon and J.M. Vargas Jr. 252 pp. USA.

2007 **Cultivation for Turfgrass Management—Its Effects and Method Using Machinery.** M. Inamori, H. Imaizumi, Y. Ushiki, K. Owata, S. Kimura, H. Tonogi, and S. Hayashi. 195 pp. Japan.

2007 **Flora of North America, Volume 24, Magnoliophyta: Commelinidae (in part): Poaceae, Part 1.** M.E. Barkworth. 939 pp. USA.

2007 **Fundamentals of Turfgrass Management.** Third rev. ed. N. Christians. 380 pp. USA.

2007 **How To Build A Sand Based Golf Green.** D.K. McIntyre, M. Parker, and D. Warwick. 250 pp. Australia.

2007 **Les Maladies Cryptogamiques des Gazons: Biologie et Contrôle (Fungal Diseases of Turfgrasses: Biology and Control).** Second rev. ed. A. Dehaye. 161 pp. France.

2007 **Manual of Grasses for North America.** M.E. Barkworth. 627 pp. USA.

2007 **STRI Disturbance Theory.** H.C. Bechelet and R.J. Windows. 80 pp. UK.

2007 **Turfgrass Establishment.** K.N. Kim. 427 pp. South Korea.

2007 **Ukrasni Travnjak (Decorative Turfgrass).** J.J. Čižek, P.M. Bošković, and N. Samardžija. 88 pp. Croatia.

2008 **All About Bowls.** Third rev. ed. J. Perris. 224 pp. UK.

2008 **Athletic Field Construction Manual.** R.W. Sheard. 73 pp. Canada.

2008 **Céspedes Ornamentales y Deportiva (Ornamentals and Sport Turf).** R. Monje-Jiménez. 527 pp. Spain.

2008 **Color Atlas of Turfgrass Weeds.** Second rev. ed. L.B. McCarty, J.W. Everest, D.W. Hall, T.R. Murphy, and F. Yelverton. 442 pp. USA.

2008 **El Swing Del Agua (Swing of the Water).** F.J. Garcia-Ircio. 92 pp. Spain.

2008 **Establishment and Management of Natural Bowling Greens in New Zealand.** Second rev. ed. D. Ormsby. 234 pp. New Zealand.

2008 **The Mathematics of Turfgrass Maintenance.** Fourth rev. ed. N.E. Christians and M.L. Agnew. 168 pp. USA.

2008 **Nutrition of Sports Turf in Australia.** J. Spencer. 206 pp. Australia.

2008 **Praxis des Sportplatzbaus (Practices in Sport Field Construction).** Second rev. ed. C.G. Matthias and W.J. Bartz. 118 pp. Germany.

2008 **Sports Field Management.** J. Gruttadaurio. 42 pp. USA.

2008 **Sports Turf Management in the Transition Zone.** M. Goatley, S. Askew, E. Ervin, D. McCall, B. Studholme, P. Schultz, and B. Horvath. 203 pp. USA.

2008 **Turfgrass Management.** Eighth rev. ed. A.J. Turgeon. 448 pp. USA.

2008 **Turfgrass Science and Management.** Fourth rev. ed. R.D. Emmons. 584 pp. USA.

2008 **Water Quality and Quantity Issues for Turfgrasses in Urban Landscapes.** J. B Beard and M.P. Kenna (eds). 318 pp. USA.

2009 **Establishment and Management of Croquet Lawns in New Zealand.** D. Ormsby. 236 pp. New Zealand.

2009 **Trávniky: Pro zahradu, kraginu a sport (Turf for Garden, Landscape and Sports).** F. Hrabě. 335 pp. The Czech Republic.

2009 **Turfgrass and Landscape Irrigation Water Quality: Assessment and Management.** R.R. Duncan, R.N. Carrow, and M.T. Huck. 485 pp. USA.

2009 **Ways to Establish and Maintain Turf Ground.** D.K. Kim. 150 pp. South Korea.

200x **Facts About Artificial Turf and Natural Grass.** Turfgrass Resource Center. 28 pp. USA.

2010 **Growing Media for Ornamental Plants and Turf.** Fourth rev. ed. K. Handreck. 559 pp. Australia.

Figure 13.1. Turfgrass Information Center at the Michigan State University Library, Overview (above) and Indexing and Abstracting Section (below). [6]*

---

* The numbers in brackets correspond to the figure source acknowledgments listed in the appendix.

CHAPTER 13

Specialty Turfgrass Books/Monographs

<small>Please see chapter 12 for details about the format of the references in this chapter.</small>

A. TURFGRASS IDENTIFICATION BULLETINS AND PUBLICATIONS

Aldous, A.E.

1932† **The Identification of Certain Native and Naturalized Grasses By Their Vegetative Characters.** Coauthor.
　　See *Copple, R.F.—1932.*

Bennett, Hugh H.

1950† **The Identification of 76 Species of Mississippi Grasses By Vegetative Morphology.*** Hugh H. Bennett, R.O. Hammons, and W.R. Weissinger. Technical Bulletin 31, Agricultural Experiment Station, Mississippi State College, State College, Mississippi, USA. 108 pp., 187 illus., 1 table, small index, soft bound in tan with black lettering, 6.1 by 8.6 inches (155 × 225 mm).
　　The publication contains detailed descriptions and drawings to use in the vegetative identification of 76 grasses. Most of the species are C_4, warm-season grasses. A small glossary and a bibliography are included in the back.

Campbell, J.A.

1950† **The Identification of Certain Native and Naturalized Grasses By Their Vegetative Characters.*** Coauthor.
　　See *Clarke, S.E.—1950.*

Carrier, Lyman

1917† **The Identification of Grasses By Their Vegetative Characters.** Lyman Carrier. Bulletin No. 461, United States Department of Agriculture, Washington, D.C., USA. 30 pp., 126 illus., no tables, no index, fold-stapled soft covers in white with black lettering, 7.8 by 9.1 inches (147 × 230 mm).
　　A pioneering small booklet with detailed vegetative descriptions and drawings for 48 seedling grasses.

Clarke, S.E.

1950† **The Identification of Certain Native and Naturalized Grasses By Their Vegetative Characters.*** S.E. Clarke, J.A. Campbell, and W. Shevkenek. Publication No. 762, Technical Bulletin No. 50, Canada Department of Agriculture, Ottawa, Canada. 129 pp., 220 illus., 1 table, indexed by both common and botanical scientific names, soft bound in tan with black lettering, 6.4 by 9.8 inches (164 × 249 mm).
　　This publication contains detailed descriptions and drawings to use in the vegetative identification of 101 grasses. Most of the species are perennial, cool-season, C_3 grasses found in Canada. A small glossary and a selected bibliography are included in the back.

Clouston, David W.A.

1962† **Identification of Grasses in Non-Flowering Condition.** David Clouston. Sports Turf Research Institute, Bingley, West Yorkshire, England, UK. 31 pp., 55 illus., 2 tables, no index, fold-stapled semisoft covers in light green with green lettering, 7.1 by 9.7 inches (181 × 246 mm).

A small booklet on the identification of grasses, primarily by means of leaf blade cross sections, which are illustrated by detailed photomicrographs. A technique for cutting transverse sections of grass leaves is presented in the appendix along with a small glossary and short bibliography. Emphasis is on cool-season turfgrasses found in the United Kingdom. A unique booklet based on the author's PhD thesis completed in 1935 at the University of Edinburgh.

Converse, James Thayer

1979† **Scotts Guide to the Identification of Grasses.** Jim Converse. O.M. Scott & Sons Company, Marysville, Ohio, USA. 82 pp., 73 illus., no tables, index in front, semisoft covers in green with gold lettering in a ring binder, 6.3 by 3.3 inches (159 × 83 mm).

A field guidebook for the identification of 60 grasses, both desirable turfgrasses and weedy grasses, commonly found in North America. It includes both written descriptions and drawings of key vegetative structures, inflorescences, and seeds. Regional vegetative identification keys are presented in the front. It is a comprehensive, quality reference guide.

n.d.† **Proturf Guide to the Identification of Grasses.** Jim Converse. O.M. Scott & Sons Company, Marysville, Ohio, USA.
circa 82 pp., 73 illus., no tables, index in front, semisoft covers in black with orange and green lettering in a ring binder, 6.3
1980s by 3.3 inches (159 × 83 mm).

This is essentially a reprint of the 1979 guide.

Copple, R.F.

1932† **The Identification of Certain Native and Naturalized Grasses by Their Vegetative Characters.** R.F. Copple and A.E. Aldous. Agricultural Experiment Station, Kansas State College of Agriculture and Applied Science, Manhattan, Kansas, USA. 73 pp., 182 illus., no tables, small index, fold-stapled covers in white with black lettering, 6.0 by 8.9 inches (152 × 226 mm).

An early, small book concerning the methods for vegetative identification of grasses and associated descriptions for 26 species. It includes a glossary and vegetative identification key.

Fischer, Walther Paul

1959† **Unsere Wichtigsten Gräser: mit Beschreibung ihrer Unterscheidungsmerkmale, ihrer Wachstumsbedingungen und ihrer Bedeutung für Landwirtschaft und Samenhandel (Our Most Important Grasses: Their importance for agriculture and the seed trade).** First edition. Walther Fischer. Verlag Mensing & Company, Hamburg, West Germany. 85 pp., 316 illus. with 247 in color, no tables, no index, glossy semisoft bound in greens and cream with black and green lettering, 6.6 by 9.4 inches (168 × 239 mm). Printed entirely in the German language.

A book concerning 34 cool-season grasses that occur in Germany, including the identifying characteristics and adaptation. Each grass is illustrated in detail in terms of growth stages and structures utilized in identification. The seed of each grass is illustrated in the back. The watercolor illustrations are by Ursla Jacobsen-Lorenzen.
Revised in 1972.

1972† **Die Wichtigsten Gräser (The Most Important Grasses).** Second revised edition. Walther Fischer and Ernst Lütke-Entrup. Mensing & Company, Hamburg-Norderstedt, West Germany. 120 pp., 90 illus. with 34 in color, 17 tables, contains an index of the grasses, cloth hardbound in light green with dark green lettering, 6.5 by 9.4 inches (166 × 238 mm). Printed entirely in the German language.

A book describing the grasses commonly used in Germany, including detailed color drawings and growth characteristics. In the back, the culture and use of the grasses, including turfgrasses, are addressed. It also contains a chapter on seed production. There are color plates of the major cool-season turfgrasses, including both vegetative and flowering plants.

Fults, Jess L.

1956† **Colorado Turfgrasses.** Jess L. Fults. Circular 2663, Extension Service, Colorado Agricultural and Mechanical College, Fort Collins, Colorado, USA. 28 pp., 159 illus., no tables, indexed, fold-stapled semisoft covers in light green with black lettering, 8.5 by 10.9 inches (215 × 278 mm).

A booklet concerning the procedures for vegetative identification of 50 grasses, both turfgrasses and weedy grasses. A key for vegetative identification and associated drawings is presented. Mainly C_3, cool-season turfgrasses are included. It also has a small glossary in the back.

Figure 13.2. Demonstration of creeping bentgrass lateral sod strength, circa 1930. [21]

Reprinted in 1959 as Circular 201-A.

Hammons, R.O.

1950† **The Identification of 76 Species of Mississippi Grasses By Vegetative Morphology.** Coauthor.
See *Bennett, Hugh H.—1950.*

Lütke-Entrup, Ernst

1972† **Die Wichtigsten Gräser (The Most Important Grasses).** Coauthor.
See *Fischer, Walther—1972.*

Phillips, C.E.

1962† **Some Grasses of the Northeast.** C.E. Phillips. Field Manual No. 2, Agricultural Experiment Station, University of Delaware, Newark, Delaware, USA. 77 pp., 300 illus., 1 table, indexed, semisoft covers in light green with black lettering in a wire ring binder, 8.5 by 10.9 inches (215 × 278 mm).
A publication on the procedures and guidelines for the identification of grasses by vegetative characters. Aids include a key and detailed drawings for each of 261 grasses. There is a glossary and a list of references at the back. It is an extensive work.

Reed, Frederick John

n.d.† **Lawn Grasses.** F.J. Reed. Ryders' Lawn Grass Seed Specialists Limited, St Albans, England, UK., 28 pp., 25 illus., 2 tables,
circa no index, soft bound in green with dark green lettering, 4.9 by 8.0 inches (125 × 203 mm).
1952 A small booklet on cool-season turfgrasses as used in the United Kingdom. It includes both vegetative and seed identification guides.

Shevkenek, W.

1950† **The Identification of Certain Native and Naturalized Grasses by Their Vegetative Characters.** Coauthor.
See *Clarke, S.E.—1950.*

Syngenta Crop Protection, Ltd.

2009† **Grass ID Guide.** Syngenta Crop Protection UK Ltd., Cambridge, Cambridgeshire, England, UK. 27 pp., 49 illus. in color, 3 tables, no index, glossy semisoft waterproof covers and pages in multicolor with black and white lettering in a wire binder, 3.1 by 7.1 inches (80 × 180 mm).
A guide to the vegetative identification of 12 turfgrasses and weedy grasses commonly found in the United Kingdom.

Weissinger, W.R.

1950† **The Identification of 76 Species of Mississippi Grasses By Vegetative Morphology.*** Coauthor.
See *Bennett, Hugh H.—1950.*

B. TURFGRASS EQUIPMENT SPECIALTY BOOKS FOR MOWING, IRRIGATION, CULTURE, AND ESTABLISHMENT

Austin, Richard L.

1990† **Lawn Sprinklers—A Do-It-Yourself Guide.*** Richard L. Austin. TAB Books, Blue Ridge Summit, Pennsylvania, USA. 159 pp., 110 illus., 13 tables, small index, textured hardbound in black with silver lettering, 7.3 by 9.1 inches (185 × 232 mm).
Guidebook for home lawn enthusiasts interested in installing an irrigation system. It includes types of systems, components, design, and installation. There is a small glossary in the back.

1990† Also printed in a glossy semisoft bound version in multicolor with white and black lettering.

Barrett, James

2003† **Golf Course Irrigation—Environmental Design and Management Practices.*** James Barrett, Brian Vinchesi, Robert Dobson, Paul Roche, and David Zoldoske. John Wiley & Sons Inc., Hoboken, New Jersey, USA. 464 pp., 180 illus., 27 tables, extensive index, glossy hardbound in multicolor with yellow, white, and black lettering, 7.8 by 9.9 inches (198 × 250 mm).
A specialty book on the design, materials, planning, installation, and maintenance of irrigation systems for golf courses plus chapters on water requirements, water sources, and pumps. Formulas and examples of computations are presented in the text. There is a small glossary in the back, along with reference tables and charts.

Bell, Brian

1991† **Machinery for Horticulture.*** First edition. Brian Bell and Stewart Cousins. Farming Press Books, Ipswich, Suffolk, England, UK. 295 pp., 351 illus. with 17 in color, 3 tables, indexed, glossy hardbound in multicolor with white and black lettering, 7.4 by 9.6 inches (187 × 244 mm).
A detailed book on the types, selection, operation, and maintenance of equipment used in professional turfgrass, landscape, and vegetable culture in the United Kingdom. There also are chapters on greenhouses and the workshop.

1995† **Fifty Years of Garden Machinery.** Brian Bell. Farming Press Books, Ipswich, Suffolk, England, UK. 235 pp., 367 illus. with 42 in color, no tables, small index, glossy hardbound in multicolor with white lettering, 7.4 by 9.8 inches (189 × 249 mm).
A specialty book concerning the history of powered gardening equipment since 1945, with a focus on UK companies and their equipment lines.

1997† **Machinery for Horticulture.*** Second revised edition. Brian Bell and Stewart Cousins. Farming Press, Ipswich, Suffolk, England, UK. 303 pp., 378 illus. with 34 in color, 2 tables, indexed, glossy hardbound in multicolor with yellow and white lettering, 7.6 by 9.4 inches (193 × 240 mm).
An update of the first edition.

Bork, Dean R.

1978† **Estimating The Feasibility of Sanitary Wastewater Effluents For Recreation Area Irrigation Projects.*** Dean R. Bork. The David Gill Corporation, St Charles, Illinois, USA. 22 pp., 9 illus., 5 tables, no index, soft bound in multicolor with black lettering, 8.8 by 10.9 inches (224 × 278 mm).
A small booklet concerning the planning and specifications involved in the use of effluent water for irrigation of turfgrass areas. It involves water requirements, inflow rates, storage capacity, and seasonal availability.

Castellano, Carmine

1975† **You Fix It: Lawn Mowers.*** Carmine C. Castellano and Clifford P. Seitz. Arco Publishing Inc., New York City, New York, USA. 256 pp.; 309 illus.; 33 tables; indexed; glossy semisoft bound in green, black and white with black, green, and white lettering; 7.8 by 10.2 inches (198 × 259 mm).
A detailed book on engine and mower problem diagnosis and repair.

Choate, Richard B.

1987† **Turf Irrigation Manual.*** Fourth revised edition. Coauthor.
See *Watkins, James A.—1987.*

Figure 13.3. Manual-push, cast iron rollers, with a width and weight ranging from 22 inches (55.9 cm) and 200 pounds (90.7 kg) to 30 inches (76.2 cm) and 350 pounds (158.8 kg), 1920s. [34]

1994† **Turf Irrigation Manual: The Complete Guide to Turf and Landscape Irrigation System.** Fifth revised edition. Richard B. Choate. Weather-matic Publication, Dallas, Texas, USA. 396 pp., 400 illus., 203 tables, indexed, glossy hardbound in multicolor with white lettering, 8.4 by 10.9 inches (214 × 278 mm).

A substantial revision of the 1987 edition. A full chapter on micro-spray and drip irrigation has been added.

Colvin, Thomas S.

1974 **Operating Tractors for Grounds Keeping and Ornamental Horticulture.** Thomas S. Colvin. American Association for Vocational Instructional Materials, Athens, Georgia, USA. 95 pp.; 261 illus.; no tables; small index; fold-stapled glossy semisoft covers in multicolor with white, black, and orange lettering; 8.4 by 9.8 inches (214 × 250 mm).

A manual for vocational students studying for employment in the grounds maintenance field.

Cousins, Stewart

1991† **Machinery for Horticulture.** Coauthor.
&   See *Bell, Brian—1991 and 1997.*
1997

Crawley, M.C.

1980† **Motor Lawnmowers—Owners Workshop Manual.** First edition. M.C. Crawley. Haynes Publishing Group, Somerset, England, UK. 164 pp.; 617 illus.; 1 table; indexed; glossy hardbound in multicolor with yellow, white, and black lettering; 8.2 by 10.7 inches (209 × 271 mm).

A book on the preventive servicing, problem diagnosis, and repair of lawn mowers, including engines. It is an extensive step-by-step pictorial format. The book is oriented to conditions and mowers manufactured in the United Kingdom.

†   *Reprinted in 1984.*

Cuthbertson, Tom

1974† **My Lawnmower Hates Me.** Tom Cuthbertson. Ten Speed Press, Berkeley, California, USA. 260 pp., 84 illus., no tables, indexed, both cloth hard and semisoft bound versions in white and black with green and brown lettering, 5.9 by 8.9 inches (150 × 227 mm).

A manual on lawnmowers, including selection, operation, maintenance, repair, and winterizing. It is oriented to mowers manufactured in North America.

Dobson, Robert

2003† **Golf Course Irrigation—Environmental Design and Management Practices.*** Coauthor.
See *Barrett, James—2003.*

Farley, Nicholas

1980† **Handbook of Garden Machinery and Equipment.*** Nicholas Farley. J M Dent & Sons Limited, London, England, UK. 272 pp., 152 illus., 24 tables, indexed, cloth hardbound in green with gold lettering and a jacket in multicolor with yellow and white lettering, 6.0 by 9.1 inches (153 × 231 mm).

A unique book on the selection, operation, and maintenance of gardening equipment, both manual and powered, including mowers, trimmers, spreaders, and irrigation equipment for turfed areas. It is oriented to equipment found in the United Kingdom.

Fitcher, Harold O.

1977† **The Complete Guide to Lawn Mowers.*** H.O. Fitcher. Menaid Press, Colorado Springs, Colorado, USA. 562 pp., 314 illus., 33 tables, small index, textured hardbound in green and black with black lettering, 8.5 by 11.0 inches (215 × 279 mm).

A comprehensive lawn mower guide, including purchase, operation, maintenance, troubleshooting, engines, and power train.

Foerste, Eleanor C.

1984† **Mowing For Money.*** Eleanor C. Foerste. Osceola County Cooperate Extension Service, Kissimmee, Florida, USA. 39 pp., 43 illus., 10 tables, no index, textured semisoft covers in light green in a 3-hole binder, 8.5 by 11.0 inches (216 × 279 mm).

A small guide to be used as a youth 4-H project. It involves lawn mowing on a fee basis.

Gasch, Rodney J.

2001† **Safety Management for Landscapers, Grounds-Care Businesses and Golf Courses.** Rodney J. Gasch. John Deere Publishing, Deere & Company, Moline, Illinois, USA. 161 pp., 351 illus., no tables, small index, glossy semisoft bound in multicolor with red and black lettering, 8.3 by 10.8 inches (211 × 274 mm).

A comprehensive book concerning the hazards and safe working practices for outdoor power equipment, compact tractors, loaders and backhoes, rotary mowers, lawn grooming equipment, chain saws, golf course equipment, chemical application equipment, utility vehicles, and trailering equipment. Also included are safety in the maintenance shop and organizing a safety program.

Hawker, Michael Frederick James

1971† **Horticultural Machinery.*** First edition. M.F.J. Hawker and J.F. Keenlyside. Macdonald Horticultural Series, Macdonald & Company (Publishers) Limited, London, UK. 175 pp., 123 illus., 3 tables, small index, semisoft bound in orange with white and black lettering, 5.5 by 8.5 inches (137 × 215 mm).

This book is designed as an introduction to machinery used in professional turfgrass, landscape, and nursery cultural practices. The basic mechanical principles are introduced.
Revised in 1977 and 1985.

1977† **Horticultural Machinery.*** Second revised edition. M.F.J. Hawker and J.F. Keenlyside. Longman Group Limited, London, England, UK. 202 pp., 151 illus., 6 tables, indexed, glossy semisoft bound in greens and black with white lettering, 5.1 by 8.4 inches (130 × 214 mm).

A revision and update of the first edition. New chapters on nursery machinery, chain saws, and glasshouse equipment have been added.
Revised in 1985.

1985† **Horticultural Machinery.*** Third revised edition. M.F.J. Hawker and J.F. Keenlyside. Longman Group Limited, London, England, UK. 190 pp., 141 illus., 2 tables, small index, glossy semisoft bound in multicolor with white and red lettering, 5.4 by 8.5 inches (137 × 216 mm).

A major update from 1977, with new sections on brakes, governors, hydrostatic transmissions, power steering, and rotary sieves.

The Irrigation Association

2000 **Advanced Head Layout.*** The Irrigation Association, Falls Church, Virginia, USA. 128 pp., 91 illus., 36 tables, no index, glossy semisoft bound in multicolor with red lettering, 8.3 by 10.7 inches (212 × 273 mm).

A manual concerning advanced design skills for planning irrigation systems for landscapes. It is oriented to experienced irrigation design specialists.

2000 **Certified Irrigation Designer Reference Manual.**[*] The Irrigation Association, Falls Church, Virginia, USA. 202 pp., 161 illus., 42 tables, indexed, glossy semisoft bound in grays with black lettering, 8.3 by 10.8 inches (212 × 274 mm).

A reference manual designed as a basic reference for the Irrigation Association Certification Examinations. Included are hydraulics, system hardware, pumps, electricity aspects, and scheduling for various types of irrigation.

2001 **Predicting and Estimating Landscape Water Use.**[*] The Irrigation Association, Falls Church, Virginia, USA. 82 pp., 36 illus., 37 tables, no index, glossy semisoft bound in multicolor with dark green lettering, 8.3 by 10.8 inches (212 × 274 mm).

A manual regarding the principles and calculations required to best estimate water usage in landscapes.

2002 **Certified Landscape Irrigation Auditor.**[*] The Irrigation Association, Falls Church, Virginia, USA. 236 pp., 39 illus., 78 tables, no index, glossy semisoft bound in multicolor with dark green lettering, 8.3 by 10.7 inches (212 × 273 mm).

A training manual for landscape irrigation auditors. Included are procedures for performing landscape irrigation audits and developing irrigation-scheduling programs. Specific system tests, work sheets for data collection, and calculation procedures are presented.

Jarrett, Albert R.

1985† **Golf Course and Grounds Irrigation and Drainage.**[*] Albert R. Jarrett. A Reston Book, Regents/Prentice-Hall Inc., Englewood Cliffs, New Jersey, USA. 246 pp., 251 illus., 97 tables, indexed, glossy hardbound in gray and greens with white and black lettering, 9.5 by 10.9 inches (214 × 277 mm).

A book on water management of golf course, sports fields, and lawn areas via irrigation and drainage systems. It encompasses planning, design, installation, and operation. It is oriented to turfgrass managers, architects, and engineers in the landscape area.

Keenlyside, John Frederick

1971, **Horticultural Machinery.**[*] Coauthor.
1977,†   See *Hawker, Michael Frederick James—1971, 1977, and 1985.*
&
1985†

Keesen, Larry

1995† **The Complete Irrigation Workbook.**[*] Larry Keesen. Franzak & Foster, Cleveland, Ohio, USA. 263 pp., 122 illus. with 99 in color, 66 tables, small index, glossy semisoft bound in multicolor with red, green, and white lettering, 8.1 by 10.5 inches (206 × 266 mm).

A quality book on lawn and landscape irrigation. It includes hydraulic principles, design basics, irrigation components, installation, problem diagnosis, system maintenance, and pumps. A set of questions is included at the back of each chapter. There is a small glossary in the back.

Figure 13.4. Horse-drawn, cast iron roller of four sizes, ranging from 4 feet wide (1.2 m) and 950 pounds (430.9 kg) to 6 feet wide (1.8 m) and 2,200 pounds (997.9 kg). [34]

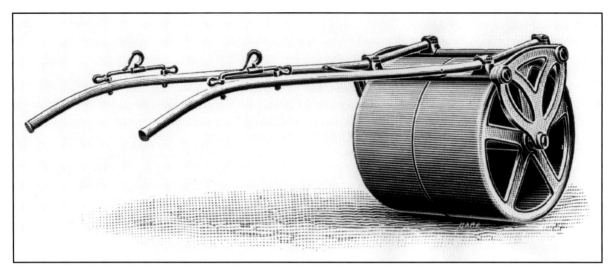

Figure 13.5. Halved, rounded-edged, 48-inch-wide (122-centimeter-wide), metal Shanks horse and pony roller, 1914. [38]

Martin, William (Bill) A.

1978 **Irrigation.** First edition. W.A. Martin and M.J. Mortimer. Management Aid No. 2, Royal Australian Institute of Parks and Recreation, Belconnen, Capital Territory, Australia. 32 pp. with 1 p. of advertising, 17 illus., 8 tables, no index, textured semisoft bound in green with black and red lettering, 6.9 by 9.75 inches (175 × 248 mm).

   A booklet on irrigation of turfgrass and landscape areas. It includes plant-water relationships, evapotranspiration, irrigation types, system components, design, drainage, and maintenance.

1988† **Turf Irrigation.** Second revised edition. Coauthor.
   See *McIntyre, David Keith—1988.*

Melby, Pete

1988† **Simplified Irrigation Design.** First edition. Pete Melby. Van Nostrand Reinhold, New York City, New York, USA. 196 pp., 272 illus., 47 tables, small index, both cloth hardbound in black with white lettering and also glossy semisoft bound in multicolor with yellow and black lettering, 8.4 by 10.8 inches (213 × 274 mm).

   This manual is oriented for those individuals who wish to be self-taught regarding the design of irrigation systems for turfgrass and landscape areas. There are sections devoted to drip irrigation in the back.
   Revised in 1995.

1995† **Simplified Irrigation Design.** Second revised edition. Pete Melby. John Wiley & Sons Inc., New York City, New York, USA. 240 pp.; 172 illus.; 80 tables; small index; glossy semisoft bound in multicolor with orange, white, red, and black lettering; 8.4 by 10.8 inches (213 × 274 mm).

   This revised edition has been expanded, especially the tables. It is described as the professional designer and installer version with measurements in both imperial and metric units.

Milne, George

1995† **Rotary & Cylinder Lawnmowers—Owners Workshop Manual.** Second revised edition. George Milne, Haynes Publishing, Somerset, England, UK. 170 pp.; 960 illus.; no tables; small index; glossy semisoft bound in multicolor with black, white, and yellow lettering; 8.2 by 10.6 inches (209 × 268 mm).

   An illustrated guidebook concerning the step-by-step procedures for the maintenance, repair, and renovation of both rotary and reel mowers. It is a revised and updated version of the 1985 edition.

Mortimer, M. John

1978 **Irrigation.** Coauthor.
   See *Martin, William (Bill) A.—1978.*

Nunn, Richard V.

1984† **Lawnmowers & Garden Equipment.** Richard V. Nunn. Creative Homeowner Press, Possair, New Jersey, USA. 160 pp.; 599 illus.; 13 tables; small index; glossy semisoft bound in multicolor with red, white, and black lettering; 8.5 by 11.0 inches (216 × 279 mm).

A guide to the repair, maintenance, and storing of mowers, lawn tractors, motors, chain saws, grass trimmers, snow throwers, garden tillers, and hand tools.

Outdoor Power Equipment Institute

1986† **American National Standard for Turf Care Equipment.*** Outdoor Power Equipment Institute and American National Standards Institute, New York City, New York, USA. 43 pp., 17 illus., no tables, no index, fold-stapled semisoft covers in olive and cream with cream and olive lettering, 8.5 by 11.0 inches (215 × 278 mm).

    This small book consists of a statement of standards for the design, manufacture, safety specifications, and operation of power lawn mowers, lawn and garden tractors, and lawn tractors. First published in 1960, these standards have been revised and republished many times.

Parker, J.M.F.

1985† **Motor Lawnmowers 2—Owners Workshop Manual.** Second revised edition. J.M.F. Parker. Nr Yeovil, Somerset, England, UK. 190 pp.; 780 illus.; 1 table; small index; glossy hardbound in multicolor with yellow, white, and black lettering; 8.2 by 10.7 inches (209 × 272 mm).

    A major revision and update of the 1980 first edition.
    See *Crawley, M.C.—1980.*
    Revised in 2000.
    See *Sparks, Andrew—2000.*

Peterson, Franklynn

1973† **Handbook of Lawn Mower Repair.*** First edition. Franklynn Peterson. Emerson Books Inc., Buchanan, New York, USA. 253 pp.; 48 illus.; 2 tables; index; cloth hardbound in light green with black lettering and a jacket in greens and white with white, black, and green lettering; 5.3 by 7.9 inches (136 × 200 mm).

    A handbook on the selection, operation, maintenance, and repair of lawnmowers. It is practically oriented to the home lawn enthusiasts and the mowers manufactured in the United States.
    Revised in 1978.

1978† **Handbook of Lawn Mower Repair.*** Second revised edition. Franklynn Peterson. Hawthorne Books Inc., New York City, New York, USA. 253 pp., 119 illus., 5 tables, indexed, glossy semisoft bound in multicolor with white and black lettering, 5.3 by 7.8 inches (135 × 198 mm).

    This book revision consists of updates and expansion of the visual presentations.

1982† Reprinted by Emerson Books Inc., Verplanck, New York, USA, in cloth hardbound in green with black lettering and a jacket in greens, black, and white with white and black lettering.

1984† Reprinted by Perigee Books, New York City, New York, USA, in glossy semisoft bound in multicolor with yellow and white lettering.

Piersol, John R.

2009† **Turf Maintenance Facility Design and Managment.*** John R. Piersol and Harry V. Smith. John Wiley and Sons, Hoboken, New Jersey, USA. 192 pp., 59 illus., 2 tables, indexed, glossy hardbound in multicolor with white and yellow lettering, 6.1 by 9.2 inches (154 × 234 mm).

    A specialty book concerning the design layout, organization, safety aspects, and operation of the maintenance facility for golf course and sports facilities. Also included are shop mechanic and manager job description, equipment features, diagnostic techniques, and record forms.

Pira, Edward S.

1963 **System Design for Golf Course Irrigation.*** First edition. Edward S. Pira. Agricultural Engineering Department, University of Massachusetts, USA. 89 pp., ~80 illus., ~40 tables, no index, soft covers in white with black lettering and stapled at side, 8.4 by 10.9 inches (214 × 279 mm).

    This pioneering instructional book is organized in a course outline for individuals studying the principles of irrigation design and operation, with emphasis on golf course turfgrasses. Two chapters are included that address subsurface drainage systems and the associated basic surveying methods. There is a glossary of terms in the appendix.
    Revised in 1968, 1973, 1975, 1979, 1982, 1989, and 1997.

1968† **Golf Course Irrigation System Design.*** Second revised edition. Edward S. Pira. Agriculture Engineering Department, University of Massachusetts, Amherst, Massachusetts, USA. 109 pp., 97 illus., 40 tables, no index, semisoft covers in light blue with dark blue lettering and stapled at side, 8.4 by 10.9 inches (214 × 279 mm).

A practical instructional book on the design of irrigation systems, especially for golf course turfgrasses. There are two parts. First is a step-by-step procedure in obtaining preliminary design data, and the second part provides guidance in applying the initial design data to the determination of the final design.
Revised in 1973, 1975, 1979, 1982, 1989, and 1997.

1973† **Golf Course Irrigation System Design.*** Third revised edition. Edward S. Pira. Food and Agricultural Engineering Department, University of Massachusetts, Amherst, Massachusetts, USA. 136 pp., 140 illus., 42 tables, no index, semi-soft covers in light green with dark green lettering in a comb binder, 8.5 by 11.0 inches (217 × 279 mm).

This third edition involves updated revisions and additions, including more drawings and table.
Revised in 1975, 1979, 1982, 1989, and 1997.

1975† **Golf Course Irrigation System Design.*** Fourth revised edition. Edward S. Pira. Food and Agricultural Engineering Department, University of Massachusetts, Amherst, Massachusetts, USA. 242 pp., 373 illus., 64 tables, no index, semi-soft covers in yellow with black lettering in a comb binder, 8.5 by 10.9 inches (215 × 278 mm).

The fourth edition is a major revision, with the addition of actual photographs. There is a glossary of selected terms in the back.
Revised in 1979, 1982, 1989, and 1997.

1979† **Golf Course Irrigation System Design.*** Fifth revised edition. Edward S. Pira. Food and Agricultural Engineering Department, University of Massachusetts, Amherst, Massachusetts, USA. 265 pp. 262 illus., 78 tables, no index, semi-soft covers in yellow with black lettering in a comb binder, 8.5 by 10.9 inches (215 × 278 mm).

This fifth edition contains minor revisions.
Revised in 1982, 1989, and 1997.

1982† **Golf Course Irrigation System Design.*** Sixth revised edition. Edward S. Pira. Food and Agricultural Engineering Department, University of Massachusetts, Amherst, Massachusetts, USA. 309 pp., 402 illus., 91 tables, no index, semi-soft covers in yellow with black lettering in a comb binder, 8.5 by 10.9 inches (215 × 278 mm).

This sixth edition is a detailed manual for students interested in the principles of irrigation system design, specifications, installation, and operation for golf course turfgrasses. It is practically oriented and extensively illustrated, with a glossary of terms and a sample contract in the back.
Revised in 1989 and 1997.

1989† **Golf Course Irrigation System Design and Drainage.*** Seventh revised edition. E.S. Pira. Food Engineering Department, University of Massachusetts, Amherst, Massachusetts, USA. 432 pp., 624 illus., 146 tables, no index, semisoft covers in green with black lettering in a comb binder, 8.5 by 10.9 inches (215 × 278 mm).

This seventh edition is an extensive manual on the design, specifications, and installation of irrigation systems for golf course turfgrasses. It includes sections on drainage systems for golf courses in the back. It is organized in an instruction format.
Revised in 1997.

1997† **A Guide to Golf Course Irrigation System Design and Drainage.*** E.S. Pira. Turfgrass Science and Practice Series, Ann Arbor Press, Chelsea, Michigan, USA. 434 pp., 650-plus illus., 120 tables, small index, glossy hardbound in multicolor with white lettering, 8.5 by 11.0 inches (216 × 279 mm).

An updated version of the 1989 edition.

Randall, Martynn

2000 **Lawn Mower Manual.*** Coauthor.
See *Sparks, Andrew—2000.*

Reader's Digest Association

1975 **The Use and Care of Lawn Mowers.*** The Reader's Digest Association Limited, London, England, UK. 48 pp.; 524 illus.; 2 tables; no index; soft bound in light green, white, and black with black and white lettering; 8.0 by 10.6 inches (204 × 269 mm).

A detailed guide to the servicing and repair of power units used on lawn mowers.

Roberts, John Mack

1988 **Irrigation Volume III of Handbook of Landscape Architectural Construction.*** Coauthor.
See *Weinberg, Scott S.—1988.*

Robinson, John

1983† **Keep Your Mower Going.*** John Robinson. Elliot Right Way Books, Kingswood, Surrey, England, UK. 158 pp., 60 illus., 7 tables, small index, soft bound in multicolor with red and black lettering, 4.3 by 7.1 inches (109 × 180 mm).

A small book on the servicing, problem diagnosis, and repair of lawn mowers. It is oriented to UK manufactured mower sources.

Robinson, Michael Richard

1990 **Turf spraying—a practical guide.*** Michael Robinson. Turfgrass Technology, Seaford, Victoria, Australia. 60 pp., 58 illus., 12 tables, no index, glossy semisoft bound in multicolor with black lettering, 8.1 by 11.6 inches (207 × 295 mm).

A guidebook on safe pesticide application, including the sprayer apparatus, sprayer calibration, pesticide calculations, sprayer operation, and sprayer cleaning and servicing.

Roche, Paul

2003† **Golf Course Irrigation—Environmental Design and Management Practices.*** Coauthor.
See *Barrett, James—2003.*

Rochester, Eugene W.

1995† **Landscape Irrigation Design.*** Eugene W. Rochester. ASAE Pub. No. 8, American Society of Agricultural Engineers, St Joseph, Michigan, USA. 217 pp., 143 illus., 60 tables, small index, glossy semisoft covers in blues with dark blue lettering in a ring binder, 8.4 by 10.9 inches (214 × 277 mm).

An introductory book on the principles, practices, and equipment involved in the design of irrigation systems for landscapes. There is a list of selected references in the back.

Sarsfield, Alio Chester

1966† **the abc's of lawn sprinkler systems.*** A.C. Sarsfield. Irrigation Technical Services, Lafayette, California, USA. 145 pp., 200-plus illus., 20 tables, no index, semisoft covers in white with black lettering in a comb binder, 10.9 by 8.5 inches (278 × 215 mm).

A small book and reference manual on the design, installation, and operation of irrigation systems for lawns. The components of an irrigation system are described and illustrated in the first section. The basic principles of design are discussed. It includes a set of pressure tables at the back.

1969† **Irrigation Technical Manual—Engineering Data.*** A.C. "Chet" Sarsfield. Irrigation Technical Services, Lafayette, California, USA. 72 pp., 2 illus., 71 tables, no index, semisoft covers in white with black lettering in a comb binder, 8.5 by 10.9 inches (215 × 278 mm).

A small manual that brings together the engineering data reference tables used in sprinkler irrigation system design for turfgrass areas.

The Scotts Company

2005† **Scotts Sprinkler & Watering Systems.*** The Scotts Company and Meredith Books, Des Moines, Iowa, USA. 192 pp.; 378 illus. in color; 25 tables; indexed; glossy semisoft bound in multicolor with orange, white, and green lettering; 8.1 by 10.8 inches (205 × 274 mm).

A guide to the planning, installation, operation, and troubleshooting of an irrigation system for a residential lawn and garden.

Smith, Harry V.

2009† **Turf Maintenance Facility Design and Management.*** Coauthor.
See *Piersol, John R.—2009.*

Smith, Stephen W.

1997† **Landscape Irrigation—Design and Management.*** Stephen W. Smith. John Wiley & Sons Inc., New York City, New York, USA. 237 pp., 130 illus., 29 tables, small index, glossy hardbound in multicolor with white and black lettering, 7.0 by 9.9 inches (177 × 250 mm).

A basic book on irrigation methods, design techniques, hydraulics, piping, control systems, and irrigation practices. There are review questions at the ends of the chapters.

Sparks, Andrew

2000† **Lawn Mower Manual.*** Third revised edition. Andrew Sparks and Martynn Randall. Haynes Publishing, Nr Yeovil, Somerset, England, UK. 192 pp., 838 illus., 1 table, no index, glossy hardbound in multicolor with white lettering, 8.2 by 10.6 inches (208 × 270 mm).

A revision and update of the second edition.
See *Parker, J.M.F.—1985.*

The Toro Company

1966† **Rainfall—Evapotranspiration Data—United States and Canada.**˙ The Toro Company, Minneapolis, Minnesota, USA. 63 pp., no illus., all tables, no index, fold-stapled soft covers in white with black lettering, 8.2 by 6.3 inches (208 × 161 mm).

   A compilation of rainfall, potential evapotranspiration, and net difference by month for 342 locations in North America.

1968† **Facts on Automatic Underground Sprinkling.**˙ The Toro Company, Riverside, California, USA. 23 pp., 51 illus. with 39 in color, no tables, no index, fold-stapled semisoft covers in multicolor with white and black lettering, 4.9 by 6.9 inches (125 × 176 mm).

   A booklet concerning lawn care focused on irrigation practices and automatic irrigation system components.

Van Leeuwen, Roger A.

1978† **Design and Install Your Own Lawn Sprinkler System.**˙ Roger A. Van Leeuwen. Dexter Publishing Company, Tulsa, Oklahoma, USA. 72 pp., 103 illus., 26 tables, no index, glossy semisoft bound in green and white with white and black lettering, 8.5 by 10.9 inches (216 × 278 mm).

   A small handbook concerning the step-by-step design procedures and subsequent installation of a lawn sprinkler system. It contains tear-out work sheets.

Vinchesi, Brian

2003† **Golf Course Irrigation—Environmental Design and Management Practices.**˙ Coauthor.

   See *Barrett, James—2003.*

Watkins, James A.

1959† **Turf Irrigation Manual.**˙ First edition. James A. Watkins. Telsco Industries, Dallas, Texas, USA. 159 pp., 40 illus., 31 tables, no index, soft bound in white with green lettering, 8.4 by 10.9 inches (214 × 277 mm).

   A source of information on materials, equipment, and design of irrigation systems for turfgrass areas. It serves as an information source on the basics as well as a reference manual. Irrigation systems for home lawns are discussed in detail.

   Revised in 1965, 1977, 1987, and 1994.

1965† **Turf Irrigation Manual.**˙ Second revised edition. James A. Watkins. Telsco Industries, Dallas, Texas, USA. 159 pp., 41 illus., 77 tables, no index, soft bound in white with green lettering, 8.4 by 10.9 inches (214 × 278 mm).

   The revision consists primarily of an update on thermoplastic pipe in the second section.

   Revised in 1977, 1987, and 1994.

1977† **Turf Irrigation Manual.**˙ Third revised edition. James A. Watkins. Weather-matic Division, Telsco Industries Publication, Dallas, Texas, USA. 353 pp.; 269 illus.; 151 tables; indexed; textured cloth hardbound in yellow with green lettering and a glossy jacket in multicolor with blue, green, black, and red lettering; 8.3 by 10.9 inches (212 × 277 mm).

   This source of information on materials, equipment, and design of irrigation systems for turfgrass and landscape areas has been updated and expanded considerably. It can serve as a book on the basics as well as a reference manual. A detailed discussion of irrigation systems for golf courses has been added.

   Revised in 1987 and 1994.

Figure 13.6. Tractor-drawn, 3-gang, cast iron roller with a 7-foot width (2.1 m), 1920s. [41]

1987† **Turf Irrigation Manual.*** Fourth revised edition. James A. Watkins and Richard B. Choate. Weather-matic Division, Telsco Industries, Dallas, Texas, USA. 363 pp., 330 illus., 110 tables, indexed, glossy hardbound in multicolor with red and blue lettering, 8.3 by 10.9 inches (211 × 277 mm).

An expanded edition in terms of automated irrigation, including microprocess controls. Water conservation concerns also are addressed.

There have been four reprintings.
Revised in 1994.

Weinberg, Scott S.

1988 **Irrigation Volume III of Handbook of Landscape Architectural Construction.*** Scott S. Weinberg and John Mack Roberts (editors). Landscape Architecture Foundation, Washington, D.C., USA. 379 pp., 283 illus., 74 tables, indexed, cloth hardbound in red with white lettering, 8.5 by 10.9 inches (215 × 277 mm).

An extensive book on landscape irrigation. It consists of 14 chapters by 12 authors. It encompasses planning, design, contracts, specifications, bidding, installation, and pumps for irrigation systems. Typical details, technical specifications, friction loss tables, evapotranspiration tables, and sample irrigation plan drawings are included in the back.

Woodson, Roger Dodge

1996 **Watering Systems for Lawn & Garden.*** Storey Communications Inc., Pownal, Vermont, USA. 144 pp.; 80 illus.; 5 tables; small index; glossy semisoft bound in multicolor with green, blue, and black lettering; 8.5 by 11.0 inches (215 × 280 mm).

A practical guide to selection and installation of an irrigation system for the home lawn and garden. It is oriented to home gardening enthusiast.

Zoldoske, David

2003† **Golf Course Irrigation—Environmental Design and Management Practices.*** Coauthor.
See *Barrett, James—2003.*

C. TURFGRASS MANAGEMENT BOOKS (MOSTLY NONAGRONOMIC)

Audubon International

1996† **A Guide To Environmental Stewardship On The Golf Course.*** First edition. Audubon International, Selkirk, New York, USA. 65 pp., 15 illus. with 8 in color, 5 tables, no index, semisoft bound in multicolor with green and black lettering, 8.5 by 11.0 inches (215 × 279 mm).

A small book that provides an overview of practical conservation management techniques for golf courses. Emphasis is on wildlife habitat, integrated pest management, and water conservation and quality.

2000† **Certification Handbook.*** Audubon International, Selkirk, New York, USA. 53 pp., 3 illus., 17 tables, no index, fold-stapled glossy semisoft covers in white and blue with black and blue lettering, 8.5 by 11.0 inches (215 × 279 mm).

A handbook of standards to qualify for the designation of Certified Audubon Cooperative Sanctuary.

2002† **A guide to environmental stewardship on the golf course.*** Second revised edition. Audubon International, Selkirk, New York, USA. 123 pp., 27 illus., 10 tables, no index, glossy semisoft bound in white, gray and blue with blue and black lettering, 8.5 by 11.0 inches (215 × 279 mm).

A major revision of a guide to aid in golf course management that functions as an ecologically valuable green space in North America.

Bartz, Wilfried J.

2008 **Praxis des Sportplatzbaus: Fehleraufdeckung und -vermeidung Mit Checkliste (Sports Field Construction Practices: Identifying and Avoiding Errors).** Coauthor
See *Matthias, Claus G.—2008.*

Bennett, Roger

1996† **Golf Facility Planning.*** Sports Turf Research Institute, Bingley, West Yorkshire, England, UK. 64 pp., 10 illus. in color, 1 table, no index, glossy semisoft covers in multicolor with white and black lettering in a wire ring binder, 8.3 by 11.7 inches (210 × 296 mm).

A guide with 32 planning steps involved in the preconstruction phases of golf facility development. It is oriented to the laws, golf entrepreneurs, and government planning authorities in the United Kingdom.

Figure 13.7. Manual-push, steel, water-ballast roller, weight adjustable via addition of water or sand, with scraper, circa 1920. [41]

Benson, Martin Erick

1992† **Golf Courses and Country Clubs: A Guide to Appraisal, Market Analysis, Development, and Financing.** *
Coauthor.
> See *Gimmy, Arthur Elston—1992.*

Brennan, Anne-Maria

1996 **Living Together—Golf And Nature In Partnership.** * Anne-Maria Brennan. The English Golf Union, Leicester, England, UK. 80 pp. plus 1 p. of advertising, 67 illus. with 66 in color, no tables, no index, glossy semisoft bound in multicolor with white lettering, 8.25 by 11.75 inches (210 × 298 mm).
> An overview booklet on the management and construction of golf courses in terms of habitats for flora and wildlife as well as environmental protection and resource conservation.

Chadwick, Ronald M.

1990† **Spon's Grounds Maintenance Contract Handbook.** * Ronald M. Chadwick. E. & F.N. Spon Limited, Chapman and Hall, London, England, UK. 157 pp., 14 illus., 5 tables, indexed, glossy hardbound in green with white and yellow lettering, 6.1 by 9.1 inches (154 × 232 mm).
> A small, practical book on the development of contracts for landscape-grounds maintenance, especially as related to government organizations. The information offered is based on the author's real-world experiences. It is oriented to the conditions and laws in the United Kingdom.

Cobham, Ralph

1990† **Amenity Landscape Management—A Resources Handbook.** * Ralph Cobham (editor). E. & F.N. Spon Limited, London, England, UK. 458 pp., 37 illus., 137 tables, indexed, glossy hardbound in multicolor with black and white lettering, 7.4 by 9.6 inches (187 × 244 mm).
> A handbook on amenity landscape requirements for practitioners involved in the preparation of plans, implementation, and management of public amenities and leisure facilities. It includes several sections on turfgrasses.

Coe, A.

2000 **Tennis Science & Technology.** Coeditor.
> See *Haake, S.J.—2000.*

Connell, Edward A.

1962† **Lawn Bowling.** * Edward A. Connell. Management Aids Bulletin No. 10, American Institute of Park Executives, Oglebay Park, Wheeling, West Virginia, USA. 27 pp., no illus., 2 tables, no index, fold-stapled glossy semisoft covers in brown and white with white and black lettering, 6.0 by 9.0 inches (152 × 228 mm).
> A booklet summarizing a survey and analysis of public operations of lawn bowling facilities in the United States.

Cook, Walter L.

1964† **Public Golf Courses.*** Walter L. Cook and Roy Holland. Management Aids Bulletin No. 33, American Institute of Park Executives Inc., Oglebay Park, Wheeling, West Virginia, USA. 37 pp., no illus., 5 tables, no index, fold-stapled glossy semisoft covers in green and white with white and black lettering, 6.0 by 8.9 inches (151 × 226 mm).

A small guide to the planning, financing, and development of a public golf course in the United States. There is a section on lighting of golf courses.

Critchley, Jacy

1987† **Good Grounds—Construction and Maintenance of Sports Fields.** Jacy Critchley. Western Region Commission Inc., Braybrook, Victoria, Australia. 50 pp. plus 1 p. of advertising, 26 illus., 2 tables, no index, glossy semisoft bound in green and black with black lettering, 3.8 by 7.6 inches (99 × 193 mm).

An introductory booklet concerning the planning, construction, and maintenance of sports fields, with emphasis on municipal facilities. It has been coordinated with a video.

Crouch, William H.

1971 **Guide to the Analysis of Golf Courses and Country Clubs.*** William H. Crouch. American Institute of Real Estate Appraisers, Chicago, Illinois, USA. 60 pp.; 2 illus.; 14 tables; no index; fold-stapled semisoft covers in cream, green, and black with black lettering; 8.4 by 10.9 inches (213 × 277 mm).

This small book addresses the valuation procedures followed in the appraisal of a golf course. Fourteen case studies are presented. There is a short selected bibliography in the back.

Davis, Spencer Harwood, Jr.

1990† **The Dictionary of Golf.*** Spencer H. Davis Jr., Steven R. Langlois, and Louis M. Vasvary. A Hearthstone Book, Carlton Press Inc., New York City, New York, USA. 79 pp.; no illus.; no tables; no index; cloth hardbound in blue with black lettering and a jacket in white, green, and black with black lettering; 5.3 by 7.9 inches (135 × 200 mm).

A small handbook of inter-mixed terms used in the game of golf and in golf course maintenance.

Dawson, Denis

1997 **Port Vale Grass Roots—From Supporter To Groundsman and Back Again.*** Denis Dawson. Chell Publications, Stoke-on-Trent, Staffordshire, England, UK. 89 pp., 28 illus., no tables, no index, glossy semisoft bound in black and white with yellow lettering, 5.9 by 8.3 inches (150 × 210 mm).

An autobiography of a soccer club groundsman at Port Vale. He relates both interpersonal experiences and soccer ground maintenance issues as a club employee.

Dodson, Ronald G.

2000† **Managing Wildlife Habitat on Golf Courses.*** Ronald G. Dodson. Ann Arbor Press, Chelsea, Michigan, USA. 185 pp., 90 illus., 5 tables, very short index, glossy hardbound in multicolor with white lettering, 7.0 by 10.0 inches (178 × 254 mm).

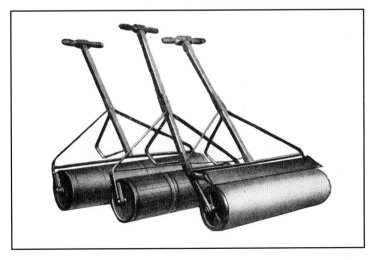

Figure 13.8. Three types of manual-push, lightweight green rollers of solid wood, spliced wood, and cast iron (left to right), each with 4-foot (1.2 m) width and 10-inch (25.4 cm) diameter. [41]

An introductory book concerning the concepts of wildlife habitat management and their practical application on golf courses. A list of selected references and a compilation of federal and state agencies involved in wildlife issues are included in the back.

2005† **Sustainable Golf Courses.*** Ronald G. Dodson. John Wiley & Sons Inc., Hoboken, New Jersey, USA. 281 pp.; 148 illus.; 12 tables; small index; glossy hardbound in multicolor with yellow, white, and blue lettering; 6.1 by 9.2 inches (155 × 233 mm).

A book on how golf course facilities can be constructed and best function in environmental enhancement, wildlife habitat, and water conservation. A case study of the Old Collier Golf Club is presented in the back. There is a foreword by Arnold Palmer.

Donovan, Richard E.

1988† **The Game of Golf and The Printed Word 1566–1985.*** Richard E. Donovan and Joseph S.F. Murdoch. Castalio Press, Endicott, New York, USA. 658 pp.; 49 illus.; no tables; indexed by author, short title, and club history; cloth hardbound in green with gold lettering and a glossy jacket in multicolor with black lettering plus a hard slip case in green; 6.0 by 9.2 inches (153 × 234 mm).

A major comprehensive bibliography of golf literature published in the English language. There are in the order of 4,800 entries. It does not include the books related to the turfgrass aspects of golf courses.

2006 **The Game of Golf and The Printed Word 1566–2005. Volume I and II.** Richard E. Donovan and Rand Jerris. Castalio Press, Endicott, New York, USA. 492 and 477 pp., 6 and 6 illus., no tables, club history index, cloth hardbound in blue and gold with imprinted gold lettering, 8.3 by 10.8 inches (212 × 274 mm).

This two-volume set is a bibliography of the golf literature represented in the English language. There are more than 15,500 entries. Turfgrass books were, for the most part, not included. There is an introductory essay on collecting golf books by Alastair J. Johnson.

Duff, Keith L.

2009† **Birds and golf courses: a guide to habitat management.** Keith L. Duff and Nigel Symes. The Royal Society for the Protection of Birds. Sandy, Bedfordshire, England, UK. 80 pp.; 202 illus. in color; 9 tables; no index; glossy semisoft bound in multicolor with white, tan, and black lettering; 8.2 by 11.4 inches (209 × 290 mm).

A small book concerning habitat management for birds on golf courses and related guidelines for the design of golf courses. Included are case studies and a list of selected references.

Dury, Peter Leslie Kennan

1998 **Grounds Maintenance: Managing Outdoor Sport and Landscape Facilities.** Peter L.K. Dury. Thorogood Limited, London, England, UK. 150 pp., 25 illus., 37 tables, no index, semisoft bound in teal with white lettering, 8.1 by 11.5 inches (205 × 191 mm).

A book covering the management aspects involved in outdoor recreation and sport facilities, with emphasis on the quality of the facility, the management plan, specifications, performance auditing, cost-effectiveness, carrying capacity, and safety. It is oriented to conditions in the United Kingdom.

The English Golf Union

The English Golf Union was formed in 1924 as the governing body of men's amateur golf in England. The objective of this non-profit organization is the development and improvement of golf facilities and players. A team of officers is available for on-site, advisory visitations. There also are a wide range of informational publications available.

1988† **Aspects of Golf Development.*** English Golf Union, Leicester, England, UK. 40 pp., no illus., no tables, indexed, semisoft covers in white, rose, and green with green lettering in a 2-hole binder, 8.3 by 11.7 inches (211 × 296 mm).

A summary of organizations affiliated with golf in the United Kingdom and their objectives, membership, and activities. Subsequently, there is an outline of planning steps to be considered in the development of a golf course and the ancillary facilities.

European Golf Association

The Ecology Unit of the European Golf Association is organized to promote the positive role of golf courses in environmental protection via their design and management. It is funded by the Royal and Ancient Golf Club of St Andrews, the PGA European Tour, and the European Golf Association.

1995† **An Environmental Strategy for Golf in Europe.*** Ecology Unit, European Golf Association, Brussels, Belgium. 48 pp., 28 illus. with 27 in color, 5 tables, no index, fold-stapled glossy semisoft covers in multicolor with white lettering, 8.2 by 11.7 inches (208 × 297 mm).

The booklet presents the aims and objectives of the European Golf Association Ecology Unit. The perceived environmental issues related to golf courses are addressed, an action plan of environmental stewardship is presented, and promotional and research approaches are proposed.

1996† **An Environmental Management Programme For Golf Courses.*** Compiled by David Stubbs. Ecology Unit, European Golf Association, Brussels, Belgium. 72 pp., 25 illus. in color, 11 tables, no index, glossy semisoft bound in multicolor with white lettering, 8.2 by 11.6 inches (208 × 295 mm).

This booklet summarizes wildlife surveys at eight European golf courses. The diversity and number of species, especially birds, were impressive. It was noted that from a turfgrass culture standpoint pesticide and fertilizer usage was minimal, but water use could be reduced.

1997† **The Committed to Green Handbook for Golf Courses.*** European Golf Association, Ecology Unit, Dorking, Surrey, UK. 40 pp., 29 illus. with 27 in color, no tables, no index, fold-stapled glossy semisoft covers in multicolor with white lettering, 8.3 by 11.7 inches (210 × 297 mm). Printed in the English, French, German, and Spanish languages.

A booklet of guidelines for maximizing environmental stewardship on golf courses and how to demonstrate the achieved benefits to the public, such as greenspace, buffer zone, wildlife habitat, and protection of water resources.

Evans, Marc

1993† **Landscape Restoration Handbook.*** Coauthor.
See *Harker, Donald F.—1993.*

Evans, Sherri

1993† **Landscape Restoration Handbook.*** Coauthor.
See *Harker, Donald F.—1993.*

Fairbrother, James

1984 **Testing the Wicket—From Trent Bridge to Lord's.** Jim Fairbrother and Reginald Moore. Pelham Books, London, England, UK. 144 pp., 25 illus., 1 table, small index, cloth hardbound in black with gold lettering, 5.4 by 8.4 inches (136 × 213 mm).

The book is organized in two parts: "I. A Groundsman Experiences Primarily at Lord's Cricket Grounds" and "II. A Procedural Overview of Turfgrass Culture and Table/Wicket Preparation Suggestions for Groundsmen."

Federazione Italiana Golf

1999† **Gli effetti ambientali delle attività ricreative sul Territorio—Il caso del golf in Italia (The environmental effects on the territory of recreation Activity—A golf case in Italy).*** P. Caggiati, S. Di Pasquale, V. Gallerani, D. Viaggi, and G. Zanni. Federazione Italiana Golf, Bologna and Roma, Italy. 129 pp., 2 illus., 46 tables, no index, semisoft bound in white with black lettering, 6.6 by 9.4 inches (168 × 240 mm). Printed in the Italian language.

This book assesses the functions of a golf course in terms of the environmental, landscape, and recreational aspects and in terms of how these functions interact with natural and socioeconomic resources. Also discussed are construction and management practices that enhance the positive benefits. There is a small glossary in the back.

Ferry, Lloyd R.

1986† **How To Start And Maintain Your Own Lawn Business in Florida.*** Lloyd R. Ferry. Ferry-Schaub Publishing Company, Englewood, Florida, USA. 64 pp., 12 illus., no tables, no index, glossy semisoft bound in red and white with white and red lettering, 5.5 by 8.2 inches (140 × 208 mm).

A small book consisting of a personal account of owning and operating a lawn and landscape maintenance business in Florida.

Fried, Gil

2005† **Managing Sport Facilities.** Gil Fried. Human Kinetics, Champaign, Illinois, USA. 375 pp.; 83 illus.; 32 tables; indexed; glossy hardbound in multicolor with red, yellow, and black lettering; 8.5 by 11.0 inches (217 × 278 mm).

A text covering the management of sports facilities, including a historical perspective, facility planning, design and construction, operation, maintenance and budgets, and event preparation and management.

Gilchrist, T. David

1983† **Trees on Golf Courses.*** T.D. Gilchrist. The Arboricultural Association, Romsey, Hampshire, England, UK. 42 pp.; 36 illus.; 2 tables; no index; semisoft bound in green, white, and black with white and black lettering; 4.0 by 8.1 inches (145 × 207 mm).

A unique booklet on trees for use on golf courses in the United Kingdom. Included are the characteristics, soil adaptation, placement relative to playing areas, and planting procedures for the various tree species. There are small sections on shrubs and on tree care.

Gillihan, Scott Wayne

2000† **Bird Conservation on Golf Courses—A Design and Management Manual.*** Scott W. Gillihan. Ann Arbor Press, Chelsea, Michigan, USA. 351 pp., 20 illus., 4 tables, no index, glossy semisoft bound in multicolor with black and white lettering, 5.9 by 9.0 (151 × 229 mm).

A manual on the design and management of golf courses for the conservation and benefit of bird species. Major sections in the appendix list breeding birds and their habitat requirements. Also included are a short glossary of selected terms and a list of selected references. The book was sponsored by the United States Golf Association Green Section and the Colorado Bird Observatory.

Gimmy, Arthur Elston

1992† **Golf Courses and Country Clubs: A Guide To Appraisal, Market Analysis, Development, and Financing.*** Arthur E. Gimmy and Martin E. Benson. Appraisal Institute, Chicago, Illinois, USA. 163 pp., 34 illus., 48 tables, indexed, glossy hardbound in multicolor with green lettering, 6.9 by 9.9 inches (176 × 252 mm).

A book on the economic characteristics of golf courses and the factors to be considered in their valuation. It is oriented to market analysts, developers, leaders, and real estate appraisers in the United States.

2003† **Analysis and Valuation of Golf Courses and Country Clubs.*** Arthur E. Gimmy and Buddie A. Johnson. Appraisal Institute, Chicago, Illinois, USA. 201 pp., 43 illus., 46 tables, no index, glossy semisoft bound in multicolor with white lettering, 7.0 by 9.9 inches (178 × 252 mm).

A major update of the 1992 book. The focus is on analysis and valuation techniques.

Golf Course Superintendents Association of America (GCSAA)

1975† **How to Find a New Position.*** Golf Course Superintendents Association of America, Laurence, Kansas, USA. 20 pp., 10 illus., no tables, no index, fold-stapled semisoft covers in green and cream with white and black lettering, 5.4 by 8.4 inches (137 × 213 mm).

A booklet concerning employment search techniques, including information sources, resume preparation, interview techniques, and contracts.

1976† **Managing Human Resources.*** Ronald C. Frame. Golf Course Superintendents Association of America, Lawrence, Kansas, USA. 38 pp., 6 illus., no tables, no index, fold-stapled semisoft covers in cream with brown lettering, 8.5 by 11.0 inches (215 × 279 mm).

A booklet on personnel management as it relates to golf course operations. Areas of emphasis include chain-of-command concepts, communications, and motivation.

n.d.† **The Budget Process On A Golf Course.*** Golf Course Superintendents Association of America, Lawrence, Kansas, circa USA. 61 pp., 31 sample forms, no tables, no index, fold-stapled semisoft covers in light green with brown lettering, 8.5 1976 by 10.9 inches (215 × 278 mm).

A booklet addressing the budget processes for the maintenance of golf courses plus a set of suggested record forms and a glossary of selected terms in the back.

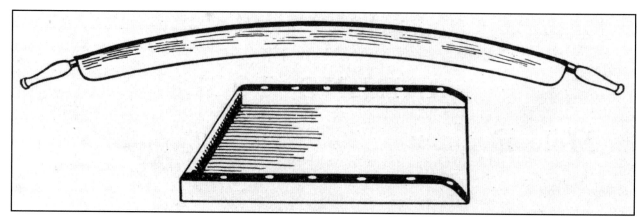

Figure 13.9. Gauge box and turf knife for cutting the soil underside of sod to a level base after placing the sod with turfgrass down into the gauge box. Needed for transplanting onto close-cut turfgrass sites. [3]

TABLE 13.1. CENTER FOR GOLF COURSE MANAGEMENT PUBLICATIONS, GOLF COURSE SUPERINTENDENTS ASSOCIATION OF AMERICA (GCSAA), SEMI–SOFT BOUND.

Year	Title	Pages	Illus.	Tables
		Number		
1990	Buying Habits of Golf Course Superintendents Report	112	0	109
1991	Pesticide and Fertilizer Usage Report	132	0	68
1991	Golf Car and Turf Utility Vehicle Report	105	0	142
1991	Mower and Maintenance Equipment Report	201	0	80
1991	Buying Habits of Golf Course Superintendents Report	130	0	82
1991	Golf Course Superintendents Report	44	0	51
1992	Golf Course Superintendents Report	51	6	53
1992	Golf Course Operations: Cost of Doing Business/Profitability	44	0	0
1992	Golf Course Operations: Maintenance Budgets & Equipment Inventories	53	0	0
1992	Golf Course Maintenance Expenditures	179	0	182
1992	Turfgrass Species and Expenditures	34	0	25
1993	Pesticide Usage Report	163	0	83
1993	Golf Car and Turf Utility Report	160	0	206
1993	Mower and Maintenance Equipment Report I	606	0	364
1993	Mower and Maintenance Equipment Report II	856	0	364
1993	Golf Course Superintendents Report	60	0	34
1993	Profiling the Golf Course Superintendent	63	48	0
1994	Golf Course Superintendents Report	93	0	69
1994	Cost of Doing Business/Profitability	51	1	37
1994	Maintenance Budgets & Equipment Inventories	62	2	37
1995	Compensation and Benefits Report	59	24	36
1996	Golf Course Superintendents Report	163	0	132
1998	Golf Course Superintendents Report	78	0	50
2000	Golf Course Superintendents Report	92	0	60
2002	Golf Course Superintendents Report	93	0	93

1993† **Golf Course Maintenance Facilities. A Guide to Planning and Design.** GCSAA Press, Golf Course Superintendents Association of America, Lawrence, Kansas, USA. 91 pp., 40 illus., no index, no tables, semisoft bound in maroon with gold lettering, 5.9 by 8.9 inches (151 × 227 mm).

A small guidebook on the basics for designing golf course operations centers. The focus is on operational efficiency and regulatory compliance.

1993† **Career Development.** Golf Course Superintendents Association of America, Laurence, Kansas, USA. 24 pp., 1 illus., no tables, no index, semisoft bound in greens with light green and black lettering, 8.5 by 10.9 inches (216 × 275 mm).

A booklet on career goal setting, current analysis, and planning. There are sample forms for job applicants and employment contracts for golf course superintendents.

1997† **GCSAA College Guide—To The Golf Course Management Profession.** First edition. Golf Course Superintendents Association of America, Laurence, Kansas, USA. 306 pp., no illus., 3 tables, no index, glossy semisoft bound in multi-color with white lettering, 6.0 by 9.0 inches (152 × 228 mm).

A guide to university and college programs offering a four-year undergraduate BSc degree and/or a two-year associate certificate in turfgrass management or golf course management education in the United States.

2000†   **GCSAA College Guide-To The Golf Course Management Profession.**[*] Second revised edition. Golf Course Superintendents Association of America, Laurence, Kansas, USA. 283 pp.; 9 illus.; 3 tables; no index; glossy semisoft bound in multicolor with black, yellow, and green lettering; 6.0 by 8.9 inches (152 × 226 mm).

       An update of the 1997 first edition.

2000†   **Sample Personnel Handbook for Golf Course Maintenance Operations.**[*] Golf Course Superintendents Association of America, Lawrence, Kansas, USA. 23 pp., no illus., no tables, no index, semisoft bound in gray with black lettering, 8.5 by 11.0 inches (216 × 279 mm).

       A model handbook for both supervisors and employees on golf courses.

Glazner, Steve

2001   **Operational Guidelines for Grounds Management.**[*] Steve Glazner (editor) with eleven contributing authors. APPA, NRPA, and PGMS, Alexandria, Virginia, USA. 173 pp., 16 illus., 8 tables, no index, glossy semisoft bound in multicolor with white and green lettering, 8.4 by 10.8 inches (213 × 274 mm).

       An organizational guide that assists grounds managers in developing facility specific operational and staffing-level plans. It also includes chapters on contracting and position descriptions. A glossary is included in the back.

Gosselin, Francis J.

1995   **Le Guide de L'Expert Tondeur (The Guide of the Expert Mower).** Francis J. Gosselin. Trois-Rivières, Quebec, Canada. 227 pp., 1 illus., 20 tables, no index, glossy semisoft covers in multicolor with white and black lettering in a ring binder, 5.4 by 8.5 inches (139 × 216 mm). Printed in the French language.

       A book concerning the start-up and operation of a residential and commercial lawn mowing service. It is oriented to eastern Canadian conditions.

Haake, S.J.

2000   **Tennis Science & Technology.**[*] S.J. Haake and A. Coe (editors). Blackwell Science Ltd., Oxford, England, UK. 476 pp., 181 illus., 57 tables, small index, glossy hardbound in multicolor with white and black lettering, 6.7 by 9.5 inches (169 × 242 mm).

       A book summarizing an extensive series of presentations concerning the equipment, playing surfaces, and player responses associated with the game of tennis.

Harker, Donald F.

1993†   **Landscape Restoration Handbook.**[*] Donald Harker, Sherri Evans, Marc Evans, and Kay Harker. Lewis Publishers, Ann Arbor, Michigan, USA. 562 pp., 75 illus. with 1 in color, 5 tables, no index, glossy hardbound in multicolor with white lettering, 6.9 by 10.0 inches (176 × 254 mm).

       A book on the principles and practices in maintaining natural landscapes as well as their restoration. A major portion of the book is devoted to the natural regions of the United States and the associated dominant ecological communities. There is a large bibliography included.

2004†   **Managing Wetlands on Golf Courses.**[*] Coauthor.

       See *Libby, Gary R.—2004.*

Harker, Kay

1993†   **Landscape Restoration Handbook.**[*] Coauthor.

       See *Harker, Donald F.—1993.*

2004†   **Managing Wetlands on Golf Courses.**[*] Coauthor.

       See *Libby, Gary R.—2004.*

Health and Safety Executive, England

1994   **Health and Safety in Golf Course Management and Maintenance.**[*] Health and Safety Executive Books, Sudbury, Suffolk, England, UK. 64 pp., 28 illus., no tables, no index, semisoft bound in multicolor with white, green, blue lettering, 6.9 by 9.8 inches (176 × 249 mm).

       A booklet on employee health and safety management on a golf course. Included are personal protective equipment, machinery operation and servicing, chemical handling, and noise and vibration aspects. There is a list of selected references in the back.

Heuer, Karla L.

1980†   **Golf Course—A Guide to Analysis and Valuation.**[*] American Institute of Real Estate Appraisers, Chicago, Illinois, USA. 138 pp., 18 illus., 40 tables, no index, cloth hardbound in green with gold lettering, 7.3 by 9.2 inches (185 × 283 mm).

A small book on criteria and procedures for the market analysis and appraisal of golf course properties, including the developmental, operational, and financial aspects. Sample analyst forms are provided in the appendix, along with a selected bibliography.

Hill, Shari

1982   **Softball Maintenance From The Ground Up.** Shari Hill. Softball Canada, Vanier, Ontario, Canada. 55 pp., 39 illus., 6 tables, no index, fold-stapled glossy semisoft covers in blues and white with red and blue lettering, 8.5 by 11.0 inches (215 × 280 mm).

A guidebook on building and care of softball fields, under the conditions in Canada. It includes site selection, soil characteristics, layout, construction, and maintenance plus ancillary equipment and facilities.

Howard, Roy

1964†   **Public Golf Courses.** Coauthor.
See *Cook, Walter L.—1964.*

Jerris, Rand

2006   **The Game of Golf and The Printed Word 1566–2005.** Coauthor.
See *Donovan, Richard E.—2006.*

Johnson, Buddie A.

2003†   **Analysis and Valuation of Golf Clubs and Country Clubs.*** Coauthor.
See *Gimmy, Arthur Elston—2003.*

Kurrein, Carol

1989†   **The Flymo Book of Garden Games and Lawn Leisure.*** Carol Kurrein. David and Charles Publishers plc., Newton Abbot, Devon, England, UK. 96 pp., 80 illus. in color, no tables, no index, hardbound in dark green with gold lettering and a jacket in multicolor with black lettering, 7.5 by 10.4 inches (192 × 264 mm).

A unique, small book covering stimulating games and healthy recreational activities enjoyed on lawns. It includes assessment criteria and development plans for these facilities as well as the actual care of the turfed lawn. There is a foreword by Alan Titchmarsh.

Lane, Dennis

2005   **The Grass is Greener.** Dennis Lane. Little Silver Publications, Matford, Exeter, England, UK. 62 pp., 13 illus., no tables, no index, fold-stapled glossy semisoft covers in multicolor with black lettering, 5.6 by 8.3 inches (143 × 210 mm).

A small autobiography concerning primarily the human relationships experienced by the author while serving as a groundsman at the Hampshire County Cricket Club and mostly at Exeter University.

Langlois, Steven Roy

1990†   **The Dictionary of Golf.** Coauthor.
See *Davis, Spencer Harwood, Jr.—1990.*

Libby, Gary R.

2004†   **Managing Wetlands on Golf Courses.*** Gary Libby, Donald F. Harker, and Kay Harker. John Wiley & Sons Inc., Hoboken, New Jersey, USA. 219 pp.; 62 illus.; 8 tables; indexed; glossy hardbound in multicolor with black, green, and white lettering; 5.9 by 8.9 inches (150 × 227 mm).

A book on wetlands organized in five parts encompassing characteristics, conservation, restoration and creation, types, and wetlands on golf courses. There is a small glossary and a list of selected references in the back.

Lilly, Sharon J.

1999†   **Golf Course Tree Management.*** Sharon Lilly. Ann Arbor Press, Chelsea, Michigan, USA. 216 pp., 133 illus., no tables, small index, glossy hardbound in multicolor with white lettering, 5.9 by 8.9 inches (151 × 227 mm).

A guidebook on tree management for golf courses. Subjects addressed include how trees grow, trees in golf course design, tree planting, and tree maintenance. There is a selected list of references in the back.

Maloney, Thomas R.

1996†   **Human Resource Management for Golf Course Superintendents.*** Coauthor.
See *Milligan, Robert Alexander—1996.*

Figure 13.10. Manual cutting and loading sod from a turfgrass nursery onto a horse-drawn wagon, circa 1905. [10]

Mascaro, Thomas Charles

1957†   **Handbook For New Green Committee Chairmen.** Tom Mascaro. West Point Products Corporation, West Point, Pennsylvania, USA. 21 pp., 1 illus., 1 table, no index, fold-stapled glossy soft covers in greens with black lettering, 5.9 by 8.9 inches (151 × 227 mm).

     A small handbook on the responsibilities and operational procedures of a green committee chairman as practiced in North America.

     See also *revisions published in 1960, 1962, and 1967.*

1960†   **Municipal Handbook for Golf Course Committees.** First edition. Tom Mascaro. Management Aids Bulletin No. 2, American Institute of Park Executives Inc., Wheeling, West Virginia, USA. 27 pp., no illus., no tables, no index, fold-stapled glossy soft covers in greens with black lettering, 6.0 by 9.0 inches (152 × 228 mm).

     An adaptation for municipal golf organizations of the 1957 **Handbook For New Green Committee Chairmen** by Tom Mascaro.

     Revised in 1962 and 1967.

1962†   **Municipal Handbook for Golf Course Committees.** Second revised edition. Tom Mascaro. Management Aids Bulletin No. 2, American Institute of Park Executives Inc., Wheeling, West Virginia, USA. 27 pp., 1 illus., 1 table, no index, fold-stapled glossy semisoft covers in greens with black lettering, 6.0 by 9.0 inches (152 × 228 mm).

     A slight revision.

     Revised in 1967.

1967†   **Municipal Handbook for Golf Course Committees.** Third revised edition. Tom Mascaro. Management Aids Bulletin No. 2, National Recreation and Park Association Inc., Washington, D.C., USA. 26 pp., 1 illus., 1 table, no index, fold-stapled glossy semisoft bound in greens and white with black lettering, 6.0 by 9.0 inches (152 × 228 mm).

     A slight revision.

Matthias, Claus G.

2002   **Praxis des Sportplatzbaus: Fehleraufdeckung und -vermeidung Mit Checkliste (Sports Field Construction Practices: Identifying and Avoiding Errors).** First edition. Claus G. Matthias. Expert-Verlag, Renningen-Maimsheim, Germany. 113 pp., 5 illus., 34 tables, no index, glossy textured semisoft bound in dark blue and red with white lettering, 5.8 by 8.2 inches (148 × 207 mm). Printed in the German language.

     A small book concerning the avoidance of errors in the design and construction of sport fields composed of turfgrass and of artificial turf.

     Revised in 2008.

2008   **Praxis des Sportplatzbaus: Fehleraufdeckung und -vermeidung Mit Checkliste (Sports Field Construction Practices: Identifying and Avoiding Errors).** Revised second edition. Claus G. Matthias and Wilfried J. Bartz. Expert Verlag, Renningen, Germany. 112 pp., 5 illus., 37 tables, no index, glossy textured semisoft bound in orange and dark blue with white lettering, 5.8 by 8.3 inches (147 × 210 mm). Printed in the German language.

     Contains some revisions.

McGonegal, Brian L.

1997 **The Microgolf Primer: Raise Golf Acres in Yards.** Brian L. McGonegal with David C. Enger. Microgolf Press, Jackson, Michigan, USA. 93 pp., no illus., no tables, no index, glossy semisoft bound in multicolor with black lettering, 4.0 by 7.0 inches (103 × 177 mm).

A small book on the construction and maintenance of a home lawn putting green plus a diversity of surrounding turfed tees.

Meyers, James

1978† **Make the Most of Your Lawn.** James Meyers. Hart Publishing Inc., New York City, New York, USA. 286 pp.; 246 illus.; 7 tables; small index; glossy semisoft bound in multicolor with white, black, and green lettering; 8.4 by 11.0 inches (214 × 279 mm).

A book on the diversity of uses for residential turfgrass lawns, including 14 lawn games such as archery, badminton, bocce, croquet, curling, paddleball, handball, horseshoe pitching, lawn bowls, platform tennis, putting games, roque, shuffleboard, tether tennis, and volleyball.

Miller, John C.

2009† **Tournament Management.** John Miller. John Wiley & Sons Inc., Hoboken, New Jersey, USA. 216 pp., 59 illus., 14 tables, indexed, glossy hardbound in multicolor with white lettering, 6.3 by 9.5 inches (160 × 241 mm).

A specialty guidebook concerning the preparation and maintenance of a golf course for a tournament competition. There is a foreword by Michael J. Hurdzan. It includes a digital version of the forms and plans in a CD-ROM format.

Milligan, Robert Alexander

1996† **Human Resource Management for Golf Course Superintendents.*** Robert A. Milligan and Thomas R. Maloney. Ann Arbor Press, Chelsea, Michigan, USA. 184 pp., 31 illus., 4 tables, small index, glossy hardbound in maroon with white lettering, 6.0 by 9.3 inches (154 × 236 mm).

A guidebook for the management of golf course green staff members. Five functional areas are discussed: planning, organizing, controlling, staffing, and directing. There is a short list of selected references at the end of each chapter.

Monty, Bryan P.

2004† **The Complete Bilingual Lawn & Landscape Training Guide (LaGuis Completa de Instrucción Bilingiie de Césped y Paisaje).*** Bryan P. Monty. Author House, Bloomington, Indiana, USA. 597 pp., no illus., 6 tables, no index, glossy semisoft bound in multicolor with black and white lettering, 6.0 by 9.0 inches (152 × 228 mm). Printed in both English and Spanish.

A simple yet detailed guide for training and use by employees of lawn and landscape companies in the United States.

Murdoch, Joseph Simpson Ferguson

1988† **The Game of Golf and The Printed Word 1566–1985.*** Coauthor.
See *Donovan, Richard E.—1988.*

National Association of Audubon Societies

n.d.
circa
192x **Golf Clubs as Bird Sanctuaries.*** National Association of Audubon Societies, New York City, New York, USA. 64 pp., 147 illus., 1 table, no index, soft bound in light brown and gray with brown lettering, 6.1 by 9.2 inches (154 × 234 mm).

This is a **very rare,** small book on the natural characteristics of a golf course that makes it a quality bird sanctuary while also functioning in insect and worm control. It includes winter bird feeding, bird houses, nesting boxes, bird baths, and plantings for roughs. It is stated that,

A fledgling robin eats 14 feet of earthworms every day.

National Golf Foundation (NGF)

1955† **Municipal Golf Course Organizing and Operating Guide.*** Verne Wickham (editor). National Golf Foundation Inc., Chicago, Illinois, USA. 120 pp.; 154 illus.; 16 tables; indexed; semisoft bound in blue, black, and white with white and black lettering; 8.9 by 11.0 inches (225 × 279 mm).

A how-to book on the planning, financing, development, operation, and staffing of a municipal golf course in the United States. It was a unique book in its time.

1957† **Golf for Industry—A Planning Guide.**˙ Ben Chlevin (editor). National Golf Foundation Inc., Chicago. Illinois, USA. 50 pp.; 96 illus.; 10 tables; no index; semisoft bound in green, tans, and white with white and dark green lettering; 8.4 by 11.0 inches (214 × 279 mm).

    The book is a practical guide on the planning of public golf courses to be sponsored by private industrial organizations.

n.d.† **Planning Information for Private Golf Clubs.**˙ National Golf Foundation Inc., Chicago, Illinois, USA. 190 pp., 58
circa illus., 33 tables, no index, semisoft covers in greens with black lettering in a comb binder, 8.5 by 11.0 inches (217 ×
1961 279 mm).

    A compilation of pamphlets and leaflets designed to assist private golf planning groups in the feasibility assessment, planning, development, and operation of private golf courses.

n.d.† **Organizing and Operation of Public Golf Course.**˙ First edition. National Golf Foundation Inc., Chicago, Illinois,
circa USA. 200 pp., 91 illus., 7 tables, no index, semisoft covers in tan with red lettering in a comb binder, 8.5 by 11.0 inches
1966 (217 × 279 mm).

    A compilation of pamphlets and leaflets designed to assist public golf course planning groups with approaches and potential problems related to the development and operation of public golf courses.

    Revised circa 1972 and 1977.

n.d.† **Planning and Building The Par-3 or Executive Golf Course Manual.** National Golf Foundation, Chicago, Illinois,
circa USA. 34 pp., 23 illus., 2 tables, no index, fold-stapled glossy semisoft covers in blacks and white with black lettering,
1970 8.6 by 11.0 inches (218 × 279 mm).

    A booklet concerning the feasibility, planning, construction, maintenance, and operation of golf short courses for public use.

n.d.† **Organizing and Operating Public Golf Courses.**˙ Second revised edition. National Golf Foundation Inc., Chicago,
circa Illinois, USA. 266 pp., 95 illus., 13 tables, no index, semisoft covers in red with black lettering in a comb binder, 8.5
1972 by 11.0 inches (217 × 279 mm).

    This revision is an update of 1966 first edition.

    Revised in 1977.

1974† **The Par 3 and Executive Golf Course Planning & Operating Manual.**˙ The National Golf Foundation Inc., Chicago, Illinois, USA. 31 pp., 32 illus., 2 tables, no index, fold-stapled semisoft covers in green with white lettering, 8.4 by 11.0 inches (213 × 207 mm).

    A small booklet outlining the site selection, design, and operation of par-3 and executive golf courses.

1977† **Organizing and Operating—Municipal Golf Courses.** Third revised edition. National Golf Foundation, North Palm Beach, Florida, USA. 414 pp., 75 illus., 23 tables, no index, textured semisoft covers in green with black lettering in a comb binder, 8.5 by 11.0 inches (216 × 278 mm).

    A revision and update of an earlier version.

Figure 13.11. Cutting and boxing of sod from an older, rear-green, wheelbarrow transport, and transplanting onto a newly constructed green in foreground, circa 1916. [35]

n.d.† **Planning Information for Private and Daily Fee Golf Clubs.** National Golf Foundation Inc., North Palm Beach,
circa   Florida, USA. 484 pp., 53 illus., 57 tables, no index, semisoft covers in yellow with black lettering in a comb binder, 8.5
1977   by 11.0 inches (217 × 279 mm).
       A compilation of pamphlets and leaflets designed to assist golf club planning groups with approaches and potential problems related to the development and operation of private and daily-fee golf organizations.

National Playing Fields Association (NPFA)

The National Playing Fields Association is a voluntary body founded in 1925 for the purpose of encouraging the provision of adequate playing fields and recreation facilities throughout England and Wales. The NPFA was incorporated by royal charter in 1933 and, in 1963, was registered as a national charity.

n.d.† **Making The Most Of School Playing Fields.*** National Playing Fields Association, London, England, UK. 24 pp.,
circa   no illus., 1 table, no index, fold-stapled soft covers in green, white and black with green and black lettering, 5.4 by 8.5
1960   inches (138 × 216 mm).
       A study report on the need and methods, during out-of-school hours, to maximize the use of school playing fields due to a shortage of public playing space in England.

1970† **Floodlighting Of Outdoor Sports Facilities.*** First edition. Christy Electrical Contractors Limited. Technical Publication, National Playing Fields Association, London, England, UK. 39 pp.; 13 illus.; no tables; no index; fold-stapled glossy semisoft covers in blue, black, and white with black lettering; 5.9 by 8.2 inches (149 × 210 mm).
       A guide to the types, selection, installation, maintenance, and operating costs of night lighting of 12 types of outdoor sports facilities. This is a significant revision which updates the improvements in lamps, fittings, and floodlighting techniques.
       Reprinted in 1976.

TABLE 13.2. NATIONAL GOLF FOUNDATION PUBLICATIONS, SEMISOFT BOUND.

Year	Title	Number		
		Pages	*Illus.*	*Tables*
1987	Golf Course Maintenance Survey	86	94	121
1989	Guidelines for Planning and Developing a Public Golf Course	64	9	12
1990	Designing, Expanding and Remodeling Golf Courses	23	0	0
1991	Golf Course Construction and Operation Contracts	38	11	0
1992	The Economic Impact of Golf Course Operations on Local, Regional & National Economics	29 (1)	1	16
1992	Golf and The Environment: A Status Report and Action Plan	127 (1)	31	12
1992	Lessons Learned from New Municipal Golf Course Developments	95 (1)	1	2
1993	Golf Ranges & Non-Regulation Facilities	125	16	23
1993	Understanding and Comparing Maintenance Budgets	47 (1)	19	4
1994	Golf Ranges, Learning Centers & Family Recreation Facilities	108	20	13
1994	Golf Course Siting and Development	135 (1)	5	0
1994	Guidelines for Financing a Golf Course	53	8	0
1995	How to Plan, Build and Operate a Successful Golf Range (Third Edition)	119	12	21
1995	Guidelines for Planning & Developing a Public Golf Course	83	6	10
1998	How to Plan, Build, and Operate a Successful Golf Range	109	6	12
1998	Guidelines for Planning and Developing a Public Golf Course (Fourth Edition)	73	5	13
1998	Golf Course Design & Construction	50 (1)	3	2
1998	Landfill Golf Courses	94 (1)	26	3
1999	Golf Facilities in the United States	86	7	32
1999	Guidelines for Financing a Golf Course	40	9	0
2000	Sample Personnel Handbook for Golf Course Maintenance Operations	23	0	1

1976† **Floodlighting Of Outdoor Sports Facilities.*** Second revised edition. Christy Electrical Contractors Limited. Technical Publication, National Playing Fields Association, London, England, UK. 40 pp., 10 illus., 3 tables, no index, fold-stapled semisoft covers in white with black lettering 6.4 by 8.5 inches (163 × 215 mm).

    This is a significant revision that updates the improvements in lamps, fittings, and floodlighting techniques. *Reprinted in 1978 and 1984.*

Nature Conservancy Council (NCC)

The Nature Conservancy Council is a government agency that controls and manages large areas of national nature reserves in Great Britain. More than 100 wardens are employed as managers.

1989† **On Course Conservation—Managing golf's natural heritage.*** Nature Conservancy Council. Peterborough, UK. 46 pp., 18 illus. in color, no tables, no index, glossy semisoft bound in multicolor with black and green lettering, 8.2 by 11.6 inches (208 × 295 mm).

    This booklet addresses the management of golf course roughs and out-of-play areas in terms of providing habitats for wildlife. Included are natural grasslands, water features, and trees. *Reprinted in 1990 and 1991.*

Nuckolls, Jonelle

1988 **Guidelines for the Design and Construction of Baseball and Softball Facilities.*** Jonelle Nuckolls and Phillip Rea. Recreation Resources Service, North Carolina State University, Raleigh, North Carolina, USA. 56 pp., 33 illus., 10 tables, no index, semisoft covers in blue with white lettering in a comb binder, 8.5 by 10.9 inches (216 × 275 mm).

    A guidebook concerning the site selection, design, site development, and installation of baseball and softball facilities. It includes turfgrass planting in North Carolina, lighting, fencing, and dugouts.

Oen, Urban Theodore

1969† **Turf Sales and Service Unit—An Individualized Learning Manual.** Urban T. Oen, Department of Agricultural Education, Michigan State University, East Lansing, Michigan, USA. 117 pp., 9 illus., 11 tables, no index, semisoft bound in light green with black lettering, 8.0 by 10.7 inches (203 × 272 mm).

    A manual for students designed for individualized learning under the guidance of an instructor. There are 10 lesson units. A selected list of books, bulletins, periodicals, films, and slide sets is presented in the back.

Peake, Malcolm

n.d.† **The Wildside of Golf.*** Coauthor.
circa See *Taylor, Robert Stephen—n.d. circa 1995.*
1995
2001 **Confessions of a Chairman of the Green.*** Malcolm Peake. Studies in Golf Course Management No. 3, The Sports Turf Research Institute, Bingley, West Yorkshire, England, UK. 76 pp., 60 illus. with 43 in color, no tables, no index, glossy semisoft bound in multicolor with white lettering, 6.75 by 8.25 inches (158 × 209 mm).

Figure 13.12. Transplanting boxed sod onto a newly constructed green, circa 1916. [35]

A small book on the personal experiences of a golfer, green committee member and chairman in renovation of Temple Golf Club in England.

Pedrotti, Robin M.

1992† **Lawn Aeration—Turn Hard Soil into Cold Cash.*** Robin M. Pedrotti. Prego Press, San Diego, California, USA. 213 pp. with 2 pp. of advertising; 45 illus.; 6 tables; small index; glossy semisoft bound in crème and green with green, black, and white lettering; 8.25 by 10.7 inches (209 × 271 mm).

A guidebook on start-up and operation of a lawn service business, with a focus on mechanical cultivation of turfgrass soils.

Penrose, Lee

n.d.† **The Wildside of Golf.*** Coauthor.
circa
1995

See *Taylor, Robert Stephen—n.d. circa 1995.*

2003† **Gorse and its management.*** Coauthor.
See *Taylor, Robert Stephen—2003.*

2003† **Bracken and its Management.** Lee Penrose, Bob Taylor, and Ian Rotherham. The Sports Turf Research Institute, Bingley, West Yorkshire, England, UK. 51 pp., 15 illus. with 13 in color, no tables, no index, glossy semisoft bound in multicolor with black lettering, 5.6 by 8.1 inches (141 × 205 mm).

A booklet concerning the biology, ecological adaptation, and culture of bracken for golf courses in the United Kingdom. There is a list of selected references in the back.

Phillips, Leonard E.

1996† **Parks: Design and Management.*** Leonard E. Phillips. McGraw-Hill Companies, New York City, New York, USA. 239 pp., 182 illus., 3 tables, small index, cloth hardbound in black with green lettering, 7.2 by 9.2 inches (184 × 234 mm).

A book on the design of parks and playgrounds plus their management. Emphasis is placed on landscape aspects and turfgrass sport and recreation areas. Lists of selected references are included at the ends of certain chapters.

Rea, Phillip

1988 **Guidelines for the Design and Construction of Baseball and Softball Facilities.*** Coauthor.
See *Nuckolls, Jonelle—1988.*

Rossi, Kevin

1994† **Lawn Care & Gardening—A Down-to-Earth Guide to the Business.*** First edition. Kevin Rossi, with illustrations by Lee Weisman. Acton Circle Publishing Company, Ukiah, California, USA. 236 pp., 51 illus., 3 tables, small index, glossy semisoft bound in greens with white and black lettering, 8.3 by 10.8 inches (211 × 275 mm).

A book on the basics for establishment and operation of a lawn-landscape service business. It includes capital investment, marketing, estimating, and bidding.
Revised in 2001.
See *Willis, Mickey—2001.*

The Royal and Ancient Golf Club of St Andrews

1994† **The Way Forward.** Royal and Ancient Golf Club of St Andrews, St Andrews, Scotland, UK. 35 pp., no illus., no tables, no index, fold-stapled glossy semisoft covers in multicolor with black lettering, 8.3 by 11.7 inches (210 × 297 mm).

A discussion document concerning a review of British golf course management and culture, including organizations, club officials, greenkeepers, golf professionals, and media.

1997† **A Course for All Seasons—A Guide to Course Management.*** The Royal and Ancient Golf Club of St Andrews and European Golf Association, St Andrews, Scotland, UK. 52 pp.; 23 illus. all in color; no tables; no index; glossy semisoft bound in multicolor with black, burgundy, and blue lettering; 8.25 by 11.9 inches (209 × 302 mm).

A small book presenting a philosophical overview of golf course development and management that is environmentally responsible and cost effective while also providing good golf.

Sachs, Paul D.

1996† **Handbook of Successful Ecological Lawn Care.** Paul D. Sachs. The Edaphic Press, Newberry, Vermont, USA. 288 pp.; 51 illus.; 9 tables; extensive index; glossy semisoft bound in multicolor with yellow, white, and black lettering; 5.4 by 8.3 inches (136 × 210 mm).

A book concerning operation of a lawn care business, with an orientation to organic philosophies of turfgrass culture.

Sanecki, Kay Naylor

1979† **Old Garden Tools.*** Kay N. Sanecki. Shire Album 41, Shire Publications Limited, Aylesbury, Buckinghamshire, England, UK. 32 pp., 46 illus., no tables, no index, fold-stapled glossy semisoft covers in multicolor with white, black, and green lettering, 5.8 by 8.2 inches (147 × 208 mm).

A small booklet documenting the early gardening and lawn tools used in England, including lawn mowers. It contains interesting early photographs and drawings.

Sayers, Philip Richard

1991† **Grounds Maintenance—A Contractor's Guide to CompetitiveTendering.*** Philip Sayers. E. & F.N. Spon Limited, London, England, UK. 208 pp.; 34 illus.; 22 tables; indexed; glossy hardbound in blues, white and greens with green, white, and blue lettering; 6.0 by 9.1 inches (153 × 232 mm).

A book on bidding for turfgrass-landscape maintenance of governmental agencies in the United Kingdom. The topics addressed are bid documents, contract specifications, time and costs, performance indicators, and quality assurance.

1991† **Managing Sport and Leisure Facilities—A Guide to Competitive Tendering.*** Philip Sayers. E. & F.N. Spon Limited, London, England, UK. 275 pp.; 30 illus.; 16 tables; indexed; glossy hardbound in greens, brown, and white with cream, white, and green lettering; 6.0 by 9.2 inches (152 × 233 mm).

A guide to the competitive management of sport and leisure facilities in the United Kingdom. Areas of emphasis include bid procedure and documents, operational management, contractor performance, and quality assurance.

Schmidgall, Raymond Stanley

1996† **Superintendent's Handbook of Financial Management.*** First edition. Raymond S. Schmidgall. Ann Arbor Press Inc., Chelsea, Michigan, USA. 156 pp., 32 illus., 3 tables, small index, glossy hardbound in green with white lettering, 7.0 by 10.0 inches (177 × 254 mm).

A handbook on the basic concepts involved in the financial management of golf course maintenance operations. Included are accounting procedures, operating budgets, financial statements, capital budgets, and leasing. There are problem sets at the end of each chapter.
Revised in 2004.

2004† **Superintendent's Handbook of Financial Management.*** Second revised edition. Raymond S. Schmidgall. John Wiley & Sons Inc., Hoboken, New Jersey, USA. 183 pp., 8 illus., 42 tables, indexed, glossy hardbound in oranges and white with white and orange lettering, 6.0 by 9.2 inches (153 × 234 mm).

A revised update of the first edition.

Scottish Golf Course Wildlife Group

2001 **Golf's Natural Heritage—An Introduction to Environmental Stewardship on the Golf Course.*** Scottish Golf Course Wildlife Group, Glasgow, Scotland, UK. 26 pp., 47 illus. in color, no tables, no index, fold-stapled semisoft covers in multicolor with blue and black lettering, 8.3 by 10.6 inches (210 × 270 mm).

A booklet providing an overview of environmental stewardship planning on golf courses in Scotland so as to ensure habitat for wildlife and protection of the environment.

The Scottish Office Environment Department

1994 **Golf Courses and Associated Developments.*** The Scottish Office Environment Department. Planning Advice Note Pan 43, The Scottish Office Environment Department, Edinburgh, Scotland, UK. 28 pp.; 21 illus.; 2 tables; indexed; semisoft covers in gray, black, and white with gray, black, and white lettering; 8.3 by 11.7 inches (210 × 297 mm).

A pamphlet concerning an overview of golf course planning and design in terms of ensuring that the policies of government authorities are addressed.

The Scottish Sports Council

The Scottish Sports Council, now Sportscotland, was formed in 1972 by royal charter to serve as the national agency for sport in Scotland. This group works in partnership with public, private, and voluntary organizations involved in sport and recreation. It is also involved in advising and implementing Scottish government policy.

1992 **Pitch & Court Markings.** The Scottish Sports Council, Edinburgh, Scotland, UK. 64 pp.; 30 illus.; 8 tables; no index; semisoft covers in multicolor with black, green, blue, and white lettering in a 2-hole binder; 5.8 by 8.7 inches (147 × 210 mm).

A quality reference guide on the specifications for the layout, color, and size of markings for sports playing areas in Scotland.

Shepherd, Matthew

2002† **Making Room for Native Pollinators—How to Create Habitat for Pollinator Insects on Golf Courses.** Matthew Shepherd. The United States Golf Association, Far Hills, New Jersey, USA. 30 pp.; 12 illus. with 10 in color; no tables; no index; semisoft bound in multicolor with white, black, and yellow lettering, 8.4 by 10.9 inches (213 × 277 mm).
  A small booklet on the biology of bees and creation of foraging habitats and nesting sites on golf courses.

Sports Council, England

The English Sports Council is now named Sport England. The objective of the group is focused on the creation of a world-leading community sport system. The government agency is accountable to Parliament and is responsible for investing National Lottery and Exchequer funds in organizations and projects that grow and sustain participation in sports.

n.d.† **A New Bias—A Report on the Future Provision for Bowls.** The Sports Council. London, England, UK. 48 pp., 51
circa illus., 15 tables, no index, fold-stapled semisoft bound in multicolor with white lettering, 8.2 by 11.5 inches (208 ×
198x 293 mm).
  A report on a study of bowls trends and needs in England. Included are participation and facility assessments for both indoor and outdoor greens.

Sportsmark Group

1980† **Groundsman's Field Handbook.** First edition. Sportsmark Group, Brentford, Middlesex, England, UK. 33 pp., 25 illus., 9 tables, no index, fold-stapled semisoft covers with waterproof pages in green and white with yellow and black lettering, 4.1 by 5.9 inches (103 × 150 mm).
  A small field handbook on specifications for the layout and marking of sports fields plus the specifications for associated field accessories. It is oriented to sports played in the United Kingdom.
  Revised in 1982, 1990, 1993, 1999, 2003, and 2004.
1982† **Groundsman's Field Handbook.** Second revised edition. Sportsmark Group, Brentford, Middlesex, England, UK. 38 pp.; 28 illus.; 5 tables; no index; fold-stapled semisoft covers with waterproof pages in green and white with yellow, black, and white lettering; 4.1 by 5.9 inches (105 × 149 mm).
  Contains some revisions.
  Reprinted at two to four year intervals.
  Revised in 1990, 1993, 1999, 2003, and 2004.
1990† **Groundsman's Field Handbooks.** Third revised edition. Sportsmark Group, Brentford, Middlesex, England, UK. 37 pp.; 36 illus.; 3 tables; no index; fold-stapled glossy semisoft covers with waterproof pages in green and white with orange, black, and white lettering; 4.1 by 5.8 inches (105 × 148 mm).
  A booklet concerning the layout dimensions and measurements, markings, and playing fixtures for a range of sports as conducted in the United Kingdom.
  Revised in 1993, 1999, 2003, and 2004.
1993† **Groundsman's Field Handbook.** Fourth revised edition. Sportsmark Group, Brentford, Middlesex, England, UK. 37 pp.; 35 illus.; 3 tables; no index; fold-stapled glossy semisoft covers with waterproof pages in green and white with orange, black, and white lettering; 4.1 by 5.8 inches (105 × 147 mm).
  Contains some revisions, primarily the field layout drawings.
  Revised in 1999, 2003, and 2004, with minor revisions.

Strawn, John

1991† **Driving The Green—The Making of a Golf Course.** John Strawn. HarperCollins Publishers, New York City, New York, USA. 344 pp., 6 illus., no tables, no index, cloth hardbound in light greens with gold lettering and a jacket in multicolor with green and black lettering, 5.6 by 9.2 inches (143 × 234 mm).
  A narrative on the development and construction of the "Ironhorse" golf course in Florida. It is a unique nontechnical book.
1992† **Driving The Green—The Making of a Golf Course.** John Strawn. Harper-Perennial, New York City, New York, USA.
  A semisoft bound version of the 1991 book in multicolor with black, white, and green lettering.
1997† **Driving The Green—The Making of a Golf Course.** John Strawn. Lyons & Burford Publishers, New York City, New York, USA. 351 pp.
  A reprinting of the 1991 edition with a new afterword.

Symes, Nigel

2009†   **Birds and golf courses: a guide to habitat management.** Coauthor.
      See *Duff, Keith L.—2009.*

Taylor, Robert Stephen

1995†   **A Practical Guide To Ecological Management Of The Golf Course.**[*] R.S. Taylor (editor). British and International
      Golf Greenkeepers Association (BIGGA), York, England, UK, and Sports Turf Research Institute (STRI), Bingley, West
      Yorkshire, England, UK. 121 pp., 36 illus. with 35 in color, 5 tables, no index, glossy semisoft bound in multicolor with
      black lettering, 5.75 by 8.2 inches (146 × 208 mm).
          A small book that presents an overview of golf course development and maintenance in terms of wildlife habitat and
      environmental protection. There is a list of selected references in the back.

n.d.†   **The Wildside of Golf.**[*] Bob Taylor, Malcolm Peake, and Lee Penrose. British and International Golf Greenkeepers
circa   Association, Aldwark, York, England, UK. 26 pp., 50 illus. in color, no tables, no index, stapled glossy semisoft covers
1995   in multicolor with black and white lettering, 11.75 by 7.9 inches (298 × 200 mm).
          A pictorial information pamphlet concerning flora and wildlife on golf courses in the United Kingdom.

2003†   **Gorse and its management.** Bob Taylor, Lee Penrose, and Ian Rotherham. The Sports Turf Research Institute, Bingley,
      West Yorkshire, England, UK. 51 pp., 32 illus. with 29 in color, 1 table, no index, glossy semisoft bound in multicolor
      with black and white lettering, 5.8 by 8.2 inches (148 × 208 mm).
          A booklet concerning the biology, ecological adaptation, and culture of gorse on golf courses in the United
      Kingdom.

2003†   **Bracken and its management.** Coauthor.
      See *Penrose, Lee—2003.*

Turfgrass Producers International (TPI) ‡

n.d.†   **Turfgrass Sod Farm Employee Handbook Models.**[*] American Sod Producers Association, Rolling Meadows, Illinois,
circa   USA. 47 pp., no illus., no tables, no index, semisoft bound in white with black lettering, 8.5 by 11.0 inches (215 × 279
1988   mm).
          A small guide to the development of an employee handbook for individual sod production operations in the United
      States.

United States Golf Association (USGA)

1988†   **How to Conduct a Competition.** United States Golf Association, Far Hills, New Jersey, USA. 61 pp., 34 illus. with 31 in color,
      2 tables, no index, fold-stapled glossy semisoft covers in multicolor with blue lettering, 6.0 by 9.0 inches (152 × 228 mm).
          A booklet concerning the planning, course preparation, course marking, course set-up, operation, and administra-
      tion of a golf competition.

1990†   **The USGA Green Section Guide.** Green Section, United States Golf Association, Far Hills, New Jersey, USA. 48 pp.;
      no illus.; no tables; no index; fold-stapled glossy semisoft covers in white, blue, and green with black, green, and blue
      lettering; 4.0 by 9.0 inches (101 × 229 mm).
          A small booklet that includes a guide to green committee members, golf course worker training and direction, and
      a dictionary of golf turfgrass terms.

1994†   **Golf and Wildlife.**[*] United States Golf Association, Far Hills, New Jersey, USA. 20 pp.; no illus.; no tables; no index;
      fold-stapled glossy semisoft covers in multicolor with green, red, blue, and black lettering; 8.4 by 10.9 inches (214 ×
      276 mm).
          A small booklet on techniques for enhancing wildlife habitat on golf courses. A set of case studies are presented.

2000†   **A Guide for Green Committee Members.**[*] United States Golf Association, Far Hills, New Jersey, USA. 51 pp., no
      illus., 1 table, no index, glossy semisoft bound in multicolor with white lettering, 5.3 by 8.4 inches (134 × 136 mm).
          A small guide concerning the organization, responsibilities, and operation of a green committee involved in oversee-
      ing golf course management. There is a glossary of terms in the back.
        Revised in 2003.

2003†   **A Guide for Green Committee Members.**[*] Second revised edition. United States Golf Association, Far Hills, New
      Jersey, USA. 43 pp., 1 illus., 2 tables, no index, fold-stapled glossy semisoft covers in multicolor with black and white
      lettering, 5.4 by 8.5 inches (138 × 215 mm).

---

‡ Prior to 1994 it was named the American Sod Producers Association (ASPA).

Figure 13.13. Manual sod kicker used for cutting sod, circa 1920.

A guide booklet concerning the activities of green committee members on golf courses. It also includes a dictionary of turfgrass terms.

2006† **Wildlife Links—Improving Golf's Environmental Game.**˙ United States Golf Association Green Section, Far Hills, New Jersey, USA. 21 pp., 32 illus. with 31 in color, 1 table, fold-stapled glossy semisoft covers in multicolor with white lettering, 8.5 by 11.0 inches (119 × 278 mm).

A small booklet on the Wildlife Links program that is a joint venture of the USGA and the National Fish and Wildlife Foundation. Best management practices are presented, along with case histories.

Vasary, Louis Michael

1990† **The Dictionary of Golf.** Coauthor.
See *Davis, Spencer Harwood, Jr.—1990.*

Webster, Frederick Annesley Michael

1940 **Sports Grounds and Buildings.**˙ Captain F.A.M. Webster. Sir Isaac Pitman 7 Sons Limited, London, England, UK. 318 pp. with 13 pp. of advertising, 144 illus., 8 tables, cloth hardbound in dark green with gold lettering, 8.5 by 10.7 inches (216 × 271 mm).

An early book on the design, layout, and construction of sports grounds, with the back one-third on sports buildings. Turfgrass facilities addressed include team sports fields, tennis courts, cricket grounds, and track & field. It contains initial chapters on sports field history and development.

Willis, Mickey

2001† **Lawn Care & Gardening—A Down-to-Earth Guide to the Business.**˙ Second revised edition. Mickey Willis. Acton Circle Publishing Company, Ukiah, California, USA. 271 pp. plus 1 p. of advertising, 91 illus., 3 tables, small index, semisoft bound in multicolor with white, yellow, black, and green lettering, 8.5 by 11.0 inches (216 × 279 mm).

A practical book on the entrepreneurial operation of a landscaping business. The 21 chapters are in four parts: about the business, business management, horticultural operations, and estimating and bidding.
See earlier edition by *Rossi, Kevin—1994.*

> *The Greens Keeper*
> Edgar A. Guest
>
> He's on the job at break of day and when the stars come out,
> There's always trouble on the course for him to fret about,
> He starts the gang to work at dawn and follows them around

Then listens to committeemen whose wisdom is profound,
They talk of "bents" and "fescues" in a way that makes him squirm
For they acquire much knowledge in one brief official term
His task is one that calls for tact, for lacking that it means
Next year there'll be another man brought on to keep the greens.

The members seldom know his name, or have a smile for him,
They only wonder why it is the course is not in trim.
And wonder why the greensman hasn't cut the rough this fall,
And when they find a cuppy lie or footprints in a trap
"The course is in a rotten shape!" declares each gloomy chap.
And yet my hat is off to him, now winter intervenes,
I want to pay my tribute to the man who keeps the greens.

He's on the job from dawn to dusk, a million pests to fight,
'Tis his to see that every green is watered well at night.
The weeds attack his finest work, the drought destroys his grass,
The rain beats down the tender shoots, but still the players pass
And still they play the game they love, a happy golfing clan
Who never stop to count the odds against a singe man.
And so I wave my hand to him, who tolls in sturdy jeans,
The best old friend all golfers have–the man who keeps the greens.

From Bulletin of Green Section of USGA, 3(1):2.

D. ARCHITECTURE AND PLANNING OF TURFGRASS FACILITIES

Encompasses books on architecture, design, and planning as related to turfgrass facilities.

Armitage, John

1996   **Six Simple Techniques To Improve Your Golf Course At The Lowest Possible Cost.** John Armitage. Green Eagle Golf Limited, Charing, Kent, England, UK. 40 pp., 9 illus., no tables, no index, fold-stapled glossy semisoft covers in multicolor with white lettering, 5.8 by 8.3 inches (147 × 210 mm).

    A small book outlining a set of simple, low-cost design improvements that can be easily implemented to upgrade individual golf holes. There is a foreword by W.F. Deeds.

Bauer, Alexander H.

1913   **Hazards, Those Essential Elements In A Golf Course Without Which The Game Would Be Tame and Uninteresting.** Aleck Bauer (editor) with six contributing authors: A. Mackenzie, Edward Ray, Donald Ross, M. Lewis Crosby, W. Laidlaw Purves, and H.S. Colt. Toby Rubovits, Chicago, Illinois, USA. 61 pp., 26 illus. with a folding course map, no tables, no index, cloth hardbound in green and quarter red with black lettering, 5.4 by 8.0 inches (138 × 203 mm).

    A **truly rare** book which is a must for collectors of historical golf books. An interesting, small booklet on the design and construction of hazards for golf courses, with emphasis on bunkers. It basically is an assemblage of articles by six authors that had been previously published in Great Britain. The chapter subject titles and contributing authors include the following:

• Introduction	• Where the British Courses Excel–Donald Ross
• Some Common Fallacies of Golf Course Construction and Green-Keeping–A. Mackenzie	• The Value of a Difficult Course–M. Lewis Crosby
• Inland Hazards–Edward Ray	• The Championship Courses–W. Laidlaw Purves
• Two Famous Water Shots	• The Construction of New Courses–H.S. Colt
	• Famous Golf Holes

    The latter portion of the book involves a discussion of English golf courses with emphasis placed on the role of bunker placement and design in the playability of individual golf holes. Each of the holes is illustrated through either photographs or drawings.
    Reprinted in 1993.

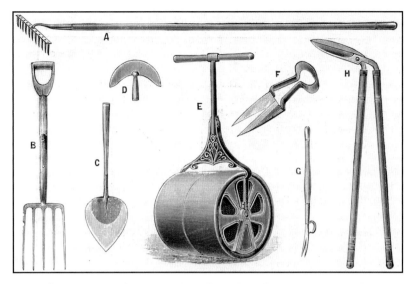

Figure 13.14. Set of manual tools used in early turfgrass maintenance. [42]

1993† **Hazards.** First reprint edition. Aleck Bauer (editor). Grant Books Limited, Droitwich, Worcestershire, England, UK. 88 pp., 39 illus., no tables, no index, cloth hardbound in green with gold lettering, 7.0 by 9.4 inches (178 × 238 mm).

A reprinting of the early 1913 book compiled by Aleck Bauer titled **Hazards**. Added were a foreword by Peter Thomson and an introduction by Fred Hawtree. Also, added in the back are articles titled "Play the Ball As It Lies" by Mr. Peter Dobereiner and "To Rake or Not to Rake Bunkers" by Philip A. Truett. There was a limited reprinting of 1,000 copies. See original printing. *Bauer, Aleck—1913.*

Butler, George Daniel

1947 **Recreation Areas—Their Design and Equipment.**˙ First edition. George D. Butler. National Recreation Association, A.S. Barnes and Company, New York City, New York, USA. 174 pp., 170 illus., 6 tables, indexed, cloth hardbound in light brown with blue lettering, 8.9 by 12.0 inches (225 × 302 mm).

A book on the design and layout of recreation facilities and sports fields. It includes a chapter on the preparation of areas for ice sports surfaces.

Revised in 1958.

1958 **Recreation Areas—Their Design and Equipment.**˙ Second revised edition. George D. Butler. National Recreation Association, The Ronald Press Company, New York City, New York, USA. 174 pp., 170 illus., 6 tables, indexed, cloth hardbound in green with black lettering, 8.9 by 12.0 inches (225 × 302 mm).

It contains minor revisions.

Carre-Riddell, Gervase

n.d.† **Practical Golf Course Design and Construction.** Gervase Carre-Riddell. Swan Hill, Victoria, Australia. 45 pp., 21
circa illus., no tables, no index, fold-stapled textured semisoft covers in greens with dark green lettering, 7.9 by 4.9 inches
1974 (200 × 124 mm).

A simplified booklet on the design, layout, and construction of golf courses. It is oriented to low budget golf developments in the Australian countryside.

Cornish, Geoffrey St. John

n.d.† **Golf Course Design—An Introduction.**˙ First edition. Geoffrey S. Cornish and William G. Robinson. Cornish and
circa Robinson, Amherst, Massachusetts, USA. 22 pp., 36 illus., no tables, no index, fold-stapled semisoft covers in tan and
1971 greens with green lettering, 8.5 × 11.0 inches (217 × 279 mm).

A small booklet outlining the basics of golf course design and planning. It is extensively illustrated by Mr. William G. Robinson.

† *Reprinted by National Golf Foundation in 1979 and in 1985.*

Revised in 1989.

1981† **The Golf Course.**˙ First edition. Geoffrey S. Cornish and Ronald E. Whitten. The Rutledge Press, New York City, New York, USA. 320 pp.; 156 illus. with 56 in color; 1 table; indexed; textured cloth hardbound in blue with gold lettering and a glossy jacket in multicolor with black, green, and white lettering; 8.4 by 10.8 inches (213 × 276 mm).

A book that documents the evolution of golf course architecture and the history of those persons contributing to the art of architecture. The second portion of the book presents the profiles of several hundred golf course architects, listed in alphabetical order. Each profile includes a list of golf courses designed. The third section contains an alphabetical listing of golf courses followed by the architect's name. Most of the sketches and diagrams are by William G. Robinson. There is a foreword by Robert Trent Jones. It is a unique book.

† *Reprinted by Ellesborough Press, London, England, UK, in 1981 as a limited edition of 200 copies.*
  Also reprinted by Windward, W.H. Smith of Leicester, England, UK.

† *Reprinted in 1982.*
  Revised in 1984 and 1987.

1984 **The Golf Course.*** Second revised edition. Geoffrey S. Cornish and Ronald E. Whitten. The Rutledge Press, New York City, New York, USA. 320 pp., 175 illus. with 52 in color, no tables, indexed, textured cloth hardbound in blue with gold lettering, 8.5 by 11.0 inches (216 × 279 mm).
  A minor revision.

1987† **The Golf Course.*** Third revised edition. Geoffrey S. Cornish and Ronald E. Whitten. The Rutledge Press, New York City, New York, USA. 367 pp., 181 illus. with 57 in color, no tables, no index, cloth hardbound in blue with gold lettering, 8.5 by 11.0 inches (217 × 278 mm).
  This is an updated version of the 1984 edition.
  Reprinted in 1988.
  Revised and expanded in 1989 as **The Classics of Golf Special Section: The Golf Course.**
  Revised and expanded under a new title in 1993.
  See The Architects of Golf—1993.

1993† **The Architects of Golf.*** Geoffrey S. Cornish and Ronald E. Whitten. Harper Collins Publishers Inc., New York City, New York, USA. 648 pp., 421 illus. with 380 being close-up photographs of golf course architects, no tables, indexed, cloth hardbound in greens with gold lettering and a glossy jacket in multicolor with yellow and white lettering, 8.4 by 10.9 inches (214 × 277 mm).
  A major revision, expansion, and updating of the author's 1981 book, **The Golf Course**. It consists of a narrative of golf course design from its beginnings to the present, followed by a unique encyclopedic listing of golf architects and their courses plus a photograph of each architect. A third section consists of a listing of 12,000-plus golf courses worldwide and the architects who designed them. A must book for collectors of golf books. There is a list of selected references in the back.

1998† **Golf Course Design.** Coeditor and contributing author.
  See Graves, Robert Muir—1998.

2002† **Classic Golf Hole Design—Using the Greatest Holes as Inspiration for Modern Courses.** Coauthor.
  See Graves, Robert Muir—2002.

2006† **Golf Course Design.** Geoffrey S. Cornish and Michael J. Hurdzan. A Grant Books Publication, Droitwich, Worcestershire, England, UK. 207 pp., 60 illus. with 52 in color, no tables, indexed plus a short book title index, cloth hardbound in yellow with gold lettering and a jacket in multicolor with white lettering, 6.9 by 9.4 inches (175 × 238 mm).
  This is a unique annotated bibliography of golf design books and related subjects. It is organized on a somewhat chronological basis. The related books include groupings on annuals and yearbooks, turfgrass science and management, drainage and irrigation, and the environment. The latter portion of the book addresses the arrangement of large collections and specific resources.

2006† A glossy semisoft bound version of the 2006 book just described.

Countryside Commission

In 1968, the Countryside Act abolished the National Parks Commission and set up the Countryside Commission. The purpose is to continually review the conservation and enhancement of landscape beauty in England and Wales as well as provide and improve the facilities of the countryside for enjoyment.

1993† **Golf Courses In The Countryside.** Countryside Commission. Cheltenham, Gloucestershire, England, UK. 48 p.; 123 illus. with 111 in color; no tables; no index; fold-stapled glossy semisoft covers in multicolor with orange, green, and black lettering; 8.2 by 11.7 inches (208 × 297 mm).
  A guidebook on the design and construction of new golf courses in England and Wales as related to the conservation of natural beauty in the countryside and informal access to these new natural beauty features. Emphasis is placed on golf course designs that blend with the topography and vegetation of the local landscape.

Daley, Paul

2003,
2003,
2005,
2008,
&
2009

Golf Architecture—A Worldwide Perspective. Vols. one, two, three, four, and five. Paul Daley (editor). Pelican Publishing, Gretna, Louisiana, USA. 254/320/368/384/351 pp., 179/228/356/363/480 illus. with 117/182/316/344/315 in color, 20 tables, small index, cloth hard-bound in black with white and lettering, 10.4 by 7.9 inches (264 × 200 mm).

A unique five-volume set on golf course architecture that brings together the diverse philosophies and opinions of 45 (volume 1), 54 (volume 2), 48 (volume 3), 43 (volume 4) and 30 (volume 5) authors from around the world who are active in golf course architecture, are active in golf writing, and/or are dedicated golfers. The volumes are well illustrated. There is a small glossary and short resume of each author in the back of each book.

Darwin, Bernard Richard Merion

1931†   **The Game of Golf.** Coauthor.
See *Wethered, Joyce—1931.*

Departments of the Army, Navy, and Air Force

1975†   **Planning and Design of Outdoor Sports Facilities.** Technical Manual 5-803-10, Navfac P-457, and Air Force Regulation 88-33, Departments of the Army, Navy, and Air Force, Washington, D.C., USA. 169 pp., 209 illus., 1 table, no index, semisoft bound in white and black with black lettering, 10.7 by 8.2 inches (271 × 207 mm).

A manual concerning the planning and design for a comprehensive range of outdoor sports. There are site specifications, layouts, and construction details.

Doak, Thomas Harry

1992†   **The Anatomy Of A Golf Course.** Tom Doak. Lyons & Burford, New York City, New York, USA. 242 pp., 78 illus. with 19 in color, 1 table, indexed, cloth hardbound in teal and black with silver lettering and a glossy jacket in multicolor with black lettering, 5.9 by 8.9 inches (151 × 227 mm).

A book on the art of golf course architecture that addresses the strategies and "why" behind the decisions of a golf architect. This young architect writes with strong convictions. There is a glossary of golf architectural terms in the back plus a listing of the author's choice of golf courses worth studying. The foreword is by Mr. Ben Crenshaw.

1996†   **The Confidential Guide to Golf Courses.** Tom Doak. Sleeping Bear Press, Chelsea, Michigan, USA. 361 pp.; 184 illus. in color; 31 tables; indexed by golf course; hardbound in light maroon with gold lettering plus a jacket in multicolor with maroon, yellow, black, and white lettering, 8.5 by 10.9 inches (215 × 278 mm).

A book in which the author evaluates the merits and undesirable aspects of those golf courses which he has personally visited around the world. He summarizes the evaluations by means of a rating from 1 to 10 with a 5 value being well above average. Included is a "Gourmet's Choice" of the 31 golf courses that are the author's favorites. Also presented is the "Doak Gazetteer," consisting of lists in which golf courses are ranked by the top 10 for a diversity of characteristics.

Douglas, Nigel B.

2004   **Golf Course Design.** Nigel B. Douglas. Nigel B. Douglas, Mary Borough, Victoria, Australia. 252 pp., 46 illus., 1 table, no index, glossy semisoft bound in multicolor with black lettering, 6.1 by 9.2 inches (154 × 234 mm).

A book concerning the design, planning, and implementation of construction of a golf course. The author presents a personal perspective and relates many of this past experiences in the Australasia region.

Fazio, Tom

2000†   **Golf Course Designs.** Tom Fazio with Cal Brown. Harry N. Abrams Inc., New York City, New York, USA. 203 pp., 85 illus. with 81 in color, 11 tables, small index, cloth hardbound in dark green with gold lettering and a jacket in multicolor with black lettering, 9.0 by 10.9 inches (228 × 278 mm).

A book on golf course design encompassing trends over many decades and factors influencing these trends, design preferences and playability, elements of design, and the design and construction processes.

Fine, Mark K.

2006†   **Bunker, Pits & Other Hazards: A Guide to the Design, Maintenance and Preservation of Golf's Essential Elements.** Coauthor.
See *Richardson, Forrest L.—2006.*

Finger, Joseph Seifter

1972† **The Business End of Building or Rebuilding A Golf Course.** Joseph S. Finger. Joseph S. Finger and Associates Inc., Houston, Texas, USA. 50 pp., 2 illus., 8 sample forms, no index, fold-stapled textured semisoft covers in light green with dark green lettering, 8.4 by 10.9 inches (214 × 278 mm).

A small book on the procedures to follow in building or rebuilding a golf course. Included are sections on initial planning, golf course architect selection, contracts, design, specifications, cost estimates, contractor selection, and cost control. There is an addendum containing sample forms. It is a unique book.

Gill, Garrett David

1977† **Golf Course Design and Construction Standards.** Garrett D. Gill. Department of Environmental Design, Texas A&M University, College Station, Texas, USA. 39 pp., 19 illus., no tables, no index, fold-stapled semisoft covers in tan with brown lettering, 8.4 by 10.9 inches (213 × 276 mm).

A small booklet on the criteria contributing to the theory of golf course design. This is followed by a discussion of the types of construction plans needed and the key components each plan should contain. Finally, there is a section outlining the bidding documents and specifications.

Golf Development Council (GDC)

1968† **Elements of Golf Course Layout and Design.** First edition. Golf Development Council, Edinburgh, Scotland, UK. 24 pp., 7 illus., 1 table, no index, fold-stapled soft covers in greens with black lettering, 9.6 by 6.0 inches (245 × 152 mm).

A small booklet summarizing the elements of golf course layout and design as practiced in the United Kingdom. *Revised in 1972 and 1980.*

1972† **Elements of Golf Course Layout and Design.** First revised edition. Golf Development Council, London, England, UK. 25 pp., 9 illus., 1 table, no index, fold-stapled semisoft covers in greens and black with black lettering, 9.7 by 6.0 inches (246 × 152 mm).

A revised second version with updates and expansion of some sections. It includes a brief section on golf course maintenance plus a list of equipment needs. *Revised in 1980.*

1980† **Elements of Golf Course Layout and Design.** Second revised edition. F.W. Hawtree. Golf Development Council, Richmond, Surrey, England. UK. 25 pp., 9 illus., 2 tables, no index, fold-stapled glossy semisoft covers in greens and black with black lettering, 9.3 by 6.0 inches (236 × 151 mm).

A minor revision of the 1972 edition.

Graves, Robert Muir

2002† **Classic Golf Hole Design—Using the Greatest Holes as Inspiration for Modern Courses.** Robert Muir Graves and Geoffrey Cornish. John Wiley & Sons Inc., Hoboken, New Jersey, USA. 323 pp., 449 illus., 11 tables, indexed, cloth hardbound in red with silver lettering, 7.9 by 9.9 inches (200 × 250 mm).

An in-depth book that describes and illustrates the classic golf holes of the world and examines the principles and composition that distinguish these classics. Also, the many variations that have evolved from these classic golf holes are discussed.

Hawtree, Frederick William

1991† **Colt & Co.—Golf Course Architects.** Fred Hawtree. Cambuc Archive, Oxford, England, UK. 214 pp., 76 illus., 4 tables, small index, cloth hardbound in green with gold lettering and a glossy jacket in multicolor with red and white lettering, 6.0 by 8.5 inches (153 × 215 mm).

A biographical study of Henry Shapland Colt (1869–1951) with his partners C.H. Alison, J.S.F. Morrison, and Dr. A. Mackenzie. The foreword is by Sir Denis Thatcher. It is a key book for golf historians. This is a limited edition of 1,000 copies.

1996† **Triple Baugé—Promenade in Medieval Golf.** Fred Hawtree. Cambuc Archive, Oxford, England, UK. 164 pp., 53 illus., 1 table, small index, cloth hardbound in gray with gold lettering, 5.8 by 8.3 inches (147 × 210 mm).

This book addresses the various potential origins of the game now known as golf. It ranges over a 900-year period.

1998† **Aspects of Golf Course Architecture I 1889–1924.** Fred Hawtree. Grant Books, Worcestershire, England, UK. 172 pp., 66 illus. with 1 in color, no tables, no index, textured cloth hardbound in yellow with imprinted gold lettering, 7.0 by 9.5 inches (177 × 241 mm).

An anthology of selected writings from 15 pioneering golf course architects of Scotland and England for the period from 1889 to 1924. A brief introduction to each article is presented by Fred Hawtree. Published as a limited edition of 675 copies, with the first 75 copies being the Woodstock edition.

2008† **Aspects of Golf Architecture II 1925–1971.** Fred Hawtree with contributions by Dr. Martin Hawtree and J. Hamilton Stutt. Grant Books, Droitwich, Worcestershire, England, UK. 208 pp., 67 illus., 2 tables, no index, cloth hardbound in pale yellow with gold lettering, 6.9 by 9.4 inches (174 × 239 mm).

An anthology of selected writings by 13 golf course architects from the British Isles. A brief resume of each architect is presented before each article. The book was published as a limited edition of 400 copies of which the first 75 copies were the Woodstock edition.

Hunter, Wiles Robert

1926† **The Links.** Robert Hunter. Charles Scribner's Sons, New York City, New York, USA. 163 pp.; 51 illus. and 10 diagrams of golf holes; 2 tables; no index; cloth hardbound in green with black lettering and a jacket in green, black, and white with gold lettering; 6.3 by 8.7 inches (160 × 222 mm).

One of the classic early American works on the architecture of golf courses. A **rare**, must book for collectors of historical golf books. The photographs of famous golf holes in North America and England are well done. The chapter subject titles include the following:

- Ante Scriptum—To Links-Land
- The Purpose of Hazards—Their Inspiration to Good Play
- Ante Scriptum—Shots—Old And New
- Things of Importance
- Placing the Hazards
- Laying Out the Course
- Constructing the Hazards
- The Climax of Golf—The Well-Placed and Well-Moulded Green
- Other Things of Importance

Regarding the question of whether roughs and hazards should be included in the tillage and planting of a golf course, Hunter stated the following:

Grassy hollows, mounds, and the banks of bunkers should be sowed with good seed and not allowed to grow wild, as otherwise they will become breeding-places for weeds. The club which can afford to do so should plough up its rough and sow it also.

1995† **The Links.** Robert Hunter. Reprinted by the United States Golf Association, Far Hills, New Jersey, USA.
1998 **The Links.** Robert Hunter. Reprinted by Ailsa Inc., Stamford, Connecticut, USA, with a foreword by Mr. Bill Coore.
1998† **The Links.** Robert Hunter. Reprinted by Sleeping Bear Press, Chelsea, Michigan, USA, following the original text, binding, sheets, and jacket format.

Hurdzan, Michael John

2003† **Building A Practical Golf Facility.** First edition. Dr. Michael J. Hurdzan. American Society of Golf Course Architects Foundation, Chicago, Illinois, USA. 120 pp.; 58 illus. with 56 in color; 4 tables; no index; glossy semisoft bound in multicolor with black, white, red, and blue lettering; 5.5 by 8.4 inches (140 × 213 mm).

A small book providing an overview concerning golf course development, with emphasis on practical golf courses. There are 21 case studies presented.
Revised in 2005.

2005† **Building A Practical Golf Facility.** Second revised edition. Dr. Michael J. Hurdzan. American Society of Golf Course Architects Foundation, Brookfield, Wisconsin, USA. 120 pp., 54 illus. with 52 in color, 4 tables, no index, glossy semi-soft bound in multicolor with white lettering, 5.4 by 8.5 inches (136 × 215 mm).

A slight revision of the 2003 first edition, with one case study added.

2006† **Golf Course Design.** Coauthor.
See *Cornish, Geoffrey St. John—2006.*

Hutchinson, Horatio Gordon

1931† **The Game of Golf.** Coauthor.
See *Wethered, Joyce—1931.*

Joffet, Robert

1947   **Traité de Construction Sportive et de Plein Air (Contracts for Sports and Leisure Construction).** Robert Joffet. Publication du Comité National des Sports, Paris, France. 322 pp., 380 illus., 18 tables, no index, semisoft bound in beige with black lettering, 8.3 by 10.6 inches (210 × 270 mm). Printed entirely in the French language.

     A **relatively rare** book on the design and technical specifications for a broad array of sports and recreation installations, including turfgrass soils. It is oriented to the soil and climatic conditions in France.

Jones, Rees Lee

1974†   **Golf Course Developments.*** Rees L. Jones and Guy L. Rando. ULI Technical Bulletin 70, Urban Land Institute, Washington, D.C., USA. 105 pp., 108 illus. with 12 in color, 6 tables, no index, glossy semisoft bound in green and blue with white and black lettering, 8.5 by 10.9 inches (215 × 277 mm).

     A small manual on golf course development oriented to real estate developers. Included are sections on site analysis, course design within a community development project, costs, construction, clubhouse, and membership. It is a unique book.

Jones, Robert Trent

n.d.
circa
1933†   **Golf Course Architecture.** Robert Trent Jones. Thompson & Jones, New York City, New York, USA. 39 pp., 15 illus., no tables, no index, semisoft bound in cream with black lettering, 5.5 by 7.7 inches (140 × 196 mm).

     A **very rare**, small booklet on golf course architecture with emphasis on strategic design concepts. A section on construction is included. Robert Jones summarized his philosophy of golf architecture as follows:

> Modern Theory of Golf Course Architecture
>
> The modern theory in golf architecture is to create a balanced hole for the various classes of golfers. In the past, the majority of players who may have been termed average golfers were punished far out of proportion to their playing skill. Traps were profusely placed in all areas of the fairway so as to catch a shot only slightly in error. Since the technique of the average golfer's swing is subject to flaws more often than the good or expert golfer's, he was constantly in trouble. As a result the game of golf lost its thrill for him.
>
>   An analysis of the situation disclosed that as far as the crack golfer was concerned, traps under 200 yards offered little or no hazard usually, whereas those same traps were constantly punishing the average golfer. It therefore was feasible that moving such traps would make the play for the average golfer less punishing without soiling the character of the course or the playing value for the expert.

Figure 13.15. A 1912 cartoon titled "Alpinisation of Courses," by Tom Wilkinson. [45]

> It was found that in the green area, the green and trap could be designed so the hole could be tightened or eased by the extent that the pin was placed behind a diagonal trap. In the old penal type of architecture, where the greens were flat and surrounded by a maze of clam-shell traps with a bottle-neck entrance, only one problem was involved—the golfer had no choice other than to hit a perfectly executed shot to the green. Since the shot required was often not in the average golfer's repertoire, he realized he was doomed before he started. With the diagonal trapping, tongue greens and alternate routing to the green, he can play a shot which he feels is within his range. This involves mental keenness as well as playing skill, as one must vary the manner in which one plays the hole on any particular day according to how well one is playing or, in the case of tournament play, according to the circumstances of the match.

Jones, Robert Trent, Jr.

1993† **Golf by Design.** Robert Trent Jones Jr. Little, Brown and Company, New York City, New York, USA. 276 pp., 177 illus. with 80 in color, no tables, no index, cloth hardbound in dark green and tan with silver lettering and a jacket in multicolor with black lettering, 8.4 by 10.9 inches (214 × 276 mm).

A book in which the design features of a golf course are discussed in relation to the playing strategies. It is well illustrated. There is a foreword by Tom Watson.

Kato, Shunsuke

1990† **What Makes a Good Golf Course Good.** Shunsuke Kato. Kato International Design Inc., Tokyo, Japan. 256 pp., 248 illus. with 215 in color, 6 tables, no index, cloth hardbound in green with gold lettering and a jacket in multicolor with gold lettering plus a slip case in blues with gold and black lettering, 8.9 by 11.9 inches (227 × 303 mm). Dual printing in both the Japanese and English languages.

A quality book on the philosophies of golf architecture by the Japanese golf course architect, Shunsuke Kato. The back half contains descriptions and excellent color photographs of the 38 golf courses he has designed.

1996 **A Good Golf Course Merges Into Nature.** Shunsuke Kato, Kato International Design Company Inc., Tokyo, Japan. 244 pp., 362 illus. with 358 in color, 1 table, no index, textured hardbound in light tan with gold lettering plus a textured slip case in tan with gold lettering, 8.9 by 11.9 inches (226 × 302 mm). Dual printing in both the Japanese and English languages.

A quality book concerning the philosophies of golf course architect Shunsuke Kato concerning golf course design. The goals of utilizing the natural terrain and harmonizing with the surrounding landscape are emphasized.

Kim, Giwi Gon

1992 **Golf Course Planning in Korea: Theory and Practice.** Givi Gon Kim, Myeong Kil Kim, Ji Deok Kim, Hwi Young Oh, Dong Geun Lee, Sang Ha Lim, and Young Kyoo Joo (authors). Chogyeong Publishing Company, Seoul South Korea. 567 pp., 365 illus., 139 tables, indexed, cloth hardbound in blue with gold lettering, 7.5 by 10.2 inches (190 × 260 mm). Printed entirely in the Korean language.

A book concerning the architectural design, planning, permitting, layout, and landscaping of golf courses under conditions in Korea. There also is a section on clubhouse construction.

Landscape Research & Development, Germany

2008 **Richtlinie für den Bau von Golf platzen—Golfplatzbaurichtlinie (Guidelines for Golf Course Construction).** Landscape Research & Development, Bonn, Germany. 60 pp., 23 illus. in color, 7 tables, no index, glossy semisoft bound in yellow with black lettering, 8.3 by 11.7 inches (210 × 296 mm). Printed in the German language.

A guidebook concerning the site development, soil and root-zone shaping, drainage, irrigation system, and planting phases of golf course construction under the conditions in Germany.

Langford, William Boice

1915 **Golf Course Architecture in The Chicago District.** William B. Langford. Chicago, Illinois, USA. 21 pp., no illus., no tables, no index, fold-stapled textured semisoft covers in dark green with gold lettering, 6.3 by 8.9 inches (160 × 226 mm).

A **very rare** book on guidance in the design and rebuilding of golf courses, with special consideration to Chicago area conditions. It is derived from a series of nine articles by the author previously published in the **Chicago Evening Post**.

Love, William Robert

1992† **An Environmental Approach to Golf Course Development.** First edition. William R. Love. American Society of Golf Course Architects, Chicago, Illinois, USA. 43 pp., 39 illus. in color, no tables, no index, fold-stapled glossy semisoft covers in multicolor with black lettering, 8.3 by 11.0 inches (210 × 279 mm).

A booklet summarizing the approaches to golf course design that contribute to a quality environment. Twelve case studies are presented. A list of selected references is included in the back.
Revised in 1999 and 2008.

1999† **An Environmental Approach to Golf Course Development.** Second revised edition. Bill Love. American Society of Golf Course Architects, Chicago, Illinois, USA. 45 pp., 45 illus. in color, no tables, no index, fold-stapled glossy semisoft covers in multicolor with black lettering, 8.5 by 11.0 inches (215 × 280 mm).

A revision of the 1992 first edition, with 16 case studies presented.
Revised in 2008.

2008† **An Environmental Approach to Golf Course Development.** Third revised edition. Bill Love with 17 contributing authors. American Society of Golf Course Architects, Brookfield, Wisconsin, USA. 58 pp., 49 illus. with 46 in color, no tables, no index, fold-stapled glossy semisoft covers in multicolor with blue and black lettering, 8.5 by 11.0 inches (215 × 280 mm).

A revised and expanded booklet summarizing the aspects of golf course design that contribute to a quality environment. Eighteen case studies are presented. A list of selected references is included in the back.

Millard, Chris

2002 **Nicklaus by Design.** Coauthor.
See *Nicklaus, Jack—2002.*

Muirhead, Gordon Desmond

1994† **Golf Course Development and Real Estate.**˙ Desmond Muirhead and Guy L. Rando with 34 contributing authors. The Urban Land Institute, Washington, D.C., USA. 191 pp., 225 illus. with 292 in color, 27 tables, no index, glossy semisoft bound in multicolor with white and black lettering, 8.3 by 11.0 inches (210 × 280 mm).

A guidebook on the development, economics, planning, design, and operation of a golf course facility. It is extensively illustrated in color. A small glossary and list of selected references are included in the back.

Nicklaus, Jack

2002 **Nicklaus by Design.** Jack Nicklaus with Chris Millard. Harry N. Abrams Inc., New York City, New York, USA. 288 pp., 163 illus. with 148 in color, no tables, indexed, hardbound in green with blue and green lettering plus a jacket in multicolor with blue and green lettering, 9.15 by 11.7 inches (240 × 296 mm).

A book concerning the architectural philosophies the author has employed in the design of golf courses, accompanied by quality pictorials. The foreword is by Mr. Pete Dye.

Peake, Malcolm

2005† **A Natural Course for Golf.** Compiled by Malcolm Peake, with 22 contributing authors. The Sports Turf Research Institute, Bingley, West Yorkshire, England, UK. 84 pp. plus 3 pp. of advertising, 65 illus. with 47 in color, no tables, no index, hardbound in multicolor with white lettering, 7.1 by 9.8 inches (180 × 248 mm).

A small book on the philosophy and issues of golf course design, construction and maintenance in terms of blending with the surrounding topography and naturalized plants versus an artificial facility. There is a "tongue in cheek" glossary in the back.

Rando, Guy L.

1974† **Golf Course Developments.** Coauthor.
See *Jones, Rees Lee—1974.*

1994† **Golf Course Development and Real Estate.** Coauthor.
See *Muirhead, Gordon Desmond—1994.*

Richardson, Forrest L.

2002† **Routing the Golf Course.** Forrest L. Richardson. John Wiley & Sons, Hoboken, New Jersey, USA. 525 pp., 337 illus., 10 tables, small index, cloth hardbound in green with silver lettering, 7.9 by 9.9 inches (200 × 250 mm).

A book on an essential component of golf course design ~ the routing. It includes a historical perspective, factors influencing routing, basic philosophy, and decision making. The foreword is by Peter Oosterhuis. There is a selected list of references in the back.

2006† **Bunkers, Pits & Other Hazards: A Guide to the Design, Maintenance and Preservation of Golf's Essential Elements.** Forrest L. Richardson and Mark K. Fine. John Wiley & Sons Inc., Hoboken, New Jersey, USA. 320 pp., 289 illus. plus 47 in color, 2 tables, small index, hardbound in multicolor with black and white lettering, 7.4 by 9.2 inches (188 × 234 mm).

A unique book on the history, design strategies, planning, construction, maintenance, and restoration of all forms of hazards. There is a chapter on psychological effects of hazards by Dr. Edward Sadalla. There is an extensive glossary and a list of selected references in the back.

Robinson, William Grieve

n.d.† circa 1975   **Golf Course Design—An Introduction.** Coauthor.
See *Cornish, Geoffrey St. John—circa 1975.*

Shackleford, Geoff

1999†   **The Golden Age of Golf Design.** Geoff Shackelford. Sleeping Bear Press, Chelsea, Michigan, USA. 315 pp., 248 illus. with 19 in color, no tables, indexed, glossy hardbound in multicolor with gold lettering and a glossy jacket in multicolor with gold and black lettering, 10.6 by 8.0 inches (268 × 202 mm).

A book concerning the pioneering golf architects and their key golf courses designed during the first four decades of the twentieth century. It includes extensive old photographs and the golf landscape paintings of Mike Miller.

2003   **Grounds for Golf.** Geoff Shackelford. Thomas Dunne Books, New York City, New York, USA. 320 pp., 149 illus. with 16 in color, no tables, indexed, cloth hardbound in green with orange lettering, 6.1 by 9.2 inches (155 × 233 mm).

A book on golf course architecture with a unique writing approach and clear statements of the author's views. The book moves chronologically from the early history, to the various "schools of golf design," then to "great golf courses." The characteristics of a golf architect and, finally, current issues and trends are discussed.

Simpson, Thomas G.

1929   **The Architectural Side of Golf.** Coauthor.
See *Wethered, Herbert Newton—1929.*

1931†   **The Game of Golf.** Coauthor.
See *Wethered, Joyce—1931.*

1952†   **Design for Golf.** Coauthor.
See *Wethered, Herbert Newton—1952.*

Sorensen, Gary L.

1976†   **The Architecture of Golf.** Gary L. Sorensen. Sorensen's Publishing Firm, College Station, Texas, USA. 106 pp., 82 illus., 3 tables, no index, soft bound in gray and browns with black lettering, 8.3 by 10.8 inches (211 × 255 mm).

A small book on golf architecture encompassing a brief history of golf and golf courses, design philosophies, architecture, and construction. It contains a selected bibliography in the back.

Thomas, George Clifford, Jr.

1927†   **Golf Architecture in America: Its Strategy and Construction.** Geo. C. Thomas Jr. The Times-Mirror Press, Los Angeles, California, USA. 342 pp., 149 illus. with 4 color plates, no tables, no index, cloth hardbound in black with green lettering and red imprint and a jacket in multicolor with black lettering, 6.0 by 8.7 inches (152 × 222 mm).

A key book that is a must for collectors of historical turf and/or golf books. It is one of the classic works on the architecture and construction of golf courses. The chapter subject topics include the following:

I.	Different Courses	VIII.	The Balance of Area
II.	The Strategy of Golf Courses	IX.	Actual Construction
III.	Something about Choice and Construction	X.	Remodeling Old Courses
IV.	The General Plan for the Property	XI.	Arbitrary Values
V.	Beauty and Utility	XII.	The Unknown Equation
VI.	Adapting the Course to the Ground	XIII.	Underdrainage
VII.	The Ability to Create		

The chapters on construction and establishment were the most comprehensive that had been written up to 1927. The photos of golf holes and the architectural drawing were well done, especially the latter. It includes a drawing for an eighteen golf course of one-shot holes in the back. George Thomas discussed the costs of constructing a golf course in California in 1927 as follows:

A representative cost to build an 18-hole golf course in California in 1927 would be US $81,000; with the minimum cost for course maintenance in California being US $25,000.

1990† **Golf Architecture in America: Its Strategy and Construction.** Geo C. Thomas. A facsimile edition by the United States Golf Association, Far Hills, New Jersey, USA. Hardbound cover in dark green with gold lettering and a light green slip case.

1997† **Golf Architecture in America: Its Strategy and Construction.** Geo. C. Thomas Jr. Reprinted by Sleeping Bear Press, Chelsea, Michigan, USA. In a similar format with a white jacket.

Tillinghast, Albert Warren

n.d.
circa
1917
Planning a Golf Course. Albert Warren Tillinghast. Privately printed, Philadelphia, Pennsylvania, USA. 24 pp., no tables, no index, saddle-stitched covers in red with black lettering, 3.5 by 8.5 inches (89 × 216 mm).

A **very rare** booklet that is basically a prospective for the Tillinghast golf course architecture firm.

1995† **The Course Beautiful–A Collection of Original Articles and Photographs on Golf Course Design.** Compiled by Richard C. Wolffe Jr., Robert S. Trebus, and Stuart F. Wolffe. Tree Wolf Productions, Short Hills, New Jersey, USA. 120 pp., 147 illus., no tables, no index, cloth hardbound in dark green with gold lettering, 8.5 by 10.9 inches (216 × 277 mm).

This interesting book is a compilation of articles written by A.W. Tillinghast between 1916 and 1936 that were originally published in various golf magazines. The old photographs are a special feature. Emphasis is on golf course design concepts, plan development, and construction supervision. Also described is the origin of the Shawnee Triple Mower of 1911.

Urban Land Institute

2001† **Golf Course Development in Residential Communities.** Urban Land Institute with nine contributing authors. Urban Land Institute, Washington, D.C., USA. 313 pp., 245 illus. with 27 in color, 25 tables, no index, glossy semisoft bound in multicolor with white and yellow lettering, 8.5 by 11.0 inches (216 × 279 mm).

A book that provides an overview of golf course development. The six chapters involve market economies, land planning, golf course design, business perspective, and legal aspects. There are 10 case studies presented.

Wethered, Herbert Newton

1929 **The Architectural Side of Golf.*** H.N. Wethered and T. Simpson. Longmans, Green Limited, London, England, UK. 211 pp., 69 illus. with 44 color plates, no tables, indexed, cloth hardbound in cream with gold lettering, 7.7 by 9.8 inches (195 × 248 mm).

A **very rare** book that is a must for collectors of historical golf and/or turfgrass books. There is a chapter on construction and maintenance of golf courses in the United Kingdom and a preface by J.C. Squire. This book also was printed as a limited edition of 50 numbered copies on large paper. The chapter subject titles include the following:

I.	The Four Ages	X.	Critics and Superstitions
II.	Attack and Defense	XI.	Two Hundred Years Ago
III.	Fairways	XII.	Going South
IV.	Putting Greens	XIII.	A Game of Confidence
V.	The Ideal Golf Course	XIV.	East Lothian
VI.	General Principles, Construction and Upkeep	XV.	Caddies We Have Met
VII.	Hazards	XVI.	The Last Chord
VIII.	Going North	XVII.	In an English Garden
IX.	The Philosopher and the Poet	XVIII.	The Last Green

The Reversible Course is discussed in the appendix:

> The Reversible Course
> To play a course backwards was an alternative that commended itself for many sound reasons almost as soon as golf courses came into being. Within living memory the Old Course at St. Andrews was habitually played in reverse, and for all we know it may still be so played. At North Berwick, and probably on most, if not all, of the older courses, the custom prevailed, in order, as it was said, to "rest the course."

In other words, this procedure helped to preserve the fairways over a given area from being unduly cut up with divot marks—a point on which opinion was then particularly stringent; it also gave the grass time to recover in the places where divots had been cut. Some method of the kind was absolutely necessary when courses were so much shorter than they are now and required the playing of a greater number of strokes within a much more limited area. In fact, iron play was rather frowned upon and discouraged as being unduly destructive of good turf. The modern practice of tearing the fairways to pieces, as well as the tees at the shorter holes, would have been regarded—and rightly so—with horror.

It will be remembered, too, that "the green," by which was meant the entire extent of the links, had very largely to look after itself. It was never so patched up or so carefully tended as it is to-day to help its wounds to heal. The course was reversed and played backwards at stated intervals to enable it to gain a little rest between whiles and recover the trueness of its surface. In addition to this, the immediate neighbourhood of the putting greens was found to be greatly benefited by a variation in the line of the approach.

1952† **Design for Golf.** New revised edition. H.N. Wethered and T. Simpson. The Sportsman's Book Club, London, England, UK. 203 pp., 50 illus., no tables, small index, cloth hardbound in green with black with versions in both gold and black lettering and jackets both in red and yellow with red and yellow lettering and in green and white with green lettering, 5.4 by 8.4 inches (136 × 213 mm).

Essentially a revised second edition of **The Architectural Side of Golf** by H. Newton Wethered, 1929, but without the color plates.

See *Wethered, Herbert Newton—1929.*

1995† **The Architectural Side of Golf.** First reprint edition. H.N. Wethered and T. Simpson. Grant Books Limited, Droitwich, Worcestershire, England, UK. 235 pp., 85 illus., no tables, short index, cloth hardbound in yellow with gold lettering, 7.0 by 9.4 inches (179 × 239 mm).

A facsimile of a classic 1929 book by architects H.N. Wethered and T. Simpson. Included is a foreword by Arthur Hills. This was a limited edition of 565 copies.

Wethered, Joyce (Lady Heathcoat-Amory)

1931† **The Game of Golf.*** First edition. Joyce Wethered, Roger Wethered, Bernard Darwin, Horace Hutchinson, and T.C. Simpson. The Lonsdale Library Volume IX, Seeley, Service & Company Limited, London, England, UK. 251 pp., 107 illus., no tables, indexed, cloth hardbound in light brown with gold lettering and a jacket with a sepia plate on a straw

Figure 13.16. A 1912 cartoon titled "Lengthening of Courses" by Tom Wilkinson. [45]

colored cover with red lettering plus a special edition of cloth and one-quarter leather with a jacket in color, 5.4 by 8.4 inches (137 × 213 mm).

This **very rare** book contains four chapters by Mr. Thomas C. Simpson on the architecture, construction, and turf-grass culture of golf courses in England. There are two chapters by Mr. Horace Hutchinson on the history and literature of golf. The chapter subject titles and contributing authors include the following:

I.	The History of the Game—Horace Hutchinson	IX.	Match & Medal Play—Bernard Darwin
II.	The Literature of the Game—Horace Hutchinson	X.	Watching for Profit—Bernard Darwin
		XI.	Ladies' Golf—Joyce Wethered
III.	Wooden Club Play—Joyce and Roger Wethered	XII.	Middle-Aged Golf—Bernard Darwin
		XIII.	Famous Courses—Bernard Darwin
IV.	The Iron Shot—Roger Wethered	XIV.	Golf Architecture—T. Simpson
V.	The Short Approach Shots—Roger Wethered	XV.	Golf Architecture (Continued)—T. Simpson
		XVI.	The Upkeep of a Golf Course—T. Simpson
VI.	Bunker Play—Roger Wethered	XVII.	The Upkeep of a Golf Course (Continued)—T. Simpson
VII.	Putting—Joyce Wethered		
VIII.	Practice—Bernard Darwin	XVIII.	Rules of Golf (Revised Edition)

Thomas Simpson took the following position concerning annual bluegrass (*Poa annua*) on putting greens in 1931:

> I do not believe in trying to maintain bents and fescues alone on putting-greens, and I am therefore opposed to the lines on which much present-day research work is tending, except perhaps from the point of view of learning what should be avoided.
>
> It does not matter in the least what kind of grass seed you sow if it be not the grass that is indigenous to the locality. In a very few years there will be little or nothing left of the imported grass which has been sown.
>
> By all means let us start our new courses with fine varieties, such as *Agrostis* and *Fescue*, in order to obtain a good sward of fine turf at the earliest possible moment. That must be insisted on from every point of view, especially in the interest of the promoters of the club, who will naturally wish to get in the first year's subscriptions as soon as possible.
>
> But after that do not let us wage an interminable and costly warfare in an endeavour to prevent our greens from being invaded with *Poa annua*, or by some other variety of grass that thrives in the locality and which, more likely that not, prefers a neutral condition of soil.
>
> I am well aware that *Poa annua* is often referred to as a weed, but, as a matter of fact, so was *Agrostis vulgaris* only a few years ago. Some weeds make an excellent surface for putting if they are kept closely cut. Yarrow, for example, is one.

†   *Reprinted in 1946 and 1948.*

1931b†   **The Game of Golf.*** American edition. Joyce Wethered, Roger Wethered, Bernard Darwin, Horace Hutchinson, and T.C. Simpson. J.B. Lippincott Company, Philadelphia, Pennsylvania, USA. 251 pp., 107 illus., no tables, indexed, cloth hardbound in light brown with gold lettering, 5.4 by 8.3 inches (137 × 211 mm).

    The same book as printed in 1931 in London, England, UK.

1951†   **The Game of Golf.*** New American edition. Joyce Wethered, Roger Wethered, Bernard Darwin, Horace Hutchinson, and T.C. Simpson. A.S. Barnes, New York City, New York, USA. 251 pp., 211 illus., no tables, indexed, cloth hardbound in light brown with gold lettering, 5.5 by 8.3 inches (140 × 211 mm).

    A reprinting of the original 1931 first edition.

Wethered, Roger Henry

1931†   **The Game of Golf.** Coauthor.

    See *Wethered, Joyce—1931.*

Whitten, Ronald Edward

1981†   **The Golf Course.** Coauthor.

    See *Cornish, Geoffrey St. John—1981.*

1993† **The Architects of Golf.** Coauthor.
See *Cornish, Geoffrey St. John—1993.*

E. PEST DIAGNOSIS FOR TURFGRASSES

Encompasses booklets emphasizing the identification of causal pest organisms usually via color photographs and documented by detailed descriptions plus management aspects.

Arkansas, USA

1981† **Weeds of Arkansas: Lawns, Turf, Roadsides, Recreation Areas—A Guide to Identification.** Ford L. Baldwin and Edwin B. Smith. Printing Service, University of Arkansas, Fayetteville, Arkansas, USA. 71 pp., 307 illus. in color, no tables, indexed, fold-stapled glossy semisoft covers in multicolor with black and white lettering, 8.5 by 10.9 inches (216 × 277 mm).
A quality guide to the identification of weeds of turfgrasses in Arkansas. It includes descriptions and color photographs of 179 weeds found in Arkansas.

California, USA

1971† **Turfgrass Pests.** First edition. Written by twelve contributing authors. Division of Agricultural Sciences Manual 41, University of California, Berkeley, California, USA. 49 pp., 93 illus. with 64 in color, 9 tables, no index, fold-stapled glossy semisoft covers in black and white with black and white lettering, 8.5 by 10.9 inches (217 × 277 mm).
A small manual on the pests of turfgrasses in California that encompasses weeds, insects, nematodes, diseases, and rodents. It includes a pest description, life history, susceptible plants, damage caused, and method of detection. There are chapters on pesticide use and safety and on area measurement, calculations, and sprayer preparation plus a glossary of terms in the back. The color photographs are very good.
Revised in 1980 and 1989.

1980† **Turfgrass Pests.** Second revised edition. W.R. Bowen (editor). Division of Agricultural Sciences Publication No. 4053, University of California, Berkeley, California, USA. 53 pp., 106 illus. with 84 in color, 7 tables, no index, glossy semisoft bound in multicolor with white lettering, 8.3 by 10.7 inches (211 × 272 mm).
A small manual on the pests of turfgrasses in California that encompasses weeds, insects, nematodes, diseases, and rodents. It includes a pest description, life history, susceptible plants, damage caused, and method of detection. There are chapters on pesticide use and safety and on area measurement, calculations, and sprayer preparation plus a glossary of terms in the back. The color photographs are very good.
Revised in 1989.

1989† **Turfgrass Pests.** Third revised edition. A.D. Ali and Clyde L. Elmore (editors) with 11 contributing authors. Cooperative Extension Publication 4053, Division of Agriculture and Natural Resources, University of California, Oakland, California, USA. 122 pp., 169 illus. with 80 in color, 7 tables, no index, glossy semisoft bound in multicolor with white lettering, 6.1 by 8.9 inches (155 × 227 mm).
A quality manual describing the procedures to identify and/or diagnose pest problems of turfgrasses found in California, with emphasis on weeds, insects, nematodes, diseases, and vertebrate pests. The manual is well illustrated. It contains a glossary in the back.

Canada

1996† **Diseases and Insects of Turfgrass in Ontario—A Handbook For Professional Turf Managers.** Pam Charbonneau (editor) with contributing authors Mark K. Sears and Tom Hsiang. Ontario Ministry of Agriculture, Food and Rural Affairs Publication 162, Ontario, Canada. 49 pp., 94 illus. with 80 in color, 3 tables, no index, glossy semisoft bound in multicolor with black and white lettering, 8.9 by 10.9 inches (226 × 277 mm).
A handbook on the identification, life cycles, and management of 17 disease and 10 insect problems that occur on turfgrasses in Ontario. There is a small glossary in the back.

1997† **Turfgrass Disease And Pest Management Guide For Professional Turfgrass Managers in B.C.** Leslie Mac Donald and Henry Gerber. Western Canada Turfgrass Association, Maple Ridge, British Columbia, Canada. 82 pp., 77 illus. with 72 in color, 5 tables, indexed, fold-stapled glossy semisoft covers in multicolor with black lettering, 8.2 by 10.9 inches (208 × 277 mm).

A guide concerning the causal pathogen, turfgrass injury diagnosis, and management of turfgrass diseases found in British Columbia plus some other insect and animal pests.

North Central USA

1963† **Lawn Weeds: Identification and Control.** J.D. Butler and F.W. Slife. Cooperative Extension Service Circular 873. University of Illinois, Urbana, Illinois, USA. 28 pp., 58 illus., 10 tables, no index, fold-stapled semisoft covers in black and white with white lettering, 6.0 by 9.0 inches (153 × 228 mm).

A guide to the identification of 28 major weeds found in turfgrasses in the Midwestern United States plus an ID key. Also included are summary tables of weed controls available.

1970† **Lawn Weeds and their Control.** Extension Publication No. 26, North Central Agricultural Extension Services, Lincoln, Nebraska, USA. 23 pp.; 56 illus. with 46 in color; 1 table; no index; fold-stapled glossy soft covers in multicolor with green, black, and red lettering; 8.5 by 1.0 inches (215 × 281 mm).

A booklet concerning the identification of 46 weeds common in turfgrasses in the north central United States plus their suggested management. There is a graph in the back showing a calendar of active growth and treatment for each weed species. *Reprinted and updated annually into the 1980s.*

1985† **Turfgrass Insect Damage, Prevention and Control—A Common Sense Approach.** Arthur H. Bruneau, David Bishop, Robert C. Shearman, and Robert E. Roselle. Nebraska Cooperative Extension Service EC 81-1238, University of Nebraska, Lincoln, Nebraska, USA. 31 pp., 20 illus. with 16 in color, no tables, no index, semisoft covers in green with white lettering in a comb binder, 7.0 by 4.5 inches (178 × 114 mm).

A guide to the diagnosis and management of insect problems on turfgrasses in Nebraska.

1997† **Integrated Turfgrass Management for the Northern Great Plains.** Frederick P. Baxendale and Roch E. Gaussoin with 10 contributing authors. Cooperative Extension EC97-1557, University of Nebraska, Lincoln, Nebraska, USA. 236 pp.; 185 illus. with 116 in color; 11 tables; indexed; glossy hard covers in multicolor with green, blue, and black lettering in a ring binder; 5.6 by 8.5 inches (141 × 217 mm).

A manual primarily devoted to the identification, management, and control of the common pests of turfgrasses in the northern Great Plains of the United States. A list of selected references is in the back.

2005† **Turf Diseases of the Great Lakes Region.** S.W. Abler and G. Yung. Cooperative Extension, University of Wisconsin, Madison, Wisconsin, USA. 25 pp., 50 illus. with 49 in color, 1 table, indexed, fold-stapled glossy semisoft bound in multicolor with blue and green lettering, 8.5 by 11.0 inches (215 × 279 mm).

A booklet concerning 25 diseases of turfgrasses that occur in the 8 states of the Great Lakes region. Included are descriptions and symptoms, grass hosts, and management strategies. There are identification keys in the front and a short glossary and calendar of occurrence in the back.

2005† **Identifying Turf and Weedy Grasses of the Northern United States.** Dianne Pedersen and Tom Voigt. Pocket ID Series, University of Illinois Extension, Urbana, Illinois, USA. 62 pp.; 129 illus. in color; no tables; no index; fold-stapled glossy semisoft covers in multicolor with green, black, and blue lettering; 4.9 by 3.5 inches (125 × 89 mm).

Figure 13.17. Manual digging of weeds from a green, circa 1930. [21]

A guide to vegetative grass identification via both close-up photographs and descriptions of the structures involved. There is a key presented in the front.

Northeastern USA

1982†   **Picture Clues to Turfgrass Problems.** A. Martin Petrovic, Maria T. Cinque, Richard W. Smiley, and Haruo Tashiro. Cornell Cooperative Extension Miscellaneous Bulletin 125, Ithaca, New York, USA. 42 pp., 47 illus. in color, no table, indexed, glossy semisoft covers in green and white with black lettering in a ring binder, 6.7 by 3.5 inches (171 × 89 mm).

A pocket field guide to aid in the diagnosis of turfgrass problems, especially insects and diseases.

2001†   **Turfgrass: Nutrient and Integrated Pest Management Manual.** Timothy M. Abbey (editor). Connecticut Agricultural Experiment Station, University of Connecticut, Storrs, Connecticut, USA. 116 pp., 218 illus., 9 tables, no index, glossy semisoft bound in multicolor with light green and white lettering, 8.5 by 11.0 inches (215 × 278 mm).

A guidebook concerning the management of turfgrasses as it influences the weed, disease, and insect problems of turfgrasses in Connecticut.

2001†   **Turfgrass Problems: Picture Clues and Management Options.** Eva Gussack and Frank S. Rossi with four contributing authors. Cooperative Extension NRAES-125, Natural Resource, Agriculture, and Engineering Service, Ithaca, New York, USA. 214 pp.; 160 illus. with 133 in color; 1 table; no index; glossy semisoft covers in multicolor with white, black, and brown lettering in a ring binder; 3.5 by 6.7 inches (88 × 170 mm).

A guide to the diagnosis of injury problems on cool-season turfgrasses, with emphasis on diseases, insects and weeds plus their management. There is a small glossary of selected terms in the back.

Pacific Northwest USA

1979†   **Diseases of Turfgrass.** First edition. Charles J. Gould, Roy L. Goss, and Ralph S. Byther. Extension Bulletin 713, Washington State University Cooperative Extension, Puyallup, Washington, USA. 40 pp., 27 illus. in color, 5 tables, no index, fold-stapled glossy soft covers in multicolor with white lettering, 8.4 by 10.9 inches (213 × 277 mm).

A booklet on the diseases of turfgrasses that occur in the Pacific Northwest. There are descriptions and color photographs of the common diseases, weather, and cultural factors influencing disease development and specific control measures. It includes selected references.

1999†   **Diseases of Turfgrass.** Second revised edition. Gary A. Chastagner, Gwen Stahnke, Ralph S. Byther, Charles J. Gould, and Roy L. Goss. Cooperative Extension EB 0713, Washington State University, Pullman, Washington, USA. 37 pp., 32 illus. with 30 in color, no tables, no index, fold-stapled glossy semisoft covers in multicolor with white and black lettering, 8.5 by 10.9 inches (216 × 278 mm).

A revised update of the 1979 first edition.

Rocky Mountain Region, USA

1996†   **Turfgrass Insects in Colorado & Northern New Mexico.** Whitney Cranshaw and Charles R. Ward. Colorado State University Cooperative Extension, Fort Collins, Colorado, USA. 40 pp.; 70 illus.; 7 tables; no index; fold-stapled glossy semisoft covers in multicolor with blue, yellow, white, green, and black lettering; 8.4 by 10.9 inches (214 × 278 mm).

A booklet concerning the appearance, habits, life cycle, and management of insects that are pests of turfgrasses in Colorado and northern New Mexico.

Southeastern USA

1967†   **Diseases of Southern Turfgrasses.** T.E. Freeman. Bulletin 713, Agricultural Experiment Stations, University of Florida, Gainesville, Florida, USA. 31 pp., 14 illus. with 7 in color, 1 table, no index, fold-stapled semisoft covers in multicolor with black lettering, 5.9 by 8.9 inches (150 × 226 mm).

A small booklet describing the disease symptoms, pathogen morphology, occurrence, and control of 24 diseases of warm-season turfgrasses. It contains a list of references in the back.

1982†   **Insect and Other Pests Associated with Turf.** James R. Baker (editor) with 13 contributing authors from 8 southeastern states. Agricultural Extension Service, North Carolina State University, Raleigh, North Carolina, USA. 108 pp., 172 illus. with 18 in color, 1 table, indexed by host plant and sub-indexed by pest, semisoft bound in cream with brown lettering, 8.5 by 10.8 inches (215 × 275 mm).

A major technical publication addressing the common insects and other pests of turfgrasses in the southeastern United States. There is a pest damage key. It contains a list of selected references and a small glossary in the back.

n.d.†   **Broadleaf Weeds in Turf: Identification and Control.** The Dow Chemical Company, Midland, Michigan, USA. 72
circa   pp.; 162 illus. in color; no tables; no index; glossy semisoft covers in maroon and gray with white, gray, and maroon
1980s   lettering in a wire ring binder, 4.2 by 8.4 inches (107 × 213 mm).

A quality pictorial reference guide for the vegetative identification of 34 broadleaf weeds that occur in turfgrass.

1985† **Turfgrass Pest Management Manual: A Guide to Major Turfgrass Pests and Turfgrasses.** First edition. Arthur H. Bruneau (editor) with six contributing authors: Arthur H. Bruneau, Joseph M. DiPaola, William B. Gilbert, William M. Lewis, Leon T. Lucas, and Robert L. Robertson. North Carolina Cooperative Extension Service Publication AG-348, North Carolina State University, Raleigh, North Carolina, USA. 64 pp., 126 illus. with 90 in color, no tables, no index, fold-stapled glossy semisoft covers in multicolor with white lettering, 8.4 by 10.8 inches (214 × 274 mm).

A guide to the identification of turfgrass species and their pests, including the weeds, diseases, and insects found in North Carolina. It contains a small glossary and set of selected references in the back.

Reprinted in 1989, 1991, and 1993.

Revised in 1989, 1991, 1997, and 2006.

n.d.† **Weeds of Southern Turfgrasses—Golf Courses—Lawns—Roadsides—Recreational Areas—Commercial Sod.** Tim R.
circa Murphy (editor) with six contributing authors: Tim R. Murphy, Daniel L. Calvin, Ray Dickens, John W. Everest, David
1992 Hall, and Lambert B. McCarty. Cooperative Extension Service, College of Agricultural and Environmental Sciences, University of Georgia, Athens, Georgia, USA. 208 pp., 444 illus. with 433 in color, no tables, indexed as to scientific and to common plant names, glossy semisoft bound in greens and white with white lettering, 5.5 × 8.4 inches (140 × 213 mm).

A quality manual on the identification and distribution of weeds associated with warm-season, perennial turfgrasses in the southeastern United States. There are good color photographs. It contains a small taxonomic glossary in the back.

Also reprinted by the Florida Cooperative Extension Service in 1992 and by the Clemson University Cooperative Extension Service in 1996.

1997† **Turfgrass Pest Management Manual.** Second revised edition. Arthur H. Bruneau (editor) with eight contributing authors. North Carolina Cooperative Extension Service Publication AG-348, North Carolina State University, Raleigh, North Carolina, USA. 75 pp.; 154 illus. with 111 in color; no tables; no index; fold-stapled glossy semisoft covers in multicolor with white, green, and black lettering; 8.5 by 10.9 inches (215 × 277 mm).

A revised update of the 1985 edition.

Revised in 2006.

2006† **Turfgrass Pest Management Manual.** Third revised edition. Arthur H. Bruneau, Gail G. Wilkerson, Bridget L. Robinson, and Emily J. Erickson (editors) with 11 contributing authors. North Carolina Cooperative Extension Publication AG-348, North Carolina State University, Raleigh, North Carolina, USA. 102 pp.; 313 illus. in color; 96 tables; no index; glossy semisoft bound in multicolor with white, red, black, and gray lettering; 8.4 by 11.0 inches (214 × 279 mm).

A major revision of the 1997 edition, especially in terms of color illustrations.

United Kingdom

1979† **Turfgrass Diseases.** First edition. Sports Turf Research Institute, Bingley, West Yorkshire, England, UK. 37 pp., 8 illus. in color, no tables, no index, fold-stapled semisoft covers in light blue with black lettering, 5.8 by 8.2 inches (148 × 209 mm).

Figure 13.18. Portable, motorized sprayer apparatus with wooden water tank, circa 1927. [2]

A booklet on the common diseases of turfgrass that occur in the British Isles. It includes descriptions, preventive strategies, and controls.

1987†   **Turfgrass Diseases.** Second revised edition. Neil A. Baldwin. Sports Turf Research Institute, Bingley, West Yorkshire, England, UK. 40 pp., 10 illus. with 8 in color, 2 tables, no index, fold-stapled glossy semisoft covers in dark blue with white lettering, 5.7 by 8.2 inches (146 × 207 mm).

An update of a practitioner-oriented booklet, which describes the diseases of turfgrass that occur in the United Kingdom plus their cultural and chemical control. The color photographs have been improved and increased in number. A small glossary has been added in the back.

Revised in 1990.

1990†   **Turfgrass Pests and Diseases.** Third revised edition. Neil A. Baldwin. Sports Turf Research Institute, Bingley, West Yorkshire, England, UK. 58 pp., 33 illus. with 32 in color, no index, 5 tables, fold-stapled glossy semisoft covers in green with white lettering, 5.7 by 8.2 inches (146 × 208 mm).

A revision of a simplified booklet on pests and diseases of turfgrasses that occur in the United Kingdom. It is organized in three parts: insect and animal pests, diseases, and pesticides. The insect part is a new addition.

1998†   **Turfgrass Diseases and Associated Disorders.** Catherine York. The Sports Turf Institute, Bingley, West Yorkshire, England, UK. 59 pp., 29 illus. with 28 in color, 5 tables, no index, glossy semisoft bound in greens and black with white and black lettering, 5.5 by 8.1 inches (140 × 206 mm).

A small book on the diseases of turfgrasses that occur in the United Kingdom under cool climatic regions. It includes the causal pathogen, turfgrass disease symptoms, habitat, and disease management.

United States

1957†   **Turf Disease Handbook.** James L. Holmes. Mallinckrodt Chemical Works, St. Louis, Missouri, USA. 27 pp., 12 illus. in color, 1 table, no index, fold-stapled textured glossy semisoft covers in green and white with green and white lettering, 5.9 by 9.0 inches (151 × 228 mm).

A small, early handbook in which 12 major diseases of turfgrasses are illustrated via both color photographs of the injury symptoms and by drawings of the causal pathogen. The text includes the injury symptoms, hosts, epidemiology, and control.

Revised in 1966, 1975, and 1993.

n.d.   **Weed Information from Scotts Research.** O.M. Scott and Sons Company, Marysville, Ohio, USA. 29 pp., 14 illus., no
circa   tables, no index, fold-stapled glossy covers in greens with black lettering, 8.4 by 11.0 inches (214 × 278 mm).
196x   A well-illustrated booklet for 14 grassy weeds commonly found in turfgrasses in the United States. The morphological structures used in identification are delineated in detail.

1966†   **Turf Pest Management Handbook.** Second revised edition. Stan Frederikson and W.A. Small. Mallinckrodt Chemical Works, St. Louis, Missouri, USA. 41 pp., 20 illus. in color, no tables, no index, fold-stapled textured glossy semisoft covers in multicolor with black and white lettering, 5.5 × 8.4 inches (139 × 214 mm).

A substantial revision of the pioneering **Turf Disease Handbook** written in 1957. It contains color illustrations and discussions of 17 diseases and 3 weedy annual grasses found in the United States. The discussion includes symptoms, grasses attacked, environment, and control.

Revised in 1975 and 1983.

1975†   **Guide to Turfgrass Pests.** Robert C. Daley (editor). National Pest Control Association Inc., Vienna, Virginia, USA. 53 pp., 77 illus. with 76 in color, no tables, no index, glossy semisoft covers in greens and white with white lettering in a comb binder, 8.6 by 11.0 inches (219 × 279 mm).

A handbook describing 19 diseases and 22 insects and other pests of turfgrasses in the United States. It is organized in a format for use in pest problem diagnosis. There is a very short glossary and a list of selected references in the back.

1975†   **Turf Pest Management Handbook.** Third revised edition. Stan Frederikson and W.A. Small. Mallinckrodt Inc., St. Louis, Missouri, USA. 59 pp., 27 illus. in color, no tables, no index, fold-stapled glossy semisoft covers in multicolor with black and white lettering, 5.4 by 8.5 inches (137 × 216 mm).

A substantial revision of the 1966 handbook. It is focused primarily on 19 common turfgrass diseases found in the United States. Included are the disease symptoms, grasses attacked, environment, and control. There are excellent color photographs of the disease symptoms plus drawings of the fungal mycelium and spores. It also includes smaller sections on broadleaf and weedy annual grass controls.

Revised in 1983.

1983†   **Turf Pest Management Handbook.** Fourth revised edition. W.A. Small. Mallinckrodt Inc., St. Louis, Missouri, USA. 49 pp., 54 illus., no tables, no index, semisoft covers in white with black and green lettering in a 6-hole binder, 5.6 by 8.3 inches (142 × 212 mm).

Figure 13.19. Manual-push, wheelbarrow-type herbicide sprayer with 15-gallon (56.8 l) tank and 4-foot-wide (1.2-meter-wide) spray boom, mid-1940s. [19]

A significant revision of the 1975 version, including the addition of four diseases. It contains quality color photographs of each disease plus mycelium and fruiting body drawings of the causal pathogens.

1986† **Diseases of Turfgrasses.** Agricultural Division, Ciba-Geigy Corporation, Greensboro, North Carolina, USA. 63 pp., 33 illus. in color, no tables, no index, glossy semisoft bound in multicolor with white and black lettering, 6.0 by 8.4 inches (152 × 213 mm).

A manual concerning the diagnosis of 23 turfgrass diseases plus guidelines for cultural controls and a seasonal occurrence calendar for six climatic regions in the United States.

Revised in 1995 and 2003.

1987† **Scotts Guide To The Identification of Turfgrass Diseases and Insects.** O.M. Scott & Sons Company, Marysville, Ohio, USA. 105 pp., 160 illus. in color, no tables, no index, glossy semisoft bound in multicolor with green and white lettering, 5.5 by 8.2 inches (140 × 209 mm).

A practical reference book on the identification and symptom diagnosis of turfgrass diseases, viruses, nematodes, insects, and mites. It includes pest descriptions, turfgrass injury symptoms, and life cycles. This comprehensive book is well illustrated in color.

198x† **Broadleaf Weeds In Turf.** Norm Thomas. The Dow Chemical Company, Midland, Michigan, USA. 72 pp.; 162 illus. in color; no tables; no index; glossy semisoft covers in maroon and gray with white, maroon, and gray lettering in a ring binder; 8.4 by 4.2 inches (213 × 107 mm).

A description and pictorial representation of 34 weeds commonly found in turfgrass. There are good quality color photographs.

1995† **Diseases of Turfgrasses.** Turf & Ornamental Products, Ciba-Geigy Corporation, Greensboro, North Carolina, USA. 35 pp., 52 illus. with 46 in color, 7 tables no index, glossy semisoft covers in multicolor with white and black lettering in a ring binder, 7.9 by 3.5 inches (200 × 89 mm).

This is a major reformatting of the 1986 manual, along with an update and the addition of five diseases.

Revised in 2003.

2001 **NPMA Field Guide to Turfgrass Pests.** National Pest Management Association, Fairfax, Virginia, USA. 176 pp., 99 illus. in color, 3 tables, no index, semisoft covers in multicolor with blue and green lettering in a 3-ring binder, 5.5 by 7.6 inches (140 × 192 mm).

A field guide to the common insects pests, nuisance pests, and diseases of turfgrasses. Included are the turfgrass injury symptoms, pest descriptions, life cycles, conditions favoring the pest, and pest control-management practices. There is a small glossary and selected list of references in the back.

2003† **Turfgrass Disease Identification Guide.** Michael Agnew, Dean Mosell, Peter Dernoeden, and Bruce Clarke. Syngenta, Greensboro, North Carolina, USA. 58 pp., 67 illus. in color, no tables, no index, glossy semisoft covers in multicolor with black and blue lettering in a ring binder, 3.5 by 6.1 inches (889 × 155 mm).

A guide to the identification and management of diseases on turfgrasses. The occurrences, both seasonally and in the United States, are illustrated for each disease.

F. PEST MANAGEMENT OF TURFGRASSES AND ORNAMENTALS

Most states in the United States have published integrated pest management (IPM) procedures for turfgrasses or turfgrasses, landscapes, and gardens. Many have pest diagnosis publications for turfgrasses. Such pest related publications are not presented herein, as they are readily obtained and tend to be repetitive.

Abbey, Timothy M.

2001† **Turfgrass: Nutrient and Integrated Pest Management Manual.** Timothy M. Abbey (editor). College of Agriculture and Natural Resources, University of Connecticut, Storrs, Connecticut, USA. 112 pp., 364 illus. with 160 in color, 10 tables, no index, glossy semisoft bound in multicolor with light green and white lettering, 8.5 by 10.8 inches (215 × 273 mm).

A manual concerning the selection, establishment, and culture of turfgrasses so as to minimize the occurrence of potentially injurious pests of turfgrasses.

Aichi-ken Prefecture

1989 **Sources of Safety Regulations of Agricultural Chemical Applications for Golf Course in Aichi Prefecture.** Aichi-ken, Japan. 174 pp., 8 illus., 31 tables, no index, textured semisoft bound in mustard shades with black lettering, 7.1 by 10.1 inches (180 × 257 mm). Printed in the Japanese language.

A government publication concerning proper safety precautions when utilizing pesticides on golf courses in Aichi Prefecture, Japan.

Balogh, James Charles

1992† **Golf Course Management and Construction: Environmental Issues.** James C. Balogh and William J. Walker (editors) with 10 contributing authors. Lewis Publishers, Chelsea, Michigan, USA. 951 pp., 23 illus., 140 tables, indexed, glossy hardbound in multicolor with white and black lettering, 7.1 by 9.2 inches (180 × 235 mm).

This lengthy book reviews the technical and scientific literature devoted to the effects of construction and management of golf courses on environmental quality. Emphasis is placed on water use, fertilization, and pesticides. Also included are chapters on wildlife and wetlands on golf courses. It is extensively referenced. Much of the environment-chemical data are from agriculture situations and do not necessarily apply to turfgrass conditions. It lacks critical assessments of reported research findings as they may or may not be applied to practical golf course turfgrass conditions.

Baxendale, Fredrick P.

1992† **Integrated Management Guide for Nebraska Turfgrass.** Fredrick P. Baxendale and Rock E. Gaussoin. E 92-1557-5, The Institute of Agriculture and Natural Resources, University of Nebraska, Lincoln, Nebraska, USA. 95 pp., 116 illus. with 57 in color, 8 tables, no index, glossy semisoft bound in multicolor with black and white lettering, 8.5 by 11.0 inches (215 × 279 mm).

A small book that addresses the specifics of an integrated pest management program for turfgrasses under the growing conditions in Nebraska.

Bebee, Charles N.

1991 **The Protection of Lawn and Turf Grasses, 1979–April 1991.** Charles N. Bebee. National Agricultural Library, Beltsville, Maryland, USA. 258 pp., no illus., no tables, author index, semisoft bound in green with black lettering, 8.5 by 11.0 inches (216 × 279 mm).

A major bibliography of citations selected from the Agricultural Online Access (AGRICOLA) database concerning weeds, diseases, insects, and environmental considerations related to turfgrasses.

Carleton, R. Milton

1957† **New Way to Kill Weeds—In Your Lawn and Garden.** R. Milton Carleton. Arco Publishing Company Inc., New York City, New York, USA. 112 pp., 200-plus illus., 1 table, no index, cloth hardbound in light green with black lettering and a jacket in multicolor with black lettering, 6.4 by 9.3 inches (162 × 237 mm).

A unique, small book on the control of weeds under conditions in the United States. It is extensively illustrated.

Carroll, Mark J.

2008† **The Fate of Nutrients and Pesticides in the Urban Environment.** Coauthor.
See *Nett, Mary—2008.*

Charbonneau, Pam

2003† **Turf IPM Manual.** Pam Charbonneau (editor). Pub. 816, Ministry of Agriculture and Food, Toronto, Ontario, Canada. 96 pp., 28 illus., 14 tables, no index, glossy semisoft bound in multicolor with white lettering, 8.4 by 11.0 inches (214 × 279 mm).

A manual concerning the use of integrated cultural practices to minimize the common pests of turfgrasses under the conditions in Ontario, Canada.

Christians, Nick Edward

2009† **Weed Control In Turf and Ornamentals.*** Coauthor.
See *Turgeon, Alfred Joseph—2009.*

Clark, J. Marshall

2000† **Fate and Management of Turfgrass Chemicals.** J. Marshall Clark and Michael P. Kenna (editors) with 91 contributing authors. ACS Symposium Series 743, American Chemical Society, Washington, D.C., USA. 477 pp., 93 illus., 107 tables, large index, glossy hardbound in greens with yellow and green lettering, 6.0 by 89 inches (162 × 226 mm).

An assemblage of 24 chapter topics, each with a list of references. It is organized in four sections: overview of the turfgrass industry and environmental issues, pesticide and nutrient fate, best chemical management practices, and biotechnology and alternative pest management.

Fogg, John Milton, Jr.

1945† **Weeds of Lawn and Garden.*** John M. Fogg. University of Pennsylvania Press, Philadelphia, Pennsylvania, USA. 215 pp., 180 illus., no tables, detailed index, cloth hardbound in black with gold lettering and a jacket in light green with dark green lettering, 5.0 by 8.1 inches (127 × 205 mm).

A unique handbook that presents descriptions, occurrence, and conditions favoring a broad array of weeds that occur in lawns and gardens. It is supported by a very good set of drawings of 175 weeds by Léonie Hagerty. It is oriented to eastern-temperate North America. There is a glossary and a bibliography in the back. Dr. Fogg offered the following scenario as to how many European origin weeds invaded North America:

> It is interesting to speculate on just what success these weeds from Europe would have had in competition with this well-established woodland vegetation, had the way not been made easy for them. Without human aid most of them would never have reached this continent. Without the transformation wrought by human activities, they might never have been given an opportunity to survive or spread. As it is, man has cut down or burned the forests, ploughed the land, built cities, railroads, and highways, altered the natural drainage; in short, he has modified the face of nature in every conceivable manner. The result is that these European adventives, instead of having to compete against a firmly entrenched and undisturbed native vegetation, have, thanks to their pioneering vigor, been able to take advantage of such artificial habitats as farms, pastures, waste places, vacant lots, lawns, and garden.

† *Reprinted in 1946, 1956, and 1963, cloth bound in green with black lettering and a jacket in color.*

French, Jackie

1997 **Organic Control of Common Weeds.** Second revised edition. Jackie French. Aird Books Pty. Ltd., Flemington, Victoria, Australia. 160 pp.; 37 illus.; no tables; small index; glossy semisoft bound in multicolor with orange, red, black, and green lettering, 5.5 by 8.,5 inches (140 × 215 mm).

A book focused on efforts to control weeds without the use of pesticides. The title indicates organic control, with the author using broad methodologies, such as manual digging and boiling water.
Reprinted in 2006.

Gaussoin, Roch E.

1992† **Integrated Management Guide for Nebraska Turfgrass.** Coauthor.
See *Baxendale, Fredrick P.—1992.*

Harrington, Kerry

2000† **Weeds of Lawns and Sports Turf.** Kerry Harrington. New Zealand Sports Turf Institute, Palmerston North, New Zealand. 55 pp. plus 2 pp. of advertising, 53 illus. in color, no tables, small index, fold-stapled glossy semisoft covers in multicolor with white and green lettering, 5.57 by 8.5 inches (146 × 209 mm).

A small book based on a series of articles that were published in the **New Zealand Turf Management Journal.** Thirty weeds common in New Zealand turfgrasses are presented in terms of their identification, habitat, and management and control.

Horgan, Brian P.

2008† **The Fate of Nutrients and Pesticides in the Urban Environment.** Coeditor.
See *Nett, Mary—2008.*

Kenna, Michael

2000† **Fate and Management of Turfgrass Chemicals.** Coeditor.
See *Clark, J. Marshall—2000.*

Kite, L. Patricia

1987† **Controlling Lawn & Garden Insects.** L. Patricia Kite. Ortho Books, Chevron Chemical Company, San Francisco, California, USA. 96 pp., 177 illus. with 159 in color, no tables, small index, glossy semisoft bound in multicolor with white lettering, 8.25 by 10.9 inches (210 × 277 mm).
A practical book on the insect pests of lawns and gardens, including diagnosis and management.

Leslie, Anne R.

1989† **Integrated Pest Management for Turfgrass and Ornamentals.** Anne R. Leslie and Robert L. Metcalf (editors) with 54 contributing authors. Office of Pesticide Programs, Environmental Protection Agency, Washington, D.C., USA. 344 pp., 32 illus., 70 tables, no index, semisoft bound in light green with green lettering, 7.8 by 10.3 inches (198 × 262 mm).
An assemblage of 25 papers presented at a symposium organized by the Integrated Pest Management Unit of the Environmental Protection Agency and held during the American Chemical Society Meetings in 1987. It focuses on the dimensions of integrated pest management, including monitoring methods, biological alternatives, and cultural practices.

1994† **Handbook of Integrated Pest Management for Turf and Ornamentals.** Anne R. Leslie (editor) with 74 contributing authors. CRC Press Inc., Boca Raton, Florida, USA. 672 pp., 248 illus., 100 tables, large index, glossy hardbound in multicolor with green and white lettering, 7.0 by 10.0 inches (178 × 254 mm).
An assemblage of 57 short chapters by a diverse range of authors that address the dimensions and experiences in integrated pest management of turfgrasses and ornamental plants.

Loewer, Peter

2001† **Solving Weed Problems.** Peter Loewer. The Lyons Press, Guilford, Connecticut, USA. 286 pp.; 56 illus.; no tables; indexed; semisoft bound in multicolor with green, black, and white lettering; 5.4 by 8.25 inches (137 × 210 mm).
A book on the identification and control of problem weeds in lawns and gardens, including weedy shrubs, trees, vines, and mosses.

Figure 13.20. Portable, motorized sprayer apparatus with Bean pump plus Fordson tractor, circa 1927. [2]

McCarty, Lambert Blanchard

2009† **Weed Control In Turf and Ornamentals.**[*] Coauthor.
    See *Turgeon, Alfred Joseph—2000.*

Metcalf, Robert L.

1989† **Integrated Pest Management for Turfgrass and Ornamentals.**[*] Coauthor.
    See *Leslie, Anne R.—1989.*

Misato, Tomomasa

1990 **A Guidebook on Agricultural Chemical Uses for Golf Courses.** Tomomasa Misato, Takao Araki, Masami Hatsukade, Makoto Konnai, Shinnkoh Gotoh, and Shozoh Kuwatsuka. Kagaku-Kogyo Nippo-Sha, Tokyo, Japan. 273 pp., 150 illus., 62 tables, indexed, glossy semisoft covers in multicolor with white and black lettering, 5.8 by 8.3 inches (147 × 210 mm). Printed in the Japanese language.
    A guidebook concerning the characteristics, conditions of occurrence, forecasting, and detailed pictures for the main pests, weeds, and diseases of turfgrasses in Japan plus the descriptions of pesticides, safety, and environmental impacts employed on turfgrass pests.

Nett, Mary

2008† **The Fate of Nutrients and Pesticides in the Urban Environment.** Mary Nett, Mark J. Carroll, Brian P. Horgan, and A. Martin Petrovic (editors). American Chemical Society, Washington, D.C., USA. 287 pp., 76 illus. with 8 in color, 46 tables, large index, glossy hardbound in multicolor with white and green lettering, 5.9 by 9.0 inches (150 × 228 mm).
    An assemblage of symposium research papers concerning chemical fate and transport that may or may not occur in turfgrass ecosystems. The influences of turfgrass cultural practices and soil-climatic aspects are addressed.

Nicholls, Jeff

2004† **Molecatcher.** Jeff Nicholls. Matador Troubador Publishing Limited, Leicester, England, UK. 82 pp.; 16 illus.; no tables; no index; glossy semisoft bound in multicolor with white, purple, red, and black lettering; 5.5 by 8.5 inches (139 × 215 mm).
    A unique book on the various methods for catching moles, with emphasis on practical approaches involving traditional means.

Owen, Mary

2000† **Protocols For An IPM System On Golf Courses.** Mary Owen with seven contributing authors. University of Massachusetts, Amherst, Massachusetts, USA. 78 pp., 3 illus., 37 tables, no index, glossy semisoft covers in multicolor with blue and white lettering in a ring binder, 10.0 by 12.0 inches (254 × 307 mm).
    A specialty manual concerning the protocols and forms for organizing and documenting an integrated pest management (IPM) system for a golf course. There is a list of selected references in the back.

2009† **Integrated Pest Management: Protocols for Turf on School Properties and Sports Fields.** Mary C. Owen, Randall G. Prostak, and Jascon D. Lanieer. University of Massachusetts, Amherst, Massachusetts, USA. 111 pp., 3 illus., 55 tables, no index, glossy semisoft covers in multicolor with black lettering in a ring binder, 8.5 by 11.0 inches (217 × 280 mm).
    A manual similar to the 2000 version, but oriented to sports grounds.

Petrovic, Martin A.

2008† **The Fate of Nutrients and Pesticides in the Urban Environment.** Coeditor.
    See *Nett, Mary—2008.*

Thompson, Patrick H.

2000† **Of Moles and Men—The Battle for the Turf.** Patrick H. Thompson. Aardvark Aventi, West Linn, Oregon, USA. 175 pp. plus 2 pp. of advertisements, 17 illus. in color, 1 table, small index, glossy semisoft bound in multicolor with blue and black lettering, 7.2 by 9.25 inches (182 × 234 mm).
    A unique, in-depth book on the biology, habitat, problems, and control of moles, especially in urban residential areas.

Turgeon, Alfred Joseph

2009† **Weed Control In Turf and Ornamentals.**[*] A.J. Turgeon, L.B. McCarty, and Nick Christians. Pearson Prentice Hall Inc., Upper Saddle River, New Jersey, USA. 312 pp.; 341 illus. with 149 in color; 33 tables; large index, glossy semisoft bound in multicolor with green, brown, and white lettering; 8.2 by 10.8 inches (208 × 275 mm).

A specialty book focused on herbicides used in the control of weeds in turfgrasses and ornamental plants. It includes herbicide action, metabolism in plants, and fate in the environment. It also includes growth regulators.

Walker, William J.

1992†   **Golf Course Management and Construction: Environmental Issues.** Coauthor.
             See *Balogh, James Charles—1992.*

G. PRIVATE COMPANY–RELATED TURFGRASS PUBLICATIONS

A number of private companies have published booklets that contain a significant amount of specific company product names and information but which also contain some turfgrass technical information. A representative group are described in this section of chapter 13.

Amchem Products, Inc.

1960†   **have a Weed Free lawn.** Amchem Products Inc., Ambler, Pennsylvania, USA. 23 pp., 35 illus., no tables, no index, fold-stapled soft covers in white and greens with green and black lettering, 5.5 by 8.3 inches (139 × 212 mm).
             A booklet on the control of weeds in lawns. It includes descriptions and drawings of 35 weeds of turfgrasses that occur in North America.
     †   Also printed in 1960 with glossy multicolor covers.

American Agricultural Chemical Company

n.d.†   **how to Have a Lastingly Beautiful Lawn.** The American Agricultural Chemical Company, New York City, New York,
circa   USA. 23 pp., 21 illus. with many as vignettes, 2 tables, no index, fold-stapled soft covers in greens and black with white
196x   and black lettering, 5.4 by 8.4 inches (138 × 214 mm).
             A booklet on the establishment and culture of lawns. It is practically oriented to the lawn enthusiasts and conditions in the United States.
             Revised circa 1960s.

n.d.†   **Your Lawn.** American Agricultural Chemical Company, New York City, New York, USA. 23 pp., 18 illus., 2 tables, no
circa   index, fold-stapled soft covers in multicolor with white lettering, 5.0 by 7.5 inches (128 × 191 mm).
196x

             The booklet contains only minor revisions of the 1960 version.

Figure 13.21. Portable, ~200-gallon Hardie sprayer, available with 9 to 12 gpm pump and 5 to 6 hp engine, circa 1934. [41]

Arthur D. Peterson Company

1936† **The Peterson Catalog and Turf Manual.** Arthur D. Peterson Company Inc., New York City, New York, USA. 48 pp., 46 illus., 5 tables, no index, fold-stapled semisoft covers of green velvet cloth with gold lettering, 7.8 by 9.8 inches (197 × 248 mm).

A booklet on the establishment and culture of cool-season turfgrasses on golf courses, sport fields, and lawns. It includes short articles on golf architecture by A.W. Tillinghast, J.R. van Kleek, H. Strong, R.T. Jones, and A.H. Tull. The back portion consists of a catalog of chemicals and equipment, including interesting photographs. The following is from the booklet:

> Each year finds a steady increase in the number of golf clubs installing fairway watering systems, and it is safe to predict that this interest will continue. The coming season will see ten public golf courses in the City of New York, as well as four State-owned courses on Long Island with full fairway irrigation, and it is only natural that the private clubs will endeavor to give their players all the advantages enjoyed by the public-links player.
>
> Many direct benefits are secured by fairway irrigation, such as more uniform playing conditions throughout the season; the avoidance of play over dusty, sun-baked ground in mid-Summer; perfect lies for brassy shots at all times; prevention of loss of grass plants during drought; better germination of new seeding; quicker healing of divots and greater benefit from applications of fertilizer. The natural beauty of the course is maintained throughout the season, where fairway irrigation has been installed.
>
> Cost is the first question that arises in the minds of the green committee, when the subject of watering fairways is suggested. A full 18-hole fairway irrigation system may cost from $10,000 to $25,000, depending upon the available water supply, texture of soil, and the layout of the course.

1939† **The Peterson Catalog and Turf Manual.** Arthur D. Peterson Company Inc., New York City, New York, USA. 39 pp., 24 photographs and numerous equipment illus., 4 tables, indexed, fold-stapled semisoft covers in green and cream with green lettering, 8.5 by 11.1 inches (216 × 283 mm).

A booklet that includes brief sections on turfgrasses, soil amendments, fertilizers, irrigation, and control of pests in the United States. The back portion is devoted to product descriptions with illustrations.

Boots Pure Drug Company, Ltd.

n.d. **Guide To Better Lawns.** Farms and Gardens Department, Boots Pure Drug Company Limited, Nottingham, England,
circa UK. 32 pp., no illus., no tables, indexed, fold-stapled semisoft bound in green with black lettering, 5.8 by 8.2 inches
19xx (147 × 208 mm).

A small booklet on the establishment and culture of lawns, with major emphasis on pest problems and their control. It is oriented to the lawn enthusiasts and cool-season conditions in the United Kingdom.

Boundary Chemical Company

n.d.† **Lawns Beautiful—How to make and keep them.** J. Lytle. The Boundary Chemical Company Limited, Liverpool,
circa England, UK. 25 pp. with 7 pp. of advertising, no illus., no tables, no index, fold-stapled semisoft covers in light brown
mid- and green with green lettering, 4.0 by 6.4 inches (102 × 162 mm).
192x A **very rare,** small booklet on the establishment and culture of lawns, with emphasis on weed and insect control. It is oriented to the lawn enthusiasts and cool-season turfgrass conditions in England.

Cambridge Soil Services, Ltd.

n.d.† **The Cambridge Sportsfield Drainage System.** Cambridge Soil Services Limited, Cambridge, England, UK. 22 pp.,
circa 13 illus., no tables, no index, fold-stapled semisoft covers in light blue and white with black lettering, 5.9 by 8.2 inches
1980 (149 × 209 mm).

A booklet on the subject of root-zone drainage, with emphasis on a procedure for mechanically inserting sand-filled vertical slits at a specified spacing into an existing sports field.

Carters Tested Seeds, Ltd.

1935 **Bowling Greens.** Reginald Beale. Carter Tested Seeds Ltd., Raynes Park, London, England, UK. 20 pp., no illus., 4 tables, no index, fold-stapled semisoft covers in green, blue, and silver with blue and silver lettering, 4.2 by 5.5 inches (107 × 139 mm).

A **rare** booklet concerning the planting and care of bowling greens, plus allied materials available. It is oriented to conditions in England.

1939† **Fertilisers Etc. For Garden and Lawn.**˙ Carters Tested Seeds Ltd., Raynes Park, London, UK. 31 pp., 12 illus., 12 tables, no index, fold-stapled covers in blacks, tans, and whites with black and white lettering, 4.1 by 5.3 inches (105 × 135 mm).

A booklet concerning fertilizers and pest control materials as properly used on turfgrass lawns and gardens in the United Kingdom.

Coastal Turf, Inc.

1996† **Champion Dwarf Bermudagrass.** Coastal Turf Inc., Bay City, Texas, USA. 29 pp., 35 illus. in color, 9 tables, no index, fold-stapled semisoft covers in white and black with black lettering in a comb binder, 8.5 by 10.9 inches (216 × 277 mm).

A small booklet in which the description of a new vertical-dwarf hybrid bermudagrass is presented based on morphological data and environmental stress assessments, including close mowing, wear, and cold.

David Miln & Company (Seedsmen), Ltd.

1960 **Descriptions and Uses of Grass and Clover Varieties—including Herbs.**˙ John W. Day. David Miln & Company (Seedsmen) Limited, Chester, Cheshire, England, UK. 38 pp., no illus., no tables, no index, fold-stapled soft covers in green with black lettering, 4.8 by 7.8 inches (123 × 199 mm).

A booklet on the characteristics of selected grass, legume, and herb cultivars available in England.

The Deming Company

n.d. **Golf and Country Club Water Supply Installation.** The Deming Company, Salem, Ohio, USA. 40 pp., 71 illus., 10 circa tables, no index, fold-stapled semisoft covers in greens with black lettering, 7.4 by 9.8 inches (182 × 241 mm).

1920 A **rare**, unique booklet concerning irrigation related equipment with emphasis on water pumps. It includes an early series of photographs principally of golf course turfs.

Diamond Shamrock Corporation

1980† **Disease and Weed Control Guide for Turf and Ornamentals.**˙ Agricultural Chemicals Division, Diamond Shamrock, Cleveland, Ohio, USA. 26 pp., 92 illus. in color, 1 table, no index, fold-stapled glossy semisoft covers in multicolor with white lettering, 8.5 × 11.0 inches (215 × 280 mm).

A small handbook in which 16 major diseases of turfgrasses are illustrated with color photographs of the damage symptoms. The text consists of discussions regarding injury symptoms, susceptible grasses, conditions favoring disease, and how the fungus survives. There also is a section on weedy grass and broadleaf weed control in the back.

D.J. Van Der Have Company

n.d.† **Amenity Grasses.** D.J. Van Der Have Company, Kapelle, the Netherlands. 44 pp., 7 illus., 1 table, no index, soft bound 19xx in multicolor with black lettering, 5.8 by 8.2 inches (147 × 209 mm).

A booklet on the turfgrass breeding and seed production operations of the company plus descriptions of the turfgrass cultivars available.

DowElanco Company

1992† **The turf manager's guide to responsible pest control.** Warm-season edition. DowElanco, Indianapolis, Indiana, USA. 43 pp.; 92 illus. with 65 in color; 8 tables; no index; fold-stapled glossy semisoft covers in multicolor with black, green, and white lettering, 8.5 by 11.0 inches (215 × 281 mm).

A small booklet on integrated pest management of weed, disease, and insect problems on warm-season turfgrasses. It includes proper pesticide handling and application.

E.I. du Pont de Nemours & Company

1957† **Turfgrass Fertilization—Its Theory and Practice.** Polychemicals Department, E.I. du Pont de Nemours & Company, Wilmington, Delaware, USA. 49 pp., 3 illus., 15 tables, no index, semisoft covers in multicolor with black lettering in a comb binder, 8.5 by 10.9 inches (217 × 278 mm).

A small manual on the culture of turfgrasses in North America, with emphasis on fertilization and especially slow-release ureaformaldehyde nitrogen carriers.

1959† **The Golf Course . . . Planning and Records.** E.I. DuPont de Nemours and Co. (Inc.), Wilmington, Delaware, USA. 62 pp., 1 illus., 42 tables, no index, semisoft covers in green and white with white and black lettering in a ring binder, 8.6 by 11.0 inches (218 × 278 mm).

A booklet with an extensive set of blank record forms encompassing the 18 holes of a golf course plus forms for weather, budget, expenditures, salaries, fuel, water, power, seed, equipment, chemicals, and supplies.

1960†   Reprinting for the 1960 year.

n.d.   **Guide to Better Turf Management.** E.I. du Pont Nemours & Company Inc., Wilmington, Delaware, USA. 42 pp.,
circa 14 illus., 4 tables, no index, fold-stapled glossy semisoft covers in white with green lettering 8.5 by 11.0 inches (215 ×
1968 279 mm).

A booklet concerning fertilization, diseases, and weed control of turfgrasses, especially for golf courses in the United States.

n.d.† **Professional Turf Manual—Diseases & Control, Fertilization, Weed Control.** Dupont Turf Products, E.I. duPont de
circa Nemours & Company, Wilmington, Delaware, USA. 33 pp., 69 illus. in color, no tables, no index, fold-stapled semisoft
early covers in multicolor with black and green lettering, 8.4 by 10.9 inches (214 × 278 mm).
1970   A small handbook in which 15 diseases of turfgrasses, plus 4 environmental stresses of turfgrasses in North America are illustrated via color photographs of the injury symptoms. There also are smaller sections on fertilization and weed control.

n.d.† **Professional Turf Manual—Diseases and Control.** DuPont Turf Products, E.I. du Pont de Nemours & Company,
circa Wilmington, Delaware, USA. 28 pp., 66 illus. in color, no tables, no index, fold-stapled glossy semisoft covers in mul-
1976 ticolor with white and yellow lettering, 8.5 by 11.0 inches (215 × 280 mm).

A small handbook in which 16 major diseases of turfgrasses in North America are illustrated via color photographs of the injury symptoms. The text is comprised of discussions regarding the causes, symptoms, and control.

En-Tout-Cas

n.d.† **The Making and Upkeep of Lawns, Grass Tennis Courts, Drives, Paths, etc.** The En-Tout-Cas Company Limited,
circa Syston, Leicester, England, UK. 20 pp., 15 illus., 3 tables, no index, fold-stapled textured semisoft covers in greens and
1938 brown with white lettering, 7.9 by 5.2 inches (201 × 133 mm).

A small brochure on seeding, sodding, fertilizers, topdressing, worm control, and bitumen use under conditions in the United Kingdom. It includes descriptions of products and prices.

F.W. Woodruff & Sons, Inc.

1932† **Something About Turf.** First edition plus eight revised editions in 1934, 1936, 1937, 1939, 1940, 1942, 1946, and
1949. W.C. Baker and et al. F.H. Woodruff & Sons, Milford, Connecticut, USA. 32 to 52 pp., 12 to 19 illus., 1 to 2 tables, no index, textured semisoft bound in colors that varied with the edition, 4.5 by 7.4 inches (114 × 187 mm).

A **rare** set of small booklets on turfgrass species, including their vegetative and seed identification plus the weed, disease, and insect problems. Emphasis is placed on cool-season turfgrass species in the United States:

> Turf is big business. The United States Government estimates that there are 1,500,000 acres of turf airfields under their supervision. Add to this the approximate 15,000,000 acres of golf courses and the enormous acreage of cemeteries, Municipal Parks and home lawns and the answer is that turf is an industry.

n.d.† **Results of Our Turf Experiments.** Bulletin No. 3, Grass Seed Division, F.H. Woodruff & Sons, Milford, Connecticut,
circa USA. 43 pp., 26 illus., 14 tables, no index, textured semisoft covers in green with black lettering in a 2-hole binder, 8.5
1939 by 0.9 inches (215 × 277 mm).

A **rare** summary of early research results obtained from 1931 through 1938 at the turfgrass field research grounds near Milford, Connecticut. The diversity of experiments ranged among cool-season turfgrass species, cultivars, and polystand assessments on various use sites such as lawn, sports field, putting green, cemetery, airfield, and shade. Also included were mowing, rolling, fertilizing, weed control, and seed timing studies.

Hartley's Pre-Sown Seeds

n.d. **Scientific Lawn Making.** Vincent Hartley. Hartley's Pre-Sown Seeds, Greenfield, Lancastershire, England, UK. 40 pp.
circa plus 8 pp. of advertising, 41 illus., 1 table, no index, fold-stapled semisoft covers in multicolor with white and black
193x lettering, 7.2 by 9.0 inches (182 × 229 mm).

A small booklet on the planting and care of lawns, with emphasis on a patented seeding method involving fixed vegetable-medium sheets with seeds attached. It is oriented to the lawn enthusiasts and cool-season turfgrass conditions in the United Kingdom.

H. Pattisson & Co., Ltd.

The company was established in 1896 by Mr. Arthur Cole. Multiple editions of this pioneering golf course equipment catalogue were published from at least the 1920s to 1965. The 1937 catalogue is described herein.

1937†   **Golf Course Requisites**. H. Pattisson & Co. Limited, Stanmore, Middlesex, England, UK. 82 pp., numerous illus., numerous tables, small index, fold-stapled semisoft covers in red and black with black and red lettering, 8.7 by 10.9 inches (220 × 276 mm).

A quality early turfgrass equipment and sports accessory catalogue by one of the pioneering English companies in golf course equipment. Contains numerous quality illustrations of equipment. A must for turfgrass history collectors.

The Humber Fishing & Fish Manure Company, Ltd.

n.d.   **The Greenkeepers Guide.** The Humber Fishing & Fish Manure Company Limited, Stoneferry, England, UK. 24 pp.
circa   with 6 pp. of advertising, 6 illus., 2 tables, no index, fold-stapled soft covers in cream and browns with brown and green
193x   lettering, 5.5 by 8.5 inches (139 × 217 mm).

A **rare**, small booklet on the culture of bowling greens, with emphasis on conditions in the United Kingdom.

Hurst Gunson Cooper Taber, Ltd.

n.d.†   **Hurst Grass Book.** Hurst Gunson Cooper Taber Limited, Witham, Essex, England, UK. 22 pp., 30 illus. with 4 in
circa   color, no tables, no index, fold-stapled glossy semisoft covers in multicolor with green, brown, and black lettering, 5.8
197x   by 8.3 inches (148 × 210 mm).

A booklet concerning cultivars of cool-season turfgrasses for use in parks, gardens, sport fields, and lawns in the United Kingdom.

Imperial Chemical Industries Limited (ICI)

The Imperial Chemical Industries was formed in 1926 by the merger of four British chemical companies, and also initiated turfgrass research in England in 1926.

n.d.†   **Improvement of Lawns, Golf Greens, and Fairways.** First edition. Imperial Chemical Industries Limited, London,
circa   England, UK. 20 pp., 1 illus., 3 tables, no index, fold-stapled semisoft covers in green with gold lettering, 5.3 by 8.4
1932   inches (134 × 214 mm).

A **rare**, small booklet on the fertilization and establishment of turfgrass for lawns, greens, fairways, and sports fields. It is oriented to the cool-season turfgrass conditions in England. ICI proposed the following fertilizer and application technique for use on turfgrass in the early 1930s as follows:

The quantities present are as

Nitrogen ....13.2 per cent

Phosphoric Acid ....2.4 per cent

Potash ....2.4 per cent

These proportions were determined after a considerable amount of experimental work and will be found suitable for almost all classes of soils. It is true that on a few soils potash may be unnecessary, but it was thought desirable to provide a fertilizer as nearly as possible suited to all requirements, and the presence of a small quantity of potash, whilst it does not add very much to the cost of the material, ensures its adaptability to soils deficient in this material and cannot do any harm to those soils which already possess a sufficiency.

A small proportion of phosphate has been included in the mixture in order that the growth of the grasses should not be checked. Investigations have shown that there is little fear that the small quantities of phosphate and potash will encourage the clovers and other leguminous species. In one particular experiment in which ammonium phosphate was applied to a lawn containing wild white clover, the clover content was reduced from 21 per cent. to 0.2 per cent. in the season, in spite of the fact that a dressing of phosphate equivalent to 1¼ tons of superphosphate per acre had been added during that time.

The fertilizer should be carried in a bucket or other convenient receptacle and distributed in small handfuls, each handful being thrown with an *upward* semicircular motion of the arm, such that the granules fall as a light spray on the turf. It is advisable to keep to a line when carrying out this operation and to keep one's back to the wind.

Revised in 1933.

n.d.† **The Improvement of Lawns, Golf Courses and Tennis Courts.** First edition. Imperial Chemical Industries Limited,
circa London, England, UK. 20 pp., no illus., 2 tables, no index, fold-stapled semisoft covers in green with gold lettering, 5.4
1932 by 8.4 inches (134 × 214 mm).

   A **rare**, small booklet on establishment and culture of turfgrasses under the conditions in England. The acid theory in turfgrass culture is refuted.

   Revised in 1936.

1933† **Improvement of Lawns, Golf Greens and Fairways**. Imperial Chemical Industries Limited, London, England, UK.
   23 pp., 1 illus., no tables, no index, fold-stapled semisoft covers in green with gold lettering, 5.4 by 8.4 inches (137 × 213 mm).

   A **rare** booklet concerning the planting and care of turfgrasses for lawns, greens, fairways, and sports grounds in the United Kingdom. Emphasis is on fertilization and weed control.

n.d.† **The Groundsman's Handbook.** Imperial Chemical Industries Limited, London, England, UK. 24 pp., 7 illus.,
circa 7 tables, no index, fold-stapled semisoft covers in black and gray with white lettering, 4.3 by 6.1 inches (110 ×
1933 156 mm).

   A **rare** booklet concerning the culture of turfgrasses on sports fields. The layout design and marking of soccer, field hockey, rugby, cricket, lawn tennis courts, bowling greens, and croquet lawns are included.

n.d.† **The Improvement of Lawns, Golf Courses and Tennis Courts**. Second revised edition. Imperial Chemical Industries
circa Limited, London, England, UK. 20 pp., no illus., no tables, no index, fold-stapled semisoft covers in green with gold
1936 lettering, 5.4 by 8.4 inches (137 × 213 mm).

   A **rare** booklet concerning the planting and care of turfgrasses for lawns, golf fairways, bowling greens, and tennis courts in the United Kingdom. There is an extended discussion of the acid theory of weed control.

n.d.† **The Groundsman's Handbook—The Care of Grass on Sports Grounds.** Imperial Chemical Industries Limited,
circa London, England, UK. 25 pp., 8 illus., 4 tables, no index, fold-stapled semisoft covers in black and white with white
1936 lettering, 4.3 by 6.3 inches (110 × 159 mm).

   A **rare,** small booklet devoted primarily to the fertilization of sports fields under the cool-season turfgrass conditions in the United Kingdom. It includes field layouts for rugby, association football (soccer), field hockey, cricket table, lawn tennis, lawn bowling, and croquet.

   Multiple reprintings.

Lebanon Chemical Corporation

n.d.† **Agronomy Journal.** Lebanon Chemical Corporation, Lebanon, Pennsylvania, USA. 39 pp., 35 illus., 7 tables,
circa no index, fold-stapled glossy semisoft covers in white and green with green lettering, 8.5 by 11.0 inches (216 ×
early 280 mm).
198x

   A booklet on the culture of lawns in North America, with emphasis on fertilization and soil pH adjustment plus insect, weed, and disease diagnoses and control.

   † *Periodic revisions and reprintings through 1999.*

1990† **Turfgrass and Ornamental Agronomy Manual.** Lebanon Chemical Corporation, Lebanon, Pennsylvania, USA. 47
   pp., 25 illus., 23 tables, no index, glossy semisoft bound in white and green with white, green and black lettering, 8.5
   by 11.0 inches (216 × 279 mm).

   An expanded booklet on the culture of lawns in North America, with emphasis on fertilization, soil pH, and maintenance calendars plus pest management.

   See also *Knoop, William Edward—1996 and 1999.*

Maxwell M. Hart, Ltd.

   A recreation ground contractor in the United Kingdom that also marketed turfgrass supplies.

n.d. **Notes on the Maintenance of Bowling Greens.** Maxwell M. Hart Limited, London, England, UK. 23 pp. with 11
circa pp. of advertising, 1 illus., no tables, no index, fold-stapled semisoft covers in green with dark blue lettering, 4.7 × 7.2
193x inches (119 × 183 mm).

   A small booklet on the culture of turfed bowling greens. It is oriented to cool-season turfgrass conditions in the United Kingdom.

Mommersteeg International

n.d.† **The Seedsman's Guide to Amenity Grasses.*** Mommersteeg International, Wellingborough, Northamptonshire, En-
circa gland, UK. 34 pp., 22 illus., 5 tables, no index, fold-stapled semisoft covers in greens with dark green lettering, 5.7 by
19xx 8.3 inches (146 × 211 mm).

>   A booklet on the cool-season turfgrass species and cultivars as used in England. It also contains a section on wild flora.

Monsanto Company

1989† **Turf Renovation Guidelines.*** Monsanto Company, St. Louis, Missouri, USA. 20 pp., 16 illus. in color, 8 tables, no index, fold-stapled glossy semisoft covers in multicolor with black and white lettering, 8.5 by 11.0 inches (216 × 279 mm).

>   A small guide on the renovation of lawns in North America, with emphasis on nonselective, systemic vegetation control.

Northrup, King & Company

1952† **Your Lawn.** Research Department, Northrup, King & Company. Minneapolis, Minnesota, USA. 34 pp.; 44 illus. and vignettes; 1 table; no index; fold-stapled semisoft covers in green, white, and black with black and white lettering; 3.8 by 8.9 inches (97 × 225 mm).

>   A small practical booklet on the planting and care of lawns. It is oriented to lawn enthusiasts and cool-climatic conditions in the United States.

Oil-Dri Corporation

1962† **Terra-Green Soil Conditioner Manual.** Woodrow A. Jaffee. Oil-Dri Corporation of America, Chicago, Illinois, USA. 36 pp., 4 illus., 9 tables, no index, fold-stapled semisoft covers in green with dark green lettering, 8.5 by 10.9 inches (215 × 277 mm).

>   A small manual on the use of fired-clay aggregates for amending root zones of golf course and sports fields in North America. A small glossary is included in the back.

Peter Henderson & Company

The company was formed by Peter Henderson in 1847 with the focus on seed sales and was actively selling turfgrass seeds in the eastern United States by 1890. This series of booklets was published annually from 1915 to 1940 under the titles listed as follows plus as **Sports Turf-Grasses and Requisites, Sport Turf-Grasses-Fertilizers-Equipment,** and **Sports Turf-Grasses, Turf Supplies and Turf Equipment.** The total pages ranged up to 64, declining to 24 as the text portion was reduced.

1925† **Sports Turf for Ideal Golf Course and Athletic Field.** Peter Henderson & Company, New York City, New York, USA. 56 pp., 115 illus., 3 tables, short index, fold-stapled textured semisoft covers in green and yellow with yellow and green lettering, 10.3 by 7.7 inches (262 × 195 mm).

>   A **rare**, small book on turfgrass history, establishment, and culture produced by an East Coast company. It includes a major section on turfgrasses for golf courses plus football, baseball, polo, lawn tennis, cricket, croquet, and parks. It contains interesting photographs of major turfgrass installations representing each of these uses. There is a section on "How to Organize a Golf Club and Planning the Ideal Course" by Mr. Seymour Dunn and on "Pointers on Golf Course Construction" by Mr. Tom Winton. Photographs and descriptions of various equipment for sale in 1925 are presented in the back. Many of these visuals are included in this book.

1929† **Sports Turf for Golf Courses and Athletic Fields.** Peter Henderson & Company, New York City, New York, USA. 64 pp.; 150 illus.; 2 tables; indexed; semisoft covers in tan, green, and black with black lettering; 9.9 by 7.6 inches (251 × 192 mm).

>   This is a significant revision, including updated photographs and descriptions of equipment for sale. A suggested approach for the construction of root zones on greens in 1929 was as follows:

> A WORD ON THE CONSTRUCTION OF GREENS, HAZARDS, THEIR CONTOURS, ETC.—The first actual work to be done on the grounds is the building, fertilizing, and seeding down of the Putting Greens. These should be irregular in outline and undulating, no two being alike, but be sure to avoid extreme ideas. Where nature has not provided the ideal undulation for a green in the location where you want it this must be built. You will frequently find that nature has partly built the undulation, and all you need to do is touch up a corner or a side here and there to complete the job. All greens should have a slight slope toward the front so that the hole itself, wherever it may be, can be seen by the player approaching it. Material can generally be found right besides the green and taken in such a way as to form a natural-looking hazard.

> Having decided on the location and undulations of the greens, the next step—a most important one is—removal of all to-soil from the green area, that is where the green is to be situated. As it is removed it should be screened and placed to one side, handy, but out of the precise green area. The same thing should be done with the top-soil from the surface of where the traps are to be. Now begin to model the undulations into the putting green, using the sub-soil from the traps as material with which to build the elevations. Be sure there are no pockets in the surface of the green where water might lay after a heavy rain. Also the entire surface of the green must be self-draining. If the soil tends towards clay, now would be the time to place the tiles (see page 19) or give the graded putting green surface from 4 to 6 inches of cinders, or finely crushed glass or porcelain. This will prevent worms from working as they do not like to bore in soil containing any material that is liable to scratch their bodies. Again, if the soil is very clayey, now would be the right time to add coarse sand to the sifted top-soil in about a 50-50 proportion, to further prevent the worms from working in the greens. This also improves the mechanical of clay soil and prevents the green from packing hard in a drought. Grasses will not grow well in a soil that packs too tightly.
>
> TEES—They should be built perfectly level and square to line of hole, preferably all on one level 60 feet by 60 feet square or in two or more terraces of equivalent area.

The Philadelphia Seed Company, Inc.

1950   **Old English Lawn Handbook.** The Philadelphia Seed Company Inc., Philadelphia, Pennsylvania, USA. 29 pp., no illus., no tables, no index, fold-stapled glossy semisoft covers in greens with black and white lettering, 3.5 × 6.2 inches (89 × 157 mm).

A small handbook concerning turfgrasses for lawns plus their planting and care in the eastern mid-Atlantic region of the United States.

Plant Protection, Ltd.

n.d.†   **How To Make and Keep A Lawn.** Plant Protection Limited, Yalding, Kent, England, UK. 21 pp., 11 illus., 1 table, no
circa   index, fold-stapled semisoft covers in black and white with green lettering, 5.5 by 8.3 inches (140 × 210 mm).
1939

A small booklet on the establishment and culture of lawns. It is practically oriented to the lawn enthusiasts and cool-season turfgrass conditions in England. A procedure for sodding and "beating turfs" was described as follows:

> Assuming that a satisfactory turf has been obtained, it is only too easy to ruin it at the outset by bad laying. First, the turves must be cut to even size and thickness, taking care to keep the edges vertical. 12 in. × 24 in. × 3 in. are suitable dimensions, the turf being trimmed down to 1½ in. to 2 in. thick before laying.
> Never lay turf on pure sand, and never lay soil on top of sand, or caking will result: the sub-stratum should be a good medium loam prepared as already described. Do not allow cut turves to lie about in heaps. If possible cut and lay out on the same day, first firming the ground by rolling.
> Leave spaces of about ½ in. between turves to allow for expansion in the subsequent beating down. In estimating the amount of turf required, allow 5 per cent of the area for these spaces. Set three to four rows only at a time, then beat with a wooden beating block or the back of a spade. On no account beat the turves too flat. Remedy the slightest defect in level at once by packing in or removing soil. Give no water during laying and stop work in heavy rain.
> When laying turf on a slope, tread the soil well into the bank and peg the turves in position if necessary. The bottom and top rows should be laid horizontally and the others vertically.

Rhone-Poulenc Ag Company

1992†   **Turf Problems and Solutions Quick Reference Guide.** Rhone-Poulenc Ag Company, Research Triangle Park, North Carolina, USA. 33 pp., 53 illus. with 35 in color, 3 tables, no index, fold-stapled glossy semisoft covers in multicolor with white and black lettering, 8.5 by 10.9 inches (215 × 278 mm).

A guide to the identification, management, and control of the major weed, disease, insect, and nematode problems occurring in turfgrasses in the United States.

Stumpp & Walter Company

1922†   **Golf Turf of High Quality.** A series of at least 16 editions, published annually in the 1930s and sometimes termed a supple-
to   ment. Also, printed in 1938 under the title of **Essentials of Producing Good Turf** and in 1945 as the **Lawn Quiz.** Stumpp
1940†   & Walter Company, New York City, New York, USA. 32 to 81 pp., many illus., a few tables, indexed, fold-stapled glossy semisoft covers in multicolor that varies with each edition with black or white lettering, 7.8 by 9.8 inches (199 × 249 mm).

A **rare** set of booklets with brief sections on the establish and care of turfgrasses for golf courses, sport fields, parks, lawn bowling, tennis courts, polo fields, airfields, and fine lawns in North America. The back one-half to two-thirds is devoted to product descriptions with illustrations and prices. It contains interesting equipment photographs. The 1937 edition lists the following book titles, descriptions, and US prices:

> **RIGHT USE OF LIME IN SOIL IMPROVEMENT.** By Alva Agee. Gives complete information on soil acidity. 20 illustrations. 104 pages ($1.25)
>
> **THE LINKS.** By Robert Hunter. A new and notable contribution to the literature of golf-course architecture. 163 pages, replete with half-tone illustrations, plans, and maps ($1.50)
>
> **GOLF SIMPLIFIED—CAUSE AND EFFECT.** By Dave Hunter of the Essex County Country Club. 43 pages. A new and simplified method of instruction in the game, whereby one idea corrects all faults common to the golfer ($1.00)
>
> **THE LAWN.** By L.S. Dickinson. Deals with the culture of turf in park, golfing, and home areas. Gives the latest practical directions which everyone may follow ($1.25)
>
> **SOIL MANAGEMENT FOR GREENKEEPERS.** By M.H. Cubbon and M.J. Markuson. An authoritative book which should be in every alert greenkeeper's library ($2.00)
>
> **TURF FOR GOLF COURSES.** By C.V. Piper and R.A. Oakley. This is an authoritative and practical treatise on the production and maintenance of grass turf ($2.50)
>
> **GOLF ARCHITECTURE IN AMERICA.** By Geo. C. Thomas, Jr. 342 pages with illustrations. Comprehensive and interesting, giving the fundamentals of golf-course construction for the benefit of the average player, as well as the expert ($5.00)

Sutton & Sons, Ltd.

n.d. circa 1912   **Turf Production from Seed.** First edition. Sutton & Sons, Reading, Berkshire, England, UK. 22 pp., 26 illus., no tables, no index, fold-stitched textured semisoft covers in tan and teal blue with impressed black lettering, 6.5 by 8.9 inches (164 × 226 mm).

A **very rare**, extensively illustrated booklet of practical illustrations of Sutton's unique method of turf formation in Britain and on the continent.

n.d. circa 1913   **Turf Production from Seed.** Second revised edition. Sutton & Sons, Reading, Berkshire, England, UK. 23 pp., 32 illus., no tables, no index, fold-stapled semisoft covers in green with dark green lettering, 6.3 by 8.9 inches (161 × 226 mm).

A dual turfgrass information and promotional booklet, with numerous photographs. A **rare** publication.

n.d.† circa 1914   **Turf Production from Seed by the Sutton System.** Third revised edition. Sutton & Sons, Reading, England, UK. 23 pp., 26 illus., no tables, no index, fold-stapled semisoft covers in green with dark green lettering, 6.5 by 8.8 inches (165 × 223 mm).

A revision of the 1913 edition.

1934†   **Sutton's Golf Course Catalogue.** Sutton & Sons Ltd., Reading, Berkshire, England, UK. 32 pp., 70 illus., 1 table, small index, fold-stapled textured semisoft covers in green with embossed dark green lettering, 7.2 by 9.8 inches (184 × 248 mm).

A dual turfgrass information and product booklet. It includes numbered drawings of equipment typically used in the 1930s. The use of mowrah meal earthworm irritant is described as follows:

> On most courses it is generally regarded as part of the regular routine to destroy worms in putting greens, and now, with the increasing use of mechanical pumping machines for applying destructive agencies, many Clubs are treating the fairways also. To keep the fairways clear of worms is no doubt a good policy, especially as players have a preference for such courses, particularly in the winter months.
>
> MOWRAH MEAL is the crushed residue of an Indian Bean known as *Bassia latifolia*. Not only is the product very efficacious in killing the worms, but it is of an organic nature and possesses fertilising properties.
>
> There are many grades of Mowrah Meal, some of which are more finely ground than others. Coarse grades are of low value for the destruction of worms, and although samples of this character are offered at cheap rates it is not economical to purchase them, for they may be only fifty per cent effective.
>
> Another point with which purchasers should bear in mind is that from certain parts of India there is sometimes exported under the name of 'Mowrah Meal' material which contains only a portion of the true bean, the remainder being from another type of bean the meal of which is somewhat similar to the genuine product. Even when this mixture is finely ground it is not properly effective.

Also published in 1935, 1936, 1937, 1938, and 1939, with some revisions, primarily product updates.

Swift & Company

1961†   **Lawn Keeping Made Easy.** Master Gardener. Swift & Company, Chicago, Illinois, USA. 29 pp., 27 illus., no tables, no index, fold-stapled glossy semisoft covers in multicolor with gray and white lettering, 5.4 by 8.3 inches (137 × 211 mm).

   A small, practical booklet on the establishment and culture of lawns. Both regional and special conditions in the United States are addressed.

Toll Lawn Dressings, Ltd.

This company published a series of booklets from 1927 to 1940, with periodic revisions. The 1936 and 1939 editions are presented herein.

1936†   **A New Way With Old Lawns and The Right Way With New Lawns.** Toll Lawn Dressings Limited. Ninth revised edition. Buxted, Sussex, England, UK. 33 pp., 4 illus., no tables, no index, fold-stapled semisoft covers in salmon with black lettering, 4.1 by 7.0 inches (103 × 179 mm).

   A **rare** booklet oriented to the use of commercially prepared topdressings for the care of turfgrasses in England. The control of weeds, diseases, and insects is emphasized, especially for lawns.

1939†   **A New Way With Old Lawns and The Right Way With New Lawns.** Twelfth revised edition in multicolor that varies. Toll Lawn Dressings Limited. Buxted, Sussex, England, UK. 42 pp., 7 illus., no tables, no index, fold-stapled semisoft covers in green with black lettering, 3.8 by 7.2 inches (97 × 182 mm).

   An expanded revision.

Vaughan's Seed Company

1957†   **Vaughan's Weed Control Chart.** Vaughan's Seed Company, Chicago, Illinois, USA. 31 pp., 23 illus., no tables, no index, soft bound in dark green, white, and black with black lettering, 3.0 by 5.4 inches (77 × 137 mm).

   A pocket-sized book containing descriptions and drawings of the common weeds that may occur in cool-season turfgrasses in the United States.

Warren's Turf Nursery

1960†   **Your Lawn in Hours.** Warren's Turf Nursery, Palos Park, Illinois, USA. 29 pp., 17 illus., 4 tables, no index, fold-stapled soft covers in greens, white, and black with black lettering, 5.5 by 8.5 inches (139 × 215 mm).

   A small booklet on the establishment of lawns by sodding plus the associated fertilizers.

   Reprinted in 1962.

Worcester Lawn Mower Company

n.d.†   **Facts On Lawn Management**. Lawrence S. Dickinson. Worcester Lawn Mower Co., Worcester Massachusetts, USA.
circa   23 pp., 8 illus., no tables, fold-stapled textured semisoft covers in light green with black lettering, 4.5 by 6.0 inches (114
1930   × 153 mm).

   A **rare**, small booklet concerning the planting and care of lawns. It is oriented to lawn enthusiasts and cool climatic conditions in the United States.

H. LAWN HISTORY AND SOCIOPOLITICAL VIEWS

Balmori, Diana

1993†   **Redesigning the American Lawn: A Search for Environmental Harmony.** Coauthor.
and
2001

   See *Bormann, F. Herbert—1993 and 2001.*

Betts, Janice Lake

1992†   **Turfgrass: Nature's Constant Benediction: The History of the American Sod Producers Association 1967–1992.** Janice Lake Betts and Wendell G. Mathews (editors). The American Sod Producers Association, Rolling Meadows, Illinois, USA. 112 pp., 405 illus., 7 tables, no index, soft bound in multicolor with white and light green lettering, 7.0 by 9.5 inches (178 × 240 mm).

   An interesting book on the 25-year history of the American Sod Producers Association (ASPA), now named Turfgrass Producers International (TPI). The early history of sod production is addressed in chapter 1 by Dr. James B Beard.

It contains many photographs plus a list of selected references in the back. A must book for the collectors of historical turfgrass books.

Revised in 2007.

See also *Mathews, Wendell Glen—2007.*

Bormann, F. Herbert

1993† **Redesigning the American Lawn: A Search for Environmental Harmony.** First edition. F. Herbert Borman, Diana Balmori, and Gordon T. Geballe. Yale University Press, New Haven, Connecticut, USA. 166 pp.; 64 illus. with 10 in color; 2 tables; small index; cloth hardbound in lavender with silver lettering and a glossy jacket in multicolor with white, black, and blue lettering and also in a semisoft bound version in multicolor with white, black, and blue lettering; 5.3 by 8.2 inches (134 × 207 mm).

This publication was developed for a course project in the School of Forestry and Environmental Studies at Yale University in 1991. The text is written by 11 graduate students and edited by 3 instructors. Following a historical review, major changes in plant species and cultural practices for lawns are proposed. The publication presents a one-sided view supported primarily by testimonials and statements of negative aspects that are not documented by sound research.

Revised in 2001.

2001† **Redesigning the American Lawn.** Second revised edition. F. Herbert Borman, Diana Balmori, and Gordon T. Geballe. Yale University Press, New Haven, Connecticut, USA. 190 pp., 89 illus. with 35 in color, 2 tables, indexed, glossy semisoft bound in blues with black lettering, 6.9 by 9.5 inches (176 × 241 mm).

A major revision of the 1993 first edition, with the addition of a chapter on sustainable lawns.

Castle, Rosemary

2001 **Liberating Lawns.** Rosemary Castle. Alternatives, Lydney, Gloucestershire, England, UK. 87 pp., 13 illus. 1 table, no index, glossy semisoft bound in multicolor with black lettering, 5.8 by 8.2 inches (148 × 209 mm).

A small book in which the author promotes the encouragement of nongrass plants or "weeds" as a dominant feature in lawns, especially as related to colorful flowers. Selected references are included, but most of it is opinion based.

Clayden, Andy

2008 **Residential Landscape Sustainability.** Coauthor.

See *Smith, Carl—2008.*

Cragg-Barber, Martin

2001 **Appreciating Lawn Weeds.** Martin Cragg-Barber with six contributing authors. That Plant's Odd, Chippenham, Wiltshire, England, UK. 40 pp., 54 illus., no tables, no index, fold-stapled glossy semisoft covers in multicolor with black lettering, 5.7 by 8.2 inches (146 × 209 mm).

Figure 13.22. Snow removal from putting green via a horse-drawn unit in 1940. [6]

A booklet oriented to the appreciation of lawn weeds from an ornamental flowering view point. Some authors focused on their opinions rather than science-based conclusions.

Dahle, Terje Nils

1985   **Rasenflächen (Turfgrass Surfaces).** Terje Nils Dahle. Literaturauslese Nr. 412, IRB Verlag, Stuttgart, Germany. 114 pp. plus 1 p. of advertising, no illus., no tables, indexed, semisoft bound in black and white with black lettering, 5.75 by 8.2 inches (146 × 208 mm). Printed entirely in the German language.

A bibliography of articles and papers published in German concerning turfgrass.

Dunnett, Nigel

2008   **Residential Landscape Sustainability.** Coauthor.
See *Smith, Carl—2008.*

Fort, Tom

2000†   **The Grass is Greener—Our Love Affair with the Lawn.** Tom Fort. Harper Collins Publishers, London, England. UK. 288 pp., 18 illus. with 2 in color, no tables, small index, cloth hardbound in green with gold lettering plus a jacket in multicolor with red and black lettering, 5.1 by 7.7 inches (129 × 195 mm).

A unique book documenting the cultural history concerning origins and evolution of the lawn and the mower in the United Kingdom. It is extensively researched in relation to the overall history of the country. The writing style contributes to enjoyable reading.

2001†   A soft-bound version of the 2000 book.

Geballe, Gordon Theodore

1993†   **Redesigning the American Lawn: A Search for Environmental Harmony.** Coauthor
and   See *Bormann, F. Herbert—1993 and 2001.*
2001†

Golf Course Superintendents Association of New Jersey

2001†   **75th Anniversary Journal.** Douglas Vogel, Golf Course Superintendents Association of New Jersey, Springfield, New Jersey, USA. 30 pp., 27 illus., no tables, no index, glossy fold-stapled cover in white and green with green lettering, 8.5 by 11.0 inches (215 × 278 mm).

A booklet summarizing the history of the Golf Course Superintendents Association of New Jersey from 1926 to 2001.

Grace, D.R.

1975†   **Ransomes of Ipswich: A History of the Firm and Guide to its Records.*** D.R. Grace and D.C. Phillips. Institute of Agricultural History, University of Reading, Reading, England, UK. 64 pp., 18 illus., no tables, no index, cloth hardbound in green with gold lettering and a jacket in multicolor with blue and black lettering, 7.2 by 10.1 inches (182 × 256 mm).

A small book consisting of a history of the Ransomes Company, a selected bibliography, and a guide to the archival collection of Ransome records.
See also *Weaver, Carol—1989.*

Haas, Eugene R.

2005†   **Caring For The Green.** Gene Haas. Wisconsin Golf Course Superintendents Association, Brookfield, Wisconsin, USA. 200 pp., 177 illus., no tables, no index, cloth hardbound in green with no lettering, plus a jacket in multicolor with white and yellow lettering, 8.0 by 8.0 inches (204 × 203 mm).

A book concerning the history of golf course superintendents in Wisconsin, as related to the Wisconsin Golf Course Superintendents Association. The focus is on the individuals and their contributions.

Halford, David G.

1982†   **Old Lawn Mowers.*** David G. Halford. Shire Album No. 91, Shire Publication Ltd., Aylesbury, Buckinghamshire/Botley, Oxford, England, UK. 32 pp.; 44 illus.; no tables; no index; fold-stapled semisoft covers in multicolor with white, red, and black lettering; 5.8 by 8.2 inches (147 × 208 mm).

A unique, interesting booklet on the origins and evolution of lawn and turfgrass mower development in the United Kingdom. It contains numerous old photographs of mowers. This booklet is a must for all collectors of historical turfgrass books.

† *Reprinted in 1993, 1999, and 2008 bound editions.*

Jacobsen, Oscar Thorkild

1977† **The Jacobsen Story.*** Oscar T. Jacobsen. Creative Printers Inc., Chapel Hill, North Carolina, USA. 68 pp., 41 illus., no tables, no index, textured semisoft bound in cream with black and orange lettering, 8.0 by 9.9 inches (202 × 252 mm).

A small book concerning the history of the Jacobsen Manufacturing Company of Racine, Wisconsin. Knud Ferdinaud Jacobsen was born in Hjoring, Denmark in 1869. He immigrated to Racine, Wisconsin, USA, in 1891 and started a pattern-making shop. It was reorganized to a tool and die company and named the Thor Machine Works in 1917. A friend A.J. Dremel proposed to Knud Jacobsen the concept of an engine-drive lawn mower, which was subsequently built and introduced for sale in 1921 under the name "4-Acre" priced at $275. By December 1920, the company name had been changed to the Jacobsen Manufacturing Company. Then in 1905, Jacobsen Manufacturing acquired the Worthington Mower Corporation of Stroudsberg, Pennsylvania, that pioneered horse- and tractor-drawn, reel, gang mowers. Jacobsen family members remained key leaders in the company until 1969, when it was merged with Allegheny Ludlum.

Jenkins, Virginia Scott

1994† **The Lawn—A History of an American Obsession.** Virginia Scott Jenkins. Smithsonian Institution Press, Washington, D.C., USA. 246 pp., 22 illus., no tables, indexed, glossy semisoft bound in black, green and pink with green, white, and black lettering, 6.0 by 9.0 inches (152 × 229 mm).

A book devoted to one side of an issue on the usefulness of lawns in America. It originated as a doctoral thesis but lacks a sound, balanced science base. The references are primarily opinion-oriented, popular articles of little scientific value.

Labbance, Robert Edwin

2002† **Keepers of the Green—A History of Golf Course Management.** Bob Labbance and Gordon Witteveen. Ann Arbor Press, Chelsea, Michigan, USA. 272 pp., 176 illus. with 29 in color, no tables, extensive index, cloth hardbound in black with gold lettering and a jacket in multicolor with gold and white lettering, 8.0 by 10.0 inches (203 × 254 mm).

Figure 13.23. Drawing of Knud F. Jacobsen (left) and a motor-powered, reel mower called Four-Acre Jacobsen (right), circa 1920s. [6, 41]

A chronological history of greenkeeping and golf course management and the individuals contributing to the changes in this profession in North America. In the back, there are short resumes for each of the presidents of the National Association of Greenkeepers of America and of the Golf Course Superintendents Association of America.

Lloyd, Clay

2005† **The History of the Northern Ohio Chapter of the Golf Course Superintendents Association of America 1923–2005.** Northern Ohio Golf Course Superintendents Association, North Olmstead, Ohio, USA. 48 pp. with 2 pp. of advertising; 44 illus.; no tables; no index; textured semisoft bound in browns, grays, and blacks with white lettering; 8.5 by 10.9 inches (215 × 277 mm).

A small book concerning the history of the Northern Ohio Golf Course Superintendents Association, originally named the Cleveland District Greenkeepers Association when formed in 1923. The early members of this group were key leaders in founding the National Association of Greenkeepers in 1926. Key events and individuals are highlighted in a horizontal format.

Mathews, Wendell Glen

1992† **Turfgrass: Nature's Constant Benediction: The History of the American Sod Producers Association.** Coeditor.
See *Betts, Janice Lake—1992.*

2007† **History of Turfgrass Producers International—40th Anniversary 1967–2007.** Dr. Wendell Mathews and Walt Pemrick (editors). Turfgrass Producers International, East Dundee, Illinois, USA. 202 pp., 657 illus., 34 tables, no index, glossy semisoft bound in multicolor with white and black lettering, 7.4 by 9.7 inches (187 × 246 mm).

An update of the 25-year history of the American Sod Producers Association (ASPA), now named Turfgrass Producers International (TPI), originally published in 1992. Much of the presentation is in chronological order. This book is a must for individuals interested in turfgrass history.

New Jersey Turfgrass Association

1980† **10th Anniversary Commemorative Book.** New Jersey Turfgrass Association, New Brunswick, New Jersey, USA. 57 pp. with 29 pp. of advertising, 19 illus., no tables, no index, fold-stapled textured semisoft covers in tans with brown lettering, 6.0 by 9.9 inches (152 × 252 mm).

A booklet that summarizes the history of turfgrass education and research at Rutgers University from 1923 onward and of the New Jersey Turfgrass Association from 1970 to 1980.

Pemrick, Walter

2007† **History of Turfgrass Producers International—40th Anniversary 1967–2007.** Coauthor.
See *Mathews, Wendell Glen—2007.*

Phillips, D.C.

1975† **Ransomes of Ipswich—A History of the Firm and Guide to its Records.*** Coauthor.
See *Grace, D.R.—1975.*

Rhode Island Golf Course Superintendents Association

2005† **75th Anniversary Volume of the Rhode Island Golf Course Superintendents Association.** Kingston, Rhode Island, USA. 227 pp., 124 illus., 1 table, no index, glossy hardbound in white with no lettering, 8.5 by 11.0 inches (215 × 279 mm).

A documentary history of the Rhode Island Golf Course Superintendents Association, originally named the Rhode Island Greenkeeper' Club when formed in 1930. Included are copies of key historical minutes, letters, papers authored by members, and photographs.

Robbins, Paul

2007† **Lawn People: How Grasses, Weeds, and Chemicals Make Us Who We Are.** Paul Robbins. Temple University Press, Philadelphia, Pennsylvania, USA. 208 pp., 18 illus., 12 tables, indexed, textured cloth hardbound in hunter green with gold lettering, 5.9 by 8.9 inches (151 × 227 mm).

A small book focused on one side of the question as to the usefulness of modern lawns. There is a lack of available peer-reviewed references presented and discussed to provide a science-based balance before drawing conclusions.

Roberts, Beverly Cruickshank

1992† **Lawns and Sports Turf History.** Beverly C. Roberts and Eliot C. Roberts. The Lawn Institute, Pleasant Hill, Tennessee, USA. 71 pp., 57 illus., no tables, no index, fold-stapled glossy semisoft covers in white and greens with green and black lettering, 8.4 by 11.1 inches (213 × 281 mm).

A small book on the history of lawns and various segments of the turfgrass industry. References are cited from various scholarly and nontechnical sources but are less than comprehensive.

1992† **Blades of Grass—Speaking Out On The Environment.** Beverly C. Roberts and Eliot C. Roberts. The Lawn Institute, Pleasant Hill, Tennessee, USA. 68 pp., 201 illus., no tables, no index, fold-stapled glossy semisoft covers in white and greens with green and black lettering, 8.3 by 11.0 inches (210 × 280 mm).

A booklet of cartoons based on two blades of grass plus a short commentary that depicts the various aspects of turfgrasses and their culture.

n.d.† **The Lawnscape—Our Most Intimate Experience With Ecology**. Beverly C. Roberts and Eliot C. Roberts. The Lawn
circa Institute. Pleasant Hill, Tennessee, USA. 44 pp., 88 illus., no tables, no index, fold-stapled glossy semisoft covers in
199x white and greens with green and black lettering, 8.4 by 11.0 inches (213 × 280 mm).

A booklet describing the biological composition of a lawn ecosystem.

Roberts, Eliot Collins

n.d.† **Lawn and Sports Turf Benefits.** Eliot C. Roberts and Beverly C. Roberts. The Lawn Institute, Pleasant Hill, Tennessee,
circa USA. 35 pp., 1 illus., no tables, no index, fold-stapled glossy semisoft covers in white and greens with green and black
199x lettering, 8.4 by 11.0 inches (213 × 279 mm).

A small book on the benefits and economic value of turfgrass. References are cited from various scholarly and nontechnical sources but are less than comprehensive.

Schroeder, Fred E.H.

1993† **Front Yard America: The Evolution and Meanings of a Vernacular Domestic Landscape.** Fred E.H. Schroeder. Bowling Green State University Popular Press, Bowling Green, Ohio, USA. 181 pp., 80 illus., no tables, no index, cloth hardbound in green with yellow lettering, 5.9 by 8.9 inches (150 × 226 mm).

A book on the historical evolution of the American front yard in terms of the domestic landscape design. There is a major chapter on the historical aspects of the lawn mower. The early photographs are worth noting. Numerous references are included in the back.

† *Also printed in semisoft bound version.*

Skirde, Werner

2011 **Entwicklung und Stand der Rasenforschung in Deutschland und Österreich (History of Turfgrass Research in Germany and Austria).** Werner Skirde, Paul Baader, and Alexander Richter. Baader Konzept, Mannheim, Germany. 144 pp., 98 illus. in color, 8 tables, no index, semisoft bound in greens and grays with black lettering, 8.2 by 11.8 inches (209 × 299 mm). Printed in the German language.

A fairly comprehensive presentation concerning the history of turfgrass research in Germany and Austria. Emphasis is on the location, university, researcher, year(s), and types of investigations.

Smith, Carl

2008 **Residential Landscape Sustainability.** Carl Smith, Andy Clayden, and Nigel Dunnett. Blackwell Publishing, Oxford, England, UK. 205 pp., 47 illus. with 34 in color, 16 tables, small index, glossy hardbound in multicolor with black and white lettering, 7.4 by 9.6 inches (187 × 245 mm).

A guidebook regarding the development of sustainable residential villages as related to both the structures and their landscapes under conditions in England. There is a residential landscape sustainable checklist in the back.

Sports Turf Research Institute

2004† **A Review of the Turfgrass Industry, past, present and future.** Dr. Mike Canaway (editor). Commemorative Bulletin, Sports Turf Research Institute, Bingley, West Yorkshire, England, UK. 101 pp., 102 illus., 2 tables, no index, glossy semisoft covers in multicolor with white lettering, 8.2 by 11.6 inches (208 × 295 mm).

An assemblage of 25 papers by authors from around the world that are oriented to the seventy-fifth anniversary of the Sports Turf Research Institute, formed in 1929 as the Board of Greenkeeping Research.

Steinberg, Theodore

2006† **American Green—The Obsessive Quest for the Perfect Lawn.** Ted Steinberg. W.W. Norton & Company Inc., New York City, New York, USA. 311 pp., 59 illus., no tables, indexed, cloth hardbound in green and white with green lettering, 6.1 by 8.2 inches (154 × 208 mm).

A book on the post–WWII sociocultural history of the American lawn. An interesting, documented approach with an entertaining presentation. Somewhat of a balance between social and science aspects is maintained.

Teyssot, George

1999† **The American Lawn.** Georges Teyssot (editor) with seven contributing authors. Princeton Architectural Press, New York City, New York, USA. 231 pp., 251 illus. with 109 in color, no tables, no index, glossy semisoft bound in multi-color with white and green lettering, 8.2 by 10.7 inches (209 × 272 mm).

An interesting collection of essays that examines the American lawn within its historical, artistic, literary, and political contexts. This collection includes "The American Lawn: Surface of Everyday Life" by Georges Teyssot, "The Saga of Grass: From the Heavenly Carpet to Fallow Fields" by Monique Mosser, "The Lawn in Early American Landscape and Garden Design" by Therese O'Malley, "Professional Pastoralism: The Writing on the Lawn, 1850–1950" by Alessandra Ponte, "Fairway Living: Lawncare and Lifestyle from Croquet to Golf Course" by Virginia Scott Jenkins, "The Lawn at War: 1941–1961" by Beatriz Colomina, "The Electric Lawn" by Mark Wigley, and "Docket" by Elizabeth Diller and Ricardo Scofidio.

Vernegaard, Lisa

1993†   **Redesigning the American Lawn: A Search for Environmental Harmony.** Coauthor.
&   See *Bormann, F. Herbert—1993 and 2001.*
2001†

Weaver, Carol

1989† **Ransomes 1789–1989—200 Years of Excellence—A Bicentennial Celebration.*** Carol Weaver and Michael Weaver. Ransomes Sims & Jefferies PLC, Ipswich, England, UK. 132 pp., 127 illus. with 19 color plates, no tables, no index, cloth hardbound in blue with gold lettering, 6.8 by 8.8 inches (172 × 223 mm).

A book on the 200-year history of equipment manufacturer, Ransomes, Sims, and Jefferies. It is based on the company's historical records that are held in the Museum of English Rural Life at the University of Reading. In 1832, Ransomes obtained a license to manufacture the first mower, which was invented by Edwin Beard Budding of Stroud. From 70 to 80 mowers were produced annually for the next 20 years.

See also *Grace, D.R.—1978.*

Weaver, Michael

1989† **Ransomes 1789–1989—200 Years of Excellence—A Bicentennial Celebration.*** Coauthor.
See *Weaver, Carol—1989.*

Weidner, Krista

2004† **The Grass Keeps Getting Greener: 75 Years of Turfgrass Research and Education at Penn State.** Krista Weidner. Pennsylvania State University, University Park, Pennsylvania, USA. 78 pp., 18 illus., no tables, indexed, semisoft covers in multicolor with black lettering, 8.5 by 11.0 inches (216 × 279 mm).

The evolution and development of the turfgrass research and education program at Penn State University are presented. Included are the origins starting in 1929, faculty and staff resumes, research facilities, and adult education programs. There is an appendix summarizing graduate students and thesis titles.

Witteveen, Gordon Cornelis

2001† **A Century of Greenkeeping.** Gordon Witteveen. Ann Arbor Press, Chelsea, Michigan, USA. 224 pp., 38 illus., no tables, no index, glossy hardbound in multicolor with brown lettering, 5.9 by 9.0 inches (151 × 228 mm).

A unique book concerning the history of golf course turfgrass culture, greenkeepers, and greenkeeper associations in the province of Ontario, Canada. The historical developments are organized chronologically. A summary list of Ontario greenkeepers plus positions and years held is presented in the back.

2002† **Keepers of the Green—A History of Golf Course Management.** Coauthor.
See *Labbance, Robert Edwin—2002.*

2008† **Keeping the Green in Canada.** Gordon Witteveen. Canadian Golf Superintendents Association, Mississauga, Ontario, Canada. 163 pp.; 140 illus.; 2 tables; no index; glossy hardbound in multicolor with green, white, and black lettering; 8.4 by 10.8 inches (214 × 273 mm).

A book concerning the history of golf course management in Canada from 1873 to 2007. Emphasis is on greenkeeping methods and the greenkeepers who were key leaders in the evolution of the profession. The history of the Canadian Golf Superintendents Association also is addressed.

I. ORNAMENTAL GRASS BOOKS

This section encompasses books that address unmowed ornamental grasses as used in landscapes.

Darke, Rick

1994† **For Your Garden: Ornamental Grasses.** Rick Darke. Little, Brown and Company, Michael Friedman Publishing Group Inc., New York City, New York, USA. 72 pp., 79 illus. in color, indexed, glossy semisoft bound in multicolor with brown lettering, 9.5 by 9.6 inches (242 × 243 mm).

A small book concerning the use of ornamental grasses in garden landscapes. Quality landscape photographs are used extensively.

2007 **The Encyclopedia of Grasses for Livable Landscapes.** Rick Darke. Timber Press, Portland, Oregon, USA. 487 pp., 1,040 illus. in color, no tables, indexed, cloth hardbound in green with gold lettering and a jacket in multicolor with white and yellow lettering, 9.9 by 10.9 inches (252 × 278 mm).

An impressive, large book regarding the unmowed ornamental grasses, sedges, rushes, restios, and cattails. Included are detailed descriptions, adaptations, and cultural requirements for both species and cultivars plus their utilization in landscape designs.

Foerster, Karl

1957 **Einzug der Gräser und Farne in die Gärten (Introduction of Grasses and Ferns into the Garden).** First edition. Karl Foerster. Neumann, Radebeul, East Germany. 221 pp., 261 illus. with 11 in color, no tables, no index, textured hardbound in greens with gold lettering, 6.7 by 9.4 inches (170 × 239 mm). Printed entirely in the German language.

A book with emphasis on perennial grasses used as unmowed ornamental plants in gardens under the conditions in Germany. Also included are ferns and selected foliage plants. There is a short bibliography in the back.

1961 Basically a reprint of the 1957 edition as a second edition by Verlag J. Neumann-Neudamm of Melsungen, West Germany, and by Neumann, Radebeul, East Germany.

1978 **Einzug der Gräser und Farne in die Gärten (Introduction of Grasses and Ferns into the Garden).** Karl Foerster and Bernhard Röllich. Verlag J. Neumann-Neudamm, Melsungen, West Germany. 256 pp., 290 illus. with 132 in color, no tables, large index, textured hardbound in green with green lettering, 6.7 by 9.4 inches (170 × 239 mm). Printed entirely in the German language.

A revision of the 1961 edition.

Basically reprinted in 1982 in both West Germany and East Germany.

1988 **Einzug der Gräser and Farne in die Gärten (Introduction of Grasses and Ferns into the Garden).** Karl Foerster and Bernhard Röllich. Neumann Verlag, Leipzig, East Germany. 255 pp., 141 illus. with 71 in color, no tables, large index, cloth hardbound in green and white with white lettering, 6.5 by 9.4 inches (165 × 238 mm). Printed entirely in the German language.

Basically a reprinting of the 1978 edition and also published by Eugen Ulmer GmbH & Co., Stuttgart, West Germany, with 238 color illustrations.

Greenlee, John

1992† **The Encyclopedia of Ornamental Grasses.** John Greenlee with photographs by Derek Fell. Michael Friedman Publishing Group Inc., New York City, New York, USA. 186 pp., 279 illus. in color, no tables, small index, cloth hardbound in blue with silver lettering and a glossy jacket in multicolor with blue and black lettering, 8.6 by 8.4 inches (176 × 214 mm).

This book gives descriptions, adaptation, landscape uses, culture, and individual close-up color photographs of more than 250 ornamental grasses. Suggested uses within various types of landscape designs also are given. A small glossary and a bibliography are found in the back.

Grounds, Roger

1979† **Ornamental Grasses.** First edition. Roger Grounds. Van Nostrand Reinhold Company, New York City, New York, USA. 216 pp., 117 illus. with 58 in color, 1 table, indexed, cloth hardbound in green with gold lettering and a jacket in multicolor with green and black lettering, 6.9 by 8.4 inches (176 × 214 mm).

A book on the true grasses plus the bamboos, sedges, rushes, and cattails used as unmowed ornamentals. It contains detailed descriptions and both botanical drawings and photographs. The numerous color photographs are very good.

Revised in 1989.

1989† **Ornamental Grasses.** Second revised edition. Roger Grounds. Christopher Helm (Publishers) Inc., Bromley, Kent, England, UK. 240 pp., 63 illus. with 20 in color, 1 table, large index, cloth hardbound in gray with gold lettering and a jacket in multicolor with green lettering, 6.1 by 9.1 inches (156 × 232 mm).

    Contains minor revisions.

1998† **The Plantfinder's Guide to Ornamental Grasses.** Roger Grounds. David & Charles Publishers, Newton Abbot, Devon, UK, and Timber Press Inc., Portland, Oregon, USA. 192 pp., 111 illus. with 106 in color, 1 table, indexed, cloth hardbound in green with gold lettering and a jacket in multicolor with white lettering, 7.5 by 10.4 inches (191 × 263 mm).

    A book that describes unmowed ornamental grasses in terms of how they can be utilized in landscape plantings and garden design themes. Emphasis is on grass species with showy flowers and/or foliage.

Holmes, Roger

1997† **Taylor's Guide to Ornamental Grasses.** Roger Holmes (editor). Houghton Mifflin Company, New York City, New York, USA. 309 pp., 188 illus. with 165 in color, 13 tables, indexed, glossy semisoft bound in multicolor with orange and white lettering, 4.3 by 8.5 inches (109 × 216 mm).

    A handbook on unmowed ornamental grasses for landscape use in the United States. There are 165-plus grass species described and shown via color photographs.

Loewer, H. Peter

1977† **Growing and Decorating with Grasses.*** H. Peter Loewer. Walker and Company, New York City, New York, USA. 128 pp., 100-plus illus., 1 table, indexed, cloth hardbound in green with gold lettering and a jacket in greens with yellow and white lettering, 8.0 by 10.9 inches (202 × 278 mm).

    A small book on annual and perennial grasses, sedges, rushes, and bamboos used as unmowed ornamentals. It contains descriptions and large botanical drawings. There is a bibliography included. It is a unique book.

1995† **Ornamental grasses—step-by-step.** Peter Loewer. Better Home and Gardens Books, Meredith Corporation, Des Moines, Iowa, USA. 132 pp.; 245 illus. in color; 2 tables; small index; glossy semisoft bound in multicolor with white, yellow, and black lettering; 9.1 by 9.7 inches (232 × 247 mm).

    A book on the use, design, planting, and care of ornamental grasses for landscapes in the United States. There are 115 grass species with color photographs and descriptions.

Macleod, Catherine

2003† **Grass Scapes: gardening with ornamental grasses**. Coauthor.

    See *Quinn, Martin—2003.*

Meyer, Mary Hockenberry

1975† **Ornamental Grasses—Decorative Plants For Home and Garden.** Mary Hockenberry Meyer. Charles Scribner's Sons, New York City, New York, USA. 136 pp., 53 illus. with 10 color plates, 14 tables, indexed, cloth hardbound in light green with silver lettering and a jacket in multicolor with black lettering, 6.0 by 8.9 inches (152 × 226 mm).

    A small book on annual and perennial grasses used as unmowed ornamentals. It contains descriptions, botanical drawings, adaptation, and cultural requirements for more than 90 grasses. There is a small glossary in the back.

Moskovitz, Mark

1989† **Ornamental Grass Gardening—Design Ideas, Functions, and Effects.** Coauthor.

    See *Reinhardt, Thomas A.—1989.*

Oakes, Albert Jackson

1990† **Ornamental Grasses and Grasslike Plants.** A.J. Oakes. Van Nostrand Reinhold, New York City, New York, USA. 614 pp., 134 illus., 17 tables, detailed index, textured cloth hardbound in green with silver lettering and a jacket in green and black with cream lettering, 6.0 by 8.9 inches (152 × 226 mm).

    An extensive, quality book on the adaptation, appearance, culture, and uses for hundreds of (a) ornamental grasses, (b) bamboo, (c) grasslike plants, and (d) rushes and sedges. The appearance aspects include form, size, color, texture, and growth habit. A glossary, a short bibliography, and quick reference summary tables are found in the back.

Ondra, Nancy J.

2002† **Grasses.** Nancy J. Ondra. Storey Publishing, North Adams, Massachusetts, USA. 143 pp., 211 illus. with 191 in color, no tables, indexed, glossy semisoft bound in multicolor with black lettering, 9.4 by 9.5 inches (239 × 241 mm).

A book concerning unmowed ornamental grasses and their use in garden landscapes. Twenty garden designs and associated plant lists are presented.

Ottesen, Carole

1989† **Ornamental Grasses—The Amber Wave.** Carole Ottesen. McGraw-Hill Publishing Company, New York City, New York, USA. 230 pp., 122 illus. with 98 in color, 6 tables, indexed, cloth hardbound in green with gold lettering and a jacket in multicolor with white and yellow lettering, 7.6 by 9.9 inches (192 × 251 mm).

A small, quality book on ornamental grasses for North America. It contains descriptions of more than 70 ornamental grass species and varieties plus sections on their care. The discussions are oriented to how these grasses can be effectively used in landscape designs.

1995† A semisoft bound version of the 1989 book.

Pesch, Barbara B.

1988† **Ornamental Grasses.** Barbara B. Pesch (editor) with 17 contributing authors. Plants and Gardens Handbook No. 117, Brooklyn Botanic Garden Record, Brooklyn, New York, USA. 104 pp., 61 illus. with 20 in color, 9 tables, no index, semisoft bound in multicolor with white lettering, 6.0 by 9.0 inches (152 × 228 mm).

A booklet concerning ornamental grasses and bamboos for use in the United States.

Quinn, Martin

2003† **Grass Scapes: Gardening with Ornamental Grasses**. Martin Quinn and Catherine Macleod. Whitecap Books Limited, North Vancouver, British Columbia, Canada. 192 pp., 281 illus. in color, 2 tables, small index, glossy semisoft bound in multicolor with white and black lettering, 7.5 by 9.2 inches (190 × 234 mm).

A book concerning the descriptions and uses of grasses, sedges and rushes as ornamentals in landscapes. Design concepts and plans are addressed.

† Also published in the United States by Ball Publishing of Batavia, Illinois, USA.

Reinhardt, Thomas A.

1989† **Ornamental Grass Gardening—Design Ideas, Functions, and Effects.** Thomas A. Reinhardt, Martina Reinhardt, and Mark Moskowitz. Michael Friedman Publishing Group Inc., New York City, New York, USA. 126 pp., 119 illus. in color, no tables, cloth hardbound in burgundy with silver lettering and a jacket in multicolor with green and burgundy lettering, 10.3 by 8.3 inches (262 × 210 mm).

A small book on ornamental grasses and their use in the landscape. Emphasis is placed on the propagation, culture, and effective utilization in landscape designs, especially the latter. There is a section on the most commonly available ornamental grasses including a description, environmental adaptation, and proper placement in landscape designs. The color photographs are of good quality.

Reinhardt, Martina

1989† **Ornamental Grass Gardening—Design Ideas, Functions, and Effects.** Coauthor.
See *Reinhardt, Thomas A.—1989.*

Taylor, Nigel J.

1992† **Ornamental Grasses, Bamboos, Rushes & Sedges.** Nigel J. Taylor. Ward Lock, London, England, UK. 96 pp., 52 illus. with 31 in color, no tables, indexed, textured cloth hardbound in green with gold lettering and a jacket in multicolor with white and orange lettering, 8.5 by 8.2 inches (218 × 207 mm).

A small book on the use of grasses, bamboos, rushes, and sedges as unmowed ornamentals in the landscape. It includes suggested garden planting plans for these species.

1994† A semisoft bound version of the 1992 book.

J. NONGRASS ARTIFICIAL SURFACE PUBLICATIONS

Australian Turfgrass Research Institute

1991† **Synthetic Surface Manual.*** There are 10 contributing authors: Ron Martin, Simon Fitzgerald, Alexandra Shakesby, Noel Turnbull, Brian Williams, Peter Wallace, John Payne, Robert Cole, Graeme Clark, and Keith Robertson. Australian

Turfgrass Research Institute, Concord West, New South Wales, Australia. 29 pp., 2 illus., 4 tables, no index, fold-stapled glossy semisoft covers in yellow with black lettering, 8.0 by 11.7 inches (203 × 297 mm).

A small manual on synthetic playing surfaces, with emphasis on synthetic grass bowling greens. There are sections on maintenance, surface evaluation, and installation.

Myer, Spencer

2007† **Synthetic Surfaced Bowling Greens.** Spencer Myer, Bob Jones, and Keith McAuliffe. The New Zealand Sports Turf Institute, Palmerston North, New Zealand. 28 pp.; 16 illus. in color; 4 tables; no index; glossy semisoft covers in multicolor with black, purple, and green lettering in a metal ring binder; 8.2 by 11.7 inches (209 × 296 mm).

A booklet concerning the history, types, maintenance, and weed control of artificial turfs as used on bowling greens. Forms for recording maintenance activities are included in the back.

National Playing Fields Association (NPFA)

1968 **Hard Porous & All-Weather Surfaces for Recreation (Outdoor).**[*] National Playing Fields Association, London, En-
to   gland, UK. 24 pp.; no illus.; 4 tables; no index; fold-stapled semisoft covers in yellow, white, and black with black
1971   lettering; 5.8 by 8.0 inches (147 × 205 mm).

This small booklet addresses the types and characteristics of both hard and porous all-weather surfaces for sport and recreational uses. It was revised periodically starting in 1968.

1976 **Hard Porous (Waterbound) Surfaces for Recreation.**[*] The National Playing Fields Association, London England, UK. 31 pp., no illus., no tables, indexed, fold-stapled glossy semisoft covers in multicolor with black lettering, 5.8 by 8.3 inches (148 × 210 mm).

A booklet concerning the characteristics, specifications, and maintenance of hard porous surfaces, as contrasted to all-weather hard surfaces.

1977 **Hard Surfaces For Play Areas.**[*] First edition. National Playing Fields Association, London, England, UK. 21 pp.; no illus.; no tables; no index; fold-stapled semisoft covers in white, black, and greens with white lettering; 5.8 by 8.3 inches (148 × 210 mm).

A small booklet of suggested specifications for the hard surfaces of game and play areas.
Revised in 1989.

1989† **Hard Porous (Waterbound) Surfaces For Recreation.**[*] National Playing Fields Association, London, England, UK. 35 pp.; 13 illus.; 1 table; no index; fold-stapled glossy semisoft covers in white, greens, and black with white and black lettering; 5.6 by 8.2 inches (146 × 208 mm).

A small booklet on the design, construction, and maintenance of nonturf, artificial surfaces for activities such as athletics, cricket, and tennis.

1989† **Hard Surfaces For Games And Play Areas.**[*] Second revised edition. National Playing Fields Association, London, England, UK. 25 pp.; no illus.; 1 table; no index; fold-stapled glossy semisoft covers in white, black, and greens with white lettering; 5.6 by 8.2 inches (146 × 208 mm).

There are minor revisions of the first edition.

Nottinghamshire County Council

1985 **To Play Like Natural Turf.**[*] Cricket Pitch Jottings, Nottinghamshire County Council, Nottingham, England, UK. 69 pp.; 203 illus.; 1 table; no index; fold-stapled semisoft covers in green, black, and white with black and white lettering; 8.2 by 11.5 inches (209 × 292 mm).

This booklet is composed of six chapters by various authors on nonturf pitches or squares for cricket play. It includes summaries of field tests conducted by Mr. Peter K.L. Dury on various types of nonturf constructions. There also is a chapter on the maintenance of nonturf surfaces.

Rhodes, David I.

1996† **The Maintenance of Artificial Sandfilled Turf Used for Sport.**[*] David I. Rhodes. Sports Surface Management, Manchester, England, UK. 93 pp. plus 6 pp. of advertising, 21 illus. with 13 in color, 4 tables, no index, glossy semisoft bound in multicolor with red and yellow lettering, 5.6 by 8.0 inches (142 × 203 mm).

A small, practical guide for the construction and maintenance of sand-filled artificial turfs, including problems with older installations.

Roberts, John R.B.

1990 **Artificial Turf Pitches for Hockey.** The Sports Council in association with the Hockey Association and the All English Women's Hockey Association, London, England, UK. 158 pp., 122 illus., 27 tables, no index, glossy hard covers in multicolor with green and black lettering in a 3-hole binder, 8.3 by 11.7 inches (210 × 297 mm).

A planning, design, construction, and management guide for artificial hockey pitches in England. It includes case studies, list of suppliers, and selected references.

Smith, Frank

1972 **Hard, Porous and All-Weather Surfaces.*** First edition. Frank Smith. SISIS Equipment Limited, Macclesfield, Cheshire, UK. 62 pp. plus 9 pp. of advertising, 37 illus., 4 tables, small index, fold-stapled semisoft covers in grays with black lettering, 5.8 by 8.2 inches (148 × 208 mm).

A small book on the selection, site preparation, construction, and maintenance of hard, porous playing areas, such as cricket wickets, tennis courts, ski slopes, and running tracks.
Revised in 1978.

1978 **Hard, Porous and Artificial Surfaces.*** Second revised edition. Frank Smith. SISIS Equipment Limited, Macclesfield, Cheshire, UK. 64 pp. plus 9 pp. of advertising, 43 illus., 1 table, small index, fold-stapled semisoft covers in red and black with red lettering, 5.8 × 8.2 inches (147 × 209 mm).

A minor revision.

Tipp, Graeme

1982† **Polymeric Surfaces for Sports and Recreation.*** G. Tipp and V.J. Watson. Applied Science Publishers Limited, Barking, Essex, England, UK. 409 pp., 104 illus., 34 tables, indexed, cloth hardbound in black with gold lettering and a jacket in green and white and black lettering, 5.5 by 8.7 inches (140 × 220 mm).

A unique book on the use of plastics and elastomers as surfaces for sports and recreational activities. Part one describes the chemistry, manufacture, and installation of the different systems. Part two deals with the degradation, maintenance, economics, and pathological aspects of the systems. Part three outlines specific testing methods for various materials.

1983† **Synthetic Surfaces.*** Coauthor.
See *Watson, Vic J.—1983.*

1987† **Impacting Absorbing Surfaces.*** Coauthor.
See *Watson, Vic J.—1987.*

1988† **Synthetic Surfaces.*** Second revised edition. Graeme Tipp and Vic Watson. Special Report No. 1, National Turfgrass Council, Bingley, West Yorkshire, England, UK. 32 pp., 8 illus., 1 table, no index, semisoft bound in yellow with black lettering, 5.8 by 8.3 inches (147 × 210 mm).

An updated, expanded revision of the 1983 edition.

Watson, Vic J.

1982† **Polymeric Surfaces for Sports and Recreation.*** Coauthor.
See *Tipp, Graeme—1982.*

1983† **Synthetic Surfaces.** First edition. Vic Watson and Graeme Tipp. Special Report No. 1, National Turfgrass Council, Bingley, West Yorkshire, England, UK. 21 pp., 4 illus., 4 tables, no index, a clear overlay with soft covers in black and white with black lettering in a comb binder, 5.8 by 8.3 inches (148 × 210 mm).

A booklet describing the types of synthetic surfaces for sports fields, including specifications and testing.

1987† **Impact-Absorbing Surfaces For Children's Playgrounds.*** V. Watson and G. Tipp. National Playing Fields Association, London, England, UK. 34 pp., 2 illus., no tables, no index, fold-stapled semisoft covers in white, black and green with black lettering, 5.8 by 8.2 inches (147 × 209 mm).

A booklet on the main types of impact absorbing surfaces, including the advantages, disadvantages, and installation.

1988† **Synthetic Surfaces.*** Coauthor.
See *Tipp, Graeme—1988.*

Wesley, Reginald

1955 **Artificial Cricket Pitches.*** Reginald Wesley. Contractors Record Limited, London, England, UK. 70 pp. with 12 pp. of advertising, 24 illus., no tables, no index, cloth hardbound in dark blue with gold lettering, 5.4 by 8.3 inches (136 × 212 mm).

A small book on artificial pitches or tables for cricket play in the United Kingdom. It contains a major chapter on concrete surfaces and their construction plus a chapter on surfacing materials. It is oriented for low-budget, local grounds situations.

K. SUMMARY OF CHINESE TURFGRASS BOOKS

1984   Cao Ping (Turf). M. Shao. 68 pp.

1984   Cao Ping: Fu Di Bei Zhi Wu (Turf: Including Ground Cover Plants). M. Shao and Z. Hu. 68 pp.

1984   Cao Ping Ji Di Bei Zhi Wu (Turf and Ground Cover Plants). Z. Hu and X. Zhao. 194 pp.

1990   Zhonggua Cao Ping Zhi Wu Zai Pei (Chinese Lawn Plants Cultivation). W. Xiao, Z. Sun, and Y. Zhao. 358 pp.

1991   Cao Ping Jian Zhi Yu Guan Li (Turf Establishment and Management). L. Han, J. Sun, and Z. Liu. 141 pp.

1991   Cao Ping Zhong Zhi (Turf Cultivation). S. Hu. 186 pp.

1991   Cao Ping Xue Ji Yin Yong Ji Shu (Turfgrass Science and Practical Techniques). S. Hu, M. Lai, and L. Dong. 198 pp.

1993   Cao Ping Zai Pei Guan Li (Turf Cultivation Management). Z. Chen. 343 pp.

1993   Cao Ping Pin Zhong Zhi Nan (Guide to Turfgrass Variety). M. Li. 155 pp.

1994   Cao Ping Guan Li Xue (Turf Management). L. Han. 337 pp.

1995   Cao Ping Di Bei Zhi Wu (Turf and Ground Cover Plants). Z. Hu and S. Liu. 236 pp.

1995   Cao Ping Xue (Turfgrass Science). J. Sun. 170 pp.

1996   Cao Ping Zhong Zhi (Turf Cultivation). S. Hu. 135 pp.

1996   Cao Ping Bing Chong Hai Ji Gi Fang Zhi (Turf Pest and Disease Control). F. Wang and H. Shang. 237 pp.

1996   Caoi Ping Di Bei Yu Ren Lei Huan Jing: Cao Ping Di Bej Jian Zhi Yuan Li Yu Shi Yong Ji Shu (Turf and Ground Cover with Human Environment). S. Zhou. 425 pp.

1997   Cao Ping Bing Chong Cao Hoi De Fa Sheng Ji Fang Zhi (Turf Pests and Pest Management). Z. Zhang, Q. Zhang, W. Wang, and S. Yang. 302 pp.

1998   Cao Ping Ke Xue Yu Yan Jiu (Turf Science and Research). F. Liu. 470 pp.

1998   Cao Ping Jian Zhi Yu Guan Li (Turf Establishment and Management). Z. Zhang. 184 pp.

1999   Cao Ping Cao Zhong Ji Qi Pin Zhong (Turfgrass Species and Cultivars). L. Han, B. Yang, and J. Deng. 203 pp.

1999   Gao Er Fu Giu Chang Cao Ping (Golf Course Turf). J. Beard, L. Han, Y. Zhang, S. Liang, M. Sun, X. Hu, and J. Zong. 457 pp.

1999   Cao Ping Za Cao (Turf Weeds). W. Cao, Z. Chen, B. Huang, S. Li, D. Liu, and L. Han. 286 pp.

1999   Cao Ping Jian Zhi (Turf Establishment). C. Chen and D. Cai. 87 pp.

1999   Cao Ping Zhi Wu Zhong Zi (Turfgrass Seeds). B. Chen, Y. Xie, and G. Xin. 174 pp.

1999   Yun Dong Chang Cao Ping (Athletic Field Turf). L. Han, B. Ding, and D. Alders. 164 pp.

1999   Cao Ping Jian Zhi Yu Guan Li Shou Ce (Turf Establishment and Management Manual). L. Han, D. Tian, and X. Mou. 309 pp.

1999   Cao Ping Ying Yang Yu Shi Fei (Turf Management: Nutrition and Fertilizer). B. Huang, W. Cao, and Z. Chen. 218 pp.

1999   Xian Dai Cao Ping Jian Zhi Yu Guan Li Ji Shu (Modern Turf Establishment and Management Techniques). F. Huang and Z. Liu. 347 pp.

1999   Cao Ping Za Cao (Turf Weeds). S. Li, D. Liu, and L. Han. 290 pp.

1999   Lu Di Cao Ping (Grounds Turf). B. Sun, M. Li, and S. Bai. 202 pp.

1999   Cao Ping Zhi Wu (Lawn Plants). J. Tan. 99 pp.

1999   Cao Ping Ji Xie (Turf Equipment). G. Yu, M. Li, and J. Sun. 230 pp.

1999   Cao Ping Chong Hai (Turf Insects). Q. Zhang, K. Liu, and G. Wang. 257 pp.

1999   Cao Ping Hai Chong (Turf Diseases and Pests). M. Zhao. 252 pp.

2000   Cao Ping Zai Pei Yu Yang Hu (Turf Cultivation and Maintenance). Z. Chen. 263 pp.

2000   Cao Ping Jian Zhi Chang Shi (Turf Establishment General Knowledge). X. He. 162 pp.

2000   Cao Ping Jian Zhi Yu Yang Hu Guan Li (Turf Establishment and Turfgrass Maintenance). Q. Lu and K. Ji. 309 pp.

2000   Cao Ping Jiang Zhi Yu Yang Hu Guan Li (Turf Establishment and Maintenance). Q. Lu, K. Ji, and M. Liu. 307 pp.

2000   Cao Ping Bing Chong Hai (Turf Pest and Disease). Office of Capital Forestry Committee. 195 pp.

2000   Cao Ping Ji Shu Zi Nan (Turf Technology Directory). J. Sun, Z. Zhang, and J. Sun. 297 pp.

2000   Cao Ping Yu Di Bei Zei Pci (Turf and Ground Cover Cultivation). J. Tan. 201 pp.

2000   Xin Bian Zhong Guo Cao Ping Yu Di Bei (New Lawns and Ground Covers in China). J. Tan and Z. Tan. 315 pp.

2000   Taiwan Cao Ping Za Cao Coi Se Tu Jian (Lawn Weeds in Taiwan). L. Xu and M. Jiang. 160 pp.

2001   Cao Ping Bao Hu Ji Shu (Turf Preservation Techniques). X. Chen. 162 pp.

2001   Cao Ping Jian Zhi Ji Shu (Turf Establishment Techniques). Z. Chen. 204 pp.

2001   Yunnan (Turf). M. Cui. 160 pp.

2001   Cao Ping Ke Xue Yu Guan Li (Turf Science and Management). L. Hu, X. Bian, and X. Yang. 459 pp.

2001   Cao Ping Di Bei Zhi Wu. Guan Shang Cao (Turf. Ground Cover Plant. Ornamental Grasses). J. Liu. 147 pp.

2001   Cao Pi Sheng Chan Ji Shu (Turf Production Technology). Z. Liu. 185 pp.

2001   Cao Ping Xin Pin Zhong Xuan Yong Ji Jian Zhi (New Turf Cultivars Selection Methods). X. Shi, Z. Zhu, and B. Jia. 155 pp.

2001   Cao Ping Yang Hu Ji Shu (Lawn Maintenance Techniques). Z. Su. 142 pp.

2001   Cao Ping Zhi Wu Zhong Zhi Ji Shu (Turf Cultivation Technology). B. Sun, G. Yi, and M. Zhang. 159 pp.

2001   Cao Ping Si Yong Ji Shu Shou Ce (Turf Practical Techniques Manual). Y. Sun, H. Zhou, and Q. Yang. 425 pp.

2001   Cao Ping Cao Xin Pin Zhong Ji Di Xuan Yu, Jian Zhi Yu Guan Li Ji Shu (New Turfgrass Cultivar Breeding and Turf Establishment and Management). B. Yang. 153 pp.

2001   YuNan Cao Ping (Turf in Yunnan). Y. Zhang and Y. Nie. 111 pp.

2001   Cao Ping Yang Mu Ji Shu (Turf Maintenance Techniques). M. Zhao, Y. Sun, and Q. Zhang. 259 pp.

2002   Cao Ping Ji Xie Shi Yong Yu Wei Hu Shou Ce (Turf Equipment Use and Maintenance Manual). C. Chen. 397 pp.

2002   Cao Ping Cao Pin Zhong Zhi Nan (Turfgrass Cultivars Guidelines). X. Liu and G. Chen. 380 pp.

2002   Cao Ping Jian Zhi Yu Yang Hu Shou Ce (Turf Establishment and Maintenance Handbook). J. Jia, D. Xi, and D. Mei. 338 pp.

2002   Cao Ping Za Cao Fang Chu Ji Shu (Turf Weed Control Technology). G. Shen, Y. He, and L. Yang. 308 pp.

2002   Cao Ping Yang Hu Wen Da 300 Li (Turf Maintenance 300 Example of Questions and Answers). X. Song. 336 pp.

2002   Cao Ping Lu Di Shi Yong Ji Shu Zhi Nan (Turf Practical Techniques Guidelines). J. Sun, L. Han, Y. Zhu, Z. Li, X. Li, Y. An, Y. Xie, Y. Bi, G. Chang, F. Jin, S. Yin, and S. Xing. 584 pp.

2002   Cao Ping Lu Di Gui Hua She Ji Yu Jian Zhi Guan Li (Ground Turf Establishment and Management). J. Sun, Z. Li, Y. Zhu, and J. Sun. 333 pp.

2002   Cao Ping Di Bei Jing Guan She Ji Yu Ying Yong (Turf and Groundcover Landscape Design and Application). J. Tan, J. Liu, and Z. Tan. 299 pp.

2002   Cao Ping Jian Zhi Yu Yang Hu (Turf Establishment and Maintenance). C. Wang. 419 pp.

2002   Cao Ping Bing Chong Hai Fang Zhi (Turf Disease and Pest Control). C. Wang. 426 pp.

2002   Cao Ping Bing Chong Hai Shi Bie Ji Gi Fang Zhi (Identification and Management of Turf Disease and Pest). F. Wang and H. Shang. 182 pp.

2002   Cao Ping Za Cao Ji Hua Xue Fang Cu Cai Se Tu Pu (Turf Weeds and Chemical Weed Control: Color Atlas). G. Xue and J. Ma. 276 pp.

2003   Cao Ping Jian Zhi Yu Guan Li (Turf Establishment and Management). Z. Chen. 307 pp.

2003   Cao Ping Zai Pei Yu Guan Li Da Quan (Compilation of Turf Cultivation and Management). Z. Chen. 487 pp.

2003   Cao Ping Yu Yuan Lin Lv Hua Ji Xie Xuan Yong Shou Ce (Turf and Landscape Equipment Selection Manual). Y. Liu, R. Shen, and Z. Gu. 316 pp.

2003   Zhongguo Cao Lei Zuo Wu Bing Li Xue Yan Jiu (Forage and Turfgrass Pathological Research in China). Z. Nan and C. Li. 258 pp.

2003   Cao Ping Xue (Turfgrass Science). J. Sun. 389 pp.

2003   Cao Ping Bing Cong Cao Hai (Turf Diseases, Pests, and Weeds). Q. Weng and D. Yu. 210 pp.

2003   Cao Ping Ji Xie (Turf Equipment). S. Yao. 188 pp.

2004   Cao Ping You Hai Sheng Wu Ji Qi Fang Zhi (Turfgrass Pests Controlling). R. Liu. 526 pp.

2004   Cao Ping Ke Xue Shi Xi Shi Yan Zhi Dao (Experiment and Trial Manual for Turfgrass Science). R. Long and T. Yao. 295 pp.

2004   Cao Ping Guan Gai Yu Pai Shui Gong Chong Xue (Turf Irrigation and Drainage Engineering). D. Su. 333 pp.

2004   Cao Ping Gong Cheng Xue (Turf Engineering). J. Sun. 255 pp.

2004   Cao Ping Yu Di Bei Zhi Wu (Turf and Ground Cover Plants). W. Wang, H. Tian, and Z. Chjen. 226 pp.

2004   Cao Ping Yuan Lin Za Cao Hua Xue Fang Chu (Chemical Weed Control for Turf and Nursery Garden). G. Xuc, J. Ma, J. Wu, B. Wei, and D. Zhu. 284 pp.

2004   Cao Ping Yang Hu Ji Xic (Turf Equipment). G. Yu. 256 pp.

2004   Cao Ping Cao Yu Shong Xue (Turfgrass Breeding). X. Zhang. 359 pp.

2004   Cao Ping Ying Yang Yu Shi Fei (Turf Nutrients and Fertilization). Z. Zhang. 297 pp.

2005   Cao Ping Xue Ji Chu (Turf Science Basis). X. Bian and X. Zhang. 207 pp.

2005   Cao Ping Cao Yu Di Bei Zhi Wu Do Xuan Ze (Turfgrass and Groundcover Plant Selection). B. Hou, Y. Chen, A. Long, J. Shen, X. Tang, L. Li, W. Liu, and X. Chen. 201 pp.

2005   Cao Ping Jian Zhi Shou Ce (Lawn Growth Handbook). X. Xian, Y. Gua, and X. Gou. 422 pp.

2005   Cao Ping Chong Hai Fang Zhi (Turf Pest Management). G. Zhang, X. Zhou, Y. Tan, and S. Li. 215 pp.

2005   Cao Ping Coa Zhong Zi Sheng Chan Ji Shu (Turfgrass Seed Production Techniques). S. Shi. 311 pp.

2005   Cao Ping Chong Hai Fang Zhi (Turfgrass Breeding). X. Zhang. 342 pp.

2006   Cao Ping Yu Di Bei Ko Xue Jin Zhan (Development of Turfgrass and Ground Cover Plant Sciences). Z. Chen and H. Zhou. 311 pp.

2006   Cao Ping Cao Sheng Wu Ji Shu Ji Ying Yong (Turfgrass Biological Technology and Application). Z. Lin, Y. Hu, X. Zhu, and X. Wang. 189 pp.

2006   Cao Ping Jian Zhi Yhu Yang Hu (Turf Establishment and Management). Z. Lu and S. Zhang. 191 pp.

2006   Cao Ping Jian Zhi Yu Yang Hu (Turf Establishment and Management). W. Pan. 190 pp.

2006   Cao Ping Bing Chon Hai Shi Bie Yu Fang Zhi (Identification and Management of Turf Pests). H. Shang and F. Wang. 190 pp.

2006   Cao Ping Gong Chong (Turf Engineering). J. Sun. 364 pp.

2006   Cao Ping Bao Hu (Turf Protection). B. Xu. 347 pp.

2006   Cao Ping Hua Xue Chu Cao (Turf Chemical Weed Control). Y. Wang, W. Zhang, and C. Wang. 173 pp.

2006   Cao Pi Sheng Chan Ji Shu (Turf Technical Manual: Sod Production Techniques). D. Zhang. 180 pp.

2007   Cao Ping Bing Chong Hai Zhen Duan Yu Fang Zhi Yuan Se Tu Pu (Primary Color Atlas of Turf Pest Identification and Pest Management). D. Yu. 165 pp.

Figure 13.24. Turfgrass genotype characterization in a glasshouse at Texas A&M University, in 1970s. [26]

CHAPTER 14

Scientific Periodicals Containing Turfgrass Papers

THIS SECTION SUMMARIZES THOSE PUBLICATIONS IN WHICH A MAJORITY OF THE SIGNIFICANT research on the basic science of turfgrass culture is first published. Most significant turfgrass research is published in peer-reviewed scientific journals. That means the papers have been subjected to critical review and assessment by research scientists at other institutions and have been found to follow proper scientific procedure, statistical analysis of data, and appropriate conclusions. The format, procedures, and terminology used in each of the individual scientific journals usually is dictated by the sponsoring scientific group.

A majority of the American scientific papers on turfgrass research are published in journals sponsored by the Crop Science Society of America (CSSA), the American Society of Agronomy (ASA), the American Society of Horticultural Science (ASHA), the American Phytopathological Society (APS), the Weed Science Society of America (WSSA), and the American Entomological Society (AES). Many of the independent, peer-reviewed scientific papers on turfgrass can be found in the **Agronomy Journal**, **Crop Science**, **HortScience**, and **Journal of American Society for Horticultural Science**. There are 16 scientific periodicals outside the United States that have been or are being published specifically in the turfgrass area. Countries involved are Australia, Germany, Japan, New Zealand, South Korea, Sweden, and the United Kingdom.

FORMAT

Publications in this category include regularly published (ranging from monthly to yearly) scientific periodicals that contain papers on turfgrass.

The citation style used in this chapter is as follows:

- Name of scientific periodical.
- Name and location of sponsor, city, state, and country.
- Years published; the first year listed is when the proceedings were first published, while the second year listed is when publication of the proceedings were terminated; if the proceedings continues to be published, this is indicated by a "+."
- The notation "peer-reviewed" indicates the paper has been assessed for sound scientific (a) research procedures, (b) data statistical analysis, and (c) conclusions by multiple, anonymous scholars and researchers who possess expertise in the particular area of investigation.
- Details about what the periodical contains and includes.
- Previous or subsequent names and dates published for the same proceedings are listed in parentheses.
- Number of times published annually for the most recent year of publication.

Agronomy Abstracts

American Society of Agronomy, Madison, Wisconsin, USA. 1955+.

No independent peer review.

Contains abstracts of papers presented at the annual meetings of the ASA, CSSA, and SSSA.

Includes abstracts of the Turfgrass Science Division C-5.

Published once annually, changed to CD-ROM in 2001 and to an electronic format in 2004.

Agronomy Journal

American Society of Agronomy, Madison, Wisconsin, USA. 1949+.

Contains independent peer-reviewed research papers of ASA members, including plant and soil research of turfgrasses.

(*formerly* Journal of American Society of Agronomy, 1913–1948).

Published six times annually.

Agronomy News

American Society of Agronomy, Crop Science Society of America, and Soil Science Society of America, Madison, Wisconsin, USA. 1956–1998.

(*see* Crop Science, Soil Science, Agronomy News, 1999).

Published 6 to 12 times annually.

Annual Experimental Report

Nishinihon Green Section Institute, Fukuoka, Wajiro, Japan. 1996–2005.

Published once annually in the Japanese language.

Annual Rutgers Turfgrass Symposium

Rutgers University, New Brunswick, New Jersey, USA. 1991+.

Applied Turfgrass Science

Plant Management Network (PMN), St. Paul, Minnesota, USA. 2004+.

Produced in an electronic format on an irregular schedule.

ASHS Newsletter

American Society for Horticultural Science, Alexandria, Virginia, USA. 1984+.

Published 12 times annually.

Asian Journal of Turfgrass Science

Turfgrass Society of Korea, Seoul, South Korea. 2011+.

Contains independent peer reviewed papers.

(*formerly* Korean Journal of Turfgrass Science, 1987–2010).

Published one to four times annually in the Korean and English languages.

Bulletin of Nishinihon Green Research Institute

Nishinihon Green Research Institute, Fukuoka, Wajiro, Japan. 1999–2000.

Published once annually in the Japanese language.

C-5 Newsletter

Turfgrass Science Division C-5, Crop Science Society of American, Madison, Wisconsin, USA. 1995+.

Published at variable times annually.

Crop Science

Crop Science Society of America, Madison, Wisconsin, USA. 1961+.

Contains independent peer-reviewed research papers of CSSA members, including research on turfgrasses by Turfgrass Science Division C-5 members.

Published six times annually.

Crop Science—Soil Science—Agronomy News

Crop Science Society of America, Soil Science Society of America, and American Society of Agronomy, Madison, Wisconsin, USA. 1999.

(*formerly* Agronomy News, 1956–1998).
(*see* CSA News, 2000+).
Published 12 times annually.

CSA News

American Society of Agronomy, Crop Science Society of America, and Soil Science Society of America, Madison, Wisconsin, USA. 2000+.
(*formerly* Agronomy News, 1956–1998, and Crop Science, Soil Science, Agronomy News, 1999).
Published 12 times annually.

European Journal of Turfgrass Science

Hortus Zeitschriften Cöllen & Bleeck/Köllen Druck +Verlag GmbHm, Bonn, Germany. 2008+.
(*formerly* Rasen: Turf-Gazon Grünflächen Begrünungen, 1976–2007).
(*also contains* Greenkeepers Journal, 2008+).
Published four times annually.

F&N Tests or Fungicide and Nematicide Tests

The American Phytopathological Society, St. Paul, Minnesota, USA. 1945–2006.
(*see* Plant Disease Management Reports, 2007+).
Published one time annually.

Gurin Kenkyu Hokokushu (Turf Research Bulletin)

Kansai Golf Union Green Section Research Center, Takarazuka, Japan. 1961+.
No external peer-review.
Contains review and research papers on turfgrasses for golf courses.
Published once annually in the Japanese language.

Half Annual Report of Nishinihon Green Research Institute

Nishinihon Green Research Institute, Fukuoka, Wajiro, Japan. 1986+.
Published twice annually in the Japanese language.

HortScience

American Society for Horticultural Science, Alexandria, Virginia, USA. 1966+.
Contains independent peer-reviewed research papers of ASHS members, including research on turfgrasses.
Published seven times annually.

HortTechnology

American Society for Horticultural Science, Alexandria, Virginia, USA. 1991+.
Published four times annually.

Journal Japanese Society of Turfgrass Science

Turfgrass Research and Development Organization of Japan, Tokyo, Japan. 1984+.
Contains review and research papers on turfgrass.
(*formerly* Journal of Japan Turfgrass Research Association, 1972–1983).
Published 12 times annually in the Japanese language with English summaries.

Journal of American Society of Agronomy

American Society of Agronomy, Geneva, New York, USA. 1913–1948.
Contains independent peer-reviewed research papers of ASA members.
(*formerly* Proceedings of the American Society of Agronomy, 1910–1912).
(*see* Agronomy Journal, 1949+).
Published six times annually.

Journal of Economic Entomology

Entomological Society of America, Lanham, Maryland, USA. 1908+.
Published six times annually.

Journal of Entomological Science
Georgia Entomological Society, Tifton/Griffin, Georgia, USA. 1985+.
Published four times annually.

Journal of Environmental Horticulture
Horticultural Research Institute Inc., Washington, DC, USA. 1983+.
Contains primarily independent peer-reviewed research papers concerning ornamental plants, with some papers on turfgrass.
Published four times annually.

Journal of Environmental Quality
American Society of Agronomy, Crop Science Society of America, and Soil Science Society of America, Madison, Wisconsin, USA. 1972+.
Contains independent peer-reviewed research and review papers of ASA, CSSA, and SSSA members, including some research on turfgrasses.
Published four times annually.

Journal of Japan Turfgrass Research Association (Shibukusa Kenkyu)
Japan Turfgrass Research Association, Matsudo-shi, Chiba-ken, Japan. 1971–1983.
Contains review and research papers on turfgrass.
(*see* Journal Japanese Society of Turfgrass Science, 1984+).
Published two times annually in the Japanese language with English summaries.

Journal of Nematology
The Society of Nematologist, College Park, Maryland, USA. 1969+.
Published four times annually.

Journal of Plant Registrations
Crop Science Society of America Inc., Madison, Wisconsin, USA. 2007+.
Published at irregular times annually.

Journal of Soil and Water Conservation
Soil and Water Conservation Society, Baltimore, Maryland/Ankeny, Iowa, USA. 1946+.
Published four to six times annually.

Journal of the American Society for Horticultural Science
American Society for Horticultural Science, Alexandria, Virginia, USA. 1969+.
Contains independent peer-reviewed research papers of ASHS members, including some research on turfgrasses.
(*formerly* Proceedings of the American Society for Horticultural Science, 1916–1968).
Published six times annually.

The Journal of the Board of Greenkeeping Research
Board of Greenkeeping Research, Bingley, Yorkshire, England, UK. 1929–1950.
Contains turfgrass research and review papers of the Board of Greenkeeping Research (BGR) staff.
No external peer review.
(*see* The Journal of the Sports Turf Research Institute, 1951–1996).
(*see also* Report of the Board of Greenkeeping Research, 1930–1950).
Published once annually, except from 1941 to 1947.

Journal of the Sports Turf Research Institute
Sports Turf Research Institute, Bingley, West Yorkshire, England, UK. 1951–1996.
Contains primarily turfgrass research papers of the STRI staff
No external peer review.
(*formerly* The Journal of the Board of Greenkeeping Research, 1929–1950).
(*see* Journal of Turfgrass Science, 1997–2002).
Published once annually.

Journal of Turfgrass and Sports Surface Science
Sports Turf Research Institute, Bingley, West Yorkshire, England, UK. 2002+.
(*formerly* Journal of Turfgrass Science, 1997–2001).
Published once annually.

Journal of Turfgrass Management
The Haworth Press Inc., Binghamton, New York, USA. 1995–1999.
Published four times annually.

Journal of Turfgrass Science
Sports Turf Research Institute, Bingley, West Yorkshire, England, UK. 1997–2001.
(*formerly* Journal of the Sports Turf Research Institute, 1951–1996).
(*see* Journal of Turfgrass and Sports Surface Science, 2002+).
Published once annually.

Korean Journal of Turfgrass Science
Turfgrass Society of Korea, Seoul, Korea. 1987–2010.
Contains independent peer-reviewed papers, including an English abstract.
(*see* Asian Journal of Turfgrass Science, 2011+).
Published one to four times annually in the Korean and English languages.

North American Grass Breeders Conference Proceedings
Grass Breeders Work Planning Conference, USA. 1971, 1973, 1975, 1980–2000.
Published biannually.

Phytopathology
The American Phytopathological Society, St. Paul, Minnesota, USA. 1911+.
This international journal contains independent peer-reviewed plant pathology research papers of APS members, including research on turfgrass diseases and causal pathogens.
Published 12 times annually.

Phytopathology News
The American Phytopathological Society, St. Paul, Minnesota, USA. 1966+.
Published 12 times annually.

Plant Disease
The American Phytopathological Society, St. Paul, Minnesota, USA. 1980+.
This international journal of applied plant pathology contains research papers of APS members on plant diseases and their control, including diseases of turfgrasses.
(*formerly* Plant Disease Reporter, 1923–1979).
Published 12 times annually.

Plant Disease Bulletin
Plant Disease Survey, Bureau of Plant Industry, United States Department of Agriculture, Beltsville, Maryland, USA. 1917–1922.
Contains research papers on plant diseases and their causal pathogens.
(*see* Plant Disease Reporter, 1923–1979).
Published at irregular times annually.

Plant Disease Management Reports
The American Phytopathological Society, St. Paul, Minnesota, USA. 2007+.
Consolidation of B&C Tests and F&N Tests.
Published one time annually.

Plant Disease Reporter
United States Department of Agriculture, Beltsville, Maryland, USA. 1923–1979.

Contains research papers on plant diseases and their causal pathogens, including diseases of turfgrasses.
(*formerly* Plant Disease Bulletin, 1917–1922).
(*see* Plant Disease, 1980+).
Published 12 times annually.

Plant Physiology
American Society of Plant Physiologists, Rockville, Maryland, USA. 1920+.
Published 12 times annually.

Proceedings of the American Society for Horticultural Science
American Society for Horticultural Science, Greensboro, North Carolina, USA. 1916–1968.
Contains independent peer-reviewed research papers of ASHS members, including some research on turfgrasses.
(*formerly* Proceedings of the Society for Horticultural Science, 1905–1915).
(*see* Journal of the American Society for Horticultural Science, 1969+).
Published two times annually.

Proceedings of the American Society of Agronomy
American Society of Agronomy, Madison, Wisconsin, USA. 1910–1912.
(*see* Journal of the American Society of Agronomy, 1913–1948).
Published one time annually.

Proceedings of the North Central Weed Control Conference
North Central Weed Control Conference Inc., Champaign, Illinois, USA. 1944–1988.
No external peer review.
Contains research papers on weeds and their control, including some on turfgrasses.
(*see* Proceedings North Central Weed Science Society Conference, 1989–2000).
Published once annually.

Proceedings North Central Weed Science Society Conference
North Central Weed Science Society, Champaign, Illinois, USA. 1989–2000.
No external peer review.
(*formerly* Proceedings North Central Weed Control Conference, 1944–1988).
Published once annually.

Proceedings of the Northeastern Weed Control Conference
Rutgers University, New Brunswick, New Jersey, USA. 1947–1970.
No external peer review.
Contains research papers on weeds and their control, including some on turfgrasses.
(*see* Proceedings of the Northeastern Weed Science Society, 1971+).
Published once annually.

Proceedings of the Northeastern Weed Science Society
Northeastern Weed Science Society, Cornell Ornamentals Research Laboratory, Farmingdale, New York, USA. 1971+.
No external peer review.
Contains research papers on weeds and their control, including some on turfgrasses.
(*formerly* Proceedings of the Northeastern Weed Control Conference, 1948–1970).
Published once annually.

Proceedings of the Southern Weed Conference
Southern Weed Conference, Knoxville, Tennessee, USA. 1947–1968.
No external peer review.
(*see* Proceedings Southern Weed Science Society, 1969–1987).
Published once annually.

Proceedings Southern Weed Science Society
Southern Weed Science Society, Champaign, Illinois, USA. 1969–2002.

No external peer review.
(*formerly* Proceedings of the Southern Weed Conference, 1947–1968).
Published once annually.

Proceedings of the Western Society of Weed Science

Western Society of Weed Science, Newark, California/Las Cruces, New Mexico, USA. 1947+.
No external peer review.
Published once annually.

Proceedings of the Society for Horticultural Science

American Society of Horticultural Science, College Park, Maryland, USA. 1903–1915.
Contains independent peer-reviewed papers of ASHS members.
(*see* Proceedings of the American Society for Horticultural Science, 1916–1968).
Published once annually.

Rasen: Turf·Gazon or Rasen—Turf·Gazon (Lawn: Turf·Sod)

Hortus-Verlag GmbH, Bonn-Bad Godesberg, West Germany. 1970–1975.
Contains research and review papers on turfgrasses.
(*see* Rasen: Turf·Gazon Grünflächen Begrünungen, 1976–2007).
Published four times annually in the German language with an English abstract.

Rasen: Turf·Gazon Grünflächen Begrünungen (Lawns: Turf·Sod Green Spaces Greening, with Greenkeepers Journal Added in 1989)

Hortus-Zeitschriften, Cöllen & Bleeck (*up to 2004*), Köllen Druck & Verlag GmbH, Bonn, Germany. 1976–2007.
Contains primarily review and science-based papers on turfgrasses. Focused on conditions in Austria, Germany, and Switzerland.
(*formerly* Rasen: Turf·Gazon, 1970–1975).
Published four times annually.

Report of Green Keeping Research

Greens Research Committee of the New Zealand Golf Association and the New Zealand Institute for Turf Culture, Palmerston North, New Zealand. 1933, 1934, 1935, 1937, 1941, 1949, 1951, and 1957.
No external peer review.

Report of the Board of Greenkeeping Research

The Board of Greenkeeping Research, St Ives Research Station, Bingley, Yorkshire, England, UK. 1930–1938.
No external peer review.
Published once annually.

Research Report

Nishinihon Green Research Institute, Fukuoka, Wajiro, Japan. 1962–1968, 1970, 1971, 1973, 1976, 1979, 1983, 1984, and 1989–1996.
Published once annually in the Japanese language.

Soil Science

Rutgers, The State University of New Jersey, New Brunswick, New Jersey, USA. 1916+.
Contains independent peer-reviewed research and review papers on soil science.
Published six times annually.

Soil Science Society of America Proceedings

Soil Science Society of America, Madison, Wisconsin, USA. 1936–1975.
Contains independent peer-reviewed research papers of SSSA members.
(*see* Soil Science Society of America Journal, 1976+).
Published six times annually.

Soil Science Society of America Journal

Soil Science Society of America, Madison, Wisconsin, USA. 1976+.

Contains independent peer-reviewed research papers of SSSA members, including some research on turfgrasses.
(*formerly* Soil Science Society of America Proceedings, 1936–1975).
Published six times annually.

Weeds

The Association of Regional Weed Control Conferences, 1951–1955, and the Weed Science Society of America, 1956–1961. Urbana, Illinois, USA. 1962–1967.
Contains independent peer-reviewed research papers of WSSA members on weeds and their control, including research on turfgrasses.
(*see* Weed Science, 1968+).
Published four times annually.

Weed Science

Weed Science Society of America, Champaign, Illinois, USA. 1968+.
Contains independent peer-reviewed research papers of WSSA members on weeds and their control, including research on turfgrasses.
(*formerly* Weeds, 1951–1967).
Published four times annually.

Weed Technology

Weed Science Society of America, Champaign, Illinois/Lawrence, Kansas, USA. 1987+.
Published four times annually.

Weibulls Gräs-tips

W. Weibull AB, Landskrona, Sweden. 1958+, with volumes 10 and 11 published in 1968, 12–14 in 1971, 15 in 1972, 16 in 1973, 17 in 1974, 18 in 1975, 19 in 1976, 20 in 1977, 21 in 1978, 22 in 1979, and 23–25 in 1982.
No external peer review.
Contains turfgrass review and research papers of the Weibullsholm staff.
Published once annually in the English and Swedish languages.

Zeitschrift für Vegetationstechnik im Landschafts- und Sportstättenbau (Journal for Vegetative Technology in Landscaping and Sport Facility Construction)

Patzer-Verlag GmbH. u. Co. KG, Berlin, Germany. 1978–1990.
Contains some review and research papers on turfgrasses.
Published four times annually in the German language with English summaries.

Figure 14.1. A turfgrass species root characterization facility in a glasshouse at Texas A&M University, in the 1980s. [26]*

---

* The numbers in brackets correspond to the figure source acknowledgments listed in the appendix.

CHAPTER 15

Turfgrass Research Publications

THERE EXISTS A MAJOR GROUP OF PUBLICATIONS ORIENTED SPECIFICALLY TO THE SUBJECT AREA OF turfgrass research. A majority of these publications have been generated from the turfgrass research programs at the American state agricultural experiment stations associated with the state universities and, to a lesser extent, from turfgrass research groups within a few private companies. Most of these annual research publications take the form of annual progress reports, annual research summaries, field day reports, and periodic research reports. The primary objective is to provide ongoing summarizations of the research activities and accomplishments on an annual basis. Some of the publications have received critical peer-review by colleagues at the sponsoring institutions. The periodicals listed herein are based on three written requests to all organizations active in turfgrass research either now or in the past. The turfgrass research reports, research summaries, and field day reports typically have not been placed in libraries or even retained as sets by the originating organizations. Thus documentation for the publications as presented herein in terms of titles and years published may not be comprehensive.

These publications should be important holdings in the libraries of turfgrass researchers and educators. The turfgrass research leadership at these institutions where the effort was made to prepare these research reports should be complimented for their diligence in processing and summarizing an extensive array of research data into meaningful presentations that serve as a permanent record for others to use.

A. Annual Turfgrass Research Progress Reports and Summaries

The first grouping described as the annual research progress reports and summaries has been published under the auspices of national organizations and state universities, primarily in the United States. These reports provide an excellent summary of the ongoing turfgrass research being conducted at these institutions at that time. Eight institutions initiated a series of research summaries in the 1960s. The first two of this type were published by Michigan State University and the University of Arizona in 1962, followed by Ohio State University and Oklahoma State University in 1965, by Rhode Island University and Rutgers University in 1966, by Virginia Tech in 1967, and by the University of Missouri in 1969. Subsequently, a series of annual research reports were initiated in the 1970s by three state universities—Illinois in 1971, Kansas State in 1975, and Nebraska in 1979 plus the USDA-Beltsville in 1976.

B. Turfgrass Research Field Day Reports

The turfgrass research field days held at state universities that conduct major field plot research programs provide an opportunity for the professional turfgrass manager to observe the ongoing research efforts on a regular basis. Especially significant is the opportunity to observe the newly developed turfgrass cultivars, fertilizers, pesticides, and cultural practices. In conjunction with this activity, a number of institutions publish a turfgrass field day report. This second section lists those turfgrass field day reports that are published on a regular basis and are sufficiently extensive to contain detailed results and conclusions for the particular year being reported. The six institutions that have published a series of field day reports prior to 1970 are summarized in this section. Included are Rhode Island in 1930, Virginia Tech in 1958, Ohio State and Oregon State in 1959, Washington State in 1961, and Michigan State in 1962.

C. Turfgrass Research Conference and Symposium Publications

The turfgrass research conferences and symposium publications are listed in the third section. Of international significance are the proceedings of the International Turfgrass Research Conference, which has been held at four-year intervals since 1969. Within the United States, there are the proceedings of the O.M. Scott & Sons Turfgrass Research Conference that consisted of a six-volume set published between 1968 and 1978 and the four proceedings of the Turfgrass Symposia sponsored by the ChemLawn Corporation that were held between 1980 and 1986.

D. Roadside Turfgrass Research Publications

The fourth grouping includes roadside research publications that have been published on an individual basis. This major thrust in roadside establishment research during the 1960s and 1970s in the United States was associated with the construction of the interstate highway network and funded primarily through the United States Public Bureau of Roads. As a result, a significant body of published work on turfgrass selection, site preparation, establishment, and culture for roadsides evolved. Most of the publications appeared between 1953 and 1977. Major state programs supported by the United States Public Bureau of Roads included: Michigan, Minnesota, New York, North Carolina, and Oklahoma.

E. Individual Major Turfgrass Research Publications

The fifth grouping includes individual turfgrass research publications, most of more than 20 pages in length, that are a one-time effort and typically summarize a range of experiments oriented to a major problem area. Of particular historical note are a series of research reports published by the experiment stations in Rhode Island in 1917, 1928, 1929, 1934, 1938, 1939, and 1944; in New Jersey in 1930, 1934, 1937, and 1946; and in Ohio in 1941, 1943, and 1949.

FORMAT

Publications in this category include (1) annual and biennial reports and (2) one-time published reports of turfgrass research that are over 20 pages in length, plus those of historical significance.

The citation style used in this chapter is as follows:
- Name of conference.
- Name and location of sponsor—city, state, and country.
- Years published; the first year listed is when the proceedings was first published, while the second year listed is when publication of the proceedings was terminated; if the proceedings continues to be published this is indicated by a "+."
- Number of times published annually—for the most recent year of publication.
- The notation "peer reviewed" indicates the paper has been assessed for sound scientific (a) research procedures, (b) data statistical analysis, and (c) conclusions by multiple, anonymous scholars and researchers who possess expertise in the particular area of investigation.
- Previous or subsequent names and dates published for the same proceedings are listed in parentheses.

A. ANNUAL TURFGRASS RESEARCH PROGRESS REPORTS AND SUMMARIES

Alberta Turfgrass Research Foundation Annual Report
Alberta Turfgrass Research Foundation, Olds College, Olds, Alberta, Canada. 1989.
(*see* Prairie Turfgrass Research Centre Annual Report, 1990+).

Annual Report Purdue University Turfgrass Science Program
Department of Agronomy, Purdue University, West Lafayette, Indiana, USA. 1997+.
(*see* Purdue University Turfgrass Research Summary, 1988–1996).

Annual Research Summary
Turfgrass Research Advisory Committee, University of California at Riverside, Riverside, California, USA. 1997–2000.

Annual Report on Green Keeping Research
Greens Research Committee of the New Zealand Golf Association, Palmerston North, New Zealand. 1934, 1935, 1937, and 1941.
(*formerly* New Zealand Annual Report on Green Keeping Research, 1933).
(*see* Report on Green Keeping Research, 1951 and 1957).

Applied Research, Demonstration and Extension Programs in Small Grains, Forage Crops, Peanuts, and Turf
Department of Entomology, North Carolina State University, Raleigh, North Carolina, USA. 1989–1990.
(*see* Research and Extension Programs, Peanut, and Turfgrass Management, 1991–1999).

Arkansas Turfgrass Report
University of Arkansas, Fayetteville, Arkansas, USA. 2007+.
Produced in an electronic format.

Berichte über die Jahrestagungen des Förderkreises "Landschafts- und Sportplatzbauliche Forschung Gießen"
(Reports on the Annual Meetings of the Promotion Association "Landscape and Sportfield Research at Giessen")
Institut für Bodenkunde und Bodenerhaltung-Landschaftsbau, Giessen, West Germany. 1981+.

The Board of Greenkeeping Research Report
St Ives Research Station, Bingley, Yorkshire, England, UK. 1931–1939.

Breeding and Evaluation of Kentucky Bluegrass and Associated Species for Turf
NE-57 and 139 Project Research Report—Preliminary, Plant Genetics and Germplasm Institute, United States Department of Agriculture, Beltsville, Maryland, USA. 1979–1981.

Chemical Control of Diseases Affecting Turf—Progress Report
Oklahoma State University, Stillwater, Oklahoma, USA. Circa 1965–1970.

Clemson University Turfgrass Program
Clemson University Public Service Activities SB671, Clemson, South Carolina, USA. 1996–2004.

Figure 15.1. Freeze stress simulator for turfgrasses as originally developed at Michigan State University, early 1960s. [26]*

Cornell Turfgrass Research Report

Cornell University, Ithaca, New York, USA. 1986–1998.

Disease Control Practices for Turfgrasses in Oklahoma Preliminary and Progress Reports

Oklahoma State University, Stillwater, Oklahoma, USA. 1970–1981.

Evaluation of Some Turfgrass Fungicides

Noel Jackson et al., Department of Plant Pathology-Entomology, University of Rhode Island, Kingston, Rhode Island, USA. 1966–1972 and 1975.

Florida Turfgrass Research

Fort Lauderdale Research and Education Center, University of Florida, Fort Lauderdale, Florida, USA. 2002–2004.

Golf Course Superintendents Report

Center for Golf Course Management, Golf Course Superintendents Association of America, Lawrence, Kansas, USA. 1990–1998.

GTI Annual Research Report

The Guelph Turfgrass Institute, Guelph, Ontario, Canada. 1986–2007.
(*formerly* Turfgrass Research Annual Report, 1981–1985).

Horticulture Studies Research Series

University of Arkansas, Fayetteville, Arkansas, USA. 1998–2003.
(*see* Arkansas Turfgrass Report, 2007+).

Illinois Turfgrass Research Summary or Report

Department of Horticulture, University of Illinois, Urbana, Illinois, USA. 1973–1983, titled as a summary, and 1984–1995, titled as a report.
Produced in an electronic format since 1995.

Iowa Turfgrass Research Report

Department of Horticulture, Iowa State University, Ames, Iowa, USA. 1984–2003.
Since 2004 has been available only on the web in electronic form.
(*formerly* Iowa Turfgrass Field Day and Equipment Show, 1981–1983).

Kansas Turfgrass Research Report or K-State Turfgrass Research

Kansas State University and Kansas Turfgrass Foundation, Manhattan, Kansas, USA. 1975+.
Produced in an electronic format since 2003.

---

* The numbers in brackets correspond to the figure source acknowledgments listed in the appendix.

Kentucky Turfgrass Research
   University of Kentucky, Lexington, Kentucky, USA. 1980–2001, 2003, and 2005.
   Produced electronically on CD since 2002.

Missouri Turfgrass Research Report
   University of Missouri Agricultural Experiment Station and Missouri Valley Turfgrass Association, Columbia, Missouri, USA. 1969–
      1995, 1998–2006, and 2009+.
   Produced in an electronic format since 2006.

MSU Turfgrass Research Reports
   Department of Crop and Soil Sciences, Michigan State University, East Lansing, Michigan, USA. 2005+.
   Produced in an electronic format.

National Turfgrass Evaluation Program Reports
   National Turfgrass Evaluation Program (NTEP), Beltsville, Maryland, USA. 1980+.
   Progress and final reports published periodically.

New Zealand Golf Green Research Committee Leaflet
   W.A. Kiely, Palmerston North, New Zealand. 1931–1932, with two leaflets annually.

New Zealand Annual Report on Green Keeping Research
   E.B. Levy, New Zealand Greenkeeping Research Committee, Palmerston North, New Zealand. 1933.
   (*see* Annual Report on Green Keeping Research, 1934, 1935, 1937, and 1940).

Pacific Turfgrass Research Program Annual Report
   The University of British Columbia, Vancouver, British Columbia, Canada. 1994–1995.

Peanut and Turfgrass Research
   Department of Entomology, North Carolina State University, Raleigh, North Carolina, USA. 2000.
   (*formerly* Research and Extension Programs, Peanut, and Turfgrass Management, 1991–1999).

Poa annua *Stress Investigations Annual Report*
   University of Minnesota, St. Paul, Minnesota, USA. 1988.

Prairie Turfgrass Research Centre Annual Report
   Alberta Turfgrass Research Foundation, Olds College, Olds, Alberta, Canada. 1990+.
   (*formerly* Alberta Turfgrass Research Foundation Annual Report, 1989).
   Also produced in an electronic format.

Purdue University Turfgrass Research Summary
   Agronomy Department, Purdue University, West Lafayette, Indiana, USA. 1988–1996
   (*formerly* Annual Report Purdue University Turfgrass Science Program, 1997+).

Rapport Annuel: CRH Secteur Gazons, Golfs, & Gazonnières
   Centre de Recherche en Horticulture, Université Laval, Québec, Canada. 1995 and 1997.

Regional Tall-Fescue Test Preliminary Report
   Southern Turfgrass Worker Group and Plant Genetics and Germplasm Institute, United States Department of Agriculture, Beltsville,
      Maryland, USA. 1980–1983.

Report of Turfgrass Research
   Florida Turfgrass Association and University of Florida, Gainesville, Florida, USA. 1980–1985.

Report on Greenkeeping Research
   New Zealand Institute of Turf Culture, Palmerston North, New Zealand. Circa 1949 and circa 1958.
   (*formerly* Annual Report on New Zealand Green Keeping Research, 1934, 1935, 1937, and 1940).

Report on Turfgrass Research

    Rutgers University, New Brunswick, New Jersey, USA. 1966–1969.

    (*see* Rutgers Turfgrass Proceedings, 1973+.).

Report on Turfgrass Research

    Turfgrass Research Committee, University of Arizona, Tucson, Arizona, USA. 1962–1970, with 1967 and 1968 published as one report.

    (*see also* Turfgrass and Ornamentals Research Summary, 1990+).

Research and Development Yearbook

    Scandinavian Turfgrass and Environmental Research Foundation, Danderyd, Sweden. 2006+.

Research and Extension Programs, Peanut, and Turfgrass Management

    Department of Entomology, North Carolina State University, Raleigh, North Carolina, USA. 1991–1999.

    (*formerly* Applied Research, Demonstration, Extension Programs in Small Grains, Forage Crops, Peanuts, and Turf, 1989–1991).

    (*see* Peanut and Turfgrass Research, 2000).

Research Projects

    Turfgrass Technology Pty. Ltd., Frankston, Victoria, Australia. 1989–1992.

Research Report

    Western Canada Turfgrass Association, Maple Ridge, British Columbia, Canada. 1989+.

Results of Research Program for Control of Weeds and Diseases of Turfgrasses and Ornamentals

    Department of Plant Pathology and Physiology, Virginia Polytechnic Institute and State University, Blacksburg, Virginia, USA. 1967–1982.

Rutgers Turfgrass Proceedings

    Rutgers University, New Brunswick, New Jersey, USA. 1973+

    (*formerly* Report on Turfgrass Research, 1966–1969).

Rutgers Turfgrass Symposium

    Rutgers University, New Brunswick, New Jersey, USA. 1991+.

    Published once annually.

Summary of Turfgrass, Forage, & Noncropland Management Research

    Crop and Soil Sciences Department, University of Georgia, Griffin, Georgia, USA. 2003–2006.

    Produced in an electronic format.

Texas Turfgrass Research Consolidated Progress Report

    Texas Agricultural Experiment Station, Texas A&M University, College Station, Texas, USA. 1976–1996.

Turf and Landscape Research Summary or Turfgrass and Ornamentals Research Summary

    Ohio Agricultural Research and Development Center, Wooster, Ohio, USA. 1965, 1968–1983.

Turf and Ornamental Weed Science Research Report

    W.L. Currey, University of Florida, Gainesville, Florida, USA. 1984.

Turf Experiments in Progress

    D.G. Sturkie, Crops, Auburn University, Auburn, Alabama, USA. 1966.

Turfgrass and Horticultural Trials Report

    Ronald C. Smith, North Dakota State University, Extension Service Center, Fargo, North Dakota, USA. 1990–2002.

Turfgrass and Ornamentals Research Summary

    Cooperative Extension and Agricultural Experiment Station, University of Arizona, Tucson, Arizona, USA. 1988–1997.

Figure 15.2. Evapotranspiration rate and modeling simulator for turfgrasses, as originally developed in the late 1970s at Texas A&M University. [26]

(*formerly* Report on Turfgrass Research, 1962–1970).
(*see* Turfgrass, Landscape, and Urban IPM Research Summary, 1988–1989, 1991, 1997, 2002, and 2006).

Turfgrass Cultivar Evaluations
Ohio State University, Columbus, Ohio, USA. 1979–1980.

Turfgrass Disease Control Trials
Houston B. Couch et al., Virginia Polytechnic Institute and State University, Blacksburg, Virginia, USA. 1985–1987 and 1989.

Turfgrass Evaluations
University of Missouri, Columbia, Missouri, USA. 1983–1984, 1987–1991, 1993, 1995–1997, and 1999–2002.

Turfgrass Investigations
Robert W. Miller, Ohio State University, Columbus, Ohio, USA. 1973.

Turfgrass, Landscape, and Urban IPM Research Summary
College of Agriculture & Life Sciences, University of Arizona, Tucson, Arizona, USA. 1988–1989, 1991, 1997, 2002, 2006, and 2009.
Produced in an electronic format since 2001.
Published at one- to three-year intervals.

Turfgrass Management Annual Research Summary Report
University of Tennessee Agricultural Experiment Station, Knoxville, Tennessee, USA. 1985–1988.

Turfgrass Pathology Research Report, University of Georgia
Department of Plant Pathology, Georgia Experiment Station, Griffin, Georgia, USA. 1990–2000.

Turfgrass Pathology Summary
Texas A&M University Research Center at Dallas, Plano, Texas, USA. 1998–2005.

Turfgrass Pathology, Weed Science, and Physiology Research Summaries
Department of Plant Science and Landscape Architecture, University of Maryland, College Park, Maryland, USA. 2005+.

Turfgrass Production and Management-Research Progress Report
Department of Agronomy, Oklahoma State University, Stillwater, Oklahoma, USA. Circa 1967–1975.

Turfgrass Program Summary
    University of Georgia, Griffin, Georgia, USA. 1995–1997.

Turfgrass Research
    Agricultural Experiment Station, Colorado State University, Fort Collins, Colorado, USA. 1958.

Turfgrass Research Annual Report
    University of Guelph, Guelph, Ontario, Canada. 1981–1985.
    (*see* GTI Annual Research Report, 1988–2007).

Turfgrass Research in Florida: A Technical Report
    Institute of Food and Agricultural Sciences, University of Florida, Gainesville, Florida, USA. 1982–1997.

Turfgrass Research—Progress Report
    Oklahoma State University, Stillwater, Oklahoma, USA. Circa 1965–1967.
    (*see* Turfgrass Production and Management, 1967–1975).

Turfgrass Research Report
    Center for Turfgrass Science, Pennsylvania University, University Park, Pennsylvania, USA. 1989, 1990, 1991, 1992, and 1998+.

Turfgrass Research Report
    Department of Horticulture, Kansas State University, Manhattan, Kansas, USA. 1976+.
    Produced in an electronic format since 2003.

Turfgrass Research Report
    Department of Plant Science, University of Connecticut, Storrs, Connecticut, USA. 2005+.

Turfgrass Research Report
    Field Crops Laboratory, United States Department of Agriculture, Beltsville, Maryland, USA. 1972–1978.

Turfgrass Research Report
    Ohio State University, Columbus, Ohio, USA. 1993, 1995, and 1998.

Turfgrass Research Report
    Soils and Crops Department, Rutgers University, New Brunswick, New Jersey, USA. 1971.

Turfgrass Research Reports
    Auburn University, Auburn, Alabama, USA. 1972–1973.

Turfgrass Research Report or Turfgrass Research & Information Report
    University of Missouri, Columbia, Missouri, USA. 1986+.

Turfgrass Research Report or Turfgrass Research Summary
    University of Nebraska, Lincoln, Nebraska, USA. 1978–1989, and 1991–2002.
    Produced in an electronic format since 1998.

Turfgrass Research Reports, Southern Turfgrass Work Group
    United States Department of Agriculture, Agricultural Research Service, Beltsville, Maryland, USA. 1976.

Turfgrass Research Summary
    Department of Crop and Soil Sciences, Michigan State University, East Lansing, Michigan, USA. 1962–1975.

Turfgrass Research Summary or Report
    Department of Horticulture, University of Illinois, Urbana, Illinois, USA. 1971–2003.

Figure 15.3. Photosynthesis assessment via turfgrass canopy gaseous exchange with temperature control under high irradiation at Michigan State University in the early 1970s. [26]

Turfgrass Research Summary
South Dakota State University, Brookings, South Dakota, USA. 1998–2000.

Turfgrass Trials
Ornamental Section, Agriculture Canada, Ottawa, Ontario, Canada. 1980–1981.

Turfgrass Variety and Experimental Selection Trials Report
Field Crops Laboratory, United States Department of Agriculture, Beltsville, Maryland, USA. 1979–19xx. Published in odd-numbered years.

Turfgrass Variety Research Report
University of California at Riverside, Riverside, California, USA. 1975.

Turfgrass Variety Trial Results
University of Maryland, College Park, Maryland, USA. 1972.

Turfgrass Variety Trials
Department of Horticultural Science, University of Guelph, Guelph, Ontario, Canada. 1968–1971.

Turfgrass Variety Trials or Turfgrass Trials
Agassiz Research Station, Canada Agriculture, British Columbia, Canada. 1970–1975.

Turf Research and Advisory Institute Report
Department of Agriculture, Victoria, Australia. 1973–1978.

Turf Research in Ontario
Ontario Turfgrass Research Foundation, Guelph, Ontario, Canada. 2007+.

Turf Research Review
United States Golf Association Green Section, Beltsville, Maryland, USA. 1950, 1951, and 1956.

Turf Variety Trials
Western Washington Research and Extension Center, Washington State University, Puyallup, Washington, USA. 1975.

University of California Turfgrass Research Releases
Turfgrass Workgroup, University of California, Riverside, California, USA. 1994.

University of Connecticut Annual Turfgrass Research Report
College of Agriculture and Natural Resources, University of Connecticut, Storrs, Connecticut, USA. 2005+.
Produced in an electronic format.

URI Turfgrass Research Review or Turfgrass Research Review
Rhode Island Agricultural Experiment Station, University of Rhode Island, Kingston, Rhode Island, USA. 1975–1990.
Published one to four issues annually.

USGA Annual Turfgrass Research Report
United States Golf Association, Far Hills, New Jersey, USA. 1985–1987.

USGA Environment Progress Report
United States Golf Association, Far Hills, New Jersey, USA. 1991–1996.
(*see* USGA Turfgrass and Environmental Progress Report, 1997–2005).

USGA Environmental Research Summary
United States Golf Association, Far Hills, New Jersey, USA. 1991–1992.
(*see* USGA Turfgrass and Environmental Research Summary, 1995+).

USGA Summary of Research Reports
United States Golf Association, Far Hills, New Jersey, USA. 1977 and 1983–1984.

USGA Turfgrass and Environmental Progress Reports
United States Golf Association, Far Hills, New Jersey, USA. 1997–2005.

USGA Turfgrass and Environmental Research Executive Summary
United States Golf Association, Far Hills, New Jersey, USA. 2001+.

USGA Turfgrass and Environmental Research Online
United States Golf Association, Far Hills, New Jersey, USA. 2002+.
Produced in an electronic format; two articles per month.

USGA Turfgrass and Environmental Research Summary
United States Golf Association, Far Hills, New Jersey, USA. 1995+.
(*combined* from USGA Turfgrass Research Summary, 1988–1994, and USGA Environmental Research Summary, 1991–1992).

USGA Turfgrass Progress Reports
United States Golf Association Green Section, Far Hills, New Jersey, USA. 1977 and 1983–1996.
(*see* USGA Turfgrass and Environmental Progress Report, 1997–2005).

USGA Turfgrass Research Summary
United States Golf Association, Far Hills, New Jersey, USA. 1983–1994.
(*see* USGA Turfgrass and Environmental Research Summary, 1995+).

Weed Control Investigations in Turfgrasses
Division of Agriculture, Forestry, and Veterinary Medicine Information Bulletin, Mississippi State University, Mississippi State, Mississippi, USA. 1989 and 1993–1998.

Wisconsin Turf Research (name changed in 1993 to Wisconsin Turfgrass Research Reports)
University of Wisconsin, Madison, Wisconsin, USA. 1983+.
Produced in an electronic format since 2000.

B. TURFGRASS RESEARCH FIELD DAY REPORTS

Arkansas Field Day Report
University of Arkansas, Fayetteville, Arkansas, USA. 2001, 2003, 2005, 2007, and 2008+.

Clemson University Turfgrass Field Day
Clemson University, Pendleton, South Carolina, USA. 1974+.

Cornell University Turfgrass Field Day
Cornell University, Ithaca, New York, USA. 1978, 1980–1983, 1985, 1998, and 2001.

Fall Field Day on Turf Culture or Field Day on Turf Culture
University of California at Los Angeles, Westwood, California, USA. 1949–1950.
(*see* Southern California Conference on Turf Culture, 1950–1951).

Golf and Grounds Field Day
Sandhills Research Station, North Carolina State University, Raleigh, North Carolina, USA. 1998.

IFAS Turfgrass Field Day or UF-IFAS Turf Field Days
Research and Education Center, Fort Lauderdale, Florida, USA. 1979 and 2003.

International Warm Season Turfgrass Research Tour
Seeds West Inc., Maricopa, Arizona, USA. 1996–2000, 2002, and 2004.

Iowa Turfgrass Field Day and Equipment Show
Iowa State University, Ames, Iowa, USA. 1981–1983.
(*see* Iowa Turfgrass Research Report, 1984–2003).

ISHS Orchard Day and Turf Field Day
Southern Illinois University, Carbondale, Illinois, USA. 2000.

Figure 15.4. Turfgrass Research field day on bentgrass green at Michigan State University in the early 1960s. [26]

Lawn and Ornamentals Days
  Ohio Agricultural Experiment Station, Wooster, Ohio, USA. 1959–1964.
  (*see* Lawn and Ornamentals Research, 1965–1971).

Lawn and Ornamentals Research or Turf and Landscape Research
  Ohio Agricultural Research and Development Center, Wooster, Ohio, USA. 1965–1971.
  (*formerly* Lawn and Ornamentals Days, 1959–1964).
  (*see* The Ohio State University Turfgrass Field Day or Turfgrass Research Field Day, 1974, 1975, and 1986–1993).

Lofts Research Report and Field Day or Lofts Research Results
  Lofts Seed Inc., Bound Brook, New Jersey, USA. Circa 1975–1994.

Maryland Turfgrass Research Field Day
  University of Maryland, College Park, Maryland, USA. Published even numbered years from 1982–2002.

Michigan State University Sod Producers Field Day Report
  Department of Crop and Soil Sciences, Michigan State University, East Lansing, Michigan, USA. 1969 and 1971.

Michigan State University Turfgrass Field Day Report
  Department of Crop and Soil Sciences, Michigan State University, East Lansing, Michigan, USA. 1962–1965 then published in odd-numbered years from 1981 and alternating with the Northern Michigan Turfgrass Field Day Report until 1980 and, subsequently, published annually as the Michigan Turfgrass Field Day Report since 1981.

Michigan Upper Peninsula Turfgrass Field Day Report
  Department of Crop and Soil Sciences, Michigan State University, Iron Mountain/East Lansing, Michigan, USA. 1969.

Minnesota Turf and Grounds Field Day Report
  University of Minnesota, St. Paul, Minnesota, USA. 2000+.
  Became available in an electronic format in 2000.

National Turf Field Day
  United States Golf Association Green Section and United States Department of Agriculture, Beltsville, Maryland, USA. 1948–1951.

North Carolina Turfgrass and Landscape Field Day Tour Summaries Guide
  North Carolina State University, Raleigh, North Carolina, USA. 1981–2003.
  (*see* North Carolina Turfgrass Field Day Guide, 2004+).

North Carolina Turfgrass Field Day Tour Summaries or Guide
  North Carolina State University, Raleigh, North Carolina, USA. 2004+.

Northern Michigan Turfgrass Field Day Report
  Department of Crop and Soil Sciences, Michigan State University, Traverse City/East Lansing, Michigan, USA. 1964–1966 and in alternate years from 1968–1980.

The Ohio State University Turfgrass Field Day or Turfgrass Research Field Day
  Department of Agronomy, Ohio State University, Columbus, Ohio, USA. 1974, 1975, 1984, 1986–1993, 1997, 2001, 2006, and 2007.
  (*formerly* Lawn and Ornamentals Research and Turf and Landscape Research, 1965–1971).

OSU Field Day
  Oregon State University, Corvallis, Oregon, USA. 1959–1995 and then at two-year intervals combined with the Washington State University (Puyallup) from 1997–2005 and at three-year intervals combined with the Oregon State University and Washington State University (Pullman) Field Days since 2008.

Program and Research Summaries of the University of Connecticut Turfgrass Field Day
  University of Connecticut, Storrs, Connecticut, USA. 2008 and 2010.

Rhode Island Turfgrass Field Day
University of Rhode Island, Kingston, Rhode Island, USA. 1930+.

Rocky Mountain Regional Turfgrass Field Day
Department of Horticulture, Colorado State University, Fort Collins, Colorado, USA. 1979–1980. (*see* Turf and Horticulture Field Day, 1988–2000).

Rutgers Turfgrass Research Field Day II, Disease Trials
Rutgers University, New Brunswick, New Jersey, USA. 1981+.

Turf and Horticulture Field Day
Horticulture Department, Colorado State University, Fort Collins, Colorado, USA. 1988–2000.

Turfgrass & Landscape Management Field Day
University of California at Riverside, Riverside, California, USA. 2001–2002 and 2008+.

Turfgrass and Landscape Field Day or Turf & Nursery Field Day
University of Illinois, Urbana, Illinois, USA. 2002, 2003, and 2005.

Turfgrass & Landscape Research Field Day or Turf and Landscape Day
Ohio Agricultural Research and Development Center, Wooster, Ohio, USA. 1974, 1975, and 1986–1993.

Turfgrass Field Day
Department of Horticulture, University of Illinois, Urbana, Illinois USA. 1972, 1973, 1975, 1976, 1978, 1980, 1982, and 1985–2001. (*see* Turfgrass and Landscape Field Day, 2002, 2003, and 2005).

Turfgrass Field Day
Washington State University, Puyallup, Washington, USA. 1960–1996 and then at two-year intervals combined with Oregon State University in 1997–2005 and at three-year intervals combined with Oregon State University and Washington State University (Pullman) Field Days since 2008.

Turfgrass Field Day for Homeowners
University of Wisconsin at Madison, Verona, Wisconsin, USA. 1996–1997.

Turfgrass Field Day or University of Georgia Turfgrass Field Day
University of Georgia at Griffin, Griffin, Georgia, USA. 1984, 1986, 1988, 1991, 1993, 1995, 1997, 1998, and 2004+. Published in even-numbered years.

Turfgrass Research Conference and Field Day
University of California at Riverside, Riverside, California, USA. 1985–2000. (*see* Turfgrass & Landscape Management Field Day, 2001, 2002, and 2008+).

Turfgrass Research Field Day
Department of Horticulture, University of Illinois, Urbana, Illinois, USA. 1983–2007.

Turfgrass Research Progress Report for Turf and Landscape Research Field Day
University of California Cooperative Extension, Santa Clara, California, USA. 1985–1999.

Turf-Seed and Pure-Seed Testing Turf Field Day
Turf-Seed Inc. and Pure Seed Testing Inc., Hubbard, Oregon/Rolesville, North Carolina, USA. 1982+. Located in alternate years at two sites starting in 1997.

University of Georgia Turfgrass and Landscape Ornamental Field Day
Georgia Agricultural Experiment Station at Griffin, Griffin, Georgia, USA. 1983, 1986, 1988, 1991, 1993, 1995, 1997, and 1998. (*See* University of Georgia Turfgrass Field Day, 2000+).

Figure 15.5. Chromotography studies of heat stress metabolism by Tom Duff at Michigan State University in the mid-1960s. [26]

University of Georgia Turfgrass Field Day
University of Georgia, Cooperative Extension Service, Griffin, Georgia, USA. 2000, 2002, 2004, 2006, 2008, and 2010+.

University of Kentucky Turf Research Field Day Report
University of Kentucky, Lexington, Kentucky, USA. 2006+.

University of Massachusetts Turf Research Field Day: Research Reports
University of Massachusetts, South Deerfield, Massachusetts, USA. 1974+.

University of Minnesota—Turf and Grounds Field Day
Agriculture Experiment Station, St. Paul, Minnesota, USA. 2001+.

University of Nebraska Turfgrass Field Day Report
Department of Horticulture, University of Nebraska, Mead, Nebraska, USA. 1976–1989.

USDA Turfgrass Research Field Day
Field Crops Laboratory, United States Department of Agriculture, Beltsville, Maryland, USA. 1950 and 1977–1981. Published in odd-numbered years.

Virginia Tech Turfgrass Field Days, Research Summary
Virginia Polytechnic and State University, Blacksburg, Virginia, USA. 1958+.

Wisconsin and Minnesota Snow Mold Field Days
Department of Plant Pathology, University of Wisconsin at Madison, Verona, Wisconsin, USA. 2005+.
(*formerly* Wisconsin Snow Mold Field Day, 2000–2004).
Produced in both hard and electronic formats.

Wisconsin Snow Mold Field Day
Department of Plant Pathology, University of Wisconsin at Madison, Verona, Wisconsin, USA. 2000–2004.
(*see* Wisconsin and Minnesota Snow Mold Field Day, 2005+).
Produced in both hard copy and electronic formats.

Wisconsin Turfgrass Field Day
University of Wisconsin at Madison, Verona, Wisconsin, USA. 1993+.

C. TURFGRASS RESEARCH CONFERENCE AND SYMPOSIUM PUBLICATIONS

ChemLawn Turfgrass Symposium Proceedings

Advances in Turfgrass Pathology
1980. P.O. Larsen and B.G. Joyner (editors).
Proceedings of a Symposium on Turfgrass Diseases of 23 papers. ChemLawn Corporation, Columbus, Ohio, USA. 1:197.

Advances in Turfgrass Entomology
1982. H.D. Niemczyk and B.G. Joyner (editors).
Proceedings of a Symposium on Turfgrass Insects of 26 papers. ChemLawn Corporation, Columbus, Ohio, USA. 2:149.

Advances in Turfgrass Weed Control
1986. B.G. Joyner and K.A. Hurto (editors).
Proceedings of a Symposium on Turfgrass Weeds of 12 papers. ChemLawn Corporation, Columbus, Ohio, USA. 3:154.

Advances in Turfgrass Fertility
1986. B.G. Joyner (editor).
Proceedings of a Symposium on Turfgrass Fertility of 12 papers. ChemLawn Corporation, Columbus, Ohio, USA. 4:199.

International Turfgrass Research Conference (ITRC) Proceedings or Journal

Most of the quadrennial proceedings and journals also have a separate supplement published that addresses the organization activities of the International Turfgrass Society.

Proceedings of the First International Turfgrass Research Conference
1970. J.R. Escritt (editor).
International Turfgrass Society & Sports Turf Research Institute, Bingley, West Yorkshire, England, UK. 1:610. 99 papers.

Proceedings of the Second International Turfgrass Research Conference
1974. E.C. Roberts (editor).
International Turfgrass Society, American Society of Agronomy, and Crop Science Society of America, Madison, Wisconsin, USA. 2:602. 71 papers.

Proceedings of the Third International Turfgrass Research Conference
1980. J. B Beard (editor).
International Turfgrass Society, American Society of Agronomy, and Crop Science Society of America, Madison, Wisconsin, USA. 3:530. 59 papers.

Proceedings of the Fourth International Turfgrass Research Conference
1981. R.W. Sheard (editor).
International Turfgrass Society and Ontario Agricultural College, University of Guelph, Guelph, Canada. 4:564. 62 papers.

Proceedings of Fifth International Turfgrass Research Conference
1985. F. Lemaire (editor).
International Turfgrass Society and Institut National de la Recherche Agronomique, Paris, France. 5:870. 84 papers.

Proceedings of the Sixth International Turfgrass Research Conference
1989. Hiroshi Takato (editor).
International Turfgrass Society and Japanese Society of Turfgrass Science, Tokyo, Japan. 6:458. 98 papers.

International Turfgrass Society Research Journal
    1993. R.N. Carrow, N.E. Christians, and R.C. Shearman (editors).
    International Turfgrass Society, Minneapolis, Minnesota, USA. 7:1021. 149 papers.

International Turfgrass Society Research Journal, Part 1 and 2
    1997. P.M. Martin and A.E. Baumann (editors).
    International Turfgrass Society, Sydney, New South Wales, Australia. 8(1):1145 and 8(2):304. 144 papers.

International Turfgrass Society Research Journal, Part 1 and 2
    2001. K. Carey (editor).
    International Turfgrass Society, Toronto, Ontario, Canada. 9(1):471 and 9(2):645. 164 papers.

International Turfgrass Society Research Journal, Part 1 and 2
    2005. Daniel Thorogood (editor).
    International Turfgrass Society, Aberystwyth, Wales, UK. 10(1):685 and 10(2):579. 174 papers.

International Turfgrass Society Annexe Technical Papers
    International Turfgrass Society, Aberystwyth, Wales, UK. 109 pages.
    These are non-peer-reviewed paper summaries.

International Turfgrass Society Research Journal, Part 1 and 2
    2009. Gwen Stahnke (editor).
    International Turgrass Society, Santiago, Chile. 11(1):730 and 11(2):555, 62 papers and 49 papers, respectively.

International Turfgrass Society Supplement Technical Papers
    International Turfgrass Society, Santiago, Chile. 70 pages.
    Includes 27 non-peer-reviewed paper summaries.

O.M. Scott & Sons Turfgrass Research Conference Proceedings

Notes of Progress, Scotts Turfgrass Research Seminar
    1968.
    O.M. Scott & Sons Co., Marysville, Ohio, USA. 95 pages.

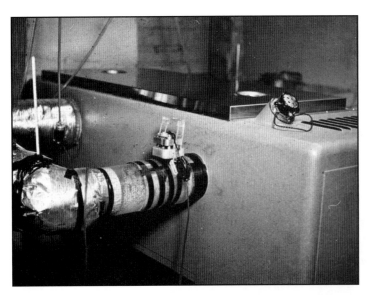 

Figure 15.6. Heat stress stimulator for whole turfgrass plant exposure (left) and close-up of plant exposure apparatus (right), developed in the mid-1960s at Michigan State University. [26]

Figure 15.7. Turfgrass microclimate apparatus and study at Texas A&M University, in the 1980s. [26]

Entomology

1969. Herbert T. Streu and Richard T. Bangs (editors).
Proceedings of Scotts Turfgrass Research Conference of 9 papers. O.M. Scott & Sons Co., Marysville, Ohio, USA. l:89.

Turfgrass Diseases

1971. Philip M. Halesky, Richard T. Bangs, and William L. Schroederer (editors).
Proceedings of Scotts Turfgrass Research Conference of 8 papers. O.M. Scott & Sons Co., Marysville, Ohio, USA. 2:99.

Weed Controls

1972. Richard E. Schmidt, Jack D. Butler, Delbert D. Hemphill, and Richard T. Bangs (editors).
Proceedings of Scotts Turfgrass Research Conference of 10 papers. O.M. Scott & Sons Co., Marysville, Ohio, USA. 3:173.

Turfgrass Breeding

1973. William R. Kneebone, William H. Daniel, Kenyon T. Payne, Richard T. Bangs, and William T. Schroederer (editors).
Proceedings of Scotts Turfgrass Research Conference of 16 papers. O.M. Scott & Sons Co., Marysville, Ohio, USA. 4:162.

Turfgrass Management

1978. Coleman Y. Ward, John R. Hall, A.J. Turgeon, A.E. Dudeck, and Richard T. Bangs (editors).
Proceedings of Scotts Turfgrass Research Conference of 10 papers. O.M. Scott & Sons Co., Marysville, Ohio, USA. 5:114.

Other Individual Turfgrass Research Conference and Symposium Publications

Amenity Grasslands, the Needs for Research

1977.
National Environmental Research Council, London, England, UK. 68 pages.

Biotechnology in Forage and Turf Grass Improvement

1998. G. Spangenberg, Z.-Y. Wang, and I. Potrykus (editors).
Monographs on Theoretical and Applied Genetics 23, Springer-Verlag, Berlin, Germany. 210 pages.

Chemical Vegetation Management

1988. John E. Kaufman and Howard E. Westerdahl (editors).
Plant Growth Regulators Society, Athens, Georgia, USA. 296 pages.

Control of Weeds, Pests, and Diseases of Cultivated Turf
1967.
Proceedings of a Symposium by the British Weed Control Council and the British Insecticide and Fungicide Council, Droitwich, Worcester, England, UK. 43 pages.

Die Kweek Van Kwaliteitsgrasblaaie Deur Omgewings-Verandering (Producing Quality Turfs Through Environmental Modification)
1974. James B Beard.
African National Parks Conference, Capetown, South Africa. 26 pages.

Establishment and Maintenance of Sporting Places Under Grass Cover
1969. F. Bures (editor) with thirty contributing authors.
Proceedings of the 1st Czechoslovak Symposium on Sportturfs, Institute of Sportturfs, University of Agriculture in Brno, Brno, Czechoslovakia. 284 pages.

European Turfgrass Society Conference
2008+. Simone Magni (editor).
European Turfgrass Society, Pisa, Italy. 192 pages.
Published every four years.

Grass Breeders Work Planning Group Proceedings
1973, 1980, 1988, 1992, 1994, 1996, 1998, and 2000.
Publication location varied with host location.

Grounds Maintenance, Dust and Erosion Control
1946.
US Army Corp of Engineers, Washington, DC, USA. 183 pages.

International Symposium on Soccer Fields
1995. Hikaru Ota (editor) with nine contributing authors.
Soft Science Publications, Tokyo, Japan. 65 pages.

Molecular Breeding of Forage Crops, 2nd International Symposium
2001. German Spangenberg (editor).
Kluwer Academic Publishers, Dordrecht, the Netherlands. 357 pages.

Molecular Breeding of Forage and Turf, 3rd International Symposium
2004. Andrew Hopkins, Zeng-Yu Wang, Rouf Mian, Mary Sledge, and Reed E. Barker (editors).
Kluwer Academic Publishers, Dordrecht, the Netherlands. 407 pages.

Molecular Breeding of Forage and Turf, 5th International Symposium
2009. Toshihido Yamada and German Spangenberg (editors).
Springer Science & Business Media LCC, New York City, New York, USA. 364 pages.

Molecular Breeding for the Genetic Improvement of Forage Crops and Turf, 4th International Symposium
2005. M.O. Humphryes (editor).
Wageningen Academic Publishers, Wageningen, the Netherlands. 286 pages.

The Next Decade in Amenity Grassland
1976. E.C. Wright (editor).
Institute of Biology and Department of Extra-Mural Studies, Queen's University, Belfast, Northern Ireland, UK. 116 pages.

Proceedings of Environmental Issues for Turf, A Symposium
1996.
Australian Turfgrass Research Institute Limited, Concord West, New South Wales, Australia. 404 pages.

Proceedings of the ATRI Turf Research Conference
1994, 1995, 1997 (third), and 1997 (fourth).
The Australian Turfgrass Research Institute Ltd., Concord West, New South Wales, Australia. 133, 131, 87, and 83 pages.

Proceedings of Australasian Turf Researchers' Conference
1989.
The Australian Turfgrass Research Institute Ltd., Concord West, New South Wales, Australia. 90 pages.

Proceedings of the First International Turfgrass Conference on Turfgrass Management and Science for Sports Fields
2004. P.A. Nektarios (editor).
Acts Horticulture 661, International Society for Horticultural Science, Leuven, Belgium. 578 pages.

Proceedings of Fate of Nitrogen from Turfgrass Fertilizer Applications
1990.
Cape Cod Cooperative Extension, Barnstable, Massachusetts, USA. 43 pages.

Proceedings of Spring Dead Spot of Bermudagrass (Cynodon spp) Workshop
1980. Leon T. Lucas (editor) with nine contributing authors.
Southern Turfgrass Workers Group Meeting, Department of Plant Pathology, North Carolina State University, Raleigh, North Carolina, USA. 46 pages.
Contains the first proposal for an integrated concept of the patch diseases.

Proceedings of the Crownvetch Symposium
1964. Guy W. McKee (editor).
Pennsylvania State University, University Park, Pennsylvania, USA. 82 pages.

Safety in American Football
1997. Earl F. Hoerner (editor).
American Society for Testing and Materials STP 1305, West Conshohocken, Pennsylvania, USA. 201 pages.

Second Crownvetch Symposium
1988. Guy W. McKee and Marvin L. Risius (editors).
Agronomy Mimeo, Pennsylvania State University, University Park, Pennsylvania, USA. 151 pages.

Figure 15.8. Dr. Jim Beard and turfgrass students being instructed in a glasshouse. [26]

Sixth Discussion Meeting of Amenity Grass Research
   1988.
   University College of Wales, Aberystwyth, Wales, UK. 270 pages.

Special Proceedings of the Turf Section Charter Meeting, Weed Society of America
   1956. Gene C. Nutter (editor) with eight contributing authors.
   West Point Products Corporation, West Point, Pennsylvania, USA. 60 pages.

D. ROADSIDE TURFGRASS RESEARCH PUBLICATIONS

Report Separates

An Investigation of Critical Problems of Establishing and Maintaining a
Satisfactory Sod Cover Along North Carolina Highways
   1967. W.B. Gilbert and D.L. Davis.
   Project ERD-110-S Final Report, Highway Research Program, North Carolina State University, Raleigh, North Carolina, USA. 100 pages.

Dispersal of Sodium Ions in Soils
   1971. F.E. Hutchinson.
   Materials and Research Technical Paper 71-10C, University of Maine, Department of Plant & Soil Sciences, in cooperation with the Maine State Highway Commission, Orono, Maine, USA. 22 pages.

L'Engazonnement Des Emprises Routieres (The Seeding of Grass Seeds on Roadsides)
   1979.
   Laboratoire Central des Ponts et Chaussées, Ministere de L'Environnement et du Cadre de Vie—Ministere des Transports, Paris, France. 72 pages.

Erosion Control Study
   circa 1969. John Beaver, Allen L. Cox, M.D. Swanner, H.T. Barr, and C.L. Mondart Jr.
   State Project No. 736-00-35, Louisiana State University, Agricultural Experiment Station, Baton Rouge, Louisiana, USA. 44 pages.

Establishment and Management of Roadside Vegetative Cover in Massachusetts
   1967. John M. Zak and Evangel J. Bredakis.
   Final Report, Bulletin No. 562, Massachusetts Agricultural Experiment Station. University of Massachusetts, Amherst, Massachusetts, USA. 28 pages.

Establishment and Use of Turf and Other Ground Covers
   1971 (final report). A.E. Dudeck and J.O. Young.
   Nebraska Highway Research Project 1, Research Study 64-1, Department of Horticulture and Forestry, University of Nebraska, Lincoln, Nebraska, USA. 165 pages.
   Annual reports in 1966, 1967, 1968, and 1969.

Establishment of Vegetation on Highway Backslopes in Iowa
   1957. Frank Benjamin Kulfinski.
   Bulletin No. 11, Iowa Agricultural Experiment Station, Iowa State College, Ames, Iowa, USA. 141 pages.

The Establishment of Vegetation on Nontopsoiled Highway Slopes in Washington
   1970. Roy L. Goss, Robert M. Blanchard, and William R. Melton.
   Final Report, Washington State Highway Commission and Washington State University Agricultural Research Center, Puyallup, Washington, USA. 29 pages.

Experiments on Seeding and Mulching for Roadside Erosion Control
1962.
Investigation No. 614 Progress Report, Materials and Research Section, Minnesota Department of Highways, St. Paul, Minnesota. 38 pages.

Heavy Duty Turf
1942. Frank V. Baircalow and Harry F. Goodloe.
Annual Report, Forage Crops Investigations, Wisconsin Agricultural Experiment Station, Madison, Wisconsin, USA. 77 pages.

Improved Establishment and Maintenance of Roadside Vegetation in Michigan
1971. James B Beard, James A. Fischer, John E. Kaufmann, and David P. Martin.
Research Report 144, Michigan Agricultural Experiment Station, Michigan State University, East Lansing, Michigan, USA. 66 pages.

Low Maintenance Turfgrass Management Systems of NC Roadsides, Final Recommendations
1999. Art Bruneau, Fred Yelverton, Rich Cooper, and Conrad Johnson.
Project Hwy 97-5, Department of Crop Science, North Carolina State University, Raleigh, North Carolina, USA. 40 pages.

Methods and Materials for the Maintenance of Turf on Highway Rights-of-Way: An Annotated Bibliography
1971. Margaret H. Smithberg and Donald B. White.
Minnesota Agricultural Experiment Station Miscellaneous Report 105, University of Minnesota, St. Paul, Minnesota, USA. 102 pages.

Methods of Controlling Erosion on Newly Seeded Highway Backslopes in Iowa
1961. Berlie Louis Schmidt.
Bulletin No. 24, Iowa Agricultural Experiment Station, Iowa State University, Ames, Iowa, USA. 51 pages.

Minnesota Highway Maintenance Costs
1970. Gerald F. Tessman.
Minnesota Department of Highways, Management Services Section, Minneapolis, Minnesota, USA. 51 pages.

Mulches for Wind and Water Erosion
1963. W.S. Chepil, N.P. Woodruff, E.H. Siddoway, and D.V. Armbrust.
Agricultural Research Service 41-84, United States Department of Agriculture, Manhattan, Kansas, USA. 23 pages.

Figure 15.9. Roadside turfgrass establishment research at Michigan State University in the early 1960s. [26]

National Bentgrass and Couchgrass Evaluation Trials
  1997. John Neylan and Michael Robinson.
  Horticultural Research and Development Commission, Gordon, New South Wales, Australia. 118 pages.

Penngift Crown Vetch for Slope Control on Pennsylvania Highways
  1954. H.B. Musser, W.L. Hottenstein, and J.P. Stanford.
  Bulletin 576, Pennsylvania State University—Agricultural Experiment Station, State College, Pennsylvania, USA. 21 pages.

Principles of Making Turf Mixtures for Roadside Seedings
  1964. R.E. Blaser.
  Number 54, Virginia Council of Highway Investigation and Research, Blacksburg, Virginia, USA. 6 pages.

Progress Report on Study of Turf Growth on Soil Mixtures Available for Highway Shoulder Construction in Michigan
  1947. J. Tyson and E.A. Finney.
  Michigan Agricultural Experiment Station, Journal Article No. 920 (N.S.), Soil Science Department, Michigan State College, East Lansing, Michigan, USA. 30 pages.

Promoting Establishment of Vegetation for Erosion Control
  1967. Wayne G. McCully and William J. Bowmer.
  Research Report Number 67-6, Texas Transportation Institute, Texas A&M University, College Station, Texas, USA. 19 pages.

Protecting Steep Construction Slopes Against Water Erosion
  1966. N.P. Swanson, A.R. Dedrick, and A.E. Dudeck.
  Progress Report of a Cooperative Study Conducted with the University of Nebraska, the Nebraska Department of Roads, and the Bureau of Public Roads, U.S. Department of Commerce, Lincoln, Nebraska, USA. 48 pages.

Public Problems in Landscape Design, Part I: Roads, Highways, Roadside Development
  1938. Paula Birner.
  University Extension Division, University of Wisconsin, Madison, Wisconsin, USA. 57 pages.

Restoration and Retention of Coastal Dune with Fences and Vegetation
  1966. J.A. Jagschitz and R.S. Bell.
  Bulletin 382, Rhode Island Agricultural Experiment Station, Kingston Rhode Island, USA. 44 pages.

Roadside Development and Erosion Control, Parts I, II, & III
  Circa 1974. Wayne W. Huffine, Lester W. Reed, and Gary W. Roach.
  Oklahoma Project No. 63-03-3, Oklahoma Agricultural Experiment Station, Oklahoma State University, Stillwater, Oklahoma, USA. 20, 87, & 108 pages plus a 74 page supplement.

Roadside Development and Erosion Control
  1987. A.D. Brede, L.M. Cargill, D.P. Montgomery, and T.J. Samples.
  Oklahoma Project No. 81-04-3, Agricultural Experiment Station, Oklahoma State University, Stillwater, Oklahoma, USA. 141 pages.

Roadside Development and Maintenance
  1970. Wesley L. Hottenstein (editor of 11 reports).
  Highway Research Record Number 335, Highway Research Board, National Academy of Sciences, Washington, DC, USA. 73 pages.

Roadside Erosion Control
  1975. Wayne W. Huffine, Lester W. Reed, and Fenton Gray.
  Oklahoma Project No. 70-01-3 Final Report, Oklahoma Agricultural Experiment Station, Oklahoma State University, Stillwater, Oklahoma, USA. 161 pages.

Roadside Erosion Control
  1977. Wayne W. Huffine, Lester W. Reed, and Fenton Gray.
  MP-102 Final Report, Oklahoma Agricultural Experiment Station, Oklahoma State University, Stillwater, Oklahoma, USA. 125 pages.

Roadside Slope Revegetation
1974. Robert M. Gallup.
Equipment Development and Test Report 7700-8, US Department of Agriculture, Forest Service Equipment Development Center, San Dimas, California, USA. 37 pages.

Roadside Vegetation Management, Final Report
1992. D.L. Martin, L.M. Cargill, and D.P. Montgomery.
MP-135, Division of Agricultural Sciences and Natural Resources, Oklahoma State University, Stillwater, Oklahoma, USA. 135 pages.

Roadside Vegetative Cover Research Project
1953. Nelson M. Wells and Harry H. Iurka.
Progress Report, Landscape Bureau, New York State Department of Public Works, Albany, New York, USA. 43 pages.

Roadside Vegetative Cover Research Project
1954. Nelson M. Wells and Harry H. Iurka.
Progress Report, Landscape Bureau, New York State Department of Public Works, Albany, New York, USA. 63 pages.

Roadside Vegetative Cover Research Project
1955. Nelson M. Wells and Harry H. Iurka.
Final Report, Landscape Bureau, New York State Department of Public Works, Albany, New York, USA. 52 pages.

Seeding Methods for Utah Roadsides
1970. C. Wayne Cook, Ingvard B. Jensen, George B. Colthorp, and Emery M. Larson.
Utah Resources Series 52, Utah Agricultural Experiment Station, Logan, Utah, USA. 23 pages.

Selection, Establishment, and Maintenance of Roadside Vegetation
1982. Wayne W. Huffine, Lester W. Reed, and Carl E. Whitcomb.
Final Report MP-110, Agricultural Experiment Station, Division of Agriculture, Oklahoma State University, Stillwater, Oklahoma, USA. 66 pages.

Selection, Establishment, and Maintenance of Vegetation Along North Carolina's Roadsides
1985. J.M. DiPaola, W.B. Gilbert, and W.M. Lewis.
Final Report 1979–1984, Center for Transportation Engineering Studies, North Carolina State University, Raleigh, North Carolina, USA. 167 pages.

A Study of Roadside Maintenance and Its Effect on Highway Design and Construction
1962. Bonner S. Coffman, Paul H. Brown, and George B. Tobey Jr.
Report No. 189-1, Engineering Experiment Station, Ohio State University, Columbus, Ohio, USA. 137 pages.

A Study of Roadside Maintenance and Its Effect on Highway Design and Construction
1963. Bonner S. Coffman and Paul H. Brown.
Report No. EES 189-2, Engineering Experiment Station, Ohio State University, Columbus, Ohio, USA. 95 pages.

Turf Establishment and Maintenance Along Highway Cuts
1962. R.E. Blaser, G.W. Thomas, C.R. Brooks, G.J. Shoop, and J.B. Martin Jr.
Number 36, Virginia Council of Highway Investigation and Research and the Agronomy Department, Virginia Polytechnic Institute, Blacksburg, Virginia, USA. 18 pages.

Turf Investigations for Airports and Highways
1953.
Engineering Experiment Station Bulletin No. 78, Purdue University, West Lafayette, Indiana, USA. 19 pages.

Utilizing Plant Growth Regulators to Develop a Cost Efficient Management System for Roadside Vegetation
1984. Michael T. McElroy, Paul E. Rieke, and Shawn L. McBurney.
Final Report, Department of Crop and Soil Sciences, Michigan State University, East Lansing, Michigan, USA. 177 pages.

Vegetation Maintenance Practices, Program, and Equipment on Minnesota Highways
1969. D.B. White and T.B. Bailey.
Investigation No. 619, Special Report, Department of Horticultural Science, University of Minnesota, St. Paul, Minnesota, USA. 69 pages.

E. INDIVIDUAL MAJOR TURFGRASS RESEARCH PUBLICATIONS

Bulletins published primarily by major universities and agricultural experiment stations in the United States are presented. Topic headings include turfgrass cultivars and botany; turfgrass soils; turfgrass nutrition and fertilization; turfgrass pest management; turfgrass culture systems; turfgrass research, general; turfgrass industry; and miscellaneous.

Turfgrasses, Cultivars and Botany

*Annual Bluegrass (*Poa annua L.*) and Its Requirements for Growth*
1937. H.B. Sprague and G.W. Burton.
Bulletin 630, New Jersey Agricultural Experiment Station, New Brunswick, New Jersey, USA. 24 pages.

*Annual Bluegrass (*Poa annua L.*), Description, Adaptation, Culture, and Control*
1978. J. B Beard, P.E. Rieke, A.J. Turgeon, and J.M. Vargas Jr.
Michigan Agricultural Experiment Station Research Report 352, East Lansing, Michigan, USA. 32 pages.

Canada Bluegrass: Its Culture and Uses
1910. R.A. Oakley.
Farmers Bulletin 402, Bureau of Plant Industry, United States Department of Agriculture, Washington, D.C., USA. 20 pages.

Cool Season Turfgrass Variety Performance in Five Western States, 1980–1981
1983. V. Gibeault and R. Autio (editors).
Western Regional Coordinating Committee-11 (Turf), University of California Cooperative Extension, University of California, Riverside, California, USA. 193 pages.

Crownvetch in West Virginia
1963. Joseph D. Ruffner and John G. Hall.
Bulletin 487, Agricultural Experiment Station, West Virginia University, Morgantown, West Virginia, USA. 19 pages.

Endophytes in Turfgrass
1996. David Aldous.
TU304 Horticultural Research and Development Corporation, Gordon, New South Wales, Australia. 32 pages.

Evaluation of Bermudagrass Varieties for General-Purpose Turf
1964. F.V. Juska and A.A. Hanson.
Agricultural Handbook No. 270, Agricultural Research Service, United States Department of Agriculture, Beltsville, Maryland, USA. 54 pages.

Evaluation of Turfgrasses for Virginia
1967. R.E. Schmidt, R.E. Blaser, and M.T. Carter.
Bulletin 12, Research Division, Virginia Polytechnic Institute, Blacksburg, Virginia, USA. 16 pages.

*Experiments with Modified Techniques for the Determination of Purity and Viability of Bluegrass Seed (*Poa Pratensis L.*)*
1938. R.H. Porter.
Research Bulletin 235, Agricultural Experiment Station, Iowa State College of Agriculture and Mechanic Arts, Ames, Iowa, USA. 22 pages.

Figure 15.10. Turfgrass cultivar plots being seeded by Dr. Jim Beard at Michigan State University in 1962. [26]

Experiments with Turf-Grasses, 1948–1955
   1958. Gunnar Weibull.
   Agri. Hort. Genetics, XVI: 209 to 220.

FloraDwarf Bermudagrass
   1997. A.E. Dudeck and C.L. Murdoch.
   Research Bulletin 901, Institute of Food and Agricultural Sciences, University of Florida, Gainesville, Florida, USA. 27 pages.

FloraTex Bermudagrass
   1994. A.E. Dudeck, J. B Beard, R.A. Reinert, and S.I. Sifers.
   Bulletin 891, Florida and Texas Agricultural Experiment Stations, Gainesville, Florida, USA. 11 pages.

Floratine St. Augustinegrass: A New Variety for Ornamental Turf
   1960. Gene C. Nutter and Robert J. Allen Jr.
   Circular S-128, University of Florida Agricultural Experiment Stations, Gainesville, Florida, USA. 12 pages.

Floratine St. Augustinegrass
   1962. Gene C. Nutter and Robert J. Allen Jr.
   Circular S-123A, University of Florida Agricultural Experiment Stations, Gainesville, Florida, USA. 14 pages.

Formation and Classification of National Ref Collection of Cultivars and Ecotypess of Couchgrass
   1996. A. Shakesby.
   Australian Turfgrass Research Institute, Concord West, New South Wales, Australia. 11 pages.

The Foundations of the Centipedegrass Seed and Hybrid Bermudagrass Industries
   1996. Wayne W. Hanna.
   The University of Georgia Coastal Plain Experiment Station USDA, Agricultural and Research Service and Abraham Baldwin Agricultural College, Tifton, Georgia, USA. 15 pages.

The Growth of Rhizomes in Kentucky Bluegrass
   1963. John E. Fisher.
   Greenhouse-Garden-Grass, Plant Research Institute, Canada Department of Agriculture, Ottawa, Ontario, Canada. 5(3):1–5.

A Guide to the Species of Cynodon (Gramineae)
1970. Jack R. Harlan, J.M.J. de Wet, Wayne W. Huffine, and John R. Deakin.
Bulletin B-673, Oklahoma State University, Agricultural Experiment Station, Stillwater, Oklahoma, USA. 37 pages.

How Kentucky Bluegrass Grows
1951. Alfred Gordon Etter.
Annals of the Missouri Botanical Garden. 38:293–375.

Kentucky Bluegrass Variety Evaluations
1970. Merle H. Niehaus and R.R. Davis.
Research Circular 177, Ohio Agricultural Research and Development Center, Wooster, Ohio, USA. 8 pages.

Lawn Grasses and Their Treatment
1912. F.G. Springs.
Agricultural Bulletin of the Federated Malay States No. 2. 76 pages.

Making a Lawn: Mixed Lawn Grass Seed Analyzed
1886. W.J. Beal.
Bulletin No. 11, Botanical Department, Agricultural College of Michigan, East Lansing, Michigan, USA. 10 pages.

National Bentgrass and Couchgrass Evaluation Trials
1997. John Neylan and Michael Robinson.
Horticultural Research and Development Commission, Gordon, New South Wales, Australia. 118 pages.

North American Species of Festuca
1906. Charles V. Piper.
United States National Herbarium Volume X, Part I, Smithsonian Institution, Washington, DC, USA. 51 pages.

*Northeastern Regional Turfgrass Evaluation of Kentucky Bluegrasses (*Poa pratensis L.*), 1968–1973*
1977. Fifteen contributing authors of the NE-57 Technical Research Committee.
Bulletin 814, Pennsylvania State University Agricultural Experiment Station, University Park, Pennsylvania, USA. 60 pages.

On-Site Evaluation of Creeping Bentgrass for Putting Greens: Progress Report 2000
National Turfgrass Evaluation Program No. 01-17, Beltsville, Maryland, USA. 34 pages.

Ornamental Grasses for the Home and Garden
1973. Mary Hockenberry Meyer and Robert G. Mower.
Information Bulletin 64, Plant Science, Floriculture and Ornamental Horticulture, New York State College of Agriculture and Life Sciences, Cornell University, Ithaca, New York, USA. 20 pages.

Penngift Crown Vetch for Slope Control on Pennsylvania Highways
1954. H.B. Musser, W.L. Hottenstein, and J.P. Stanford.
Bulletin 576, Agricultural Experiment Station, Pennsylvania State University, State College, Pennsylvania, USA. 21 pages.

A Preliminary Report on Carpet Grass
1924. J. Lambourne.
Malayan Agricultural Journal. 12(12):402–403.

Rhode Island Bent and Related Grasses
1918. Charles V. Piper.
Part I, Bulletin No. 692, Bureau of Plant Industry, United States Department of Agriculture, Washington, DC, USA. 12 pages.

Royal Cape Bermudagrass: A New Variety for Lawn Use in the Imperial Valley
1960. A.A. Hanson.
Crops Research Division, ARS-USDA, Beltsville, Maryland, USA. 2 pages.

The Seeds of Redtop and Other Bent Grasses
  1918. F.H. Hillman.
  Part II, Bulletin No. 692, Bureau of Plant Industry, United States Department of Agriculture, Washington, DC, USA. 11 pages.

The Seeds of the Bluegrasses
  1905. Edgar Brown and F.H. Hillman.
  Bulletin No. 84, Bureau of Plant Industry, United States Department of Agriculture, Washington, DC, USA. 38 pages.

Status Report on the Investigation of a Nonstructural Carbohydrate Extraction Procedure for Turfgrasses
  1989. Bill Richie, Fernando Gonzales, and Robert Green.
  Department of Soil and Crop Sciences, Texas A&M University, College Station, Texas, USA. 50 pages.

Studies on the Horticultural Classification and Development of Japanese Lawn Grasses
  1970. Fumio Kitamura.
  Bulletin 3, Kemigawa Arboretum, University of Tokyo, Tokyo, Japan. 60 pages.

A Study of Factors Affecting the Germination, Establishment, and Competition of the Turfgrass Species Red Fescue
(Festuca ruba L. *spp.* litoralis Vasey*), Perennial Ryegrass* (Lolium perenne L.*), and Kentucky Bluegrass* (Poa pratensis L.*)*
  2004. Soren Ugilt Larsen.
  Forest & Landscape Research No. 34-2004, Danish Centre for Forest, Landscape and Planning, Horsholm, Denmark, 141 pages.

A Study of the Root Systems of Some Important Sod-Forming Grasses
  1930. A.S. Laird.
  Bulletin 211, Agricultural Experiment Station, University of Florida, Gainesville, Florida, USA. 27 pages.

Temperature and Other Factors Applied to Turf on the Fescue Seed
  1939. Vivian Kerns and E.H. Toole.
  Technical Bulletin No. 638, United States Department of Agriculture, Washington, DC, USA. 35 pages.

Turfgrass Varieties in New Jersey
  n.d. circa 196x. C. Reed Funk, Ralph E. Engel, Philip M. Halisky, and Henry W. Indyk.
  Rutgers, State University, New Brunswick, New Jersey, USA. 21 pages.

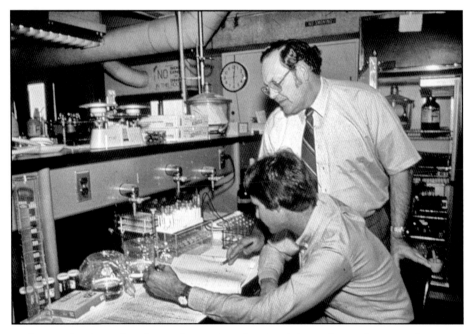

Figure 15.11. Laboratory biochemical studies involving polyacrylamide gel electrophoresis for genetic characterizations. [26]

US Turfgrass Variety List
  1981, 1983, 1988, 1990, and 1992. William A. Meyer, Crystal Rose, and Jack Murray.
  Western Regional Turfgrass Research Coordinating Committee and Beltsville Agricultural Research Center, Hubbard, Oregon, USA.

Turfgrass Soils

Aerification Studies in Progress and Planned
  Circa 1950.
  West Point Lawn Products, West Point, Pennsylvania, USA. 19 pages.

Amenity Grass, Drainage Review
  1983. C.J. Ward.
  Sports Turf Research Institute, Bingley, West Yorkshire, England, UK. 134 pages.

Athletic Field Systems Study 2000–2003: An Evaluation and Comparison of Naturally and Artificially Enhanced Athletic Field Sand Textured Root Zones
  2003. J.J. Henderson, J.N. Rogers, III, and J.R. Crumb.
  Final Report, Department of Crop and Soil Sciences, Michigan State University, East Lansing, Michigan, USA. 59 pages.

Drainage Problems on Playing Fields
  1971. V.I. Stewart.
  National Playing Fields Association, London, England, UK. 14 pages.

The Effect of Nutrients and Pesticides Applied To Turf on the Quality of Runoff and Percolating Water
  1989. Thomas L. Watschke, Ralph O. Mumma, Scott A. Harrison, and George W. Hamilton Jr.
  Report ER8904, Environmental Resources Research Institute, Pennsylvania State University, University Park, Pennsylvania, USA. 66 pages.

The Establishment and Maintenance of Turf on Stabilized Granular Material
  1949. W.H. Skrdla.
  Department of Agronomy, Purdue University, West Lafayette, Indiana, USA. 162 pages.

Geotextiles as an Intermediate Layer in USGA and USGA-Type Greens
  2001. L.M. Callahan, R.S. Freeland, J.M. Parham, A.M. Saxton, R.D. von Bernuth, D.P. Shepard, and J.M. Garrison.
  Bulletin 699, Department of Ornamental Horticulture and Landscape Design, Tennessee Agricultural Experiment Station, University of Tennessee, Knoxville, Tennessee, USA. 67 pages.

Golf Green Topdressing: Physical and Chemical Properties of Commercial Mixes
  1994. A.S. McNitt, P.J. Landshoot, B.F. Hoyland, and D.V. Waddington.
  Bulletin 872, Agricultural Experiment Station, Penn State College of Agricultural Sciences, University Park, Pennsylvania, USA. 18 pages.

A Guide to Evaluating Sands and Amendments Used for High Trafficked Turfgrass
  1970. William B. Davis, J.L. Paul, John H. Madison, and Leon Y. George.
  AXT-n 113, University of California Agricultural Extension, Davis, California, USA. 94 pages.

Leveling Tests on Soil Surfaces under Turf Cover
  1951.
  United States Air Force, Chanute Air Force Base, Rantoul, Illinois, USA. 23 pages.

The Persistence of Lawn and Other Grasses as Influenced Especially by the Effect of Manures on the Degree of Soil Acidity
  1917. Burt L. Hartwell and S.C. Damon.
  Bulletin 170, Agricultural Experiment Station of the Rhode Island State College, Kingston, Rhode Island, USA. 28 pages.

Relationship between Athletic Field Hardness and Traction, Vegetation, Soil Properties, and Maintenance Practices
1988. J.N. Rogers III, D.V. Waddington, and J.C. Harper II.
Progress Report 393, Agricultural Experiment Station, Pennsylvania State University, University Park, Pennsylvania, USA. 16 pages.

Sintered Fly Ash as a Soil Modifier
1968. J.C. Patterson, P.R. Henderlong, and L.M. Adams.
Department of Agronomy and Genetics, West Virginia University, Morgantown, West Virginia, USA. 21 pages.

Soil Construction, Drainage, and Maintenance for Swedish Grassed Parks and Sports Fields
1988. Ingrid M. Karrlsson.
Supplement 26, Acta Agriculturae Scandinavica, Stockholm, Sweden. 99 pages.

Soil Modification during Nursery Sod Production
1978. R.W. Sheard and M. Van Patter.
Department of Land Resource Science, University of Guelph, Guelph, Ontario, Canada. 69 pages.

Soil Modification for Turfgrass Areas 1: Physical Properties of Physically Amended Soils
1974. D.V. Waddington, T.L. Zimmerman, G.J. Shoop, L.T. Kardos, and J.M. Duich.
Pennsylvania State University, University Park, Pennsylvania, USA. 96 pages.

A Soil Technology Study on Effectuating and Maintaining Adequate Playing Conditions of Grass Sports Fields
1980. A.L.M. van Wijk.
Agricultural Research Report 903, Institute for Land and Water Management Research, Wageningen, the Netherlands. 124 pages.

Stabilization and Enhancement of Sand-Modified Root Zones for High Traffic Sports Turfs with Mesh Elements
1993. James B Beard and Samuel I. Sifers.
B-1710, Texas Agricultural Experiment Station, College Station, Texas, USA. 40 pages. Printed in English and Japanese.

Figure 15.12. Nitrogen fertilizer carrier treatments being applied by Dr. Paul Rieke of Michigan State University in the mid-1960s at Traverse City turfgrass research plots. [26]

A Study of the Growth of Grass in "Surface-Stabilized" Soil
> n.d. circa 1945. Martin A.F. Sutton, T.F.N. Alexander, and F.C. West.
> Sutton & Sons Limited, Reading, England, UK. 21 pages.

Testing of Sands for Golf Green Construction and Bunkers
> n.d. circa 1989. J. Neylan and P. Pierce.
> Turf Research and Advisory Institute, Frankston, Victoria, Australia. 13 pages.

Turfgrass Nutrition and Fertilization

A Comparison of Nitrogen Carriers for Bent Grass Fertilization
> 1939. J.A. DeFrance and T.E. Odland.
> Contribution No. 567, Rhode Island Agricultural Experiment Station, Kingston, Rhode Island, USA. 7 pages.

Availability of Urea-Formaldehyde Reaction Products as Shown by Pot Tests
> 1947. W.H. Armiger, F.O. Lundstrom, K.G. Clark, and M.S. Anderson.
> Division of Soils, Fertilizers and Irrigation, ARS-USDA, Beltsville, Maryland, USA. 14 pages.

Evaluation of Fertilizers Containing Composted Refuse
> 1972. D.W. Waddington, J.M. Duich, and E.L. Moberg.
> Progress Report 327, Agricultural Experiment Station, Pennsylvania State University, University Park, Pennsylvania, USA. 12 pages.

Evaluation of Turfgrass Fertilizers on Kentucky Bluegrass
> 1985. D.V. Waddington, T.R. Turner, and J.M. Duich.
> Progress Report 386, Agricultural Experiment Station, Pennsylvania State University, University Park, Pennsylvania, USA. 9 pages.

Fertilizing Grasses in South Africa
> 1948. D.B.D. Meredith.
> University of Witwatersrand and the Agricultural Advisory Section, African Explosive & Chemicals Industries Limited, Johannesburg, South Africa. 187 pages.

Fertilizing "Tifgreen" Bermudagrass Golf Greens in Louisiana
> 1982. E.P. Barrios, L.G. Jones, K.L. Koonce, and L.P. Leger.
> Bulletin No. 734, Louisiana State University, Agricultural Experiment Station, Baton Rouge, Louisiana, USA. 34 pages.

Influence of Certain Nitrogen Fertilizers on Quality and Growth of Lawngrasses in Louisiana
> 1982. E.P. Barrios, L.G. Jones, and L.P. Leger.
> Bulletin No. 739, Louisiana State University, Agricultural Experiment Station, Baton Rouge, Louisiana, USA. 21 pages.

Influence of Fertilizers on the Accumulation of Roots from Closely Clipped Bent Grasses and on the Quality of the Turf
> 1944. Robert S. Bell and J.A. DeFrance.
> Contribution No. 638, Rhode Island Agricultural Experiment Station, Kingston, Rhode Island, USA. 7 pages.

Lawn Fertilizer Test
> 1969. D.V. Waddington, J.M. Duich, and E.L. Moberg.
> Progress Report 296, Pennsylvania State University, College of Agriculture, Agriculture Experiment Station, University Park, Pennsylvania, USA. 57 pages.

A Literature Review on Sewage Utilization for Turfgrass Purposes with Annotated Bibliography
> 1979. A.E. Dudeck, E.C. Donoho, A.J. Turgeon, P.L. Heinzen, and G.E. Stout.
> Miscellaneous Publication 1979-1, Department of Ornamental Horticulture, University of Florida, Gainesville, Florida, USA. 92 pages.

Mineral Deficiency Symptoms on Turfgrass I: Major and Secondary Nutrient Elements
1962. James E. Love.
Wisconsin Academy of Sciences, Arts and Letters, Wisconsin, USA. 51:135–140.

Nutrient Retention in a USGA Green Rootzone over Seven Years
2003. L.M. Callahan, J.M. Parham, A.M. Saxton, R.W. Freeland, R.D. van Bernuth, D.P. Shepard, and J.M. Garrison.
Tennessee Agricultural Experiment Station Bulletin 701, University of Tennessee, Knoxville, Tennessee, USA. 16 pages.

The Persistence of Certain Lawn Grasses as Affected by Fertilization and Competition
1929. E.S. Garner and S.C. Damon.
Bulletin 217, Agricultural Experiment Station of the Rhode Island State College, Kingston, Rhode Island, USA. 22 pages.

Response of Kentucky Bluegrass, Creeping Red Fescue, and Bentgrass to Nitrogen Fertilizers
1960. J.M. Duich and H.B. Musser.
Progress Report 214, Agricultural Experiment Station, Pennsylvania State University, University Park, Pennsylvania, USA. 20 pages.

Studies of Several Long-Lasting Nitrogen Sources on Several Floral and Nursery Crops and Turfgrasses
1958. O.R. Lunt, J. Stark, R.H. Sciaroni, and T. Bryne.
Agricultural Experiment Station, University of California, Los Angeles, California, USA. 38 pages.

Urea-Formaldehyde and Other Nitrogenous Fertilizers for Use on Turf
1951. H.B. Musser, J.R. Watson Jr., J.P. Stanford, and J.C. Harper II.
Bulletin 542, Pennsylvania Agricultural Experiment Station, Pennsylvania State College, State College Pennsylvania, USA. 16 pages.

*Utilization of Nutrients by Colonial Bentgrass (*Agrostis tenuis*) and Kentucky Bluegrass (*Poa pratensis*)*
1934. H.B. Sprague and E.E. Evaul.
Bulletin 570, New Jersey Agricultural Experiment Station, New Brunswick, New Jersey, USA. 20 pages.

Yellowing of Centipede Grass and Its Control
1933. O.C. Bryan.
Bulletin 450, Florida Agricultural Experiment Station Press, Gainesville, Florida, USA. 2 pages. Reprinted in 1936 and 1941.

The Yield and Chemical Composition of Clippings from a Bermudagrass Green
1956. O.J. Noer and J.E. Hamner.
Turf Service Bureau, Milwaukee Sewerage Commission, Milwaukee, Wisconsin, USA. 10 pages.

Turfgrass Pest Management

Abstract of Literature on Turf Treatments
1964.
Abstract Nos. 1-81, Velsicol Chemical Corporation, Chicago, Illinois, USA. 84 pages.

An Annotated Bibliography and Literature Review on the Potential Impacts of Golf Courses on Freshwater Environments
1992. M.K. Brewin.
Trutta Environments and Management, Cochrane, Alberta, Canada. 146 pages.

The Biology and Control of Turf Grubs
1959. Joseph B. Polinka.
Research Bulletin 829, Ohio Agricultural Experiment Station, Wooster, Ohio, USA. 30 pages.

Bionomical Observations and Control of Ataenuis spretulus *in West Virginia*
1978. Joseph E. Weaver and J. Douglas Hacker.
Current Report 72, West Virginia University Agricultural and Forestry Experiment Station, Morgantown, West Virginia, USA. 16 pages.

Figure 15.13. Early postemergence experimental crabgrass control studies at Purdue University in 1957. [26]

Control of Ants in Turf and Soil
1948. John C. Shread and Gordon C. Chapman.
Bulletin 515, Connecticut Agricultural Experiment Station, New Haven, Connecticut, USA. 23 pages.

Control of Large Crabgrass and Goosegrass in Warm-Season Turfgrasses
1987. B.J. Johnson and T.R. Murphy.
Research Bulletin 364, Georgia Agricultural Experiment Station, University of Georgia, Griffin, Georgia, USA. 29 pages.

Control of Lawn Weeds and the Renovation of Lawns
1941. F.A. Welton and J.C. Carroll.
Bulletin 619, Ohio Agricultural Experiment Station, Wooster, Ohio, USA. 85 pages.

Control of Selected Weeds with Herbicides in Cool-Season Turfgrasses in the Mountain Region of Georgia
1977. B.J. Johnson.
Research Bulletin 199, Agricultural Experiment Station, University of Georgia, Experiment, Georgia, USA. 30 pages.

Control of the Japanese Beetle and Its Grub in Home Yards
1938. W.E. Fleming and F.W. Metzger.
Circular No. 401, United States Department of Agriculture, Washington, DC, USA. 14 pages.

Controlling Goosegrass in Bermudagrass Turf with Herbicides
1980. B.J. Johnson.
Bulletin 261, University of Georgia College of Agriculture Experiment Stations, Experiment, Georgia, USA. 24 pages.

Dandelion Control with 2,4-D
1946. Dayton Klingman.
Bulletin No. 274, University of Wyoming Agricultural Experiment Station, Laramie, Wyoming, USA. 12 pages.

Destroying Lawn Weeds with 2,4-D
1946. Gilbert H. Ahlgren and Herbert R. Cox.
Bulletin 725, New Jersey Agricultural Experiment Station, Rutgers University, New Brunswick, New Jersey, USA. 11 pages.

Development of Ecologically Based Management Procedures for Controlling the Invasion of Bent Grass by Couch
1996. P.M. Martin.
HRDC Final Report, Australian Turfgrass Research Institute, Concord West, New South Wales, Australia. 79 pages.

The Effect of Air Temperature on the Pathogenicity of Rhizoctonia solani *Parasitizing Grasses on Putting-Green Turf*
1930. Lawrence S. Dickinson.
Contribution No. 105, Massachusetts Agricultural Experiment Station, Amherst, Massachusetts, USA. 13 pages.

Experiments on the Control of Brown-Patch with Chlorophenol Mercury
1925. George H. Godfrey.
Professional Paper Number 1, Boyce Thompson Institute for Plant Research, Yonkers, New York, USA. 7 pages.

Herbicide and Cultural Practices for Bermudagrass Greens
1987. B.J. Johnson.
Research Bulletin 358, Georgia Agricultural Experiment Station, University of Georgia, Griffin, Georgia, USA. 18 pages.

Herbicides for Weed Control During Establishment of Warm Season Turfgrasses
1975. B.J. Johnson.
Agricultural Experiment Station, University of Georgia, Experiment, Georgia, USA. 27 pages.

Management of Herbicides for Weed Control in Bermudagrass Turf
1984. B.J. Johnson.
Research Bulletin 313, Agricultural Experiment Station, University of Georgia, Experiment, Georgia, USA. 47 pages.

Maximizing the Effectiveness of Fungicides
1988. Houston B. Couch.
Milliken Chemical, Spartanburg, South Carolina. 11 pages.

Nematode Types and Levels in Bowling Greens in Victoria
1996. J. Neylan.
HRDC Final Report TU201, Turfgrass Technology, Victoria, Australia. 37 pages.

Pesticide Use and Pest Management Practices by the Indiana Lawn Care Industry
1992. Timothy Gibb, David Kellam, Fred Whitford, Clifford Sadof, Clark Throssell, and Jay Johnson.
Bulletin No. 715, Department of Entomology, Purdue University, West Lafayette, Indiana, USA. 41 pages.

Postemergence Herbicide Activity on Broadleaf Winter Weeds, Wild Garlic, and Bermudagrass Turf
1987. B.J. Johnson.
Research Bulletin 359, Georgia Agricultural Experiment Station, University of Georgia, Griffin, Georgia, USA. 27 pages.

Results of Area Campaigns Against Japanese Beetles in Ohio
1962. J.B. Polinka.
Research Circular 108, Ohio Agricultural Experiment Station, Wooster, Ohio, USA. 16 pages.

Some Common Lawn Mushrooms
1964. J. Walton Groves.
Greenhouse-Garden-Grass, Plant Research Institute, Research Branch, Canada Department of Agriculture, Ottawa, Ontario, Canada. 4(3):18–22.

Special Bulletin on Turf Diseases Including Brown Patch-Scald-Snow Mold
n.d. circa 1920.
O.M. Scott & Sons Co., Marysville, Ohio, USA. 7 pages.

Figure 15.14. First successful Typhula blight control study with a granular fungicide conducted at Traverse City, Michigan, USA, 1966. [26]

Spraying Lawns with Iron Sulfate to Eradicate Dandelions
1919. M.T. Munn.
Bulletin No. 466, New York Agricultural Experiment Station, Geneva, New York, USA. 39 pages.

A Survey of Pesticide Use and Bird Activity on Selected Golf Courses in British Columbia
1992. Ian E. Moul and John E. Elliott.
Technical Report Series No. 163, Canadian Wildlife Service, Delta, British Columbia, Canada. 111 pages.

A Survey of Pesticide Use on Ohio Golf Courses
1978. Philip O. Larsen and Richard L. Miller.
OCES Bulletin 682 and OARDC Research Bulletin 1127, Cooperative Extension Service, Ohio State University, Columbus, Ohio, USA. 17 pages.

To Identify, Collate, and Assess Research on the Management and Control
of the Main Pests and Diseases on European Golf Courses
2004. Ruth Mann.
Document No. 2112/1, Sports Turf Research Institute, Bingley, West Yorkshire, England, UK. 80 pages.

Turf Disease Control in 1950
1950. Eric G. Sharvelle.
Mimeo BP 41, Indiana Agricultural Experiment Station, Purdue University, West Lafayette, Indiana, USA. 19 pages.

Weed-Free Compost and Seedbeds for Turf
1952. J.A. DeFrance.
Miscellaneous Publication No. 31, Rhode Island Agricultural Experiment Station, Kingston, Rhode Island, USA. 15 pages.

Winter-Hardiness and Overwintering Diseases of Amenity Turfgrasses
With Special Reference to the Canadian Prairies
1987. J.D. Smith.
Technical Bulletin 1987-12E, Research Branch, Agriculture Canada, Saskatoon, Saskatchewan, Canada. 193 pages.

Turfgrass Cultural Systems

An Assessment of Turf Species Composition and Management of Grass Racetracks
1997. John Neylan.
RIRDC Research Paper Series No. 97/39, Turfgrass Technology Pty. Limited, Sandringham, Victoria, Australia. 78 pages.

Bentgrass Maintenance for Putting Greens
  1996. John Neylan and Michael Robinson.
  TU202, Horticultural Research and Development Corporation, Gordon, New South Wales, Australia. 64 pages.

Clipping Disposal Investigations with Rotary Lawn Mowers
  1975. J. B Beard and R.L. Yoder.
  Department of Crop and Soil Sciences, Michigan State University, East Lansing, Michigan, USA. 11 pages.

Commercial Turfgrass: Sod Production in Alabama
  1981. J.L. Adrian, J.A. Yates, and R. Dickens.
  Bulletin 529, Alabama Agricultural Experiment Station, Auburn University, Auburn, Alabama, USA. 48 pages.

A Comparative Cost Analysis of Golf Course Turf Maintenance in Georgia
  1972. M.G. LaPlante.
  Research Bulletin 108, College of Agriculture Experiment Station, University of Georgia, Athens, Georgia, USA. 33 pages.

The Effect of Conservation Programs on the Quality of Urban Lawns
  1986. A.S. Winje and J.E. Flack.
  Completion Report No. 142, Colorado Water Resources Research Institute, Fort Collins, Colorado, USA. 71 pages.

Fall Establishment of Lawn Grasses
  1964. W.E. Cordukes.
  Greenhouse-Garden-Grass, Plant Research Institute, Research Branch, Canada Department of Agriculture, Ottawa, Ontario, Canada. 4(3):12–17.

Heathland/Moorland Management
  1996. Robert Stephen Taylor.
  Studies in Golf Course Management No. 1, Sports Turf Research Institute, Bingley, West Yorkshire, England, UK. 44 pages.

Optimal Water Use of Turf Grass
  1997. J.D. Jansen van Vuuren.
  WRC Report No. 417/1/97, Department of Plant and Soil Sciences, Potchefstroom University, Patchefstroom, Transvaal, South Africa. 110 pages.

The Phytotoxicitic Effects of Three Soil Wetting Agents Applied to a Bentgrass (Agrostis spp.) Green
  1992. M.G. Wallis.
  The Australian Turfgrass Research Institute, Concord West, New South Wales, Australia. 18 pages.

Putting Green Grasses and Their Management
  1934. H.F.A. North and T.E. Odland.
  Bulletin 245, Agricultural Experiment Station of the Rhode Island State College, Kingston, Rhode Island, USA. 44 pages.

Rasenspielplalzuntersuchung (Investigation on Sports Grounds)
  1969.
  Deel 1 and 2, Nederlandse Sport Federatie, Arnhem, the Netherlands. 87 and 51 pages.

Turfgrass Water Use, Drought Resistance, and Rooting Patterns in the Southeast
  1991. Robert N. Carrow.
  ERC 01-91, Department of Agronomy, University of Georgia, Griffin, Georgia, USA. 63 pages.

Water Requirements and Application Rates for Lawns
  1978. Larry O. Pochop, John Borrelli, John R. Barnes, and Patrick K. O'Neill.
  Water Resources Series No. 71, Water Resources Research Institute, University of Wyoming, Laramie, Wyoming, USA. 37 pages.

Turfgrass Research, General

Amenity Grasslands, the Needs for Research
1977.
Publications Series 'C' No. 19, Natural Environment Research Council, London, England, UK. 68 pages.

American Beachgrass (Establishment, Fertilization, Seeding)
1966. J.A. Jagschitz and R.S. Bell.
Agricultural Experiment Station Bulletin 383, University of Rhode Island, Kingston, Rhode Island, USA. 43 pages.

An Analytical Study of the Putting Greens of Rhode Island Golf Courses
1928. Basil E. Gilbert.
Bulletin 212, Agricultural Experiment Station of the Rhode Island State College, Kingston, Rhode Island, USA. 15 pages.

*Annual Bluegrass (*Poa annua L.*) Tolerance to Subfreezing Temperatures and Impermeable Covers*
2000. Yves Castonguay, Philippe Rockette, and Julie Dionne.
Soil and Crops Research and Development Centre, Agriculture and Agri Foods Canada, Sainte-Foy, Quebec, Canada. 57 pages.

Athletic Field Quality Studies
1966. Tobias Grether and Juergen Gramckow.
Cal-Turf Inc., Camarillo, California, USA. 49 pages.

Artificial Grass Surfaces for Association Football
n.d. circa 1985. Walter Winterbottom.
Summary Report, Sports Development Unit, Sports Council, London, England, UK. 73 pages.

Australian Racecourses National Track Standards, Monitoring, Assessment, and Design
2007.
Publication No. 07/159, Rural Industries Research and Development Corp., Australia. 44 pp.

Better Turf through Research: Experiments with Cynodon dactylon *and Other Species*
1948.
South African Turf Research Station, Frankenwald, South Africa. 90 pages.

Building and Stabilizing Coastal Dunes with Vegetation
1982. S.W. Brome, E.D. Seneca, and W.W. Woodhouse Jr.
UNC Sea Grant Publication No. USC-SG-82-05, North Carolina State University, Raleigh, North Carolina, USA. 20 pages.

Comparison of Natural Turf and Synthetic Bowling Greens
n.d. circa 1996. M. Robinson.
TU 112 Final Report, Turfgrass Technology, Frankston, Victoria, Australia. 37 pages.

Controlling Drifting Sand Dunes on Cape Cod
1967. John M. Zak.
Bulletin No. 563, Massachusetts Agricultural Experiment Station, University of Massachusetts, Amherst, Massachusetts, USA. 15 pages.

Current and Planned Playing Surfaces Research in UK
1986. Martin Hawtree (editor).
Research Bulletin No. 1, National Turfgrass Council, Bingley, West Yorkshire, England, UK. 33 pages.

The Effect of Nutrients and Pesticides Applied to Turf on the Quality of Runoff and Percolating Water
n.d. circa 198x. Thomas L. Watschke, Ralph O. Mumma, Scott A Harrison, and George W. Hamilton Jr.
ER 8904 Final Report, Environmental Resources Research Institute, Pennsylvania State University, University Park, Pennsylvania, USA. 66 pages.

Figure 15.15. Turfgrass wear stress simulator for studying internal plant characteristics contributing to wear tolerance. [26]

Effects of Drought, Temperature, and Nitrogen on Turf Grasses
1943. J.C. Carroll.
Plant Physiology, Rockville, Maryland, USA. 13:19–36.

Experiments with Turf Grasses in New Jersey
1930. Howard B. Sprague and E.E. Evaul.
Bulletin 497, New Jersey Agricultural Experiment Station, New Brunswick, New Jersey, USA. 55 pages.

Garden Lawns and Playing Greens Establishment and Maintenance
1935. E.A. Madden.
Bulletin No. 165, New Zealand Department of Agriculture. 14 pages.

Grass Surface Standards
1989. John Shildrick (editor).
National Turfgrass Council Workshop Report No. 10, Bingley, West Yorkshire, England, UK. 73 pp.

Guideline Specifications to Turfgrass Sodding
1972 and periodically through 2006.
Turfgrass Producers International, Rolling Meadows, Illinois, USA. 16 pages.

Guide to Vermont Ski Trail Construction and Management
circa 1972. Glenn M. Wood, Paul Winkelaar, and John J. Lindsay.
Pamphlet 39, Agricultural Experiment Station, University of Vermont, Burlington, Vermont, USA. 27 pages.

How to Build and Save Beaches and Dunes
1971. John A. Jagschitz and Robert C. Wakfield.
Marine Leaflet Series Number 4, University of Rhode Island, Kingston, Rhode Island, USA. 12 pages.
Reprinted in 1976.

Indoor Domed Stadia Natural Turf Feasibility Study, Final Report
1991. James B Beard, Arthur J.M. Milberger, and Cornelius H.M. van Bavel.
Department of Soil and Crop Science, Texas A&M University, College Station, Texas, USA. 44 pages.

Lawn Experiments
1949. K. Welton and J.C. Carroll.
Ohio Agricultural Experiment Station, Wooster, Ohio, USA. 45 pages.

Lawn Grasses and Their Management
    1938. H.F.A. North, T.E. Odland, and J.A. DeFrance.
    Bulletin 264, Agricultural Experiment Station of the Rhode Island State College, Kingston, Rhode Island, USA. 36 pages.

The Lawn Problem in the South
    1915. W.C. Coker.
    Journal of the Elisha Mitchell Scientific Society, Chapel Hill, North Carolina, USA. 31(3):162–165.

Nordic Research on Limitation of Winter Damage to Golfgreens
    1996. L. Johansson.
    Swedish University of Agricultural Sciences, Uppsala, Sweden. 49 pages.

Observations on the Lawns of Chapel Hill
    1915. W.C. Coker and E.O. Randolph.
    Journal of the Elisha Mitchell Scientific Society, Chapel Hill, North Carolina, USA. 31(2):113–119.

Playing Surfaces, Final Report
    1989.
    Commission of Enquiry, The Football League, London, England, UK. 200 pp.

Post 1976 Turfgrass Industry Challenges in Research, Teaching, and Continuing Education
    1979.
    A Bicentennial Symposium of the Turfgrass Science Division C-5, Crop Science Society of American, Madison, Wisconsin, USA. 33 pages.

Racetrack Design and Performance
    2004. A.K. Stubbs et al.
    Publication No. 04/039, Rural Industries Research and Development Corporation, Australian Capital Territory, Australia. 56 pages.

A Research Study and Review of Intensively Managed Amenity Turfgrass in the UK
    1987. Brian M.C. Symes.
    Special Report No. 2, National Turfgrass Council, Bingley, West Yorkshire, England, UK. 40 pages.

A Review of Playing Surfaces Research
    1983. J.P. Shildrick and A.L. Dye.
    National Turfgrass Council, Bingley, Yorkshire, England, UK. 33 pages.

Sod and Turf: Michigan's $350 Million Carpet
    1969. J. B Beard (editor).
    Michigan Science in Action, Michigan State University Agricultural Experiment Station, East Lansing, Michigan, USA. 24 pages.

Standards of Playing Quality for Natural Turf
    1987. G. Holmes and M.J. Bell.
    Sports Turf Research Institute, Bingley, West Yorkshire, England, UK. 80 pages.

Strategy for Establishment and Maintenance of Fescue on Danish Golf Courses
    2003. Bente Mortensen.
    North Sealand Experience Working Group, North Sealand, Denmark. 37 pages.

A Study of Factors Affecting the Growth of Lawn Grasses
    1943. Walter S. Lapp.
    Proceedings of the Pennsylvania Academy of Science, Pennsylvania, USA. XVII:117–148.

A Study of Front-Throw Lawn Mowers on the Hybrid Bermudagrasses
    1968.
    Cal-Turf, Camarillo, California, USA. 36 pages.

To Play Like Grass
1985. Peter Dury (Junior) and Peter L.K. Dury.
Nottinghamshire County Council, Nottinghamshire, England, UK. 77 pp.

Trees and Shrubs for Noise Abatement
1971. David I. Cook and David F. Van Hoverbeke.
Research Bulletin 246, USDA Forest Service and University of Nebraska, Lincoln, Nebraska, USA. 77 pages.

Turfgrass Investigations, Compilation of Research Data
1963.
Texas Agricultural Experiment Station, Texas A&M University, College Station, Texas, USA.

Turfgrass Water Conservation Species, Cultivars, and Cultural Strategies
1974. James B Beard.
Australia Irrigation Exposition, Sydney, New South Wales, Australia. 19 pages.

Turf Industry in Ontario
1976. J.L. Eggens (editor).
Notes on Agriculture, University of Guelph, Guelph, Ontario, Canada. 21 pages.

Vegetative Erosion Control Studies, Tennessee-Tombigbee Waterway
1981. Jeffrey V. Krans, Clifford Trammel, Richard Harrod, and Veronica Henning.
Final Report, Mississippi Agriculture and Forestry Experiment Station, Mississippi State University, Mississippi, USA. 134 pages.

Water Quality, Pesticide Occurrence, and Effects of Irrigation with Reclaimed Water at Golf Courses in Florida
1996. Amy Swancar.
Water-Resources Investigations Report 95-4250, U.S. Geological Survey, Tallahassee, Florida, USA. 90 pages.

Winter Grasses of Chapel Hill
1915. W.C. Coker.
Journal of the Elisha Mitchell Scientific Society, Chapel Hill, North Carolina, USA. 31(3):156–161.

Figure 15.16. Joe DiPaola in original turfgrass rhizotron at Texas A&M University in 1976. [26]

Turfgrass Industry

Alabama's Turfgrass-Sod Industry
1991. Robert W. White, John L. Adrian, and Ray Dickens.
Bulletin 610, Alabama Agricultural Experiment Station, Auburn University, Auburn, Alabama, USA. 55 pages.

Allocating Resources for Golf Course Maintenance: An Economic Analysis
1974. W.W. Wood Jr. and John Van Dam
Agricultural Extension MA-73, University of California, USA. 17 pages.

British Columbia Turfgrass Industry Profile
1999.
Western Canada Turfgrass Association Industry Profile Research Study, British Columbia, Canada. 181 pages.

Contribution of the Golf Course Industry to the State Economy: South Carolina, 1994
1995. David L. Barkley, Mark S. Henry, and Michalann G. Evatt.
Extension Economics Report 159, Department of Agricultural and Applied Economics, Clemson University, South Carolina, USA. 25 pages.

Contribution of the Golf Industry to the Arizona Economy
1989. David L. Barkley and Larry Simmons.
Technical Bulletin 263, Department of Agricultural Economics, College of Agriculture, University of Arizona. Tucson, Arizona, USA. 40 pages.

Contributions of the Ornamental Horticulture and Turfgrass Industries to the State Economy, South Carolina, 1994
1995. P. James Rathwell, Mark S. Henry, David L. Barkley, and Michalann G. Evatt.
Extension Economic Report 160, Department of Agricultural and Applied Economics, Clemson University, Clemson, South Carolina, USA. 40 pages.

Contributions of the Ornamental Horticulture and Turfgrass Industry to the South Carolina Economy, 1999
2001. P. James Rothwell, Michalann G. Evatt, and Mark S. Henry.
Department of Agricultural and Applied Economics EER 194, Clemson University, Clemson, South Carolina, USA. 41 pages.

Contribution of the Turfgrass Industry to Florida's Economy, 1991–92: A Value-Added Approach
1994. A.W. Hodges, J.J. Haydu, P.J. van Blokland, and A.P. Ball.
Economics Report ER94-1, Food and Resource Economics Department, Institute of Food and Agricultural Sciences, University of Florida, Gainesville, Florida, USA. 68 pages.

The Costs of Large Scale Turf Maintenance
n.d. circa 1966.
Economics Research Associates, Los Angeles, California, USA. 36 pages.

Economic Analysis of the Australian Turfgrass Industry
2007. David E. Aldous, John J. Haydu, and Loretta N. Satterthwaite.
University of Melbourne and University of Florida. 38 pages.

1993 Economic Contribution of Colorado's Green Industry
1997.
Department of Agricultural and Resource Economics, Colorado State University, Fort Collins, Colorado, USA. 88 pages.

Economic Feasibility of Turfgrass-Sod Production
1995. John L. Adrian Jr., William M. Lloyd, and Patricia A. Duffy.
Bulletin 625, Alabama Agricultural Experiment Station, Auburn University, Auburn, Alabama, USA. 30 pages.

An Economic and Agronomic Analysis of Mississippi Turfgrass Sod Farms
1988. Charles R. Hall, Lennie G. Kizer, Jeffrey V. Krans, Travis D. Phillips, and G. Euel Coats.
Departments of Agricultural Economics and of Agronomy, Mississippi State University, Mississippi State, Mississippi, USA. 69 pages.

The Economic Impact of the Texas Turfgrass Industry
1996. Curtis F. Lard, Charles R. Hall, and Rebecca K. Berry.
Horticultural (Turfgrass) Economics Research Report #96-9, College of Agriculture and Life Sciences, Texas A&M University, College Station, Texas, USA. 130 pages.

An Economic Survey of New Jersey Turfgrasses, 1983
1985. A. Turner Price, Daniel Rossi, Pritam Dhillon, and McKinely Hailey.
New Jersey Agricultural Experiment Station P-02530-1-85, Cook College, Rutgers, State University of New Jersey, New Brunswick, New Jersey, USA. 62 pages.

Florida Turfgrass Survey, 1974
1976.
Florida Department of Agriculture & Consumer Services, Division of Marketing, Tallahassee, Florida, USA. 33 pages.

Golf in Colorado: An Independent Study of the 2002 Economic Impact and Environmental Aspects of Golf in Colorado
2005.
Colorado State University and THK Consulting, Fort Collins, Colorado, USA. 52 pages.

A Gyepgazdálkodásról (Turfgrass Industry and Statistics)
1982. Moharos László and József Peigelbeck.
Központi Statisztakai Hivatal, Budapest, Hungary. 127 pages.

The Illinois Green Industry: Economic Impact, Structure, Characteristics
2001.
Dept. Report Series 2001-01, Department of Natural Resources and Environmental Sciences, University of Illinois, Champaign, Illinois. 115 pages.

The Impact of the Turfgrass Industry on Mississippi's Economy
1996. Michael D. Richard, J. Robert B. Field, J. Michael Goatley Jr., Victoria P. Richard, and J. Patrick Sneed.
Bulletin 1062, Mississippi Agricultural & Forestry Experiment Station, Mississippi State University, Mississippi State University, Mississippi, USA. 24 pages.

Iowa's Turfgrass Industry
2001.
Iowa Agricultural Statistics Service, Des Moines, Iowa, USA. 44 pages.

Kansas Turfgrass Survey
1994.
Kansas Agricultural Statistics, Manhattan, Kansas, USA. 20 pages.

Kentucky Turfgrass Survey
1989.
Kentucky Agricultural Statistics Service, Louisville, Kentucky, USA. 60 pages.

Maintenance Costs of Public Golf Courses
1942. Laurie Davidson Cox and Rhodell E. Owens.
Bulletin No. 22, New York State College of Forestry at Syracuse University, Syracuse, New York, USA. 38 pages.

Maryland Turfgrass Report: An Economic Impact Study
1981. James L. Frey.
Maryland Department of Agriculture Publication #201, Maryland, USA. 20 pages.

Michigan Sod Statistics
1971. Charles A. Hines and Kenneth A. Tanner.
Michigan and US Department of Agriculture, Lansing, Michigan, USA. 13 pages.

Michigan Turfgrass Industry Report
1988.
Michigan Turfgrass Foundation, Lansing, Michigan, USA. 72 pages.

The New Mexico Turfgrass Industry: An Empirical Analysis of the Demand for Turfgrass Sod
1986. Joel A. Diemer and Dino Francescutti.
Research Report 582, Agricultural Experiment Station, New Mexico State University, Las Cruces, New Mexico, USA. 16 pages.

New Mexico Turfgrass Survey
1980. Mike Fishburn.
New Mexico State University, Las Cruces, New Mexico, USA. 9 pages.

New York Turfgrass Survey
2003.
New York Agricultural Statistics Service, Albany, New York, USA. 80 pages.

North Carolina Turfgrass Industry
n.d. circa 2000.
Turfgrass Council of North Carolina, North Carolina, USA. 6 pages.

1994 North Carolina Turfgrass Survey
1995. Jeff Chaffin, Teresa Bunch, and David Luckenbach.
Number 183, North Carolina Agricultural Statistics Service, North Carolina Department of Agriculture, Raleigh, North Carolina, USA. 56 pages.

North Carolina Turfgrass Survey
1999. Kathy Neas and Holly Smith.
North Carolina Agricultural Statistics, Raleigh, North Carolina. 64 pages.

North Carolina Turfgrass Survey Detailed Summary Report, 1986
1987. Jeff Chaffin, Teresa Bunch, and David Lukenback.
North Carolina Crop and Livestock Reporting Service, Raleigh, North Carolina, USA. 76 pages.

The 1989 Ohio Turfgrass Survey
1989. Thomas L. Sporleder, Deborah L. Snyder, and William E. Distad.
Ohio State University and Ohio Turfgrass Foundation, Columbus, Ohio, USA. 150 pages.

Figure 15.17. Transplant rooting studies via glass-faced root observation boxes conducted at Michigan State University in the late 1960s. [26]

1978 Oklahoma Turfgrass Survey
1979.
Bulletin MP-105, Agricultural Experiment Station, Oklahoma State University, Stillwater, Oklahoma, USA. 16 pages.

1987 Oklahoma Turfgrass Survey
1990. Glenn E. Martin.
Master's thesis, Oklahoma State University, Stillwater, Oklahoma, USA. 69 pages.

Pennsylvania Turfgrass Survey
n.d. circa 1990. Peter J. Landschoot and Thomas L. Watschke.
Pennsylvania Turfgrass Council, Pennsylvania, USA. 24 pages.

Pennsylvania 1989 Turfgrass Survey
1989.
Pennsylvania Agricultural Statistics Service, Harrisburg, Pennsylvania, USA. 92 pages.

A Research Study and Review of Intensively Managed Amenity Turfgrass in the UK
1987. Brian M.C. Symes.
National Turfgrass Council Special Report No. 2, Bingley, West Yorkshire, England, UK. 40 pages.

1994 South Carolina Golf Course Study
1995.
Extension Economic Report 159, Department of Agricultural and Applied Economics, Clemson University, Clemson, South Carolina, USA. 25 pages.

The 1986 South Carolina Ornamentals and Turf Survey
1988. Janice Tuten, J.S. Lytle, and P.J. Rathwell.
No. 452, South Carolina Agricultural Experiment Station, Clemson University, Clemson, South Carolina. 63 pages.

1994 South Carolina Ornamental and Turfgrass Study
1995.
Extension Economic Report 160, Department of Agricultural and Applied Economics, Clemson University, Clemson, South Carolina, USA. 36 pages.

Summary Results: The 1989 Ohio Turfgrass Survey
1991.
Ohio Turfgrass Foundation, Columbus, Ohio, USA. 11 pages.

Turfgrass Enterprise Cost Analysis
1987. William Givan.
Miscellaneous Publication No. 280, Cooperative Extension Service, College of Agriculture, University of Georgia, Athens, Georgia, USA. 13 pages.

Turfgrass-Sod Marketing in Alabama
1985. John Adrian, Charles Lokey, and Ray Dickens.
Bulletin 571, Alabama Agricultural Experiment Station, Auburn, Alabama, USA. 47 pages.

1966 Turfgrass Survey
1967.
CRS-42, Pennsylvania Crop Reporting Service, Pennsylvania Department of Agriculture, Harrisburg, Pennsylvania. 38 pages.

Turfgrass Survey, Los Angeles County, California
1954. James Beutel and Fred Roewekamp.
Southern California Golf Association, Los Angeles, California, USA. 12 pages.

Turf Survey in Southern California
1950. Charles K. Hallowell.
University of California, Los Angeles, California, USA. 17 pages.

The Value of the Ornamental Plants and Turfgrass Industries to the Louisiana Economy
1997. David W. Hughes and Roger A. Hinson.
Department of Agricultural Economics Research Report No. 708, Louisiana State University Agricultural Center, Baton Rouge, Louisiana, USA. 47 pages.

Virginia's Turfgrass Industry
2000.
Virginia Agricultural Statistics Service, Richmond, Virginia, USA. 80 pages.

Virginia Turfgrass Survey
1983.
Virginia Crop Reporting Service, Richmond, Virginia, USA. 47 pages.

1973 Virginia Turfgrass Survey
1973.
Virginia Cooperative Crop Reporting Service, Richmond, Virginia. 40 pages.

Washington Turfgrass Survey, 1967
1969. Roy L. Goss, Emery C. Wilcox, and Alvin G. Law.
Cooperative Extension Service, College of Agriculture, Washington State University, Pullman, Washington, USA. 33 pages.

Wisconsin Turfgrass Industry Survey
1966.
University of Wisconsin, Madison, Wisconsin, USA. 18 pages.

Wisconsin Turfgrass Industry Survey
1999.
University of Wisconsin and Wisconsin Agricultural Statistics Service, Madison, Wisconsin, USA. 16 pages.

Miscellaneous

An Engineering-Economic Analysis of Systems Utilizing Aquifer Storage for the Irrigation of Parks and Golf Courses with Reclaimed Wastewater
1967.
Hydrology and Water Resources Publication No. 5, Center for Water Resources, Desert Research Institute, University of Nevada System at Reno, Reno, Nevada, USA. 131 pages.

Golf Courses and Wildlife: A Literature Review
2000.
Müstakis Institute for the Rockies, University of Calgary, Calgary, Canada. 77 pages.

Performance Standards for Surface Levels of Artificial Turf Sports Surfaces
1988. J.B. Murphy and S.W. Baker.
Sports Turf Research Institute, Bingley, West Yorkshire, England, UK. 62 pages.

Racetrack Design & Performance
2004. A.K. Stubbs.
Rural Industries Research and Development Corporation Publication No. 04/039, Barton, Australian Capital Territory, Australia. 56 pages.

Racetrack Management
> 2002. A.K. Stubbs and J.J. Neylan.
> Rural Industries Research and Development Corporation Publication No. 02/002, Kingston, Australian Capital Territory, Australia.
>> 135 pages.

A Review of Playing Surfaces Research
> 198x. J.P. Shildrick and A.L. Dye.
> National Turfgrass Council, Bingley, Yorkshire, England, UK. 33 pages.

Stress Physiology and Microclimate of Turfgrasses
> 1984. James B Beard.
> Department of Soil and Crop Sciences, Texas A&M University, College Station, Texas, USA. 37 pages.

A Survey of Synthetic Turf Kickabout Areas
> 1988. J.P. Murphy and S.W. Baker.
> Sports Turf Research Institute, Bingley, West Yorkshire, England, UK. 12 pages.

Figure 15.18. Cold hardiness poststress survival assessments under glasshouse conditions, in the 1960s. [26]

CHAPTER 16

Turfgrass Conference Proceedings

THE NATIONAL, REGIONAL, AND STATE TURFGRASS CONFERENCES HAVE SERVED AS A MAJOR MEANS of educating turfgrass professionals in the United States. Typically, they were sponsored jointly by a state land-grant university in cooperation with a state or regional turfgrass association. The resultant conference proceedings contain papers by professional turfgrass managers plus practically oriented interpretations of research results by turfgrass scientists associated with state universities and private companies. These turfgrass conference proceedings provide one of the more interesting documented sources concerning the evolution of turfgrass science from the late 1940s to the 1980s. Note that the years when conference proceedings were published do not necessarily correspond with the years when actual conferences were held. Most turfgrass conference proceedings ceased to be published by the 1990s. Their role as a basic information source had been replaced by turfgrass books and trade periodicals, while advances in turfgrass research were communicated via annual turfgrass research summaries and reports and field day reports from universities.

United States Turfgrass Conference Proceedings

The pioneering initiation of turfgrass conferences in most states originated with key golf course superintendents, turfgrass agronomists employed by private companies, and academic agronomists. Most of the early turfgrass conferences were organized with leadership provided by the state university and college turfgrass educators, and eventually, leaders from state turfgrass organizations assumed an increasing role in the planning and implementation. A summary of the years when state or regional turfgrass conferences were published is shown in table 16.1. These data are based on known holdings primarily held at the Michigan State University Library and from surviving records of some sponsoring organizations.

The three oldest known, continuously published annual state turfgrass conference proceedings were (a) the **Northwest Turfgrass Conference Proceedings** initiated by Washington State University in 1946 and continued through 1997, a total of 52 years; (b) the **Midwest Regional Turfgrass Conference Proceedings** published by Purdue University for 38 years from 1948 to 1986; and (c) the **Florida Turf Conference Proceedings** published initially in 1949 by the University of Florida and continued through 1994, a total of 45 years.

Many of the older turfgrass educational conference proceedings have been discarded without recognizing their historical significance. Thus the James B Beard Collection at the Michigan State University Library serves as a very important depository for holding collections, especially for the earlier conference

TABLE 16.1. SUMMARY OF YEARS WHEN KNOWN US STATE AND REGIONAL TURFGRASS CONFERENCE PROCEEDINGS WERE PUBLISHED.

Conference proceedings	Year(s) conference proceedings published	Total year(s)
Alabama (Auburn Turfgrass Short Course)	1960–1968	8
Arizona	1953–1960, 1972–1985	22
California, Northern (Davis)*	1950–1966	17
California, Southern (LA/Riverside)*	1950–1954, 1959–1960, 1963, 1970–2002	40
Colorado (Rocky Mountain Regional)	1954–1968, 1972–1981, 1989	26
Florida*	1949–1994	45
Georgia (Southeastern)	1951, 1955–1970	17
Hawaii	1965–1970	6
Illinois	1960–1986	27
Indiana/Midwest Regional (Purdue)	1948–1986	38
Kansas (Central Plains)*	1950–1951, 1961, 1963	4
Maine	1963–1970	8
Maryland (Mid-Atlantic)*	1949–1951, 1953–1966, 1968–1976	16
Massachusetts*	1956–1985	30
Michigan	1972–2005	34
Missouri (Lawn & Turf)	1960–1967, 1970–1984	23
Nebraska	1962–1964, 1976–1981	9
Nevada	1986	1
New Hampshire	1980–1985	6
New Jersey (Rutgers)*	1949–1960, 1963–1964, 1966, 1969–2010	56
New Mexico (Southwest)*	1947–1948, 1951, 1955–1962, 1982–2001	31
New York	1977–1983	7
North Carolina	1981, 1985, 1987, 1989	4
Ohio	1967–1999	33
Oklahoma	1953–1963	10
Oklahoma-Texas	1949–1950	2
Pennsylvania	1949–1956	8
Short Course on Roadside Dev. (Ohio)	1953–1972	20
South Carolina*	1974–1977, 1986–1990	9
Southeastern New York	1950	1
Southern	1960, 1963–1965, 1967–1968, 1970–1974	11
Tennessee*	1966–1981, 1985–1990	22
Texas*	1951–1980	30
Turfgrass Sprinkler Irrigation (California)	1963–1977, 1980, 1982	17
Utah (Intermountain)	1954, 1963–1964	3
Virginia	1961, 1963, 1969–1970, 1973–1994	26
Washington (Northwest)*	1946–1997	52
West Virginia	1967–1968	2
Wisconsin*	1962–1966, 1973, 1980, 1981, 1983	9

*Published under several different names.

proceedings of the 1940s, 50s, and 60s. Many of the state and regional conference proceedings published during the 1950s were sponsored by the West Point Products Corporation under the leadership of Thomas Mascaro. This was a key contribution in the initial education of turfgrass managers across the United States.

International Turfgrass Conference Proceedings

At the national and international level, the oldest educational turfgrass conference has been sponsored by the Golf Course Superintendent Association of America (GCSAA). Proceedings have been published intermittently for ~51 years from 1938 to 1998. The conference itself originated in 1926. The New Zealand Turf Conference is the second oldest. The **New Zealand Turf Conference Proceedings**, with various earlier titles, was first published in 1939 and published subsequently in intermittent years, for a total of 17. The third oldest international conference is the Canadian National Turfgrass Conference, which in its earlier years was sponsored by the Royal Canadian Golf Association. Proceedings of this educational conference were published from 1950 through 1988, a total of 39 years.

Two more recent international conferences include the conference sponsored by the Australian Golf Course Turf Management Association, whose proceedings have been published since 1969 on a biennial basis, and the National Seminar on Turf sponsored by the Royal Australian Institute of Parks and Recreation, with proceedings first published in 1980.

TABLE 16.2. SUMMARY OF YEARS WHEN KNOWN INTERNATIONAL TURFGRASS CONFERENCE PROCEEDINGS WERE PUBLISHED.

Conference proceedings name	Year(s) conference proceedings published	Total years conference proceedings published
American Sod Producers	1981–1986	6
ATRI Turfgrass (New South Wales, Australia)	1980–1997	18
Australian Turfgrass[*]	1969, 1972–2002[**]	30
Canadian National Turfgrass[*]	1950–1989	39
National/International Turfgrass, Golf Course Superintendents Assn. of America.[*]	1930–1933, 1936–1937, 1949–1950, 1952–1953, 1954–1963, 1966–1975, 1977–1998	51
National Seminar (P&R, Australian Capital Territory)[*]	1980–1983, 1988, 1990, 1999, 2001	8
National Turfgrass (England)[*]	1982–1985	4
New Zealand[*]	1939, 1951–1955, 1980, 1982, 1985, 1990, 1991, 1994, 1998, 2001, 2003, 2005, 2007	17
Science and Golf (WSCG, Scotland, UK)	1990, 1994, 1998, 2002, 2008	5
Trávniky (Czech Republic)	1998	1
Univ. Melbourne (Victoria, Australia)[*]	1983, 2002–2003	3
Western Canada (British Columbia, Canada)[*]	1971, 1982	2
Western Turf (Alberta, Canada)	1953	1

[*]Published under several different names.
[**]Alternate years 1972 to 1990.

FORMAT

Publications in this category are printed at regular intervals, usually annually. The listings are grouped as (1) United States National Turfgrass Conference Proceedings, (2) United States Regional and State Turfgrass Conference Proceedings, and (3) International Turfgrass Conference Proceedings. The citation style used in this chapter is as follows:

- Name of conference.
- Name and location of sponsor—city, state, and country.
- Years published; the first year listed is when the proceedings was first published, while the second year listed is when publication of the proceedings was terminated; if the proceedings continues to be published, this is indicated by a "+."
- Previous or subsequent names and dates published for the same proceedings are listed in parentheses.

A. UNITED STATES NATIONAL TURFGRASS CONFERENCE PROCEEDINGS

American Sod Producers Association Midwinter Conference
American Sod Producers Association, Hastings, Nebraska/Rolling Hills, Illinois, USA. 1981–1986.

Annual Conference, in the Golf Course Reporter
Golf Course Superintendents Association of America, Jacksonville Beach, Florida, USA. 1960.
(*formerly* National Turfgrass Conference, in the Golf Course Reporter, 1954–1959).
(*see* Summary of Papers Presented at the International Turfgrass Conference, 1961–1963).

Annual Conference of NAGA
National Association of Greenkeepers of America, Chicago, Illinois, USA. 1936–1937.
(*formerly* Annual Conference, 1930–1932).
(*see* National Turf Conference, 1949–1950).

Annual Convention, in the National Greenkeeper
National Association of Greenkeepers of America, Cleveland, Ohio, USA. 1933.
(*formerly* National Greenkeepers Conference, 1930–1932).
(*see* Annual Conference of NAGA, 1936–1937).

Conference Summary, in The Golf Course Reporter
Golf Course Superintendents Association of America, St Charles, Illinois, USA. 1952–1953.
(*formerly* National Turf Conference, 1949–1950).
(*see* National Turfgrass Conference, in the Golf Course Reporter, 1954–1959).

Golf Environments of the 21st Century: Integrating Wildlife Conservation and Golf Courses
United States Golf Association and National Fish and Wildlife Foundation, Far Hills, New Jersey, USA. 1997.

International Turfgrass Conference
Golf Course Superintendents Association of America, Des Plaines, Illinois/Lawrence, Kansas, USA. 1966–1975 and 1977–1984.
(*formerly* Summary of Papers Presented, 1961–1963).
(*see* International Golf Course Conference, 1985–1998).

International Golf Course Conference
Golf Course Superintendents Association of America, Lawrence, Kansas, USA. 1985–1998.
(*formerly* International Turfgrass Conference, 1977–1984).

National Greenkeepers Convention
National Association of Greenkeepers of America, published by the O.M. Scott and Sons Co., Marysville, Ohio, USA. 1930–1932.
(*see* Annual Convention, 1933).

National Turf Conference
The Greenkeeping Superintendents Association, Chicago, Illinois, USA. 1949–1950.

(*formerly* Annual Conference of NAGA, 1936–1937).
(*see* Conference Summary, 1952–1953).

National Turfgrass Conference, in the Golf Course Reporter
Golf Course Superintendents Association of America, St Charles, Illinois, USA. 1954–1959.
(*formerly* Conference Summary, in the Golf Course Reporter, 1952–1953).
(*see* Summary of Papers Presented at the International Turf-Grass Conference, 1961–1963).

Summary of Papers Presented at the International Turf-Grass Conference
Golf Course Superintendents Association of America, Jacksonville Beach, Florida, USA. 1961–1963.
(*formerly* Annual Conference of GCSAA, 1960).
(*see* International Turfgrass Conference, 1966–1975).

Wastewater Conference
Jointly ASGCA Foundation, USGA Green Section, NGF, and GCSAA. 1978.

B. UNITED STATES REGIONAL AND STATE TURFGRASS CONFERENCE PROCEEDINGS

Arizona Turfgrass Conference or Arizona Turf and Landscape Conference or Arizona Turf, Landscape, and Irrigation Conference
University of Arizona, Tucson, Arizona, USA. 1953–1960, and 1972–1985.

Bluegrass Symposium
Oregon State University, Tualatin, Oregon, USA. 1977.

California Golf Course Superintendents Institute
California Golf Course Superintendents Association and University of California at Davis, Pacific Grove, California, USA. 1973–1981.

Central Plains Turf Conference
Central Plains Turf Foundation and Kansas State College, Manhattan, Kansas, USA. 1950–1962.

Clemson University Turfgrass Conference
Clemson University, Clemson, South Carolina, USA. 1974–1977.

Desert Turfgrass Conference
University of Nevada, Reno, Nevada, USA. 1986.

Florida Turf Conference
University of Florida, Gainesville, Florida, USA. 1953–1954.
(*formerly* Turf Management Conference, 1949–1950).
(*see* University of Florida Turf Management Conference, 1955–1958).

Florida Turf-Grass Conference or Florida Turfgrass Conference
Florida Turf-Grass Association, Orlando, Florida, USA. 1981–1994.
(*formerly* Florida Turf-Grass Management Conference, 1967–1980).

Florida Turf-Grass Management Conference
Florida Turf-Grass Association, Orlando, Florida, USA. 1967–1980.
(*formerly* University of Florida Turf-Grass Management Conference, 1959–1966).
(*see* Florida Turfgrass Conference, 1981–1994).

Ground Maintenance Conference
South Carolina Landscape and Turfgrass Association and Clemson University, Clemson, South Carolina, USA. 1986–1990.

Illinois Turfgrass Conference
Department of Horticulture, University of Illinois, Urbana, Illinois, USA. 1960–1986.

Intermountain Turf Association Conference
Intermountain Turf Association, Salt Lake City, Utah, USA. 1963–1964.

Large Turf Irrigation Systems: Design and Management Update
Center for Irrigation Technology, California State University, Fresno, California, USA. 1987.

Lawn and Turf Conference or Missouri Lawn & Turf Conference or Missouri Green Industry Conference
Department of Horticulture, University of Missouri, Columbia, Missouri, USA. 1960–1967, 1970–1984, and 1987+.

Maine Turf Conference or University of Maine Turf Conference or Maine Winter Turf Conference
University of Maine, Orono, Maine, USA. 1963–1970.

Maryland Sod Conference
University of Maryland, College Park, Maryland, USA. 1973 and 1975.

Massachusetts Turfgrass Conference
University of Massachusetts, Amherst, Massachusetts, USA. 1981–1985.
(*formerly* Turf Clippings, 1956–1980).

Michigan Turfgrass Conference
Michigan Turfgrass Foundation and Department of Crop and Soil Sciences, Michigan State University, East Lansing, Michigan, USA. 1972–2005.
Both 2002 and 2003 are on CD-ROM and 2004 and 2005 are in an electronic format.

Mid-Atlantic Association of Golf Course Superintendents Conference
Mid-Atlantic Golf Course Superintendents Association and Department of Agronomy, University of Maryland, College Park, Maryland, USA. 1949–1951, 1953–1966, and 1968–1976.

Midwest Regional Turf Conference or Turf Conference
Midwest Regional Turf Foundation and Department of Agronomy, Purdue University, West Lafayette, Indiana, USA. 1948–1986, except for 1978 during energy crisis.

Nebraska Turfgrass Conference
Department of Horticulture, University of Nebraska, Lincoln, Nebraska, USA. 1962–1964 and 1976–1985.

New Hampshire Turf Conference
New Hampshire Golf Course Superintendents Association and Cooperative Extension Service, University of New Hampshire, Manchester, New Hampshire, USA. 1980–1985.

New Mexico Turfgrass Conference
New Mexico A&M College and New Mexico State University, University Park, New Mexico, USA. 1955–1962.

New York State Turfgrass Conference
New York State Turfgrass Association and Cornell University, Ithaca, New York, USA. 1977–1983.

North Carolina Turfgrass Conference
Turfgrass Council of North Carolina and Department of Crop Science, North Carolina State University, Raleigh, North Carolina, USA. 1980–1992.

Northern California Turf Conference
University of California, Davis, California, USA. 1950–1958.

(*see* Northern California Turfgrass Institute, 1959–1963).

Northern California Golf Course Superintendents Institute
University of California Cooperative Extension, Northern California Golf Association, and Golf Course Superintendents Association of Northern California, Davis, California, USA. 1973–1981.

Northern California Professional Turf and Landscape Exposition Education Seminar
Northern California Turfgrass Council and University of California Cooperative Extension, Haywood, California, USA. 1983–1985.

Northern California Turfgrass Institute
University of California, Davis, California, USA. 1959–1963.
(*formerly* Northern California Turf Conference, 1950–1958).
(*see* Turf, Nursery, and Landscape Tree Conference, 1964–1966).

Northwest Turf Conference
Northwest Turfgrass Association and Department of Agronomy, Washington State University, Pullman, Washington, USA. 1952–1961.
(*see* Northwest Turfgrass Conference, 1962–1997).

Northwest Turfgrass Conference
Northwest Turfgrass Association and Washington State University, Puyallup, Washington, USA. 1962–1997.
(*formerly* Northwest Turf Conference, 1946–1961).

Ohio Turfgrass Conference
Ohio Turfgrass Foundation and Department of Agronomy, Ohio State University, Columbus, Ohio, USA. 1967–1999.

Oklahoma-Texas Turf Conference
Texas A&M College, College Station, Texas, USA. 1949–1950.
(*see* Texas Turfgrass Conference, 1951–1978, and Oklahoma Turfgrass Conference, 1952–1963).

Oklahoma Turfgrass Conference
Oklahoma Turfgrass Association and Oklahoma A&M College and State University, Stillwater, Oklahoma, USA. 1953–1963 with 1960 and 1961 as one proceeding.

Pennsylvania State University Turfgrass Conference
Department of Agronomy, Pennsylvania State University, University Park, Pennsylvania, USA. 1949–1956.

Rocky Mountain Regional Turfgrass Conference
Department of Horticulture, Colorado A and M College and Colorado State University, Fort Collins, Colorado, USA. 1954–1968, 1972–1981, and 1989.

Rutgers Turfgrass Proceedings
Rutgers Center for Turfgrass Science, Cook College-Rutgers University, New Brunswick, New Jersey, USA. 1973+.

Rutgers University Short Course in Turf Management or Rutgers Turf Short Course
Rutgers University, New Brunswick, New Jersey, USA. 1949–1960, 1963–1964, 1966, and 1969–1972.

Short Course on Roadside Development
Ohio State University and Ohio Department of Highways, Columbus, Ohio, USA. 1953–1972.

Southeastern New York Turf School
New York State Turfgrass Association, New York City, New York, USA. 1950.

Southeastern NC Professional Turfgrass Conference
North Carolina State University, Raleigh, North Carolina, USA. 1991–1999.

Southeastern Turfgrass Conference
University of Georgia Coastal Plain Experiment Station, Tifton, Georgia, USA. 1951 and 1955–1970.

Southern California Conference on Turf Culture
University of California at Los Angeles, Westwood, California, USA. 1950–1951.
(*see* Southern California Turfgrass Institute, 1959–1960 and 1963).

Southern California Turf Conference or Southern California Turfgrass Conference
University of California at Los Angeles, Westwood, California, USA. 1952–1954.
(*see* Southern California Turfgrass Institute, 1955–1963).

Southern California Turfgrass Institute
University of California, Riverside, California, USA. 1955–1963.
(*formerly* Southern California Turfgrass Conference, 1952–1954).
(*see* Turf and Landscape Horticulture Institute, 1970–1989).

Southwest Turfgrass Conference
Southwest Turfgrass Association, New Mexico State University, Las Cruces, New Mexico, USA. 1947–1948, 1951, and 1982–2001.

Tennessee Turfgrass Conference
Tennessee Turfgrass Association, Franklin, Tennessee, USA. 1967–1981 and 1985–1990.
(*formerly* Turf and Landscape Clinic, 1966).

Texas Turf Conference or Texas Turfgrass Conference
Texas Turfgrass Association and Texas A&M College/University, College Station, Texas, USA. 1951–1980.
(*formerly* Oklahoma-Texas Turf Conference, 1949–1950).

Turf and Landscape Clinic
Tennessee Turfgrass Association, Franklin, Tennessee, USA. 1966.
(*see* Tennessee Turfgrass Conference, 1967–1990).

Turf and Landscape Horticulture Institute or Turf and Landscape Institute
University of California, Riverside, California, USA. 1970–1989.
(*formerly* Southern California Turfgrass Institute, 1952–1963).

Turf Clippings and Massachusetts Turfgrass Conference
Stockbridge School of Agriculture, University of Massachusetts, Amherst, Massachusetts, USA. 1956–1980.
(*see* Massachusetts Turfgrass Conference, 1981–1985).

Turf Conference
Southern Turfgrass Association, Memphis, Tennessee, USA. 1960, 1963–1965, 1967–1968, and 1970–1974.

Turfgrass Management Conference
University of Hawaii, Cooperative Extension Service, Honolulu, Hawaii, USA. 1965–1970.

Turfgrass Sprinkler Irrigation Conference
University of California Cooperative Extension, Lake Arrowhead, California, USA. 1963–1979.

Turf Management Conference
Florida Greenkeeping Superintendents Association and Southeastern Turf Conference, Gainesville, Florida, USA. 1949–1950.
(*see* Florida Turf Conference, 1953–1954).

Turf, Nursery, and Landscape Tree Conference
University of California, Davis, California, USA. 1964–1966.
(*formerly* Northern California Turf Institute, 1959–1963).

Turfgrass Short Course
Auburn University, Auburn, Alabama, USA. 1960–1968.

Turfgrass Sprinkler Irrigation Conference
Sprinkler Irrigation Association and University of California Cooperative Extension, Lake Arrowhead, California, USA. 1963–1979. (*see* The Irrigation Association Annual Technical Conference, 1980 and 1982).

University of Florida Turf-Grass Management Conference
Department of Ornamental Horticulture, University of Florida, Gainesville, Florida, USA. 1959–1966.
(*formerly* University of Florida Turf Management Conference, 1955–1958).
(*see* Florida Turf-Grass Management Conference, 1967–1980).

University of Florida Turf Management Conference
Department of Ornamental Horticulture, University of Florida, Gainesville, Florida, USA. 1955–1958.
(*formerly* Florida Turf Conference, 1953–1954).
(*see* University of Florida Turf-Grass Management Conference, 1959–1966).

Utah Turfgrass Conference
West Point Products Corporation and Utah Golf Course Superintendents, Salt Lake City, Utah, USA. 1954.

Virginia Turfgrass Conference or Virginia Turf Conference or Virginia Turfgrass and Landscape Conference
Virginia Turfgrass Council and Virginia Polytechnic Institute, Blacksburg, Virginia, USA. 1961, 1963, 1969, 1970, and 1973–1994.

Washington Turf Conference
Department of Agronomy, Washington State University, Pullman, Washington, USA. 1948–1951.
(*see* Northwest Turf Conference, 1952–1961).

West Virginia Turfgrass Conference
West Virginia University, Morgantown, West Virginia, USA. 1967–1968.

Wisconsin Golf Turf Symposium
Wisconsin Golf Course Superintendents Association and the Milorganite Division of the Milwaukee Metropolitan Sewerage District, Milwaukee, Wisconsin, USA. 1966, 1973, 1980, 1981, and 1983.

Wisconsin Turfgrass Conference
Department of Horticulture, University of Wisconsin, Madison, Wisconsin, USA. 1962–1966.

C. INTERNATIONAL TURFGRASS CONFERENCE PROCEEDINGS

Annual General Meeting and Conference of the New Zealand Greenkeepers Association
New Zealand Institute of Turf Culture, Palmerston North, New Zealand. 1953–1955.

ATRI Turfgrass Management Seminar and Symposium
The Australian Turfgrass Research Institute Ltd., Concord West, New South Wales, Australia. 1980–1997, usually with multiple site events annually totaling 34.

Australian National Turfgrass Conference
Australian Golf Course Turf Management Association, Sydney, New South Wales/Clayton, Victoria, Australia. 1969, 1972, 1974, 1976, 1978, 1980, 1982, 1984, 1986, 1988, 1990, 1992, 1994, 1996, 1998, and 2000+.
The conference was organized by various state golf course superintendent associations up until 1978.

Australian Turfgrass Seminar
Australian Golf Course Superintendents Association, Clayton, Victoria, Australia. 1993, 1995, 1997, and 1999.

Better Results from Sports Turf
Technical Bulletin No. 5, Sports Council, East Midland Region, Nottingham, Nottinghamshire, England, UK. 1978.

British Columbia Turfgrass Conference
  Western Canada Turfgrass Conference, British Columbia, Canada, 1971.

Canadian Turfgrass Conference and Show
  Canadian Golf Superintendents Association, Toronto, Canada. 1975–1988.
  (*formerly* RCGA National Turfgrass Conference, 1964–1974).

Conference of Bowling Greenkeepers and Greenkeepers
  New Zealand Institute of Turf Culture, Palmerston North, New Zealand. 1951–1952.

The Government Grounds Turf Seminar
  New Zealand Soil Bureau, Department of Scientific and Industrial Research, Wellington, New Zealand. 1984.

Institute of Groundsmanship Annual Conference
  Institute of Groundsmanship, Milton Keynes, England, UK. 1994.

National Conference on Urban Irrigation
  Irrigation Association of Australia Limited, Canberra, Australian Capital Territory, Australia. 1986.

National Seminar on Turf Management
  Royal Australian Institute of Parks and Recreation, Canberra, Australian Capital Territory, Australia. 1980, 1981, 1982, 1983, 1988, 1990, and 1999.

National Turfgrass Conference
  National Turfgrass Council, Bingley, West Yorkshire, England, UK. 1982–1985.

National Turf Grass Conference
  New Zealand Golf Course Superintendents Association Inc., Auckland, New Zealand. 1991.

New Zealand Sports Turf Conference
  New Zealand Sports Turf Institute, Palmerston North, New Zealand. 2001, 2003, and 2005.
  (*see* New Zealand Turf Conference, 2007+).
  Published every two years.

New Zealand Sports Turf Convention
  New Zealand Sports Turf Institute, Palmerston North, New Zealand. 1980, 1982, 1985, 1990, 1994, and 1998.
  (*see* New Zealand Sports Turf Conference, 2001, 2003, and 2005).
  Published every three to five years.

New Zealand Turf Conference
  New Zealand Sports Turf Institute, Palmerston North, New Zealand. 2007+.
  (*formerly* New Zealand Sports Turf Conference, 2001, 2003, and 2005).
  Published every two years.

The Next Decade in Amenity Grassland
  Agriculture and Food Science, Queen's University of Belfast, Northern Ireland, UK. 1976.

NSW Bowling Greenkeepers Association Conference
  New South Wales Bowling Greenkeepers Association, Sydney, New South Wales, Australia. 1992, 1998, 1999, 2000, 2001, and 2003.

On Course for Change
  The Royal and Ancient Golf Club of St Andrews, St Andrews, Fife, Scotland, UK. 2000.

PGA European Tour Greenkeepers Conference
  European Professional Golfers Association, England, UK. 1989 and 1995.
  Portfolio of presentations.

RCGA National Turfgrass Conference
    Green Section of the Royal Canadian Golf Association, Toronto, Ontario, Canada. 1965–1974.
    (*formerly* School of Soils, Fertilization, and Turf Maintenance, 1951–1954, and RCGA Sports Turfgrass Conference, 1950–1964).
    (*see* Canadian Turfgrass Conference, 1975–1988).

RCGA Sports Turfgrass Conference
    Ontario Agricultural College and the Green Section of the Royal Canadian Golf Association, Guelph/Toronto, Ontario, Canada.
        1950–1965.
    (*formerly* School of Soils, Fertilization, and Turf Maintenance, 1951–1954).
    (*see* RCGA National Turfgrass Conference, 1965–1974).

Report on State Tour
    NSW Golf Course Curators' Association, Sydney, New South Wales, Australia. 1954, 1957, 1960, and 1964.

School of Soils, Fertilization, and Turf Maintenance
    Royal Canadian Golf Association Green Section, Toronto, Ontario, Canada. 1951–1954.
    (*see* RCGA Sports Turfgrass Conference, 1950–1964).

Science and Golf
    World Scientific Congress of Golf, St Andrews Fife, Scotland, UK. Vol. 1, 2, 3, 4, and 5. 1990, 1994, 1998, 2002, and 2008.

Seminar
    Turfgrass Association of Australia (ACT Region), Canberra, Australian Capital Territory, Australia. 2000+.

Seventh Annual General Meeting and Conference
    New Zealand Greenkeepers Association, Palmerston, North, New Zealand. 1939.

Trávniky (Turfgrasses)
    Mendel University of Agriculture and Forestry, Brno/Agentura Bonus, Hrdejovice, Czech Republic. 1998.

Preparation and Maintenance of Turf Cricket Wickets Seminar
    Australian Instituted of Parks and Recreation, Tasmanian Division, Hobart, Tasmania, Australia. 1976.

Turfgrass Seminar
    Turfgrass Association of Australia (New South Wales Region) and Sports Turf Association NSW, Sydney, New South Wales, Australia. 2004, 2006, and 2011.

Turfgrass Seminar—Disease, Annual Bluegrass/Microorganisms
    Victorian College of Agriculture and Horticulture, Burnley, and the Turf Research Institute, Frankston, Victoria, Australia. 1983, 2002, and 2003.

Turf Seminar
    Australian Bowling Greenkeepers Federation, Hobart, Tasmania, Australia. 1999.

Western Canada Turfgrass Association Conference
    Western Canada Turfgrass Association, British Columbia, Canada. 1982.

Western Turf Conference
    Green Section of the Royal Canadian Golf Association and University of Alberta, Edmonton, Alberta, Canada. 1953.

Figure 16.1. Evapotranspiration minilysimeters being acclimated for field experiments under glasshouse conditions, in the 1980s. [26]*

---

* The numbers in brackets correspond to the figure source acknowledgments listed in the appendix.

CHAPTER 17

Turfgrass Technical/Practitioner/ Trade Periodicals

PUBLICATIONS IN THIS CATEGORY ENCOMPASS REGULARLY PUBLISHED TITLES DEVOTED PRIMARILY to turfgrass related subjects. Included are turfgrass trade and association publications of a semitechnical nature. The listings are grouped in (a) United States National Turfgrass Technical Periodicals; (b) United States Regional, State, and Area Turfgrass Periodicals; and (c) International Turfgrass Technical Periodicals.

From 1892 to 1926, or for 34 years, the turfgrass technical information was disseminated in the United States primarily via books. Then between 1926 and 1928, the first five turfgrass periodical publications appeared:

(a) The **Bulletin** of the United States Golf Association in 1921.
(b) **Clubhouse and Fairway** of M.B. Smith Publishing in 1926.
(c) The **National Greenkeeper** of the National Association of Greenkeepers of America in 1927.
(d) **Golfdom** of Herb & Joe Graffis Publishers in 1927.
(e) **Lawn Care** of the O.M. Scott & Sons Company in 1928.

Lawn Care was the first turfgrass periodical focused on residential lawns. It was first published in 1928 and continued, with issues numbers 1 to 47, for 10 years, with 5 per year (February, March, April, August, and September). During this period, collected, and sometimes selected, issues also were reprinted and released in a binder format. In some cases, earlier numbers were updated and reissued as well as new numbers being issued. It was converted to a digest-format beginning circa 1948 and was widely distributed in this collected format, "revised and brought up-to-date annually" until circa 1952. In a related offshoot, the digests were converted to small-format books beginning in 1955. In addition, regional distributions of the serial form began in 1955, with five regions and a frequency varying from four to six per year. Twenty-two regional editions were produced by 1959 with a seasonal emphasis and varying frequencies up to eight per year, including the first "special editions." Both the nature of the regionals and their frequency varied throughout, including some highly specialized ones, such as a Japanese-language Central California issue. The regional approach, though more restrained in diversity, continued until circa 1983, when the periodical became oriented to product promotion. Thereafter, an annual spring issue may have been the only one produced until circa 1997 and the eventual conversion to an electronic e-mail format.

FORMAT

The citation style used in this chapter is as follows:
- Name of publication.
- Name and location of publisher—city, state, and country.
- Years published; the first year listed is when the periodical was first published, while the second year listed is when publication of the periodical was terminated; if the periodical continues to be published this is indicated by a "+."
- Previous or subsequent names and dates published for the same periodical are listed in parentheses.

A. UNITED STATES NATIONAL TURFGRASS TECHNICAL PERIODICALS

ALA

GIE Inc., Cleveland, Ohio, USA. 1984–1988.
(*combined* from American Lawn Applicator, 1980–1982, and Lawn Care Professional, 1981–1984).
(*see* ALA: Lawn and Landscape Maintenance, 1986–1988).
Published eight times annually.

ALA: American Lawn Applicator

GIE Inc., Cleveland, Ohio, USA. 1982–1984.
(*formerly* American Lawn Applicator, 1980–1982).
(*see* ALA, 1984–1988).
Published eight times annually.

ALA: Lawn and Landscape Maintenance

GIE Inc., Cleveland, Ohio, USA. 1986–1988.
(*formerly* ALA, 1984–1988).
(*see* Lawn and Landscape Maintenance, 1989–1995).
Published 12 times annually.

American Lawn Applicator or American Lawnapplicator

American Lawn Applicator Inc., Farmington, Michigan, USA. 1980–1982.
(*see* ALA, 1984–1988).
Published six times annually.

ASPA Bulletin

American Sod Producers Association, Hastings, Nebraska, USA. 1973–1976.
(*see* TurfNews, 1977+).
Published two times annually.

Athletic Turf

Advanstar Communications Inc. (changed to Questex Media Group Inc. in 2005), Cleveland, Ohio, USA. 2001+.
(*formerly* Athletic Turf: Maintenance & Technology, 1998–2001).
Produced in an electronic format since 2003.
Published six times annually.

Athletic Turf: Maintenance & Technology

Advanstar Communications Inc., Cleveland, Ohio, USA. 1998–2001.
(*formerly* Athletic Turf Maintenance Technology, 1988–1997).
(*see* Athletic Turf, 2001+).
Published 12 times annually.

Athletic Turf Maintenance Technology

Advanstar Communications Inc., Cleveland, Ohio, USA. 1988–1997.
(*see* Athletic Turf: Maintenance & Technology, 1998–2001).
Published six times annually.

Athletic Turf News

> Harcourt Brace Jovanovich Publications (changed to Questex Media Group Inc. to 2005), Cleveland, Ohio, USA. 2003+.
> Produced 12 times annually in an electronic format.
> Published 12 times annually.

Bayer Golf Advantage

> Bayer Environmental Science, Montvale, New Jersey, USA. 2004–2008.
> Published three times annually.

Bioturf News

> Advanstar Communications Inc., Cleveland, Ohio, USA. 1993–1994.
> (An insert in Landscape Management, 1992–1993).
> Published six times annually.

The Bulletin or Bulletin

> United States Golf Association Green Section, Washington, DC, USA. 1921–1933.
> A classic, early periodical containing a combination of turfgrass research summaries and field experience of practitioners.
> (*see* Turf Culture, 1939–1942).
> Published 12 times annually.

Business Management Newsletter or Business Management

> American Sod Producers Association (changed to Turfgrass Producers International in 1994), Hastings, Nebraska/Hillsdale, Rolling
> Meadows, and East Dundee, Illinois, USA. 1982+.
> Produced in an electronic format only since 2009.
> Published 6 to 12 times annually.

Clippings

> Golf Course Mechanics Association, Weston, Massachusetts, USA. 1995.
> Published at variable times annually.

Clubhouse and Fairway

> M.B. Smith Publishing Co., Chicago/Mount Morris, Illinois, USA. 1926–1927.
> Published 12 times annually.

Crittenden Turf News or Crittenden Turf

> Crittenden Turf Inc., Coronado/Escondido, California, USA. 1995–2004.
> Published 12 times annually.

The Crowning Touch

> Advanta Seeds West, Albany, Oregon, USA. 1996–1998.
> Published two to four times annually.

Dakota Turf Talk

> Dakota Peat & Equipment, Grand Forks, North Dakota, USA. 1997–2000.
> Published at variable times annually.

Earth-Shaping News

> Golf Course Builders Association of America, Lincoln, Nebraska, USA. 1991+.
> Published four times annually.

Fore Front

> Golf Course Superintendents Association of America, Lawrence, Kansas, USA. 1974–1981.
> Published six times annually.

The Forum

> Professional Grounds Management Society, Baltimore, Maryland, USA. 1977+.

(*formerly* Grounds Management Forum, 1970–1976).
Produced in an electronic format since 2004.
Published six times annually.

Foster Turf News
Palm Desert, California, USA. 1987–1989.
Published at variable times annually.

Foundation
GCSAA Scholarship & Research, Lawrence, Kansas, USA. 1993–1998.
(*see* IMPACT, 1999–2008).
Published two times annually.

GCM
Golf Course Superintendents Association of America, Lawrence, Kansas, USA. 2007+.
(*formerly* Golf Course Management, 1979–2006).
Also became available electronically in 2009 as Digital GCM.
Published 12 times annually.

GCM NewsWeekly
Golf Course Superintendents Association of America, Lawrence, Kansas, USA. 2008–2009.
(*formerly* GCSA NewsWeekly, 1999–2008).
Produced 51 times annually in an electronic format.

The GCS Bulletin
The Golf Collectors Society Inc., published in various locations where the editor was resident, USA. 1970+.
Published at variable times annually.

GCSAA Industry Spotlight
Golf Course Superintendents Association of America, Lawrence, Kansas, USA. 2010+.
Produced 52 times annually in an electronic format.

GCSAA Newsletter
Golf Course Superintendents Association of America, Lawrence, Kansas, USA. 1976–1981.
(*see* Newsline, 1982–2006).
Published four times annually.

GCSAA NewsWeekly
Golf Course Superintendents Association of America, Lawrence, Kansas, USA. 1999–2008.
(*see* GCM NewsWeekly, 2008–2009).
Published 51 times annually.

GCSAA This Week
Golf Course Superintendents Association of America, Lawrence, Kansas, USA. 2006+.
Produced 52 times annually in an electronic format.

Golf
Josiah Newman Publisher (changed to Harper & Brothers in 1900), New York City, New York, USA. 1897–1917.
(*became* Official Bulletin of the United States Golf Association in 1899).
(*absorbed* Golfing in 1899 and American Golf in 1902).
Published 12 times annually.

Golf Advantage
Bayer Environmental Science, Research Triangle Park, North Carolina, USA. 2004.
Published at variable times annually.

Golf & sportsTURF
Sports Turf Managers Association and Gold Trade Publishers (changed to STMA and Adams Publishing in 1991), Van Nuys, California, USA. 1990–1992.
(*formerly* sportsTURF, 1985–1989).
(*see* SportsTURF, 1992+).
Published 12 times annually

Golf and Turf Annual
Golf & Turf Annual Inc., North Miami, Florida, USA. 1990–1992.
Published once annually.

Golf Business
The Harvest Publishing Co., Cleveland, Ohio, USA. 1976–1981.
(*formerly* Golfdom, 1974–1976).
Published 12 times annually.

The Golf Course
Peterson, Sinclaire, & Miller Inc. and Carters Tested Seeds Inc., New York City, New York, USA. 1916–1918 and 1921.
An early turfgrass periodical.
Published at variable times annually.

Golf Course Industry
GIE Media Inc., Richfield, Ohio, USA. 2007+.
(*formerly* Golf Course News, 1988–2006).
Produced in an electronic format since 2007.
Published 12 times annually.

Golf Course Irrigation (changed in 1997 to Golf Course and Irrigation)
Adams/Green Industry Publishing, Cathedral City, California, USA. 1993–1997.
(*changed* to Golf Course and Irrigation in 1997).
Published six times annually.

Golf Course Management
Golf Course Superintendents Association of America, Lawrence, Kansas, USA. 1979–2006.
(*see* GCM, 2007+).
(*formerly* The National Greenkeeper, 1927–1933; The National Greenkeeper and Turf Culture, 1933; The Greenkeepers' Bulletin, 1933;
The Greenkeepers' Reporter, 1933–1951; The Golf Course Reporter, 1951–1965; and The Golf Superintendent, 1966–1978).
Published 12 times annually.

Golf Course News
United Publications Inc., Yarmouth, Maine, USA (changed to GIE Media Inc., Cleveland, Ohio, USA, in 2004). 1988–2006.
(*see* Golf Course Industry, 2007+).
Published 6 to 12 times annually.

Golf Course Reporter
Golf Course Superintendents Association of America, Chicago, Illinois/Jacksonville Beach, Florida, USA. 1951–1965.
(*formerly* The National Greenkeeper, 1927–1933; The National Greenkeeper and Turf Culture, 1933; The Greenkeepers' Bulletin,
1933; The Greenkeepers' Reporter, 1933–1951).
(*see* The Golf Superintendent, 1966–1978; Golf Course Management, 1979–2006; and GCM, 2007+).
Published six times annually through 1952 and then eight times annually.

Golf Course Superintendents Association of America Newsletter
Golf Course Superintendents Association of America, Jacksonville Beach, Florida/Des Plaines, Illinois/Lawrence, Kansas, USA.
1962–1978.
Published four times annually.

The Golf Course Trades

The Trades Publishing Co., Crossville, Tennessee, USA. 1985+.
Produced in an electronic format since 2008.
Published 12 times annually.

Golfdom

Herb & Joe Graffis Publishers, Chicago, Illinois, USA. 1927–1965.
A pioneering golf course management periodical.
(*see* Golfdom, 1965–1972).
Published 10 to 12 times annually.

Golfdom

Advanstar Communications Inc., Cleveland, Ohio, USA. 1999–2004.
(*formerly* Golfdom, 1974–1976).
(*see* Golfdom, 2004+).
Published 12 times annually.

Golfdom

The Harvest Publishing Co., Cleveland, Ohio, USA. 1974–1976.
(*formerly* Golfdom, 1972–1974).
(*see* Golf Business, 1976–1981).
Published 10 times annually.

Golfdom

Popular Science Publishing Co./Times Mirror Magazine, New York City, New York, USA. 1972–1974.
(*formerly* Golfdom, 1965–1972).
(*see* Golfdom, 1974–1976).
Published 10 times annually.

Golfdom

Universal Publishing and Distributing Corp./UPD Publishing Corp., New York City, New York, USA. 1965–1972.
(*formerly* Golfdom, 1927–1965).
(*see* Golfdom, 1972–1974).
Published 10 times annually.

Golfdom

Questex Media Inc., Cleveland, Ohio, USA. 2004+.
(*formerly* Golfdom, 1999–2004).
Published 12 times annually.

The Golf Superintendent

Golf Course Superintendents Association of America, Des Plaines, Illinois/Lawrence, Kansas, USA. 1966–1978.
(*formerly* The Golf Course Reporter, 1951–1965).
(*see* Golf Course Management, 1979–2006, and GCM, 2007+).
Published 12 times annually.

Golfweek's Superintendent News (changed in 2003 to Golfweek's SuperNews)

The Golfweek Group, Turnstile Publishing Co., Orlando, Florida, USA. 1999–2007.
(*changed* to Golfweek's SuperNews in 2003).
(*see* TurfNet, the Magazine, 2007+).
Published 1 to 24 times annually.

Government Relations Advocacy

Golf Course Superintendents Association of America, Lawrence, Kansas, USA. 1999–2000.
Published 12 times annually as part of Newsline.

Government Relations Briefax
Golf Course Superintendents Association of America, Lawrence, Kansas, USA. 1996–1998.
Published 12 times annually and distributed by fax.

Government Relations Briefing
Golf Course Superintendents Association of America, Lawrence, Kansas, USA. 1989–1995.
(*see* Government Relations Briefax, 1996–1998).
(*incorporated* into Newsline in 1996).
Published 12 times annually.

Grass Blade Review
Jacklin Seed Co., Post Falls, Idaho, USA. 2006+.
Published four times annually.

Grass Clippings
Jacklin Seed Co., Post Falls, Idaho, USA. 1988–1994.
Published four times annually.

The Greenkeepers' Bulletin
The National Association of Greenkeepers of America, Cleveland, Ohio, USA. July 1933.
(*changed* in the second issue of August, 1933, to The Greenkeepers' Reporter, 1933–1951).
(*formerly* The National Greenkeeper, 1929–1933).

The Greenkeepers' Reporter
The National Association of Greenkeepers of America, Chicago, Illinois, USA. 1933–1951.
(*formerly* The Greenkeepers' Bulletin, 1933).
(*see* The Golf Course Reporter, 1951–1965; The Golf Superintendent, 1966–1978; Golf Course Management, 1979–2006; and GCM, 2007+).
Published five times annually.
The 1934 educational program of the National Association of Greenkeepers of America held at the William Penn Hotel in Pittsburgh, Pennsylvania, USA (from page 12 of the January 1934 issue [vol. 2, no. 1] of The Greenkeepers' Reporter):

1934
GREENKEEPER'S EDUCATIONAL PROGRAM

Wednesday, January 31
Address of Welcome—By The Honorable John S. Fisher, President of the Pittsburgh Chamber of Commerce.
Getting Acquainted with the Soil—By Professor J.W. White, Penn State College, Penn State, Pa.
Golf Course Maintenance—By John McNamara, Pittsburgh Field Club, Aspinwall, Pa.
Irrigation—By Wendell P. Miller, 33 W. 60th St., New York, N.Y.
What the Depression Taught Greenkeepers About Sound Economy—By Charles Nuttall, Fox Chapel G. & C.C., Aspinwall, Pa.

Thursday, February 1
Trouble Shooting on Golf Courses in the Metropolitan Area—By Dr. Edw. E. Evaul, Agricultural Experimental Station, New Brunswick, N.J.
Lubrication—By J.H. Schenck, Gulf Oil Co. of Pittsburgh, Pittsburgh, Pa.
The Value of District Turf Gardens to Golf Clubs—By John Anderson, Crestmont C.C., West Orange, N.J.
Grubs and Webworms—By Professor L.B. Smith, Penn State College, Penn State, Pa.
Our Attitude Toward Golf and Our Clubs—By Oscar B. Fitts, Columbia C.C., Chevy Chase, Md.

Friday, February 2
Golf Course Fertilization—By O.J. Noer, Milwaukee, Wis.
Saving Money in Irrigation System Operation—By Robert Duguid, Evanston G.C., Evanston, Ill.
Economics in Course Maintenance—By Dr. John Monteith Jr., U.S.G.A., Washington, D.C.
Selling Ourselves to our Officials—By W.J. Sansom, Toronto Golf Club, Long Branch, Ontario, Canada.

Greens and Grass Roots
  Golf Course Superintendents Association of America, Lawrence, Kansas, USA. 1996–2005.
  Published 2 to 24 times annually.

Ground Rules
  CIBA Turf & Ornamental Products, Greensboro, North Carolina, USA. 1987–1995.
  Published four times annually.

Grounds Maintenance
  Intertec Publishing Corporation, Overland Park, Kansas, USA. 1966–2006.
  Published 12 times annually.

Grounds Maintenance: Golf Edition
  Intertec Publishing Corporation, Overland Park, Kansas, USA. 2000–2002.
  Published 12 times annually.

Grounds Maintenance Management Guidelines
  Professional Grounds Management Society, Pikesville/Hunt Valley/Baltimore, Maryland, USA. 1983, 1986, 1988, 1991, 1997, 1998, and 1999.
  Published periodically.

Grounds Management Forum
  Professional Grounds Management Society, Pikesville/Hunt Valley, Maryland, USA. 1970–1976.
  (*see* The Forum, 1977+).
  Published 12 times annually.

IMPACT
  The GCSAA Foundation (changed to Environmental Institute for Golf in 2003), Golf Course Superintendents Association of America, Lawrence, Kansas, USA. 2001–2008.
  Also produced in an electronic format since 2001.
  Published at variable times annually.

Irrigation & Green Industry
  ISG Communications Inc., Los Angeles/Reseda, California, USA. 1998+.
  Published six times annually.

Jacklin Research Newsletter
  Jacklin Seed and Simplot Turf & Horticulture, Post Falls, Idaho, USA. 2006+.
  Published six times annually.

Jacobsen Newsreel
  Jacobsen Division of Textron Inc., Racine, Wisconsin, USA. 1989–1990.
  Published at variable times annually.

Journal of Environmental Turfgrass (changed in 1990 from Turfgrass Environment)
  American Sod Producers Association, Rolling Meadows, Illinois, USA. 1989–1993.
  (*changed* from Turfgrass Environment in 1990).
  Published once annually.

Lakeshore News
  Lakeshore Equipment & Supply Co., Elyria, Ohio, USA. 1967–1985.
  (*see* LESCO News, 1986–2003).
  Published at variable times annually.

Lawn & Landscape
  GIE Media Inc., Cleveland, Ohio, USA. 1995+.

(*formerly* Lawn and Landscape Maintenance, 1989–1995).
Published 12 times annually.

Lawn & Landscape Digest
Lawn Institute, Marietta, Georgia, USA. 1996–1997.
Published four times annually.

Lawn and Landscape Maintenance
GIE Inc. Publishers, Cleveland, Ohio, USA. 1989–1995.
(*formerly* ALA: Lawn and Landscape Maintenance, 1988–1989).
(*see* Lawn and Landscape, 1995+).
Produced in an electronic format since 2006.
Published 12 times annually.

Lawn Care
O.M. Scott & Sons Co., Marysville, Ohio, USA. 1928–1955.
Regional editions initiated in 1955.
A pioneering turfgrass periodical for lawn enthusiasts.
Published five times annually.

Lawn Care Industry
Harvest Publishing Co. (changed to Harcourt Brace Jovanovich Publications in 1987), Cleveland, Ohio, USA. 1977–1991.
(*incorporated* into Landscape Management in 1991).
Published 12 times annually.

Lawn Care Professional
GIE Inc., Cleveland, Ohio, USA. 1981–1984.
(*see* ALA, 1984–1988).
Published six times annually.

Lawn Care (Professional Edition; changed in 1961 to The Lawn Pro)
O.M. Scott & Sons Company, Marysville, Ohio, USA. 1961–1962.
(*changed* to The Lawn Pro in 1961).
Published at variable times annually.

Lawn Care Report
Dow Chemical, Midland, Michigan (changed to DowElanco in 1989), Indianapolis, Indiana, USA. 1986–1990.
Published four times annually.

The Lawn Institute Digest
Lawn Institute, Marietta, Georgia, USA. 1992–1996.
(*formerly* Lawn Institute Harvests, 1982–1992).
(*see* Lawn & Landscape Digest, 1996–1997).
Published four times annually.

Lawn Institute Harvests
The Lawn and Turf Institute, Pleasant Hill, Tennessee, USA. 1983–1992.
Contains reviews of recently published papers.
(*see* The Lawn Institute Digest, 1992–1996).
Published four times annually.

Lawn Institute Special Topic Sheets
Lawn Institute, Pleasant Hill, Tennessee, USA. 1987–1991.
Published at variable times annually.

Lawn Servicing
Intertec Publishing Corporation, Overland Park, Kansas, USA. 1984–1989.
Published 6 to 10 times annually.

Lawn Weed Bulletin
O.M. Scott & Sons Company, Marysville, Ohio, USA. 1968–1969.
Published at variable times annually.

Leaderboard
Golf Course Superintendents Association of America, Lawrence, Kansas, USA. 1992–2005.
Published six times annually.

LESCO News
LESCO Inc., Rocky River, Ohio, USA. 1986–2003.
(*formerly* Lakeshore News, 1967–1985).
Produced in an electronic format until 2007.
Published 10 times annually.

LM Direct! Enews
Questex Media Group Inc., Cleveland, Ohio, USA. 2005+.
(*formerly* LM Week in Review, 2001–2005).
Produced 24 times annually in an electronic format.

LM Week in Review
Questex Media Group Inc., Cleveland, Ohio, USA. 2001–2005.
(*see* LM Direct! Enews, 2005+).
Published 12 times annually.

Lofts Research Update
Lofts Pedigreed Seed Inc., Bound Brook, New Jersey, USA. 1979–1983.
Published two times annually.

Managers Memo
Professional Grounds Management Society, McLean, Maryland, USA. 1973–1979.
(*formerly* The Professional Gardener, 1949–1972).
(*see* Grounds Management Forum, 1977+).
Published 12 times annually.

The Medalist
Medalist American Turfgrass Seed Co., Albany, Oregon, USA. 1992–1996.
(*formerly* Turf News Segment, 1989–1992).
Published four times annually.

The National Greenkeeper
The National Association of Greenkeepers of America, Cleveland, Ohio, USA. 1927–1933.
A pioneering golf course turfgrass periodical.
(*see* The National Greenkeeper and Turf Culture, 1933; The Greenkeepers' Bulletin, 1933; The Greenkeepers' Reporter, 1933–1951; The
    Golf Course Reporter, 1951–1965; The Golf Superintendent, 1966–1978; Golf Course Management, 1979–2006; and GCM, 2007+).
Published 12 times annually.

The National Greenkeeper and Turf Culture
The National Association of Greenkeepers of America, Cleveland, Ohio, USA. 1933.
(*formerly* The National Greenkeeper, 1927–1933).
(*see* The Greenkeepers' Bulletin, 1933; The Greenkeeper's Reporter, 1933–1951; The Golf Course Reporter, 1951–1965; The Golf
    Superintendent, 1966–1978; Golf Course Management, 1979–2006; and GCM, 2007+).
Published 12 times annually.

Newsline
Golf Course Superintendents Association of America, Lawrence, Kansas, USA. 1982–2006.
(*formerly* GCSAA Newsletter, 1976–1981).
Converted to an electronic format in 2006.
Published 12 times annually.

Northern Turf and Landscape Press
Argus Agronomics, Clarksdale, Mississippi, USA. 1994–1995.
(*formerly* Northern Turf Management, 1990–1994).
Published 12 times annually.

Northern Turf Management
Farm Press Publications/Argus Agronomics, Clarksdale, Mississippi, USA. 1990–1994.
(*see* Northern Turf and Landscape Press, 1994–1995).
Published 12 times annually.

NTEP Comings and Goings
National Turfgrass Evaluation Program, Beltsville, Maryland, USA. 2010+.
Produced at variable times annually in an electronic format.

NTEP Newsline
National Turfgrass Evaluation Program, Beltsville, Maryland, USA. 1998–2001.
Became all electronic in 2001.
Produced variable times annually.

On Course
O.M. Scott & Sons Co., Marysville, Ohio, USA. 1991–1992.
Published six times annually.

On the Green
Jacklin Seed Co., Post Falls, Idaho, USA. 1990–1997.
Published four times annually.

PACE Highlights
PACE Turfgrass Research Institute, San Diego, California, USA. 2005+.
Produced in an electronic format since 2010.
Published 12 times annually.

PACE Insights
PACE Turf, San Diego, California, USA. 1995+.
(*see* also PACE Highlights, 2005+).
Published at variable times annually.

PACE Turf Updates
PACE Turf, San Diego, California, USA. 1993+.
Produced 52 times annually, originally sent via fax and now in an electronic format.

Park & Grounds Management
Madisen Publishing Division, Appleton, Wisconsin, USA. 1991–1999.
(*formerly* Park Maintenance and Grounds Management, 1978–1990).
Published 6 to 12 times annually.

Park Maintenance
Madisen Publishing Division, Appleton, Wisconsin, USA. 1947–1977.
(*see* Park Maintenance and Grounds Management, 1978–1990).
Published 12 times annually.

Park Maintenance and Grounds Management
  Madisen Publishing Division, Appleton, Wisconsin, USA. 1978–1990.
  (*formerly* Park Maintenance, 1947–1977).
  (*see* Park & Grounds Management, 1991–1999).
  Published 12 times annually.

Pocket Seed Guide
  Harcourt Brace Jovanovich Publications, Cleveland, Ohio, USA. 1989–2006.
  Published once annually.

ProSource
  Professional Lawn Care Association of America, Marietta, Georgia, USA. 1980–1997.
  Published four to six times annually.

Proturf/ProTurf Magazine
  Proturf Division, O.M. Scott & Sons Co., Marysville, Ohio, USA. 1971–1990.
  Turf and Grounds Manager combined with ProTurf Magazine in 1984.
  (*changed* to ProTurf Magazine in 1984).
  Published two to five times annually.

Research News Flash
  Jacklin Seed Co., Post Falls, Idaho, USA. 1998+.
  Published six times annually.

Sod Grower Newsletter
  Proturf Division, O.M. Scott & Sons Co., Marysville, Ohio, USA. 1972–1976.
  Published three times annually.

Southern Golf
  Brantwood Publications Inc., Elm Grove, Wisconsin/Clearwater, Florida, USA. 1976–1984.
  (*formerly* Southern Golf Course Operations for the Bermudagrass Belt, 1970–1976).
  Published five times annually.

Southern Golf Design, Construction, & Operations
  Brantwood Publication Inc., Clearwater, Florida, USA. 1992–2000.
  (*formerly* Southern Golf Landscape and Resort Management, 1984–1992).
  Published six times annually.

Southern Golf Landscape and Resort Management
  Brantwood Publications Inc., Tampa, Florida, USA. 1984–1992.
  (*formerly* Southern Golf Course Operations, 1969–1981).
  (*see* Southern Golf Design, Construction, & Operations, 1992–2000).
  Published six times annually.

Southern Golf Course Operations for the Bermudagrass Belt
  Brantwood Publications Inc., Elm Grove, Wisconsin, USA. 1970–1976.
  (*see* Southern Golf, 1976–2000).
  Published four times annually.

Southern Golf/Southern Golf: Design, Construction & Operations
  Brantwood Publications Inc., Clearwater, Florida, USA. 1976–2000.
  (*formerly* Southern Golf Course Operations for the Bermudagrass Belt, 1970–1976).
  (*incorporated* Southern Landscape and Turf in 1984).
  (*changed* to Southern Golf: Design, Construction, and Operations in 1992).
  Published six times annually.

Southern Landscape and Turf
Brantwood Publications Inc., Elm Grove, Wisconsin/Tampa, Florida, USA. 1983–1984.
(*formerly* Landscape and Turf, 1981–1982).
(*see* Southern Golf, 1984–2000).
Published seven times annually.

Southern Turf and Landscape Press
Argus Agronomics, Clarksdale, Mississippi, USA. 1994–1995.
(*formerly* Southern Turf Management, 1990–1994).
Published 12 times annually.

Southern Turf Management
Farm Press Publications, Clarksdale, Mississippi, USA. 1990–1994.
(*see* Southern Turf and Landscape Press, 1994–1995).
Published 12 times annually.

Southern Turf Newsletter
Southern Turf Nurseries, Tifton, Georgia, USA. 1963–1976.
Published at variable times annually.

SpecTalk
Irrigation Division, The Toro Co., Riverside, California, USA. 1983–1987.
Published at variable times annually.

SportsField Management
Moose River Publishing/Moose River Media, St. Johnsbury, Vermont, USA. 2005+.
Published 10 times annually plus produced since 2011 in an electronic format.

sportsTURF
Gold Trade Publishers Inc., Van Nuys/Encino, California, USA. 1985–1989.
(*see* Golf & sportsTURF, 1990–1992).
Became the official publication of the Sports Turf Managers Association in 1988.
Published 12 times annually.

SportsTURF
Sports Turf Managers Association, Lawrence, Kansas, USA, and Adams/Green Industry Publishing, Bev-Al Communications/M2 MEDIA360, Park Ridge, Illinois, USA. 1992+.
(*formerly* Golf & sportsTurf, 1990–1992).
Published 12 times annually.

Sports Turf Manager
Sports Turf Managers Association, Council Bluffs, Iowa/Chicago, Illinois, USA. 1987–2003.
(*see* STMA News Online, 2004+).
Now available in an electronic format.
Published four to six times annually.

Sports Turf News/Sports Turf Newsletter
Sports Turf Managers Association, Ontario, California, USA. 1985–1988.
(*see* sports TURF, 1988–1989).
(*changed* to Sports Turf Newsletter in 1985).
Published four times annually.

Sports Turf Topics
Sports Turf Managers Association, Council Bluffs, Iowa/Chicago, Illinois, USA. 1992–2002.
Compendium of primarily STMA articles.
Published once annually.

STMA News Online

Sports Turf Managers Association, Lawrence, Kansas, USA. 2004+.

(*formerly* Sports Turf Manager, 1982–2003).

Produced 12 times annually in an electronic format; for the first year, it was called E-Digest.

Student Links

Golf Course Superintendents Association of America, Lawrence, Kansas, USA. 1995–circa 2001.

Published at variable times annually.

Superintendent

Moose River Media, St. Johnsbury, Vermont, USA. 2001+.

(*formerly* an insert in Turf).

Published 4 to 12 times annually.

The Sward

Turfgrass Information Center, Michigan State University Library, East Lansing, Michigan, USA. 1989+.

Produced in an electronic format since 2006.

Published at variable times annually.

T&O Service Tech

GIE Inc. Publishers, Cleveland, Ohio, USA. 1995–1998.

Published six times annually.

Timely Turf Topics

United States Golf Association, Washington, DC, USA. 1940–1948.

(*see* USGA Journal and Turf Management, 1948–1963, and USGA Green Section Record, 1963+).

Published six times annually.

Turf (North, South, Central, West, and Extra editions)

NEF Publishing Co., Moose River Publishing/Moose River Media, St. Johnsbury, Vermont, USA. 1987+.

Initiation date for North, 1988; South, 1989; Central, 1990; West, 1991; and Extra, 2003.

Published 12 times annually.

Turf and Grounds Manager or T & GM

O.M. Scott & Sons Co., Marysville, Ohio, USA. 1976–1983.

Combined with **ProTurf Magazine** in 1984.

Published four times annually.

Turf and Landscape Digest

Argus Agronomics, Clarksdale, Mississippi, USA. 1995.

(*formerly* Turf Management Digest, 1991–1994).

Published once annually.

Turf & Landscape Digest

Intertec Publishing, Overland Park, Kansas, USA. 1991–1997.

Published once annually.

Turfax

International Sports Turf Institute Inc. and James B Beard Publisher, College Station, Texas, USA. 1993–2002.

Published six times annually and distributed by fax.

Turf Comm

Douglas T. Hawes Publisher, Plano, Texas, USA. 1985–2005.

Published six to nine times annually.

Turf Culture
United States Golf Association Green Section, Washington, DC, USA. 1939–1942.
(*see* Timely Turf Topics, 1940–1948).
Published 12 times annually.

Turfgrass Producers International E-Newsletter
Turfgrass Producers International, East Dundee, Illinois, USA. 2008+.
Produced 12 or more times annually in an electronic format.

Turf-Grass Times
Turf-Grass Publications Inc., Jacksonville Beach, Florida (changed to Brantwood Publications Inc., Elm Grove, Wisconsin, in 1973), USA. 1965–1980.
(*incorporated* Golf Course Operations in 1966).
(*merged* to become Landscape & Turf Industry, 1980–1981).
Published 4 to 7 to 12 times annually.

Turfgrass TRENDS or Turfgrass trends
Turf Information Group Inc., Wilmington, Delaware, USA. 1992–1997.
(*publication* assumed by Advanstar Communications Inc., Cleveland, Ohio, USA, in 1997 and integrated within Golfdom in 2003).
Published 12 times annually.

The Turf Letter
W. Lee Berndt, Jupiter, Florida, USA. 1996.
Published four times annually.

Turf Management Digest
Farm Press Publications, Clarksdale, Mississippi, USA. 1991–1994.
(*see* Turf and Landscape Digest, 1995).
Published once annually.

TurfNet, the Magazine/TurfNet and TurfNetSports
Turnstile Publishing Co./Turnstile Media Group, Orlando, Florida, USA. 2007+.
(*formerly* Golfweek's SuperNEWS, 1999–2007).
Produced 12 to 24 times annually.

TurfNet Monthly/TurfNet-the Newsletter
Turnstile Publishing Co., Orlando, Florida, USA. 1994+.
(*changed* to TurfNet the Newsletter in 2007).
Produced 12 times annually in an electronic format.

TurfNews/TPI Turf News
American Sod Producers Association (changed to Turfgrass Producers International in 1994), Hastings, Nebraska/Hillsdale, Rolling Meadows, and East Dundee, Illinois, USA. 1977+.
(*formerly* ASPA Bulletin, 1973–1976).
(*changed* to TPI Turf News in 1994).
Published six times annually.

Turf News Segment
Medalist Division, Northrup, King & Co./NK Lawn and Garden, Minneapolis, Minnesota, USA. 1989–1992.
(*see* The Medalist, 1992–1996).
Published four times annually.

The Turf Professional
Warren's Turf Nursery Inc., Crystal Lake, Illinois, USA. 1987–1994.
Published three times annually.

The Turf Survey—A Topical Textbook of Turf
    G.A. Farley Publisher, Cleveland, Ohio, USA. 1936.
    Published 12 times annually.

Turf Talks
    O.M. Scott & Sons Co., Marysville, Ohio, USA. 1966–1974.
    Published 3 and 12 times annually.

Turf Talks
    Professional Lawn Care Association of America, Marietta, Georgia, USA. Circa 1980s.
    Published three times annually.

Turf Tech
    Turf-Seed Inc., Hubbard, Oregon, USA. Circa 1976–1988.
    Published four times annually.

Turf Tips
    Minnesota Toro Inc. and Agri-Chemicals Division of United States Steel, Minneapolis, Minnesota, USA. Circa 1977–1978.
    Published four times annually.

Turf Tips
    Sewerage Commission of the City of Milwaukee, Milwaukee, Wisconsin, USA. 1962–1967.
    Published at variable times annually.

Turf Trader
    WMR Media Inc., Perry, Georgia, USA. 2009+.
    Published six times annually.

Turf Weekly
    Golf Course Superintendents Association of America, Lawrence, Kansas, USA. 2013+.
    Produced 50 times annually in an electronic format.

The USGA Clippings
    Southeastern Region, United States Golf Association Green Section, Griffin, Georgia, USA. 1994–1998.
    Published four times annually.

USGA Green Section Record
    United States Golf Association, Far Hills, New Jersey, USA. 1963–2010.
    (*formerly* The Bulletin, 1926–1933; Turf Culture, 1939–1942; and USGA Journal and Turf Management, 1948–1963).
    Published six times annually then, in 2010, changed to a weekly electronic format.

USGA Green Section Weekly Update
    United States Golf Association Green Section, Far Hills, New Jersey, USA. 2004+.
    Produced 52 times annually in an electronic format.

USGA Journal and Turf Management
    United States Golf Association, New York City, New York, USA. 1948–1963.
    (*combined* from USGA Journal and Timely Turf Topics in 1948).
    (*see* USGA Green Section Record, 1963+).
    Published seven times annually.

Weeds and Turf: Pest Control/Weeds and Turf
    Trade Magazine Inc., The Harvest Publishing Co./Harcourt Brace Jovanovich Publications, Cleveland, Ohio, USA. 1962–1964.
    (*see* Weeds, Trees, and Turf, 1963–1987).
    (*changed* to Weeds and Turf in 1963).
    Published 12 times annually.

Weeds, Trees, and Turf

The Harvest Publishing Co./Harcourt Brace Jovanovich Publications, Cleveland, Ohio, USA. 1963–1987.
(*formerly* Weeds and Turf, 1964).
(*see* Landscape Management, 1987+).
Published 12 times annually.

Weedwatch

Questex Media Inc. and Dow Agro Sciences, Cleveland, Ohio, USA. 2011+.
Produced four times annually in an electronic format.

Western Turf and Landscape Press

Argus Agronomics, Clarksdale, Mississippi, USA. 1994–1995.
(*formerly* Western Turf Management, 1990–1994).
Published 12 times annually.

Western Turf Management

Farm Press Publications/Argus Agronomics, Clarksdale, Mississippi, USA. 1990–1994.
(*see* Western Turf and Landscape Press, 1994–1995).
Published 12 times annually.

West Pointers

West Point Lawn Products, West Point, Pennsylvania, USA. 1946–1965.
Published at variable times annually.

B. UNITED STATES REGIONAL, STATE, AND AREA TURFGRASS PERIODICALS

The Alabama Green Journal/Alabama Green

Alabama Golf Course Superintendents Association, Birmingham, Alabama, USA. 1994+.
(*changed* to Alabama Green in 2004).
Published five times annually.

Alabama Turfgrass Association Newsletter

Auburn University, Auburn, Alabama, USA. 1981–1983.
(*see* The ATA Scribe, 1984–1988).
Published four times annually.

AL Green

Alabama Golf Course Superintendents Association, Birmingham, Alabama, USA. 1987–1993.
(*see* The Alabama Green Journal, 1994+).
Published four times annually.

Arkansas Turf Tips

Department of Horticulture, University of Arkansas, Fayetteville, Arkansas, USA. 2007+.
Produced 12 times annually in an electronic format.

The ATA Scribe

Alabama Turfgrass Association, Auburn, Alabama, USA. 1984–1988.
(*formerly* Alabama Turfgrass Association Newsletter, 1981–1983).
Published four times annually.

The Ballmark

Central Illinois Golf Course Superintendents Association, Jacksonville/Auburn, Illinois, USA. 1984–1990.
(*formerly* Turfgrass News, 1981–1983).
Published four times annually.

Better Turf thru Agronomics
    Turfgrass Research Advisory Committee, University of California, Riverside, California, USA. 1996–2002.
    Published at variable times annually.

Between the Lines
    Keystone Athletic Field Managers Organization, Dauphin, Pennsylvania, USA. 1999+.
    Produced in an electronic format since 2008.
    Published three times annually.

The Blade
    New England Sports Turf Managers Association, Westwood, Massachusetts, USA. 1991+.
    Published three to four times annually.

The Bluegrass Blade
    Bluegrass Golf Course Superintendents Association, Versailles, Kentucky, USA. 1975+.
    Published 12 times annually.

The Bonnie Greensward
    Philadelphia Association of Golf Course Superintendents, Gladwyne/Glenside, Pennsylvania, USA. Circa 1950s+.
    Produced in an electronic format since 2008.
    Published nine times annually.

Buckeyeturf
    Ohio State University, Columbus, Ohio, USA. 2001+.
    (*includes* SportsNotes and International SportsNotes).
    Produced 35 to 50 times annually in an electronic format.

Bulletin
    Inland Empire Golf Course Superintendents Association, Spokane, Washington, USA. 1991–1996.
    (*formerly* Newsletter, 1988–1990).
    (*see* Newsletter, 1997+).
    Published at variable times annually.

The Bulletin
    Nebraska Turfgrass Association, Lincoln, Nebraska, USA. 2003+.
    (*formerly* NTF Turfgrass Bulletin, 1980–2002).
    Produced in an electronic format since 2011.
    Published up to four times annually.

The Bull Sheet
    Midwest Association of Golf Course Superintendents, Glen Ellyn/Highland Park/Carpenterville, Illinois, USA. 1947–1996.
    A pioneering local golf course superintendents organization periodical.
    (*see* On Course, 1996+).
    Published 12 times annually.

Cactus Clippings
    Cooperative Extension, University of Arizona, Tucson, Arizona, USA. 1992–2003.
    Published at variable times annually.

Cactus Clippings
    Cactus and Pine Golf Superintendents Association, Sun City/Scottsdale, Arizona, USA. 1987+.
    Produced in an electronic format since 2008.
    Published 6 to 12 times annually.

California Fairways
    California Golf Course Superintendents Association/Adams Publishing Corporation, Cathedral City, California, USA. 1992–1995.

(*formerly* Golden State Fairways, 1989–1991).
(*see* California GCSA E-Magazine, 1995+).
Published six times annually.

California GCSA E-Magazine
California Golf Course Superintendents Association, Reedly, California, USA. 1995+.
Produced 12 times annually in an electronic format.

California Turfgrass Culture
University of California, Riverside, California, USA. 1960–2003.
A pioneering state turfgrass educational periodical.
(*formerly* Southern California Turfgrass Culture, 1951–1959).
All issues digitized and available online.
Published four times annually.

Carolinas Newsletter/Carolinas Green
Carolinas Golf Course Superintendents Association, Clemson/Liberty, South Carolina, USA. 1965+.
(*changed* to Carolinas Green in 1992).
Published six times annually.

Central Illinois Golf Course Superintendents Newsletter
Central Illinois Golf Course Superintendents Association, Jacksonville, Illinois, USA. 1975–1980.
(*see* Turfgrass News, 1981–1983).
Published four times annually.

Central Plains Turfgrass Foundation Newsletter/CPTF Newsletter
Central Plains Turfgrass Foundation and Horticulture Department, Kansas State University, Manhattan, Kansas, USA. 1950–1986.
(*see* TurfNews, 1987+).
(*changed* to CPTF Newsletter in 1986).
Published four times annually.

Chips & Putts
Pocono Turfgrass Association, Shavertown/Clarks Summit/Harding, Pennsylvania, USA. 1971+.
Published nine times annually.

Clippings
Minnesota Turf and Grounds Foundation, Wayzata, Minnesota, USA. 1993+.
Published two times annually.

Clippings
New Jersey Turfgrass Association, Milltown, New Jersey, USA. 1984–2002.
(*formerly* Green World, 1971–1983).
(*see* Clippings & Green World, 2002+).
Published three times annually.

Clippings
Western Washington Golf Course Superintendents Association, Tacoma, Washington, USA. 1995–1997.
Published four times annually.

Clippings & Green World
New Jersey Turfgrass Association, Milltown, New Jersey, USA. 2002+.
(*see* Green World, 1971–1983, and Clippings, 1984–2002).
Published three times annually.

Clippings from the Glades
Everglades Golf Course Superintendent Association, Naples, Florida, USA. Circa 1997–2000.
Published 12 times annually.

Commonwealth Crier

Virginia Golf Course Superintendents Association, Haymarket/Glen Allen/Draper, Virginia, USA. 1999+.
Also produced in an electronic format since 2007.
Published four times annually.

The Conn Clippings/Connecticut Clippings/Connecticut Clippings Newsletter/Connecticut Clippings

Connecticut Association of Golf Course Superintendents, Groton/Stamford/Woodbridge, Connecticut, USA. 1968+.
(*changed* to Connecticut Clippings in 1982, *changed* to Connecticut Clippings Newsletter in 1988, *changed* to Connecticut Clippings in 1992).
Produced in an electronic format since 2006.
Published four times annually.

Course Conditions

Michigan Golf Course Superintendents Association, Lansing/St. Johns, Michigan, USA. 2008+.
(*consolidation* of A Patch of Green, 1971–2003; Turf Times, 1988–2004; and Western Views, 1982–2002).
Published four times annually.

CUTT

Cornell University Turfgrass Times and New York Turfgrass Association, Ithaca/Latham, New York, USA. 1990+.
Published four times annually.

The Divot News

Golf Course Superintendents Association of Southern California, Los Angeles/Buena Park/Newhall, California, USA. 1963–2005.
(*see* Sea to Sand Magazine, 2006+).
Published 12 times annually.

Divots

Miami Valley Golf Course Superintendents Association, Middletown/Fairborn, Ohio, USA. 1950–2003.
Published ~10 times annually.

Eastern Turfletter

United States Golf Association Green Section, Eastern District, New Brunswick, New Jersey, USA. 1957–1963.
(*formerly* Northeastern Turfletter, 1954–1957).
Published six times annually.

eVerdure

Chicagoland Association of Golf Course Superintendents, Wheeling, Illinois, USA. 2009+.
(*formerly* Verdure Newsletter, 1975–2009).
Produced at variable times annually in an electronic format.

The Florida Green

Florida Golf Course Superintendents Association, Lake Worth/Jensen Beach, Florida, USA. 1980+.
(*formerly* South Florida Green, 1973–1980).
Published four times annually.

Florida Turf

Florida Turf-grass Association, Orlando, Florida, USA. 1968–1983.
(*formerly* Florida Turf-Grass Association Bulletin, 1954–1967).
(*see* Florida Turf Digest, 1984+).
Produced in an electronic format since 2007.
Published four times annually.

Florida Turf Digest

Florida Turfgrass Association, Orlando/Lakeland, Florida, USA. 1984+.
(*formerly* Florida Turf, 1968–1983).
Published six times annually.

Florida Turf-Grass Association Bulletin
Florida Turf-Grass Association, Gainesville, Florida, USA. 1954–1967.
(*see* Florida Turf, 1968–1983, and Florida Turf Digest, 1984+).
Published four times annually.

Fore!
Quad-State Turfgrass Association, Princeton/Paducah, Kentucky, USA. 1985–1991.
Published 12 times annually.

Fore
San Diego Golf Course Superintendents Association, San Diego, California, USA. 1984–1990.
(*see* Informer, 1990+).
Published 12 times annually.

Fore Your Information
Sierra Nevada Golf Course Superintendents Association, Clovis, California, USA. 1975+.
Produced in an electronic format since 2009.
Published six times annually.

Front Nine News
The Northwestern Pennsylvania Golf Course Superintendents Association, Girard/Erie/Forestville, Pennsylvania, USA. 1996+.
Produced in an electronic format since 2008.
Published four times annually.

FTGA Newsletter
Florida Turf-Grass Association, Orlando, Florida, USA. 2009+.
Published 12 times annually

Gateway Green
Mississippi Valley Golf Course Superintendents Association, St. Louis, Missouri, USA. 1975+.
Published four times annually.

GCSANJ Newsletter
New Jersey Golf Course Superintendents Association, New Jersey, USA. 1978–1979.
(*formerly* New Jersey Golf Course Report, 1968–1978).
(*see* The Greenerside, 1977+).
Published six times annually.

Georgia Sod Producers Association News
Georgia Sod Producers Association, Griffin, Georgia, USA. 1991+.
Produced in an electronic format since 2001.
Published three times annually.

Georgia Turfgrass Foundation News/Georgia Turfgrass Association News
Georgia Turfgrass Foundation and Georgia Turfgrass Association, Norcross/Athens/Acworth, Georgia, USA. 1985–1992.
(*see* GTA Today, 1993+).
(*changed* to Georgia Turfgrass Association News in 1989).
Published five times annually.

The Georgia Turfgrass News
Georgia Golf Course Superintendents Association, Athens, Georgia, USA. 1968–1989.
(*see* Through the Green, 1989+).
Published six times annually.

Golden State Fairways
California Golf Course Superintendents Association/RK Communications Group Inc., Las Vegas, Nevada, USA. 1989–1991.

(*see* California Fairways, 1991–1995).
Published four times annually.

Golf Course Superintendents Association of Arkansas Newsletter
Golf Course Superintendents Association of Arkansas, Bryant, Arkansas, USA. 2000+.
Produced in an electronic format since 2008.
Published four times annually.

Grass Clippings
New Jersey Golf Course Superintendents Association, Closter, New Jersey, USA. 1953–1964.
(*see* Turf Clippings, 1965–1967).
Published 12 times annually.

Grass Clippings
Southern Idaho Golf Course Superintendents Association/Idaho Golf Course Superintendents Association, Boise, Idaho/Shelby/Lolo, Montana, USA. 1998+.
Published three times annually.

Grass Root News
Central New York Golf Course Superintendents Association, Camillus, New York, USA. 1978–1985.
Published four times annually.

The Grass Roots
Wisconsin Golf Course Superintendents Association, Madison, Wisconsin, USA. 1975+.
(*formerly* Super Talk, 1966–1968).
Published six times annually.

The Green Breeze
Greater Cincinnati Golf Course Superintendents Association, Cincinnati/Lebanon, Ohio, USA. 1948+.
Produced in an electronic format since 2007.
Published eight times annually.

The Greenerside
Golf Course Superintendents Association in New Jersey, New Brunswick/Springfield, New Jersey, USA. 1977+.
Published six times annually.

Green Examiner
Intermountain Golf Course Superintendents Association, Centerville/South Ogden, Utah, USA. 1985+.
Produced in an electronic format since 2008.
Published four to six times annually.

Green Mountain Greens
Vermont Golf Course Superintendents Association, Hartford/Poultney, Vermont, USA. Circa 198x+.
Produced in electronic format since 2009.
Published six times annually.

The Green Sheet
Central Pennsylvania Golf Course Superintendents Association, Lebanon, Pennsylvania, USA. 1993+.
Produced in an electronic format since 2009.
Published seven times annually.

The Green Sheet
Florida Golf Course Superintendents Association, Boca Raton/Stuart/Jensen Beach, Florida, USA. 1985+.
Published six times annually.

Greens Talk
Texas Golf Coast Superintendents Association, Harlingen/Corpus Christi, Texas, USA. 1992+.
Published six times annually.

Greensward
Hoosier Turfgrass Association and Hoosier Golf Course Superintendents Association, Fort Wayne, Indiana, USA. 1958+.
Produced in an electronic format since 2008.
Published seven to nine times annually.

Green World
New Jersey Turfgrass Association, New Brunswick, New Jersey, USA. 1971–1983.
(*see* Clippings, 1984–2001).
Published three times annually.

GTA Today
Georgia Turfgrass Association, Acworth/Commerce, Georgia, USA. 1993+.
(*see* Georgia Turfgrass Association News, 1985–1992).
Produced in an electronic format since 2007.
Published six times annually.

Gulf Coast Gazette
Gulf Coast Golf Course Superintendents Association, Auburn, Alabama, USA. 2003–2008.
(*see* Gulf Coast GCSA Chapter e-Newsletter 2009+).
Published two times annually.

Gulf Coast GCSA Chapter e-Newsletter
Gulf Coast Golf Course Superintendents Association, Auburn, Alabama, USA. 2009+.
(*formerly* Gulf Coast Gazette, 2003–2008).
Produced 12 times annually in an electronic format.

Heart Beat (plus starting in 1990 Heart Beat Newsletter)
Heart of America Golf Course Superintendents Association, Kansas City, Missouri, USA. 1969+.
Each published in alternating months, a total of 12 times annually.

The Hole Nine Yards
Long Island Golf Course Superintendents Association, West Hempstead, New York, USA. 1987+.
Produced in an electronic format since 2008.
Published six times annually.

Hole Notes
Minnesota Golf Course Superintendents Association, Edina/Wayzata, Minnesota, USA. 1969+.
Published 10 to 12 times annually.

Hudson Valley Foreground/Foreground
Hudson Valley Golf Course Superintendents Association, Newburgh/Accord/Elmsford, New York, USA. 1969+.
(*formerly* Hudson Valley Golf Course Superintendents Association Newsletter, 1964–1968).
(*changed* to Foreground in 1989).
Produced in an electronic format since 2008.
Published 6 to 12 times annually.

Hudson Valley Golf Course Superintendents Association Newsletter
Hudson Valley Golf Course Superintendents Association, Newburgh, New York, USA. 1964–1968.
(*see* Hudson Valley Foreground, 1969+).
Produced four times annually.

Iagreen Turfgrass Management Newsletter
> Department of Horticulture, Iowa State University, Ames, Iowa, USA. 1962–1966.
> Published four times annually.

The IGCSA Newsletter
> Indiana Golf Course Superintendents Association, Fishers, Indiana, USA. 1975–1996.
> (*see* Indiana Superintendent News, 1988–2007).
> Published 12 times annually.

Illinois Turfgrass Update
> Illinois Turfgrass Foundation, Urbana/Chicago/Lemont, Illinois, USA. 1981+.
> Produced in an electronic format since 2000.
> Published at variable times annually.

Indiana Chapter GCSAA Newsletter
> Indiana Golf Course Superintendent Association, Indianapolis, Indiana, USA. 2008+.
> (*formerly* Indiana Superintendent News, 1988–2007).
> Published four times annually.

Indiana Superintendent News
> Indiana Golf Course Superintendents Association, Indianapolis, Indiana, USA. 1988–2007.
> (*formerly* IGCSA Newsletter, 1975–1987).
> (*see* Indiana Chapter GCSAA Newsletter, 2008+).
> Published 12 times annually.

Informer
> San Diego Golf Course Superintendents Association, San Diego, California, USA. 1990+.
> (*formerly* Fore, 1984–1990).
> Became electronic and biweekly in 2009.
> Published six times annually.

Inland Empire Golf Course Superintendents Association Newsletter
> Inland Empire Golf Course Superintendents Association, Spokane, Washington/Lolo, Montana, USA. 1997+.
> (*formerly* Bulletin, 19991–1996).
> Published three times annually.

Iowa GCSA Reporter
> Iowa Golf Course Superintendents Association, Urbandale, Iowa, USA. 1995+.
> (*formerly* The Reporter, 1968–1995).
> Produced in an electronic format since 2010.
> Published 10 times annually.

Iowa Turfgrass Grower
> Department of Horticulture, Iowa State University, Ames, Iowa, USA. 1977–1992.
> (*see* Ornamental & Turf Newsletter, 1992–1995).
> Published two times annually.

Kansas Grassroots Newsletter
> Kansas Golf Course Superintendents Association, Winfield/Manhattan, Kansas, USA. 1972+.
> Produced in an electronic format since 2009.
> Published six times annually.

Kentuckiana Klippings
> Kentuckiana Golf Course Superintendents Association, Louisville, Kentucky, USA. 1958+.
> Produced in an electronic format since 2008.
> Published 6 to 12 times annually.

Kentucky Turfgrass News/Kentucky Turfgrass Council Newsletter
Eastern Kentucky University, Richmond, Kentucky, USA. 1972–2000.
(*see* The State of Grass, 2000+).
(*changed* to Kentucky Turfgrass Council Newsletter in 1983).
Published two to four to six times annually.

The Keynoter
The Pennsylvania Turfgrass Council Inc., University Park/Lemont/State College, Pennsylvania, USA. 1971–2003.
Published four to five times annually.

LIGCSA Newsletter or LIGCSA Supernews
Long Island Golf Course Superintendents Association, Oyster Bay/Huntington, New York, USA. 1973–1982.
Published 10 times annually.

Lone Star Links
Lone Star Golf Course Superintendents Association, Kilgore/Kingsland/Carrollton, Texas, USA. 1988+.
Produced in an electronic format since 2003.
Published four times annually.

The Louisiana Turfgrass Bulletin
Louisiana Turfgrass Association, Baton Rouge, Louisiana, USA. 1964+.
Published six times annually.

Mainely Green
Maine Golf Course Superintendents Association, Yarmouth, Maine, USA. 1972+.
Published four times annually.

Maryland Turfgrass Council News
Maryland Turfgrass Council Inc., College Park/Elderburg/Churchville, Maryland, USA. 1981–2009.
(*see* MTC Turf News, 2009+).
Published four times annually.

Michigan Turfgrass Foundation Newsnotes
Michigan Turfgrass Foundation, Saginaw/Lansing, Michigan, USA. 1978–2000.
(*see* NewsNotes, 2000+).
Published four times annually.

Michigan Turfgrass Report
Department of Crop and Soil Sciences, Michigan State University, East Lansing, Michigan, USA. 1963–1971.
Published two to three times annually.

Mid-Atlantic News Letter
Mid-Atlantic Association of Golf Course Superintendents, Rockville, Maryland/Jeffersonton/Springfield, Virginia, USA. 1950–1983.
(*see* Newsletter, 1984–1989).
Published 12 times annually.

Mid-Atlantic Turfletter
United States Golf Association Green Section, Mid-Atlantic District, Beltsville, Maryland, USA. 1955–1957.
Published six times annually.

Mid-Continent News
United States Golf Association Green Section, Waco, Texas, USA. 1992–1993.
Published at variable times annually.

Mid-Continent Turfletter
>   United States Golf Association Green Section, Midwestern District, Chicago, Illinois/College Station, Texas, USA. 1957–1963.
>   Published six times annually.

Midwest Factsheet
>   Midwest Regional Turf Foundation, Purdue University, West Lafayette, Indiana, USA. 2002+.
>   Published at variable times annually.

Midwest Memo
>   Midwest Regional Turf Foundation, Purdue University, West Lafayette, Indiana, USA. 1988+.
>   Published six times annually.

Midwest Regional Turf Foundation Newsletter or Midwest Turf Newsletter
>   Midwest Regional Turf Foundation, Purdue University, West Lafayette, Indiana, USA. 1971 and 1989–1991.
>   Published at variable times annually.

Midwest Turf-News and Research or News and Research
>   Midwest Regional Turf Foundation, Purdue University, West Lafayette, Indiana, USA. 1946–1974.
>   A pioneering regional turfgrass organization periodical.
>   Published four to variable times annually.

MiSTMA Newsletter/MiSTMA
>   Michigan Sports Turf Managers Association, Lansing, Michigan, USA. 2001–2005 and 2008–2010.
>   (*changed* to MiSTMA in 2008).
>   Published four times annually.

Mountain State Greensletter or Mountain State Greenletter
>   West Virginia Chapter Golf Course Superintendents Association, Bridgeport/Williamston, West Virginia, USA. 1979+.
>   Published nine times annually.

Mountain West Turf
>   Harris Publishing Inc., Idaho Falls, Idaho, USA. 2001+.
>   Published four times annually.

MTC Turf News
>   Maryland Turfgrass Council Inc., College Park, Maryland, USA. 2009+.
>   (*formerly* Maryland Turfgrass Council NEWS, 1981–2009).
>   Published four times annually.

MVTA
>   Missouri Valley Turfgrass Association Inc., University of Missouri, Columbia, Missouri, USA. 1978+.
>   Published at variable times annually.

NCTGA Newsletter
>   North Central Turfgrass Association, Bismarck, North Dakota, USA. 1986–1991.
>   (*see* Talking Turf, 1991+).
>   Published at variable times annually.

New England Turf News
>   New England Regional Turfgrass Foundation, Newport, Rhode Island, USA. 2003+.
>   Published once annually.

New Hampshire Turf Talk
>   New Hampshire Golf Course Superintendents Association, Manchester, New Hampshire, USA. 1971–1993.

(*see* Turf Talk, 1994+).
Published six times annually.

New Jersey Golf Course Report

New Jersey Golf Course Superintendents Association, New Jersey, USA. 1968–1978.
(*see* GCSANJ Newsletter, 1978–1979).
Published six times annually.

The New Leaf News

Oregon Golf Course Superintendents Association, Gresham, Oregon, USA. Circa 1970–1982.
(*see* OGCSA Newsletter, 1982–1997).
Published one to two times annually.

Newsletter

Chicagoland Golf Course Superintendents Association, Highland Park/Lamont, Illinois, USA. 1968–1970.
(*see* Verdure, 1970–2009).
Published at variable times annually.

Newsletter

The Greenkeepers Club of New England, Fall River/Walpole, Massachusetts, USA. 1929–1937.
(*see* The Newsletter of the Golf Course Superintendents Association of New England Inc., 1970+).
Published 12 times annually.

Newsletter

Inland Empire Golf Course Superintendents Association, Spokane, Washington, USA. 1988–1990.
(*see* Bulletin, 1991–1996).
Published three to four times annually.

Newsletter

Mid-Atlantic Association of Golf Course Superintendents, Arlington, Virginia, USA. 1984–1989.
(*formerly* Mid-Atlantic News Letter, 1950–1983).
(*see* Turf Talk, 1989–1990).
Published 10 times annually.

Newsletter

Midwest Regional Turfgrass Foundation, Purdue University, West Lafayette, Indiana, USA. 1966–1989.
Published at variable times annually.

Newsletter

North Central Turf Grass Association, Bismarck, North Dakota, USA. 1986–1991.
Published four times annually.

The Newsletter

Peaks & Prairies Golf Course Superintendents Association, Bozeman, Montana, USA. 1988–1989.
(*see* The Perfect Lie, 1989+).
Published four times annually.

The Newsletter of the Golf Course Superintendents Association of New England Inc.

Golf Course Superintendents Association of New England Inc., Newberry/Weston/Norton, Massachusetts, USA. 1970+.
(*formerly* Newsletter of the Greenkeepers Club of New England, 1929–1937).
Produced in an electronic format since 2006.
Published 12 times annually.

Newsletter of the Northern Michigan Turf Managers Association

Northern Michigan Turf Managers Association, Traverse City, Michigan, USA. 1972–1988.

(*see* Turf Times, 1988–2004).
Published four times annually.

Newsletter of the Rhode Island Golf Course Superintendents Association

Rhode Island Golf Course Superintendents Association, Riverside, Rhode Island, USA. 1986.
(*formerly* The Superintendents News, 1968–1987).
(*see* Surf 'n' Turf, 1987+).
Published six times annually.

Newsnotes or News Notes

Michigan Turfgrass Foundation and Michigan State University, Saginaw/Lansing/East Lansing, Michigan, USA. 2000+.
(*formerly* Michigan Turfgrass Foundation Newsnotes, 1978–2000).
Produced in an electronic format since 2005.
Published at variable times annually.

New York State Turf Association Bulletin

New York State Turf Association and Cornell University, Lakeview/Syracuse/Liverpool/Ithaca, New York, USA. 1949–1959.
A pioneering state turfgrass educational periodical.
(*see* New York State Turfgrass Association Bulletin, 1959–1974).
Published at variable times annually.

New York State Turfgrass Association Bulletin or The Bulletin

New York State Turfgrass Association, Massapequa/Latham, New York, USA. 1959–1974, 1977–1979, and 1984+.
Published at variable times annually, with certain years missed.

North Carolina Turfgrass

The Turfgrass Council of North Carolina Inc., Raleigh/Southern Pines/Cary, North Carolina, USA. 1983+.
Produced in an electronic format since 2004.
Published four to six times annually.

Northeastern Turfletter

United States Golf Association Green Section, North Eastern Office, New Brunswick, New Jersey, USA. 1954–1957.
(*see* Eastern Turfletter, 1957–1963).
Published five to six times annually.

Northern Ohio Turfgrass News or Northern Ohio Turf

Northern Ohio Chapter, Golf Course Superintendents Association, Canton/Bedford/Westfield Center, Ohio, USA. 1958+.
Published 6 to 9 to 10 to 12 times annually.

North Texas News

North Texas Golf Course Superintendents Association, Hurst/Fort Worth, Texas, USA. 1994+.
Published at variable times annually.

Northwest Turfgrass Topics

Northwest Turfgrass Association and Washington State University, Puyallup/Olympia, Washington, USA. 1959–1997.
(*formerly* Pacific Northwest Turf Association Newsletter, 1953–1957).
(*see* Turfgrass Management in the Pacific Northwest, 1998–2008).
Published three times annually.

NTF Turfgrass Bulletin

Nebraska Turfgrass Foundation, Hastings/Lincoln, Nebraska, USA. 1980–2002.
(*see* The Bulletin, 2003+).
Published four times annually.

OGCSA Newsletter

Oregon Golf Course Superintendents Association, Salem, Oregon/Vancouver, Washington/Sisters, Oregon, USA. 1982–1997.

(*formerly* The New Leaf News, circa 1970–1981).
(*see* Turfgrass Management in the Pacific Northwest, 1998–2008).
Published four times annually.

Ohio Turfgrass Council Newsletter

The Ohio Turfgrass Council, Wooster, Ohio, USA. 1963–1966.
(*see* Ohio Turfgrass Foundation Newsletter, 1967–1991).
Published two times annually.

Ohio Turfgrass Foundation Newsletter

The Ohio Turfgrass Foundation, Columbus, Ohio, USA. 1967–1991.
(*formerly* Ohio Turfgrass Council Newsletter, 1963–1966).
(*see* OTF Newsletter or OTF Turf News, 1990+).
Published at variable times annually.

Oklahoma Turf

Oklahoma Turfgrass Research Foundation and Oklahoma State University, Stillwater, Oklahoma, USA. 1983–1995.
(*see* Oklahoma Turfgrass Notes, 1995+).
Published one to four times annually.

Oklahoma Turfgrass Notes

Oklahoma State University and Oklahoma Turfgrass Research Foundation, Stillwater, Oklahoma, USA. 1995+.
(*formerly* Oklahoma Turf, 1983–1995).
Produced from 2 to 35 times annually in an electronic format.

The Old Dominion Spreadsheet

Old Dominion Golf Course Superintendents Association, Chesterfield, Virginia, USA. 2004–2006.
Published four times annually.

On Course

Midwest Association of Golf Course Superintendents, Carpentersville/Lemont, Illinois, USA. 1996+.
(*formerly* The Bull Sheet, 1947–1996).
Produced in an electronic format since 2008.
Published 12 times annually.

On the Road—Quarterly Bulletin

Northern California Golf Association, Pebble Beach, California, USA. 2002–2006.
Published three to four times annually.

Ornamental & Turf Newsletter

Department of Horticulture, Iowa State University, Ames, Iowa, USA. 1992–1995.
(*formerly* Iowa Turfgrass Grower, 1977–1992).
Published four times annually.

OTF Newsletter or OTF Turf News

Ohio Turfgrass Foundation, Columbus/Zanesville, Ohio, USA. 1988+.
(*formerly* Ohio Turfgrass Council Newsletter, 1963–1966, and Ohio Turfgrass Foundation Newsletter, 1967–1991).
Produced in an electronic format since 2001.
Published four to five to six times annually.

Our Collaborator

The Northeastern Golf Course Superintendents Association, Newtonsville/Rexford, New York, USA. 1954+.
Published seven to eight times annually.

Ozark Turf Association Newsletter

Ozark Turf Association, Ozark, Missouri, USA. 1985+.

Produced in an electronic format since 2006.
Published eight times annually.

The Pacific Greenkeeper

The Southern California Greenkeepers Association, San Gabril, California, USA. Circa 1920–circa 1932.
Published 12 times annually.

Pacific Northwest Turf Association Newsletter

Northwest Turf Association, Pullman/Puyallup, Washington, USA. 1953–1957.
(*see* Northwest Turfgrass Topics, 1959–1997).
Published at variable times annually.

The Pallet

Turfgrass Producers of Texas, Wharton, Texas, USA. 2002+.
Published four times annually.

A Patch of Green

Michigan & Border Cities Golf Course Superintendents Association/Greater Detroit Golf Course Superintendents Association, Birmingham/Troy, Michigan, USA. 1971–2003.
(*see* Course Conditions, 2008+).
Published 12 times annually.

The Perfect Lie

Wy-Mont Golf Course Superintendents Association/Peaks & Prairies Golf Course Superintendents Association, Bozeman/Shelby/Lolo, Montana, USA. 1989+.
(*formerly* Turf Talk, 1976–1988, and The Newsletter, 1988–1989).
Published four times annually.

Plain Turf

Nebraska Golf Course Superintendents Association, Omaha, Nebraska, USA. 2001–2007.
Published eight times annually.

Profile: The Newsletter of the Eastern Shore Association of Golf Course Superintendents

Eastern Shore Association of Golf Course Superintendents, Berlin, Maryland, USA. 1993–2002.
Changed to an electronic version in 2003.
Published nine times annually.

Quarterly News Bulletin of the Southern Turfgrass Association

Mississippi State, Mississippi/College Station, Texas, USA. 1965–1988.
Published four times annually.

The Reporter

Iowa Golf Course Superintendents Association, Cedar Rapids/Ames, Iowa, USA. 1968–1995.
(*see* Iowa GCSA Reporter, 1995+).
Published 10 times annually.

The Reporter or Reporter

Rocky Mountain Golf Course Superintendents Association, Denver, Colorado, USA. 1992+.
(*formerly* Rocky Mountain Reporter, 1965–1992).
Published eight times annually.

RGGCSA Newsletter

Rio Grande Golf Course Superintendents Association, Rio Rancho, New Mexico, USA. 2009+.
(*formerly* The Super News, 1998–2009).

Produced in an electronic format since 2009.
Published five to six times annually.

Rocky Mountain Reporter

Rocky Mountain Golf Course Superintendents Association, Aurora, Colorado, USA. 1965–1992.
(*formerly* The Poop Sheet, 1966, and The Scoope Sheet, 1967).
(*see* The Reporter, 1992+).
Published eight times annually.

Rub of the Green

Hi-Lo Desert Golf Course Superintendents Association, Palm Desert/Lomita, California, USA. 1972–2005.
(*see* Sea to Sand Magazine, 2006+).
Published 12 times annually.

The Sand Blaster

Cape Cod Turf Managers Association/Golf Course Superintendents Association of Cape Cod, Hyannisport and Centerville, Massachusetts/North Scituate, Rhode Island, USA. 1988+.
Produced in an electronic format since 2007.
Published six times annually.

Sea to Sand Magazine

Golf Course Superintendents Association of Southern California, Los Angeles, California, USA. 2006+.
(*formerly* Rub of the Green, 1972–2005, and The Divot News, 1963–2005).
Published 12 times annually.

Short Cuttings

Ohio State University, Columbus, Ohio, USA. 1990–1992.
Published two times annually.

SoCal STMA Newsletter

Southern California STMA, San Diego, California, USA. 2000+.
Produced in an electronic format since 2002.
Published four times annually.

The Sod Grower

Sod Growers Association of Michigan, Sterling Heights/Williamston, Michigan, USA. 1971.
Published four times annually.

South Dakota Turf Manager

South Dakota Turf Foundation and Department of Horticulture, Forestry, Landscape and Parks, South Dakota State University, Brookings, South Dakota, USA. 1999–2001.
Published three to four times annually.

South Dakota Turf Prose

Golf Course Superintendents Association of South Dakota and South Dakota Turf Research Foundation, Aberdeen, South Dakota, USA. 1990+.
Published four times annually.

Southeastern Turfletter

United States Golf Association Green Section, Southeastern District, Tifton, Georgia, USA. 1954–1956.
Published three times annually.

Southern California Turf Culture

College of Agriculture, University of California at Los Angeles, Westwood, California, USA. 1951–1954.

A pioneering state turfgrass educational periodical.
(*see* Southern California Turfgrass Culture, 1954–1959, and California Turfgrass Culture, 1960–2003).
Published four times annually.

Southern California Turfgrass Culture
University of California, Riverside, California, USA. 1954–1959.
A pioneering state turfgrass educational periodical.
(*formerly* Southern California Turf Culture, 1951–1954).
(*see* California Turfgrass Culture, 1960–2003).
Published four times annually.

Southern Turf Foundation Bulletin
Southern Turf Foundation, Tifton, Georgia/Orangeburg, South Carolina, USA. 1950–1954.
Published two to four times annually.

Southern Turfletter
United States Golf Association Green Section, College Station, Texas/Beltsville, Maryland, USA. 1957–1963.
Published six times annually.

The South Florida Green
South Florida Golf Course Superintendents Association, Miami, Florida, USA. 1973–1980.
(*see* The Florida Green, 1980+).
Published four times annually.

Southwestern Turfletter
United States Golf Association Green Section, Southwestern District, College Station, Texas, USA. 1955–1957.
Published six times annually.

The State of Grass
The Kentucky Turfgrass Council, Richmond, Kentucky, USA. 2000+.
(*formerly* Kentucky Turfgrass Council Newsletter, 1972–2000).
Published six times annually.

Suncoast Newsletter
Suncoast Golf Coast Golf Course Superintendents Association, Laurel, Florida, USA. 1980–2005.
Produced in an electronic format since 2003.
Published 12 times annually.

The Superintendent News
Rhode Island Golf Course Superintendents Association, 1968–1987.
(*see* Surf 'n' Turf, 1987+).
Published six times annually.

The Super News
Rio Grande Golf Course Superintendents Association, Bernalillo, New Mexico, USA. 1998–2009.
(*see* RGGCSA Newsletter, 2009+).
Published five to six times annually.

Super Talk
Wisconsin Golf Course Superintendents Association, Madison, Wisconsin, USA. 1966–1968.
(*see* The Grass Roots, 1975+).
Published seven times annually.

Surf 'n' Turf
Rhode Island Golf Course Superintendents Association, Riverside/North Scituate, Rhode Island, USA. 1987+.

(*formerly* The Superintendent News, 1968–1987).
Produced in an electronic format as a web newsletter since 2008.
Published six times annually.

Talking Turf
North Central Turfgrass Association, Fargo, North Dakota, USA. 1991+.
(*formerly* NCTGA Newsletter, 1986–1991).
Published four times annually.

Talkin' Turf
Northern California Golf Association, Pebble Beach, California, USA. 2000+.
Published two to four times annually.

Talk Turf
Wisconsin Sports Turf Managers Association, Verona, Wisconsin, USA. 2009+.
(*formerly* WSTMA NEWS, 1998–2008).
Placed on an electronic website from 2002 onward.
Published four times annually.

Tech Turf Topics
Virginia Polytechnic and State University, Blacksburg, Virginia, USA. 1977–1983.
Published four times annually.

Tee Off!/Tee-Off News
Golf Course Superintendents Association of Central California, Fresno, California, USA. 1988–1998.
(*changed* to Tee-Off News in 1996).
Published 12 times annually.

Tee to Green
Louisiana-Mississippi Golf Course Superintendents Association, New Orleans and Monroe, Louisiana/Starkville, Mississippi, USA. 1971+.
Published four times annually.

Tee to Green
Metropolitan Golf Course Superintendents Association, New Rochelle/Rye, New York, USA. 1962+.
Published 6 to 12 times annually.

Tee to Green
South Texas Golf Course Superintendents Association Inc., Fulshear/Houston, Texas, USA. Circa 1970+.
Published nine times annually.

Tennessee Turfgrass
Tennessee Turfgrass Association and Tennessee Valley Sports Turf Managers Association/Leading Edge Communication, Franklin, Tennessee, USA. 1987+.
Produced in an electronic format since 2011.
Published six times annually.

Texas Turfgrass
Texas Turfgrass Association Inc., Bryan/College Station, Texas, USA. 1981+.
(*formerly* Turf News of Texas, 1949–1980).
Published four to six times annually.

Three Rivers Green
Greater Pittsburg Golf Course Superintendents Association, Allison Park, Pennsylvania, USA. 1985+.
Published four times annually.

Through the Green

Georgia Golf Course Superintendent Association, Watkinsville/Hartwell/Athens, Georgia, USA. 1989+.
(*formerly* The Georgia Turfgrass News, 1968–1989).
Published six times annually.

Thru the Green

Golf Course Superintendents Association of Northern California, Diamond Springs, California, USA. 1997+.
Published 12 times annually.

Tidewater Turf Topics or Shortgrass

Tidewater Turfgrass Association, Norfolk, Virginia, USA. 1985+.
Published 12 times annually.

Treasure Coast Tide

Treasure Coast Golf Course Superintendents Association, Vero Beach, Florida, USA. 1995–2005.
Published six times annually.

The Tri-Stater News

Tri-State Golf Course Superintendents Association, Louisville, Kentucky/Evansville and Mount Vernon, Indiana, USA. Late 1960s+.
Published four times annually.

Turf and Grounds Maintenance Newsletter

Agricultural Extension Service, University of Tennessee Institute of Agriculture, Knoxville, Tennessee, USA. 1984–1988.
Published four times annually.

Turf Bulletin

Massachusetts Turf and Lawn Grass Council Association Inc. and University of Massachusetts, Amherst/Hadley, Massachusetts, USA. 1962–1985.
Published at variable times annually.

Turf Clippings

New Jersey Golf Course Superintendents Association, New Jersey, USA. 1965–1967.
(*see* New Jersey Golf Course Report, 1968–1978).
Published six times annually.

Turfgrass Bulletin

Nebraska Turfgrass Foundation and Nebraska Turfgrass Association, Hastings, Nebraska, USA. 1980–2002.
(*see* The Bulletin, 2003+).
Published at variable times annually.

Turfgrass Management in the Pacific Northwest

Northwest Turfgrass Association, Inland Empire, Oregon, and Western Washington Golf Course Superintendents Associations, Sisters, Oregon, USA. 1998–2008.
Changed to an online electronic format in 2008 produced from Olympia, Washington.
Published at variable times annually.

Turfgrass Management Newsletter

Department of Crop and Soil Sciences, Michigan State University, East Lansing, Michigan, USA. 1970–1987.
Published at variable times annually.

Turfgrass Matters

Mid-Atlantic Association of Golf Course Superintendents, Lutherville/Rockville/Abington, Maryland, USA. 1990+.
(*formerly* Turf Talk, 1989–1990).
Published five to six times annually.

Turfgrass News
 Central Illinois Golf Course Superintendents Association, Jacksonville, Illinois, USA. 1981–1983.
 (*formerly* Central Illinois Golf Course Superintendents Newsletter, 1975–1980).
 (*see* The Ballmark, 1984–1990).
 Published four times annually.

Turfgrass News
 Northern Ohio Golf Course Superintendents Association, Cleveland, Ohio, USA. 1959+.
 Produced in an electronic format since 2009.
 Published nine times annually.

Turfgrass Times
 Northern Illinois Golf Course Superintendents Association and Northwestern Illinois Golf Course Superintendents Association,
  Savanna/Hanover/Rockford, Illinois, USA. 1985+
 Produced in an electronic format since 2003.
 Published six times annually.

Turfgrass Topics
 Cooperative Extension Service, University of Georgia, Athens, Georgia, USA. 1981–1990.
 Published four times annually.

TurfNews
 Kansas Turfgrass Foundation and Kansas State University, Manhattan, Kansas, USA. 1987+.
 (*formerly* Central Plains Turfgrass Foundation Newsletter or CPTF Newsletter, 1950–1986).
 Produced in an electronic format since 2007.
 Published four times annually.

Turf Newsletter for Golf Course Superintendents
 North Carolina Agriculture Extension Service, Raleigh, North Carolina, USA. Circa 1970–1984.
 Published three times annually.

Turf News of Texas
 Texas Turfgrass Association, College Station, Texas, USA. 1949–1980.
 (*see* Texas Turfgrass, 1981+).
 Published four times annually.

Turf Notes
 New England Cooperative Extension Turf Program, University of Massachusetts, Amherst/Worchester, Massachusetts, USA.
  1989–2005.
 Published at variable times annually.

Turf Related
 Great Lakes Chapter of the Ohio Sports Turf Managers Association, Sylvania, Ohio, USA. 1997+.
 Published four times annually.

Turf Tales
 The Southern California Turfgrass Council, Van Nuys, California, USA. 1994–2002.
 Published four times annually.

Turf Talk
 Cooperative Extension Turf Program, University of Massachusetts, Amherst, Massachusetts, USA. 1998+.
 Produced at variable times annually in an electronic format.

Turf Talk
 Mid-Atlantic Golf Course Superintendents Association, Rockville, Maryland, USA. 1989–1990.

(*formerly* Newsletter, 1984–1989).
(*see* Turfgrass Matters, 1990+).
Published five times annually.

Turf Talk

New Hampshire Golf Course Superintendents Association, Londonderry, New Hampshire, USA. 1994+.
(*formerly* New Hampshire Turf Talk, 1971–1993).
Published five times annually.

Turf Talk

Western New York Golf Course Superintendents Association, Hamburg, New York, USA. 1980–1982.
Published six times annually.

Turf Talk

West Texas Golf Course Superintendents Association, Amarillo/Andrews, Texas, USA. 1987+.
Produced in an electronic format since 2005.
Published six times annually.

Turf Talk

Wy-Mont Golf Course Superintendents Association, Bozeman, Montana, USA. 1976–1988.
(*see* The Perfect Lie, 1989+).
Published five to six times annually.

Turf Times

Alabama Turfgrass Association, Auburn, Alabama, USA. 2001+.
(*formerly* The ATA Scribe, 1984–1988).
Produced in an electronic format since 2008.
Published four times annually.

Turf Times

Northern Michigan Turf Managers Association, Traverse City/Lansing, Michigan, USA. 1988–2004.
(*formerly* Newsletter of the Northern Michigan Turf Managers Association, 1972–1988).
(*see* Course Conditions, 2008+).
Published four to six times annually.

Turf Tips

Department of Horticulture, University of Arkansas, Fayetteville, Arkansas, USA. 2007+.
Produced 12 times annually in an electronic format.

Turf Tips

Midwest Regional Turf Foundation, West Lafayette, Indiana, USA. 1998+.
Published at variable times annually.

Turf Tips

University of Arizona, Cooperative Extension, Tucson, Arizona, USA. 1994–2003.
Published 10 to 12 times annually.

Turf Topics

Northwest Turfgrass Association and Washington State University, Puyallup, Washington, USA. 1959–1997.
(*see* Turfgrass Management in the Pacific Northwest, 1998+).
Published four times annually.

Turfuture

Midwest Regional Turf Foundation, Purdue University, West Lafayette, Indiana, USA. 1989.

UCR Turf News
University of California, Riverside, California, USA. 2003+.
Published four times annually.

UMass Turftalk
University of Massachusetts, Amherst, Massachusetts, USA. 1998+.
Produced variable times annually in an electronic format.

Update
Sports Field Managers Association of New Jersey, Pennsville, New Jersey, USA. 2001+.
Published four to six times annually.

Verdure/Verdure Newsletter
Chicagoland Association of Golf Course Superintendents, Chicago/Wheeling/Lemont, Illinois, USA. 1970–2009.
(*formerly* Newsletter, 1968–1970).
(*changed* to Verdure Newsletter in 1975).
(*see* eVerdure, 2009+).
Published at variable times annually.

Virginia Turfgrass Quarterly Journal/Virginia Turfgrass Journal
Virginia Turfgrass Council Inc., Virginia Beach, Virginia, USA/Harvest Publishing Co., Winter Haven, Florida, USA. 1996+.
(*formerly* Virginia Turfgrass News, 1974–1996).
(*changed* to Virginia Turfgrass Journal in 2001).
Also produced in an electronic format since 2008.
Published four to six times annually.

Virginia Turfgrass News
Virginia Turfgrass Council Inc. and Virginia Polytechnic Institute, Blacksburg/Richmond/Virginia Beach, Virginia, USA. 1974–1996.
(*see* Virginia Turfgrass Journal, 1996+).
Published two to four times annually.

West Coast Wind
Florida West Coast Golf Course Superintendents Association, Pinnellas Park/St. Petersburg/Wesley Chapel, Florida, USA. 2003+.
Produced in an electronic format since 2004.
Published 6 to 12 times annually.

Western Turfletter
United States Golf Association Green Section Western District, Garden Grove, California, USA. 1957–1962.
Published six times annually.

Western Views
Western Michigan Golf Course Superintendents Association, Kalamazoo, Michigan, USA. 1982–2002.
(*see* Course Conditions, 2008+).
Published four times annually.

Wisconsin Turfgrass News
Wisconsin Turfgrass Association, Verona, Wisconsin, USA. 1981+.
Last 15 years available on an electronic website.
Published two to three to four times annually.

WSTMA News
Wisconsin Sports Turf Managers Association, Verona, Wisconsin, USA. 1998–2008.
(*see* Talk Turf, 2009+).
Published four times annually.

C. INTERNATIONAL TURFGRASS TECHNICAL PERIODICALS

Accent on Grass
> Oseco Inc., Brampton, Ontario, Canada. 1973–1985.
> Published at variable times annually.

AGCSA Action
> Australian Golf Course Superintendents Association, Clayton, Victoria, Australia. 1994+.
> Produced four to six times annually in an electronic format, currently in alternate months to the Australian Turfgrass Management Journal.

Agrostologist—A Bulletin of Turf Management Forming the Link between Theory and Practice
> T.W. Evans and I.G. Lewis Publishers, London, England, UK. 1936–1940.
> Published three times annually.

AGSA Turf News
> Atlantic Golf Superintendents Association, Sackville, Nova Scotia, Canada. 1987–2004.
> (*formerly* Atlantic Turf News, 1983–1986).
> (*see* The Turf News, 2004+).
> Produced in an electronic format since 2003.
> Published four times annually.

Alberta Newsletter
> Alberta Golf Superintendents Association, Airdrie, Alberta, Canada. 1987–1998.
> (*see* A Touch of Green, 1999+).
> Published six times annually.

Amenity Management
> ACP Publishers Limited, Oxon, England, UK/Haymarket Trade and Leisure Publications Ltd., Teddington, Middlesex, England, UK. 1989–1996.
> (*see* Turf & Amenity Management, 1997–2004).
> Published 12 times annually.

Around the Green
> British and International Golf Greenkeepers Association, York, North Yorkshire, England, UK. 1988.
> Published at variable times annually.

Atlantic Turf News
> Atlantic Golf Superintendents Association, St Johns, New Brunswick, Canada. 1983–1986.
> (*see* AGSA Turf News, 1987–2004).
> Published two times annually.

ATRI Turf Notes
> Australian Turfgrass Research Institute Ltd., Concord West, New South Wales, Australia. 1982–1997.
> (*includes* ATRI Research Reports).
> Published four times annually.

Australasian Turf Care
> Advantage Publishing, Mentone, Victoria, Australia. 1981.
> (*changed* title to TurfN Tree by 1982).

The Australian Golf Course Review
> Printcraft Press, Brookvale, New South Wales, Australia. 1965–1966.
> Published four times annually.

The Australian Golf Superintendent
> Robert Dransfield Publisher, Sydney, New South Wales, Australia. 1982–1984.
> Published three to four times annually.

The Australian Greenkeeper
   The Australian Greenkeeper, Sydney, New South Wales, Australia. 1936–1941.
   Published four times annually.

The Australian Greenkeeper Year Book
   Austral International Publications, Sydney, New South Wales, Australia. 1964–1965 and 1967–1968.
   Published at variable times.

Australian Turf Management Journal or Australian Turfgrass Management Journal
   Australian Golf Course Superintendents Association, Clayton, Victoria, Australia. 1999+.
   (*formerly* AGCSA was with Golf and Sports Turf Australia, 1993–1999).
   Published six times annually.

Australian Turfgrass Management Offshoot
   Australian Golf Course Superintendents Association and Australian Turfgrass Management, Clayton, Victoria, Australia.
      2005–2008.
   Published six times annually.

BCGSA Newsletter or The Dogwood
   British Columbia Golf Superintendents Association, Lake Cowichan, British Columbia, Canada. 1997+.
   Published two to three times annually.

The Bowling Greenkeeper
   New South Wales Bowling Greenkeepers' Association, Ryde/Sydney, New South Wales, Australia. 1963+.
   Published six times annually.

Breaking Golf Course Management News
   Ikki Publishing Co. Limited, Taito-ku, Tokyo, Japan. 1999+.
   Published 12 times annually.

The British Golf Greenkeeper (New Series)
   The British Golf Greenkeepers' Association and Media Promotion Ltd., London/Croydon, Surrey, England, UK. Circa 1945–1976.
   (*formerly* Journal of the Golf Greenkeepers' Association, London, England, UK, circa 1913–1936).
   (*see* Golf Greenkeeping and Course Maintenance, 1979–1989).
   Published 12 times annually.

The Bulletin
   Grass Research Bureau Limited, West Ryde, New South Wales, Australia. 1958–1961 and New Series from 1965–1967.
   (*see* Grass Research, 1955–1957).
   Published at variable times annually.

Bulletin
   Queensland Board of Greenkeeping Research, Brisbane, Queensland, Australia. 1935–1936.
   Published at variable times annually.

Bulletin
   Victoria Golf Association Research Section, Melbourne, Victoria, Australia. 1935–1936.
   Published at variable times annually.

The Bulletin of the New South Wales Golf Council Green Research Committee
   New South Wales Golf Council, Sydney, New South Wales, Australia. 1932–1933.
   Published at variable times annually.

Buying Turfgrass Seed
   Sports Turf Research Institute, Bingley, West Yorkshire, England, UK. 1975–1997.
   (*see* Turfgrass Seed, 1998+).
   Published once annually.

Césped Deportivo

La Asociación Española de Técnicos en Mantenimiento de Campos de Golf, Malaga, Spain. 1999–2001.
(*see* Greenkeepers, 2002+).
Published one to two times annually.

Clippings

Ontario Golf Superintendents' Association, Guelph, Ontario, Canada. 2008+.
Produced 51 times annually in an electronic format.

Courses for Horses

Australian Racecourse Managers Association, Greensborough, Victoria, Australia. 2010+.
Published four times annually.

The Cut

Australian Golf Course Superintendents Association, Clayton, Victoria, Australia. 2005+.
Produced weekly in an electronic format.

Euro Turf and Landscape Management

Farm Press Publications, Clarksdale, Mississippi, USA. 1993.
Published six times annually.

Fieldmanager

NWST NeWSTories bv, Nijmegen, the Netherlands. 2005+.
Published six times annually.

Fieldmanager Nieuwsbrief

NWST NeWSTories bv, Nijmegen, the Netherlands. 2005+.
Produced 52 times annually in an electronic format.

Golf & Groundscare

Nelson Communications Limited, Salisbury, Wiltshire, England, UK. 1997–2000.
(*added* Parks, Golf Courses, and Sports Grounds).
(*see* Turf Professional, 2000+).
Published six times annually.

Golf & Sports Turf Australia

Glenvale Publications, Mount Waverly, Victoria, Australia. 1993+.
(*incorporated* as the official magazine of the Australian Golf Course Superintendents Association from 1993 to 1999).
Published six times annually.

The Golf Course

Park View Publications Limited/Fore Golf Publications Limited, Harwich, Essex, England, UK. 1987–1990; with British and International Golf Greenkeepers Association from 1987–1989.
(*added* Golf Greenkeeping in 1989).
(*see* Greenkeeper International, 1991+).
Published 10 times annually.

Golf Course Management & Maintenance Magazine

Golf Digest, Tokyo, Japan. 1968+.
Published 12 times annually.

Golf Course Management Information

Korean Turfgrass Research Institute, Seongnam, Kyonggi Province, South Korea. 1990+.
Published four times annually.

Golf Course News International
Market Drayton, Shropshire, England, UK,/Golf Course News International, Cellardyke, Fife, Scotland, UK. 1993–2006.
Published six times annually.

Golfdom Europe
Questex Media Inc., Cleveland, Ohio, USA. 2007–2008.
Published four times annually.

Golf Greenkeeping
British Golf Greenkeepers' Association/Wharfedale Publications Limited, West Yorkshire, England, UK. 1987–1988.
(*formerly* Golf Greenkeeping and Course Maintenance, 1981–1987).
(*see* The Golf Course, 1987–1990).
Published 10 times annually.

Golf Greenkeeping and Course Maintenance
British Golf Greenkeepers' Association, Great Bookham, Surrey, England, UK. 1981–1987.
(*formerly* British Golf Greenkeeper, 1945–1976).
(*see* Golf Greenkeeping, 1987–1988).
Published 10 times annually.

Golf Industry Research
Ikki Publishing Co. Limited, Tokyo, Japan. 1980–1989.
(*see* Monthly Golf Management, 1989+).
Published four times annually.

Grass Research
Grass Research Bureau, West Ryde, New South Wales, Australia. 1955–1957.
(*see* The Bulletin, 1958–1965).
Published at variable times annually.

Grassroots
Golf Course Superintendents Association of Malaysia, Selangor, Malaysia. 2000–2005.
Published at variable times annually.

Greenbladet
Swedish Greenkeepers Association, Höllviken, Sweden. 1987+.
(*formerly* Gröna Bladet, 1984–1987).
Published four times annually.

Green Is Beautiful
Ontario Golf Superintendents' Association, Etobicoke/Guelph, Ontario, Canada. 1975+.
(*formerly* OGSA Newsletter, 1967–1975).
Produced in an electronic format since 2010.
Published four times annually.

Greenkeeper
A. Quick & Co. Limited, Harwich, Essex, England, UK. 1981–1984; with Scottish and International Golf Greenkeepers' Association from 1982–1984 and European and International Greenkeeper Association from 1983 to 1984.
(*see* Greenkeeper & The International Greenkeeper, 1984–1989).
Published 10 times annually.

Greenkeeper or Green-Keeper
Association Francaise des Personnels d'Entreten de Golf/French Association of Greenkeepers, Biarritz, France. 1986–2002.
(*see* Green Magazine, 2004+).
Published four times annually.

Greenkeeper & The International Greenkeeper

A. Quick & Co. Limited (changed in 1986 to Fore Golf Publications Limited), Harwich, Essex, England, UK. 1987–1989; with Scottish and International Golf Greenkeepers Association and European and International Greenkeepers Association from 1984–1986.

(*formerly* Greenkeeper 1980–1984).

(*incorporated* The International Greenkeeper in 1984).

Published 10 times annually.

Greenkeeper International

British and International Golf Greenkeepers Association, Alne, York, North Yorkshire, England, UK. 1991+.

(*formerly* The Golf Course, 1987–1990).

Produced in an electronic format since 2007.

Published 12 times annually.

Greenkeeper Nieuwsbrief

NWST NEWS Toriés bv, Nijmegen, the Netherlands. 1990+.

Produced six times annually in an electronic format.

Greenkeepers

La Asociación Española de Greenkeepers, Malaga, Spain. 2002+.

(*formerly* Césped Deportivo, 1999–2001).

Published one to four times annually.

Greenkeepers Journal (Greenkeeper Verband)

Greenkeeper Verband Deutschland and HORTUS-Zeitschriften, Cöllen & Bleeck GbR/Köllen Druck & Verlag GmbH, Bonn, Germany. 2008+.

(*now an insert* in European Journal of Turfgrass Science, 2008+).

(*formerly an insert* in Rasen: Turf·Gazon, 1989–2007).

Published four times annually.

Greenkeeper (Vakblad voor Greenkeeping)

Nederlandse Golf Federatie en de Nederlandse Greenkeeepers Associatie, Nijmegen, the Netherlands, 1990+.

Published two to six times annually.

Greenkeeping

Gornal Wood, Dudley, West Midlands, UK/Union Press Limited, London, England, UK. 2002+.

Published six times annually.

Greenkeeping Management

British and International Golf Greenkeepers Association, York, Yorkshire, England, UK. 1989–1990.

(*formerly* Greenkeeper & the International Greenkeeper, 1984–1989).

(*see* Greenkeeper International, 1991+).

Published 12 times annually.

Green Letter

Kansai Golf Union, Takarazuka, Hyogo-ken, Japan. 2003+.

Published 12 times annually.

Green Magazine

Association Francaise des Personnels d'Entreten de Golf or French Association of Greenkeepers, Biarritz, France. 2004+.

(*formerly* Green-Keeper, 1986–2002).

Published four times annually.

The GreenMaster

Canadian Golf Superintendents Association, Toronto, Ontario, Canada. 1965–1980.

First three issues published under the title News Bulletin.

(*see* GreenMaster, 1981+).
Published six times annually.

GreenMaster
Canadian Golf Superintendents Association and Kenilworth Publishing Inc., Toronto/Richmond Hill, Ontario, Canada. 1981+.
(*formerly* The GreenMaster, 1965–1980).
Published 6 to 12 times annually.

Greenside
Golf Course Superintendents Association of Ireland, Bainbridge, Ireland. 1996+.
During the first year, it was published as a newsletter, also issued online.
Published four times annually.

GreenTechino
Golf, Estudio, Asesoraments, Servicious Agrupados, Madrid, Spain. 1994–1998.
Published at variable times annually.

Gröna Bladet
Swedish Greenkeepers Association, Höllviken, Sweden. 1984–1987.
(*see* Greenbladet, 1987+).
Published one time annually, progressing to four times annually.

The Groundsman or Groundsman
The Institute of Groundsmanship, Wolverton Mill East, Milton Keynes, Buckinghamshire, England, UK. 1952+.
(*formerly* Journal of National Association of Groundsman, 1948–1952).
Published 6 to 12 times annually.

Grounds Maintenance—Japanese Edition
Intertec Publishing Corporation, Overland Park, Kansas, USA. 1991–1992.
Published at variable times annually.

gsi (Golf Science International)
World Scientific Congress of Golf Trust, St Andrews, Fife, Scotland, UK. 1998–2000.
Published four to six times annually.

The GTI Advisor
Guelph Turfgrass Institute, Guelph, Ontario, Canada. 1996+.
Produced 10 to 20 times annually in an electronic format.

The International Greenkeeper
International Greenkeepers Association, Babette Haradine, Caslano, Switzerland. 1970–1989.
(*see* Greenkeeper & The International Greenkeeper, 1984–1987).
Published three times annually.

International Turfgrass
International Turfgrass Society. 1993+.
Published at variable times annually.

International Turfgrass Bulletin
Sports Turf Research Institute Limited, Bingley, West Yorkshire, England, UK. 1996+.
(*formerly* Sport Turf Bulletin, 1951–1996).
Published four times annually.

Journal of National Association of Groundsmen
National Association of Groundsmen, New Madden, Surrey, England, UK. 1948–1952.

(*see* The Groundsman, 1952+).
Published 12 times annually.

Journal of the Golf Greenkeepers' Association

Golf Greenkeepers' Association, London, England, UK. Circa 1913–1936.
(*see* Official Organ of the GGA, 1936, and The British Golf Greenkeeper [New Series], 1945–1976).

Keeping It Green

Australian Golf Course Superintendents Association and Golf Australia, Clayton, Victoria, Australia. 2008+.
Produced at variable times annually in an electronic format.

Kenkyu Seiseki Hokokusho (Bulletin of West Japan Turf Research Institute)

Nishi Nihon Gurin Kenkyujo, Fukuoka-Ken, Japan. 1962–circa 1990s.
Published two times annually.

Monthly Golf Management

Ikki Publishing Co. Limited, Taito-ku, Tokyo, Japan. 1989+.
(*formerly* Golf Industry Research, 1980–1989).
Published 12 times annually.

National Turfgrass Council Turfgrass Reports

National Turfgrass Council, Bingley, West Yorkshire, England, UK. 1983–1991.
Fifty-one variably published reports from workshops.

Neue Landschaft: Fachzeitschrift für Garten-, Landschafts-, Spiel- und Sportplatzbau (New Landscape: Professional Journal for Garden, Landscape, Play, and Sport Stadiums)

Patzer, Berlin, Germany. 1995+.
Published 12 times annually.

NEWS

National Turfgrass Council, Bingley, West Yorkshire, England, UK. 1989–1993.
Published at variable times annually.

Newsletter of the New Zealand Institute for Turf Culture or Newsletter of the New Zealand Turf Culture Institute

New Zealand Institute for Turf Culture and New Zealand Turf Culture Institute, Palmerston North, New Zealand. 1960–1976.
(*see* Sports Turf Review, 1976+).
Published six times annually.

Newsline or Toro Newslines

Toro, St Neots, Huntington, Cambridgeshire, England, UK. 1994+.
Published at variable times annually.

New Zealand Turf Management Journal

New Zealand Turf Sports Institute, Palmerston North, New Zealand. 1986+.
(*formerly* Sports Turf Review, 1976–1986).
Published four times annually.

Notes from St Ives Research Station

Board of Greenkeeping Research, Bingley, West Yorkshire, England, UK. 1948–1950.
(*see* Sports Turf Bulletin, 1951–1996).
Published 12 times annually.

NTF Newsletter

National Turfgrass Council, Bingley, West Yorkshire, England, UK. 1988–1994.
(*formerly* an insert in Parks, Golf Course, and Sports Grounds).
Published six times annually.

NSWGCSA
New South Wales Golf Course Superintendents Association, Sydney, New South Wales, Australia. 2000+.
Published four times annually.

NSW News
Turf Grass Association of Australia, Sydney, New South Wales, Australia. 2002–2008.
(*see* On Common Ground, 2008+).
Published three times annually.

OGSA Newsletter
Ontario Golf Course Superintendents Association, Etobicoke, Ontario, Canada. 1967–1975.
(*see* Green Is Beautiful, 1975+).
Published six times annually.

On Common Ground
Sports Turf Association, Sydney, New South Wales, Australia. 2008+.
Published three times annually.

On Course—Greenkeepers Training Committee (GTC) Newsletter
Greenkeepers Training Committee Limited, Alne, York, Yorkshire, England, UK. 2001+.
Published four times annually.

Parks and Grounds
Avonwold Publishing Co., Saxonwold, South Africa. 1977–2004.
Published four times annually.

Parks and Sports Grounds/Parks & Sports Grounds
Clark and Hunter Limited, London, England, UK. 1943.
(*see* Parks, Golf Courses, and Sports Grounds, 1935–1999).
Published 12 times annually.

Parks, Golf Courses, and Sports Grounds/Parks, Golf Courses, & Sports Grounds.
Clark & Hunter Limited, Staines, Middlesex, England, UK. 1935–1999.
(*formerly* Parks and Sports Grounds, 1943).
One of the older turfgrass trade periodicals in the United Kingdom.
Published 12 times annually.

Pitchcare or Pitchcare the Magazine or PC pitchcare
Pitchcare.com Limited, Wolverhampton, West Midlands, England, UK. 2005+.
Produced six times annually in an electronic format.

The Professional & Greenkeeper
Sports Press, London, England, UK. 1910–1917.
Published 12 times annually.

The Queensland Greenkeeper
Queensland Greenkeepers Association, Brisbane, Queensland, Australia. 2001–2004.
Published four times annually.

South Easterner
South East Region Association of Groundsman, London, England, UK. 1954–1963.
Published four times annually.

Sportowe Nawierzchnie Trawiaste (Sport Turf Fields)
Wszelkie Prawa Zastrzezone, Pabianice, Poland. 2004–2006.
Produced six times annually in an electronic format.

Sports Field Forum NZ Quarterly Newsletter
> Sports Field Forum NZ Inc., Palmerston North, New Zealand. 2006+.
> Produced four to six times annually in an electronic format.

Sports Turf, Amenity, & Leisure
> Institute of Groundsmanship, Wolverton Mill East, Milton Keynes, Buckinghamshire, England, UK. 2009+.
> Published six times annually.

Sports Turf Bulletin
> Sports Turf Research Institute, Bingley, West Yorkshire, England, UK. 1951–1996.
> (*formerly* Notes from the St Ives Research Station, 1948–1950).
> (*see* International Turfgrass Bulletin, 1996+).
> Published 4 to 12 times annually.

Sports Turf Manager
> The Sports Turf Managers Association, Guelph, Ontario, Canada. 1995+.
> (*formerly* Sports Turf Newsletter, 1987–1994).
> Produced in an electronic format since 2010.
> Published four times annually.

Sports Turf Newsletter
> Sports Turf Managers Association, Guelph, Ontario, Canada. 1987–1994.
> (*see* Sports Turf Manager, 1995+).
> Published four times annually.

Sports Turf Review
> New Zealand Turf Culture Institute Inc., Palmerston North, New Zealand. 1976–1986.
> (*formerly* Newsletter of the New Zealand Culture Institute, 1974–1976).
> (*see* New Zealand Turf Management Journal, 1986+).
> Published six times annually.

Sutton's Golf Journal
> Sutton & Sons, Reading, Berkshire, England, UK. 1922–1923.
> Published at variable times.

Talking Turf
> Federation of European Golf Greenkeepers Associations, Warwrickshire, England, UK. 2000+.
> Published two times annually.

TGA News
> Turfgrass Growers Association, Woodbridge, Suffolk, England, UK. 2000.
> Published two times annually.

TGM
> Campos Deportivos y Espacios Verdes, Buenos Aires, Argentina. 1992+.
> Published four to six times annually.

TGM Newsletter online
> Campos Deportivos y Espacios Verdes, Buenos Aires, Argentina. 200x+.
> Produced six times annually in an electronic format.

A Touch of Green
> Alberta Golf Superintendents Association, Airdrie, Alberta, Canada. 1999+.
> (*formerly* Alberta Newsletter, 1987–1998).
> Published four times annually.

Trávniky, Sport, Zeleň, Ekologie (Greens, Sports, Greenery, Ecology)
Gramina, Brno, Czechoslovakia. 1991–1992.
Published four times annually.

Turf & Amenity Management
Haymarket Business Publication Limited, London, England, UK. 1997–1998.
Nexus Media Communications Limited, Swanley, Kent, England, UK. 1999–2004.
(*combined* from Amenity Management, 1989–1996, and Turf Management, 1982–1998).
Published 12 times annually.

Turf & Recreation
Turf & Recreation Publishing Inc., Surrey, British Columbia/Delhi, Ontario, Canada/MRK Publishing, Vancouver, British Columbia, Canada. 1988+.
Published 7 to 10 times annually.

TURFCRAFT
Scottish Golf Greenkeepers Association (changed to Scottish and International Golf Greenkeepers Association in 1972), Scotland, UK. Circa 1967–1974.
Published four times annually.

TurfCraft Aust.
Claim Pty. Limited, Hampton, Victoria, Australia. 1987–1994.
(*incorporated* the official journal of the Australian Golf Course Superintendents Association from 1987–1992).
(*see* TurfCraft International, 1994+).
Published five times annually.

TurfCraft International
Rural Press, Agricultural Publishers Pty. Limited, Port Melbourne, Victoria, Australia. 1994+.
(*formerly* TurfCraft Aust., 1987–1994).
Published six times annually.

Turf for Sport
Sutton & Sons Limited, Reading, England, UK. 1947–1964.
Published four times annually.

Turfgrass Seed
Sports Turf Research Institute, Bingley, West Yorkshire, England, UK. 1998+.
(*formerly* Buying Turfgrass Seed, 1975–1997).
Published once annually.

turfguide
Turf Australia, Turf Producers Australia Ltd., Cleveland, Queensland, Australia. 2004 and 2011.

The Turf Line News
Western Canada Turfgrass Association, Maple Ridge/Hope, British Columbia, Canada. 1972+.
Published four to six times annually.

Turf Management
Haymarket Business Publications Limited, Teddington, Middlesex, England, UK. 1982–1998.
(*see* Turf & Amenity Management, 1997–2004).
Published 12 times annually.

Turf Management for Sport and Leisure
Golf World Ltd., London, England, UK. 1982–1986.
(*incorporated* to Landscape Management in 1985).
Published 12 times annually.

The Turf News
Atlantic Golf Superintendents Association, Sackville, Nova Scotia, Canada. 2004+.
(*formerly* AGSA Turf News, 1987–2004).
Produced four times annually in an electronic format.

Turf News
Kansai Golf Union, Green Section Research Center, Takarazuka, Hyogo-ken, Japan. 1967+.
Published three to four times annually.

Turf Notes
Horticulture Department, University of Guelph, and Guelph Turfgrass Institute, Guelph, Ontario, Canada. 1989–1995.
Published four times annually.

Turf Professional
Nelson Communications Limited/Nelson Professional Limited, Salisbury, Wiltshire, England, UK. 2000+.
(*formerly* Golf & Groundscare, 1997–2000).
Published six times annually.

TURF Talk or TalkingTURF
Syngenta Seeds, Cambridge, Cambridgeshire, England, UK. 2007+.
Published one to two times annually.

Turf Talk
Jacobsen Turf Equipment, Northamptonshire, England, UK. 1993–1998.
Published 12 times annually.

Turf Topics Newsletter
Turf Research & Advisory Institute, Department of Agriculture (changed to Turfgrass Technology Pty. Ltd. in 1989), Frankston, Victoria, Australia. 1974–1993.
Published one to four times annually.

Weibulls Gräs-tips or Weibulls Gräs Tips
W. Weibull AB, Landskrona, Sweden. 1958–1983.
Published at variable times annually (vol. 10 and 11 in 1968, vol. 12–14 in 1971, vol. 15–22 in 1972–1979, and vol. 23–25 in 1983).

Zahrada, Park, Krajina, Trávniky (Garden, Park, Landscape, Turfgrass)
Společnost Pro Zuhrodni a Krajinnou Tvorbu, Brno, Czech Republic. 1993–1994.
Published four times annually.

CHAPTER 18

Supporting Technical/Practitioner Periodicals

PUBLICATIONS IN THIS CATEGORY INCLUDE REGULARLY PUBLISHED (MONTHLY TO YEARLY) TECHNI-
cal periodicals that are devoted primarily to nonturfgrass subjects but that do include turfgrass-related
articles regularly. The listings are grouped as (1) United States Supporting Technical Periodicals and (2)
International Supporting Technical Periodicals.

A. UNITED STATES SUPPORTING TECHNICAL PERIODICALS

Center for Grassland Studies
University of Nebraska, Lincoln, Nebraska, USA. 1995+
Published four to three times annually.

Crittenden Golf Inc. or Golf Inc.
Crittenden Golf Magazine, San Diego, California, USA. 1992–2007.
Published 12 times annually..

Erosion Control
International Erosion Control Association and Forester Communications Inc., Santa Barbara, California, USA. 1994+.
Published nine times annually.

e-Times
Irrigation Association, Falls Church, Virginia, USA. 2008+.
Produced 12 times annually in an electronic format.

Golf Business
National Golf Course Owners Association, Mt Pleasant, South Carolina, USA. 1995+.
Published 12 times annually.

Golf Business Today
Golf Course Superintendents Association of America, Lawrence, Kansas, USA. 1993+.
Published 12 times annually.

Golf Industry Report
National Golf Foundation, Jupiter, Florida, USA. 2001–2009.
Published two to four times annually.

Highway Research News

Highway Research Board, National Academy of Sciences, Washington, DC, USA. 1963–1974.
Published four times annually.

Industrial Vegetation, Turf, and Pest Management

Agricultural Products Department, Dow Chemical Co., Midland, Michigan, USA. 1979–1980.
Published at variable times annually.

Landscape

South Carolina Landscape & Turfgrass Association, Clemson/Columbia, South Carolina, USA. 1983+.
Produced in an electronic format since 2009.
Published four times annually.

Landscape & Hardscape Construction

Moose River Media, St Johnsbury, Vermont, USA. 2008+.
(*formerly* Landscape Construction, 2003–2008).
Published 12 times annually.

Landscape & Irrigation

Gold Trade Publications Inc., Van Nuys/Adams Business Media Inc., Cathedral City, California, USA. 1983+.
(*formerly* Landscape West and Irrigation News, 1977–1982).
Published 12 times annually.

Landscape and Turf

Brantwood Publications, Elm Grove, Wisconsin/Tampa, Florida, USA. 1981–1983.
(*formerly* Landscape Industry, 1969–1980).
(*see* Southern Landscape and Turf, 1983–1984).
Published seven times annually.

Landscape & Turf Industry

Brantwood Publications Inc., Elm Grove, Wisconsin, USA. 1980–1981.
(*formed* by merging Landscape Industry and Turf-Grass Times in 1980).
(*see* Landscape and Turf, 1981–1983).
Published seven times annually.

Landscape Architecture

American Society of Landscape Architects, Louisville, Kentucky, USA. 1910+.
Contains occasional articles on turfgrass.
Published six times annually.

Landscape Construction

Moose River Publishing, St Johnsbury, Vermont, USA. 2003–2008.
Published 12 times annually.

Landscape Design and Construction

Brantwood Publications Inc., Elm Grove, Wisconsin, USA. 1961–1968.
(*formerly* Landscaping, 1955–1961).
(*see* Landscape Industry, 1969–1980).
Published 12 times annually.

Landscape Design/Build

Advanstar Landscape Group, Cleveland, Ohio, USA. 2003–2006.
Published 12 times annually.

Landscape Industry
Brantwood Publications Inc., Elm Grove, Wisconsin, USA. 1969–1980.
(*formerly* Landscape Design and Construction, 1961–1968).
(*see* Landscape & Turf Industry, 1980–1981).
Published six times annually.

Landscape Management
Harcourt Brace Jovanovich Publications/Advanstar Communications Inc./Questex Media Publications, Cleveland, Ohio, USA. 1987+.
(*incorporated* Lawn Care Industry in 1991).
(*formerly* Weeds, Trees, and Turf, 1963–1987).
Also produced in an electronic format since 2010 as i-news.
Published 12 times annually.

Landscape Superintendent and Maintenance Professional
Landscape Communications, Tustin, California, USA. 2004+.
Published 10 to 12 times annually.

Landscape West & Irrigation News
Gold Trade Publications, Van Nuys, California, USA. 1977–1982.
(*see* Landscape & Irrigation, 1983+).
Published 12 times annually.

Landscaping
Automotive Press, Los Angeles, California, USA. 1955–1961.
(*see* Landscape Design and Construction, 1961–1968).
Contains occasional articles on turfgrass.
Published 12 times annually.

Lawn & Garden Marketing
Intertec Publishing Corp., Kansas City, Missouri, USA. 1961–1991.
Published 10 times annually.

The Modern Cemetery
O.H. Sample, Rockford, Illinois, USA. 1933–1951.
(*formerly* Park and Cemetery and Landscape Gardening, 1890–1932, and Park and Cemetery, 1932).
Contains occasional articles on turfgrass.
Published 12 times annually.

Ornamentals South
Publications South Inc., Atlanta, Georgia, USA. 1978+.
Contains occasional articles on turfgrass.
Published six times annually.

Park and Cemetery
Cemetery Publishing Co., Madison, Wisconsin, USA. 1932.
(*formerly* Park and Cemetery and Landscape Gardening, 1890–1932).
(*see* Modern Cemetery, 1933–1951).
Contains occasional articles on turfgrass.
Published 12 times annually.

Park and Cemetery and Landscape Gardening
R.J. Haight Publisher, Chicago, Illinois, USA. 1890–1932.
Contains occasional articles on turfgrass.
(*see* Park and Cemetery, 1932).
Published 12 times annually.

The Park and Recreation Trades
The Trades Publishing Co., Crossville, Tennessee, USA. 1993+.
Produced in an electronic format starting in 2009.
Published 12 times annually.

Parks & Recreation
American Institute of Park Executives, Wheeling, West Virginia, USA. 1917+.
Published 12 times annually.

Parks & Recreation
National Recreation and Park Association, New York City, New York/Ashburn, Virginia, USA. 1966+.
Published 12 times annually

Parks & Recreation Business or PRB
Northstar Publishing, Medina, Ohio, USA. 2002+.
Published 12 times annually.

The Professional Gardener
Professional Grounds Management Society and National Association of Gardeners, New York City, New York, USA. 1949–1972.
(*see* Manager's Memo, 1973–1979).
Published 12 times annually.

Professional Landscaper and Groundsman
Albatross Publications, Horsham, West Sussex, England, UK. 1986+.
Published two to six times annually.

Southwest Lawn & Landscape
RK Communications Group Inc., Las Vegas, Nevada, USA. 1986–1992.
Published 12 times annually.

Southwest Trees and Turf
Stone Peak Services, Las Vegas, Nevada, USA. 1996+.
Published 11 to 12 times annually.

Superintendent's Videomagazine
EPIC of Wisconsin Inc., West Bend, Wisconsin, USA. 2007.
Published at variable times annually.

Western Landscaping News/Western Landscaping
William/Lawrence Co. (changed to Hester Communications Inc. in 1979), Irvine, California, USA. 1961–1983.
Contains articles oriented to the western United States.
(*changed* to Western Landscaping in 1984).
Published 12 times annually.

B. INTERNATIONAL SUPPORTING TECHNICAL PERIODICALS

Australasian Parks and Leisure
Leisure Australia and New Zealand Recreation Association, Bendigo, Victoria, Australia. 2002–2005.
(*formerly* Australian Parks and Leisure, 1998–2001).
Published four times annually.

Australian Parks and Leisure
Leisure Australia, Bendigo, Victoria, Australia. 1998–2001.

(*see* Australasian Parks and Leisure, 2002–2005).
Published four times annually.

***Das Gartenamt: Stadt und Grün: Organ der Ständigen Konferenz der Gartenbauamtsleiter beim
Deutschen Städtetag (Journal for Design of Environmental, Landscape, and Sport Facilities)***
Patzer-Verlag Gmbh. u. Co. KG., Hanover, West Germany. 1952–1995.
Contains occasional articles on turfgrass.
Published 12 times annually.

GCA Journal
The Japanese Society of Golf Course Architects, Tokyo, Japan. 1993+.
Published one time annually.

GCA Newsletter
Japanese Society of Golf Course Architects, Tokyo, Japan. 1993–1999.
Published at variable times annually.

Golf Architecture
Society of Australian Golf Course Architects, South Melbourne, Victoria, Australia. 2002+.
(*formerly* The Journal of the Society of Australian Golf Course Architects, 1997–2001).
Published once annually.

Golf Architecture Newsletter
Sports Turf Research Institute, Bingley, West Yorkshire, England, UK. 2002+.
Published once annually.

Golf Course Architecture
Tudor Rose, Leicester, England, UK. 2005+.
Published four times annually.

***Havenkunst: Kobenhaven: Nordisk Tidssbkrift for Planlaegning af Have og Landskal
(Garden Art: Scandinavian Review for Garden and Landscape Planning)***
Copenhagen, Denmark. 1920–1968.
Contains occasional articles on turfgrass.
(*see* Landskap, 1969–1981).
Published eight times annually.

The Journal of the Society of Australian Golf Course Architects
Society of Australian Golf Course Architects, South Melbourne, Victoria, Australia. 1997–2001.
(*see* Golf Architecture, 2002+).
Published once annually.

Landmark
Consolidated Communications, Calgary, Alberta, Canada. 1989+.
Published six times annually.

Landskap (Landscape)
Scandinavian Review for Garden and Landscape Planning, Copenhagen, Denmark. 1969–1981.
Contains occasional articles on turfgrass.
(*formerly* Havenkunst, 1920–1968).
Published eight times annually.

Panstadia International Quarterly Report
Panstadia Publishing Co. Ltd., Castle Donington, Derbyshire, England, UK. 1994+.
Published four times annually.

Parks and Recreation
Institute of Park and Recreation Administration, Reading, England, UK. 1970–2007.
Contains occasional articles on turfgrass.
Published 12 times annually.

Paysage Actualities (Landscape News)
Monitor Group, Paris, France. 1976+.
(*incorporated* Espaces Verts in 1982).
Published 12 times annually.

Figure 18.1. Turfgrass shade adaptation studies being established at Michigan State University, in the 1960s. [26]*

---

* The numbers in brackets correspond to the figure source acknowledgments listed in the appendix.

CHAPTER 19

Post-1950 General Lawn
and Landscape Books

PLEASE SEE CHAPTER 12 FOR DETAILS ABOUT THE FORMAT OF THE REFERENCES IN THIS CHAPTER.

A. POST-1950 GENERAL LAWN BOOKS

Encompasses general lawn books that were less than 100 pages and/or published after 1950, usually by journalist authors. They are oriented primarily to residential lawn enthusiasts.

American Potash Institute

1959† **You Can Grow A Good Lawn.** American Potash Institute, composed of six sections by contributing authors J.F. Reed, S.E. Younts, R.W. Schery, W.H. Daniel, W.L. Nelson, and J.M. Duich. The American Potash Institute Inc., Washington, DC, USA. 33 pp., 36 illus., 4 tables, no index, fold-stapled glossy semisoft covers in multicolor with white and black lettering, 5.9 by 8.9 inches (151 × 226 mm).

A small booklet on lawn planting and care, with emphasis on fertilization practices. It is practically oriented to lawn enthusiasts and conditions in North America.

Atkinson, Robert E.

1964† **All about Dichondra.** Robert E. Atkinson (editor) with eight contributing authors: Theodore Payne, V.B. Youngner, S.E. Spaulding, H. Hamilton Williams, Wayne C. Morgan, Robert E. Atkinson, R.N. Jefferson, and A.S. Deal. Lasca Leaves XIV (3), California Arboretum Foundation Inc., Arcadia, California, USA. 25 pp., 11 illus., no tables, no index, fold-stapled glossy semisoft covers in multicolor with black lettering, 6.0 by 9.0 inches (152 × 228 mm).

A small booklet on dichondra (*Dichondra repens*) as a vegetative cover in southern California. Topics include a description, seed production, physiology, culture, weed control, and pests of dichondra. There are references in the backs of some sections. The only publication available that is focused on dichondra as used for lawns.

Australian Turfgrass Research Institute

1991† **On Your Home Turf—A guide to the establishment and care of the home lawn.** David S. Drane (editor) with four contributing authors: Gary W. Beehag, Ian C. McIver, Alexandra Shakesby, and Murray G. Wallis. Australian Turfgrass Research Institute, Concord West, New South Wales, Australia. 51 pp., 10 illus. plus 3 pp. of advertising, 6 tables, no index, fold-stapled glossy semisoft covers in greens and white with black and green lettering, 8.5 by 11.5 inches (215 × 292 mm).

A small book concerning site preparation, turfgrass selection, planting, care, and pest management for lawns. It is practically oriented to lawn enthusiasts and conditions in Australia.

† *Reprinted in 1992 and 1993, with minor updates.*

Bagshaw, F.A.

1966   **The Home Lawn.** Second revised edition. F.A. Bagshaw (editor). Bulletin P226, New South Wales Department of Agriculture, New South Wales, Australia. 39 pp., 30 illus., no tables, no index, semisoft stapled covers in green and black with black lettering, 5.3 by 8.3 inches (135 × 210 mm).

A booklet concerning the planting and care of lawns. It is oriented to lawn enthusiasts and conditions in New South Wales.

Baker, Jerry

1972†   **Make Friends With Your Lawn.** Jerry Baker. The Benjamin Company Inc., New York City, New York, USA. 73 pp., 14 illus., 5 tables, no index, fold-stapled glossy semisoft covers in multicolor with black lettering, 5.0 by 8.0 inches (127 × 204 mm).

A small booklet on the care of lawns in North America. It contains some questionable recommendations in terms of sound, beneficial turfgrass responses.

1973†   **Make Friends With Your Lawn.** Jerry Baker. Simon and Schuster, New York City, New York, USA. 95 pp., 134 illus., 2 tables, no index, glossy semi–soft bound in multicolor with black and green lettering, 8.3 by 10.8 inches (210 × 275 mm).

A small book on lawn care plus sections on seeding and sodding. It contains some questionable recommendations in terms of sound, beneficial turfgrass responses.

1976†   **Make Friends With Your Lawn.** Jerry Baker. Pocket Books, New York City, New York, USA. 176 pp.; 140 illus.; 2 tables; no index; semi–soft bound in multicolor with yellow, white, and black lettering; 4.5 by 6.9 inches (115 × 175 mm).

A smaller pocket-book version of a 1972 booklet with more illustrations. It contains some questionable recommendations in terms of sound, beneficial turfgrass responses.

1987†   **Jerry Baker's Lawn Book.** Jerry Baker. Ballantine Books, Random House Inc., New York City, New York, USA. 203 pp.; 76 line-drawing illus.; 90 tables; indexed; glossy semi–soft bound in multicolor with green, yellow, and white lettering; 7.1 by 9.0 inches (179 × 229 mm).

A book on the planting and care of lawns in North America, with emphasis on culture and pest control. There is a question and answer section.

1987†   **The Impatient Gardener's Lawn Book.** Jerry Baker. Ballantine Books, New York City, New York, USA. 203 pp.; 81 illus.; 108 tables; small index; glossy semi–soft bound in multicolor with green, yellow, and white lettering; 7.2 by 9.2 inches (184 × 233 mm).

The book is a reissue from an earlier 1987 book with a variant title. It was also printed in 1987 with the same title but with a different cover in multicolor with white lettering.

2001†   **Green Grass Magic.** Jerry Baker. American Master Products Inc./Storey Communications Inc., Pownal, Vermont, USA. 368 pp.; 648 illus.; 11 tables; indexed; glossy hardbound in multicolor with white, yellow, and black lettering; 7.1 by 8.9 inches (180 × 226 mm).

A book representing the author's perceptions and humor about the planting and care of lawns.

Ball, Jeff Norman

1994†   **Smart Yard—60-Minute Lawn Care.** Jeff Ball and Liz Ball. Fulcrum Publishing, Golden, Colorado, USA. 215 pp., 119 illus., 10 tables, small index, glossy semi–soft bound in multicolor with green and black lettering, 7.0 by 9.0 inches (177 × 228 mm).

A practical book on the planting and care of lawns. It is oriented to lawn enthusiasts and cool-season conditions of northern North America.

1996†   *Reprinted in 1996* with very minor changes.

Ball, Liz Geigle

1994†   **Smart Yard—60-Minute Lawn Care.** Coauthor.
&   See *Ball, Jeff Norman—1994 and 1996.*
1996†

Beckett, Kenneth A.

1985†   **Lawns.** First edition. Kenneth A. Beckett. Aura Garden Handbooks, Greenford, Middlesex, England, UK. 48 pp., 77 illus. in color, no tables, small index, fold-stapled glossy semisoft covers in multicolor with black lettering, 5.8 by 8.2 inches (147 × 210 mm).

A booklet on the planting and care of turfgrasses for lawns. It is practically oriented to lawn enthusiasts and cool-season conditions in the United Kingdom.

Revised in 1988.

1988† **Lawns.** Second revised edition. Kenneth A. Beckett. Collins Aura Garden Handbooks, William Collins Sons & Company Limited, London, England, UK. 48 pp., 85 illus. in color, 2 tables, small index, glossy semi–soft bound in multicolor with gray and black lettering, 5.8 by 8.2 inches (147 × 208 mm).

The booklet contains minor revisions.

Bernard, John D.

1954† **Lawns.** John D. Bernard. Rinehart's Garden Library, Rinehart & Company Inc., New York City, New York, USA. 94 pp., 17 illus., 4 tables, no index, cloth hardbound in multicolor with yellow and white lettering, 5.4 by 8.5 inches (137 × 215 mm).

A small book on lawn planting and care. It is practically oriented to home lawn enthusiasts and conditions in North America.

1955† **Lawns.** John D. Bernard. The Small Garden Library, William Heinemann Limited, London, England, UK. 64 pp., 11 illus., no tables, no index, cloth hardbound in multicolor with yellow and white lettering, 5.4 by 8.4 inches (137 × 214 mm).

A small book on lawn planting and care. It is practically oriented to home lawn enthusiasts and cool-season conditions in the United Kingdom.

Better Homes and Gardens

1993† **Lawns.** The Gardener's Collection, Meredith Books, Des Moines, Iowa, USA. 64 pp., 39 illus. with 34 in color, no tables, small index, glossy semi–soft bound in multicolor with black and white lettering, 7.4 by 9.8 inches (188 × 250 mm).

A small book concerning the planting, care, and pests of lawns. The back third contains a directory of lawn plants, including dicotyledons. It is practically oriented to lawn enthusiasts and conditions in the United States and Canada.

Bonar, Ann

1965† **Lawns.** A. Bonar. Amateur Gardening Picture Book Number 13, W.H. & L. Collingridge Limited, London, England, UK. 60 pp., 116 illus., no tables, no index, cloth hardbound in light green with silver lettering and a glossy jacket in multicolor with red and black lettering, 5.8 by 8.1 inches (148 × 205 mm).

A small book on the planting and care of lawns. It is practically oriented to lawn enthusiasts and cool-season conditions in England. It contains extensive photographs.

Böswirth, Daniel

2002 **Rasenprobleme: Erkennen und Beheben (Lawn Problems: Diagnosis and Correction).** Daniel Böswirth and Alice Thinschmidt. Verlag Eugen Ulmer, Österreichischer Agrarverlag, Stuttgart, Germany. 95 pp., 90 illus. in color, 5 tables, no index, glossy hard bond in multicolor with white and light green lettering, 6.5 by 9.0 inches (164 × 228 mm). Printed entirely in the German language.

A small book concerning solving problems that occur during the planting and care of lawns composed of turfgrasses and, occasionally, flowers. It is practically oriented to gardening enthusiasts under the cool-season conditions in Germany.

Bowker, Mike

1991† **Caring For Lawns.** Mike Bowker. NK Lawn and Garden Company, Avon Books, The Hearst Corporation, New York City, New York, USA. 80 pp., 180 illus., 3 tables, no index, glossy semi–soft bound in multicolor with white lettering, 8.0 by 9.9 inches (202 × 252 mm).

A small book on lawn care. It is practically oriented to lawn enthusiasts and conditions in the United States.

See also *Zeman, Anne M.—1993.*

Bradfield, Tom

1982† **How to grow lawns.** Tom Bradfield (editor). Hathercliff Limited, London, England, UK. 31 pp., 8 illus. in color, no tables, no index, glossy semi–soft bound in multicolor with black lettering, 4.4 by 6.4 inches (113 × 163 mm).

A small, simplified booklet on the planting and care of lawns. It is practically oriented to lawn enthusiasts and cool-season conditions in England.

Bradshaw, John

n.d.†
circa
early
195x
All About Lawns—Guide to Better Gardening. John Bradshaw. Leland Publishing Company Limited, Toronto, Ontario, Canada. 65 pp., 75 illus. in color, 1 table, no index, soft bound in multicolor with red and black lettering, 6.9 by 10.0 inches (176 × 253 mm).

A booklet on lawn planting and care. It is practically oriented to lawn enthusiasts and cool-season turfgrass conditions of North America.

1961†
All About Lawns. John Bradshaw. Book 1 of the 16 Volume Complete Guide to Better Gardening, Leland Publishing Company Limited, New York City, New York, USA. 96 pp., 119 illus. with 92 in color, 1 table, no index, glossy hardbound in multicolor with white, red, blue, and black lettering, 6.9 by 9.9 inches (175 × 252 mm).

A small book on the planting and care of lawns in the northern, cool-climatic regions of North America. It is practically oriented to the home lawn enthusiasts.

1965†
Better Lawns. John Bradshaw. Bolens Division, FMC Corporation, Wisconsin, USA. 47 pp., 45 illus. plus 18 color plates, 1 table, no index, fold-stapled glossy semisoft covers in multicolor with red and black lettering, 6.0 by 9.0 inches (154 × 229 mm).

A booklet on the planting and care of lawns in the northern United States. It is practically oriented to home lawn enthusiasts.

1982†
The Lawn Book: How to Grow the Perfect Lawn. John Bradshaw. McClelland and Stewart Limited, Toronto, Ontario, Canada. 96 pp., 64 illus. with 49 in color, no tables, small index, glossy semi–soft bound in multicolor with yellow, white, and black lettering, 5.9 by 9.3 inches (150 × 236 mm).

A small book on the planting and care of cool-season turfgrasses for lawns that is organized on a month-by-month calendar basis. It is practically oriented to home lawn enthusiasts in northern North America.

Brochard, Daniel

1980†
Réussir un Gazon (To Have a Successful Turf). First edition. Daniel Brochard. Collection LaVie en Vert, Dargaud Editeur, Paris, France. 95 pp., 25 illus., 1 table, small index, glossy semi–soft bound in multicolor with white, black, and yellow lettering, 6.4 by 8.5 inches (163 × 216 mm). Printed entirely in the French language.

A small book on the planting and care of lawns. It is practically oriented to lawn enthusiasts and conditions in France.

Reprinted multiple times, including in Quebec, Canada.

1995
Réussir un Gazon (To Have a Successful Turf). Second revised edition. Daniel Brochard. Dargaud Editeur, Paris, France. 96 pp. including 2 pp. of advertising, 65 illus., 1 table, small index, glossy semi–soft bound in multicolor with yellow and blue lettering, 6.2 by 8.7 inches (158 × 220 mm). Printed entirely in the French language.

Some revisions, with a number of new illustrations.

1998
Le Gazon Avec Succés (The Successful Lawn). First edition. Daniel Brochard. Éditions Rustica, Paris, France. 63 pp., 121 illus. in color, 1 table, small index, hardbound in multicolor with white and green lettering, 8.0 by 10.0 inches (202 × 254 mm). Printed in the French language.

A small book on the planting and care of lawns. It is practically oriented to lawn enthusiasts and conditions in France.

2000
Le Gazon Avec Succés (The Successful Lawn). Second revised edition. Daniel Brochard. Éditions Rustica, Paris, France. 128 pp., illus. in color, 1 table, indexed, hardbound in multicolor with white lettering. Printed in the French language.

An expanded revision of the first edition.

Carleton, R. Milton

1959†
Your Lawn—How To Make It And Keep It. First edition. R. Milton Carleton. D. Van Nostrand Company Inc., Princeton, New Jersey, USA. 165 pp., 23 illus., no tables, indexed, cloth hardbound in green and gray with green lettering with a jacket in greens and white with black and white lettering, 5.4 by 8.2 inches (138 × 208 mm).

This book addresses the basics of turfgrass planting and care within a how-to approach. It is practically oriented to home lawn enthusiasts and conditions in the United States.

† *Reprinted in 1960 and 1967.*
Revised in 1971.

1971†
Your Lawn: How to Make It and Keep It. Second revised edition. R. Milton Carleton. Van Nostrand Reinhold Company, New York City, New York, USA. 127 pp.; 56 illus.; no tables; indexed; cloth hardbound in light yellow with brown lettering and a glossy jacket in multicolor with green, blue, and black lettering; 7.9 by 7.9 inches (200 × 200 mm).

A major revision of the book on the basics of turfgrass planting and care. It is practically oriented to home lawn enthusiasts and conditions in the United States. The photographs have been improved substantially.

Carr, David

1981† **Lawns—A Green Fingers Guide.** David Carr. Ebury Press, London, England, UK. 63 pp.; 160 illus.; 3 tables; no index, glossy semi–soft bound in multicolor with black, green, and blue lettering; 6.6 by 9.4 inches (168 × 240 mm).

A small booklet on the planting and care of lawns. It is practically oriented to lawn enthusiasts and cool-season conditions in the United Kingdom.

Carson, Sean McBirney

1954† **Lawns.** S.McB. Carson and R.W. Colton. Help Yourself Series No. 34, Burke Publishing Company Limited, London, England, UK. 64 pp.; 17 illus.; 4 tables; no index; glossy hardbound in red, black, and white with white, black, and red lettering; 4.2 by 6.5 inches (108 × 165 mm).

A small book on the construction, planting, and care of lawns. It is practically oriented to lawn enthusiasts and cool-season conditions of England. Included is a chapter on paths and verges.

Chatard, Jean

1978† **Le Gazon à l'Anglaise (English Turf).** Jean Chatard. La Maison Rustique, Paris, France. 31 pp., 18 illus. with 7 in color, no tables, no index, fold-stapled glossy semisoft covers in multicolor with black and white lettering, 5.3 by 7.0 inches (134 × 178 mm). Printed entirely in the French language.

A small booklet on the planting and care of lawns. It is practically oriented to home lawn enthusiasts and conditions in France.

Chessmore, Roy A.

1961† **Lawn Improvement in Southern Oklahoma.** Roy A. Chessmore. Miscellaneous Publication-2, The Samuel Roberts Noble Foundation, Ardmore, Oklahoma, USA. 24 pp.; 3 illus.; no tables; no index; fold-stapled semisoft covers in green, blacks, and white with black lettering; 5.4 by 8.1 inches (137 × 206 mm).

A booklet concerning turfgrass selection, planting, and care of lawns under the conditions in southern Oklahoma. *Reprinted in late 1961.*

Chicago Park District

1938 **You Can Have A Good Lawn.** Chicago Park District, Burnham Park, Chicago, Illinois, USA. 29 pp., 39 illus., no tables, no index, fold-stapled semisoft covers in multicolor with black lettering, 5.3 by 8.1 inches (135 × 206 mm).

A booklet about the planting and care of turfgrass for lawns in the Chicago area. It is practically oriented to lawn enthusiasts.

Clayton, B.C.

1962† **Die Rasenfibel (The Turf Primer).** Coauthor.
See *Dawson, Robert Brian—1962.*

Colton, Ronald William

1954† **Lawns.** Coauthor.
See *Carson, Sean McBirney—1954.*

Cornish, Geoffrey St. John

1951† **Your Lawn.** Geoffrey S. Cornish. Authentic Publications Inc., New York City, New York, USA. 48 pp.; 122 illus.; 1 table; no index; fold-stapled soft covers in cream, green, and blue with blue and white lettering; 8.1 by 10.6 inches (207 × 270 mm).

A small book on lawn planting and care. It is practically oriented to lawn enthusiasts and conditions in North America.

1952† **Your Guide to a Greener Lawn.** Geoffrey S. Cornish. The Massachusetts Horticultural Society, Boston, Massachusetts, USA. 63 pp.; 74 illus.; 2 tables; indexed; fold-stapled semisoft covers in yellow, dark green, and burgundy with green lettering; 6.0 by 9.1 inches (152 × 231 mm).

A small handbook on lawn planting and care. It is practically oriented to home lawn enthusiasts and cool-season conditions in northeastern United States.

† *Reprinted in 1953.*

Courtier, Jane

2001† **Lawn and Lawncare.** Jane Courtier. Time-Life Books, Alexandria, Virginia, USA. 111 pp.; 50 illus. in color; no tables; small index; semisoft covers in multicolor with white, green, blue, and black lettering in a spiral binder; 5.1 by 8.8 inches (130 × 224 mm).

A small book concerning the planting and care of lawns. It is practically oriented to lawn enthusiasts and conditions in the United States.

2001   **Lawns & Lawn Care.** Jane Courtier. Marshall Editions Developments Ltd., London, England, UK. 102 pp.; 149 illus. in color; 7 tables; small index; glossy semisoft covers in multicolor with green, blue, black, orange, and white lettering in a wire spiral binder; 5.7 by 8.5 inches (144 × 215 mm).

A small, practical book about the planting, care, and renovation of lawns under the conditions in England.

Dawson, Robert Brian

1954†   **Lawns.** R.B. Dawson. Amateur Gardening' Handbook No. 11, W.H. & L. Collingridge Limited, London, England, UK. 92 pp., 18 illus., 2 tables, no index, glossy hardbound in multicolor with black and red lettering, 4.5 by 6.4 inches (116 × 162 mm).

A small book on the planting and care of lawns in England. It is practically oriented to home lawn enthusiasts. There is a short section on lawn tennis courts in the back.

†   *Reprinted in 1958.*

n.d.†   **Lawns: Essentials of Establishment and Maintenance.** R.B. Dawson. The Royal Horticultural Society, London, England, UK. 23 pp., 8 illus., 8 tables, no index, fold-stapled semisoft covers in light green with dark green lettering, 5.6 by 8.7 inches (142 × 221 mm).

circa
1956

A small booklet on lawn care that is practically oriented to lawn enthusiasts and cool-season conditions in the United Kingdom.

†   *Reprinted through 1966.*

1962†   **Die Rasenfibel (The Turf Primer).** R.B. Dawson and B.C. Clayton, as translated by Dr. Bernhard von Limburger. Münstertor-Verlag, Stuttgart, West Germany. 91 pp. with 5 pp. of advertising, no illus., 4 tables, no index, semi–soft bound in tan and a jacket in green and white with black lettering, 5.0 by 7.4 inches (128 × 188 mm). Printed entirely in the German language.

A small book on the planting and care of lawns. It is practically oriented to home lawn enthusiasts.

Dean, Michael D.

1986†   **In Search of the Perfect Lawn.** Michael Dean. Black Moss Press, Windsor, Ontario, Canada. 64 pp., 3 illus., no tables, no index, glossy semi–soft bound in multicolor with black lettering, 5.4 by 8.7 inches (137 × 221 mm).

A small, humorous book on lawns and their care. It is an assemblage of writings that had been published in the Poetry Toronto Newsletter, Five On Fiche, and Wordfest 983.

Dobbs, Steve

2002†   **The Perfect Texas Lawn—Attaining and Maintaining the Lawn You Want.** Steve Dobbs. One of a state series. Cool Springs Press, Nashville, Tennessee, USA. 160 pp.; 115 illus. with 79 in color; 9 tables; indexed; glossy semi–soft bound in multicolor with green, red, white, and black lettering; 6.0 by 9.0 inches (152 × 228 mm).

A book on the planting and care of turfgrasses for lawns. It is practically oriented to lawn enthusiasts and conditions in the state of Texas. There is a small glossary in the back.

NOTE: This book is representative of a series of similar books by the same author and publisher all with the same length and format but changed somewhat to other states such as Alabama, Arkansas, Georgia, Louisiana, Mississippi, North Carolina, Oklahoma, South Carolina, and Tennessee.

See also *Myers, Melinda—2003; MacCubbin, Tom—2004.*

Drane, Davis S.

1991†   **On Your Home Turf.** David S. Drane (editor) with four contributing authors: Gary W. Beehag, Ian C. McIver, Alexandra Shakesby, and Murray G. Wallis. Australian Turfgrass Research Institute, Concord West, New South Wales, Australia. 50 pp.; 10 illus.; 6 tables; no index; glossy semi–soft bound in multicolor with white, black, and green lettering; 8.3 by 11.6 inches (210 × 294 mm).

A practical guide to the planting and care of lawns. It is practically oriented to lawn enthusiasts and conditions in Australia.

Edwards, Ray

1980†   **Lawns: Easy Making and Maintenance.** Ray Edwards. David & Charles Limited, Newton Abbot, Devon, England, UK. 48 pp., 53 illus., no tables, no index, textured semi–soft bound in multicolor with yellow and white lettering, 5.8 by 8.2 inches (147 × 209 mm).

A small book on the planting and care of lawns. It is practically oriented to home lawn enthusiasts and cool-season conditions in England.

Ellis, Barbara

1997† **Safe & Easy Lawn Care.** Barbara Ellis (editor). Taylor's Weekend Gardening Guides, Houghton Mifflin Company, New York City, New York, USA. 127 pp.; 111 illus. with 70 in color; 1 table; small index; glossy semi–soft bound in multicolor with green, yellow, and black lettering; 7.5 by 9.4 inches (190 × 238 mm).

A book on the planting and care of turfgrasses for lawns, with emphasis on organic methods. It is practically oriented to home owners under conditions in the United States.

Escritt, John Robert

1979† **Lawns.** J.R. Escritt. The World of the Garden Series, Hodder and Stoughton Paperbacks, Sevenoaks, Kent, England, UK. 113 pp.; 33 illus.; 8 tables; indexed; semi–soft bound in multicolor with black, red, and white lettering; 5.0 by 7.8 inches (127 × 198 mm).

A small book on the planting and care of lawn turfgrasses. It is practically oriented to lawn enthusiasts and cool-season conditions in England. There is a small glossary of selected terms in the back.

† US edition by David McKay & Company Inc., New York City, New York, USA.

Evans, Eva Knox

1947 **Let's Plant Grass.** Second revised edition. Eva Knox Evans and John W. Barnes. Hinds, Hayden & Eldrege Inc., New York City, New York, USA. 31 pp., 22 illus., no tables, no index, semi–soft bound in multicolor with white lettering, 7.1 by 9.6 inches (181 × 243 mm).

A unique children's book about the activity of planting and care of grass for a lawn in a school yard. Characters from Winnie the Pooh are included. Learning activities involve arithmetic and spelling.

Faust, Joan Lee

1957† **The New York Times Book of Lawn Care.** Joan Lee Faust (editor) with seven contributing authors: John R. Rebham, J.M. Duich, Marvin H. Ferguson, John F. Cornman, Alexander M. Radko, Ralph E. Engel, and Bernard Gladstone. Alfred A. Knopf Inc., New York City, New York, USA. 88 pp.; 49 illus.; 1 table; indexed; cloth hardbound in blue with gold lettering and a jacket in green, yellow, and white with white, yellow, black lettering; 6.1 by 8.9 inches (155 × 226 mm).

A small book on the planting and care of lawns. It is practically oriented to the home lawn enthusiasts and turfgrass conditions in the United States.

† *Reprinted in 1958, 1959, 1960, 1961, 1962, 1963, 1964, 1965, and 1972.*

Fech, John

2002† **Taunton's Lawn Guide.** John Fech. The Taunton Press Inc., Newtown, Connecticut, USA. 144 pp.; 80 illus. in color; 23 tables; small index; glossy semi–soft bound in multicolor with green, yellow, and black lettering; 8.5 by 10.9 inches (215 × 276 mm).

A book concerning the site preparation, planting, and care of turfgrasses for lawns. There is a seasonal calendar of lawn care practices grouped by the turfgrass climatic regions of the United States.

Feldman, Fran

1993† **Lawns.** Fran Feldman (editor). Sunset Publishing Corporation, Menlo Park, California, USA. 80 pp., 128 illus., 2 tables, small index, glossy semi–soft bound in multicolor with yellow and white lettering, 8.2 by 10.7 inches (208 × 271 mm).

A small book on the planting and care of lawns. It is practically oriented to home lawn enthusiasts and conditions in North America.

† *Reprinted in 1994 and 1997.*

Fish, Martin

2005† **Lawns—Grow & Maintain Healthy Grass.** Martin Fish. Harper Collins Publishers, London, England, UK. 144 pp., 277 illus., 3 tables, small index, glossy semi–soft bound in multicolor with white lettering, 7.4 by 9.7 inches (188 × 246 mm).

A practical book about the planting and care of lawns. It is oriented to lawn enthusiasts under the conditions in England.

Fisons Horticulture Limited

James Fison's began as a flour mill and bakery in the late 1700s in Birmingham, England. James Fison and Sons was formed in 1808. By the 1840s, the company entered the fertilizer business and moved their headquarters to Ipswich, England. The company became the dominant fertilizer producer in the United Kingdom and, subsequently, entered the pest control market in the 1950s. The Fison's Horticulture Division was organized out of the Agrochemical Division in 1977.

n.d.† circa 195x — **Fisons Lawn Book.** First edition. Horticulture Department, Fisons Limited, Felixstowe, Suffolk, England, UK. 24 pp.; 3 illus.; no tables; no index; fold-stapled semisoft covers in green, white, and brown with brown lettering; 4.8 by 7.1 inches (122 × 181 mm).

A small booklet on the planting and care of lawns, with emphasis on fertilization and pest control. It is practically oriented to lawn enthusiasts and cool-season conditions in the United Kingdom.

Revised in 1960 and circa 1963.

1960† — **Fisons Lawn Book.** Second revised edition. Fisons Horticulture Limited, Felixstowe, Suffolk, England, UK. 32 pp., 30 illus., no tables, no index, fold-stapled glossy semisoft covers in multicolor with white and black lettering, 5.5 by 8.7 inches (142 × 222 mm).

A small book on the planting and care of lawns. It is practically oriented to lawn enthusiasts and cool-season conditions in the United Kingdom.

Revised circa 1963.

n.d.† circa 1963 — **Fisons Lawn Book.** Third revised edition. Fisons Horticulture Limited, Felixstowe, Suffolk, England, UK. 32 pp., 37 illus., no tables, no index, fold-stapled glossy semisoft covers in multicolor with black lettering, 5.6 by 8.7 inches (141 × 220 mm).

There are small revisions, primarily to the drawings.

Flynn, Joseph F.

1951† — **How to Grow and Keep a Better Lawn.** Joseph F. Flynn. Simon and Schuster Inc., New York City, New York, USA. 76 pp. plus 4 pp. of advertising, 36 illus., no tables, no index, cloth hardbound in dark blue with white lettering, 5.7 by 8.7 inches (146 × 222 mm).

A small booklet on lawn planting and care. It is practically oriented to home lawn enthusiasts and cool-season conditions in the northeastern United States. In 1951, Mr. Flynn's suggested a high phosphorus fertilizer analysis for lawns as follows:

> A complete fertilizer replenishes *all* the elements in quantities necessary for grass, namely, nitrogen, phosphoric acid, and potash. Nitrogen is the key to a good lawn, so you want a fertilizer with a high nitrogen content. I recommend a formula such as 8-6-4 or 6-10-4.

1951† — **How to Grow and Keep a Better Lawn.** Joseph F. Flynn. Simon and Schuster Publishers and Hart Publishing Company Inc., New York City, New York, USA. 89 pp.; 41 illus.; 9 tables; no index; glossy semi–soft bound in multicolor with light green, white, and black lettering; 5.3 by 7.9 inches (134 × 201 mm).

This is a semi–soft bound version of the book previously described.

Galaktionov, Ivan Innokent'evich

1963 — **Mnogoletnie Gazony Srednei Polosy RSFSR (Perennial Lawns of the Anxral Part of Russia).** I.I. Galaktionov. Izd-vo Ministaratva kommunal'nogo khoziaistva RSFSR, Moscow, Russia, USSR. 41 pp.; 6 illus.; 5 tables; no index; fold-stapled semisoft covers in green, yellow, and white with black lettering; 5.0 by 7.0 inches (130 × 199 mm). Printed entirely in the Russian language.

A booklet concerning the planting and care of lawns under the conditions in the middle zone of Russia. There is a list of selected references in the back.

Gandert, Klaus-Dietrich

1967† — **Rasen im Garten (Lawn in the Garden).** First edition. Klaus Gandert. VEB Deutscher Landwirtschaftsverlag, Berlin, East Germany. 96 pp., 21 illus., 7 tables, no index, glossy semi–soft bound in multicolor with white and black lettering, 4.3 by 7.0 inches (109 × 178 mm). Printed entirely in the German language.

A small book on turfgrasses, soil preparation, planting, and care of lawns. A practically oriented book for cool-season conditions in East Germany.

Revised in 1970, 1975, 1982, and 1988.

1970   **Rasen im Garten (Lawn in the Garden).** Second revised edition. Klaus-Dietrich Gandert. Deutscher Landwirtschafts-verlag. Berlin, East Germany. 96 pp., 29 illus., 12 tables, no index, glossy semi–soft bound in multicolor with white and black lettering, 4.2 by 7.0 inches (107 × 177 mm). Printed entirely in the German language.

   A minor revision.

   Revised somewhat in 1975, 1982, and 1988.

1975   **Rasen im Garten (Lawn in the Garden).** Third revised edition. Klaus-Dietrich Gandert. Deutscher Landwirtschafts-verlag VEB, Berlin, Germany. 104 pp., 30 illus., 12 tables, indexed, glossy semi–soft bound in green with dark green and white lettering, 4.5 by 6.5 inches (110 × 165 mm). Printed in the German language.

   A small revision.

   Revised in 1982 and 1988.

1982   **Rasen im Garten (Lawn in the Garden).** Fourth revised edition. Klaus-Dietrich Gandert. VEB Deutscher Landwirtschaftsverlag, Berlin, Germany. 96 pp., 20 illus., 7 tables, small index, semi–soft bound in multicolor with black lettering, 4.2 by 6.9 inches (107 × 176 mm). Printed in the German language.

   A minor revision.

   Revised in 1988.

1988   **Rasen im Garten (Lawn in the Garden).** Fifth revised edition. Klaus-Dietrich Gandert. VEB Deutscher. Landwirtschaftsverlag, Berlin, Germany. 94 pp., 21 illus., 7 tables, small index, semi–soft bound in multicolor with black lettering, 4.1 by 6.9 inches (105 × 177 mm). Printed in the German language.

   A minor revision.

Gardiner, Ruth

1973†   **The Perfect Lawn.** Ruth Gardiner and Roger Grounds (editors). Concorde Books, Ward Lock Limited, and Peter Way Ltd., London, England, UK. 94 pp., 104 illus. with 39 in color, 4 tables, small index, both cloth hardbound and soft bound versions in multicolor with white lettering, 5.7 by 7.8 inches (146 × 197 mm).

   A small book on lawn planting and care. It is practically oriented to home lawn enthusiasts and cool-season conditions in England. It is well illustrated.

1973   **The Perfect Lawn—50 Ways to Raise a Glorious Green.** Ruth Gardiner (editor). Peter Way Ltd., Covent Garden, London, England, UK. 33 pp.; 107 illus.; 1 table; indexed; fold-stapled glossy semisoft covers in multicolor with light green, pink, and white lettering; 7.7 by 11.2 inches (196 × 284 mm).

   A small, practical guide concerning the planting and care of lawns, with the subject titles organized alphabetically. It is oriented to lawn enthusiasts in England.

1977   **Het ideale gazon (The ideal lawn).** Ruth Gardiner and C. Den Engelsen. B.V.W.J. Thieme & Cie-Zutphen, the Netherlands. 95 pp., 108 illus. with 74 in color, 1 table, small index, glossy semi–soft bound in multicolor with yellow and black lettering, 5.6 by 7.8 inches (143 × 197 mm). Printed in the Dutch language.

   A small book concerning the planting and care of turfgrasses for lawns. It is practically oriented to lawn enthusiasts under the conditions in the Netherlands.

1978†   **So wird der Rasen perfekt (The Perfect Lawn).** Ruth Gardiner, as translated by Walter Büring. Verlag Paul Parey, Berlin/Hamburg, West Germany. 94 pp., 131 illus. with 92 in color, 1 table, indexed, glossy semi–soft bound in multicolor with white and black lettering, 5.8 by 7.8 inches (148 × 197 mm). Printed entirely in the German language.

   A small, practically oriented book on the design of gardens, turfgrass types, planting, and care of lawns in West Germany.

Gil Rodriguez, Nuria

2005   **Césped: Instalación y Mantenimiento (Turfgrass: Installation & Maintenance).** Nuria Gil Rodriguez. Salamanca, Rio Mondego-Vistahermosa, Spain. 81 pp., illus., no tables, no index. glossy semi–soft bound in multicolor, 8.3 by 11.8 inches (210 × 300 mm). Printed in the Spanish language.

   A practical booklet concerning the planting and care of lawns under the conditions in Spain.

Gilmer, Maureen

1999†   **The Complete Idiot's Guide to a Beautiful Lawn.** Maureen Gilmer, Alpha Books, Macmillan Company, New York City, New York, USA. 416 pp.; 300-plus illus. with 43 in color; 5 tables; indexed; glossy semi–soft bound in multicolor with blue, green, and black lettering; 7.5 by 9.1 inches (191 × 231 mm).

   A book on turfgrass selection, planting, care, renovation, and pest management for lawns. It is oriented to lawn enthusiasts under conditions in the United States.

Goldstein, Alrica

2006† **Scotts Lawn.** First edition. Alrica Goldstein (editor) with writers Dr. Nick E. Christians and Ashton Ritchie. The Scotts Miracle-Gro Company, Marysville, Ohio, USA. 95 pp.; 149 illus.; no tables; indexed; glossy semisoft covers in multi-color with white, yellow, and black lettering in a wire ring binder; 5.0 by 8.2 inches (127 × 208 mm).

   A guide to the planting and care of turfgrasses. It is practically oriented to lawn enthusiasts under conditions in the United States.

Golovach, Aleksandr Grigor'evich.

1955 **Gazony, ikh ustroislov i soderzhanie (Lawns: Their Construction and Composition).** A.G. Golovach. Izd-vo Akademii nauk SSSR, Moscow, Russia, USSR. 339 pp., 86 illus., 51 tables, no index, textured cloth hardbound in green with yellow lettering, 6.0 by 10.0 inches (169 × 258 mm). Printed entirely in the Russian language.

   A quality early Russian book concerning site preparation, turfgrass selection, planting, and care of lawns under the conditions in Russia. There is a list of selected references in the back.

Grayson, Esther Catherine

1956† **The Complete Book of Lawns.** Coauthor.
   See *Rockwell, Frederick Frye—1956.*

Green, Douglas

2001† **The Everything Lawn Care Book.** Douglas Green. Adams Media Corporation, Holbrook, Massachusetts, USA. 285 pp. plus 3 pp. of advertising; 52 illus. with 18 in color plus numerous vignettes; 21 tables; indexed; glossy semi–soft bound in multicolor with green, red, and blue lettering; 7.9 by 9.2 inches (202 × 234 mm).

   A book on the planting and care of lawns. It is practically oriented to lawn enthusiasts and conditions in the United States.

Grounds, Roger

1973† **The Perfect Lawn.** Coauthor.
   See *Gardiner, Ruth—1973.*

1974 **The Perfect Lawn.** Roger Grounds (editor). Lard Lock Limited, London, England, UK. 94 pp., 108 illus. with 74 in color, 1 table, small index, cloth hardbound in green with gold lettering, 5.9 by 7.8 inches (151 × 197 mm).

   A book concerning the planting and care of turfgrasses. It is practically oriented to lawn enthusiasts and conditions in the United Kingdom.

Guérin, Jean-Paul

2007† **Un Beau Gazon: Création, Entretien, Problèms et Solutions (A Beautiful Lawn: Establishment, Maintenance, and Troubleshooting).** Jean-Paul Guérin. Editions Eugen Ulmer, Paris, France. 66 pp., 121 illus. with 112 in color, 3 tables, no index, semi–soft bound in multicolor with green and white lettering, 6.7 by 8.3 inches (170 × 212 mm). Printed in the French language.

   A small book on the planting and care of lawns. It is oriented to lawn enthusiasts under the conditions in France.

2008 **Erste Hilfe für den Rasen (First Aid for the Lawn).** Jean-Paul Guérin, with translation by Christiane Schoelzel. Eugen Ulmer KG (Hohenheim) Stuttgart, Germany. 66 pp.; 111 illus. with 101 in color; 2 tables; no index; glossy semi–soft bound in multicolor with white, yellow, black, and green lettering; 6.6 by 8.3 inches (168 × 211 mm). Printed entirely in the German language.

   A practical small book concerning the planting and care of lawns under the cool conditions in Germany. Emphasis is placed on pest and disease problems.

Gurie, Albert

n.d.† **Lawns—Their Construction and Upkeep.** Albert Gurie. News Chronicle Publications, London, England, UK. 65 pp. circa plus 7 pp. of advertising; 21 illus.; 3 tables; no index; glossy semi–soft bound in multicolor with green, yellow, and blue 195x lettering; 4.8 by 7.1 inches (123 × 180 mm).

   A small book on the planting and care of lawns. It is practically oriented to lawn enthusiasts and cool-season conditions in the United Kingdom.

Haas, Cathy

1996† **Lawns.** Cathy Haas. Ortho Easy-Step Books, Ortho Books, San Ramon, California, USA. 64 pp.; 146 illus. with 142 in color; 6 tables; no index; glossy semi–soft bound in multicolor with white, black, and green lettering; 5.0 by 9.0 inches (127 × 229 mm).

   A small book on the planting and care of lawns. It has a practically oriented, simplified approach for novices.

n.d.† **Césped (Turf).** Cathy Haas. Libros Ortho en pasos fáciles, Ortho Books, San Ramon, California, USA. 64 pp.; 145
circa illus. in color; 10 tables; no index; glossy hardbound in multicolor with white, black, and green lettering; 5.1 by 8.9
1996 inches (130 × 226 mm). Printed entirely in the Spanish language.

A small book on the planting and care of lawns. It is practically oriented to Spanish lawn enthusiasts.

Habegger, Ernst

1985 **Der Rasen (The Lawn).** Ernst Habegger. Hallwag Verlag AG, Bern, Switzerland. 140 pp. plus 4 pp. of advertising, 183
illus. with 111 in color, 16 tables, small index, glossy semi–soft bound in multicolor with white lettering, 4.1 by 5.7
inches (104 × 146 mm). Printed in the German language.

A small book on the planting and care of lawns. It is oriented to lawn enthusiasts and conditions in Switzerland.

Håbjörg, Atle

1980 **Hageselkapets Plenbok (The Lawn Book of the Garden Society).** Atle Håbjörg. Gröndahl & Sön Forlag A/S, Oslo,
Norway. 103 pp., 39 illus. with 12 in color, no tables, no index, soft bound in multicolor with white lettering, 5.9 by
7.2 inches (151 × 132 mm). Printed entirely in the Norwegian language.

A small book on the planting and care of lawns. It is practically oriented to lawn enthusiasts and cool-season conditions in Norway.

Hansen, Waldenar

1948 **Græsplæner (Lawns).** Waldenar Hansen. Praktiske Håndbøger, udgivet af Alm. dansk Gartnerforening, Copenhagen,
Denmark. 78 pp., 63 illus., no tables, no index, fold stapled soft covers in green and black with black lettering, 5.9 by
8.2 inches (150 × 208 mm). Printed entirely in the Danish language.

A small book on the planting and care of lawns. It is practically oriented to lawn enthusiasts and cool-season conditions in Denmark.

Hastings, Chris

2000† **Southern Lawns.** Chris Hastings. Longstreet Press Inc., Atlanta, Georgia, USA. 223 pp., 60 illus. with 59 in
color, 22 tables, indexed, glossy semi–soft bound in multicolor with white lettering, 8.0 by 10.0 inches (203 ×
245 mm).

A quality book on the planting and care of lawns, with chapters organized around seven turfgrass species. It
is practically oriented to home owners and conditions in southeastern United States. There is a small glossary in
the back.

Hayman, Steven

2008† **The Lawn Guide.** Coauthor.
See *Sharples, Philip—2008.*

Hedger, George

1960† **Your Lawn: Establishing, Maintaining, Improving.** George Hedger. Agricultural Division, The Samuel Roberts
Noble Foundation, Ardmore, Oklahoma, USA. 22 pp., 8 illus., 2 tables, no index, fold-stapled soft covers in white with
black lettering, 5.4 by 8.5 inches (136 × 215 mm).

A small booklet concerning the turfgrasses and their planting and care. It is oriented to lawn enthusiasts and warm
condition of Oklahoma.

Hellyer, Arthur George Lee

1970† **Your Lawn.** First edition. Arthur Hellyer. Hamlyn Publishing Group Limited, London, England, UK. 64 pp., 132 illus.
with 81 in color, no tables, small index, glossy semi–soft bound in multicolor with green and black lettering; 5.7 by 8.7
inches (146 × 222 mm).

A small book on the planting and care of lawns. It is practically oriented to home lawn enthusiasts and cool-season
conditions in the United Kingdom. This booklet is well illustrated with good color photographs.
Revised in 1986.

1986† **Lawn Care.** Second revised edition. Arthur Hellyer. Hamlyn Gardening Guide, Hamlyn Publishing Group Limited,
Twickenham, Middlesex, England, UK. 64 pp.; 88 illus. in color; no tables; small index; glossy semi–soft bound in
multicolor with green, brown, and black lettering, 5.7 by 8.7 inches (145 × 222 mm).

A small booklet on lawn planting and care. It is practically oriented to lawn enthusiasts and cool-season conditions
in the United Kingdom.

Hertel, Fritz

n.d.
circa
1954 **Rasenanlage und -Pflege (Turf Establishment and Care).** Fritz Hertel. Lehrmeister-Bucherei Nr. 304, Philler-Verlaag, Minden, West Germany. 64 pp., 12 illus., no tables, no index, semi–soft bound in multicolor with blue and white lettering, 4.0 by 6.0 inches (117 × 164 mm). Printed in the German language.

A small book concerning the planting and care of lawns under the conditions in West Germany. There are chapters on equipment and sport fields in the back.

Herwig, Rob

1967 **Uw Gazon: Het aanleggen en het onderhouden van het betere gazon (Your Turf: To start and maintain the best turf).** Rob Herwig. L.J. Veen Uitgeversmij N.V., Amsterdam, the Netherlands. 112 pp. plus 8 pp. of advertising, 43 illus. with 12 in color, 3 tables, no index, soft bound in multicolor with black lettering, 5.3 by 8.2 inches (133 × 208 mm). Printed entirely in the Dutch language.

A book on site development, turfgrass selection, planting, and care of cool-season turfgrass lawns under conditions in the Netherlands.

Hessayon, David Gerald

1961† **Be Your Own Lawn Expert.** First edition. D.G. Hessayon. Pan Britannica Industries Limited, Waltham Abbey, Essex, England, UK. 35 pp., 150-plus illus. with more than one-half in color, 7 tables, no index, soft bound in multicolor with black and red lettering, 7.2 by 9.3 inches (182 × 236 mm).

A small booklet on lawn planting and care. It is practically oriented to home lawn enthusiasts and cool-season conditions in England. It is extensively illustrated in color.
Revised in 1975.

1961† **Be Your Own Lawn Expert.** First edition. D.G. Hessayon. Pan Britannica Industries Ltd., Watham Abbey, Essex, England, UK. 35 pp.; 139 illus. with 88 in color; no tables; no index; fold-stapled glossy semisoft covers in multicolor with black, red, and yellow lettering; 7.4 by 9.5 inches (188 × 240 mm).

A small booklet on the planting and care of lawns. It is oriented to lawn enthusiasts and conditions in the United Kingdom.
Revised in 1967 and 1975.

1967† **Be Your Own Lawn Expert.** Second revised edition. D.G. Hessayon. Pan Britannica Industries Limited, Waltham Cross, Hertfordshire, England, UK. 35 pp.; 210 illus. in color; 4 tables; no index; glossy semi–soft bound in multicolor with black, gold, and white lettering; 7.2 by 9.4 inches (184 × 238 mm).

A slight revision of the 1961 edition in full color.
Revised in 1975.

1975† **Be Your Own Lawn Expert.** Second revised edition. D.G. Hessayon. pbi Publications, Pan Britannica Industries, Waltham Cross, Hertfordshire, England, UK. 34 pp.; 140 illus. in color; 6 tables; no index; glossy semi–soft bound in multicolor with black, gold, and white lettering; 7.2 by 9.4 inches (184 × 238 mm).

A revision of a small extensively illustrated booklet on lawn planting and care. It is oriented to lawn enthusiasts and cool-season conditions in England.
Numerous reprintings.

1982† **The Lawn Expert.** First edition. D.G. Hessayon. pbi Publications, Waltham Cross, Hertfordshire, England, UK. 104 pp., 300-plus illus. in color, 14 tables, small index, glossy semi–soft bound in multicolor with green and black lettering, 7.2 by 9.3 inches (182 × 238 mm).

A small, extensively illustrated book on the planting and care of lawns. It is an expansion of an earlier booklet titled **Be Your Own Lawn Expert.** It is practically oriented for home lawn enthusiasts and cool-season conditions in England.

† *Numerous reprintings.*
Revised in 1988.

1983 **Rasen (Lawn).** D.G. Hessayon. A German translation of his 1982 book.

1986 **Césped: Manual de Cultivo y Conservación (Lawn: Manual of Culture and Conservation).** First edition translation. Dr. D.G. Hessayon, as translated by Luisa Moysset. Leopold Blume, Barcelona, Spain. 104 pp., 289 illus. in color, 10 tables, small index, glossy semi–soft bound in multicolor with green lettering, 7.3 by 9.3 inches (184 × 235 mm). Printed entirely in the Spanish language.

Basically a translation of the **The Lawn Expert**, ninth edition.
Reprinted in 1994 and 1996.
Revised edition translation in 1998.

1988† **The Lawn Expert.** Second revised edition. D.G. Hessayon. pbi Publications, Waltham, Hertfordshire, England, UK. 104 pp., 300-plus illus. in color, 9 tables, small index, glossy semi–soft bound in multicolor with light-green and black lettering, 7.4 by 9.4 inches (189 × 238 mm).

A minor update of the 1982 first edition.

There are three slightly revised editions: third, fourth, and fifth.

1991† **The Lawn Expert.** Sixth revised edition. D.G. Hessayon. pbi Publications, Waltham Cross, Hertfordshire, England, UK. 104 pp., 350+ illus. in color, 18 tables, very small index, glossy semi–soft bound in multicolor with green and black lettering, 7.5 by 9.4 inches (189 × 238 mm).

A slight revision of an extensively illustrated book on lawn planting and care.

1997† **The New Lawn Expert.** D.G. Hessayon. Expert Books, Transworld Publishers Limited, London, England, UK. 128 pp., 400-plus illus. in color, 6 tables, small index, glossy semi–soft bound in multicolor with white and light green lettering, 7.4 by 9.4 inches (189 × 238 mm).

A revision and expansion of an earlier edition titled **The Lawn Expert.** A section on nongrass covers has been added.

Reprinted in 1999, with very minor revisions.

1998 **Césped: Manual de Cultivo y Conservación (Lawn: Manual of Culture and Conservation).** Revised edition. Dr. D.G. Hessayon. Leopold Blume, Barcelona, Spain. 128 pp., 400-plus illus. in color, 7 tables, small index, glossy semi–soft bound in multicolor with light-green and black bound in multicolor with light green and black lettering, 7.2 by 9.3 inches (184 × 235 mm). Printed entirely in the Spanish language.

There are minor revisions of an earlier translation.

Reprinted in 2000 and 2001.

2000† **The Lawn Expert.** D.G. Hessayon. Expert Books, Transworld Publishers Limited, London, England, UK. 128 pp., 400+ illus. in color, 10 tables, small index, glossy semi–soft bound in multicolor with white and black lettering, 7.2 by 9.3 inches (184 × 235 mm).

Minor revisions.

Reprinted in 2002 with very minor revisions.

2008† **The Lawn Expert.** Dr. G. Hessayon. Expert Books, Transworld Publishers, London, England, UK. 128 pp., 400-plus illus. in color, 8 tables, small index, glossy, semi–soft bound in multicolor with white lettering, 7.2 by 9.3 inches (184 × 237 mm).

A minor revision of the 2000 edition.

Himmelhuber, Peter

2001 **Rasen: richtig anlegen und pflegen (Lawns: Correct installation and care).** Peter Himmelhuber. Verlagsunion Pabel Moewig KG, Rastatt, Germany. 96 pp., 129 illus. in color, 2 tables, small index, glossy semi–soft bound in multicolor with yellow lettering, 6.0 by 7.4 inches (15 3 × 189 mm). Printed entirely in the German language.

A small book concerning the planting and care of turfgrass lawns under the cool-conditions in Germany. There also a section on flower-grass meadows.

Home Garden Magazine

1971† **Book of Lawn Care.** Home Garden Magazine. Award Books, Universal Publishing and Distributing Corporation, New York City, New York, USA. 192 pp.; 85 illus.; 17 tables; indexed; glossy semi–soft bound in multicolor with green, brown, and black lettering; 4.1 by 6.9 inches (105 × 174 mm).

A book on the planting and care of lawns. There are informational tables concerning trees and shrubs in the back portion. It is oriented to lawn and landscape enthusiasts and conditions in the United States.

The Horticultural Trade Association

1972 **Lawns.** The Horticultural Trades Association. KTG Know the Garden Series, EP Publishing Company, Wakefield, Yorkshire, England, UK. 40 pp., 36 illus., 4 tables, no index, semi–soft bound in multicolor with black lettering, 8.0 by 5.3 inches (202 × 134 mm).

A small book on the planting, renovation and care of lawns. It is practically oriented to lawn enthusiasts and cool-season conditions in the United Kingdom.

Imbrigiotta, Bob

2002† **Journey to a bulletproof Lawn.** Bob Imbrigiotta. Bulletproof Lawn Company, Fort Lauderdale, Florida, USA. 64 pp.; 264 illus. with 257 in color; 6 tables; indexed; glossy semi–soft bound in multicolor with black, gray, green, and yellow lettering; 8.4 by 11.0 inches (213 × 279 mm).

A small guide book on the planting and care of St Augustinegrass lawns. It is practically oriented to conditions in Florida. It is extensively illustrated.

James, E.B.

n.d.† **Better Lawns—Their Establishment and Maintenance.** E.B. James. Ray Hay Publications, Haslemere, Surrey, England, circa UK. 23 pp. plus 8 pp. of advertising, 6 illus., no tables, no index, fold-stapled semisoft covers in green and black with cream 194x lettering, 6.0 by 9.4 inches (152 × 239 mm).

A short manual on the planting and care of lawns. It is practically oriented to home lawn enthusiasts and cool-season conditions in England. Mr. James described the concerns and approaches to sodding lawns in England in 1950:

> When a lawn is required for almost immediate use, turfing is probably the speediest and most fool-proof method to adopt, but also the most expensive.
>
> The chief advantage is that the turfing season stretches over several months, when pressure of work is slackest, whereas seeding is confined to a few weeks in Spring and Autumn when there are so many other jobs to be done in the garden that time for sowing can ill be spared.
>
> One of the principal disadvantages of turfing (apart from cost) is the acute shortage of really good sods, and when purchasing any appreciable amount, inspection of the actual site from which they are to be cut will enable the gardener to determine whether the turf will suit his requirements.
>
> Old park or downland turf is generally the most suitable as the finer Bent and Fescue grasses usually predominate—avoid at all costs turf that consists of coarse species; it is infinitely better to buy turves containing fine grass and weeds, the majority of which are susceptible to selective weedkillers, than to purchase weed-free turf composed of coarse grasses. Another good tip worth remembering is to choose turf from a similar type of soil to that of the site on which it is to be laid.

Johns, Rodney

2006† **Tips & Traps for Growing and Maintaining the Perfect Lawn.** Rodney Johns. McGraw-Hill Companies Inc., New York City, New York, USA. 216 pp.; 60 illus.; 4 tables; small index; glossy semi–soft bound in multicolor with blue, yellow, white, and black lettering; 5.9 by 9.0 inches (150 × 228 mm).

A practical book concerning the planting and care of lawns. It is oriented to lawn enthusiasts under conditions in the United States.

Kahl, Erich

1971† **Anlage und Pflege des Rasens (Establishment and Maintenance of Lawns).** Coauthor.

See *Woess, Friedrich—1971.*

Knoop, William Edward

1986† **The Complete Guide To Texas Lawn Care.** Dr. William E. Knoop. TG Press, Waco, Texas, USA. 146 pp., 103 illus. with 67 in color, 29 tables, small index, cloth hardbound in green with gold lettering, 7.0 by 10.1 inches (177 × 257 mm).

A book on site preparation and planting and care of lawns. It is practically oriented to warm-season lawn enthusiasts under conditions in Texas. There is a small glossary of selected terms in the book.

Also printed with glossy semi–soft bound covers in multicolor with white lettering.

Kupfer, Karl-Heinz

1967 **Immer Grüner Rasengarten (Permanent Green Lawn Garden).** Karl-Heinz Kupfer. WOLF Garten Center GmbH, Lexika-Verlag, Döffingen, West Germany. 139 pp., 65 illus. with 95 in color, 3 tables, indexed, semi–soft bound in multicolor with black lettering, 6.2 by 8.6 inches (158 × 218 mm). Printed entirely in the German language.

A small, well-illustrated book on the design, planting, and care of lawns and ornamental plants in West Germany. It contains a glossary in the back.

Laptev, Alekseĭ Alekseevich

1955 **Gazony ustroĭstvo i ukhod za nimi (Lawns: Arrangement and Care).** A.A. Laptev. Akademii a arkhitektury Ukrainskoĭ SSR, Kiev, Ukraine, USSR. 76 pp., 40 illus., 17 tables, no index, soft bound in multicolor with maroon and white lettering, 5.0 by 8.0 inches (149 × 220 mm). Printed entirely in the Russian language.

An early Russian book concerning lawn grasses and legumes plus their care. It is oriented to cool-season grasses under conditions in the Ukraine. There is a selected list of references in the back.

Lawfield, Wilfred Norman

1959† **Lawns and Sportsgreens.** W.N. Lawfield. W.H. & L. Collingridge Limited, London, England, UK. 84 pp.; 43 illus.; 8 tables; small index; cloth hardbound in green with white lettering and a jacket in multicolor with white, red, and black lettering; 5.0 by 7.7 inches (128 × 196 mm).

A small book on lawn planting and care. It is practically oriented to home lawn enthusiasts and cool-season conditions in England. It includes one short chapter on tennis courts, bowling greens, hockey fields, and croquet courts.

Lindsay, Pax

1976† **The Victa Lawn Book.** First edition. Pax Lindsay. A.H. & A.W. Reed, Pty. Limited, Sydney, New South Wales, Australia. 96 pp.; 73 illus. with 30 in color; 3 tables; no index; cloth hardbound in red with gold lettering and semi–soft bound in multicolor with green, white, and black lettering; 5.25 by 8.3 inches (133 × 210 mm).

A small book on site preparations, planting, and care of lawns plus a section on mower maintenance. It is practically oriented to lawn enthusiasts under conditions in Australia and New Zealand.

† *Reprinted in 1977.*

1981 **The Australian Gardener's Guide to Lawn Care.** P. Lindsay. A.H. & A.W. Reed Pty Limited, Sydney, New South Wales, Australia. 87 pp., 35 illus., 3 tables, no index, glossy semi–soft bound in multicolor with white and black lettering, 5.0 by 7.9 inches (127 × 200 mm).

A small book on the planting and care of lawns. It is practically oriented to lawn enthusiasts under conditions in Australia and New Zealand.

Lorence, Harry E.

1973† **"Hay, How's Your Lawn?"** Harry E. Lorence. Thomson Publications, Indianapolis, Indiana, USA. 60 pp., 35 illus., no tables, no index, semisoft covers in multicolor with black lettering in a comb binder, 8.5 by 10.9 inches (215 × 278 mm).

A small book on the planting and care of lawns. It is practically oriented to home lawn enthusiasts and conditions in the United States.

MacCaskey, Michael Robert

1979† **All About Lawns.** Michael MacCaskey. Ortho Books, Chevron Chemical Company, San Francisco, California, USA. 97 pp., 150-plus illus. in color, 17 tables, indexed, glossy semi–soft bound in multicolor with white lettering, 28.2 by 10.9 inches (208 × 278 mm).

A small book on the planting and care of lawns. It is practically oriented to home lawn enthusiasts and conditions in North America. It includes editions for the Midwest/Northeast, South, and West within a similar format.

1982† **Lawns.** Michael MacCaskey. Illustrated Encyclopedia of Gardening, The American Horticultural Society, Mount Vernon, Virginia, USA. 144 pp., 267 illus. in color, 99 tables, indexed, glossy hardbound in greens with white lettering, 8.1 by 10.9 inches (206 × 278 mm).

A book on the planting and care of lawns. It is oriented to lawn enthusiasts and conditions in the United States, including regional guidelines. The book is extensively illustrated in color.

1985† **All About Lawns.** Revised edition. Michael MacCaskey. Ortho Books, Chevron Chemical Company, San Francisco, California, USA. 112 pp., 221 illus. in color, 20 tables, indexed, glossy semi–soft bound in multicolor with white and black lettering, 8.2 by 10.9 inches (208 × 276 mm).

A major revision involving a consolidation of the three regional editions published in 1979.

Revised in 1994.

1994† **All About Lawns.** Second revised edition. Cathy Haas and Michael MacCaskey. Ortho Books, The Solaris Group, San Ramon, California, USA. 112 pp., 167 illus. in color, 6 tables, indexed, glossy semi–soft bound in multicolor with white lettering, 8.2 by 10.9 inches (209 × 278 mm).

Some revisions, especially among the photographs.

See *Rogers, Marilyn—1999.*

MacCubbin, Tom

2003† **The Perfect Florida Lawn.** Tom MacCubbin. Cool Springs Press, Nashville, Tennessee, USA. 175 pp.; 118 illus. with 109 in color; 13 tables; indexed; glossy semi–soft bound in multicolor with green, red, white, black lettering; 6.0 by 9.0 inches (152 × 227 mm).

A book on the planting and care of turfgrasses for lawns. It is practically oriented to lawn enthusiasts and conditions in the state of Texas. There is a glossary of selected terms in the back.

See also *Dobbs, Steve—2002*; *Myers, Melinda—2003.*

Machielse, P.L.

1966† **Lawnmakers' Handbook.** P.L. Machielse. Farmer's Weekly Supplement, Bloemfontein, Orange Free State, South Africa. 32 pp., 25 illus., 1 table, no index, fold-stapled soft covers in tan with red lettering, 6.9 by 9.8 inches (175 × 249 mm).

A booklet on the planting and care of lawns under conditions in South Africa. It includes an interesting section on early turfgrass selections used in South Africa.

Mann, David

1981† **Adrian Bloom's Guide to Lawns.** David Mann. Book 9, A Jarrold Garden Series. Jarrold Colour Publications, Norwich, England, UK. 32 pp., 38 illus. in color, no tables, no index, glossy semi–soft bound in multicolor with yellow and white lettering, 5.2 by 7.5 inches (132 × 190 mm).

A small booklet on the planting and care of lawns. It is practically oriented to home lawn enthusiasts and cool-season conditions in England.

Manz, Inge

1971† **Der Rasen um unser Haus (Grasses around Our House).** Inge Manz. Falken-Verlag Erich Sicker, Wiesbaden, West Germany. 96 pp.; 67 illus. with 8 in color; no tables; indexed; soft bound in multicolor with green, teal, and black lettering; 5.8 by 8.1 inches (147 × 207 mm). Printed entirely in the German language.

A small book on lawn planting and care. It is practically oriented to home lawn enthusiasts and cool-season conditions in Germany.

McDonald, David K.

1999 **Ecologically Sound Lawn Care for the Pacific Northwest.** David K. McDonald. Seattle Public Utilities, Seattle, Washington, USA. 87 pp., no illus., 1 table, no index, fold-stapled semisoft covers in black with black lettering, 8.3 by 11.0 inches (210 × 279 mm).

A small manual with suggested cultural practices for lawns in the Pacific Northwest based on a review of selected turfgrass and pesticide literature.

McGourty, Frederick, Jr.

1973† **The Home Lawn Handbook.** First edition. Frederick McGourty Jr. (editor) with 16 contributing authors—Robert W. Schery, R.B. Alderfer, C.G. Wilson, C.R. Skogley, Fred V. Grau, W.H. Daniel, James B Beard, Ralph E. Engel, Herbert Cole Jr., Herbert T. Streu, Henry W. Indyk, J.M. Duich, C.R. Funk, Glenn W. Burton, Victor B. Youngner, and J.R. Watson. Brooklyn Botanic Garden, Brooklyn, New York, USA. 90 pp.; 82 illus. with 10 in color; 15 tables; no index; fold-stapled glossy semisoft covers in green, black, and white with black lettering; 5.9 by 9.0 inches (150 × 228 mm).

A handbook on lawn planting and care in the United States. It contains good illustrations, including color plates of the common diseases of turfgrasses. It is a special printing from **Plants and Gardens** 29(1).

Earlier versions edited by P.K. Nelson were published in 1956 and 1963.

† *Reprinted in 1974, 1979, and 1981.*

Revised in 1998.

1988† **Home Lawn Handbook.** Fred McGourty and Marjorie J. Dietz. Brooklyn Botanic Garden, Brooklyn, New York, USA. 84 pp., 82 illus. with 10 in color, 14 tables, no index, fold-stapled glossy semisoft covers in multicolor with white and black lettering, 5.9 by 9.0 inches (150 × 228 mm).

A minor revision. It is a special revised printing from **Plants and Gardens** 29(1).

McKenzie, Gary

2004 **Turf Sustain.** Coauthor.

See *Ruscoe, Peter—2004.*

McKinley, Michael

2005† **Ortho's All About Lawns.** Michael McKinley (editor). Meredith Books, Des Moines, Iowa, USA. 128 pp., 285 illus. in color, 6 tables, indexed, glossy semi–soft bound in multicolor with white and yellow lettering, 8.1 × 10.6 inches (206 × 275 mm).

A practical book concerning turfgrass selection, planting, and care of lawns in the United States.

Melady, John Hayes

1952† **Better Lawns—For Your Home.** John Hayes Melady. The Melady Garden Books, Grosset & Dunlap Publishers, New York City, New York, USA. 130 pp., 109 illus. with 4 color plates, 18 tables, indexed, cloth hardbound in green with dark green lettering and a glossy jacket in multicolor with white and green lettering, 5.0 by 7.3 inches (127 × 185 mm).

A small book on lawn planting and care. It is practically oriented to home lawn enthusiasts and cool-season conditions of northern United States. It contains a section on "judging lawns in competition." Illustrations are by Eva Melady. J.H. Melady prefaced his book as follows:

> Our individual worlds are shrinking in this modern era. We are thinking less and doing less, while more of our planning is being done for us. We have come a long way from the sturdy independence of the pioneers who built their homes and made their own utensils and furniture.

No longer do we practice piano scales by the hour, for our twenty-dollar radio brings us more canned music than we can use; films of our comedies and dramas reach our community on the morning train in steel boxes. Or we buy furniture and fashions so that we may view the films in appropriate style at home by television. No longer need we go to boxing matches and wrestling bouts, to baseball, football or roller-derbies: they come to us over the air so long as we can keep up our payments and turn a button. Modern man is becoming one who watches rather than participates.

But there is one frontier remaining for our individualism—the home garden and especially the home lawn. The lawn calls for rugged effort, and the more we put into it the better it will become: our property will increase in value, our neighbors will imitate us, and our town will be a better place to live in.

Mellor, David Reynold

2003†   **The Lawn Bible.** David R. Mellor. Hyperion, New York City, New York, USA. 286 pp.; 81 illus.; 4 tables; indexed; glossy semi–soft bound in green and white with orange, green, and white lettering; 8.5 by 10.9 inches (215 × 276 mm).

A book on the planting and care of lawns. It is practically oriented to lawn enthusiasts and conditions in the United States.

Mioulane, Patrick

1984   **Comment Réussir Votre Pelouse (How to Have a Successful Lawn).** Patrick Mioulane. Guide Mon Jardin et Ma Maison, Le Port Marly, France. 35 pp., 48 illus., no index, soft bound in multicolor with black and white lettering, 7.1 by 9.5 inches (180 × 240 mm). Printed entirely in the French language.

A booklet on the planting and care of lawns. It is practically oriented to lawn enthusiasts and conditions in France.

1987   **Votre Gazon (Your Turf).** Patrick Mioulane and Michel Rocher. Numéro hors-série "Mon Jardin, Ma Maison," Boulogne-Billancourt, France. 98 pp.; 171 illus.; 2 tables; no index; soft bound in multicolor with red, black, and white lettering; 8.3 by 11.0 inches (210 × 280 mm). Printed entirely in the French language.

A small book on the planting and care of lawns. It is practically oriented to lawn enthusiasts and conditions in France.

Myers, Melinda

2003†   **The Perfect Ohio Lawn—Attaining and Maintaining the Lawn You Want.** Melinda Myers. One of a state series. Cool Springs Press, Nashville, Tennessee, USA. 157 pp.; 108 illus. with 80 in color; 6 tables; indexed; glossy semi–soft bound in multicolor with green, red, white, and black lettering; 6.0 by 9.0 inches (152 × 228 mm).

A book on the planting and care of turfgrasses for lawns. It is practically oriented to lawn enthusiasts and conditions in the state of Ohio. There is a small glossary in the back.

NOTE: This book is representative of a series of similar books by the same author and publisher all with a similar length and format but changed somewhat for other states, such as Indiana, Iowa, Michigan, Minnesota, Missouri, and Wisconsin.

See also *Dobbs, Steve—2002*; *MacCubbin, Tom—2003*.

Nelson, Peter K.

1956†   **Handbook on Lawns.** First edition. Peter K. Nelson (editor) with 16 contributing authors. Brooklyn Botanic Garden, Brooklyn, New York, USA. 93 pp. plus 4 pp. of advertising, 125 illus. with 8 in color, 5 tables, no index, fold-stapled semisoft covers in white and black with black lettering, 5.9 by 8.8 inches (149 × 223 mm).

An interesting, small handbook on lawn planting and care in North America. It is well illustrated with photographs, including color plates of the common turfgrass diseases. It is a special printing from **Plants and Gardens** New Series 12(2).
Revised in 1963, 1973, and 1988.

1963†   **Handbook on Lawns.** Second revised edition. Peter K. Nelson (editor) with 15 contributing authors. Brooklyn Botanic Garden, Brooklyn, New York, USA. 89 pp., 116 plus 8 in color, 7 tables, no index, fold-stapled semisoft covers in grays and white with black lettering, 6.0 by 8.9 inches (151 × 226 mm).

An update and minor revision of the first edition. It is a revised special printing from **Plants and Gardens** 12(2).
Revised in 1973 and 1988.
See *McGourty, Frederick Jr.—1973 and 1988.*

Noble Foundation

n.d.†   **Your Lawn.** The Samuel Roberts Noble Foundation Inc., Ardmore, Oklahoma, USA. 22 pp., 8 illus., 3 tables, no index,
circa   soft bound in multicolor with black lettering, 5.0 by 8.0 inches (138 × 211 mm).
196x

A booklet on the planting and care of lawns in the Oklahoma area. It is practically oriented to lawn enthusiasts.

Nonn, Harald

2003   **Rasen: schnell & einfach (Lawn: Quick & easy).** Harald Nonn. Gräfs und Unzer, München, Germany. 62 pp. plus 2 pp. of advertising; 81 illus. in color; no tables; small index; glossy semi–soft bound in multicolor with yellow, white, and black lettering; 6.5 by 7.3 inches (164 × 199 mm). Printed entirely in the German language.

A practical small book concerning the planting and care of lawns. It is oriented to lawn enthusiasts under cool-season conditions in Germany.

Reprinted annually through at least 2006.

Ogren, Tom

2004†   **What the "Experts" May Not Tell You About . . . Growing the Perfect Lawn.** Tom Ogren. Warner Books, New York City, New York, USA. 219 pp.; no illus.; 12 tables; indexed; glossy semi–soft bound in multicolor with yellow, white, maroon, and black lettering; 5.1 by 8.0 inches (130 × 203 mm).

A book on the care of turfgrasses for lawns in the United States. It is practically oriented to lawn enthusiasts.

Oklahoma Turfgrass Research Foundation

n.d.†
circa
197x
    Lawns for Oklahoma. Oklahoma Turfgrass Research Foundation, Stillwater, Oklahoma, USA. 27 pp., 6 illus., 5 tables, no index, soft bound in multicolor with black lettering, 5.8 by 9.0 inches (147 × 228 mm).

A small booklet on the planting and care of lawns in Oklahoma. It is practically oriented to lawn enthusiasts. There is a glossary of terms in the back.

O.M. Scott and Sons Company

1953†   **Lawn Care-How to Have a Better Lawn.** First edition. O.M. Scott and Sons Company, Marysville, Ohio, USA. 65 pp., 51 illus. with 13 in color, no tables, indexed, glossy semi–soft bound in multicolor with black lettering, 4.4 by 6.7 inches (111 × 170 mm).

A small, practical booklet on the planting and care of turfgrass lawns in the United States. It was published for the employees of Western Electric.

1953†   **Lawn Care—How to Have a Better Lawn.** Second revised edition. O.M. Scott and Sons Company, Marysville, Ohio, USA. 132 pp., 82 illus., 3 tables, indexed in front, glossy semi–soft bound in multicolor with black and orange lettering, 4.3 by 6.9 inches (108 × 174 mm).

A book composed of two parts. Part one contains reprints of current issue Nos. 113 to 128 of **Lawn Care.** Part two is a condensed digest of early issues in a bound set of 13 chapters. It is practically oriented to lawn enthusiasts and conditions in North America.

1955†   **Lawn Care—Building and Maintaining.** First edition. O.M. Scott and Sons Company, Marysville, Ohio, USA. 95 pp., 87 illus., 1 table, small index, fold-stapled soft covers in green and white with white and black lettering, 5.4 by 8.2 inches (138 × 208 mm).

A small handbook on lawn planting and care in the cool-season climatic region. It is practically oriented to lawn enthusiasts and conditions in North America.

Revised in 1956.

1955†   **Lawn Care—Building and Maintaining—The Amateur's Guide.** First edition. O.M. Scott and Sons Company, Marysville, Ohio, USA. 107 pp.; 103 illus.; 1 table; small index; cloth hardbound in greens, black, orange, and white with black and white lettering; 5.4 by 8.3 inches (138 × 210 mm).

A small book on site preparation, turfgrass planting, care, and renovation of lawns. It is practically oriented to the lawn enthusiasts and conditions in North America.

1955†   **Lawn Care—Building and Maintaining—The Amateur's Guide.** Second revised edition. O.M. Scott and Sons Company, Marysville, Ohio, USA. 95 pp., 79 illus. with 28 in color, 1 table, small index, glossy semisoft, found in multicolor and white lettering, 4.2 by 6.9 inches (109 × 174 mm).

A small handbook on lawn planting and care in the cool-season turfgrass region. A practical book oriented to home lawn enthusiasts.

n.d.†
circa
1955
    Lawn Care—Building and Maintaining—The Amateur's Guide. Third revised edition. O.M. Scott and Sons Company, Marysville, Ohio, USA. 95 pp., 83 illus. with 37 in color, 1 table, small index, glossy semi–soft bound in multicolor with black and white lettering, 4.2 by 6.9 inches (106 × 175 mm).

A small revision of the second edition.

1955†   **Lawn Care—Building and Maintaining—The Amateur's Guide.** First edition. O.M. Scott and Sons Company, Marysville, Ohio, USA. 49 pp.; 34 illus. with 14 in color; 1 table; indexed; fold-stapled soft covers in greens, white, and black with black and white lettering; 4.2 by 6.9 inches (107 × 175 mm).

A practical booklet on the planting and care of turfgrass lawns in the United States, with emphasis on pest problems.

† A special edition in the same format and content also was published in 1955 but with white and green lettering.

1955† **Lawn Care—Building and Maintaining—The Amateur's Guide.** O.M. Scott and Sons Company, Marysville, Ohio, USA. 33 pp.; 25 illus. with 14 in color; 1 table; no index; semi–soft bound in greens, white, and black with white and green lettering; 4.3 by 6.9 inches (109 × 176 mm).

A small, practical booklet on the planting and care of turfgrass lawns in the United States. It was published for the General Motors employees.

1955† **Lawn Care—Building and Maintaining—The Amateur's Guide.** Special edition. O.M. Scott & Sons Company, Marysville, Ohio, USA. 47 pp.; 43 illus.; 1 table; no index; fold-stapled soft covers in greens, white, and black with white and green lettering; 4.3 by 6.8 inches (108 × 173 mm).

A small pocket book on the site development, planting, and care of lawn turfgrasses. It is practically oriented to lawn enthusiasts.

1955† **Lawn Care—Building and Maintaining—The Amateur's Guide.** Special edition for General Motors personnel staff. O.M. Scott & Sons Company, Marysville, Ohio, USA. 33 pp.; 30 illus.; 1 table; no index; semi–soft bound in greens, white, and black with white and light green lettering; 4.2 by 6.8 inches (107 × 174 mm).

A small pocket handbook on lawn planting and care. It is practically oriented to lawn enthusiasts and cool-season climatic conditions in North America.

1956† **Lawn Care—Building and Maintaining.** Second book edition. O.M. Scott and Sons Company, Marysville, Ohio, USA. 109 pp., 80 illus., 1 table, small index, fold-stapled glossy semisoft covers in multicolor with black lettering, 5.2 by 8.3 inches (132 × 210 mm).

A small handbook on lawn planting and care in the cool-season turfgrass region. A practical book oriented to home lawn enthusiasts and conditions in North America.

1966† **The 1966 Lawn Care Book.** Joseph E. Howland (editor). O.M. Scott and Sons Company. Marysville, Ohio, USA. 22 pp., 39 illus. in color, no tables, no index, fold-stapled soft covers in multicolor with green and white lettering, 9.0 by 6.0 inches (230 × 153 mm).

A practical booklet on weed identification, weed control, and seeding of cool-season turfgrasses in the United States.

† *Reprinted in 1967* with a different cover and introduction.

Ozawa, Tomoo

1967 **Lawn Garden Construction and Care.** Tomoo Ozawa. Bunka-shuppankyoku, Tokyo, Japan. 116 pp., 177 illus. with 49 in color, 22 tables, indexed, semi–soft bound in multicolor. Printed entirely in the Japanese language.

A small book concerning the planting and care of lawns under the conditions in Japan.

Palin, Robert

1985† **The Master Gardener's Guide to Lawn Care.** Robert Palin. Salamander Books Limited, London, England, UK. 77 pp.; 129 illus. in color; 3 tables; small index; glossy hardbound in multicolor with yellow, white, and black lettering; 4.5 by 8.5 inches (114 × 216 mm).

A small book on the construction, planting, and care of lawns, including turfgrass pests. It also includes a section on equipment for lawns. It is practically oriented to lawn enthusiasts and cool-season conditions in the United Kingdom.
Reprinted in 1987.

1986 **Gazonverzorging: Alles wat u moet weten over aanleg en onderhoud van een gazon (Lawn Care: Everything You Have to Know about Construction and Maintenance of a Lawn).** Robert Palin and Son Tyberg. Publiboek, Antwerpen, the Netherlands. 77 pp., 99 illus. in color, 3 tables, indexed, glossy hardbound in light green with yellow and white lettering, 4.8 by 8.8 inches (120 × 222 mm). Printed in the Dutch language.

A small, practical book concerning the planting and care of a lawn. It is oriented to lawn enthusiasts and conditions in the Netherlands.

1987 **The Master Gardener's Guide to Lawn Care.** Robert Palin. Tetra Press.

Essentially a reprinting of the 1985 book with different front and back covers and printed again in London, England, UK, and in Morris Plains, New Jersey, USA.

1991 **Gazon zonder mos en onkruid: aanleg, verzorging (A Lawn without Moss and Weeds: Construction and Care).** Robert Palin and Son Tyberg. Zuid-Hollandsche U.M., Weert, the Netherlands. 77 pp., 100 illus. in color, 3 tables, indexed, glossy hardbound in green with white lettering, 4.8 by 8.8 inches (122 × 224 mm). Printed in the Dutch language.

A small practical book about turfgrass lawn planting and care. It is oriented to lawn enthusiasts under the cool climatic conditions in the Netherlands.

Parfitt, David

2007† **Lawnscapes: Mowing Patterns To Make Your Yard A Work Of Art.** David Parfitt. Quirk Books, Philadelphia, Pennsylvania, USA. 80 pp.; 96 illus. in color; no tables; small index; glossy hardbound in multicolor with a five-millimeter-high artificial turf front and black, red, green, and white lettering; 7.6 by 5.7 inches (194 × 144 mm).

A unique small book concerning the use of a lawn mower to form stripping and a diverse array of ornamental designs in a turfgrass area. It includes how-to instructions.

2008† **The Art of the Lawn.** A UK version of the 2007 book printed under a different title by Ivy Press, Newton Abbot, Devon, England, UK.

Peigelbeck, Will

1957† **Get Rid of Crab Grass.** Will Peigelbeck. MACO Magazine Corporation, New York City, New York, USA. 48 pp.; 67 illus.; 4 tables; no index; fold-stapled soft covers in multicolor with black, red, and green lettering; 8.6 by 10.9 inches (217 × 278 mm).

A booklet that emphasizes the procedures for the control of crabgrass in lawns. It includes planting, renovation, and preventive care methods. It is practically oriented to lawn enthusiasts and conditions in the United States.

Pessey, Christian

1983 **Les Pelouses (Lawns).** Christian Pessey. Collection Bricolage et Maison de l'Encyclopédie Intégrate de la Consommation, La Nouvelle Librairie, Paris, France. 64 pp.; 50 illus.; 7 tables; indexed; semi–soft bound in multicolor with white, red, and yellow lettering; 4.3 by 9.0 inches (110 × 229 mm). Printed entirely in the French language.

A booklet on the planting and care of lawns. It is practically oriented to lawn enthusiasts and conditions in France.

Peterson, Chris

2011† **The Complete Guide to a Better Lawn.** Chris Peterson. Creative Publishing International, Minneapolis, Minnesota, USA. 224 pp., 446 illus. in color, 4 tables, small index, glossy semi–soft bound in multicolor with white lettering 8.3 by 10.8 inches (210 × 275 mm).

A book concerning the planting and care of lawns. It is practically oriented to lawn enthusiasts and conditions in the United States.

Porter, Wesley R.

1986 **Green Side Up—Growing a Perfect Lawn in Canada.** Wesley R. Porter. Fitzhenry & Whiteside Limited, Markham, Ontario, Canada. 108 pp.; 10 illus. as vignettes; 7 tables; no index; glossy semi–soft bound in multicolor with red, yellow, and black lettering; 6.0 by 9.0 inches (151 × 228 mm).

A small, practical book on turfgrasses for lawns plus their planting and care under the cool-season conditions in Canada.

1987 **Pour une pelouse parfaite: ensemencement, entretien, équipement (For a Perfect Lawn: Sowing, Maintenance, Equipment).*** Wesley R. Porter, as translated to French by Jean Prévost. Transmonde, Montreal, Quebec, Canada. 154 pp.; 1 illus.; 13 tables; no index; glossy hardbound in multicolor with white, red, and black lettering; 5.2 by 8.0 inches (132 × 203 mm). Printed in the French language.

A practical book on the planting and care of turfgrasses for lawns. It is oriented to lawn enthusiasts and climatic conditions in Canada.

1999 **Green Side Up—Growing a Perfect Lawn in Northern Climates.*** Wes R. Porter. Fitzhenry & Whiteside, Markham, Ontario, Canada. 140 pp., 56 illus., 71 tables, small index, glossy semi–soft bound in greens with white and black lettering, 6.0 by 8.9 inches (152 × 226 mm).

An update and expansion of the 1986 Canadian edition.

The 1999 book was first published in the United States in 2000.

Pouliot, Paul

1975 **Préparer, Entretenir, Embellir Votre Pelouse (Prepare, Maintain, Beautify Your Lawn).** Paul Pouliot. Les Éditions de l'Homme Ltée, Montreal, Quebec, Canada. 279 pp., 213 illus., 2 tables, no index, semi–soft bound in multicolor with black and green lettering, 5.4 by 7.9 inches (136 × 201 mm). Printed entirely in the French language.

A book on the planting and care of lawns. It is practically oriented to lawn enthusiasts and the cool-season turfgrass conditions in Canada.

1976† **Caring for Your Lawn.** Paul Pouliot. Habitex Series, Les Éditions de l'Homme, Montreal, Quebec, Canada. 275 pp. plus 6 pp. of advertising; 323 illus.; 3 tables; no index; glossy semi–soft bound in multicolor with green, blue, and black lettering; 5.3 by 8.0 inches (134 × 202 mm).

An English version of the 1975 French edition.

Provey, Joseph R.

1999† **Better Lawns—Step by Step.** First edition. Joe Provey and Kris Robinson. Creative Homeowner Press, Upper Saddle River, New Jersey, USA. 160 pp. plus 1 p. of advertising; 424 illus. with 408 in color; 8 tables; indexed; glossy semi–soft bound in multicolor with white, red, yellow, green, and black lettering; 9.0 by 9.9 inches (288 × 252 mm).

A book on the planting and care of lawns. It is practically oriented to lawn enthusiasts and conditions in the United States. There is a small glossary in the back.

Revised in 2008.

2008 **Toro Expert Guide to Lawns.** Second revised edition. Joseph R. Provey and Kris Robinson with Van Cline. Creative Homeowner, Upper Saddle River, New Jersey, USA. 175 pp. plus 1 p. of advertising; 478 illus. in color; 8 tables; small index; glossy semi–soft bound in multicolor with yellow, white, and green lettering; 8.5 by 10.8 inches (217 × 275 mm).

Some revision of the 1999 first edition.

Pycraft, David

1972† **Lawns.** David Pycraft. Wisley Handbook 4, The Royal Horticultural Society, London, England, UK. 42 pp., 18 illus., 2 tables, no index, fold-stapled glossy semisoft covers in multicolor with black and white lettering, 5.7 by 8.3 inches (146 × 210 mm).

A small booklet on lawn planting and care. It is practically oriented to home lawn enthusiasts and cool-season conditions in England.

† *Reprinted in 1980.*

Radloff, Holger

2004 **Alles Zum Thema Endlich Schöner Rasen (Everything about & Achieving Beautiful Lawns).** Holger Radloff. Gruner und Jahr, Hamburg, Germany. 98 pp., 177 illus. in color, 1 table, no index, glossy semi–soft bound in multicolor with yellow and white lettering, 5.1 by 7.3 inches (129 × 189 mm). Printed in the German language.

A practical book concerning the design, planting, and care of lawns. It is oriented to lawn enthusiasts and cool-season conditions in Germany.

Raymond, Dick

1993† **Down-To-Earth Natural Lawn Care.** Dick Raymond. A Storey Publishing Book, Storey Communications Inc., Pownal, Vermont, USA. 154 pp., 130 illus., no tables, indexed, glossy semi–soft bound in multicolor with black and green lettering with a jacket in multicolor with black and green lettering, 8.5 by 10.8 inches (215 × 275 mm).

A small book on lawn planting and care in the United States. It contains a chapter on ground covers and ornamental grasses in the back.

Riov, Claude

1979 **Les Secrets des Beaux Gazons (Secrets of Beautiful Turf).** Claude Riov. Collection LeBerger Vert, Berger-Levrault Éditeur, Paris, France. 63 pp., 33 illus., no tables, no index, semi–soft bound in multicolor with black and white lettering, 4.8 by 6.9 inches (123 × 176 mm). Printed entirely in the French language.

A small book on the planting and care of lawns. It is practically oriented to lawn enthusiasts and conditions in France.

Ripa, A.

1982 **Zaliéni (Lawns).** A. Ripa, H. Ranka, and I. Holms. Avots, Riga, Latvia. 88 pp., 21 illus., 13 tables, no index, semi–soft bound in multicolor with white lettering, 5.1 by 7.9 inches (129 × 201 mm).

A small, practical book on lawn planting and care of turfgrasses. It is oriented to conditions in Latvia.

Robey, Melvin John

1972† **Home Lawn Care—145 Questions and Answers.** Melvin J. Robey. Melvin Robey Publisher, West Lafayette, Indiana, USA. 42 pp., 45 illus., 15 tables, no index, fold-stapled glossy semisoft covers in greens and white with white lettering; 5.9 by 9.0 inches (151 × 226 mm).

A small book on the planting and care of lawns. It is practically oriented to lawn enthusiasts and conditions of North America.

Robinson, Kris

1999 **Better Lawns—Step by Step.** Coauthor.

See *Provey, Joseph R.—1999.*

2008 **Toro Expert Guide to Lawns.** Coauthor.
See *Provey, Joseph R.—2008.*

Rocher, Michel

1987 **Votre Gazon (Your turf).** Coauthor.
See *Mioulane, Patrick—1987.*

Rockwell, Frederick Frye

1956† **The Complete Book of Lawns.** F.F. Rockwell and Esther C. Grayson. The American Garden Guild and Doubleday and Company Inc., Garden City, New York, USA. 190 pp., 84 illus. plus 10 color plates, 13 tables, good index, cloth hardbound in green with yellow lettering and a jacket in multicolor with green and black lettering, 5.5 by 8.2 inches (140 × 208 mm).

A book on the basics of lawn planting and care within a how-to approach. Emphasis is placed on lawn turfgrasses and conditions in the United States.

Rodale, Inc.

2000† **Lawns.** Rodale Organic Gardening Basics, Rodale Inc., Emmaus, Pennsylvania, USA. 112 pp., 100 illus. with 85 in color, 1 table, indexed, glossy semi–soft bound in multicolor with white and green lettering, 5.8 by 8.8 inches (147 × 224 mm).

A small book concerning the planting and care of turfgrass lawns, with an orientation to organic approaches.

Rogers, Marilyn

1999† **Ortho's All About Lawns.** Marilyn Rogers (editor). Meredith Corporation, Des Moines, Iowa, USA. 96 pp., 250 illus. in color, 7 tables, small index, glossy semi–soft bound in multicolor with white and yellow lettering, 8.1 by 10.9 inches (206 × 276 mm).

A practical book on the planting and care of lawns. It is oriented to lawn enthusiasts under conditions in the United States and Canada.

Reprinted in 2000 by the Scotts Company.

2002† **Scotts Lawns: Your Guide to a Beautiful Yard.** First edition. Marilyn Rogers (editor) with contributing writers Dr. Nick E. Christians and Ashton Ritchie. The Scotts Company, Marysville, Ohio, USA. 192 pp.; 370 illus. in color; 3 tables; small index; glossy semi–soft bound in multicolor with white, yellow, and black lettering; 8.0 by 10.9 inches (205 × 276 mm).

A guide book concerning the planting and care of lawns. It is practically oriented to lawn enthusiasts and environmental conditions in the United States.

Revised in 2007.

2007† **Lawns 1-2-3.** Marilyn Rogers (editor). The Home Depot, Meredith Books, Des Moines, Iowa, USA. 191 pp., 423 illus. in color, 10 tables, small index, glossy hardbound in multicolor with white and yellow lettering, 8.0 by 10.8 inches (204 × 273 mm).

A well-illustrated book regarding the planting and care of turfgrass lawns. It is practically oriented to the home lawn enthusiasts and conditions in North America.

2007† **Scotts Lawns: Your Guide to a Beautiful Yard.** Second revised edition. Marilyn Rogers (editor) with contributing writers Nick Christians, David R. Mellor, and Ashton Ritchie. Scotts Miracle-Gro Company, Marysville, Ohio, USA. 233 pp.; 367 illus. with 365 in color; 8 tables; small index; glossy semi–soft bound in multicolor with yellow, white, and red lettering; 8.1 by 10.8 inches (205 × 274 mm).

A revision and update of the 2002 edition.

2008† **Scotts Southern Lawns.** Marilyn Rogers (editor) with contributing writers Nick Christians, David R. Mellor, and Ashton Ritchie. The Scotts Miracle-Gro Company, Marysville, Ohio, USA. 223 pp.; 322 illus. in color; no tables; indexed; glossy semi–soft bound in multicolor with yellow, white, red, and black lettering; 8.1 by 10.9 inches (206 × 276 mm).

A guidebook concerning the planting and care of lawns. It is practically oriented to lawn enthusiasts and warm-season environmental conditions in the southern United States.

Rogers, Trey

2007† **Lawn Geek.** Trey Rogers. New American Library, Penguin Group, New York City, New York, USA. 349 pp.; 44 illus.; 16 tables; indexed; glossy semi–soft bound in multicolor with yellow, light green, and white lettering; 5.9 by 9.0 inches (151 × 228 mm).

A book on the planting and care of lawns. It is oriented to home lawn enthusiasts under conditions in the United States.

Ruscoe, Peter

2004 **Turf Sustain.** Peter Ruscoe, Ken Johnston, and Gary McKenzie. Sports Turf Technology, West Australia, Australia. 90 pp. with 2 pp. of advertising, 361 illus. with 179 in color, 5 tables, no index, glossy semi–soft bound in multicolor with blues and black lettering, 7.0 by 9.8 inches (178 × 249 mm).

A small book concerning soil management, plant selection, planting, and care of turfgrasses. It is practically oriented to conditions in Western Australia.

Sasias, Gérard

2008† **Pelouses et Gazons (Lawns and Grass).** Gérard Sasias. Les Clefs du Jardinage, Artémes Éditions, France. 96 pp., 118 illus. in color, 3 tables, indexed, glossy hard covers in multicolor with white and black lettering in a ring binder, 5.7 by 7.7 inches (145 × 195 mm). Printed entirely in the French language.

A practical book concerning the planting and care of lawns. It is oriented to lawn enthusiasts under both cool and warm climatic conditions in France.

Scherer, Hans

1966† **Schöner Rasen—Aber wie? (Beautiful Lawns—But How?).** * Hans Scherer. Verlag Stichnote Gmb.H, Darmstadt, West Germany. 144 pp. with 15 pp. of advertising, 64 illus. with 3 color photos, 3 tables, indexed, hardbound in green and red with white and black lettering, 4.8 by 8.0 inches (122 × 202 mm). Printed entirely in the German language.

A small book on the planting and care of lawns under West German conditions. It contains a chapter on sports field turfgrasses in the back.

Schery, Robert Walter

1960† **Lawn Establishment and Care.** Robert W. Schery. Bulletin No. 7, Missouri Botanical Garden, St. Louis, Missouri, USA. 21 pp., 14 illus., no tables, no index, semi–soft bound in black and white with black lettering, 5.0 by 8.0 inches (151 × 227 mm).

A booklet on the planting and care of lawns. It is practically oriented to lawn enthusiasts and conditions in the United States.

1962† **Selecting Lawn Grasses.** Dr. Robert W. Schery. T.F.H. Publications Inc., Jersey City, New Jersey, USA. 32 pp., 33 illus., no tables, no index, fold-stapled glossy semisoft covers in multicolor with black and yellow lettering, 5.4 by 8.5 inches (137 × 216 mm).

A small booklet on the selection of turfgrasses for lawns in both the northern and southern climatic regions of the United States.

1962† **10 Frequent Lawn Problems—And What To Do About Them.** Dr. Robert W. Schery. T.F.H. Publications Inc., Jersey City, New Jersey, USA. 32 pp., 42 illus., no tables, no index, fold-stapled glossy semisoft covers in multicolor with green and black lettering, 5.5 by 8.5 inches (140 × 216 mm).

A small booklet on the common problems encountered in the planting and care lawns. It is practically oriented to home lawn enthusiasts and conditions in the United States.

1976† **Better Lawns.** Countryside Books. A.B. Morse Company, Barrington, Illinois, USA. 48 pp., 62 illus. with 34 in color, no tables, no index, glossy semi–soft bound in multicolor with black and green lettering, 5.5 by 8.0 inches (140 × 203 mm).

A small booklet that emphasizes primarily the selection of adapted turfgrass species and their establishment in lawns. It is practically oriented to home lawn enthusiasts and conditions in the United States. The booklet is authored by Dr. Robert W. Schery, although not acknowledged in the text.

Schultz, Warren

1989† **The Chemical-Free Lawn.** Warren Schultz. Rodale Press, Emmaus, Pennsylvania, USA. 194 pp., 106 illus., 20 tables, indexed, glossy semi–soft bound in multicolor with green and orange lettering, 7.5 by 9.1 inches (191 × 232 mm).

A book on organic methods for the planting and care of lawns under conditions in the United States. Much of the book is oriented to nonpesticide approaches to control pests of turfgrasses.

1999† **A Man's Turf—The Perfect Lawn.** Warren Schultz. Three River Press, New York City, New York, USA. 180 pp., 185 illus. with 148 in color, 2 tables, small index, glossy semi–soft bound in multicolor with green and white lettering, 9.9 by 9.9 inches (252 × 252 mm).

This book is an intermix of historical aspects and current lawn care practices with emphasis on organic approaches. It is oriented to conditions in the United States. The photographs are by Mr. Roger Foley.

Sharples, Philip

2008† **The Lawn Guide.** Philip Sharples and Steven Hayman. S&H Publishing, London, England, UK. 99 pp.; 91 illus. in color; 13 tables; small index; glossy semi–soft bound in multicolor with green, black, and white lettering; 7.4 by 9.8 inches (187 × 250 mm).

A practical, small book concerning the planting and care of lawns. It is oriented to lawn enthusiasts in the cool climatic regions of the United States.

Shiels, George Roger

1984† **The Lawn.** George R. Shiels. City and Guilds Leisurecraft Books, Worlds Work Limited, Tudworth, Surrey, England, UK. 72 pp., 107 illus. with 33 in color, 24 tables, indexed, glossy semi–soft bound in multicolor with orange and black lettering, 7.9 by 9.9 inches (202 × 252 mm).

A small book on the planting and care of home lawns. It is practically oriented to cool-season conditions in England. The book is designed to complement a study by readers for associated national examinations.

Sigalov, Boris

1955 **Dekorativnye gazony (Ornamental Lawns).** Boris Sigalov. Ministerstvo Kommunal'nogo Khoziaistva RSFSR, Moscow, Russia, USSR. 64 pp., 16 illus., 10 tables, no index, semi–soft bound in greens and white with black lettering, 5.0 by 8.0 inches (144 × 219 mm). Printed entirely in the Russian language.

A small book on the planting and care of lawns. It is oriented to lawn enthusiasts and conditions in Russia.

Solly, Cecil

1942 **A Perfect Turf: How to Plant a Lawn and Maintain a Perfect Turf in the Pacific Northwest.** First edition. Cecil Solly. The Garden Notebook Series—Volume I, Seattle, Washington, USA. 17 pp., 22 illus., no tables, no index, fold-stapled semisoft covers bound in multicolor with black lettering, 5.0 by 6.8 inches (127 × 173 mm).

A practical booklet concerning the planting and care of lawns. It is oriented to lawn enthusiasts and cool-season conditions in the Pacific Northwest.

Revised in 1944, 1945, and 1948.

1944 **The Perfect Turf: How to Plant a Lawn and Maintain a Perfect Turf in the Pacific Northwest.** Revised edition. Cecil Solly. The Garden Notebook Series—Volume I, Seattle, Washington, USA. 24 pp., 27 illus., no tables, no index, fold-stapled semisoft covers in multicolor with black lettering, 5.0 by 6.8 inches (127 × 173 mm).

A minor revision of the 1942 edition.

Revised in 1945 and 1948.

1945 **A Perfect Turf: How to Plant a Lawn and Maintain a Perfect Turf in the Pacific Northwest.** Second revised edition. Cecil Solly. The Garden Notebook Series—Volume I, Seattle, Washington, USA. 32 pp., 34 illus., no tables, no index, fold-stapled glossy semisoft covers in multicolor with black lettering, 5.0 by 6.8 inches (127 × 173 mm).

A booklet on the planting and care of lawns. It is practically oriented to lawn enthusiasts and conditions in the Pacific Northwest region of the United States.

Revised in 1948.

1948 **A Perfect Turf: How to Plant and Maintain a Perfect Turf in the Pacific Northwest.** Revised edition. Cecil Solly. The Garden Notebook Series—Volume I, Seattle, Washington, USA. 33 pp., 31 illus., no tables, no index, fold-stapled glossy semisoft covers in multicolor with black lettering, 5.0 by 6.8 inches (127 × 173 mm).

A slightly revised version.

Soltys, Karen Costello

2000† **Rodale Organic Gardening Basics—Lawns.** Karen Costello Soltys (editor). Rodale Inc., Emmaus, Pennsylvania, USA. 112 pp., 103 illus. with 102 in color, 3 tables, small index, glossy semi–soft bound in multicolor with green and white lettering, 5.8 by 8.8 inches (147 × 223 mm).

A small book on the planting and care of lawns employing organic methods.

Stella, Anselmo

1994† **Impianto e Cura del Tappeto Verde (Construction and Maintenance of Turf).** Anselmo Stella. Edagricole-Edizione Agricole, Bologna, Italy. 80 pp., 79 illus. with 54 in color, no tables, no index, glossy semi–soft bound in multicolor with white and black lettering, 5.3 by 8.3 inches (135 × 210 mm). Printed entirely in the Italian language.

A small book on the planting and care of lawns. It is practically oriented to lawn enthusiasts and conditions in Italy.

Stern, Michael

1998   **Der Gepflegte Rasen: Zierrasen, Blumenrasen und Spielrasen (The Well-Kept Lawn: Ornamental Grass, Flowering Grass, and Recreation Grass).** Michael Stern. Natur Book Verlag GmbH, Augsburg, Germany. 64 pp., 53 illus. in color, 1 table, small index, glossy semi–soft bound in multicolor with white and yellow lettering, 6.3 by 8.4 inches (159 × 214 mm). Printed entirely in the German language.

A practical, small book concerning grasses and their planting and care for lawns. It is oriented to cool-season grasses under the conditions in Germany.

Stevenson, Tom

1965   **Lawn Guide.** Tom Stevenson. Robert B. Luce Inc., Washington, DC, USA. 76 pp., 14 illus., no tables, no index, cloth hardbound in multicolor with white and black lettering, 4.8 by 7.4 inches (121 × 188 mm).

A small book on the planting and care of lawns. It is practically oriented to home lawn enthusiasts in the United States.

St. Remy Media

2002   **Smart Lawn Care.** St. Remy Media Inc., Montreal, Quebec, Canada. 22 pp.; 79 illus. in color; 4 tables; no index; glossy semisoft covers in multicolor with yellow, blue, and white lettering in a spiral ring binder; 7.0 by 3.8 inches (177 × 96 mm).

A field guide to diagnosing lawn problems under Canadian conditions.

Templeton, Matt

1976†   **Lawns.** Matt Templeton. David and Charles (Publishers) Limited, Newton Abbot, England, UK. 144 pp.; 47 illus.; 5 tables; no index; cloth hardbound in both salmon and in light green with gold lettering and a jacket in green, white, and black with green, white, and black lettering; 5.3 by 8.4 inches (135 × 215 mm).

A small book on lawn planting and care that is organized by subject in alphabetical order. It is practically oriented to home lawn enthusiasts and cool-season conditions in England.

Thinschmidt, Alice

2002   **Rasenprobleme: Erkennen und Beheben (Lawn Problems: Diagnosis and Correction).** Coauthor.
See *Böswirth, Daniel—2002.*

Titchmarch, Alan

1984†   **Lawns.** Alan Titchmarsh's Gardening Guides. The Hamlyn Publishing Group Limited, London, England, UK. 32 pp., 19 illus. with 7 in color, 1 table, indexed, fold-stapled glossy semisoft covers in multicolor with white and black lettering, 3.9 by 5.9 inches (100 × 150 mm).

A small booklet on the planting and care of lawns. There is a section on aromatic lawns of chamomile and thyme. It is practically oriented to lawn enthusiasts and cool-season conditions of the United Kingdom.

Toogood, Alan R.

1993†   **Lawn Craft.** Alan Toogood. Ward Lock Master Gardener, Ward Lock Limited, London, England, UK. 96 pp.; 103 illus. in color; 13 tables; small index; glossy semi–soft bound in multicolor with green, white, and black lettering; 8.5 by 8.2 inches (218 × 208 mm).

A small book on the planting and care of lawns. It is practically oriented to lawn enthusiasts and cool-season condition in the United Kingdom.

Touwen L.

1968   **Het Gazon: Aanleg en Onderhoud (Lawns: Construction and Maintenance).** L. Touwen. Zomer & Keunings N.V. Gebr., Wageningen, the Netherlands. 88 pp., 41 illus., 4 tables, small index, glossy semi–soft bound in multicolor with yellow and black lettering, 5.0 by 7.5 inches (128 × 190 mm). Printed entirely in the Dutch language.

A small book on the construction, planting, and care of cool-season lawns under conditions in the Netherlands. There is a monthly calendar of cultural practices and a very short bibliography in the back.

Tukey, Paul Boardway

2007†   **The Organic Lawn Care Manual: An All-Natural, Low-Maintenance System For A Beautiful, Safe Lawn.** Paul Tukey. Storey Publishing, North Adams, Massachusetts, USA. 271 pp.; 274 illus. with 272 in color; 13 tables; indexed; glossy semi–soft bound in multicolor with black, gray, red, and white lettering; 8.5 by 10.9 inches (215 × 276 mm).

A book concerning lawn care under conditions in the United States and Canada utilizing primarily organic methods. There is a section on ground covers and a small glossary in the back.

Tyberg, Son

1986   **Gazonverzorging: Alles wat u moet weten over aanleg en onderhoud van een gazon (Lawn Care: Everything You Have to Know about Construction and Maintenance of a Lawn).** Coauthor.
    See *Palin, Robert—1986.*

1991   **Gazon zonder mos en onkruid: aanleg, verzorging (A Lawn Without Moss and Weeds: Construction and Care).** Coauthor.
    See *Palin, Robert—1991.*

Vavassori, Angelo

1993†   **Progetto, impianto e cura del Prato (Design, Construction and Maintenance of the Lawn).** Angelo Vavassori and Daniela Vavassori. Demetra S.r.l., Verona, Italy. 78 pp., 79 illus. with 57 in color, 13 tables, no index, glossy semi–soft bound in multicolor with green and black lettering, 6.6 by 9.4 inches (168 × 238 mm). Printed entirely in the Italian language.
    A small book on the planting and care of turfgrasses for lawns. It is practically oriented to residential lawn owners under conditions in Italy.
    †   Also published by Mistral, Demetra S.r.l., Sommacampagna, Italy, as a semi–soft bound version in multicolor with yellow and white lettering.

Vavassori, Daniela

1993†   **Progetto, impianto e cura del Prato (Design, Construction and Maintenance of the Lawn).** Coauthor.
    See *Vavassori, Angelo—1993.*

Walheim, Lance

1998†   **Lawn Care for Dummies.** Lance Walheim and editors of the National Gardening Association. Wiley Publishing Inc., New York City, New York, USA. 360 pp. plus 2 pp. of advertising; 158 illus. with 24 in color; 11 tables; large index; glossy semi–soft bound in multicolor with yellow, white, black, and maroon lettering; 7.4 by 9.2 inches (188 × 233 mm).
    An introductory book for novices in lawn planting and care. It is practically oriented to lawn enthusiasts and conditions in the United States. There is a selected list of references in the back.
    Also published by IDG Books Worldwide Inc., Foster City, California, USA, in 1998.

Warren, Benedict Orlando, Jr.

1955†   **Green Grows Your Lawn: If You'll Let It.** Ben Warren. Warren's Turf Nursery, Palos Park, Illinois, USA. 31 pp., 12 illus., 3 tables, no index, fold-stapled semisoft covers in light green with dark green lettering, 5.5 by 8.2 inches (140 × 209 mm).
    A small booklet concerning lawn planting with sod and subsequent care in the northern United States.

1959†   **Green Grows Your Lawn.** Ben Warren. Warren's Turf Nursery, Palos Park, Illinois, USA. 31 pp., 13 illus., 3 tables, no index, fold-stapled semisoft covers in light green with dark green lettering, 5.5 by 8.2 inches (140 × 209 mm).
    A small booklet on the planting from sod and care of lawns in the northern United States. There is a small section on ground covers in the back.

1960†   **Green Grows Your Lawn.** First edition. Ben Warren. Warren's Turf Nursery, Palos Park, Illinois, USA. 65 p., 24 illus., 4 tables, no index, fold-stapled textured semisoft covers in greens with dark green lettering, 5.5 × 8.3 inches (139 × 212 mm).
    A small handbook on the planting and care of lawns in the northern United States. Sod production and transplanting receive special emphasis. There is an interesting pictorial series on the evolution of sod handling equipment.
    Revised in 1961, 1962, 1966, and 1969.

1961†   **Green Grows Your Lawn.** Second revised edition. Ben Warren. Warren's Turf Nursery, Palos Park, Illinois, USA. 63 pp., 24 illus., 4 tables, no index, fold-stapled textured semisoft covers in greens with white lettering, 5.4 by 8.4 inches (137 × 213 mm).
    A major revision. It contains interesting photographs of early sod harvesting equipment.
    Revised in 1962, 1966, and 1969.

1962†   **Green Grows Your Lawn.** Third revised edition. Ben Warren. Warren's Turf Nursery, Palos Park, Illinois, USA. 53 pp., 32 illus., 4 tables, no index, fold-stapled textured semisoft covers in greens with white lettering, 5.4 by 8.4 inches (137 × 213 mm).

A significant revision. It contains interesting historical photographs showing the evolution of sod harvesting and handling equipment representative of the early years in the United States.
Revised in 1966 and 1969.

1966† **Green Grows Your Lawn.** Fourth revised edition. Ben Warren. Warren's Turf Nursery, Palos Park, Illinois, USA. 53 pp., 35 illus., 4 tables, no index, fold-stapled semisoft covers in greens with white lettering, 5.4 by 8.4 inches (137 × 213 mm). Minor revisions.
Revised in 1969.

1969† **Green Grows Your Lawn.** Fifth revised edition. Ben Warren. Warren's Turf Nursery, Palos Park, Illinois, USA. 63 pp., 40 illus., 3 tables, no index, fold-stapled semisoft covers in greens with white lettering, 5.4 by 8.4 inches (137 × 213 mm). Minor revisions.

Weibull, Gunnar

1957† **Gräsmattor (Lawns).** Gunnar Weibull. Ljus SMA Handböcker, Ljus Förlag, Stockholm, Sweden. 76 pp., 20 illus., 4 tables, no index, semi–soft bound in cream with green and black lettering, 4.8 by 7.4 inches (121 × 188 mm). Printed entirely in the Swedish language.

A small book on lawn planting and care. It is practically oriented to home lawn enthusiasts and cool-season conditions in Sweden.

1961† **Den Gröna Mattan (The Green Carpet).** Gunnar Weibull. P.A. Norstedt & Söners Förlag, Stockholm, Sweden. 96 pp., 33 illus., 4 tables, no index, soft bound in multicolor with black lettering, 4.7 by 7.4 inches (121 × 108 mm). Printed entirely in the Swedish language.

A small book on lawn planting and care. It is practically oriented to home lawn enthusiasts and cool-season conditions in Sweden.
Reprinted in 1962.

Whitehead, Stanley Bamford

1955† **Garden Lawns.** Stanley B. Whitehead. Foyle's Handbooks, W. & G. Foyle Limited, London, England, UK. 96 pp.; 42 illus.; no tables; small index; glossy semi–hardbound in light green, black, and white with white and black lettering; 4.8 by 7.2 inches (123 × 182 mm).

A small book on lawn planting and care. It is practically oriented to lawn enthusiasts and cool-season conditions in the United Kingdom. The book contains a chapter on chamomile and flower lawns.

Whitney Seed Company

1955† **abc's of Lawn Preparation and Maintenance.** Whitney Seed Co., Inc., and Geoffrey S. Cornish. Whitney Seed Co., Inc., Buffalo, New York, USA. 79 pp., 71 illus., 7 tables, no index, semisoft covers in black and white with green lettering plus clear overlays in a comb binder, 6.1 by 8.9 inches (156 × 227 mm).

A practical, small book on the planting and care of lawns. It is oriented to lawn enthusiasts in the cool climatic regions of the United States.

Williamson, Don

2005 **Lawns for Canada: Natural and Organic.** Don Williamson. Lone Pine Publishing, Edmonton, Alberta, Canada. 160 pp., 190 illus. with 156 in color, 6 tables, small index, glossy textured semi–soft bound in multicolor with white and yellow lettering, 5.4 by 8.5 inches (138 × 215 mm).

A book on the planting and care of lawns involving a philosophy of organic culture without the use of pesticides. It is oriented to lawn enthusiasts under the cool climatic regions of Canada.

2006 **Lawns Natural and Organic.** Don Williamson. Lone Pine Publishing International Inc., Auburn, Washington, USA. 158 pp., 189 illus. with 155 in color, 6 tables, small index, glossy semi–soft bound in multicolor with white and yellow lettering, 5.4 by 8.5 inches (138 × 215 mm).

A minor revision of the 2005 book **Lawns for Canada: Natural and Organic.** An attempt has been made to broaden the coverage to warm-season grasses across the United States, but only zoysiagrass and bermudagrass descriptions were added.

Winward, Lane L.

1992† **The Healthy Lawn Handbook.** Lane L. Winward. Lyone & Burford Publishers, New York City, New York, USA. 136 pp.; 84 illus.; no tables; indexed; glossy semi–soft bound in white, green, and blue with white, black, and blue lettering; 7.0 by 9.2 inches (177 × 234 mm).

A practical book on the planting and care of lawns. It is oriented to lawn enthusiasts and conditions in the United States.

Woess, Friedrich

1971† **Anlage und Pflege des Rasens (Establishment and Maintenance of Lawns)**. Friedrich Woess and Erich Kahl. Moldavia-Verlag, Wien, Austria. 96 pp., 11 illus., 10 tables, indexed, glossy semi–soft bound in multicolor with white and black lettering, 3.9 by 7.4 inches (99 × 189 mm). Printed entirely in the German language.

A small booklet on the planting, care, and renovation of lawns. It is practically oriented to home lawn enthusiasts and cool-season conditions in Austria.

Zeman, Anne M.

1993† **Installing & Renewing Lawns.** Anne M. Zeman. NK Lawn & Garden Company, Avon Books, The Hearst Corporation, New York City, New York, USA. 80 pp., 159 illus. in color, 1 table, indexed, glossy semi–soft bound in multicolor with white lettering, 8.0 by 9.9 inches (202 × 252 mm).

A small book on the planting and care of lawns. It is practically oriented to lawn enthusiasts and conditions in North America.

B. DUAL LAWN AND LANDSCAPE BOOKS

Consists of general, practically oriented landscape books of over 20 pages that contain significant sections on turfgrass, ground covers, flowers, shrub, trees, and/or gardens. Many are written by gardening journalists.

Aldous, David Earnest

1991† **Lawn Care and Lawn Alternatives.** David E. Aldous. Lothian Australian Garden Series, Lothian Publishing Company, Port Melbourne, Victoria, Australia. 64 pp.; 106 illus. with 18 in color; 11 tables; short index; glossy semi–soft bound in multicolor with white, black, and yellow lettering; 8.2 by 10.8 inches (209 × 274 mm).

A small book on the planting and care of lawns plus sections on ornamentals and ground covers. It is practically oriented to lawn enthusiasts and conditions in Australia.

American Chemical Paint Company

1956† **Better Lawns and Gardens.** American Chemical Paint Company, Ambler, Pennsylvania, USA. 63 pp.; 67 illus.; 2 tables; small index; fold-stapled soft covers in multicolor with black, white, and orange lettering; 5.4 by 8.2 inches (137 × 208 mm).

A small book on the planting and care of lawns, flowers, and vegetables in the United States.

Arthurs, Kathryn L.

1979 **Lawns & Ground Covers.** Kathryn L. Arthurs (editor). Lane Publishing Co., Menlo Park, California, USA. 96 pp., 239 illus. in color, 1 table, small index, glossy semi–soft bound in multicolor with white and black lettering, 8.2 by 10.9 inches (209 × 277 mm).

A practical book concerning the planting and care of lawns and ground covers. It is oriented to landscape enthusiasts in the United States.

Asseray, Phileppe

2004† **Réussir Votre Gazon (Your Successful Turf).** Phileppe Asseray (editor). Les Guides No 105, Mon Jardin & Ma Maison, Boulogne-Billancourt, France. 66 pp.; 127 illus. in color; 2 tables; very small index; fold-stapled glossy semisoft covers in multicolor with white, green, and black lettering; 7.0 by 9.5 inches (180 × 240 mm). Printed entirely in the French language.

A practical, small book concerning the planting and care of lawns. It is oriented to landscape enthusiasts under conditions in France.

Bahir, Zvi

1971† **Efficient Maintenance and Irrigation of Gardens.** Z. Bahir and M. Gil (editors) with 14 contributing authors: J. Palgi, Z. Bahir, E. Berniker, J. Ben-Arav, J. Silberman, B. Rabinovitch, R. Dekel, J. Keini, M. Gil, U. Tal, J. Seligman, J. Apel, M. Menachem, and R. Weissman. Gan Vanof (Garden & Landscape), Ministry of Agriculture, Tel Aviv, Israel.

57 pp., 45 illus., 16 tables, no index, fold-stapled semisoft covers in yellow with black lettering, 6.7 by 9.3 inches (170 × 236 mm). Printed entirely in the Hebrew language.

A small handbook on the efficient maintenance of landscape gardens in Israel. It includes a section on lawns. Emphasis is placed on water conservation.

Ball, Liz

2000† **Yard Care—Step-by-Step.**[*] Liz Ball. Better Homes and Gardens, Meredith Corporation, Des Moines, Iowa, USA. 169 pp., 466 illus. in color, 8 tables, small index, glossy semi–soft bound in multicolor with yellow, white, and black lettering, 9.2 by 9.7 inches (233 × 247 mm).

A practical book on the planting and care of lawns, ground covers, trees, shrubs, and vines. It is practically oriented to conditions in the United States.

Bilbrey, C. Robert

1957† **Lawn and Garden Book.**[*] Robert C. Bilbrey (editor) in collaboration with James Burdett. Popular Mechanics Press, Chicago, Illinois, USA. 192 pp.; 233 illus.; 2 tables; indexed; cloth hardbound in green with black lettering and a jacket in green, white, and black with black and white lettering; 6.3 by 9.1 inches (160 × 231 mm).

A book on landscape and vegetable gardening, with a small section on home lawns. It is practically oriented to gardening enthusiasts in the cool-climatic regions of North America.

See earlier version *Throm, Edward Louis.—1951.*

Bourne, Jeffrey A.

n.d.† **Grounds Maintenance Management Guidelines.**[*] First edition. Jeffrey A. Bourne (editor). Professional Grounds Management Society, Pikesville, Maryland, USA. 27 pp., no illus., 4 tables, no index, semi–soft bound in browns with dark brown lettering, 8.3 by 10.7 inches (210 × 273 mm).
circa
1983

This booklet contains guidelines concerning the basic minimum standards for grounds maintenance operations in the United States.

Breschke, Joachim

1996† **Rasen und Wiese—Rasenanlage. Rasenpflege. Naturwiese. Blumenwiese (Lawns and Meadows—Making a lawn. Maintaining a lawn. Natural meadows. Flowering meadows.).**[*] Joachim Breschke. Freude am eigenen Garten (Pleasure in your Garden series), Verlageunion Erich Pabel-Arthur Moewig KG, Rastatt, Germany. 126 pp.; 17 illus. with 12 in color; 6 tables; no index; semi–soft bound in multicolor with light green, white, and black lettering; 4.5 by 7.0 inches (115 × 179 mm). Printed entirely in the German language.

A small book on the planting and care of lawn and flower meadows. It is practically oriented to lawn-garden enthusiasts and conditions in Germany.

Bruning, Walter

1963† **Minimum Maintenance Landscaping.**[*] Walter Bruning. Jacobsen Manufacturing Company, Racine, Wisconsin, USA. 33 pp.; 28 illus. plus numerous vignettes; no tables; no index; glossy semi–soft bound in gray and black with yellow, white, and black lettering; 5.2 by 8.2 inches (131 × 209 mm).

A booklet concerning landscape layouts that facilitate more efficient maintenance, such as mowing.

Butler, Jack Dean

1988† **Landscape Management: Planting and Maintenance of Trees, Shrubs, and Turfgrasses.**[*] Coauthor.
See *Feuchet, James R.—1988.*

Cassiday, Bruce

1976† **Home Guide to Lawns and Landscaping.**[*] Bruce Cassiday. A Popular Science Skill Book, Harper & Row, Times Mirror Magazines Inc., New York City, New York, USA. 210 pp., 291 illus., 22 tables, indexed, cloth hardbound in greens with light green lettering and a jacket in multicolor with white and black lettering, 6.0 by 9.0 inches (152 × 228 mm).

A simplified book on the planting and care of both turfgrass lawns and landscape plants. It is practically oriented to gardening enthusiasts and conditions in the United States.

Chas. H. Lilly Co.

1959 **Better Lawn Plan.**[*] Chas. H. Lilly Co., Seattle, Washington, USA. 23 pp., 2 illus., 2 tables, no index, fold-stapled soft covers in multicolor with green and white lettering, 5.3 by 8.4 inches (135 × 213 mm).

A booklet with primary emphasis on the planting and care of lawns. It is practically oriented to lawn enthusiasts in the Pacific Northwest, United States.
Reprinted with small revisions in 1963.

Chono, Hidesato

1967†   **Turfgrass Establishment and Care of the Home Lawn—Including Home Garden Design.*** Coauthor.
See *Ehara, Kaoru—1967.*

Cochrane, Karen

2001   **Rasen & Bodendecker (Lawns and Groundcover).*** Coauthor.
See *Stebbings, Geoff—2001.*

Colborn, Nigel

2000†   **The Garden Floor.*** Nigel Colborn. Trafalgar Square Publishing, North Pomfret, Vermont, USA. 127 pp., 169 illus. with 163 in color, no tables, indexed, glossy hardbound in multicolor with white lettering and a jacket in multicolor with white lettering, 8.8 by 10.1 inches (224 × 257 mm).
A book concerning the diversity of plant and hard surfaces available for the design and utilization of open spaces in a garden.

Collins, Donald N.

1983†   **Turf and Garden Fertilizer Handbook.*** Donald N. Collins (editor) with eight contributing authors: Robert C. Rund, William H. Daniel, R.P. Freeborg, John R. Hall, III, O.A. Lorenz, John C. Harper II, Harold Davidson, and Fred P. Miller. The Fertilizer Institute, Washington, DC, USA. 126 pp., 49 illus. with 5 in color, 16 tables, small index, glossy hardbound in green with white lettering, 4.0 by 7.0 inches (101 × 178 mm).
A small handbook on fertilizers and fertilization strategies for lawns, landscape plants, and vegetable gardens in the United States. It contains a glossary in the back.

Corbett, Lee Cleveland

1904   **Beautifying the Home Grounds.*** Lee Cleveland Corbett. Farmers Bulletin 185, United States Department of Agriculture, Washington, DC, USA. 24 pp., 8 illus., no tables, no index, soft bound in white with black lettering, 6.2 by 9.5 inches (158 × 240 mm).
A pioneering booklet on landscaping procedures for the home lawn. It is practically oriented to lawn enthusiasts and conditions in the United States.

Crockett, James Underwood

1971†   **Lawns And Ground Covers.*** First edition. James Underwood Crockett. Time-Life Books, New York City, New York, USA. 159 pp., 291 illus., 4 tables, indexed, cloth hardbound in multicolor with white lettering, 7.8 by 10.7 inches (199 × 272 mm).
A small book covering the planting and care of lawns and ground covers, with emphasis on the how-to aspects. It contains an interesting introductory chapter on "The Lore of Lawns" and a large chapter in the back on ground covers. It is well illustrated in color.
Revised in 1973.

1973†   **Lawns And Ground Covers.*** Second revised edition. James Underwood Crockett. Time-Life Books, New York City, New York, USA. 159 pp., 291 illus., 4 tables, indexed, cloth hardbound in multicolor with white lettering, 7.8 by 10.7 inches (199 × 272 mm).
Slight revisions.

Crotta, Carol A.

2000†   **Lawns.*** Carol A. Crotta. Black & Decker Outdoor Home, Creative Publishing International Inc., Minnetonka, Minnesota, USA. 112 pp.; 230 illus. in color; no tables; indexed; glossy semi–soft bound in multicolor with white, orange, green, and black lettering; 8.2 by 10.0 inches (208 × 274 mm).
A book on the planting and care of lawns and ground covers. It is practically oriented to landscape enthusiasts and conditions in North America.

Mr. Cuthbert

1954   **Mr Cuthbert's Guide to Growing Lawns & Hedges.*** Mr. Cuthbert. Cassell & Company Limited, London, UK. 31 pp., 9 illus., no tables, no index, fold-stapled glossy semisoft covers in white with black lettering, 3.75 by 5.1 inches (95 × 130 mm).

A small booklet on the planting and care of lawns and hedges. It is practically oriented to landscape enthusiasts and conditions in England.

Daniel, William Hugh

1971† **Basic Grounds Maintenance.**˙ Coauthor.
See *Haskell, Theodore James—1971.*

Daniels, Stevie

1995† **the Wild Lawn handbook.**˙ Stevie Daniels. Macmillan Inc., New York City, New York, USA. 223 pp.; 17 illus.; 6 tables; detailed index; hardbound in light green and dark green with gold lettering and a glossy jacket in multicolor with white, yellow, and black lettering; 7.3 by 9.1 inches (185 × 232 mm).

A book on unmowed vegetation to replace mowed lawns. Much of the content consists of testimonials by individuals about their conversion activities, but there is a lack of long-term follow-up on the ultimate result and the costs. It contains numerous undocumented statements and broad conclusions that lack validation.

Davis, Mary

1977† **Lawns and Groundcovers.**˙ Coauthor.
See *Pittendrigh, Stuart—1977.*

DiNella, Glenn R.

2004† **Scotts Lawnscaping.**˙ Glenn R. DiNella. The Scotts Company, Marysville, Ohio, USA. 192 pp., 400 illus., 1 table, indexed, glossy semi–soft bound in multicolor with pink, white, and green lettering, 8.2 by 10.9 inches (207 × 276 mm).

A book concerning integrating landscaping with lawns, including lawn history, design strategies and styles, planting, and care. It is oriented to conditions in North America.

Doxon, Lynn Ellen

1999† **High Desert Yards and Gardens.**˙ Lynn Ellen Doxon. University of New Mexico Press, Albuquerque, New Mexico, USA. 242 pp.; 33 illus.; 2 tables; index; glossy hardbound in multicolor with green, red, and brown lettering; 6.0 by 8.0 inches (152 × 203 mm).

A book concerning the planting and care of landscape gardens, with the focus on high desert conditions of New Mexico.

Duble, Richard Lee

1977† **Southern Lawns and Groundcovers.**˙ Richard Duble and James Carroll Kell. Pacesetter Press, Gulf Publishing Company, Houston, Texas, USA. 91 pp., 55 illus., 21 tables, glossy semi–soft bound in multicolor with white and black lettering, 8.2 by 10.7 inches (208 × 271 mm).

A small book on the planting and care of lawns and ground covers. It is strictly oriented to gardening enthusiasts in the warm-climatic regions of the United States.

Ehara, Kaoru

1967† **Turfgrass Establishment and Care of the Home Lawn-Including Home Garden Design.**˙ Kaoru Ehara, Yoshisuke Maki, and Hidesato Chono (editors). Seibundo Shinkosha Inc., Tokyo, Japan. 237 pp., 216 illus. with 55 designs of lawn gardens, 3 tables, no index, cloth hardbound in scarlet and black with scarlet lettering and a glossy jacket in multicolor with black lettering, 5.8 by 7.9 inches (147 × 201 mm). Printed entirely in the Japanese language.

A manual on lawn planting and lawn-garden development oriented to home owners in Japan. Subjects included are lawn culture, lawn landscape gardening, 28 models of lawn garden layouts, arrangement of flowers and trees in the lawn garden, flower garden development, and ground cover plants.

Everett, Thomas H.

1956† **Lawns and Landscaping Handbook.**˙ Thomas H. Everett. A Fawcett How-to Book No. 302, Fawcett Publications Inc., Greenwich, Connecticut, USA, and then in Do-it-yourself Series, Arco Publishing Company, New York City, New York, USA. 144 pp.; 100-plus illus.; no tables; no index; both cloth hardbound in greens with black lettering and soft bound versions in multicolor with yellow, white, and red lettering; 6.6 by 9.2 inches (167 × 234 mm).

A small book on lawn planting and care. Smaller sections in the back discuss landscape plants. It is practically oriented to home gardening enthusiasts and conditions in North America.

1975† **Lawns and Landscaping.**ˈ T.H. Everett. Grosset & Dunlop Publishers, New York City, New York, USA. 95 pp., 95 illus., no tables, indexed, both cloth hardbound and glossy semi–soft bound versions in multicolor with white lettering, 7.7 by 10.7 inches (196 × 271 mm).

A small book on lawn, ground cover, and flower establishment and culture. It is practically oriented to home garden enthusiasts and conditions in North America. It contains sections on soils and site development.

Feuchet, James R.

1988† **Landscape Management—Planting and Maintenance of Trees, Shrubs, and Turfgrasses.**ˈ James R. Feuchet and Jack D. Butler. Van Nostrand Reinhold Company, New York City, New York, USA. 179 pp., 128 illus., 27 tables, indexed, glossy semi–soft bound in greens with white lettering, 8.1 by 9.2 inches (206 × 233 mm).

A guide on landscape planting and care, with two small chapters on turfgrasses. It is oriented to landscape practitioners of North America. It includes chapters on plant growth and development, soils and plant fertility, irrigation practices, pesticide safety, and pesticide application equipment.

Florida Green Industries

2002† **Best Management Practices For Protection Of Water Resources In Florida.**ˈ Florida Green Industries and Department of Environmental Protection, Tallahassee, Florida, USA. 63 pp., 25 illus., 9 tables, no index, glossy semi–soft bound in multicolor with brown and green lettering, 8.4 by 11.0 inches (214 × 279 mm).

A guidebook on best cultural practices for use on landscapes in Florida, with emphasis on water quality.

Funakoshi, Ryoji

2006 **Lawn and Groundcover.**ˈ Ryoji Funakoshi. Shufu no Tomosha, Tokyo, Japan. 128 pp., 390 illus. in color, 9 tables, indexed, glossy semi–soft bound in green with light green and white lettering, 7.3 by 8.3 inches (184 × 210 mm). Printed in the Japanese language.

A book on the planting and care of lawns and ground covers under conditions in Japan. There is a section on garden designs.

Garden Life Hen

1976 **Turf and Turf Gardens.**ˈ Garden Life Hen. Seibundo Shinkosha, Tokyo, Japan. 200 pp., 370 illus. with 32 in color, 26 tables, no index, textured semisoft bound in gray with red lettering, 7.3 by 9.0 inches (186 × 228 mm).

A book concerning the design of landscaped lawns involving turfgrasses and ground covers plus sections on planting and care of lawns.

Gil, M.

1971† **Efficient Maintenance and Irrigation of Gardens.**ˈ Coauthor.
See *Bahir, Zvi—1971.*

Gilmer, Maureen

1994† **Easy Lawn and Garden Care.**ˈ Maureen Gilmer. The Homeowner's Library, Consumer Reports Books, Consumers Union, Yonkers, New York, USA. 239 pp., 62 illus., 20 tables, indexed, glossy semi–soft bound in multicolor with white and green lettering, 8.3 by 10.8 inches (212 × 275 mm).

A book on the planting and care of lawn grasses and landscape plants in the United States. At times, the approaches presented lack a sound factual basis.

Green, Douglas

2001† **The Everything Lawn Care Book.**ˈ Douglas Green. Adams Media Corporation, Holbrook, Massachusetts, USA. 285 pp. plus 3 pp. of advertising; 52 illus. with 18 in color; 24 tables; indexed; glossy semi–soft bound in multicolor with blue, red, green, and black lettering; 8.0 by 9.2 inches (203 × 284 mm).

A practical book concerning the planting and care of lawns and ground covers, with emphasis on organic, low-maintenance aspects.

Griffith, Roger M.

1989 **Lawns and Landscaping.**ˈ Roger M. Griffith and the editors of Garden Way. Doubleday Book & Music Clubs Inc., New York City, New York, USA. 159 pp.; 131 illus.; 8 tables; indexed; glossy hardbound in multicolor with green, white, and black lettering; 8.3 by 11.0 inches (210 × 278 mm).

A book on the planting and care of lawns and landscapes. It is practically oriented to landscape enthusiasts in the United States.

Hampshire, Kristen

2007† **John Deere Landscaping & Lawn Care—The Complete Guide To A Beautiful Yard Year-Round.**˙ Kristen Hampshire. Quarry Books, Quayside Publishing Group, Gloucester, Massachusetts, USA. 175 pp.; 200 illus. in color; 1 table; small index; glossy semi–soft bound in multicolor with black, green, and yellow lettering; 8.4 by 10.8 inches (214 × 275 mm).

A practical book on turfgrass planting and care for lawns in the United States plus other landscape features—paths, lighting, ponds, ground covers, and plant beds.

Haskell, Theodore James

1971† **Basic Grounds Maintenance.**˙ Theodore J. Haskell, Ira B. Lykes, and William H. Daniel. Madisen Publishing Division, Appleton, Wisconsin, USA. 125 pp., 34 illus., 17 tables, no index, soft bound in tan with black lettering, 8.4 by 10.8 inches (214 × 275 mm).

This small book contains selected topics on tree care, turfgrass culture, and facility operations.

Hendy, Jenny

2007 **Creative Ideas for Lawns, Patios, Decks & Paths.**˙ Jenny Hendy. Anness Publishing Ltd., London, England, UK. 64 pp., 111 illus. in color, no tables, small index, glossy semi–soft bound in multicolor with maroon lettering, 7.9 by 9.3 inches (200 × 235 mm).

A small, practical book about the design and the types of materials and selection for English garden surfaces and floor. It includes grass, stone, wood, brick, tile, and gravel.

Hessayon, D.G.

2007† **The Lawn Expert.**˙ Dr. D.G. Hessayon. Expert Books, Transworld Publishers, Random House Group Ltd., London, England, UK. 128 pp., 400-plus illus. in color, 9 tables, small index, glossy semi–soft bound in multicolor with white lettering, 7.2 by 9.3 inches (184 × 235 mm).

A major revision of the 2002 edition, with a section devoted to alternate ground covers, including meadows, wild flower meadows, synthetic lawns, nongrass lawns, hard landscapes, and ground cover plants.

† *Printed in 2008,* with very minor revisions.

Hill, Lewis

1995† **Lawns, Grasses and Groundcovers.**˙ Lewis Hill and Nancy Hill. Rodale's Successful Organic Gardening, Rodale Press, Emmaus, Pennsylvania, USA. 160 pp., 42 illus. with 310 in color, 1 table, indexed, glossy semi–soft bound in multicolor with black and cream lettering, 8.0 by 10.9 inches (204 × 276 mm).

A guidebook on the use of organic techniques for the planting and care of lawns, ornamental grasses, and ground covers in the home landscape. It is practically oriented to conditions in the United States.

2000† **The Lawn & Garden Owner's Manual.** Lewis Hill and Nancy Hill. Storey Publishing, North Adams, Massachusetts, USA. 191 pp.; 93 illus. in color; 34 tables; indexed; glossy semi–soft bound in multicolor with blue, green, and black lettering; 8.4 by 10.9 inches (214 × 276 mm).

A manual concerning the planting and care of lawns and landscape plants. It is practically oriented to gardening enthusiast under conditions in the United States.

Hill, Nancy

1995† **Lawns, Grasses and Groundcovers.**˙ Coauthor.
    See *Hill, Lewis—1995.*

2000† **The Lawn & Garden Owner's Manual.** Coauthor.
    See *Hill, Lewis—2000.*

James, Ronald

1961 **Lawns, Trees, and Shrubs of Central Africa.**˙ Ronald James. Purnell & Sons (S.A.) Pty. Limited, Capetown & Johannesburg, South Africa. 174 pp., 48 illus., 12 tables, indexed, cloth hardbound in green with gold lettering, 3.3 by 8.5 inches (135 × 216 mm).

This book addresses the basics of selection, management, and placement of trees, shrubs, and lawns for the development of beautiful landscape gardens in Rhodesia and Nyasaland, Central Africa. There is a foreword by Lord Malvern.

Jedicke, Eckhard

1986 **Blumenwiese oder Rasen?: Anlage u. Pflege (Flower Meadow or Turfgrass? Installation and Care).**˙ Eckhard Jedicke. Franckh, Stuttgart, Germany. 79 pp., 31 illus. with 26 in color, 8 tables, indexed, glossy semi–soft bound in gray with black and white lettering, 6.3 by 7.7 inches (159 × 195 mm). Printed in the German language.

A small book concerning the planting and care of grass-flower plant communities. It is oriented to conditions in Germany.

Jindal, S.L.

1982   **Lawns and Gardens.**˙ S.L. Jindal. Publications Division, Ministry of Information and Broadcasting, Government of India, New Delhi, India. 217 pp., 94 illus. with 24 in color, no tables, indexed, cloth hardbound in white with green lettering, 6.3 by 9.5 inches (160 × 241 mm).

A book about gardening under the conditions in India. There is a small section concerning the planting and care of lawns.

Jönsson-Rose N.

1897†   **Lawns and Gardens.**˙ N. Jönsson Rose. G.C. Putnam's Sons, New York City, New York, USA. 426 pp., 172 illus., no tables, large index of common and scientific plant names, textured cloth hardbound in cream and green with gold lettering, 7.1 by 9.8 inches (180 × 249 mm).

A major early book on planning, site preparation, planting, and care of landscapes for the home. It includes the use a lawns and turfgrasses, trees, shrubs, hedges, vines, herbaceous plants, flowers, rocks, and ponds.

Kanameishi, Betty M.

1951†   **how to grow the best Lawn and Garden in your neighborhood.**˙ Coauthor.
    See *Throm, Edward Louis—1951.*

Kell, James Carrol

1977†   **Southern Lawns and Groundcovers.**˙ Coauthor.
    See *Duble, Richard Lee—1977.*

Kim, Ho-Jun

1997   **Compendium of Turfgrass and Tree Diseases and Insects with Color Plates.**˙ Ho-Jun Kim, Gyu Yul Shin, Seung-Weon Yang, and Jin Won Kim. Korean Turfgrass Research Institute, Seoul, South Korea. 200 pp., 271 illus. in color, 5 tables, indexed, semi–soft bound in purple with gold lettering, 6.1 by 8.7 inches (155 × 220 mm). Printed in the Korean language.

A book concerning the diagnosis and control procedures for the major pests of turfgrasses and trees under the conditions in South Korea. It includes 28 diseases and 19 insects and small animal pests of turfgrasses and 18 diseases and 36 insect pests of trees.

Kim, Jin Won

1997   **Compendium of Turfgrass and Tree Diseases and Insects with Color Plates.**˙ Coauthor.
    See *Kim, Ho-Jun—1997.*

Knoop, William Edward

1996†   **The Landscape Handbook.**˙ William E. Knoop (editor). Advanstar Communications, Duluth, Minnesota, USA. 179 pp., 96 illus., 35 tables, indexed, glossy multicolor in orange and green lettering, 8.0 by 10.9 inches (203 × 276 mm).

A book on soil preparation, planting, and care of turfgrasses, woody ornamentals, and trees. A major portion of the book addresses weed, disease, and insect problems. There also is a small section on ponds. A glossary of landscape terms is included in the book.

1999†   **The Turf and Landscape Management Handbook.**˙ William E. Knoop (editor). Advanstar Communications, Cleveland, Ohio, USA. 93 pp. with 6 pp. of advertising; 69 illus.; 25 tables; no index; glossy semi–soft bound in multicolor with white, green, and black lettering; 8.0 by 10.7 inches (204 × 272 mm).

A revision of the earlier book, with most of the coverage focused on turfgrasses.

Kramer, Richard

2003†   **Lawn & Landscape Technician's Handbook.**˙ Dr. Richard Kramer. G.I.E. Media Inc., Cleveland, Ohio, USA. 238 pp.; 96 illus. with 56 in color; no tables; no index; glossy semi–soft bound in multicolor with red, green, and white lettering; 4.2 by 7.1 inches (107 × 181 mm).

A field guide to the identification of insect pests of turfgrass and ornamental plants plus their management. It is oriented to service technicians in the lawn and landscape industry. There is a small glossary in the back.

Lacey, Stephen

1991†   **Lawns & Ground Covers.**˙ Stephen Lacey (editor). The National Trust Guide, Trafalgar Square Publishing, North Pomfret, Vermont, USA, and Pavilion Books Limited, London, England, UK. 110 pp., 86 illus. with 53 in color, no tables, indexed, cloth hardbound in green with gold lettering and a glossy jacket in multicolor with black lettering, 9.0 by 9.0 inches (228 × 228 mm).

A book on the design, installation, and care of lawns, ground covers, and paved areas as used in gardens. It is practically oriented to conditions in England.

Lamerz-Beckschäfer, Birgit

2004   **Rasen & Bodendecker (Lawns and Groundcovers).**˙ Coauthor.
See *Stebbings, Geoff—2001.*

Lütke-Entrup, Ernst

1959†   **Die Wichtigsten Gräser (The Most Important Grasses).**˙ Coauthor.
&   See *Fischer, Walther—1959 and 1972.*
1972†

Lykes, Ira B.

1971†   **Basic Grounds Maintenance.**˙ Coauthor.
See *Haskell, Theodore James—1971.*

MacCaskey, Michael

1982†   **Lawns and Ground Covers—How to Select, Grow and Enjoy.**˙ Michael MacCaskey. Horticultural Publishing Company Inc., Tucson, Arizona, USA. 176 pp.; 243 illus. in color; 22 tables; indexed; semi–soft bound in multicolor with yellow, white, and black lettering, 8.4 by 10.8 inches (214 × 274 mm).

A small book on the planting and care of ground covers and lawns. The front one-third is devoted to turfgrasses. It is practically oriented to home gardening enthusiasts and conditions in North America.

†   *There are several reprintings over subsequent years.*

Maki, Yoshisuke

1967†   **Turfgrass Establishment and Care of the Home Lawn—Including Home Garden Design.**˙ Coauthor.
See *Ehara, Kaoru—1967.*

Martinelli, Janet

1993†   **The Natural Lawn & Alternatives.**˙ Janet Martinelli (editor) with 10 contributing authors. Brooklyn Botanic Garden, Brooklyn, New York, USA. 96 pp., 86 illus., 3 tables, small index, glossy semi–soft bound in multicolor with black and white lettering, 6.0 by 8.9 inches (152 × 225 mm).

A booklet on lawns and conversion to alternative vegetation such as native plants, ground covers, moss, and sedge under conditions in the United States.
Revised in 1999.

1999†   **Easy Lawns.**˙ Janet Martinelli (editor) with 10 contributing authors. Bulletin #160, Brooklyn Botanic Garden, Brooklyn, New York, USA. 111 pp.; 95 illus. with 56 in color; no tables; small index; glossy semi–soft bound in multicolor with blue, white, and green lettering; 6.0 by 9.0 inches (152 × 228 mm).

A major revised version of the 1993 booklet, with more emphasis on plant species utilization by regions within the United States.

McDonald, Elvin

1971†   **The Low-Upkeep Book of Lawns and Landscape.**˙ Elvin McDonald and Lawrence Power. Hawthorn Books Inc., New York City, New York, USA. 152 pp., 102 illus., 7 tables, indexed, textured cloth hardbound in brown and gray with brown lettering and a jacket in multicolor with black lettering, 6.0 by 9.1 inches (152 × 231 mm).

A book on planning, planting, and establishing lawn-garden landscapes for home grounds. Emphasis is placed on low maintenance designs and plant selection. It is practically oriented to garden enthusiasts and conditions in the United States.

1977† **Making Your Lawn & Garden Grow.**ˑ Elvin McDonald. A Melnor Guide, Dorison House Publishers Inc., New York City, New York, USA. 144 pp., 48 illus., 10 tables, small index, glossy hardbound in multicolor with green and black lettering, 6.9 by 8.9 inches (175 × 226 mm).

    A book on a broad array of gardening activities, with emphasis on plant selection. It is practically oriented to landscape enthusiasts and conditions in the United States.

McGraw-Hill Book Company

1980† **Landscape & Lawn Care.**ˑ McGraw-Hill Book Company, New York City, New York, USA. 94 pp., 123 illus., 11 tables, no index, glossy semi–soft bound in multicolor with white and black lettering, 8.5 by 10.9 inches (216 × 277 mm).

    A shorter revision of a 1975 book on a broad array of gardening activities.

    See also *Minnesota Mining and Manufacturing Company—1975.*

McHoy, Peter

1989 **The Garden Floor—The Design and Maintenance of Lawns and Their Alternatives.**ˑ Peter McHoy. Headline Book Publishing PLC, London, England, UK. 120 pp., 141 illus. in color, 3 tables, small index, cloth hardbound in green with gold lettering and a glossy jacket in multicolor with black lettering, 10.1 by 8.2 inches (257 × 208 mm).

    A book on the design of large garden areas that encompasses turfgrasses, hard surfaces, and ground covers. It is practically oriented to lawn enthusiasts and conditions in the United Kingdom.

McIndoe, Andrew

2006† **Living Lawns.**ˑ Andrew McIndoe. David & Charles/OutHouse Publishing, Winchester, Hampshire, England, UK. 80 pp., 97 illus., no tables, small index, glossy semi–soft bound in multicolor with white lettering, 5.8 by 8.2 inches (148 × 209 mm).

    A practical book on the design, planting, care, and renovation of lawns plus smaller sections on trees and ornamental plants. It is oriented to garden enthusiasts in the United Kingdom.

Minnesota Mining and Manufacturing Company

1975 **The Home Pro Landscape and Lawn Care Guide.**ˑ Automotive and Hardware Trades Division, Minnesota Mining and Manufacturing Company, Minneapolis, Minnesota, USA. 166 pp.; 239 illus.; 11 tables; no index; glossy semi–soft bound in multicolor with green, black, and orange lettering; 8.0 by 5.0 inches (203 × 129 mm).

    A how-to book on gardening including landscaping lawns, ground covers, trees, shrubs, and gardens. It is practically oriented to home gardening enthusiasts and conditions in the United States.

    Revised 1980.

    See *McGraw-Hill Book Company—1980.*

Oldale, Adrienne

1969† **Lawns, Hedges, and Fences.**ˑ Adrienne Oldale and Peter Oldale. Volume Two, Collins Step-by-Step Guides, Collins, Glasgow, Scotland, UK. 95 pp., 201 illus., no tables, no index, hardbound in multicolor with black and blue lettering, 5.4 by 8.3 inches (137 × 210 mm).

    A small, practical book on the initial phases of garden development, such as the lawn, boundary fences and hedges, gates, and shrubs. It is oriented to garden enthusiasts and conditions in the United Kingdom.

Oldale, Peter

1969† **Lawns, Hedges, and Fences.**ˑ Coauthor.

    See *Oldale, Adrienne—1969.*

O.M. Scott and Sons Company

1995† **See & Do Solutions—Lawns and Groundcovers.**ˑ The Scotts Company, Marysville, Ohio, USA. 128 pp.; 124 illus. with 116 in color; 6 tables; indexed; glossy hard covers in multicolor with white, black, red, and green lettering in a wire ring binder; 7.5 by 10.0 inches (190 × 254 mm).

    A guidebook on the design, plant selection, and planting for a diverse range of landscape and use conditions in residential areas.

1976† **Scotts Information Manual for Lawns and Gardens.**ˑ O.M. Scotts and Sons, Marysville, Ohio, USA. 155 pp., 152 illus. with 12 in color, 12 tables, no index, textured glossy semisoft covers in green with white lettering in a ring binder, 8.4 by 11.0 inches (214 × 279 mm).

    A manual mainly focused on lawn planting and care in the United States. Emphasis is placed on the identification, characteristics, and occurrence of weeds, diseases, and insects. There are small sections on ornamentals and vegetables plus a glossary of terms in the back.

Pittendrigh, Stuart

1977† **Lawns and Groundcovers.**˙ Stuart Pittendrigh and Mary Davis. Ure Smith, Sydney, Australia. 80 pp.; 100 illus.; 3 tables; indexed; semi–soft bound in multicolor with orange, white, and black lettering; 8.0 by 10.9 inches (202 × 276 mm).

A small book on the planting and care of lawns and ground covers. It is practically oriented to home grounds enthusiasts and conditions in Australia.

Reprinted in 1979.

Power, Lawrence

1971† **The Low-Upkeep Book of Lawns and Landscape.**˙ Coauthor.

See *McDonald, Elvin—1971.*

Pycraft, David

1980† **Lawns, Ground Cover and Weed Control.**˙ First edition. David Pycraft. Mitchell Beazley Publishers Limited, London, England, UK. 96 pp., 300+ illus., 6 tables, indexed, glossy semisoft covers in green and white with green and black lettering in a ring binder, 11.3 by 9.1 inches (287 × 230 mm).

A small handbook on the planting and care of turfgrasses and ground covers; plus a major section on weeds and their control. It is practically oriented to gardening enthusiasts and conditions in England.

† *Reprinted in 1982, 1983, 1984, 1986, 1988, and 1992 as semi–soft bound versions in multicolor with white, green, and black lettering.*

1985† **Il Prato-tappeti erbosi, piante tappezzanti, infestanti (Lawns, Ground Covers and Weed Control).**˙ David Pycraft. Serie di giardinaggio a cura della Royal Horticultural Society, Zanichelli Editore S. p. A., Bologna, Italy. 190 pp., 366 illus., 8 tables, indexed, glossy semi–soft bound in multicolor with white lettering, 6.0 by 9.4 inches (153 × 238 mm). Printed entirely in the Italian language.

A translation of the 1980 book by David Pycraft.

1992† **Lawns, Weeds & Ground Cover.**˙ Second revised edition. David Pycraft. The Royal Horticultural Society Encyclopedia of Practical Gardening, Mitchell Beazley International Limited, London, England, UK. 192 pp., 351 illus., 9 tables, indexed, glossy semi–soft bound in multicolor with white and brown lettering, 6.3 by 9.1 inches (160 × 230 mm).

A major revision of the handbook on the planting and care of turfgrasses and ground covers, with emphasis on weed control. It is practically oriented to gardening enthusiasts and conditions in the United Kingdom.

† *Reprinted in 1993.*

Rao, Balakrishna

1992† **Landscape Problem Management.**˙ Balakrishna Rao. Landscape Management, Cleveland, Ohio, USA. 153 pp., 2 illus., 3 tables, no index, soft bound in multicolor with white and black lettering, 5.9 by 8.9 inches (151 × 227 mm).

A book which is a categorized compilation of questions-and-answers previously published in **Landscape Management** magazine.

Roby, Melvin John

1977† **Lawns.**˙ Melvin J. Robey. David McKay Company Inc., New York City, New York, USA. 224 pp., 49 illus., 40 tables, indexed, cloth hardbound in green with light green lettering, 5.4 by 8.2 inches (138 × 208 mm).

A book on the planting and care of lawns and ground covers. It is oriented to lawn enthusiasts under conditions in the United States.

Rosenfeld, Richard

2001 **A Gardener's Guide To Hedges, Lawns & Groundcovers.**˙ Richard Rosenfeld (editor). Murdock Books UK Ltd., London, England UK. 112 pp., 220 illus., no tables, small index, glossy semi–soft bound in multicolor with white lettering, 8.4 by 10.8 inches (212 × 275 mm).

A practical guidebook about the selection, planting, and care of landscape plants, including hedges and ground covers. It includes a short section on turfgrasses.

Roth, Susan A.

1987† **Controlling Lawn & Garden Insects.**˙ Susan A. Roth. Ortho Books, Chevron Chemical Company, San Francisco, California, USA. 96 pp., 186 illus., no tables, small index, glossy semi–soft bound in multicolor with white lettering, 8.3 by 11.0 inches (211 × 279 mm).

A book on the identification, biology, and control of problem insects in lawns and gardens. It is oriented to conditions in the United States.

Rubin, Carole

1989† **How to get your lawn & garden off drugs.** Carole Rubin. Friends of the Earth, Ottawa, Ontario, Canada. 97 pp., 18 illus., 1 table, no index, glossy semisoft covers in multicolor with black lettering in a ring binder, 5.4 by 8.4 inches (137 × 214 mm).
A small book on suggested organic methods of lawn and garden care.

Sanchez, Janet H.

1995† **Lawns, Ground Covers & Vines.** Janet H. Sanchez. Better Home and Gardens Books, Des Moines, Iowa, USA. 132 pp., 387 illus. in color, no tables, small index, glossy hard covers in multicolor with white and black lettering in a coiled wire binder, 9.0 by 9.7 inches (228 × 247 mm).
A book concerning garden design and the planting and care of lawns, ground covers, and vines under the conditions in the United States and Canada.

Schery, Robert Walker

1963† **The Householder's Guide to Outdoor Beauty.** Robert W. Schery. Pocket Books Inc., New York City, New York, USA. 337 pp.; 100-plus illus.; 7 tables; no index; glossy semi–soft bound in multicolor with white, yellow, and black lettering; 4.2 by 7.1 inches (106 × 180 mm).
A manual on the planting and care of lawns with the back portion covering landscape plants. It is practically oriented to home gardening enthusiasts and conditions in North America.

Seale, Allan

1991† **Allan Seale's Garden Book of Lawns & Ground Cover.** Allan Seale. Treasure Press, Port Melbourne, Victoria, Australia. 79 pp., 61 illus., 2 tables, no index, glossy semi–soft bound in multicolor with green and white lettering, 4.5 by 8.2 inches (114 × 209 mm).
A small handbook on the planting and care of lawns and ground covers. It is practically oriented to gardening enthusiasts and conditions in Australia and New Zealand.

Sharon Publications

1976† **Lawns and Landscaping.** Sharon Publications, Fort Lee, New Jersey, USA. 92 pp., 98 illus., 1 table, indexed, cloth hardbound in multicolor with yellow and white lettering, 7.6 by 10.7 inches (194 × 271 mm).
A small book on the how-to aspects of turfgrass, tree, shrub, and ground cover care in the home grounds. It is practically oriented to home gardening enthusiasts and conditions of North America. The book contains a chapter on landscaping principles in the back.

Shell Canada Limited

n.d.† **Quick-Action Guide to Lawn & Garden Care.** Shell Canada Limited, Time-Life Custom Publishing, USA. 44 pp., 84
circa illus., 2 tables, no index, hardbound in multicolor with white lettering, 8.5 by 5.5 inches (215 × 140 mm).
19xx A small, quick-reference booklet on the care of lawns, gardens, landscape plants, and trees.

Shewell-Cooper, W.E.

1959 **The A.B.C. of Soils—Including Manuring, Composts and Lawns.** W.E. Shewell-Cooper. The English Universities Press Ltd., London, England, UK. 160 pp., 42 illus., 4 tables, small index, cloth hardbound in light green with black lettering, 4.8 by 7.2 inches (122 × 183 mm).
A practitioner's book concerning the philosophy of organic soil management plus a back portion devoted to planting and care of lawns. Chamomile, clover, and alpine lawns are included.

Shim, Gyu Yul

1997 **Compendium of Turfgrass and Tree Diseases and Insects with Color Plates.** Coauthor.
See *Kim, Ho-Jun—1977.*

Siems, H.B.

1931† **For Better Lawns and Gardens.** H.B. Siems. Swift & Company, Chicago, Illinois, USA. 58 pp., 24 illus., 3 tables, no index, soft bound in multicolor with black lettering, 5.9 by 8.5 inches (150 × 216 mm).

A booklet on gardening that includes lawns, flowers, shrubs, trees, vegetable gardens, and house plants. It is practically oriented to gardening enthusiasts and conditions in the United States, with emphasis on fertilizing and soils management. It contains two chapters on lawns. Dr. Siems introduced the book as follows:

> Plants have contributed to our comfort and pleasure since time immemorial, but not until a few years ago did their usefulness in beautifying our homes become so universally appreciated.
>
> History records few changes in our mode of living that have equalled in growth and extent the present-day method of home beautification. It has spread in to every city, town, and hamlet. Few changes have done so much to improve our standard of living or contributed more to our health and welfare.
>
> Everywhere the old-fashioned front lawns have given way to beautiful open lawns with graceful shrubs softening the corners and walls. The back yard of yesterday has been converted into an "outdoor living room," carpeted with rich-green velvety grass, with walls of shrubs and trees, tapering down to border of flowers arranged in groups to form as many different scenes as the imagination can inspire—living scenes that change with the hours, the days, and the seasons.
>
> Here, away from the hazards of the street, children may play in safety. Here each member of the family finds an alluring retreat that only such an environment could provide, and the pleasure and satisfaction one gets from creating and maintaining this new beauty is almost equal to that which he receives from it.

Simpson, A.G. William

1992† **Growing Lawns, including Lawn Alternatives and Ground Covers.** A.G.W. Simpson. Kangaroo Press Pty. Ltd., Kenthurst, New South Wales, Australia. 144 pp., 204 illus. with 96 in color, 5 tables, small index, glossy semi–soft bound in multicolor with black and white lettering, 7.2 by 9.5 inches (182 × 240 mm).

A small book on the planting and care of lawns in Australia and New Zealand. One third of the book is devoted to lawn alternatives, including ground covers.

Sombke, Laurence

1991† **The Environmental Gardener.** Laurence Sombke. Master Media Limited, New York City, New York, USA. 168 pp., 5 illus., no tables, no index, glossy hardbound in orange and yellow with black and white lettering, 5.1 by 8.0 inches (129 × 203 mm).

A book concerning care of the lawn and garden via organic approaches, including composting that is combined with integrated pest management.

1994† **Beautiful Easy Lawns and Landscapes.** Laurence Sombke. The Glade Pequot Press, Old Saybrook, Connecticut, USA. 182 pp., 42 illus., 3 tables, indexed, glossy semi–soft bound in multicolor with green and black lettering, 6.9 by 9.9 inches (175 × 251 mm).

A book on the care of lawns and landscape plants. One-third of the book is devoted to lawn care calendars for the four major climatic regions of the United States. Some of the author's presentation lacks a sound basis in factual information.

South Australia, Education Department

1979† **Grounds Maintenance Handbook For Schools.** Education Department of South Australia, Adelaide, South Australia, Australia. 37 pp., 40 illus., 8 tables, no index, soft bound in white and black with black lettering, 8.2 by 11.6 inches (208 × 294 mm).

A booklet concerning the care of turfgrasses, trees, and shrubs on grounds surrounding schools. It is oriented to school councils, principals, and groundspersons under conditions in South Australia.

Stebbings, Geoff

1999† **Lawns and Ground Cover.** First edition. Geoff Stebbings. American Horticultural Society, DK Publishing Inc., New York City, New York, USA. 80 pp.; 266 illus. in color; 2 tables; small index; glossy semi–soft bound in multicolor with gold, green, white, and black lettering; 5.8 by 8.4 inches (147 × 213 mm).

A small book on the planting and care of turfgrasses for lawns and for ground covers. It is oriented to gardening enthusiasts under conditions in the United States.

The previous 1999 book also was published in Spanish as **Césped y Otras Alternativas** by Albatros, Buenos Aires, Argentina, and in Dutch as **Gazons en Bodembedekkers** by Terra, Tielt: Lannoo, Warnsveld, the Netherlands plus in English as **Lawns & Ground Cover** by Dorling Kindersley, London, England, UK, and by Cavendish Books, Vancouver, British Columbia, Canada.

Reprinted in UK in 2003.

2001 **Rasen & Bodendecker (Lawns and Groundcovers).** First edition. Geoff Stebbings. Darling Kindersley Verlag GmbH, Starnberg, Germany. 80 pp., 87 illus., 11 tables, small index, glossy hardbound in multicolor with white and yellow lettering, 6.7 by 9.4 inches (171 × 231 mm). Printed in the German language.

A German translation of the 1999 book.

2001 A French translation of the 1999 book **Lawns and Ground Covers**.

2002 **Lawns and Ground Covers.** Second revised edition. Geoff Stebbings. Dorling Kindersley Limited, London, England, UK. 80 pp., 239 illus. in color, 4 tables, indexed, semi–soft bound in multicolor with white lettering, 5.5 × 8.3 inches (140 × 210 mm).

2002 A Chinese translation of the 1999 book **Lawns and Ground Covers** titled **Cao Ping Jian Zhi Pu Yang Hu Cal Se Tu Shuo**.

2004 **Rasen & Bodendecker: Einfache Lösungen Für Einen Schönen Rasen (Lawns and Groundcovers: Simple Solutions for a Beautiful Lawn).** Second revised edition. Geoff Stebbings and Birgit Lamerz-Beckschäfer. Weltbild, Augsburg, Germany.

A German translation of the 2002 second revised translation of **Lawns and Ground Covers.**

Stevens, W.J.

1908 **Lawns.** W.J. Stevens. One & All Garden Books Volume II, London, England, UK. 20 pp., 27 illus., no tables, no index, semi–soft bound in green with black lettering, 5.4 by 8.3 inches (138 × 210 mm).

A **rare**, early small booklet oriented primarily to vegetables, fruits, and flowers, with one chapter on lawns.

Sulzberger, Robert

2004 **Rasen Blumenwiese: Anlegen und Pflegen (Grass-Blooming Flowers: Creating and Maintaining.)** Robert Sulzberger. BLV Verlagsgesellschaft mbH, München, Germany. 95 pp.; 128 illus. in color; 8 tables; small index; glossy semi–soft bound in multicolor with white, blue, and red lettering; 6.5 by 8.3 inches (164 × 211 mm). Printed entirely in the German language.

A booklet about the planting and care of lawns, both turfgrass and flowering meadows. It is oriented to lawn enthusiasts and conditions in Germany.

2005 **Alles Over Gazon en Bloemenweide (Everything about Lawns and Meadow Flowers).** Robert Sulzberger and Emmy Middelbeek-van der Ven. Deltas, Aartselaar (Oosterhout), Belgium. 95 pp., 126 illus. in color, 8 tables, small index, glossy semi–soft bound in multicolor with red and black lettering, 6.5 by 8.3 inches (164 × 211 mm). Printed in the Dutch language.

A small book concerning the planting and care of lawns composed of turfgrasses or flowers and turfgrasses. It is practically oriented to gardening enthusiasts under the cool conditions in the Netherlands.

Sunset Books

1955† **How to Install and Care for Your Lawn.** A Sunset Book. Lane Publishing Company, Menlo Park, California, USA. 64 pp., 114 illus., 1 table, no index, semi–soft bound in multicolor with yellow and dark green lettering, 8.3 by 10.4 inches (211 × 264 mm).

A small book on lawn planting and care. It is practically oriented to home lawn enthusiasts, especially those in the western United States. The book contains a brief section on trees in the back.

Revised in 1960, 1964, 1979, and 1989.

1960† **Sunset Lawn and Ground Cover Book.** Second revised edition. Joseph F. Williamson (editor). Lane Book Company, Menlo Park, California, USA. 96 pp., 120 illus., 6 tables, no index, soft bound in green, brown and white with white and black lettering, 8.2 by 10.7 inches (209 × 272 mm).

A small book on lawn planting and care. It is practically oriented to home gardening enthusiasts, especially in the western United States. There is a section on ground covers.

The first edition was published under the title **How To Install and Care For Your Lawn.**

Revised in 1964, 1979, and 1989.

1964† **Lawns and Ground Covers—A Sunset Book.** Third revised edition. Joseph F. Williamson (editor). Lane Magazine and Book Company, Menlo Park, California, USA. 112 pp.; 166 illus.; 6 tables; indexed; glossy semi–soft bound in multicolor with white, green, and black lettering; 8.2 by 10.6 inches (209 × 270 mm).

Revised and updated, especially the section on ground covers.

Revised in 1979 and 1989.

1979† **Sunset Lawns and Ground Covers.** Fourth revised edition. Kathryn L. Arthurs (editor). Lane Publishing Company, Menlo Park, California, USA. 96 pp.; 193 illus. with 106 in color; 4 tables; indexed; glossy semi–soft bound in multicolor with white, green, and black lettering; 8.2 by 10.6 inches (209 × 270 mm).

A major revision, with the lawn sections substantially reduced in size and the ground cover sections expanded. Color photographs have been added, primarily on ground covers.

† *Reprinted in 1980.*
Revised in 1989.

1989† **Sunset Lawns & Ground Covers.**˙ Fifth revised edition. Fran Feldman (editor). Sunset Publishing Corporation, Menlo Park, California, USA. 160 pp.; 225 illus. in color; 11 tables; indexed; glossy semi–soft bound in multicolor with white, orange, and black lettering; 8.2 by 10.7 inches (208 × 271 mm).

A major revision of a book on the planting and care of lawns and ground covers. It is practically oriented to the gardening enthusiasts and conditions in the United States. It has much-improved color photographs.

† *Reprinted in 1990 and 1991.*

Taloumis, George

1976† **Winterize Your Yard and Garden.**˙ George Taloumis. J.P. Lippincott Company, Philadelphia, Pennsylvania, USA. 288 pp., 273 illus. with 12 in color, no tables, indexed, cloth hardbound in red with silver lettering and a jacket in multicolor with yellow and white lettering, 6.0 by 9.2 inches (153 × 233 mm).

A book on autumn, winter, and spring gardening activities in the United States, including small sections on lawns. It is practically oriented to the home gardening enthusiasts and conditions in the United States.

Throm, Edward Louis

1951† **How to grow the best Lawn and Garden in your neighborhood.**˙ Edward L. Throm and Betty M. Kanameishi (editors). Popular Mechanics, Chicago, Illinois, USA. 176 pp., 284 illus., 6 tables, indexed, cloth hardbound in green with yellow lettering and a jacket in light green and white with black and green lettering, 6.4 by 9.2 inches (162 × 235 mm).

A small book on landscape gardening, with a section on home lawns. It is practically oriented to gardening enthusiasts and conditions in the cool-climatic regions of North America.

Revised in 1957.
See *Bilbrey, C. Robert—1957.*

Time-Life Books

1989† **Lawns And Ground Covers.**˙ The Time-Life Gardener's Guide, Alexandra, Virginia, USA. 158 pp., 349 illus. in color, 3 tables, index, glossy hardbound in multicolor with white lettering, 10.8 by 9.2 inches (274 × 234 mm).

An extensively illustrated book on the planting and care of turfgrass lawns and ground covers. It is practically oriented to landscape enthusiasts and conditions in the United States and Canada.

n.d.† **Quick-Action Guide to Lawn & Garden Care.**˙ Vol. 10. Time-Life Books and Shell Canada Limited, Alexandra, Vir-
circa ginia, USA. 44 pp., 84 illus., 1 table, no index, glossy hardbound in multicolor with white lettering, 5.4 by 8.5 inches
199x (138 × 216 mm). Printed in both the English and French languages.

A booklet on preventive practices and problem solving for lawns, flowers, shrubs, and trees in the garden landscape.

Time Publication Ventures

2000† **Essential Yard Care and Landscaping Projects.**˙ Time Publication Ventures, New York City, New York, USA. 192 pp.; 343 illus. with 317 in color; no tables; no index; glossy semi–soft bound in black with blue, yellow, and white lettering; 9.0 by 10.0 inches (229 × 254 mm).

A general book concerning the planting and care of lawns and trees plus the numerous associated aspects, such as decks, fences, walks, walls, patios, and lighting.

Titchmarsh, Alan

2009† **how to garden—Lawns, Paths and Patios.**˙ Alan Titchmarsh. BBC Books, London, England, UK. 132 pp., 254 illus. in color, 1 table, indexed, glossy semi–soft bound in multicolor with white and green lettering, 7.4 by 9.2 inches (187 × 234 mm).

A book concerning the construction and care of lawns, trees, shrubs, ground covers, and hard surfaces. It is oriented to gardening enthusiasts and conditions in the United Kingdom.

Toronto Public Health

2004 **Pesticide Free . . . A Guide to Natural Lawn and Garden Care.**˙ Toronto Public Health, Toronto, Canada. 43 pp.; 19 illus.; 3 tables; no index; semisoft covers in multicolor with white, black, and blue lettering in a ring binder; 8.5 by 10.9 inches (216 × 278 mm).

A guide for the use of nonchemical or natural means that are proposed to solve lawn, tree, and garden problems, including weeds, diseases, and insects.

Trumble, H.C.

1946† **Blades of Grass.**˙ H.C. Trumble. Georgian House, Melbourne, Australia. 294 pp., 62 illus., no tables, no index, cloth hardbound in green with gold lettering, 5.4 by 8.3 inches (136 × 211 mm).

An individual's personal observations on forage and turfgrass use and culture as experienced over several decades, primarily in Australia. It contains a list of references in the back.

Tsitsin, Nikolai Vasiljevich

1977 **Gazony: Osnovy Semenovodstva i Ratonirovaniya (Lawns: Foundation of Seed Growing and Division).**˙ Nikolai Vasiljevich Tsitsin. USSR Akademiya Nauk SSSR, Sovet Botanicheskikh Sadov SSSR. Maskva: Nauka, Moscow, Russia, USSR. 248 pp., 9 illus., 78 tables, small index, hardbound in light green and white with black lettering, 5.5 by 8.3 inches (140 × 212 mm). Printed entirely in the Russian language.

A book on the planting and care of turfgrass lawns and ground covers. It is oriented to turfgrass conditions in the Soviet Union. There is a selected, short bibliography in the back.

van der Ven, Emmy Middelbeek

2005 **Alles Over Gazon en Bloemenweide (Everything about Lawns and Meadow Flowers).**˙ Coauthor.
See *Sulzberger, Robert—2005.*

Webb, David A.

1985 **Practical Landscaping & Lawn Care.**˙ David A. Webb. TAB Books, Blue Ridge Summit, Pennsylvania, USA. 240 pp., 85 illus. with 8 in color, no tables, small index, cloth hardbound in multicolor with green and black lettering, 7.3 by 9.2 inches (186 × 233 mm).

A practical book concerning the planting and care of landscape plants and lawns. It is oriented to gardening enthusiasts in the United States.

Wohlaschlager, Josef

1984 **Rasen und Blumenwiese (Turf and Flower Meadow).**˙ First edition. Josef Wohlschlager. Eugen Ulmer Gmbh & Co., Stuttgart, West Germany. 128 pp., 116 illus. with 75 in color, no tables, indexed, glossy semi–soft bound in multicolor with black lettering, 4.9 by 7.5 inches (124 × 190 mm). Printed entirely in the German language.

A book on the planting and care of lawns and flowers. It is practically oriented to lawn and garden enthusiasts under conditions in Germany.
Revised in 1996.

1996 **Rasen und Blumenwiese (Turf and Flower Meadow).**˙ Second revised edition. Josef Wohlschlager. Verlag Engen Ulmer GmbH & Co., Stuttgart (Hohenheim), Germany. 128 pp., 109 illus. with 96 in color, no tables, small index, glossy semi–soft bound in multicolor with black lettering, 5.1 by 7.4 inches (129 × 189 mm).

A revised update of the 1984 first edition, especially the color photographs.

Yang, Seung-Weon

1997 **Compendium of Turfgrass and Tree Diseases and Insects with Color Plates.**˙ Coauthor.
See *Kim, Ho-Jun—1997.*

Figure Sources

1. African Explosives and Chemical Industries and South African Turf Research Fund. 1948. *Better Turf through Research.* African Explosives and Chem. Industries and South African Turf Res. Fund, Johannesburg, South Africa.

2. Archives, Golf Course Superintendents Association of America, Lawrence, Kansas, USA.

3. Archives, Sports Turf Research Institute, Bingley, West Yorkshire, England, UK.

4. Archives, Tom Mascaro, Tallahassee, Florida, USA.

5. Archives, The Toro Co., Bloomington, Minnesota, USA.

6. Archives, Turfgrass Collection, Michigan State University Library, East Lansing, Michigan, USA.

7. Archives, United States Golf Association Green Section, Far Hills, New Jersey, USA.

8. Archives, University of Rhode Island, Kingston, Rhode Island, USA.

9. Arthur D. Peterson Co. Ltd. 1936. *The Peterson Catalog and Turf Manual.* Arthur D. Peterson Co. Ltd., New York City, New York, USA.

10. Barron, L. 1906. *Lawns and How to Make Them.* Doubleday, Page & Co., New York, New York, USA.

11. Beale, R. 1906. *Golf.* James Carter & Co., London, England, UK.

12. Beale, R. Circa 1910. *The Practical Greenkeeper.* James Carter & Co., Raynes Park, London, England, UK.

13. Beale, R. Circa 1913. *The Practical Greenkeeper.* James Carter & Co., Raynes Park, London, England, UK.

14. Beale, R. 1924. *Lawns for Sports—Materials, Tools and Fittings. Supplement.* James Carter & Co., Raynes Park, London, UK.

15. Beale, R. 1924. *Lawns for Sports—Their Construction and Upkeep.* James Carter & Co., Raynes Park, London, England, UK.

16. Beale, R. 1939. *The Practical Greenkeeper.* Carters Tested Seeds, London, England, UK.

17. Board of Greenkeeping Research. 1929. *J. of the Board of Greenkeeping Res. 1.*

18. Carters Tested Seeds Ltd. 1939. *Fertilizers etc. for Garden and Lawn.* Carters Tested Seeds Ltd., Raynes Park, London, England, UK.

19. Davis, F.F. 1947. *Turf Weed Control with 2,4-D.* National Park Service, United States Department of Interior, Washington, DC, USA.

20. Early Local History Group. 2006. *Sutton Seeds: A History 1806–2006.* Early Local History Group, Reading, England, UK.

21. Farley, G.A. 1930. *Golf Course Commonsense.* Farley Libraries, Cleveland Heights, Ohio, USA.

22. Gault, W.K. Circa 1912. *Practical Golf Greenkeeping.* The Golf Printing and Publ. Co., London, England, UK.

23. The Greens Research Committee of the New Zealand Golf Association. 1934. *Annual Report on Green Keeping Research.* The Greens Research Committee of the New Zealand Golf Association. Palmerston North, New Zealand.

24. Hall, T.D., D. Meredith, R.E. Altona, and N.J. Nentz. Circa 193x. *Essentials for Good Lawns.* African Explosives and Chem. Industries Ltd., Johannesburg, South Africa.

25. Hughes, H.D., M.E. Heath, and D.S. Metcalfe, editors. 1955. Forages: The Science of Grassland Agriculture. 2nd ed. Iowa State College Press, Ames, Iowa, USA.

26. James B Beard Collection, College Station, Texas, USA.

27. J.M. Thornburn & Co. 1908. The Seeding and Preservation of Golf Links. J.M. Thornburn & Co., New York City, New York, USA.

28. MacDonald, J. 1923. Lawns, Links & Sportsfields. Country Life Ltd. and George Newnes Ltd., London, England, UK.

29. Mascaro, T. 1960. Fall Renovation of Greens and Fairway. West Point Products Corp., West Point, Pennsylvania, USA.

30. Messrs. Maxwell M. Hart Ltd. In: P.W. Smith, 1950, The Planning, Construction and Maintenance of Playing Fields. Oxford Univ. Press, London, England, UK.

31. O.M. Scott & Sons Co. 1928. Bent Lawns. O.M. Scott & Sons Co., Marysville, Ohio, USA.

32. O.M. Scott & Sons Co. O.M. Scott & Sons Co., Marysville, Ohio, USA.

33. Paul, W. Circa 1920. The Care and Upkeep of Bowling Greens. Paisley, Scotland, UK.

34. Peter Henderson & Co. 1920s and 1930s. Sports Turf for the Ideal Golf Course and Athletic Field. Peter Henderson & Co., New York, New York, USA.

35. Piper, C.V., and R.A. Oakley. 1917. Turf for Golf Courses. The Macmillan Co., New York, New York, USA.

36. Ricci, James B. Reel Lawn Mower History & Preservation Project. Haydenville, Massachusetts, USA.

37. Sanders, T.W. 1911. Lawns and Greens: Their Formation and Management. W.H. & L Collingridge, London, England, USA.

38. Shanks Collection of James Wallace, Chapelton of Boysack, Arbroath, Scotland, UK, via Graeme Forbes of Stewart, Dalkeith, Scotland, UK.

39. The Sports Press. 1910–1917. The Professional and Greenkeeper. London, England, UK.

40. Stewart & Co. Circa 1913. Formation and Treatment of Links and Lawn Turf. Stewart & Co., Edinburgh, Scotland, UK.

41. Stumpp & Walter Co. 1920 and 1930s. Golf Turf Supplement. Stumpp & Walter Co., New York City, New York, USA.

42. Sutton & Sons. Circa 1903. Garden Lawns, Tennis Lawns, Bowling Greens, Croquet Grounds, Putting Greens, Cricket Grounds. Sutton & Sons, Reading, England, UK.

43. Sutton, M.A.F., editor. 1933. Golf Courses: Design, Construction and Upkeep. Simpkin Marshall Ltd., London, England, UK.

44. Sutton, M.A.F., editor. 1950. Golf Courses: Design, Construction and Upkeep. Sutton & Sons, Reading, England, UK.

45. Sutton, M.H.F., editor. 1912. The Book of the Links. W.H. Smith & Son, London, England, UK.

46. Sutton, M.H.F. Circa 191x. The Laying Out and Upkeep of Golf Courses and Putting Greens. Sutton & Sons, Reading, England, UK.

47. University of St Andrews Library, St Andrews, Scotland, UK.

48. Warren, B. 1962. Green Grows Your Lawn. Warren's Turf Nursery, Palos Park, Illinois, USA.